Quiz Yourself

Quiz Yourself problems allow you to immediately apply the concepts and skills you have just learned. Use them to check your understanding of the material that has just been covered while you are reading the text or to review key ideas before an exam.

EXAMPLE 1 *Using the APR Table*

Hector has agreed to pay off a $3,500 loan by making 24 monthly payments. If the total finance charge on his loan is $460, what is the APR he is being charged?

SOLUTION: The finance charge per $100 financed is

$$\frac{\text{finance charge}}{\text{amount borrowed}} \times 100 = \frac{460}{3,500} \times 100 \approx 0.1314 \times 100 \approx \$13.14.$$

Because Hector is making 24 monthly payments, we use the row in Table 9.6 for 24 payments, as we show in Figure 9.5. Reading across this line, we find that the closest amount to $13.14 is $12.98. The top of this column shows the approximate APR for Hector's loan, which is 12%.

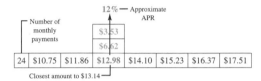

Number of monthly payments			12% ← Approximate APR				
			$3.53				
			$6.62				
24	$10.75	$11.86	$12.98	$14.10	$15.23	$16.37	$17.51

Closest amount to $13.14 ⎯

FIGURE 9.5 Using Table 9.6 to find the APR for Hector's loan.

Now try Exercises 5 to 12. ✳ **21**

We can find the APR if we know the number and size of payments on a loan.

Quiz Yourself **21**

Assume that Jason will repay a loan for $11,250 by making 36 payments. Assume that the finance charge is $1,998.

a) What is the finance charge per $100 financed? $17.76

b) What is the APR? about 11%

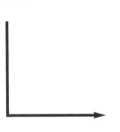

Math in Your Life

Can Ants Do Mathematics?

Yes, believe it or not, ants can do mathematics! Of course, they do not use tiny pencils and paper or minicalculators, but in their own way, when you see ants scurrying around their nest, they are actually working on finding the shortest routes to food sources in a way that is similar to the TSP problem discussed in this section.

In more recent years, scientists have applied mathematics to study how ants, bees, and other social insects cooperate using "swarm intelligence" to solve problems. This research has paid off in applications such as finding improved telephone routes in congested lines and helping robots work cooperatively. Eric Bonabeau, a researcher in ant algorithms, foresees a world in which computer "chips are embedded in every object, from envelopes to trashcans to heads of lettuce" and believes that ant algorithms will be necessary to allow these chips to communicate.

Math In Your Life

Math in Your Life links the mathematics in this course to the world around you, making the math more interesting to learn and easier to understand.

Looking Deeper

Looking Deeper sections at the conclusion of each chapter go more in depth on a particularly interesting topic from the chapter.

Looking Deeper

11.5 Fair Division

Objectives

1. Understand the difference between discrete and continuous division of items.
2. Use the method of sealed bids to make a fair division of items.

How often do you agree to a deal only to find out after it is done that you will receive more than you expected? You will now learn how to divide any group of items among several interested parties that does exactly that. This method works in dividing an inheritance among the heirs of an estate, awarding property to a couple getting divorced, or dissolving a business among partners.

Fair Division

There are various ways to solve these problems, which are called **fair division** problems. We will now explain one method for handling *discrete* fair division problems as opposed to *continuous* fair division problems.

> **DEFINITIONS** **Discrete fair division** means that the items being distributed cannot be subdivided into fractional parts. **Continuous fair division** means that the items being distributed can be subdivided into parts and the division can be done as finely as we want.

Mathematics
All Around

Thomas L. Pirnot

Kutztown University of Pennsylvania

Addison-Wesley

Boston • New York • San Francisco

London • Toronto • Sydney • Tokyo • Singapore • Madrid

Mexico City • Munich • Paris • Cape Town • Hong Kong • Montreal

Executive Editor:	Anne Kelly
Acquisitions Editor:	Marnie Greenhut
Senior Project Editor:	Rachel S. Reeve
Project Manager:	Sandi Goldstein
Assistant Editor:	Leah Goldberg
Editorial Assistant:	Leah Driska
Senior Managing Editor:	Karen Wernholm
Senior Production Supervisor:	Sheila Spinney
Senior Designer:	Barbara T. Atkinson
Digital Assets Manager:	Marianne Groth
Media Producer:	Nathaniel Koven
Software Development:	Eileen Moore and Mary Durnwald
Executive Marketing Manager:	Becky Anderson
Marketing Coordinator:	Bonnie Gill
Senior Manufacturing Buyer:	Carol Melville
Cover Design:	Beth Paquin
Text Design:	Carolyn Deacy
Production Coordination, Composition, and Illustrations:	Progressive Information Technologies
Illustrations:	Karen Heyt
Cover photo:	Telephone © Shutterstock; Photo Enhancements by Beth Paquin

Library of Congress Cataloging-in-Publication Data
Pirnot, Thomas L.
     Mathematics all around / Thomas L. Pirnot.—4th ed.
       p. cm.
     Includes index.
     ISBN 10: 0-321-56797-8 (student edition)
     ISBN 13: 978-0-321-56797-0 (student edition)
       1. Mathematics—Textbooks.   I. Title.
     QA39.3.P57 2008
     510—dc22
                                        2008028327

For permission to use copyrighted material, grateful acknowledgment is made to the copyright holders on page p-1, which is hereby made part of this copyright page.

Many of the designations used by manufacturers and sellers to distinguish their products are claimed as trademarks. Where those designations appear in this book, and Addison-Wesley was aware of a trademark claim, the designations have been printed in initial caps or all caps.

Copyright © 2010, 2007, 2004, 2001 Pearson Education, Inc.

All rights reserved. No part of this publication may be reproduced, stored in a retrieval system, or transmitted, in any form or by any means, electronic, mechanical, photocopying, recording, or otherwise, without the prior written permission of the publisher. Printed in the United States of America. For information on obtaining permission for use of material in this work, please submit a written request to Pearson Education, Inc., Rights and Contracts Department, 501 Boylston Street, Boston, MA 02116, fax your request to 617-848-7047, or e-mail at www.pearsoned.com/legal/permissions.htm.

1 2 3 4 5 6 7 8 9 10—QWT—10 09 08

Addison-Wesley
is an imprint of

ISBN 10: 0-321-56797-8
ISBN 13: 978-0-321-56797-0

DEDICATION

*To my wife Ann, whose loving support and encouragement
continue to be the wind beneath my wings*

Contents

Preface   xi

1   Problem Solving

Strategies and Principles   1

1.1   Problem Solving   2
1.2   Inductive and Deductive Reasoning   18
1.3   Estimation   26
    Chapter Summary   35
    Chapter Review Exercises   35
    Chapter Test   36

2   Set Theory

Using Mathematics to Classify Objects   38

2.1   The Language of Sets   39
2.2   Comparing Sets   47
2.3   Set Operations   54
2.4   Survey Problems   64
2.5   **Looking Deeper:** Infinite Sets   73
    Chapter Summary   78
    Chapter Review Exercises   78
    Chapter Test   79

3   Logic

The Study of What's True or False or Somewhere in Between   81

3.1   Statements, Connectives, and Quantifiers   82
3.2   Truth Tables   91
3.3   The Conditional and Biconditional   102
3.4   Verifying Arguments   111
3.5   Using Euler Diagrams to Verify Syllogisms   120
3.6   **Looking Deeper:** Fuzzy Logic   127
    Chapter Summary   134
    Chapter Review Exercises   135
    Chapter Test   136

4 Graph Theory (Networks)
The Mathematics of Relationships   138

4.1   Graphs, Puzzles, and Map Coloring   139
4.2   The Traveling Salesperson Problem   153
4.3   Directed Graphs   164
4.4   **Looking Deeper:** Scheduling Projects Using PERT   171
Chapter Summary   180
Chapter Review Exercises   181
Chapter Test   182

5 Numeration Systems
Does It Matter How We Name Numbers?   185

5.1   The Evolution of Numeration Systems   186
5.2   Place Value Systems   194
5.3   Calculating in Other Bases   203
5.4   **Looking Deeper:** Modular Systems   215
Chapter Summary   224
Chapter Review Exercises   225
Chapter Test   225

6 Number Theory and the Real Number System
Understanding the Numbers All Around Us   227

6.1   Number Theory   228
6.2   The Integers   239
6.3   The Rational Numbers   247
6.4   The Real Number System   259
6.5   Exponents and Scientific Notation   270
6.6   **Looking Deeper:** Sequences   280
Chapter Summary   291
Chapter Review Exercises   293
Chapter Test   294

7 Algebraic Models
How Do We Approximate Reality?   296

7.1   Linear Equations   297
7.2   Modeling with Linear Equations   308
7.3   Modeling with Quadratic Equations   316
7.4   Exponential Equations and Growth   325
7.5   Proportions and Variations   336
7.6   Functions   343
7.7   **Looking Deeper:** Dynamical Systems   352
Chapter Summary   359
Chapter Review Exercises   360
Chapter Test   362

8 Modeling with Systems of Linear Equations and Inequalities

What's the Best Way to Do It?   *364*

8.1   Systems of Linear Equations   365
8.2   Systems of Linear Inequalities   376
8.3   **Looking Deeper:** Linear Programming   384
     Chapter Summary   **392**
     Chapter Review Exercises   **393**
     Chapter Test   **393**

9 Consumer Mathematics

The Mathematics of Everyday Life   *395*

9.1   Percent Change and Taxes   396
9.2   Interest   404
9.3   Consumer Loans   414
9.4   Annuities   422
9.5   Amortization   430
9.6   **Looking Deeper:** Annual Percentage Rate   439
     Chapter Summary   **445**
     Chapter Review Exercises   **447**
     Chapter Test   **448**

10 Geometry

Ancient and Modern Mathematics Embrace   *450*

10.1   Lines, Angles, and Circles   451
10.2   Polygons   460
10.3   Perimeter and Area   470
10.4   Volume and Surface Area   482
10.5   The Metric System and Dimensional Analysis   490
10.6   Geometric Symmetry and Tessellations   500
10.7   **Looking Deeper:** Fractals   513
     Chapter Summary   **523**
     Chapter Review Exercises   **525**
     Chapter Test   **527**

11 Apportionment

How Do We Measure Fairness?   *530*

11.1   Understanding Apportionment   531
11.2   The Huntington–Hill Apportionment Principle   539
11.3   Applications of the Apportionment Principle   545
11.4   Other Paradoxes and Apportionment Methods   552
11.5   **Looking Deeper:** Fair Division   567
     Chapter Summary   **575**
     Chapter Review Exercises   **576**
     Chapter Test   **577**

12 Voting
Using Mathematics to Make Choices   579

12.1 Voting Methods   580
12.2 Defects in Voting Methods   590
12.3 Weighted Voting Systems   601
12.4 **Looking Deeper:** The Shapley-Shubik Index   610
  Chapter Summary   617
  Chapter Review Exercises   618
  Chapter Test   619

13 Counting
Just How Many Are There?   622

13.1 Introduction to Counting Methods   623
13.2 The Fundamental Counting Principle   631
13.3 Permutations and Combinations   638
13.4 **Looking Deeper:** Counting and Gambling   650
  Chapter Summary   655
  Chapter Review Exercises   656
  Chapter Test   656

14 Probability
What Are the Chances?   658

14.1 The Basics of Probability Theory   659
14.2 Complements and Unions of Events   673
14.3 Conditional Probability and Intersections of Events   681
14.4 Expected Value   695
14.5 **Looking Deeper:** Binomial Experiments   703
  Chapter Summary   710
  Chapter Review Exercises   711
  Chapter Test   712

15 Descriptive Statistics
What a Data Set Tells Us   714

15.1 Organizing and Visualizing Data   715
15.2 Measures of Central Tendency   727
15.3 Measures of Dispersion   740
15.4 The Normal Distribution   750
15.5 **Looking Deeper:** Linear Correlation   763
  Chapter Summary   772
  Chapter Review Exercises   773
  Chapter Test   774

Appendix   776

Answers to Quiz Yourself Problems   778

Answers to Exercises   784

Photo Credits   P1

Index   I1

Preface

In the fourth edition, I have continued the approach and philosophy that have made *Mathematics All Around* a readable, accessible, interesting, and relevant textbook for *today's* liberal arts mathematics students.

Because students now live in an increasingly complex world where both information and, unfortunately, misinformation are growing rapidly, it is more important than ever that students learn to think clearly and become comfortable in the numerical world all around them. The writing continues to reflect my belief that students are most successful when they focus on understanding over memorization.

Throughout the text, I emphasize fundamental principles and intuitive thinking to help students understand and remember the material more easily. Based on user and reviewer suggestions, I have enhanced many portions of the text by expanding explanations, adding numerous diagrams and annotations, and using color to highlight important points. As a result, students will find that the many interesting topics in the text are more understandable and easier to master.

As with previous editions, I have written this edition to provide students majoring in the liberal arts, the social sciences, education, business, and other nonscientific areas with an understanding and appreciation of mathematics, the contributions of diverse cultures, and its many fascinating applications. *Mathematics All Around* is particularly appropriate for students who need to satisfy a one- or two-course requirement in mathematics for graduation or to transfer to another institution.

Content Changes

- To emphasize problem solving, I have moved the topics of **problem solving**, **estimation**, and **inductive and deductive reasoning** into a separate chapter. The problem-solving techniques and principles along with other topics in Chapter 1 form a common thread that is woven through all chapters of the book.

- As a result, Chapter 2 is now totally focused on **set theory**.

- In addition, every chapter after the first has been reorganized so that:

  a. **Chapter Objectives** have been rewritten and coordinated with section headings and also with the summary subdivisions.

  b. The previous **Of Further Interest** is now called **Looking Deeper** and has been moved to the main part of the chapter. Exercises for these sections now appear in the **Chapter Review Exercises** and the **Chapter Test**.

  c. A selection of **Group Projects** follow each **Chapter Test**.

  d. We have converted the old statement headings into **Key Points** marginal notes.

- A number of chapter openers and section openers have been rewritten to make them more current and more relevant.

- Many examples have been rewritten to make the overall flow of ideas better and to make the material more student friendly.

- Many exercises in Chapter 9 on consumer mathematics have been rewritten to make the notation consistent with that used by the financial functions available on the TI-83 Plus and TI-84 Plus graphing calculators. Not only in Chapter 9, but throughout the text numerous screens are given to illustrate how to use graphing calculators to enhance problem solving. TI-83 Plus and TI-84 Plus tutorials will be available to adopters.

- Over 600 exercises are either new, updated, or changed in some way. We have reduced the amount of duplication in the exercises and increased the variety.

New and Continuing Pedagogical Features

The many strong pedagogical features of *Mathematics All Around* make this text a useful tool for the student.

ENHANCED!

Format

The format makes this text easy to use and makes learning mathematics easier for the student. I introduce ideas conversationally and then illustrate them with clear, simple examples. I have included numerous **visual explanations** to help students "see" the mathematics that they are learning. Carefully annotated diagrams and the effective use of color help reinforce the written explanations of numerous topics in this edition. After reinforcing highlighted points with more examples, students test their understanding by working through the **Quiz Yourself** problems. More detailed examples occur later in a section after the ideas have been clearly discussed. Summary boxes help students locate information when doing exercises.

UPDATED!

Motivation with Emphasis on Applications

Applications throughout the text motivate the discussion of mathematics and increase students' interest in the material. Naturally, I have updated and revised the applications to include more current and fresh ideas so they are of interest to today's students. In addition, each chapter opener emphasizes a problem in a realistic situation and gives a broad overview of the chapter material. Toward the end of this brief section, I frequently present problems to motivate the mathematics we will be discussing. After developing the necessary mathematical tools throughout the chapter, I return to solve the problem, reinforcing the value of the concepts studied in the chapter.

Objectives

Objectives appear at the beginning of each section to give students a clear idea of what they are going to learn in each section. In addition, reviewing these objectives after reading the section is an excellent way of checking comprehension of the key ideas just covered.

Problem Solving

Problem solving continues to be one of the main themes of this text. Section 1.1 discusses strategies and principles that help the student understand and attack problems more effectively. In the **Problem-Solving** boxes placed throughout the text, I then remind students of particular problem-solving methods that are appropriate to the example they are reviewing. These boxes relate back to the problem-solving material introduced in Section 1.1 for further review.

Some Good Advice

In addition to the problem-solving boxes, there are numerous boxes titled **Some Good Advice,** which point out common mistakes, provide further advice on problem solving, and make connections between different areas of mathematics.

Math in Your Life

Each chapter contains one or more **Math in Your Life** boxes that connect the mathematics in this course to the various majors that these students are studying, as well as to their everyday lives. These connections make the mathematics more interesting to learn and easier to understand contextually, as well as highlighting the usefulness of mathematics in everyday life.

Quiz Yourself

Each section contains numerous short quizzes called **Quiz Yourself,** which students can use to check their understanding of the material immediately preceding the quiz. These quizzes can be used as a break in the flow of the lecture material and to encourage active learning. A special symbol, **10**, indicating when students should try a **Quiz Yourself** exercise, has been included at the end of appropriate examples to provide students with further assistance while studying the material.

UPDATED!

Highlights

Each chapter contains **Highlight** boxes that discuss the history and practical applications of the topics being presented. These highlights help students understand the material that they are learning in a broader context. A listing of the historical figures mentioned in the text is located at the end of the book and includes individuals such as Aristotle, George Polya, Sophie Germain, and Georg Cantor.

New highlights, titled **Between the Numbers**, encourage students to think carefully about interpreting relevant numerical information in their lives. Some examples of these involve the reliability of crowd estimates at a rally, the soundness of the Social Security system, reading what a graph is really telling you (or perhaps not telling you), the dangers of adjustable rate mortgages, and runaway credit card debt.

Also in the **Highlight** boxes, I encourage students to take advantage of available technology by explaining how to use tools such as graphing calculators and spreadsheets effectively to solve problems. Although the text does not depend on technology, instructors can download pertinent technology from the Instructor's Resource Center at www.pearsonhighered.com/irc and from the related MyMathLab® course. Contact your local Pearson representative for more information about MyMathLab and to learn how to access it.

To more completely integrate the material covered in the Highlights, questions about them are included in the Looking Back exercises at the end of each section.

NEW!

Looking Deeper

The last section of each chapter is titled **Looking Deeper**, which discusses an enrichment topic that goes beyond the main body of material in the chapter. This material is included in the Chapter Summary, and tested in the Chapter Review Exercises and the Chapter Test. Topics include fuzzy logic, linear programming, fractals, and counting and gambling.

UPDATED!

Exercises

Exercise sets have been updated to reduce repetition, enhance variety, and emphasize current applications and have been regrouped into new categories. **Looking Back** exercises encourage students to read and understand the examples in the section before trying the remaining exercises. In order to encourage the student to read the numerous Highlights, I have included questions about the Highlights in this group. Students can use **Sharpening Your Skills** exercises to hone their newly acquired skills. **Applying What You've Learned** exercises are application oriented, with many using real, current data. **Communicating Mathematics** asks the student to write about the mathematics that he or she is learning. Throughout the text, I encourage students to use the Internet and graphing calculators to enhance their understanding of the material through the **Using Technology to Investigate Mathematics** exercises. More challenging exercises are grouped in the **For Extra Credit** category. Additionally, many examples are cross-referenced with exercises to help students in doing homework.

UPDATED!

Summary, Chapter Review Exercises, and Chapter Test

Each chapter has a summary followed by **Chapter Review Exercises** and a **Chapter Test**. I have tightly coordinated the **Chapter Summary** with the chapter objectives and include cross-references in the Chapter Summary to help students locate relevant examples for review more quickly.

The **Chapter Review Exercises** are also cross-referenced to the sections that cover the topics being tested. This enables a student who is not comfortable with a particular concept to return to the section covering that material. The **Chapter Test** then checks students' knowledge of the chapter material by providing a variety of questions that are not referenced to specific sections. There is now a greater variety of problems in the Chapter Review Exercises and Chapter Tests. Chapter Test problems are presented in a random order to better test students' understanding and ability to do these problems. Several group exercises follow each chapter test.

UPDATED!   ## TI Calculator Help

I have increased the emphasis on TI graphing calculators and have included numerous screens illustrating how to use this technology effectively. Tutorials on using the TI graphing calculators are available to adopters of *Mathematics All Around* at the Instructor's Resource Center at www.pearsonhighered.com/irc and from the related MyMathLab course. Contact your local Pearson representative for more information about MyMathLab and how to access it.

Supplements

Supplements for the Instructor

Annotated Instructor's Edition
ISBN 0-321-56505-3 / 978-0-321-56505-1

The Annotated Instructor's Edition contains answers to most exercises on the page where they occur.

Instructor's Solutions Manual
James Lapp
ISBN 0-321-57582-2 / 978-0-321-57582-1

This manual contains detailed, worked-out solutions to all the exercises in the text.

Instructor's Testing Manual
Paul Ache, Kutztown University of Pennsylvania
ISBN 0-321-57587-3 / 978-0-321-57587-6

The Testing Manual includes three alternative tests per chapter. These items may be used as actual tests or as references for creating tests.

Insider's Guide
ISBN 0-321-57595-4 / 978-0-321-57595-1

This supplement includes resources to assist faculty with course preparation and classroom management. Included are helpful teaching tips correlated to each section of the text, as well as additional resources for classroom enrichment.

TestGen®

TestGen (**www.pearsonhighered.com/testgen**) enables instructors to build, edit, print, and administer tests using a computerized bank of questions developed to cover all the objectives of the text. TestGen is algorithmically based, allowing instructors to create multiple but equivalent versions of the same question or test with the click of a button. Instructors also can modify test bank questions or add new questions. Tests can be printed or administered online. The software and test bank are available for download from Pearson Education's online catalog.

PowerPoint® Lecture Slides

Available through **www.pearsonhighered/irc** or MyMath-Lab, these lecture slides include definitions, key concepts and examples for use in a lecture setting and are available for each section of the text.

Pearson Math Adjunct Support Center

The Pearson Math Adjunct Support Center (www.pearson tutorservices.com/math-adjunct.html) is staffed by qualified instructors with more than 50 years of combined experience at both the community college and university levels. Assistance is provided for faculty in the following areas:

- Suggested syllabus consultation
- Tips on using materials bundled with your book
- Book-specific content assistance

- Teaching suggestions, including advice on classroom strategies

Supplements for the Student

Student Edition
ISBN 0-321-56797-8 / 978-0-321-56797-0

Student's Solutions Manual
James Lapp
ISBN 0-321-57584-9 / 978-0-321-57584-5

This manual contains detailed, worked-out solutions to all the odd-numbered section exercises and to all Chapter Review and Chapter Test exercises. Students will find this manual very helpful.

Video Lectures on DVD with Optional Subtitles
ISBN 0-321-57593-8 / 978-0-321-57593-7

The Video Lectures for this text are on DVD, making it easy and convenient for students to watch the videos from a computer at home or on campus. The complete digitized video set, affordable and portable for students, is ideal for distance learning, or supplemental instruction. Videos include an optional window to display subtitles in both English and Spanish.

Media Supplements

MathXL

MathXL is a powerful online homework, tutorial, and assessment system that accompanies Pearson's textbooks in mathematics or statistics. With MathXL, instructors can create, edit, and assign online homework and tests using algorithmically generated exercises correlated at the objective level to the textbook. They also can create and assign their own online exercises and import TestGen tests for added flexibility. All student work is tracked in MathXL's online gradebook. Students can take chapter tests in MathXL and receive personalized study plans based on their test results. The study plan diagnoses weaknesses and links students directly to tutorial exercises for the objectives they need to study and retest. Students can also access supplemental animations and video clips directly from selected exercises. MathXL is available to qualified adopters. For more information, visit our Web site at **www.mathxl.com**, or contact your sales representative.

MyMathLab

MyMathLab is a series of text-specific, easily customizable online courses for Pearson's textbooks in mathematics and statistics. Powered by CourseCompass™ (our online teaching and learning environment) and MathXL (our online homework, tutorial, and assessment system), MyMathLab gives you the tools you need to deliver all or a portion of your course online, whether your students are in a lab setting or working from home. MyMathLab provides a rich and flexible set of course materials, featuring free-response exercises that are algorithmically generated for unlimited practice and mastery. Students can also use online tools, such as video lectures, animations, and a multimedia textbook, to independently improve their understanding and performance. Instructors can use MyMathLab's homework and test managers to select and assign online exercises correlated directly to the textbook, and they also can create and assign their own online exercises and import TestGen tests for added flexibility. MyMathLab's online gradebook—designed specifically for mathematics and statistics—automatically tracks students' homework and test results and gives the instructor control over how to calculate final grades. Instructors also can add offline (paper-and-pencil) grades to the gradebook. MyMathLab also includes access to the **Pearson Tutor Center** (**www.pearsontutorservices.com**). The Tutor Center is staffed by qualified mathematics instructors who provide textbook-specific tutoring for students via toll-free phone, fax, e-mail, and interactive Web sessions. MyMathLab is available to qualified adopters. For more information, visit our Web site at **www.mymathlab.com** or contact your sales representative.

Acknowledgments

It is a pleasure to thank the many wonderfully talented people who have helped me in so many ways in writing this edition. In particular, I thank Marnie Greenhut, acquisitions editor at Pearson Addison-Wesley, who oversaw the project; my project editor Rachel Reeve; and my project manager Sandi Goldstein who all gave wonderful advice and encouragement to reshape this edition and bring it to fruition. I am grateful for Leah Goldberg's and Leah Driska's careful attention in handling the myriad of administrative details for this revision. I appreciate the work of Sheila Spinney who supervised production of the text and Barbara Atkinson, senior designer, who has produced yet another beautiful design. Thanks also to Becky Anderson and Bonnie Gill for continuing to spread the message of *Mathematics All Around*. Nathaniel Koven has done a wonderful job producing and managing all of the fine media supplements for the text, including MyMathLab.

I also greatly appreciate Greg Tobin's continuing confidence in me and support for my books over the years.

I would like to thank James Lapp, who wrote the solutions manuals and corrected my errors for this edition. I am grateful for Paul Ache's thoughtful work in updating the testing manual. I appreciate the careful work of Cathy Ferrer and John Samons in checking for accuracy in the text and exercise answers. Susan Whalen was a great help to me in updating many examples and exercises and making the text fresher and more current in numerous ways.

My wife Ann and my children Matt, Tony, Joanna, and Mike continue to be a great source of encouragement and support for me in my writing.

Finally, last, but certainly not least, I thank all my reviewers and users of previous editions for helping me to continue improving *Mathematics All Around*, and reshaping it into a better text. I always listen carefully to your generous advice and appreciate your many constructive suggestions. The following is a list of reviewers of all editions (reviewers of the fourth edition are marked with an "*").

Gisela Acosta, *Valencia Community College*
Randall Allbritton, *Daytona Beach Community College*
Seth Armstrong, *Southern Utah University*
James Arnold, *University of Wisconsin*
Carmen Q. Artino, *The College of Saint Rose*
Bruce Atkinson, *Samford University*
John Atkinson, *Missouri Western State College*
René Barrientos, *Miami Dade Community College–Kendall*
Thomas Bartlow, *Villanova University*
Linda Barton, *Ball State University*
Kathleen Bavelas, *Manchester Community Technical College*
Kari Beaty, *Midlands Technical College–Airport*
David Behrman, *Somertset Community College*
Norma Biscula, *University of Maine–Augusta*
Terence R. Blows, *Northern Arizona University*
Warren S. Butler, *Daytona Beach Community College*
Marc Campbell, *Daytona Beach Community College*
David Cochener, *Austin Peay State University*
Linda F. Crabtree, *Metropolitan Community College*
Daniel Cronin, *New Hampshire Technical College*
Debra Curtis, *Bloomfield College*
Walter Czarnec, *Framingham State College*

Greg Dietrich, *Florida Community College at Jacksonville*
Margaret M. Donlan, *University of Delaware*
* Cameron B. Douthitt, *University of Colorado at Boulder*
* Gina Poore Dunn, *Lander University*
John Emert, *Ball State University*
Catherine Ferrer, *Valencia Community College*
Tom Foley, *Glendale Community College*
Olivia Garcia, *University of Brownsville*
John Gaudio, *Waubonsee Community College*
John Gimbel, *University of Alaska–Fairbanks*
Gail Gonyo, *Adirondack Community College*
Donald R. Goral, *Northern Virginia Community College*
Glen Granzow, *Idaho State University*
Tom Greenwood, *Bakersfield College*
Joshua Guest, *University of Mississippi*
Philip Gustafson, *Mesa State College*
* Cliff A. Harris, *University of New Mexico*
Ward Heilman, *Bridgewater State College*
Nyeita Irish-Schult, *St. Petersburg Junior College*
Stacey Jones, *Benedict College*
Anne Jowsey, *Niagara County Community College*
Vicky Kauffman, *University of New Mexico*

* Danny T. Lau, *Gainesville State College*
Kathryn Lavelle, *Westchester Community College*
Ana Leon, *Lexington Community College*
Maita Levine, *University of Cincinnati*
Rhonda MacLeod, *Florida State University*
* Mahdi Majidi-Zolbanin, *University of Illinois at Springfield*
Rafael Marino, *Nassau Community College*
* Anne Martincic, *McHenry County College*
Karen L. McLaren, *University of Maryland–College Park*
Kenneth Myers, *Bloomfield College*
Bill Naegele, *South Suburban College*
Nancy H. Olson, *Johnson County Community College*
John L. Orr, *University of Nebraska–Lincoln*
Joanne Peeples, *El Paso Community College*
Scott Reed, *College of Lake County*
Nancy Ressler, *Oakton Community College*
Nolan Rice, *College of Southern Idaho*
Len Ruth, *Sinclair Community College*
Gene Schlereth, *University of Tennessee–Chattanooga*
* Linda Schultz, *McHenry County College*
Mark Sherrin, *Flagler College*
Daniel Seabold, *Hofstra University*
Mary Lee Seitz, *Erie Community College–City Campus*

Kara Shavo, *Mercer County Community College*
Marguerite Smith, *Merced College*
Jamal Tartir, *Catholic University*
Jamal Tartir, *Youngstown State University*
Beverly Taylor, *Valencia Community College*
William Thistleton, *State University of New York–Utica/Rome*
Mark Tom, *College of the Sequoias*
Robert R. Urbanski, *Middlesex County College*
* Rebecca Walls, *West Texas A & M University*
Sheela L. Whelan, *Westchester Community College*
Carol S. White, *Flagler College*
John L. Wisthoff, *Anne Arundel Community College*
Robert Woods, *Broome Community College*
Rob Wylie, *Carl Albert State University*
Marvin Zeman, *Southern Illinois University*

High School Reviewers

Courtney Carpenter, *Albemarle High School*
David Emanuel, *Prunell Swett High School*
Cathy Hicks, *Wakefield High School*
Scott Maxwell, *Athens Drive High School*

I welcome comments and suggestions from users of this book and if you wish to contact me, my e-mail address is pirnot@kutztown.edu.

Thomas Pirnot
Kutztown, PA

To The Student: How To Succeed At Mathematics

I would like to suggest to you some things that you can do before, during, and after class to become a more successful mathematics student.

Prepare ahead.

Although it takes discipline to do so, you will get much more out of your mathematics class if you are able to read the material (even briefly) before it is covered in class. Reading ahead will make lectures more meaningful and questions might occur to you that you would not think to ask had you not read the section in advance. You should use the Key Points that are highlighted in the margins at the beginning of each subsection to get an overview of the topics covered before trying to read the section more carefully.

Read slowly and carefully.

Also remember that reading a math book is very different from reading a textbook in other subjects, such as history, sociology, or music. Because mathematical language is very compressed, a few symbols or words usually contain a lot of information. By paying careful attention to the exact meaning of the definitions and notation being used, you will increase your understanding of mathematics. As you read your book, always try to be actively involved with the material. After reading an example, cover the solution and then try to rework the example yourself. If you cannot work all of the way through, restudy the example to see what points you are missing and try again.

Try to understand the big picture.

I cannot emphasize strongly enough that mathematics is much more than a list of facts, formulas, and equations to be memorized. The more that you can see the big picture of how the material is developed, the better your overall understanding of mathematics will be. If your instructor makes comments in class about the reasons for doing a particular example, or how one topic relates to another, be sure to get those comments in your notes.

Remember how to remember.

Although it is important to remember the basic material for a test, you will find that memorizing without understanding is not very useful. Throughout this text, we will use many diagrams, analogies, and examples to help you understand mathematical terminology, formulas, equations, and solution methods. If you work at understanding the intuitive meaning behind mathematical concepts, those ideas will become more permanent in your memory. We call this process "Remembering How to Remember," and you will find this method of studying an effective way to retain information and hone your math skills.

Do homework intelligently.

When doing exercises, first work on those labeled Looking Back that begin each exercise set. These exercises closely follow the order of the topics presented in the section and will reinforce your understanding of the overall plan of the section.

Students sometimes tell me that although they understand the lectures and do the homework, they still are not successful on exams. I believe that this occurs because they are too book-dependent; that is, they can only do a certain type of problem when they know where it is placed in the book. I have helped many such students improve their grade dramatically by encouraging them to make 3×5 flash cards. To make these cards, choose typical problems from each section. Put a question on one side of a card and its solution on the back. Then when studying, shuffle these cards and give yourself a test so that the questions are coming randomly. This is a great help in finding out whether you can do problems on your own or can only do them when you are looking at a particular section of the text.

Study with a friend.

It is a great study aid if you can team up with other members in your class to study before quizzes and exams. Other students in your class are hearing the same lectures and will understand your instructor's point of view on the material. You can quiz each other by using the flash cards and the Looking Back exercises. I have seen many students' grades improve once they started working in study groups.

I sincerely hope that you enjoy reading this book as much as I have enjoyed writing it. Good luck in your mathematical studies!

Problem Solving
Strategies and Principles

"Poetry Professor Dies from Overdose of Chromium"

Years ago I sometimes worried that if I were careless with my problem solving, I would wake one morning and, while enjoying my morning cup of coffee, read a headline similar to the above.

My friend and colleague, Dr. Al, was a professor of English, specializing in poetry, and, paradoxically, also coach of our school's national championship powerlifting team. Although he never asked for my advice on poetry, Dr. Al sometimes did seek my mathematical advice when he was experimenting with various nutrients to increase his strength. He would need to make conversions from ounces to milligrams, or perhaps milligrams to micrograms, and would ask me to do the calculations for him. Because an overdose of some of these substances could be fatal, I was always very careful to make sure that my advice to him was correct.

In addition to Dr. Al, over the years, other friends and colleagues have asked me to help them solve a variety of real problems such as the following:

"I think that my oil deliveryman is cheating me. Can you help me?"

"Our youth group is building a large star for the top of our church and we need some help with trigonometry."

"Can you tell me how much concrete I'll need for my son's basketball court?"

"Do you think that I am better off retiring now or after we get the new contract?"

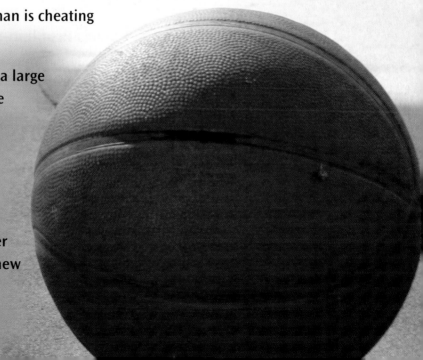

(continued)

"Should I take the free financing on the car, or the $2,000 rebate?"

And so on, and so on . . . .

You will find that the many real problems you will encounter in life are not as easily stated as those you studied in your high school algebra textbook. However, I do believe that by studying the methods presented in this chapter, you will increase your ability to understand and to solve problems in your life. ●

1.1 Problem Solving

Objectives

1. Understand Polya's problem-solving method.
2. State and apply fundamental problem-solving strategies.
3. Apply basic mathematical principles to problem solving.
4. Use the Three-Way Principle to learn mathematical ideas.

My goal in writing this section is to introduce you to some practical techniques and principles that will help you to solve many personal and professional problems in your life—such as whether to buy or lease a car, borrow money for graduate school, or organize a large class project. You will find that although real-life problems are often more complex than those found in this text, by mastering the techniques presented here, you will increase your ability to solve problems throughout your life. It is important to remember that you cannot rush becoming a good problem solver. Like anything else in life, the more you practice problem solving, the better you become at it.

Much of the advice presented in this section is based on a problem-solving process developed by the eminent Hungarian mathematician George Polya (see the historical highlight at the end of this section). We will now outline Polya's method.

 **KEY POINT**

George Polya developed a four-step problem-solving method.

George Polya's Problem-Solving Method

Step 1: Understand the problem. It would seem unnecessary to state this obvious advice, but yet in my years of teaching, I have seen many students try to solve a problem before they completely understand it. The techniques that we will explain shortly will help you to avoid this critical mistake.

Step 2: Devise a plan. Your plan may be to set up an algebraic equation, draw a geometric figure, or use some other area of mathematics that you will learn in this text. The plan you choose may involve a little creativity because not all problems succumb to the same approach.

Step 3: Carry out your plan. Here you do what many often think as "doing mathematics." However, realize that steps 1 and 2 are at least as important as the mechanical process of manipulating numbers and symbols to get an answer.

Step 4: Check your answer. Once you think that you have solved a problem, go back and determine if your answer fits the conditions originally stated in the problem. For example, if you are to find the number of snowboarders who participate

Math in Your Life

Problem Solving and Your Career

It may seem to you that some of the problems that we ask you to solve in this book are artificial and of no practical use to you. However, consider the following question that was asked of a prospective employee during a job interview. "How many quarters—placed one on top of the other—would it take to reach the top of the Empire State Building?" This question may strike you as strange and unrelated to your qualifications for getting a job. However, according to an article at monster.com, such puzzle-type questions can play a critical role in determining who

is hired for an attractive, well-paying job and who is not. What is important about this question is not the answer, but rather the applicant's ability to display creative problem-solving techniques in a pressure situation.

If you are interested in learning more about how your ability to think creatively in solving puzzles can affect your future job prospects, see the book *How Would You Move Mount Fuji? Microsoft's Cult of the Puzzle—How the World's Smartest Company Selects the Most Creative Thinkers* by William Poundstone.

---

in the Winter X Games, $19\frac{1}{2}$ people is not an acceptable answer. Or, in an investment problem, it is highly unlikely that your deposit of $1,000 would earn $334 interest in a bank account. If your solution is not reasonable, then look for the source of your error. Maybe you have misunderstood one of the conditions of the problem, or perhaps you made a simple computational or algebraic mistake.

KEY POINT

Problem solving relies on several basic strategies.

Problem-Solving Strategies

Problem solving is more of an art than a science. We will now suggest some useful strategies; however, just as we cannot list a set of rules describing how to write a novel, we cannot specify a series of steps that will enable you to solve every problem. Artists, composers, and writers make creative decisions as to how to use their tools, and so you also must be creative in using your mathematical tools.

Mathematics is not as rigid as you may believe from your past experiences. It is important to use the strategies in this section to keep your focus on understanding concepts rather then memorizing formulas. If you do this, you may be surprised to find that a given problem can be solved in several different ways.

STRATEGY
Draw Pictures

Problems usually contain several conditions that must be satisfied. You will find it useful to draw pictures to understand these conditions before trying to solve the problem.

EXAMPLE 1   *Visualizing a Condition in a Word Problem*

Four architects are meeting for lunch to discuss preliminary plans for a new performing arts center on your campus. Each will shake hands with all of the others. Draw a picture to illustrate this condition, and determine the number of handshakes.

Quiz Yourself ❶*

How might your diagram change in Example 1 if we were counting the ways the architects could send text messages to each other? Realize now that a message sent from A to B is not the same as a message sent from B to A.
Hint: Consider putting arrowheads on the edges.

SOLUTION: We will use points labeled A, B, C, and D, respectively, to represent the people, and join these points with lines representing the handshakes, as in Figure 1.1.

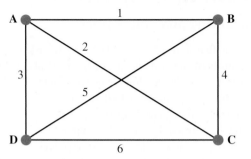

FIGURE 1.1 Visualizing handshakes.

If we represent the handshake between A and B by AB, then we see that there are six handshakes; namely, AB, AC, AD, BC, BD, and CD.

Now try Exercises 7 to 10. ❋ ❶

In later chapters, you will often be interested in determining all the possibilities that can occur when a series of things are occurring. For example, in Chapter 14, you will be solving probability problems involving the flipping of coins and the rolling of dice. The next example illustrates one way to visualize such situations.

EXAMPLE 2 *Drawing a Tree Diagram*

Draw a diagram to illustrate the different ways that you can flip three coins.

SOLUTION: We can draw the following diagram (Figure 1.2), which is called a *tree diagram*, to show the different possibilities. To keep straight in our mind what each coin is

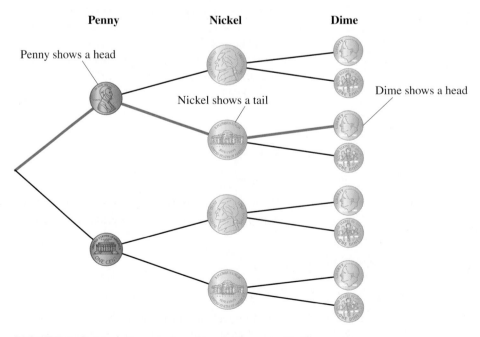

FIGURE 1.2 A tree diagram shows the eight ways to flip three coins.

---

doing, we will assume that the three coins are a penny, a nickel, and a dime. The third branch of the tree (shown in red) illustrates that one possibility is that the penny shows a head, the nickel shows a tail, and the dime shows a head. By tracing through this diagram, you can see that there are eight different ways that the three coins can be flipped.

Now try Exercises 19 and 20. ✳

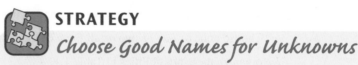

STRATEGY
Choose Good Names for Unknowns

It is a good practice to name the objects in a problem so you can remember their meaning easily.

Example 3 combines good naming with the drawing strategy mentioned earlier.

EXAMPLE 3   *Combining the Naming Strategy and the Drawing Strategy*

Assume that one group of dance students is taking swing lessons and another is taking Latin dance lessons. Choose good names for these groups, and represent this situation with a diagram.

SOLUTION: In Figure 1.3, the region labeled S represents students taking swing dance lessons and region L represents the students taking Latin dance lessons.

Quiz Yourself ❷

Describe the students who are represented by region r_3. What about r_4?

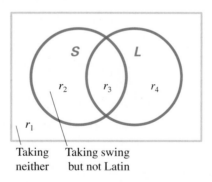

FIGURE 1.3 Representing several groups of people in a diagram.

Quiz Yourself ❸

Choose meaningful names for the objects mentioned in the following situation.

Two amounts are invested— one at a high interest rate, the other at a low rate.

As you can see from Figure 1.3, the region marked r_2 indicates students who are taking swing lessons, but are not taking Latin dance lessons. Region r_1 represents students who are taking neither type of lesson.

Now try Exercises 25 and 26. ✳ ❷ ❸

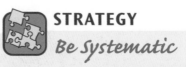

STRATEGY
Be Systematic

If you approach a situation in an organized, systematic way, frequently you will gain insight into the problem.

EXAMPLE 4   *Systematically Listing Options*

Javier is buying an iPhone and is considering which optional features to include with his purchase. He has narrowed it down to three choices: extended-life battery, deluxe ear buds, and 8 gigabytes of memory. Depending on price, he will decide how many of these options he can afford. In how many ways can he make his decision?

SOLUTION: In making his decision, we see that there are four cases.

Choose none *or* one *or* two *or* all three of the options.

We organize these possibilities in Table 1.1.

	Battery	**Ear Buds**	**8 Gigabytes**
choose none {	No	No	No
choose one {	Yes	No	No
	No	Yes	No
	?	?	?
choose two {	Yes	Yes	No
	Yes	No	Yes
	?	?	?
choose three {	Yes	Yes	Yes

TABLE 1.1 Systematic listing of choices of iPhone options.

Quiz Yourself ❹

Complete Table 1.1.

We see that there are eight ways that Javier can make his decision.

Now try Exercises 17, 18, 21, and 22. ✺ ❹

STRATEGY

Look for Patterns

If you can recognize a pattern in a situation you are studying, you can often use it to answer questions about that situation.

EXAMPLE 5   *Finding Patterns in Pascal's Triangle*

You will encounter the pattern that we show in Figure 1.4, called *Pascal's triangle*, in later chapters.

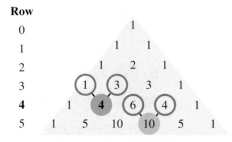

FIGURE 1.4 Pascal's triangle.

Notice how each number is the sum of the two numbers immediately above it, which are a little to the right and a little to the left. Suppose we want to find the total of all the numbers that will be in the ninth row of this diagram.

SOLUTION: (It will be convenient when we discuss this diagram in later chapters to begin numbering rows with 0 instead of 1.)

Notice that in the zeroth row, the total is 1; in the first row, the total is 2; in the second row, the total is 4; and in the third row, it is 8. We continue this pattern in Table 1.2.

Quiz Yourself **5**

a) List the numbers in the sixth and seventh rows of Pascal's triangle.

b) What are the first two numbers in the 100th row of Pascal's triangle?

Row	0	1	2	3	4	5	6	7	8	9
Total	1	2	4	8	16	32	64	128	256	512

TABLE 1.2 The sum of the numbers in each row of Pascal's triangle.

We now easily see that the desired total is 512. ❀ **5**

STRATEGY
Try a Simpler Version of the Problem

You can begin to understand a complex problem by solving some scaled-down versions of the problem. Once you recognize a pattern in the way you are solving the simpler problems, then you can carry over this insight to attack the full-blown problem.

In these days of identity theft, it is important when you send personal information, such as your Social Security number or bank account numbers, to another party that the information cannot be intercepted and your identity stolen. In Example 6, we consider a problem similar to the handshake problem you saw earlier.

Recall that in Example 1, by drawing a diagram of the possible handshakes among four people, you could see that there were six possibilities. If we had asked the same question for 12 people, it would be very cumbersome to draw a picture to count the handshakes, so we would have to have used another technique.

EXAMPLE 6 *Secure Communication Links*

Suppose that 12 branches of Bank of America need secure communication links among them so that financial transfers can be made safely. How many links are necessary?

SOLUTION: Instead of considering all 12 branches, we will look at much smaller numbers of bank branches, count the links, and see if we recognize a pattern. Let's call the branches A, B, C, D, E, F, G, H, I, J, K, and L. In Table 1.3, we will write AB for the link between A and B, AC for the link between A and C, etc.

From looking at these smaller examples, we see an emerging pattern. We notice that as we add new branches, the number of links increases first by 1, then by 2, then by 3, and then by 4, etc.

You can easily see why this is the case if you imagine establishing the branch offices of the banks one at a time. First A is established, so no links are required. Then when B is

Number of Branches	Branches	Links	Number of Links	
1	A	None	0	
2	A, B	AB	1	⟵ add 1 link
3	A, B, C	AB, AC, BC	3	⟵ add 2 links
4	A, B, C, D	AB, AC, AD, BC, BD, CD	6	⟵ add 3 links
5	A, B, C, D, E	AB, AC, AD, AE, BC, BD, BE, CD, CE, DE	10	⟵ add 4 links

TABLE 1.3 Looking for a pattern in the links between Bank of America branches.

built, one link is required between A and B. When C is built, two additional links are needed, namely, AC and BC. When branch D is added, we will need three additional links—AD, BD, and CD. If we continue this pattern as in Table 1.4, we can solve the original problem.

Number of Branches	1	2	3	4	5	6	7	8	9	10	11	12
Number of Links	0	1	3	6	10	15	21	28	36	45	55	66

TABLE 1.4 Counting the links.

We now see that for 12 branches, there will be 66 links.

Now try Exercises 33 to 38. ❋ **6**

Quiz Yourself **6**

Determine the number of ways e-mail can be sent between the various branches. Notice now that e-mail from A to B, written AB, is not the same as e-mail from B to A, written BA.

STRATEGY

Guessing Is OK

One of the difficulties in solving word problems is that you can be afraid to say something that may be wrong and consequently sit staring at a problem, writing nothing until you have the full-blown solution. Making guesses, even incorrect guesses, is not a bad way to begin. It may give you some understanding of the problem. Once you make a guess, evaluate it to see how close you are to meeting all the conditions of the problem.

Imagine that in a few years, when you have graduated, you decide to purchase a brand-new house. At first you are happy with your decision, but then one day when you open your mail, you are shocked to find a school real estate tax bill for $5,200. You are not alone, because in some areas of the country,* both young homeowners and senior citizens are struggling with excessive property taxes. As a result, taxpayer groups have urged politicians to consider other ways of funding public education. We consider a hypothetical case in our next example.

---

*As I was writing this fourth edition, a bitter controversy was going on in Pennsylvania, where property owners both young and old alike were losing their homes, in large part, due to rising property taxes.

EXAMPLE 7 *Solving a Word Problem by Guessing*

Suppose that you own a house in a school district that has a yearly budget of $100 million. In order to reduce the portion of the budget borne by property owners, your local taxpayers' organization has negotiated the following political agreement:

1. The amount funded by the state income tax will be three times the amount funded by property taxes.
2. The amount funded by the state sales tax will be $15 million more than the amount funded by property taxes.

How much of the budget will be funded by property taxes?

SOLUTION: In a later chapter, we will discuss how to solve problems algebraically, but for now, all we want to do is make several educated guesses and then keep adjusting them until we get an acceptable answer. Let us call the amount of the budget due to property taxes p, the amount due to income taxes i, and the amount due to sales taxes s. We will organize our guesses in the following chart.

Guesses for p, i, s (in millions of $)	Evaluation of Guess	
	Good Points	**Weak Points**
20, 20, 60	Total is 100.	Amounts are not different.
20, 30, 60	Amounts are different.	Total is not 100. The amount i is not three times p. The amount s is not 15 more than p.
20, 60, 35	The amount i is three times p. The amount s is 15 more than p.	The total is greater than 100, so we have to reduce the amount p.
18, 54, 33	The amount i is three times p. The amount s is 15 more than p.	The total is still more than 100.
17, 51, 32	We have a solution. The amount from property taxes is $17 million.	

Now try Exercises 65 to 74. ❋

You may not believe that the guessing approach to problem solving that we suggested in Example 7 is doing mathematics. However, if you are making intelligent guesses and systematically refining them, you probably have some solid, intuitive, underlying logical reasons for what you are doing. And, as a result of this thinking, you are doing legitimate mathematics. The problem with guessing is that it can be inefficient and, if the answer is complex, you probably will not be able to find it without doing some algebra. **7**

Quiz Yourself **7**

The Field Museum in Chicago, which houses Sue, the largest preserved *T. Rex*, has acquired three more mechanical dinosaurs for a new exhibit. The combined weight of the three new dinosaurs is 50 tons. If the weight of the Apatosaurus is seven times the weight of the Duckbill, and the Diplodocus is 14 tons more than the Duckbill, what are the weights of each dinosaur?

 **STRATEGY**

Convert a New Problem to an Older One

An effective technique in solving a new problem is to try to connect it with a problem you have solved earlier. It is often possible to rewrite a condition so that the problem becomes exactly like one you have seen before.

EXAMPLE 8   *Converting a New Problem to an Older One*

Assume that your dear Uncle Trump has died and bequeathed you a large sum of money and you are considering putting it in the following possible investments: stocks, bonds, mutual funds, real estate, certificates of deposit, collectibles, and precious metals. How many ways can you invest your money using these options if you can choose from none to all seven of these options?

SOLUTION: We could solve this problem by doing systematic listing or considering simpler examples as we have done earlier; however, we will solve this problem by recognizing that it is a problem that we have solved before. We are just using different language. Recall that in Example 2 when we were counting the ways to flip three different coins we used a tree diagram to list the possibilities. If instead of thinking "coins" we think "investment options," and instead of "heads or tails" we think "yes of no," then we see that we can draw a similar tree. So consider the tree diagram in Figure 1.5 illustrating the ways that you can select none to all three of the options: stocks, *s*, bonds, *b*, or mutual funds, *m*. A "yes" in the tree means that you are taking that option; a "no" means that you are not taking it. The branch highlighted in red indicates that one possibility is to invest in stocks and mutual funds, but not bonds.

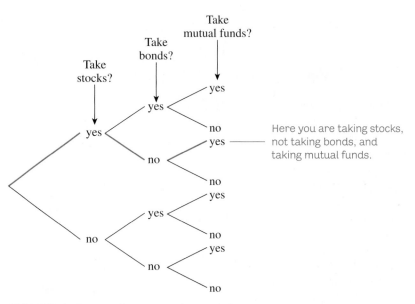

FIGURE 1.5 A tree diagram showing the different ways to choose none to all three investments.

Quiz Yourself ⑧

How many ways could you choose to invest your inheritance if you had 10 different investments to choose from?

From this diagram you can see that there are eight ways to choose among three investments. If we were to add a fourth investment, such as certificates of deposit, we would add two branches "yes or no" to the existing eight, to give us 16. So each time we add an investment, we double the number of possible ways to select investments. Table 1.5 shows us how to continue this pattern to solve the problem. ❋ ⑧

Number of Investments	1	2	3	4	5	6	7
Number of Selections	2	4	8	16	32	64	128

TABLE 1.5 Counting possibilities for investing.

 **KEY POINT**

Remembering several fundamental principles helps us with problem solving.

Some Mathematical Principles

We will now discuss some basic mathematical principles that we will refer to frequently throughout this text.

 The Always Principle

When we say a statement is true in mathematics, we are saying that the statement is true *100% of the time.* One of the great strengths of mathematics is that we do not deal with statements that are "sometimes true" or "usually true."

EXAMPLE 9 *An Algebraic Statement That Is Always True*

Find an example to illustrate that the statement

$$(x + y)^2 = x^2 + 2xy + y^2$$

is true.*

SOLUTION: In algebra, we prove that the statement $(x + y)^2 = x^2 + 2xy + y^2$ is true for all numbers x and y. This means that any number we substitute for x and y should make the statement true.

Suppose that $x = 3$ and $y = 4$. Then, substituting in the given statement we get

$$(3 + 4)^2 = 3^2 + 2 \cdot 3 \cdot 4 + 4^2 \quad \text{or} \quad 49 = 9 + 24 + 16. \;❀\; \boxed{9}$$

where $(x+y)^2$, $2 \cdot x \cdot y$, x^2, y^2 are labeled.

Quiz Yourself ❾

Find another example to show that

$$(x + y)^2 = x^2 + 2xy + y^2.$$

In a similar way, we accept a mathematical argument only if it holds up under every set of conditions. If we can think of even a single situation in which the argument fails, then the argument is not acceptable.

 The Counterexample Principle

An example that shows that a mathematical statement fails to be true is called a *counterexample.* Keep in mind that if you want to use a mathematical property and someone can find a counterexample, then the property you are trying to use is not allowable. A hundred examples in which a statement is true do not prove it to be *always* true, yet a single example in which a statement fails makes it a false statement. Be careful to understand that when we say a statement is false, we are not saying that it is always false. We are only saying that the statement is *not always true.* That is, we can find at least one instance in which it is false.

*For a review of the order of operations in arithmetic and algebra, see Appendix A.

EXAMPLE 10 *Counterexamples to False Algebra Statements*

Although the following two statements are false, those who are not secure in working with algebra often believe them to be true.

a) $(x+y)^2 = x^2 + y^2$ b) $\sqrt{x+y} = \sqrt{x} + \sqrt{y}$

Provide a counterexample for each statement.

SOLUTION:

a) Assume that $x = 3$ and $y = 4$. Then,

$$(x+y)^2 = (3+4)^2 = 7^2 = 49,$$

which is not equal to

$$x^2 + y^2 = 3^2 + 4^2 = 9 + 16 = 25.$$

b) Let $x = 16$ and $y = 9$. Then,

$$\sqrt{x+y} = \sqrt{16+9} = \sqrt{25} = 5.$$

However,

$$\sqrt{x} + \sqrt{y} = \sqrt{16} + \sqrt{9} = 4 + 3 = 7.$$

Now try Exercises 39 to 48. ❋ **10**

Quiz Yourself **10**

Find a counterexample to the following statement:

$$\frac{a+b}{a+c} = \frac{b}{c}.$$

The Order Principle

When you read mathematical notation, pay careful attention to the *order* in which the operations must be performed. The order in which we do things in mathematics is as important as it is in everyday life. When getting dressed in the morning, it makes a difference whether you first put on your socks and then your shoes, or first put on your shoes and then your socks. Although the difference may not seem as dramatic, reversing the order of mathematical operations can also give unacceptable results. Note that we are not saying that it is *always* wrong to reverse the order of mathematical operations; we are saying that if you reverse the order of operations, you *may* accidentally change the meaning of your calculations.

Quiz Yourself **11**

Consider the equation

$$\sqrt{x+y} = \sqrt{x} + \sqrt{y}.$$

a) On the left side of the equation we are performing two operations. What are these operations and what is their order?

b) What is the order of the operations performed on the right side of the equation?

c) Summarize what is wrong with this equation, as we did in Example 11(a).

EXAMPLE 11 *Reversing the Order of Operations Can Change the Result of Mathematical Computations*

Explain how Example 10 illustrates the order principle.

SOLUTION: In Example 10, the problem with statements a) and b) is that in each case we carelessly changed the order in which we performed the operations.

a) In the equation $(x+y)^2 = x^2 + y^2$, the left side of the equation tells us to *first* add x and y and *then* square the resulting sum. The right side of the equation tells us to *first* square x and y separately and *then* add these two squares. To say this more succinctly, we can say that to "first add and then square" is not the same as to "first square and then add."

b) The problem here is similar. Test your understanding of the order principle by solving Quiz Yourself 11.

Now try Exercises 49 to 52. ❋ **11**

Understanding the order principle will be important when you work with set theory in Chapter 2 and also when you study logic in Chapter 3.

The Splitting-Hairs Principle

You should "split hairs" when reading mathematical terminology. If two terms are similar but sound slightly different, they usually do not mean exactly the same thing. In everyday English, we may use the words *equal* and *equivalent* interchangeably; however, in mathematics they do not mean the same thing. The same is true for notation. When you encounter different-looking notation or terminology, work hard to get a clear idea of exactly what the difference is. Representing your ideas precisely is part of good problem solving.

Quiz Yourself 12

In your own words, explain the differences you see in each pair of symbols. Do not be concerned if you do not know what these symbols mean.

a) $\supset$ and $\supseteq$

b) $\{\varnothing\}$ and $\varnothing$

c) $\cup$ and $\vee$

d) $\{0\}$ and 0

EXAMPLE 12 *Identifying Differences in Notation*

Notice the difference in the following pairs of symbols. It is not important that you know what these symbols mean at this time; all we want you to do is recognize that there are slight differences in the notations.

a) $<$ and $\leq$. There is an extra line under the $<$ symbol on the right.

b) $\cap$ and $\wedge$. The symbol on the left is rounded; the symbol on the right is pointed.

c) $\in$ and $\subset$. The symbol on the left has an extra line.

d) $\varnothing$ and 0. The symbol on the right is the number 0; the symbol on the left is something else, not a number. ✹ **12**

The Analogies Principle

Much of the formal terminology that we use in mathematics sounds like words that we use in everyday life. This is not a coincidence. Whenever you can associate ideas from real life with mathematical concepts, you will better understand the meaning behind the mathematics you are learning.

EXAMPLE 13 *Relating Mathematical Terms to Everyday Language*

In Table 1.6, the left column contains formal mathematical terminology. The right column contains ordinary English words that will help you remember the mathematical concepts.

Quiz Yourself 13

Which English words are you reminded of by the following mathematical terminology?

a) intersection

b) simultaneous (as in simultaneous equations)

Mathematical Concept	Related English Ideas
Union	Labor union, marriage union, united
Complement	Complete
Equivalent	Equivalent (in some way the same)
Slope	Ski slope, slope of a roof

TABLE 1.6 Mathematical terms related to words in everyday language.

Now try Exercises 61 to 64. ✹ **13**

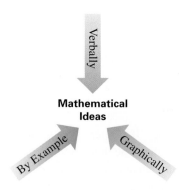

FIGURE 1.6 The Three-Way Principle.

The Three-Way Principle

We conclude this section with a method for approaching mathematical concepts that we illustrate in Figure 1.6.

Whether you are learning a new concept or trying to gain insight into a problem, it is helpful to use the ideas we have discussed in this chapter to approach mathematical situations in three ways.

- *Verbally*—Make analogies. State the problem in your own words. Compare it with situations you have seen in other areas of mathematics.
- *Graphically*—Draw a graph. Draw a diagram.
- *By example*—Make numerical or other kinds of examples to illustrate the situation.

Not every one of these three approaches fits every situation. However, if you get in the habit of using a verbal-graphical-example approach to doing mathematics, you will find that mathematics is more meaningful and less dependent on rote memorization. If you practice approaching mathematics using the strategies and principles that we have discussed, you will find eventually that you are more comfortable and more successful in your mathematical studies.

✸ ✸ ✸ HISTORICAL HIGHLIGHT

George Polya

We could argue that George Polya (1887–1985) is the father of problem solving as we teach it in so many of today's mathematics textbooks (including this one). As a youth, Polya decided to study law but because of his dislike for memorization, he found it tedious and eventually chose a career in mathematics. While working as a mathematics tutor, he began to develop a problem-solving method that led him to write a book called *How to Solve It*. This book, which has sold more than 1 million copies, discusses many of the approaches to problem solving that you will learn in this section.

Polya loved to solve problems, and it seemed that he could find them everywhere. One day while he and his wife-to-be, Stella, were walking in a garden in Switzerland, they encountered another young couple six times. Polya wondered what the likelihood was that they would meet the same couple so many times on the same walk. His attempts to answer that question eventually led him to publish research papers on the topic of the random walk problem.

In the 1940s, due to their concern about the increasing influence of Nazism in Europe, George and Stella emigrated to the United States. Polya eventually accepted a position at Stanford University in California, where he conducted research and worked on problem solving until he was well into his nineties.

Exercises 1.1

Looking Back*

These exercises follow the general outline of the topics presented in this section and will give you a good overview of the material that you have just studied.

1. List the four steps in Polya's problem-solving method.

2. Name the seven problem-solving strategies that we have presented in this section.

3. Give an example of a situation in which you might use each problem-solving strategy that you mentioned in Exercise 2.

4. List the six principles that we discussed in this section. Give an example of how each principle is used.

5. What well-known book on problem solving did Polya write?

*Before doing these exercises, you may find it useful to review the note *How to Succeed at Mathematics* on page xix.

6. What is the point of the *Mathematics in Your Life* highlight on page 3?

Sharpening Your Skills

In Exercises 7–10, draw a picture to illustrate each situation. You are not being asked to solve any problems—just draw a picture.

7. Five liters of a 10% sugar solution are mixed with pure water to get a 5% solution.

8. Six people clink each other's glasses in a toast.

9. Messages are being sent among three spies: (A)ustin Powers, James (B)ond, and (C)hloe O'Brien.

10. The Democratic, Republican, and Green political parties have no members in common.

In Exercises 11–16, choose names that would be meaningful for each item. Again, you are not being asked to solve the problem—just make a recommendation for some good names.

11. A solar panel is three times as long as it is wide.

12. In order to reduce dependence on foreign oil, a presidential commission is considering increasing funding for research in the development of hybrid automobiles, windmill turbines, and solar energy.

13. Meredith, Christina, Izzie, and Alex are planning a party for Derek.

14. Two pathways in the brain cross.

15. A person makes two investments—one in stocks, the other in bonds.

16. Lance Armstrong wants to include calcium and protein in his diet.

In Exercises 17 and 18, list the items mentioned. Try to organize your list in a systematic way.

17. List the different combinations of heads and tails that can occur when a penny and a nickel are flipped.

18. Using the numbers 1, 2, and 3, form as many ordered pairs of numbers as you can. For example, (2, 3) is one such pair; (3, 2) is a different one. You are allowed to repeat numbers, so (1, 1) is an allowable pair.

19. If you draw a tree, as we did in Example 2, to show the number of ways to flip four coins, how many possibilities would there be?

20. If you draw a tree, as we did in Example 2, to show the number of ways to roll two dice (one red and one green), how many possibilities would there be? (*Hint:* Draw the branches for the roll of the first die and then attach more branches to correspond to the roll of the second die.)

21. Choose two recording artists to host the Grammy Awards. Your choices are Kanye West, Beyonce Knowles, Christina Aguilera, and Carrie Underwood. (Represent the artists by the first letter of their last name. One pair would be Kanye West and Beyonce Knowles, which you would represent by WK. Note that KW is the same as WK.)

22. Follow the arrows in the given diagram to travel from "Begin" to "End."

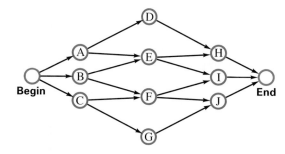

23. Some role playing games, such as *Dungeons and Dragons*, have dice with other than six sides. Assume that you are rolling two four-sided dice—with faces numbered 1, 2, 3, and 4. Draw a tree diagram and then list all of the possible ordered pairs of numbers that can be obtained when the two dice are rolled.

24. Repeat the previous problem, but now only list pairs of numbers that are different.

25. Consider the following diagram, which represents various groups of people. *G* is the group of people who are good singers, and *A* is the group of people who have appeared as contestants on *American Idol*. Describe the people who are in regions r_3 and r_4.

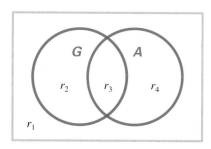

26. Redraw the diagram in Exercise 25, except now label the two groups of people as *W*, those who are working to reduce global warming, and *H*, those who are striving to reduce world hunger. Describe the people who are in regions r_2 and r_4.

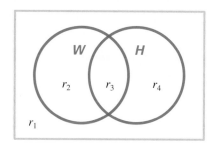

In Exercises 27–32, continue the pattern for five more items in the list. (There may be more than one way to continue the pattern. We will provide only one solution.)

27. 7, 14, 21, 28, . . .

28. 1, 3, 9, 27, . . .

29. *ab, ac, ad, ae, bc, bd, be,* . . .

30. (1, 1), (1, 2), (1, 3), (1, 4), (1, 5), (1, 6), (2, 1), (2, 2), (2, 3), . . .

31. 1, 1, 2, 3, 5, 8, 13, . . .

32. 2, 3, 5, 7, 11, 13, 17, . . .

In Exercises 33–38, we state a problem. Instead of trying to solve that problem, state a simpler problem and solve it instead.

33. Ten people are being honored for their work in reducing pollution. In how many ways can we line up these people for a picture?

34. If you guess at 10 true–false questions, how many different ways can you fill in the 10 answers?

35. Using all the letters of the alphabet, how many two-letter codes can we form if we are allowed to use the same letter twice? For example, *ah, yy, yo,* and *bg* are all allowable pairs. (*Note: bg* is different from *gb*.)

36. A family has seven children. If we list the genders of the children (for example, *bbggbgb*, where *b* is boy and *g* is girl), how many different lists are possible?

37. An electric-blue Ferrari comes with seven options: run flat tires, front heated seats, polished rims, front parking sensor, carbon interior trim, fender shields, and a custom tailored cover. You may buy the car with any combination of these options (including none). How many different choices do you have?

38. In printing your résumé, you have 10 different colors of paper to use and 20 different font styles. How many different ways can you print your résumé?

In Exercises 39–48, determine whether each statement is true or false. If a statement is true, give two examples to illustrate it. If it is false, give a single counterexample.

39. Months of the year have 31 days.

40. All past presidents of the United States are deceased.

41. $\dfrac{a}{b} + \dfrac{c}{d} = \dfrac{a+c}{b+d}$

42. If $a < b$, then $a + c < b + c$.

43. If A is the father of B and B is the father of C, then A is the father of C.

44. If X is acquainted with Y and Y is acquainted with Z, then X is acquainted with Z.

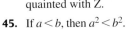

45. If $a < b$, then $a^2 < b^2$.

46. $x + y - z = x + z - y$

47. If the lengths of the sides of a square are doubled, then the area of the square is also doubled.

48. If the price on a Blu-ray player is increased by 10% and then later reduced by 10%, the price will be the same as the original price.

In Exercises 49–52, decide whether the two sequences of operations will give the same results.

49. Squaring a number, then adding 5 to it; adding 5 to a number and then squaring that sum.

50. Squaring two numbers *x* and *y*, then subtracting the results; subtracting *x* and *y*, then squaring the difference.

51. Adding two numbers *x* and *y*, then dividing the result by 3; dividing *x* and *y* by 3, then adding the results.

52. Multiplying two numbers *x* and *y* by 5, then adding the results; adding *x* and *y*, then multiplying the result by 5.

In Exercises 53–60, explain the differences you see in each pair of symbols.

53. 5 and {5}

54. A and A′

55. U and u

56. (1, 2) and [1, 2]

57. {1, 2} and (1, 2)

58. (2, 3) and (3, 2)

59. ∅ and 0

60. { } and ∅

In later chapters, we will use the following mathematical terminology. Of what common English usage do these terms remind you? Explain.

61. path, circuit, bridge, directed

62. dimension, reflection, translation, transform

63. intercept, best fit, compounding

64. median, deviation, central tendency, correlation

Applying What You've Learned

In Exercises 65–74, do not try to solve each problem algebraically. Instead, make a guess that satisfies one or more conditions of the problem and then evaluate your guess, as we did in Example 7. Keep adjusting your guesses until you have a solution that fits all the conditions of the problem.

65. The local historical society wants to preserve two buildings. The total age of the buildings is 321 years. If one building is twice as old as the other, what are the ages of the two buildings?

66. To celebrate Dunder Mifflin's 40th anniversary, Michael Scott has made a 40-foot-long sandwich. The sandwich is cut into three unequal pieces. The longest piece is three times as long as the middle piece, and the shortest piece is 5 feet shorter than the middle piece. What are the lengths of the three pieces?

67. Janine worked 15 hours last week. One job as a clerk in a sporting goods store paid her $7.25 per hour, while her job giving piano lessons paid $12 per hour. If she earned $137.25 between the two jobs, how many hours did she work at each job?

68. Vince worked 18 hours last week. Part of the time he worked in a fast-food restaurant and part of the time he worked tutoring a high school student in mathematics. He was paid $7.25 per hour in the restaurant and $15 per hour tutoring. If he earned $161.50 total, how many hours did he work at each job?

69. Last season, LaDainian Tomlinson carried the football two times more than Steven Jackson and 68 times less than Larry Johnson. If the three running backs carried the ball 1,110 times, how many times did each carry the ball?

70. In 2007, Alex Rodriguez of the New York Yankees and Magglio Ordoñez of the Detroit Tigers had a total of 295 runs batted in. If Rodriguez had 7 more than Ordoñez, how many did each have?

71. Heather has divided $8,000 between two investments—one paying 8%, the other 6%. If the return on her investment is $550, how much does she have in each investment?

72. Carlos has $9,000 in two mutual funds. One fund pays 11% interest and the other fund pays 8%. If his income from the two funds last year was $936, how much did he invest in each fund?

73. The administration at Center City Community College has formed a 26-person planning committee. There are five times as many administrators as there are students and five more faculty members than students. How many students are there on the committee?

74. Last week Minxia made 55 phone calls to gather support for the reelection of her local state representative. She contacted three times as many senior citizens as she did young adults. The number of middle-age adults was one-half the number of senior citizens. How many senior citizens did she contact?

Communicating Mathematics

75. What is the benefit of choosing good names for objects in a mathematical discussion?

76. Explain the advantage of listing items systematically in a discussion rather than in a random fashion.

77. How does the technique of considering simpler examples of a problem relate to the technique of looking for patterns in solving a problem?

78. Why is guessing a good way to gain insight into a problem?

79. Students often ignore the first and last step of Polya's problem-solving method. Why do you think that this is so? What are the dangers of ignoring these steps?

80. Explain in your own words the Three-Way Principle.

81. Does using the Three-Way Principle in learning mathematics seem different from the ways you have approached learning mathematics in the past? Explain.

82. The French mathematician Jacques Hadamard wrote a book, *An Essay on the Psychology of Invention in the Mathematical Field* (Dover 1954), that gives insights into how real mathematicians use the subconscious mind to solve problems. Write a brief report on this book.

Using Technology to Investigate Mathematics

83. Search the Internet for the term *problem solving*, and write a brief report on one of the sites that you find interesting.

84. Search the Internet for the terms *creativity* and *mathematics*. Write a brief report on an interesting site that you find.

85. Use a calculator to find the last digit in the number 2^{100}. If you try to do this directly, you will get an answer that looks something like 1.2676506 E 30, which is the number written in what is called *scientific notation* that we discuss later in Chapter 6. That is not what we are asking you to do. If you instead consider the sequence of computations $2, 2 \times 2, 2 \times 2 \times 2, 2 \times 2 \times 2 \times 2$, etc. until you see a pattern, that will allow you to predict the last digit in 2^{100}.

86. Proceed as in Exercise 85 and use a calculator to find the last digit in the number 7^{100}.

For Extra Credit

87. How many ordered triples are possible if we roll three 6-sided dice?

88. How many ordered triples are possible if we roll three 12-sided dice?

89. Draw a diagram to show all of the possible routes that a sales representative for a company could take by starting at Los Angeles, visiting Chicago, Houston, and Philadelphia in some order, and then returning home. (Describe each route by listing the first letter of each city visited. For example, LHPCL is one route.)

90. Reconsider Exercise 89, but now assume you have 10 different cities, including Los Angeles. By looking at simpler examples, find a pattern to determine the total number of routes that would be possible. Assume that you are always starting at Los Angeles.

91. Continue the following sequence of pairs of numbers by listing the next two pairs in the sequence: (3, 5), (5, 7), (11, 13), (17, 19), (29, 31), ...

92. Continue the following sequence of pairs of numbers by listing the next two pairs in the sequence: (5, 11), (7, 13), (11, 17), (13, 19), (17, 23), (19, 29), ...

93. We will call the following figure a 5-by-5 square. In this figure, we have highlighted a 1-by-1 square in red and a 3-by-3 square in blue. Find the number of all possible squares in the figure. Try to be systematic in solving this problem by considering all 5×5 squares, then all 4×4 squares, etc.

94. Consider in the following map how many different ways there are to go directly from the Hard Rock Cafe to The Cheesecake Factory.

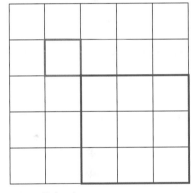

Hard Rock Cafe

The Cheesecake Factory

1.2 Inductive and Deductive Reasoning

Objectives

1. Understand how inductive reasoning leads to making conjectures.
2. Give examples of correct and incorrect inductive reasoning.
3. Understand the difference between inductive and deductive reasoning.

We will begin this section with three apparently unrelated ramblings:

1. As I was writing this section of this book, my house was being invaded by ants. I noticed that on sunny days, I had an abundance of ants in my bathroom and kitchen, whereas on rainy days there was hardly an ant to be seen. Today I woke up, saw that it was raining, and concluded that I probably wouldn't see many ants.

2. About a week prior to spring break, our telephone rang and my wife said, "I'll bet that's Karen calling. She probably wants to get together over the break."

3. You notice that a new technology stock, TechniCom, has doubled in price for each of the last three weeks. You decide to invest in TechniCom because you want to double your money.

What do these situations about the ants, our friend Karen, and your TechniCom stock all have in common? In each case, we were using inductive reasoning. We were looking at recurring patterns and drawing general conclusions. Inductive reasoning is one type of reasoning that we use in mathematics, science, and everyday life.

Inductive Reasoning

> **DEFINITION** **Inductive reasoning** is the process of drawing a general conclusion by observing a pattern in specific instances. This conclusion is called a **hypothesis** or **conjecture.**

Mathematicians and scientists often make conjectures based on their observations. Mathematicians try to prove that conjectures are true by using the laws of mathematics. Example 1 explains a famous conjecture that was made hundreds of years ago, but mathematicians have not yet been able to prove it.

EXAMPLE 1 *Goldbach's Conjecture*

In 1742, Christian Goldbach made the conjecture that we could write every even integer greater than 2 as the sum of two prime numbers.* A prime number is a number such as 3, 5, and 7 that can be divided evenly only by itself and 1. Illustrate Goldbach's conjecture, by expressing the even integers 20, 48, and 100 as the sum of two prime numbers.

SOLUTION: Write $20 = 13 + 7$, $48 = 11 + 37$, and $100 = 41 + 59$.

Now try Exercises 31 to 34. ❋ **14**

Although no one has yet been able to prove Goldbach's conjecture, many believe it to be true based on inductive reasoning.

In Example 2, we ask you to make a conjecture that mathematicians have been able to prove is true.

---

*The primes do not have to be different.

KEY POINT

In inductive reasoning, we begin with specific examples.

Quiz Yourself **14**

Verify Goldbach's conjecture for a) 38 and b) 46.

HISTORICAL HIGHLIGHT ✸ ✸ ✸

IQ Tests and Inductive Reasoning*

Has it ever occurred to you how much of your future is dependent on inductive reasoning? Many intelligence tests, such as college entrance exams and placement tests in the armed services, contain many questions that involve pattern recognition and inductive reasoning. Why should that one ability play such a large role in determining your future? Maybe it shouldn't!

One way of measuring intelligence is the IQ, or intelligence quotient, which is credited to German psychologist Wilhelm Stern. In 1912, Stern defined a child's IQ to be his mental age divided by his chronological age times 100. For example, a 10-year-old child with a mental age of 11, would have an IQ of $11/10 = 1.10 = 110\%$, or 110.

In more recent years, as IQ tests have begun to fall out of favor, many have accepted Howard Gardner's alternative notion of multiple intelligences such as logical-mathematical, musical, bodily kinesthetic, and interpersonal. Others say that we should not test at all and only measure achievement itself.

EXAMPLE 2   *A Divisibility Test for 9*

Consider the numbers a) 72, b) 963, c) 10,854, and d) 7,236,261, which are each evenly divisible by 9. (Verify this.) Add the digits in each number. Do you see any pattern? Make a conjecture.

SOLUTION:

a) $7 + 2 = 9$ b) $9 + 6 + 3 = 18$

c) $1 + 0 + 8 + 5 + 4 = 18$ d) $7 + 2 + 3 + 6 + 2 + 6 + 1 = 27$

Notice that in each case we get a sum that is evenly divisible by 9. Our conjecture is that in order for 9 to divide evenly into a number, 9 must divide the sum of the digits of the number. ✳

 PROBLEM SOLVING

The Always Principle

Keep in mind that in doing inductive reasoning, you are only making an educated guess. You cannot be sure that your conclusion is true. Recall our discussion of the Always Principle and the Counterexample Principle in Section 1.1.

In Chapter 3, we will discuss an important problem, called the Traveling Salesperson Problem[†] that has many real-world applications, such as routing calls through a telephone network, scheduling airline flights, and sending e-mails and text messages through a network of computers. This problem, simply stated, is: What is the best way to have a salesperson travel through a collection of cities with minimal cost? We will not get into the complete solution here as we do in Chapter 3, but in the next example, we will use inductive reasoning to determine how many different routes are possible.

EXAMPLE 3   *Determining the Number of Routes for a Salesperson*

Suppose that Sharifa is going to visit branch offices of her medical supply company based in Atlanta with branch offices in Boston, New York, Cincinnati, Miami, Detroit, Portland, Los Angeles, Houston, and Kansas City. How many different ways can she begin in Atlanta, visit all branches, and return home?

---

*This highlight is based on the essay *The History of IQ Testing* that can be found at www.ivillagehealth.com.
[†]This problem is known historically as the Traveling Salesman Problem.

SOLUTION: We will look at smaller examples until we see a pattern. (In this discussion, we will represent cities by the first letter of their names.) Suppose there was only the one office in Atlanta. Then Sharifa would make no visits. If there were one other city—say, Boston—then she would visit Boston and return home. We will represent her one possible trip by a tree diagram as in Figure 1.7(a). The diagram shows that there is only one possible trip: she would travel from A to B. Now suppose that there are two branch offices, Boston and New York. The diagram would now look like Figure 1.7(b). Here we see that there are two possible trips: ABN and ANB, and then return home.

With three branch offices—Boston, New York, and Cincinnati—the diagram would look like Figure 1.7(c).

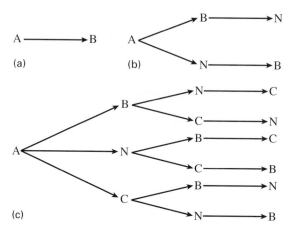

FIGURE 1.7 (a) One branch office, Boston. (b) Two branch offices, Boston and New York. (c) Three branch offices—Boston, New York, and Cincinnati.

We begin to see a pattern that we explain in Table 1.7. We will indicate a route by listing the first letter of the cities in the order that they are visited.

Number of Branch Offices	Branches	Routes	Number of Routes	
0	None	None	0	
1	B	AB	1	
2	B, N	ABN ANB	$2 = 2 \times 1$	⟵ Twice as many routes. She can first go to B or N and then has only one choice for the last city.
3	B, N, C	ABNC ABCN ANBC ANCB ACBN ACNB	$6 = 3 \times 2 \times 1$	⟵ Three times as many routes. Starting at A, she can first go to either B or N or C, then she can visit the remaining two cities in two ways.
4	B, N, C, M	ABNCM ABNMC ABMNC ABMCN ANBCM ANBMC Etc.	$24 = 4 \times 3 \times 2 \times 1$	⟵ Four times as many routes. Now she can start at one of four cities, and then visit the remaining three cities in six ways.

TABLE 1.7 Systematically listing routes as more cities are added.

Math in Your Life

Is Everybody in Iceland Tall?

Although inductive reasoning is important, misusing it can be dangerous. It is easy to jump to a wrong conclusion by making a judgment based on a few isolated examples. For example, suppose that in your psychology class you were to meet an exchange student from Iceland named Fridrik Stefansson, who is 6 feet 9 inches tall. And then, in your history class you meet Fannar Olafsson and Kristinn Jonasson—both from Iceland. Fannar is 6 feet 8 inches tall and Kristinn is 6 feet 9. From these experiences, can you conclude that everyone in Iceland is tall? No—because the students that I have just mentioned are members of the 2007 Iceland National Basketball Team and are certainly not typical Icelanders.

Assuming general characteristics of a certain group of people, based on a small sample, leads to stereotyping. As to how this incorrect form of inductive reasoning relates to you, you might find it interesting to read about how stereotyping can determine your future in the corporate world.*

Quiz Yourself ⑮

In Example 3, conjecture how many routes there will be if we have 12 branch offices.

If there were five cities, Sharifa could first visit one of the five cities and then travel to the other four cities in $4 \times 3 \times 2 \times 1$ different ways. Continuing this pattern for nine cities, the number of possible routes would be an amazing $9 \times 8 \times 7 \times 6 \times 5 \times 4 \times 3 \times 2 \times 1 = 362,880$ ways. ✳ ⑮

Incorrect Inductive Reasoning

Sometimes inductive reasoning can mislead us into thinking that something is true that is not, as we see in Example 4.

EXAMPLE 4   *False Inductive Reasoning*

We want to divide a circle into regions by selecting points on its circumference and drawing line segments from each point to each other point. Figure 1.8 shows the greatest number of regions that we get if we have one point (no line segment is possible for this case), two, three, and four points.

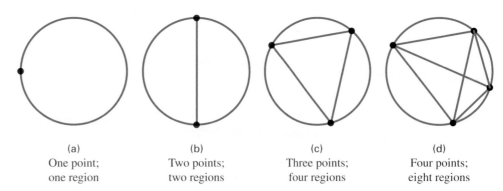

(a)	(b)	(c)	(d)
One point; one region	Two points; two regions	Three points; four regions	Four points; eight regions

FIGURE 1.8 Dividing a circle into regions.

Use inductive reasoning to find the greatest number of regions we would get if we had six points on the edge of the circle.

SOLUTION: This problem seems somewhat similar to what we did in Example 3. Notice that it appears that each time we add another point, we double the number of regions. It is

---

*See James M. Henslin *Sociology* (Boston: Allyn & Bacon, 2007), pp. 186–187.

natural to conjecture that if we have five points, there would be 16 regions, and with six points, we would get 32 regions. However, this is not true. You can try it for yourself by drawing a large circle and picking six points in different ways on the circle. The largest number of regions that you will find is 31, not 32. ✽

 *KEY POINT*

Deductive reasoning begins with accepted facts and principles.

Deductive Reasoning

In inductive reasoning, we draw a general conclusion by considering a number of specific examples. In a sense, deductive reasoning is a reverse way of reasoning from inductive reasoning.

> **DEFINITION** In **deductive reasoning,** we use accepted facts and general principles to arrive at a specific conclusion.

In mathematics, we often use general mathematical principles to prove some specific fact. In Example 5, we will show you how to explain what at first seems to be a mystifying number trick.

EXAMPLE 5   *Explaining a Number Trick by Using Deductive Reasoning*

Consider the following number trick:*

1. How many days a week do you eat out?
2. Multiply this number by 2.
3. Add 5 to the number you got in step 2.
4. Multiply the number you obtained in step 3 by 50.
5. If you have already had your birthday this year, add 1,760; if you haven't, add 1,759.
6. Subtract the four-digit year that you were born.
7. You should now have a three-digit number. The first digit is the number of times that you eat out each week. (There is more to this trick that we will investigate in Exercises 67 and 68.)

Use deductive reasoning to explain why this trick works.

SOLUTION: We will use general algebra principles and deductive reasoning to explain this trick. Let's begin by calling the number of days that you eat out n and go through the trick step-by-step.

Step 2: Multiplying the number by 2 will give us $2n$.

Step 3: Adding 5, we get $2n + 5$.

Step 4: Multiplying by 50 gives us the expression $(2n + 5)50$.

Step 5: Let's assume that you haven't had your birthday yet, so we will add 1,759 to get $(2n + 5)50 + 1,759$.

---

*Thanks to Dr. Robert Voytas of the Kutztown University Psychology Department for asking me to explain why this trick works.

Step 6: Let's assume that you were born in 1991, so we will subtract 1,991 to get $(2n + 5)50 + 1{,}759 - 1{,}991$.

When we simplify the expression in step 6, we get

$$(2n + 5)50 + 1{,}759 - 1{,}991 = 100n + 250 + 1{,}759 - 1{,}991$$
$$= 100n + 2{,}009 - 1{,}991$$
$$= 100n + 18.$$

Notice that when we look at this expression in this form, the hundred's digit is n, which is the number of days that you eat out each week. Notice that the difference $2{,}009 - 1{,}991$ has no effect on the hundred's digit. We will have more to say about this difference in the Further Exercises. ✳

Exercises 1.2

Looking Back*

These exercises follow the general outline of the topics presented in this section and will give you a good overview of the material that you have just studied.

1. Find an example of inductive reasoning in this section that leads to a correct conclusion.

2. Find an example of inductive reasoning in this section that leads to an incorrect conclusion.

3. Why is Example 5 an example of deductive reasoning rather than inductive reasoning?

4. In solving a problem, how would you arrange these steps in the order in which you would use them? a) deductive reasoning, b) inductive reasoning, c) making a conjecture

5. Do you agree with the emphasis put on inductive reasoning in intelligence tests that we mentioned in the Historical Highlight on page 19?

6. How is stereotyping an example of inductive reasoning?

Sharpening Your Skills

Is each of the following situations, an example of inductive or deductive reasoning?

7. It has rained the past three weekends, canceling your softball game. You expect that next Saturday it will rain again.

8. Carla is calculating her income taxes to determine if she will get a refund this year.

9. As you read *Gone Baby Gone*, you are keeping track of the clues that the author has given to predict who the killer will be.

10. You tell your friend Jay to be ready 15 minutes before you actually intend to pick him up because Jay is always late for his appointments.

11. Luis has noticed that the stock market has gone up on the Friday before each of the last three holidays. He plans to buy stock on the Friday before Labor Day to cash in on this trend.

12. Marianne is using the rules of algebra to solve a word problem on a quiz.

13. Brett is calculating his expenses for next year to determine how large his student loan should be.

14. Latisha noticed that on every true–false quiz so far this semester, her instructor has given twice as many false questions as true questions. On the next quiz, if she is not sure of an answer, she will guess "false."

15. The American Conference team has won the Super Bowl the past four times. You expect that the American Conference team will win again this year.

16. Emily estimates that if she can average 50 miles per hour, she will reach San Diego in $5\frac{1}{2}$ hours.

---

*Before doing these exercises, you may find it useful to review the note *How to Succeed at Mathematics* on page xix.

In Exercises 17–24, use inductive reasoning to predict the next term in the sequence of numbers.

17. 1, 4, 7, 10, 13, ?

18. 2, 8, 14, 20, 26, ?

19. 3, 6, 12, 24, 48, ?

20. 5, 15, 45, 135, 405, ?

21. $\dfrac{1}{2}, \dfrac{1}{4}, \dfrac{1}{8}, \dfrac{1}{16}, \dfrac{1}{32}$, ?

22. $\dfrac{1}{2}, \dfrac{2}{3}, \dfrac{3}{4}, \dfrac{4}{5}, \dfrac{5}{6}$, ?

23. 1, 1, 2, 3, 5, 8, 13, ?

24. 0.1, 0.10, 0.101, 0.1010, 0.10101, ?

In Exercises 25–28, use inductive reasoning to draw the next figure in the pattern. There may be several correct answers.

25.

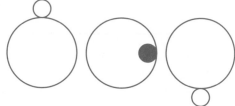

26.

27.

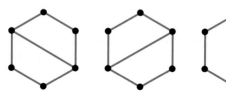

28.

In Exercises 29 and 30, place the next X in a box to continue the pattern.

29.

| | X | X | | X | | X | | | | X | | | | | |

30.

| X | X | X | | X | | | X | | | | | X | | |

Illustrate Goldbach's conjecture for each of the following numbers.

31. 16

32. 18

33. 20

34. 26

Applying What You've Learned

35. In preparation for his induction into the baseball hall of fame, Cal Ripken, Jr. autographed a stack of baseballs to give away to his fans. How many baseballs are in the stack?

36. If a stack of baseballs similar to those shown in Exercise 35 was six layers high, how many baseballs would be in the stack?

A magic square is a square arrangement of numbers such that if you add the numbers in any row, column, or diagonal, you always get the same sum. In the magic square shown below, called a 3-by-3 square, the sum of any row, column, or diagonal is always 15.

8	1	6
3	5	7
4	9	2

In Exercises 37 and 38, the magic squares use each of the integers from 1 to 16 exactly once. Use deductive reasoning to

a) *Determine the total of all the numbers in the square.*

b) *Determine the total for each row, column, and diagonal.*

c) *Complete the magic squares.*

Explain your reasoning.

37.

7			9
			8
13		2	
4	1		14

38.

16	3		13
5			8
	6		12
4		14	

In Exercises 39–42, follow the instructions for each "trick" starting with several different numbers of your choice. Make a conjecture as to what result you get in each case. Use algebra and deductive reasoning, as we did in Example 5, to explain why your conjecture is correct.

39. a. Choose any natural number.

    b. Multiply the number by 3.

    c. Add 9 to the product you just found.

    d. Divide the results of part (c) by 3.

    e. Subtract the number that you started with.

40. a. Choose any natural number.

    b. Multiply the number by 5.

    c. Add 20 to the product you just found.

    d. Divide the results of part (c) by 5.

    e. Subtract 4.

41. a. Choose any natural number.

    b. Multiply the number by 8.

    c. Add 12 to the product you just found.

    d. Divide the results of part (c) by 4.

    e. Subtract 3.

42. a. Choose any natural number.

    b. Multiply the number by 15.

    c. Add 20 to the product you just found.

    d. Divide the results of part (c) by 5.

    e. Subtract 4.

In Exercises 43 and 44, draw the next figure in the sequence.

43.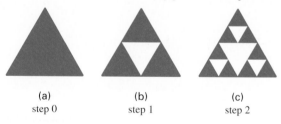

    (a)   (b)   (c)

   step 0   step 1   step 2

44.

Communicating Mathematics

45. Explain the difference between inductive reasoning and deductive reasoning.

46. In what sense are inductive reasoning and deductive reasoning "opposite" processes?

47. In Example 3, explain why the number of routes for Sharifa increased as it did. For example, if you found that the number of routes for five cities is 120, how would the total number of routes increase for six cities?

48. Show the conjecture that we made in Example 4 is true for five points. Draw a figure that has 31 regions if we are using six points. How was Example 4 an example of false inductive reasoning?

Exercises 49 to 52 are actual questions taken from an IQ test. Use inductive reasoning to solve them. Explain your thinking as to how you found a solution.

49. GGAGLLGA is to 46336466 as LLGAAGGL is to what number?

50. Assume that NASA scientists have analyzed alien signals from outer space and have deduced the following:

    "Foofrug Merduc Lilit" means: Where is your leader?

    "Niurus Tuume Gazist" Means: Our planet is distant.

    "Foofrug Merduc Gazist" means: Where is your planet?

    What is the best translation for the word "Lilit?"

51. Committee is to Etimoc as 367768899 is to what number?

52. What is the next number in the following series: 8 – 14 – 12 – 18 – 16 – 22?

Using Technology to Investigate Mathematics

53. Perform an Internet search for intelligence tests and identify questions that require inductive reasoning.

54. Search the Web for the history of intelligence testing.

In each of the next four exercises, we give you a series of numerical equations. Make a conjecture as to what the next two equations in the pattern are. Check your conjecture with a calculator.

55. a. $1 + 2 = \dfrac{2 \times 3}{2}$

    b. $1 + 2 + 3 = \dfrac{3 \times 4}{2}$

    c. $1 + 2 + 3 + 4 = \dfrac{4 \times 5}{2}$

56. a. $2 + 4 = 2 \times 3$

    b. $2 + 4 + 6 = 3 \times 4$

    c. $2 + 4 + 6 + 8 = 4 \times 5$

57. a. $1 + 3 = 4$

    b. $1 + 3 + 5 = 9$

    c. $1 + 3 + 5 + 7 = 16$

58. a. $\dfrac{1}{1 \times 2} + \dfrac{1}{2 \times 3} = \dfrac{2}{3}$

    b. $\dfrac{1}{1 \times 2} + \dfrac{1}{2 \times 3} + \dfrac{1}{3 \times 4} = \dfrac{3}{4}$

    c. $\dfrac{1}{1 \times 2} + \dfrac{1}{2 \times 3} + \dfrac{1}{3 \times 4} + \dfrac{1}{4 \times 5} = \dfrac{4}{5}$

For Extra Credit

59. We will call the rectangle below, which has three rows and four columns, a 3-by-4 rectangle. How many squares of various sizes can you find in this rectangle?

60. Repeat Exercise 59 for a 6-by-4 rectangle.

61. a) Repeat Exercise 59, but now count rectangles of all types, including squares. b) Repeat Exercise 60, now counting rectangles of all types, including squares. It is important when doing this that you are systematic. Count rectangles of sizes 1 by 1, 1 by 2, 1 by 3, 1 by 4, 2 by 1, 2 by 2, and so on.

62. Can you find some general pattern that would enable you to count rectangles of all various sizes in a 10-by-6 rectangle without actually drawing the rectangle and counting? Explain your thinking.

63. Stacking baseballs. If a stack of baseballs (similar to the one shown in Exercise 35) has a rectangular base that is six balls long and four balls wide, and is stacked with as many balls as possible, how many balls will be in the stack? Explain your reasoning. We can only place a ball on the stack if it is resting on four other balls.

64. Stacking baseballs. Redo Exercise 63, but now assume the base is seven balls long and five balls wide.

65. Make up a 3-by-3 magic square of your own using the natural numbers from 1 to 9. Explain how you constructed the magic square.

66. Make up a 4-by-4 magic square of your own using the natural numbers from 1 to 16. Explain how you constructed the magic square.

67. Explaining a trick. In Example 5, we mentioned that there is more to the trick than what we explained. The rest of the trick states that the last two digits of the three-digit number you obtain is your age. However, this part of the trick will only work in the year 2010. Explain why this is so.

68. Explaining a trick. How would you adjust the trick in Exercise 67 so that it will work in the year 2012? Explain how you came up with your adjustment.

1.3 Estimation

Objectives

1. Use rounding to make estimates.
2. Approximate calculations by using compatible numbers.
3. Make estimates based on information presented in graphs, maps, and photographs.

"That can't be right!" Do you ever find yourself saying that to a clerk in a convenience store? On occasion, I do. As I collect my purchases, I often do a rough mental estimate of what I expect to pay when the bill is presented to me. For another example, if you want to leave a tip of roughly 15% for a $37.86 bill in a restaurant, what should you leave? Chances are that you do not have a calculator in your pocket, so I guess that you have to wing it. What's a quick way to decide on the tip?

Unfortunately, in spite of the many forms of technology that we have available to do our calculations for us, chances are there will be times in real life when we have to make a quick, rough estimate, which is the topic of this section. Estimation is also an important part of effective problem solving in mathematics. To check your work, often you will find it useful to be able to make a reliable estimate to decide if your answer is reasonable.

✎ *KEY POINT*

Rounding is an effective way to make an estimate.

Rounding to Make Estimates

We will now examine how to use rounding to do estimation. In rounding, we consider the digit to the right of the digit being rounded.

If that digit to the right is 5 or more, we round up; otherwise, we round down.

For example, in Figure 1.9 we see that if we want to round 12,435 to the nearest hundred, we look at the digit in the tens place, which is 3. Because 3 is less than 5, we round

We want to round to the hundreds place. ———┐

12,435

└——— 3 is less than 5, so round down.

We want to round to the thousands place. ———┐

7,864

└——— 8 is greater than 5, so round up.

FIGURE 1.9 Rounding to hundreds and thousands.

Quiz Yourself **16**

Round each of the following numbers to the nearest thousand:

a) 635,421   b) 723,562

12,435 down to 12,400. On the other hand, in rounding 7,864 to the nearest thousand, because the hundreds digit is 8, we round 7,864 up to get 8,000. **16**

Example 1 illustrates a familiar situation in which rounding is useful.

EXAMPLE 1   *Estimating a Grocery Bill*

On your way home from work, you stopped at the market to pick up the items listed in Table 1.8. You also would like to buy a half gallon of ice cream for $3.59, but you remember that you have only $20 in your wallet and you don't want to be caught in the checkout line without enough money. Use rounding to the nearest 10 cents to

a) Estimate the total cost of your purchases.

b) Decide if it is safe to put the ice cream in your cart.

Item	Cost ($)	Cost Rounded to Nearest 10 Cents
Cereal	4.29	4.30
Milk	2.41	2.40
Bread	1.89	1.90
Lunch meat	3.36	3.40
Pickles	2.37	2.40
Dishwashing liquid	2.87	2.90
	Total: $17.19	Total: $17.30

TABLE 1.8 Estimating the cost of grocery items.

SOLUTION:

a) You can add the rounded prices mentally to get $17.30.

b) You decide that the half gallon is probably too expensive and decide to buy a quart of ice cream instead. ❋

Notice in Example 1 that if you had rounded to the nearest dollar, your estimate would have been $16, and you would have taken the half gallon, which would have put you over $20.

 *Some Good Advice*

In Example 1, you do not need a calculator to add the estimated prices. For example, to add 4.30 and 2.40, first add the dollars to get $2 + 4 = 6$. Then, add the cents to get $30 + 40 = 70$ cents, so the total is 6.70. You can easily add in the other numbers one at a time to get the total. Doing this kind of mental arithmetic will make you a stronger mathematics student.

KEY POINT

Using compatible numbers simplifies approximate calculations.

KEY POINT

Estimation can help interpret graphical data.

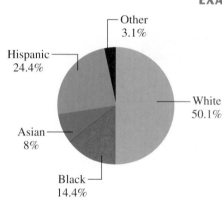

—Other
3.1%

Hispanic
24.4%

—White
50.1%

Asian—
8%

Black—
14.4%

FIGURE 1.10 The U.S. population in the year 2050.

Quiz Yourself **17**

Use compatible numbers and Figure 1.10 to estimate the number of blacks there will be in the United States in 2050.

Using Compatible Numbers

Another way to make a quick estimate that is slightly different than rounding is to use compatible numbers. In using compatible numbers, instead of using the given numbers in a problem, we substitute other numbers that are easier to work with. For example, instead of dividing 298 by 14, we might divide 300 by 15 to get 20. Or, instead of multiplying 11 times 73, we multiply 10 times 73 to get 730.

EXAMPLE 2   *Using Compatible Numbers to Estimate a Bill*

Marcia is buying 19 Blu-ray disks at $12.89 each. Use compatible numbers to estimate the cost of the Blu-ray disks.

SOLUTION: We can replace $12.89 by $13 and 19 by 20 to get $20 \times 13 = 260$. Because we have replaced the $12.89 and 19 by two larger numbers, we know that Marcia's actual bill will be somewhat less than $260.

Now try Exercises 23 to 36. ❋

Estimating Graphical Data

Often, you can use estimation to summarize information that is given to you graphically.

EXAMPLE 3   *Using Compatible Numbers to Estimate a Population*

According to the U.S. Bureau of the Census, by the year 2050, the population of the United States will be 419,854,000. Use the graph in Figure 1.10 to estimate the number of Hispanics who will be living in the United States in 2050.

SOLUTION: We can replace the percentage* of Hispanics in 2050, which is estimated to be 24.4%, by the simpler number 25%. Recall that 25% is equal to the fraction $\frac{1}{4}$. Also, we can replace the total U.S. population of 419,854,000 by 400,000,000. Then, multiplying we get

$$\text{Number of Hispanics} = \frac{1}{4} \times 400,000,000$$
$$= 100,000,000.$$

Now try Exercises 49 to 56. ❋ **17**

EXAMPLE 4   *Who's Doing the Housework?*

A sociology researcher,[†] studying the division of household labor between married couples who are both working, found the data given in the graph in Figure 1.11.
Use this information to estimate answers to the following questions.

a) How many more hours per week do women spend on cooking and cleaning up than men do?

b) How many more hours per week do men spend on outdoor chores, repairs, and paying bills than women do?

---

*If you need to review how to work with percents, see Appendix A.
[†]See James M. Henslin, *Sociology* (Boston: Allyn & Bacon, 2007), p. 454, for a reference to the actual research article.

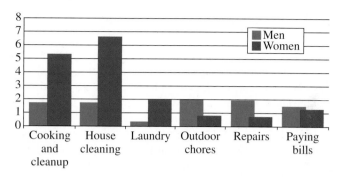

FIGURE 1.11 Household labor done by married couples.

c) How many more total hours per week do women spend on household chores than men do?

d) If this weekly discrepancy is constant throughout one year, how many more 8-hour days of work per year is the typical wife in this study putting in at housework than her husband does?

SOLUTION: We will estimate the amounts shown in the graph and summarize our estimates in Table 1.9.

	Men	Women
Cooking and cleanup	1.8	5.2
House cleaning	1.8	6.8
Laundry	0.2	2.0
Outdoor chores	2.0	0.8
Repairs	2.0	0.6
Paying bills	1.5	1.3
Total	9.3	16.7

—— Women spend 3.4 more hours here.

—— Men spend 2.8 more hours here.

TABLE 1.9 Estimates of household labor done by married couples.

a) It appears that women spend $5.2 - 1.8 = 3.4$ hours more per week on cooking and cleanup.

b) From our estimates, the men spend

$$(2.0 + 2.0 + 1.5) - (0.8 + 0.6 + 1.3) = 5.5 - 2.7 = 2.8$$

more hours per week on the three chores mentioned.

c) For total chores, from the graph, it appears that women spend $16.7 - 9.3 = 7.4$ hours more per week on these chores than men do.

d) If we multiply 7.4 by 52 weeks in a year, we get $7.4 \times 52 \approx 385$ more hours per year. Dividing this by 8 hours in a typical work day, we get $\frac{385}{8} = 48.1$ more days of work for women per year.

Now try Exercises 37 to 40 and 45 to 48. ❋

In Examples 5 and 6, we show you some other ways to make estimates.

✺ ✺ ✺ HISTORICAL HIGHLIGHT

An Amazing Estimate*

The ancient Greek mathematician Eratosthenes (276–194 BC) devised a clever way to estimate Earth's circumference. He believed that the cities Alexandria and Syene (today called Aswan) in Egypt were on the same great circle of Earth. He also knew the distance between the cities because it had been measured by a surveyor called a *bematistes*, who was trained to walk in equal steps. The bematistes, using a unit of measurement called a *stadium*, which was 516.73 feet, found the cities to be 5,000 stadia apart.

Using simple but clever geometry calculations, which we explain in Example 3 of Section 10.1, Eratosthenes determined the distance from Alexandria to Syene to be equal to $\frac{1}{50}$ of Earth's circumference. He then used this information to estimate Earth's circumference to be 24,662 miles, which is just 245 miles less than its true value.

FIGURE 1.12 Map of Alaska.

EXAMPLE 5   *Estimating Distances from a Map*

In the Great Alaska Challenge, boats will race from the starting point at Prudhoe Bay (A), to the finish line at Kodiak (F), (see Figure 1.12).

a) Estimate the length of the race course. Assume that at any point in the race, boats will take the shortest path.

b) If the boat *The Inyuk* travels at 35 miles per hour, how long will it take to complete the course?

SOLUTION:

a) We have marked points A, B, C, and so on, on the map to approximate the total length of the race. If we measure the total length of these distances on the map, we get $2\frac{1}{4}$ inches. According to the scale† of the map, 1 inch equals 286 miles; thus the total length of the race course is approximately 644 miles.

b) If *The Inyuk* travels at 35 miles per hour, then the race will take $\frac{644}{35} = 18.4$ hours, or about 18 hours and 24 minutes. ✺

⊚ *Some Good Advice*

When making an estimate, it is a good idea to "estimate your estimate." By that, we mean you should have a rough idea of whether your estimate is too large or too small.

Usually after a large political rally takes place, the news media report an estimate of the size of the crowd. Of course, if the crowd is in the hundreds of thousands, no one has actually counted the people. Example 6 explains a technique that can be used for making this type of estimate. However, the highlight that follows cautions us to be careful in accepting the reported numbers.

EXAMPLE 6   *Estimating the Size of a Large Number of Objects*

Estimate the number of M&Ms that are visible in Figure 1.13.

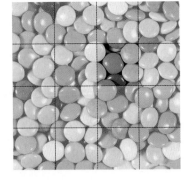

FIGURE 1.13 Estimating number of M&Ms.

---

*This highlight is based on material from David M. Burton, *The History of Mathematics: An Introduction* (New York: McGraw-Hill, 1999), pp. 179–180.
†The scale shows that $\frac{7}{8}$ inch = 250 miles, so 1 inch = $\frac{8}{7} \times 250 \approx 286$ miles.

SOLUTION: In Figure 1.13, we have divided the photograph into 16 equally sized rectangles. It appears that each rectangle has approximately the same number of candies. We see that the rectangle we have highlighted contains 11 candies (we will count any candy that is partially visible). Therefore, it is reasonable to estimate that there are $11 \times 16 = 176$ candies in the photograph. ✳

HIGHLIGHT ✿ ✿ ✿

Between the Numbers—Whom Can You Believe?

War Protestors Estimated to Be Somewhere Between 400,000 and Two Million People

If you were to come across this headline on an article on your favorite news Web site, you might think that the people making the estimate really didn't know the size of the crowd. Of course, a news source probably would not give you both estimates. The organizers of the rally would report the crowd to be very large, while those who oppose the rally may want it to seem small. Sometimes, however, a poor crowd estimate is not due to politics, but rather to a lack of proper methods and perhaps using the wrong equipment.

Such was the case in the fall of 1995, when the Million Man March, organized by Afro-American political leader

and activist Louis Farrakhan, was reported by U.S. Park Police to be 400,000. Dismayed by the low estimate of crowd size, Farrakhan, who thought the number to be between 1.5 and 2 million, threatened to sue the National Park Service.

The law suit was avoided however, when three days after the march, Dr. Farouk El-Baz, director of The Center for Remote Sensing at Boston University, used enlarged aerial photographs to obtain a more accurate estimate of 870,000, which was acceptable to both Farrakhan and the Park Service. Because of the controversy generated by this bad estimate, the Park Police have since been barred by Congress from issuing crowd estimates.

Exercises 1.3

Looking Back*

These exercises follow the general outline of the topics presented in this section and will give you a good overview of the material that you have just studied.

1. Give an example of how you round up a number to the hundreds place.

2. Give an example of how you round down a number to the thousandths place.

3. What do we mean when we say that we are using compatible numbers to make an estimate? Give an example of this.

4. List three ways that estimation can be useful in your life.

5. Why did we mention Eratosthenes in this section?

6. What is the point that we were making in the "Between the Numbers" Highlight?

Sharpening Your Skills

Round each of the following numbers to the nearest hundred.

7. 37,264

8. 6,835

9. 8,327

10. 14,872

Estimate the answers to the following problems. Use either rounding or compatible numbers. In some cases, your answers may differ from ours depending on what method you use.

11. $21.6 + 38.93 + 191 + 42.5$

12. $17.18 + 27.2 + 10.31 + 87.6$

13. $34.6 - 15.3$

14. $107.28 - 68.49$

15. 4.75×16.3

16. $1,028 \times 14.8$

17. $17.4/3.31$

18. $2,068/72.4$

19. 0.091×785

20. $0.0008 \times 4,026.3$

21. $8.7\% \times 1,024$

22. $18\% \times 683$

Applying What You've Learned

Estimate each of the following answers. Explain how you made your estimate and, where possible, state whether your estimate is larger or smaller than the exact answer. We will provide one possible answer, and yours may differ from ours depending on how you make your estimate.

---

*Before doing these exercises, you may find it useful to review the note *How to Succeed at Mathematics* on page xix.

23. **Training for cross country.** Mike is training for the cross country team. He will run 3.7 miles per day, five days a week. How many miles will he run during the next six weeks?

24. **Purchasing supplies.** Julia is purchasing 18 packs of pencils at $0.94 per pack. What is her bill?

25. **Travel time.** Olive's father is averaging 47.5 miles per hour on the Hoover's trip to Redondo Beach, CA, for a beauty pageant. If he still has 325 miles to travel and it is now 1:00 PM, about what time should they arrive?

26. **The price of gasoline.** Ramon spent $45.85 last month for 13.5 gallons of gasoline. What was the price per gallon?

27. **Calculating a tip.** The Sopranos went out to dinner and their bill was $82.45. If they want to leave a 15% tip, how much should they leave?

28. **Sharing apartment expenses.** Ted, Lily, and Marshall share an apartment. Their total utilities bill for last month was $76.38. What is each person's share of the bill?

29. **Buying plants.** Emily is buying plants for spring planting. She bought three packs of petunias for $2.95 per pack, four packs of vincas for $1.39 per flat, and a package of potting soil for $2.79. What is her total bill?

30. **Buying a computer.** Shandra bought a new computer for $1,389. If the sales tax is 6%, what is her total bill?

31. **Capacity of an elevator.** An elevator has a capacity of 2,300 pounds. Alicia and her fifth-grade class of 21 students want to crowd into the elevator. Do you think it is safe? Explain.

32. **Capacity of an elevator.** Would it be safe for eight linemen for the NFL Washington Redskins to crowd into the elevator mentioned in Exercise 31? Explain.

33. **Calculating taxes.** Dwight's taxable income is $37,840. If the state income tax rate is 2.4% and the county wage tax is 1.1%, what is the total that he will pay for those two taxes?

34. **Estimating a tire warranty.** Chuck bought new tires for the Nerd Herd company car that are guaranteed for 42,000 miles. If he drives 11.5 miles each way to work five days a week and then an additional 14 miles each weekend, how many weeks will it be before his warranty expires?

35. **Calculating tax deductions.** Mary Rose works at home as a financial consultant. She wants to claim some of her yearly expenses as tax deductions. Her average monthly basic phone bill is $13.75, electricity is $68.45 per month, and water and sewer is $12.80 per month. If she can claim 1/7 of these expenses as deductions, how much will her annual deduction be?

36. **Calculating tax deductions.** Ben bought a new car for $19,880 and paid a 2.5% sales tax. He can deduct this amount on his federal income tax return. If he is in the 21% bracket, how much will this deduction save him in taxes?

During the past 30 years, women have made great progress in earning professional degrees in areas such as medicine, dentistry, law, and theology. The following graph shows the increase in women completing degrees in medicine, dentistry, and law between 1980 and 2004. Use this graph to solve Exercises 37–40.

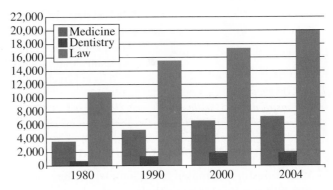

37. In which of the three areas have women made the largest percentage growth? (*Hint:* In one of the areas, there are two and one-half as many women graduating in 2004 as there were in 1980.)

38. In which of the three areas have women made the smallest percentage growth?

39. Did women make a larger gain in the number of law degrees between 1980 and 1990 or between 1990 and 2000?

40. Approximately how many women earned degrees in dentistry in 1990?

In December of 2006, the Harris Organization polled 2,309 adult viewers of YouTube to determine whether they would change their viewing habits if short commercials were included before every clip. Use the following pie chart, which summarizes the results of the poll, to solve Exercises 41–44.

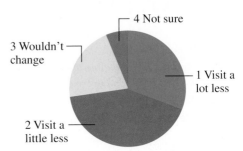

41. Which of the four categories contains the most viewers?

42. Which is the best estimate of the number of viewers that were polled that gave response 3: 500, 300, 1,000, or 1,200?

43. Was response 1 or 2 given more frequently?

44. Estimate how many viewers gave response 4.

The following graph shows the change in the sales of electronic devices from 2004 to 2007 (The New York Times Almanac, 2008). Use this graph to estimate the answers for Exercises 45–48. We will give the exact answers in the back of the text.

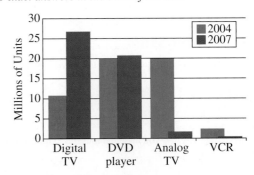

45. How many DVD players were sold in 2007?

46. How many VCRs were sold in 2007?

47. How many more digital TVs were sold in 2007 than in 2004?

48. What device showed the biggest drop in sales?

The following pie chart shows revenues of the federal government in billions of dollars for the fiscal year 2007. Use this graph to estimate answers to Exercises 49–52. The total revenue is $2,407 billion.

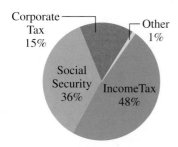

49. How much revenue is derived from Social Security taxes?

50. How much revenue is derived from income taxes?

51. How much revenue is derived from corporate taxes?

52. How much revenue is derived from Social Security taxes and income taxes combined?

The following pie chart shows a distribution of immigration into the United States in a recent year. Use this graph to estimate the answers to Exercises 53–56. The total number of immigrants was 705,361.

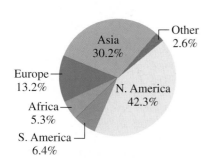

53. How many people immigrated to the United States from Asia?

54. How many people immigrated to the United States from Europe?

55. If 130,661 people immigrated from Mexico, what percentage was that of the total number of immigrants?

56. If 41,034 people immigrated from China, what percentage was that of the total number of immigrants?

Communicating Mathematics

57. What do we mean by "estimating your estimate?"

58. Is rounding the same as using compatible numbers?

59. Argue for or against this statement: "With all of the technology that we have available for doing computations, it is really not important to be able to make good estimates."

60. What dangers can result in doing estimation? Give specific examples of the danger of overestimation and underestimation.

61. What is the relationship between estimation and being able to do mental computations?

62. Criticize the estimation that we made of the number of M&Ms in Figure 1.13.

Using Technology to Investigate Mathematics

63. Use the Internet to search for data that is of interest to you. Present it as a graph, as in Examples 3 or 4. Make some estimations, and write a concise report of your findings. (Some good sources of data are the U.S. Bureau of the Census, various online almanacs, Harris and Gallup polling organizations, and *USA Today.*)

64. Do an Internet search on "estimation" to find an interesting article similar to the "Math in Your Life" highlight in this section. Write a brief report of your findings.

65. Use your calculator to check the accuracy of your estimates for Exercises 33 and 35. Using the true value for a base, calculate the percent* of error that you made in your estimates.

66. Repeat the instructions in Exercise 65 for Exercise 42.

For Extra Credit

67. **Buying fertilizer.** The Martinez's yard is 96 feet by 169 feet, and a diagram of their yard is given in Figure 1.14.

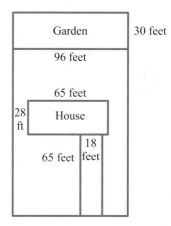

FIGURE 1.14 The Martinez's yard.

The dimensions of the house, driveway, and garden are also given. The rest of the yard is grass. Mr. Martinez wants to fertilize his lawn. A bag of fertilizer covers 5,000 square feet. Estimate how many bags of fertilizer he will need. Explain how you made your estimate.

68. **Purchasing paint.** Heidi and Spencer are painting their living room, which is 18.5 feet long and 11 feet wide. The room is 7.75 feet high. If they want to put two coats of paint on

---

*For a review of percent, see Section 9.1.

the walls and a gallon of paint will cover 200 square feet, how many gallons of paint should they buy? Explain how you arrived at your estimate.

69. **Estimating Earth's circumference.** Use a map of Egypt to estimate the distance between Alexandria and Aswan. Then multiply your estimate by 50. When Eratosthenes made his estimate of Earth's circumference, he also added an additional 2,000 stadia to his estimate. Do this also. You need to know that there are 5,280 feet in a mile to find the extra miles in your estimate. How does your estimate compare with Eratosthenes' estimate?

70. Ask a parent or some other relative or friend what they think estimation has to do with each of the following situations or some other that they may want to discuss. Compare your results, and make a short report based on your findings.

Budgets

Cooking

Planning for a party

Income taxes

Planning for college

Planning a wedding

Repaying student loans

Anticipating bonuses based on company revenue

71. In constructing the graph in Exercise 37, the actual data that we used said there were 3,486 women graduates of medical school in 1980 and 7,169 in 2004.

   a. Use this information to calculate the average increase in women graduates from medical school over the 24-year period from 1980 to 2004.

   b. Assuming the same average rate of increase until 2014, estimate the number of women graduates from medical school in 2014.

72. Repeat Exercise 71, but now estimate the number of women graduating from dental school in 2014, assuming that there were 700 graduates in 1980 and 1803 in 2004.

CHAPTER SUMMARY

SECTION	SUMMARY	EXAMPLE
SECTION 1.1	The four steps in Polya's problem-solving method are: 1. Understand the problem. 2. Devise a plan. 3. Carry out your plan. 4. Check your answer.	Discussion, p. 2
	The following strategies are helpful in solving problems:	
	Draw pictures.	Examples 1 and 2, pp. 3, 4
	Choose good names for unknowns.	Example 3, p. 5
	Be systematic.	Example 4, p. 6
	Look for patterns.	Example 5, p. 6
	Try a simpler version of the problem.	Example 6, p. 7
	Guessing is OK.	Example 7, p. 9
	Convert a new problem to an older one.	Example 8, p. 10
	The following principles are useful in problem solving:	
	The **Always** Principle	Example 9, p. 11
	The **Counterexample** Principle	Example 10, p. 12
	The **Order** Principle	Example 11, p. 12
	The **Splitting-Hairs** Principle	Example 12, p. 13
	The **Analogies** Principle	Example 13, p. 13
	The **Three-Way Principle** tells us that we can understand mathematical ideas verbally, graphically, and by making examples.	Discussion, p. 14
SECTION 1.2	**Inductive reasoning** is the process of drawing a general conclusion by observing a pattern in specific instances.	Examples 1–3, pp. 18, 19
	Inductive reasoning can lead to a false conclusion.	Example 4, p. 21
	In **deductive reasoning**, we use accepted facts and general principles to arrive at a specific conclusion.	Example 5, p. 22
SECTION 1.3	In **rounding**, we consider the digit to the right of the digit being rounded. If that digit to the right is 5 or more, we round up; otherwise, we round down.	Example 1, p. 27
	In using **compatible numbers**, instead of using the numbers given in a problem, we substitute other numbersthat are easier to work with.	Examples 2 and 3, p. 28
	Estimation can help us understand information that is presented to us in maps, graphs, and photographs.	Examples 5 and 6, p. 30

CHAPTER REVIEW EXERCISES

Section 1.1

1. List the four steps in Polya's problem-solving method.

2. What is a counterexample?

3. Dr. House's Fellowship applicants: "Thirteen," Lawrence, Chris, Amber, and Travis are working on diagnosing a patient. In how many different ways can we choose two of these students to present their results? Realize that the order in which we choose the students is not important.

4. At a T.G.I. Friday's, you have 8 appetizers, 20 entrées, and 10 desserts. How many different meals can you choose if you select one appetizer, one entrée, and one dessert? Do not solve this problem, but state a simpler version and solve it instead.

5. Picaboo worked 20 hours last week. Part of the time she worked as a stockperson for $5.65 per hour. The rest of the time she worked as a ski instructor for $8 per hour. If she earned $141.20, how many hours did she work at each job? Solve this problem by making and evaluating guesses until you find the answer.

6. Is the following statement true or false?

$$\frac{a}{b} + \frac{c}{d} = \frac{a+c}{b+d}$$

If it is true, give two examples. If it is false, give a counterexample.

7. Explain the Three-Way Principle.

Section 1.2

8. Explain the difference between inductive and deductive reasoning.

9. Do the following situations illustrate inductive or deductive reasoning?

   a. You are following a list of clues in *The DaVinci Code* as Robert Langdon searches for the Holy Grail.

   b. J.K. Rowling's last four books have sold over 10 million copies. If she writes another book, it will sell over 10 million copies.

10. Use inductive reasoning to predict the next term in the following sequences: a) 2, 7, 12, 17, 22, . . . b) 3, 4, 7, 11, 18, 29, ...

11. Use inductive reasoning to draw the next figure in the pattern.

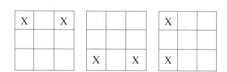

12. Illustrate Goldbach's conjecture for the number 48.

13. Follow the instructions for this "trick" starting with several different numbers of your choice. Make a conjecture as to what result you get in each case. Use algebra and deductive reasoning to explain why your conjecture is correct.

   a. Choose any natural number.

   b. Multiply the number by 8.

   c. Add 12 to the product you just found.

   d. Divide the results of part (c) by 4.

   e. Subtract 3.

Section 1.3

14. Round each of the following numbers to the nearest thousand.

   a. 46,358

   b. 27,541

15. Use compatible numbers to estimate the answers to the following problems. Your answers may differ from ours.

   a. $209.35 - 61.19$

   b. 5.85×15.64

16. Juana is averaging 52.4 miles per hour on her trip to Miami. If she still has 156 miles to travel and it is now 4:00 PM, about what time should she arrive?

17. Use the following graph to estimate a) the average earnings for women with a bachelor's degree and b) the difference in earnings between men and women who are high school graduates with no further education.

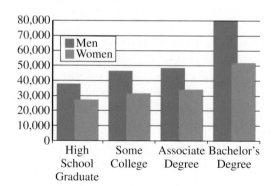

CHAPTER TEST

1. List three problem-solving techniques that we discussed in Section 1.1.

2. Identify which of the following statements is false and give a counterexample.

   a. $\dfrac{a+b}{c} = \dfrac{a}{c} + \dfrac{b}{c}$ **b.** $\dfrac{a}{b+c} = \dfrac{a}{b} + \dfrac{a}{c}$

3. Solve the following problem by making a series of guesses. In August of 2007, three of the top-selling video game players were Nintendo's Wii, Microsoft's Xbox 360, and Sony's PS3. The total

sales (in thousands) among the three were 812 units. The Xbox sold 15 more than twice the PS3, while Wii sold 4 less than Xbox and PS3 combined. How many Wiis were sold?

4. According to *USA Today*, NASA is tracking 12,000 objects the size of a grapefruit or larger that are orbiting around Earth. The given graph shows the distribution of owners of this space debris.

   a. Estimate the number of objects owned by Russia.

   b. Estimate the number of objects owned by China.

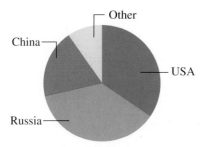

Figure for Exercise 4

5. Round 36,478 a) to the nearest thousand and b) to the nearest hundred.

6. What is The Splitting-Hairs Principle?

7. Explain the difference between inductive and deductive reasoning.

8. What type of reasoning do the following situations illustrate?

   a. Earlier you called your friend Carla and left her a message to call you back. Since Carla is faithful in returning your calls, and has not called back yet, you assume that she has not returned home yet.

   b. You use Google Maps' estimated time for a trip from Chicago to Boston and calculate that to make that time you will have to drive over 63 miles per hour.

9. State the Three-Way Principle.

10. What is the next likely term in the following sequence of numbers?

$$1, 2, 6, 15, 31, 56, 92, \ldots$$

11. Continue the following pattern for three more items in the list:

$$\text{abc, abd, abe, bcd, bce, bcf, } \ldots$$

12. What is the likely next figure in the following sequence?

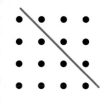

13. Illustrate Goldbach's Conjecture for 60.

14. Determine if the following statement is true or false. If it is false, give a counterexample. If the price of a laptop computer is decreased by 10% and then increased by 10%, the price will be the same as the original.

15. Follow the instructions for the following "trick" by starting with several different values for n. Make a conjecture as to what you will get for an arbitrary n. Prove your conjecture using algebra.

   a. Choose any natural number.

   b. Multiply the number by 4.

   c. Add 40 to the product you just found.

   d. Divide by 2.

   e. Subtract 20.

GROUP PROJECTS

1. The ancient Greeks were intrigued with certain numbers that they called figurate numbers because they could illustrate the numbers with geometric figures, as we show in Figure 1.15. We represent the first triangular number by T_1, the second by T_2, the third by T_3, etc. Similarly, we represent square numbers by $S_1, S_2, S_3 \ldots$, and the pentagonal numbers by $P_1, P_2, P_3 \ldots$.

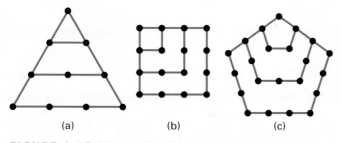

FIGURE 1.15 (a) Triangular. (b) Square. (c) Pentagonal.

   a. Figure 1.15(a) shows the first four triangular numbers to be 1, 3, 6, and 10. Calculate the next three triangular numbers.

   b. It is known that the nth triangular number is given by the equation $T_n = \frac{n(n+1)}{2}$. Verify this equation for $n = 1, 2, 3, 4, 5, 6, 7$.

   c. Find T_{10} and T_{20}.

   d. The first four square numbers are 1, 4, 9, and 16. Find the next three square numbers.

   e. The nth square number is given by the equation $S_n = n^2$. Verify this equation for $n = 1, 2, 3, 4, 5, 6, 7$.

   f. Consider the following diagram and explain how it shows a relationship between the square numbers and triangular numbers. Fill in the question marks in the equation $T_? + T_? = S_?$. Consider other square numbers, look for similar patterns, and express them as similar equations.

   g. Complete the following equation: $T_? + T_? = S_n$.

   h. Search the Internet for a formula for P_n and compute the next three pentagonal numbers.

2. Get a photo of a crowd scene. Have each member of the group make an estimate independently from the rest of the group. Compare your estimates. Average your estimates, and compare your average with the averages of the other groups in the class.

3. Use Google Earth and have each member of the group estimate the distance between two cities. Compare your estimates. Average your estimates, and compare your average with the averages of the other groups in the class.

4. Research the word *sudoku* and explain how solving a sudoku puzzle illustrates deductive reasoning.

Set Theory
Using Mathematics to Classify Objects

2

Have you ever gone shopping online at a site like itunes.com and ordered a book or music video and the site suggested another item that you might like? I had that experience several years ago when shopping for music CDs as gifts for one of my sons. He had given me a list of five suggestions and after I placed my order for one of the CDs, the site popped up with a list of several others that I might consider purchasing. To my amazement, three of the suggestions made were also on the list that my son had given me. It was a little weird. How did itunes know so much about my son's likes and dislikes?

A similar experience also happens to me when I am shopping in a supermarket or a drugstore. When I am given my checkout receipt, there is a list of discount coupons attached suggesting other items that I might

(continued)

want to buy. Many stores have "Preferred Customer Cards" that are scanned when you make a purchase. The information that is gathered is organized, using the same set theory that you will learn in this chapter, to construct a database of information about you and the other customers.

Not only stores but also search engines, such as Google, use set theory to organize data so that you can retrieve huge amounts of information in a matter of seconds. As you study this chapter, you will understand these applications and learn how to use set theory to solve problems by organizing large amounts of data. ●

2.1 The Language of Sets

Objectives

1. Specify sets using both listing and set-builder notation.
2. Understand when sets are well defined.
3. Use the element symbol properly.
4. Find the cardinal number of sets.

Did you Google today? If you are savvy about using the Internet, you know that I am asking you whether you used the popular search engine Google to look for information. As I began revising this section, I did a search for "Japanese luxury automobiles" by first specifying each of the individual words separately. When I searched for "Japanese," I received a list of 78,400,000 Web sites, including the Japanese language, cancer research in Japan, Japanese museums, Japanese gardens, as well as many, many other topics. The search for "automobiles" returned 9,080,000 sites, whereas searching for "luxury" returned 34,000,000 sites, including luxury real estate, clothes, and hotels. Finally, searching for "luxury Japanese automobiles," I received a considerably shorter list of "only" 53,000 sites.

In the first three searches, Google was searching through its huge collection of Web addresses to identify a subcollection that contained the word I specified, such as "Japanese." In the last search, we found the sites that had the words, "Japanese," "luxury," and "automobiles" in common.

Mathematicians frequently use this notion of grouping together objects with common characteristics so they can treat that collection as a single mathematical object.

In mathematical terms, a collection of objects is called a **set** and the individual objects in this collection are called the **elements** or **members** of the set.

We generally use capital letters to name sets and lowercase letters to denote elements in a set. For example, we may label the set of Olympic gold medal winners by the letter G and denote elements of that set of people by $x, a, p,$ or d. It is important to choose a name for a set to help remember what it is. We could call a group of Republicans X and a group of Democrats Y, but it is easier to work with these sets if we name the set of Republicans R and the Democrats D.

✎ *KEY POINT*

We represent sets by listing elements or by using set-builder notation.

Representing Sets

We often write a set by listing its elements within braces. For example, consider the set of seasons of the year to be the set S. Then, we may write $S = \{$spring, summer, fall, winter$\}$. Although it may be possible to list all the elements of a set, it is sometimes inconvenient to do so. If B were the set of all natural numbers* from 1 to 1,000 inclusive, listing all its elements would be cumbersome. Instead, we could write $B = \{1, 2, 3, \ldots , 1,000\}$. The first few

---

*We describe some common sets of numbers on page 40.

❀ ❀ ❀ HIGHLIGHT

Sets of Numbers Commonly Used in Mathematics

$N = \{x : x \text{ is a natural number}\} = \{1, 2, 3, \ldots\}$
$W = \{x : x \text{ is a whole number}\} = \{0, 1, 2, 3, \ldots\}$
$I = \{x : x \text{ is an integer}\} = \{\ldots -2, -1, 0, 1, 2, 3, \ldots\}$
$Q = \{x : x \text{ is a rational number}\} = \{x : x \text{ is of the } form \; \frac{a}{b},$
  where a, b are integers and $b \neq 0\}$

$R = \{x : x \text{ is a real number}\} = \{x : x \text{ has a decimal expansion}\}$
The set N is also sometimes called the set of counting numbers and denoted by C.

---

elements of B are written to establish a pattern. The dots, called an ellipsis, indicate that the list continues in the same manner up to the last number in the set, which is 1,000. We would write the set W, consisting of all whole numbers, in a similar fashion as $W = \{0, 1, 2, 3, \ldots\}$. Because we do not put a number after the ellipsis, this means that the list does not end.

If the elements of a set all share some common characteristics that are satisfied by no other object, then we can use **set-builder notation** to represent the set. For example, suppose C is the set of all carnivorous (meat-eating) animals. Using set-builder notation, we write

$$C = \{x : x \text{ is a carnivorous animal}\}.$$

As shown in Figure 2.1, we read this equation as

"C is the set of all x such that x is a carnivorous animal."

$C = \{x : x \text{ is a carnivorous animal}\}$

"C"
"the set"   "such that"
"equals"   "x is a carnivorous animal"
or "is"
"of all x"

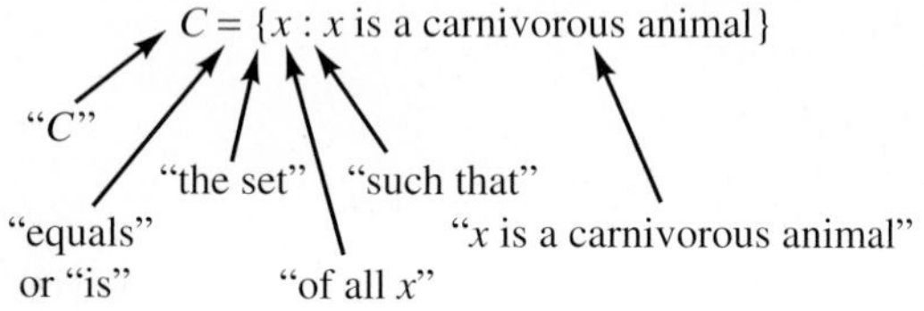

FIGURE 2.1 Reading set-builder notation.

Clearly, *lion* is one of the elements of C, but *lamb* is not.

We can write a set represented by the listing method by using set-builder notation, and vice versa.

 Some Good Advice

Using terminology and symbols properly will improve your work in mathematics. Notice that the set equation in Figure 2.1 sounds like a grammatically correct English sentence when you read it out loud.

Quiz Yourself ➊ *

Express each set using an alternative method.

a) $A = \{y : y \text{ is a day of the week}\}$

b) $B = \{1, 2, 3, \ldots, 60\}$

EXAMPLE 1   *Using Set Notation*

Use an alternative method to write each set.

a) $T = \{A^-, A^+, B^-, B^+, AB^-, AB^+, O^-, O^+\}$

b) $B = \{y : y \text{ is a color of the American flag}\}$

c) $A = \{a : a \text{ is a counting number less than 20 and is evenly divisible by 3}\}$

SOLUTION: a) We can use set-builder notation to write

$$T = \{x : x \text{ is a blood type}\}.$$

b) B can be written using the listing method as $B = \{\text{red, white, blue}\}$.

c) We can write $A = \{3, 6, 9, 12, 15, 18\}$.

Now try Exercises 19 to 30. ❀ ➊

Well-Defined Sets

✎ **KEY POINT**

Sets must be well defined.

A set is **well defined** if we are able to tell whether any particular object is an element of that set. Example 2 illustrates this idea.

Quiz Yourself ②

Which sets are well defined?

a) {x : x is a mountain over 10,000 feet high}

b) {y : y is a scary movie}

EXAMPLE 2 *Determining Whether a Set Is Well Defined*

Which sets are well defined?

a) $A = \{x : x \text{ is a winner of an Academy Award}\}$ b) $T = \{x : x \text{ is tall}\}$

SOLUTION:

a) This set is well defined because we can always determine whether a person belongs to A. Academy award winners such as Denzel Washington, Hilary Swank, and George Clooney would be members of A. Hillary Clinton, Barney, and Barry Bonds would not be members of A because they have never won an Oscar.

b) Whether or not a person belongs to this set is a matter of how we interpret *tall*; therefore, T is not well defined. Can you think of one situation in which a person who is 6 feet tall would be considered tall and a different situation in which that same person would be considered short?

Now try Exercises 31 to 38. ✳ ②

In using set-builder notation, it is possible to have a condition that no object satisfies. For example, the set

$$M = \{m : m \text{ is in your math class and is also a star on the Sopranos.}\}$$

has no elements.

> **DEFINITION** The set that contains no elements is called the **empty set** or **null set.** This set is labeled by the symbol ∅. Another notation for the empty set is { }.

The set M, that we just described, is the empty set. We can write this as $M = \varnothing$.

EXAMPLE 3 *Using Similar Notations Precisely*

a) Does {∅} have the same meaning as ∅?   b) Do {∅} and {0} mean the same thing?

SOLUTION:

a) Note that {∅} is not the same as ∅. To make this more clear, you might think of a set as a paper bag that you might get at a supermarket. Then, the empty set ∅ corresponds to an empty bag ⌷, whereas the set {∅} could be visualized as one bag containing a

second bag, which is empty.

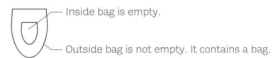

— Inside bag is empty.

— Outside bag is not empty. It contains a bag.

b) Similarly, {0} is not the same as {∅}. If we make bag drawings, then we see that {∅} corresponds to a bag containing an empty bag, whereas {0} corresponds to a bag containing the number zero.

Now try Exercises 57 to 60. ✳

Another set that we frequently use is called the universal set.

> **DEFINITION** The **universal set** is the set of all elements under consideration in a given discussion. We often denote the universal set by the capital letter *U*.

For example, in a certain problem we may want to use only the counting numbers from 1 to 10. For this discussion, the universal set would be $U = \{1, 2, 3, \ldots, 10\}$. In another situation we might consider only female consumers living in the United States. In this case, the universal set would be $U = \{x : x \text{ is a female consumer living in the United States}\}$.

Math in Your Life

How Do Babies Learn?

From the time when you were an infant, you needed to make sense out of a world full of new objects, people, sights, sounds, and emotions. How did you ever learn it all? Psychologists who study cognitive development are interested in answering exactly that question. One of the methods by which scientists investigate learning is by presenting subjects with a collection of different objects and trying to understand how the subjects learn to identify the characteristics that are common to various sets of the objects.

For example, in one experiment a researcher was interested in teaching a chimp named Sarah to recognize the similarities and differences in a set of objects. The scientist had plastic symbols for actions such as give and take, other symbols for colors, fruits, etc. One day the scientist wanted Sarah to give him a particular fruit for which he had no plastic symbol. When he showed her the symbols for "give," "apple," and the color "orange," Sarah responded by handing him an orange. As you read this chapter, you will recognize that Sarah was dealing with some of the same basic set theory ideas that you are learning in this chapter.*

 KEY POINT

The element symbol expresses that an object is a member of a set.

The Element Symbol

We use the symbol $\in$ to stand for the phrase *is an element of.* Although $\in$ looks somewhat like the letter *e*, these two symbols are *not the same* and should not be confused. The notation "$3 \in A$" expresses that 3 is an element of the set A. If 3 is not an element of set A, we write "$3 \notin A$."

EXAMPLE 4   *Using Set Element Notation*

Replace the symbol # in each statement by either $\in$ or $\notin$.

a) $3 \,\#\, \{2, 3, 4, 5\}$ b) $\{5\} \,\#\, \{2, 3, 4, 5\}$

c) Bill Gates # $\{x : x$ is a billionaire$\}$

d) jogging # $\{y : y$ is an aerobic exercise$\}$

e) the ace of hearts # $\{f : f$ is a face card in a standard 52-card deck$\}$

SOLUTION: a) $3 \in \{2, 3, 4, 5\}$

b) $\{5\} \notin \{2, 3, 4, 5\}$. Notice that 5 is a number and is not the same as $\{5\}$, which is a set.

c) Bill Gates $\in \{x : x$ is a billionaire$\}$

d) jogging $\in \{y : y$ is an aerobic exercise$\}$

e) the ace of hearts $\notin \{f : f$ is a face card in a standard 52-card deck$\}$

Now try Exercises 39 to 50. ❈ **3**

 **3**

Decide whether each statement is true or false.

a) $3 \in \{x : x$ is an odd counting number$\}$

b) $2 \notin \varnothing$

c) chocolate $\in \{x : x$ is a vitamin$\}$

 KEY POINT

The cardinal number of a set indicates its size.

Cardinal Number

In solving set theory problems, we often need to know the number of elements in a set.

> **DEFINITIONS** The number of elements in set A is called the **cardinal number** of set A and is denoted $n(A)$. A set is **finite** if its cardinal number is a whole number. An **infinite** set is one that is not finite.

---

*If you are interested in learning more about cognitive development, you can consult a general psychology text such as *Psychology* (9th edition) by Brannon and Lefton, or *Cognitive Psychology: Applying the Science of the Mind* (2nd edition) by Robinson-Riegler & Robinson-Riegler, both published by Allyn & Bacon. For an interesting Web site discussing some current research in cognitive psychology by Dr. Liz Brannon at Duke University, see www.mind.duke.edu/faculty/brannon/.

─────── **HIGHLIGHT** ❀ ❀ ❀ ─────

Between the Numbers—Is Pluto a Planet or Isn't It?

In 2006, it was common to see a headline proclaiming that our beloved little planet Pluto was no longer to be considered a planet. After 8 days of debate by the International Astronomical Union (IAU), according to the new definition of planet, Pluto had been demoted to a dwarf planet. Of course, what we are dealing with is the importance of a set being well defined. Depending on how we define the set of planets, Pluto either is one—or it isn't. Controversy remains. Alan Stern, the leader of NASA's mission to Pluto, says, "The definition stinks . . . ." Stern called it "absurd" that only 424 astronomers were allowed to vote, out of some 10,000 professional astronomers around the globe, and many astronomers are trying to get the definition overturned.

The following diagram will help you remember the meaning of the notation $n(A)$.

$$n(A)$$

The *n* reminds us of the word "number."   Capital *A* reminds us that we are dealing with a set.

EXAMPLE 5   *Finding the Cardinal Number of a Set*

State whether each set is finite or infinite. If it is finite, state its cardinal number using $n(A)$ notation.

a) $P = \{x : x$ is a planet in our solar system$\}$

b) $N = \{1, 2, 3, \ldots\}$

c) $A = \{y : y$ is a person living in the United States who is not a citizen$\}$

d) $\varnothing$

e) $X = \{\{1, 2, 3\}, \{1, 4, 5\}, \{3\}\}$

SOLUTION:

a) There are eight planets*; therefore, this is a finite set. $n(P) = 8$.

b) The set of counting numbers is an infinite set.

c) There are a finite number of people living in the United States who are not citizens; however, we probably do not know $n(A)$.

d) The empty set has no elements, so it is a finite set. Thus, $n(\varnothing) = 0$.

e) If you consider the given bag diagram, you can easily see that the set X contains three objects: the set $\{1, 2, 3\}$, the set $\{1, 4, 5\}$, and the set $\{3\}$. Therefore, $n(X) = 3$.

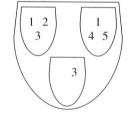

Now try Exercises 51 to 56. ✳ **4**

Quiz Yourself **4**

Find the cardinal number of each set.

a) $\{2, 4, \ldots, 20\}$

b) $\{\{1, 2\}, \{1, 3, 4\}\}$

c) $\{s : s$ is one of the United States$\}$

Exercises | 2.1 |

Looking Back†

These exercises follow the general outline of the topics presented in this section and will give you a good overview of the material that you have just studied.

1. Give an example of a set by using the listing method and then by using set-builder notation.

2. Give an example of a set that is not well defined, and explain why your set is not well defined.

3. Make two separate examples involving the set $\{1, 3, 5, 7, 8\}$, showing how to use the symbols $\in$ and $\notin$ properly.

4. Make an example of a set A such that $n(A) = 5$.

5. Explain what the Highlight "How Do Babies Learn?" has to do with set theory.

─────────

*Pluto is no longer considered a planet. See the Highlight above.

†Before doing these exercises, you may find it useful to review the note *How to Succeed at Mathematics* on page xix.

6. What set theory concept is being illustrated in the highlight regarding Pluto?

Sharpening Your Skills

In Exercises 7–18, use set notation to list all the elements of each set.

7. The natural numbers from 10 to 15 inclusive

8. The letters of the alphabet that follow *e* and come before *k*

9. $\{17, 18, \ldots, 25\}$

10. The integers between -5 and 5, not inclusive

11. The natural numbers less than 30 that are divisible by 4

12. $\{y : y$ is an odd natural number between 6 and 20$\}$

13. The days of the week

14. $\{x : x$ is one of the 50 U.S. states that begins with the letter "A," "B," or "C"$\}$

15. $\{y : y$ is a natural number less than 0$\}$

16. $\{11, 13, 15, \ldots, 25\}$

17. The set of U.S. presidents who served after Bill Clinton but before Richard Nixon

18. $\{x : x$ is one of the 50 U.S. states that borders no other state$\}$

In Exercises 19–30, use an alternative method to express each set. There may be several acceptable answers.

19. $\{3, 6, 9, 12\}$

20. $\{t : t$ is a letter in the word *number*$\}$

21. $\{m, u, s, i, c\}$

22. $\{t : t$ is a digit in the number 173,268$\}$

23. $\{z : z$ is a negative integer$\}$

24. $\{y : y$ is a month of the year$\}$

25. $\{y : y$ is a letter of the word *set* and also a letter of the word *ring*$\}$

26. $\{$Aries, Taurus, Gemini, Cancer, $\ldots$, Aquarius, Pisces$\}$

27. $\{y : y$ is a natural number greater than 100$\}$

28. $\{2, 4, 6, 8, 10, \ldots\}$

29. $\{2, 4, 6, 8, 10, \ldots, 100\}$

30. $\{x : x$ is a natural number that is evenly divisible by 3$\}$

In Exercises 31–38, determine whether each set is well defined.

31. $\{x : x$ lives in Michigan$\}$

32. $\{2, 4, 6, 8, 10, \ldots\}$

33. $\{y : y$ has an interesting job$\}$

34. $\{t : t$ has traveled much$\}$

35. $\{x : x$ is a ferocious animal$\}$

36. $\{y : y$ is a mammal$\}$

37. $\{1, -3, 5, -7, 9, -11, \ldots\}$

38. $\{y : y$ is an easy cell phone number to remember$\}$

In Exercises 39–50, replace each # with either $\in$ or $\notin$ to express a true statement.

39. $3 \# \{2, 4, 6, 8\}$

40. $3 \# \{x : x$ is a whole number$\}$

41. Franklin Roosevelt # $\{x : x$ is a past president of the United States$\}$

42. Albert Einstein # $\{y : y$ is a living American poet$\}$

43. IBM # $\{w : w$ is a manufacturer of computers$\}$

44. Tiger Woods # $\{a : a$ is a professional ice skater$\}$

45. $5 \# \{x : x$ is a rational number$\}$

46. $-5 \# \{y : y$ is a real number$\}$

47. $0 \# \varnothing$

48. $\varnothing \# \{0\}$

49. Florida # $\{x : x$ is a state south of Pennsylvania$\}$

50. $\{$Florida$\} \# \{x : x$ is a state east of the Mississippi$\}$

Find n(A) for each of the following sets A.

51. $\{1, 3, 5, 7, \ldots, 11\}$

52. $\{3, 4, 5, \ldots, 13\}$

53. $\{x : x$ is a living American president born before 1900$\}$

54. $\{x : x$ is one of the continental United States$\}$

55. $\{x : x$ is a letter in the word Mississippi$\}$

56. $\{x : x$ is a vowel in the alphabet$\}$

In Exercises 57–60, draw a "bag diagram" similar to what we did in Example 5 on page 43 to illustrate each set, A, before finding its cardinal number.

57. $\{\{1, 2\}, \{1, 2, 3\}\}$

58. $\{\{1\}, \varnothing, 0, \{0\}\}$

59. $\{\{\{\varnothing\}\}\}$

60. $\{\{1\}, \{2\}, \{3\}, \{1, 2, 3\}\}$

Describe each of the following sets as either finite or infinite.

61. $\{x : x$ is a word written by Shakespeare$\}$

62. $\{y : y$ is the number of people who have walked on the moon$\}$

63. $\{y : y$ is a number between 4 and 5$\}$

64. $\{x : x$ is an element of the empty set$\}$

In Exercises 65–72, find an element of set A that is not an element of set B. There are many correct answers.

65. $A = \{y : y$ is a number between 4 and 10$\}$

$B = \{y : y$ is a natural number between 4 and 10$\}$

66. $A = \{y : y$ is a member of the human race$\}$

$B = \{y : y$ is a citizen of the United States$\}$

67. $A = \{y : y$ is a manufacturer of electronic products$\}$

$B = \{y : y$ is a company based in the United States$\}$

68. $A = \{y : y$ is an animal$\}$

$B = \{y : y$ is covered with fur$\}$

69. $A = \{y : y$ is a world political leader$\}$

$B = \{y : y$ is European$\}$

70. $A = \{y : y$ is a car manufacturer$\}$

$B = \{y : y$ is a company based in the United States$\}$

71. $A = \{y : y$ is a day of the week$\}$

$B = \{y : y$ is a weekday$\}$

72. $A = \{y : y$ is a state whose name begins with the letter "A," "B," or "C"$\}$

$B = \{y : y$ is a state whose name begins with the letter "A" or "B"$\}$

Applying What You've Learned

In Exercises 73–76, use the table of general education electives to describe each set in an alternative way.

	Humanities	Writing	World Culture	Cultural Diversity
History012	Yes	Yes	Yes	No
History223	Yes	Yes	Yes	Yes
English010	Yes	Yes	No	No
English220	Yes	Yes	No	No
Psychology200	No	Yes	No	No
Geography115	No	No	Yes	Yes
Anthropology111	Yes	No	Yes	Yes

73. {History012, History223, English010, English220, Anthropology111}

74. {English010, English220, Psychology200}

75. $\{x : x$ satisfies a world culture requirement$\}$

76. $\{x : x$ does not satisfy a cultural diversity requirement$\}$

In October of 2007, the average price of regular gasoline in the United States was $2.77, or 277 cents, per gallon. Use this information and the given graph to answer Exercises 77–80.

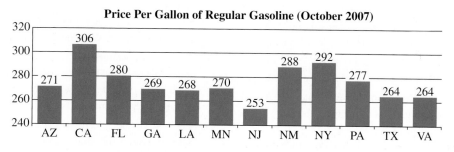

Price Per Gallon of Regular Gasoline (October 2007)

77. Use the listing method to list the set of states L that had regular gasoline at a price less than the national average.

78. Use the listing method to list the set of states G that had regular gasoline at a price greater than the national average.

79. Use set-builder notation to describe the following set of states: {CA, NY}.

80. Use set-builder notation to describe the following set of states: {GA, LA, NJ, TX, VA}.

Use the given graph, which shows the top 10 Internet sites in millions of visitors according to the New York Times Almanac (2007), *to answer Exercises 81–84.*

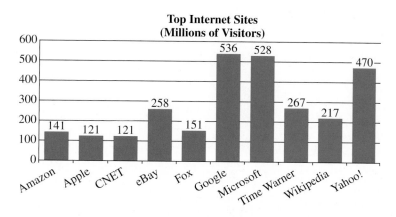

Top Internet Sites (Millions of Visitors)

81. Use the listing method to list the set of sites L that had less than 300 million visitors.

82. Use the listing method to list the set of sites M that had more visitors than Time Warner.

83. Use set-builder notation to describe the following set of sites: {Amazon, Apple, CNET, Fox}.

84. Use set-builder notation to describe the following set of sites: {Apple, CNET}.

Communicating Mathematics

85. When would it be appropriate to use set-builder notation instead of the listing method to describe a set?

86. What does it mean to say that a set is well defined?

87. Give a careful explanation of the difference in the meanings of $\{\varnothing\}$ and $\varnothing$.

88. How do you "remember how to remember" the meaning of the notation $n(A)$?

Using Technology to Investigate Mathematics

89. In the section opener, we discussed doing an Internet search. Draw a diagram illustrating the sets of pages containing "luxury Japanese automobiles" that is similar to the one we drew in Example 3 on page 5 showing dance students who were taking swing and Latin American dance lessons. There should be eight different regions in your diagram. Explain what pages are in each of the eight regions.

90. Repeat Exercise 89, doing a search on some other item that is described by three characteristics, and draw a diagram similar to the one we asked you to draw in Exercise 89.

91. Search the Internet for applications of set theory in medicine.

92. Use the Internet to find an explanation of what we mean by fuzzy sets. Find some applications of fuzzy set theory.

For Extra Credit

93. Let U = {Shaquille O'Neal, Babe Ruth, Beethoven, Bach, Leonardo da Vinci, J K Rowling, Bart Simpson, LeBron James, Hillary Clinton, Napoleon, Eli Manning, Rembrandt, Winston Churchill, Julius Caesar, Shakespeare, Bill Clinton, Beyonce, Barney}. From U we can choose a set of elements that all share some common characteristic and can give this set using both the listing method and set-builder notation—for example, {LeBron James, Shaquille O'Neal} = $\{x : x$ is a basketball player$\}$. Find as many such sets as you can and give them using both the listing method and set-builder notation.

94. **Sets of common objects.** Consider the universal set

    U = {apple, flat-screen TV, hat, Sirius radio, fish, sofa, washing machine, shoe, dog, automobile, potato chip, toenail clipper, bread, banana, vacuum cleaner, hammer, bed, pizza}.

    From U we can choose a set of elements that all share some common characteristic and can give this set using both the listing method and set-builder notation—for example, {flat screen TV, Sirius radio, washing machine, vacuum cleaner} = $\{x : x$ is an electrical appliance$\}$. Find as many such sets as you can and give them using both the listing method and set-builder notation.

We mentioned earlier that serious paradoxes arose in mathematics. We will define a paradox as a statement that contradicts itself, or a statement that can be proved to be both true and false at the same time. Use this definition in thinking about Exercises 95–97.

95. **A paradox.** A small town has only one barber, who is male. Furthermore, this barber shaves those men, and only those men, who do not shave themselves. Who shaves the barber? One of two things can happen—either the barber shaves himself or he does not. First assume that the barber shaves himself. What must you conclude? Now assume that the barber does not shave himself. What must you conclude?

96. Consider the statement, "This sentence is false." Is this sentence true or false? First assume that the sentence is true. What must you conclude? Now assume that the sentence is false. What must you conclude?

97. Let A be the set $\{1, 2, 3\}$. Clearly, $A \notin A$. Now think of the set of all such sets that are not elements of themselves and call that set S. That is, $S = \{X : X$ is a set and $X \notin X\}$. Now we ask the question, "Is $S \in S$?" First assume that $S \in S$. What must you conclude? Now assume that $S \notin S$. What must you conclude?

2.2 Comparing Sets

Objectives

1. Determine when sets are equal.
2. Know the difference between the relations of subset and proper subset.
3. Use Venn diagrams to illustrate set relationships.
4. Determine the number of subsets of a given set.
5. Distinguish between the ideas of "equal" and "equivalent" sets.

From the time you were very young, you have been making comparisons like the following: "I can run faster than you." "She's smarter than her brother." "These are my most comfortable shoes." Mathematicians also make many comparisons. In an algebra course, we compare numbers and algebraic expressions, and in geometry, we compare geometric figures. In this section, we will compare sets using methods that in some ways are similar to how you have compared quantities in the past, and in other ways, are different.

 KEY POINT

Equal sets contain the same members.

Set Equality

One of the most fundamental things we need to know about two sets is when do we consider them to be the same.

> **DEFINITION** Two sets A and B are **equal** if they have exactly the same members. In this case, we write $A = B$. If A and B are not equal, we write $A \neq B$.

This definition says that for sets A and B to be equal, every element of A must also be a member of B and every element of B must also be in A.

 Some Good Advice

When reading a definition, be careful not to assume conditions that are not specifically stated. The set equality definition says nothing about the order in which elements are listed in the sets or whether elements are repeated.

EXAMPLE 1 *Set Equality*

Which pairs of sets are equal?

a) $\{A^-, A^+, B^-, B^+, AB^-, AB^+, O^-, O^+\}$ and $\{A^-, B^-, AB^-, O^-, A^+, B^+, AB^+, O^+\}$
b) $A = \{x : x$ is a citizen of the United States$\}$ and $B = \{y : y$ was born in the United States$\}$

SOLUTION:

a) Notice that the left-hand set and the right-hand set contain exactly the same elements. The order of the elements is not important; therefore, the two sets are equal.

b) Because Arnold Schwarzenegger is an element of set A, but is not an element of set B, the sets are not equal. Can you think of any other elements of A that are not in B?

Now try Exercises 5 to 14. ✿ **5**

Quiz Yourself **5**

Determine whether each statement is true or false.

a) {Socrates, Shakespeare, Beethoven} = {Shakespeare, Beethoven, Socrates}

b) {tiger, gray whale, giant panda} = {y : y is an endangered species}

 *KEY POINT*

One set is a subset of another if all its elements are found in the other set.

Subsets

Another way we compare sets is to determine whether one set is part of another set.

> **DEFINITION** The set A is a **subset** of the set B if every element of A is also an element of B. We indicate this relationship by writing $A \subseteq B$. If A is not a subset of B, then we write $A \nsubseteq B$.

In order to show that $A \subseteq B$, we must show that every element of A also occurs as an element of B. To show that A is not a subset of B, all we have to do is find one element of A that is not in B.

 ## PROBLEM SOLVING
*The Analogies Principle**

Similarities in notation and terminology often reflect corresponding similarities in the ideas being represented. Because the notation for "is a subset of," $\subseteq$, reminds us of the notation for "less than or equal to," $\leq$, we might expect that both ideas share similar properties.

EXAMPLE 2 *Identifying Subsets*

Determine whether either set is a subset of the other.

a) $A = \{1, 2, 3\}$ and $B = \{1, 2, 3, 4\}$

b) $A = \{$Terry O'Quinn, America Ferrera, James Spader, Sally Field$\}$ and $E = \{x : x$ has won an Emmy Award since 2007$\}$

SOLUTION:

a) Every member of A is also in B, so we can say $A \subseteq B$. Because there is an element of B that is not in A, we write $B \nsubseteq A$.

b) If you check an almanac, you will find that every person mentioned in set A has won an Emmy since 2007 and so we can say $A \subseteq E$. The actor Robert Duvall won an Emmy for *Broken Trail* in 2007 but is not a member of A, so E is not a subset of A. ✻

 *KEY POINT*

Venn diagrams represent set relationships graphically.

If A is any set, then $A \subseteq A$ because clearly each element of A is an element of A. Also, the empty set is a subset of every set. For example, $\emptyset \subseteq \{1, 2, 3\}$. Even though this sounds strange, it is true that every element of the empty set is also an element of $\{1, 2, 3\}$. According to the Counterexample Principle from Section 1.1, to show that this statement is false, *you must find a counterexample*. That is, you must find an element of $\emptyset$ that is not in $\{1, 2, 3\}$. Because this is impossible, our statement is true.

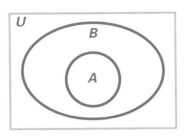

FIGURE 2.2 A Venn diagram illustrating that A is a subset of B.

Venn Diagrams

Drawings called **Venn diagrams** are used to visualize relationships among sets. Figure 2.2 is a Venn diagram that illustrates A is a subset of B.

We represent the universal set by the rectangular region, labeled U. The region labeled A is completely contained in region B, indicating that all A's elements are also in B.

> **DEFINITION** The set A is a **proper subset** of the set B if $A \subseteq B$ but $A \neq B$. We write this as $A \subset B$. If A is not a proper subset of B, then we write $A \not\subset B$.

---

*See Section 1.1.

From this definition, we see that $\{2, 4, 6, \ldots\} \subseteq \{1, 2, 3, \ldots\}$. Also, $\{2, 4, 6, \ldots\} \subset \{1, 2, 3, \ldots\}$ because $\{1, 2, 3, \ldots\}$ contains elements that are not members of $\{2, 4, 6, \ldots\}$. Also, in Example 2(b), the set $A = \{$Terry O'Quinn, America Ferrera, James Spader, Sally Field$\}$ is a proper subset of E because, again, Robert Duvall is not a member of A.

Some Good Advice

Distinguish between the notation for subset and proper subset. The lower half of the subset symbol in the notation "$A \subseteq B$" looks somewhat like an equal sign because it is *possible* that sets A and B are equal. Of course, the sets do not have to be equal. When we write $A \subset B$, the lower line is missing to remind us that the sets *cannot* be equal.

EXAMPLE 3 *Identifying Subsets*

Consider the following table of information regarding medalists in the 2004 Summer Olympic Games in Athens, Greece.

	Event	**Medal**	**Country**
Catalina Ponor	Balance beam (Gymnastics)	Gold	Romania
Veronica Campbell	200-meter sprint	Gold	Jamaica
Carly Patterson	All-around (Gymnastics)	Gold	United States
Kirsty Coventry	100-meter backstroke	Silver	Zimbabwe
Justin Gatlin	100-meter sprint	Gold	United States
Lishi Lao	Platform diving	Silver	China
Joachim Olsen	Shot put	Bronze	Denmark
Shawn Crawford	200-meter sprint	Gold	United States
Pyrros Dimas	Weightlifting (85 kg)	Bronze	Greece
Mi Ran Jang	Weightlifting (+75 kg)	Silver	South Korea

Assume that this set is the universal set, and define the following sets:

$$T = \text{the set of 200-meter winners}$$
$$A = \text{the set of U.S. athletes}$$
$$G = \text{the set of gold medal winners}$$

Which statements are true?

a) $T \subseteq G$ b) $T \subset G$ c) $G \subseteq A$

SOLUTION:

a) This is true because every element of T, which equals the set $\{$Campbell, Crawford$\}$, is also an element of G, which equals the set $\{$Ponor, Campbell, Patterson, Gatlin, Crawford$\}$.

b) This is also true. We already know that T is a subset of G. Because G contains an element that is not in T—for example, Ponor—this means that T is a proper subset of G.

c) This is false. Set G contains Campbell, which is not an element of A.

Now try Exercises 15 to 24. ✸ **6**

Quiz Yourself **6**

Decide whether each statement is true or false.

a) $\{2, 4, 6, 8, \ldots\} \subseteq \{1, 2, 3, 4, \ldots\}$

b) In the table in Example 3, $T \subseteq A$.

When one of our cars was totaled in an accident, my wife and I decided to buy a new vehicle. Before we signed on the dotted line, the salesperson asked us if we wanted to consider adding an extended warranty, OnStar navigation system, an XM radio, or extra paint protection. Suppose you were faced with such a situation. How might you consider the number of ways to make this decision? You might begin by randomly listing different sets of options. We will use w for warranty, o for OnStar, x for XM radio, and p for the paint protection. Then some of your choices might be $\{w, o\}$, $\{x, p\}$, $\{o\}$, $\{w, o, p\}$, $\{w, x, p\}$, and so on.

The trouble with this random approach is that after identifying about 11 or 12 subsets, it becomes harder and harder to think of new ones, and you might stop without creating a complete list. As we pointed out in Section 1.1, it is important to attack a problem systematically. We illustrate an organized solution to a similar type of problem in Example 4.

EXAMPLE 4   *Finding All Subsets of a Set Systematically*

Find all subsets of the set $\{1, 2, 3, 4\}$.

SOLUTION: We can organize this problem by considering subsets according to their size, going from 0 to 4. This method is illustrated in the following table.

Size of Subset	Subsets of This Size	Number of Subsets of This Size
0	∅	1
1	$\{1\}, \{2\}, \{3\}, \{4\}$	4
2	$\{1, 2\}, \{1, 3\}, \{1, 4\}, \{2, 3\}, \{2, 4\}, \{3, 4\}$	6
3	$\{1, 2, 3\}, \{1, 2, 4\}, \{1, 3, 4\}, \{2, 3, 4\}$	4
4	$\{1, 2, 3, 4\}$	1
		Total = 16

Quiz Yourself ❼

Find all subsets of the set $\{1, 2, 3\}$.

Now try Exercises 35 to 38. ✳ ❼

If we can generalize the solution to one problem, we can often use this knowledge to solve other, related problems. Let us try to generalize the solution that we found in Example 4. To discover a pattern, we look at further examples having different size sets S.

S	All Subsets of S	Number of Subsets of S
∅	∅	1
$\{1\}$	∅, $\{1\}$	2
$\{1, 2\}$	∅, $\{1\}$, $\{2\}$, $\{1, 2\}$	4
$\{1, 2, 3\}$	See Quiz Yourself 7	8
$\{1, 2, 3, 4\}$	See Example 4	16

Quiz Yourself ❽

a) If a set has five elements, how many subsets will it have?

b) How many subsets can be formed using letters of the alphabet?

We see that each time we add an element to S, the number of subsets of S doubles. The pattern $1 = 2^0$, $2 = 2^1$, $4 = 2^2$, $8 = 2^3$, and $16 = 2^4$ gives us the general relationship that we seek. ❽

> **THE NUMBER OF SUBSETS OF A SET** A set that has k elements has 2^k subsets.

To get a better idea of why this formula for counting subsets holds, read the discussion of tree diagrams in Exercises 61 and 62.

 **KEY POINT**

Equivalent sets have the same number of elements.

Equivalent Sets

Another relationship that can occur between two sets is that the elements of one set may match up with the elements in the other set.

> **DEFINITION** Sets *A* and *B* are **equivalent,** or in **one-to-one correspondence,** if $n(A) = n(B)$. Another way of saying this is that two sets are equivalent if they have the same number of elements.*

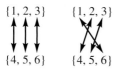

FIGURE 2.3 Two ways to match *A* and *B* in a one-to-one correspondence.

The sets {1, 2, 3} and {4, 5, 6} are equivalent because they have the same number of elements. One way of thinking about equivalent sets is that each element of one set can be matched with exactly one element of the other in a one-to-one correspondence. Figure 2.3 shows two ways to match the elements of {1, 2, 3} and {4, 5, 6}. Can you think of any others? Exercise 77 considers the question of exactly how many one-to-one correspondences there are between two sets.

In Example 3, the set of gold medal winners was {Ponor, Campbell, Patterson, Gatlin, Crawford}, which is not equivalent to the set of U.S. athletes {Patterson, Gatlin, Crawford}.

 Some Good Advice

Recall the Splitting Hairs Principle. The words *equal* and *equivalent* sound similar but do not have the same meanings. *Be careful not to use them interchangeably.*

Exercises

Looking Back[†]

These exercises follow the general outline of the topics presented in this section and will give you a good overview of the material that you have just studied.

1. What does it mean when we say that two sets are equal?

2. Explain the difference between the notation $A \subseteq B$ and $A \subset B$.

3. How does the discussion on page 50 following Example 4 help you "remember how to remember" the number of subsets of a *k*-element set?

4. Give an example of two equivalent sets that are not equal.

Sharpening Your Skills

In Exercises 5–14, decide whether each pair of sets is equal. Justify your answer.

5. {1, 3, 5, 7, 9} and {1, 5, 9, 3, 7}

6. {*a, e, i, o, u*} and {*a, b, . . . , u*}

7. {*x* : *x* is a counting number between 5 and 19 inclusive} and {*y* : *y* is a rational number between 5 and 19 inclusive}

8. {*x* : *x* is a counting number between 3 and 10 inclusive} and {*y* : *y* is a whole number between 3 and 10 inclusive}

9. {1, 3, 5, . . . , 99} and {*x* : *x* is an odd counting number between 0 and 100}

10. {3, 6, 9, 12, 15} and {*x* : *x* is a counting number that is a multiple of 3}

11. {*z* : *z* is a month of the year} and {January, February, March, . . . , December}

12. {*y* : *y* is a weekday} and {Monday, Tuesday, . . . , Friday}

13. ∅ and {*x* : *x* is a living American born before 1800}

14. ∅ and {∅}

In Exercises 15–24, decide whether each statement is true or false. Justify your answer.

15. {rose, daisy, sunflower, orchid} ⊆ {rose, orchid, sunflower, daisy, carnation}

16. {rose, daisy, sunflower, orchid} ⊂ {rose, orchid, sunflower, daisy, carnation}

17. {collie, poodle, beagle, chihuahua, bulldog} ⊂ {bulldog, chihuahua, beagle, poodle, collie}

---

*If you are interested in reading about one-to-one correspondences between infinite sets, see Section 2.5.

[†]Before doing these exercises, you may find it useful to review the note *How to Succeed at Mathematics* on page xix.

18. {collie, poodle, beagle, chihuahua, bulldog} ⊆ {bulldog, chihuahua, beagle, poodle, collie}

19. {$x : x$ is a letter in the word *happy*} ⊆ {$y : y$ is a letter in the word *happiness*}

20. {$t : t$ is a letter in the word *Ruth*} ⊂ {$z : z$ is a letter in the word *truth*}

21. ∅ ⊆ {1, 3, 5}   22. ∅ ⊂ ∅

23. {∅} ⊆ {0}   24. {0} ⊆ ∅

In Exercises 25–34, decide whether each pair of sets is equivalent. Justify your answer.

25. {1, 2, 3, 4, 5} and {*a, e, i, o, u*}

26. {2, 4, 6, 8, 10, 12} and {2, 3, 4, . . . , 12}

27. {$x : x$ is a letter in the word *song*} and {$x : x$ is a letter in the word *songs*}

28. {$x : x$ is a letter in the word *tenacity*} and {$x : x$ is a letter in the word *resolve*}

29. ∅ and {∅}

30. {∅} and {0}

31. {1, 3, 5, 7, . . . , 15} and {4, 6, 8, 10, . . . , 18}

32. {a, b, c, d, e, . . . , z} and {3, 4, 5, 6, . . . , 26}

33. {$x : x$ is a day in the year 2012} and {$y : y$ is a day in the year 2011}

34. {$x : x$ was in the starting lineup of the New England Patriots in the 2008 Super Bowl} and {$x : x$ was in the starting lineup of the New York Giants in the 2008 Super Bowl}

35. List all the two-element subsets of the set {1, 2, 3}.

36. List all the two-element subsets of the set {1, 2, 3, 4}.

37. List all the three-element subsets of the set {1, 2, 3, 4}.

38. List all the three-element subsets of the set {1, 2, 3, 4, 5}.

39. If set *A* has five elements, how many subsets does *A* have? How many of these subsets are proper?

40. If set *A* has seven elements, how many subsets does *A* have? How many of these subsets are proper?

Applying What You've Learned

Use the following table to answer Exercises 41–44.

	Major	Class Rank	GPA	Activities
Allen	Music	Freshman	1.9	Drama
Belinda	Art	Senior	2.8	Newspaper
Carmen	English	Freshman	3.1	Baseball
Dana	History	Freshman	2.9	Drama
Elston	Art	Senior	2.8	Band
Frank	Sociology	Sophomore	3.1	Football
Gina	Chemistry	Junior	2.6	Newspaper
Hector	Physics	Freshman	2.2	Band
Ivana	English	Junior	3.5	Basketball
James	English	Sophomore	2.9	Newspaper

In Exercises 41–44, consider the following sets: U (upperclassmen), L (lowerclassmen), S (science majors), V (GPA above 3.0), A (art majors), T (athletes), and D (involved in drama).

41. Find a set that is equal to *V*.

42. Find a set that is equivalent to *S*, but not equal to *S*.

43. Find a set whose cardinal number is the largest of all the sets.

44. Find a set whose cardinal number is the smallest of all the sets.

45. Domino's Pizza advertises that you can order your pizza plain, or with extra cheese, or with any combination of the toppings peppers, pepperoni, onion, sausage, anchovies, or olives. In how many different ways can you order your pizza?

46. If Domino's Pizza wants to advertise that there are over 500 ways to order your pizza, how many different toppings must they have available?

47. Burger King advertises that you can "Have it your way." If their burger can be purchased with from none to eight toppings, such as pickle, onion, tomato, etc., in how many different ways can you order your burger?

48. Burger King wishes to outdo Domino's Pizza in Exercise 46 and wants to advertise that you have over a thousand different ways to order your burger. What is the minimum number of different toppings that they must have available?

49. The owners of the Phoenix Flames football team own different amounts of stock in the franchise, so when they vote on an issue they have different amounts of voting power. Assume that Alvarez' and Cianci's votes each will count twice, Belardo's will be counted three times, Devlin's four times, and Espinoza's once. In order for a motion to be passed, the vote must be a total voting weight of at least nine. How many different subsets of this group of owners have a voting weight of at least nine?

50. Five Internet companies are merging so that they can compete with the larger companies. The companies own different stock and so their voting power is weighted accordingly. ComCore's vote will count as four votes, AvantNet's and NanoWeb's will each count as three, eNet's will count as two, and MicroNet's will count only once. For any policy to be passed, a total voting weight of 10 or more is needed. How many different subsets of these companies have a total weight of 10 or more?

Communicating Mathematics

51. List two conditions that are not important when considering set equality. Give examples.

52. Which of the following statements are incorrect? Explain the mistake that is being made.

    a. {1} ∈ {1, 2, 3}   **b.** 3 ∈ {1, 2, 3}

    c. {2} ⊆ {1, 2, 3}   **d.** 3 ⊆ {1, 2, 3}

53. What does the Splitting Hairs Principle that we explained on page 13 have to do with understanding the notations ∈, ⊆, and ⊂?

54. Give an example to show that if $n(A) = n(B)$, then we can match the elements of A and B in a one-to-one correspondence.

55. How would you help a friend remember the meaning of $n(A)$?

56. If $A \subseteq B$, can we conclude that $A \subset B$? Explain. If $A \subset B$, can we conclude that $A \subseteq B$? Explain.

On an exam, Pete said that the set B has 25 subsets. Use this information to answer Exercises 57–60.

57. What is wrong with Pete's answer?

58. What mistake do you think Pete made?

59. How many elements do you think that B had? Why?

60. How many subsets do you think that B had? Why?

61. How does the following diagram explain why the set $A = \{1, 2\}$ has four subsets? Draw a similar diagram for $B = \{1, 2, 3\}$.

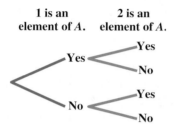

62. If a set has three elements, then it has 8 subsets. Use a tree diagram as in Exercise 61 to give an explanation as to why you would then expect a four-element set to have 16 subsets.

Using Technology to Investigate Mathematics

When mathematicians find a solution to a problem, they are interested in the practicality of being able to actually compute the solution. You have seen in the box following Example 4, that a set having k elements will have 2^k subsets. We will pursue this idea in Exercises 63–66.

63. Find the largest value for k on your calculator that when you try to compute 2^k you do not get an overflow. (On the calculator that I was using, the largest k for which I could get an answer was $k = 332$.)

64. If a set has 30 elements, then there will be $2^{30} = 1{,}073{,}741{,}824$ subsets. If you could write one subset every second, how long would it take you to list all 2^{30} subsets? (*Hint:* There are $365 \times 24 \times 60 \times 60$ seconds in a year—we are ignoring leap years.)

65. Find by trial and error the largest k that you could use to list 2^k subsets in 100 years. Again, you are listing one subset per second.

66. If you were using a computer that could calculate one billion subsets per second, approximately how many years would it take to list all of the subsets in a 100-element set? (*Note:* Your calculator will not give you an exact answer. The answer will be in scientific notation—see Section 6.5.)

67. Do an Internet search to find more information about Pascal's triangle (see Exercises 69–72). Describe other patterns that are found in Pascal's triangle and another application of it other than set theory. Write a short report about your findings.

68. Search the Internet for the history of set theory.

For Extra Credit

In Exercises 69–72, recall that in Section 1.1 we introduced the following arrangement of numbers, which is called Pascal's triangle.

Notice that the fourth line* of this triangle contains the numbers 1, 4, 6, 4, 1, which, as we saw in Example 4, are precisely the counts of the number of subsets of a four-element set with 0, 1, 2, 3, and 4 elements, respectively.

69. With this observation in mind, how do you interpret the fifth line of Pascal's triangle?

70. How do you interpret the sixth line of Pascal's triangle?

71. Tyra Banks is choosing three of the remaining nine contestants in the running to be America's Next Top Model to appear in *Cover Girl Magazine*. In how many ways can this be done? (*Hint:* Use Pascal's triangle as you did in Exercises 69 and 70.)

72. Donald Trump has 10 contestants left on *The Apprentice* and wishes to consider four of them as potential project managers for the next task. In how many ways can this be done? (*Hint:* Use Pascal's triangle as you did in Exercises 69 and 70.)

73. Add the numbers across each row of Pascal's triangle. Do the results look familiar?

74. Notice that the arrangement of numbers in each row is symmetric. See how the 4s are balanced in the fourth row, the 5s are balanced in the fifth row, and so on. Can you give a set theory explanation to account for this?

We mentioned that the subset notation, $\subseteq$, and the notation for "less than or equal to," $\leq$, appear to be similar. In Exercises 75–76, for each property of $\leq$, state the corresponding property of $\subseteq$. Next, convince yourself that the newly stated property is indeed a valid set theory property.

75. If $a \leq b$ and $b \leq c$, then $a \leq c$.

76. If $a \leq b$ and $b \leq a$, then $a = b$.

77. By considering examples, find a general pattern that will tell you how many different one-to-one correspondences you can list for a set having n elements.

78. Discuss why it would be impossible with finite sets to have $A \subset B$ and also have a one-to-one correspondence between A and B. Give an example of how this is possible with infinite sets. (*Hint:* See the Of Further Interest section of this chapter.)

---

*We start counting these lines with 0, not 1.

2.3 Set Operations

Objectives

1. Perform the set operations of union, intersection, complement, and difference.
2. Understand the order in which to perform set operations.
3. Know how to apply DeMorgan's laws in set theory.
4. Use Venn diagrams to prove or disprove set theory statements.
5. Calculate the cardinal number of the union of two sets.

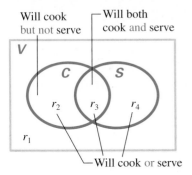

Will cook but not serve — Will both cook and serve

— Will cook or serve

FIGURE 2.4 A Venn diagram illustrating the volunteers at the soup kitchen.

Assume a community service club on your campus is asking for volunteers to cook and serve food, as well as to perform various other tasks at a nearby soup kitchen. In the Venn diagram in Figure 2.4, V is the set of all the volunteers, C is the set of people who cook, and S is the set of people who serve.

There are various subsets of V in this diagram. For example, if we join sets C and S together to form a larger set, then we have the set of people who will either cook or serve the food. We can represent this set by regions r_2, r_3, and r_4.

The set of people who will both cook and serve food is shown in Figure 2.4 as region r_3. You can think of this set as the place where sets C and S overlap.

Finally, we can find the set of people who will only cook, shown as r_2, by starting with set C and removing the elements of set S. **9**

You could think of what we were doing in Figure 2.4 as treating the various sets as mathematical objects, and then performing informal operations on them such as "joining," "overlapping," and "removing" to form new sets. We will now give these informal operations the more precise mathematical names of "union," "intersection," and "set difference," and then show you how to do computations with these set operations.

 **KEY POINT**

We form the union of sets by joining sets together.

Union of Sets

Many technical words that we use in mathematics sound exactly like words we use in English. Such is the case with the set theory operation of "union." If you think of how we use *union* in everyday life, you might think of a labor union, the European Union, the United States, a marriage union, etc. In each case, you see the idea of things coming

Math in Your Life

What's in Your Database?

"We appear to be on the brink of a post-September 11 surveillance society."* In an effort to search out and identify terrorists, the United States is employing a dazzling array of technological weapons, including iris scans, voiceprints, and facial recognition software. In addition to these technologies, both the government and private corporations maintain large databases (sets) of information with intricate cross-references to search for relationships that will provide information about who the terrorists are and how they are operating.

These technologies offer us the long-term promise of not only protecting us from terrorists but also the

elimination of identity theft, shorter security lines at airports, and faster credit approvals. However, some are concerned that as others gather and catalog information about the details of our lives, we are trading our security for our privacy and eventually our freedom.

One private company, called ChoicePoint, has gathered more than 250,000,000,000,000 bytes of data on 220 million people—a very large set indeed. If all the information in this database were printed out, it would stretch from Earth to the moon and back 77 times!

*This note is based on the article, "They're Watching You," which appeared in *BusinessWeek*, January 24, 2005.

Quiz Yourself ❾

a) Describe the set of people in Figure 2.4 if we remove regions r_2, r_3, and r_4.

b) What regions do you get if you remove all of *C* from *V* in Figure 2.4? What set of people do these regions represent?

together to make something larger. In forming a set union (see below), we are joining sets together to form a larger set. For example, $\{1, 3, 4, 5\} \cup \{2, 4, 6\} = \{1, 2, 3, 4, 5, 6\}$. Notice that although 4 is an element of both sets, it is not necessary to list it twice.

DEFINITION The **union** of sets *A* and *B*, written $A \cup B$, is the set of elements that are members of either *A* or *B* (or both). Using set-builder notation,

$$A \cup B = \{x : x \text{ is a member of } A \text{ or } x \text{ is a member of } B\}.$$

The union of more than two sets is the set of all elements belonging to at least one of the sets.

We will now introduce a situation that we will use in several of the examples of this section.

Before deciding on an activity for an exercise campaign, you might consider how convenient it would be to perform the activity and also whether it would require any cost. With this in mind, the given table compares the features of several fitness activities.

Activity	Requires Special Location	Requires Special Equipment
Hot room yoga (yo)	Yes	No
Resistance training (rt)	No	Yes
Bicycling (bi)	No	Yes
Calisthenics (ca)	No	No
Hiking (hi)	No	No
Jogging (jo)	No	No
Elliptical machine (el)	Yes	Yes
Tennis (te)	Yes	Yes

These activities form a universal set

$$U = \{\text{yo, rt, bi, ca, hi, jo, el, te}\},$$

where we represent each element by a two-letter abbreviation. Let's define the following subsets of *U* that we will use in Examples 1 and 2:

E = the set of activities that need special equipment = {rt, bi, el, te}

L = the set of activities that must be done in a special location = {yo, el, te}

EXAMPLE 1 *Finding the Union of Sets*

Find the union of the following pairs of sets.

a) $A = \{1, 3, 5, 6, 8\}$, $B = \{2, 3, 6, 7, 9\}$

b) The sets of fitness activities *E* and *L* described previously.

SOLUTION:

a) $A \cup B = \{1, 3, 5, 6, 8\} \cup \{2, 3, 6, 7, 9\}$ = the set of elements that are in *A* or *B* or both = $\{1, 3, 5, 6, 8, 2, 3, 6, 7, 9\} = \{1, 2, 3, 5, 6, 7, 8, 9\}$.

Notice in our final answer how we listed the elements *in order* and *did not list duplicate elements* because doing so does not affect set equality.

b) $E \cup L = \{\text{rt, bi, el, te}\} \cup \{\text{yo, el, te}\}$ = the set of elements in *E* or *L* or both = {yo, rt, bi, el, te} ✸

It is good to practice reading the statements in Example 1 aloud to become comfortable with set notation and terminology.

KEY POINT

The intersection of sets is the set of elements they have in common.

Intersection of Sets

Another set operation is intersection (see below), which corresponds to our earlier informal notion of overlapping sets. Again, if you think of how we use the word *intersection* in other situations, such as the intersection of streets or the intersection of lines in geometry, you see the essential idea of intersection is a region that is common to both sets. Generally, the intersection of two sets is smaller than either set whereas the union is generally larger.

> **DEFINITIONS** The **intersection** of sets A and B, written $A \cap B$, is the set of elements common to both A and B. Using set-builder notation,
>
> $$A \cap B = \{x : x \text{ is a member of } A \text{ and } x \text{ is a member of } B\}.$$
>
> The intersection of more than two sets is the set of elements that belong to each of the sets. If $A \cap B = \varnothing$, then we say that A and B are **disjoint.**

Quiz Yourself ⑩

Let $M = \{x : x$ is a letter in the word *mathematics*$\}$ and let $B = \{y : y$ is a letter in the word *beauty*$\}$. Find $M \cup B$ and $M \cap B$.

EXAMPLE 2 *Finding the Intersection of Sets*

We will find the intersection of the sets discussed in Example 1.

a) $A \cap B = \{1, 3, 5, 6, 8\} \cap \{2, 3, 6, 7, 9\} =$ the set of elements that are in both A and $B = \{3, 6\}$

b) $E \cap L = \{$rt, bi, el, te$\} \cap \{$yo, el, te$\} =$ the set of elements in E and $L = \{$el, te$\}$.

Now try Exercises 9 to 26 that involve only union and intersection. ❋ ⑩

Often you will understand a set theory problem better if you represent it graphically. Figure 2.5 shows how to visualize the union and intersection of two sets.

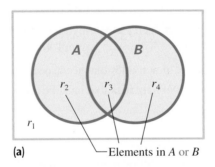

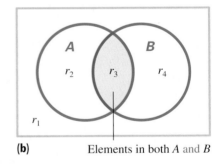

(a)   Elements in A or B (b)   Elements in both A and B

FIGURE 2.5 (a) Venn diagram of $A \cup B$. **(b)** Venn diagram of $A \cap B$.

If sets A and B are disjoint, we draw the Venn diagram so that A and B do not overlap.

KEY POINT

The elements not in a set form its complement.

Set Complement

The third set operation that we consider is complementation. When thinking of the complement of a set, think of what elements you need to "complete the set" to get the whole universal set.

> **DEFINITION** If *A* is a subset of the universal set *U*, the **complement** of *A* is the set of elements of *U* that are *not* elements of *A*. This set is denoted by *A*′. Using set-builder notation,
>
> $$A' = \{x : x \in U \text{ but } x \notin A\}.$$

EXAMPLE 3 *Finding the Complement of Sets*

Find the complement of each set. We have stated a universal set for each set.

a) $U = \{1, 2, 3, \ldots, 10\}$ and $A = \{1, 3, 5, 7, 9\}$.

b) *U* is the set of people living in the United States, and *M* is the set of people who have pages on MySpace.

c) *U* is the set of cards in a standard 52-card deck, and *F* is the set of face cards.

SOLUTION:

a) *A*′ is the set of elements in *U* that are not in *A*, so $A' = \{2, 4, 6, 8, 10\}$.

b) *M*′ is the set of people living in the United States who do not have pages on MySpace.

c) *F*′ is the set of nonface cards.

The complement of *A* is the shaded region in Figure 2.6. ❋

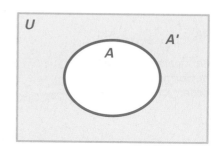

FIGURE 2.6 Venn diagram of the complement of *A*.

KEY POINT

To form a set difference, begin with one set and remove all elements that appear in a second set.

Set Difference

The last set operation that we discuss in this section is set difference.

> **DEFINITION** The **difference** of sets *B* and *A* is the set of elements that are in *B* but not in *A*. This set is denoted by *B* − *A*. Using set-builder notation,
>
> $$B - A = \{x : x \text{ is a member of } B \text{ and } x \text{ is not a member of } A\}.$$

The set difference symbol may remind you of the symbol we use for subtracting numbers—and there is some similarity. When I ask my students how they first learned about subtraction, they usually mention "taking away objects." For example, if you start with five apples and take away two, the result is three. So, a good way to remember how to calculate *B* − *A* is to start by listing the elements of set *B* and then "taking away" the elements that are in set *A*. The difference of sets *B* and *A* is shaded in Figure 2.7.

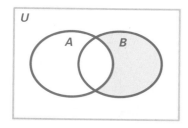

FIGURE 2.7 Venn diagram of *B* − *A*.

EXAMPLE 4 *Finding the Difference of Sets*

a) Find $\{3, 6, 9, 12\} - \{x : x \text{ is an odd integer}\}$.

b) Recall the sets of physical activities $E = \{\text{rt, bi, el, te}\}$ and $L = \{\text{yo, el, te}\}$ from Examples 1 and 2. Find *E* − *L* and *L* − *E*.

SOLUTION:

a) We start with $\{3, 6, 9, 12\}$ and remove all the odd integers to get $\{6, 12\}$.

  Remove elements that are odd.

b) The set $E - L = \{\text{rt, bi, el, te}\} - \{\text{yo, el, te}\} = \{\text{rt, bi}\}$.

  Remove elements that are found in *L*.

Notice that the set $L - E = \{\text{yo, el, te}\} - \{\text{rt, bi, el, te}\} = \{\text{yo}\}$. As you might expect, changing the order of the way we calculate the set difference affects our final answer. ❋ **11**

Quiz Yourself **11**

a) Let $U = \{a, c, e, g, i, k, m, o, q\}$ and $C = \{x : x \text{ is a letter in the word } game\}$. Find *C*′.

b) Let $A = \{1, 3, 4, 5, 6\}$ and $B = \{2, 3, 4, 6, 7, 8\}$. Find *B* − *A* and *A* − *B*.

PROBLEM SOLVING

*The Analogies Principle**

It is important to remember the analogies that we have mentioned between set operations and ordinary English words. For *union*, think of joining sets together to get a larger set. *Intersection* reminds you of overlapping streets—you generally get a smaller set. In computing a *complement*, think about finding the elements needed to complete the set to obtain the universal set. In finding a set difference, remember the *take away* model of subtraction. However, it is important to keep in mind that *these informal analogies are not the definitions*—if your instructor asks you for a definition on an exam, use the formal definitions given in the definition boxes.

✏ *KEY POINT*

Set operations must be performed in the correct order.

Order of Set Operations

Just as we must perform arithmetic operations in a certain order,[†] set notation specifies the order in which we perform set operations.

EXAMPLE 5   *Order of Set Operations*

Let $U = \{1, 2, 3, \ldots, 10\}$, $E = \{x : x \text{ is even}\}$, $B = \{1, 3, 4, 5, 8\}$, and $A = \{1, 2, 4, 7, 8\}$. Find $(A \cup B)' \cap (E' \cup A)$.

SOLUTION: The notation indicates that some of the operations in this expression must be done before others. For example, in considering $(A \cup B)'$, the parentheses tell us to form the union before taking the complement. In the expression $(E' \cup A)$, we first must find E' before calculating the union. The following is one way to do these calculations.

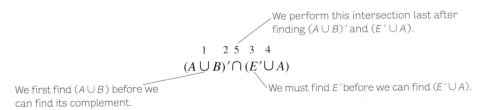

We perform this intersection last after finding $(A \cup B)'$ and $(E' \cup A)$.

$$\overset{1 \quad 2\,5 \quad 3 \quad 4}{(A \cup B)' \cap (E' \cup A)}$$

We first find $(A \cup B)$ before we can find its complement.

We must find E' before we can find $(E' \cup A)$.

We follow these steps in order:

1. $(A \cup B) = \{1, 2, 3, 4, 5, 7, 8\}$
2. $(A \cup B)' = \{6, 9, 10\}$
3. $E' = \{1, 3, 5, 7, 9\}$
4. $(E' \cup A) = \{1, 2, 3, 4, 5, 7, 8, 9\}$
5. $(A \cup B)' \cap (E' \cup A) = \{9\}$

   Now try Exercises 9 to 20. ✻

Sometimes notation leads us into thinking that something is true when it is not. For example, in algebra we know that $(x + y)^2 \neq x^2 + y^2$, even though it looks like it might be so. Similarly, in set theory if we do not think carefully about the statement $(A \cup B)' = A' \cup B'$, it appears as though it could be true. Example 6 shows that you cannot change the order in which you calculate unions and complements; that is, $(A \cup B)' \neq A' \cup B'$.

---

*See Section 1.1.
[†]For a review of the order of operations in arithmetic, see Appendix A. See also the Order Principle in Section 1.1.

HISTORICAL HIGHLIGHT ✺ ✺ ✺

DeMorgan and Boole

Just as biologists classify animals and flowers according to their various characteristics, mathematicians classify mathematical systems according to the properties present in the systems. Set theory is an example of a system called a Boolean algebra, named after the British mathematician George Boole (1815–1864). The son of a shopkeeper, Boole was not able to attend the schools of the more privileged students and had to teach himself Latin, Greek, and mathematics. His book *Investigation of the Laws of*

Thought, published in 1854, is one of the classics in the history of mathematics.

Boole, together with Augustus DeMorgan (1806–1871) and others, developed a method of performing logical computations similar to the way in which we do algebra. Their work formed the mathematical basis for all the digital devices that are so prevalent today, such as MP3 players, TiVo, and digital cameras.

Quiz Yourself 12

Let $A = \{1, 2, 5, 7, 8, 9\}$ and $B = \{2, 3, 5, 6, 7\}$ be subsets of the universal set $U = \{1, 2, 3, \ldots, 10\}$. Find the following.

a) $(A \cap B)'$ b) A'

c) B' d) $A' \cup B'$

EXAMPLE 6 *Order of Calculating Unions and Complements*

Let $U = \{1, 2, 3, 4, 5\}$, $A = \{1, 3, 5\}$, and $B = \{1, 2, 3\}$.

a) Find $(A \cup B)'$.   b) Find $A' \cup B'$.

SOLUTION:

a) This notation tells us to *first* find the union and *then* take the complement. Hence, $(A \cup B)' = \{1, 2, 3, 5\}' = \{4\}$.

b) Here, we first take complements and then find the union. This gives us

$$A' \cup B' = \{2, 4\} \cup \{4, 5\} = \{2, 4, 5\}.$$

Thus, we see that $(A \cup B)' \neq A' \cup B'$. ✿ 12

If you look carefully at Example 6, you will see that $(A \cup B)' = A' \cap B'$. This is an example of one of DeMorgan's laws in set theory. Quiz Yourself 12 illustrates the second of DeMorgan's laws.

DEMORGAN'S LAWS FOR SET THEORY
If A and B are sets, then $(A \cup B)' = A' \cap B'$ and $(A \cap B)' = A' \cup B'$.

Viewing set theory as a mathematical system, we can ask whether set operations satisfy some of the same properties that are present in number systems. For example, we know that multiplication is distributive over addition—for all integers a, b, and c, $a \times (b + c) = (a \times b) + (a \times c)$. Example 7 investigates whether intersection is distributive over union.

We are shading regions that are in B, or C, or both.

EXAMPLE 7 *Intersection Distributes over Union*

Let A, B, and C be sets in a universal set U. Does intersection distribute over union? That is, does

$$A \cap (B \cup C) = (A \cap B) \cup (A \cap C)?$$

SOLUTION: We will answer this question by drawing a Venn diagram for the set on the left-hand side of the equation and another for the set on the right-hand side.

In shading the left-hand set, the parentheses tell us that we must first take the union before considering the intersection. In Figure 2.8, we have shaded $B \cup C$, which consists of regions r_3, r_4, r_5, r_6, r_7, and r_8.

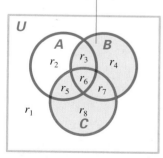

FIGURE 2.8 Venn diagram of $B \cup C$.

We are shading where
A "overlaps" B ∪ C.

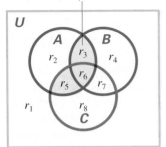

FIGURE 2.9 Venn diagram
of $A \cap (B \cup C)$.

When we intersect $B \cup C$ with set A, we see that of the shaded regions in Figure 2.8, only regions r_3, r_6, and r_5 are also in A. This gives us the Venn diagram of $A \cap (B \cup C)$ shown in Figure 2.9.

Next, we look at the right-hand side of our equation, $(A \cap B) \cup (A \cap C)$. The parentheses tell us to first consider $(A \cap B)$ and $(A \cap C)$ and then take their union. We shade these sets in Figure 2.10.

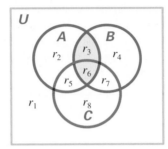 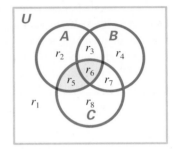

FIGURE 2.10 Venn diagrams of $A \cap B$ and $A \cap C$.

Therefore, $(A \cap B) \cup (A \cap C)$ is made up of regions r_3, r_6, and r_5, which is the same set we shaded in Figure 2.9. Because these two Venn diagrams are the same, this shows that $A \cap (B \cup C) = (A \cap B) \cup (A \cap C)$.

Now try Exercises 27 to 42. ❈

Prior to 1900, physicians were puzzled as to why some patients were saved and others died after receiving blood transfusions. In 1901, Dr. Karl Landsteiner, researcher at the University of Vienna, solved the mystery with his discovery of blood types, for which he later received the Nobel Prize. The next example is based on Landsteiner's classification of blood types, which involves two antigens* labeled A and B. There are four possibilities: A person can have only A, only B, both A and B, or neither of these antigens. The numbers in the next example are based on Red Cross data.

EXAMPLE 8   *Incorrectly Counting the Union of Blood Types*

Suppose we test a group of 100 people and find that 40 people have the A antigen and 11 have the B antigen. If we designate the first set of people by A and the second set by B, what is wrong with the following equation: $n(A \cup B) = n(A) + n(B) = 40 + 11 = 51$?

SOLUTION: In counting the people, we have ignored the fact that some people might possess both the A and B antigens, and so were counted twice. According to the Red Cross, in our group of 100, we can assume that there are 4 people who have both the A and B antigens. So, we have to subtract 4 from our total. That is,

$$n(A \cup B) = n(A) + n(B) - n(A \cap B) = 40 + 11 - 4 = 47.$$

4 people are in both A and B and have been counted twice.

(We will pursue the example of blood typing in greater depth in Section 2.4.) ❈

THE CARDINAL NUMBER OF THE UNION OF TWO SETS
If A and B are sets, then $n(A \cup B) = n(A) + n(B) - n(A \cap B)$.

We must subtract $n(A \cap B)$ so we do not count these elements twice.

---

*Antigens are proteins that stimulate the body's immune systems.

Exercises 2.3

Looking Back*

These exercises follow the general outline of the topics presented in this section and will give you a good overview of the material that you have just studied.

1. Complete the following definition of the union of sets A and B:
$$A \cup B = \{x : \quad \}$$

2. How would your definition change in Exercise 1 if we asked for intersection instead of union?

3. Give a careful definition of what we mean by the complement of set A.

4. Draw a Venn diagram illustrating the set $B - A$.

5. State one of DeMorgan's laws for set theory, and explain how it relates to the Order Principle that we explained in Section 1.1.

6. Why is the equation $n(A \cup B) = n(A) + n(B)$ generally incorrect?

7. Explain how the Analogy Principle can help you remember set theory terminology.

8. What contribution did Boole and DeMorgan make to modern society?

Sharpening Your Skills

In Exercises 9–20, let $U = \{1, 2, 3, \ldots, 10\}$, $A = \{1, 3, 5, 7, 9\}$, $B = \{1, 2, 3, 4, 5, 6\}$, *and* $C = \{2, 4, 6, 7, 8\}$. *Perform the indicated operations.*

9. $A \cap B$
10. $A \cup B$
11. $B \cup C$
12. $B \cap C$
13. $A \cup \varnothing$
14. $A \cap \varnothing$
15. $A \cup U$
16. $A \cap U$
17. $A \cap (B \cup C)$
18. $A' \cap (B \cup C')$
19. $(A - B) \cap (A - C)$
20. $A - (B \cup C)$

In Exercises 21–26, consider the universal set $U = \{$apple, flat-screen TV, hat, satellite radio, fish, sofa, hybrid automobile, potato chip, bread, banana, hammer, pizza$\}$.

Let $M = \{x : x$ *is human-made*$\}$, $E = \{y : y$ *is edible*$\}$, $G = \{t : t$ *grows on a plant*$\}$. *Find each set.*

21. $M \cap E$
22. $M - E$
23. $E - M$
24. E'
25. $M' \cap G'$
26. $G \cap (M' \cap E)$

In Exercises 27–34, represent each set using a Venn diagram.

27. $A - (B \cup C)$
28. $A \cap (B - C)$
29. $(A \cap B) - C$
30. $(A \cup B) - C$
31. $A \cup (B - C)$
32. $A \cup (B \cup C)$
33. $(A \cup (B \cup C))'$
34. $(A \cap (B \cap C))'$

In Exercises 35–42, describe the shaded region using set theory notation.

35.

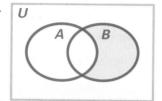

36.

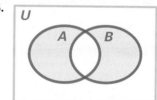

37.

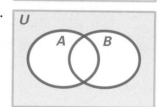

38.

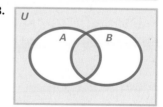

39.

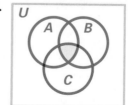

40.

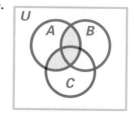

---

*Before doing these exercises, you may find it useful to review the note *How to Succeed at Mathematics* on page xix.

41.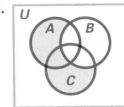

42.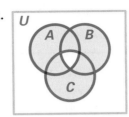

We have indicated the number of elements in each region of the Venn diagram to the right. In Exercises 43–50, find the following.

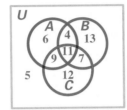

43. $n(A)$

44. $n(A \cup B)$

45. $n(C')$

46. $n(A - C)$

47. $n(A \cap C)$

48. $n(A \cap B \cap C)$

49. $n((A \cup B) \cap C)$

50. $n((A \cup C) - (B \cup C))$

In Exercises 51–54, assume $A \subseteq B$. Express each set in a simpler way.

51. $A \cap B$

52. $A - B$

53. $A \cup B$

54. $A' \cap B'$

Applying What You've Learned

In the following table, we have listed various features that a person might consider when buying a new car.

	Cost	Size	Warranty (years)	Safety Rating	Antitheft Package
a	$17,800	Subcompact	3	Good	Yes
b	$19,500	Midrange	2	Good	No
c	$17,300	Subcompact	3	Moderate	Yes
d	$21,500	Compact	3	Poor	No
e	$17,200	Subcompact	2	Poor	No
f	$18,700	Compact	3	Poor	No
g	$20,000	Compact	4	Moderate	Yes
h	$21,700	Midrange	4	Good	No

In the universal set $U = \{a, b, c, \ldots, h\}$, define subsets of cars with the following characteristics:

P = price is above $19,000
C = is compact
G = has good safety rating
A = has antitheft package
W = warranty is at least three years

In Exercises 55–62, first describe each set in words and then find the set.

55. $P \cap C$

56. $A \cup G$

57. $W \cap G'$

58. $G - A$

59. $P \cap (G \cup W)$

60. $G' \cap C'$

61. $P - (G \cup A)$

62. $P' - (G \cup C)$

In the following table, we have given nutritional information for the following items available at a fast-food restaurant:* Monster (m), Monster with Cheese (mc), Bacon Cheeseburger (bc), Hamburger (h), Cheeseburger (c), Fish Sandwich (fs), and Ham Cheesy (hc).

	% of Minimum Daily Allowance (MDA)			
	Protein	Vitamin A	Vitamin B1	Calcium
m	42	14	25	8
mc	49	21	25	21
bc	49	8	20	17
h	23	3	16	4
c	27	7	16	10
fs	29	—	18	5
hc	37	15	58	19

Let P = {x : x is an item that provides at least 30% of the MDA for protein}, C = {x : x is an item that provides at least 10% of the MDA for calcium}, B = {x : x is an item that provides at least 25% of the MDA for vitamin B1}, A = {x : x is an item that provides at least 15% of the MDA for vitamin A}

In Exercises 63–66, find each set.

63. $P \cap (B \cup A)$

64. $(P \cup C) \cap (B \cup A)$

65. $P \cup C \cup B$

66. $P \cap C \cap B$

*This information is based on menu items at a well-known fast-food restaurant.

According to the New York Times Almanac (2007), when adjusted for inflation, the 10 top-grossing movies* of all time are those listed in the table below. Define the following sets:

E = the set of movies that earned more than $800 million
O = the set of movies that won an Oscar
B = made before 1960

Movie	Year	Inflation-Adjusted Gross ($ millions)	Won Oscar
a. *Birth of a Nation*	1915	789	No
b. *Snow White and the Seven Dwarfs*	1937	698	No
c. *Gone With the Wind*	1939	1,293	Yes
d. *The Ten Commandments*	1956	838	No
e. *101 Dalmatians*	1961	639	No
f. *The Sound of Music*	1965	911	Yes
g. *Star Wars*	1977	1,140	No
h. *E.T.*	1982	908	No
i. *Titanic*	1997	827	Yes

Describe each of the following sets in words and then list its elements as a set. For example B = {a, b, c, d}.

67. $E \cap B$ **68.** E' **69.** $O \cap B'$

70. $B - O$ **71.** $E \cap B \cap O$ **72.** $(E \cup B)'$

Communicating Mathematics

73. How does the usage of the terms *union* and *intersection* in the English language help you "remember how to remember" their meaning in set theory?

74. What does the notion of set difference have to do with the idea of removing objects?

75. Students often misstate one of DeMorgan's laws as $(A \cup B)' = A' \cup B'$? Why do you think this is so?

76. Give some examples in this section showing how the Three-Way Principle from Section 1.1 can help you understand set theory concepts.

In Exercises 77–82, decide whether each statement is always true. Explain your answer by considering appropriate examples or by drawing a Venn diagram. If you think that a statement is not always true, provide a counterexample. Assume that all sets are finite.

77. If $A \subseteq B$, then $n(A) < n(B)$.

78. If $A - B = \varnothing$, then $A \subseteq B$.

79. If $n(A) = n(B)$ and $A \subseteq B$, then $A = B$.

80. If $A \cup B = A \cap B$, then $A = B$.

81. $n(X - Y) = n(X) - n(Y)$

82. If $X \subseteq Y$, then $n(Y - X) = n(Y) - n(X)$.

Using Technology to Investigate Mathematics

In Exercises 83 and 84, consider the following. When using an Internet search engine, the instructions for the search are given by specifying keywords. For example, you can ask for a search to locate sites that discuss music by specifying the keyword "music." If you want to make a more complex search, you can join keywords using "OR" or "AND." If you want to search for sites discussing music or art, you can specify "music OR art." If you specify "music AND alternative," the search engine identifies sites dealing with both "music" and "alternative," that is, alternative music.

83. When you use AND, what set operation are you asking the search engine to perform? What about OR?

84. If we specify "music AND American OR British," it is not clear how the search engine would group the words specified. Using parentheses, we can group as follows:

(music AND American) OR British

or

music AND (American OR British).

Are both requests the same? What does this have to do with set theory? Discuss. If you have access to a search engine, see whether you can learn how it handles compound requests for searches.

85. Applets are interactive programs written in the Java programming language that are often used to illustrate various concepts in mathematics in a dynamic way. Do an Internet search for "set theory and applets" to find applets illustrating some of the concepts that we have discussed in this section. After you run an applet, write a brief description of what it does, and explain how the applet illustrates some concepts that you have learned about set theory.

86. Search the Internet for "fuzzy set theory" and write a brief report on your findings.

For Extra Credit

In comparing number systems with set theory, it is useful to notice that union is in some ways similar to addition and that intersection is in some ways similar to multiplication. In Exercises 87–90, for each familiar property of the real number system,

a) State the corresponding property for set theory.

b) Determine whether the property stated in (a) is true.

87. $a \cdot b = b \cdot a$

88. $a + b = b + a$

89. $(a \cdot b) \cdot c = a \cdot (b \cdot c)$

90. $(a + b) + c = a + (b + c)$

91. We know that multiplication is distributive over addition, and we have shown in Example 7 that intersection distributes over union. Show that addition does not distribute over multiplication by providing an appropriate counterexample.

92. Does union distribute over intersection? Use Venn diagrams to confirm your answer.

---

*I found it interesting when researching this example to compare the earnings of these older classics with the earnings (in millions) of more recent "blockbusters" such as *Spider-Man 3* ($336.5), *Transformers* ($319), and *Pirates of the Caribbean: At World's End* ($309).

2.4 Survey Problems

Objectives

1. Label sets in Venn diagrams with various names.
2. Use Venn diagrams to solve survey problems.
3. Understand how to handle contradictory information in survey problems.

If you were to buy a sleek, new Razr2 cell phone, the manufacturer might ask you to fill out a survey asking, "Will you use this phone to download videos? What is your age? Will you use this phone to share photos? What features are most important to you?" The company would organize your responses in a database, and then use set theory principles to extract information that it can use in designing and advertising new products.

Governmental agencies also gather information about you and your family to design legislation to help keep college affordable and to ensure that Social Security benefits will still be available when your parents retire.

In this section, we will show how to use set theory to solve survey problems, but first we need to sharpen our skills in working with Venn diagrams.

KEY POINT

Regions of a Venn diagram can have many different names.

Naming Venn Diagrams

In everyday life, it is common for a person to be known by many names. For example, a woman might be known at work as Ms. Smith, at home by Mom or Honey, by her sister as Sis, by her Internet provider as ASmith123, and by the IRS as 178-34-7886. Similarly, the same set can have many different names. For example, consider the four regions shown in the Venn diagram in Figure 2.11.

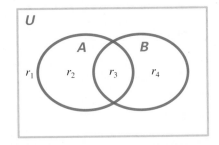

FIGURE 2.11 A Venn diagram with two sets.

We can use set notation to name these regions. For example, region r_4 is $B - A$, and region r_1 is $(A \cup B)'$. These names are not unique. Using one of DeMorgan's laws, region r_1 could also be called $A' \cap B'$.

As you can see, we can represent the same set in different ways, just as we often express numbers in different ways. Depending on what calculations we are doing, we may write 6 one time as $3 + 3$ and another time as $5 + 1$. These ideas carry over to Venn diagrams with more than two sets.

EXAMPLE 1 *Naming Regions of a Three-Set Venn Diagram*

Use Figure 2.12(a) on page 65 to answer each question.

a) Name the set we get by combining regions r_6 and r_7.

b) What is a set name for region r_2?

c) Express $A - B$ as the union of two sets.

SOLUTION:

a) Regions r_6 and r_7 are precisely the regions common to B and C; therefore, this set is $B \cap C$. See Figure 2.12(b) on page 65.

b) Region r_2 is clearly *part* of set A; however, elements in region r_2 are not elements of B, nor are they elements of C. Therefore, one name for this set is $A \cap B' \cap C'$. See Figure 2.12(c) on page 65.

Notice that another way you can view this set is to start with set A (regions $r_2, r_3, r_5,$ r_6) and remove those regions that are in $B \cup C$, namely, $r_3, r_5,$ and r_6. Thus, we also can name region r_2 as $A - (B \cup C)$.

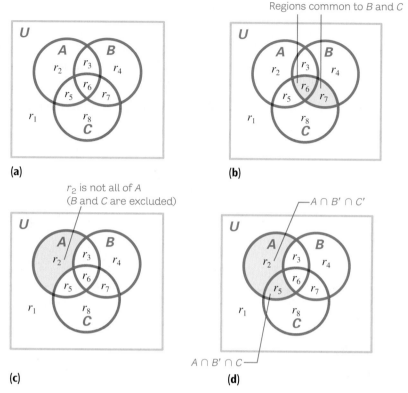

FIGURE 2.12 (a) A Venn diagram with three sets. (b) Venn diagram of $B \cap C$. (c) Venn diagram of $A \cap B' \cap C'$. (d) Venn diagram of $A - B$.

Quiz Yourself **13**

Use Figure 2.12 to answer each question.

a) What is a set name for region r_8?

b) Express $B - C$ as the union of two sets.

c) $A - B$ consists of regions r_2 and r_5, which have set names $A \cap B' \cap C'$ and $A \cap B' \cap C$. We can write $A - B = (A \cap B' \cap C') \cup (A \cap B' \cap C)$. See Figure 2.12(d) above. Now try Exercises 7 to 18. ❋ **13**

🖉 **KEY POINT**

Venn diagrams can organize information in a survey problem.

Survey Problems

When we collect data and organize it into sets, we then usually want to analyze the information to answer questions about those sets. These types of problems are often called **survey problems.** Example 2 is an example of a survey problem.

EXAMPLE 2 *A Survey Problem Involving Solutions for Global Warming*

The National Resource Defense Council[*] believes that by using technology properly, we can cut U.S. global warming pollution by half. Three of the solutions proposed by the NDRC are using energy-efficient appliances, driving energy-efficient cars, and using renewable energy sources. Assume that you surveyed 100 members of Congress to determine which solutions they favored funding and obtained the following results:

a) 12 favored funding the increased use of renewable energy sources only.

b) 20 recommended funding both energy-efficient appliances and renewable energy sources.

---

*The NDRC has 1.2 million members and was instrumental in drafting and sponsoring California's Global Warming Solutions Act.

c) 22 favored funding both energy-efficient cars and increased use of renewable energy sources.

d) 14 want to fund all three areas.

From this information, determine the total number who favored increased funding for renewable energy.

SOLUTION: We will represent this information by defining the following set:

$$A = \{x : x \text{ favors funding energy-efficient appliances}\}$$
$$C = \{x : x \text{ favors funding energy-efficient cars}\}$$
$$R = \{x : x \text{ favors funding renewable energy sources}\}$$

Then conditions (a) – (d) can be rewritten using set notation as:

a) $n(A' \cap C' \cap R) = 12$ b) $n(A \cap R) = 20$ c) $n(C \cap R) = 22$ d) $n(A \cap C \cap R) = 14$

favor R, not A, not C   favor A and R   favor C and R   favor A, C, and R

We will display this information in a Venn diagram. For example, you can see in Figure 2.13(a) the fact that $n(A \cap C \cap R) = 14$ and $n(A' \cap C' \cap R) = 12$.

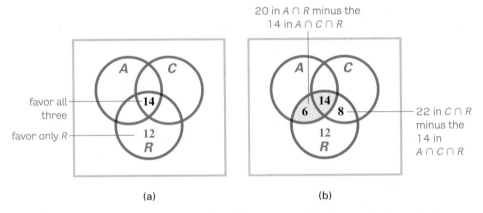

FIGURE 2.13 (a) Venn diagram showing conditions a) and d). (b) Venn diagram showing conditions b) and c).

We have shaded $A \cap R$ in Figure 2.13(b). Because we have already accounted for 14 in $A \cap C \cap R$, there are 6 remaining in $(A \cap R) - C$. Similarly, we can use the fact that $n(C \cap R) = 22$ to get 8 in $(C \cap R) - A$. Adding these numbers, we see that R contains $6 + 14 + 8 + 12 = 40$ members of Congress. ❁

Example 3 analyzes a survey problem that requires a Venn diagram with three sets.

EXAMPLE 3   *A Survey of TV Viewer Preferences*

A television network conducted a market survey to determine the evening viewing preferences of people in the 18–25 age bracket. The following information was obtained:

a) 3 prefer a reality show early on weekdays.

b) 14 want to watch TV early on weekdays.

c) 21 want to see reality shows early.

d) 8 want reality shows on weekdays.

e) 31 want to watch TV on weekdays.

f) 36 want to watch TV early.

g) 40 want to see reality shows.

h) 13 prefer late, weekend shows that are not reality shows.

From this information, determine how many people do not want to see reality shows and how many prefer to watch TV on the weekend.

SOLUTION: The set of people surveyed is a universal set containing three subsets.

$$W = \text{those who prefer to watch TV on weekdays}$$
$$E = \text{those who desire early programming}$$
$$R = \text{those who want to see reality shows}$$

We draw these sets in Figure 2.14. By condition a), we know that there are 3 people in $W \cap E \cap R$. Because condition b) states there are 14 people in $W \cap E$, and we have counted 3 of them, there must be 11 more in this region. We have recorded this information in Figure 2.14(a).

By condition d), 8 people want to see weekday reality shows, so we can deduce that 5 have not yet been counted in $W \cap R$. Similarly, from condition c), we see that 18 people have not been counted in $R \cap E$. This is shown in Figure 2.14(b).

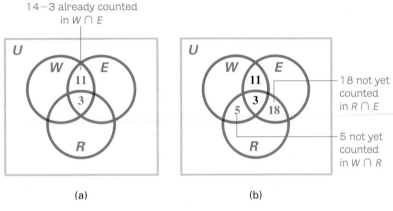

FIGURE 2.14 Counting elements in $W \cap E$, $W \cap R$, and $R \cap E$.

Condition e) states that 31 people desire weekday programming, but we have already counted 19. Therefore, there are 12 more in set W. Similarly, there are 4 more people to be counted in set E and 14 more in set R. Condition h) tells us that there are 13 people outside sets W, E, and R. We summarize this information in Figure 2.15.

With this information, we can answer the original questions. The people wanting to see nonreality shows fall outside R, so they total $12 + 11 + 4 + 13 = 40$. Those who prefer weekend programming are outside W. This count is $14 + 18 + 4 + 13 = 49$.

Now try Exercises 33 to 44. ❈

Contradictions in Survey Problems

It is possible to list a set of conditions on sets that cannot all hold at the same time.

EXAMPLE 4   *Inconsistent Survey Data*

Suppose that an Internet blog states the following information concerning which Web browsers its readers use:

a) 316 use WebMagic.

b) 478 use iBrowse.

c) 104 use both WebMagic and iBrowse.

d) 567 use only one of the Web browsers.

Find the inconsistency in these data.

FIGURE 2.15 The number of elements in each region of the Venn diagram.

✏️ **KEY POINT**

A survey problem may contain contradictory information.

SOLUTION: We begin by drawing a Venn diagram in Figure 2.16(a). Let W be the set of WebMagic users and let I be the set of those who use iBrowse.

Figure 2.16(a) shows the 104 people who use both Web browsers. Because W contains 316 elements, this means that $316 - 104 = 212$ elements are in $W - I$. Similarly, $478 - 104 = 374$ people use only iBrowse, so $I - W$ contains 374 elements. We show this new information in Figure 2.16(b).

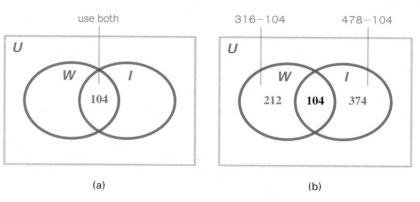

(a)   (b)

FIGURE 2.16 (a) 104 use both WebMagic and iBrowse. (b) 212 use only WebMagic; 374 use only iBrowse.

According to Figure 2.16(b), $212 + 374 = 586$ people use only one of the two Web browsers. This directly contradicts condition d). Therefore, these data are inconsistent. ✵

Information may be organized in a computer's database in table form. We analyze such a situation in Example 5.

EXAMPLE 5   *Surveying Media Users*

A media company that produces Pilates videos surveyed its customers to determine whether they would prefer to purchase the videos by buying an access code to allow them to download from a Web site, purchase hard copies on a DVD, or access the videos via a podcast. The company has extracted the following table of information from its customer database:

	Access Code (*A*)	Podcast (*P*)	DVD (*D*)	Total
Under 41 (*Y*)oung	20	15	9	44
41 to 55 (*M*)iddle-aged	44	**34**	8	86
Over 55 (*S*)enior	31	14	5	50
Total	95	63	22	180

a) Find the number of elements in $M \cup D$.

b) Find the number of elements in A'.

SOLUTION:

a) The number of elements in $M \cup D$ is the number of people who are middle-aged or prefer DVDs. Thus,

$$n(M \cup D) = n(M) + n(D) - \overset{\text{counted twice}}{n(M \cap D)} = 86 + 22 - 8 = 100.$$

b) A' is the set of people who do not want to buy an access code, so

$$n(A') = 63 + 22 = 85.$$

Now try Exercises 45 to 48. ✵ **14**

Quiz Yourself **14**

Use the table in Example 5 to find $n(Y \cup D)$.

HISTORICAL HIGHLIGHT ✺ ✺ ✺

Georg Cantor and Set Theory

It is unlikely that Georg Cantor, the nineteenth-century German mathematician who invented many of the set theory ideas that you have been studying, would have foreseen how his theory would be used so commonly in today's society. It was surely not Cantor's intent to develop mathematics that would be used to determine consumer preferences, help a government keep track of its citizens, or organize information about suspected terrorists in a database.

As you will see in the next section, Cantor discovered some remarkable ideas regarding infinite sets—ideas that were not accepted by his contemporaries. His former teacher

Leopold Kronecker called Cantor's methods "a dangerous type of mathematical insanity" and the distinguished Frenchman Henri Poincaré proclaimed, "Later generations will regard [Cantor's set theory] as a disease from which one has recovered."

Today, however, most mathematicians have accepted Cantor's results, and set theory now provides a foundation for virtually all of mathematics. The eminent twentieth-century mathematician David Hilbert said that set theory is ". . . one of the highest achievements of man's intellectual processes."

Exercises 2.4

Looking Back*

These exercises follow the general outline of the topics presented in this section and will give you a good overview of the material that you have just studied.

1. Express $A \cup B$ in Figure 2.11 as the union of three *disjoint* sets. That is, fill in the equation $A \cup B = (set\ 1\quad) \cup (set\ 2\quad) \cup (set\ 3\quad)$.

2. Express A in Figure 2.12(a) as the union of four disjoint sets.

3. Explain how we obtained the 8 elements for the region shown in Figure 2.13.

4. In constructing Example 4, we first drew a Venn diagram, put numbers in the various regions, and then wrote the conditions for the problem. Do the same thing, but use different numbers to make another example with similar contradictory information.

5. How were Cantor's ideas regarded when they were first introduced?

6. According to the Historical Highlight box on Cantor, what are some of the ways that set theory is applied in our everyday life?

Sharpening Your Skills

In Exercises 7–12, determine which numbered regions make up the indicated set.

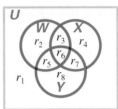

7. X

8. Y

9. $X \cap Y$

10. $X \cup Y$

11. $X - Y$

12. $Y - X$

In Exercises 13–18, describe each set by referring to the numbered regions, as in Example 1.

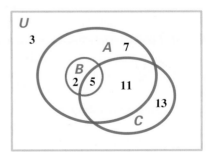

13. W 14. $W \cap Y$

15. $X - W$ 16. $W - (X \cup Y)$

17. $X \cap Y \cap W'$ 18. $X \cap Y \cap W$

The numbers in the regions of the given Venn diagram indicate the number of elements in each region. Use this diagram to solve Exercises 19–24.

19. $n(A)$ 20. $n(C')$

21. $n(A - B)$ 22. $B \cup C$

23. $n(A \cap C)$ 24. $n(B - A)$

---

*Before doing these exercises, you may find it useful to review the note *How to Succeed at Mathematics* on page xix.

The numbers in the regions of the given Venn diagram indicate the number of elements in each region. Use this diagram to solve Exercises 25–28.

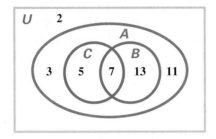

25. $n(B)$

26. $n(C')$

27. $n(A - B)$

28. $(B \cup C)'$

In Exercises 29–32, find, if possible, the number of elements in sets A, B, and C using the given information. If there is an inconsistency in the information, state where it occurs.

29. $n(A \cap B) = 5$, $n(A \cap B \cap C) = 2$, $n(B \cap C) = 6$, $n(B - A) = 10$, $n(B \cup C) = 23$, $n(A \cap C) = 7$, $n(A \cup B \cup C) = 31$.

30. $n(B - C) = 4$, $n(C - B) = 9$, $n(A \cap B \cap C) = 3$, $n(B \cup C) = 22$, $n(A - C) = 7$, $n(A \cap B) = 7$, $n(A \cap C) = 5$.

31. $A \cap C = \varnothing$, $n(A \cap B) = 3$, $n(C - B) = 2$, $n(B - C) = 7$, $n(B \cup C) = 16$, $n(A \cup B) = 16$.

32. $A \subset B$, $A \cap C = \varnothing$, $n(C - A) = 8$, $n(A \cup C) = 12$, $n(C - B) = 3$, $n(B \cup C) = 17$.

Applying What You've Learned

33. Automobile accidents. An investigation of a number of automobile accidents revealed the following information:

a. 18 accidents involved alcohol and excessive speed.

b. 26 accidents involved alcohol.

c. 12 accidents involved excessive speed but not alcohol.

d. 21 of the accidents involved neither alcohol nor excessive speed.

How many accidents were investigated?

34. Perceptions of students. The political science department at a university surveyed a number of politicians regarding their perceptions of college students and gathered the following information:

a. 15 saw students as hardworking and economically advantaged.

b. 25 saw students as hardworking.

c. 11 believed students were economically advantaged but not hardworking.

d. 16 believed students were neither hardworking nor economically advantaged.

How many surveyed believed students were economically advantaged?

35. There are 82 people protesting the destruction of the rain forests. There are 34 males, 23 of whom are students, and 20 female nonstudents. How many female students are protesting?

36. There are 95 students who have applied for a scholarship. There are 18 minorities, 7 of whom are athletes. If there are 20 athletes applying, how many applicants are neither an athlete nor a minority?

37. Survey of vacationers. A survey is taken of 100 people who vacationed at a dude ranch. The following information was obtained:

a. 17 took horseback-riding lessons, attended the Saturday night barbecue, and purchased a tour guide.

b. 28 attended the Saturday night barbecue and purchased a tour guide.

c. 24 took horseback-riding lessons and purchased a tour guide.

d. 42 took horseback-riding lessons but didn't attend the barbecue.

e. 86 took horseback-riding lessons or purchased a tour guide.

f. 14 purchased only a tour guide.

g. 14 did none of these three things.

How many attended the barbecue? How many purchased a tour guide?

38. Repeat Exercise 37, given that

a. 19 took horseback-riding lessons, attended the Saturday night barbecue, and purchased a tour guide.

b. 34 attended the Saturday night barbecue and purchased a tour guide.

c. 30 took horseback-riding lessons and purchased a tour guide.

d. 33 took horseback-riding lessons but did not attend the barbecue.

e. 86 took horseback-riding lessons or purchased a tour guide.

f. 8 purchased only a tour guide.

g. 3 did none of these three things.

How many attended the barbecue or purchased the tour guide? How many did not purchase a tour guide?

39. Media survey. A survey of young adults was taken to determine which of the various sources they use to obtain news. Of the 36 people who use the Internet, 13 use the Internet only to learn the news. Of the 48 people who use the newspaper, 11 use the newspaper only for news coverage. There are 23 people who use newspapers, the Internet, and television for news coverage. From this information, determine how many use both the Internet and television to learn the news.

40. Mass transit survey. A survey was made of 200 city residents to study their use of mass transit facilities. According to the survey:

a. 83 did not use mass transit.

b. 68 used the bus.

c. 44 used only the subway.

d. 28 used both the bus and subway.

e. 59 used the train.

Explain how you can use this information to deduce that some residents must use both the bus and train.

41. Pandora.com surveyed a group of subscribers regarding which online music channels they use on a regular basis. The following information summarizes their answers:

7 listened to rap, heavy metal, and alternative rock.

10 listened to rap and heavy metal.

13 listened to heavy metal and alternative rock.

12 listened to rap and alternative rock.

17 listened to rap.

24 listened to heavy metal.

22 listened to alternative rock.

9 listened to none of these three channels.

a. How many people were surveyed?

b. How many people listened to either rap or alternative rock?

c. How many listened to heavy metal only?

42. *Personal Fitness Magazine* surveyed a group of young adults regarding their exercise programs and the following results were obtained:

3 were using resistance training, Tae Bo, and Pilates to improve their fitness.

5 were using resistance training and Tae Bo.

12 were using Tae Bo and Pilates.

8 were using resistance training and Pilates.

15 were using resistance training only.

30 were using Tae Bo.

17 were using Pilates but not Tae Bo.

14 were using something other than these three types of workouts.

a. How many people were surveyed?

b. How many people were using only Tae Bo?

c. How many people were using Pilates but not resistance training?

43. The Dean of Academic Services surveyed a group of students about which support services they were using to help them improve their academic performance and found the following results:

5 were using office hours, tutoring, and online study groups to improve their grades.

16 were using office hours and tutoring.

28 were using tutoring.

14 were using tutoring and online study groups.

8 were using office hours and online study groups but not tutoring.

23 were using office hours but not tutoring.

18 were using only online study groups.

37 were using none of these services.

a. How many students were surveyed?

b. How many students were using only office hours?

c. How many students were using online study groups?

44. A group of young adults who were asked which issues would be important during the next decade gave the following answers:

13 believed that nuclear war, terrorism, and environmental concerns would be important.

43 believed that nuclear war would be important.

17 believed that nuclear war and terrorism would be important.

23 believed that nuclear war or terrorism but not the environment would be important.

28 believed that nuclear war and the environment would be important.

18 believed that terrorism but not nuclear war would be important.

7 believed that only the environment would be a serious issue.

6 believed that none of these issues would be important.

a. How many people were surveyed?

b. How many people believed the environment would be an important issue?

c. How many people believed only terrorism would be an important issue?

In Exercises 45–48, use the following table from Example 5 to find the cardinal number of each set.

	Access Code (A)	Podcast (P)	DVD (D)	Total
Under 41 (Y)oung	20	15	9	44
41 to 55 (M)iddle-aged	44	34	8	86
Over 55 (S)enior	31	14	5	50
Total	95	63	22	180

45. $A \cup P$ **46.** $S \cap D$ **47.** $A - (M \cup S)$ **48.** $Y \cap (A \cup D)$

In Example 8 of Section 2.3, we discussed Nobel laureate Dr. Karl Landsteiner's discovery of the antigens A and B in human blood. According to which antigens are present, a person is classified as A, B, AB (has both), or O (has neither). It was later discovered that a person may or may not have the Rh (Rhesus) factor. A person having the Rh factor is classified as Rh positive (+) and those lacking the factor Rh negative (−). For example, if a person has the A antigen, lacks the B antigen, and has the Rh factor, her blood would be classified as A⁺. A person with AB⁻ blood has both the A and B antigens, but lacks the Rh factor. The Venn diagram below shows the eight possible blood types that are based on this classification. In Exercises 49–52 describe the blood type(s) in each region.

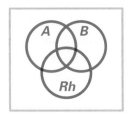

49. $A - (B \cup Rh)$ **50.** $(A \cup B \cup Rh)'$

51. $(A \cup B) - Rh$ **52.** $(B \cap Rh) - A$

Continuing our discussion of blood types, a person can safely receive a transfusion from another person provided the receiver has all of the A, B, and Rh factors of the donor. So, a person with B⁺ blood can receive blood from persons with types B⁺, B⁻, O⁺, and O⁻, but not from types A⁺, A⁻, AB⁺, or AB⁻. Use this information to answer Exercises 53 and 54.

53. Describe using set notation the set of blood types of persons that can receive blood from anyone (universal receivers).

54. Describe using set notation the set of blood types of persons that can donate blood to anyone (universal donors).

Communicating Mathematics

55. In Figure 2.11, students often call region r_2 by the name A. What is wrong with this thinking?

56. In Figure 2.12(a), students often call region r_3 by the name $A \cap B$. What is wrong with this thinking?

Using Technology to Investigate Mathematics

57. Search the Internet to find an area of interest and gather information to write a survey problem similar to Exercises 41–44. Write the survey problem and have your classmates solve it. Try to be creative in the ways that you give the clues.

58. Search the Internet for sites that explain how to construct Venn diagrams and report on an interesting site that you find.

For Extra Credit

59. Notice that if a Venn diagram has one set A, then it divides the universal set into two regions, A and A'. If the diagram has two sets A and B, then there will be four regions: $A \cap B$, $A \cap B'$, $A' \cap B$, and $A' \cap B'$. What is the maximum number of regions for a three-set Venn diagram? Name the regions in a manner similar to what we have just done to one- and two-set Venn diagrams.

60. Thinking along the lines of Exercise 59, what do you expect to be the maximum number of regions possible for a Venn diagram with four sets? Name them as in Exercise 59. Which of these regions are missing in the following Venn diagram with four sets?

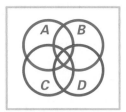

61. Communicating Mathematics. Make up a survey question whose conditions will satisfy the given Venn diagram.

62. Survey of aerobic activities. Draw a Venn diagram that summarizes numerically the information in the following table regarding sets C (burns more than 600 calories per hour), A (can be done alone), and W (can be done in any kind of weather).

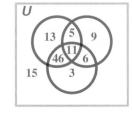

Activity	Calories Burned per Hour	Can Be Done Alone	Can Be Done in Any Kind of Weather
Ballroom dancing (ba)	300	No	Yes
Bench stepping (be)	610	Yes	Yes
Bicycling (bi)	415	Yes	No
Calisthenics (ca)	300	Yes	Yes
Hiking (hi)	400	Yes	No
Jogging (jo)	655	Yes	No
Elliptical machine (el)	655	Yes	Yes
Tennis (te)	425	No	No

Looking Deeper
Infinite Sets

Objectives

1. Understand the definition of infinite sets.
2. Show that the set of natural numbers and the set of integers are countable.
3. Recognize that the set of rational numbers is countable.
4. Recognize that not all infinite cardinal numbers are the same size.

In the late nineteenth century, the great German mathematician Georg Cantor* made the remarkable discovery that infinite sets can have different sizes. In this section, we will explain some of Cantor's revolutionary ideas; but first we will use the notion of one-to-one correspondence to make the notion of cardinal number more precise.

> **DEFINITION** Two sets A and B are in **one-to-one correspondence** if we can pair the elements of the two sets so that every element of A is paired with exactly one element of B and every element of B is paired with exactly one element of A.

For example, the set $T = \{$Carreras, Domingo, Pavarotti$\}^{\dagger}$, a set of famous tenors, is in one-to-one correspondence with $F = \{$Beyonce, Aguilera, Clarkson$\}$, a set of popular female vocalists because we can pair each tenor in T with exactly one singer in F as follows:

$$\begin{array}{ccc} \text{Carreras} & \text{Domingo} & \text{Pavarotti} \\ \updownarrow & \updownarrow & \updownarrow \\ \text{Beyonce} & \text{Aguilera} & \text{Clarkson} \end{array}$$

Quiz Yourself 15

Show another one-to-one correspondence between T and F.

Of course, there are several other ways of pairing the elements of T with the elements of F in a one-to-one correspondence, and we will investigate this further in the exercises. **15**

We can now say that a set A has cardinal number n, or is finite, if we can place A in a one-to-one correspondence with the set $\{1, 2, 3, \ldots, n\}$. For example, if $A = \{x : x$ is a letter of the alphabet$\}$, then $n(A) = 26$ because we have the following one-to-one correspondence between A and $\{1, 2, 3, \ldots, 26\}$:

$$\begin{array}{cccc} a & b & c & \cdots & z \\ \updownarrow & \updownarrow & \updownarrow & & \updownarrow \\ 1 & 2 & 3 & \cdots & 26 \end{array}$$

Infinite Sets

Notice that we cannot place a finite set in a one-to-one correspondence with one of its proper subsets. Cantor used this simple observation to define infinite sets.

> **DEFINITION** A set is **infinite** if we can place it in a one-to-one correspondence with a proper subset of itself.

---

*See the Historical Highlight on page 69.
†The classical album *Carreras—Domingo—Pavarotti: The Three Tenors in Concert (1990)* holds the Guinness world record for the best-selling classical album of all time.

Quiz Yourself **16**

Find another one-to-one correspondence between N and one of its proper subsets. Be sure to include the general pattern for the correspondence.

EXAMPLE 1   *The Set of Natural Numbers Is Infinite*

Show that the set of natural numbers is infinite.

SOLUTION: Let $N = \{1, 2, 3, \ldots\}$ and $E = \{2, 4, 6, \ldots\}$, then the following is a one-to-one correspondence between N and its proper subset E:

$$
\begin{array}{ccccc}
1 & 2 & 3 & \cdots & n & \cdots \\
\updownarrow & \updownarrow & \updownarrow & & \updownarrow & \\
2 & 4 & 6 & \cdots & 2n & \cdots
\end{array}
$$

✤ **16**

Countable Sets

Instead of simply saying that the set of natural numbers is infinite, we can now be more precise.

> **DEFINITION** Cantor called the cardinal number for the set of natural numbers $\aleph_0$. The symbol $\aleph$, aleph, is the first letter of the Hebrew alphabet. We pronounce $\aleph_0$ as "Al-luf null" or "Al-luf naught." Therefore, we can write $n(N) = \aleph_0$. Any set that is either finite or that is in a one-to-one correspondence with N, and therefore has cardinal number $\aleph_0$, is called a **countable set.**

EXAMPLE 2   *The Set of Integers Is Countable*

Show that I, the set of integers, can be put in a one-to-one correspondence with the set of natural numbers.

SOLUTION: The following matching shows the correspondence. (In the correspondence, we are assuming n represents a positive integer.)

$$
\begin{array}{ccccccccccc}
0 & 1 & -1 & 2 & -2 & 3 & -3 & \cdots & n & -n & \cdots \\
\updownarrow & \updownarrow & \updownarrow & \updownarrow & \updownarrow & \updownarrow & \updownarrow & & \updownarrow & \updownarrow & \\
1 & 2 & 3 & 4 & 5 & 6 & 7 & \cdots & 2n & 2n+1 & \cdots
\end{array}
$$

Therefore, $n(I) = \aleph_0$, and I is a countable set. ✤

When we work with infinite sets, we find some surprising results. It would seem that there are so many more rational numbers than there are integers, that we might expect that the cardinal number of the rational numbers is larger than $\aleph_0$. As Example 3 shows, this is not the case.

```
1/1  2/1  3/1  4/1  5/1 ...
1/2  2/2  3/2  4/2  5/2 ...
1/3  2/3  3/3  4/3  5/3 ...
1/4  2/4  3/4  4/4  5/4 ...
1/5  2/5  3/5  4/5  5/5 ...
```

FIGURE 2.17 Listing the positive rational numbers.

EXAMPLE 3   *The Set of Rational Numbers Is Countable*

Show that the set of rational numbers is countable.

SOLUTION: First consider the positive rational numbers as we list them in Figure 2.17. The first row of the arrangement has all positive rational numbers with denominator 1. The second row has all denominators 2, the third row has denominators of 3, and so on. We then trace through the arrangement following the line in Figure 2.17, skipping over numbers that we have encountered earlier.

We will follow the path through the rational numbers in Figure 2.17 and list those numbers in a straight line, matching them with the natural numbers as follows:

$$
\begin{array}{cccccccccccccc}
1/1, & 2/1, & 1/2, & 1/3, & 2/2, & 3/1, & 4/1, & 3/2, & 2/3, & 1/4, & 1/5, & 2/4, & 3/3, & 4/2, & 5/1 \cdots \\
\updownarrow & \updownarrow & \updownarrow & \updownarrow & \text{skip} & \updownarrow & \updownarrow & \updownarrow & \updownarrow & \updownarrow & \updownarrow & & \text{skip} & & \updownarrow \\
1, & 2, & 3, & 4, & & 5, & 6, & 7, & 8, & 9, & 10, & & & & 11 \cdots
\end{array}
$$

Notice that in our matching, we have skipped over rational numbers that we have encountered earlier, such as 2/2, 2/4, 3/3, and 4/2.

Quiz Yourself ⓱

a) List the next three numbers after 5 that we would encounter if we were to continue the listing of the positive rational numbers that we began in Example 3.

b) What is the next number that we would skip after 4/2?

When we look at the positive rational numbers in this way, we see that there is a first number, then a second, then a third, and so on, and we are listing each number exactly once. Therefore, there is a one-to-one correspondence between *N* and the positive rational numbers.

To show that the entire set of rational numbers is countable, we would argue as we did in Example 2 to account for zero and the negative rationals; however, we will not go into the details of how to do it. Therefore, the set of rational numbers has the same cardinal number as the natural numbers, which is $\aleph_0$. ✳ ⓱

The Cardinal Number *c*

So far, the infinite sets that you have studied have all been countable. We will next show you a set that has a cardinal number greater than $\aleph_0$. Because the argument that we will show you is a little sophisticated, we will look at a simpler version of our reasoning before going on to the real thing.

When I introduce infinite sets to my students, I often have a class of about 35 students in a room with about 42 chairs. I ask them to tell me as quickly as they can whether there are more students or more chairs in the room. Instantly, a student responds "more chairs." I ask the student how she counted so quickly and she tells me, "I didn't." I ask, "Then how did you know that there are more chairs?" She explains that every student is seated and yet there are empty chairs left over. You will see this same type of reasoning in Example 4.

EXAMPLE 4   *A Cardinal Number Greater Than* $\aleph_0$

We will reproduce Cantor's argument that the set of real numbers between 0 and 1 has cardinal number greater than $\aleph_0$. We begin by assuming that we *can* put the set of numbers between 0 and 1 in a one-to-one correspondence with the natural numbers and show that no matter how we try, there will always be some number that we could not have listed.

Although we would not actually know what the listing would be, for the sake of argument, let us assume that we had listed all the numbers between 0 and 1 as follows:

$$1 \leftrightarrow 0.6348291347 \ldots$$
$$2 \leftrightarrow 0.2373261008 \ldots$$
$$3 \leftrightarrow 0.4821063391 \ldots$$
$$4 \leftrightarrow 0.6824537128 \ldots$$
$$5 \leftrightarrow 0.4657189233 \ldots$$
$$. \qquad .$$
$$. \qquad .$$
$$. \qquad .$$

Although we have assumed that all numbers between 0 and 1 are listed, we will now show you how to construct a number *x* between 0 and 1 that is not on this list. We want *x* to be different from the first number on the list, so we will begin the decimal expansion of *x* with a digit other than a 6 in the tenths place, say $x = 0.5 \ldots$. Because we don't want *x* to equal the second number on the list, we make the hundredths place not equal to 3, say 4; so far, $x = 0.54 \ldots$.

Continuing this pattern, we make sure that *x* is different from the third number in the third decimal place, say 3, and different from the fourth number in the fourth decimal place, make it 5, and so on. At this point, $x = 0.5435 \ldots$. By constructing *x* in this fashion, it cannot be the first number on the list, or the second, or the third, and so on. In fact, *x* will differ from every number on the list in at least one decimal place, so it cannot be any of the numbers on the list.

This means that our assumption that we were able to match the numbers between 0 and 1 with the natural numbers is wrong. So the cardinal number of this set is not $\aleph_0$. Cantor used the number *c*, for the word *continuum,* for this cardinal number. Because there is a number between 0 and 1 that cannot be matched with a natural number, we can argue as we did in our discussion of the seats and the students that the cardinal number *c* is greater than the cardinal number $\aleph_0$. ✳

❋ ❋ ❋ HISTORICAL HIGHLIGHT

The Continuum Hypothesis

Perhaps you believe that mathematics is pretty much "cut-and-dried" in the sense that if you know certain rules and formulas agreed on by all mathematicians, then that is all there is to mathematics. Nothing could be farther from the truth. If you were to go more deeply into mathematics, you would find that at certain critical spots, some mathematicians go one way, whereas others go in a completely opposite direction.* This is certainly the case with Cantor's set theory. Although Cantor proved what many considered to be remarkable results, others rejected his methods completely.

Among Cantor's discoveries was that there is an infinite number of infinite cardinal numbers. That is, there is an infinite number of sizes of infinity. The smallest infinite cardinal number he called $\aleph_0$ which as you have seen is the cardinal number of the natural numbers. He proved that the set of all subsets of the natural numbers has another cardinal number $\aleph_1$ that is greater than $\aleph_0$. Cantor believed that perhaps $\aleph_1$ was equal to c. Although he labored for years to prove this conjecture, called the *continuum hypothesis*, Cantor was unable to prove it.

In 1900, the noted German mathematician David Hilbert said that the continuum hypothesis was one of the great unanswered questions in mathematics, and for the first half of the twentieth century, mathematicians could not answer this question. However, in 1963, a mathematician at Princeton University named Paul Cohen made the remarkable discovery that *it is impossible to either prove or disprove the continuum hypothesis*. That is, you can either accept it or reject it. If you accept it, then you are doing Cantorian set theory. If you reject it, then you are doing non-Cantorian set theory. Both are equally consistent but different areas of mathematics.

Exercises 2.5

Looking Back†

These exercises follow the general outline of the topics presented in this section and will give you a good overview of the material that you have just studied.

1. Give another example of a one-to-one correspondence between the sets T and F on page 73 than the one we gave.

2. In Example 1, how did we show that the set of natural numbers is infinite?

3. What did Cantor prove about the number of infinite cardinal numbers?

4. What is Cantor's continuum hypothesis?

Sharpening Your Skills

In Exercises 5–12, show that each set has cardinal number $\aleph_0$ by showing a one-to-one correspondence between the natural numbers and the given set. Be sure to indicate the general correspondence.

5. $\{4, 8, 12, 16, 20, \ldots\}$
6. $\{5, 10, 15, 20, 25, \ldots\}$
7. $\{8, 11, 14, 17, 20, \ldots\}$
8. $\{7, 11, 15, 19, 23, \ldots\}$
9. $\{2, 4, 8, 16, 32, \ldots\}$
10. $\{3, 9, 27, 81, 243, \ldots\}$
11. $\{1, 1/2, 1/3, 1/4, 1/5, \ldots\}$
12. $\{1/2, 2/3, 3/4, 4/5, 5/6, \ldots\}$

In Exercises 13–16, we give an expression describing the number that corresponds to the natural number n. Use this expression to describe a one-to-one correspondence between the natural numbers and one of its subsets. For example, if we gave you the expression 2n, you would write the following correspondence that we gave in Example 1:

$$
\begin{array}{ccccccc}
1 & 2 & 3 & \ldots & n & \ldots \\
\updownarrow & \updownarrow & \updownarrow & & \updownarrow & \\
2 & 4 & 6 & \ldots & 2n & \ldots
\end{array}
$$

13. $3n$
14. $2n + 3$
15. $3n - 2$
16. $4n + 5$

In Exercises 17–26, describe a one-to-one correspondence between the given set and one of its proper subsets. For example, if we gave you the set $\{3, 5, 7, 9, 11, \ldots\}$, the nth term is $2n + 1$. You could then write the correspondence by matching the elements of $\{3, 5, 7, 9, 11, \ldots\}$ with the elements of the subset $\{5, 7, 9, 11, 13, \ldots\}$. The general correspondence would match $2n + 1$ with $2n + 3$.

17. $\{2, 4, 6, 8, 10, \ldots\}$
18. $\{5, 10, 15, 20, 25, \ldots\}$
19. $\{7, 10, 13, 16, 19, \ldots\}$
20. $\{6, 9, 12, 15, 18, \ldots\}$
21. $\{2, 4, 8, 16, 32, \ldots\}$
22. $\{3, 9, 27, 81, 243, \ldots\}$
23. $\{1, 1/2, 1/3, 1/4, 1/5, \ldots\}$
24. $\{1/2, 2/3, 3/4, 4/5, 5/6, \ldots\}$
25. $\{1/2, 1/4, 1/6, 1/8, 1/10, \ldots\}$
26. $\{1/2, 1/4, 1/8, 1/16, 1/32, \ldots\}$

---

*For a good example of this divergence in thinking, see the extensive highlights on non-Euclidean geometry in Chapter 10.

†Before doing these exercises, you may find it useful to review the note *How to Succeed at Mathematics* on page xix.

In Example 3, we showed you how to match the natural numbers with the positive rational numbers in a one-to-one correspondence. Use that matching to answer the following questions.

27. What rational number corresponds to the natural number 12?

28. What rational number corresponds to the natural number 15?

29. What natural number is matched with the rational number 4/5?

30. What natural number is matched with the rational number 1/6?

Communicating Mathematics

31. In Example 3, what did we do to show that the set of positive rational numbers is countable?

32. What was the essence of the argument in Example 4 that showed the cardinal number of the set of numbers between 0 and 1 is greater than $\aleph_0$?

33. How would you convince someone that the set {1, 2, 3, 4, 5} is not infinite?

34. How would you convince someone that the set {2, 4, 6, 8, . . .} is infinite?

35. In Example, 3, why did we ignore rational numbers written as 2/2, 2/4, 3/3, and 4/2?

36. In constructing the number x in Example 4, how would you decide what to put in the 99th place?

Using Technology to Investigate Mathematics

37. Do an Internet search on the term *infinite sets*. Write a report on your findings.

38. Search the Internet for biographical information on Georg Cantor. Report on your findings.

For Extra Credit

39. How many one-to-one correspondences are there between the set {1, 2, 3} and the set {4, 5, 6}?

40. How many one-to-one correspondences are there between two four-element sets?

41. The arithmetic of infinite cardinal numbers has some peculiar properties. For example, it is a fact that $1 + \aleph_0 = \aleph_0$. Show this by forming the union of a set whose cardinal number is 1 with a disjoint set whose cardinal number is $\aleph_0$ and then showing that the union of the sets is in one-to-one correspondence with the natural numbers.

42. Another strange infinite arithmetic fact is that $\aleph_0 + \aleph_0 = \aleph_0$. Show why this is true by considering the union of two disjoint sets having cardinal number $\aleph_0$.

CHAPTER SUMMARY

SECTION	SUMMARY	EXAMPLE
SECTION 2.1	We can represent sets by **listing** elements or by using **set-builder notation**.	Example 1, p. 40
	Sets must be **well defined**.	Example 2, p. 41
	The **element symbol** expresses that an object is a member of a set.	Example 4, p. 42
	The **cardinal number** of a set indicates its size.	Example 5, p. 43
SECTION 2.2	Two sets are **equal** if they have exactly the same members.	Example 1, p. 47
	The set A is a **subset** of set B, written $A \subseteq B$, if every element of A is also found in B. If $A \neq B$, then we say that A is a **proper subset** of B and write $A \subset B$.	Example 2, p. 48
	We can use **Venn diagrams** to illustrate subset relationships.	Discussion, p. 48
	A set that has k elements has 2^k subsets.	Example 4, p. 50
	Equivalent sets have the same number of elements.	Discussion, p. 51
SECTION 2.3	a) The **union** of sets A and B, written $A \cup B$, is the set of elements that are elements of either A or B, or both.	Example 1, p. 55
	b) The **intersection** of sets A and B, written $A \cap B$, is the set of elements common to both A and B.	Example 2, p. 56
	c) The **complement** of set A, written A', is the set of elements in the universal set that are not elements of A.	Example 3, p. 57
	d) The **difference** of sets B and A, written $B - A$, is the set of elements that are in B but not in A.	Example 4, p. 57
	We must perform set operations in a proper order.	Example 5, p. 58
	DeMorgan's laws state that $$(A \cup B)' = A' \cap B' \quad \text{and} \quad (A \cap B)' = A' \cup B'.$$	Example 6, p. 59
	We can use **Venn diagrams** to prove or disprove set theory statements.	Example 7, p. 59
	If A and B are sets, then $$n(A \cup B) = n(A) + n(B) - n(A \cap B).$$	Example 8, p. 60
SECTION 2.4	Sets in Venn diagrams can have various names.	Example 1, p. 64
	We can use Venn diagrams to solve survey problems.	Examples 2 and 3, pp. 65–67
	A survey problem can have contradictory information.	Example 4, p. 67
SECTION 2.5	An **infinite set** can be put in a one-to-one correspondence with a proper subset of itself.	Example 1, p. 74
	The natural numbers and the set of integers are countable.	Examples 1 and 2, p. 74
	The set of rational numbers is countable.	Example 3, p. 74
	The set of real numbers between 0 and 1 is not countable.	Example 4, p. 75

CHAPTER REVIEW EXERCISES

Section 2.1

1. Use an alternative method to express each set.

    a. {2, 4, 6, 8, . . . , 18}

    b. {Alabama, Alaska, Arizona, . . . , Wisconsin, Wyoming}

    c. {$x : x$ is a letter in the word *stupendous*}

    d. {$y : y$ is a person in your math class and also a former president of the United States}

2. Explain why $\varnothing \neq \{\varnothing\}$.

3. Make up a "bag diagram" to illustrate the set {3, {∅}, {{1, 2}, {1, 2, 3}}}.

4. Find the cardinal number of each of these sets.

    a. {3, 6, 9, . . . , 18, 21}

    b. $\varnothing$

    c. {$x : x$ is one of the states in the United States}

Section 2.2

5. Decide whether each pair of sets is equal. Justify your answer.

   a. {1, 3, 5, 7, 9} and {9, 7, 5, 3, 1}

   b. {2, 3, 4, 5, 4, 6, 4, 7, 8, 9, 10, 9, 11, 12} and {2, 3, 4, . . . , 12}

   c. {2, 4, 6, 8, . . .} and {2, 4, 6, . . . , 1,000}

6. Decide whether each statement is true or false. Justify your answer.

   a. {daffodil, daisy, rose, carnation} $\subset$ {$f : f$ is a flower}

   b. {daffodil, daisy, rose, carnation} $\subseteq$ {$f : f$ is a flower}

   c. {$x : x$ is a letter in the word *plenty*} $\subseteq$ {$x : x$ is a letter in the word *plentiful*}

   d. $\varnothing \subseteq$ {1, 2, 3}

   e. $\varnothing \subset \varnothing$

7. Which pairs of sets are equivalent?

   a. {baseball, football, basketball, field hockey, volleyball} and {b, f, b, f, v}

   b. {1, 3, 5, 7, 9, . . . , 99} and {2, 4, 6, 8, 10, . . . , 100}

   c. {$x : x$ is a letter in the word *tulips*} and {$x : x$ is a letter in the word *flower*}

   d. {$\varnothing$} and {0}

8. **a.** List all of the subsets of the set {a, b, c}

   b. How many subsets does the set {3, 5, 8, 9, 12, 15, 17} have?

Section 2.3

9. Let $U =$ {1, 2, 3, . . . , 10} and let $A =$ {2, 5, 7, 8, 9}, $B =$ {3, 4, 5, 7, 9, 10}, and $C =$ {5, 3, 8, 9, 2}. Find the following sets:

   a. $A \cap B$ **b.** $B \cup C$

   c. C' **d.** $A - C$

10. Using the same sets as in Exercise 9, find the following sets:

   a. $(A \cup B)'$ **b.** $(A - C) \cup (A - B)$ **c.** $A' \cap (B' \cup C)$

11. Represent each set using a Venn diagram.

   a. $A \cup B$ **b.** $B \cap C$

   c. $A' \cap B \cap C$ **d.** $(B \cup C) - A$

12. Use DeMorgan's laws to represent $(A \cup B)'$ in a different way.

13. **a.** List three algebraic properties satisfied by the set operation of union.

   b. List three algebraic properties satisfied by the set operation of intersection.

   c. What algebraic property relates union and intersection?

Section 2.4

14. Use the following information to answer the given questions.

$n(A \cap B) = 4$, $n(A \cap B \cap C) = 1$, $n(B \cap C) = 8$, $n(B - A) = 9$, $n(B \cup C) = 23$, $n(A \cap C) = 6$, $n(A \cup B \cup C) = 35$, $n(B') = 32$.

   a. How many elements are there in $C - B$?

   b. How many elements are there in A'?

15. A survey was taken of college freshmen regarding the factors that they considered important in choosing a college.

   a. 82 said cost.

   b. 15 said cost but not academics.

   c. 70 said social life.

   d. 48 said all three.

   e. 56 said cost and social life.

   f. 25 said academics but not social life.

   g. 16 said academics but not cost.

How many said academics in the survey? How many said both academics and social life?

Section 2.5

16. What is the definition of an infinite set?

17. Show that the set of natural numbers is infinite.

18. In matching the rational numbers with the natural numbers as we did in Example 3 in Section 2.5, what rational number matches with 7?

19. In creating the number x in Example 4 in Section 2.5, where we proved that the set of real numbers between 0 and 1 cannot be put in a one-to-one correspondence with the natural numbers, how did we decide what to put in the third decimal place?

CHAPTER TEST

1. Use an alternative method to express each set.

   a. {101, 102, 103, 104, . . .}

   b. {$x : x$ is a month in the year}

   c. {$y : y$ is a person in your math class and also more than 100 years old}

2. Decide whether each pair of sets is equal. Justify your answer.

   a. {a, b, c, d, e} and {e, a, b, d, c}

   b. {{1}, {2}, {3}} and {1, 2, 3}

   c. {2, 4, 6, 8, . . .} and {$x : x$ is an even natural number greater than 1}

3. Which pairs of sets are equivalent?

   a. {Beyonce, Oprah, Sting, Dr. Phil} and {rose, daisy, collie, Mars, chocolate}

   b. {1, 3, 5, 7, 9, . . . , 99} and {2, 4, 6, 8, 10, . . . , 100}

   c. {$\varnothing$} and {0}

4. Let $U =$ {1, 2, 3, . . . , 10} and let $A =$ {1, 2, 5, 6, 9}, $B =$ {2, 3, 4, 5}, and $C =$ {2, 4, 6, 8, 10}. Find the following sets:

   a. C' **b.** $B - C$

   c. $(A \cap B)'$ **d.** $(A' \cap B') \cup C$

5. Explain why $\varnothing \neq$ {$\varnothing$}.

6. Find the cardinal number of each of these sets.

   a. $\{2, 4, 8, 16, 32, 64, 128, 256\}$ **b.** $\{\varnothing\}$

7. How many subsets does the set $\{1, 2, 3, \ldots, 8\}$ have?

8. Make up a "bag diagram" to illustrate the set $\{\{2\}, \varnothing, \{\{1,2\}, \{1,2,3\}\}\}$.

9. Decide whether each statement is true or false. Justify your answer.

   a. $\{\text{boxer, poodle, chihuahua, collie}\} \subseteq \{d : d \text{ is a dog}\}$

   b. $\{x : x \text{ is a letter in the word } love\} \subset \{x : x \text{ is a letter in the word } lovely\}$

   c. $\varnothing \in \{1, 2, 3\}$

10. Use DeMorgan's laws to represent $(A \cap B)'$ in a different way.

11. Represent each set using a Venn diagram.

   a. $A \cap C$ **b.** $(A \cup C) - B$

12. Use the following information to answer the given questions.

   $n(A \cap C) = 11$, $n(A \cap B \cap C) = 5$, $n(B \cap C) = 8$, $n(C - A) = 10$, $n(B \cup C) = 34$, $n(A \cap B) = 10$, $n(A \cup B \cup C) = 54$, $n(C') = 64$.

   a. How many elements are there in $C - A$?

   b. How many elements are there in B'?

13. A survey was taken of drivers regarding the factors that they considered important in buying a new car.

   a. 84 said cost.

   b. 15 said cost but not gas mileage.

   c. 72 said safety.

   d. 48 said all three.

   e. 56 said cost and safety.

   f. 25 said gas mileage but not safety.

   g. 20 said gas mileage but not cost.

   How many said gas mileage in the survey? How many said both gas mileage and safety?

14. What is the definition of an infinite set?

15. Show that the set of natural numbers is infinite.

16. In matching the rational numbers with the natural numbers as we did in Example 3 in Section 2.5, what rational number matches with 9?

17. In creating the number x in Example 4 in Section 2.5, where we proved that the set of real numbers between 0 and 1 cannot be put in a one-to-one correspondence with the natural numbers, how did we decide what to put in the fifth decimal place?

GROUP PROJECTS

1. Draw a Venn diagram with four sets A, B, C, and D showing all 16 regions. (*Hint:* You cannot do it with four circles, but you can do it with four ovals.) Number the regions 1 to 16 and then give the set theory name for each region. For example, one region will be named $A \cap B' \cap C \cap D'$. Try to draw a Venn diagram for five sets having 32 distinct regions. If you cannot do this, then search the Internet for examples of such a diagram and present your results.

2. In set theory as you have studied it, an element x is either in a set A or it is not. That is, we can say $x \in A$ or $x \notin A$. In **fuzzy set theory**, membership in sets is not so cut and dried. We will start with a set of people $X = \{\text{Lily, Marshall, Ted, Robin, Barney}\}$ and define the fuzzy set T as the set of tall people as follows:

$$T = \{(\text{Ted}, 0.8), (\text{Marshall}, 0.4), (\text{Lily}, 0.7),$$
$$(\text{Robin}, 0.3), (\text{Barney}, 0.5)\}.$$

Ted, Marshall, Lily, Robin, and Barney are members of T, but they have different **degrees of membership**, depending on how tall they are. For example, Ted's degree of membership is 0.8, whereas Robin's is 0.3. This means that Ted has the property of "being tall" to a much higher degree than Robin does. Degrees of membership are always between 0 and 1 inclusive. Two fuzzy sets are **equal** if they have the same members, and the degrees of membership of the elements of the two sets are the same.

To compute the **complement** of T, we replace each set membership degree with 1 minus the degree of membership, as follows:

$$T' = \{(\text{Ted}, 0.2), (\text{Marshall}, 0.6), (\text{Lily}, 0.3),$$
$$(\text{Robin}, 0.7), (\text{Barney}, 0.5)\}.$$

Let us now define S, the fuzzy set of strong people:

$$S = \{(\text{Ted}, 0.5), (\text{Marshall}, 0.6), (\text{Lily}, 0.4),$$
$$(\text{Robin}, 0.4), (\text{Barney}, 0.6)\}.$$

We compute the **intersection** of T and S by associating with each member of X the smaller degree of membership, as follows:

$$S \cap T = \{(\text{Ted}, 0.5), (\text{Marshall}, 0.4), (\text{Lily}, 0.4),$$
$$(\text{Robin}, 0.3), (\text{Barney}, 0.5)\}.$$

The **union** of sets T and S is similar, except now we take the larger of the set memberships for each member of X. So,

$$S \cup T = \{(\text{Ted}, 0.8), (\text{Marshall}, 0.6), (\text{Lily}, 0.7),$$
$$(\text{Robin}, 0.4), (\text{Barney}, 0.6)\}.$$

   a. Define some other fuzzy sets using X above.

   b. Do several computations using S, T, and the new fuzzy sets that you have defined in (a).

   c. Verify that some of the rules you have learned earlier are satisfied in fuzzy set theory. For example, show by computing examples that DeMorgan's laws hold. Show that intersection distributes over union, etc.

3. Research the Internet to find applications of fuzzy set theory. You should be able to find many interesting applications in medicine. Write a report on your findings.

Logic

3

The Study of What's True or False or Somewhere in Between

*I*f you needed an operation, how would you feel about having the operation done by a robot? Or, how about an even scarier scenario—what if the doctor controlling the robot was actually in another country and was doing the surgery over the Internet? This is not some fanciful science fiction, but is now a scientific reality.

Scientists at Kyung Hee University in Korea have shown that surgery can indeed be performed over the Internet by demonstrating a simulated operation on a set of pig intestines. Remote-controlled surgery is just another of the rapidly growing list of digital technologies, such as the Internet, satellite TV, digital video recorders, and cell phones, that are based on the logical principles that you will be studying in this chapter.

You will learn how to perform logical computations to determine when statements are true or false and when arguments

(continued)

are valid. Next you will learn how to analyze arguments called *syllogisms,* which is the way Aristotle approached logic more than 2,000 years ago. Finally, we will show you a modern form of logic called *fuzzy logic* that is being used today in a variety of real-life, practical applications. ●

Statements, Connectives, and Quantifiers

3.1

Objectives

1. Identify statements in logic.
2. Represent statements symbolically using five connectives.
3. Understand the difference between the universal and existential quantifiers.
4. Write the negations of quantified statements.

"You just don't understand what I'm saying!" Have you ever blurted that out while arguing with a friend or relative? At such times, wouldn't it be nice if you both could type your arguments into your "logic calculators," hit return, and, in a blink, determine who is correct?

In a sense, that was what George Boole* was trying to do when he invented *symbolic logic* in the nineteenth century. Boole believed that he could calculate with logical ideas symbolically—similar to the way we calculate with numerical quantities in algebra. You may recall from algebra that $(x + y)^2 = x^2 + 2xy + y^2$, which is a general statement that is true for any numbers x and y. Similarly, Boole wanted to develop methods for doing routine logical calculations to determine when statements were true or false and whether logical arguments were valid. Although symbolic logic never achieved the success that Boole envisioned, as you will see in the exercise set in Section 3.2, it is important today because it forms the mathematical basis for digital devices that you use every day, such as MP3 players, digital video recorders, satellite radio and TV, and cell phones.

In studying algebra, you learned that to solve problems, the first step is to represent verbal expressions with symbols. For example, instead of using the sentence "The area of a rectangle equals the product of its length times its width," we write the more concise statement "$A = l \times w$." Similarly, in this section, you will learn to represent English statements symbolically; in Section 3.2, we will determine when these statements are true or false.

Statements in Logic

🖉 **KEY POINT**

In symbolic logic, we only care whether statements are true or false—not their content.

As you begin your study of symbolic logic, keep in mind that *we are only concerned with the truth or falsity of the sentences we analyze and not their content.* As you will see in Section 3.4, an argument can seem ridiculous and yet have a proper logical form. On the other hand, an argument that seems believable may have an unacceptable logical form.

> **DEFINITION** A **statement** in logic is a declarative sentence that is either true or false. We represent statements by lowercase letters such as *p*, *q*, or *r*.

The following are examples of statements. Remember that to be a statement, the sentence must be either true or false; however, *we do not need to know which it is.*

a) AIDS is a leading killer of women.

b) If you eat less and exercise more, you will lose weight.

---

*See the Historical Highlight on George Boole in Section 2.3.

c) In the last 10 years, we have reduced by 25% the amount of greenhouse gases in the atmosphere that produce global warming.

d) Oprah Winfrey or Steven Spielberg had the highest earnings in the entertainment industry in 2007.

The following are not statements:

e) You're fired! (This is not a declarative sentence.)

f) When did dinosaurs become extinct? (This is not a declarative sentence.)

g) This statement is false. (This is a paradox. It cannot be either true or false. If we assume that this sentence is true, then we must conclude that it is false. On the other hand, if we assume it is false, then we conclude that it must be true.)

Notice that statements a)–d) fall into two categories. Statements a) and c) each express a single idea; if we were to remove any part of these sentences, they would no longer make sense. Such sentences are called *simple statements*. In contrast, statements b) and d) clearly contain several ideas connected to make a more complex sentence. These sentences are called *compound statements*. Notice that words such as *if . . . then* and *or* connect the ideas to form the compound sentences. ①

Quiz Yourself ① *

Identify each sentence as simple or compound.

a) The 2007 Red Sox were the best team in the history of baseball.

b) If you break your lease, then you forfeit your deposit.

c) The world's tallest skyscrapers are in Kuala Lumpur and Chicago.

> **DEFINITIONS** A **simple statement** contains a single idea. A **compound statement** contains several ideas combined together. The words used to join the ideas of a compound sentence are called **connectives**.

✎ **KEY POINT**

In logic, we connect ideas using *not, and, or, if . . . then,* and *if and only if.*

Connectives

Because the English language is so rich, we can use many words to connect ideas. However, the connectives we use in logic generally fall into five categories: *negation, conjunction, disjunction, conditional,* and *biconditional.* We will now discuss each connective individually.

> **DEFINITION Negation**
> **Negation** is a statement expressing the idea that something is not true. We represent negation by the symbol ∼.[†]

EXAMPLE 1 *Negating Sentences*

Negate each statement:

a) *p*: The high-definition screen is included in the price of your MacBook.

b) *b*: The blue whale is the largest living creature.

SOLUTION:

a) The high-definition screen is *not* included in the price of your MacBook. We write this negation symbolically as ∼*p*.

b) The blue whale is *not* the largest living creature. The symbols ∼*b* represent this statement. ❋

---

*Quiz Yourself answers begin on page 778.

[†]Even though a negation does not combine two separate ideas, we will consider a negation to be a compound statement.

> **DEFINITION** **Conjunction**
> A **conjunction** expresses the idea of *and*. We use the symbol ∧ to represent a conjunction.

The next example illustrates some statements that you might encounter when signing a lease for your new apartment.

EXAMPLE 2 *Joining Statements Using And*

Consider the following statements:

p: The tenant pays utilities. d: A $150 deposit is required.

a) Express the statement "It is not true that: the tenant pays utilities and a $150 deposit is required" symbolically.

b) Write the statement $\sim p \wedge \sim d$ in English.

SOLUTION:

a) This sentence has the form $\sim(p \wedge d)$.

b) The tenant does not pay utilities and a $150 deposit is not required. ❊

 PROBLEM SOLVING

The Order Principle

Notice the difference in the form of the statements in Example 2 that we show in this diagram:

First use "and."   First negate.

a) $\sim(p \wedge d)$ b) $\sim p \wedge \sim d$

Then negate.   Then use "and."

These statements sound similar but do not say the same thing. The Order Principle from Section 1.1 warns that changing the order of mathematical operations *may* change the meaning. In Section 3.2, we will explain the difference between these statements.

> **DEFINITION** **Disjunction**
> A **disjunction** conveys the notion of *or*. We use the symbol ∨ to represent a disjunction.

EXAMPLE 3 *Joining Ideas Using Or*

Consider the following statements:

h: We will build more hybrid cars. f: We will use more foreign oil.

a) Write the following statement symbolically:

    We will not build more hybrid cars or we will use more foreign oil.

b) Write the symbolic form $\sim(h \vee f)$ in English.

SOLUTION:

a) The symbolic form for this sentence is $(\sim h) \vee f$.

b) Translating into English, we get:

    It is not true that: we will build more hybrid cars or we will use more foreign oil.

❊ **2**

Quiz Yourself **2**

Consider the following statements:

  d: I will buy a Blu-ray disc player.

  i: I will buy an iPod touch.

Write each statement in symbolic form.

a) I will not buy a Blu-ray disc player or I will not buy an iPod touch.

b) I will not buy a Blu-ray disc player and I will buy an iPod touch.

HISTORICAL HIGHLIGHT ✺ ✺ ✺

The Evolution of Logic

Over the centuries, numerous Greek philosophers, Buddhist monks, European theologians, and American scholars have studied logic, yet many believe that the Greek Aristotle was the greatest logician of all time. In fact, in this chapter you will learn some of the same principles that Aristotle taught his students thousands of years ago.

Although many early logicians studied logic to discover religious truths and lead more perfect lives, gradually the emphasis changed and by the seventeenth century logic had become an area of mathematics. In fact, the German mathematician Gottfried Leibniz, believing that

the invention of symbolic logic would make philosophy unnecessary, wrote:

> . . . an alphabet of human thought could be devised, and that everything could be discovered and distinguished by the combination of the letters of this alphabet . . . discussions between two philosophers will not be any longer necessary. It will be enough to take pen in hand [and] say to one another: We calculate!

Unfortunately, logic has not achieved Leibniz's expectations; however, it is a powerful tool in modern mathematics with numerous applications.

> **DEFINITION Conditional**
> A **conditional** expresses the notion of *if . . . then*. We use an arrow, →, to represent a conditional.

EXAMPLE 4 *Connecting Ideas Using If . . . Then*

Suppose that p represents "The Phillies win the World Series" and k represents "Kiefer will win an Emmy."

a) We would read the statement $p \rightarrow k$ as "If the Phillies win the World Series, then Kiefer will win an Emmy."

b) We would write the statement "If the Phillies do not win the World Series, then Kiefer will not win an Emmy" symbolically as $\sim p \rightarrow \sim k$. ❋

Again, you must be particular about the order in which you use connectives. It would be incorrect to change the order of the negations and the conditional to write the answer for b) as $\sim(p \rightarrow s)$.*

Although mathematicians use the next connective frequently, it is not as common in ordinary language.

> **DEFINITION Biconditional**
> A **biconditional** represents the idea of *if and only if*. Its symbol is a double arrow, ↔.

EXAMPLE 5 *Connecting Ideas with If and Only If*

Write each biconditional statement in symbolic form:

a) A polygon has three sides if and only if it is a triangle.

b) The software can be returned if and only if the seal is not broken.

SOLUTION:

a) Let p represent the statement "A polygon has three sides" and let t represent the statement "The polygon is a triangle." Then this statement has the form $p \leftrightarrow t$.

---

*You will understand the difference in meaning between $\sim(p \rightarrow s)$ and $\sim p \rightarrow \sim s$ when we discuss logically equivalent statements in Section 3.2.

Consider the following statements:

f: I fly to Houston.

q: I will qualify for frequent-flyer miles.

Write each statement in symbolic form.

a) If I do not fly to Houston, then I will not qualify for frequent-flyer miles.

b) I fly to Houston if and only if I will qualify for frequent-flyer miles.

b) Let *r* represent the statement "The software can be returned" and let *b* represent the statement "The seal is broken." Then this statement has the form $r \leftrightarrow \sim b$. ✤ **3**

In working with connectives we will often use the informal terms *not, and, or, if . . . then,* and *if and only if* instead of the more formal terms *negation, conjunction, disjunction,* and so on. By combining these five connectives, we can form extremely complex sentences. Consider these simple statements:

t: Today is Tuesday.

r: It is raining.

s: It is springtime.

h: You are hungry.

Suppose we make a compound statement:

If today is not Tuesday and it is raining, then either you are hungry or it is not springtime.

If you are hungry for lunch on a rainy Wednesday in April, is the previous statement true or false? What if you are not hungry on a sunny Tuesday in March? As you can see, it is often difficult to have a precise understanding of what an English sentence is saying. Notice that the form of this statement is $(\sim t \wedge r) \rightarrow (h \vee \sim s)$. In Section 3.2, we will develop rules for computing with connectives so that we can then easily answer such questions.

 Some Good Advice

You can use commas to group the components of a compound statement. For example, you would translate "Stock prices will rise and inflation will fall, or bond prices will decrease" as $(s \wedge i) \vee b$. You would write the statement $s \wedge (i \vee b)$ into English as "Stock prices will rise, and inflation will fall or bond prices will decrease."

✏ *KEY POINT*

Quantifiers state how many objects satisfy a given property.

Quantifiers

In addition to connectives, there are other special words called *quantifiers* that you need to understand in analyzing sentences. **Quantifiers** tell us "how many" and fall into two categories: universal quantifiers and existential quantifiers.

> **DEFINITION** **Universal quantifiers** are words such as *all* and *every* that state that all objects of a certain type satisfy a given property.

The following statements contain universal quantifiers:

All citizens over age 18 have the right to vote.

Every triangle has an interior angle sum of 180°.

Each Nascar driver must register for the Daytona 500 by August 1.

> **DEFINITION** **Existential quantifiers** are words such as *some, there exists,* and *there is at least one* that state that there are one or more objects that satisfy a given property.

The following statements contain existential quantifiers:

Some drivers qualify for reduced insurance rates.

There is a number whose square is 25.

There exists a bird that cannot fly.

Negating Quantifiers

Suppose we want to negate the statement "All professional athletes are wealthy." In order to do this, we must understand how to negate a statement that contains quantifiers. To see how to do this, consider the diagrams* in Figure 3.1. We illustrate the original statement to be true in Figure 3.1(a) by drawing the set of all athletes as contained within the set of all wealthy people. In Figure 3.1(b), we move the circle representing all athletes slightly so that part of the circle lies outside the circle labeled *wealthy*. We have shown that the statement "Not all athletes are wealthy" means the same thing as "Some athletes are not wealthy."

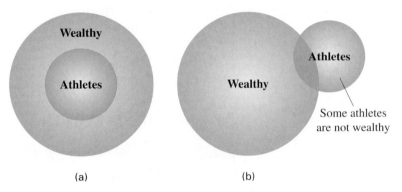

(a)   (b)

FIGURE 3.1 (a) All athletes are wealthy. (b) Not all athletes are wealthy (some athletes are not wealthy).

Let us now investigate how to negate a statement with an existential quantifier. Consider the statement "Some students will get a scholarship," which is illustrated in Figure 3.2(a). To say "It is false that some students will get a scholarship" means that the circle representing the students can have nothing in common with those getting a scholarship. This is shown in Figure 3.2(b). Thus, we can state the negation of the original statement as "No students will get a scholarship."

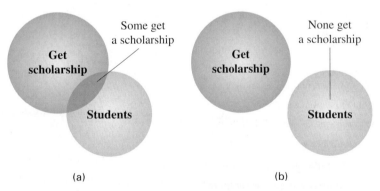

(a)   (b)

FIGURE 3.2 (a) Some students get a scholarship. (b) It is false that some students get a scholarship (none get a scholarship).

We now summarize this discussion.

NEGATING STATEMENTS WITH QUANTIFIERS
The phrase *Not all are* has the same meaning as *At least one is not*.
The phrase *Not some are* has the same meaning as *None are*.

---

*We will consider such diagrams, called *Euler diagrams*, in greater depth in Section 3.5.

EXAMPLE 6   *Negating Quantified Statements*

Negate each quantified statement and rewrite it in English.

a)   All customers will get a free dessert.

b)   Some computers have a two-year warranty.

SOLUTION:

Quiz Yourself

Negate each quantified statement and rewrite it in English.

a)   All professors are friendly.

b)   Some dogs bite.

a)   We want to rewrite the statement "Not all customers will get a free dessert." Because this statement has the form *Not all are*, we can rewrite it as *At least one is not.* In English we would say "At least one customer will not get a free dessert."

b)   The negation of the statement is "It is not true that some computers have a two-year warranty." Because this statement has the form *Not some have*, we can rewrite it as *None have.* In English we would say "No computers have a two-year warranty."

Now try Exercises 41 to 50.

 PROBLEM SOLVING

The Three-Way Principle

The Three-Way Principle in Section 1.1 reminds us that drawing diagrams of quantified statements will help us understand better how to write their negations.

Exercises 3.1

Looking Back*

These exercises follow the general outline of the topics presented in this section and will give you a good overview of the material that you have just studied.

1.   List five connectives that we use in logic. Give both their informal and formal names.

2.   What is the meaning of each of the following symbols? $\sim$, $\rightarrow$, $\wedge$, $\leftrightarrow$, $\vee$

3.   What do we mean by a universal quantifier?

4.   What is an existential quantifier?

5.   How do you negate a statement of the form "All are?"

6.   How do you negate a statement of the form "Some are?"

7.   Why did early philosophers study logic?

8.   What was Leibniz's expectation for logic?

Sharpening Your Skills

In Exercises 9–18, determine which sentences are statements.

9.   More than half the households in the country have digital cameras.

10.   The research plane, the X-15, could fly at a speed of over 4,500 miles per hour.

11.   Come skiing with me this weekend in Tahoe.

12.   Are you going to the Dave Matthews Band concert?

13.   Where did you buy your 180s?

14.   Water quality has steadily improved during the past 10 years.

15.   You will be a millionaire by the time you are 30.

16.   *The Simpsons* is a popular TV show about lawyers.

17.   Will you still love me when I'm 64?

18.   Romeo, Romeo, wherefore art thou Romeo?

In Exercises 19–28, identify each statement as simple or compound. If a statement is compound, identify the connectives used.

19.   If you want to ride, you must be at least 48 inches tall and have no food, drinks, or Kryptonite.†

20.   Travolta is satisfied with his performance in the musical.

21.   LeBron or Kobe will lead the conference in scoring.

22.   If Hillary supports environmental issues, she will succeed in politics.

23.   Alvaro likes opera.

24.   Amira did not buy the online tutoring package for her math book.

25.   If *Pirates of the Caribbean: Dead Man's Chest* grossed $1 million the fastest of any movie, then either *Spider-Man* or *Shrek the Third* was second.

---

*Before doing these exercises, you may find it useful to review the note *How to Succeed at Mathematics* on page xix.
†Regulations for riding the Superman the Escape roller coaster at Six Flags Magic Mountain in Los Angeles, California.

26. Jena is going to Europe this summer.

27. Red sky in morning, sailor take warning.

28. I have a dream.

Consider the following statements: g: Global warming will increase. a: We will develop alternate energy sources. c: Congress will pass new energy legislation.

In Exercises 29–34, write each statement in symbolic form.

29. Global warming will increase or Congress will not pass new energy legislation.

30. We will not develop alternative energy sources and global warming will increase.

31. If global warming does not increase then: Congress will not pass new energy legislation and we will not develop alternative energy sources.

32. If Congress will not pass new energy legislation then: global warming will increase or we will develop alternative energy sources.

33. Global warming will increase if and only if we do not develop alternative energy sources.

34. Congress will not pass new energy legislation or we will develop alternative energy sources, if and only if global warming will not increase.

Consider the following statements: t: The radial tires are included. s: The sunroof is extra. w: Power windows are optional.

In Exercises 35–40, translate each statement into words.

35. $t \vee (\sim s)$ **36.** $\sim s \wedge \sim w$

37. $\sim(s \wedge t)$ **38.** $\sim(w \wedge t)$

39. $t \rightarrow (s \vee \sim w)$ **40.** $(s \wedge t) \rightarrow \sim w$

Communicating Mathematics

In Exercises 41–48, negate each quantified statement and then rewrite it in English in an alternative way. Draw diagrams to illustrate your thinking. Explain what your diagrams tell you.

41. All snakes are poisonous.

42. Some auto mechanics are incompetent.

43. Some personal items are not covered by this insurance policy.

44. All married couples must file a joint tax return.

45. Some scientists believe that an asteroid collision led to the extinction of the dinosaurs.

46. All polygons have four sides.

47. All modern art is difficult to understand.

48. Some factories emit toxic wastes.

Applying What You've Learned

Use the following table of student information to determine whether the statements in Exercises 49–54 are true or false. If a statement is false, explain why it is false using the rules for quantifiers that we have developed. For example, the statement "All sophomores are on scholarship" is false because "There exists a sophomore (Stephen) who is not on scholarship."

Name	Year	Scholarship	Athlete	Commuter
Lennox	Freshman	Yes	Yes	No
Marilu	Freshman	No	No	Yes
Stephen	Sophomore	No	Yes	Yes
Omarosa	Junior	Yes	No	No
Tito	Sophomore	Yes	Yes	Yes
Nadia	Sophomore	Yes	Yes	No
Piers	Junior	Yes	No	Yes

49. All sophomores are commuters.

50. There exists an athlete who is not on scholarship.

51. There exists a sophomore who is not an athlete.

52. All freshmen are athletes.

53. No freshmen are on scholarship.

54. No athletes are commuters.

In Exercises 55–60, examine each statement to identify which connectives are being used. Exercises 55–58 are based on the 1040 federal income tax form.

55. If you were a household employee who did not receive a Form W-2 because your employer paid you less than $1,000, be sure to include the amount you were paid on Form 1040-A.

56. If you received a Form 1099 showing federal income tax withheld on dividends to interest income, include the amount withheld in the total on line 29a.

57. You were divorced and you made joint estimated tax payments with your former spouse or you changed your name and you made estimated tax payments using your former name.

58. If you were covered by a plan at work, your IRA deduction may be reduced or eliminated, but you can still make contributions to an IRA even if you cannot deduct them.

59. When you turn on the DVD player or replace a disk, the player returns to the original default setting.

60. If you press the MEMORY or RETURN button while the display appears on the screen, the display disappears.

Because the English language is so complex, it is often difficult to see clearly the structure of compound statements. In Exercises 61–66, examine each statement to identify which connectives are being used. You may need to rephrase the sentence to see its structure more clearly.

61. "There was one of two things I had a right to, liberty or death; if I could not have one, I would have the other . . . ." (Harriet Tubman, conductor-in-chief of the Underground Railroad)

62. "I have found the best way to give advice to your children is to find out what they want and then advise them to do it." (Harry S. Truman)

63. "The challenge is to practice politics as the art of making what appears to be impossible, possible." (Hillary Clinton)

64. "The growing and dying of the moon reminds us of our ignorance, which comes and goes. . . ." (Black Elk, Oglala Sioux holy man)

65. "Political division, based on color, is entirely artificial; and when it disappears, so will the domination of one color group by another." (Nelson Mandela)

66. "Woman must not depend upon the protection of man, but must be taught to protect herself." (Susan B. Anthony)

In Exercises 67–72, determine if the following quantified statements are true or false. If a statement is false, negate it to change it to a correct statement. (See page 40 for a review of common sets of numbers.)

67. All integers are rational numbers.

68. There exists a rational number that is not an integer.

69. All whole numbers are natural numbers.

70. All real numbers are rational.

71. There exists an integer that is not a whole number.

72. There exists a whole number that is not an integer.

Communicating Mathematics

73. In Example 2, what is the significant difference between the statements $\sim(p \wedge d)$ and $\sim p \wedge \sim d$?

74. In Example 4, what is the significant difference between the statements $\sim(p \rightarrow d)$ and $\sim p \rightarrow \sim d$?

75. Show with diagrams how you can "remember how to remember" the negation of a statement having a universal quantifier.

76. Show with diagrams how you can "remember how to remember" the negation of a statement having an existential quantifier.

Using Technology to Investigate Mathematics

77. Many logic puzzles give pieces of information about a group of people, and by putting all the information together, you are to identify the people being described. For example, a puzzle may start by saying that there are five friends and you are to determine who has the following occupations: pilot, bassoon player, stand-up comic, wine taster, and baker. You are given various pieces of information such as: (a) Bree attended the play but not with Lynette or the wine taster. (b) Edie, Bree, and the pilot saw the play last night. (c) Neither Lynette nor Susan is a musician. And so forth.

One way to solve such a puzzle is to make a table and place Xs at places in the table where you can rule out a person's occupation. For example, statement (b) allows us to eliminate Bree and Edie as the pilot.

	Pilot	Bassoon Player	Comic	Wine Taster	Baker
Bree	X				
Gabrielle					
Lynette					
Susan					
Edie	X				

Perform a search on the Internet for similar logic puzzles and solve some. Then try to make up a puzzle of your own.

78. Find several interesting sites on the Internet that discuss logic—perhaps regarding the history of logic, applications of logic, and so on. Write a brief report on the results of your search. If you search for "Fuzzy Logic," you will find some interesting applications.

If you have a calculator, look at your manual to see if it has the logical connectives that we have discussed. For example, on my calculator I used the "or" connective to see if the statement $8 > 4$ or $2 = 3$ is true (see accompanying copy of calculator screen). The calculator returned the number 1, which stands for "true." If my statement had been false, then the calculator would have returned a 0.

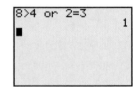

79. Learn how to use the logical connectives on your calculator and test the following statements to see if they are true or false.

a. $(5 \geq 6/3)$ *and* $1 < 4$ **b.** $6 = 2*3$ *or* $8 = 7*5$

80. Repeat Exercise 79 with an Excel spreadsheet.

For Extra Credit

81–86. In symbolic logic, the form of statements is more important than their contents. However, as you saw in Exercises 61–66, writers and orators are concerned with the eloquence of their words. Rewrite each statement in Exercises 61–66 using the connectives you learned in this section to make the form of the statements more clear. First do your rewrites separately, and then, as a group, compare your answers to determine which rewrite is truest to the meaning of the original quote.

87. Think of a real-life situation that you might want to analyze using symbolic logic. What difficulties do you encounter in doing this?

88. Provide arguments for or against the view that "Symbolic logic has nothing to do with real life."

Truth Tables

Objectives

1. Learn the truth tables for the five fundamental connectives.
2. Compute truth tables for compound statements.
3. Determine when statements are logically equivalent.
4. State and apply DeMorgan's laws.

Imagine that as you're finishing lunch with your friends Charlie and Tony, the server brings you the check. Charlie suggests, "We've basically had the same meals, so why don't we just split this three ways?" Agreeing, you decide that a total of $36 is fair. Dividing by three, you each leave $12 on the table. There is no argument here because the rules of arithmetic and algebra that you used in settling your bill have existed for hundreds of years.

In the nineteenth century, George Boole and others developed rules to try to make logical reasoning as easy as this. We will now introduce you to some of these rules and show you how to do some simple logical computations.

In algebra, after you understood how to represent numerical quantities with symbols such as x and y, you learned rules for computing with these quantities. Frequently you were interested in whether you could consider two different-looking expressions to be the same. You know that although $(x + y)^2$ and $x^2 + y^2$ look similar, they do not always represent the same quantity. For example, $(3 + 4)^2 \neq 3^2 + 4^2$.

In logic, we want to know whether or not a pair of statements with similar forms, such as $\sim(p \wedge q)$ and $\sim p \wedge \sim q$, convey exactly the same meaning. We will now show you how to compute truth tables in order to determine when compound statements are true and when they are false, and whether or not they have the same meaning.

 Some Good Advice

In logic, the truth values for statements using connectives such as *not, and, or,* and *if . . . then* are *sometimes* slightly different from everyday usage. It is important that you understand these differences.

 **KEY POINT**

Negating a statement reverses its truth value.

Truth Tables

Negation works in logic exactly as it does in everyday language. If p is a true statement such as "$2 + 2 = 4$," then its negation $\sim p$ is false. Also, if q is a false statement such as "George Washington was the king of England," then its negation $\sim q$ is true. We summarize the behavior of negation in Figure 3.3, which is an example of a truth table.

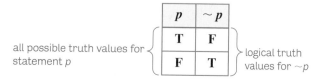

p	$\sim p$
T	F
F	T

all possible truth values for statement p · logical truth values for $\sim p$

FIGURE 3.3 The truth table for negation.

The left column indicates *all the possibilities* for statement p, namely, that it can be either true or false. The right column indicates what logical values $\sim p$ takes on for the distinct possibilities for p. If p is true, the truth table tells us that $\sim p$ is false. If p is false, then $\sim p$ is true.

EXAMPLE 1 *Finding the Truth Value of Negations*

Determine whether each statement is true or false and then state its negation. Notice how the negation of each statement has the opposite truth value of the original statement.

a) The Dallas Cowboys are a well-known football team.

b) Mexico is a prosperous country in Asia.

SOLUTION:

a) This is true. Therefore, its negation, "The Dallas Cowboys are *not* a well-known football team," is false.

b) This is false. So, "Mexico is *not* a prosperous country in Asia" is true. ❋

The *and* connective also works in logic as it does in everyday life. Suppose you make the following statement to a friend:

"I am a basketball fan and I like rock music."

This statement is of the form $p \wedge q$, where p represents "I am a basketball fan" and q represents "I like rock music." The two left columns in Figure 3.4 list all the possible combinations of truth values for p and q. The right column shows the truth values for $p \wedge q$.

If you think carefully about this, you realize that the only way you are telling the truth is if you are a basketball fan and you also like rock music, that is, if *both p and q* are true. Take a moment to interpret lines 2, 3, and 4 of Figure 3.4 to convince yourself that in each case, the original conjunction is false.

 KEY POINT

A conjunction is true only when both of its component parts are true.

p	q	$p \wedge q$
T	T	T
T	F	F
F	T	F
F	F	F

all possible ways that p and q can be true or false

FIGURE 3.4 A conjunction is true *only* when both p and q are true.

EXAMPLE 2 *Finding the Truth Values of Conjunctions*

Determine whether each statement is true or false.

a) J. K. Rowling is a well-known writer and U2 is a famous rock band.

b) Shaquille O'Neal played first base for the Atlanta Braves and he refuses to do TV commercials.

c) When you are 50 years old, you will be a billionaire and Maine is larger than California.

SOLUTION: These statements are of the form $p \wedge q$, so we can use Figure 3.4 to determine their logical values.

a) In this case, both components of the conjunction are true, so by line 1 of Figure 3.4, the entire statement is true.

b) Because both component statements are false, using the fourth line of Figure 3.4, we see that the statement is false.

c) Of course, we don't know if you will be a billionaire by age 50, so p is either true or false, but we don't know which. However, we do know that Maine is not larger than California. Therefore, we are dealing with either line 2 or 4 of Figure 3.4. In either case, the statement is false. ❋

 **KEY POINT**

A disjunction is false only when both of its component parts are false.

p	q	$p \vee q$
T	T	T — inclusive or
T	F	T
F	T	T
F	F	F

all possible ways that p and q can be true or false

FIGURE 3.5 A disjunction is false *only* when both p and q are false.

Mathematicians use the word *or* slightly differently than we use *or* in everyday life. Suppose that your mathematics instructor announces that there will be a quiz next Tuesday or Thursday. You come to class on Tuesday, have a quiz, and, not expecting another quiz, you cut Thursday's class. When you return the following week, you find out that you missed a quiz on Thursday. You would probably feel that your instructor had misled you. The problem is that you misinterpreted your instructor's use of *or*. In everyday conversation, we use what is called the **exclusive or**. Even though we do not explicitly say it, we understand this version of *or* to mean "one or the other, *but not both.*" In mathematics and logic, we use the **inclusive or**, which is defined in Figure 3.5. Notice that on the first line of the table, when p is true and q is true, the statement $p \lor q$ is true, which differs from the way *or* is usually used in everyday life.

Some Good Advice

If you learn the truth tables for *and* and *or* by remembering the one different line in each table, it will help you do logic calculations more quickly.

EXAMPLE 3 *Analyzing the Logic of an Advertisement*

Consider the following help-wanted ad:

> "Management trainee wanted. Applicant must have four-year degree in accounting or three years of experience working in a financial institution."

Which applicants should be considered for the position if we interpret *or* according to Figure 3.5?

a) Aya has a four-year degree in accounting and has worked two years for a loan company.

b) Heidi has studied accounting in college (but did not graduate) and has worked for five years selling electronics.

c) Monte earned a four-year degree in accounting and has five years of experience working for a credit card company.

SOLUTION:

a) Aya's background corresponds to line 2 of Figure 3.5, so she should be considered for the job.

b) Heidi's credentials are reflected in line 4 of Figure 3.5, so she is not qualified.

c) Monte's background corresponds to line 1 of Figure 3.5, so he should also be considered. ✳ **5**

As we mentioned in Section 1.1, similar notations often represent similar mathematical ideas. Notice that the *and* symbol, ∧, in logic looks somewhat like the intersection symbol, ∩, in set theory, and the *or* symbol, ∨, is similar to the union symbol, ∪. This is not a coincidence because mathematicians classify set theory and logic as examples of Boolean algebras. Because of their similar structures, there are many parallel ideas that occur in logic and set theory. The table shows some of these parallels.

Logic		Set Theory	
and	∧	∩	intersection
or	∨	∪	union
not	~	′	complement
if . . . then	→	⊆	subset

Similarities between set theory and logic.

Often if a result is true for set theory, then a similar result also holds for logic (and vice versa). As you read this chapter, think about how the ideas discussed here remind you of corresponding ideas in set theory.

Quiz Yourself **5**

Determine whether each statement is true or false.

a) Exercise strengthens your bones or milk is a good source of calcium.

b) $10^2 < 0$ and $(5-3)^2 = 4$.

c) Chocolate is a vitamin or a mineral.

d) You will live to be 70 years old or Earth has a single moon.

KEY POINT

We use truth tables to find the logical values of complex statements.

Compound Statements

When calculating truth tables of compound statements, you also have to pay attention to the order in which you perform logical operations.

EXAMPLE 4   *Finding the Truth Table for a Compound Statement*

Compute a truth table for a statement of the form $(\sim p \wedge q) \vee (p \wedge q)$.

SOLUTION: The numbers in the tables below specify the order of the logical computations. We begin by copying the values for p and q from the left columns beneath all instances of p and q *in the table.* We also compute $\sim p$, which is the first step in completing this table.

		negate			copy q		copy p		copy q
				copy p					
		1		**2**		**4**		**3**	
p	**q**	**(~**	**p**	**∧**	**q)**	**∨**	**(p**	**∧**	**q)**
T	T	F	T		T		T		T
T	F	F	T		F		T		F
F	T	T	F		T		F		T
F	F	T	F		F		F		F

Next, calculate $\sim p \wedge q$ in step 2 and $p \wedge q$ in step 3.

		1		**2**		**4**		**3**	
p	**q**	**(~**	**p**	**∧**	**q)**	**∨**	**(p**	**∧**	**q)**
T	T	F	T	F	T		T	T	T
T	F	F	T	F	F		T	F	F
F	T	T	F	T	T		F	F	T
F	F	T	F	F	F		F	F	F

Use these columns to compute step 2. ⟍   Use these columns to compute step 3.

Finally, we use columns 2 and 3 (highlighted) to compute step 4.

		1		**2**		**4**		**3**	
p	**q**	**(~**	**p**	**∧**	**q)**	**∨**	**(p**	**∧**	**q)**
T	T	F	T	F	T	T	T	T	T
T	F	F	T	F	F	F	T	F	F
F	T	T	F	T	T	T	F	F	T
F	F	T	F	F	F	F	F	F	F

└— final column

Quiz Yourself ⑥

Construct a truth table for the statement
$(p \wedge \sim q) \vee (\sim p \vee q)$.

You can now use this truth table to find values for statements of the form $(\sim p \wedge q) \vee (p \wedge q)$. For example, if p is false and q is true, then the third line of the table tells you that $(\sim p \wedge q) \vee (p \wedge q)$ is true.

Now try Exercises 25 to 32. ❊ ⑥

We could have saved time in step 1 of the table in Example 4 if we directly wrote the truth values for ~*p* instead of first writing the values for *p* and then negating them.

In the final column for the truth table in Quiz Yourself 6, you got all Ts. A statement that is always true is called a **tautology**.

Because we compute truth tables so frequently, it is important to set them up efficiently. As you have seen, if a statement has a single variable such as *p*, then the truth table will have two lines because *p* can be either T or F. If a statement has two variables, there will be four lines, corresponding to the cases TT, TF, FT, and FF.

The tree diagram in Figure 3.6 shows why a statement having three variables *p*, *q*, and *r* has eight lines in its truth table.

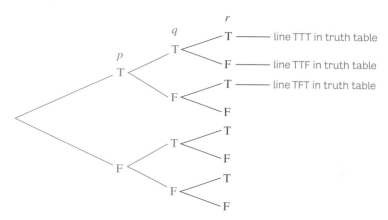

FIGURE 3.6 A statement with three variables has eight lines in its truth table.

Notice how each time we add one variable, each branch in the previous tree splits into two branches—one for T and one for F. In general, the following pattern is true.

> **THE NUMBER OF LINES IN A TRUTH TABLE** If the statement has *k* variables, then its truth table will have 2^k lines.

A statement with the variables *p*, *q*, and *r* will have $2^3 = 8$ lines. To fill in the columns quickly under *p*, *q*, and *r*, first place 4 Ts and then 4 Fs under *p*. Next place 2 Ts, 2 Fs, 2 Ts, and 2 Fs under *q*. Finally, place TFTFTFTF under *r*. You will see this pattern in Example 5.

EXAMPLE 5 *Constructing a Truth Table for a Statement with Three Variables*

Construct a truth table for $(\sim p \wedge q) \vee (\sim r)$.

SOLUTION: We perform the following steps in order:

Step 1: To speed the process, place values for ~*p* directly in the table.

Step 2: Use values for ~*p* and *q* to compute values for ∧.

Step 3: Write values for ~*r* in the table.

Step 4: Use values from steps 2 and 3 (highlighted) to compute values for ∨.

			negate p ⌐───────┐		copy q	negate r	
			1	**2**		**4**	**3**
p	**q**	**r**	**(~p**	**∧**	**q)**	**∨**	**~r**
T	T	T	F	F	T	F	F
T	T	F	F	F	T	T	T
T	F	T	F	F	F	F	F
T	F	F	F	F	F	T	T
F	T	T	T	T	T	T	F
F	T	F	T	T	T	T	T
F	F	T	T	F	F	F	F
F	F	F	T	F	F	T	T

└ final column

KEY POINT

Logically equivalent statements express the same meaning.

Logically Equivalent Statements

It is easy to confuse statements that seem similar but may, in fact, mean very different things. Therefore, it is important for us to know when statements that look different in logic express exactly the same information and when they do not. We use the following criterion to decide whether two statements mean the same thing.

> **DEFINITION** Two statements are **logically equivalent** if they have the same variables and when their truth tables are computed, the final columns in the tables are identical.

EXAMPLE 6 *Determining When Statements Mean the Same Thing*

Assume that you have an entertainment book containing discount coupons for movies, restaurants, and other leisure activities. You are considering eating at either the Pasta Bar or the Deli. Do the following two statements say the same thing?

a) It is not true that: the Pasta Bar accepts coupons and the Deli accepts coupons.

b) The Pasta Bar does not accept coupons or the Deli does not accept coupons.

✺ ✺ ✺ HISTORICAL HIGHLIGHT

The Death of Leibniz's Dream

For hundreds of years, mathematicians pursued Leibniz's dream* of developing a logical theory that could be used to build all of mathematics. In the 1920s, two British mathematicians, Bertrand Russell and Alfred North Whitehead, tried to develop such a plan and were unsuccessful. All hopes for success were dashed in 1931 when Kurt Gödel, a Princeton mathematician, proved his famous *Incompleteness Theorem*, which showed in a mathematical theory such as elementary arithmetic that there will always exist statements that are "true" but cannot be proved using the theory! Leibniz's dream was finally dead.

*See Historical Highlight on page 85.

SOLUTION: If we let p represent "The Pasta Bar accepts coupons" and let d represent "The Deli accepts coupons," then we can write a) symbolically as

$$\sim(p \wedge d)$$

and b) as

$$(\sim p) \vee (\sim d).$$

To decide whether these statements say the same thing, all we have to do is make two truth tables.

First compute "and" then "negate."

			2		1
p	d	$\sim$	$(p$	$\wedge$	$d)$
T	T	**F**	T	T	T
T	F	**T**	T	F	F
F	T	**T**	F	F	T
F	F	**T**	F	F	F

First "negate" then compute "or."

1	3	2
$(\sim p)$	$\vee$	$\sim d$
F	**F**	F
F	**T**	T
T	**T**	F
T	**T**	T

Quiz Yourself 7

Prove that $\sim(p \vee q)$ is logically equivalent to $(\sim p) \wedge (\sim q)$.

Because the final columns (highlighted) in the two truth tables are identical, the two statements are logically equivalent and therefore express exactly the same information.

Now try Exercises 49 to 56. ❋ 7

DeMorgan's Laws

It is easier to remember mathematical facts if we make connections between them and see analogies. Notice that in Example 6, the logical equivalence of $\sim(p \wedge d)$ and $(\sim p) \vee (\sim d)$ looks very similar to one of DeMorgan's laws in set theory, namely $(A \cap B)' = A' \cup B'$. You should expect to see such a parallel because logic and set theory are both Boolean algebras.

> **DEMORGAN'S LAWS FOR LOGIC** If p and q are statements, then
>
> a) $\sim(p \wedge q)$ is logically equivalent to $(\sim p) \vee (\sim q)$.
>
> b) $\sim(p \vee q)$ is logically equivalent to $(\sim p) \wedge (\sim q)$.

By defining logical equivalence, we have taken a large step toward treating logic algebraically. In algebra, when we say $2(x + y) = 2x + 2y$, we are making a general statement that holds for an infinite number of specific cases. For example, we can substitute $x = 3$ and $y = 5$ to get the true statement $2(3 + 5) = 2 \cdot 3 + 2 \cdot 5$. Similarly, when we say that $\sim(p \wedge q)$ is logically equivalent to $(\sim p) \vee (\sim q)$, we are saying that an infinite number of pairs of English language statements are equivalent. For example, if p is the statement "Today is Tuesday" and q is the statement "It is raining," then the following two statements have the same meaning:

"It is not true that: today is Tuesday and it is raining." $\sim(t \wedge r)$

"Today is not Tuesday or it is not raining." $\sim t \vee \sim r$

We can now reason about the meaning of complex statements based on their form rather than their content. Documents such as apartment leases, warranties, car loan applications, college admission forms, and income tax instructions often are written in a legalistic style using the connectives you have been studying.

Quiz Yourself ⑧

Use one of DeMorgan's laws to rewrite (in English) the following statement taken from a car warranty:

The car is not more than five years old and has not been driven over 50,000 miles.

EXAMPLE 7 *Applying Logic to Legal Documents*

Use DeMorgan's laws to rewrite the following statement, which is based on instructions for filing Form 1040-A with the U.S. Internal Revenue Service.

It is false that: you received interest from a seller-financed mortgage and the buyer used the property as a personal residence.

SOLUTION: It is easy to rewrite this statement if first we represent it in symbolic form. Let *r* represent "You received interest from a seller-financed mortgage" and let *b* represent "The buyer used the property as a personal residence." This statement has the form $\sim(r \wedge b)$.

By DeMorgan's laws this statement is equivalent to $(\sim r) \vee (\sim b)$. We can now rewrite this in English as "You did not receive interest from a seller-financed mortgage or the buyer did not use the property as a personal residence."

Now try Exercises 41 to 48. ✷ ⑧

 Some Good Advice

When you use logic to rewrite a statement, the result can sound awkward. You may want to smooth out the grammar so that the sentence sounds better. *This is usually a mistake!* Unless you are quite careful, you can easily change the meaning of a sentence by rewriting it. In logic, the form of a statement is more important than its literary style.

Some people prefer another method to construct truth tables, which we will now explain.

🖉 **KEY POINT**

There is an alternative way to construct truth tables.

EXAMPLE 8 *Using an Alternative Method to Construct Truth Tables*

a) We will reconstruct the truth table for $(\sim p \wedge q) \vee (p \wedge q)$ that you saw in Example 4. Recall that we constructed that table by performing the following steps:

Step 1: Calculate $\sim p$.

Step 2: Compute values for $\sim p \wedge q$.

Step 3: Find values for $p \wedge q$.

Step 4: Calculate values for $(\sim p \wedge q) \vee (p \wedge q)$.

In the following table, we make separate columns for each of these steps to obtain the same table as we did in Example 4.

		1	2	3	4
p	**q**	**~p**	**~p ∧ q**	**p ∧ q**	**(~p ∧ q) ∨ (p ∧ q)**
T	T	F	F	T	T
T	F	F	F	F	F
F	T	T	T	F	T
F	F	T	F	F	F

b) We will construct a truth table for $(\sim p \wedge q) \vee (\sim r)$ that we explained in Example 5. We constructed that table by performing the following steps:

Step 1: Calculate $\sim p$.

Step 2: Find values for $\sim p \wedge q$.

Step 3: Compute values for $\sim r$.

Step 4: Calculate values for $(\sim p \wedge q) \vee (\sim r)$.

Again, we make separate columns for each step. Notice that the final column in this table is exactly the same as the final column in Example 5.

			1	2	3	4
p	q	r	~p	~p ∧ q	~r	(~p ∧ q) ∨ (~r)
T	T	T	F	F	F	F
T	T	F	F	F	T	T
T	F	T	F	F	F	F
T	F	F	F	F	T	T
F	T	T	T	T	F	T
F	T	F	T	T	T	T
F	F	T	T	F	F	F
F	F	F	T	F	T	T

Exercises 3.2

Looking Back*

These exercises follow the general outline of the topics presented in this section and will give you a good overview of the material that you have just studied.

1. What is a quick way to remember the truth tables for the ∧ and ∨ connectives?

2. How did we show that the statements in Example 6 were logically equivalent?

3. How do DeMorgan's laws in logic illustrate the order principle that we discussed in Section 1.1?

4. In using logic to rewrite an English sentence, we use a three-step process that we could describe briefly as: a) Convert to symbols. b) Replace the symbols. c) Convert back to English. Explain how this applies to Example 7.

5. Who showed that Leibniz's dream was impossible?

6. What was the significance of Kurt Gödel's Incompleteness Theorem?

Sharpening Your Skills

In Exercises 7–10, use numbers to specify the order in which you would perform the logical operations for each statement.

7. $\sim(p \lor \sim q)$

8. $\sim(\sim p \land q)$

9. $p \land \sim(p \lor \sim q)$

10. $(p \land \sim q) \lor \sim p$

For Exercises 11–18, fill in the missing values in the following truth table.

				2	1	4		3	
p	q	r	(p	∧	~r)	∨	(q	∨	r)
T	T	T	T	F	F	17.____	T	T	T
T	T	F	T	11.____	T	T	T	T	F
T	F	T	T	F	16.____	T	F	T	T
T	F	F	T	T	T	T	F	12.____	F
F	T	T	F	F	F	T	T	14.____	T
F	T	F	F	15.____	T	T	T	T	F
F	F	T	F	F	F	T	F	13.____	T
F	F	F	F	F	T	18.____	F	F	F

---

*Before doing these exercises, you may find it useful to review the note *How to Succeed at Mathematics* on page xix.

In Exercises 19 and 20, determine how many lines will be in the truth table for each statement.

19. $(p \lor q) \land (r \lor s) \land t$

20. $p \land q \land r \land s \land t \land u$

In Exercises 21–24, state whether each number is a possibility for the number of the lines in a truth table.

21. 16 **22.** 36

23. 9 **24.** 32

In Exercises 25–36, construct a truth table for each statement.

25. $p \land {\sim}q$ **26.** ${\sim}p \land q$

27. ${\sim}({\sim}p \land q)$ **28.** ${\sim}(p \lor {\sim}q)$

29. ${\sim}(p \lor q) \land {\sim}(p \land q)$ **30.** ${\sim}(p \land q) \lor {\sim}(p \lor q)$

31. $(p \lor r) \land (p \land {\sim}q)$ **32.** $(p \land r) \lor (p \land {\sim}q)$

33. $(p \lor q) \land (p \lor r)$ **34.** $(p \land q) \lor (p \land r)$

35. ${\sim}(p \lor {\sim}q) \land r$ **36.** $(p \land {\sim}q) \lor {\sim}r$

In Exercises 37–40, determine whether we are using the inclusive or *or the* exclusive or *in each sentence.*

37. Pay me now or pay me later.

38. You will earn a tax rebate if your income is less than \$23,500 or if you are older than 65.

39. The warranty is void if the appliance has been abused or improperly maintained.

40. The penalty is a \$500 fine or 40 hours of community service.

In Exercises 41–48, use DeMorgan's laws to rewrite the negation of each statement.

41. Bill is tall and thin.

42. Yorrel will go to law school or pursue an MBA.

43. Christian will apply for either a loan or work study.

44. Joanna will quit her job and join the Peace Corps.

45. Ken qualifies for a rebate or a reduced interest rate.

46. Pei Li got a larger disk drive and the extended warranty with her computer.

47. The number x is not equal to 5 and s is not odd.

48. The number y divides 10, but it is not even.

In Exercises 49–56, determine whether the pairs of statements are logically equivalent.

49. ${\sim}(p \land {\sim}q),$ $({\sim}p) \lor q$

50. ${\sim}(p \lor {\sim}q),$ $({\sim}p) \land q$

51. ${\sim}(p \lor {\sim}q) \land {\sim}(p \lor q),$ $p \lor (p \land q)$

52. ${\sim}({\sim}p \lor {\sim}q),$ $p \lor q$

53. ${\sim}(p \lor {\sim}q) \land {\sim}(p \lor q),$ $({\sim}p \land q) \land ({\sim}p \land {\sim}q)$

54. $(p \land {\sim}q) \lor {\sim}(p \land q),$ ${\sim}(p \lor {\sim}q) \lor ({\sim}p \lor {\sim}q)$

55. $p \lor ({\sim}q \land r),$ $(p \lor ({\sim}q)) \land (p \lor r)$

56. $p \land ({\sim}q \lor {\sim}r),$ $(p \land ({\sim}q)) \lor (p \land {\sim}r)$

Applying What You've Learned

Use the following graph that summarizes the status of couples five years after they decide to live together* to find the truth values of the statements in Exercises 57–62. Let p *represent the statement "Couples who considered living together as a step towards marriage were most likely to be married after five years." Let* q *represent the statement "Couples who considered living together as a substitute for marriage were most likely to be split up five years later." Let* r *represent "More couples who used living together as a substitute for marriage were married five years later than those who were living together but just dating."*

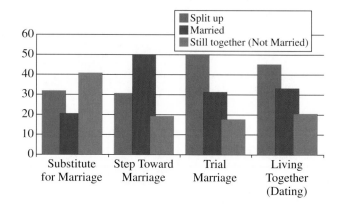

57. $({\sim}p) \lor ({\sim}q)$ **58.** ${\sim}(p \land q)$

59. $({\sim}p \lor q) \land r$ **60.** $(p \land q) \lor {\sim}r$

61. $p \lor ({\sim}q \land r)$ **62.** $r \land {\sim}(q \lor r)$

The statements in Exercises 63–66 are excerpts from the instructions for filing the Form 1040-A with the Internal Revenue Service. Rewrite them as we did in Example 7.

63. The earned income tax did not reduce the tax you owe or did not give you a refund.

64. You cannot claim yourself or your spouse as a dependent.

65. You are not single and not the head of a household.

66. It is not true that: you are filing a joint return and are covered by a retirement plan at work.

In 1937, Claude Shannon showed that computer scientists could use symbolic logic to design computer circuits by using the following approach. Electricity passes through a switch when it is closed and does not flow when the switch is open.

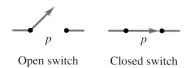

Open switch Closed switch

As shown in the following diagram, electricity flows through a series circuit only when switches p *and* q *are both closed. A **series** circuit corresponds to a conjunction,* p $\land$ q, *in logic. A **parallel** circuit*

---

corresponds to a disjunction, p ∨ q. *Electricity will flow through a parallel circuit if either* p *or* q *is closed.*

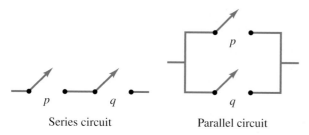

Series circuit   Parallel circuit

We can build more complicated circuits by combining series and parallel circuits. Such circuits can be represented by more complex logical forms. For example, we can represent the circuit

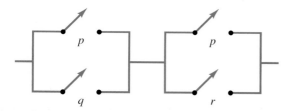

by the logical form (p ∨ q) ∧ (p ∨ r). *Notice that this form is equivalent to the form* p ∨ (q ∧ r) *which can be represented by the circuit.*

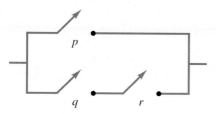

In Exercises 67–70, represent each circuit by a logical form and then rewrite that logical form in an equivalent form. Use truth tables to prove that the second form is equivalent to the first. Draw a circuit that corresponds to the second logical form. If possible, try to choose the second form so that the corresponding circuit has fewer switches than the original circuit.

67.

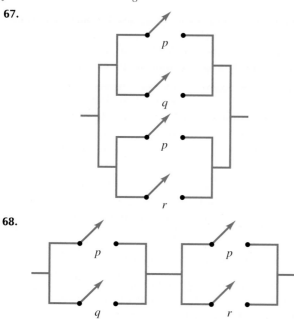

68.

69.

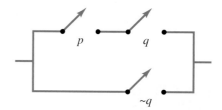

70.

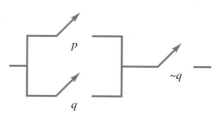

Communicating Mathematics

71. In logic, we use the inclusive *or* as opposed to the exclusive *or* we use in everyday life. What line in Figure 3.5 tells us that we are using the inclusive *or*?

72. Give a concrete example of when you would use the exclusive *or* in everyday life.

73. Give a concrete example of when you would use the inclusive *or* in everyday life.

74. What advantage do you see in using truth tables to determine logical equivalence rather than using our "common sense" and knowledge of the English language?

75. Explain the purpose of having eight lines in the truth table for a statement having three variables *p*, *q*, and *r*.

76. Why does the number of lines in a truth table double if we increase the number of variables in the statement by 1?

Using Technology to Investigate Mathematics

77. Research the life of George Boole and write a brief report on your findings.

78. Research the life of Augustus DeMorgan and write a brief report on your findings.

79. Use your calculator to make examples similar to the one we gave in the exercise set for Section 3.1 to verify the truth tables that we gave for *and* and *or* in this section.

80. Search the Internet for applets that allow you to construct truth tables and report on your findings.

For Extra Credit

81. In set theory, we showed that a set with *k* elements has 2^k subsets. If a logical statement has variables *p*, *q*, and *r*, explain how the eight subsets of {*p*, *q*, *r*} correspond to the eight lines of the truth table for this statement.

82. Make a truth table for $(p \vee q) \wedge (p \vee r)$. Show a correspondence between the subsets of {*p*, *q*, *r*} and the lines of the truth table.

83. Show that the *and* connective is unnecessary in the sense that any statement of the form $(p \wedge q)$ can be written in an equivalent form that does not use ∧.

84. Show that the *or* connective is unnecessary in the sense that any statement of the form ($p \lor q$) can be written in an equivalent form that does not use $\lor$.

The stroke connective has the following truth table:

p	q	$p \mid q$
T	T	F
T	F	T
F	T	T
F	F	T

85. The stroke connective is sometimes called NAND (*not and*). Explain why this is so.

86. Prove that $\sim p$ is logically equivalent to $p \mid p$.

87. Prove that $p \lor q$ is logically equivalent to $(p \mid p) \mid (q \mid q)$.

88. Give a form using only the stroke connective that is equivalent to $p \land q$.

3.3 The Conditional and Biconditional

Objectives

1. Construct truth tables for conditional statements.
2. Identify logically equivalent forms of a conditional.
3. Use alternative wording to write conditionals.
4. Construct truth tables for biconditional statements.

"If it ain't broke, don't fix it."

How many times have you heard that sound advice? Is this folk wisdom also telling us,

"If it is broke, then fix it?"

Maybe—or maybe not. You'll understand exactly what is being said here and, *more important, what is not being said* after you finish studying the conditional in this section.

Conditional statements are common in everyday life—they are literally "all around us." The manufacturer of our DVD-burning software tells us, "If the seal on the package is broken, then the software cannot be returned." As a young child at the amusement park, you were no doubt disappointed when you saw a sign warning, "If you are not 48 inches tall, you cannot ride this roller coaster."

In this section, we will use truth tables for conditionals and biconditionals to compute truth values of compound statements. You will notice that sometimes we use conditionals slightly differently in mathematics than in everyday language.

 **KEY POINT**

There is only one way a conditional can be false.

The Conditional

In order to understand when a conditional statement is true or false, consider this example. Mr. Gates, the owner of a small factory, has a rush order that must be filled by next Monday and he approaches you with this generous offer:

If you work for me on Saturday, then I'll give you a $100 bonus.

If we let w represent "You work for me on Saturday" and b represent "I'll give you a $100 bonus," then this statement has the form $w \rightarrow b$. We must examine four cases to determine exactly when Mr. Gates is telling the truth and when he is not.

Case 1 (*w* is true and *b* is true.):

You come to work and you receive the bonus.

In this case, Mr. Gates certainly made a truthful statement.

Case 2 (*w* is true and *b* is false.):

You come to work and you don't receive the bonus.

Mr. Gates has gone back on his promise, so he has made a false statement.

Case 3 (*w* is false and *b* is true.):

You don't come to work, but Mr. Gates gives you the bonus anyway.

Think carefully about this case because it differs from the way we tend to use *if . . . then* in everyday language. When you listen to Mr. Gates, do not read more into his statement than he actually said. You do not expect to get the bonus if you did not come to work because that is your experience in everyday life. However, Mr. Gates never said that. *You are assuming this condition.* Remember that in logic, a statement is either true or false. Therefore, because Mr. Gates did not say something false, he has told the truth.

Case 4 (*w* is false and *b* is false.):

You don't come to work and you don't receive the bonus.

In this case, Mr. Gates is telling the truth for exactly the same reason as in Case 3. Because you did not come to work, Mr. Gates can give you the bonus or not give you the bonus. In either case, he has not told a falsehood and therefore is telling the truth.

This discussion explains the truth table for the *if . . . then* connective in Figure 3.7.

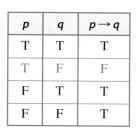

p	*q*	*p* → *q*
T	T	T
T	F	F
F	T	T
F	F	T

FIGURE 3.7
A conditional is false *only* when the hypothesis is true and the conclusion is false.

DEFINITIONS For a conditional *p* → *q*, statement *p* is called the *hypothesis* and *q* is called the *conclusion.**

Some Good Advice

A good way to remember the truth table for the conditional is that the *only way a conditional can be false* is if it has a true hypothesis and a false conclusion.

We can build truth tables for statements that combine the conditional with the connectives you have studied earlier.

EXAMPLE 1   *Constructing a Truth Table for a Conditional Statement*

Construct a truth table for the statement $(\sim p \lor q) \rightarrow (\sim p \land \sim q)$.

SOLUTION: Because there are two variables, *p* and *q*, the truth table has four lines. After carrying out steps 1 through 5, the values for the hypothesis of the conditional are under the *or* symbol of step 2 (highlighted) and the values for the conclusion are under the *and* symbol of step 5 (highlighted).

In step 6, lines 1 and 3 are the only cases with a true hypothesis and a false conclusion, so we put an F on those lines. The final truth table then looks like this:

---

*Some call the hypothesis the *antecedent* and the conclusion the *consequent*.

		1		2		6 ∘	3	5	4	
p	**q**	**(~p**	**∨**	**q)**	**→**	**(~p**	**∧**	**~q)**	
T	T	F	T	T	F	F	F	F	
T	F	F	F	F	T	F	F	T	
F	T	T	T	T	F	T	F	F	
F	F	T	T	F	T	T	T	T	

hypothesis · · · conclusion

True hypothesis and false conclusion occur on these lines.

Quiz Yourself ❾

Construct a truth table for the statement
$(p \wedge \sim q) \rightarrow (\sim p \vee q)$.

Now try Exercises 7 to 22. ❋ ❾

Many formal and informal contracts use the *if . . . then* connective. We can understand a condition in an agreement by writing it in a symbolic form.

EXAMPLE 2 *Analyzing a Refund Policy*

Emilio has just purchased a new solar-powered music system for his iPod from overstock.com that has the following return policy:

> *If* you return the system within 30 days in its original packaging, *then* you will be given a full refund on your purchase.

Dissatisfied with his system, Emilio returns it in its original packaging but is denied a refund. What can you deduce from this information?

SOLUTION: If we let *r*, *p*, and *f* represent the following statements:

r: Emilio returned the music system within 30 days.

p: The system was in its original packaging.

f: He is given a full refund.

we see that the store's policy has the form $(r \wedge p) \rightarrow f$. We know that *f* is false because Emilio did not get a refund. If we assume that the store is truthful in its policy, then the hypothesis $r \wedge p$ cannot be true because a conditional of the form $T \rightarrow F$ is always false. However, we do know that *p* is true, because Emilio used the original packaging. If *r* were also true, that would make $r \wedge p$ true, which we just said cannot be. So we conclude that *r* is false—the system was not returned within 30 days.

Now try Exercises 79–80. ❋

❋ ❋ ❋ HIGHLIGHT

Between the Numbers—Always Read the Fine Print!

Have you ever been tempted to buy a large flat-screen TV with the come-on

Buy Now—No Interest Payments for One Year!

Be careful when you see such an ad. Usually the fine print states something like *if* you continue to meet certain conditions of the contract, *then* the 0% interest rate stays in effect. But, as you saw in the section opener with Mr. Gates, if the hypothesis of this conditional is false—in other words, if you are even one day late in a payment, or perhaps carry

a balance past the one-year period for what is called the "tickle rate" of the loan—then the penalties can be huge. You may have to suffer paying late fees, and, now that the agreement has been violated on your part, the interest rates could go as high as 30%. Lenders who make such tempting loans are counting on the fact that you will not be able to pay off your balance by the end of the 0% interest period or that you will slip up in some other way, violating the contract and thus allowing them to have their way with you financially.

KEY POINT

The converse, inverse, and contrapositive are three derived forms of a conditional.

Derived Forms of a Conditional

There are three typical ways that you might want to rephrase a conditional; however, always remember that when you rewrite a statement in logic, you may also be changing its meaning.

> **DEFINITIONS** We can derive the following statements from the conditional $p \rightarrow q$:
>
> **The converse has the form $q \rightarrow p$.**
> **The inverse has the form $\sim p \rightarrow \sim q$.**
> **The contrapositive has the form $\sim q \rightarrow \sim p$.**

The following table will help you remember how to construct these derived forms of a conditional:

Name of Derived Form	How It Is Constructed
Conditional	$p \rightarrow q$
Converse	Switch p and q.
Inverse	Negate both p and q. Don't switch.
Contrapositive	Negate both p and q and also switch.

 Some Good Advice

In order to understand a statement's form, you should first represent it symbolically before rewriting it in English.

EXAMPLE 3   *Rewriting the Converse, Inverse, and Contrapositive of Statements in Words*

Write, in words, the converse, inverse, and contrapositive of the statement:

"If marijuana is legalized, then drug abuse will increase."

SOLUTION: We can write this statement symbolically as $m \rightarrow d$, where

m stands for "Marijuana is legalized."

and

d stands for "Drug abuse will increase."

We first write each of the derived forms symbolically.

To form the converse, we *switch* d and m to get the form $d \rightarrow m$. Translating this into words gives us

"If drug abuse will increase, then marijuana is legalized."

To form the inverse, we *negate* both m and d to get the form $\sim m \rightarrow \sim d$, which we can write as

"If marijuana is not legalized, then drug abuse will not increase."

To form the contrapositive, we *negate* both m and d and also *switch* to get the form $\sim d \rightarrow \sim m$. We write this as

"If drug abuse will not increase, then marijuana is not legalized."

Now try Exercises 27 to 34. ❋ ⑩

Quiz Yourself ⑩

Write, in words, the inverse and contrapositive of the statement:

If the price of downloading videos increases, then people will copy them illegally.

You must be careful when you rephrase a conditional because the derived statement may not be logically equivalent to the original. We now investigate which derived forms of a conditional are logically equivalent.

EXAMPLE 4   *Equivalence of the Derived Forms of a Conditional*

Which of the statements $p \rightarrow q$, $q \rightarrow p$, $\sim p \rightarrow \sim q$, and $\sim q \rightarrow \sim p$ are logically equivalent?

SOLUTION: We will answer this question by comparing truth tables for these statements.

		Conditional	Converse			Inverse			Contrapositive		
p	q	$p \rightarrow q$	q	$\rightarrow$	p	$\sim p$	$\rightarrow$	$\sim q$	$\sim q$	$\rightarrow$	$\sim p$
T	T	T	T	T	T	F	T	F	F	T	F
T	F	F	F	T	T	F	T	T	T	F	F
F	T	T	T	F	F	T	F	F	F	T	T
F	F	T	F	T	F	T	T	T	T	T	T

logically equivalent (Conditional — Contrapositive)

logically equivalent (Converse — Inverse)

Because the final columns under the conditional and contrapositive are identical, these statements are logically equivalent. We also see that neither the converse nor the inverse are equivalent to the original conditional; however, they are equivalent to each other. ✳

Example 5 shows how to rephrase the types of statements found in leases, contracts, warranties, and tax forms without changing their meaning.

EXAMPLE 5   *Rewriting Legal Statements*

Use the contrapositive to rewrite the following statements:

a) If you pay a deposit, then we will hold your concert tickets.

b) If you do not itemize, then you take the standard deduction.

SOLUTION: The following diagrams show how to write the contrapositives by negating both the hypothesis and conclusion and then interchanging them.

a) Let p represent "You pay your deposit" and h represent "We will hold your concert tickets."

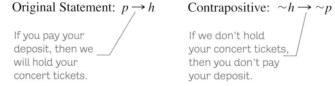

Original Statement: $p \rightarrow h$
If you pay your deposit, then we will hold your concert tickets.

Contrapositive: $\sim h \rightarrow \sim p$
If we don't hold your concert tickets, then you don't pay your deposit.

b) Let i represent "You itemize deductions" and s represent "You take the standard deduction."

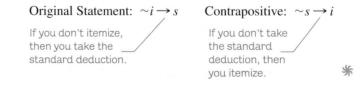

Original Statement: $\sim i \rightarrow s$
If you don't itemize, then you take the standard deduction.

Contrapositive: $\sim s \rightarrow i$
If you don't take the standard deduction, then you itemize. ✳

Alternate Wording of Conditionals

There are many ways to express a conditional without using the words *if . . . then*. Each of the following forms expresses the conditional "if *p* then *q*":

q if *p*	Here the *if* still is associated with *p* even though it occurs later in the sentence.
p only if *q*	Recognize that *only if* does not say the same thing as *if*. The *if* condition is the hypothesis; the *only if* condition is the conclusion.
p is sufficient for *q*.	The sufficient condition is the hypothesis.
q is necessary for *p*.	The necessary condition is the conclusion.

The following diagram will help you remember the information above:

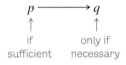

EXAMPLE 6 *Rewriting Statements in If . . . Then Form*

Write each statement in *if . . . then* form.

a) Your driver's license will be suspended if you are convicted of driving under the influence of alcohol.

b) You will graduate only if you have a 2.5 grade point average.

c) To hold your reservation, it is sufficient to give us your credit card number.

d) To qualify for a discount on your airline tickets, it is necessary to pay for them two weeks in advance.

SOLUTION:

a) Because the *if* goes with the clause "you are convicted of driving . . . ," that clause is the hypothesis. We can rewrite this sentence as

> "If you are convicted of driving under the influence of alcohol, then your driver's license will be suspended."

b) The *only if* goes with the clause "you have a 2.5 grade point average"; therefore, this is the conclusion. We write this sentence as

> "If you graduate, then you have a 2.5 grade point average."

c) The phrase *it is sufficient* identifies the hypothesis. We write the sentence as

> "If you give us your credit card number, then [we will] hold your reservation."

d) The necessary condition is the conclusion. Rewriting this sentence, we get

> "If you qualify for a discount on your airline tickets, then you pay for them two weeks in advance."

Now try Exercises 49 to 56. ✳ ⑪

Quiz Yourself ⑪

Write each statement using the *if . . . then* connective.

a) In order to travel on the Amazon, it is necessary to update your immunization.

b) You will increase your cardiovascular fitness only if you exercise three times a week.

✐ **KEY POINT**

The biconditional means that two statements say the same thing.

The Biconditional

The biconditional (*if and only if*) indicates that two statements mean the same thing. For example, we could say that "Today is Tuesday if and only if tomorrow is Wednesday." In algebra we say "$x + 3 = 7$ if and only if $x = 4$." The notation for the biconditional, which is $p \leftrightarrow q$, suggests that we are saying $p \rightarrow q$ and $q \rightarrow p$ at the same time. We show the truth table for the biconditional in Figure 3.8.

Math in Your Life

Thinking About How You Think

You and I think all the time, but do we ever think about how we think? Scientists who study artificial intelligence use formal logic to try to understand human thinking in order to build machines and write programs that duplicate the way we reason.

How does this apply to you and me? If you go to your doctor with a mysterious ailment, your doctor might contact an artificial intelligence program to analyze your symptoms and suggest a diagnosis. Engineers apply logic to devise intelligent climate controls for large buildings that use "common-sense" rules to save energy. To make our drive home from work safer, automobile designers build artificial intelligence into our cars' systems to enable them to adjust to driving conditions.

Although researchers are still far from understanding the complexities of human thinking, artificial intelligence has become a multibillion-dollar industry that now has many practical applications.

p	q	$p \leftrightarrow q$
T	T	T
T	F	F
F	T	F
F	F	T

FIGURE 3.8 The biconditional is true when *both* p and q have the same value.

You should use truth tables to verify that the biconditional $p \leftrightarrow q$ is logically equivalent to the statement $(p \rightarrow q) \wedge (q \rightarrow p)$.

EXAMPLE 7 *Computing a Truth Table for a Complex Biconditional*

Construct a truth table for the statement $\sim(p \vee q) \leftrightarrow (\sim q \wedge p)$.

SOLUTION:

p	q	($\sim$	$(p \vee q)$) 2,1	$\leftrightarrow$ 5	($\sim q$ 3	$\wedge$ 4	p)
T	T	F	T	T	F	F	T
T	F	F	T	F	T	T	T
F	T	F	T	T	F	F	F
F	F	T	F	F	T	F	F

Another way to verify that two statements are logically equivalent is to join them with a biconditional and see whether this new statement is a tautology. We will investigate this further in the exercises.

Exercises 3.3

Looking Back*

These exercises follow the general outline of the topics presented in this section and will give you a good overview of the material that you have just studied.

1. What is a quick way to remember the truth table for the $\rightarrow$ connective?

2. How does the story about working for Mr. Gates help you remember the third and fourth lines of the conditional's truth table?

3. When using logic to rewrite an English sentence, we use a three-step process that we could describe briefly as a) Convert to symbols. b) Replace the symbols. c) Convert back to English. Explain how this applies to Example 5.

4. What cases in the biconditional truth table result in a value of true?

5. What does the Highlight "Always Read the Fine Print" have to do with the conditional in logic?

6. What is artificial intelligence?

---

*Before doing these exercises, you may find it useful to review the note *How to Succeed at Mathematics* on page xix.

Sharpening Your Skills

In Exercises 7–14, assume that p *represents a true statement,* q *a false statement, and* r *a true statement. Determine the truth value of each statement.**

7. $\sim(p \lor q) \to \sim p$ **8.** $(p \land \sim q) \to q$

9. $(p \land q) \to (q \lor r)$ **10.** $(p \lor \sim q) \to r$

11. $(\sim p \lor \sim q) \to r$ **12.** $r \to (\sim p \land q)$

13. $\sim(\sim p \land q) \to \sim r$ **14.** $\sim(\sim p \lor r) \to \sim q$

In Exercises 15–26, construct a truth table for each statement.

15. $p \to \sim q$ **16.** $\sim p \to q$

17. $\sim(p \to q)$ **18.** $\sim(q \to p)$

19. $(p \lor r) \to (p \land \sim q)$ **20.** $(p \land q) \to (p \land \sim r)$

21. $\sim(p \lor r) \to \sim(p \land q)$ **22.** $\sim(p \land r) \to \sim(p \lor q)$

23. $(p \lor q) \leftrightarrow (p \lor r)$ **24.** $(p \land q) \leftrightarrow (p \land r)$

25. $(\sim p \to q) \leftrightarrow (\sim q \to p)$

26. $(p \to \sim q) \leftrightarrow (q \to \sim p)$

In Exercises 27–34, write in words the converse, inverse, or contrapositive as indicated for each statement.

27. If it rains, it pours. (converse)

28. If this appliance fails within 30 days, then it will be fixed free of charge. (converse)

29. If you buy the all-weather radial tires, then they will last for 80,000 miles. (inverse)

30. If you are older than 18, then you must register with Selective Service. (inverse)

31. If a geometric figure is an equilateral triangle, then its sides are all equal in length. (contrapositive)

32. If a geometric figure is a quadrilateral, then the sum of its interior angles is 180 degrees. (contrapositive)

33. If x evenly divides 6, then x evenly divides 9. (inverse)

34. If x is an even prime number, then x is divisible by 2. (inverse)

Assume that you begin with a statement of the form p → q. *In Exercises 35–40, describe each of the given forms in a simpler way.*

35. The converse of the inverse

36. The inverse of the converse

37. The converse of the converse

38. The contrapositive of the inverse

39. The contrapositive of the contrapositive

40. The inverse of the contrapositive

In Exercises 41–44, write the converse, inverse, and contrapositive of each statement in symbolic form.

41. $(\sim p) \to q$ **42.** $p \to \sim q$

43. $(\sim p) \to \sim(q \land r)$ **44.** $\sim(p \lor r) \to q$

In Exercises 45–48, determine which pairs of statements are equivalent. (It is helpful to first write the statements in symbolic form.)

45. If you activate your cell phone before October 1, then you receive 100 free minutes. If you do not receive 100 free minutes, then you do not activate your cell phone before October 1.

46. If you don't register for the LSATs before August 1, then you must pay a $30 late fee. If you register for the LSATs before August 1, then you do not pay a $30 late fee.

47. If it is raining, then use your headlights. If you use your headlights, then it is raining.

48. If Gretchen Bleiles does not score high in the Winter Olympics snowboarding competition, then she will not win a medal. If Gretchen Bleiles qualifies for a medal, then she scores high in the Winter Olympics snowboarding competition.

In Exercises 49–56, rewrite each statement using the words if . . . then.

49. I'll take a break if I finish my workout.

50. You can return the iPod touch only if you have not opened the package.

51. To qualify for this deduction, it is necessary for you to complete Form 3093.

52. To reserve a campsite, it is sufficient that you pay a small deposit.

53. You will receive a free cell phone only if you sign up before March 1.

54. I'll go to Florida if I can save $850.

55. To get a reduction on your auto insurance, it is sufficient that you remain accident free for three years.

56. To graduate this semester, it is necessary that you complete 18 credits.

Applying What You've Learned

In Exercises 57–60, write the converse, inverse, or contrapositive of each conditional statement based on instructions for filing Form 1040-A with the Internal Revenue Service.

57. Interpreting tax forms. If your gross income is over $2,250, then you cannot be claimed by someone else as a dependent. (contrapositive)

58. Interpreting tax forms. If you are filing a joint return, then include your spouse's income. (inverse)

59. Interpreting tax forms. If the amount you overpaid is large, then decrease the amount being withheld from your pay. (converse)

---

*In interpreting these expressions, assume that the negation symbol affects as little of the expression as possible. For example, you should interpret $\sim p \land q$ as meaning $(\sim p) \land q$ rather than $\sim(p \land q)$.

60. **Interpreting tax forms.** If you are a nonresident alien, then you cannot claim an earned income tax credit. (contrapositive)

Use this graph, which shows the results of a February 2005 Gallup poll of 18- to 29-year-olds regarding who they think is the greatest U.S. president of all time, to determine if the statements in Exercises 61–64 are true or false.

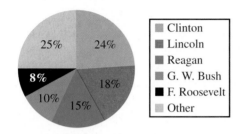

- ☐ Clinton
- ☐ Lincoln
- ☐ Reagan
- ☐ G. W. Bush
- ☐ F. Roosevelt
- ☐ Other

61. If more than half of those surveyed chose Clinton or Lincoln, then more than 10% chose Reagan.

62. If 10% chose Bush, then 10% also chose Roosevelt.

63. If Roosevelt was not ranked higher than Clinton, then Lincoln was not the most preferred.

64. If Bush was not preferred over Lincoln, then Reagan was not preferred by more than 10%.

65. Prove that the biconditional is logically equivalent to two conditionals.

66. Prove that $p \leftrightarrow q$ is logically equivalent to $\sim p \leftrightarrow \sim q$.

Perhaps you have heard the term "helicopter parents"—parents who "hover" over their children, protecting them from making mistakes, particularly in school situations. The following graph* summarizes some activities in which parents are very involved or even take over for their children when they are applying for college. Use this information to determine if the statements in Exercises 67–70 are true or false.

Percent of Parents Highly Involved with Students' College Application Process

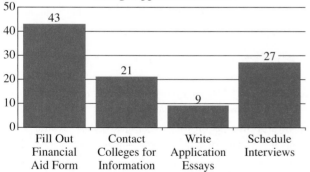

67. If less than 10% of parents help write application essays, then more than 40% fill out financial aid forms.

68. If more than 20% of parents contact colleges for information, then either 27% schedule interviews or less than 40% fill out financial aid forms.

69. If more than 20% are highly involved in scheduling interviews, then more than 10% are involved with writing application essays and contacting colleges for information.

70. If less than 20% schedule interviews, then more than 50% fill out financial aid forms.

Communicating Mathematics

71. Explain how you form the converse, inverse, and contrapositive of a conditional. (For example, sometimes we switch clauses, sometimes we negate, and so on.)

72. Give an example of a true conditional whose converse is false.

73. Give an example of a true conditional whose inverse is false.

74. Is it possible to have a false conditional whose converse is also false? Explain.

75. Explain why it is reasonable to expect that the forms $p \to q$ and $\sim p \vee q$ are logically equivalent. Use truth tables to verify that these forms are logically equivalent.

76. Why is it reasonable to expect that the form $p \wedge \sim q$ is logically equivalent to the form $\sim(p \to q)$? Use truth tables to verify that these forms are logically equivalent.

77. If you assume that the result in Exercise 75 is true, how would DeMorgan's laws lead you to expect the result in Exercise 76?

78. Use Exercise 75 to rewrite the following statement:

If you are older than 18, then you must register with Selective Service.

In Exercises 79 and 80, assume that a credit card company has the following policy:

If you have a platinum credit card with an outstanding balance of more than \$1,000, or you have been a member for at least 10 years, then you qualify for the discount loan rate.

Represent the conditions stated in this policy as follows:

p = *You have a platinum credit card.*

b = *You have a balance of more than \$1,000.*

m = *You have been a member for at least 10 years.*

d = *You qualify for the discount loan rate.*

Based on the partial information given, what can you deduce about each situation? Explain your thinking. (Hint: Recognize that the policy can be written in symbolic form as $((p \wedge b) \vee m) \to d$.)

79. Jamie is a platinum credit card member and has an outstanding balance of \$750. He does not qualify for the discount loan rate.

80. Carla has a balance of \$1,245 on her platinum credit card and has been a member for 12 years.

For Extra Credit

In Exercises 81 and 82, use your knowledge of logic to rewrite each statement in a simpler, equivalent form using fewer connectives.

81. $(\sim p \vee \sim q) \to \sim r$

82. $\sim q \to (\sim r \wedge p)$

---

*Complete survey available at collegeboard.com.

In Exercises 83 and 84, construct truth tables for each statement. Study these tables carefully and see if you can recognize how to write these statements in simpler, equivalent forms.

83. $p \vee (r \wedge (p \wedge q))$

84. $p \wedge (r \vee (p \vee q))$

Exercises 85–88 are based on the exercise sets in earlier sections in which we discussed the correspondence between electrical circuits and logical forms. Draw a circuit that corresponds to each form.

85. $p \rightarrow q$ (*Hint:* Consider a form that is equivalent to this conditional that uses other connectives.)

86. $p \leftrightarrow q$

87. $(\sim p) \rightarrow \sim (q \wedge r)$

88. $p \leftrightarrow (q \wedge \sim r)$

3.4 Verifying Arguments

Objectives

1. Use truth tables to show an argument to be valid.
2. Understand how a truth table can show an argument to be invalid.
3. Recognize some common valid argument forms.
4. Recognize some common fallacies.

In a famous Monty Python skit called *The Argument Clinic,** Michael Palin, playing a character called "Man," arranges to pay to have an argument. After paying his fee for a five-minute argument, he is directed to a room where John Cleese, playing the character of "Mr. Vibrating," awaits. The man knocks on the door and Mr. Vibrating speaks first.

Mr. Vibrating: Come in.   **Man:** Is this the right room for an argument?
Mr. Vibrating: I've told you *once.*   **Man:** No you haven't.
Mr. Vibrating: Yes I have.   **Man:** When?
Mr. Vibrating: Just now!   **Man:** No you didn't.
Mr. Vibrating: Yes I did!   **Man:** Didn't.
Mr. Vibrating: Did.   *And so the skit continues.*

When you hear the word *argument,* you may think of an altercation or verbal fight such as this. However, as you will learn in this section, an argument in logic is a series of statements that have a precise form whose validity we can analyze using truth tables.

Verifying Arguments

In this section, we will consider when a collection of statements produces a logical conclusion. Instead of expressing an argument in sentence form, we will represent it symbolically and use truth tables to determine whether it is valid. The following is a simple example of an argument:

1. If Mahala passed the bar exam, then she is qualified to practice law.
2. Mahala passed the bar exam.

   _____

3. Therefore, Mahala is qualified to practice law.

This argument begins with statements 1 and 2, called *premises* and ends with the final statement 3, called a *conclusion.* If the two premises are true, then the conclusion must also be true. Another way of saying this is that *the conclusion follows from the premises.*

 KEY POINT

We use truth tables to determine whether arguments are valid.

_____

*For the entire script of this skit, go to www.montypythonpages.com.

> **DEFINITIONS** An **argument** is a series of statements called **premises** followed by a single statement called the **conclusion**. An argument is **valid** if whenever all the premises are true, then the conclusion must also be true.

We will use a truth table to determine whether the preceding argument is valid. Let us represent the statement "Mahala passed the bar exam" by p and represent the statement "She is qualified to practice law" by q. Then the argument has the following symbolic form. (The symbol $\therefore$ means *therefore* in the conclusion of an argument.)

$$\left. \begin{array}{l} p \to q \\ p \end{array} \right\} \text{premises}$$

$$\overline{}$$

$$\therefore q \} \text{conclusion}$$

We can view this argument as a conditional statement that has the following form:

If the first premise is true

and

the second premise is true

then the conclusion is true.

This argument then has the symbolic form $[(p \to q) \land p] \to q$. We can verify that this argument is valid by showing that this conditional statement is *always* true, so we will consider its truth table.

		premises					conclusion	
		1			**3**	**2**	**5**	**4**
p	**q**	**[(p**	**→**	**q)**	**∧**	**p]**	**→**	**q**
T	T	T	T	T	T	T	T	T
T	F	T	F	F	F	T	T	F
F	T	F	T	T	F	F	T	T
F	F	F	T	F	F	F	T	F

All Ts show argument to be valid.

Because this statement is a tautology, the argument is valid. The form $[(p \to q) \land p] \to q$ is called the **law of detachment**. Note that not only have we proved that this particular argument is valid but also that *any argument having this form is valid*.

> **VERIFYING AN ARGUMENT** We verify an argument by performing the following steps:
>
> 1. Write the argument symbolically.
> 2. Join the premises together using the *and* connective.
> 3. Form a conditional statement using the conjunction from step 2 for the hypothesis and the conclusion of the argument as the conclusion of the conditional.
> 4. If the statement you form in step 3 is a tautology, then the argument is valid. If there are any Fs in the final column, then the argument is not valid.

EXAMPLE 1 *Determining the Validity of an Argument*

Determine if the following argument is valid:

If you subscribe to the most popular Netflix DVD rental plan, then you have unlimited rentals per month.

You do not have unlimited rentals per month.

∴ You do not have the most popular Netflix DVD rental plan.

SOLUTION: Let p stand for "You subscribe to the most popular Netflix rental plan" and let u represent "You have unlimited rentals per month." Then this argument has the following form:

$$p \rightarrow u$$
$$\sim u$$
$$\overline{}$$
$$\therefore \sim p$$

We now consider the truth table for the conditional $[(p \rightarrow u) \wedge (\sim u)] \rightarrow (\sim p)$.

		premises					conclusion	
		1			**3**	**2**	**5**	**4**
p	u	$[(p$	$\rightarrow$	$u)$	$\wedge$	$\sim u]$	$\rightarrow$	$\sim p$
T	T	T	T	T	F	F	T	F
T	F	T	F	F	F	T	T	F
F	T	F	T	T	F	F	T	T
F	F	F	T	F	T	T	T	T

Argument is valid.

Because the final column of the truth table contains all true values, we conclude that this argument is valid. In essence, we have shown that whenever both of the premises are true, the conclusion is also true. ❁

The argument in Example 1 is an example of the **law of contraposition**.

Invalid Arguments

KEY POINT

If the truth table for an argument has even a single F, then the argument is invalid.

EXAMPLE 2 *Showing That an Argument Is Invalid*

Determine whether the following argument is valid:

If I have the new calling plan, then I can text my friends for free.

I can text my friends for free.

Therefore, I have the new calling plan.

SOLUTION: If we let c represent "I have the new calling plan" and let f represent "I can text my friends for free," then we can write the argument symbolically as follows:

$$c \rightarrow f$$
$$f$$
$$\overline{}$$
$$\therefore c$$

We next consider a truth table for the statement $[(c \rightarrow f) \wedge f] \rightarrow c$.

		premises					conclusion	
		1				**2**	**4**	**3**
c	f	$[(c$	$\rightarrow$	$f)$	$\wedge$	$f]$	$\rightarrow$	c
T	T	T	T	T	T	T	**T**	T
T	F	T	F	F	F	F	**T**	T
F	T	F	T	T	T	T	**F**	F
F	F	F	T	F	F	F	**T**	F

False value shows argument to be invalid.

Math in Your Life

Do We All Reason the Same Way?

You might think that although we may approach problem solving differently, in the end, we all basically reason the same way. After all, "logic is logic." Not so according to research in cultural psychology.*

The different cultural traditions of the East and the West affect the way people reason, and researchers have found that the reasoning tendencies of people in China are different from those found in the Western world. For example, when forced to reason about contradictory statements, Western Europeans and Americans prefer

noncompromising solutions, whereas the Chinese prefer compromises.

Chinese tradition recognizes contradictions in life and embraces them—all things are connected. Western culture views reality analytically—reality is precise and constant—things are isolated and independent of one another. Psychologists find that these different worldviews affect how people in the East and the West approach and solve problems.

Quiz Yourself 12

Determine whether the following argument is valid:

If you install the new antivirus software, then your computer will be protected.

You do not install the new antivirus software.

Therefore, your computer will not be protected.

Because an F appears in the third row of the final column of the table, the argument is not valid. ✳ 12

Example 2 illustrates a common invalid argument form called the **fallacy of the converse.**

 Some Good Advice

You may feel intuitively that the argument in Example 2 is valid because of the way calling plans work. It is very important when analyzing arguments that you do not bring outside information or opinions into the discussion. *The form of the argument is more important than the content of the statements we are making.*

Valid Argument Forms and Fallacies

 **KEY POINT**

Arguments are often one of several standard forms.

We list some common forms of valid and invalid arguments.

Valid Arguments

Law of Detachment	Law of Contraposition	Law of Syllogism	Disjunctive Syllogism
$p \to q$	$p \to q$	$p \to q$	$p \lor q$
p	$\sim q$	$q \to r$	$\sim p$
$\therefore q$	$\therefore \sim p$	$\therefore p \to r$	$\therefore q$

Invalid Arguments

Fallacy of the Converse	Fallacy of the Inverse
$p \to q$	$p \to q$
q	$\sim p$
$\therefore p$	$\therefore \sim q$

*See the article "Culture, Dialectics, and Reasoning About Contradiction," by Peng and Nisbett in the *American Psychologist* 54, (1999), pp. 741–754.

EXAMPLE 3   *Identifying the Form of an Argument*

Identify the form of each argument and state whether it is valid or invalid.

a) If you want to improve your cardiovascular fitness, then take up cross-country skiing.

   You take up cross-country skiing.

   ───────────────────────────────

   Therefore, you want to improve your cardiovascular fitness.

b) If you sleep through your morning math class, then you will be well rested.

   If you are well rested, then you will do well on your math test.

   ───────────────────────────────

   Therefore, if you sleep through your morning math class, then you will do well on your math test.

SOLUTION:

a) Let *i* represent "You want to improve your cardiovascular fitness" and let *c* represent "You take up cross-country skiing." This argument has the following form:

$$i \rightarrow c$$
$$c$$
$$\overline{}$$
$$\therefore i$$

This is the fallacy of the converse, which is an invalid argument form.

b) Let *s* represent "You sleep through your morning math classes," let *r* represent "You will be well rested," and let *w* represent "You will do well on your math test." The argument then has the following form:

$$s \rightarrow r$$
$$r \rightarrow w$$
$$\overline{}$$
$$\therefore s \rightarrow w$$

This is a syllogism and is therefore valid, even though the reasoning seems to make no sense. Now try Exercises 7 to 24. ✳ **13**

You may have been surprised in Example 3(b) that the argument was valid. Keep in mind that *the symbolic form, not the content,* determines the validity of an argument.

We can use logic to analyze more complex arguments than those we have studied so far.

EXAMPLE 4   *Analyzing a Complex Argument with Three Variables*

Determine whether this argument is valid or invalid:

   If we balance the budget or reduce taxes, then there will be more money available to fight pollution.

   If we do not balance the budget, then we will not reduce taxes.

   We will not reduce taxes.

   ───────────────────────────────

   Therefore, there will be more money to fight pollution.

SOLUTION: Consider the statements *b*, "We balance the budget"; *r*, "We reduce taxes"; and *m*, "There will be more money available to fight pollution." This argument then has the form

$$[((b \vee r) \rightarrow m) \wedge (\sim b \rightarrow \sim r) \wedge (\sim r)] \rightarrow m,$$

and, as usual, we construct a truth table to investigate the argument's validity.

Quiz Yourself **13**

Identify the form of the argument and state whether it is valid:

If more is spent on research, then medical science will advance.

More is not spent on research.

Therefore, medical science will not advance.

b	r	m	[((b ∨ r)	→	m)	∧	(~b	→	~r)	∧	(~r)]	→	m
			1	**2**		**6**	**3**	**5**	**4**	**8**	**7**	**9**	
T	T	T	T	T	T	T	F	T	F	F	F	**T**	T
T	T	F	T	F	F	F	F	T	F	F	F	**T**	F
T	F	T	T	T	T	T	F	T	T	T	T	**T**	T
T	F	F	T	F	F	F	F	T	T	F	T	**T**	F
F	T	T	T	T	T	F	T	F	F	F	F	**T**	T
F	T	F	T	F	F	F	T	F	F	F	F	**T**	F
F	F	T	F	T	T	T	T	T	T	T	T	**T**	T
F	F	F	F	T	F	T	T	T	T	T	T	**F**	F

The F in the eighth row of the final column shows that this argument is invalid.

Now try Exercises 25 to 36. ✳

Even though the truth table in Example 4 has Ts in seven of eight rows, by the Always Principle, we must say that the argument is invalid.

✳ ✳ ✳ HIGHLIGHT

Logic and National Defense

Successful Missile Defense Intercept Test Takes Place Near Hawaii

In October 2007, the Missile Defense Agency, a subdepartment of the Department of Defense, posted this news release, touting the most recent success of the Ballistic Missile Defense System, an offspring of the "Star Wars" missile defense system proposed by President Reagan over a quarter of a century ago.

Because the system uses complex computer programs to control radar, lasers, and missiles to destroy incoming nuclear warheads, some concerned scientists believe that flawed computer code may cause a failure in the event of a real nuclear attack.

Verifying program correctness is similar to verifying that a logical argument is valid—but on a much larger scale, because these programs may contain millions of lines of code. Although research on program verification continues, computer scientists have not completely solved the problem of determining when a complex program is completely error free.

Exercises 3.4

Looking Back*

These exercises follow the general outline of the topics presented in this section and will give you a good overview of the material that you have just studied.

1. Express an argument with premises p_1 and p_2 and conclusion c, as a conditional statement.

2. What does it mean to say that the conclusion of an argument follows from the premises?

3. How did we conclude in Example 1 that the argument was valid?

4. Why did we conclude in Example 2 that the argument was not valid?

5. How useful would it be to you if you were doing business in a foreign country to realize that not all cultures reason in the same way as we do in Western culture?

6. Why are some scientists concerned about the Ballistic Missile Defense System?

---

*Before doing these exercises, you may find it useful to review the note *How to Succeed at Mathematics* on page xix.

Sharpening Your Skills

In Exercises 7–24, identify the form of each argument and state whether the argument is valid.

7. If a car has air bags, then it is safe.

This car has air bags.

_____

Therefore, this car is safe.

8. If news on inflation is good, then stock prices will increase.

News on inflation is good.

_____

Therefore, stock prices will increase.

9. If a movie is exciting, then it will gross a lot of money.

This movie grossed a lot of money.

_____

Therefore, it is exciting.

10. If we develop alternative fuels, then we will use less foreign oil.

We are using less foreign oil.

_____

Therefore, we are developing alternative fuels.

11. This laptop has the enhanced video card or an optical disc drive.

This laptop does not have the enhanced video card.

_____

Therefore, this laptop has an optical disc drive.

12. Either my MP3 player is defective or this download is corrupted.

My player is not defective.

_____

Therefore, this download is corrupted.

13. If you pay your tuition late, then you will pay a late penalty.

You do not pay your tuition late.

_____

Therefore, you will not pay a late penalty.

14. If you perform maintenance on your PC, then you violate your warranty.

You do not perform maintenance on your PC.

_____

Therefore, you do not violate your warranty.

15. If you watch *The Apprentice*, then you will succeed in business.

If you succeed in business, then you will have a skyscraper named after you.

_____

Therefore, if you watch *The Apprentice*, you will have a skyscraper named after you.

16. If June 1 is Monday, then June 2 is Friday.

If June 2 is Friday, then June 5 is Wednesday.

_____

Therefore, if June 1 is Monday, then June 5 is Wednesday.

17. If you do not buy the sports package, then you will not get the leather seats.

You do get the leather seats.

_____

Therefore, you did buy the sports package.

18. If you love me, then you will do everything I ask.

You do not do everything I ask.

_____

Therefore, you do not love me.

19. If Phillipe joins the basketball team, then he will not be able to work part-time.

Phillipe did not join the basketball team.

_____

Therefore, he will be able to work part-time.

20. If Carrie gets a raise, then she will be able to afford a bigger apartment.

Carrie gets a raise.

_____

Therefore, Carrie will be able to afford a bigger apartment.

21. If Malik buys a satellite dish, then he will get 128 TV channels.

Malik does not get 128 TV channels.

_____

Therefore, he did not buy a satellite dish.

22. If you cut the amount of fat in your diet, then you will have more energy.

You don't have more energy.

_____

Therefore, you did not cut the amount of fat in your diet.

23. Minxia took the job in Boston or San Diego.

Minxia did not take the job in Boston.

_____

Therefore, Minxia did take the job in San Diego.

24. Rob will take the $1,000 rebate or the interest-free loan.

Rob will not take the $1,000 rebate.

_____

Therefore, Rob will take the interest-free loan.

In Exercises 25–36, determine whether each form represents a valid argument.

25. p

$q \rightarrow \sim p$

_____

$\therefore \sim q$

26. p

$\sim q \rightarrow p$

_____

$\therefore \sim p \vee q$

27. $\sim r$

$r \rightarrow q$

∴ $\sim q \wedge r$

28. p

$\sim q \rightarrow \sim p$

∴ q

29. $\sim q \rightarrow p$

$r \rightarrow \sim q$

∴ $\sim p \rightarrow r$

30. p

$\sim q \rightarrow \sim p$

$(p \wedge q) \rightarrow r$

∴ $q \rightarrow r$

31. p

$q \rightarrow \sim p$

$q \rightarrow (r \wedge p)$

∴ r

32. p

$\sim p \rightarrow \sim q$

$q \rightarrow r$

∴ r

33. r

$r \rightarrow \sim q$

$p \vee q$

∴ p

34. r

$r \rightarrow q$

$\sim p \vee \sim q$

∴ $\sim p$

35. $q \rightarrow \sim p$

$r \rightarrow \sim q$

∴ $\sim p \rightarrow r$

36. p

$\sim p \rightarrow \sim q$

$(p \wedge q) \rightarrow r$

∴ $\sim q \rightarrow r$

Applying What You've Learned

37. If the product has a lower price, then the product does not have quality.

If the product does not have a lower price or does not have quality, then the product is not reliable.

The product has a lower price.

Therefore, the product is reliable.

38. If the team wins this game, then it will qualify for the playoffs.

The team will play in a tournament or will not qualify for the playoffs.

The team wins this game.

Therefore, the team will play in a tournament and will qualify for the playoffs.

39. If you buy from a reputable breeder, then your labradoodle will not require shots.

If you do not buy from a reputable breeder, then it will cost you extra.

It will not cost you extra.

Therefore, your labradoodle will not require shots.

40. If Dave is alone tonight, then he will not come to the party.

If Dave is not alone tonight or does not come to the party, then he will work on his term paper.

Dave will not work on his term paper tonight.

Therefore, Dave is not alone tonight.

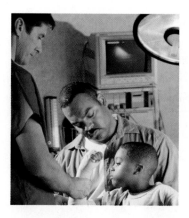

41. If health care is not improved, then the quality of life will not be high.

If health care is improved and the quality of life is high, then the incumbents will be reelected.

The quality of life is high.

Therefore, the incumbents will be reelected.

42. Raul has an IRA.

If Raul has an IRA, then he will not withdraw from his CD.

Raul will withdraw from his CD or invest in bonds.

Therefore, Raul will invest in bonds.

43. Jamie is fluent in Spanish.

If Jamie is fluent in Spanish, then she will work in Madrid.

She will not visit Mexico or she will not work in Madrid.

Therefore, she will visit Mexico.

44. Christian is going to Cancun.

If Christian does not go to Cancun, then he will not go on spring break.

If Christian goes to Cancun on spring break, then his friends will be envious.

Therefore, if Christian goes an spring break, then his friends will be envious.

Communicating Mathematics

In the spirit of "remembering how to remember," in Exercises 45–50, give an intuitive reason as to why the following laws and fallacies are named as they are. For example, what does the law of detachment have to do with detaching?

45. Law of detachment

46. Law of contraposition

47. Law of disjunctive syllogism

48. Fallacy of the converse

49. Fallacy of the inverse

50. What do we mean when we say that the content of an argument is not important?

We have emphasized that the form of a logical argument is more important than its content. Recall in Example 2 that you saw an argument that sounds reasonable, yet has an invalid form. In Exercises 51–54, refer to the tables on page 114 that show forms of some standard valid and invalid arguments. In each exercise, explain your thinking. Try to make your examples interesting by referring to topics currently in the news.

51. Write an argument that sounds reasonable and also has a valid form.

52. Write an argument that sounds reasonable but has an invalid form.

53. Write an argument that sounds unreasonable but has a valid form.

54. Write an argument that sounds unreasonable and also has an invalid form.

Using Technology to Communicate Mathematics

55. Go to www.montypythonpages.com and read the script of *The Argument Clinic*. State the definition of an argument that occurs in that skit.

56. Find a site on the Internet that allows you to download a program that generates crossword puzzles. Make a crossword puzzle using key words of logic. Try to be clever in giving your clues.

57. Use the Internet to find examples of arguments written by Lewis Carroll. Write a brief report on one that you find particularly amusing.

58. Research the Internet for "false arguments." Explain one type of false argument and make up an example of it.

For Extra Credit

In a complicated argument with many variables, it is not practical to use truth tables because of their size. We can, however, use valid argument forms to reason without using truth tables. For example, consider the following argument:

$$p$$
$$p \rightarrow q$$
$$p \land q \rightarrow r$$
$$\sim s \rightarrow \sim r$$
$$\overline{}$$
$$\therefore s$$

We assume that all the premises are true, and we reason like this to prove that the argument is valid:

1. *We assumed that p and p → q are both true, therefore by the law of detachment, we have that q is true.*

2. *Now both p and q are true, so p ∧ q is also true.*

3. *By the law of detachment again, p ∧ q true and p ∧ q → r true force r to be true.*

4. *Because the statement ~s → ~r is equivalent to its contrapositive, we then know that r → s is true.*

5. *Knowing that r and r → s are both true, we conclude that s is also true.*

Therefore, by assuming that all the premises are true, we were able to reason that the conclusion s also must be true.

This means that the argument is valid. In Exercises 59 and 60, reason similarly to prove that each argument is valid.

59. $a \land b$
$b \rightarrow c$
$d \rightarrow \sim c$
$\overline{}$
$\therefore \sim d$

60. $p \rightarrow \sim q$
$p \rightarrow r$
$\sim s \rightarrow q$
p
$\overline{}$
$\therefore r \land s$

In addition to the argument forms that you studied in this section, there are many situations when a person will use a flawed form of an argument to make his or her point. In Exercises 61–64, we describe some well-known flawed argument forms. In each case, write another flawed argument of the same form.

61. Circular Reasoning. In circular reasoning, we support a statement by simply repeating the statement in different or stronger terms. Example: "Arnold Schwarzenegger is a successful governor because he is the best governor our state's ever had."

62. Slippery Slope. Slippery slope arguments falsely assume that one thing must lead to some exaggerated other thing. Example: "If we don't fight this tuition increase now, soon nobody will be able to afford college."

63. Biased Sample. We draw a conclusion about a population based on a prejudiced sample. Example: You interview 20 students at your library and find that the average student on campus studies more than 25 hours per week.

64. False Analogy. Two objects are assumed to be similar and because one object has a certain property, then the other must also. Example: "Running a college is like running a business. In both, the most important thing is the bottom line."

3.5 Using Euler Diagrams to Verify Syllogisms

Objectives

1. Use Euler diagrams to identify a valid syllogism.
2. Use an Euler diagram to identify an invalid syllogism.

In this section, we will analyze arguments called syllogisms. Like the arguments you studied in Section 3.4, a **syllogism** consists of a set of statements called *premises* followed by a statement called a *conclusion*. However, syllogisms differ from earlier arguments in that the premises and conclusion of a syllogism may contain quantifiers such as *all*, *some*, and *none*. The arguments in Section 3.4 did not.

Valid Syllogisms

A syllogism is **valid** if whenever its premises are all true, then the conclusion is also true. If the conclusion of a syllogism can be false even though all the premises are true, then the syllogism is **invalid.** Perhaps the most famous syllogism is

> All people are mortal.
>
> Socrates is a person.
> _____
>
> Therefore, Socrates is mortal.

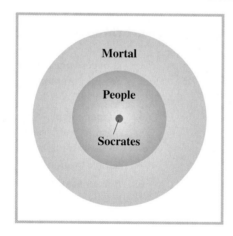

FIGURE 3.9 Euler diagram for the Socrates syllogism.

We can determine whether this syllogism is valid by drawing an **Euler*** (pronounced "oiler") **diagram**. In Figure 3.9, we represent the first premise "All people are mortal," by drawing a small circle labeled "People" inside a larger circle labeled "Mortal." We next place a dot labeled "Socrates" within the circle labeled "People" to indicate "Socrates is a person." Doing this forces the dot "Socrates" to be in the circle "Mortal," so we can see that drawing the two premises in Figure 3.9 forces the conclusion to occur in the diagram. Therefore, the syllogism is valid.

We will often use ∴ for the word *therefore* in the conclusion of a syllogism.

EXAMPLE 1   *Determining the Validity of a Syllogism*

Use an Euler diagram to determine whether the syllogism is valid.

> All poets are good spellers.
>
> Dante is not a good speller.
> _____
>
> ∴ Dante is not a poet.

---

*See the Historical Highlight on page 141 for an article about the famous Swiss mathematician Leonhard Euler.

— HISTORICAL HIGHLIGHT ✻ ✻ ✻

Polish Logicians

Jan Lukasiewicz, in his papers on Aristotle,* made the fundamental ideas of logic available in the Polish language, which led to a school of logic scholars that flourished in Poland in the early part of the twentieth century. Among these was Alfred Tarski, who is considered to be one of the principal logicians of the twentieth century, and Father Salamucha, who formed a group in Krakow to use mathematical logic to modernize Catholic dogma.

Sadly, World War II ended the golden age of Polish logic. During this period, Tarski went to America and Salamucha and a number of Jewish mathematicians were murdered. Lukasiewicz and the others who survived the war chose not to return to Poland.

SOLUTION: Consider the Euler diagrams in Figure 3.10.

(a) All poets are good spellers.

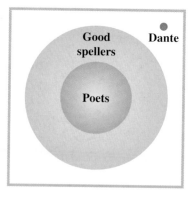

(b) Dante is not a good speller.

FIGURE 3.10

We see from Figure 3.10(b) that if Dante is not a good speller, then he cannot be a poet. Therefore, the conclusion is valid. ✳

Although you may believe that it is possible for a person to be a poet without being a good speller, your feeling about this premise is not relevant to the validity of the argument. Remember: *It is possible for individual premises or the conclusion to be false and yet the syllogism can be valid.* You must rely on Euler diagrams rather than your intuition when deciding whether syllogisms are valid.

Quiz Yourself 14

Draw an Euler diagram to determine whether the syllogism is valid:

All credit cards are made of plastic.

This card is not a credit card.

∴ This card is not made of plastic.

Invalid Syllogisms

EXAMPLE 2 *Using an Euler Diagram to Show That a Syllogism Is Invalid*

Use an Euler diagram to show that the syllogism is invalid.

All tigers are meat eaters.

Simba is a meat eater.

∴ Simba is a tiger.

*You can find many sites discussing Aristotle's syllogisms on the Internet.

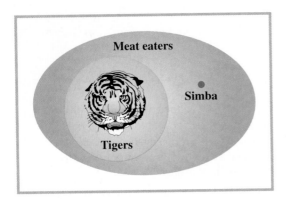

FIGURE 3.11 Simba is a meat eater but not a tiger.

SOLUTION: Figure 3.11 illustrates the two premises. We see in Figure 3.11 that it is possible for Simba to be a meat eater without being a tiger. The argument is therefore invalid because the premises do not force the conclusion to hold. ✿ **14**

Example 3 shows how to handle a premise containing the quantifier *none are* or *no*.

EXAMPLE 3 *Analyzing a Syllogism That Contains the Quantifier No*

Assume that a syllogism begins with the following two premises:

No iMacs get viruses.

My computer does not get viruses.

a) Can we conclude that my computer is an iMac?

b) Can we conclude that my computer is not an iMac?

SOLUTION:

a) We need to determine the validity of the syllogism

No iMacs get viruses.

My computer does not get viruses.

∴ My computer is an iMac.

FIGURE 3.12 Because the conclusion holds in this diagram, you may mistakenly think that the syllogism is valid.

Figure 3.12 shows one possible way to draw the first two premises in an Euler diagram. We illustrate the first premise by drawing two disjoint circles to represent iMacs and computers that get viruses. The rectangle represents the set of all computers.

From Figure 3.12, you might conclude that my computer is an iMac. *That would be a mistake.* As we show in Figure 3.13, there is another way to draw the premise "My computer does not get viruses."

From Figure 3.13 we see that the syllogism is invalid because the given premises do not force the conclusion to hold. Therefore, you cannot conclude that my computer is an iMac.

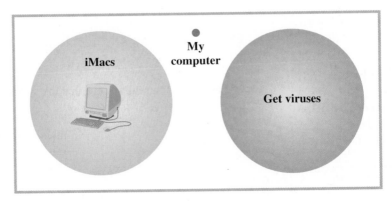

FIGURE 3.13 The conclusion does not hold in this diagram, so the syllogism *cannot* be valid.

b) Figure 3.12 shows that the premises can hold and yet we cannot conclude that my computer is not an iMac. Therefore, based on the given premises, we cannot conclude that my computer is not an iMac. ✸

The quantifier *some* can be misleading in syllogisms because there are several different ways to represent *some* in Euler diagrams. In Figure 3.14, each diagram represents the premise "Some As are Bs."

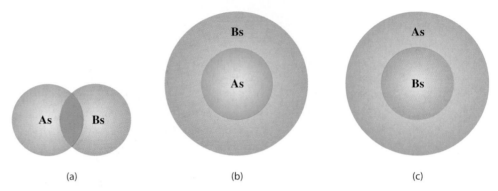

(a)   (b)   (c)

FIGURE 3.14 Euler diagrams illustrating that "Some As are Bs."

Notice in Figure 3.14(b) that the fact that "All As are Bs" does not rule out the possibility that "Some As are Bs."

EXAMPLE 4   *Analyzing a Syllogism Having the Quantifier Some*

Is the syllogism valid?

All cars manufactured after 2007 have driver-side air bags.

Some cars with driver-side air bags have passenger-side air bags.

———————————————————————————————

∴ Some cars with passenger-side air bags were manufactured after 2007.

SOLUTION: Because a single diagram can show a syllogism to be invalid, our strategy will be to find such a diagram. After looking at several diagrams, if we feel that it is impossible to show the argument invalid, then we will say that it is valid. Figure 3.15 shows three ways to illustrate the two premises.

By luck, the first diagram we drew, Figure 3.15(a), shows that *it is possible* that no cars manufactured after 2007 have passenger-side air bags. Therefore, the conclusion does not follow from the premises, so the syllogism is invalid. ✸ **15**

Quiz Yourself **15**

Determine whether the syllogism is valid:

All firefighters are brave.

Some women are firefighters.

———————————————

∴ Some women are brave.

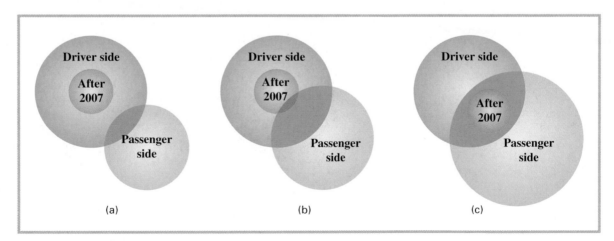

(a)   (b)   (c)

FIGURE 3.15 Three ways to illustrate the premises of the air bag syllogism.

In Example 4, we cannot conclude that no cars manufactured after 2007 have passenger-side air bags. Clearly the diagrams in Figures 3.15(b) and 3.15(c) show that *it is possible* for some cars manufactured after 2007 to have passenger-side air bags. All we are able to say in Example 4 is that even though the conclusion sounds reasonable, it does not follow from the premises.

PROBLEM SOLVING
The Counterexample Principle

From the Counterexample Principle in Section 1.1 you know that one Euler diagram can show a syllogism to be invalid. In testing the validity of a syllogism, it is a good strategy to draw several diagrams, with the idea that one of them may show the syllogism to be invalid. If you are convinced that no diagram can show the syllogism to be invalid, then you can conclude that the syllogism is valid.

The problem with drawing Euler diagrams, particularly for syllogisms using the quantifier *some*, is that whatever diagram you draw will show extra conditions that were not stated as a premise. These extra conditions can mislead you in determining the validity of a syllogism.

EXAMPLE 5 *Extra Conditions Are Always Present in an Euler Diagram*

Each Euler diagram in Figure 3.16 illustrates the following premises:

> All As are Bs.
> Some Bs are Cs.

State a condition that is present in each diagram that has not been stated as a premise.

SOLUTION: There are many possible conditions. We will give one for each diagram.

a) Because the circles containing As and Cs do not intersect, we have drawn the extra condition that "No As are Cs."

b) Because the circles containing As and Cs do intersect, we have drawn the condition that "Some As are Cs."

c) Because we drew the circle containing As inside the circle containing Cs, we have the additional condition that "All As are Cs." ✳ **16**

Quiz Yourself **16**

Find one additional condition in Figure 3.16, other than those given in Example 5, that was not stated as a premise.

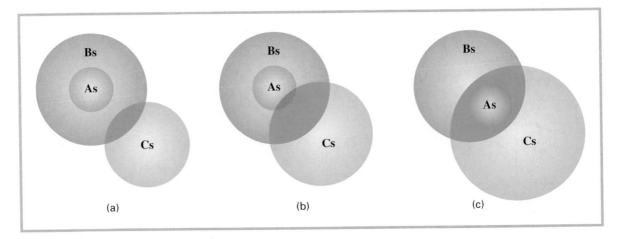

(a) (b) (c)

FIGURE 3.16 Euler diagrams illustrating "All As are Bs" and "Some Bs are Cs."

In Figure 3.16, several other extra conditions appear in the diagrams that were not stated as premises. It may be hard to show that a syllogism is valid because by the Always Principle, the conclusion must hold in every Euler diagram that illustrates the premises. You must be careful to consider a variety of diagrams before you accept a syllogism as valid.

Exercises 3.5

Looking Back*

These exercises follow the general outline of the topics presented in this section and will give you a good overview of the material that you have just studied.

1. When is a syllogism valid?

2. When is a syllogism invalid?

3. What was the point of Example 5?

4. How many Euler diagrams must you draw to prove that a syllogism is not valid?

5. Which Polish mathematician introduced logic to Polish scholars?

6. What ended the golden age of Polish logic?

Sharpening Your Skills

In Exercises 7–22, determine whether each syllogism is valid or invalid.

7. All original parts are under warranty.

This part is an original part.

∴ This part is under warranty.

8. All college students are ambitious.

Cameron is ambitious.

∴ Cameron is a college student.

9. All imports are subject to a surcharge.

This jacket is subject to a surcharge.

∴ This jacket is an import.

10. All documentarians are biased.

Michael Moore is a documentarian.

∴ Michael Moore is biased.

11. All vitamins are healthful.

Milk is healthful.

∴ Milk is a vitamin.

12. All aerobic activities are healthful.

Blogging is an aerobic activity.

∴ Blogging is healthful.

13. All athletes are fit.

Julio is not an athlete.

∴ Julio is not fit.

14. All athletes are fit.

Sheena is not fit.

∴ Sheena is not an athlete.

15. All politicians are honest.

Walt is not honest.

∴ Walt is not a politician.

16. All salespeople are sincere.

Emily is not sincere.

∴ Emily is not a salesperson.

17. Some large corporations are concerned about the environment.

General Motors is not a large corporation.

∴ General Motors is not concerned about the environment.

18. Some children love cookies.

Dani does not love cookies.

∴ Dani is not a child.

19. Some biologists believe the Loch Ness monster exists.

All who believe the Loch Ness monster exists are irrational.

Antawn is not irrational.

∴ Antawn is not a biologist.

20. Some investors are wealthy.

All wealthy people are happy.

∴ Some investors are happy.

21. Some mammals are large.

All large animals are dangerous.

∴ Some mammals are dangerous.

*Before doing these exercises, you may find it useful to review the note *How to Succeed at Mathematics* on page xix.

22. Some mammals are large.

Some dangerous animals are large.

∴ Some mammals are dangerous.

Applying What You've Learned

In Exercises 23–26, complete each syllogism so that it is valid and the conclusion is true. There may be several correct answers.

23. Some taxes are unfair.

All unfair taxes should be abolished.

∴

24. All honest politicians should be supported.

Marika should not be supported.

∴

25. No team that plays in a domed stadium has won the Super Bowl.

Some teams that wear red uniforms have won the Super Bowl.

∴

26. Some mathematicians are fine musicians.

All fine musicians are intelligent.

∴

Communicating Mathematics

27. Draw an Euler diagram for the statements "All As are Bs" and "Some Bs are Cs."

28. Draw an Euler diagram for the statements "Some As are Bs" and "Some Bs are Cs."

29. Draw an Euler diagram for the statements "No As are Bs" and "All Bs are Cs."

30. In each of your drawings for Exercises 27–29, state a condition in your drawing that is not present in the statement that you are trying to illustrate.

31. Give an example of a valid syllogism that has a false statement for its conclusion.

32. Give an example of an invalid syllogism that has a true statement for its conclusion.

Using Technology to Investigate Mathematics

33. Search on the Internet at a history of mathematics site to find a biography of Leonhard Euler. Write a brief report on your findings.

34. The author Charles Dodgson, who wrote under the pseudonym Lewis Carroll, is well known for his works *Alice in Wonderland* and *Through the Looking Glass*. He was also a logician who wrote many clever syllogisms. Perform a search on the Internet to find some of his syllogisms and then solve them using Euler diagrams.

For Extra Credit

In Exercises 35–38, write a valid syllogism that would be illustrated by each Euler diagram. For example, the Socrates syllogism would be illustrated by a diagram similar to the one in Exercise 35.

35.

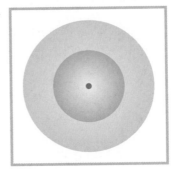

36.

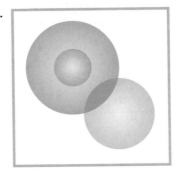

37.

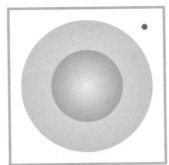

38.

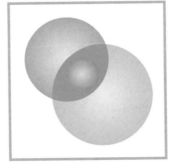

Looking Deeper

3.6 Fuzzy Logic

Objectives

1. Understand the definition of statements in fuzzy logic.
2. Be able to work with connectives in fuzzy logic.
3. Use fuzzy logic to make decisions.

"Aristotle Is Out and Buddha Is In"*

You have been learning Aristotelian logic in this chapter, but increasingly some researchers believe that we would be better off without a right or wrong, true or false approach to problem solving—we should, instead, be using fuzzy thinking to solve our problems.

In fact, by using *fuzzy logic*, researchers have been able to come to your rescue and put an end to some of life's little annoyances. Fuzzy air conditioners make your home more comfortable, with a fuzzy shower controller you can avoid the shock of extreme water temperature changes when a toilet is flushed, and subways operated by fuzzy computers will give you a smoother ride with no need to hang onto a strap during your morning commute.

In this section, we will teach you how to use fuzzy logic to analyze situations when what is being said is not black or white, but rather some shade of gray.

For example, to say "It is warm today" may not mean the same thing on the first day of March as it does in the middle of August. Although it may be warm today, perhaps it is not as warm as it was two weeks ago. To simply say "It is warm today" ignores exactly how warm it is. The strict condition that statements must be either true or false makes symbolic logic unsuitable for representing many real-life situations.

✏ KEY POINT

Statements in fuzzy logic have truth values between 0 and 1.

Fuzzy Statements

Mathematicians developed *fuzzy logic* in order to apply the techniques of logic more widely. This is a perfectly acceptable, well-developed area of mathematics. Once we define the meaning of fuzzy statements and explain how fuzzy connectives behave, we can work with them as we did for other (nonfuzzy[†]) statements.

> **DEFINITIONS** In **fuzzy logic**, a **statement** is a declarative sentence that has an associated **truth value** between 0 and 1 inclusive.

EXAMPLE 1 *The Truth Values of Statements in Fuzzy Logic*

Here are some examples of statements in fuzzy logic:

a) "I like peach ice cream," with a truth value of 0.9.
b) "Tiger Woods is a great golfer," with a truth value of 0.92.
c) "Maine is a large state," with a truth value of 0.45.
d) "*Gone with the Wind* is a great movie," with a truth value of 0.73.
e) "You find mathematics interesting," with a truth value of 0.7. ✳

---

*Quoted from a book review on fuzzy thinking, by Kirkus Reviews, Kirkus Associates, LP, amazon.com (2007).
[†]Some use the term *crisp statements* to describe the statements we studied earlier.

The way we assigned truth values in Example 1 is somewhat arbitrary. You may want to assign a truth value of 0.85 to the statement "*Gone with the Wind* is a great movie" because you believe that it is a greater movie than we do. *This is perfectly OK and is really the point of fuzzy logic.* If we could not differ in our assignment of truth values, we would then both be forced to say that *Gone with the Wind* was a great movie, but we could not express the difference of our opinions of exactly how great the movie was.

✎ **KEY POINT**

Negation, conjunction, and disjunction in fuzzy logic are defined differently than the corresponding connectives in nonfuzzy logic.

Fuzzy Connectives

We will now consider how connectives work in fuzzy logic. In Example 1, we assigned a truth value of 0.9 to "I like peach ice cream." Because this expresses the degree to which I like peach ice cream, it also says that $1 - 0.9 = 0.1$ indicates the degree to which I *do not like* peach ice cream. The next definition explains how to calculate the truth value of negations in fuzzy logic.

> **DEFINITION** If p is a statement in fuzzy logic, then the truth value of the **negation** of p, written $\sim p$, is
>
> $$1 - \text{the truth value of } p.$$

EXAMPLE 2   *Calculating the Truth Values of Negations in Fuzzy Logic*

Calculate the truth values for the negations of statements b) and c) in Example 1.

SOLUTION:

b) "Tiger Woods is not a great golfer" has a truth value of $1 - 0.92 = 0.08$.

c) "Maine is not a large state" has a truth value of $1 - 0.45 = 0.55$.

    Now try Exercises 13 to 16. ❀ **17**

It is easy to compute conjunctions and disjunctions in fuzzy logic.

Quiz Yourself **17**

Calculate the truth value for the negations of statements d) and e) in Example 1.

> **DEFINITIONS** Suppose that p and q are two statements in fuzzy logic.
>
> a) The truth value of the **disjunction** of p and q, written $p \lor q$, is the *maximum* of the truth values for p and q.
>
> b) The truth value of the **conjunction** of p and q, written $p \land q$, is the *minimum* of the truth values for p and q.

EXAMPLE 3   *Calculating Disjunctions and Conjunctions in Fuzzy Logic*

Let e represent the statement "iMacs are easier to use than PCs" and let c represent "iMacs cost more than PCs." Assume that e has a truth value of 0.85 and c has a truth value of 0.73. Find the truth values of the following statements:

a) "iMacs are easier to use than PCs *and* cost more than PCs."

b) "iMacs are easier to use than PCs *or* cost more than PCs."

Quiz Yourself 18

Let *l* represent the statement "I like living in a large city" with a truth value of 0.82 and let *e* represent the statement "I like living on the East Coast" with a truth value of 0.45. Find the truth value of each statement.

a) $l \wedge e$

b) $l \vee e$

SOLUTION:

a) The statement "iMacs are easier to use than PCs *and* cost more than PCs" has the form $e \wedge c$, so its truth value is the minimum of 0.85 and 0.73, which is 0.73.

b) The statement "iMacs are easier to use than PCs *or* cost more than PCs" has the form $e \vee c$, so its truth value is the maximum of 0.85 and 0.73, which is 0.85.

Now try Exercises 17 to 20. ✸ 18

We can use our knowledge of fuzzy connectives to analyze more complex statements.

EXAMPLE 4 *Computing the Truth Value of a Statement with Three Variables*

Let *h*, *s*, and *c* be defined as follows:

    h: "Foreign cars have high resale value," with truth value 0.85.

    s: "Foreign cars are safer than domestic cars," with truth value 0.73.

    c: "Foreign cars cost more than domestic cars," with truth value 0.6.

Find the truth value of the statement

    "Foreign-made cars have high resale value, but it is not true that they are safer or cost more than domestic cars."

SOLUTION: This statement has the following form:

$$h \wedge \sim (s \vee c)$$
$$0.85 \quad\quad 0.73 \;\; 0.6$$

Quiz Yourself 19

Let *p* have a truth value of 0.65, let *q* have a truth value of 0.80, and let *r* have a truth value of 0.55. Find the truth value of each statement.

a) $\sim(p \wedge q) \vee (\sim r)$

b) $((\sim p) \wedge q) \vee r$

We have written the truth values of *h*, *s*, and *c* to make the calculations easier to follow.

    The truth value of $s \vee c$ is the maximum of 0.73 and 0.6, which is 0.73. The truth value of $\sim(s \vee c)$ is then $1 - 0.73 = 0.27$. The truth value of the conjunction of $\sim(s \vee c)$ and *h* is the minimum of 0.27 and 0.85, namely 0.27. ✸ 19

    It is awkward to always talk about the minimum of the truth values of *h* and *s*, the maximum of the truth values of *h* and *s*, and so on, so for simplicity's sake, in calculating the truth value of $h \vee s$, we will write $0.85 \vee 0.73 = 0.85$ and the truth value of $h \wedge s$ as $0.85 \wedge 0.73 = 0.73$. We will handle negation similarly. With this agreement in mind, we can calculate the truth value of $\sim(h \vee \sim(s \wedge \sim h))$:

$$\sim(0.85 \vee \sim(0.73 \wedge \sim 0.85)) = (0.85 \vee \sim(0.73 \wedge 0.15))$$
$$= \sim(0.85 \vee \sim 0.15)$$
$$= \sim(0.85 \vee 0.85)$$
$$= \sim 0.85 = 0.15$$

KEY POINT

We use an equivalent form of the conditional in ordinary logic to define the conditional in fuzzy logic.

    It is easy to use truth tables to prove that the conditional $p \rightarrow q$ is logically equivalent to $(\sim p) \vee q$. We use this result in defining how to compute the truth value of conditionals in fuzzy logic.

    DEFINITION If *p* and *q* are two statements in fuzzy logic, then the truth value of the **conditional**, $p \rightarrow q$, is defined to be the truth value of $(\sim p) \vee q$.

EXAMPLE 5 *Computing the Truth Values of Conditionals*

Let *p* have a truth value of 0.75 and let *q* have a truth value of 0.46. Find the truth value of each statement.

a) $p \rightarrow q$ b) $q \rightarrow p$ c) $\sim q \rightarrow \sim p$

Quiz Yourself **20**

Let p have a truth value of 0.65, let q have a truth value of 0.80, and let r have a truth value of 0.55. Find the truth value of each statement.

a) $p \rightarrow q$

b) $(p \wedge q) \rightarrow r$

c) $\sim(p \vee q) \rightarrow \sim r$

SOLUTION:

a) The truth value of $p \rightarrow q$ is the same as the truth value of $(\sim p) \vee q$, which is $(\sim 0.75) \vee 0.46 = 0.25 \vee 0.46 = 0.46$.

b) To calculate the truth value of $q \rightarrow p$, instead we calculate the truth value of $(\sim q) \vee p$, which is $(\sim 0.46) \vee 0.75 = 0.54 \vee 0.75 = 0.75$.

c) The truth value of $\sim q \rightarrow \sim p$ is found by calculating the truth value of $\sim(\sim q) \vee (\sim p)$. We simplify this by replacing $\sim(\sim q)$ with q. The truth value of $q \vee (\sim p)$ is $0.46 \vee (\sim 0.75) = 0.46 \vee 0.25 = 0.46$. It should not surprise you that this is the same truth value that $p \rightarrow q$ had. Why?

Now try Exercises 21 to 28. ✤ **20**

Fuzzy Decision Making

We will now show how to use fuzzy logic to make decisions. Deciding on what to do when your car needs an expensive repair is certainly a situation in which the factors you consider are not clear-cut. It is a perfect example of a problem that we can analyze using fuzzy logic.

EXAMPLE 6* *Making Decisions Using Fuzzy Logic*

Assume that your car has broken down and you are trying to decide whether to repair it, buy a newer used car, or buy a brand-new car. In Table 3.1,[†] we list the characteristics of a car that are important to you and assign a number between 0 and 1 to indicate how important each characteristic is to you. Table 3.2 lists the degree to which each option satisfies these characteristics.

Characteristic	Importance
Cost	0.70
Reliability	0.60
Size (large enough)	0.45
Good gas mileage	0.65
Safety	0.75

TABLE 3.1 Fuzzy truth values describe how important each characteristic is to you.

	Repair	Newer Used Car	New Car
Cost	0.80	0.65	0.30
Reliability	0.40	0.60	0.95
Size (large enough)	0.90	0.75	0.50
Good gas mileage	0.55	0.80	0.85
Safety	0.45	0.60	0.75

TABLE 3.2 Fuzzy truth values describe how well each option satisfies the characteristics that are important to you.

The ideal option for you is the one that scores the highest in the characteristics that are important to you. In other words, we want to consider fuzzy statements of the following type:

"If cost is important to you, then repairing your car satisfies this condition."

"If reliability is important to you, then buying a new car satisfies this condition."

We thus want truth values for fuzzy statements of the form $i \rightarrow c$, where i is a statement that a characteristic is important to you and c is a statement that an option possesses that characteristic. Recall that $i \rightarrow c$ has the same truth value as $(\sim i) \vee c$. Let us look at the truth value for the statement

"If cost is important to you, then repairing your car satisfies this condition."

Here i is the statement "Cost is important to you," which from Table 3.1 has a truth value of 0.70. The statement c is "Repairing your car satisfies this condition (cost)," which, from

---

*The method used in this example is based on Ronald R. Yager, "Concepts, Theory and Techniques: A New Methodology for Ordinal Multiobjective Decisions Based on Fuzzy Sets," *Decision Sciences* 12(4) (October 1981), pp. 589–600.

[†]In Tables 3.1 and 3.2, the assignment of values between 0 and 1 is somewhat arbitrary. Another person might assign these values differently.

Table 3.2, has a truth value of 0.80. The truth value of $(\sim i) \vee c$ is thus $(\sim 0.70) \vee 0.80 = 0.30 \vee 0.80 = 0.80$.

We will now construct a truth table for all the statements $(\sim i) \vee c$ for each option: repair, buy a used car, and buy a new car.

	i	$\sim i$	Repair c	Repair $(\sim i) \vee c$	Used c	Used $(\sim i) \vee c$	New c	New $(\sim i) \vee c$
Cost	0.70	0.30	0.80	0.80	0.65	0.65	0.30	0.30
Reliability	0.60	0.40	0.40	0.40	0.60	0.60	0.95	0.95
Size	0.45	0.55	0.90	0.90	0.75	0.75	0.50	0.55
Gas mileage	0.65	0.35	0.55	0.55	0.80	0.80	0.85	0.85
Safety	0.75	0.25	0.45	0.45	0.60	0.60	0.75	0.75
Conjunction of all $(\sim i) \vee c$				**0.40**		**0.60**		**0.30**

Quiz Yourself 21

Calculate the truth value of each statement.

a) "If reliability is important to you, then buying a new car satisfies this condition."

b) "If being large enough is important to you, then buying a newer used car satisfies this condition."

For an option to be satisfactory, you want to have all the conditions satisfied as well as possible. Therefore, you are interested in cost AND reliability AND size AND gas mileage AND safety. Thus, to finish making your decision, you calculate the *conjunction* of all the expressions $(\sim i) \vee c$ for each option. The conjunction of all the expressions $(\sim i) \vee c$ for repairing your car is the minimum of the truth values in this column, which is 0.40 (highlighted). Similarly we obtain a value of 0.60 (highlighted) for buying a newer used car and 0.30 (highlighted) for buying a new car. Because buying a newer used car has the highest final rating, the best choice for you *according to this rating method* is to buy a newer used car.

Now try Exercises 29 to 32. ❋ 21

The method we used in Example 6 is a real way to use fuzzy logic to make decisions. If you do not agree totally with this process for making the final decision, we invite you to criticize this method and to suggest alternatives in the exercises.

Exercises 3.6

Looking Back*

These exercises follow the general outline of the topics presented in this section and will give you a good overview of the material that you have just studied.

1. What truth values are possible for a fuzzy statement?
2. If p is a fuzzy statement, what is the truth value of $\sim p$?
3. If p and q are fuzzy statements, what is the truth value of $p \wedge q$?
4. If p and q are fuzzy statements, what is the truth value of $p \vee q$?

Sharpening Your Skills

In Exercises 5–12, assign a truth value between 0 and 1 to each fuzzy statement. Of course, the assignments will vary from student to student.

5. *The Simpsons* is a television show intended for adults.
6. Physicians earn high salaries.
7. Madonna is a good singer.
8. Abraham Lincoln was our best president.
9. The United States should increase its foreign aid.
10. We should spend more money to shelter the homeless.
11. Money brings happiness.
12. Baseball is a difficult sport.

In Exercises 13–16, calculate the truth value of the negation of each fuzzy statement. The truth value of each fuzzy statement is given in parentheses.

13. Picasso was a famous artist. (0.95)
14. Ryan Howard is known mainly for his prowess at pitching. (0.15)
15. Learning a musical instrument can teach self-discipline. (0.75)
16. Space research is vital to national defense. (0.40)

*Before doing these exercises, you may find it useful to review the note *How to Succeed at Mathematics* on page xix.

In Exercises 17–20, consider the following fuzzy statements:

p: *Students choose a college because of its location. (0.75)*

q: *Students choose a college because of its cost. (0.85)*

Determine the truth value of each statement.

17. Students choose a college because of its location or cost.

18. Students choose a college because of its location and cost.

19. Students choose a college because of its location and not its cost.

20. It is not true that students choose a college because of its location or its cost.

In Exercises 21–28, assume that p *has a truth value of 0.27,* q *has a truth value of 0.64, and* r *has a truth value of 0.71. Find the truth value of each statement.*

21. $p \wedge \sim q$

22. $\sim(p \vee q)$

23. $\sim(p \vee \sim q)$

24. $(p \vee r) \wedge (p \wedge \sim q)$

25. $\sim(p \vee \sim q) \wedge \sim r$

26. $p \rightarrow r$

27. $q \rightarrow (p \vee r)$

28. $\sim(p \vee q) \rightarrow \sim(q \wedge r)$

Applying What You've Learned

In Exercises 29–32, use the method described in Example 6 to evaluate each situation.

29. You are trying to decide which of two job offers to accept. The first job is a retail sales trainee for a large consumer products company and the second is a marketing data analyst for a communications company. In the first table, we list the characteristics of a job that are important to you; in the second table, we list the degree to which each job satisfies these characteristics. Which job should you take?

Characteristic	Importance
Salary	0.70
Interesting	0.80
Work with people	0.60
Flexible hours	0.75

	Sales Trainee	Data Analyst
Salary	0.60	0.65
Interesting	0.50	0.80
Work with people	0.90	0.30
Flexible hours	0.80	0.60

30. You are trying to decide whether to buy a house or continue to rent an apartment. In the first table, we list the characteristics that are important to you; in the second table, we list the degree to which each situation satisfies these characteristics. What should you do?

Characteristic	Importance
Cost	0.70
Close to job	0.60
Adequate space	0.55
Near city attractions	0.65

	House	Apartment
Cost	0.80	0.65
Close to job	0.40	0.80
Adequate space	0.90	0.75
Near city attractions	0.55	0.90

31. Your younger sister is having a difficult time deciding which of three colleges to attend: Little Small College (LSC), Good Old State (GOS), and Mom's Favorite University* (MFU). In the first table, we list features of each school that are important to your sister and rank these qualities; in the second table, we list the degree to which each school satisfies these characteristics. What should she do?

Characteristic	Importance
Size (not too large)	0.60
Cost	0.80
Academics	0.75
Social life	0.90
Close to home	0.30

	LSC	GOS	MFU
Size (not too large)	0.95	0.60	0.40
Cost	0.40	0.80	0.30
Academics	0.80	0.70	0.65
Social life	0.55	0.75	0.70
Close to home	0.70	0.65	0.95

32. You are trying to decide which of three investments to make: (A)llied InterServe, (B)itLogic, or (C)omQual. In the first table, we list the characteristics of investing that are important to you; in the second table, we list the degree to which each choice satisfies these characteristics. Which investment is your best choice?

*The reason that this is Mom's favorite university is that Mom would like to see your sister go to school close to home.

Characteristic	Importance
Past performance	0.85
Safety	0.60
Liquidity	0.75
Minimum amount to invest	0.35
Management fee	0.55

	A	B	C
Past performance	0.80	0.65	0.90
Safety	0.45	0.60	0.80
Liquidity	0.60	0.75	0.75
Minimum amount to invest	0.85	0.60	0.65
Management fee	0.45	0.80	0.75

Communicating Mathematics

33. How are the rules for computing the truth tables for connectives that you learned for nonfuzzy logic consistent with the rules for computing truth values of compound statements in fuzzy logic?

34. Explain the fuzzy decision-making process that we illustrated in Example 6. What do you do first? Second?

Using Technology to Investigate Mathematics

35. Search the Internet to find some applications of fuzzy logic at various universities and commercial sites and report on what you find.

36. Search the Internet for some applications of fuzzy logic in designing appliances, and cars, and in medicine.

37. From your instructor, obtain a spreadsheet for doing fuzzy logic decision making. Perform the calculations of Example 6 using this spreadsheet.

For Extra Credit

38. Choose a situation you will face in which you must make a decision, and make the decision using the method in Example 6. Begin by identifying characteristics that are important to you and assign them levels of importance between 0 and 1, inclusive. Next, specify to what degree each choice satisfies each characteristic. Finally, mimic the calculations in Example 6 to make your decision.

39. Do you have any criticisms of the decision-making method that we used in Example 6? Are you comfortable with the way in which decisions are made by this method? Can you suggest any changes? If so, try to implement these changes and then reconsider the examples and exercises of this section to see whether your method results in different decisions being made.

CHAPTER SUMMARY*

SECTION	SUMMARY	EXAMPLE
SECTION 3.1	A **statement** in logic is a declarative sentence that is either true or false.	Discussion, p. 82
	A **simple statement** contains a single idea. A **compound statement** contains several ideas combined together with words called **connectives**.	Discussion, p. 83
	Negation, represented by ~, expresses the idea that something is *not* true.	Example 1, p. 83
	Conjunction, represented by ∧, expresses the idea of *and*.	Example 2, p. 84
	Disjunction, represented by ∨, expresses the idea of *or*.	Example 3, p. 84
	The **conditional**, represented by →, expresses the notion of *if . . . then*.	Example 4, p. 85
	The **biconditional**, represented by ↔, expresses the idea of *if and only if*.	Example 5, p. 85
	Universal quantifiers are words such as *all* and *every* which state that all objects of a certain type satisfy a given property.	Discussion, p. 86
	Existential quantifiers are words such as *some*, *there exists*, and *there is at least one* which state that there are one or more objects that satisfy a given property.	
	To negate quantified statements we remember that:	Example 6, p. 88
	The phrase *Not all* has the same meaning as *At least one is not*.	
	The phrase *Not some are* has the same meaning as *None are*.	
SECTION 3.2	The **negation** (~) of a statement reverses its truth value.	Truth table, p. 91
		Example 1, p. 92
	A **conjunction** (∧) is true only when both component parts are true.	Truth table, p. 92
		Example 2, p. 92
	A **disjunction** (∨) is false only when both component parts are false.	Truth table, p. 92
		Example 3, p. 93
	If a statement has k variables, then its truth table has 2^k lines.	Discussion, p. 95
	Statements are **logically equivalent** if they have the same variables and their truth tables are identical.	Example 6, p. 96
	DeMorgan's laws state:	Example 7, p. 98
	a) $\sim(p \wedge q)$ is logically equivalent to $(\sim p) \vee (\sim q)$.	
	b) $\sim(p \vee q)$ is logically equivalent to $(\sim p) \wedge (\sim q)$.	
SECTION 3.3	A **conditional** (→) is false only if its hypothesis is true and its conclusion is false.	Truth table, p. 103
		Example 1, p. 103
	Derived forms of the **conditional** $p \rightarrow q$ are:	Discussion, p. 105
	Converse: $q \rightarrow p$	Example 3, p. 105
	Inverse: $\sim p \rightarrow \sim q$	
	Contrapositive: $\sim q \rightarrow \sim p$	
	A conditional and its contrapositive are logically equivalent.	Example 4, p. 106
	A **biconditional** (↔) is true only when both component parts have the same value (both Ts or both Fs).	Truth table, p. 108
		Example 7, p. 108
SECTION 3.4	An **argument is valid** if whenever the premises are true, then the conclusion is true.	Discussion, p. 112
		Example 1, p. 112
	An **argument is invalid** if when we compute its truth table, there is even one F.	Example 2, p. 113
	There are several standard **valid argument** forms:	Example 3, p. 115

Law of Detachment	Law of Contraposition	Law of Syllogism	Disjunctive Syllogism
$p \rightarrow q$	$p \rightarrow q$	$p \rightarrow q$	$p \vee q$
p	$\sim q$	$q \rightarrow r$	$\sim p$
$\therefore q$	$\therefore \sim p$	$\therefore p \rightarrow r$	$\therefore q$

*Before studying this chapter's material, it would be useful to reread the note *How to Succeed at Mathematics* on page xix.

| | There are several standard **fallacies**: | Example 3, p. 115 |

Fallacy of the Converse	**Fallacy of the Inverse**
$p \rightarrow q$	$p \rightarrow q$
q	$\sim p$
$\therefore p$	$\therefore \sim q$

SECTION 3.5	A **syllogism** consists of a set of statements called **premises** followed by a statement called a **conclusion**. The premises and conclusion of the syllogism may contain quantifiers such as *all*, *some*, and *none*.	Discussion, p. 120
	A syllogism is **valid** if whenever its premises are all true, then the conclusion is also true.	Example 1, p. 120
	If the conclusion of a syllogism can be false even though all the premises are true, then the syllogism is **invalid**.	Example 2, p. 121
SECTION 3.6	**Statements** in fuzzy logic have truth values between 0 and 1 inclusive.	Example 1, p. 127
	In fuzzy logic, the value of the **negation** of p, written $\sim p$, is $1 -$ the value of p.	Example 2, p. 128
	The truth value of the **disjunction** of p and q, written $p \vee q$, is the *maximum* of the truth values for p and q. The truth value of the **conjunction** of p and q, written $p \wedge q$, is the *minimum* of the truth values for p and q.	Example 3, p. 128
	We can use fuzzy logic to make **decisions**.	Example 6, p. 130

CHAPTER REVIEW EXERCISES

Section 3.1

1. Which of the following are statements? Explain your answers.

   a. Why do these things always happen to me?

   b. The distance from Los Angeles to New York City is 2,000 miles.

   c. Bring me back a pizza.

2. Let b represent the statement "I will buy an iMac" and let s represent the statement "I will sell my old computer." Write each statement in symbolic form.

   a. I will not buy an iMac or I will sell my old computer.

   b. It is not true that: I will buy an iMac and not sell my old computer.

3. Let f represent "Antonio is fluent in Spanish" and let l represent "Antonio has lived in Spain for a semester." Write each statement in English.

   a. $\sim(f \wedge \sim l)$ b. $\sim f \vee \sim l$

4. Negate each quantified statement and then rewrite it in English in an alternate way.

   a. All writers are passionate.

   b. Some graduates received several job offers.

Section 3.2

5. Let p represent some true statement, q represent some false statement, and r represent some false statement. What is the truth value of the following statements?

   a. $p \wedge (\sim q)$ b. $r \vee (\sim p \wedge q)$ c. $\sim(p \vee q) \wedge \sim r$

6. How many rows will be in the truth table for each statement?

   a. $\sim(p \vee q) \wedge \sim(r \vee p)$ b. $(p \vee q) \wedge (r \vee s) \wedge t$

7. Construct a truth table for each statement.

   a. $\sim(p \vee \sim q)$ b. $\sim(p \vee \sim q) \wedge \sim r$

8. Negate each statement and then rewrite the negation using DeMorgan's laws.

   a. Either the deluxe monitor or the enhanced video card is included with your purchase.

   b. I will not sign the lease or I will not accept the housing agreement.

9. Which pairs of statements are logically equivalent?

   a. $\sim(p \wedge \sim q), (\sim p) \vee q$

   b. $\sim(p \vee \sim q) \wedge \sim(p \vee q), p \vee (p \wedge q)$

Section 3.3

10. Assume that p represents a true statement, q a false statement, and r a true statement. What is the truth value of each statement?

    a. $\sim(p \vee q) \rightarrow \sim p$ b. $(p \wedge q) \leftrightarrow (q \vee r)$

    c. $((\sim p) \vee (\sim q)) \rightarrow r$

11. Construct a truth table for each statement.

    a. $\sim p \rightarrow q$ b. $\sim(p \wedge r) \leftrightarrow \sim(p \vee q)$

12. Write in words the converse, inverse, and contrapositive for the statement "If the disk drive has been damaged, then we cannot recover the data."

13. Rewrite each statement using the words *if . . . then*.

    a. The Knicks will get to the finals only if they beat the Lakers.

    b. To be an astronaut, it is necessary to have a pilot's license.

Section 3.4

14. Identify the form of each argument.

a. If you make lots of money, then you will be happy.

You do make lots of money.

∴ You are happy.

b. If Felicia enjoys spicy food, then she will enjoy this Cajun chicken.

Felicia does not enjoy this Cajun chicken.

Therefore, Felicia does not enjoy spicy food.

15. Determine whether the form represents a valid argument:

$$\sim p$$
$$q \to p$$
$$(p \lor q) \to r$$
$$\overline{}$$
$$\therefore r$$

16. Use a truth table to determine whether the argument is valid.

If you pay more for your phone plan, then you will have more calling minutes.

If you pay more for you phone plan or have more calling minutes, then you will call your mother more often.

You paid more for your phone plan.

Therefore, you will call your mother more often.

Section 3.5

In Exercises 17 and 18, use Euler diagrams to determine whether each syllogism is valid or invalid.

17. Some used cars are expensive.

All expensive cars are safe.

This car is not safe.

∴ This is not a used car.

18. All professors are wealthy.

Some professors are absent-minded.

All wealthy people are happy.

∴ Some absent-minded people are happy.

Section 3.6

19. Assume that p and q are fuzzy statements having truth values of 0.47 and 0.82, respectively. Compute the truth values for the following statements:

a. $p \land \sim q$ **b.** $\sim(p \lor \sim q)$ **c.** $p \to \sim q$

CHAPTER TEST

1. Which of the following are statements?

a. New York City is the largest city in North America.

b. When did the Red Sox last win the pennant?

2. Negate each quantified statement and then rewrite it in English in an alternate way.

a. All rock stars are fine musicians.

b. Some dogs are aggressive.

3. Let p represent the statement "I will pass my lifeguard test" and let f represent "I will have fun this summer." Write each statement in symbolic form.

a. I will pass my lifeguard test or I will not have fun this summer.

b. It is not true that: I will not pass my lifeguard test and have fun this summer.

4. Let m represent "The Mets will win the series" and let p represent "Pedro will win the Cy Young Award." Write each statement in English.

a. $\sim(m \lor \sim p)$ **b.** $\sim m \land \sim p$

5. How many rows will there be in a truth table for a logical statement having six variables?

6. If p is false and q is true and r is false, what is the truth value of each statement?

a. $\sim(p \lor \sim q)$ **b.** $\sim(p \lor q) \land \sim r$

c. $\sim(\sim r \lor \sim p) \land r$

7. Assume that p, q, and r are fuzzy statements having truth values of 0.65, 0.38, and 0.75, respectively. Find the truth values of the following fuzzy statements:

a. $\sim(p \lor r)$ **b.** $(\sim r) \lor \sim(p \land q)$

8. Construct a truth table for each statement.

a. $\sim(p \land \sim q)$ **b.** $(\sim p \lor \sim q) \land r$

9. Write each statement using the words *if . . . then*.

a. To get enough sources for your research term paper, it is sufficient to go to Wikipedia.

b. Ticketmaster will mail the concert tickets only if you pay a fee.

10. Negate each statement and then rewrite the negation using DeMorgan's laws.

a. You can take the final exam or write a term paper.

b. I will not finish the painting or I will not show it at the gallery.

11. Determine whether the following pairs of statements are logically equivalent.

    a. $\sim(p \vee \sim q)$, $\sim p \wedge q$

    b. $(\sim p \vee \sim q) \wedge (\sim p \vee q)$, $\sim p \wedge q$

12. Write in words the converse, inverse, and contrapositive for the statement "If it glitters, then it is gold."

13. If p is true, q is false, and r is true, what is the truth value of each statement?

    a. $(p \vee \sim q) \rightarrow \sim q$ **b.** $(\sim p \wedge q) \rightarrow \sim r$

    c. $(p \vee \sim q \leftrightarrow \sim (p \wedge q)$

14. Construct a truth table for each statement.

    a. $(\sim p \vee q) \rightarrow \sim (p \wedge q)$

    b. $(\sim p \vee \sim q) \leftrightarrow r$

15. Determine whether the form represents a valid argument.

$$p$$
$$\sim q \rightarrow \sim r$$
$$(q \vee r) \rightarrow \sim p$$
$$\overline{\therefore \sim r}$$

16. Identify the form of each argument.

    a. If it ain't broke, then don't fix it.

    It is broke.

    ―――――――――――

    Therefore, fix it.

    b. I'll major in music or art history.

    I am not majoring in music.

    ―――――――――――

    Therefore, I am majoring in art history.

17. In fuzzy logic, we replaced the conditional $p \rightarrow q$ by what logical form?

18. Use a truth table to determine if the argument is valid.

If you go to eBay, then you will find a bargain or you will waste your money.

If you do not waste your money, then you will find a bargain.

You will not go to eBay or you will not find a bargain.

―――――――――――

Therefore, you will waste your money.

19. Use an Euler diagram to determine whether the syllogism is valid or invalid.

Some poets are sensitive.

No sensitive people are selfish.

Brittany is a poet.

―――――――――――

Therefore, Brittany is not selfish.

GROUP PROJECTS

1. In Sections 3.1 to 3.4, you studied a two-valued logic. Statements were either true or false. In the early part of the twentieth century, the Polish logician Jan Lukasiewicz* and others invented three-valued logic. Instead of true or false, we include a third option of "maybe."

    a. Truth tables for this three-valued logic are similar to truth tables for two-valued logic except now for k variables we would have 3^k lines. We give the truth table for the "and" connective; you construct the truth tables for "not," "or," and "if . . . then." Recall that in two-valued logic, the statement $p \rightarrow q$ is logically equivalent to $\sim p \vee q$.

p	q	p ∧ q
T	T	T
T	M	M
T	F	F
M	T	M
M	M	M
M	F	F
F	T	F
F	M	F
F	F	F

    b. Compute truth tables for some compound statements using two variables p and q, as we did in Section 3.2.

    c. Do DeMorgan's laws hold in three-valued logic? Explain.

    d. Investigate equivalent and nonequivalent derived forms of the conditional as we did in Section 3.3.

2. Recall the stroke connective, written |, that we introduced in the exercise set in Section 3.2. Try writing some English sentences using only the word *stroke* for connectives. For example, we know that the statements $\sim p$ and $p \mid p$ are logically equivalent, so in a weather report, instead of saying "It will not rain today," we could say "It will rain today *stroke* it will rain today" and avoid the use of the negation. As another example, consider how a weatherperson would report "If the clouds break, then it will be warm and humid this afternoon," using only the stroke connective.

―――――――――――

*See the Historical Highlight on page 121.

Graph Theory (Networks)
The Mathematics of Relationships

4

The best method to control something is to understand how it works.

J. Doyne Farmer,* Santa Fe Institute

Would it surprise you to know that the mathematics we use to describe the interaction of cancer cells, efficiently schedule snow plows and garbage collection in New York City, and minimize the cost of an executive's business trip all involves a similar type of analysis?

Although these situations seem to be unrelated, all of them require us to under-stand the relationships among objects in a complex system. How do the cancer cells interact with each other? How are the streets of New York City related? How is travel from one city to the next related in terms of the cost of plane fare? *(continued)*

---

*Dr. Farmer is a researcher at the Santa Fe Institute who applies math-ematics to study complex systems.

In this chapter, we will show you how to use an area of mathematics called **graph theory**, to model many situations to solve a variety of realistic problems. Researchers in such diverse areas as sociology, biology, political science, and urban studies use graph theory to solve their problems. In Section 4.4, you will also learn how we can use graph theory to schedule tasks in a complex project, such as building a space colony. By using these schedules, you will see how to allocate resources efficiently to complete the project. ●

Graphs, Puzzles, and Map Coloring

4.1

Objectives

1. Understand graph terminology.
2. Apply Euler's theorem to graph tracing.
3. Understand when to use graphs as models.
4. Use Fleury's theorem to find Euler circuits.
5. Utilize graph coloring to simplify a problem.

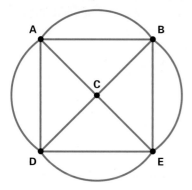

FIGURE 4.1 Puzzle tracing.

 **KEY POINT**

Graphs consist of vertices and edges.

When my oldest son was about 5 years old, I gave him the puzzle that you see drawn in Figure 4.1 and that you might want to try. Place your pencil on any dot and trace the figure completely without lifting your pencil and without tracing any part of any line twice. Can you do it?

After a few minutes of struggling, my son came to me and said, "I can't do it. . . . Nobody can do it." "Why do you say that?" I asked. He replied, "There are too many places to get stuck." Did you also notice that? Several hundred years ago, the brilliant Swiss mathematician Leonhard Euler did and, after learning a little graph theory, you will also see why this problem cannot be done.

The mathematics that you study in this chapter is probably different from the mathematics that you have studied before, and at first you may think that graph theory is only concerned with children's games. However, this field has many practical applications, as you will see later in the chapter.

We will begin by introducing some basic terminology.

Graph Terminology

> **DEFINITIONS** A **graph** consists of a finite set of points, called **vertices**, and lines, called **edges**, that join pairs of vertices.*

Figure 4.1 is an example of a graph. Its vertices are the points A, B, C, D, and E, and the 12 connecting lines are the edges. Generally, we use capital letters to designate vertices. If there is only one edge joining a pair of vertices, then we label that edge by the vertices it connects. For example, in Figure 4.1 we can refer to the edge joining vertices A and C as AC or CA; the order is not important. To refer to edge AB would be confusing because there are two edges joining vertices A and B. In this case, we might label one edge e_1 and the other e_2.

---

*Certain types of graphs are called *networks*.

The graph in Figure 4.2(a) has five vertices: A, B, C, D, and E. Although edges BC and AD intersect, the point of intersection is not a vertex. We will not consider a point of intersection of a pair of edges to be a vertex, unless we place a solid dot there and label it as a vertex.

In drawing a graph, the important information is what vertices we connect by edges. The placement of the vertices and the shape of the edges are unimportant. We could have also drawn Figure 4.2(a) as in Figure 4.2(b), because this graph contains exactly the same vertices and edges. If a graph looks awkward the first time you draw it, you should redraw it to present the information more clearly.

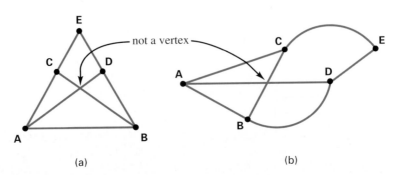

(a)   (b)

FIGURE 4.2 Two ways to draw the same graph.

KEY POINT

Graphs represent
relationships among objects.

Graph Tracing

We will now introduce several situations that we can model by graphs.

One of the most famous and yet elementary applications of graph theory originated in Koenigsberg, Prussia, during the eighteenth century. The Pregel River divided Koenigsberg into four distinct sections, as shown on the map in Figure 4.3.

Seven bridges connected the four portions of Koenigsberg. It was a popular pastime for the citizens of Koenigsberg to start in one section of the city and take a walk visiting all sections of the city, trying to cross each bridge exactly once and to return to the original starting point. This problem is called the **Koenigsberg bridge problem**.

It might not be apparent that this is a graph theory problem. However, as you see in Figure 4.4, we can model Koenigsberg by using vertices A, B, C, and D to represent the bodies of land and seven edges to represent the bridges.

In 1735, Leonhard Euler discovered a simple way to determine when a graph can be traced.* To **trace** a graph means to begin at some vertex and draw the entire graph without lifting your pencil and without going over any edge more than once. Before giving the solution to this graph tracing problem, we need a few more definitions.

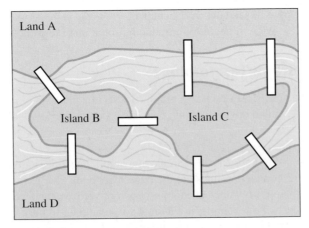

FIGURE 4.3 Map of Koenigsberg.

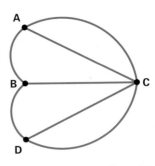

FIGURE 4.4 Graph model of Koenigsberg.

---

*A very readable translation of Euler's original paper is J. R. Newman, Ed., "Leonhard Euler and Koenigsberg Bridges," *Scientific American* (July 1953), pp. 66–70.

HISTORICAL HIGHLIGHT ❋ ❋ ❋

Pioneers in Graph Theory

In 1735, the Swiss mathematician Leonhard Euler became the first person to work in graph theory by solving the Koenigsburg bridge problem. Euler was a prolific mathmatician, writing almost 900 papers. He produced mathematics so effortlessly that his biographer Arago said he calculated without effort: "as men breathe, or as eagles sustain themselves in the wind." Incredibly, he did about half of his work during the last 17 years of his life while totally blind. One afternoon, while working on a problem, Euler suffered a stroke and with the words, "I die," he ceased calculating.

Later, in the nineteenth century, the English mathematician Arthur Cayley became interested in the four-color problem (page 147) and subsequently wrote several papers applying graph theory to chemistry.

> **DEFINITIONS** A graph is **connected*** if it is possible to travel from any vertex to any other vertex of the graph by moving along successive edges. A **bridge** in a connected graph is an edge such that if it were removed the graph is no longer connected.

Figure 4.5 illustrates these definitions.

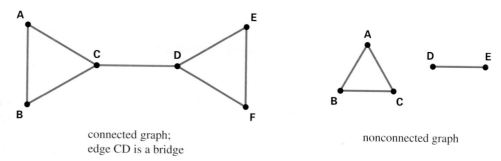

connected graph;
edge CD is a bridge

nonconnected graph

FIGURE 4.5 Examples of a connected graph and a nonconnected graph.

> **DEFINITIONS** A vertex of a graph is **odd** if it is an endpoint of an odd number of edges of the graph. Similarly, a vertex is **even** if it is an endpoint of an even number of edges.

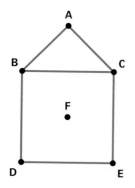

FIGURE 4.6 Odd and even vertices.

EXAMPLE 1 *Odd and Even Vertices in a Graph*

In Figure 4.6, determine which vertices are odd and which are even.

SOLUTION: Because B and C are endpoints of three edges, they are both odd. Vertices A, D, and E are endpoints of two edges, so they are even. Vertex F is also even because it is the endpoint of zero edges. ❋ **1**

Euler's Theorem

We will now make two observations that tell us when a graph can be traced.

Observation 1: If, while tracing a graph, we neither begin nor end with vertex A, then A must be an even vertex.

---

*Connected graphs are sometimes called *networks*.

Quiz Yourself ❶*

Is this graph connected? List its odd and even vertices. Are any edges bridges?

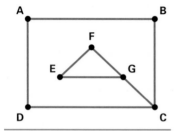

This is easy to see. Assume that we are tracing a graph and that vertex A is neither a beginning nor ending vertex. Eventually we must come into A by means of one edge, call it e_1, and then must leave by another edge, call it e_2. See Figure 4.7(a). If no other edges are joined to A, then A is the endpoint of exactly two edges, so it is even.

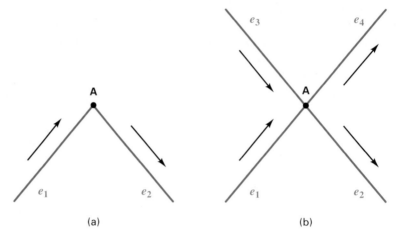

FIGURE 4.7 Every time we come into A by one edge, we must leave by another.

If more than two edges are joined to vertex A, then as we continue tracing, we will come into A again by a third edge, call it e_3, and leave by a fourth edge, e_4. See Figure 4.7(b). If A is the endpoint of more than four vertices, continuing this line of thinking, you can see that every time we come into vertex A by one edge, we must leave by another, so A must be the endpoint of an even number of edges.

From this we conclude that an odd vertex can only be used as either a starting point or ending point when tracing a graph.

Observation 2: If a graph can be traced, then it can have at most two odd vertices.

This follows directly from Observation 1. In tracing the graph, one odd vertex could be the starting vertex and another could be the ending vertex. Because no other vertices could be the starting or ending vertex, all other vertices must be even.

We now paraphrase Euler's theorem, which tells us when a graph can be traced.

KEY POINT

Euler's theorem tells when a graph can be traced.

> **EULER'S THEOREM**   A graph can be traced if it is connected and has zero or two odd vertices.

If a graph has two odd vertices, the tracing must begin at one of these and end at the other. If all the vertices are even, then the graph tracing must begin and end at the same vertex. It does not matter at which vertex this occurs.

EXAMPLE 2   *Tracing Graphs*

Which of the graphs in Figure 4.8 can be traced? Recognize that Figure 4.8(a) is the puzzle graph in Figure 4.1, and Figure 4.8(b) is the graph of Koenigsberg and its bridges.

---

*Quiz Yourself answers begin on page 778.

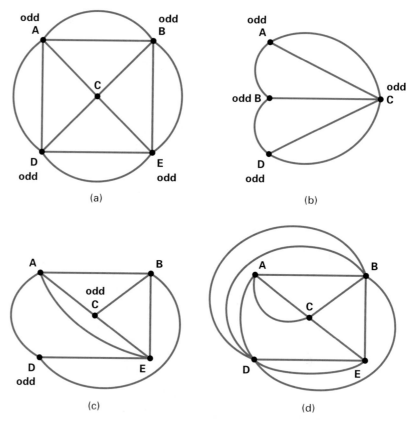

FIGURE 4.8 (a) Puzzle graph. (b) Koenigsberg graph. Only (c) and (d) can be traced by Euler's theorem.

Quiz Yourself ②

Using Euler's theorem, state which of the following graphs can be traced. For a graph that can be traced, list a sequence of vertices that describes how to trace the graph. For a graph that cannot be traced, state which part of Euler's theorem fails.

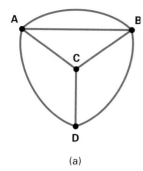

(a)

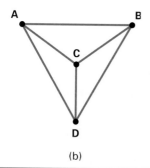

(b)

SOLUTION:

a) Notice that vertices A, B, D, and E are all odd; therefore, by Euler's theorem, this graph cannot be traced.

b) All vertices are odd; this means that the Koenigsberg bridge graph cannot be traced.

c) This graph has two odd vertices, C and D. One way to trace this graph is to begin at D and follow this sequence of vertices:

   D, A, B, D, E, B, C, A, E, C

d) All vertices in this graph are even. Can you find a sequence of vertices describing how to trace this graph?

Now try Exercises 23 to 30. ✳ ②

 PROBLEM SOLVING

Draw a Picture

The Koenigsburg bridge problem has two features that are present in every graph model:

1. A set of objects—the four bodies of land

2. A relationship among the objects—they were connected by bridges

If these two features are present in a problem, you can model it with a graph as follows:

First, represent the objects by vertices having good names.

Second, join the vertices representing related objects with edges.

We can model many diverse relationships with graphs. Countries might be related because they share a common border. Houses in a town could be related if a fiber-optic cable connects them. Two tasks may be related if one must be done before the other.

Before discussing further applications of Euler's theorem, we will introduce more terminology.

> **DEFINITIONS** A **path** in a graph is a series of consecutive edges in which no edge is repeated. The number of edges in a path is called its **length**. A path containing all the edges of a graph is called an **Euler path**. An Euler path that begins and ends at the same vertex is called an **Euler circuit**. A graph with all even vertices contains an Euler circuit and is called an **Eulerian graph**.

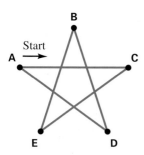

FIGURE 4.9 A graph containing an Euler circuit.

EXAMPLE 3   *An Euler Circuit*

Find some paths in the graph shown in Figure 4.9.

SOLUTION: There are many paths in this graph. For example, the path ACEB from A to B has three edges and therefore has length 3. Also, the path ACEBDA is an Euler path of length 5 that is also an Euler circuit because it begins and ends at the same vertex. ✳ ❸

Quiz Yourself ❸

a) Find an Euler path in this graph.

b) Is there an Euler circuit? Explain.

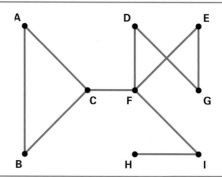

Math in Your Life

It's a Small World After All

The term "six degrees of separation," first mentioned in a 1929 novel by the Hungarian writer Frigyes Karinthy, states that any two people on Earth are connected by a chain of no more than six associations.*

The sociologist Stanley Milgram tested this theory, called "the small-world problem," by randomly selecting people in the Midwest to send packages to a stranger in Massachusetts. They were to send the package to a person they knew who they thought was most likely to know the target. Then that person would do the same, and so on, until the package was delivered. Milgram found that, on the average, the package was delivered after passing through the hands of five to seven people. In similar experiments, other researchers have found that the average length of a path of e-mails between a random sender and receiver was six.

While discussing this with my wife, we realized that because of my appearance on a national television amateur show in my teens, my degree of separation with the Beatles is three and as a result of a graduate course that she took, her degree of separation from Mother Theresa of Calcutta is two.

So what does this have to do with you? In your activities at school—fraternities, clubs, musical organizations, and sports teams—you are creating links in these personal chains that you may be able to use in later life to your benefit. In fact, companies like Linked In, develop software that will provide you with chains of contacts that can help you in the corporate world. Scientists also use "small worlds" to analyze electrical power grids, study disease transmission, and explore the organization of computer circuits.

*Perhaps you have played the trivia game, "Six Degrees of Kevin Bacon."

 **KEY POINT**

We use Fleury's algorithm to find Euler circuits.

Fleury's Algorithm

Although we can use Euler's theorem to find out whether Euler circuits exist for a given graph, it does not tell us how to find them. Until now, we have used trial and error; however, in a large graph, this is not efficient. A method called **Fleury's algorithm** provides a systematic technique for finding Euler circuits. An **algorithm** is a series of steps that we follow to accomplish something. You might think of an algorithm as a recipe to do some mathematical task.

> **FLEURY'S ALGORITHM** If a connected graph has all even vertices, we can find an Euler circuit for it by beginning at any vertex and traveling over consecutive edges according to these rules:
>
> 1. After you have traveled over an edge, erase it. If all the edges for a particular vertex have been erased, then erase that vertex also.
>
> 2. Travel over an edge that is a bridge only if there is no alternative.

EXAMPLE 4 *Using Fleury's Algorithm to Find an Efficient Route*

Assume you are doing maintenance work along pathways joining locations A, B, and so on in a theme park, as shown in Figure 4.10. Find an Euler circuit in this graph to make your job efficient by not retracing pathways. Assume that you leave from and return to building C.

SOLUTION: We will use Fleury's algorithm to find an Euler circuit in this graph.

Step 1: We begin at vertex C and traverse edge CJ, next JK and KI, and then IF. We numbered these edges 1, 2, 3, and 4 in Figure 4.10, indicating the order in which we will travel over them. We erase these edges and also vertices J and K because they no longer have any edges joined to them. This gives us the graph in Figure 4.11.

Step 2: We are now at vertex F. We cannot traverse FC because it is a bridge. So we traverse FG, GI, IH, and HF (marked 5, 6, 7, and 8), erasing these edges and vertices G, I, and H. The graph now looks like Figure 4.12(a).

Step 3: We have no choice now but to traverse FC (edge 9). We follow this with CA and AB (marked 10 and 11). After erasing appropriate edges and vertices, we have the graph in Figure 4.12(b).

We can now finish the circuit by traveling over BE, EC, CD, DB, and BC (marked 12, 13, 14, 15, and 16). The final circuit is CJKIFGIHFCABECDBC. Notice that we have traversed every edge exactly once and ended at our starting point, vertex C. If you were to follow this circuit, you would cover each path exactly once in performing maintenance on the paths in the park.

Now try Exercises 31 to 36. ✳

FIGURE 4.10 Graph showing paths in a theme park.

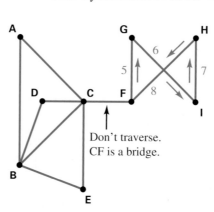

FIGURE 4.11 Step 1.

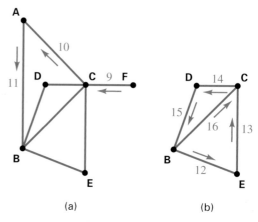

(a)   (b)

FIGURE 4.12 (a) Step 2. (b) Step 3.

Some of the decisions we made in Example 4 were arbitrary, and there are certainly many different ways to construct an Euler circuit for this graph.

Eulerizing a Graph

✏️ *KEY POINT*

We can add edges to convert a non-Eulerian graph to an Eulerian graph.

Have you ever been on a duck tour? Cities such as Washington, D.C., Miami, Seattle, Chicago, and many others offer guided tours in authentic, renovated World War II amphibious landing vehicles called DUCKs. On a DUCK tour in Boston, Massachusetts, our tour guide, called a "ConDUCKtor," explained that although we were allowed to "quack" at people on the street, certain areas were zoned by law to be "nonquacking zones," and in these zones, quacking was forbidden. In order to minimize traffic where possible, a tour would not go over the same street twice. In the next example, you will see how to use graph theory to design an optimal DUCK tour.

To solve our problem, we will have to add some edges to a non-Eulerian graph (a graph having odd vertices) so that the new graph is Eulerian. This technique is called **Eulerizing a graph.**

EXAMPLE 5   *Designing a DUCK tour*

The Boston DUCK tour company wants to design a DUCK route in a historic area of Boston shown in the map in Figure 4.13. We want to begin and end the tour at the same location and minimize traveling over any street more than once. Find such a route.

SOLUTION: We can model this map with the graph shown in Figure 4.14. We represent each intersection by a vertex and each section of street joining two intersections by an edge.

We labeled the *odd* vertices in this graph A, B, and so on. To Eulerize this graph, we duplicate some edges so that the new graph has only even vertices. We show this in Figure 4.15.

If we begin our route at the upper right corner of the graph and follow the edges as they are numbered, we will traverse all the edges of the graph and return to our starting point. Notice that the five pairs of duplicate edges— AB, CG, and so on—represent streets that must be traveled twice. Although we will not prove it, there is not a better way to Eulerize the original graph to reduce the number of streets that are traveled twice.

Now try Exercises 37 to 40. ❋ ④

When Eulerizing a graph, you can only duplicate edges that are *already present* in the graph. In Example 5, you are not allowed to insert a *single* edge from F to H to Eulerize the graph.

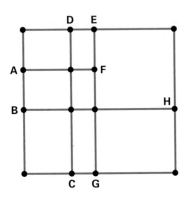

FIGURE 4.13 Map of Boston historic area.

FIGURE 4.14 Graph representing Boston historic area.

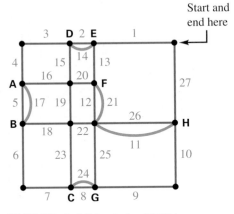

FIGURE 4.15 Route for DUCK tour.

Quiz Yourself ④

Add edges to this graph to make it Eulerian.

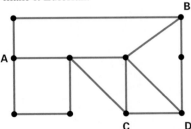

KEY POINT

The four-color problem is another application of graph theory.

Map Coloring

We now turn our attention to an interesting problem called the **four-color problem**. Although we state this as a puzzle, like the Koenigsberg bridge problem, it has interesting real-life applications. In 1852, Francis Guthrie, a student at University College, London, first posed this famous question to his mathematics professor, Augustus DeMorgan.

> **THE FOUR-COLOR PROBLEM** Using at most four colors, is it always possible to color a map so that any two regions sharing a common border receive different colors?

FIGURE 4.16 Map of South America.

Unable to find an answer, DeMorgan communicated the problem to his friend Sir William R. Hamilton at Trinity College, Dublin, Ireland. For more than a hundred years, this problem remained unsolved. However, in 1976, Professors Kenneth Appel and Wolfgang Haken of the University of Illinois announced that they had solved this problem by proving that it is possible to color any map using no more than four colors. However, their proof was treated with some skepticism, because the proof was not done in the traditional method—by hand, where each step could be checked for validity. Instead Appel and Haken programmed a computer to do the proof; the completion of the proof required 1,200 hours of computer time!

For an illustration of the statement of this problem, consider the map of South America in Figure 4.16. Using at most four colors, we want to color this map so that we use a different color for any two countries having a common border. For example, we cannot use the same color for Colombia and Peru; however, we can use the same color for Paraguay and Uraguay.

EXAMPLE 6 *Solving the Four-Color Problem for South America*

Model the map of South America by a graph and use this graph to color the map using at most four colors.

SOLUTION: In this problem, we have a set of countries, some of which are related in that they share a common border. Therefore, we can model this situation by a graph.

We will represent each country by a vertex; if two countries share a common border, we draw an edge between the corresponding vertices. This graph appears in Figure 4.17.

Note that we connect the vertices representing Peru and Colombia with an edge because they share a common boundary. We do not connect the

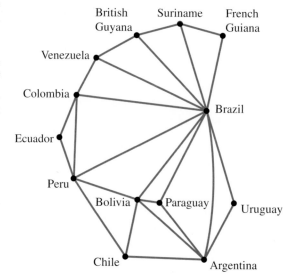

FIGURE 4.17 Graph model of map of South America.

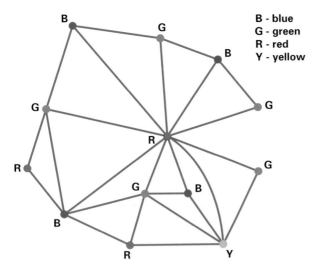

B - blue
G - green
R - red
Y - yellow

FIGURE 4.18 Coloring of graph of South America.

vertices representing Argentina and Peru, because they have no boundary in common.

We can rephrase the map-coloring question now as follows: Using four or fewer colors, can we color the vertices of a graph so that no two vertices of the same edge receive the same color? It is easier to think about coloring a graph than it is to think about coloring the original map.

There are several ways to color the graph in Figure 4.17. However, unlike tracing graphs, there is no particular procedure for accomplishing this coloring except trial and error. One possible coloring for the graph appears in Figure 4.18. You may want to color this graph in a different way; however, notice that you cannot do it using fewer than four colors.

Now try Exercises 41 to 44. ✳ **5**

Example 7 shows how we can use map coloring for a practical purpose.

Quiz Yourself **5**

Color the following "map" by first modeling it with a graph and then coloring the graph, as we did in Example 6.

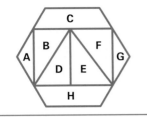

EXAMPLE 7 *Using Graph Theory to Schedule Committees*

Each member of a city council usually serves on several committees to oversee the operation of various aspects of city government. Assume that council members serve on the following committees: police, parks, sanitation, finance, development, streets, fire department, and public relations.

Use Table 4.1, which lists committees having common members, to determine a conflict-free schedule for the meetings. We do not duplicate information in Table 4.1. That is, because police conflicts with fire department, we do not also list that fire department conflicts with police.

SOLUTION: We first will model the information in this table with the graph in Figure 4.19. We join two committees by an edge provided they have a conflict. This problem is similar to the map-coloring problem. If we color this graph, then all vertices having the same color represent committees that can meet at the same time. We show one possible coloring of the graph in Figure 4.19.

From Figure 4.19, we see that the police, streets, and sanitation committees have no common members and therefore can meet at the same time. Public relations, development, and the fire department can meet at a second time. Finance and parks can meet at a third time. ✳

Committee	Has Members in Common With
Police	Public relations, fire department
Parks	Streets, development
Sanitation	Fire department, parks
Finance	Police, public relations
Development	Streets
Streets	Fire department, public relations
Fire department	Finance

TABLE 4.1 Committees that have common members.

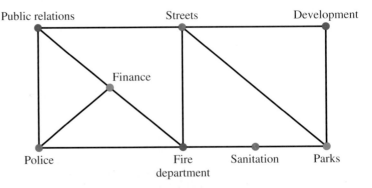

FIGURE 4.19 Graph model of committees with common members.

Exercises 4.1

Looking Back*

These exercises follow the general outline of the topics presented in this section and will give you a good overview of the material that you have just studied.

1. What is the difference between an odd vertex and an even vertex?

2. How many odd vertices can a connected graph have if it can be traced?

3. In step 2 of the solution of Example 4, why did we not traverse the edge FC immediately after traversing edges 1, 2, 3, and 4?

4. In Example 5, what was the purpose of putting the extra edges in the graph?

5. Name two early mathematicians who helped develop graph theory.

6. Of what practical importance does the "six degrees of separation" theory have to your life?

Sharpening Your Skills

In Exercises 7–14, determine whether the graph is connected. Which vertices are odd? Which vertices are even?

7.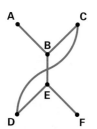

8.

9.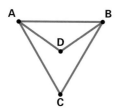

10.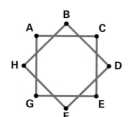

11.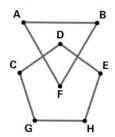

12.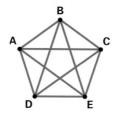

13.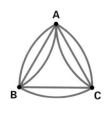

14.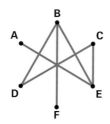

In Exercises 15–22, try to give an example of each graph that we describe. If, after several tries, you cannot find the graph that we have requested, explain why you think that it may be impossible to find that example. (In the answer key, we will give examples that are as simple as possible in the sense that they have the fewest number of vertices and edges.) The degree *of a vertex is the number of edges that are joined to that vertex.*

15. A graph with four even vertices

16. A graph with four odd vertices

17. A graph with three odd vertices

18. A graph with four vertices of degree two and two vertices of degree three

19. A connected graph with one even vertex and four odd vertices

20. A graph with one odd vertex

21. A graph with six vertices of degree three

22. A graph with five vertices and the largest number of edges so that no edges are repeated. That is, there will be only one edge joining A and B, only one edge joining A and C, and so on.

In Exercises 23–30, use Euler's theorem to decide whether the specified graph can be traced. If the graph cannot be traced, tell which condition of the theorem fails.

23. The graph in Exercise 7

24. The graph in Exercise 8

25. The graph in Exercise 9

26. The graph in Exercise 10

27. The graph in Exercise 11

28. The graph in Exercise 12

29. The graph in Exercise 13

30. The graph in Exercise 14

In Exercises 31–36, if the given graph is Eulerian, find an Euler circuit in it. If the graph is not Eulerian, first Eulerize it and then find an Euler circuit. Write your answer as a sequence of vertices, as we did in Example 4. There are many possible correct answers to these exercises. We will provide only one in the answer key.

31.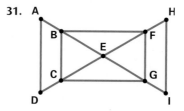

32.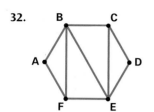

33.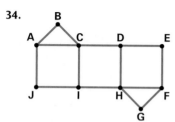

34.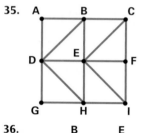

35.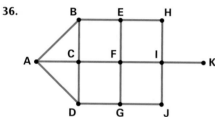

36.

Applying What You've Learned

37. Finding an efficient route. A taxi driver wants to travel over each of the streets indicated in the following map, but does not want to travel over any part of the route more than once. Can this be done? Explain your answer.

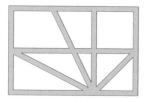

38. Finding an efficient route. Repeat Exercise 37 for the following map.

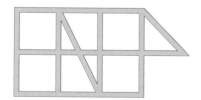

Exercises 39 and 40 are similar to the DUCK tour problem in Example 5. Model each map by a graph and then Eulerize it to design a route that has a minimal number of streets more than once.

39. 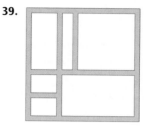   **40.**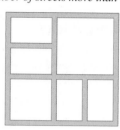

In Exercises 41–44, use the map of the continental United States to represent each group of states with a graph as we did in Example 6. Recall that vertices representing two states will be joined by an edge if and only if the states share a common border.

Figure for Exercises 41–44

41. Map coloring. Washington, Oregon, Idaho, Montana, Wyoming

42. Map coloring. Ohio, Kentucky, Tennessee, Arkansas, Missouri, Illinois, Indiana, Iowa, Minnesota, Wisconsin

43. Map coloring. Texas, Oklahoma, Arkansas, Louisiana, Mississippi, Alabama, Tennessee, North Carolina, South Carolina, Georgia, Florida

44. Map coloring. California, Nevada, Utah, Arizona, New Mexico, Colorado, Kansas, Oklahoma, Arkansas, Louisiana

In Exercises 45–48, we give you a group of states. Is it possible to begin in one of the states in the group and travel through all the states without ever crossing the same boundary between two states twice? (Hint: Think of the graphs that you drew in Exercises 41–44.)

45. Use the states that we listed in Exercise 41.

46. Use the states that we listed in Exercise 42.

47. Use the states that we listed in Exercise 43.

48. Use the states that we listed in Exercise 44.

49. Finding an efficient route. Because of Michael's escape from Fox River State Penitentiary, security procedures are being reexamined. In the following floor plan of a section of the prison, if all the doors are open, is it possible for a guard to enter this section from the hallway, pass through each door locking it behind him, and then exit without ever having to open a door that has been previously locked? (*Hint:* Model this with a graph where you consider the set of objects to be the rooms and the hallway and that any two rooms are related if they are connected by an open door.)

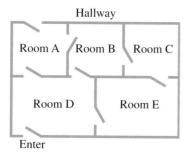

Hallway

Figure for Exercise 49

50. **Finding an efficient route.** The following is a floor plan for another section of Fox River. The situation is the same as in Exercise 49, except there is now only one door by which the guard can enter the room. Place an exit door in one of the rooms A, B, or C so that the guard can enter by the door marked "Enter," pass through each door and lock it behind him, and then exit by the door you've placed in the plan. Explain why this is the only possible place to locate the door.

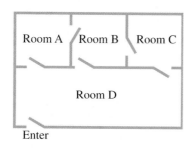

Use the technique presented in Example 7 to answer Exercises 51–54. We do not list duplicates in the tables of information.

51. **Avoiding Conflicts.** The Griffins are looking forward to a "Family Guy" wedding, but there is concern about the impending rehearsal dinner because certain people invited to the dinner just don't get along with each other. Therefore, it is important that people who are not friendly be seated at different tables. Use the information in the table to determine a satisfactory seating arrangement for the dinner using as few tables as possible.

Dinner Guest	Is Not Friendly With
Peter	Carter
Carter	Glenn, Chris
Glenn	Tom
Lois	Steve
Barbara	Tom, Dianne
Tom	Dianne, Meg
Dianne	Cleveland
Meg	Cleveland

52. **Avoiding conflicts.** The designer of the new Jungle World entertainment complex wants to include several large enclosures in which wild animals can roam freely. If one animal can harm another, then those two animals cannot share the same enclosure. (See table below.) Use this information to determine the smallest number of enclosures necessary to contain the animals. Also, state how you could assign the animals.

Animal	Cannot Be Placed With
Tiger	Zebra, leopard, rhinoceros, giraffe, antelope, ostrich
Leopard	Zebra, boar, antelope, giraffe, ostrich
Crocodile	Ostrich, heron
Boar	Tiger, crocodile, zebra

53. **Scheduling meetings.** A college's student government has a number of committees that meet Tuesdays between 11:00 and 12:00. To avoid conflicts, it is important not to schedule two committee meetings at the same time if the two committees have students in common. Use the following table, which lists possible conflicts, to determine an acceptable schedule for the meetings.

Committee	Has Members in Common With
Academic standards	Academic exceptions, scholarship, faculty union
Computer use	University advancement, event scheduling
Campus beautification	Curriculum, faculty union, event scheduling
Affirmative action	Academic exceptions, scholarship
University advancement	Parking, curriculum, academic standards
Parking	Academic standards, affirmative action
Faculty union	Computer use, event scheduling
Scholarship	Campus beautification

54. Scheduling meetings. Repeat Exercise 53 using the following table.

Committee	Has Members in Common With
Academic standards	Scholarship, university advancement
Computer use	University advancement, affirmative action
Campus beautification	Curriculum, academic festival, faculty union, event scheduling, scholarship
Affirmative action	Scholarship
University advancement	Curriculum, academic festival, faculty union
Faculty union	Computer use, event scheduling, academic festival
Long-range planning	Campus beautification, computer use

Communicating Mathematics

55. If in tracing a graph, we neither begin nor end at vertex A, why must A be even?

56. In what kind of situations do we use graphs as models?

57. Explain Fleury's algorithm in your own words.

58. What is the four-color problem?

59. Make up several examples of graphs. By examining these graphs, explain why the sum of the degrees of all the vertices in a graph must be an even number.

60. Use Exercise 59 to explain why a graph cannot have an odd number of odd vertices.

Using Technology to Investigate Mathematics

61. There are many Internet sites that have interactive programs called *applets* that illustrate graph theory concepts. Find an interesting site, run the applets, and report on your findings.

62. Perform an Internet search to find map coloring applets. Run the applets and report on your findings.

63. Many people think of mathematics as a static subject that was invented years ago. However, that is not the case. Mathematicians are constantly reexamining old mathematics and inventing new mathematics. Perform an Internet search to find discussions regarding controversies of the proofs of the solution of the four-color problem and report on your findings.

While researching topics for exercises for this section, I came across many fascinating applications of graph theory. In Exercises 64–68, search the Internet for applications of graph theory in the given categories. You should be able to find some very interesting graphs. Write a brief report of your findings.

64. Social networking   **65.** E-mail traffic

66. Google   **67.** Cancer

68. Music

For Extra Credit

69. Draw a graph that can be colored with only two colors.

70. Draw a graph that cannot be colored with two colors but can be colored with three. Can you state what configuration of vertices will force you to use at least three colors in coloring a graph?

71. Draw a graph that cannot be colored with three colors but can be colored with four.

72. Can you state what configuration of vertices will force you to use at least four colors in coloring a graph?

73. Different notes on a trumpet are obtained by moving its three valves up and down. The table below indicates the eight possible positions for these three valves and a note that would sound if the valves were in the indicated position. Is it possible to play one of these notes and then, by changing only one valve at a time, to play all the other notes without repeating a note twice? (*Hint:* Consider the notes as the objects of a set and that two notes are related if one can be obtained from the other by moving only one valve.)

Valve 1	Valve 2	Valve 3	Note
Up	Up	Up	C
Up	Up	Down	A
Up	Down	Up	B
Up	Down	Down	E♭
Down	Up	Up	F
Down	Up	Down	D
Down	Down	Up	E
Down	Down	Down	C#

74. If an instrument has four valves, there are sixteen possible positions for the valves. Repeat Exercise 73 for this situation.

75. Assume that the registrar at your school is building a final exam schedule and because the following courses are known to be particularly difficult, the registrar does not want any student to take finals for two of them on the same day. The courses are Anth215, Bio325, Chem264, Mat311, Phy212, Fin323, and ComD265. The asterisks in the given table indicate which courses have students in common and therefore should have their finals on different days. Assume that two courses are related if they have any students in common. Model this set of courses with a graph and use it to design a final exam schedule.

	Anth215	Bio325	Chem264	Mat311	Phy212	Fin323	ComD265
Anth215		*	*	*		*	*
Bio325	*		*			*	
Chem264	*	*		*	*		
Mat311	*		*		*	*	*
Phy212			*	*			
Fin323	*	*		*			
ComD265	*			*			

Table for Exercise 75

76. Make up a scheduling problem that would be of some interest to you similar to what we did in Exercise 75. Use graph theory to develop a solution.

The Traveling Salesperson Problem

4.2

Objectives

1. Understand how to solve the traveling salesperson problem using Hamilton circuits.
2. Determine all Hamilton circuits in a complete graph.
3. Solve the traveling salesperson problem using the brute force algorithm.
4. Solve the traveling salesperson problem using the nearest neighbor algorithm.
5. Solve the traveling salesperson problem using the best edge algorithm.

Some problems in mathematics are so simple to state that a child can understand them, yet the greatest mathematicians in the world cannot solve them. This is the case with a famous and difficult problem in graph theory called **the traveling salesperson problem (TSP)**. The TSP gets its name from the problem of determining the most efficient way for a salesperson to schedule a trip to a series of cities and then return home.

For an example of a TSP, suppose that Danielle, who is regional sales manager for a publishing company, lives in Philadelphia and must make visits next week to branch offices in New York City, Cleveland, Atlanta, and Memphis (see Figure 4.20). To determine which would be her cheapest trip, she has obtained prices of flights between each pair of cities.* Danielle could start at Philadelphia and fly first to New York, then to

FIGURE 4.20 Cities Danielle must visit.

*We will assume that direction is not important here and that a one-way flight between two cities costs the same in either direction.

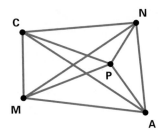

FIGURE 4.21 Graph representing cities Danielle will visit.

Cleveland, Atlanta, and finally to Memphis before returning home. Or, she could first fly to Cleveland, then Atlanta, Memphis, New York, and then home. If Danielle wants to find the cheapest trip, she will have to spend some effort in considering all possible trips through the four cities and then return home.

In keeping with the spirit of this chapter, we model this situation with a graph. We represent each city by a vertex and join two vertices by an edge if Danielle might fly from one to the other. In this graph, every pair of vertices will be joined by an edge. For clarity, we will represent each city by a letter—P for Philadelphia, N for New York City, and so on. This gives us the graph in Figure 4.21.

Hamilton Paths

We now model Danielle's possible trips by paths in this graph. For instance, we can represent the trip (Philadelphia, New York, Cleveland, Atlanta, Memphis, home) by the path PNCAMP, and the trip (Philadelphia, Cleveland, Atlanta, Memphis, New York, home) by the path PCAMNP. If we knew the prices of all the possible flights, we could then solve Danielle's problem. But first we need to develop some more graph theory.

> **DEFINITIONS** A path that passes through all the vertices of a graph exactly once is called a **Hamilton path**. If a Hamilton path begins and ends at the same vertex, then it is called a **Hamilton circuit**. If a graph has a Hamilton circuit we will say it is **Hamiltonian**.

 PROBLEM SOLVING

The Splitting-Hairs Principle

Recall the Splitting-Hairs Principle in Section 1.1. Although the definitions of *Hamilton path* and *Euler path* sound similar, they are not the same. In producing a Hamilton path, you do not have to trace every edge, as with an Euler path.

EXAMPLE 1 *Hamilton Paths and Circuits*

Find a Hamilton path in each graph shown in Figure 4.22.

SOLUTION:

a) In Figure 4.22(a), the path ABCFDGE (in red) is a Hamilton path. If we return to vertex A, then ABCFDGEA is a Hamilton circuit.

b) The graph in Figure 4.22(b) has no Hamilton paths or circuits. If you start at any vertex, you will see that you cannot follow edges to visit every vertex exactly once. For example, if you start at vertex A and then go to vertices B and C, you are in trouble. If you go to D, then you cannot get to E without passing through C a second time. Similarly, if you go to E, you cannot get to D. You may start your path at other vertices, but you will find that no matter how you try you cannot find a Hamilton path.

Now try Exercises 7 to 12. ✳

FIGURE 4.22 (a) ABCFDGE is a Hamilton path. (b) No Hamilton path exists.

Unlike the situation with Euler's theorem for graph tracing, we will not give a rule for determining when a graph has a Hamilton path.

KEY POINT

Tree diagrams help us find Hamilton circuits systematically.

Finding Hamilton Circuits

If you think about it, the reason we could not find a Hamilton path in the graph in Example 1(b) was that there were not enough edges. When we went to vertex D, we were stuck. If we had more edges to use, we might have avoided going back through C a second time to get to vertex E. Often we will be dealing with graphs that have every possible edge.

> **DEFINITION** A **complete graph** is one in which every pair of vertices is joined by an edge. A complete graph with n vertices is denoted by K_n.*

EXAMPLE 2 *Complete Graphs*

Each graph shown in Figure 4.23 is complete.

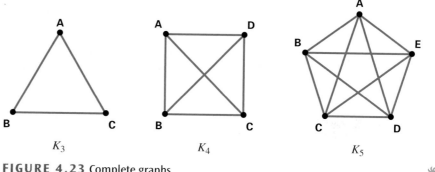

FIGURE 4.23 Complete graphs.

Often we need to find all the Hamilton circuits in a graph. It is easy to do this in complete graphs.

EXAMPLE 3 *Finding Hamilton Circuits in K_4*

Find all Hamilton circuits in K_4.

SOLUTION: Path ABCDA shown in Figure 4.24 is one of the Hamilton circuits in K_4. We will consider path BCDAB to be the same path because it passes through the same vertices in the same order and the only difference is that we are beginning and ending at vertex B rather than vertex A. In light of this remark, we will assume that all Hamilton circuits in this example begin at vertex A.

The *tree diagram* in Figure 4.25, shows how to list all Hamilton circuits in K_4 systematically. Begin at vertex A, then go to either B, C, or D. If we decide to go to B, we can then

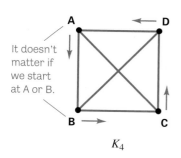

It doesn't matter if we start at A or B.

FIGURE 4.24 Circuit ABCDA is the same as BCDAB.

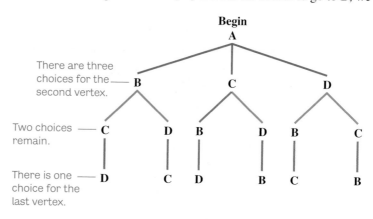

FIGURE 4.25 Finding Hamilton circuits in K_4.

---

*The notation K for complete graphs is chosen to honor the twentieth-century Polish mathematician Kasimir Kuratowski, who discovered several important theorems in graph theory.

Quiz Yourself **6**

a)Write the circuit ACBDA, but begin with vertex B.

b)What is the reversal of ACBDA?

choose *either* C *or* D, and so on. The red path, highlighted in Figure 4.25, shows how to construct the Hamilton path ACBD and then, by returning to A, get the Hamilton circuit ACBDA.

Tracing through the six branches of the tree, we see that K_4 has six Hamilton circuits:

Notice in Example 3 that the Hamilton circuits occur in pairs. Consider the circuit ABCDA. If we list these vertices in reverse order, we get the circuit ADCBA. In any graph, once we have found one Hamilton circuit, we automatically have another one by listing the vertices in reverse order.

Let's generalize what we saw in Figure 4.25. We always began the Hamilton paths at vertex A, which left us with three choices for the second vertex of the path. After choosing the second vertex, there were only two vertices remaining, which gave us two new branches for each of the existing three branches. The tree now had six branches. We now had only one vertex left to finish the Hamilton path, so the final tree had $3 \times 2 \times 1 = 6$ branches.

The following diagram shows how we would calculate the number of Hamilton paths in K_5:

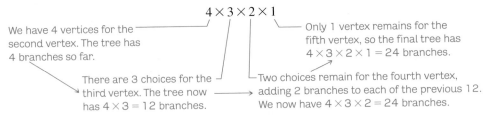

Therefore, there are $4 \times 3 \times 2 \times 1 = 24$ Hamilton paths in K_5.

You can see now that if we were solving this problem for K_n, we would start each Hamilton path at vertex A. We would then draw $n - 1$ branches corresponding to the $n - 1$ remaining vertices. We would then draw $n - 2$ branches from each of these $n - 1$ branches, giving us $(n - 1)(n - 2)$ branches. Continuing in this fashion, we would get a final tree having $(n - 1)(n - 2)(n - 3)(n - 4) \cdots 3 \times 2 \times 1$ branches.

> **THE NUMBER OF HAMILTON CIRCUITS IN K_n** K_n has $(n - 1)(n - 2)(n - 3)$ $(n - 4) \cdots 3 \times 2 \times 1$ Hamilton circuits. This number is written $(n - 1)!$ and is called $(n - 1)$ *factorial.*

Table 4.2 shows the number of Hamilton circuits in K_n for various values of n.

n	Number of Hamilton Circuits in K_n
3	$2! = 2 \cdot 1 = 2$
4	$3! = 3 \cdot 2 \cdot 1 = 6$
5	$4! = 4 \cdot 3 \cdot 2 \cdot 1 = 24$
10	$9! = 362,880$
15	$14! = 87,178,291,200$
20	$19! = 121,645,100,408,832,000$

TABLE 4.2 Number of Hamilton circuits in K_n.

Quiz Yourself **7**

How many Hamilton circuits are there in K_7?

As you can see, as n increases, the number of Hamilton circuits in K_n grows at an incredible rate. **7**

— HISTORICAL HIGHLIGHT 🌟 🌟 🌟 —

William Rowan Hamilton*

Hamiltonian circuits are named after perhaps the greatest Irish mathematician of all time, William Rowan Hamilton, who was born in Dublin, Ireland in 1805. In his early teens, he took part in an arithmetic contest with an American youth named Zerah Colburn, who was known as a "lightning calculator." When Hamilton came in second best, he dedicated himself fiercely to studying mathematics.

Hamilton was very skilled at learning languages and could read the works of Euclid in the original Greek, Isaac Newton in Latin, and the eminent mathematician Pierre Simon Laplace in French. At age 17, he sent an error he found in Laplace's work to the Royal Irish Academy, whereby the president of the academy declared him to be a first-rate mathematician.

Hamilton attended Trinity College in Dublin, where he distinguished himself in both mathematics and poetry— saying that although he was a mathematician by profession, he was a poet by heart. His work in optics was so impressive that he was appointed a professor of astronomy while still an undergraduate student!

Although his work seemed theoretical at the time, Hamilton made contributions to many areas of mathematics that today have numerous practical applications.

 **KEY POINT**

In solving a TSP problem by brute force, we consider all possible Hamilton circuits.

Solving the TSP by Brute Force

Now that we have introduced some basic ideas regarding Hamilton circuits, we are almost ready to solve the TSP posed earlier. Recall that Danielle was deciding which order for visiting her branch offices would be the cheapest. In Figure 4.26, we redraw the graph that models her possible trips, except now on each edge we have indicated the cost of traveling between the cities represented by the vertices of the edge.

> **DEFINITIONS** When we assign numbers to the edges of a graph, the graph is called a **weighted graph** and the numbers on the edges are called **weights**. The **weight of a path** in a weighted graph is the sum of the weights of the edges of the path.

In Danielle's problem, the weights represent money; in other problems, the weights might represent distance or time.

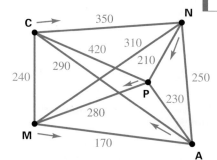

FIGURE 4.26 Graph model of Danielle's possible trips. Weights on edges denote cost of travel between cities.

EXAMPLE 4   *Solving a TSP Using Brute Force*

Use Figure 4.26 to find the sequence of cities for Danielle to visit that will minimize her total travel cost.

SOLUTION: Another way to state this problem is that we want to find the Hamilton circuit that has the smallest weight. For example, if Danielle chooses the circuit PMACNP, the cost of her trip will be

$$280 \quad + \quad 170 \quad + \quad 290 \quad + \quad 350 \quad + \quad 210 \quad = \quad \$1,300.$$

| Philadelphia to Memphis | Memphis to Atlanta | Atlanta to Cleveland | Cleveland to New York | New York to Philadelphia | Total |

Our plan for a solution is simple but tedious. We will calculate the weight of each Hamilton circuit in the graph. The circuit with the smallest weight will be our solution.

---

*This highlight is based on material from David M. Burton, *The History of Mathematics: An Introduction* (New York: McGraw-Hill, 1999).

Hamilton Circuit	Weight ($)
PACMNP	1,280
PACNMP	1,460
PAMCNP	1,200
PAMNCP	1,480
PANCMP	1,350
PANMCP	1,450
PCAMNP	1,400
PCANMP	1,550
PCMANP	1,290
PCNAMP	1,470
PMACNP	1,300
PMCANP	1,270

TABLE 4.3 Hamilton circuits and their weights.

Because this graph has five vertices there will be 4! = 24 Hamilton circuits. However, in Table 4.3 we have listed only 12 of these circuits because clearly each circuit has the same weight as its reverse.

From Table 4.3, we see that the smallest weight is $1,200, which corresponds to the circuit PAMCNP. Thus, Danielle should start at Philadelphia and then travel to Atlanta, Memphis, Cleveland, and New York City in that order and then return home.

Now try Exercises 23 to 26. ✳

In Example 4, our solution was not clever or sophisticated. We simply considered every possible Hamilton circuit and chose the best one. This approach to the TSP is called **brute force.** Realize how impractical this method would have been if Danielle visited 15 cities. In that case, from Table 4.2 we see that we would have had to consider 14! = 87,178,291,200 Hamilton circuits.

Although this problem is very large, you might think that a fast computer could solve it fairly quickly. Let's suppose that we used a computer that could examine 100 circuits per second. There are 31,536,000 seconds in a year, so the computer could examine $31,536,000 \times 100 = 3,153,600,000$ Hamilton circuits per year. Thus, it would still take our computer

$$\frac{14!}{3,153,600,000} = \frac{87,178,291,200}{3,153,600,000} \approx 28 \text{ years}$$

to complete the solution!

Recall that an algorithm is a series of steps that finds a solution for a problem. We will formalize what we did in Example 4. **8**

Quiz Yourself **8**

Use the brute force algorithm to find a Hamilton circuit with minimum weight in the graph.

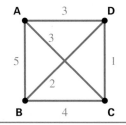

THE BRUTE FORCE ALGORITHM FOR SOLVING THE TSP

Step 1: List all Hamilton circuits in the graph.

Step 2: Find the weight of each circuit found in step 1.

Step 3: The circuits with the smallest weights tell us the solution to the TSP.

The Nearest Neighbor Algorithm

It would be nice if there were an algorithm that would solve a TSP much faster than the brute force algorithm. Unfortunately, at present there is no such algorithm and, in fact, mathematicians do not know whether it is even possible to find such an algorithm.

In a situation such as this, mathematicians ask a different question: "Is there an algorithm that, although it may not find the best solution, does find a pretty good approximation to the best solution?" In fact, a number of algorithms will do this. One rather intuitively simple approach is, when constructing a Hamilton circuit, always choose the next edge that has the smallest weight. This method is called the **nearest neighbor algorithm.**

KEY POINT

There are algorithms that give good approximations to solutions to the TSP.

THE NEAREST NEIGHBOR ALGORITHM FOR SOLVING THE TSP

Step 1: Start at any vertex X.

Step 2: Of all the edges connected to X, choose any one that has the smallest weight. (There may be several with smallest weight.) Select the vertex at the other end of this edge. This vertex is called the *nearest neighbor* of X.

Step 3: Choose subsequent *new* vertices as you did in step 2. When choosing the next vertex in the circuit, choose one whose edge with the current vertex has the smallest weight.

Step 4: After all vertices have been chosen, close the circuit by returning to the starting vertex.

Math in Your Life

Can Ants Do Mathematics?

Yes, believe it or not, ants can do mathematics! Of course, they do not use tiny pencils and paper or minicalculators, but in their own way, when you see ants scurrying around their nest, they are actually working on finding the shortest routes to food sources in a way that is similar to the TSP problem discussed in this section.

In more recent years, scientists have applied mathematics to study how ants, bees, and other social insects cooperate using "swarm intelligence" to solve problems. This research has paid off in applications such as finding improved telephone routes in congested lines and helping robots work cooperatively. Eric Bonabeau, a researcher in ant algorithms, foresees a world in which computer "chips are embedded in every object, from envelopes to trashcans to heads of lettuce" and believes that ant algorithms will be necessary to allow these chips to communicate.

EXAMPLE 5 *Solving a TSP Using the Nearest Neighbor Algorithm*

Use the nearest neighbor algorithm to schedule Danielle's trip.

SOLUTION: The graph in Figure 4.27 shows the possibilities for her trip. Danielle will begin in Philadelphia, so we start the Hamilton circuit with vertex P. The four edges joined to P (with their weights) are PA (230), PC (420), PM (280), and PN (210). Since edge PN has the smallest weight, we choose N as the second vertex in our circuit.

Because we do not want to return to P, we consider the three new edges joined to N, namely, NA (250), NC (350), and NM (310). Edge NA has the smallest weight, so we add vertex A to our circuit.

We are now at vertex A and must choose either M or C for the fourth vertex. Edge AM (170) has a smaller weight than AC (290), so the fourth vertex in the circuit is M. This means that we complete the circuit by adding vertex C and then returning to P. The Hamilton path we have constructed is therefore PNAMCP, whose weight is

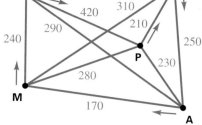

FIGURE 4.27 Graph model of Danielle's possible trips.

Quiz Yourself **9**

Redo Example 5, but this time assume that Danielle is based in Memphis and must visit each of the other four cities. Again, use the nearest neighbor algorithm to find a Hamilton circuit and state its weight.

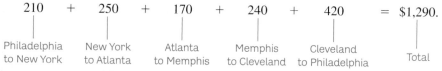

$$210 \quad + \quad 250 \quad + \quad 170 \quad + \quad 240 \quad + \quad 420 \quad = \$1{,}290.$$

| Philadelphia to New York | New York to Atlanta | Atlanta to Memphis | Memphis to Cleveland | Cleveland to Philadelphia | Total |

Now try Exercises 27 to 30. ❉ **9**

Notice from Examples 4 and 5 that although the nearest neighbor algorithm did not find the best solution to Danielle's problem, it did find a reasonably inexpensive way to schedule her trip. In Quiz Yourself 9, the nearest neighbor algorithm did find the best solution. Of course, we cannot predict when the nearest neighbor algorithm will find the best solution to the TSP.

The Best Edge Algorithm

We will investigate another way to find an approximate solution to the TSP. In looking at Figure 4.27, it seems a mistake not to try to use edges like AM (weight 170) and PN (weight 210). If we focus our attention on always choosing the "best edge," rather than trying to construct the circuit vertex by vertex, we might be able to find a pretty good approximation to the best solution to the TSP.

THE BEST EDGE ALGORITHM FOR SOLVING THE TSP

Step 1: Begin by choosing any edge with the smallest weight.

Step 2: Choose any remaining edge in the graph with the smallest weight.

Step 3: Keep repeating step 2; however, do not allow a circuit to form until all vertices have been used. Also, because the final Hamilton circuit cannot have three edges joined to the same vertex, never allow this to happen during the construction of the circuit.

EXAMPLE 6 *Solving a TSP Using the Best Edge Algorithm*

Use the best edge algorithm to schedule Danielle's trip.

SOLUTION: We begin by selecting edge AM, which has the smallest weight, 170. Next choose edges NP (weight 210), AP (weight 230), and CM (weight 240). We have highlighted the edges we have already selected in Figure 4.28.

The edge with the next smallest weight is AN; however, selecting it forms a circuit, so we do not choose it. We do not choose MP, AC, or MN for the same reason. Selecting the last remaining edge, CN, forms a Hamilton circuit. Notice that it has a weight of 1,200, which also makes it the best solution to Danielle's problem.

Now try Exercises 31 to 34. ✳

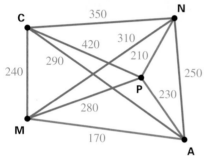

FIGURE 4.28 First four edges selected for Danielle's trip using the best edge algorithm.

Although the best edge algorithm found the solution for Danielle's TSP in Example 6, there is no guarantee that this will happen for other TSPs.

Exercises 4.2

Looking Back*

These exercises follow the general outline of the topics presented in this section and will give you a good overview of the material that you have just studied.

1. On page 156, in computing the number of Hamilton circuits in K_5, after we had chosen the product 4×3, how many vertices remained for us to choose in completing a Hamilton circuit?

2. Using the brute force algorithm to solve the TSP in a complete graph with six vertices, how many Hamilton circuits would you have to consider if you don't count reversals of circuits?

3. In using the nearest neighbor algorithm, after choosing a vertex, how do you decide what vertex to choose next?

4. In Example 6, why did we not choose the edge AN?

5. What event inspired Hamilton to study mathematics more intensely?

6. Mention two applications of "swarm intelligence" in solving real-world problems.

Sharpening Your Skills

For many of these exercises, there may be several possible correct answers. We will provide only one in the answer key.

In Exercises 7–10, find a Hamilton circuit in each graph that begins with the specified edge.

7. **a.** AD   **b.** EB   8. **a.** AC   **b.** BD

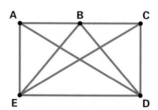

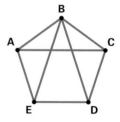

---

*Before doing these exercises, you may find it useful to review the note *How to Succeed at Mathematics* on page xix.

9. a. AD   **b.** EA

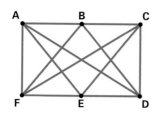

10. a. AE   **b.** FD

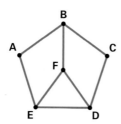

In Exercises 11–14, it helps organize your solution if you construct a tree similar to the one shown in Figure 4.25.

11. Find all the Hamilton circuits that begin with edge AB in the graph shown in Exercise 7.

12. Find all the Hamilton circuits that begin with edge AF in the given graph.

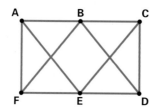

13. How many Hamilton circuits are in K_7?

14. How many Hamilton circuits are in K_8?

15. Draw K_6.   **16.** Draw K_7.

There are several theorems which will tell you when a graph is Hamiltonian.

> **Dirac's Theorem:** *If a simple* graph has n vertices where n ≥ 3 and each vertex has at least $\frac{n}{2}$ edges joined to it, then the graph is Hamiltonian.*

> **Ore's Theorem:** *If a graph has n vertices where n ≥ 2 and for any two nonadjacent vertices the sum of their degrees is at least n, then the graph is Hamiltonian.*

For the graphs in Exercises 17–20, determine: a) Are the conditions of Dirac's theorem satisfied? b) Are the conditions of Ore's theorem satisfied? c) Is the graph Hamiltonian?

17.

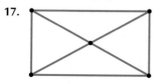

18.

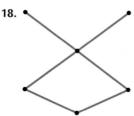

19.

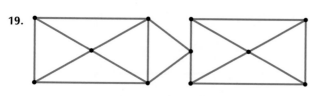

20.

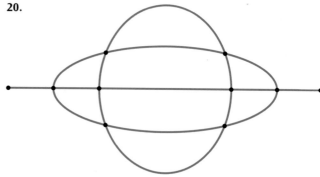

21. Find the weight of the specified paths in the following graph.

   a. AGEDCB   **b.** ABCDGEF

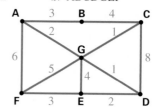

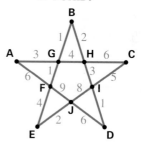

22. Find the weight of the specified paths in the following graph.

   a. AGBHCI   **b.** FGHIDJ

In Exercises 23–26, use the brute force algorithm to find a Hamilton circuit that has minimal weight. Recall that we found all Hamilton circuits in K_4 in Example 3.

23.

24.

25.

26.

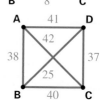

In Exercises 27–30, use the nearest neighbor algorithm to find a Hamilton circuit that begins at vertex A in each graph.

27.

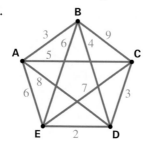

---

*Simple means that any two vertices in the graph have at most one edge between them and no vertex has an edge to itself.

28.

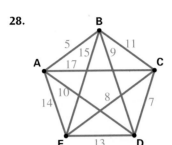

29.

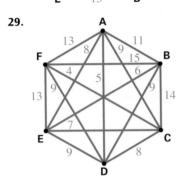

30.

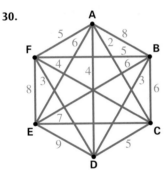

In Exercises 31–34, use the best edge algorithm to find a Hamilton circuit in each graph. List the circuit beginning at vertex A.

31. The graph of Exercise 27

32. The graph of Exercise 28

33. The graph of Exercise 29

34. The graph of Exercise 30

In Exercises 35–38, explain why each graph cannot have any Hamilton circuits.

35.

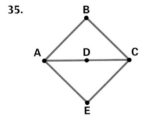

36.

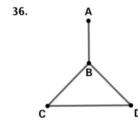

37.

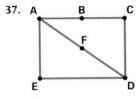

38.

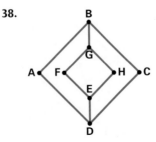

Applying What You've Learned

In Exercises 39–40, we have added some cities to Danielle's territory. Instead of drawing a graph, we have listed the cost between pairs of cities in a table. Again, Danielle will start at Philadelphia, visit all the cities, and then return home. Recall that the original cities in this problem were (P)hiladelphia, (N)ew York, (A)tlanta, (M)emphis, and (C)leveland.

39. Finding the cheapest route. Assume that (R)aleigh and (B)oston are to be added to the trip.

	P	N	A	M	C	R	B
P	0	210	230	280	420	240	430
N	210	0	250	310	350	320	180
A	230	250	0	170	290	90	510
M	280	310	170	0	240	120	500
C	420	350	290	240	0	270	380
R	240	320	90	120	270	0	290
B	430	180	510	500	380	290	0

a. If you were to draw a graph as we did in Example 4 and then use the brute force algorithm, how many Hamilton circuits would you have to consider?

b. If you could examine one Hamilton circuit per minute, how many hours would it take you to solve this TSP using the brute force algorithm?

c. Use the nearest neighbor algorithm to find a Hamilton circuit beginning at Philadelphia.

d. Use the best edge algorithm to find a Hamilton circuit beginning at Philadelphia.

40. Finding the cheapest route. Redo Exercise 39; however, in addition to Raleigh and Boston, assume that we have also added (D)allas to the trip. The cost of travel between Dallas and each of the other cities is given in the following table.

	P	N	A	M	C	R	B
D	510	530	410	370	450	260	560

Use the map to solve Exercises 41 and 42. Assume that all blocks are the same length.

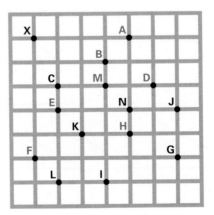

Figure for Exercises 41 and 42

41. Finding the most efficient route. Papa John's Pizza is located at the position marked A on the map. Assume that deliveries must be made to locations B, D, E, H, F, and M, and then the delivery truck must return to the store.

    a. Use the nearest neighbor algorithm to devise a delivery route that will minimize the number of blocks traveled. You may assume that all streets are two-way streets. (*Hint:* First make a table of the distances between the various locations, similar to what we did in Exercise 39.)

    b. Repeat part a) using the best edge algorithm.

42. Finding the most efficient route. Repeat Exercise 41, but now assume that the deliveries must be made to B, E, H, I, J, and L.

Use the map for Exercises 43 and 44.

43. Finding the most efficient route. FedEx is located at the position marked X on the map.

    a. Use the nearest neighbor algorithm to devise an efficient route to make deliveries to locations A, B, C, D, E, F, and P. The route starts and ends at X.

    b. Repeat part a) using the best edge algorithm.

44. Finding the most efficient route. Repeat Exercise 43, but now deliveries must be made to locations G, H, I, J, K, L, and N.

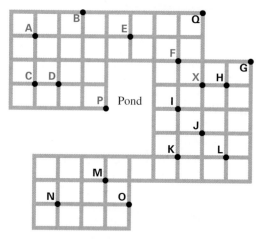

Figure for Exercises 43 and 44

Communicating Mathematics

45. What is the difference between a Hamilton circuit and an Euler circuit?

46. What is the disadvantage of using the brute force algorithm for large graphs?

47. What is the advantage of using the nearest neighbor and the best edge algorithms in solving the TSP?

48. What is the disadvantage of using the nearest neighbor and the best edge algorithms in solving the TSP?

49. Explain why the extended nearest neighbor algorithm may give you a better solution to a TSP than the nearest neighbor algorithm.

50. Explain why there must always exist Hamilton circuits in K_n.

51. When asked to find all Hamilton circuits in a complete graph, Michael found 15. How do we know he has made a mistake?

52. Explain why there are $(n - 1)!$ Hamilton circuits in K_n.

Using Technology to Investigate Mathematics

53. On my calculator, the highest factorial that I could compute was 69!, which was displayed in scientific notation as 1.711 E 98. Find the largest factorial that your calculator can compute.

54. People who work with very large mathematical problems are interested in how rapidly the size of a problem grows. In Chapter 1, we showed that a set having n elements will have 2^n subsets. In Chapter 3, we saw that a logical statement having n variables will have 2^n lines in its truth table. Compare the size of 2^n with $n!$ for several values of n. What conclusion do you draw?

There are many wonderful Internet sites devoted to the TSP; however, traditionally this problem is called "The Traveling Salesman Problem." You can search the Internet to find interesting sites devoted to this problem to solve Exercises 55–57. (There is a particularly good site, with many links, at Georgia Tech University.)

55. Find some of the famous TSP problems and explain what they are. For example, what is the Sweden TSP? Find some others. Report on your findings.

56. Some sites on the Internet have beautiful artistic renderings of particular solutions to the TSP. Find such a site and download the art.

57. Do an Internet search to find 10 or more interesting applications of the traveling salesman problem.

For Extra Credit

58. If a computer can examine 1,000 Hamilton circuits per second, how long would it take to examine all the Hamilton circuits in K_{10}?

59. If a computer can examine 1,000,000 Hamilton circuits per second, how long would it take to examine all the Hamilton circuits in K_{20}?

If a weighted graph has vertices A, B, C, and so on, a variation of the nearest neighbor algorithm is to use the algorithm to find a Hamilton circuit starting at A, then find another one starting at B, then find another one starting at C, and so on. Then pick the Hamilton circuit from among these that has the smallest weight. We will call this algorithm the extended nearest neighbor *algorithm.*

60. Redo Exercise 27 using the extended nearest neighbor algorithm.

61. Redo Exercise 29 using the extended nearest neighbor algorithm.

62. Draw a weighted graph in which the brute force, nearest neighbor, and best edge algorithms all give different Hamilton circuits.

4.3 Directed Graphs

Objectives

1. Understand how directed graphs model relationships that go in only one direction.
2. Use directed graphs to model influence.
3. Model the spread of a disease by a directed graph.

In real life, you are related to people in many different ways. You are friends with some people, smarter than others, and live on the same street as still others. "Living on the same street" is an example of a relationship that has a symmetry to it. If you live on the same street as Angelina Jolie, then Angelina Jolie lives on the same street as you do.

Other relationships lack this symmetry. "Love" is such a relationship. You may love Chris, but tragically Chris may not love you (sigh). As another example, consider the "mother of" relationship. If Carla is the mother of Izzy, then Izzy is certainly not the mother of Carla. Relationships that are nonsymmetric are all around us, and to model them with graphs, we assign directions to the edges of the graphs.

Directed Graphs

When an edge has a direction it is called a **directed edge**. A graph in which all edges are directed is called a **directed graph**. One application of directed graphs is to model the flow of information within a group of people or within an organization.

✏️ **KEY POINT**

Directed graphs model the flow of information.

EXAMPLE 1 *Modeling the Spread of Rumors*

We want to model how rumors are spread among four people: Lauren, Heidi, Whitney, and Audrina. Assume that we have gathered the following information:

1. If Lauren hears a rumor, she will communicate it to Heidi, but Heidi will not tell rumors she hears to Lauren.
2. Heidi and Whitney will tell any rumors that they hear to each other.
3. Heidi will tell rumors she hears to Audrina, but Audrina does not relay rumors she hears to Heidi.

Model this situation with a directed graph.

SOLUTION: Because rumors do not always flow in both directions, we can model this situation with the directed graph shown in Figure 4.29. The arrows on the edges show the direction in which rumors travel.

For example, the directed edge from L to H shows that Lauren communicates rumors to Heidi. Likewise, the absence of a directed edge from H to L indicates that Heidi never tells rumors to Lauren. ❈

To obtain information from a directed graph, it is often useful to identify certain paths in the graph.

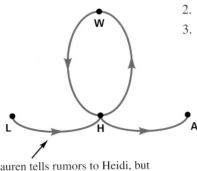

Lauren tells rumors to Heidi, but Heidi does not tell rumors to Lauren.

FIGURE 4.29 A directed graph modeling the spread of rumors.

DEFINITIONS Suppose that X and Y are vertices in a directed graph. If it is possible to begin at X, follow a sequence of edges in the directions indicated, and end at Y, then we call the sequence of edges encountered a **directed path from X to Y**. We will denote a directed path by the sequence of vertices encountered along the path. The **length** of a directed path is the number of edges along that path. (Recall that edges cannot be repeated in a path.)

Quiz Yourself ❿

Use the graph to answer the following questions:

a) Can you find two directed paths from A to C?

b) Are there any directed paths from C to A?

c) What is the length of the directed path ABCDB?

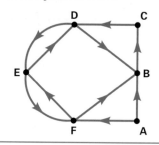

EXAMPLE 2 *Finding Paths in a Directed Graph*

Consider the directed graph shown in Figure 4.30.

a) What is the length of the directed path ACDE?

b) Is ABCE a directed path?

SOLUTION:

a) ACDE is a directed path of length 3 from A to E.

b) ABCE is *not* a directed path from A to E because CE has the wrong direction.

Now try Exercises 7 to 10. ❈ ❿

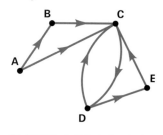

FIGURE 4.30 A directed graph.

Modeling Influence

An interesting application of directed graphs is to model how members of a set influence each other. Because influence is usually not exerted equally in both directions, we use directed graphs as models in these situations.

EXAMPLE 3 *Modeling Influence*

BigMart wants to build a shopping center and is receiving opposition from the residents of Peacefultown. BigMart's lawyers suggest that the best approach to get approval for the project is to gather the support of the most influential members of the township zoning board. After a careful investigation, BigMart's lawyers determine the following patterns of influence:

✎ **KEY POINT**

Directed graphs model how members of a set exert influence.

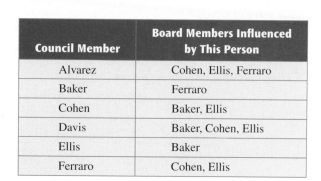

Council Member	Board Members Influenced by This Person
Alvarez	Cohen, Ellis, Ferraro
Baker	Ferraro
Cohen	Baker, Ellis
Davis	Baker, Cohen, Ellis
Ellis	Baker
Ferraro	Cohen, Ellis

TABLE 4.4 Influence on zoning board members.

Use Table 4.4 to determine which board member has the most influence.

SOLUTION: Because influence is not exerted in both directions (Alvarez influences Cohen, but Cohen does not influence Alvarez), we use the directed graph shown in Figure 4.31 to model this relationship among zoning board members.

The directed edges in the graph indicate who influences whom. For instance, the directed edge from A to E reflects that Alvarez has influence over Ellis. Because Alvarez and Davis each exert direct influence over three people, we might say that both are equally influential.

However, let us consider what we will call two-stage influence. We observe that Alvarez influences Cohen, who in turn influences Ellis. Therefore, we will say that Alvarez has two-stage influence over Ellis. In terms of our model, if there is a directed path of length 2 from vertex X to vertex Y, then board member X exerts two-stage influence over Y.

Let us now determine the number of ways that Alvarez can influence Ellis, either directly or in two stages. In Figure 4.31, in addition to the directed edge AE, there are also the directed paths ACE and AFE. Therefore, there are three ways in which Alvarez can influence Ellis in one or two stages.

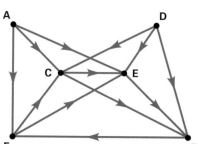

FIGURE 4.31 Directed graph modeling influence on zoning board.

Table 4.5 shows the amount of one- and two-stage influence for each pair of board members. For example, the 3 in row A, column E, of Table 4.5 indicates that there are three paths of length 1 or 2 from A to E. The entry of row C, column B, is 2 because there are two paths of length 1 or 2 from C to B—namely, CB and CEB. In general, the entry in row X, column Y, in the table tells us the number of paths of length 1 or 2 from vertex X to vertex Y in the graph.

two paths of length 1 or 2 from C to B

three paths of length 1 or 2 from A to E

Alvarez's total influence

	A	B	C	D	E	F	**Total Direct and Two-Stage Influence**
A	0	2	2	0	3	1	8
B	0	0	1	0	1	1	3
C	0	2	0	0	1	1	4
D	0	3	1	0	2	1	7
E	0	1	0	0	0	1	2
F	0	2	1	0	2	0	5

From (label for rows)

TABLE 4.5 Total one- and two-stage influence exerted by board members.

Most Influential

Alvarez
Davis
Ferraro
Cohen
Baker
Ellis

Least Influential

FIGURE 4.32 Ranking of zoning board members according to one- and two-stage influence.

Now that we have determined the amount of one- or two-stage influence, we can rank the board members. Although Alvarez and Davis influence Ellis directly and also by way of Cohen, we see that Alvarez also influences Ferraro, who influences Ellis. There are three ways in which Alvarez exerts influence over Ellis but only two ways in which Davis can. Therefore, it is reasonable to say that Alvarez exerts more influence over Ellis than Davis can.

We can now rank board members according to their one- and two-stage influence. Adding the entries in row A, we get

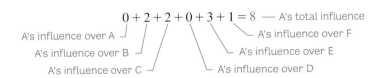

$$0 + 2 + 2 + 0 + 3 + 1 = 8 \quad \text{— A's total influence}$$

A's influence over A
A's influence over B
A's influence over C
A's influence over D
A's influence over E
A's influence over F

which means that there are eight ways Alvarez exerts influence over other board members in one or two stages. Adding the entries for each of the other rows, we obtain the last column in Table 4.5. According to the way we are measuring influence, Figure 4.32 shows the ranking of the board members.

Now try Exercises 21 to 26. ❈ **11**

Quiz Yourself **11**

List in a table the number of directed paths of length 1 or 2 between each pair of vertices in the given graph.

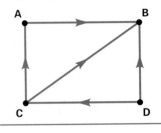

As in Example 3, we often want to rank objects with respect to some property. For example, we could rank football teams by using their number of victories. Assume Ohio State and Nebraska have played the same number of games, but Ohio State has four victories and Nebraska has three. We could then rank Ohio State above Nebraska. Suppose instead that each team has four victories. In order to break a tie in the ranking order, we could consider the notion of two-stage dominance; that is, if Ohio State has beaten Michigan and Michigan has defeated Indiana, then Ohio has demonstrated two-stage dominance over Indiana. If Ohio State has more two-stage dominances over other

Math in Your Life

You'll Have That Tomorrow

Isn't it marvelous how you can drop off a package and within 24 hours it reaches its destination? Recently, I received a tracking report that went roughly like this: Pickup 2:41 PM—Reading, PA 7:01 PM—Philadelphia, PA 10:52 PM—Chicago, IL 4:10 AM—Delivered 10:27 AM. This report describes how my package traveled over a path in a huge directed graph that models the delivery company's operations.

Scientists, working in the area of **operations research**, frequently use graph theory to organize the flow of tens of millions of packages each day, coordinate millions of phone calls, and help hoards of airline passengers arrive safely at their destinations.

teams than Nebraska has, then we could rank Ohio State above Nebraska. If the two teams are still tied with regard to the number of two-stage dominances, we might then consider the number of three-stage dominances, and so forth.

Keep this example in mind when solving the exercises at the end of this section in which you are asked to rank objects of a set.

 Some Good Advice

Realize that when building mathematical models, *you* decide what features you feel are important in the model. You may not agree with the way a model is constructed. For example, should two-stage dominance count the same as one-stage dominance in the last discussion? If not, then you may want to change the method of building the model.

KEY POINT

Directed graphs model disease transmission.

Modeling Disease

We now use directed graphs to answer a different type of question.

EXAMPLE 4 *Modeling the Spread of a Disease*

Army virologist Lieutenant Colonel Robert Neville has quarantined eight New Yorkers who have contracted a deadly virus and hopes that no others have contracted this virus. He believes that one person introduced the virus and communicated it to the others in the group. Use the information in Table 4.6 to determine if Neville's hypothesis is correct.

SOLUTION: We can easily model this situation using the directed graph in Figure 4.33. We represent each of the eight patients with a vertex and draw a directed edge from vertex X to vertex Y, if X could have transmitted the virus to Y.

Because Dustin could have given the virus to Caterina, we draw a directed edge from D to C. Similarly, because Brian could have transmitted the virus to Frank, there is a directed edge drawn from B to F. The directed paths in the graph show how the infection might have spread within the group. For example, the directed path ADCF shows that it

Patient	Others Within the Group Who Could Have Contracted the Virus from This Patient
Amanda	Dustin, Jackson
Brian	Caterina, Frank, Ina
Caterina	Frank
Dustin	Caterina
Frank	Louisa
Ina	Brian, Frank
Jackson	Amanda, Caterina, Frank
Louisa	Caterina

TABLE 4.6 How a virus might have spread in a town.

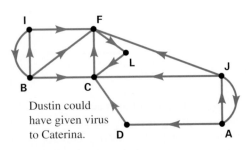

Dustin could have given virus to Caterina.

FIGURE 4.33 Directed graph representing spread of a virus.

was possible for the virus to travel from Amanda to Dustin to Caterina, and finally to Frank. It is important to realize that if the virus spread from X within the group to Y, then there must be a directed path in the graph from vertex X to vertex Y.

We see that it was impossible for the virus to have started with Amanda and spread within the group to Brian because there is no directed path from A to B. In fact, checking each of the eight people, we see that it was impossible for the virus to start with any one of them and then spread to all the others. Therefore, if our information and assumption are correct, there is at least one other person in the city who has the virus but who has not yet been identified.

Now try Exercises 29 to 32. ✤

Exercises 4.3

Looking Back*

These exercises follow the general outline of the topics presented in this section and will give you a good overview of the material that you have just studied.

1. Why did we put arrows on the edges of the graph in Figure 4.29?

2. In Example 3, why is FCEA not a directed path from vertex F to vertex A?

3. What do the numbers in the right column of Table 4.5 represent? What do these numbers have to do with directed edges and directed paths?

4. Explain in Example 4 why the virus could not have originated with Dustin and spread to the rest of the group.

5. What does graph theory have to do with overnight shipping companies?

6. Mention two applications of graph theory to everyday life.

Sharpening Your Skills

For many of these exercises there may be several correct answers. We will provide only one in the answer key.

In Exercises 7–10, use each graph to find the requested items, if it is possible. If it is not possible to find a requested item, explain why not.

7. **a.** Two different directed paths from A to E

   b. A directed path from A to C

   c. A directed path of length 3 from A to E

   d. A directed path of length 2 from A to E

   e. A directed path of length 5 from A to A

8. **a.** Two different directed paths from A to E

   b. A directed path from A to C

   c. A directed path of length 3 from A to E

   d. A directed path of length 2 from A to E

   e. A directed path of length 5 from A to A

9.

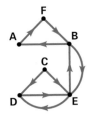

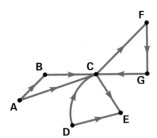

   a. A directed path from B to E

   b. A directed path from A to G

   c. A directed path of length 5 from A to E

   d. A directed path from F to D

   e. A directed path of length 5 from A to C

10.

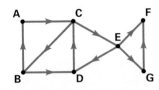

   a. Two different directed paths from D to E

   b. A directed path from B to F

   c. A directed path from F to C

   d. A directed path of length 5 from A to E

   e. A directed path of length 5 from A to A

11. Construct a table that displays the number of directed paths of length 1 or 2 between each pair of vertices in the graph of Exercise 7.

---

*Before doing these exercises, you may find it useful to review the note *How to Succeed at Mathematics* on page xix.

12. Construct a table that displays the number of directed paths of length 1 or 2 between each pair of vertices in the graph of Exercise 8.

13. Construct a table that displays the number of directed paths of length 1 or 2 between each pair of vertices in the graph of Exercise 9.

14. Construct a table that displays the number of directed paths of length 1 or 2 between each pair of vertices in the graph of Exercise 10.

Applying What You've Learned

15. Modeling the spread of rumors. Ryan, Dwight, Pam, Jim, Angela, and Kevin work in the same office. A rumor has been circulating that their company is being taken over by a Singapore-based firm in a hostile takeover. The given directed graph shows how rumors travel among these six employees.

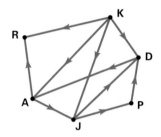

a. Are there any people among these six who could have initiated the rumor?

b. Change the direction of as few edges as possible to make Pam the person who started the rumor.

16. Modeling the spread of classified information. Several news organizations have made public a secret report on how to improve the U.S. diplomatic stature in the Middle East. Based on sources (which we cannot reveal), the given graph indicates how the information could have passed among these organizations.

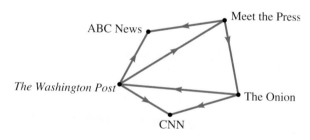

a. Determine which organizations could have first obtained this information.

b. Change the direction of only one edge in the graph so that only the *Washington Post* could have obtained the information first.

17. Modeling the flow of paperwork through a bureaucracy. In a large corporation, five signatures are needed on a form. It is known that certain managers will not give their approval before others. This situation is depicted in the following directed graph. A directed edge from vertex A to vertex B indicates that A must sign the form before B. If a secretary must hand-carry the form from office to office, in which sequence can all the signatures be obtained?

Figure for Exercise 17

18. Modeling the spread of disease. The following directed graph models the spread of a motovirus among a group of passengers on a cruise ship.

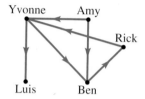

a. Are there any people in this group who could have introduced the virus to the group? Explain.

b. Change the direction of one edge in the graph so that no one could have introduced the virus to the group.

19. Modeling a food chain. The African hare and gazelle are vegetarians that feed primarily on grass. Gazelles and hares are eaten by lions, cheetahs, and humans. Draw a directed graph to model this food chain.

20. Modeling a communications network. An AMBER Alert emergency broadcast system has been set to transmit messages among several cities. The following table indicates how this system has been set up. Draw a directed graph modeling this emergency network. Can a message originate in one city and spread throughout the network? Explain.

City	Can Broadcast To
Philadelphia	New York, Detroit
New York	Philadelphia, Boston
Boston	Philadelphia, New York
Dallas	Los Angeles, Phoenix
Detroit	Philadelphia, Dallas
Los Angeles	Dallas, Phoenix
Phoenix	Dallas, Los Angeles

21. Ranking football teams. The given graph shows the results among several football teams in the Southeast Conference. Rank these teams using one- and two-stage dominance.

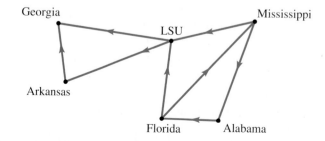

22. **Ranking American Gladiators.** The graph below shows (hypothetical) partial results after several rounds of head-to-head competition among the American Gladiators in the Hit and Run Contest. Use one- and two-stage dominance to rank these competitors. An arrow from vertex X to vertex Y indicates that X defeated Y.

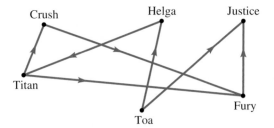

23. **Modeling influence.** A designer, who wishes to obtain the contract for creating a new advertising campaign for Center City Community College, wants to determine the most influential members of the selection committee. The table below lists the influences that have been observed among committee members.

Member	Lee	Feinstein	Johnson	Cordaro	Murciano
Influences	No one	Cordaro, Murciano	Lee, Cordaro	Lee	Lee, Johnson

Model this committee with a directed graph and use one- and two-stage influence to determine the most influential committee members.

24. **Ranking teams.** In a round-robin singles pool tournament, each of six competitors plays each other once. The results of the tournament are shown in the following table:

Player	Defeated
Carla	Rob, Sara, Orlando, Matt
Rob	Tanya, Orlando, Matt
Tanya	Carla, Sara, Matt
Sara	Rob, Matt
Orlando	Tanya, Sara
Matt	Orlando

Use a directed graph to model this situation and determine a ranking order for the players using one- and two-stage dominance.

25. **Modeling drink preferences.** In a bottled water "paired comparison" test, college students were asked to compare the drinks Propel, Aquafina, Fuji, Dasani, and Vitaminwater. Use the table below to rank the drinks using one- and two-stage preference.

Brand	Propel	Aquafina	Fuji	Dasani	Vitaminwater
Preferred Over	Aquafina, Fuji	Vitaminwater	Aquafina, Dasani, Vitaminwater	Propel, Aquafina, Vitaminwater	Propel

26. **Modeling drink preferences.** Redo Exercise 25 using the following table:

Brand	Propel	Aquafina	Fuji	Dasani	Vitaminwater
Preferred Over	Aquafina, Dasani	Vitaminwater	Aquafina, Dasani, Vitaminwater	Aquafina, Vitaminwater	Propel

Communicating Mathematics

27. When is a directed graph a better model for a relationship than a nondirected graph?

28. In Figure 4.31, how do we model two-stage influence versus one-stage influence?

For Exercises 29–32, use Figure 4.33 in Example 4, which models the spread of the Hanta virus.

29. What do you see in the graph that tells you the virus could not have been introduced by either I or B?

30. What do you see in the graph that tells you the virus could not have been introduced by J, D, or A?

31. What would have to be present in the graph for it to be possible that the virus was introduced by F? Is there one edge whose direction you could change to make this happen?

32. Is there one edge whose direction you could change so that the graph would indicate that the virus could have been introduced by D? Explain.

33. In modeling influence, we counted two-stage influence as equal with one-stage influence. Criticize this approach. Can you suggest another method? How would this affect our tables that summarize one- and two-stage influence? Explain.

34. Think of a situation that we have not discussed in this section in which you feel directed graphs could be used as models. Describe the application and explain the reasons why you think it can be represented by a directed graph.

Using Technology to Investigate Mathematics

35. Perform an Internet search on the words *directed*, *graphs*, and *applications* to find applications of directed graphs. Write a brief report on your findings.

36. Directed graphs are often used to model the flow of some commodity through a network. Perform an Internet search on the words *directed*, *graphs*, *flow*, and *applications*. Find an interesting application and write a brief report on your findings.

37. There are many beautiful artistic renderings of the Internet as multicolored directed graphs. Find several such images and, if possible, make a colored hard copy of them to show to the class.

38. From your instructor, obtain the chapter on matrices. The "Of Further Interest" section will show you how to do the calculations in this section in a very efficient way by using matrices. In fact, one of the highlights in that section shows how to use a computer algebra system to perform directed graph calculations. Solve some of the problems in this section using matrices.

For Extra Credit

39. Modeling consumer preferences. When purchasing food products, a consumer would consider ease of preparation, nutritional value, price, and taste. Conduct a consumer survey by asking five people you know to complete a copy of the following ballot.

For each of the following pairs, circle the quality that is more important to you:

Ease of preparation	Nutritional value
Ease of preparation	Price
Ease of preparation	Taste
Nutritional value	Price
Nutritional value	Taste
Price	Taste

On the third line of the ballot, if more people chose taste over ease of preparation, then we can say that taste is preferred over ease of preparation. Use the results of your survey to determine which quality is preferred for each pair, and then summarize your results by means of a directed graph.

40. Modeling car feature preferences. When shopping for an automobile, a buyer may consider each of the following features: economy, safety, comfort, style, manufacturer, and availability of service. Assume you are a car buyer and compare each possible pair of features, indicating which of the two is more important to you. After completing this paired-comparison test, determine a ranking order by importance of the six features.

Looking Deeper

4.4 Scheduling Projects Using PERT

Objectives

1. Understand the language of PERT diagrams.
2. Use PERT diagrams to schedule projects.

Imagine that you had been hired as Donald Trump's apprentice to oversee the construction of his latest skyscraper. Or, what if NASA asked you to take charge of building the next space shuttle? Would your first reaction be

"I just don't know where to begin?"

 **KEY POINT**

PERT diagrams help organize large projects.

PERT Diagrams

Many people are overwhelmed when facing a job that has a large number of subtasks. There is a scheduling system called PERT that helps organize a large project. In order to give you an understanding of this method, let's consider the following hypothetical situation.

Imagine that you are serving on a committee to develop a timetable for constructing and populating an orbiting space colony. Your goal is to schedule the 10 major tasks listed in Table 4.7 that are needed to complete the project. Also, you are to determine if reducing the amount of time needed to build the life support systems will shorten the total time for the project.

It is tempting to simply add the time required for each task and conclude that the project requires 75 months. However, because some tasks can be done simultaneously, the total time needed will be less than 75 months. The last column of Table 4.7 shows which tasks must precede others.

Task	Time Required (months)	Preceding Tasks
1. Train construction workers	6	None
2. Build shell	8	None
3. Build life support systems	14	None
4. Recruit colonists	12	None
5. Assemble shell	10	1, 2
6. Train colonists	10	2, 3, 4
7. Install life support systems	4	1, 2, 3, 5
8. Install solar-energy systems	3	1, 2, 5
9. Test life support and energy systems	4	1, 2, 3, 5, 7, 8
10. Bring colonists to the colony	4	1, 2, 3, 4, 5, 6, 7, 8, 9

TABLE 4.7 Time and precedence of the space colony tasks.

We can display the data in Table 4.7 in a special type of directed graph called a **PERT diagram**, shown in Figure 4.34. We represent each task by a vertex containing the number of months needed for that task and assume that the beginning and end of the project require no time.

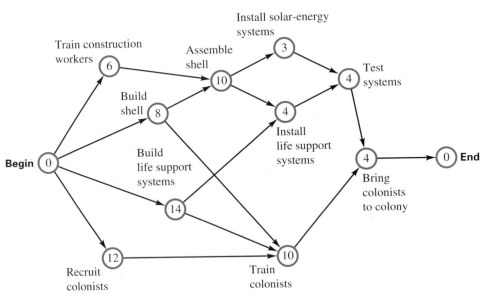

FIGURE 4.34 A PERT diagram for the space colony project.

If a task X *immediately* precedes a task Y, we draw a directed edge from vertex X to vertex Y in the graph. For example, from Table 4.7 we see that training construction workers *immediately* precedes assembling the shell in space, so we have drawn a directed edge from the vertex "Train construction workers" to the vertex "Assemble shell." Notice that because building the shell on Earth does not immediately precede installing the life support systems, we do not draw a directed edge between these two vertices. We will return to the solution of this problem after Quiz Yourself 12. **12**

Some Good Advice

If your first attempt at drawing a PERT diagram does not show the relationship between the tasks clearly, take the time to redraw it so that the information stands out.

Quiz Yourself 12

The following table lists the times required to complete the tasks of a particular project and the dependency among these tasks. Draw a PERT diagram for this project.

Task	Days Required	Preceding Tasks
A	3	None
B	4	A
C	6	A
D	2	A, B, C
E	5	A, B, C, D
F	7	A, B, C, D, E

We now return to our space colony scheduling problem.

EXAMPLE 1   *Determining When to Install the Life Support Systems in the Space Colony*

Use the PERT diagram in Figure 4.34 to determine the earliest time we could install life support systems.

SOLUTION: We see from the diagram that several different sequences of tasks must precede installing the life support systems. For example, the sequence

Begin, Train construction workers, Assemble shell

must be completed before we can install the life support systems. This sequence requires $0 + 6 + 10 = 16$ months to complete.

By examining the diagram, we see that the most time-consuming sequence of tasks preceding the installation of life support systems is the sequence

Begin, Build shell, Assemble shell,

which requires $0 + 8 + 10 = 18$ months. This means that we can begin installing the life support systems in the nineteenth month of the project. ❋

 **KEY POINT**

We can use PERT diagrams to schedule tasks efficiently.

Efficient Scheduling

Note that an efficient schedule would have construction of the shell and the life support systems taking place simultaneously because neither of these tasks depends on the other. We can schedule other tasks using reasoning similar to that in Example 1. To simplify our discussion, we introduce the following definition:

DEFINITION Suppose T is a task in a PERT diagram. Let us consider all directed paths from "Begin" to T. If we add the time along each of these paths, any path requiring the most time to complete is called a **critical path** for task T.

Using this new terminology, we can say that a critical path for "Install life support systems" is

Begin, Build shell, Assemble shell, Install life support systems.

> **SCHEDULING A TASK IN A PERT DIAGRAM** To determine when to schedule a task T in a PERT diagram, do the following:
>
> 1. Find a critical path for task T.
> 2. Add all times along this critical path, with the exception of the time required for task T. This sum gives us the time to be allowed before scheduling T.

EXAMPLE 2 *Scheduling the Testing of the Colony's Systems*

Use the preceding scheduling procedure to determine when to start the testing of the colony's systems.

SOLUTION: From our PERT diagram in Figure 4.34, we see that a critical path for this task is

Begin, Build shell, Assemble shell, Install life support systems, Test systems.

Examining this path, we see that $0 + 8 + 10 + 4 = 22$ months must be allowed *before* the testing of the colony's systems can begin. Therefore, this task should be scheduled to begin in the twenty-third month. ❄

Table 4.8 lists a schedule for all the tasks of the space colony project.

Task	Month Task Begins
1. Train construction workers	1st
2. Build shell	1st
3. Build life support systems	1st
4. Recruit colonists	1st
5. Assemble shell	9th
6. Train colonists	15th
7. Install life support systems	19th
8. Install solar-energy systems	19th
9. Test life support and energy systems	23rd
10. Bring colonists to the colony	27th

TABLE 4.8 Schedule for the space colony tasks.

EXAMPLE 3 *Scheduling the Space Colony Project*

Determine the time needed for the entire space colony project.

SOLUTION: To do this, we must find a critical path for the vertex "End." Such a path is

Begin, Build shell, Assemble shell, Install life support systems, Test systems, Bring colonists to the colony, End,

which is highlighted by the red edges in Figure 4.35. From this diagram, we find that we need $0 + 8 + 10 + 4 + 4 + 4 + 0 = 30$ months to complete the project. ❄ **13**

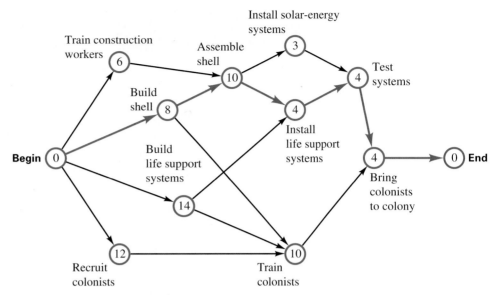

FIGURE 4.35 Critical path for the space colony project.

Quiz Yourself 🔞

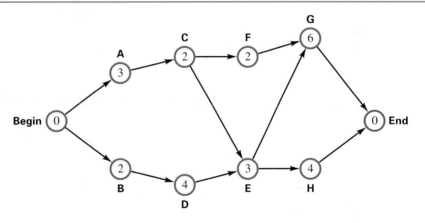

a) Find a critical path for G.

b) What is a critical path for "End"?

c) Assuming that the numbers in the vertices represent days, when should task H be scheduled?

d) How long will the whole project take?

Let us now consider whether it is possible to shorten the total length of the project by reducing the time spent in building the life support systems. Referring to Figure 4.35, we see that the task of building the life support systems *does not lie on a critical path* for the vertex "End." Therefore, reducing this time would in no way decrease the total length of the project.

EXAMPLE 4 *Using PERT to Organize a Concert*

Assume that you are in charge of organizing a concert to raise money to aid victims of a recent earthquake. Your job is to develop a schedule to complete the project in the shortest possible time. The project tasks and their dependencies are given in Table 4.9 and are displayed in the PERT diagram in Figure 4.36.

Task	Time Required (weeks)	Preceding Tasks
1. Obtain permits from the city	2	None
2. Raise funds from local agencies	1	None
3. Canvass local merchants for advertising support	4	None
4. Hire performers	3	Obtain permits, raise funds, canvass merchants
5. Rent an auditorium	2	Obtain permits
6. Print the program	1	All of the above
7. Advertise the concert	2	All of the above except print programs

TABLE 4.9 Tasks for concert project.

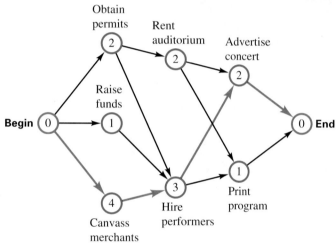

FIGURE 4.36 PERT diagram for concert project.

SOLUTION: By finding critical paths for each of the vertices in the PERT diagram in Figure 4.36, we easily obtain the following schedule of the tasks:

Task	Week Task Begins
1. Obtain permits	1
2. Raise funds	1
3. Canvass merchants	1
4. Rent auditorium	3
5. Hire performers	5
6. Advertise concert	8
7. Print program	8

We determine the shortest time period to complete this project by finding a critical path for the vertex "End." Such a path (marked in red) is

Begin, Canvass merchants, Hire performers, Advertise concert, End.

Thus, the entire concert project can be completed in 9 weeks, which is the sum of the times along this path. ✳

Exercises 4.4

Looking Back*

These exercises follow the general outline of the topics presented in this section and will give you a good overview of the material that you have just studied.

1. What does the number 8 in the vertex "Build shell" in Figure 4.34 represent?

2. In Figure 4.34, why did we not draw a directed edge from the vertex "Build shell" to the vertex "Install life support systems?"

Sharpening Your Skills

Some of these exercises may have several correct answers. We will give only one in the answer key.

In Exercises 3–6, assume that the time is measured in days.

3. Use the following PERT diagram to answer the following questions.

*Before doing these exercises, you may find it useful to review the note *How to Succeed at Mathematics* on page xix.

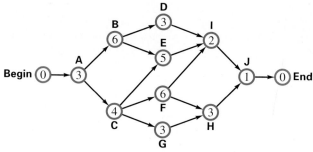

Figure for Exercise 3

a. Find a critical path for task I.

b. Find a critical path for task E.

c. On what day will task H begin?

d. When will task I be completed?

e. What is the least number of days needed to complete this project?

f. Find a critical path for the vertex "End."

4. Use the following PERT diagram to answer the following questions.

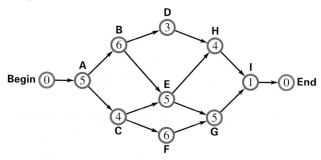

a. Find a critical path for task H.

b. Find a critical path for task G.

c. On what day will task G begin?

d. When will task H be completed?

e. What is the least number of days needed to complete this project?

f. Find a critical path for the vertex "End."

5. Use the following PERT diagram to answer the following questions.

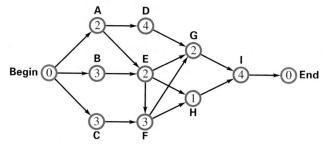

a. Find a critical path for task G.

b. Find a critical path for task H.

c. What is a critical path for "End"?

d. When should task F be scheduled?

e. When should we schedule task G?

f. How many days are required for the whole project?

6. Use the following PERT diagram to answer the following questions.

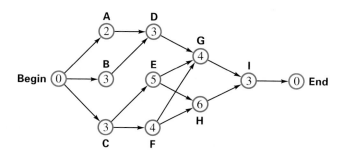

a. Find a critical path for task G.

b. Find a critical path for task H.

c. What is a critical path for "End"?

d. When should task H be scheduled?

e. When should we schedule task G?

f. How many days are required for the whole project?

In Exercises 7–10, use the given PERT diagrams to schedule the tasks so each task is completed in the least possible amount of time. Assume the numbers in the vertices refer to days.

7.

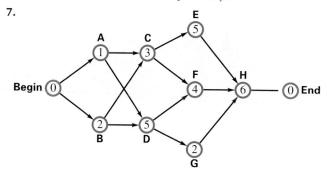

8.

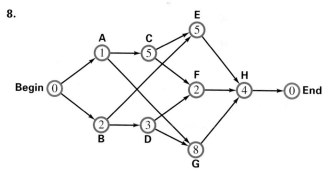

9.

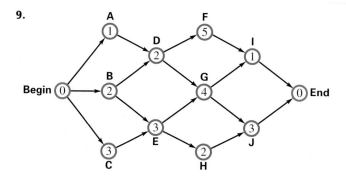

10.

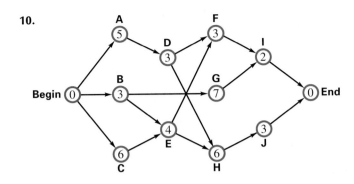

Applying What You've Learned

11. Planning a festival. The Earth Day Committee has begun planning for an Earth Day festival. We list the subtasks for this project and the dependencies among them in the following table. Draw a PERT diagram for this project and construct a schedule for the tasks.

Task	Preceding Tasks	Time Required (weeks)
1. Get funding	none	2
2. Get permits	none	1
3. Decide on program	1	2
4. Rent tents	1, 2, 3	2
5. Arrange for speakers, entertainers, etc.	1, 2, 3, 4	4
6. Advertise	1, 3, 5	2
7. Set up tents, booths, etc.	1, 2, 3, 4	1
8. Set up festival	1, 2, 3, 4, 5, 6, 7	1

12. Organizing a project. Assume that you need to complete a senior group project in order to graduate. Because you have a heavy workload, it is important that you plan carefully so you will graduate on time. You need to choose an advisor for this project and a group of students to work with. The subtasks for this project and the dependencies among them are listed in the following table. Draw a PERT diagram for this project and give a schedule for the tasks.

Task	Preceding Tasks	Time Required (weeks)
1. Decide on group	none	2
2. Choose advisor	1	1
3. Choose project	1, 2	4
4. Divide responsibilities for project among group	1, 2, 3	2
5. Do research	1, 2, 3	4
6. Develop rough outline of project	1, 2, 3	2
7. Refine outline	1, 2, 3, 4, 5, 6	2
8. Complete project	1, 2, 3, 4, 5, 6, 7	4
9. Arrange for presentation	1, 2, 3, 4, 5, 6, 7, 8	1

13. Building a student union building. Waldenville Community College is planning to build a new student union building. The subtasks for this project and their dependencies are listed in the following table. Draw a PERT diagram for this project and give a schedule for the tasks.

Task	Preceding Tasks	Time Required (months)
1. Get funding	none	3
2. Choose contractor	1	2
3. Draw plans	1	4
4. Grade land	1, 2, 3	1
5. Lay underground utilities, sewers, etc.	1, 2, 3, 4	1
6. Build building	1, 2, 3, 4, 5	8
7. Install utilities in building	1, 2, 3, 4, 5, 6	2
8. Install computer network in building	1, 2, 3, 4, 5, 6	1
9. Purchase furniture	1, 2, 3, 4, 5, 6	2
10. Inspect building	1, 2, 3, 4, 5, 6, 7, 8, 9	1

14. **Organizing a health program.** A developing nation plans to improve available health care for its citizens. Use the following table to draw a PERT diagram for this project and then schedule the tasks so the project can be completed as efficiently as possible.

Task	Preceding Tasks	Time Required (months)
1. Appropriate money	None	3
2. Build health clinics	1	8
3. Construct hospitals	1	18
4. Educate citizens in health practices	1	12
5. Recruit students	1	8
6. Train doctors	1, 3, 5	30
7. Train aides	1, 2, 5	15
8. Inoculate	1, 2, 6, 7	4
9. Conduct follow-up study	All of the above	2

15. **Organizing an advertising campaign.** The American Legacy Foundation together with MTV is developing an advertising campaign, called *Truth*, to reduce smoking in young adults. A series of newspaper ads, billboards, and TV and radio commercials are to be produced. Use the following table to draw a PERT diagram for this project and then schedule the tasks so the project can be completed efficiently.

Task	Preceding Tasks	Time Required (months)
1. Conduct survey	None	3
2. Develop budget	1	1
3. Hire PR firm	1	6
4. Set up production schedule	1, 2, 3	1
5. Produce ads	1, 2, 3	8
6. Disseminate ads	1, 2, 3, 5	2
7. Evaluate results	1, 2, 3, 4, 5, 6	6

Communicating Mathematics

16. The task "Build life support systems," which takes 14 weeks, is the most time-consuming task in the space colony project. Would it shorten the total length of the project if we could shorten this time to 12 weeks? Explain.

17. How do we determine when to schedule a task in a PERT diagram?

18. How do we determine the time required for an entire project that is represented by a PERT diagram?

19. You saw in Example 3 that the total length of the space colony project was 30 months. If you could reduce the time for "Assemble shell" by 4 months, would that shorten the length of the project to 26 months?

20. Write a brief report on a real-life project that can be organized using PERT.*

Using Technology to Investigate Mathematics

21. Research the history of PERT using the Internet. What was the first project that used PERT? Report on your findings.

22. Search for "Gantt charts" on the Internet. How are they similar to PERT diagrams?

23. Try to find applets on the Internet that allow you to build and analyze PERT diagrams. Report on your findings.

For Extra Credit

24. **Planning an innovative house.** A local electric company wants to build an experimental house that will conserve energy. Company officials believe the project breaks down naturally into the following tasks (the time required for each task is given in parentheses):

    1. Draw plans for house (2 months)

    2. Design energy systems for house (6 months)

    3. Develop new insulation techniques (3 months)

    4. Purchase land (4 months)

    5. Build conventional shell of house (6 months)

    6. Install new energy systems (2 months)

    7. Install new insulation (1 month)

    8. Finish the remaining conventional parts of the house (3 months)

    9. Landscape (1 month)

    10. Perform tests to determine energy usage in house (8 months)

Determine a reasonable list of dependencies for these tasks, then draw a PERT diagram and develop a schedule for these tasks.

25. Select a project that interests you. Identify the subtasks and draw a PERT diagram to develop a schedule for the project.

---

*Books on operations research often contain sections that discuss PERT. PERT is sometimes called the *critical path method,* or CPM.

CHAPTER SUMMARY

SECTION	SUMMARY	EXAMPLE
SECTION 4.1	A **graph** is a set of points, called **vertices**, and lines, called **edges**, that join pairs of vertices.	Definitions, p. 139
	A graph is **connected** if it is possible to travel from any vertex to any other vertex of the graph by moving along successive edges. A **bridge** in a connected graph is an edge such that if it were removed, the graph would no longer be connected. A vertex is **odd** if it is an endpoint of an odd number of edges. A vertex is **even** if it is the endpoint of an even number of edges.	Definitions, p. 141 Example 1, p. 141
	Euler's theorem states that a graph can be traced if it is connected and has zero or two odd vertices.	Discussion, p. 142 Example 2, p. 142
	A graph can be used to **model** a set of objects that have some relationship among them.	Problem Solving, p. 143
	We can use **Fleury's algorithm** to find Euler circuits.	Example 4, p. 145
	It is always possible to **color a map** with at most four colors so that any two regions sharing a common border will have different colors.	Example 6, p. 147
SECTION 4.2	We use Hamilton circuits to solve the **Traveling Salesperson Problem (TSP)**, which asks: What is the most efficient way for a salesperson to schedule a trip to a series of cities and then return home? A **Hamilton path** is a path that passes through all the vertices of a graph exactly once. A **Hamilton circuit** is a Hamilton path that begins and ends at the same vertex.	Discussion, p. 153 Example 1, p. 154
	In a **complete graph**, every pair of vertices is joined by an edge. A complete graph with n vertices is denoted by K_n and has $(n-1)!$ Hamilton circuits. A **weighted graph** has numbers assigned to its edges. The **weight of a path** is the sum of the weights of the edges of the path.	Example 2, p. 155 Discussion, p. 156 Discussion, p. 157
	The **brute force algorithm** for solving the TSP is as follows: Step 1. List all the Hamilton circuits in the graph. Step 2. Find the weight of each circuit found in step 1. Step 3. The circuits with the smallest weights are solutions to the TSP.	Example 4, p. 157
	The **nearest neighbor algorithm** for solving the TSP is as follows: Step 1. Start at any vertex X. Step 2. Choose the edge with the smallest weight that is connected to X. (There may be several edges with the smallest weight.) Select the vertex at the other end of this edge. This vertex is called the *nearest neighbor* of X. Step 3. Choose subsequent *new* vertices as you did in step 2, by considering the weights of the edges connected to the current vertex. Step 4. After all vertices have been chosen, return to the starting vertex.	Example 5, p. 159
	The **best edge algorithm** for solving the TSP is as follows: Step 1. Begin by choosing the edge with the smallest weight. Step 2. Keep choosing *new* edges as in step 1. However, do not let a circuit to be formed until all vertices have been used, and never allow three edges to be joined to the same vertex.	Example 6, p. 160
SECTION 4.3	An edge in a graph that is given a direction is called a **directed edge**. A graph with all directed edges is called a **directed graph**.	Discussion, p. 164 Example 1, p. 164
	If we can begin at vertex X in a directed graph and follow a sequence of edges in the direction indicated to reach vertex Y, that path is called a **directed path**.	Example 2, p. 165
	We use directed graphs as **models** when the relationship between objects is not necessarily exerted in both directions. This is the case with influence, dominance, and disease transmission.	Example 3, p. 165 Example 4, p. 167
SECTION 4.4	We can use a directed graph called a **PERT diagram** to schedule the order of tasks in a complex project.	Discussion, p. 171
	A **critical path** for task T is a path from "Begin" to T that requires the most time. The time required before we can begin task T tells us when to schedule T.	Examples 1 and 2, pp. 173, 174

CHAPTER REVIEW EXERCISES

Section 4.1

1.

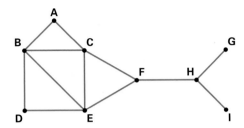

Use the preceding graph to answer the following questions.

a. How many edges does the graph have?

b. Which vertices are odd? Which are even?

c. Is the graph connected?

d. Does the graph have any bridges?

2. Explain how graphs are used to model a collection of objects in which some of the objects are related to each other. Give an example.

3. Which of the following graphs can be traced? Explain your answer by referring to Euler's theorem.

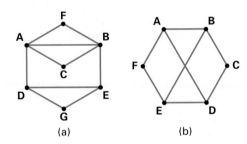

4. Use Fleury's algorithm to find an Euler circuit in the following graph. Describe the circuit by listing the vertices on the path.

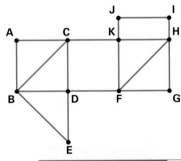

5. Model the following street map with a graph and design an efficient way of traveling through all the streets so a minimal number of streets are traveled more than once.

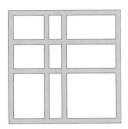

6. Model the following "map" of countries A, B, C, . . . , M by a graph and then devise a way of coloring the map using a minimal number of colors.

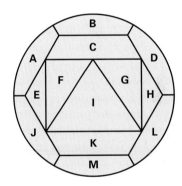

7. Allison, Branden, Colin, Donny, Erica, Jim, Kami, Lance, Marshall, and Nicole are competing on a whitewater rafting trip in the Amazing Race and will be traveling in several canoes. In considering a variety of factors such as weight, skill level, etc., the trip leaders have decided that certain people should not be in the same canoe as certain other people. The table below shows who should not travel together. We will abbreviate using the first letter of each name and will not list duplicate information; for example, if A cannot travel with B, we will not also list that B cannot travel with A. How many canoes are needed? Develop a plan as to who can travel together.

Person	A	B	C	D	E	K	L	M
Cannot Travel With	B, J, D	D, J	K, L, E	J	K, L	D, L	M, N	N

Section 4.2

8. Find all Hamilton circuits that begin at vertex A in K_5 and pass next through vertex B.

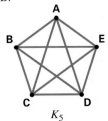

K_5

Use the following weighted graph to answer Exercises 9–11. There may be several possible correct answers. We will provide only one in the answer key.

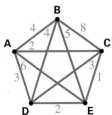

9. Use the brute force algorithm to find a Hamilton circuit that has minimal weight.

10. Use the nearest neighbor algorithm to find a Hamilton circuit that begins at vertex A.

11. Use the best edge algorithm to find a Hamilton circuit that begins at vertex A.

Section 4.3

12. Use the following directed graph to find the requested items, if possible. If it is not possible to find an item, explain why not.

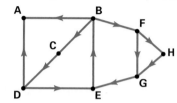

    a. A directed path from C to B

    b. A directed path of length greater than 5 from C to B

    c. A directed path from H to F

    d. Two different directed paths from F to B

13. When are directed graphs rather than nondirected graphs used as models?

14. A student action committee has been formed on your campus to lobby local legislators on social justice issues such as poverty, discrimination, environmentalism, taxation, diversity, etc. The influence exerted among committee members is listed in the following table. Use a graph to determine one- and two-stage influence among these people and determine who is the most influential committee member.

Committee Member	Committee Members Influenced by This Person
Pham	Stein, Vaccaro, Alvarez
Vaccaro	Alvarez
Stein	Vaccaro, Bartkowski
Robinson	Stein, Alvarez
Bartkowski	Vaccaro
Alvarez	Stein, Bartkowski

Section 4.4

15. Use the following PERT diagram to answer the following questions.

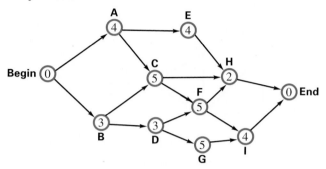

    a. Find a critical path for task H.

    b. What is a critical path for "End"?

    c. When should task F be scheduled?

    d. How many days are required for the whole project?

16. Congratulations! You are planning your wedding. The tasks and their dependencies for this happy event are listed in the following table. Develop a schedule for these tasks.

Task	Preceding Tasks	Time Required
1. Decide who will finance wedding	none	2
2. Decide on date for wedding	none	2
3. Hire wedding planner	1, 2	3
4. Decide on wedding party	1, 2	2
5. Create guest list	1, 2	3
6. Decide on places for ceremony and reception	1, 3	2
7. Decide on music, florist, pictures	1, 3	3
8. Decide on menu	1, 3, 6	2

CHAPTER TEST

1.

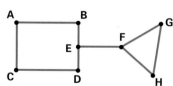

Use the preceding graph to answer the following questions.

    a. How many edges does the graph have?

    b. Which vertices are odd? Which are even?

    c. Is the graph connected?

    d. Does the graph have any bridges?

2. Which of the following graphs can be traced? If a graph cannot be traced, explain what part of Euler's theorem fails.

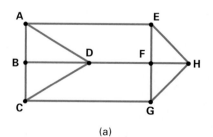

(a)

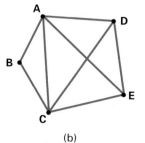

(b)

3. Use Fleury's algorithm to find an Euler circuit in the following graph beginning at vertex A. Describe the circuit by listing its vertices.

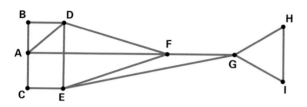

4. Find all Hamilton circuits that begin at vertex A in K_5 and pass next through vertex D.

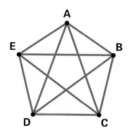

5. Model the following street map with a graph and design an efficient way of traveling through all the streets so that a minimal number of streets are traveled more than once.

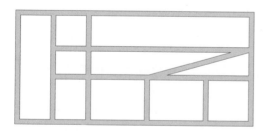

6. Model the following "map" of countries A, B, C, . . . , H by a graph and then devise a way of coloring the graph using a minimal number of colors.

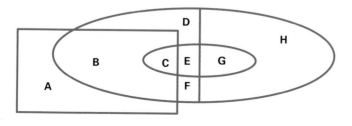

Use the following weighted graph to answer Exercises 7–9. Begin all circuits at vertex A. There may be several possible correct answers. We will provide only one in the answer key.

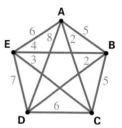

7. Use the brute force algorithm to find a Hamilton circuit that has minimal weight.

8. Use the nearest neighbor algorithm to find a Hamilton circuit that has minimal weight.

9. Use the best edge algorithm to find a Hamilton circuit that has minimal weight.

10. Use the given directed graph to find the following items. If it is not possible, explain why not.

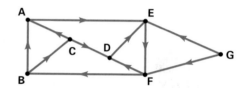

  a. A directed path from B to F

  b. A directed path from C to G

11. Use the given PERT diagram to answer the following questions.

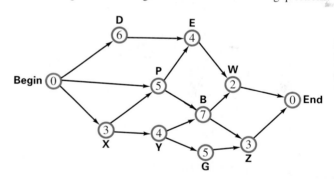

  a. Find a critical path for task Z.

  b. What is a critical path for "End"?

  c. When should task W be scheduled?

  d. How many days are required for the whole project?

12. The following graph models one- and two-stage influence among a group of people: A, B, C, . . . , F. Find the person who has the most influence.

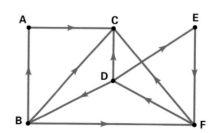

GROUP PROJECTS

1. As a group, pick five locations either on campus or in town and measure the distance between each pair of locations. If the locations are within walking distance, you can use a pedometer to measure the distance. If the locations are more distant, you can use the odometer on your car. Then solve the traveling salesman problem using the brute force, the nearest neighbor, and the best edge algorithms.

2. As a group, discuss the idea of "six degrees of separation."

   a. Have a contest to see who has the smallest degrees of separation from well-known personalities.

   b. Try to determine the degrees of separation among different members of the class (other than the obvious one, because you all are taking the same class). For example, perhaps you know someone outside of class, Jerome, who plays in the band and another band member, Jamie, is in your class. That would make your degree of separation two from Jamie.

3. Perform a search of the Internet for "traveling salesman problem applets." You will be able to find sites with interactive programs devoted to the TSP. You may want to search separately or as a group to find a site that you can use easily. Some sites, although they have programs that will give solutions, are difficult to understand. As a group, you may make more sense out of what they are doing than if you work alone. Some sites will allow you to vary the algorithms that are used to solve the problem. Finally, agree on which is the best site and write a brief explanation of what the site does. You may want to present your findings to your class.

*W*e begin this chapter by listening in on a conversation between Bart and Lisa Simpson . . .

Lisa: Remember, Bart, mom said you can't watch Krusty until you learn your number facts. I'll bet you haven't even started yet.

Bart: Lighten up Lis—I'm already done.

Lisa: Really? Prove it.

Bart: Really, Sis! Ask me anything.

Lisa: OK. What's $7 + 5$?

Bart: No Problemo—14. Gimme another.

Lisa: OK, so you got one right. What's 4×7?

Bart: Another easy one—34. . . .

It's hard to believe, but Bart did study and he got both answers right. You see, Bart lives in Springfield, where cartoon characters have

(continued)

four fingers on each hand, so it's likely that their arithmetic would be based on counting with eight fingers rather than ten, as ours is. Surprisingly, eight-finger arithmetic and several of its cousins play a large role in designing today's digital devices such as iPhones, satellite radios, and DVD players.

In addition to learning how to do arithmetic Bart's way, you will travel back in time to study Babylonian, Chinese, and Greek mathematics, where you will see the beginnings of elegant ideas that thousands of years later have blossomed into the Hindu-Arabic system that works so beautifully for us today. By studying the Hindu-Arabic system, you will understand why the arithmetic that you learned as a child works as well as it does.

Finally, we will show you how to do arithmetic on strangely numbered clocks, which has serious applications in situations where we need to send secure coded messages, such as in banking and the military. ●

5.1 The Evolution of Numeration Systems

Objectives

1. Represent numbers and compute in the Egyptian system.
2. Understand the Roman numeration system.
3. Understand the Chinese numeration system.

We can easily take a brilliant idea for granted because of its inherent simplicity. Such is the case with our familiar system for representing numbers that is the result of thousands of years of development culminating in the Hindu-Arabic system that we use today. In this section, we will examine some of the early attempts made by various cultures to develop practical numeration systems.

Primitive societies needed only simple counting systems. An early sheepherder might have scratched tally marks in the dirt, tied knots in a vine, cut notches in a tally stick, or matched a pile of small pebbles with his sheep in order to keep track of them. Such early counting methods as these eventually led to the invention of the abstract concept of number. The number 10 could represent 10 sheep, 10 pebbles, or 10 people.

> **DEFINITIONS** A **number** tells us how many objects we are counting; a **numeral** is a symbol which represents a number.

Although the *number* ten would mean the same amount in each culture, the *numeral* for ten (the symbols we use to write the number) varied greatly. For example, Babylonians used the symbol ❬, Egyptians used ∩, Greeks used the letter iota ɩ, Romans used X, and we, of course, represent ten by the symbols 10.

We will first discuss the simple grouping systems used by the Egyptians and Romans and then explain the multiplicative grouping system of the Chinese, which has some similarities to our current Hindu-Arabic system that we discuss in Section 5.2.

The Egyptian Numeration System

The Egyptian hieroglyphic numeration system, which is more than 5,500 years old, is an example of a **simple grouping** system. In the Egyptian system, we form numerals by

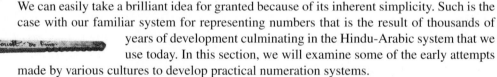

✏️ **KEY POINT**

In a simple grouping system, we represent a number as the sum of the values of its numerals.

Number	Symbol	Name
1	l	Stroke
10	∩	Heel bone
100	9	Scroll
1,000	⌠	Lotus flower
10,000	⟋	Pointing finger
100,000	⌒	Fish or tadpole
1,000,000	𝓧	Astonished person

TABLE 5.1 Egyptian numeration symbols.

combining the symbols that represent various powers of 10, as shown in Table 5.1. The value of the numeral is the sum of the values of the numerals. For example, we would write 325 as follows:

999∩∩lllll.

/ | \
3   2   5
hundreds tens ones

Because the order of the symbols is not important, we could also write 325 as

ll99∩9∩lll.

EXAMPLE 1   *Converting between Egyptian and Hindu-Arabic Notation*

a) Convert ⌠⌠⌠9999∩∩lllll to Hindu-Arabic notation.

b) Write 1,230,041 in Egyptian notation.

SOLUTION:

a) We have 3 thousands, 4 hundreds, 2 tens, and 5 ones, so this Egyptian numeral represents

$$3,000 + 400 + 20 + 5 = 3,425.$$

b) We will represent 1,000,000 by 𝓧, 200,000 by ⌒⌒, 30,000 by ⟋⟋⟋, 40 by ∩∩∩∩ and 1 by l, to get

𝓧⌒⌒⟋⟋⟋∩∩∩∩l. ✸ ❶

Quiz Yourself ❶ *

a) Write the Egyptian numeral
⟋⟋⌠999∩∩∩∩∩llllll in
Hindu-Arabic notation.

b) Write the number 241,536
using Egyptian notation.

As you can see, the Egyptian system is not an efficient way to represent numbers. The Egyptian numeral lllllllll requires nine times as many symbols as we use to represent nine using the Hindu-Arabic system. The number 68 requires 14 symbols, ∩∩∩∩∩∩llllllll. Imagine how awkward it would be to balance your checkbook or total your restaurant bill using this system.

It is straightforward to add and subtract in the Egyptian hieroglyphic system.

EXAMPLE 2   *Adding in the Egyptian System*

Add 999∩∩∩∩∩∩∩ll and 99∩∩∩∩∩lll.

SOLUTION: We add these two numbers by simply grouping all the symbols together (the addition sign is not part of the Egyptian notation). After obtaining the sum, we regroup 10 heel bones as one scroll.

999∩∩∩∩∩∩∩ll
+ 99∩∩∩∩∩lll
───────────────
99999∩∩∩∩∩∩∩∩∩∩∩∩lllll

⌐ Rewrite 10 heel bones as one scroll.

We can now rewrite this answer as

999999∩∩lllll. ✸

✳ ✳ ✳ HISTORICAL HIGHLIGHT

Solving the Mystery of Egyptian Hieroglyphics*

In 1798, the French emperor Napoleon sailed with a large army to conquer Egypt and disrupt the lucrative trade routes to India. Although he was severely defeated, this military disaster turned out to be a scientific triumph for Europe. Napoleon had taken scholars with him to study Egypt's culture, and when they returned, they brought back a wealth of information about this ancient civilization. Unfortunately, much of the material was written in a form of hieroglyphics called **demotic script**, which no one was able to translate.

Fortunately, Napoleon also brought back with him the key to solving this puzzle—a polished stone, called the **Rosetta stone** that contained writing in Greek, demotic script, and ancient hieroglyphics. Scholars believed that the stone contained three versions of the exact same text and set about using their knowledge of Greek to translate the other two cryptic sections.

After returning from Egypt with Napoleon, the French mathematician Jean-Baptiste Fourier showed some

hieroglyphics to an 11-year-old boy named Jean Francois Champollion. When Fourier stated that no one could read hieroglyphics, the boy replied boldly, "I will do it when I am older," and from then on he dedicated his life to translating hieroglyphics. Legend has it that when Champollion finally solved the mystery of hieroglyphics, he exclaimed, "I've got it," and fainted.

Much of our knowledge of early Egyptian mathematics comes from the Rhind papyrus, named after a Scotsman named A. Henry Rhind, who purchased it in 1858. The scroll was written by a scribe named Ahmes in 1650 BC who claimed that the document contained "a thorough study of all things, insights into all that exists, [and] knowledge of all obscure secrets." This was an exaggerated promise, because when the papyrus was translated, it was found to contain only a collection of mathematical exercises and rules for doing multiplication and division.

Quiz Yourself ②

Perform the following operations using Egyptian notation:

a) 999∩∩∩∩||||| +
   99999999∩∩|||||

b) 𓏤𓏤𓏤𓏤𓏤∩∩∩∩|| −
   𓏤𓏤9999∩∩|||||||||

Power of 2	Value
2^0	1
2^1	2
2^2	4
2^3	8
2^4	16
2^5	32
2^6	64

TABLE 5.2 Powers of 2.

EXAMPLE 3   *Subtracting Using Egyptian Notation*

Subtract ∩∩∩∩∩∩|| from 99∩∩∩||||| using Egyptian notation.

SOLUTION: We have the reverse problem from that of Example 1. Although we can subtract two strokes from four strokes, we cannot subtract six heel bones from three heel bones unless we "borrow." That is, we convert one scroll to 10 heel bones.

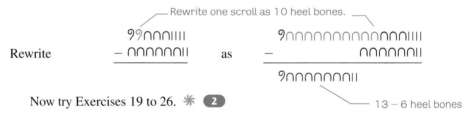

Rewrite one scroll as 10 heel bones.

Rewrite   99∩∩||||| − ∩∩∩∩∩∩|| as   9∩∩∩∩∩∩∩∩∩∩∩∩∩||||| − ∩∩∩∩∩∩||

9∩∩∩∩∩∩∩||

13 − 6 heel bones

Now try Exercises 19 to 26. ✳ ②

Notice in Examples 2 and 3 how the Egyptian calculations remind us of the way we carry and borrow in performing addition and subtraction using Hindu-Arabic notation.

⊙ *Some Good Advice*

You can check your work in performing Egyptian arithmetic by converting all the numbers to Hindu-Arabic notation and then redoing the computations.

The Egyptians also had a method for doing multiplication, as we show in Example 4.[†] This method is based on the fact that any positive integer can be written as the sum of powers of 2 (see Table 5.2). For example, $19 = 1 + 2 + 16$ and $81 = 1 + 16 + 64$.

---

*This highlight is based on David M. Burton, *The History of Mathematics: An Introduction* (New York: WCB-McGraw Hill, 1999), pp. 31–35.
[†]In explaining this doubling method, we will use Hindu-Arabic numerals rather than the cumbersome Egyptian numerals.

EXAMPLE 4   *Computing Area Using the Egyptian Method of Doubling*

A craftsman is restoring a rectangular wall of the Temple of Queen Hatshepsut (the fifth pharaoh of the Eighteenth Dynasty of Ancient Egypt) with tiles. If the wall measures 13 feet by 21 feet, use the Egyptian method of doubling to determine how many square feet must be covered.

SOLUTION: First we write 13 as $1 + 4 + 8$. To find the area, we will calculate 13×21 as follows:

$$13 \times 21 = (1 + 4 + 8) \times 21* = 1 \times 21 + 4 \times 21 + 8 \times 21 = 21 + 84 + 168 = 273.$$

We find these products by repeatedly doubling 21 in Table 5.3.

So, the craftsman must use 273 square feet of tile for the temple wall.

Now try Exercises 27 to 32. ✳ ③

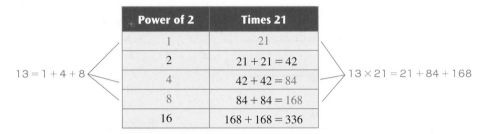

$13 = 1 + 4 + 8$

Power of 2	Times 21
1	21
2	$21 + 21 = 42$
4	$42 + 42 = 84$
8	$84 + 84 = 168$
16	$168 + 168 = 336$

$13 \times 21 = 21 + 84 + 168$

TABLE 5.3 Powers of 2 times 21.

Quiz Yourself ③

a) Represent 25 as the sum of powers of 2.

b) Use the Egyptian method of doubling to calculate 25×43.

 **KEY POINT**

The Roman numeration system is a more sophisticated simple-grouping numeration system.

Number	Roman Numeral
1	I
5	V
10	X
50	L
100	C
500	D
1,000	M

TABLE 5.4 Roman numerals.

The Roman Numeration System

The Roman numeration system, which was developed between 500 BC and 100 AD, has several improvements over the Egyptian system. The Romans used letters of the alphabet as numerals to represent certain numbers, as we show in Table 5.4.

The first advantage that the Roman system had over the Egyptian system was that the Romans used a subtraction principle that allowed them to represent numbers more concisely than the Egyptians could.

EVALUATING ROMAN NUMERALS In Roman notation, we add the values of the numerals from left to right, provided we never have a numeral with a smaller value than the numeral to its right.

For example, DCLXXVIII has the value

$$500 + 100 + 50 + 10 + 10 + 5 + 1 + 1 + 1 = 678.$$

Notice that the values of the numerals either stay the same or decrease, but never increase.

However, *if the value of a numeral is ever less than the value of the numeral to its right,* then the value of the left numeral is subtracted from the value of the numeral to its right. For example,

IV represents $5 - 1 = 4$,

IX represents $10 - 1 = 9$,

XL represents $50 - 10 = 40$,

and CM represents $1,000 - 100 = 900$.

---

*We are using the property that multiplication distributes over addition here.

✻ ✻ ✻ HISTORICAL HIGHLIGHT

The Algorists versus the Abacists*

In 1299, the rulers of Florence, Italy, passed a law forbidding bankers from using Hindu-Arabic numerals (our commonly used numerals 1, 2, 3, . . .). Although the more efficient Hindu-Arabic numerals had been introduced to Europe 500 years earlier, they were still outlawed as late as the fourteenth century.

To understand why such a law might be necessary, we first must understand how computations were done using Roman numerals. Archaeologists have found stone and marble counting boards called abaci† (plural of *abacus*), which were tablets with four grooves cut into them, as we show in Figure 5.1.

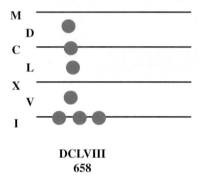

FIGURE 5.1 Roman abacus Showing 658.

The grooves corresponded to 1, 10, 100, and 1,000, and the spaces represented the intermediate values, 5, 50, and 500. The **abacist** (person computing with an abacus) would place small stones or metal counters in the grooves and spaces to represent Roman numerals, as we show in Figure 5.1. Then computations would be done by sliding stones back and forth, replacing two intermediate values with one higher value. For example, the abacist would replace two counters in the 5 space with one counter in the 10 groove. It was easy to understand these calculations in which counters were moved about and regrouped on the board.

In contrast, **algorists** did Hindu-Arabic calculations with pen and paper—but people were suspicious of this new method in which one ink mark could represent two, three, or even four counters. To make matters worse, algorists had a symbol that stood for no counters at all! In medieval times, the notion of zero was confusing to many people and not necessary to those who used Roman numerals. Furthermore, unscrupulous merchants and bankers could cheat uneducated customers by changing a Hindu-Arabic numeral such as a 0 to a 6 or a 1 to a 4.

Eventually, because of the efficiency of Hindu-Arabic notation, the algorists won the battle, which is probably why you do not calculate with counting boards and pebbles today.

There are two restrictions on this subtraction principle:

1. We can only subtract the numerals I, X, C, and M. For example, we cannot use VL to represent 45.

2. We can only subtract numerals from the next two higher numerals. For instance, we can only subtract I from V and X; therefore, we cannot use IC to represent 99.

Notice that the subtraction principle allows us to write multiples of 4 and 9 more efficiently than we could using Egyptian notation. For example, instead of writing 99 as LXXXXVIIII, we can write it as XCIX, saving six symbols.

EXAMPLE 5 *Translating between Roman Numerals and Hindu-Arabic Notation*

a) Convert MCMXLIII to Hindu-Arabic notation.

b) Write 492 in Roman numerals.

SOLUTION:

a) The following diagram shows how to interpret this numeral. Remember, every time we see a smaller numeral to the left of a larger numeral, we must subtract.

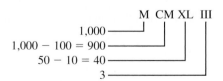

---

*This highlight is based on the article by Barbara E. Reynolds, "The Algorists vs. the Abacists: An Ancient Controversy on the Use of Calculators," *The College Mathematics Journal*, Vol. 24, No. 3, May 1993, pp. 218–223.

†Chinese and Japanese abaci were invented much later and are much different from the Roman ones. We will discuss these other abaci in Section 5.2.

Quiz Yourself ④

a) Identify two errors made in writing the Roman numeral LDIL.

b) Convert DXLVIII to Hindu-Arabic notation.

c) Convert |IX| and $\overline{V}$ to Hindu-Arabic notation.

Thus, MCMXLIII represents $1{,}000 + 900 + 40 + 3 = 1{,}943$.

b) You can write 400 as CD, 90 as XC, and 2 as II to express 492 as CDXCII.

A second advantage the Roman system had over the Egyptian system was that it used a multiplication principle that is similar to those found in more advanced numeration systems, such as the Chinese system that you will study next and also our Hindu-Arabic system.

In the Roman system, a bar above a symbol means to multiply the value of the symbol by 1,000. Also, bracketing a symbol by two vertical lines multiplies the value of the symbol by 100.* Thus, $\overline{X}$ represents $10 \times 1{,}000 = 10{,}000$, |V| represents $5 \times 100 = 500$, and $|\overline{L}|$ represents $50 \times 1{,}000 \times 100 = 5{,}000{,}000$.

Now try Exercises 33 to 54. ✳ ④

The Chinese Numeration System

The traditional Chinese numeration system uses multiples of powers of 10 such as 10, 100, and 1,000 to represent numbers and is based on the notation shown in Table 5.5. These symbols originated during the Han dynasty, which extended from 206 BC to 220 AD. Thus, these symbols have not changed much during the past 2,000 years.†

This system is an example of a **multiplicative system** because we form numerals by writing products of integers between 1 and 9, inclusive, and powers of 10. Traditionally, the ancient Chinese wrote numerals vertically. For example, to write 300, they would write

 ⟸ ← 3
 百 ← times 100,

and they would write 5,000 as

 ﾋ ← 5
 千 ← times 1,000.

However, the Chinese now write these symbols horizontally, and we will also follow this modern practice. (There are also other features of modern Chinese notation that we will not go into here but will discuss in the exercises.) Therefore, we will write 300 as ⟸百 and 5,000 as ﾋ千.

KEY POINT

The Chinese numeration system is a multiplicative numeration system.

Chinese Numeral	Value
～	1
⟝	2
⟸	3
⊖	4
ﾋ	5
六	6
七	7
入	8
ﾉ	9
十	10
百	100
千	1,000

TABLE 5.5 Chinese numerals.

EXAMPLE 6   *Translating from Chinese to Hindu-Arabic Notation*

Express each of the following in Hindu-Arabic notation.

a) ﾉ百⊖十⟝   b) ﾋ千⟝十入

SOLUTION:

a) As you can see in Figure 5.2(a), the first two symbols represent the product of 9 and 100, or 900. The second two symbols represent 4 times 10, or 40, and the last figure represents 2. Thus, the number we are representing is 942.

Notice how we write the units numeral without multiplying by a power of 10.

ﾉ百   ⊖十   ⟝   ﾋ千   ⟝十   入
9×100 4×10 2 $5 \times 1{,}000$ 2×10 8
(a)   (b)

FIGURE 5.2 Translating Chinese numerals.

---

*See Jan Gullberg, *Mathematics: From the Birth of Numbers* (New York: W. W. Norton & Company, 1997), p. 39.
†The Chinese numeration system that we describe here was used for representing numbers (much as we use Roman numerals to number the Super Bowls) but *not* for doing calculations. Another Chinese system, called a "rod system," which we will describe in the group project exercises in the Chapter Test, was actually used as early as the thirteenth or fourteenth centuries to do computations. Some believe that this early rod system was adopted by the Arabs and Indians and led to our current Hindu-Arabic numeration system.

b) As we see from Figure 5.2(b), this numeral represents 5 times 1,000, plus 2 times 10, plus 8, or 5,028.

Now try Exercises 55 to 60. ❀

EXAMPLE 7   *Translating from Hindu-Arabic to Chinese Notation*

Write the following numbers using Chinese notation:

a) 694   b) 9,056

SOLUTION:

a) To express this number, we need 6 hundreds, 9 tens, and 4 units. We write this as

六百 九十 四

$$6 \times 100 + 9 \times 10 + 4.$$

Again, in using this notation, we do not multiply the units symbol, 4, by a power of 10. Also, we have separated the groups of symbols for emphasis.

b) Here, we need 9 thousands, 5 tens, and 6 units. We write this as

九千五十六.

Now try Exercises 61 to 66. ❀

The Chinese did not have a symbol for zero in the traditional notation, nor did they use the concept of place value, which we will discuss in Section 5.2. Therefore, when the Chinese wrote 六 (6), they had to specify whether they meant 6 tens, 6 hundreds, or 6 thousands, etc., which required them to use an extra symbol. As you will see in Section 5.2, the concept of place value allows us to avoid these extra symbols and express numbers more efficiently.

Exercises 5.1

Looking Back*

These exercises follow the general outline of the topics presented in this section and will give you a good overview of the material that you have just studied.

1. List four different numerals that represent the number ten.

2. How did we perform what we call "carrying" in addition in Example 2?

3. How did we perform what we call "borrowing" in subtraction in Example 3?

4. How does Example 6 illustrate that the Chinese system is a multiplicative system?

5. What document gives us much of our knowledge about Egyptian mathematics?

6. Why were people in Europe at first suspicious of Hindu-Arabic numerals?

Sharpening Your Skills

Write the Egyptian numerals using Hindu-Arabic numerals.

7. 999∩∩IIII

8. ∩9999∩∩II

9. ℓℓ𐤊999∩∩IIII

10. ⌒⌒ℓℓℓ𐤊𐤊9∩∩III

11. 𝍩𝍩⌒ℓℓ𐤊99∩

12. 𝍩⌒⌒ℓℓℓ99∩III

Write each Hindu-Arabic numeral using Egyptian numerals.

13. 245   14. 362   15. 3,245

16. 23,416   17. 245,310   18. 2,036,042

Perform each of the following addition problems using Egyptian notation.

19. 𓍢𓍢𓍢𓎆𓎆𓏺𓏺𓏺𓏺𓏺 + 𓍢𓍢𓍢𓍢𓍢𓍢𓍢𓍢𓎆𓎆𓏺𓏺𓏺𓏺𓏺𓏺𓏺𓏺𓏺

20. 𓍢𓎆𓎆𓎆𓎆𓎆𓎆𓎆𓎆𓏺𓏺𓏺𓏺𓏺𓏺 + 𓍢𓍢𓍢𓍢𓍢𓍢𓎆𓎆𓎆𓎆𓏺𓏺𓏺𓏺𓏺𓏺𓏺

21. 𓁨𓂭𓂭𓂭𓂭𓆼𓍢𓍢𓎆 + 𓁨𓆐𓂭𓂭𓂭𓂭𓂭𓂭𓂭𓍢

22. 𓆼𓆼𓆼𓆼𓆼𓍢𓍢𓍢𓍢𓍢𓍢 + 𓂭𓆼𓆼𓆼𓆼𓆼𓆼𓍢𓍢𓍢𓍢𓍢𓎆

Perform each of the following subtraction problems using Egyptian notation.

23. 𓍢𓍢𓍢𓍢𓍢𓍢𓎆𓎆𓏺𓏺𓏺𓏺𓏺𓏺𓏺𓏺 − 𓍢𓍢𓍢𓎆𓎆𓎆𓎆𓎆𓏺𓏺𓏺𓏺

24. 𓆼𓍢𓍢𓎆𓎆𓎆𓏺𓏺𓏺 − 𓍢𓍢𓍢𓎆𓎆𓏺𓏺𓏺𓏺

25. 𓂭𓂭𓂭𓍢𓍢𓍢𓍢𓍢 𓂭𓂭𓆼𓆼𓆼𓆼𓍢𓍢𓍢𓎆𓎆𓎆𓎆𓎆𓏺𓏺𓏺𓏺

26. 𓁨𓆐𓂭𓂭𓂭 − 𓁨𓂭𓂭𓆼

Use the Egyptian method of doubling to calculate the following products.

27. 14 × 43 **28.** 11 × 57 **29.** 12 × 107

30. 21 × 126 **31.** 35 × 121 **32.** 64 × 103

Write each Roman numeral using Hindu-Arabic numerals.

33. LIX **34.** LXI

35. DLXIV **36.** CLXIX

37. MCMLXIII **38.** MDCXXXVI

39. $\overline{\text{V}}$MMDXLIV **40.** $\overline{\text{X}}$MMMCDLIV

41. |D|CCLXII **42.** |M|DLVII

43. |$\overline{\text{V}}$|MCDXX **44.** |$\overline{\text{L}}$|MMMDX

Write each numeral in Roman notation. (There may be several correct answers.)

45. 44 **46.** 96

47. 278 **48.** 947

49. 444 **50.** 999

51. 4,795 **52.** 3,247

53. 89,423 **54.** 98,546

Write each Chinese numeral as a Hindu-Arabic numeral. (Recall that we are writing Chinese numerals horizontally rather than in the traditional vertical form.)

55. 四百三十六 **56.** 八千五百二十五

57. 五千六十七 **58.** 四千三百四

59. 九千九百九十九 **60.** 四千九十七

Write each numeral using Chinese numerals.

61. 495 **62.** 726

63. 2,805 **64.** 3,926

65. 9,846 **66.** 8,054

Applying What You've Learned

67. The great pyramid at Giza was completed in 𓆼𓆼𓍢𓍢𓍢𓍢𓍢𓍢𓎆𓎆𓎆𓎆𓎆 BC. Write this date using Hindu-Arabic notation.

68. Cheops, the builder of the Great Pyramid at Giza. died in 𓆼𓆼𓍢𓍢𓍢𓍢𓍢𓎆𓎆𓏺𓏺𓏺𓏺𓏺𓏺𓏺𓏺 BC. Write this date using Hindu-Arabic notation.

69. An Egyptian merchant has a warehouse that contains 𓆐𓆐𓂭𓂭𓂭𓆼𓆼𓆼𓆼𓍢𓍢𓍢𓍢𓍢𓎆𓎆𓎆𓎆 square feet of storage. If he purchases another warehouse containing 𓆐𓆼𓆼𓎆𓎆𓎆𓎆𓎆𓎆𓎆𓎆 square feet of storage, what is the total square feet of storage that he now has? Do the calculation using Egyptian notation and double-check your answer using Hindu-Arabic notation.

70. An ancient Egyptian merchant had on hand 𓂭𓆼𓆼𓆼𓆼𓍢𓍢𓍢𓎆𓎆𓎆𓎆𓎆 bushels of wheat and sold 𓂭𓆼𓆼𓍢𓍢𓍢𓍢𓍢𓎆𓎆𓎆 to another merchant. How many bushels of wheat does he have remaining? Do the calculation using Egyptian notation and double-check your answer using Hindu-Arabic notation.

Frequently, Roman numerals are used today in movie credits to specify the date that a movie was released. Translate each of the following movie dates into Roman notation.

71. *Gone with the Wind* (1939)

72. *On the Waterfront* (1954)

73. *Forrest Gump* (1994)

74. *The Bourne Ultimatum* (2007)

75. The oldest discovery of Chinese written numerals is from the Yin dynasty (1523–1027 BC). Express these dates using Chinese numerals.

76. When Marco Polo visited China in 1274, he was impressed by the grandeur of the court of Kublai Khan. Express this date in Chinese numerals.

Communicating Mathematics

77. What is the difference between "number" and "numeral?"

78. Explain two advantages of the Roman numeration system over the Egyptian system.

79. Why is there no need for zero in the traditional Chinese system?

80. The modern Chinese numeration system has a symbol for zero. Give an example of how that would allow the Chinese to write some of their numerals more efficiently.

81. Research the Ionic Greek numeration system, which is an example of a ciphered numeration system. Write a paragraph explaining how this system works. What is one advantage and one disadvantage of this system?

Using Technology to Investigate Mathematics

82. Experiment with the Microsoft Excel spreadsheet ROMAN function which converts Hindu-Arabic numerals to Roman numerals. For example, ROMAN(499,0) equals CDXCIX. If you change the 0 to 1, 2, 3, or 4, the form of the Roman numeral becomes increasingly simplified. For example, ROMAN(499,2) equals XDIX and ROMAN(499,4) equals ID. Practice using this function to convert several numerals using different forms.

83. What are some of the ways that options 1, 2, 3, and 4 for the ROMAN function violate the classical rules for writing Roman numerals?

84. Search the Internet to find a calculator that allows you to compute using Egyptian numerals. Use it to verify some of the computations in this section and report on your findings.

85. Search the Internet to find a Roman numeral calculator. Use it to verify some of the computations in this section and report on your findings.

86. Search the Internet to find a Web site that has an interactive abacus and use it to do some simple additions and subtractions.

For Extra Credit

87. In the Egyptian numeration system, whenever we have 10 of one symbol we regroup using a single symbol representing the next highest power of 10. For example, we replace 10 ∩s by one ૧. If we follow this rule, what is the largest number that you can express using the Egyptian notation that we discussed in this section? Write your answer using Hindu-Arabic notation.

88. Suppose that Egyptian numeration was based on 5 rather than 10. That is, ∩ would represent five strokes and ૧ would represent five ∩s, etc. In this case, what would be the largest number that you could express in this system? Write your answer in Hindu-Arabic notation.

89. Invent an Egyptian type of numeration system using the following symbols: √, ⊗, ∇, ∝, ≈, ↑, ◇, where these symbols represent units, tens, hundreds, and so on, respectively.

   a. Write 3,142 and 7,203 using this notation.

   b. Find 3,142 + 7,203 similar to the way we added in Example 2.

   c. Find 7,203 − 3,142 similar to the way we subtracted in Example 3.

90. Write the number 1,999 in Roman numerals as many ways as you can. (*Hint:* Realize that it is not mandatory to use the subtraction principle in writing Roman numerals.)

Place Value Systems

Objectives

1. Understand the Babylonian place value system.
2. Understand the Mayan numeration system.
3. Understand and compute in the Hindu-Arabic system.
4. Use galleys and Napier's rods to calculate.

What is the most important invention in the history of mathematics and science?

Before reading further, think about how you would answer that question. . . . When I ask my classes that question, I get a variety of answers, but never the one that I am looking for—the simple numeration system that we all learned in elementary school. We probably don't notice it because it works so well.

> **DEFINITION** In a **place value system**, also called a **positional system**, the placement of the symbols in a numeral determines the value of the symbols.

For example, in the numeral 35, the 3 represents three tens, or 30, and the 5 represents five ones, or 5. However, in the numeral 53, the 3 now represents 3 and the 5 represents 50.

If the Chinese numeration system that we discussed in Section 5.1 had used the concept of place value,* then it would have been possible to write numbers more

---

*The ancient Chinese also had another numeration system, called a **rod numeral** system, that was based on 10 and used place value. We discuss this in the group project exercises Chapter Test.

efficiently. For example, the number 543 in Figure 5.3(a) could have been written as in Figure 5.3(b).

5 × 100   +   4 × 10   +   3 × 1   100s   10s   1s

(a)   (b)

FIGURE 5.3 (a) Chinese notation (b) with the concept of place value.

With this modification, the rightmost symbol represents units, the symbol to its left represents tens, the next symbol to the left represents hundreds, etc. As you will see later, in order to have a true place value system, it also would have been necessary to invent a symbol for zero.

We will now discuss several place value systems, beginning with the ancient Babylonian system.

The Babylonian Numeration System

The Babylonians* had a primitive place value system that was based on powers of 60 and hence is called a *sexagesimal* system. There were two symbols in the Babylonian system: **❙**, which represented 1, and **❬**, which represented 10. These symbols were written in wet clay with wedge-shaped sticks, and when the clay hardened, there remained a permanent record of the calculations. To write small numbers, this system worked much like the Egyptian system.

For example, the number 23 could be written as **❬❬❙❙❙**. However, to represent larger numbers, the Babylonians used several groups of these symbols, separated by spaces, and multiplied the value of these groups by increasing powers of 60, as we illustrate in Example 1.

KEY POINT

The Babylonians developed an early example of a place value system.

EXAMPLE 1   *Converting Babylonian Numerals to Hindu-Arabic Numerals*

Convert **❙❙   ❬❬❬❙   ❬❬❙❙❙** to Hindu-Arabic notation.

SOLUTION: We interpret the three groups of numerals as follows: the right group represents units, the middle group represents 60s, and the left group represents 60^2s, as shown in Figure 5.4.

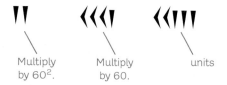

Multiply by 60^2.   Multiply by 60.   units

FIGURE 5.4 Babylonian notation.

Thus, this notation represents the number

$$2 \times 60^2 + 31 \times 60 + 23 = 2 \times 3{,}600 + 31 \times 60 + 23 = 9{,}083.$$ ❋

Because the early Babylonians did not have a symbol for zero, it could be hard to tell exactly how many spaces were between groups of symbols. For example, it could be difficult to determine whether **❙   ❬❙❙❙❙** represents $1 \times 60 + 14$ or $1 \times 60^2 + 14$.[†]

The Babylonians used the symbol **❡** to indicate subtraction. Thus, the numeral **❬❬❡❙❙❙** represents $20 - 3 = 17$.

---

*Ruins of ancient Babylon are located almost 60 miles south of modern-day Baghdad in Iraq. The Babylonian civilization lasted from 2000 BC to about 600 BC.

[†]A much later Babylonian numeration system did have a symbol for zero that avoids this problem.

To convert Hindu-Arabic numerals to Babylonian numerals, we must divide by powers of 60, similar to the way that we convert seconds to hours and minutes. For example, to convert 7,717 seconds to hours and minutes, we first divide by 3,600 (there are 3,600 seconds in 1 hour) to get the number of whole hours.

$$
\begin{array}{r}
2 \text{ hours} \\
3{,}600\overline{)7{,}717} \\
\underline{7{,}200} \\
517 \text{ seconds}
\end{array}
$$

Thus, we have 2 whole hours with 517 seconds left over. Now, we divide 517 by 60 to find the whole number of minutes.

$$
\begin{array}{r}
8 \text{ minutes} \\
60\overline{)517} \\
\underline{480} \\
37 \text{ seconds}
\end{array}
$$

So, we see that 7,717 seconds corresponds to 2 hours, 8 minutes, and 37 seconds.

EXAMPLE 2   *Converting from Hindu-Arabic to Babylonian Notation*

Write 12,221 as a Babylonian numeral.

SOLUTION: First divide by $3{,}600 = 60^2$, which is the largest power of 60 that will divide into 12,221.

$$
\begin{array}{r}
3 \text{ times } 60^2 \\
3{,}600\overline{)12{,}221} \\
\underline{10{,}800} \\
1{,}421 \text{ units left over}
\end{array}
$$

The quotient 3 tells us how many 60^2s are present in the number. Next, divide 1,421 by 60.

$$
\begin{array}{r}
23 \text{ times } 60 \\
60\overline{)1{,}421} \\
\underline{1{,}380} \\
41 \text{ units left over}
\end{array}
$$

The quotient 23 tells us that there are 23 60s in the number, and the remainder 41 tells us that there are 41 units left over. The number 12,221 can now be written as $3 \times 60^2 + 23 \times 60 + 41$. We write this in Babylonian notation as

𝌆𝌆𝌆   ≪𝌆𝌆𝌆𝌆   ≪≪≪≪𝌆

Now try Exercises 7 to 24. ❈ **5**

Quiz Yourself **5**

a) Convert **𝌆𝌆𝌆 ≪≪𝌆𝌆𝌆𝌆 ≪≪≪𝌆𝌆** to a Hindu-Arabic numeral.

b) Convert 7,573 to Babylonian notation.

You might wonder why the Babylonians chose such a strange base as 60 for their mathematical system. Some speculate that it was due to the fact that they did fraction calculations by combining unit fractions such as $\frac{1}{2}, \frac{1}{3}, \frac{1}{4}, \frac{1}{5}$, and so on. For example, they would write $\frac{7}{12}$ as $\frac{1}{3} + \frac{1}{4}$. The number 60 was a convenient number to use in these situations because it has many different divisors. In computing a sum such as $\frac{1}{12} + \frac{1}{5} + \frac{1}{10} + \frac{1}{15} + \frac{1}{15} = \frac{31}{60}$, the denominator 60 would frequently arise. Thus, using the base 60 made it easier for them to do fractional computations. You can see the influence of a base-60 system today in that we measure an hour as 60 minutes, or $3{,}600 = 60^2$ seconds. Also, in geometry, we have 360 degrees in a circle.

The Hindu-Arabic Numeration System

✎ **KEY POINT**

The Hindu-Arabic numeration system is a place value system based on 10.

The oldest known examples of Hindu-Arabic numerals were found on a stone column in India and are believed to have been written about 250 BC. That Hindu system had no symbol for zero, and scholars are not certain as to when our current Hindu-Arabic system became fully developed. The Persian mathematician al-Khowarizmi became aware of Hindu notation, and in 825, he wrote a book whose title translated into English means "The Book of al-Khowarizmi on Hindu Number." After this, Hindu notation was adopted by the Arabs, and in 1202, the Italian mathematician Leonardo Fibonacci, after studying in the Middle East, wrote a book on arithmetic and algebra, titled *Liber Abaci* (a Book on the Abacus), which helped spread Hindu-Arabic numerals throughout Europe.

The Hindu-Arabic system is a place value system based on 10, unlike some of the earlier systems you have studied. One feature of this numeration system is that we can write all numerals using only the digits* 0, 1, 2, . . . , 9. Thus we do not need special symbols for 10, 100, 1,000, etc., as did some of the earlier numeration systems. Also, the invention of zero, used as a place holder, makes it easy to distinguish between numbers such as 5,001, 501, and 51. Contrast this approach with that of the early Babylonians, who used spaces between groups of symbols.

We can write Hindu-Arabic numerals in *expanded form* to show explicitly how each digit is multiplied by a power of 10. For example,

$$6{,}582 = 6 \times 10^3 + 5 \times 10^2 + 8 \times 10^1 + 2 \times 10^0.$$

(Recall that $10^0 = 1$, $10^1 = 10$, $10^2 = 10 \times 10 = 100$, $10^3 = 10 \times 10 \times 10 = 1{,}000$, etc.)

EXAMPLE 3 *Writing Hindu-Arabic Numbers in Expanded Form*

a) Write 53,024 in expanded form.

b) Write $4 \times 10^3 + 0 \times 10^2 + 2 \times 10^1 + 5 \times 10^0$ using Hindu-Arabic notation.

SOLUTION:

a) $53{,}024 = 5 \times 10^4 + 3 \times 10^3 + 0 \times 10^2 + 2 \times 10^1 + 4 \times 10^0$

b) $4 \times 10^3 + 0 \times 10^2 + 2 \times 10^1 + 5 \times 10^0 = 4 \text{ thousands} + 0 \text{ hundreds} + 2 \text{ tens} + 5 \text{ ones}$
$= 4{,}025$

Now try Exercises 37 to 48. ❋

We can use expanded notation to explain the algorithms to perform the arithmetic operations that you learned in elementary school.

EXAMPLE 4 *Using Expanded Notation to Explain the Addition Algorithm*

Calculate $4{,}625 + 814$ using expanded notation.

SOLUTION: Using expanded notation, we can write this problem as

$$
\begin{array}{rl}
4{,}625 = & 4 \times 10^3 + 6 \times 10^2 + 2 \times 10^1 + 5 \times 10^0 \\
+\ \ 814 = +\ & \underline{\hspace{2.5cm} 8 \times 10^2 + 1 \times 10^1 + 4 \times 10^0} \\
5{,}439 = & 4 \times 10^3 + 14 \times 10^2 + 3 \times 10^1 + 9 \times 10^0
\end{array}
$$

We take 10 of the 10^2s and represent them as one 10^3 in the column to the left.

---

*The word *digit* is the Latin word for *finger*. Because we have 10 fingers, it is not surprising that many numeration systems are based on the number 10.

✸ ✸ ✸ HISTORICAL HIGHLIGHT

Mayan Mathematics and Astronomy

The Maya Indians lived on the Yucatan Peninsula in Central America from about 200 BC to 1540 AD. During the height of their culture, from about 290 to 925 AD, the Mayans made remarkable contributions to mathematics, astronomy, and art. They invented a numeration system based on the number 20. The system used bars and dots, with each bar representing five dots. The Mayans would count as shown in Figure 5.5.

After 19, the Mayans used positioning of symbols to represent larger numbers. They wrote numerals vertically, with the lowest position representing units, the next higher representing 20s, the next position representing 20×18, the next representing $20 \times 18 \times 20$, as shown in Figure 5.6.* They used the symbol ⬭, which looked somewhat like a seashell, to represent zero.

You may think it's strange that in Figure 5.6 we multiplied 14 by 20×18 rather than 20 squared. However, the Mayans had a good reason for doing this because their calendar consisted of 18 *uinals* ("months") of 20 days each, plus 5 extra days added on at the end of the year.

They had a remarkably good estimate of the length of a year—365.242000 days versus our current estimate of

•	$1 \times 20 \times 18 \times 20^2$	$= 144{,}000$
⬭	$0 \times 20 \times 18 \times 20$	$= \quad 0$
•••• ̿	$14 \times 20 \times 18$	$= \quad 5{,}040$
••• —	8×20	$= \quad 160$
☰	15	$= \quad 15$
		$\overline{\quad 149{,}215}$

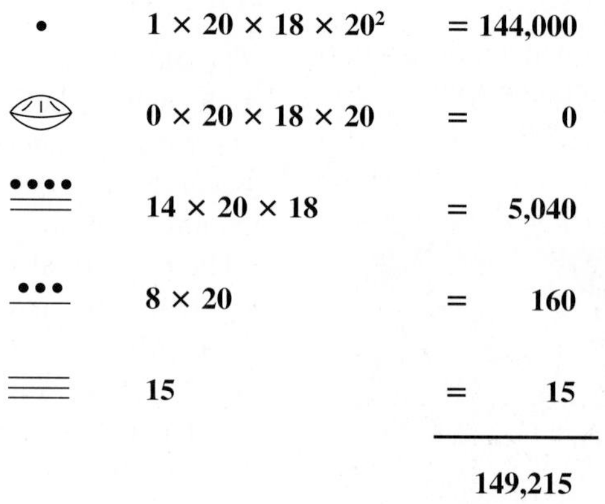

FIGURE 5.6 Representing a large number with Mayan notation.

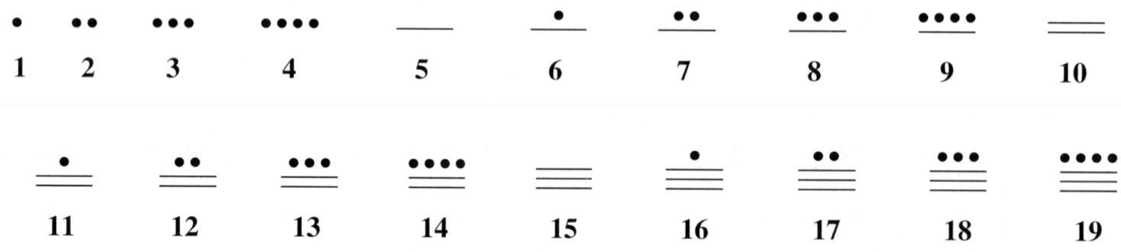

365.242198 days. Mayan astronomers also observed the movements of the sun, moon, and planets and accurately predicted solar eclipses. Their accomplishments were truly remarkable considering that the Mayans did not know how to make glass, so they did not have the advantage of telescopes as did later astronomers.

FIGURE 5.5 Counting with Mayan numerals.

The computations in the 10^0 and 10^1 places are straightforward; however, when we add 6×10^2 and 8×10^2, we get 14×10^2. We cannot express 14 using a single place, so instead, we think of 14 as $10 + 4$. Then we can write $14 \times 10^2 = (10 + 4) \times 10^2 = 10 \times 10^2 + 4 \times 10^2$. (We are using the fact that multiplication distributes over addition here.) This gives us 4 ten squareds and an additional ten cubed, making 5 ten cubeds. In expanded notation, we can write this as

$$5 \times 10^3 + 4 \times 10^2 + 3 \times 10^1 + 9 \times 10^0,$$

which equals 5,439.

Now try Exercises 49 to 52. ✸

The way we handled the 14 in the answer in Example 4 explains why you were taught to "put down the 4 and carry the 1 to the next column to the left" in performing this addition.

---

*Exercises 25 to 32 will give you more practice in working with Mayan notation.

KEY POINT

Hindu-Arabic notation allows us to simplify computations.

EXAMPLE 5 *Using Expanded Notation to Explain the Subtraction Algorithm*

Calculate $728 - 243$ using expanded notation.

SOLUTION: Using expanded notation, we can write this problem as

$$728 = \quad 7 \times 10^2 + 2 \times 10^1 + 8 \times 10^0$$
$$- 243 = - 2 \times 10^2 + 4 \times 10^1 + 3 \times 10^0.$$

The computations in the 10^2 and 10^0 places would be straightforward, except for the fact that we cannot directly subtract 4×10^1 from 2×10^1.

To remedy this, we express one of the ten squareds as 10×10^1. This gives us 12 tens and 6 ten squareds. We can now rewrite the subtraction as

We took one 10^2 and treated it as ten 10^1s.

$$728 = \quad 6 \times 10^2 + 12 \times 10^1 + 8 \times 10^0$$
$$\underline{- 243 = - 2 \times 10^2 + \quad 4 \times 10^1 + 3 \times 10^0}$$
$$485 \qquad 4 \times 10^2 + \quad 8 \times 10^1 + 5 \times 10^0$$

We now can write $4 \times 10^2 + 8 \times 10^1 + 5 \times 10^0 = 485$.

Now try Exercises 53 to 56. ✳

When you learned to do subtraction, you probably were taught to do this same "borrowing," but in a less drawn-out fashion and without using expanded notation.

An important benefit of Hindu-Arabic notation is that we can do basic numerical calculations very simply using pencil and paper rather than an abacus or counting board. In Example 6, we explain the galley method for doing multiplication, which you will see is a forerunner of the multiplication method that we use today. This method was popular in Italy in the fifteenth century; however, because printers found it cumbersome to typeset galleys (particularly when doing division), this method eventually gave way to our more modern way of doing multiplication.

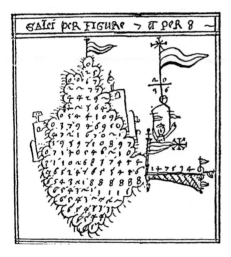

An example of using the galley method to do division.*

EXAMPLE 6 *Multiplying Using the Galley Method*

Multiply 685 and 49 using the galley method.

SOLUTION: We begin by constructing the rectangle, divided into triangles, which is called a galley,† in Figure 5.7(a). Then we compute the partial products in each box of the galley as shown in Figure 5.7(b). For example, we put the 2 and 4 in the triangles of the upper left-hand box because the product of 6 and 4 is 24.

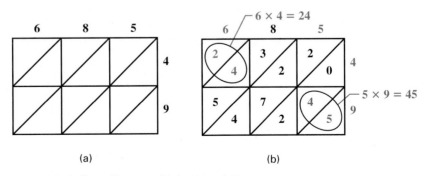

(a)   (b)

FIGURE 5.7 A galley to multiply 685 and 49.

---

*From Frank J. Swetz (Ed.), *Five Fingers to Infinity* (Chicago: Open Court Pub. Co., 1994).
†It is also possible to do division using the galley method, and when doing so, the pattern of numbers resembles a ship, or galley. This method is also sometimes called the *gelosia* method. *Gelosia* in Italian means "window."

✷ ✷ ✷ HIGHLIGHT

John Napier*

In the sixteenth century, long before computers were invented, Napier's rods were an important calculating device. To perform a lengthy multiplication, all that one had to do was line up the rods, read off the partial products, and then use simple addition to compute the product.

Napier, the eighth Baron of Merchiston, Scotland, was also an amateur theologian. He reasoned that the Roman pope was the anti-Christ, and from studying the Bible's Book of Revelation, he deduced that God would destroy the world by 1700. Napier mistakenly considered these religious "discoveries" to be his greatest contributions to mankind.

Although he probably would have disagreed, his greatest work was in mathematics. In addition to his rods, he spent 20 years developing tables of logarithms,[†] which greatly simplified scientific computations. This discovery led to the invention of the slide rule, one of the principal tools of scientists and engineers during the twentieth century.

The noted French mathematician Pierre de Laplace said that by shortening the time it took to do lengthy calculations, Napier had doubled the life of astronomers.

Next we add the numbers along at the diagonals, starting at the bottom right, as shown in Figure 5.8. If the sum along a diagonal is greater than 9, place the units digit at the end of the diagonal and carry the 1 to the channel to the left.

Follow the numbers as indicated by the red arrow to get the final product of 33,565.

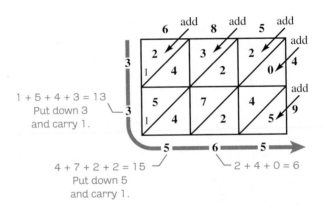

FIGURE 5.8 Computing the product by adding partial products.

Now try Exercises 57 to 64. ✽ **6**

Multiply 328×39 using the galley method. Show the galley as well as your final answer.

The galley method is much like the multiplication algorithm that we use today. The channels correspond to the columns in the multiplication below with the numbers slightly rearranged.

$$
\begin{array}{r}
685 \\
\times 49 \\
\hline
45 \\
720 \\
5{,}400 \\
200 \\
3{,}200 \\
+24{,}000 \\
\hline
33{,}565
\end{array}
$$

9×685

40×685

---

*This highlight is based on David M. Burton, *The History of Mathematics: An Introduction* (New York: WCB-McGraw Hill, 1999), pp. 325–330.

[†]Your calculator probably has keys labeled "log" or "ln," which are today's electronic equivalents of Napier's logarithms tables.

 **KEY POINT**

Napier's rods are a variation on the galley method.

In the seventeenth century, the English mathematician John Napier developed a device called **Napier's rods** or **Napier's bones** for doing multiplication. This device consists of a series of strips, each labeled at the top with a digit, 0, 1, 2, . . . , 9. The remainder of each strip lists all multiples of the label at the top, as shown in Figure 5.9. There is an additional strip called the *index*, which is also shown in Figure 5.9.

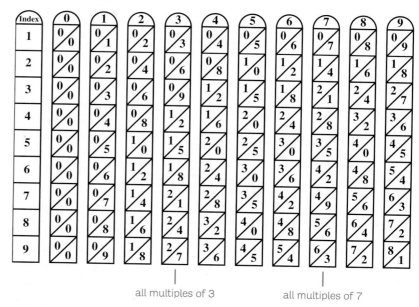

all multiples of 3   all multiples of 7

FIGURE 5.9 Napier's rods.

To calculate a product such as 6 × 325, you select strips for 3, 2, and 5 and place them side by side next to the index, as shown in Figure 5.10. We then use the small galley, formed by the three boxes next to the 6 in the index, to compute the product, 1,950, as we did in Example 6. See Figure 5.11.

We will investigate how to use Napier's rods to do more complex multiplications in Exercises 71 and 72.

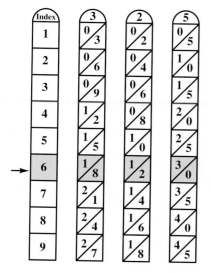

FIGURE 5.10 Using Napier's rods to compute 6 × 325.

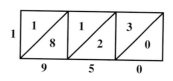

FIGURE 5.11 6 × 325 = 1,950.

Exercises 5.2

Looking Back*

These exercises follow the general outline of the topics presented in this section and will give you a good overview of the material that you have just studied.

1. Name four place value systems that we discussed in this section.

2. What are two differences between the Babylonian numeration system and the Roman system?

3. Name one feature that the Mayan numeration system has that is lacking in the Babylonian system.

4. Why is zero important in a place value system?

5. What was the base for the Mayan numeration system?

6. Name two of John Napier's contributions to mathematics.

Sharpening Your Skills

Write the following Babylonian numerals as Hindu-Arabic numerals.

7. ⟨⟨⟨▌▌   8. ⟨⟨⟨⟨⟨▌▌▌▌▌▌

*Before doing these exercises, you may find it useful to review the note *How to Succeed at Mathematics* on page xix.

9. 〈〈 〈   **10.** 〈 〈〈〈〈

11. 〈〈 〈〈〈〈 〈〈〈〈〈〈〈

12. 〈〈 〈〈〈 〈〈〈〈〈

Perform the following operations using the Babylonian numeration system. If you can, try to think using the Babylonian notation instead of converting to Hindu-Arabic notation.

13. 〈〈〈〈〈 + 〈〈〈〈〈

14. 〈〈〈〈〈 + 〈〈〈

15. 〈〈〈 〈〈 − 〈〈〈 〈〈〈〈

16. 〈〈〈〈 〈〈 − 〈〈〈 〈〈〈

Write each number using Babylonian notation.

17. 8,235   **18.** 7,331

19. 18,397   **20.** 26,411

21. 123,485   **22.** 227,597

23. 188,289   **24.** 173,596

Translate each of the following Mayan numerals to Hindu-Arabic notation.

25.   **26.**   **27.**   **28.**

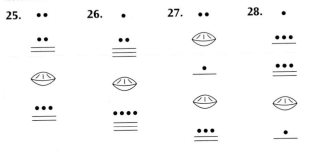

In Exercises 29–32, write the numeral that precedes the given Mayan numeral.

29.   **30.**   **31.**   **32.**

   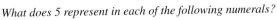

What does 5 represent in each of the following numerals?

33. 37,521   **34.** 53,184

35. 105,000   **36.** 5,023,671

Write each number using expanded notation.

37. 25,389   **38.** 37,248

39. 278,063   **40.** 820,634

41. 1,200,045   **42.** 3,002,608

Write each number using standard Hindu-Arabic notation.

43. $5 \times 10^3 + 3 \times 10^2 + 6 \times 10^1 + 8 \times 10^0$

44. $8 \times 10^3 + 2 \times 10^2 + 1 \times 10^1 + 4 \times 10^0$

45. $3 \times 10^5 + 7 \times 10^4 + 0 \times 10^3 + 0 \times 10^2 + 8 \times 10^1 + 2 \times 10^0$

46. $6 \times 10^5 + 0 \times 10^4 + 8 \times 10^3 + 2 \times 10^2 + 0 \times 10^1 + 4 \times 10^0$

47. $3 \times 10^6 + 7 \times 10^4 + 5 \times 10^2 + 2 \times 10^0$

48. $6 \times 10^6 + 3 \times 10^2 + 8 \times 10^1 + 2 \times 10^0$

Perform the following additions and subtractions using expanded notation as we did in Examples 4 and 5.

49. 2,863 + 425   **50.** 5,264 + 583

51. 3,482 + 2,756   **52.** 7,843 + 1,692

53. 926 − 784   **54.** 835 − 362

55. 5,238 − 1,583   **56.** 3,417 − 2,651

Applying What You've Learned

In Exercises 57–62, a) Use the galley method to perform the multiplication, and b) rewrite the calculations in a more conventional fashion, as we did in the discussion following Example 6.

57. 4×235 **58.** 6×382

59. 23×876 **60.** 56×371

61. 293×465 **62.** 473×628

Use the partially completed galleys to determine the numbers that are being multiplied.

63.

64.

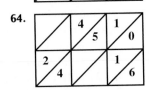

Use Napier's rods to perform each multiplication.

65. 3×628 **66.** 7×346

67. 8×492 **68.** 5×728

69. 6×924 **70.** 4×834

71. Devise a method for using Napier's rods to multiply numbers having several digits, such as 324×615. (*Hint:* Think of 324 as $300 + 20 + 4$.) Use this method to multiply 324×615.

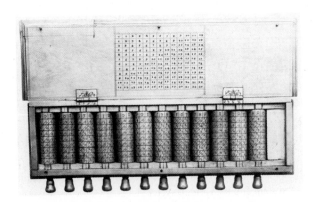

72. Use the method that you invented in Exercise 71 to multiply 426×853.

Communicating Mathematics

73. What are some of the advantages of a place value system over the earlier nonplace value systems?

74. Explain why we do carrying and borrowing when we calculate in the Hindu-Arabic system.

75. How is multiplying using the galley system similar to the way that we do multiplication today? Give an example.

76. How is multiplying using Napier's rods similar to the galley method?

77. Why might we confuse 65 and 3,605 when using Babylonian notation?

78. Give an example of two numbers, other than 65 and 3,605, that might be confused when using Babylonian notation. Why might they be confused?

79. What difficulty does the ancient Chinese numeration system have in representing very large numbers? How does the Hindu-Arabic system resolve this problem?

80. Place the following types of numeration systems in order beginning with the earliest.

   a. place value system

   b. multiplicative grouping

   c. tally marks

   d. simple grouping

Give an example of each type of system. Explain why you decided to order the systems as you did.

Using Technology to Investigate Mathematics

81. Search the Internet for more information about the Babylonian and Mayan numeration systems. Write a report on your findings.

82. Search the Internet for information about John Napier and his work in mathematics. Write a report on your findings.

For Extra Credit

The modern Chinese numeration system is a base-10 place value system that uses the symbol ○ for zero. Rewrite each of the following numerals using this modern notation. Realize that now you do not need special symbols for 10, 100, 1,000, etc.

83. 四百三十六

84. 入十五百二十五

85. 五千六十七

86. 四十三百四

87. 九千九

88. 四千九十

5.3 Calculating in Other Bases

Objectives

1. Count in non-base-10 systems.
2. Understand the similarity of notation in a base-5 and a base-10 system.
3. Perform arithmetic operations in a non-base-10 system.
4. Understand the relationship between binary, octal, and hexadecimal systems.

Imagine that you could listen in on a romantic conversation between two computers who had fallen in love. As the one computer looks lovingly at the other, he says longingly:

"1001000 1001111 1010111 1000100 1001111 1001001 1001100 1001111
1010110 1000101 1011001 1001111 1010101–1001100 1000101 1010100
1001101 1000101 1000011 1001111 1010101 1001110 1010100 1010100
1001000 1000101 1010111 1000001 1011001 1010011".*

This string of 0s and 1s is meaningless to us, but, in fact, that is the way the first computer might profess his love to the other in what is called binary, or base-2, notation. (In Exercises 77 to 80, you will learn how to translate this binary code.)

You have already seen the Babylonian numeration system, which is based on 60, and the Mayan system that uses 20, as well as our Hindu-Arabic base-10 system. Now, you

---

*This introduction may sound silly, but at one time or another, you may have accidentally dialed a fax machine, or some other electronic device and heard a variety of hissing, screeching, and beeping sounds as you listened in on a binary conversation. The translation of this statement is: "HOWDOILOVEYOU—LETMECOUNTTHEWAYS."

will learn how to work not only in a base-2 system but also in several other systems based on 5, 8, and 16. As you will see shortly, we can build numeration systems using any integer greater than 1. These systems are not just mathematical curiosities, because various societies around the world use other bases, such as 2, 3, 4, 5, or 8, for their numeration systems.

Although we will emphasize the base-5 numeration system in this section, you will find that the principles we explain are easy to carry over to systems having other bases.

Non-Base-10 Systems

KEY POINT

The principles of the Hindu-Arabic numeration system carry over to systems that use other bases.

We will now explain how to count using bases other than 10, but first, we must examine carefully how our Hindu-Arabic base-10 system works. We begin counting 1, 2, 3, etc. until we reach 9 and then, instead of writing a single symbol for 10, we write "one zero." It is important to remember that the 1 represents 1 times the base, and the zero represents 0 units, as shown below.

$$1 \times 10^1 \quad \underset{\displaystyle 1 \ 0}{\rule{0pt}{0pt}} \quad 0 \times 10^0$$

Similarly, in any base, b, we follow the same pattern—the numeral 10 represents 1 times the base plus 0 units, as seen in the following diagram.

$$1 \times b^1 \quad \underset{\displaystyle 1 \ 0}{\rule{0pt}{0pt}} \quad 0 \times b^0$$

Keeping this in mind, we can now count in base 5, remembering that every time we get a 5 in any position, we "put down a zero and carry a one." We start counting 1, 2, 3, 4, and because the base 5 is next, we write 10_5. The subscript 5 tells us that the base is 5 instead of 10. Continuing this pattern, when we reach 14_5, the next number would require us to write 5 in the units place, which we cannot do—so we write 20_5 instead. Therefore, we count in base 5 as follows:

⌐ We do not use a single symbol to represent five.
$$1, 2, 3, 4, 10_5, 11_5, 12_5, 13_5, 14_5, 20_5, 21_5, 22_5, \ldots, 44_5.$$
└ We cannot count any more objects in the 5^0 place or the 5^1 place.

If we add 1 more to 44_5, we would have 5 in the units place, so we write a 0 and carry the 1. However, we now have 5 in the 5's place, so we again write 0 and carry a 1 to the five-squared place. Thus the number following 44_5 is 100_5.

In reading a base-5 numeral such as 23_5, do not say "twenty-three base five" because "twenty" will remind you of base 10 and perhaps confuse you when you are doing computations. Say "two three base five" instead.

In a base-8 system, called an **octal** system, we would count

$$1, 2, 3, 4, 5, 6, 7, 10_8, 11_8, 12_8, \ldots, 17_8, 20_8, 21_8, \ldots, 77_8, 100_8, \ldots.$$

decimal: 8 9 10   15 16 17   63 64

However, in counting in a base-16 system, called a **hexadecimal** system, we run into a problem. If we start counting in the usual way—1, 2, 3, 4, 5, 6, 7, 8, 9—we cannot write 10_{16} next because 10_{16} represents $1 \times 16 + 0 \times 1$, which is decimal 16—not 10. To solve this problem, it is customary to use the symbols A, B, C, D, E, and F to represent the decimal numbers 10, 11, 12, 13, 14, and 15. Thus, in base 16, we would count

$$1, 2, 3, \ldots, 9, A, B, C, D, E, F, 10_{16}, 11_{16}, \ldots, 1F_{16}, 20_{16}, 21_{16}, \ldots, FF_{16}, 100_{16}, \ldots.$$

decimal: 10 11 12 13 14 15 16   17   31 32 33   255 256

Quiz Yourself ⑦

Count to the decimal number 10 using a base-3 numeration system.

In designing digital devices such as satellite radios and MP3 players, engineers use a base-2, or **binary**, system. This system uses only the digits 0 and 1, so we count as follows:

$$1, 10_2, 11_2, 100_2, 101_2, 110_2, 111_2, 1000_2, 1001_2, \ldots.$$

$$\uparrow \quad \uparrow \quad \uparrow \quad \uparrow \quad \uparrow \quad \uparrow \quad \uparrow \quad \uparrow \quad \uparrow$$

decimal: 1   2   3   4   5   6   7   8   9   ⑦

Expanded notation works in non-base-10 systems very much the way it works in the Hindu-Arabic system. For example, we write the numeral $2{,}304_5$ in expanded notation as

$$2 \times 5^3 + 3 \times 5^2 + 0 \times 5^1 + 4 \times 5^0.$$

 Some Good Advice

It will help you to work in different bases if you always keep two things in mind.

1. In base b, you only use the numerals $0, 1, 2, \ldots, b-1$. For example, in base 8 you can only use 0, 1, 2, 3, 4, 5, 6, and 7. You cannot use 8 or 9.
2. Just as in base 10, the places in the numeral represent multiples of powers of the base.

We will now use expanded notation to convert numerals from non-base-10 systems to Hindu-Arabic numerals.

EXAMPLE 1 *Converting to Decimal Notation*

a) Convert $40{,}312_5$ to decimal (base-10) notation.

b) Convert $7{,}036_8$ to decimal notation.

SOLUTION:

a) To convert to decimal notation, we write $40{,}312_5$ in expanded notation as

$$\begin{aligned}
40{,}312_5 &= 4 \times 5^4 + 0 \times 5^3 + 3 \times 5^2 + 1 \times 5^1 + 2 \times 5^0 \\
&= 4 \times 625 + 0 \times 125 + 3 \times 25 + 1 \times 5 + 2 \times 1 \\
&= 2{,}500 + 0 + 75 + 5 + 2 \\
&= 2{,}582.
\end{aligned}$$

b) In base 8, we use the digits, $0, 1, 2, \ldots, 7$ and express numbers in expanded notation using powers of 8. Thus,

$$\begin{aligned}
7{,}036_8 &= 7 \times 8^3 + 0 \times 8^2 + 3 \times 8^1 + 6 \times 8^0 \\
&= 7 \times 512 + 0 \times 64 + 3 \times 8 + 6 \times 1 \\
&= 3{,}584 + 0 + 24 + 6 \\
&= 3{,}614.
\end{aligned}$$

Quiz Yourself ⑧

Convert 354_6 to decimal notation.

Now try Exercises 7 to 28. ✳ ⑧

We will now explain how to convert a decimal numeral to a base-5 numeral. In this process, which we explain in Example 2, we first determine the number of units, next the number of 5s, then the number of 25s, and so on.

Math in Your Life

What Do Numbers Sound Like*

You probably don't realize it, but when you download a song, it is transmitted as a string of base-2 numbers called bits. How many numbers? I'm glad you asked.

To record music digitally, it is first changed into electrical waves that are then sampled at a rate of 44,100 times per second. Next, each sample is coded using 16 bits. To record in stereo, this coding must be done for each ear. So, for 1 second of stereo music, you must gather $44,100 \times 16 \times 2 = 1,411,200$ bits of information. If you were to listen to a 3-minute song, you would really be listening to a string of roughly one-fourth of a billion 0s and 1s. Printing out all of these bits would require over 76,000 pages, which, if laid end-to-end, would stretch over 13 miles!

EXAMPLE 2   *Converting from Decimal to Base-5 Notation*

Write 384 as a base-5 numeral.

SOLUTION: The following division shows that when dividing 5 into 384, we get a quotient of 76 with 4 units left over.

$$
\begin{array}{r}
76 \\
5\overline{)384} \\
380 \\
\hline
\end{array}
$$
$$4 \text{ units left over}$$

We next divide 76 by 5. This is equivalent to dividing the original number by 25, which is 5 squared. From this division, we see that after dividing by 25, there is 1 five left over. This means that in the base-5 numeral, there will be a 1 in the fives place.

$$
\begin{array}{r}
15 \\
5\overline{)76} \\
75 \\
\hline
\end{array}
$$
$$1 \text{ five left over}$$

We continue this process and express the divisions more compactly, as in Figure 5.12.

Remainders

5 ⌊ 384	4 ——	5^0 = **units**
5 ⌊ 76	1 ——	5^1 = **fives**
5 ⌊ 15	0 ——	5^2 = **twenty-fives**
5 ⌊ 3	3 ——	5^3 = **one-twenty-fives**
0		

Read remainders in reverse order.

Stop when quotient is zero.

FIGURE 5.12 Repeatedly dividing by 5 to find a base-5 numeral.

Quiz Yourself ❾

Convert the decimal number 113 to base 5.

In Figure 5.12, we continued dividing each quotient by 5 until we obtained a quotient of 0. Reading the remainders in reverse order will give us the base-5 numeral for 384, which is $3,014_5$.

Now try Exercises 29 to 42. ✳ ❾

It is important to remember, in Example 2, that we stopped the process when we reached *a quotient of zero*—not a remainder of zero.

---

*This example is based on "The Math of Online Music Trading" (February 2002) by Keith Devlin. See www.maa.org/devlin/devlin_02_02.html.

PROBLEM SOLVING
The Analogies Principle

In Section 1.1, we recommended that to understand mathematical ideas, it is useful to make analogies with situations that you have seen before. If you thoroughly understand the concept of place value in our Hindu-Arabic numeration system, you can easily modify the techniques you have used before to understand how to compute in other place value systems.

KEY POINT

Arithmetic operations in other bases are similar to base-10 operations.

Arithmetic in Non-Base-10 Systems

Imagine that you had grown up in a base-5 world. As a small child, perhaps you learned to count by watching *Sesame Street*. Instead of hearing repeatedly the ditty

"one, two, three, four, five, six, seven, eight . . . nine . . . ten,"

you may have learned to count

"one, two, three, four, fen, fenone, fentwo, thirfeen, fourfeen, twenfy, twenfy one, etc."

It is doubtful that you would have used the cumbersome name "one zero base five" for 5 or "one one base five" for 6, etc. We have invented the shorter, fanciful names "fen, fenone, fentwo" for numbers such as 10_5, 11_5, and 12_5. Do not bother to memorize these make-believe names.

Your early days of elementary school would have been much easier because you would have had fewer number facts to memorize. You would not have had to learn that $7 + 8 = 15$ because problems using the digits 7 and 8 would never arise. Also, some number facts would be written differently in a base-5 world. For example, you would write the decimal fact that $4 + 4 = 8$ as $4 + 4 = 13_5$. Using our previous made-up terminology, you would say

"four plus four equals thirfeen."

You would have spent a good deal of time memorizing the addition facts that we list in Table 5.6. For example, in Table 5.6, we see that $3 + 4 = 12_5$. ⑩

Quiz Yourself ⑩

Find $2 + 4$ and $2 + 3$ in Table 5.6.

+	0	1	2	3	4
0	0	1	2	3	4
1	1	2	3	4	10_5
2	2	3	4	10_5	11_5
3	3	4	10_5	11_5	12_5	—— $3 + 4 = 12_5$
4	4	10_5	11_5	12_5	13_5

TABLE 5.6 Base-5 addition facts.

We will use the addition facts in Table 5.6 to do the addition in Example 3.

EXAMPLE 3   *Adding in Base 5*

Add $\begin{array}{r} 342_5 \\ + \ 223_5 \end{array}$.

SOLUTION: We add much like we do in performing decimal addition. First we add the units, $2 + 3 = 5$ decimal, which we express as 10_5. We place the 0 in the units place and carry the 1 to the fives place, as we show in Figure 5.13(a).

$$
\begin{array}{r}
1 \\
342_5 \\
+\ 223_5 \\
\hline
0_5
\end{array}
\quad
\begin{array}{l}
2 + 3 = 10_5 \\
\text{write 0, carry 1.}
\end{array}
\qquad
\begin{array}{r}
11 \\
342_5 \\
+\ 223_5 \\
\hline
20_5
\end{array}
\quad
\begin{array}{l}
1 + 4 + 2 = 12_5 \\
\text{write 2, carry 1.}
\end{array}
$$

(a)   (b)

FIGURE 5.13 Adding in base 5.

We next add the numerals in the fives place: $1 + 4 + 2 = 12_5$ as we show in Figure 5.13(b). Note how we write down the 2 and carry the 1 to the five-squared place. We finish by adding the numerals in the five-squared place, carrying as necessary.

$$
\begin{array}{r}
111 \\
342_5 \\
+\ \ \ 223_5 \\
\hline
1120_5
\end{array}
$$

$1 + 3 + 2 = 11$, write 1, carry 1. ✽

EXAMPLE 4   *Subtracting in Base 5*

Subtract $\begin{array}{r} 424_5 \\ -\ 143_5 \end{array}$.

SOLUTION: Subtraction in base 5 is also much like base-10 subtraction. We begin by subtracting three units from four units to get one unit.

In the fives place, we cannot subtract 4 fives from 2 fives, so we borrow 1 twenty-five and write it as 5 fives. Thus, we now have 7 fives in the fives place, which we have written as 12_5, as shown in Figure 5.14. We complete the subtraction by subtracting 1 twenty-five from the 3 remaining twenty-fives.

$$
\begin{array}{r}
3 \\
\cancel{4}^{\,1}2\ 4_5 \\
-\ 1\,4\,3_5 \\
\hline
2\,3\,1_5
\end{array}
$$

Borrowing 1 twenty-five
we now have 12_5, or decimal 7
fives. Subtracting, we get 3 fives.

FIGURE 5.14 Subtracting in base 5 may require borrowing.

Now try Exercises 43 to 56. ✽ **11**

Quiz Yourself **11**

a) Add $315_6 + 524_6$.
b) Subtract $325_6 - 131_6$.

Some Good Advice

Although it is useful to check base-5 computations by converting all numbers to base 10, you will increase your skill in working in other bases if you try to do all computations by thinking in base 5 even though it may seem a little difficult at first.

Quiz Yourself **12**

Find 2×4 and 4×4 in Table 5.7.

In order to understand how to multiply in other bases, we first need to know our multiplication table. The multiplication facts will look different from what you are used to seeing in base 10. For example, in Table 5.7, we see that $3 \times 4 = 22_5$, which is the base-5 representation of decimal 12. You should verify the other base-5 multiplication facts given in Table 5.7. **12**

×	0	1	2	3	4
0	0	0	0	0	0
1	0	1	2	3	4
2	0	2	4	11_5	13_5
3	0	3	11_5	14_5	22_5
4	0	4	13_5	22_5	31_5

$3 \times 4 = 22_5$

TABLE 5.7 Base-5 multiplication facts.

Before giving an example of multiplication, it is useful to review the discussion following Example 6 of Section 5.2, explaining the meaning of galley multiplication in base 10. Recall that we wrote all the partial products in multiplying 685 and 49 as follows:

$$
\begin{array}{r}
685 \\
\times 49 \\
\hline
45 \\
720 \\
5{,}400 \\
200 \\
3{,}200 \\
+24{,}000 \\
\hline
33{,}565
\end{array}
$$

45 — units times units
720 — units times tens
5,400 — units times hundreds — } Combine into first partial product.
200 — tens times units
3,200 — tens times tens
+24,000 — tens times hundreds — } Combine into second partial product.

Notice how

$$\text{units times units} = \text{units,}$$

$$\text{units times tens} = \text{tens,}$$

$$\text{tens times tens} = \text{hundreds, and so on.}$$

Also, when we do this multiplication in the usual way, we combine the first three products into a single partial product and the second three products into a second partial product. We will follow this pattern in doing base-5 multiplication.

EXAMPLE 5 *Multiplying in Base 5*

Multiply

$$
\begin{array}{r}
134_5 \\
\times\ 32_5
\end{array}
$$

SOLUTION: We first multiply 2 units times 4 units to get 8 units in base 10, or 13_5 units. We write down 3 units and carry 1 to the fives place.

$$
\begin{array}{r}
1 \\
134_5 \\
\times\ 32_5 \\
\hline
3_5
\end{array}
$$

1 — 5 units expressed as 1 five.

3 units

We next multiply 2 units times 3 fives to get 6 fives, plus the 1 five we carried, gives 7 fives, which we write as 12_5. We write the 2 and carry the 1 to the five-squared column.

$$
\begin{array}{r}
11 \\
134_5 \\
\times\ 32_5 \\
\hline
23_5
\end{array}
$$

5 fives expressed as 1 five squared.

2 fives

We complete the first partial product by multiplying the 2 units times 1 five squared and add the 1 five squared, which was carried, to get 3 five squareds.

$$
\begin{array}{r}
134_5 \\
\times\ 32_5 \\
\hline
323_5
\end{array}
$$

We compute the second partial product in a similar way. We begin by multiplying 3 fives times 4 units to get 22_5 fives. We write 2 fives and carry a 2 to the five-squared column.

$$
\begin{array}{r}
2 \\
134_5 \\
\times\ 32_5 \\
\hline
323_5 \\
2
\end{array}
$$
— We indent one place because we are multiplying by 3, which represents 3×5.

We complete the second partial product as follows:

$$
\begin{array}{r}
22 \\
134_5 \\
\times\ 32_5 \\
\hline
323_5 \\
1012
\end{array}
$$

We then add the partial products to get the final product.

$$
\begin{array}{r}
134_5 \\
\times\ 32_5 \\
\hline
323_5 \\
1012 \\
\hline
10{,}443_5
\end{array}
$$
— Adding in base 5.

Quiz Yourself ⓭

Multiply:
a) $34_5 \times 42_5$
b) $56_8 \times 47_8$

Now try Exercises 57 to 60. ❈ ⓭

Before doing division, you must have a very good understanding of multiplication.* For example, consider how we do the first step in the following decimal division:

$$
\begin{array}{r}
2 \\
18\overline{)4721} \\
36 \\
\hline
11
\end{array}
$$
— Experience in base 10 tells us that 18 divides into 47 twice.

Based on years of experience in working with the decimal system and our ability to make good estimations, we see that 18 divides into 47 twice, but not three times. In dividing,

$$
23_5\overline{)4132_5}
$$
— We lack experience in base 5 here, so we make a table of multiples of 23_5.

*Several years ago, while waiting in the parents' lounge for my son to finish his saxophone lesson, a parent asked me why I thought children in her daughter's grade always had trouble with learning division. She told me the problem was so serious that for the past several years, the teacher in the previous year cut short teaching multiplication to give the children a head start in learning division. I told her, "I think that's the problem. Children cannot learn to divide unless they are comfortable with multiplication."

we do not have the same deep background that enables us to find the trial quotient, so we have to use another approach, as we show in Example 6.

EXAMPLE 6 *Dividing in Base 5*

Perform the division

$$23_5 \overline{)4132_5}.$$

SOLUTION: To be able to estimate trial quotients, we have to know the multiples of 23_5, which we list below. (You should verify these facts.)

$$23_5 \times 0 = 0$$
$$23_5 \times 1 = 23_5$$
$$23_5 \times 2 = 101_5$$
$$23_5 \times 3 = 124_5$$
$$23_5 \times 4 = 202_5$$

From this list, we see that 23_5 divides into 41_5 once but not twice. So we start the division by placing a 1 in the quotient.

$$
\begin{array}{r}
1 \\
23_5 \overline{)4132_5} \\
23 \\
\hline
13
\end{array}
$$
— Base-5 subtraction

Remember that when subtracting $41_5 - 23_5$, we are doing a base-5 subtraction. We next bring down the digit 3 and add another digit to the quotient. From the previous list, we see that 23_5 divides into 133_5 three times with a remainder of 4.

$$
\begin{array}{r}
13 \\
23_5 \overline{)4132_5} \\
23 \\
\hline
133 \\
124 \\
\hline
4
\end{array}
$$
— Base-5 subtraction

Next, we bring down the 2 and complete the division.

$$
\begin{array}{r}
131_5 \\
23_5 \overline{)4132_5} \\
23 \\
\hline
133 \\
124 \\
\hline
42 \\
23 \\
\hline
14_5
\end{array}
$$
— Base-5 subtraction

Thus, 23_5 divides into $4{,}132_5$, with a quotient of 131_5 and a remainder of 14_5. Now try Exercises 61 to 64. ❀

Some Good Advice

To check to see if we have properly done a division, such as the one we did in Example 6, you can multiply the divisor, 23_5, by the quotient, 131_5, and then add the remainder, 14_5. The result should be the dividend, 4132_5.

🖉 *KEY POINT*

Binary, octal, and hexadecimal notation are closely related.

Binary		Octal
0 0 0	=	0
0 0 1	=	1
0 1 0	=	2
0 1 1	=	3
1 0 0	=	4
1 0 1	=	5
1 1 0	=	6
1 1 1	=	7

FIGURE 5.15 Three binary places correspond to one octal place.

Binary, Octal, and Hexadecimal Systems

As we have mentioned, digital devices such as computers, MP3 players, and iPhones use the binary system. You might think of the 1s and 0s as representing "on" and "off" commands controlling microscopic switches on electronic chips. The values 0 and 1 are called **bits**, which is a contraction of the words **binary digit**. A video card in your PC, for example, may use the 16-bit binary pattern 1011001101011110 as the command to change the pen color to red in a drawing package. Because it is tedious to remember such long strings of bits, often we rewrite these commands in a form that is easier for humans to remember.

Figure 5.15 shows how we can represent three binary places efficiently as one octal place. This means that we can express the previous 16-bit command using fewer octal numerals, as we show in Example 7.

EXAMPLE 7 *Converting from Binary to Octal and Hexadecimal*

a) Write the binary command 1011001101011110 using octal notation.

b) Write the binary command 1011001101011110 using hexadecimal notation.

SOLUTION:

a) We begin by arranging the bits in 1011001101011110 in groups of three *beginning from the right*. We then interpret each group of three bits as an octal number, as shown here. (For the sake of efficiency, we often omit the subscripts that indicate the base.)

⌐ Begin grouping
from the right.

binary	⟶	1	011	001	101	011	110
octal	⟶	1	3	1	5	3	6

We see that we can represent the binary command 1011001101011110 more succinctly as the octal command 131536. Notice how much easier it is to remember the octal form of the command rather than the binary form.

Binary		Hexadecimal
0 0 0 0	=	0
0 0 0 1	=	1
0 0 1 0	=	2
0 0 1 1	=	3
.		.
.		.
.		.
1 1 1 0	=	E
1 1 1 1	=	F

FIGURE 5.16 Four binary places correspond to one hexadecimal place.

b) Figure 5.16 shows that we can represent four binary places as one hexadecimal place. Thus, we first group the bits in 1011001101011110 in groups of four, again beginning from the right. We then interpret each group of four bits as a hexadecimal number, as we show here.

binary	⟶	1011	0011	0101	1110
hexadecimal	⟶	B(11)	3	5	E(14)

We can now represent the binary command 1011001101011110 more concisely as the hexadecimal command B35E.

Now try Exercises 65 to 68. ✳

HISTORICAL HIGHLIGHT ❈ ❈ ❈

From the Abacus to the Computer

The earliest mechanical calculating device is the abacus, which dates back to 300 BC. Early abaci, similar to the Roman abacus we described in the Historical Highlight on page 190, were widely used in Europe prior to the adoption of written Hindu-Arabic numerals. Some believe that Christians introduced the abacus to China beginning about 1200 AD and later to Japan and Korea.

We show a typical Chinese abacus, or suan-pan, in Figure 5.17. The suan-pan consists of beads that slide on wires that represent powers of 10, as we show in Figure 5.17. Beads below the horizontal bar represent 1; beads above the bar represent 5. Beads next to the bar are active and are used to represent the number. The suan-pan in Figure 5.17 shows how we would represent 9,073.

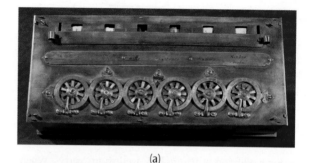

(a)

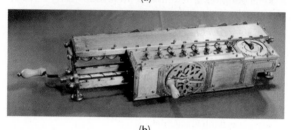

(b)

FIGURE 5.18 (a) Pascal's and (b) Leibniz's calculators.

believed that there would be a worldwide demand for only about a half-dozen of these costly machines. Little did they imagine that computers would become so small and cheap and such a commonplace part of modern-day life.

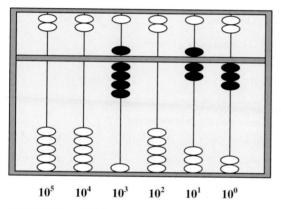

$$10^5 \quad 10^4 \quad 10^3 \quad 10^2 \quad 10^1 \quad 10^0$$

FIGURE 5.17 A Chinese suan-pan representing 9,073.

In 1642, the renowned French mathematician Blaise Pascal invented an adding machine, developing principles that would be used in later calculators (see Figure 5.18(a)). Later in the century, the German Gottfried Leibniz improved on Pascal's design (see Figure 5.18(b)). The British mathematician Charles Babbage invented a complicated calculating machine called an analytical engine in 1826 (see Figure 5.19). Although Babbage was unable to actually construct his machines,* his designs laid the foundation for today's modern computers.

The first electronic computer was constructed by J. Presper Eckert and John Mauchly at the University of Pennsylvania in the early 1940s. The computer was so large (more than 30 tons) and so expensive (about a half million dollars) that some of the leading businessmen of the time

FIGURE 5.19 Babbage's analytic engine.

---

*Many years later, IBM constructed perfectly working models of Babbage's machines using his designs.

Exercises 5.3

Looking Back*

These exercises follow the general outline of the topics presented in this section and will give you a good overview of the material that you have just studied.

1. What is the largest numeral that we use in a base-5 numeration system?

2. How do we represent 8 in a base-8 system? How do we represent 16 in a base-16 system?

3. In doing addition in Example 5, we had to "carry." When we carried the 1 to a column, how many objects did the carried 1 represent in the column to its right?

4. In Example 6, why did we make the table of multiples of 23_5?

5. How is the suan-pan similar to Hindu-Arabic notation?

6. Name three mathematicians who were instrumental in developing early mechanical calculators.

Sharpening Your Skills

List the numbers that precede and follow each of the given numbers in the given base.

7. 24_5

8. 500_6

9. 1011_2

10. 77_8

11. EF_{16}

12. 100_{16}

Write each number as a base-10 numeral.

13. 432_5

14. 243_5

15. 504_6

16. 555_6

17. $100,111_2$

18. $100,101_2$

19. $1,110,101_2$

20. $1,100,111_2$

21. 267_8

22. 137_8

23. 704_8

24. 561_8

25. $2F4_{16}$

26. $18E_{16}$

27. $D08_{16}$

28. $C3B_{16}$

Convert each decimal number to a numeral in the given base.

29. 334 base 5

30. 1,298 base 5

31. 1,838 base 6

32. 3,968 base 6

33. 103 base 2

34. 51 base 2

35. 94 base 2

36. 107 base 2

37. 3,403 base 8

38. 2,297 base 8

39. 2,792 base 16

40. 2,219 base 16

41. 3,562 base 16

42. 3,827 base 16

Perform each addition or subtraction.

43. $3,412_5 + 231_5$

44. $3,215_6 + 423_6$

45. $4,013_5 + 1,242_5$

46. $2,067_8 + 2,443_8$

47. $2A18_{16} + 43B_{16}$

48. $BF2E_{16} + A35_{16}$

49. $11,011_2 + 10,101_2$

50. $100,111_2 + 10,111_2$

51. $2,412_5 - 321_5$

52. $1,325_6 - 453_6$

53. $A83_{16} - 43B_{16}$

54. $6C2E_{16} - A35_{16}$

55. $111,011_2 - 10,101_2$

56. $100,101_2 - 10,011_2$

Perform each multiplication or division.

57. $41_5 \times 23_5$

58. $24_5 \times 31_5$

59. $302_5 \times 43_5$

60. $413_5 \times 34_5$

61. $3,412_5 \div 24_5$

62. $2,143_5 \div 32_5$

63. $4,132_5 \div 42_5$

64. $4,402_5 \div 14_5$

Write each binary number first as an octal number and then as a hexadecimal number.

65. 1011101101_2

66. 1011110111_2

67. 1111101001_2

68. 1010100101_2

Write each number as a binary number.

69. 246_8

70. 573_8

71. $A3E_{16}$

72. $B8C_{16}$

73. Convert 3524_8 to hexadecimal.

74. Convert 6235_8 to hexadecimal.

75. Convert $3AC_{16}$ to octal.

76. Convert $D7B_{16}$ to octal.

Applying What You've Learned

Many electronic devices store and communicate information using the ASCII coding system, which is an acronym for American Standard Code for Information Interchange. *The ASCII coding system uses 7-bit binary codes to represent characters. For example, the binary equivalents of the numbers 65 to 90 represent the capital letters A to Z. Using ASCII coding, we represent A by 1000001, B by 1000010, C by 1000011, and so forth. In Exercises 77–80, translate each binary string into English. (First group the bits into 7-bit groups.)*

77. 1000011100000110011101000100101011001

78. 1001000100010110011001001100100111111

79. 100110010011111010110110000101

80. 1010100101001010101011010100101001000

81. What base would you be using to change seconds to hours and minutes? What about changing pennies to nickels and quarters?

82. Think of other situations similar to those we mentioned in Exercise 81 in which you change one unit of measurement into another unit of measurement. State what base you are using when doing the conversion.

---

*Before doing these exercises, you may find it useful to review the note *How to Succeed at Mathematics* on page xix.

Communicating Mathematics

83. How many addition and multiplication facts would you have to memorize in a base-8 system? In a base-16 system? Why is that the case?

84. If $3{,}231_b = 11{,}453_c$, which base is larger? Explain your answer.

85. Why did we use A in a hexadecimal system to represent 10? Why could we not use 10?

86. Three binary places correspond to how many octal places? Explain.

87. Locate a book or article that discusses the history of numeration systems and give a brief report on societies that use bases other than 10.

88. Research the ancient Chinese bar numeration system. Write a brief report on your findings and give several illustrations of how you would compute with bar numerals.

Using Technology to Investigate Mathematics

89. Your instructor can provide you with programs that will do conversions from one base to another and that will also do arithmetic in various bases. Use these programs to reproduce some of the computations in this section.

90. Search the Internet to find several sites that discuss the role mathematics plays in creating digital music. You might search "digital," "music," and "mathematics." Report on your findings.

For Extra Credit

Make the following conversions.

91. Convert 201221_3 to base 5.

92. Convert 4523_7 to base 8.

93. Convert $B05_{16}$ to base 9.

94. Convert 365_8 to base 4.

95. Some are proponents of a base-12, or duodecimal, system.* Invent a base-12 numeration system. Give careful thought to what additional symbols you will need in this system. Count from 1 to 25 using your system.

Consider a base-6 numeration system that uses the symbols $\mp$, $\oplus$, $\neq$, \$, @, and %, corresponding to 0, 1, 2, 3, 4, and 5, respectively.

96. Count to decimal 20 using this system.

97. Translate $\oplus \mp \neq \$$ to a decimal number.

98. Translate the decimal number 30 to this system.

99. Add $\oplus \neq \mp \%$ and $\% \$ @$ in this system.

---

*You may find it interesting to research the Duodecimal Society.

Looking Deeper

5.4 Modular Systems

Objectives

1. Understand modular systems.
2. Perform operations in modular systems.
3. Solve congruences in modular systems.
4. Understand some applications of modular systems.

As a collector of antique cars, you've come across a deal that seems almost too good to be true—a 1981 Ford Thunderbird with only 47,000 miles on it. Or, let's think about it, maybe it is really 147,000 miles—or perhaps 247,000 miles. The odometer on your car is an example of a mathematical system in which counting is done in a cyclic fashion where the same numbers are used over and over again. Your bedroom clock also ticks off the hours in the same way—1, 2, 3, . . . , 12, and then recycles these numbers.

Both systems are an example of a modulo-*m* system, which we will study next. These systems have many properties in common with the familiar system of integers. As you will see, in modulo-*m* systems, we can count, perform arithmetic operations, and solve equations. In addition, modulo-*m* systems have interesting, real-life applications.

 *KEY POINT*

We use a clock to visualize the operations in a modulo-*m* system.

Modulo-*m* Systems

> **DEFINITIONS** If *m* is an integer greater than 1, then a **modulo-*m* system*** consists of the numbers 0, 1, 2, . . . , *m* − 1. Counting and arithmetic operations are performed in a manner corresponding to movements on an *m*-hour clock. The number *m* is called the **modulus** of the system.

FIGURE 5.20 A 12-hour clock.

In order to understand a modulo-12 system, we draw a 12-hour clock as shown in Figure 5.20. Notice how we have replaced the 12 on the clock with the number 0—the reason for doing this will become clear shortly. The numbers 0, 1, 2, . . . , 11 now form a modulo-12 system. If we were to begin at 0 and count to 53 in this system, at what number would we stop? Of course, we could begin counting on the clock, 1, 2, 3, 4, 5, 6, 7, 8, 9, 10, 11, 0, 1, 2, and so on until we counted 53 numbers. However, there is an easier way. Notice that if we count on the clock to multiples of 12, such as 12, 24, 36, and then 48, we return to the 0 position (see Figure 5.21). If we now count an additional 5, we reach 53, which is position 5 on this 12-hour clock. Thus, counting to 53 in a modulo-12 system will give us 5.

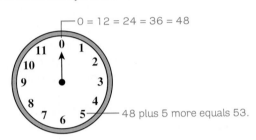

FIGURE 5.21 Counting to 53 on a 12-hour clock.

 Some Good Advice

Because multiples of 12 are the same as 0 on a 12-hour clock, a quicker way to determine the 5 would have been to divide 53 by 12, and keep the remainder of 5.

We can visualize the days of the week, Monday (1), Tuesday (2), Wednesday (3), Thursday (4), Friday (5), Saturday (6), and Sunday (7) on the 7-hour clock shown in Figure 5.22. Notice that we replace the 7 by 0 on this clock. The numbers 0, 1, . . . , 6 form a modulo-7 system.

If we count to 45 on this clock, we see that we reach 0 at each multiple of 7, namely 7, 14, 21, 28, 35, and 42. If we now count 3 more, we reach 45. This means that 45 in a modulo-7 system is the same as 3. Notice again, that dividing 45 by 7 and keeping the remainder 3 is a quicker way to solve this problem. **14**

We can express the fact that 53 occupies the same position as 5 on a 12-hour clock more precisely.

┌─────────────────────────────────────┐
│ 0 = 7 = 14 = 28 = 35 = 42 │
│   0   │
│   6   1   │
│   5   2   │
│   4   3 — 42 plus 3 more equals 45. │
└─────────────────────────────────────┘

FIGURE 5.22 Counting to 45 on a 7-hour clock.

Quiz Yourself **14**

a) Count to 75 on a 12-hour clock.

b) Count to 41 on a 7-hour clock.

> **DEFINITION** We say that *a* **is congruent to** *b* **modulo** *m*, written *a* ≡ *b* (mod *m*), provided that *m* evenly divides *a* − *b*. Note, if you find it more convenient to divide *m* into *b* − *a*, this is also acceptable.

---

*Modulo-*m* systems are also called **modular arithmetic systems**, or simply **modular systems**.

Because 12 divides $53 - 5 = 48$ evenly, this means that $53 \equiv 5 \pmod{12}$. We read this as, 53 is congruent to 5 modulo 12. Similarly, because $45 - 3 = 42$ is a multiple of 7, we write $45 \equiv 3 \pmod 7$.

PROBLEM SOLVING

The Splitting-Hairs Principle

Remember the Splitting-Hairs Principle from Section 1.1. The three lines $\equiv$ look somewhat like the equal sign but do not mean exactly the same thing. When we write $a \equiv b \pmod m$, we are not saying that a and b are equal in the usual sense, but in terms of an m-hour clock, we can consider them to be the same.

EXAMPLE 1　*Determining When Numbers Are Congruent*

Determine which statements are true.

a) $39 \equiv 15 \pmod{12}$

b) $33 \equiv 19 \pmod 8$

c) $25 \equiv 48 \pmod 7$

d) $11 \equiv 35 \pmod 6$

SOLUTION:

a) $39 - 15 = 24$, which is evenly divisible by 12. Therefore, $39 \equiv 15 \pmod{12}$ is a true statement.

b) $33 - 19 = 14$, which is not evenly divisible by 8. Thus, $33 \equiv 19 \pmod 8$ is false.

c) Instead of dividing 7 into $25 - 48$, we consider $48 - 25 = 23$. Because 7 does not divide evenly into 23, this statement is false.

d) Because 6 divides evenly into $35 - 11 = 24$, this statement is true.

Now try Exercises 9 to 14. ❈ **15**

Quiz Yourself **15**

Determine which statements are true.

a) $11 \equiv 33 \pmod{12}$

b) $27 \equiv 59 \pmod 8$

You might wonder why we have not been discussing negative numbers. This is because in a modulo-m system, we do not need negative numbers. As you can see on the 5-hour clock in Figure 5.23, the number -2 corresponds to moving 2 in a counterclockwise direction. If we do this, we stop at 3. This means that -2 is congruent to 3 and so -2 and 3 can be used interchangeably.

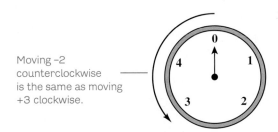

Moving -2 counterclockwise is the same as moving $+3$ clockwise.

FIGURE 5.23 -2 is congruent to 3 modulo 5.

When doing calculations in a modulo-m system, you can use any two congruent numbers interchangeably. Therefore, if you are calculating in a modulo-5 system and get a result of 19, you can replace it by 4 because $19 \equiv 4 \pmod 5$. We will use this principle frequently when we are performing arithmetic operations in modular systems.

❋ ❋ ❋ **HIGHLIGHT**

Are You Really Who You Think You Are?

But he that filches from me my good name/Robs me of that which not enriches him/And makes me poor indeed.

—Shakespeare, *Othello*, Act iii, Scene 3.

According to the results of a survey released by the Federal Trade Commission in November of 2007, it was estimated that about 8.3 million people in the United States have been the victims of identity theft. By "shoulder surfing" (looking over your shoulder at your ATM), eavesdropping as you give out sensitive information on your cell phone, or "dumpster diving" at your apartment's community trash bin, criminals can gather enough information to in effect become you— buying a new Wii game console with "your" new credit card, withdrawing money from your bank account, and then, after running up enough debt, declaring bankruptcy in your name.

The magnitude of identity theft warns us that we have to be careful as to how we give out information such as Social Security numbers, credit card numbers, and other personal financial information. When sensitive information is sent electronically, such as by using the Internet, it is critical to code the information so that it is not vulnerable to hacking. One popular method for encrypting messages based on modular arithmetic is the **RSA algorithm**, named after its inventors Ronald Rivest, Adi Shamir, and Leonard Adelman, which we will discuss in Exercises 61 to 64.

For more information about identity theft, you and your family might want to visit the following sites: www.ftc.gov/os/2007/11/SynovateFinalReportIDTheft2006 .pdf; www.usdoj.gov/criminal/fraud/websites/idtheft.html.

 **KEY POINT**

Operations in a modulo-m system are closely related to operations on the integers.

Operations in Modulo-m Systems

It is easy to add, subtract, and multiply in a modulo-m system. For example, in a modulo-7 system, $6 + 4 = 10$. However, because $10 \equiv 3 \pmod 7$, we can write $6 + 4 \equiv 3 \pmod 7$. It helps to see this computation if you draw the clock in Figure 5.24. If we begin at 0 and count to 6 and then count 4 more, we see that we stop at 3. All modulo-m operations are done in a similar way.

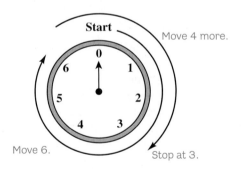

FIGURE 5.24 A 7-hour clock showing $6 + 4 = 3 \pmod 7$.

PERFORMING ARITHMETIC OPERATIONS IN A MODULO-*M* SYSTEM

 To add, subtract, and multiply* in a modulo-m system:

1. Perform the operation as usual.
2. Replace the result in (1) by one of the numbers $0, 1, 2, \ldots, m - 1$, which is congruent to the result in part (1).

---

*We do division differently. See Exercise 65.

EXAMPLE 2   *Adding in a Modulo-m System*

a) Find $10 + 9$ (mod 12).   b) Find $7 + 3$ (mod 8).

SOLUTION:

a) To find $10 + 9$ (mod 12) is asking us in a succinct way to compute $10 + 9$ in a modulo-12 system. Now, $10 + 9 = 19$; however, $19 \equiv 7$ (mod 12). Therefore, we can write $10 + 9 \equiv 7$ (mod 12).

b) Here, $7 + 3 = 10$; but $10 \equiv 2$ (mod 8), so $7 + 3 \equiv 2$ (mod 8).

Now try Exercises 15 to 18. ✻

Doing subtraction requires a slight variation of our approach, as we see in Example 3.

EXAMPLE 3   *Subtracting in a Modulo - m System*

a) Find $2 - 5$ (mod 12).   b) Find $3 - 7$ (mod 8).

SOLUTION:

-3 is the same as 9.

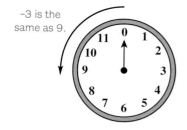

FIGURE 5.25 A 12-hour clock showing $2 - 5 = -3 = 9$.

a) We begin by calculating $2 - 5 = -3$. However, as we pointed out earlier, we do not need negative numbers in a modulo-*m* system. On a 12-hour clock, to represent -3, we think of beginning at 0 and moving counterclockwise three numbers, as shown in Figure 5.25. Because we stop at 9, we can say $2 - 5 \equiv 9$ (mod 12).

b) First calculate $3 - 7 = -4$. If you visualize an 8-hour clock, you see that -4 is the same as 4. Therefore, $3 - 7 \equiv 4$ (mod 8).

Now try Exercises 19 to 22. ✻

 PROBLEM SOLVING

Verify Your Answers

To check a subtraction problem in the integers, such as $8 - 5 = 3$, you can perform the addition $3 + 5 = 8$. Similarly, to check subtraction in a modulo-*m* system, we do a corresponding addition. In Example 3, we can check that $2 - 5 \equiv 9$ (mod 12) by showing $9 + 5 \equiv 2$ (mod 12).

Multiplication is straightforward in a modulo-*m* system.

EXAMPLE 4   *Multiplying in a Modulo m System*

a) Find 9×7 (mod 12).   b) Find 5×8 (mod 9).

SOLUTION:

a) We first calculate $9 \times 7 = 63$. Next, we divide 63 by 12 to get a remainder of 3. Therefore, $9 \times 7 \equiv 3$ (mod 12).

b) We begin by calculating $5 \times 8 = 40$. Dividing 40 by 9, we get a remainder of 4. Thus, $40 \equiv 4$ (mod 9), so $5 \times 8 \equiv 4$ (mod 9).

Now try Exercises 23 to 26. ✻ **16**

Quiz Yourself **16**

Perform the following operations.

a) $4 + 6$ (mod 8)

b) $3 - 8$ (mod 12)

c) 4×5 (mod 6)

 **KEY POINT**

We can use trial and error to solve congruences.

Solving Congruences

In a modulo-*m* system, we solve congruences instead of equations. As you will see in Example 5, we can solve congruences by trial and error.

EXAMPLE 5   *Solving Congruences Using Trial and Error*

a)   Solve $4 + x \equiv 2$ (mod 5).   b)   Solve $2x - 3 \equiv 3$ (mod 6).   c)   Solve $2x \equiv 3$ (mod 6).

SOLUTION:

a)   We will test each of the numbers 0, 1, 2, 3, and 4 to see which ones solve the congruence.

$$4 + 0 \equiv 2 \text{ (mod 5)} \qquad \text{(False; not a solution.)}$$
$$4 + 1 \equiv 2 \text{ (mod 5)} \qquad \text{(False; not a solution.)}$$
$$4 + 2 \equiv 2 \text{ (mod 5)} \qquad \text{(False; not a solution.)}$$
$$4 + 3 \equiv 2 \text{ (mod 5)} \qquad \text{(True; 3 is a solution.)}$$
$$4 + 4 \equiv 2 \text{ (mod 5)} \qquad \text{(False; not a solution.)}$$

Thus we find that 3 is the only solution of this congruence.

b)   When we test each number from 0 to 5, we find that both 0 and 3 are solutions.

$$2 \times 0 - 3 \equiv 3 \text{ (mod 6)} \qquad \text{(True; 0 is a solution.)}$$
$$2 \times 1 - 3 \equiv 3 \text{ (mod 6)}$$
$$2 \times 2 - 3 \equiv 3 \text{ (mod 6)}$$
$$2 \times 3 - 3 \equiv 3 \text{ (mod 6)} \qquad \text{(True; 3 is a solution.)}$$
$$2 \times 4 - 3 \equiv 3 \text{ (mod 6)}$$
$$2 \times 5 - 3 \equiv 3 \text{ (mod 6)}$$

c)   If you substitute 0, 1, 2, . . . , 5 for x in the congruence, you will find that none of these numbers makes the congruence true. (Verify this.) Therefore, this congruence has no solutions.

Now try Exercises 27 to 38. ✻ **17**

In Example 6, you must solve a pair of congruences simultaneously.

Quiz Yourself **17**

Solve the following congruences.

a) $4 - x \equiv 7$ (mod 8)

b) $3x \equiv 0$ (mod 9)

EXAMPLE 6   *Solving a Pair of Congruences*

Lauren and Whitney are arranging a collection of designer clothes for display in a Teen Vogue fashion show. If they arrange the clothes in stacks of five, they have three left over. If they arrange the clothes in stacks of four, they have two left over. What is the smallest pile of clothes that they can have?

SOLUTION: Let us say that x is the number of clothes. When Lauren and Whitney stack the clothes five at a time, they have three left over, which tells us that if we divide x by 5, the remainder is 3. In terms of congruences, this means that $x \equiv 3$ (mod 5). We conclude that x is one of the following numbers:

$$3, 8, 13, \mathbf{18}, 23, 28, 33, 38, \ldots .$$

Similarly, when they stack the clothes four at a time, they have two left over, which tells us that $x \equiv 2$ (mod 4). This means that x must be in the following list:

$$2, 6, 10, 14, \mathbf{18}, 22, 26, 30, \ldots .$$

The smallest number that is found in both of these lists is 18, so the smallest pile of clothes Lauren and Whitney have to display is 18.

Now try Exercises 39 to 42. ✻ **18**

Quiz Yourself **18**

Find the smallest positive integer that satisfies $x \equiv 5$ (mod 7) and $x \equiv 6$ (mod 9).

An interesting application of modular arithmetic appears on many of the products that we use each day. Consider the 12-digit Universal Product Code (UPC) found underneath the bar code on a box of Honey Nut Cheerios, shown in Figure 5.26. The first digit, 0, identifies the product. The next five digits, 16000, identify the manufacturer. The next group of digits, 66590, provides information about the product. The last digit, in this case 3, is called a *check digit*. The check digit provides a way of determining if a UPC is a valid code.

FIGURE 5.26 Honey Nut Cheerios product code.

Example 7 explains how to compute the check digit for a UPC.

EXAMPLE 7 *Using Modular Arithmetic to Find Check Digits*

If the 12-digit UPC is of the form $a_1a_2a_3a_4a_5a_6a_7a_8a_9a_{10}a_{11}a_{12}$,* then we compute the check digit in two steps.

1. First we compute the expression

$$3a_1 + a_2 + 3a_3 + a_4 + 3a_5 + a_6 + 3a_7 + a_8 + 3a_9 + a_{10} + 3a_{11}.$$

   That is, we multiply the first digit in the UPC by 3; add the second digit; then add three times the third digit; add the fourth digit, and so on.

2. Choose the check digit, a_{12}, so that the sum in step (1) plus a_{12} is congruent to 0 modulo 10.

   a) Verify that the UPC for Honey Nut Cheerios in Figure 5.26 is a valid code.

   b) Suppose that in the middle of typing this UPC you accidentally typed the pattern 655 instead of 665. Show why this would not be a valid UPC.

SOLUTION:

a) In the Honey Nut Cheerios UPC, which is 016000665903, the first digit, a_1, is 0, the second digit, a_2, is 1, the third digit, a_3, is 6, and so on. Therefore,

$$3a_1 + a_2 + 3a_3 + a_4 + 3a_5 + a_6 + 3a_7 + a_8 + 3a_9 + a_{10} + 3a_{11}$$
$$= 3 \times 0 + 1 + 3 \times 6 + 0 + 3 \times 0 + 0 + 3 \times 6 + 6 + 3 \times 5 + 9 + 3 \times 0 = 67.$$

To find the check digit, a_{12}, we must find a number such that $67 + a_{12} \equiv 0$ (mod 10). It is easy to see that if $a_{12} = 3$, then $67 + 3 \equiv 0$ (mod 10). Therefore, the check digit for this UPC is 3, so the UPC is a valid code.

b) If you had typed 016000655903 for the UPC, then in computing $3a_1 + a_2 + 3a_3 + a_4 + 3a_5 + a_6 + 3a_7 + a_8 + 3a_9 + a_{10} + 3a_{11}$ you would calculate $3 \times 0 + 1 + 3 \times 6 + 0 + 3 \times 0 + 0 + 3 \times 6 + 5 + 3 \times 5 + 9 + 3 \times 0 = 66$. This would mean that to have $66 + a_{12} \equiv 0$ (mod 10), the check digit would have to be 4, not 3. This means that you typed an invalid UPC. ✳

Identification numbers occur in many other places, such as on bank checks, U.S. postal money orders, driver's licenses, express shipping labels, airline tickets, and ISBN numbers on books. Although there are many different ways to calculate check digits other than the method that we explained in Example 6, most methods are based on modular arithmetic. We will discuss several of these other methods in the exercises.

---

*Don't let this notation scare you. All we are saying is that the UPC consists of 12 numbers. The first number is called a_1, the second is called a_2, and so forth.

Exercises 5.4

Looking Back*

These exercises follow the general outline of the topics presented in this section and will give you a good overview of the material that you have just studied.

1. In discussing arithmetic on a 12-hour clock, why did we replace 12 with 0?

2. What was a slow way to determine what 53 would be on a 12-hour clock? What would be a quicker way?

3. Give an example on a 12-hour clock to show why we don't need negative numbers when we do modular arithmetic.

4. What is the purpose of the RSA algorithm?

Sharpening Your Skills

Count to the number that is specified on the clock that is indicated.

5. 43 on a 12-hour clock

6. 21 on a 6-hour clock

7. 39 on a 7-hour clock

8. 69 on an 8-hour clock

Determine which of the following statements are true.

9. $35 \equiv 59 \pmod{12}$
10. $31 \equiv 14 \pmod 7$
11. $11 \equiv 43 \pmod 8$
12. $29 \equiv 18 \pmod 6$
13. $46 \equiv 78 \pmod{10}$
14. $53 \equiv 75 \pmod{11}$

Perform the following operations.

15. $8 + 9 \pmod{12}$
16. $4 + 7 \pmod 8$
17. $3 + 4 \pmod 5$
18. $6 + 8 \pmod{11}$
19. $4 - 9 \pmod{12}$
20. $5 - 6 \pmod 8$
21. $3 - 4 \pmod 5$
22. $3 - 8 \pmod{11}$
23. $4 \times 9 \pmod{12}$
24. $5 \times 6 \pmod 8$
25. $2 \times 5 \pmod 7$
26. $9 \times 3 \pmod{14}$

Solve the following congruences.

27. $3 + x \equiv 1 \pmod{12}$
28. $9 + x \equiv 6 \pmod{12}$
29. $6 + x \equiv 4 \pmod 8$
30. $5 + x \equiv 2 \pmod 6$
31. $3 - x \equiv 4 \pmod 7$
32. $5 - x \equiv 7 \pmod{10}$
33. $2 - x \equiv 8 \pmod 9$
34. $3 - x \equiv 4 \pmod 5$
35. $3x \equiv 5 \pmod 7$
36. $5x \equiv 3 \pmod 8$
37. $4x \equiv 2 \pmod{10}$
38. $6x \equiv 4 \pmod 8$

Find the smallest positive integer that solves all of the given congruences.

39. $x \equiv 3 \pmod 7, x \equiv 4 \pmod 5$
40. $x \equiv 1 \pmod 6, x \equiv 7 \pmod 8$
41. $x \equiv 2 \pmod{10}, x \equiv 2 \pmod 8, x \equiv 0 \pmod 6$
42. $x \equiv 5 \pmod 9, x \equiv 1 \pmod{11}, x \equiv 7 \pmod 8$

Applying What You've Learned

43. In a Jules Verne novel, Phileas Fogg took an imaginary trip around the world in 80 days. If he were counting the days of his trip on a 7-hour clock, what would the clock show when he landed?

44. Columbus's first voyage to America took 71 days. If he were counting the days of his voyage on a 7-hour clock, what would the clock show when he landed in America?

45. **Packaging candy.** A package designer for a chocolate manufacturer is designing boxes to hold a certain number of candies. When arranging the pieces six in a row, there will be four left over. When arranging the pieces five in a row, there will be three left over. What is the smallest number of pieces that the manufacturer might be considering to place in the box?

46. **Arranging a marching band.** A marching band is considering different configurations for its upcoming half-time show. When the members are arranged eight or 10 in a row, there are two members left over. If they are arranged 12 in a row, there are six left over. What is the smallest number of members that the band can have?

While taking a class on the culture of China, you have learned about the Chinese zodiac in which people fall into one of 12 categories, depending on the year of their birth. The categories, numbered 0 to 11, correspond to the following animals: (0) monkey, (1) rooster, (2) dog, (3) pig, (4) rat, (5) ox, (6) tiger, (7) rabbit, (8) dragon, (9) snake, (10) horse, and (11) goat. Those who believe in this zodiac think that the year of a person's birth influences both their personality and fortune in life.

To find your zodiac sign, divide the year of your birth by 12. The remainder then determines your sign. The remainder, when we divide 1985 by 12, is 5; therefore, a person born in 1985 is an ox according to the Chinese zodiac. Use this information to solve Exercises 47 and 48.

47. **Estimating a person's age.** You want to date Chris who is in your Chinese culture class. However, before asking for a date, you want to find out Chris's age without asking directly. Chris is at least 18 and seems to be younger than 30 years old. If according to the Chinese zodiac Chris is a dog, then in what year was Chris born?

48. **Estimating a person's age.** A person who is between 40 and 50 years old in the year 2003, and who according to the Chinese zodiac is a pig, was born in what year?

---

*Before doing these exercises, you may find it useful to review the note *How to Succeed at Mathematics* on page xix.*

If you look at the back cover of the student edition of this textbook, you will see a 10-digit number called an ISBN (International Standard Book Number). This book's ISBN is 0-321-56797-8. The first digit, 0, indicates that the book is from an English-speaking country. The digits 321 identify the publisher (in this case, Pearson Addison-Wesley), and the third group of digits 56797 identifies the book. The last digit, 8, is a check digit. Finding check digits for ISBNs is more complicated than finding check digits for UPCs. We calculate an ISBN check digit in three steps.

1. *Multiply the first digit by 10, the second digit by 9, the third digit by 8, continuing until you multiply the ninth digit by 2, as in Table 5.8. Then find the total of these products. In this example, the total is 179.*

$10 \times 0 =$	0
$9 \times 3 =$	27
$8 \times 2 =$	16
$7 \times 1 =$	7
$6 \times 5 =$	30
$5 \times 6 =$	30
$4 \times 7 =$	28
$3 \times 9 =$	27
$2 \times 7 =$	14
Total	179

TABLE 5.8 Table for calculating ISBN numbers.

2. *Divide the total by 11, keep the remainder, which is 3, and call it* r.
3. *Find the check digit* c *such that* r + c = 11. *In this case* c = 8.

Find the missing digit d *in each of the following ISBNs.*

49. *A Thousand Splendid Suns*—07435d4450
50. *Finding the Next Starbucks*—15918d1348
51. *Time and Materials*—0061349d07
52. *Breathe My Name*—1595140d48

Communicating Mathematics

53. Explain in your own words how to add, subtract, and multiply in a modulo-m system.
54. How would you solve a congruence in a modulo-12 system?
55. How would you solve a pair of congruences?

56. What is the purpose of the check digit in Example 7?
57. Browse the book, *Identification Numbers and Check Digit Schemes*, by Joseph Kirtland, (Mathematical Association of America, 2001) and report on a check digit scheme that we have not discussed in this section.

Using Technology to Investigate Mathematics

58. Your instructor can provide you with technology that will do modular arithmetic. Use that technology to reproduce some of the computations in this section.
59. Search the Internet to find several sites that discuss the RSA algorithm. Report on the details of how the algorithm works. For example, how do we choose n and e? Report on some applications of the RSA algorithm.
60. Find some Web sites that discuss modular systems and art, and report on your findings.

For Extra Credit

The RSA algorithm uses the equation c = m^e (mod n) *to encode a message. Here,* c *is the coded form of the letter* m. *We choose the numbers* n* *and* e *according to certain rules that we will not explain. To illustrate this method, suppose that we want to encode the letter C, which we will represent by the number 3. Let* n = 35 *and* e = 5. *(Do not be concerned how we picked 35 and 5.) Then to encode 3, we compute* c = 3^5 (mod 35) = 243 (mod 35) = 33. *So 33 is the coded form of the letter C. Use this method to encode each letter in Exercises 61–64.*

61. L   **62.** O
63. V   **64.** E

65. When we do usual division of integers, to say that $12/4 = 3$ is the same as saying that $12 = 4 \cdot 3$. More generally, if $a/b = x$, then $a = b \cdot x$. Use this idea to devise a way to perform division in a modulo-m system. That is, how would you explain how to solve a congruence such as $4/6 \equiv x$ (mod 8)? Use the method you devise to find
  a. 5/3 (mod 8)   **b.** 2/6 (mod 8)
66. Solve the congruence $2/x \equiv 5$ (mod 6).

---

*In practice, when using the RSA algorithm, the number *n* is a number having more than 500 digits, and unless you know how to factor *n* (see Section 6.1), it is impossible to break the code.

CHAPTER SUMMARY*

You will learn the items listed in this summary more thoroughly if you keep the following advice in mind:

1. Focus on "remembering how to remember" the ideas. What pictures, word analogies, and examples help you to remember these ideas?
2. Practice writing each item without looking at the book.
3. Make 3 × 5 flash cards to break your dependence on the book. Use these cards to give yourself practice tests.

SECTION	SUMMARY	EXAMPLE
SECTION 5.1	A **number** tells us how many objects we are counting; a numeral is a symbol that represents a **number**.	Discussion, p. 186
	In a **simple grouping system**, such as the Egyptian numeration system, the value of a number is the sum of the values of its numerals.	Example 1, p. 187
	Addition and **subtraction** in the Egyptian system are based on grouping and regrouping its basic symbols.	Examples 2 and 3, pp. 187–188
	The Egyptians performed multiplication using a "**doubling method**," which is based on the fact that we can write any positive integer as a sum of powers of 2.	Example 4, p. 189
	The **Roman numeration system** is more sophisticated than the Egyptian's system. It has a *subtraction principle* that we use to write numbers more concisely and a *multiplication principle* that makes it easier to write large numbers.	Example 5, p. 190 and discussion p. 190
	The **Chinese numeration system** is a *multiplicative system*. We form Chinese numerals by writing integers between 1 and 9, inclusive, multiplied by powers of 10.	Examples 6 and 7, pp. 191–192
SECTION 5.2	In a **place value system**, also called a **positional system**, the placement of the symbols in a numeral determines the value of the symbols.	Discussion 1, p. 194
	The **Babylonian numeration system** was a primitive place value system based on powers of 60 and is called a *sexagesimal system*.	Example 1, p. 195 Example 2, p. 196
	The **Hindu-Arabic numeration system** is a place value system based on 10. We write Hindu-Arabic numerals in **expanded notation** to show how each digit in the numeral is multiplied by a power of 10.	Discussion, p. 197 Example 3, p. 197
	The **Mayans** had a positional numeration system based on the number 20.	Historical Highlight, p. 198
	In the Hindu-Arabic system, expanded notation helps us to understand the algorithms to perform **arithmetic operations**.	Examples 4 and 5, pp. 197–199
	In the **galley method**, we compute partial products in each box of the galley and then add along diagonals to compute the final product.	Example 6, p. 199
	Napier's rods are a variation of the galley method that enables us to compute partial products quickly.	Discussion, p. 201
SECTION 5.3	**Counting** in a non-base-10 numeration system is very similar to counting in a base-10 system.	
	A **base-5 system** uses the digits 0, 1, 2, 3, and 4. In expanded notation, we use powers of 5 rather than powers of 10.	Example 1, p. 205
	Arithmetic operations in a non-base-10 system are done much like the operations in a base-10 system.	Examples 3, 4, and 5, pp. 207–209
	The **binary** system has a base of 2; the **octal** system has a base of 8; the **hexadecimal** system has a base of 16.	Example 7, p. 212
SECTION 5.4	A **modulo-*m* system** consists of the numbers 0, 1, 2, 3, . . . , $m - 1$. Counting and arithmetic operations are performed in a manner corresponding to movements on an *m*-hour clock. The number *m* is called the **modulus** of the system.	Discussion, p. 216
	We say that *a* **is congruent to** *b* **modulo** *m* if *m* evenly divides $a - b$.	Example 1, p. 217
	To **add, subtract, or multiply in a modular system**, we perform the operation as usual, divide the result by *m*, and keep the remainder.	Examples 2, 3, and 4, p. 219
	We **solve congruences** by using trial and error.	Examples 5 and 6, pp. 220–221

*Before reviewing this chapter, you will find it useful to reread the note *How to Succeed at Mathematics* on page xix.

CHAPTER REVIEW EXERCISES

Section 5.1

1. Write ![Egyptian numerals] using Hindu-Arabic numerals.

2. Find ![Egyptian numerals] using Egyptian notation.

3. Use the Egyptian method of doubling to calculate 37×53.

4. Write 4,795 in Roman notation.

5. Write ![Chinese numerals] as a Hindu-Arabic numeral.

6. Do the words *number* and *numeral* mean the same thing? Explain.

7. Explain two advantages of the Roman numeration system over the Egyptian system.

8. Why were people in Europe at first suspicious of Hindu-Arabic numerals?

Section 5.2

9. Write 11,292 using Babylonian notation.

10. Subtract $4,237 - 2,673$ using expanded notation.

11. Use the galley method to multiply 46×103.

12. Why is zero important in a place value system?

Section 5.3

13. Write the following Mayan numerals in Hindu-Arabic notation:

    a.

    b.

14. Write 342_5 and $B3D_{16}$ as base-10 numerals.

15. Write decimal 3,403 in base-8 notation.

16. Add $10,111_2 + 11,001_2$.

17. Divide $4,312_5 \div 23_5$.

18. Write $1,011,100,010_2$ first as an octal number and then as a hexadecimal number.

19. Convert 463_7 to base-5 notation.

Section 5.4

20. Count to 76 on an 8-hour clock.

21. Determine which of the following statements are true:

    a. $54 = 72 \ (\text{mod } 6)$ b. $29 = 75 \ (\text{mod } 11)$

22. Perform the indicated operations.

    a. $8 + 11 \ (\text{mod } 12)$ b. $7 - 9 \ (\text{mod } 11)$

    c. $4 \times 8 \ (\text{mod } 9)$

23. Solve the following congruences:

    a. $3x \equiv 9 \ (\text{mod } 12)$ b. $4x \equiv 3 \ (\text{mod } 8)$

CHAPTER TEST

1. Write 3,685 in Roman notation.

2. Give two advantages of the Hindu-Arabic numeration system over the Chinese numeration system.

3. Write 264_7 and $A3E_{16}$ as base-10 numerals.

4. Write ![Egyptian numerals] using Hindu-Arabic numerals.

5. Write the following Mayan numerals in Hindu-Arabic notation:

    a.

    b.

6. Determine which of the following statements are true:

    a. $43 \equiv 57 \ (\text{mod } 8)$ b. $16 \equiv 52 \ (\text{mod } 6)$

7. Find ![Egyptian numerals] − ![Egyptian numerals] using Egyptian notation.

8. Write ![Chinese numerals] as a Hindu-Arabic numeral.

9. Count to 57 on a 6-hour clock.

10. Give an example of three different numerals that represent the same number.

11. Add $101,101_2 + 110,101_2$.

12. Write 10,937 using Babylonian notation.

13. Use the galley method to multiply 67×238.

14. Write decimal 2,305 as a base-8 numeral.

15. Perform the indicated operations.

    a. $9 + 12 \ (\text{mod } 13)$ b. $6 - 10 \ (\text{mod } 12)$

    c. $5 \times 4 \ (\text{mod } 6)$

16. Add $1,738 + 526$ using expanded notation.

17. What problem did the lack of zero cause in the ancient Chinese system?

18. Solve the following congruences:

    a. $4x \equiv 8 \ (\text{mod } 12)$

    b. $4x \equiv 7 \ (\text{mod } 10)$

19. Find $3,142_5 \div 24_5$.

20. Use the Egyptian method of doubling to calculate 26×59.

21. Write $110,101,011,110_2$ first as an octal number and then as a hexadecimal number.

22. Find $42_5 \times 34_5$.

GROUP PROJECTS

1. *In Example 7, in Section 5.3 we discussed a shortcut method to make conversions between binary, octal, and hexadecimal. These conversions worked because 8 and 16 are powers of 2. With this in mind, you would expect to be able to make similar types of conversions between bases where one base is a power of the other—such as between 3 and 9, 2 and 4, and 4 and 16. Use this notion to invent conversion techniques to solve Exercises a)–f).*

   a. Convert $201{,}221_3$ to base 9.

   b. Convert $101{,}101{,}011{,}101_2$ to base 4.

   c. Convert $A0B5_{16}$ to base 4.

   d. Convert 768_9 to base 3.

   e. Convert $301{,}233_4$ to base 16.

   f. Convert $312{,}032_4$ to base 16.

2. Research the ancient Chinese rod numerals and answer the following questions*:

   a. What are the two ways to write numerals in this system?

   b. What is the base for this system?

   c. How did this system handle zero?

   d. How did this system represent decimals?

   e. When was this system created and what other countries adopted this system?

3. There are various methods based on modular arithmetic that can be used to calculate the day of the week for any date in the past or the future. Investigate the three methods below and calculate the day of the week for several dates. To make sure that you understand a method, calculate the day using the date on the present day.

   a. Zeller's congruence (In order to understand this method, you have to understand the "floor function." We use the notation $\lfloor x \rfloor$ to indicate the floor of x, which means the greatest integer that is less than or equal to x. To calculate the floor for positive numbers, we simply drop the decimal part of a number. For example $\lfloor 13.63 \rfloor = 13$, $\lfloor 2.34 \rfloor = 2$, and $\lfloor 8 \rfloor = 8$.)

   b. The Doomsday rule

   c. Babwani's formula

---

*We will not provide answers for this exercise, but you should be able to find answers by searching the Internet.

Number Theory and the Real Number System

6

Understanding the Numbers All Around Us

Atoms are very tiny things. In fact, they are so small that it would take about 10 million of them to make a length that is the width of the capital "T" in the word "Theory" in the title of this chapter.

If we were able to line up all the atoms in your body and stretch them out in a straight line, how far do you think they would reach? Would they stretch from New York City to Chicago? Could they stretch around the world? We will answer this question later in this chapter, and the answer will astound you.

Have you ever thought about what it would be like to take a trip around our solar system? If a rocket were to travel at 25,000 miles per hour, how long would it take to travel to a distant planet such as Neptune?

(continued)

On a less fanciful (and somewhat scarier) note, do you have any understanding of how the national debt has changed in the last 50 years? Since 1960, the U.S. population has almost doubled, but the national debt has grown from $284 billion to over $9 trillion. How can we make such large numbers understandable?

In this chapter, you will first work with different common number systems and then will study lists of numbers called *sequences*. Some of the sequences that you will study have interesting applications in mathematics and also in nature.

The author John Paulos wrote a book about a condition that he calls *innumeracy*, which is the mathematical equivalent of illiteracy. After studying this chapter, you can certainly expect to be more "numerate." ●

6.1 Number Theory

Objectives

1. Determine when numbers are prime.
2. Apply divisibility tests to factor numbers into products of prime factors.
3. Find the GCD and LCM of two natural numbers.
4. Use the GCD and LCM to solve applied problems.

From the time you were a small child, you have known about a set of numbers that has fascinated the greatest minds in the history of mathematics for thousands of years. The numbers that we use for counting, {1, 2, 3, . . .}, are called the **natural numbers**, or **counting numbers**. This simple set of numbers has elegant properties and patterns that have intrigued mathematicians from ancient times to the present. The study of the natural numbers and their properties is called **number theory**.

 **KEY POINT**

Prime numbers are the building blocks of the integers.

Prime Numbers

One question that people have studied since the time of the ancient Greeks is exactly which natural numbers can be written as a product of *other* natural numbers and which cannot. For example, $30 = 2 \cdot 3 \cdot 5$, whereas $17 = 17 \cdot 1$. The fact that we cannot write 17 as a product of natural numbers without using 17 greatly interests mathematicians. In order to investigate this question, we have to explain what it means for one natural number to divide another.

> **DEFINITIONS** If a and b are natural numbers, then we say a **divides** b, written $a|b$, provided there is a natural number q such that $b = qa$. Other ways of stating this is that a is a **divisor** of b, a is a **factor** of b, or that b is a **multiple** of a. A way of testing to see if a divides b is to actually divide b by a and see if you get a zero remainder.

For example, we can say that 5 divides 30, because there is a natural number, namely 6, such that $30 = 5 \cdot 6$. We can write this fact more concisely as $5|30$. Also, $17|17$ because $17 = 17 \cdot 1$. When we write a natural number as a product of natural numbers, we say that we have **factored** the number. Several factorizations of 72 are $8 \cdot 9$ and $2 \cdot 2 \cdot 2 \cdot 3 \cdot 3$.

Numbers like 17, which have only trivial factorizations, are particularly important in number theory.

The TI-84 screen below shows how you can use a calculator to test for divisibility. Because 5 divides 30, no digits follow the decimal point. Because 7 does not divide 45, you see several digits after the decimal point.

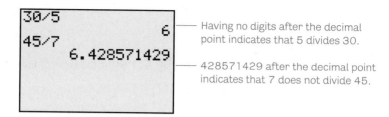

Having no digits after the decimal point indicates that 5 divides 30.

428571429 after the decimal point indicates that 7 does not divide 45.

> **DEFINITIONS** A natural number, greater than 1,* which has only itself and 1 as factors, is called a **prime number**. A natural number, greater than 1, that is not prime is called **composite**.

Numbers such as 2, 5, 17, and 29 are prime numbers because their only divisors are themselves and 1. On the other hand, numbers such as 4, 33, 87, and 102 are composite because you can find divisors for them that are other than 1 and the number itself.

One very well-known way to generate a list of primes is called the **Sieve of Eratosthenes**, which is named after the Greek mathematician Eratosthenes of Alexandria, whom we discuss in the Historical Highlight on page 233. We explain his process in Example 1.

EXAMPLE 1   *Finding a List of Prime Numbers*

Use the Sieve of Eratosthenes to find all prime numbers less than 50.

SOLUTION: First we list all natural numbers from 1 to 50 in Figure 6.1 and then systematically cross off all nonprimes according to the following steps:

1. Cross off 1 because it is not prime.
2. Circle 2 because it is prime and then cross off all other multiples of 2, such as 4, 6, 8, . . . , 50.
3. The next number in the list that has not been crossed off is 3, which is prime. Circle it and cross off all remaining multiples of 3, namely 9, 15, 21, . . . , 45.
4. Next circle 5, a prime, and cross off its remaining multiples, 25 and 35.
5. Finally circle 7 and cross off its one remaining multiple, 49.
6. Because the next prime, which is 11, is greater than the square root of 50, we can stop looking for composites and circle all remaining numbers in the list. If you think about it, you will see that it is impossible to have a composite number of the form $a \times b$ where both a and b are 11 or greater and yet the product is less than 50.

It would be a good idea for you to reproduce the above steps† to produce Figure 6.1 from scratch.

Now try Exercises 15 and 16. ✳

---

*There are technical reasons, which we will not get into, why mathematicians consider 1 to be neither prime nor composite.

†If you wanted to find all primes between 1 and 500, it would take a long time to create the list. You can find interactive Java applets on the Internet that will allow you to use this sieve without the tedium of writing a long list of numbers.

After crossing off multiples of 2, 3, 5, and 7, the numbers remaining in the table are prime.

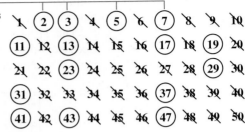

FIGURE 6.1 Using the Sieve of Eratosthenes.

 Some Good Advice

When doing problems, you will frequently need to know some of the smaller prime numbers. It is a good idea to memorize the following primes: 2, 3, 5, 7, 11, 13, 17, 19, 23, 29, 31, 37, 41, 43, 47.

EXAMPLE 2 *Identifying Prime Numbers*

Determine whether the following numbers are prime or composite.

a) 83   b) 87   c) 397

SOLUTION:

a) To see if 83 has any factors, we start dividing it by primes such as 2, then 3, then 5, and so on. *We don't need to check to see if any composites divide 83, because if a composite like 6 divides 83, then the primes 2 and 3 would have divided 83.* We keep checking for divisibility by primes until we reach the square root of 83. Notice that

$$9 < \sqrt{83} < 10.$$

$9^2 = 81$, so 9 is less than the square root of 83. $10^2 = 100$, so 10 is greater than the square root of 83.

As in Example 1, if you do not find a prime divisor of 83 by the time you reach the square root of 83, you will not find another divisor until you reach 83 because multiplying two integers greater than 9 will give a product that is at least 100. Because 2, 3, 5, and 7 do not divide 83, we conclude that 83 is prime.

b) The number 87 factors as $3 \cdot 29$. Again, in testing 87, you only need to check primes up to the square root of 87, which is between 9 and 10.

c) Because $\sqrt{397} \approx 19.92$, to determine if 397 is prime, we must check all primes less than or equal to 19 to see if they divide 397. You can do this by hand, or to speed the process, you can use your calculator, as we show in the accompanying TI-84 screen. The nonzero digits after the decimal point show that neither 2 nor 3 divides 397 evenly. If you do the division, you will find that no primes from 2 to 19 divide 397; therefore, 397 is prime.

```
397/2
              198.5
397/3
         132.3333333
```

Quiz Yourself ❶ *

Identify the following numbers as either prime or composite.

a) 89   b) 187
c) 143   d) 211

Now try Exercises 21 to 28. ✳ ❶

Divisibility Tests and Factoring

In checking larger numbers for primality, it is useful to have some quick tests for divisibility so that we don't actually have to do a lengthy division. For example, it would be nice to be able to look at the number 21,021 and have a quick test to see whether 3 divides into it without actually having to do the division. Table 6.1 lists some divisibility tests. We will examine why these tests work in the exercises.

Although there are divisibility tests for 7 and also 11, they are hard to remember, and you are really better off simply dividing by 7 or 11 rather than using the test.

Number Is Divisible by	Test	Example
2	The last digit of the number is divisible by 2.	2 divides 13,578 because 2 divides 8.
3	The sum of the digits is divisible by 3.	3 divides 21,021 because 3 divides $2 + 1 + 0 + 2 + 1 = 6$.
4	The number formed by the last two digits is divisible by 4.	102,736 is divisible by 4 because 4 divides 36.
5	The last digit is 0 or 5.	607,895 is divisible by 5.
6	The number is divisible by both 2 and 3.	802,674 is divisible by both 2 and 3, so it is divisible by 6.
8	The number formed by the last three digits is divisible by 8.	8 divides 230,264 because 8 divides 264.
9	The sum of the digits is divisible by 9.	2,081,763 is divisible by 9 because 9 divides $2 + 0 + 8 + 1 + 7 + 6 + 3 = 27$.
10	The number ends in 0.	12,865,890 is divisible by 10.

TABLE 6.1 Divisibility tests for some small numbers.

Math in Your Life

What's the Buzz on Prime Numbers?*

During the late spring of 2004, billions of mysterious insects, called *cicadas*, emerged from the ground in the eastern United States to live briefly in the fresh air and sunshine, mate, and then die. You might wonder, "What does this incredible swarm of insects have to do with mathematics?" Well, periodical cicadas, as they are called, emerge every 13 or 17 years—a curious choice of numbers. Why would the cicadas settle in on prime numbers for their life cycles?

Paleontologist Stephen Gould proposed that by emerging every 13 or 17 years, cicadas minimize their chance of synchronizing with predators who will destroy them. For example, a 17-year cycle for the cicadas would synchronize with a 5-year predator only every 85 years. Mathematician Glenn Webb, working at Vanderbilt University, has created mathematical models to investigate Gould's theory and has found that indeed by emerging every 13 or 17 years, cicadas do improve their chances of survival. Other scientists, however, are skeptical and believe that it is only a coincidence that the cicada's life cycles happen to be prime numbers. Webb admits, "I don't know if there will ever be a satisfying scientific resolution."

---

*This note is based on "Mathematicians Explore Cicada's Mysterious Link with Primes," *The Baltimore Sun*, May 10, 2004 p. 10.A.

EXAMPLE 3   *Applying Divisibility Tests*

Test the number 11,352 for divisibility by

a) 2   b) 3   c) 4   d) 5

e) 6   f) 8   g) 9   h) 10

SOLUTION:

a) The last digit of 11,352 is divisible by 2, so 2 divides the number.

b) The sum of the digits is $1 + 1 + 3 + 5 + 2 = 12$, which is divisible by 3, so 11,352 is also divisible by 3.

c) The last two digits of 11,352 form the number 52, and 52 is divisible by 4; therefore, 11,352 is also divisible by 4.

d) The last digit is neither 0 nor 5, so 5 does not divide this number.

e) Both 2 and 3 divide 11,352; therefore, 6 also divides this number.

f) The number formed by the last three digits is 352, which is divisible by 8, so 8 divides 11,352.

g) In part b), we saw that the sum of the digits of 11,352 is 12. Because 9 does not divide 12, 9 does not divide 11,352.

h) The number does not end in 0, so 10 does not divide the number.

Now try Exercises 29 to 36. ❈ **2**

Quiz Yourself **2**

Test the number 30,690 for divisibility by

a) 2   b) 3   c) 4

d) 5   e) 6   f) 8

g) 9   h) 10

In chemistry, we form compounds by combining simpler objects called atoms. For example, a molecule of table salt is formed by combining one sodium and one chlorine atom. Similarly, in mathematics, objects are built from simpler objects. For example, $120 = 2 \cdot 2 \cdot 2 \cdot 3 \cdot 5$ and $84 = 2 \cdot 2 \cdot 3 \cdot 7$. The fundamental theorem of arithmetic states that every natural number greater than 1 is "built" by multiplying a unique combination of prime numbers. This theorem appeared in another way in Euclid's *Elements* and was stated in its present form by the great German mathematician Karl Friedrich Gauss.

> **THE FUNDAMENTAL THEOREM OF ARITHMETIC** Every natural number greater than 1 is a unique product of prime numbers, except for the order of the factors. (Product could mean a single prime number.)

Example 4 shows us one way to find the prime factorization of a natural number by using factor trees.

EXAMPLE 4   *Factoring a Natural Number*

Factor 4,620.

SOLUTION: To factor 4,620, we first try to think of a way to write it as a product of two smaller numbers. For example, $4,620 = 462 \cdot 10$. It really doesn't matter how you write it as a product, *what is important is that you somehow express 4,620 as a product of smaller, simpler numbers.* We can represent this preliminary factorization graphically by the diagram in Figure 6.2.

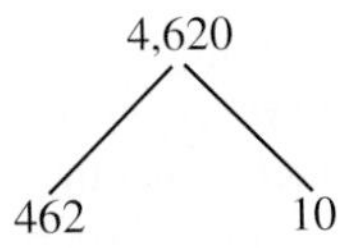

We call this a **factor tree**. As we add more branches, you will see that this diagram looks like an upside-down tree. Next, we factor 462 and 10 and add new branches to the tree, as in Figure 6.3(a). Using the divisibility test, we see that 3 divides 231.

If we now multiply all the primes at the ends of the branches (shown in red) in Figure 6.3(b), we see that $4,620 = 2 \cdot 3 \cdot 7 \cdot 11 \cdot 2 \cdot 5 = 2^2 \cdot 3 \cdot 5 \cdot 7 \cdot 11$.

FIGURE 6.2
A first factorization of 4,620.

HISTORICAL HIGHLIGHT ✵ ✵ ✵

Eratosthenes

One of the Greeks who made great contributions to the field of number theory was Eratosthenes (276–194 BC). He studied at Plato's academy in Athens and was considered to be one of the most learned men of antiquity. Eratosthenes was a scholar in many areas, including geography, philosophy, history, astronomy, mathematics, and literary criticism. In fact, he had so many varied interests that his friends called him Pentathis, a name given to a champion in five athletic events. His enemies, however, called him Beta, the second letter of the Greek alphabet, charging that although he worked in many fields, he was outstanding in none.

Many would argue that Eratosthenes was certainly first-rate in both mathematics and geography. He was the first to try to put geography on a sound scientific basis. In his work *Geographia*, he argued that the world was round—1,700 years before Columbus made his famous voyage in 1492. He also produced the most accurate map of the known world of his time and was the first to use a grid of lines of longitude and latitude that is still used in maps today. Although writers of mathematics textbooks remember him for his prime sieve, he is probably more famous for devising a method for measuring the circumference of Earth.*

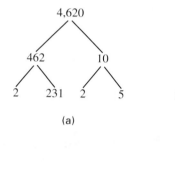

FIGURE 6.3 Completing the factorization of 4,620, using a factor tree.

Quiz Yourself ③

a) Draw a factor tree to factor 1,560.

b) Write the factorization.

Now try Exercises 41 to 48. ✳ ③

If you were factoring a number like 2,400 and recognized that it is the product of 24 and 10 and 10, then you could begin drawing your tree with three branches, as in Figure 6.4. As we said, it really doesn't matter how you start your tree, the important thing is to express 2,400 as a product of smaller natural numbers.

It is important to understand that the fundamental theorem of arithmetic tells us that although we may factor a natural number several different ways, we will always end up with the same factorization when we are finished.

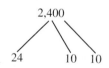

FIGURE 6.4
A possible first step in factoring 2,400.

Greatest Common Divisors and Least Common Multiples

In many applications of number theory, we need to find the greatest common divisor and least common multiple of natural numbers.

✎ **KEY POINT**

We use prime factorization to find the GCD and LCM of numbers.

> **DEFINITION** The **greatest common divisor** or **GCD** of two natural numbers is the largest natural number that divides both numbers.

For small numbers, you can find the GCD by looking at the numbers and identifying the largest natural number that divides both numbers. For example, to find the GCD of

---

*The method we explain in Example 3 of Section 10.1 is essentially the one that Eratosthenes invented.

24 and 40, we list the divisors of both numbers. The common divisors are highlighted in red.

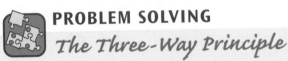

Divisors of 24: 1, 2, 3, 4, 6, 8, 12, 24

Divisors of 40: 1, 2, 4, 5, 8, 10, 20, 40

From these lists, we see that the largest number dividing both 24 and 40 is 8.

For larger numbers, it is tedious to list all divisors of both numbers, so we use the approach shown in Example 5, which involves factoring both numbers.

EXAMPLE 5   *Using Prime Factorization to Find the GCD*

Find the GCD of 600 and 540.

SOLUTION: We first factor both numbers using factor trees to get

One 3 divides both. ⌐   ⌐ One 5 divides both.
$$600 = 2 \cdot 2 \cdot 2 \cdot 3 \cdot 5 \cdot 5 \text{ and } 540 = 2 \cdot 2 \cdot 3 \cdot 3 \cdot 3 \cdot 5.$$
                                              └─ Two 2s divide both.

If we think about what "greatest common divisor" means, we recognize that we need to find as many primes as possible that divide *both* numbers at the same time. We see that two 2s will divide both 600 and 540; however, three 2s will not. Similarly, only one 3 divides both numbers and only one 5 divides both numbers. Therefore, the GCD of 600 and 540 is $2 \cdot 2 \cdot 3 \cdot 5 = 60$. ✿ ④

Quiz Yourself ④

Find the GCD of 126 and 588.

Another way of thinking about what we did in Example 5 is to write $600 = 2^3 \cdot 3^1 \cdot 5^2$ and $540 = 2^2 \cdot 3^3 \cdot 5^1$. Then in forming the GCD, we multiplied the 2^2, the 3^1, and 5^1, which were the *smallest powers* of the primes that divide both numbers. So the GCD is $2^2 \cdot 3^1 \cdot 5^1$.

PROBLEM SOLVING

The Three-Way Principle

The meaning of "greatest common divisor" tells you how to compute the GCD of two numbers. The word *greatest* tells you to find the *largest* natural number that divides evenly into *both* numbers. That is why, for instance, we used 3^1, rather than 3^3, in constructing the GCD of 600 and 540, because 3^1 divides both numbers; however, 3^3 does not divide 600.

Next we will discuss the least common multiple of two numbers.

DEFINITION The **least common multiple** or **LCM** of two natural numbers is the smallest natural number that is a multiple of both numbers.

To find the LCM of small numbers such as 8 and 6, we could list multiples of both numbers and then choose the smallest common multiple as follows:

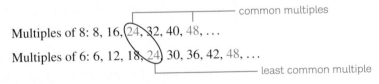

Multiples of 8: 8, 16, 24, 32, 40, 48, ...

Multiples of 6: 6, 12, 18, 24, 30, 36, 42, 48, ...

However, for larger numbers, we can use a prime factorization approach to find the LCM, similar to what we did in Example 5.

EXAMPLE 6 *Using Prime Factorization to Find the LCM*

Find the LCM of 600 and 540.

SOLUTION: As in Example 5, we start by factoring both numbers. Recall that

600 requires three 2s.

$$600 = 2 \cdot 2 \cdot 2 \cdot 3 \cdot 5 \cdot 5 \text{ and } 540 = 2 \cdot 2 \cdot 3 \cdot 3 \cdot 3 \cdot 5.$$

540 requires three 3s.

600 requires two 5s.

Now, think about the words *least common multiple*. Here we recognize that we need to find the *smallest* natural number that is *divisible by all the prime factors* of 600 and 540. However, we don't want to use any more primes in our number than we absolutely have to. The LCM will have to have three 2s, three 3s, and two 5s. So, the LCM of 600 and 540 is

$$2 \cdot 2 \cdot 2 \cdot 3 \cdot 3 \cdot 3 \cdot 5 \cdot 5 = 5{,}400.$$

Although this number seems large, realize that we cannot omit any of the prime factors. For example, if we omit one of the 3s, then 540 will not divide into the number. If we omit one of the 5s, then 600 will not divide into the number.

Now try Exercises 49 to 56. ❈

Another way of thinking about what we did in Example 6 is to write $600 = 2^3 \cdot 3^1 \cdot 5^2$ and $540 = 2^2 \cdot 3^3 \cdot 5^1$. Then in forming the LCM, we multiplied the 2^3, the 3^3, and 5^2, which were the *highest* powers of the primes that divide either number. So, the LCM is $2^3 \cdot 3^3 \cdot 5^2$.

> **FINDING THE GCD AND LCM BY USING FACTORIZATION** To find the GCD and LCM of two numbers, do this:
>
> 1. Factor both numbers, and write each as a product of powers of primes.
> 2. To calculate the GCD, multiply the *smallest* powers of any primes that are common to both numbers.
> 3. To calculate the LCM, multiply the *largest* powers of all primes that occur in either number.

If we look at what happened in Examples 5 and 6 carefully, we see the following pattern:

$$600 = \boxed{2^3} \enspace \textcircled{3^1} \enspace \boxed{5^2}$$
$$540 = \textcircled{2^2} \enspace \boxed{3^3} \enspace \textcircled{5^1}$$

You see that when we multiply the circled factors we get the GCD, and when we multiply the boxed factors we get the LCM. This means that the GCD times the LCM gives us the product of 600 and 540. Another way of saying this is that once you have found the factors that make up the GCD, the product of the remaining factors is the LCM.

Applying the GCD and LCM

We find applications of the GCD in situations where we want to take several larger objects and represent them as a collection of smaller objects of the same size. For example, a clerk in a store may want to pile different types of dishes in piles of the same size, or a cabinet maker might want to cut several boards of different lengths to make smaller boards of the same size.

The LCM occurs in situations where we have two types of objects and we want to take some of each to get a larger object of the same size. For example, one person has every fifth day off and another may have every sixth day off, and we ask when will both be off together.

For over 40 years, Japan has been a leader in developing high-speed trains that can travel at over 300 miles per hour with remarkable reliability and safety records. The next example discusses a problem of scheduling the so-called "bullet trains" leaving the Tokyo station.

❋ ❋ ❋ HIGHLIGHT

There's Big Money in Big Numbers

Mathematicians are fascinated with finding certain types of very large prime numbers. A prime p is called a Sophie Germain prime (see Historical Highlight on page 254) if $2p + 1$ is also prime. For example, 2, 3, 5, and 11 are Sophie Germain primes because $2 \cdot 2 + 1 = 5$, $2 \cdot 3 + 1 = 7$, $2 \cdot 5 + 1 = 11$, and $2 \cdot 11 + 1 = 23$ are prime numbers. Thirteen is not a Sophie Germain prime because $2 \cdot 13 + 1 = 27$, which is not prime. It is conjectured that there are infinitely many Sophie Germain primes, but this has not been proven. At the time of writing this Highlight, the largest known Sophie Germain prime was $48047305725 \times 2^{172403} - 1$, which, when written out, has 51,910 decimal digits. Although this is a large number, it is nowhere close to being the largest known prime. In 2006, Nayan Hajratwala won a $50,000 prize for finding a Mersenne prime (a prime of the form $2^p - 1$) that had over 9 million digits. The Electronic Frontier Foundation will award $100,000 to the first person who finds a prime number with 10 million digits and $250,000 for a prime with 1 billion digits.*

EXAMPLE 7

Assume that bullet trains have just departed from Tokyo to Osaka, Niigata, and Akita. If a train to Osaka departs every 90 minutes, a train to Niigata departs every 120 minutes, and a train to Akita departs every 80 minutes, when will all three trains again depart at the same time?

SOLUTION: Trains to Osaka will leave in 90, 180, 270, . . . minutes; trains to Niigata leave in 120, 240, 360, . . . minutes; trains to Akita leave in 80, 160, 240, . . . minutes. We want to know the smallest number of minutes until all three trains again leave at the same time. That is, we want to find the LCM of 90, 80, and 120. We can factor these numbers as follows:

greatest powers of 3 and 5 in any of the three numbers

greatest power of 2 in any of the three numbers

$$90 = 2 \times 3^2 \times 5 \qquad 80 = 2^4 \times 5 \qquad 120 = 2^3 \times 3 \times 5$$

Thus, the LCM of these three numbers is $2^4 \times 3^2 \times 5 = 720$ minutes, or, dividing by 60 minutes, we get $\frac{720}{60} = 12$ hours.

Now try Exercises 67 to 73. ❋

Exercises ⟨6.1⟩

Looking Back†

These exercises follow the general outline of the topics presented in this section and will give you a good overview of the material that you have just studied.

1. In determining that 83 was prime in Example 2, why did we stop trying to find factors after 7?

2. How are the divisibility tests for 2, 4, and 8 similar? How are the divisibility tests for 3 and 9 similar?

3. When did we stop adding branches to the factor tree in Example 4?

4. What is the result if we multiply the GCD of a and b by the LCM of a and b?

5. Mention a contribution that Eratosthenes made to science outside of mathematics.

6. What is one theory as to how it benefits cicadas to emerge every 13 or 17 years?

---

*You can join in on the search for this number by going to www.mersenne.org.
†Before doing these exercises, you may find it useful to review the note *How to Succeed at Mathematics* on page xix.

Sharpening Your Skills

Determine whether each of the following statements is true.

7. 8 divides 56.

8. 21 is a multiple of 2.

9. 27 is a multiple of 6.

10. 5 is a divisor of 35.

11. 9 is a factor of 96.

12. 14 divides 42.

13. 7 is a divisor of 63.

14. 6 is a factor of 76.

15. Use the Sieve of Eratosthenes to find all prime numbers between 51 and 100, inclusive.

16. Use the Sieve of Eratosthenes to find all prime numbers between 101 and 120, inclusive.

Estimate the square root of each of the following numbers, n, by finding a natural number a such that $a < \sqrt{n} < a + 1$.

17. 95

18. 138

19. 153

20. 229

Identify the following numbers as either prime or composite. If a number is composite, factor it as a product of two smaller natural numbers (not necessarily prime numbers).

21. 231

22. 89

23. 113

24. 153

25. 197

26. 143

27. 119

28. 137

Check each of the following numbers for divisibility by each of these numbers: 2, 3, 4, 5, 6, 8, 9, 10. State the numbers that divide the given number.

29. 141,270

30. 18,036

31. 47,385

32. 476,376

Find the smallest natural number that is divisible by

33. 2, 3, 5, and 6.

34. 2, 4, 8, and 5.

35. 4, 5, and 10.

36. 2, 3, 4, 5, 6, and 10.

Provide a counterexample to show that each of the following statements is false.

37. If 2 and 4 divide *a*, then 8 divides *a*.

38. If 3 and 6 divide *a*, then 18 divides *a*.

39. If 10 and 4 divide *a*, then 40 divides *a*.

40. If 4 and 6 divide *a*, then 24 divides *a*.

Factor each of the following natural numbers.

41. 980

42. 396

43. 9,900

44. 2,106

45. 621

46. 805

47. 319

48. 403

Find the GCD and LCM of each of the following pairs of natural numbers.

49. 20, 24

50. 60, 72

51. 56, 70

52. 66, 110

53. 216, 288

54. 675, 1,125

55. 147, 567

56. 275, 363

There is another way to find the GCD of two natural numbers called the **Euclidean algorithm**. *The Euclidean algorithm is as follows:*

Suppose that we want to find the GCD of two numbers such as 24 and 88.

First we divide the smaller number, 24, into the larger number, 88, as shown in Figure 6.5.

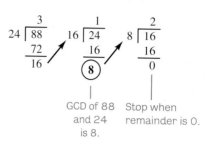

FIGURE 6.5 The Euclidean algorithm.

Next we divide the remainder, 16, into the previous divisor, 24, to get a new remainder, 8. We continue this process, always dividing the remainder into the previous divisor. So, next we divide 8 into 16. At this point, the remainder is 0, so we stop. The previous remainder, 8, is the GCD of 24 and 88.

Use the Euclidean algorithm to find the GCD of each of the following pairs of natural numbers.

57. 280, 588

58. 84, 1,200

59. 495, 1,575

60. 99, 1,155

It can be proved that if you multiply the GCD and LCM of two natural numbers a and b, the product is the same as the product a · b. In Exercises 61–64, first find the GCD of the two numbers and then divide the GCD into the product of the two numbers to find the LCM. To find the GCD, you can use the Euclidean algorithm.

61. 12, 27

62. 16, 56

63. 90, 120

64. 17, 178

Applying What You've Learned

65. Cicadas. In addition to avoiding their predators, another advantage of the 13- and 17-year emergence pattern for cicadas is that the two species rarely emerge during the same year to compete for food. Suppose that type A cicada has a 13-year emergence cycle and type B cicada has a 17-year cycle. If both type A and type B cicadas emerge during 2010, when is the next year that both cicadas will emerge during the same year?

66. Cicadas. Continuing the discussion in Exercise 65, suppose that a predator of both type A and type B cicadas emerges every 5 years. If the predator also emerged in 2010, when will both types of cicadas and the predator all emerge together again?

67. Scheduling flights. Assume that Jet Blue flights for Miami leave every 35 minutes and flights for Dallas leave every 20 minutes. If flights to Miami and Dallas have just departed, how many minutes will it be before this will happen again?

68. Scheduling nurses' shifts. Carla and Laverne are nurses who occasionally work in the emergency room. Carla works in the emergency room every 6 weeks and Laverne works in the emergency room every 8 weeks. If Carla and Laverne have just worked together in the emergency room, in how many weeks will they work in the emergency room together again?

69. Storing medical supplies. In a medical supply room, there are 36 packages of type O positive blood and 30 packages of type AB negative blood. It is important not to confuse these two types of blood. If we want to stack piles of each type of blood on a shelf, with only one type of blood in a pile, what is the largest number of packets of blood that we can have in each pile if each pile is of the same size?

70. Displaying store merchandise. A sporting goods store has 20 instructional DVDs on skiing and 12 DVDs on snowboarding. The owner of the store wants to display the DVDs on a shelf with stacks of the same size and each stack consisting of only one type of DVD. What is the most number of DVDs in each stack that will accomplish this?

71. Tiling a museum floor. In the Central America room of the Ancient Civilization Museum, we want to tile the floor with replicas of ancient Aztec tiles. If the floor measures 33 feet by 21 feet, what is the largest-size square tile that we can use to tile the floor without having to cut any tiles?

72. Scheduling lawn service. At the Berkshire Country Club, the lawn service cuts the grass every 8 days and the pest service sprays for insects every 30 days. If both services have just been to the country club, in how many days will both be there again?

73. Monitoring pollution. The Environmental Protection Agency is monitoring the Lackawanna River to determine the cause of a fish kill. A steel mill releases hot water into the river every 48 hours, and a plastics plant discharges pollutants into the river every 54 hours. If both hot water and pollutants from the plastics plant have just been discharged, in how many hours will both be released into the river again?

74. Movie showings. At the 24/7 Monster MoviePlex, *Harry Potter* shows every 120 minutes, *The Dark Knight* shows every 140 minutes, and *Indiana Jones* shows every 90 minutes. If all three movies have just begun, in how many minutes will they again all begin at the same time?

Communicating Mathematics

75. If you were using the Sieve of Eratosthenes to find the primes up to 300, what is the largest prime whose multiples you would cross off before you knew that the remaining numbers in the table were prime?

76. Explain how you would use the prime factorization method to find the GCD of two natural numbers.

77. Explain how you would use the prime factorization method to find the LCM of two natural numbers.

78. If we were to give you two numbers and their GCD, what is a quick way to find the LCM without doing any factoring?

79. Explain why the divisibility test for 4 works by considering the number 36,824. It helps to think of 36,824 as 36,800 + 24.

80. Explain why the divisibility test for 3 works by considering the number 5,712. It helps to think of the following equations:

$$5,712 = 5(1,000) + 7(100) + 1(10) + 2$$
$$= 5(999 + 1) + 7(99 + 1) + 1(9 + 1) + 2$$
$$= 5 \cdot 999 + 7 \cdot 99 + 1 \cdot 9 + (5 + 7 + 1 + 2)$$

81. Explain why the rules that we stated for divisibility in Exercises 37 to 40 failed.

82. Explain how you would apply the Euclidean algorithm to find the GCD of three numbers.

Using Technology to Investigate Mathematics

Use your calculator to test the following numbers to see if they are prime. To do this,

1. *Find the square root of the number,* n, *that you are checking.*

2. *Divide* n *by all prime numbers up to and including* $\sqrt{n}$. *If none of these primes divides* n, *then* n *is a prime number.*

83. 493   **84.** 577   **85.** 677   **86.** 713

87. A Mersenne prime is a prime of the form $2^n - 1$, where *n* is a prime number. For example, $7 = 2^3 - 1$ is a Mersenne prime. As of January 4, 2008, the largest known Mersenne prime was $2^{32582657} - 1$, which we will call *x*. If *x* were written out, it would have 9,808,358 digits. To put the size of this number in perspective, assume that a typical page in your word processor has 46 lines with 90 characters per line. Do the following calculations:

a. Divide 9,808,358 by the number of characters per page to find the number of pages that would be required to print out *x*.

b. Take the number of pages that you found in part a) and multiply it by 11 inches to determine the length of the paper (in inches) required to print *x*. Next, divide this number by 12 to find the length of the paper in feet.

c. Divide the number that you found in part b) by 5,280 (the number of feet in 1 mile) to determine the length of the paper in miles.

88. Research on the Internet to find the current largest Mersenne prime, and if it is different from *x*, then repeat the calculations in Exercise 87 for this larger Mersenne prime.

For Extra Credit

There are many famous conjectures* regarding prime numbers. In Exercises 89 and 90, we investigate two of the most famous conjectures.

89. A pair of **twin primes** is a pair of numbers that differ by 2 that are both prime. For example, 17 and 19 are a pair of twin primes. Also 41 and 43 are another pair of twin primes. The **twin prime conjecture** states that there are an infinite number of twin primes. Find the next three twin primes that are greater than 43.

90. **Goldbach's conjecture** states that any even number greater than 2 can be written as the sum of two primes. For example, $16 = 11 + 5$, $26 = 13 + 13$, and $40 = 3 + 37$. Write each of the following as a sum of two primes:†

   a. 80   **b.** 100   **c.** 200

---

*You can think of a conjecture as an educated guess that has not yet been proved.
†There may be several ways to do this.

91. If both p and q divide the natural number n, then does $p \cdot q$ divide the number n? Either explain why this must be true or else provide a counterexample to this statement.

92. The divisibility test for 6 is that the number must be divisible by both 2 and 3. Explain why the divisibility test for 8 is *not* that the number must be divisible by both 2 and 4.

93. Devise a divisibility test for 15.

94. Research the technique for checking addition called "casting out nines." How is it similar to our test for divisibility by 9?

6.2 The Integers

Objectives

1. Model the rules for adding and subtracting integers.
2. Understand the rules for multiplying and dividing integers.
3. Perform computations using integers.

Although the natural numbers were known in various forms to many ancient peoples, zero did not appear until many years later. The Hindus are credited with inventing zero, which later became part of our Hindu-Arabic system of notation.

> **DEFINITION** We adjoin zero to the set of natural numbers to form the set $\{0, 1, 2, 3, \ldots\}$, called the set of **whole numbers**.

Many mathematicians were slow to accept zero, arguing that we could not have a number that represented nothing. The Greek mathematician Diophantus said that to solve the equation $3x + 20 = 8$ and get a solution of -4 was absurd. As late as the sixteenth century, many European mathematicians would use negative numbers in doing calculations but would not accept zero or a negative number as an answer to a problem. Now, of course, once we learn the rules for calculating with negative numbers, we can calculate with them as easily as we do with positive numbers.

> **DEFINITION** The set of **integers** is the set $\{\ldots, -3, -2, -1, 0, 1, 2, 3, \ldots\}$.

In this section, we will investigate the rules for adding, subtracting, multiplying, and dividing integers.

 Some Good Advice

Keep in mind that if you understand *why* a rule works, that will help you to remember *what* the rule is telling you to do.

✎ *KEY POINT*

We can explain addition rules by considering movement on a number line.

Adding and Subtracting Integers

A common model that we can use to explain the addition of integers is movement on the number line. In this model, we think of integers as points on the number line, as shown in Figure 6.6.

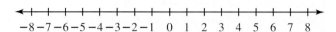

FIGURE 6.6 Representing integers on the number line.

Then, beginning at 0, we interpret positive numbers as movement to the right and negative numbers as representing movement to the left. We illustrate this in Example 1.

EXAMPLE 1   *Representing Integer Addition as Movement on the Number Line*

Perform the following additions:

a) $(+3) + (+5)$ b) $(+9) + (-12)$ c) $(-149) + (137)$

SOLUTION:

a) Starting at 0, think of $+3$ as a movement 3 places to the right. Then think of $+5$ as an additional movement of 5 places to the right, as in Figure 6.7. Thus, we see that $(+3) + (+5) = +8$.

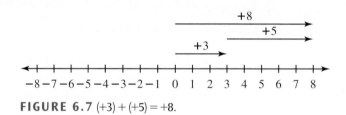

FIGURE 6.7 $(+3) + (+5) = +8$.

b) To calculate $(+9) + (-12)$, we first move 9 places to the right and then move 12 places to the left, as we show in Figure 6.8.

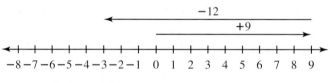

FIGURE 6.8 $(+9) + (-12) = -3$.

Quiz Yourself ⑤

Find:

a) $(+4) + (6)$

b) $(+8) + (-13)$

c) $(-3) + (-7)$

d) $(+145) + (-123)$

c) It is not practical to draw a number line when calculating with such large numbers; however, you still can imagine movements along the number line. If you visualize moving 149 units to the left and then 137 units to the right, you see that the net effect is that you have moved $149 - 137 = 12$ units to the left, so $(-149) + (+137) = -12$.

Now try Exercises 7 to 14. ❋ ⑤

Mathematicians consider the set of integers as an abstract system having only two operations—addition and multiplication. At first, they do not mention subtraction and division, considering these operations to be somewhat secondary operations.

Now we will define subtraction in terms of addition, and later we will define division in terms of multiplication. However, first, we need a definition.

✎ *KEY POINT*

We define subtraction of integers in terms of addition.

HISTORICAL HIGHLIGHT ✾ ✾ ✾

Negative Numbers*

It took mathematicians a long time to accept the notion of negative numbers. Historians tell us that there is no evidence of negative numbers in the mathematics of ancient civilizations such as the Babylonians and the Greeks. In fact, Greek mathematics was so heavily based on geometry, which involved the measuring of lengths and distances, that they had no need for negative numbers.

By the seventh century, Indian bookkeepers used negative numbers to represent debt, and Indian mathematicians such as Brahmagupta showed that they had a clear idea of how to use negative numbers in doing scientific calculations.

During the Renaissance, the Italian mathematician Girolamo Cardano used negative numbers in his 1545 work *Ars Magna,* which was a text on algebraic equations. However, a century later, other European mathematicians were still slow to accept negative numbers. The Frenchman René Descartes called negative numbers "false roots" and his contemporary, Blaise Pascal, was convinced that numbers "less than zero" could not exist. It was not until the eighteenth century that negative numbers were generally accepted.

> **DEFINITION** Two integers x and y are **opposites**[†] if $x + y = 0$.

For example, -8 is the opposite of $+8$ because $(-8) + (+8) = 0$. Also because $(-13) + (+13) = 0$, -13 and $+13$ are opposites. Because $0 + 0 = 0$, we say that 0 is its own opposite. As you can see in Figure 6.9, opposites are balanced on opposite sides of 0.

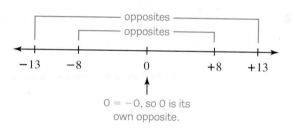

FIGURE 6.9 Opposite numbers.

> **DEFINITION** If a and b are integers, then $a - b = a + (-b)$.

This definition means to compute $3 - 8$, you change the subtraction to an addition, as in the following diagram:

$$3 - 8 = 3 + (-8) = -5$$
Add the opposite of 8. —/   \— Result is -5.

We compute $(-6) - (-19)$ as follows:

$$(-6) - (-19) = (-6) + (+19) = +13$$
Add the opposite of -19. —/   \— Result is $+13$.

Now try Exercises 15 to 22. **6**

 **6**

Convert the following subtraction problems to addition problems and find the answers:

a) $(+5) - (+12)$

b) $(-3) - (-9)$

 Some Good Advice

You may have heard in the past that "two negatives make a positive." This is not a rule in mathematics, but rather a memory device that sometimes works and sometimes does not, depending on the situation in which you try to apply it. For example, $(-6) + (-2)$ does not make a positive. Rather than thinking of "two negatives make a positive," it is better to think of $-(-6)$ as the opposite of -6, or $+6$, and think of $3 - (-5)$ as $3 + (+5)$.

---

*This highlight is based on material in Jan Gullberg, *Mathematics: From the Birth of Numbers* (New York: W. W. Norton, 1997), pp. 72–73.
[†]Numbers that are opposites are also called *additive inverses* of each other.

KEY POINT

We can use a money model to explain integer multiplication.

Multiplying and Dividing Integers

You can understand the rules for multiplying integers, if you consider the following model. Call today day zero, and assume that you have $100 in your wallet. Also assume that each day one of two things happens, either you receive $5 in the mail from your rich uncle or you spend $5 for a pizza. We assume that you are making no other changes in your finances.

We will now consider the product of integers a and b. The integer a will represent changes in days—a positive a represents days in the future, a negative a represents days in the past. The number b will represent changes in your finances—a positive b means you receive $5, a negative b means you spend $5.

For example, we interpret $(+3)(-5)$ to mean that for the next 3 days, you spend $5 each day on pizza. The product $(-4)(+5)$ means for the past 4 days, you have been receiving $5 from your uncle. With this in mind, consider Table 6.2.

Interpretation of $a \cdot b$		Multiplication Rule
$(+3)(+5) = +15$ —— You will gain $15 in 3 days. For the next 3 days, you receive $5.		A positive times a positive is a positive.
$(+3)(-5) = -15$ —— You will lose $15 in 3 days. For the next 3 days, you spend $5.		A positive times a negative is a negative.
$(-3)(+5) = -15$ —— You had $15 less 3 days ago. For the past 3 days, you received $5.		A negative times a positive is a negative.
$(-3)(-5) = +15$ —— You had $15 more 3 days ago. For the past 3 days, you spent $5.		A negative times a negative is a positive.

TABLE 6.2 Rules for multiplying integers.

We can state these rules for multiplying integers more succinctly.

> **RULES FOR MULTIPLYING INTEGERS** If a and b are integers, then:
>
> a) If a and b have the same sign, then $a \cdot b$ is positive.
>
> b) If a and b have opposite signs, then $a \cdot b$ is negative.

EXAMPLE 2 *Applying the Rules for Multiplying Integers to the Stock Market*

a) Assume that stock for Itech, an international technology company, has been losing value at $3 per day for each of the last 8 days. How much more was the stock worth 8 days ago?

b) If the stock continues to lose its value at $3 per day, how much less will the stock be worth in 4 days?

SOLUTION:

a) To find the worth 8 days ago, we represent "8 days ago" by -8 and "losing value at $3 per day" by -3. We then calculate the product $(-8)(-3) = +24$. So, the stock was worth $24 more 8 days ago.

b) The change in value in 4 days will be $(+4)(-3) = -12$, so the stock will be worth $12 less.

Now try Exercises 23 to 34. ✻ **7**

Quiz Yourself **7**

Find the following products:

a) $(-4)(-6)$

b) $(+8)(-7)$

KEY POINT

We define division in terms of multiplication.

Like subtraction, we consider division to be somewhat of a secondary operation, and we define division in terms of multiplication.

> **DEFINITION** If a, b, and c are integers, then $a/b = c$ means that $a = b \cdot c$.

Try to imagine that you did not know any division facts but were comfortable with doing multiplication. If you knew the definition of division, to compute 8/2 you would consider the equation $\frac{8}{2} = c$, which according to the definition of division means that you should consider $8 = 2 \cdot c$ instead. From your knowledge of multiplication, you would know that 4 is a solution for this equation; therefore, $\frac{8}{2} = 4$. In Example 3, we will use the definition of division to explain the rules for dividing signed numbers.

EXAMPLE 3 *Deriving the Rules for Dividing Signed Numbers*

Use the definition of division to find each of the following quotients:

a) $\dfrac{+6}{+2}$ b) $\dfrac{+12}{-6}$ c) $\dfrac{-15}{+3}$ d) $\dfrac{-20}{-5}$

SOLUTION: In each case, we will rewrite the division problem as a corresponding multiplication problem.

a) To solve $\frac{+6}{+2} = c$, we solve $+6 = (+2) \cdot c$ instead. The solution is clearly $+3$, so we can say $\frac{+6}{+2} = +3$. This illustrates that a *positive divided by a positive is a positive*.

b) Instead of solving $\frac{+12}{-6} = c$, we solve $+12 = (-6) \cdot c$. We see that $c = -2$. Thus, $\frac{+12}{-6} = -2$. This shows that a *positive divided by a negative is a negative*.

c) To compute $\frac{-15}{+3}$, we must solve the equation $-15 = (+3) \cdot c$. In this case, $c = -5$, so $\frac{-15}{+3} = -5$. Thus, a *negative divided by a positive is a negative*.

d) Last, to find $\frac{-20}{-5}$, we solve $-20 = (-5) \cdot c$. Because $c = +4$, we can say $\frac{-20}{-5} = +4$. This illustrates that a *negative divided by a negative is a positive*.

Now try Exercises 35 to 50. ✳

We now state the rules for dividing integers that we derived in Example 3 more concisely. Notice the similarity between these rules and the rules for multiplying integers.

> **RULES FOR DIVIDING INTEGERS** If a and b are integers and $b \neq 0$, then:
>
> a)   If a and b have the *same sign*, then $\frac{a}{b}$ is positive.
>
> b)   If a and b have *opposite signs*, then $\frac{a}{b}$ is negative.

Notice that it is permissible to divide *into* zero. For example, $\frac{0}{5} = 0$. However, we do not allow you to divide *by* zero. For an explanation of why we cannot divide by zero, see Exercises 89 and 90.

We can use a combination of the rules for working with signed integers to simplify a complex expression.

EXAMPLE 4 *Calculating with Integers*

Calculate $\left(-4 - (-9)\right) \cdot \left(\dfrac{-6 + (-2)}{9 - 5}\right)$.

SOLUTION: We will begin by working inside the left parentheses.

$$\left(-4-(-9)\right)\cdot\left(\frac{-6+(-2)}{9-5}\right)=\left(-4+(+9)\right)\cdot\left(\frac{-6+(-2)}{9-5}\right)=5\cdot\left(\frac{-6+(-2)}{9-5}\right)$$

└─── changed subtraction to an addition

Next, we simplify the expression inside the remaining parentheses.

$$5\cdot\left(\frac{-6+(-2)}{9+(-5)}\right)=5\cdot\left(\frac{-8}{4}\right)=5\cdot(-2)=-10.$$

└─── changed subtraction to an addition

Now try Exercises 51 to 58. ✻ **8**

Quiz Yourself **8**

Simplify

$$\left(-3+(-9)\right)\cdot\left(\frac{-5-(-9)}{4-2}\right).$$

 PROBLEM SOLVING

The Analogies Principle

Recall the Analogies Principle in Section 1.1 that says it is important to make connections between mathematical ideas. You should keep in mind that once you know your multiplication rules, the division rules are exactly the same. If you think about this, you will realize that this is not a coincidence because we derived the division rules from the rules for multiplication.

Exercises 6.2

Looking Back*

These exercises follow the general outline of the topics presented in this section and will give you a good overview of the material that you have just studied.

1. How do we visualize opposite numbers?

2. In the discussion following the definition of subtraction on page 241, explain how we used the definition of subtraction, as well as the notion of movement along a number line, to calculate $(-6)-(-19)$.

3. In Table 6.2, what was our explanation that a negative times a negative is a positive?

4. In Example 3, what was our explanation that $\frac{+12}{-6}=-2$?

5. Why were negative numbers not necessary to the ancient Greeks?

6. How many centuries earlier did Indian mathematicians use negative numbers before they were accepted in Europe?

Sharpening Your Skills

Calculate each of these sums using movements on the number line as we did in Example 1.

7. $(+8)+(-5)$
8. $(-9)+(-4)$
9. $(+4)+(+6)$
11. $(-3)+(+7)$
11. $(-128)+(+137)$
12. $(+47)+(+32)$
13. $(+57)+(-38)$
14. $(-38)+(-53)$

Rewrite the following subtraction problems as addition problems and then find the answers.

15. $(+18)-(-5)$
16. $(-8)-(-13)$
17. $(+6)-(+19)$
18. $(-3)-(+14)$
19. $(-28)-(+37)$
20. $(+41)-(+13)$
21. $(+32)-(-18)$
22. $(-28)-(-23)$

Find the following products.

23. $(-5)(-7)$
24. $(+5)(+3)$
25. $(-7)(+8)$
26. $(+3)(-8)$
27. $(+8)(+6)$
28. $(-3)(-9)$
29. $(+19)(-2)$
30. $(-6)(+7)$
31. $(-9)(+7)$
32. $(+5)(+9)$
33. $(-8)(-9)$
34. $(+7)(-3)$

Write each of the following division problems as a corresponding multiplication problem and then solve it as we did in Example 3.

35. $\frac{+16}{-2}$
36. $\frac{+24}{+6}$
37. $\frac{-14}{-2}$
38. $\frac{-26}{+2}$

Perform the following divisions.

39. $\frac{-24}{+8}$
40. $\frac{+20}{+5}$
41. $\frac{-30}{-2}$
42. $\frac{+28}{-4}$
43. $\frac{+15}{+3}$
44. $\frac{-14}{+7}$
45. $\frac{+30}{-5}$
46. $\frac{-22}{-11}$

*Before doing these exercises, you may find it useful to review the note *How to Succeed at Mathematics* on page xix.

47. $\dfrac{-18}{-3}$ **48.** $\dfrac{+20}{+5}$ **49.** $\dfrac{-40}{+10}$ **50.** $\dfrac{+25}{-5}$

Perform each of the following calculations.

51. $(-2)(-3) + (+4)(-8)$

52. $(-2)(+5) + (-4)(+7)$

53. $(-4)(-2) + (+4)(-5)$

54. $(+8)(-2) + (-3)(-5)$

Simplify each of the following expressions.

55. $(-6 - (-14)) \cdot \left(\dfrac{-12 + (-6)}{7 - 4} \right)$

56. $(5 + (-9)) \cdot \left(\dfrac{10 - (-4)}{-3 - 4} \right)$

57. $\left(\dfrac{-1 - (-5)}{4 - 6} \right) \cdot \left(\dfrac{-4 + (-2)}{-5 + 3} \right)$

58. $\left(\dfrac{3 + 9}{-1 - 2} \right) \cdot \left(\dfrac{8 - (-4)}{7 - (+9)} \right)$

Determine whether the following statements are true or false. Remember the Always Principle from Section 1.1 when deciding on your answer. If a statement is false, then provide a counterexample.

59. The product of two negative integers is a positive integer.

60. The product of two whole numbers is a positive integer.

61. The sum of a negative and a positive integer is a negative integer.

62. The product of a negative integer and a positive integer is negative.

63. The quotient of two negative integers is negative.

64. The quotient of a negative integer and a positive integer is negative.

65. The product of a natural number and a positive integer is positive.

66. The quotient of two natural numbers is a natural number.

Applying What You've Learned

The following chart gives some heights of mountains and depths of ocean trenches. Use this information to answer Exercises 67–70.

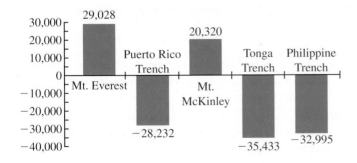

67. Comparing extreme distances in geography. How much higher is the top of Mount Everest than the bottom of the Puerto Rico Trench?

68. Comparing extreme distances in geography. How much higher is the top of Mount McKinley than the bottom of the Tonga Trench?

69. Comparing extreme distances in geography. How much higher is the bottom of the Puerto Rico Trench than the bottom of the Tonga Trench?

70. Comparing extreme distances in geography. How much higher is the bottom of the Philippine Trench than the bottom of the Tonga Trench?

71. Comparing extreme temperatures. According to the *World Almanac*, the highest temperature ever recorded in the world was 136 degrees Fahrenheit in Libya on September 13, 1922, and the lowest temperature recorded was −129 degrees Fahrenheit in Antarctica on July 21, 1983. What is the difference between these two records?

72. Comparing extreme temperatures. According to the *World Almanac*, the highest temperature recorded in California was 134 degrees Fahrenheit on July 10, 1913, and the lowest temperature recorded in California was −45 degrees Fahrenheit on January 20, 1937. What is the difference between these two records?

73. The distance an elevator travels. The ground floor of a dormitory is labeled floor 0. However, there are floors below labeled −1, −2, and so on. If an elevator goes from floor −7 to floor 59, how many floors has the elevator traveled?

74. Temperature change in a laboratory. The temperature in a cryogenics laboratory chamber has dropped from 14 degrees Fahrenheit to −63 degrees Fahrenheit. How much has the temperature dropped?

75. Change in historical dates. Confucius was born in 551 BC, and the Ming Dynasty began in China in 1368 AD. How far apart did these two events occur? (*Warning*: In doing this calculation, there is no year zero.)

76. Change in historical dates. Euclid was born in 323 BC and Karl Friedrich Gauss was born in 1777 AD. How far apart were their birthdates? (*Warning*: In doing this calculation, there is no year zero.)

77. Temperature change. The present temperature is 65 degrees Fahrenheit and the weather bureau is predicting that as a cold front comes in, the temperature will drop 7 degrees per hour over the next four hours. What will the temperature be at that time?

78. **Tracking the stock market.** Assume that the Dow Jones Average for the stock market began the week at 12,782 points. On Monday the market went down 102 points, then on Tuesday rose 78 points, rose 11 points on Wednesday, declined 14 points on Thursday, and rose 12 points on Friday. What was the Dow Jones Average at the close on Friday?

Communicating Mathematics

79. On page 240, we mention that subtraction is "somewhat of a secondary operation" on the integers. What did we mean by that? (*Hint:* Consider the definition of subtraction.)

80. Why do we discourage memorizing an artificial rule such as "two negatives make a positive"? Give an example of when this "rule" doesn't work.

81. Can you combine the rules for multiplying and dividing signed integers into a single rule?

82. Use the definition of division on page 243 to explain why $\frac{-20}{-5} = +4$.

If it is possible, give an example of each of the following. If it is not possible, explain why it is not possible. There may be many possible answers.

83. An integer that is not a whole number

84. A negative integer that is also a whole number

85. A whole number that is not a natural number

86. An integer that is not a natural number

87. A natural number that is not a whole number

88. A natural number that is not an integer

89. Use the definition of division in terms of multiplication to explain why we cannot divide 8 by 0. What does it mean to say $\frac{8}{0} = x$?

90. Use the definition of division in terms of multiplication to explain why we cannot divide 0 by 0. What does it mean to say $\frac{0}{0} = x$?

Explain each of the following calculations using a money model similar to the ones given in Table 6.2. For example, $(-2)(+7)$ would mean for the past 2 days, you have been receiving \$7. Explain what the change in your finances would be in each instance.

91. $(+4)(-6)$ 92. $(+5)(+8)$
93. $(-3)(+7)$ 94. $(-6)(-3)$

Using Technology to Investigate Mathematics

95. Use the Internet to research the history of zero. Write a brief report on your findings.

96. Use the Internet to research the history of negative numbers. Write a brief report on your findings.

Notice that this magic square uses the digits 1 to 9. A 4-by-4 magic square would customarily use the natural numbers from 1 to 16. In our exercises, we will relax this rule so that we only have to use consecutive integers.

For Extra Credit

A magic square* is a square arrangement of numbers as shown here. Because this square has three rows and three columns, it is called a 3-by-3 square.

8	1	6
3	5	7
4	9	2

In a magic square, if you add the numbers in any row, column, or diagonal, you always get the same sum. For example, in this magic square, the sum of any row, column, or diagonal is always 15.

Complete the entries in the following magic squares.

97.

−9	5	4	e
a	−4	b	−1
−2	0	1	−5
3	c	d	6

98.

6	a	−8	3
−5	b	1	−2
−1	c	−3	2
−6	5	d	e

Use the following block of numbers to answer Exercises 99–100.

1	2	3	4	5	6	7	8
9	10	11	12	13	14	15	16
17	18	19	20	21	22	23	24
25	26	27	28	29	30	31	32
33	34	35	36	37	38	39	40
41	42	43	44	45	46	47	48
49	50	51	52	53	54	55	56
57	58	59	60	61	62	63	64

99. What is the relationship of the sum of the numbers in the solid box, which we will call a 3-by-3 box, to the number in the center of the box?

100. Does the relationship you saw in Exercise 99 hold for other boxes of the same size? Can you explain why this does or does not happen?

101. Choose a box of numbers of size 4 by 4, and determine a pattern for finding the sum of the numbers in the box.

102. Without doing any calculations, state how to find the sum of numbers in a 5-by-5 box. Do the same for a 6-by-6 box.

6.3 The Rational Numbers*

Objectives

1. Determine when two rational numbers are equal.
2. Add and subtract rational numbers.
3. Multiply and divide rational numbers.
4. Perform operations using mixed numbers.
5. Represent rational numbers as repeating decimals.

There was a time when you were very young and knew only about the natural numbers—1, 2, 3, ... Then, eventually you learned about zero and then negative numbers. When you were very young, you had no need for fractions. You might wonder why people decide that they need different types of numbers. One way of thinking about how number systems evolve is that while working in one system, we pose a problem that we cannot solve using the numbers in that system. So we have to invent new numbers.

For example, suppose that you knew only about the natural numbers and were asked to solve the equations $x + 2 = 6$ and $x + 6 = 2$. On the surface, they seem to be the same problem. However, the first equation has a solution within the natural numbers, namely $x = 4$, but the second equation does not. If you want to solve the second equation, you need to invent a new number, -4. The need for these new numbers leads us to develop the system of integers that you studied in Section 6.2.

Similarly, if, while working in the system of integers, you want to solve the equation $6x = -2$, you find that there are no solutions. So, you have to invent another number, $\frac{-2}{6} = -\frac{1}{3}$. To solve such equations, we need the system of rational numbers.

> **DEFINITIONS** The set of **rational numbers**, which we will denote by Q, is the set of all numbers that can be written in the form $\frac{a}{b}$, where the a and b are integers and $b \neq 0$. The top number a is called the **numerator**, and the bottom number b is called the **denominator**.

Numbers such as $\frac{1}{2}, \frac{7}{13}, \frac{-4}{5}$, and $\frac{-9}{-20}$ are examples of rational numbers. Also, integers such as 5, -3, and 0 are rational numbers because they *can be* written in the form $\frac{5}{1}, \frac{-3}{1}$, and $\frac{0}{1}$. Also, as you will see later, many decimal numbers are also rational numbers.

 KEY POINT

We cross multiply to test for equality of rational numbers.

Equality of Rational Numbers

One of the most important things to know in any mathematical system is when two different-looking objects are the same. There is a straightforward way to determine if two rational numbers are equal.

> **DEFINITION** We define **equality of rational numbers** as follows:
>
> $$\frac{a}{b} = \frac{c}{d}^{\dagger} \quad \text{if and only if } ad = bc$$

FIGURE 6.10
$ad = bc$.

Computing the products in the equation $ad = bc$ is called **cross multiplying** or calculating the **cross product**. Figure 6.10 helps us remember this equation.

---

*In order to work with rational numbers, you need to be familiar with some of the basic properties of common number systems. For a review of these properties, see Appendix A.
†For the remainder of this section, assume that any expression such as $\frac{a}{b}$ is a rational number.

Some Good Advice

Although there are other methods for determining when rational numbers are equal, it is best to learn this one method well and use it all the time. Many of the rules that we use in working with rational numbers are based on this definition of equality.

EXAMPLE 1   *Testing for Equality of Rational Numbers*

Determine which of the following pairs of rational numbers are equal:

a) $\dfrac{3}{8} = \dfrac{12}{32}$ b) $\dfrac{39}{116} = \dfrac{35}{108}$ c) $\dfrac{1,419}{1,892} = \dfrac{165}{220}$

SOLUTION:

a) Calculating the cross product, we get $3 \cdot 32 = 8 \cdot 12$, or $96 = 96$, so the two rational numbers are equal.

b) When we cross multiply, we get $39 \cdot 108 = 4,212$ and $116 \cdot 35 = 4,060$, so the two rational numbers are not equal.

c) It is unlikely that you would ever have to deal with such awkward numbers; however, we gave this example to make the point that the cross-multiplication method works even for "bad" numbers.

　　Cross multiplying, we get $1,419 \cdot 220 = 312,180$ and $1,892 \cdot 165 = 312,180$. Thus, the two rational numbers are equal.

Now try Exercises 7 to 14. ✳

We use the following rule to simplify rational numbers.

 Quiz Yourself ⑨

Determine whether the following pairs of rational numbers are equal:

a) $\dfrac{51}{187}, \dfrac{45}{165}$ b) $\dfrac{31}{91}, \dfrac{63}{147}$

✎ **KEY POINT**

Reducing fractions is based on the notion of equality.

> **CANCELING COMMON FACTORS**
>
> $$\frac{a \cdot c}{b \cdot c} = \frac{a}{b}$$

This rule says that we can cancel the same *factor* from both the numerator and denominator of a rational number. This process of canceling common factors is called **reducing the rational number**. When all common factors have been canceled, we say that the number is in **lowest terms**.

　　When you are working with awkward rational numbers, it is sometimes useful to use your calculator. The accompanying screens show how to use the "Frac" command to test the two fractions from Example 1 for equality. On my calculator I get the "Frac" command by first pressing the "Math" key and then selecting "Frac" from the menu shown in Screen 1. Screen 2 shows that when we reduce $\frac{1,419}{1,892}$ and $\frac{165}{220}$, in both cases we get $\frac{3}{4}$, so these numbers are equal.

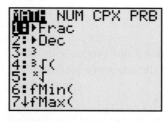

SCREEN 1

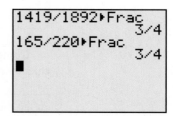

SCREEN 2

EXAMPLE 2   *Reducing Rational Numbers*

Reduce the following rational numbers:

a) $\dfrac{48}{72}$ b) $\dfrac{1{,}848}{2{,}112}$

SOLUTION:

a) There are several ways to do this. For example, using the divisibility rules, we recognize that 4 divides both the numerator and the denominator and then we cancel 4 to get the number $\frac{12}{18}$, as we show in the first step of Figure 6.11. Next, we see that 6 is a common factor of both 12 and 18 and cancel again to get $\frac{2}{3}$. Notice that you stop this reducing process when you see that the numerator and denominator have no factors in common.

$$\frac{48}{72} = \frac{\cancel{4} \cdot 12}{\cancel{4} \cdot 18} = \frac{12}{18} = \frac{\cancel{6} \cdot 2}{\cancel{6} \cdot 3} = \frac{2}{3}$$

FIGURE 6.11 Reducing $\frac{48}{72}$ by canceling common factors.

b) Here it is difficult to see all the common factors of the numerator and the denominator, so we do the cancellation in stages. First, we recognize that 4 is a factor that we can cancel, and then we can proceed as you see in the following steps.

4 divides 48 so 4 divides 1,848. Similarly, 4 divides 2,112.

Cancel 4.

Numerator and denominator are divisible by both 2 and 3, and therefore, 6.

$$\frac{1{,}848}{2{,}112} = \frac{462 \cdot 4}{528 \cdot 4} = \frac{462}{528} = \frac{77 \cdot 6}{88 \cdot 6} = \frac{77}{88} = \frac{7}{8}$$

Cancel 6.   Cancel 11.

We have written $\dfrac{1{,}848}{2{,}112}$ in lowest terms as $\dfrac{7}{8}$.

Now try Exercises 15 to 22. ✳ **10**

Quiz Yourself **10**

Reduce $\frac{252}{336}$.

$a \cdot c \cdot b = b \cdot c \cdot a$

FIGURE 6.12
The cancellation rule is valid because of the definition of equality of rational numbers.

It is one thing to *know* how to use a rule, it is another to understand *why* we can use the rule. The cancellation rule states that $\frac{a \cdot c}{b \cdot c} = \frac{a}{b}$. Why is this equation true? If we use the definition of equality and cross multiply, we see that we get an equality, as we show in Figure 6.12.

🧩 **PROBLEM SOLVING**

The Three-Way Principle

When extending this cancellation rule to algebra, students sometimes mistakenly cancel terms rather than factors. For example, it is tempting to cancel 5 from the numerator and denominator of a quotient such as $\frac{x+5}{y+5}$ to get $\frac{x+\cancel{5}}{y+\cancel{5}} = \frac{x}{y}$. You can convince yourself that this is not valid by considering an example as the Three-Way Principle in Section 1.1 suggests. If you substitute 2 for x and 3 for y, you see that the equation $\frac{x+5}{y+5} = \frac{x}{y}$ becomes the equation $\frac{2+5}{3+5} = \frac{2}{3}$, which is false. Therefore, this cancellation of terms is not valid.

We will next consider operations on rational numbers.

Adding and Subtracting Rational Numbers

 **KEY POINT**

We can add or subtract rational numbers that have the same denominators.

When we add rational numbers that have the same denominator, such as $\frac{3}{5} + \frac{8}{5}$, we can think that "three of one thing, plus eight of that same thing is eleven of that thing." So, $\frac{3}{5} + \frac{8}{5} = \frac{3+8}{5} = \frac{11}{5}$.

When adding rational numbers with different denominators, the addition is more complicated because we can't, so to speak, add apples and oranges. However, we can add rational numbers with different denominators if first we rewrite both fractions in a form where they have common denominators. For example, we could add $\frac{1}{6} + \frac{3}{4}$ as follows:

$$\frac{1}{6} + \frac{3}{4} = \frac{1 \cdot 4}{6 \cdot 4} + \frac{6 \cdot 3}{6 \cdot 4} = \frac{4}{24} + \frac{18}{24} = \frac{22}{24} = \frac{11}{12}$$

We need the same   Now the denominators are
denominators.   the same, so we can add.

This simple example will help you remember the following rule for adding rational numbers. The rule for subtracting is essentially the same.

> ### ADDING AND SUBTRACTING RATIONAL NUMBERS
>
> $$\frac{a}{b} + \frac{c}{d} = \frac{a \cdot d + b \cdot c}{b \cdot d} \qquad \frac{a}{b} - \frac{c}{d} = \frac{a \cdot d - b \cdot c}{b \cdot d}$$

We illustrate these rules in Example 3.

EXAMPLE 3   *Adding and Subtracting Rational Numbers*

a) Find $\frac{5}{6} + \frac{3}{4}$.   b) Find $\frac{11}{18} - \frac{3}{8}$.

SOLUTION:

a) We apply the rule for adding rational numbers.

$$\frac{5}{6} + \frac{3}{4} = \frac{5 \cdot 4 + 6 \cdot 3}{6 \cdot 4} = \frac{20 + 18}{24} = \frac{38}{24} = \frac{19}{12}.$$

b) We apply the rule for subtracting rational numbers.

$$\frac{11}{18} - \frac{3}{8} = \frac{11 \cdot 8 - 18 \cdot 3}{18 \cdot 8} = \frac{34}{144} = \frac{17}{72}.$$

Quiz Yourself **11**

Find $\frac{3}{8} + \frac{5}{6}$ and $\frac{13}{16} - \frac{11}{24}$.

Now try Exercises 23 to 34. ❀ **11**

If you can recognize the LCM of the two denominators, you may find it quicker to add rational numbers by rewriting the numbers as quotients having a common denominator. In Example 3(a), if you recognize that the LCM of 6 and 4 is 12, then you could write

$$\frac{5}{6} + \frac{3}{4} = \frac{5 \cdot 2}{6 \cdot 2} + \frac{3 \cdot 3}{4 \cdot 3} = \frac{10}{12} + \frac{9}{12} = \frac{19}{12}.$$

It is up to you to decide in each situation which method is the fastest and most reliable for you.

If you want to add three numbers, you can add the first two numbers, get an answer, and then add the third. To compute an expression like $x - y - z$, first subtract $x - y$ and then subtract z from that result.

◎ *Some Good Advice*

Why is $\frac{a}{b} + \frac{c}{d} = \frac{a + c}{b + d}$ an incorrect way to do rational number addition?

Recall that the Always Principle from Section 1.1 states that for a mathematical statement to be true, it must always be true 100% of the time, *without exception*. If we perform the addition $\frac{1}{2} + \frac{1}{2}$ incorrectly, as stated previously, we get

$$\frac{1}{2} + \frac{1}{2} = \frac{1 + 1}{2 + 2} = \frac{2}{4} = \frac{1}{2},$$

which is clearly false.

✎ **KEY POINT**

To multiply rational numbers, we multiply numerators and denominators.

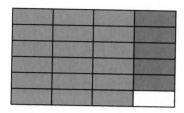

FIGURE 6.14 Representation of $\frac{3}{4} \times \frac{5}{6} = \frac{15}{24}$.

Multiplying and Dividing Rational Numbers

We will now explain the multiplication rule for rational numbers. In Figure 6.13(a), we visualize the fraction $\frac{3}{4}$, and in Figure 6.13(b), we represent $\frac{5}{6}$.

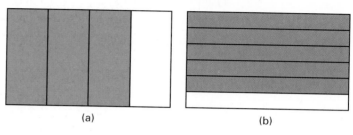

(a)   (b)

FIGURE 6.13 (a) Representation of $\frac{3}{4}$; (b) representation of $\frac{5}{6}$.

If we overlap these two figures, we get Figure 6.14, which shows that if we take $\frac{5}{6}$ of the $\frac{3}{4}$, we get 15 of the 24 sections, or $\frac{15}{24}$, which equals $\frac{5}{8}$.

Another way of looking at Figure 6.14 is $\frac{5}{6} \cdot \frac{3}{4} = \frac{5 \cdot 3}{6 \cdot 4} = \frac{15}{24} = \frac{5}{8}$. This helps us remember the multiplication rule for rational numbers.

> **MULTIPLYING RATIONAL NUMBERS**
>
> $$\frac{a}{b} \cdot \frac{c}{d} = \frac{a \cdot c}{b \cdot d}$$

EXAMPLE 4 *Multiplying Rational Numbers*

Find a) $\frac{4}{3} \cdot \frac{3}{15}$ b) $\frac{18}{25} \cdot \frac{10}{81}$ c) $\left(\frac{9}{4}\right) \cdot \left(-\frac{12}{5}\right)$

SOLUTION:

a) Applying the multiplication rule, we get $\frac{4}{3} \cdot \frac{3}{15} = \frac{4 \cdot 3}{3 \cdot 15} = \frac{12}{45} = \frac{4}{15}$.

b) You can use the multiplication rule to get $\frac{18}{25} \cdot \frac{10}{81} = \frac{180}{2,025}$ and then reduce this number; however, this is a lot of unnecessary work.

    If you recognize that eventually you are going to cancel common factors, it is easier to do that first before multiplying the numerators and denominators. You can cancel 9 from the first numerator and second denominator, and then cancel 5 from the first denominator and the second numerator. (You can also do the cancellation in one step if you prefer.) Finally, multiply the simpler numbers $\frac{2}{5}$ and $\frac{2}{9}$ to get $\frac{4}{45}$, as we show in Figure 6.15.

$$\frac{\cancel{18}^{2}}{25} \cdot \frac{10}{\cancel{81}_{9}} = \frac{2}{\cancel{25}_{5}} \cdot \frac{\cancel{10}^{2}}{9} = \frac{2}{5} \cdot \frac{2}{9} = \frac{4}{45}$$

Cancel 9.   Cancel 5.   Now multiply.

FIGURE 6.15 Canceling before multiplying saves work.

Quiz Yourself ⑫

Find:

a) $\dfrac{24}{9} \cdot \dfrac{21}{20}$

b) $\left(-\dfrac{8}{15}\right) \cdot \left(-\dfrac{25}{12}\right)$

c) *The rules for multiplying signed rational numbers are the same as those for multiplying integers,* so the product of a positive number and a negative number is a negative number. We will first multiply without regard to signs, remembering that the final product is negative.

$$\frac{9}{\cancel{4}_{1}} \cdot \frac{\cancel{12}^{3}}{5} = \frac{9}{1} \cdot \frac{3}{5} = \frac{27}{5}$$

So the answer is $-\frac{27}{5}$. Although you may recall that $-\frac{27}{5}$, $\frac{-27}{5}$, and $\frac{27}{-5}$ are all equal, we tend to write the answer as $-\frac{27}{5}$. ❋ ⑫

✎ *KEY POINT*

Division of rational numbers is based on multiplication.

To understand the rule for dividing rational numbers, consider how we might think about dividing $\dfrac{\frac{3}{4}}{\frac{5}{8}}$. What makes this division complicated is that we have $\frac{5}{8}$ in the denominator.

One way to get rid of the $\frac{5}{8}$ is to multiply it by its reciprocal, $\frac{8}{5}$. However, if we multiply the denominator by $\frac{8}{5}$, we must also multiply the numerator by $\frac{8}{5}$. Multiplying by $\frac{8}{5}$ gives us a 1 in the denominator and the product $\frac{3}{4} \cdot \frac{8}{5}$ in the numerator, as we see here.

$$\frac{\frac{3}{4}}{\frac{5}{8}} = \frac{\frac{3}{4} \times \frac{8}{5}}{\frac{5}{8} \times \frac{8}{5}} = \frac{\frac{3}{4} \times \frac{8}{5}}{1} = \frac{3}{\overset{1}{\cancel{4}}} \times \frac{\overset{2}{\cancel{8}}}{5} = \frac{6}{5}$$

Instead of going through this lengthy process each time we divide, we use the following rule, which tells us in effect to "invert the denominator and multiply."

DIVIDING RATIONAL NUMBERS

$$\frac{\dfrac{a}{b}}{\dfrac{c}{d}} = \frac{a}{b} \cdot \frac{d}{c}$$

EXAMPLE 5 *Dividing Rational Numbers*

Perform the following divisions:

a) $\dfrac{\frac{25}{12}}{\frac{10}{3}}$.   b) $\dfrac{-\frac{11}{6}}{\frac{7}{9}}$

SOLUTION:

a) Applying the division rule,* we invert the denominator and multiply.

$$\frac{\frac{25}{12}}{\frac{10}{3}} = \frac{\overset{5}{\cancel{25}}}{\underset{4}{\cancel{12}}} \cdot \frac{\overset{1}{\cancel{3}}}{\underset{2}{\cancel{10}}} = \frac{5}{4} \cdot \frac{1}{2} = \frac{5}{8}$$

Quiz Yourself ⑬

Find: a) $\dfrac{\frac{5}{28}}{\frac{11}{14}}$ b) $\dfrac{-\frac{33}{14}}{-\frac{11}{8}}$

b) *The rules for division of rational numbers are the same as the rules for division of integers*, so a negative number divided by a positive will result in a negative number. We will remember that fact and divide without regard to signs.

Again, we invert and multiply to get $\dfrac{\frac{11}{6}}{\frac{7}{9}} = \frac{11}{\underset{2}{\cancel{6}}} \cdot \frac{\overset{3}{\cancel{9}}}{7} = \frac{11}{2} \cdot \frac{3}{7} = \frac{33}{14}$, so the answer is $-\frac{33}{14}$.

Now try Exercises 35 to 52. ✲ ⑬

Note that you have to be careful how you interpret a quotient such as $\dfrac{\frac{4}{8}}{16}$. In this case, the longer horizontal bar tells us that we are dividing $\frac{4}{8}$ by 16, so you can think of $\dfrac{\frac{4}{8}}{16}$ as $\dfrac{\frac{4}{8}}{\frac{16}{1}}$. You might find it enlightening to compute $\dfrac{\frac{4}{8}}{16}$ and $\dfrac{4}{\frac{8}{16}}$. You will see that you do not get the same result.

Example 6 requires us to use several operations on rational numbers at the same time.

---

*When multiplying and dividing rational numbers with your calculator, you should put parentheses around the operands as follows: $\left(\frac{4}{3}\right) \cdot \left(\frac{3}{15}\right)$ and $\left(\frac{25}{12}\right) \cdot \left(\frac{10}{3}\right)$; otherwise, the calculator results may be incorrect.

EXAMPLE 6   *Using the Scale Function in a Graphics Program*

Suppose you are drawing a landscaping plan using a graphics package, such as Adobe CS3: Master Collection, and have drawn a rectangle measuring $\frac{8}{3}$ inches long by $\frac{5}{6}$ inches wide representing a rectangular Japanese rock garden.

a) If you scale down the length by a factor of $\frac{3}{4}$ and increase the width by a factor of $\frac{9}{8}$, what are the new dimensions of the sides of the garden in your drawing?

b) How does the area of the new rectangle compare with the area of the original rectangle in your drawing?

SOLUTION:

a) We will multiply the length of the original rectangle by the scaling factor of $\frac{3}{4}$ to get a length of $\frac{8}{3} \cdot \frac{3}{4} = \frac{24}{12} = 2$ inches. We multiply the width of the original rectangle by the scaling factor $\frac{9}{8}$ to get a new width of $\frac{5}{6} \cdot \frac{9}{8} = \frac{45}{48} = \frac{15}{16}$ inch.

b) To answer this question, we will divide the area of the new rectangle by the area of the old rectangle. The area of the new rectangle is $2 \cdot \frac{15}{16} = \frac{30}{16} = \frac{15}{8}$ square inches. The area of the old rectangle is $\frac{8}{3} \cdot \frac{5}{6} = \frac{40}{18} = \frac{20}{9}$ square inches. Dividing the new area by the old area, we get

$$\frac{\frac{15}{8}}{\frac{20}{9}} = \frac{\overset{3}{\cancel{15}}}{8} \cdot \frac{9}{\underset{4}{\cancel{20}}} = \frac{27}{32}.$$

So, the area of the new garden in the drawing is $\frac{27}{32}$ the area of the old garden in the drawing. ❋

KEY POINT

Mixed numbers help us understand the size of rational numbers.

$$8\overline{)123}\begin{array}{r}15\\ \\ 120\\ \hline 3\end{array}$$ — Quotient tells us that 8 divides into 123 fifteen whole times.

— Remainder tells us that 3 eighths are left over.

FIGURE 6.16 Converting $\frac{123}{8}$ to $15\frac{3}{8}$.

Mixed Numbers

In working with rational numbers, you could get an answer such as $\frac{123}{8}$. A rational number such as this, in which the numerator is greater than the denominator, is called an **improper fraction.***

Although the answer may be correct in its present form, we have to think a little to understand the size of this number. However, if we divide 123 by 8 to get the quotient 15 and remainder 3 (Figure 6.16), then we see that $\frac{123}{8} = 15\frac{3}{8}$, which is not quite 15 and $\frac{1}{2}$.

We can use this example to give a general rule for writing an improper fraction as a mixed number.

CONVERTING AN IMPROPER FRACTION TO A MIXED NUMBER

To convert the improper fraction $\frac{a}{b}$ to a mixed number, perform the division

$$b\overline{)a}\begin{array}{r}q\\ \\ \vdots\\ \hline r\end{array}$$

Then, write $\frac{a}{b}$ as $q + \frac{r}{b} = q\frac{r}{b}$.

---

*Rational numbers with a positive numerator and a positive denominator are often called **fractions**. In this discussion, we will talk only about fractions.

✾ ✾ ✾ HISTORICAL HIGHLIGHT

Sophie Germain

Sophie Germain was born in France in 1776, at a time when women were discouraged from studying mathematics. Because women were barred from enrolling at the Ecole Polytechnique in Paris, she secretly obtained lecture notes of various professors at this prestigious university. Students were allowed to make written comments on the lectures of their professors and, even though she was not a student, she submitted her commentaries using the pseudonym

M. Leblanc. Unaware that his "student" was a woman, Joseph Lagrange, one of the eminent mathematicians of the time, had high praise for the work of M. Leblanc.

When Lagrange discovered Leblanc's true identity, he extolled Germain as one of the promising young mathematicians of the time. In 1816, she was awarded a prize from the French Academy for her paper on the mathematics of elastic surfaces.

EXAMPLE 7 *Converting an Improper Fraction to a Mixed Number*

Convert the following improper fractions to mixed numbers:

a) $\frac{45}{6}$ b) $\frac{133}{8}$

SOLUTION:

a) As you can see in the accompanying figure, when we divide 45 by 6, we get a quotient of 7 and a remainder of 3. Thus, $\frac{45}{6} = 7\frac{3}{6} = 7\frac{1}{2}$.

b) Similarly, in (b), $\frac{133}{8} = 16\frac{5}{8}$. ✳

We can also convert mixed numbers into improper fractions. For example, we think of $5\frac{3}{4}$ as $5 + \frac{3}{4} = \frac{5 \cdot 4}{4} + \frac{3}{4} = \frac{23}{4}$. This calculation is the basis for the following conversion rule. Now try Exercises 57 to 64.

> **CONVERTING A MIXED NUMBER TO AN IMPROPER FRACTION**
>
> The mixed number $q\frac{r}{b}$ equals the improper fraction $\frac{q \cdot b + r}{b}$.

Don't let this formula intimidate you. All we are saying is you multiply q times b and add r to get the numerator. Then use b for the denominator.

EXAMPLE 8 *Converting a Mixed Number to an Improper Fraction*

Convert the following mixed numbers to improper fractions:

a) $5\frac{2}{7}$ b) $8\frac{3}{11}$

SOLUTION:

a) $5\frac{2}{7} = \frac{5 \cdot 7 + 2}{7} = \frac{37}{7}$ b) $8\frac{3}{11} = \frac{8 \cdot 11 + 3}{11} = \frac{91}{11}$. ✳ **14**

In calculating with mixed numbers, you will often find it useful to first convert the numbers to improper fractions before doing the computation. For example, to compute $3\frac{1}{2} \times 2\frac{2}{5}$, rewrite it as $\frac{7}{2} \times \frac{12}{5} = \frac{84}{10} = \frac{42}{5} = 8\frac{2}{5}$.

Quiz Yourself **14**

a) Convert $\frac{23}{4}$ to a mixed number.

b) Convert $2\frac{3}{5}$ to an improper fraction.

KEY POINT

Rational numbers have repeating decimal representations.

Repeating Decimals

If you were to divide 3/16 on your calculator, the answer would look something like 0.1875000. We will now discuss how to write rational numbers in decimal form and then how to write a decimal number as the quotient of two integers.

EXAMPLE 9 *Writing a Rational Number in Decimal Form*

Write each of the following rational numbers as a decimal number:

a) $\frac{5}{8}$ b) $\frac{7}{16}$ c) $\frac{82}{111}$

SOLUTION:

a) If you divide the numerator by the denominator, you will get $\frac{5}{8} = 0.625$.

b) Dividing, we find that $\frac{7}{16} = 0.4375$.

c) Although this seems like a strange choice for an example, as you will see, it is very interesting. If you divide $\frac{82}{111}$ on your calculator, you might get an answer that looks something like 0.7387387387. This is not quite correct.

   If you do the division by hand, as in Figure 6.17, you will see that the digits in the quotient will repeat indefinitely. Once you have calculated a quotient of 0.738 and get a remainder of 82, the calculations that you just did will repeat all over again to give you the quotient 0.738738.

$$\begin{array}{r} .738 \\ 111\overline{\smash{)}82.000000} \\ \underline{777} \\ 430 \\ \underline{333} \\ 970 \\ \underline{888} \\ 820 \end{array}$$

We are again dividing 111 into 820, so the digits in the quotient repeat all over again.

FIGURE 6.17 $\frac{82}{111}$ has a repeating decimal expansion.

The true decimal expansion of $\frac{82}{111}$ is actually the infinite repeating decimal 0.738738738738 . . . . We usually write the part of a decimal that repeats with a bar over it, so we would write $\frac{82}{111}$ as $0.\overline{738}$.

Now try Exercises 65 to 72. ❋ **15**

Quiz Yourself **15**

Write $\frac{21}{33}$ as a decimal number.

When dividing a rational number to get a decimal expansion, we can have only a finite number of possible remainders, so at some point, the type of repetition that we saw in Example 9(c) must always occur.

> **DECIMAL EXPANSIONS OF RATIONAL NUMBERS**
>
> When writing a rational number in decimal form, we always get either a terminating expansion, as in a) and b) in Example 9, or an infinite repeating expansion, as in c). (We could think of a terminating expansion as a repeating expansion in which 0 repeats from some point on.)

0.1 2 4

tenths ⌐

hundredths ⌐

thousandths ⌐

FIGURE 6.18
Reading 0.124.

We will now explain how to write decimal numbers as quotients of integers. If a number has a terminating expansion such as 0.124, we remember that we read this number as 124 thousandths (see Figure 6.18).

So we can write 0.124 as

$$\frac{124}{1,000} = \frac{31 \cdot \cancel{4}}{250 \cdot \cancel{4}} = \frac{31}{250}.$$

└─ 4 divides both numerator and denominator.

To write a repeating decimal such as $0.\overline{36}$ as a quotient of integers is a little more complicated, as we explain in Example 10.

EXAMPLE 10 *Writing a Repeating Decimal as a Quotient of Integers*

Write $x = 0.\overline{36}$ as a quotient of integers.

SOLUTION: The problem in writing 0.36363636363636 . . . as a quotient of integers is that we have to deal with the infinite "tail" of 363636 . . . . The technique that we use is to create another number with the same tail that x has. We then subtract the two numbers to get a number without a repeating infinite tail.

Consider the number $100 \cdot x = 36.36363636$ . . . . If we subtract $100 \cdot x - x$, we get

$$\begin{array}{r} 100 \cdot x = 36.36363636 \ldots \\ -\quad x = -0.36363636 \ldots, \\ \hline 99 \cdot x = 36 \end{array}$$

so the infinite tails have disappeared. Solving $99 \cdot x = 36$, we find that $x = \frac{36}{99} = \frac{4}{11}$. Thus, $0.36363636363636 \ldots = \frac{4}{11}$. *

Now try Exercises 73 to 84. ✸ **16**

If in Example 10 we were dealing with a number such as $x = 0.\overline{634}$, then we would subtract $1,000 \cdot x - x$ to get rid of the infinite tail.

A number with a nonrepeating decimal expansion is not a rational number, and we will discuss such numbers in the next section.

Quiz Yourself **16**

a) Write 0.2548 as a quotient of integers.

b) Write $x = 0.\overline{54}$ as a decimal number.

Exercises 6.3

Looking Back†

These exercises follow the general outline of the topics presented in this section and will give you a good overview of the material that you have just studied.

1. What is a good way to check if two rational numbers are equal?

2. How did we reduce the rational numbers in Example 2?

3. Before we can add two rational numbers, we rewrite them having what the same?

4. In our discussion prior to Example 5, we simplified $\dfrac{\frac{3}{4}}{\frac{5}{8}}$ by multiplying both the numerator and denominator by $\frac{8}{5}$. What is our memory device for remembering this computation?

5. What does the "Frac" command do?

6. Who was M. Leblanc?

Sharpening Your Skills

Write all answers in this exercise set in lowest terms.

Which of the following pairs of rational numbers are equal?

7. $\dfrac{2}{3}, \dfrac{8}{12}$ 8. $\dfrac{5}{6}, \dfrac{10}{18}$ 9. $\dfrac{12}{14}, \dfrac{14}{16}$

10. $\dfrac{11}{6}, \dfrac{18}{20}$ 11. $\dfrac{22}{14}, \dfrac{30}{21}$ 12. $\dfrac{5}{16}, \dfrac{10}{32}$

13. $\dfrac{5}{14}, \dfrac{15}{42}$ 14. $\dfrac{9}{7}, \dfrac{54}{42}$

Reduce each fraction.

15. $\dfrac{15}{35}$ 16. $-\dfrac{30}{48}$ 17. $\dfrac{-24}{72}$

18. $\dfrac{60}{135}$ 19. $\dfrac{225}{350}$ 20. $-\dfrac{132}{96}$

21. $\dfrac{143}{154}$ 22. $\dfrac{-120}{216}$

---

*Notice that $0.\overline{36} = \frac{36}{99}$. You would find that $0.\overline{634} = \frac{634}{999}$; in general, you can often make this type of conversion quickly by following this pattern.
†Before doing these exercises, you may find it useful to review the note *How to Succeed at Mathematics* on page xix.

Perform the following operations. Express your answer as a positive or negative quotient of two integers in reduced form.

23. $\dfrac{2}{3} + \dfrac{1}{2}$ **24.** $\dfrac{3}{4} + \dfrac{1}{6}$ **25.** $\dfrac{1}{6} - \dfrac{1}{2}$

26. $\dfrac{13}{16} - \dfrac{5}{8}$ **27.** $\dfrac{7}{16} - \dfrac{1}{3}$ **28.** $\dfrac{2}{9} - \dfrac{3}{8}$

29. $\dfrac{5}{24} + \dfrac{7}{18}$ **30.** $\dfrac{5}{12} + \dfrac{3}{14}$ **31.** $\dfrac{3}{4} + \dfrac{5}{6} + \dfrac{7}{8}$

32. $\dfrac{1}{3} + \dfrac{2}{5} + \dfrac{5}{6}$ **33.** $\dfrac{1}{8} - \dfrac{2}{3} + \dfrac{1}{2}$ **34.** $\dfrac{2}{9} - \dfrac{2}{27} + \dfrac{1}{4}$

Perform the following operations. Express your answer as a positive or negative quotient of two integers in reduced form.

35. $\dfrac{2}{3} \cdot \dfrac{1}{2}$ **36.** $\dfrac{5}{16} \cdot \dfrac{4}{15}$

37. $\dfrac{1}{6} \div \dfrac{1}{2}$ **38.** $\dfrac{5}{16} \div \dfrac{3}{8}$

39. $\dfrac{7}{8} \div \left(-\dfrac{5}{24}\right)$ **40.** $\dfrac{2}{33} \div \dfrac{6}{55}$

41. $\dfrac{7}{18} \cdot \left(-\dfrac{3}{4}\right)$ **42.** $-\dfrac{7}{32} \cdot \dfrac{8}{35}$

43. $\left(\dfrac{7}{18} \cdot \left(-\dfrac{3}{4}\right)\right) \div \left(\dfrac{7}{9}\right)$ **44.** $\left(\dfrac{14}{25} \div \dfrac{4}{5}\right) \cdot \left(\dfrac{10}{3}\right)$

45. $\left(\dfrac{11}{30} \div \left(-\dfrac{1}{6}\right)\right) \cdot \left(\dfrac{15}{4}\right)$ **46.** $\left(\dfrac{7}{4} \cdot \dfrac{8}{21}\right) \div \left(\dfrac{2}{5}\right)$

Perform the following operations. Express your answer as a positive or negative quotient of two integers in reduced form.

47. $\dfrac{5}{3} \cdot \left(\dfrac{8}{15} + \dfrac{2}{3}\right)$ **48.** $\dfrac{8}{9} \cdot \left(\dfrac{5}{6} + \dfrac{1}{4}\right)$

49. $\dfrac{22}{27} \cdot \left(\dfrac{3}{11} + \dfrac{2}{3}\right)$ **50.** $\dfrac{10}{9} \cdot \left(\dfrac{2}{5} - \dfrac{2}{3}\right)$

51. $\dfrac{7}{30} \div \left(\dfrac{1}{6} - \dfrac{3}{14}\right)$ **52.** $\dfrac{11}{40} \div \left(\dfrac{3}{5} - \dfrac{1}{4}\right)$

Each of the following divisions represents a conversion of an improper fraction to a mixed number. Write that conversion as an equation with the improper fraction on the left and the mixed number on the right.

53.
$$\begin{array}{r} 15 \\ 5\overline{)79} \\ \underline{75} \\ 4 \end{array}$$

54.
$$\begin{array}{r} 4 \\ 8\overline{)35} \\ \underline{32} \\ 3 \end{array}$$

55.
$$\begin{array}{r} 9 \\ 3\overline{)29} \\ \underline{27} \\ 2 \end{array}$$

56.
$$\begin{array}{r} 17 \\ 8\overline{)143} \\ \underline{136} \\ 7 \end{array}$$

Convert each improper fraction to a mixed number.

57. $\dfrac{27}{4}$ **58.** $\dfrac{139}{8}$ **59.** $\dfrac{121}{15}$ **60.** $\dfrac{214}{12}$

Convert each mixed number to an improper fraction.

61. $2\dfrac{3}{4}$ **62.** $5\dfrac{3}{8}$ **63.** $9\dfrac{1}{6}$ **64.** $4\dfrac{5}{12}$

Write each rational number as a terminating or repeating decimal.

65. $\dfrac{3}{4}$ **66.** $\dfrac{5}{8}$ **67.** $\dfrac{3}{16}$

68. $\dfrac{27}{32}$ **69.** $\dfrac{9}{11}$ **70.** $\dfrac{16}{33}$

71. $\dfrac{4}{13}$ **72.** $\dfrac{4}{7}$

Write each decimal as a quotient of two integers in reduced form.

73. 0.64   **74.** 0.075   **75.** 0.836

76. 0.345   **77.** 12.2   **78.** 4.068

79. $0.\overline{4}$ **80.** $0.3\overline{8}$ **81.** $0.\overline{189}$

82. $0.\overline{21}$ **83.** $0.3\overline{18}$ **84.** $0.\overline{384615}$

Applying What You've Learned

85. Living expenses. Andre spends $\frac{1}{3}$ of his paycheck on rent, $\frac{1}{4}$ on food, and $\frac{1}{6}$ on utilities. What fractional part of his paycheck does he have left for other expenses?

86. Scholarships allocation. At Central State College, $\frac{1}{8}$ of the students have athletic scholarships and $\frac{1}{3}$ of the students have academic scholarships. (Assume that no students have both.) What fractional part of the student body has one or the other of these scholarships?

87. Dividing the purse in a tournament. Assume that in a Texas Hold 'em tournament, the winner receives $\frac{1}{3}$ of the purse, the person who comes in second receives $\frac{1}{4}$, and the remainder of the purse is split among four other players. What fractional part of the purse does each of these four players receive?

88. Advertising budget. The Great Downhill Ski Company spends $\frac{1}{3}$ of its advertising budget on print media, $\frac{2}{5}$ of the budget on TV ads, and $\frac{1}{6}$ of the budget on the radio. What part of the budget remains for other types of advertising?

Use the following graph, which describes the proportion of women officers in four branches of the armed services in 2007 to answer Exercises 89 and 90 (Source: Department of Defense).

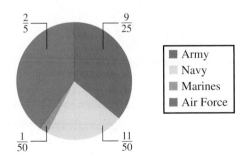

89. What fractional part of women officers were in either the Army or the Navy?

90. What fractional part of women officers were in neither the Army nor the Air Force?

91. Tiling walls of a museum. The director of the New Central American Art Museum is redecorating the walls of a room with replicas of ancient Inca tiles. The square tiles are $8\frac{1}{2}$ inches. If the room is 17 feet high, how many $8\frac{1}{2}$-inch tiles will be needed vertically? Will any of the tiles have to be cut?

92. Buying art supplies. Banksy is an artist who purchases spray paint in $2\frac{3}{8}$-liter cuysans. If he b the larger can, which is $1\frac{1}{3}$ times as large as the smaller can, how many liters will it contain?

93. Modifying a recipe. In *The Fanny Farmer* cookbook, the recipe for Hungarian Goulash, which serves four, calls for (among other ingredients) $1\frac{1}{2}$ tablespoons of lemon juice. If you want to increase this recipe to serve 10, how much lemon juice should you use?

94. Modifying a recipe. In *The Fanny Farmer* cookbook, the recipe for 1 cup of horseradish cream sauce (to accompany roast beef) calls for $\frac{3}{4}$ cup of heavy cream. If you want to make $1\frac{1}{2}$ cups of this sauce, how many cups of heavy cream should you use?

95. Unit pricing.* Milan can buy a $5\frac{1}{4}$-ounce tube of toothpaste for $2.60 or a $7\frac{1}{8}$-ounce tube of the same toothpaste for $3.60. Which size is the better buy? Explain. (*Hint:* Convert the mixed numbers to decimals and then calculate the price per ounce for each size.)

96. Unit pricing. Janita can buy a $37\frac{1}{2}$-ounce bottle of orange juice for $1.75 or a $44\frac{3}{4}$-ounce bottle of the same juice for $2.70. Which size is the better buy? Explain. (*Hint:* Convert the mixed numbers to decimals and then calculate the price per ounce for each size.)

Use the given graph, which gives a projection for the racial-ethnic makeup of the United States in 2025 to answer Exercises 97 and 98. (Source: Sociology *by James Henson, Allyn & Bacon, 2007)*

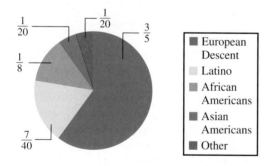

97. What fractional part of the projected U.S. population is expected to be of neither European descent nor Latino?

98. What fractional part of the projected U.S. population is expected to be African American, Latino, or Asian American?

99. Minimizing waste on a remodeling project. Ty Pennington is placing $3\frac{3}{8}$-foot strips of bamboo flooring on the floor of a house for the show *Extreme Makeover: Home Edition* and wants to minimize the waste caused by having leftover pieces that are too short. How many $3\frac{3}{8}$-foot strips can he cut from a 12-foot strip? If Ty buys the strips in 15-foot lengths, will there be more or less waste per 15-foot strip than for a 12-foot strip?

100. Estimating a painting job. Paige is estimating the cost of a painting job. A room measures $10\frac{1}{4}$ feet on two walls and $13\frac{1}{2}$ feet on the other two walls. Also, the room is $8\frac{3}{4}$ feet high. What is the total area of the four walls of the room?

Communicating Mathematics

101. In multiplying two rational numbers, what might we do first before multiplying the numerators and denominators?

102. Give a counterexample to show that it is incorrect to add rational numbers as follows:

$$\frac{a}{b} + \frac{c}{d} = \frac{a+c}{b+d}$$

103. Give an example of how you would convert an improper fraction to a mixed number.

104. Give an example of how you would convert a mixed number to an improper fraction. When is it useful to make this conversion?

105. Explain why the decimal expansion of a rational number has to repeat.

106. Draw a diagram to illustrate the product $\frac{3}{8} \cdot \frac{2}{3}$ similar to the discussion that preceded Example 4. Explain how your diagram illustrates the product.

Perform each of the following divisions using the technique that we used to calculate $\dfrac{\frac{3}{4}}{\frac{5}{8}}$ in the discussion prior to Example 5.

107. $\dfrac{\frac{3}{2}}{\frac{1}{4}}$ **108.** $\dfrac{\frac{5}{6}}{\frac{2}{3}}$ **109.** $\dfrac{\frac{5}{12}}{\frac{3}{8}}$ **110.** $\dfrac{\frac{3}{10}}{\frac{2}{5}}$

Using Technology to Investigate Mathematics

111. Use the Internet to find Java applets that illustrate the concepts we have discussed in this section. For example, you can find many interactive programs that illustrate the notion of equivalent rational numbers and operations on rational numbers. Report on your findings.

---

*You may find it interesting to go to a market and find examples of products where the smaller size costs less per unit than does the larger "economy" size.

112. Search the Internet to find information on the history of irrational numbers. Write a brief report on your findings.

For Extra Credit

113. Making a bookshelf. Antonio is making a student bookshelf from boards and cinder blocks. He has a piece of weathered pine that is $10\frac{7}{8}$ feet long and wants to cut it into four pieces of equal length. If we ignore the width of the saw's cut, how long will each shelf be? Express your answer in terms of feet.

114. Making a bookshelf. Redo Exercise 113 taking into account that each saw cut will be $\frac{1}{8}$ of an inch wide.

115. Hanging a mirror. Suppose that you want to hang a mirror that is $40\frac{1}{2}$ inches wide on a wall that is 6 feet (72 inches) wide. You want to center the mirror and put two hooks into the wall to support the mirror at $\frac{1}{3}$ and $\frac{2}{3}$ distance from one side of the mirror to the other side (see Figure 6.19). How far should the hooks be placed from the edges of the wall?

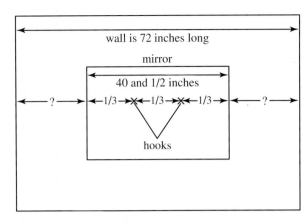

FIGURE 6.19 Hanging a mirror.

116. Publishing a textbook. A publisher wants to increase the content in a new edition of a textbook without increasing the number of pages. The current printed area on each page is $5\frac{1}{4}$ inches wide and inches long. If we increase the width of the printed area to $6\frac{1}{8}$ inches and increase the length to $8\frac{1}{2}$ inches, by how many square inches will the printed area increase per page?

6.4 The Real Number System

Objectives

1. Understand that irrational numbers have nonrepeating decimal expansions.
2. Compute using radicals.
3. Understand the properties of common number systems.

When you think of the word *irrational,* what words come to mind? Nonsensical? Unreasonable? Crazy? In order to understand why mathematicians would use such an unflattering term to describe certain numbers, we need to discuss a little history.

About 500 BC, a school of ancient Greek mathematicians led by Pythagoras of Samos (see Historical Highlight on page 264) discovered an amazing number, the square root of 2, which we write as $\sqrt{2}$. Their discovery was based on the Pythagorean theorem that we illustrate in Figure 6.20(a). This theorem states that for a right triangle, the square of the length of the hypotenuse (the longest side) is equal to the sum of the squares of the lengths of the other two sides. Or, $c^2 = a^2 + b^2$. If the right triangle has two sides of length 1, as in Figure 6.20(b), then $c^2 = 1^2 + 1^2 = 2$, or $c = \sqrt{2}$.

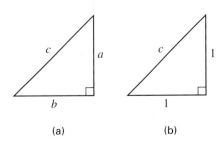

FIGURE 6.20 (a) $c^2 = a^2 + b^2$;
(b) $c^2 = 1^2 + 1^2 = 2$.

What is amazing about this discovery is that we cannot write $\sqrt{2}$ as the quotient of two integers, and therefore, it is not a rational number. Prior to that time, the Greeks believed that all numbers were rational. So, when the Greeks discovered that $\sqrt{2}$ was not a quotient of integers, it didn't make sense—it was irrational. We will give Pythagoras' proof that $\sqrt{2}$ is not rational in Exercise 101.

✏️ **KEY POINT**

Irrational numbers have nonrepeating decimal expansions.

Irrational Numbers

You saw in Section 6.3 that the rational numbers are precisely those numbers that have repeating* decimal expansions. This means that any number that is not rational, such as $\sqrt{2}$, must have a nonrepeating decimal expansion.

> **DEFINITION** An **irrational number** is a number that is not a rational number and therefore has a nonrepeating decimal expansion.

A number such as 5.12112111211112 . . . is an example of an irrational number. Although there is a pattern to the digits in this expansion, there is no single block of numbers that repeats from some point on. Other examples of irrational numbers would be 15.1234567891011121314 . . . and 0.10200300040000500000 . . . .

If you use your calculator to find $\sqrt{2}$, you may get an answer such as $\sqrt{2} = 1.414213562$. Because this is a terminating (and hence repeating) decimal, what your calculator is telling you is not quite true. In reality, your calculator gave you a very close rational number approximation, which is a little less than $\sqrt{2}$. *In fact, any number of the form $\sqrt{n}$, where n is not a perfect square, will be an irrational number.*

Although you probably have had experience with many more rational numbers than irrational numbers, you may find it surprising that there are more irrational numbers than there are rational numbers.† We will discuss several famous irrational numbers: π (pi), ϕ (phi), and e.

The number $\pi = 3.141592654$ . . . frequently occurs in geometry and is defined as the ratio of the circumference of a circle to its diameter. For example a circle with diameter 2, will have a circumference of 2π. See Figure 6.21.

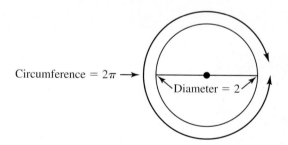

FIGURE 6.21 A circle with diameter 2 has a circumference of 2π.

The Greeks thought that the number $\phi = \dfrac{\sqrt{5}+1}{2} = 1.618033989$ . . . described the ratio of the length to the width of the most beautiful rectangle. The architecture of both the Greek Parthenon and the United Nations building (see page 287), as well as many works of art, are based on this ratio. The number $e = 2.718281828$ . . . , named after the famous Swiss mathematician Leonhard Euler (pronounced "oiler"), has many important applications in science and engineering. It is probably the only number on your calculator keypad that is named after a person.

---

*Recall that we consider terminating decimals to be repeating decimals with the repeated digit being zero.
†See Section 2.6, which discusses infinite sets.

EXAMPLE 1 *Approximating π, φ, and e*

Although $\pi = 3.141592654\ldots$, $\phi = 1.618033989\ldots$, and $e = 2.718281828\ldots$ are irrational numbers and therefore have nonrepeating decimal expansions, there are methods for computing good approximations to these numbers.

a) Approximate π by using the first 20 terms of the following series:

$$4 - \frac{4}{3} + \frac{4}{5} - \frac{4}{7} + \frac{4}{9} - \frac{4}{11} + \frac{4}{13}\ldots.$$

b) Use your calculator to approximate ϕ as follows:

Step 1: Enter 1.

Step 2: Press the $\boxed{1/x}$ or $\boxed{x^{-1}}$ key.

Step 3: Add 1 to the result in step 2.

Step 4 and beyond: Keep repeating steps 2 and 3—that is, invert your answer and add 1 repeatedly until you get the first five decimals after the decimal point in the decimal expansion of ϕ.

c) Approximate e by evaluating the expression $(1 + 1/n)^n$ for $n = 10$, 100, 1,000, and 10,000.

SOLUTION:

a) If you compute $4 - \frac{4}{3} + \frac{4}{5} - \frac{4}{7} + \frac{4}{9} - \frac{4}{11} + \frac{4}{13} - \cdots - \frac{4}{39}$, you should get the result 3.091623807. As you add more terms to this expression, you will get values closer to π. I did this (using a computer) for 100 terms and got 3.131592904. For 10,000 terms, the approximation was 3.141492654.

b) The following calculator screens show the beginning of the process and then the calculations at a much later stage for evaluating ϕ:

```
1⁻¹+1
                    2
Ans⁻¹+1
                  1.5
Ans⁻¹+1
          1.666666667
```
```
            1.617977528
Ans⁻¹+1
            1.618055556
Ans⁻¹+1
            1.618025751
Ans⁻¹+1
            1.618037135
■
```

c) Using my calculator, I obtained the following results:

$$(1 + 1/10)^{10} = 2.59374246$$
$$(1 + 1/100)^{100} = 2.704813829$$
$$(1 + 1/1{,}000)^{1{,}000} = 2.716923932$$
$$(1 + 1/10{,}000)^{10{,}000} = 2.718145927 \text{ ❊}$$

Computing with Radicals

In order to do exact calculations with irrational numbers, we often express them using radicals (roots).

> **DEFINITIONS** A number, such as $\sqrt{x}$, is called a **radical**. The symbol $\sqrt{}$ is called a **radical sign** and x is called the **radicand**.

Because scientific and mathematical formulas often contain radicals, you should understand the following rules that are essential to working with radicals.

> **MULTIPLYING AND DIVIDING RADICALS** If $a \geq 0$ and $b \geq 0$, then
> $$\sqrt{ab} = \sqrt{a}\sqrt{b}, \text{ and if } b \neq 0, \text{ then } \sqrt{\frac{a}{b}} = \frac{\sqrt{a}}{\sqrt{b}}.$$

```
√(6*5)
        5.477225575
√(6)*√(5)
        5.477225575
■
```

EXAMPLE 2 *Multiplying and Dividing Radicals*

Use the rules for multiplying and dividing radicals to rewrite each of the following. Use a calculator to verify your answers. (Remember that your calculator is only giving you approximations instead of exact results.)

a) $\sqrt{6 \cdot 5}$ b) $\sqrt{8}\sqrt{3}$ c) $\sqrt{\frac{36}{12}}$ d) $\frac{\sqrt{7}}{\sqrt{15}}$

SOLUTION:

a) $\sqrt{6 \cdot 5} = \sqrt{6}\sqrt{5}$. Using a calculator, $\sqrt{6 \cdot 5} = \sqrt{30} = 5.477225575$, and $\sqrt{6}\sqrt{5} = 5.477225575$. Note when calculating $\sqrt{6}\sqrt{5}$, calculate this product* by first calculating $\sqrt{6}$; do not clear the calculator. Then press the multiplication key. Next calculate $\sqrt{5}$ and then press "Enter" ("=" key).

b) $\sqrt{8}\sqrt{3} = \sqrt{8 \cdot 3} = \sqrt{24}$. Both sides of the equation have a decimal approximation of 4.898979486.

c) $\sqrt{\frac{36}{12}} = \frac{\sqrt{36}}{\sqrt{12}} = \frac{6}{\sqrt{12}}$. Both sides of the equation have an approximate decimal value of 1.732050808.

d) $\frac{\sqrt{7}}{\sqrt{15}} = \sqrt{\frac{7}{15}}$. Calculating both sides, we get 0.6831300511. ❋

We can use the product and quotient rules for radicals to simplify radicals. In simplifying a radical, we want to eliminate all perfect squares (numbers such as $2 \cdot 2 = 4$, $3 \cdot 3 = 9$, $5 \cdot 5 = 25$, and so on) under the radical. There are usually several different ways to do calculations that will give you the same simplification.

Math in Your Life

When Is "Close" Not Good Enough?

You might argue that in this day of powerful computers, there is really no reason to work with a number such as $\sqrt{2}$ when the approximation 1.414213562 would do just as well. Realize that when doing lengthy computations, eventually the small errors introduced by using approximations can amount to a larger, intolerable error in the final result. If precision is critical, it is important to hold off using approximations as long as we can in obtaining a final result.

For example, at the University of Maryland Medical Center, doctors perform a relatively new type of non-invasive brain surgery called Gamma Knife®. In this surgery, no incision is made; instead, more than 200 beams of relatively weak radiation are all precisely focused to provide an intense amount of radiation at a specific spot in the brain. As you can imagine, to achieve such precision, it is important that the calculations are as exact as is mathematically possible.

---

*Practice duplicating our calculations until you are confident in working with your calculator.

EXAMPLE 3 *Simplifying Radicals*

Simplify each of the following:

a) $\sqrt{18}$ b) $\sqrt{15}\sqrt{5}$ c) $\sqrt{\dfrac{54}{12}}$ d) $\dfrac{\sqrt{20}}{\sqrt{15}}$

SOLUTION:

a) We recognize that 9 is a factor of 18 and 9 is a perfect square. So, $\sqrt{18} = \sqrt{2\cdot 9} = \sqrt{2}\sqrt{9} = \sqrt{2}\cdot 3 = 3\sqrt{2}.$ *

b) We notice that 5 is a factor of 15 and 5 also appears in the second radical sign. So, if we combine the two radicals into one radical, then we can simplify the 5 squared in the single radical. Thus,

$$\sqrt{15}\sqrt{5} = \sqrt{15\cdot 5} = \sqrt{3\cdot 5\cdot 5} = \sqrt{3\cdot 25} = \sqrt{3}\sqrt{25} = \sqrt{3}\cdot 5 = 5\sqrt{3}.$$

c) There are several ways to do this. We will begin by using the quotient rule for radicals. Thus,

— Identify perfect squares in 54 and 12.

$$\sqrt{\frac{54}{12}} = \frac{\sqrt{54}}{\sqrt{12}} = \frac{\sqrt{6\cdot 9}}{\sqrt{3\cdot 4}} = \frac{\sqrt{6}\cdot\sqrt{9}}{\sqrt{3}\cdot\sqrt{4}} = \frac{\sqrt{6}\cdot 3}{\sqrt{3}\cdot 2} = \sqrt{\frac{6}{3}}\cdot\frac{3}{2} = \sqrt{2}\cdot\frac{3}{2} = \frac{3}{2}\sqrt{2}.$$

d) Here we see that if we combine the two radicals into a single radical, then we can cancel the 5 from the numerator and denominator. So,

— Notice that 20 and 15 have a common factor of 5.

$$\frac{\sqrt{20}}{\sqrt{15}} = \sqrt{\frac{20}{15}} = \sqrt{\frac{4\cdot 5}{3\cdot 5}} = \frac{\sqrt{4}}{\sqrt{3}} = \frac{2}{\sqrt{3}}.$$

— Cancel common factor.

Now try Exercises 17 to 24 and 33 to 44. ✳ **17**

Quiz Yourself **17**

Simplify the following:

a) $\sqrt{45}$

b) $\sqrt{\dfrac{15}{48}}$

Prior to the invention of calculators, people had to use pencil and paper computations to make rational number estimates of numbers such as $\frac{2}{\sqrt{3}}$. It was easier to do such calculations if quotients did not have radicals in the denominator. Shortly, we will discuss how to rewrite quotients such as $\frac{2}{\sqrt{3}}$ in this preferred form.

Some Good Advice

In doing a calculation, you will find it to your advantage to try to work with smaller numbers whenever possible. For example, if you write $\sqrt{72}\sqrt{48}$ as $\sqrt{72\cdot 48} = \sqrt{3{,}456}$, then you have the problem of finding the factors in 3,456.

On the other hand, if you recognize that $\sqrt{72} = \sqrt{2\cdot 36} = 6\sqrt{2}$ and $\sqrt{48} = \sqrt{3\cdot 16} = 4\sqrt{3}$, then you can do this calculation quickly as $\sqrt{72}\sqrt{48} = 6\sqrt{2}\cdot 4\sqrt{3} = 24\sqrt{6}.$

✎ **KEY POINT**

Rationalizing the denominator makes a quotient more understandable.

Sometimes when we calculate with radicals, we get an answer such as $\frac{2}{\sqrt{3}}$. We have no difficulty evaluating this expression with modern calculators; however, without using a calculator, it is a little difficult to estimate the size of this number. If we write the number in another way, without a radical in the denominator, we can get a better intuitive understanding of the number. The process of writing a quotient so that it has no radicals in the denominator is called **rationalizing the denominator**. We illustrate this technique in Example 4.

—————

*Although we will not do so, you can use a calculator to verify that $\sqrt{18} = 3\sqrt{2}$.

✸ ✸ ✸ HISTORICAL HIGHLIGHT

The Pythagoreans*

Shortly before 500 BC, Pythagoras founded a school in the Greek settlement of Crotona, Italy, that was to have a profound influence on mathematics for the next 2,500 years. The school, which had about 300 students (including about 30 women), was somewhat like a secret society or fraternity. All students studied the same subjects, or *mathemata*—number theory, music, geometry, and astronomy. In addition, they studied logic, grammar, and rhetoric.

Beginners, called *acoustici*, listened silently as Pythagoras lectured them from behind a curtain. After three years of obedient study, they became *mathematici*, and the secrets of the society were revealed to them. Many

of these "secrets" are the mathematical theorems that we still study today.

The Pythagoreans would not eat beans or drink wine. They believed that one's soul could leave the body and live in another person or animal. Accordingly, they would not eat meat or fish so that they would not consume the residence of another's soul.

Around 500 BC, there was a political revolt in Crotona. The school was burned, and Pythagoras was killed. There are conflicting reports of his death. One romantic account tells that as he fled for his life, Pythagoras stopped at the edge of a sacred field of beans, allowing his enemies to kill him rather than trample his sacred plants.

EXAMPLE 4 *Rationalizing Denominators*

Rationalize the denominator in the following expressions:

a) $\dfrac{2}{\sqrt{3}}$ b) $\dfrac{7}{\sqrt{12}}$

SOLUTION:

a) We want to write this quotient so that it has a perfect square in the radical in the denominator. So we multiply both the denominator and numerator by $\sqrt{3}$ to get

$$\frac{2}{\sqrt{3}} = \frac{2 \cdot \sqrt{3}}{\sqrt{3} \cdot \sqrt{3}} = \frac{2 \cdot \sqrt{3}}{\sqrt{3 \cdot 3}} = \frac{2 \cdot \sqrt{3}}{\sqrt{9}} = \frac{2 \cdot \sqrt{3}}{3} = \frac{2}{3}\sqrt{3}.$$

b) If we factor 12, we find that $12 = 2 \cdot 2 \cdot 3$. The product $2 \cdot 2$ is already a perfect square, so we need another factor of 3 to have the radical contain a perfect square. Thus, we will multiply the numerator and denominator by $\sqrt{3}$. So we have

$$\frac{7}{\sqrt{12}} = \frac{7 \cdot \sqrt{3}}{\sqrt{12} \cdot \sqrt{3}} = \frac{7 \cdot \sqrt{3}}{\sqrt{36}} = \frac{7 \cdot \sqrt{3}}{6} = \frac{7}{6}\sqrt{3}.$$

Now try Exercises 45 to 52. ✳ **18**

Quiz Yourself **18**

Rationalize the denominator in $\dfrac{3}{\sqrt{15}}$.

KEY POINT

We add and subtract radicals with the same radicand.

It is a little more complicated to add and subtract radicals. If we add $5\sqrt{3} + 4\sqrt{3}$, we can think "five of something plus four of something equals nine of something." Thus, $5\sqrt{3} + 4\sqrt{3} = 9\sqrt{3}$. However, if we want to add $5\sqrt{2} + 4\sqrt{7}$, we are, so to speak, adding apples and oranges. We cannot add five of one thing to four of another thing. In order to add (or subtract) expressions with radicals, all the radicals must have the same radicand. We illustrate this in Example 5.

EXAMPLE 5 *Adding Expressions Containing Radicals*

Perform the operations.

a) $3\sqrt{2} + 6\sqrt{2}$ b) $7\sqrt{5} - 9\sqrt{5}$ c) $5\sqrt{3} + 8\sqrt{12}$ d) $\sqrt{45} + \sqrt{12}$

SOLUTION:

a) This addition is straightforward because both radicands are the same. So, $3\sqrt{2} + 6\sqrt{2} = 9\sqrt{2}$.

---

*This highlight is based on material from David Burton, *The History of Mathematics* (New York: WCB McGraw-Hill, 1999), pp. 86–90.

b) Similarly, because the radicands are the same, the subtraction is quite simple. Thus we get $7\sqrt{5} - 9\sqrt{5} = -2\sqrt{5}$.

c) Here the radicands are different. However, we see that 12 has a factor of 4, so we can simplify $\sqrt{12}$. Therefore,

$$5\sqrt{3} + 8\sqrt{12} = 5\sqrt{3} + 8\sqrt{4 \cdot 3} = 5\sqrt{3} + 8\sqrt{4}\sqrt{3}$$
$$= 5\sqrt{3} + 8 \cdot 2\sqrt{3} = 5\sqrt{3} + 16\sqrt{3} = 21\sqrt{3}.$$

d) We rewrite $\sqrt{45} + \sqrt{12}$ as $3\sqrt{5} + 2\sqrt{3}$ and see that because the radicands are different, we cannot simplify this expression any further.

Now try Exercises 25 to 32. ✶ ⓳

Find:
a) $3\sqrt{5} + 8\sqrt{5}$
b) $3\sqrt{20} - 2\sqrt{45}$

 **KEY POINT**

Number systems share many common properties.

Properties of the Real Numbers

Figure 6.22 summarizes the relationships between the number systems that you have studied in this chapter.

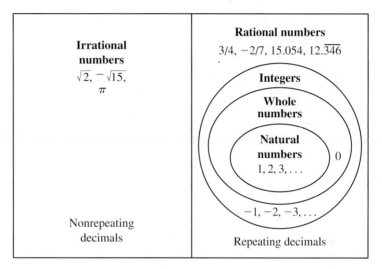

FIGURE 6.22 The set of real numbers.

If we combine the set of rational numbers with the set of irrational numbers, we get the set of real numbers. The set of real numbers, together with the operations of addition and multiplication,* form a mathematical system that has the properties listed in Table 6.3 on the next page.

EXAMPLE 6 *The Closure Property*

Which of the following sets are closed with respect to a given operation?

a) The rational numbers under addition.
b) The natural numbers under subtraction.
c) The irrational numbers under multiplication.

SOLUTION:

a) The sum of two rational numbers is a rational number, so the rational numbers are closed under addition.

---

*Recall that in Section 6.2 we said subtraction and division are not independent operations, but are based on addition and multiplication.

Property	Definition	Examples
Closure property for addition and multiplication	If a and b are real numbers, then $a + b$ and $a \cdot b$ are also real numbers.	3 and π are both real numbers, so $3 + \pi$ is a real number as well. $\frac{1}{2}$ and $\sqrt{3}$ are real numbers, so $\frac{1}{2} \cdot \sqrt{3}$ is also a real number.
Commutative property for addition and multiplication	If a and b are real numbers, then $a + b = b + a$ and $a \cdot b = b \cdot a$. (The *order* in which we combine the numbers is not important.)	$3 + \sqrt{2} = \sqrt{2} + 3$ and $(-5) \cdot \pi = \pi \cdot (-5)$
Associative property for addition and multiplication	If a, b, and c are real numbers, then $(a + b) + c = a + (b + c)$ and $(a \cdot b) \cdot c = a \cdot (b \cdot c)$. (The way we *group* the numbers is not important.)	$(3 + \sqrt{5}) + \sqrt{7} = 3 + (\sqrt{5} + \sqrt{7})$ and $\left((-6) \cdot \frac{1}{2} \right) \cdot \pi = (-6) \cdot \left(\frac{1}{2} \cdot \pi \right)$
Zero is an **identity element for addition**, and 1 is an **identity element for multiplication**.	If a is any real number, then $0 + a = a + 0 = a$ and $1 \cdot a = a \cdot 1 = a$.	$0 + 5 = 5 + 0 = 5$ and $1 \cdot \sqrt{10} = \sqrt{10} \cdot 1 = \sqrt{10}$
If x is a real number, then $-x$ is an **additive inverse** for x. If $x \neq 0$, then $\frac{1}{x}$ is a **multiplicative inverse** for x.	For any real number x, $x + (-x) = (-x) + x = 0$ and $x \cdot \left(\frac{1}{x} \right) = \left(\frac{1}{x} \right) \cdot x = 1$.	$3 + (-3) = (-3) + 3 = 0$ and $\pi \cdot \frac{1}{\pi} = \frac{1}{\pi} \cdot \pi = 1$
Multiplication **distributes** over addition.	If a, b, and c are real numbers, then $a \cdot (b + c) = (a \cdot b) + (a \cdot c)$.	$\frac{1}{2} \cdot (4 + 6) = \left(\frac{1}{2} \cdot 4 \right) + \left(\frac{1}{2} \cdot 6 \right) = 5$

TABLE 6.3 Properties of the real numbers.

Quiz Yourself 20

Give an example to show that closure *does not hold* for the operations on the given sets.

a) Irrational numbers, subtraction

b) Negative integers, multiplication

b) If we compute the difference, $5 - 8 = -3$, we do not get a natural number. Thus, the natural numbers are not closed under subtraction.

c) Be careful! The product $\sqrt{2} \cdot \sqrt{2} = 2$, so the product of two irrational numbers need not be irrational. Therefore, the set of irrational numbers is not closed under multiplication. ❋ 20

PROBLEM SOLVING

The Always Principle

In checking whether a property such as closure holds for an operation, remember the Always Principle from Section 1.1. If we say that the property holds, then it must hold for *every* possibility *without even one exception*. For example, the set of rational numbers is not closed under division because we cannot divide by the rational number 0.

In our next example, we will investigate the commutative and associative properties. Although students often get these properties confused, it is easy to remember the difference if you think how we use similar words in English. "Commutative" reminds us of the word *commuter*, who is a person who goes back and forth between two places. So the commutative property involves numbers changing places. Similarly, the associative property means that we are *reassociating*, or regrouping, numbers.

HIGHLIGHT ✳ ✳ ✳

Between the Numbers—A Million-Dollar Comma

Just as we use the associative property in mathematics to regroup numbers, we use commas in English to regroup ideas. Although punctuation may strike you as an unimportant and tedious topic, in fact, changing punctuation often changes the meaning of a sentence.*

An article in the *New York Times* business section on October 25, 2006 describes how an extra comma in a

contract allowed a Canadian phone company to pull out of a deal earlier than expected, thereby saving the phone company one million dollars. Although the intent of the agreement seemed clear, the regulator ruled that both parties had to live by what was written in the contract, rather than what they thought they had agreed upon.

EXAMPLE 7 *Investigating Commutativity and Associativity*

Determine whether the given statements are true or false. If a statement is true, give an example to illustrate it. If a statement is false, provide a counterexample.

a) Multiplication is commutative on the set of rational numbers.

b) Subtraction is associative on the set of integers.

c) Division is associative on the set of integers.

d) Addition is associative on the set of whole numbers.

SOLUTION:

a) This is true. For example, $\frac{2}{3} \cdot \frac{4}{5} = \frac{2 \cdot 4}{3 \cdot 5} = \frac{4 \cdot 2}{5 \cdot 3} = \frac{4}{5} \cdot \frac{2}{3}$.

b) This is false. A counterexample would be $8 - (4 - 2) = 8 - 2 = 6$, but regrouping we get

$(8 - 4) - 2 = 4 - 2 = 2$, not 6.

c) This is false. One counterexample would be $27 \div (9 \div 3) = 27 \div 3 = 9$. However, if we regroup, we get $(27 \div 9) \div 3 = 3 \div 3 = 1$, not 9.

d) This is true. For example, $3 + (0 + 7) = (3 + 0) + 7$. ✳ **21**

Quiz Yourself **21**

Give examples to illustrate each of the following facts:

a) Addition is associative on the set of rational numbers.

b) Multiplication is commutative on the set of integers.

EXAMPLE 8 *Finding Inverses in a Mathematical System*

Find the additive and multiplicative inverses for each of the following numbers:

a) 5   b) −8   c) 0   d) $\frac{3}{4}$ e) $\sqrt{2}$ f) π

SOLUTION:

a) The additive inverse for 5 is a number that when added to 5 gives us 0. Clearly, −5 is the additive inverse for 5. The multiplicative inverse for 5 is a number that when multiplied by 5 gives us 1. Because $\frac{1}{5} \cdot 5 = 1$, we see that $\frac{1}{5}$ is the multiplicative inverse for 5.

b) Because $8 + (-8) = 0$, we have that 8 is the additive inverse for −8. The reciprocal of −8, namely $\frac{1}{-8} = -\frac{1}{8}$, is the multiplicative inverse for −8 because $-8 \cdot \left(-\frac{1}{8}\right) = 1$.

c) Zero is the additive inverse for zero because $0 + 0 = 0$. Zero has no multiplicative inverse. We cannot find a number, call it a, such that $a \cdot 0 = 1$, because any number times 0 will be 0.

---

*See Exercise 104 regarding "A panda eats shoots and leaves" versus "A panda eats, shoots, and leaves."

d) Because $\frac{3}{4} + \left(-\frac{3}{4}\right) = 0$, $-\frac{3}{4}$ is the additive inverse for $\frac{3}{4}$. Also, because $\frac{3}{4} \cdot \frac{4}{3} = 1$, we have that $\frac{4}{3}$ is the multiplicative inverse for $\frac{3}{4}$.

e) The number $-\sqrt{2}$ is the additive inverse for $\sqrt{2}$ and $\frac{1}{\sqrt{2}}$ is the multiplicative inverse for $\sqrt{2}$.

f) Clearly, $-\pi$ is the additive inverse for π and $\frac{1}{\pi}$ is the multiplicative inverse for π.
Now try Exercises 65 to 74. ✸

We will now look at some examples of distributivity, which relates addition and multiplication.

EXAMPLE 9 *Using Distributivity*

Use distributivity to rewrite each of the following expressions. Then evaluate the expression.

a) $2 \cdot (3 + 4)$ b) $5 \cdot 3 + 5 \cdot 4$ c) $2 \cdot (3 - 5)$

SOLUTION:

a) $2 \cdot (3 + 4) = 2 \cdot 3 + 2 \cdot 4 = 6 + 8 = 14$

b) $5 \cdot 3 + 5 \cdot 4 = 5 \cdot (3 + 4) = 5 \cdot 7 = 35$

c) $2 \cdot (3 - 5) = 2 \cdot (3 + (-5)) = 2 \cdot 3 + 2 \cdot (-5) = 6 - 10 = -4$ ✸

Exercises 6.4

Looking Back*

These exercises follow the general outline of the topics presented in this section and will give you a good overview of the material that you have just studied.

1. In terms of decimal expansions, how is an irrational number different from a rational number?

2. When will $\sqrt{n}$ be an irrational number?

3. Complete the following equations: $\sqrt{ab} =$, $\sqrt{\frac{a}{b}} =$.

4. In simplifying $\sqrt{72}\sqrt{48}$, why is it not a good idea to write this as $\sqrt{72 \cdot 48} = \sqrt{3{,}456}$? What should you do instead?

5. Why is it sometimes important to use exact numbers, such as $\sqrt{2}$, rather than numerical approximations?

6. When did Pythagoras live?

Sharpening Your Skills

Which of the following numbers are rational, and which are irrational?

7. $\frac{3}{8}$

8. 5.0136

9. $1.23456789101112\ldots$

10. $\sqrt{10}$

11. 3.1416

12. 0.10110111

13. $0.101101110\ldots$

14. $\sqrt{81}$

15. Give an example to show that $\sqrt{n}$ can be a rational number.

16. Give an example to show that $\sqrt{n}$ can be an irrational number.

Simplify the given radicals.

17. $\sqrt{18}$ **18.** $\sqrt{27}$ **19.** $\sqrt{75}$ **20.** $\sqrt{50}$

21. $\sqrt{48}$ **22.** $\sqrt{24}$ **23.** $\sqrt{189}$ **24.** $\sqrt{240}$

If possible, combine the radicals into a single radical.

25. $3\sqrt{5} + 8\sqrt{5}$ **26.** $2\sqrt{3} - 4\sqrt{2}$

27. $5\sqrt{7} - 3\sqrt{5}$ **28.** $3\sqrt{7} + 2\sqrt{7}$

29. $\sqrt{20} + 6\sqrt{5}$ **30.** $5\sqrt{12} - 4\sqrt{3}$

31. $\sqrt{50} + 2\sqrt{75}$ **32.** $\sqrt{28} + 2\sqrt{63}$

Perform the indicated operation, and simplify if possible.

33. $\sqrt{18}\sqrt{2}$ **34.** $\sqrt{15}\sqrt{5}$ **35.** $\sqrt{12}\sqrt{15}$

36. $\sqrt{72}\sqrt{10}$ **37.** $\sqrt{28}\sqrt{21}$ **38.** $\sqrt{27}\sqrt{33}$

39. $\frac{\sqrt{24}}{\sqrt{6}}$ **40.** $\frac{\sqrt{54}}{\sqrt{6}}$ **41.** $\frac{\sqrt{32}}{\sqrt{18}}$

42. $\frac{\sqrt{45}}{\sqrt{20}}$ **43.** $\frac{\sqrt{96}}{\sqrt{72}}$ **44.** $\frac{\sqrt{63}}{\sqrt{112}}$

Rationalize the denominator and simplify.

45. $\frac{3}{\sqrt{5}}$ **46.** $\frac{4}{\sqrt{3}}$ **47.** $\frac{12}{\sqrt{6}}$

---

*Before doing these exercises, you may find it useful to review the note *How to Succeed at Mathematics* on page xix.

48. $\dfrac{21}{\sqrt{14}}$ **49.** $\dfrac{10}{\sqrt{22}}$ **50.** $\dfrac{8}{\sqrt{10}}$

51. $\dfrac{3\sqrt{10}}{\sqrt{72}}$ **52.** $\dfrac{3\sqrt{5}}{\sqrt{135}}$

Find a) a rational number and b) an irrational number between the two given numbers. Give your answers in decimal form. There are many correct answers. (Hint: Write all numbers in decimal form before answering the questions.)

53. 0.43 and 0.44   **54.** 1.245 and 1.246

55. $0.\overline{4578}$ and $0.45\overline{78}$ **56.** $0.\overline{123}$ and $0.123\overline{1}$

57. $\frac{4}{7}$ and $\frac{5}{7}$ **58.** $\frac{5}{8}$ and $\frac{3}{4}$

59. 0.12112111211112 . . . and $0.\overline{12}$

60. $0.\overline{123}$ and 0.1234567891011121314 . . .

Put the numbers in order from smallest to largest.

61. a. 0.345345   **b.** $0.\overline{345}$ **c.** $0.3\overline{45}$ **d.** 0.34534534

62. a. $0.\overline{261}$ **b.** $0.26\overline{1}$ **c.** $0.2\overline{6}$ **d.** 0.2626

63. a. $\dfrac{4}{9}$ **b.** $\dfrac{5}{9}$ **c.** $0.4\overline{54}$ **d.** $0.5\overline{54}$

64. a. $\dfrac{5}{13}$ **b.** $0.38\overline{4}$ **c.** $\dfrac{4}{9}$ **d.** $0.3\overline{8}$

State which property of the real numbers we are illustrating.

65. $3(4+5) = 3 \cdot 4 + 3 \cdot 5$

66. $6(8+2) = (8+2)6$

67. $3 + (6+8) = (6+8) + 3$

68. $5(3 \cdot 2) = (5 \cdot 3)2$

69. $3 + (6+8) = (3+6) + 8$

70. $5(3 \cdot 2) = (3 \cdot 2)5$ **71.** $7 + 0 = 7$

72. $8 + (-8) = 0$

73. $4 + (2 + (-1)) = 4 + ((-1) + 2)$

74. $7(8-2) = 7 \cdot 8 - 7 \cdot 2$

Applying What You've Learned

In Exercises 75–78, first express your answer as a radical expression and then as a decimal rounded to the nearest tenth.

75. We can use the equation $D = \sqrt{2H}$ to approximate the distance D, in miles, that you can see to the horizon at a height of H feet. How many miles could a forest ranger see to the horizon if his eyes are 6 feet above ground level and he is standing on a fire tower that is 42 feet high?

76. Using the equation in Exercise 75, how far to the horizon could you see from a hot air balloon that is 160 feet above the ground?

77. Police investigating an accident can use the equation $v = 2\sqrt{5L}$ to estimate the speed at which a vehicle was traveling when the driver hit the brakes. Here, v is the speed of the vehicle in miles per hour and L is the length of the skid marks in feet. If a car leaves a skid mark of 120 feet, how fast was the car going when the driver hit the brakes?

78. Repeat Exercise 77, but now assume that the skid mark is 160 feet.

79. The time for a pendulum to swing all the way forward and then back (called its **period**) depends on its length. If a pendulum is L feet long, then we use the formula $T = 2\pi\sqrt{\dfrac{L}{32}}$ to calculate T, the time in seconds of its period. If a child is swinging from a 20-foot rope attached to a tree branch hanging over a pond, how long would it take for him to swing all the way out and then back again?

80. Repeat Exercise 79 for a 30-foot rope.

81. Marius Pudzianowski, a frequent champion in the World's Strongest Man competition, is competing in a contest in which forces of 120 and 160 pounds are applied to two ropes that form a right angle, as shown in the figure. Marius must hold a third rope, attached to the vertex of the right angle, for as long as he can. Find the force that Marius is resisting, which corresponds to length of the hypotenuse in the right triangles in the figure.

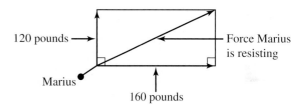

82. Repeat Exercise 81 for weights of 180 and 240 pounds.

83. The equation $t = \dfrac{\sqrt{d}}{4}$ estimates the time t in seconds that it takes an object to fall a distance of d feet. Assume that you are on an observation deck 1,600 feet above the ground on the Taipei 101 skyscraper in Taiwan, China. If your MP3 player slips out of your hand, how long will it take to hit the ground?

84. Repeat Exercise 83 assuming you are now on an observation deck that is 1,000 feet above the ground on the Empire State Building.

Communicating Mathematics

85. In simplifying a radical such as $\sqrt{20}$, what kind of factors are you looking for inside the radical sign?

86. What is the purpose of rationalizing the denominator in an expression such as $\dfrac{3}{\sqrt{2}}$? How do we do this rationalizing?

87. When we are showing that a property holds in a number system, what principle are we using from Section 1.1?

88. How do we show that a property fails in a number system?

Decide whether the following statements are true or false. Explain your answers.

89. $\sqrt{2} = 1.414213562$.

90. Every irrational number has a nonterminating decimal expansion.

91. If n is a natural number, then $\sqrt{n}$ is irrational.

92. A real number cannot be both rational and irrational.

93. $\sqrt{3}\sqrt{5}$ is irrational.

94. $\frac{\sqrt{72}}{\sqrt{8}}$ is irrational.

95. The sum of a rational number and an irrational number is an irrational number.

96. The product of two irrational numbers is always irrational.

Using Technology to Investigate Mathematics

97. Continue the calculation of π in Example 1 part a) for 20 more terms. How does this approximation compare with the one we obtained in Example 1?

98. Continue the calculation of ϕ in Example 1 part b) until your approximation is correct to five places after the decimal.

99. There are many interesting Web sites devoted to the irrational number pi. Have a contest to find the site that reports on computing the most digits in the decimal expansion of pi.

100. A mnemonic (a memory device) for memorizing the digits in the decimal expansion of pi is a sentence such as "Can I have a small container of coffee?" If you count the letters in each word of this sentence, you will get 3, 1, 4, 1, 5, 9, 2, and 6, which are the first eight digits in the infinite decimal expansion of pi. Search the Internet for other mnemonics, and then have a contest to see who can write the longest mnemonic for memorizing pi. (When I did this search, I found mnemonics in Bulgarian, Dutch, German, French, and Polish, as well as in many other languages.)

For Extra Credit

101. We will now give a proof that $\sqrt{2}$ is irrational. Explain why each of the steps in the proof is valid.

    a. Assume that $\sqrt{2}$ can be written as a quotient of integers $\frac{a}{b}$, where the quotient is in reduced form. So, $\sqrt{2} = \frac{a}{b}$.

    b. $2 = \dfrac{a^2}{b^2}$ **c.** $2 \cdot b^2 = a^2$

    d. a^2 is even, that is, divisible by 2.

    e. a cannot be odd, so it must be even.

    f. $a = 2 \cdot k$ for some integer k.

    g. $2 \cdot b^2 = (2k)^2$.   **h.** $2 \cdot b^2 = 4k^2$.

    i. $b^2 = 2k^2$.   **j.** b^2 is even.

    k. b is even.

    l. We have a contradiction, so our assumption that $\sqrt{2}$ can be written as a quotient of integers $\frac{a}{b}$ is false. Therefore, $\sqrt{2}$ is not rational.

102. Do a similar proof to show that $\sqrt{3}$ is irrational.

103. Find the length of line segment AB in the given figure.

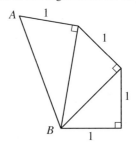

104. We mentioned in the Highlight on page 267 that changing the placement of commas can change the meaning of a sentence. For example:

    A panda walks into a cafe and orders a bamboo sandwich. After finishing his meal, he pulls out a gun and shoots the waiter. The police arresting him ask, "Why did you do that?" The panda responds, "I have to because it says in my encyclopedia that a panda eats, shoots, and leaves."

The panda has misplaced his commas and changed the meaning of the sentence. It should read " . . . that a panda eats shoots and leaves," meaning that his diet consists of shoots and leaves.

    a. Rewrite Sam I Am's statement "I like green, eggs, and ham" to read correctly.

    b. My daughter recently said to her husband, "We have cream cheese and bagels in the freezer." What do you think she really said?

105. The triple of natural numbers 4, 5 is called a Pythagorean triple. Why do you think that this is so? (*Hint:* Consider right triangles.)

106. Find another triple of natural numbers that is a Pythagorean triple.

6.5 Exponents and Scientific Notation

Objectives

1. Understand and apply the basic rules of exponents.
2. Perform operations using scientific notation.
3. Solve problems using scientific notation.

In this section, we will be able to answer the question that we posed at the very beginning of this chapter. If we were able to unravel all the atoms in your body, as though you were a great big ball of thread, and then stretch out all those atoms in a straight line, how far would they reach? To be able to represent both the tiny size of an atom, as well as the vastness of the size of the universe, we have to represent real numbers in a new way, which we will call *scientific notation*. We will introduce this notation after you study some rules of exponents.

KEY POINT

All exponent rules are based on the definition of exponents.

Exponents

In order to write such very large and very small numbers efficiently, we will use exponential notation. Some examples of exponential notation are, $10^3 = 10 \cdot 10 \cdot 10 = 1,000$ and $2^5 = 2 \cdot 2 \cdot 2 \cdot 2 \cdot 2 = 32$. In general, we have the following definition.

> **DEFINITION** If a is any real number and n is a counting number, then
>
> $$a^n = \underbrace{a \cdot a \cdot a \cdots \cdot a}_{\substack{\text{product} \\ \text{of } n \text{ } a\text{'s}}}.$$
>
> The number a is called the **base**, and the number n is called the **exponent**.

This simple definition is the basis for all the rules for exponents that we will discuss in this section.

EXAMPLE 1   *Evaluating Expressions with Exponents*

Write each of the following expressions in another way using the definition of exponents, then evaluate the expression.

a) $3 \cdot 3 \cdot 3 \cdot 3 \cdot 3$ b) 2^4 c) 0^6 d) $(-2)^4$ e) -2^4 f) 5^1

SOLUTION:

a) $3 \cdot 3 \cdot 3 \cdot 3 \cdot 3 = 3^5 = 243$

b) $2^4 = 2 \cdot 2 \cdot 2 \cdot 2 = 16$

c) $0^6 = 0 \cdot 0 \cdot 0 \cdot 0 \cdot 0 \cdot 0 = 0$

d) $(-2)^4 = (-2) \cdot (-2) \cdot (-2) \cdot (-2) = +16$

e) This is not the same as d). Notice that here the exponent, 4, takes precedence* over the negative sign. So, first we multiply the four 2s and then insert the negative sign. Thus,

First raise 2 to the fourth power.

$$-2^4 = -(2 \cdot 2 \cdot 2 \cdot 2) = -16.$$

f) $5^1 = 5$ ✳

Suppose in a physics class you encountered the expression $x^3 \cdot x^4$. You could simplify this expression by using only the definition of exponents as follows:

$$x^3 \cdot x^4 = \underbrace{(x \cdot x \cdot x)}_{3 \text{ } x\text{'s}} \cdot \underbrace{(x \cdot x \cdot x \cdot x)}_{4 \text{ } x\text{'s}} = \underbrace{x \cdot x \cdot x \cdot x \cdot x \cdot x \cdot x}_{7 \text{ } x\text{'s}} = x^7.$$

Thus, $x^3 \cdot x^4 = x^7$.

By thinking of a simple example such as this, you can remember the following rule for rewriting certain products:

> **PRODUCT RULE FOR EXPONENTS** If x is a real number and m and n are natural numbers, then
>
> $$x^m x^n = x^{m+n}.$$

*For a review of the order of operations, see Appendix A.

Notice in this rule that all the bases are the same. This rule would not apply to an expression such as x^3y^5 because the bases are different.

EXAMPLE 2   *Applying the Product Rule for Exponents*

Use the product rule for exponents to rewrite each of the following expressions, if possible:

a) $2^5 \cdot 2^9$ b) $3^2 \cdot 5^4$

SOLUTION:

a) $2^5 \cdot 2^9 = 2^{5+9} = 2^{14}$

Quiz Yourself 22

Use the product rule to rewrite $3^4 \cdot 3^2$.

b) We cannot apply the product rule here. The product $3^2 \cdot 5^4$ contains two 3s and four 5s as factors and we cannot combine these six factors into a single expression. We could, however, rewrite this as $3 \cdot 3 \cdot 5 \cdot 5 \cdot 5 \cdot 5 = 5{,}625$. ✳ 22

Suppose that you did not recall the rule for simplifying the expression $(y^3)^4$. You could think of $(y^3)^4$ as (y^3) multiplied by itself four times as follows:*

$$(y^3)^4 = (y^3)(y^3)(y^3)(y^3) = (y \cdot y \cdot y)(y \cdot y \cdot y)(y \cdot y \cdot y)(y \cdot y \cdot y) = y^{12}.$$

You see from this example that we simplified the expression by multiplying the exponents, which leads us to the following rule:*

> **POWER RULE FOR EXPONENTS** If x is a real number and m and n are natural numbers, then
>
> $$(x^m)^n = x^{m \cdot n}.$$

EXAMPLE 3   *Applying the Power Rule for Exponents*

Use the power rule for exponents to simplify each of the following:

a) $(2^3)^2$ b) $((-2)^3)^4$

SOLUTION:

a) $(2^3)^2 = 2^{3 \cdot 2} = 2^6 = 64$

b) $((-2)^3)^4 = (-2)^{3 \cdot 4} = (-2)^{12} = 4{,}096$ ✳ 23

Quiz Yourself 23

Use the power rule for exponents to simplify each of the following:

a) $(5^4)^3$

b) $((-2)^2)^4$

PROBLEM SOLVING
The Three-Way Principle

The Three-Way Principle in Section 1.1 suggests that a good way to remember what exponent rule to use in a particular situation is to think of simple examples. Then use the definition of exponents to recall how we derived the rule. For example, to remember whether you add or multiply exponents in simplifying the expression $(a^8)^7$, think what you would do with $(x^3)^2$. If you write this expression as $x^3 \cdot x^3 = x \cdot x \cdot x \cdot x \cdot x \cdot x = x^6$, then you recall that you are supposed to multiply exponents in this situation.

We often have to divide expressions containing exponents. For example, suppose that we want to simplify the expression $\frac{x^7}{x^3}$. Using the definition of exponents, we can rewrite the numerator and denominator as follows:

$$\frac{x^7}{x^3} = \frac{x \cdot x \cdot x \cdot x \cdot \cancel{x} \cdot \cancel{x} \cdot \cancel{x}}{\cancel{x} \cdot \cancel{x} \cdot \cancel{x}} = \frac{x \cdot x \cdot x \cdot x}{1} = x \cdot x \cdot x \cdot x = x^4.$$

---

*We will explain some other power rules for exponents in the exercises.

This example leads us to the following quotient rule for exponents:

> **QUOTIENT RULE FOR EXPONENTS** If x is a nonzero real number and both m and n are natural numbers, then
> $$\frac{x^m}{x^n} = x^{m-n}.$$

This rule is fine provided m is greater than n. However, if m equals n, then $m - n = 0$ and if m is less than n, then $m - n < 0$. To understand what to do in these cases, let's look at some examples. Suppose that $m = n = 3$, then $\dfrac{x^m}{x^n} = \dfrac{x^3}{x^3} = \dfrac{\cancel{x} \cdot \cancel{x} \cdot \cancel{x}}{\cancel{x} \cdot \cancel{x} \cdot \cancel{x}} = 1$. Also, if $m = 3$ and $n = 5$, we get $\dfrac{x^m}{x^n} = \dfrac{x^3}{x^5} = \dfrac{\cancel{x} \cdot \cancel{x} \cdot \cancel{x}}{\cancel{x} \cdot \cancel{x} \cdot \cancel{x} \cdot x \cdot x} = \dfrac{1}{x \cdot x} = \dfrac{1}{x^2}$. These examples will help you to remember the following definitions:

> **DEFINITIONS** If $a \neq 0$, then $a^0 = 1$, and if n is a natural number, then $a^{-n} = \dfrac{1}{a^n}$.

EXAMPLE 4 *Applying the Quotient Rule for Exponents*

Use the quotient rule to simplify the following expressions and write your answer as a single number:

a) $\dfrac{3^7}{3^5}$ b) $\dfrac{7^5}{7^8}$ c) $\dfrac{17^9}{17^9}$

SOLUTION:

a) $\dfrac{3^7}{3^5} = 3^{7-5} = 3^2 = 9$ b) $\dfrac{7^5}{7^8} = 7^{5-8} = 7^{-3} = \dfrac{1}{7^3} = \dfrac{1}{343}$

c) $\dfrac{17^9}{17^9} = 17^0 = 1$ ✽ **24**

We explained the rules for working with exponents by using very simple examples in which the exponents m and n were both natural numbers. Without proving it, we will simply state that the *rules don't change* when you are working with other types of exponents (see Figure 6.23).

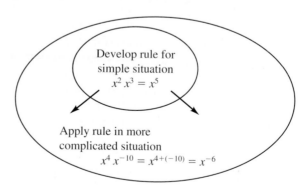

FIGURE 6.23 The rules don't change as we work in more complicated situations.

For example, if you are dealing with the expression $3^{-4} \cdot 3^6$, you can rewrite it as $3^{-4} \cdot 3^6 = 3^{-4+6} = 3^2 = 9$. In other mathematics courses, you may see rational exponents, or even irrational exponents! However, if you remember that the rules are the same

Quiz Yourself **24**

Simplify and evaluate these expressions.

a) $\dfrac{6^8}{6^{10}}$ b) $\dfrac{(-8)^{11}}{(-8)^9}$

c) $\dfrac{11^0}{11^0}$

when dealing with these more complicated examples, you will be able to do your computations confidently.

✎ **KEY POINT**

We can extend the exponent rules to all integer exponents.

EXAMPLE 5 *Using Exponent Rules When the Exponents Are Not Positive*

Assume that the exponent rules extend to these expressions and then simplify as in Example 4.

a) $5^{-4} \cdot 5^7$ b) $(2^{-3})^2$ c) $\dfrac{(-3)^{-2}}{(-3)^{-7}}$ d) $\dfrac{12^{-3}}{12^{-3}}$

SOLUTION: *

a) We apply the product rule to get $5^{-4} \cdot 5^7 = 5^{-4+7} = 5^3 = 125$.

b) Here, we apply the power rule:

$$(2^{-3})^2 = 2^{-3 \cdot 2} = 2^{-6} = \frac{1}{2^6} = \frac{1}{64}.$$

c) Now, we use the quotient rule to get

$$\frac{(-3)^{-2}}{(-3)^{-7}} = (-3)^{-2-(-7)} = (-3)^{-2+7} = (-3)^5 = -243.$$

d) We'll use the definition of a zero exponent to simplify

$$\frac{12^{-3}}{12^{-3}} = 12^{-3-(-3)} = 12^{-3+3} = 12^0 = 1.$$

Now try Exercises 7 to 38. ✳ **25**

Quiz Yourself **25**

Simplify:
a) $3^8 \cdot 3^{-6}$ b) $(2^{-2})^{-1}$

Scientific Notation

As we mentioned at the beginning of this section, scientists often have to deal with very large and very small numbers. In order to be able to understand and calculate with such extreme quantities, mathematicians have developed scientific notation. Before giving the formal definition, we will illustrate scientific notation with several examples.

Astronomers often measure distances in light-years, which is the distance that light travels in one year. A light-year is approximately 5,865,696,000,000 miles. To convert such a very large number to scientific notation, we keep dividing the number by 10 until the number is between 1 and 10. We do this by moving the decimal point to the left and keeping track of how many places we have moved the decimal point. In this case, we need to move the decimal point 12 places to the left (Figure 6.24(a)). Because moving the decimal point to the left makes the number smaller, we must multiply by 10^{12} to restore the size of the number (Figure 6.24(b)).

In physics, the electric charge on an electron is 0.00000000000000000016 coulomb. To write this number in scientific notation, we move the decimal point 19 places to the right to get 1.6, which, in effect, multiplies the number by 10^{19}. To restore the number to its small size, we must divide by 10^{19} or, what is equivalent, multiply by 10^{-19}. Thus, $0.00000000000000000016 = 1.6 \times 10^{-19}$.

✎ **KEY POINT**

We use scientific notation to represent very large and very small numbers.

5,865,696,000,000. 5.865696×10^{12}

Moving the decimal point 12 places to the left makes the number smaller.

Multiplying by 10^{12} restores the size of the number.

(a)   (b)

FIGURE 6.24 (a) Moving the decimal point to the left. (b) Multiplying by 10^{12}.

> **DEFINITION** A number is written in **scientific notation** if it is of the form $a \times 10^n$, where $1 \leq a < 10$ and n is any integer.

---

*Be aware that we are showing you only one way to apply the rules of exponents to get these solutions. There are other valid ways to solve these problems.

Remember that for a number of the form $a \times 10^n$ to be in scientific notation, the a must be greater than or equal to 1 and less than 10. The two numbers in the following diagram are not in scientific notation:

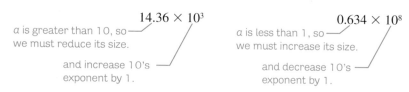

14.36×10^3

a is greater than 10, so we must reduce its size.

and increase 10's exponent by 1.

0.634×10^8

a is less than 1, so we must increase its size.

and decrease 10's exponent by 1.

We can write these numbers in scientific notation as 1.436×10^4 and 6.34×10^7.

If your calculator has the capability to express numbers in scientific notation, it probably has one of the following keys:

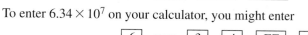

| EE | EEx | EXP |

To enter 6.34×10^7 on your calculator, you might enter

| 6 | . | 3 | 4 | EE | 7 |

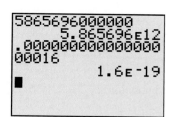

Calculators also display scientific notation in different ways. One popular way to indicate 6.34×10^7 is to display 6.34 E 7. The E 7 tells us that the exponent of 10 is 7 in scientific notation. Another popular method uses a space instead of the E to display 6.34×10^7 as 6.34 7.

RULES FOR CONVERTING A DECIMAL NUMBER TO SCIENTIFIC NOTATION

If the number x is very large,

a) Make the number smaller by moving the decimal point to the left until the number is between 1 and 10.

b) Restore the size of x by multiplying the number you obtained in part a) by 10 raised to the power, which is the number of places that you moved the decimal point to the left. (If you move the decimal point three places to the left, multiply by 10^3.)

If x is very small,

c) Make the number larger by moving the decimal point to the right until the number is between 1 and 10.

d) Restore the size of x by multiplying the number you obtained in part c) by 10 raised to the power, which is the *negative* of the number of places that you moved the decimal point to the right. (If you move the decimal point eight places to the right, multiply by 10^{-8}.)

Math in Your Life

Exponents Are All Around Us

What do your credit card balance, the number of people coming down with the latest flu virus, the population of China, and the sale of Botox all have in common? All of these quantities follow a basic pattern known as exponential growth—which we will study in detail in Section 7.4. The fact that credit card debt grows so rapidly is the same reason that a flu virus spreads so quickly through a population. We can use the same mathematics of exponents to describe the growth of China's population, as well as the increase in sales of a new beauty product such as Botox.

Scientists also use exponential models to explain the growth of greenhouse gas emissions, determine the age of dinosaur bones, and describe the metabolism of drugs in your bloodstream.

EXAMPLE 6   *Converting from Standard Notation to Scientific Notation*

Rewrite the numbers in the following statements in scientific notation:

a) The length of Manhattan Island is approximately 708,660 inches.

b) The MacBook Air takes 0.0000000034 of a second to perform an operation.

SOLUTION:

a) Move the decimal point in 708,660. to the left five places to get 7.08660. Now multiply this number by 10^5 to get 7.08660×10^5.

b) Move the decimal point in 0.0000000034 to the right nine places to get 3.4. Now multiply this number by 10^{-9} to get 3.4×10^{-9}. ✳

EXAMPLE 7   *Converting from Scientific Notation to Standard Notation*

Rewrite the numbers in the following statements in standard notation:

a) The precise length of a year is 3.1556925511×10^7 seconds.

b) The diameter of an atom is about 2.0×10^{-10} meter.

SOLUTION:

a) We move the decimal point in 3.1556925511 to the right seven places to get 31,556,925.511 seconds.

b) Move the decimal point in 2.0 ten places to the left to get 0.0000000002 meter.

Now try Exercises 39 to 56. ✳ **26**

Quiz Yourself **26**

a) Convert 53,728.41 to scientific notation.

b) Convert 2.45×10^{-3} to standard notation.

✏ *KEY POINT*

We use rules of exponents to multiply and divide numbers in scientific notation.

It is easy to multiply and divide numbers written in scientific notation. For example,

$$(3 \times 10^2) \cdot (4 \times 10^5)$$
$$= (3 \times 4) \cdot (10^2 \times 10^5) \quad \text{Regroup factors.}$$
$$= 12 \times 10^7. \quad \text{Add exponents.}$$

Also,

$$\frac{8 \times 10^6}{2 \times 10^4}$$
$$= \frac{8}{2} \times \frac{10^6}{10^4} \quad \text{Multiplication of rational numbers.}$$
$$= 4 \times 10^{6-4} \quad \text{Subtract exponents when dividing.}$$
$$= 4 \times 10^2. \quad \text{Simplify exponent.}$$

PROBLEM SOLVING

The Analogies Principle

Notice that in multiplying and dividing numbers written in scientific notation, you are not doing anything new. You are simply applying properties such as associativity and commutativity, as well as the rules for working with rational numbers that you have learned earlier.

Applications of Scientific Notation

We will now use scientific notation to solve some applied problems.

EXAMPLE 8 *Comparing the Storage Capacity of Personal Computers*

Carlos's new hard drive has 1 terabyte of storage, which is 1.0×10^{12} bytes. His previous drive had 1.8 gigabytes of storage, which is 1.8×10^{9}. How many times larger is the capacity of his new drive than the capacity of his old drive?

SOLUTION: We must divide 1.0×10^{12} by 1.8×10^{9} as follows:

$$\frac{1.0 \times 10^{12}}{1.8 \times 10^{9}} = \frac{1.0 \times 10^{12-9}}{1.8} = \frac{10^{3}}{1.8} = \frac{1,000}{1.8} \approx 556.$$

Therefore, Carlos's new drive has about 556 times the storage of his previous one. ❋

EXAMPLE 9 *Finding the Total Length of All the Atoms in a Person's Body*

If all the atoms in a person weighing 175 pounds were lined up in a straight line, how long would that line be? Compare this length with the distance to the nearest star, Proxima Centauri. Use the following information* to answer the question:

The person's body contains 3.4×10^{27} atoms.

The diameter of an atom is 2×10^{-10} meter (a meter is slightly more than 39 inches).

The distance to the nearest star, Proxima Centauri, is 4.6 light-years.

A light-year is 9.46×10^{15} meters.

SOLUTION:

a) We first multiply the number of atoms in the person's body by the diameter of a single atom.

$$(3.4 \times 10^{27})(2 \times 10^{-10}) = (3.4 \times 2)(10^{27} \times 10^{-10})$$

$$= 6.8 \times 10^{27+(-10)} = 6.8 \times 10^{17}$$

Thus, we find that total length of the atoms in the body is 6.8×10^{17} meters.

b) Next, we find the distance, in meters, to Proxima Centauri. To do this, we multiply $4.6 \times 9.46 \times 10^{15}$. This gives us

— *a* must be between 1 and 10.

$$4.6 \times 9.46 \times 10^{15} = 43.516 \times 10^{15} = 4.3516 \times 10^{16}.$$

c) Finally, we divide the length we found in part a) by the distance we found in part b), to get

$$\frac{6.8 \times 10^{17}}{4.3516 \times 10^{16}} = \frac{6.8}{4.3516} \times \frac{10^{17}}{10^{16}} \approx 1.56 \times 10 = 15.6$$

Thus, we have found that the total length of the atoms in the person's body is roughly 16 times the distance to the nearest star! ❋

Generally, we will not be giving you exercises as complicated as Example 9; however, we just wanted to show you the power of using the rules of exponents and scientific notation.

---

*This information is taken from Cesare Emiliani, *The Scientific Companion* (New York: Wiley, 1988), pp. 1–10.

Exercises ⟨6.5⟩

Looking Back*

These exercises follow the general outline of the topics presented in this section and will give you a good overview of the material that you have just studied.

1. Give an example to show how you can remember how to simplify $x^m x^n$.

2. Give an example to show how you can remember how to simplify $(x^m)^n$.

3. Give an example to show how you can remember how to simplify $\dfrac{x^m}{x^n}$.

4. What is the point of our diagram in Figure 6.23?

5. Mention some situations in which we might find exponential growth.

6. How does a calculator display scientific notation?

Sharpening Your Skills

Evaluate each expression.

7. $2 \cdot 2 \cdot 2 \cdot 2 \cdot 2$	**8.** 5^3	**9.** -2^4
10. 0^3	**11.** -3^2	**12.** 5^{-2}
13. 9^1	**14.** 3^{-4}	**15.** 3^0
16. 5^1	**17.** 0^6	**18.** $(-5)^2$

Use the rules for exponents to first rewrite each of the following expressions and then evaluate the new expression.

19. $3^2 \cdot 3^4$	**20.** $(-2)^3 \cdot (-2)^5$
21. $(7^2)^3$	**22.** $8^0 \cdot 8^2$
23. $5^4 \cdot 5^{-6}$	**24.** $4^2 \cdot 4^3$
25. $(3^2)^{-3}$	**26.** $2^{-4} \cdot 2^{-3}$
27. $(-3)^{-2} (-3)^3$	**28.** $(3^2)^4$
29. $\dfrac{5^9}{5^7}$	**30.** $(7^{-1})^{-3}$
31. $\dfrac{6^{-2}}{6^{-4}}$	**32.** $\dfrac{(-3)^6}{(-3)^9}$
33. $\dfrac{-3^6}{-3^9}$	**34.** $(2^{-3})^{-2}$
35. $(3^{-4})^0$	**36.** $(7)^{-2} \cdot (7)^6$

37. $\dfrac{2^9}{4^3}$ (*Hint:* Rewrite 4.)

38. $9^3 \cdot 27^{-2}$ (*Hint:* Rewrite 9 and 27.)

Rewrite each of the following numbers in scientific notation.

39. 4,356,000	**40.** 3,200,000,000
41. 783	**42.** 0.000258

43. 0.0024	**44.** 28
45. 0.008	**46.** 8,056

Rewrite each of the following numbers in standard notation.

47. 3.25×10^4	**48.** 4.7×10^8
49. 1.78×10^{-3}	**50.** 7.41×10^{-8}
51. 6.3×10^1	**52.** 9.7×10^1
53. 4.5×10^{-7}	**54.** 8×10^{-7}
55. 1×10^6	**56.** 1×10^{-5}

Each of the following numbers is not written in scientific notation. Rewrite them so that they are in scientific notation.

57. 23.81×10^6	**58.** 426.5×10^5
59. 0.84×10^3	**60.** 0.03×10^8

Use scientific notation to perform the following operations. Leave your answer in scientific notation form.

61. $(3 \times 10^6)(2 \times 10^5)$

62. $(4 \times 10^3)(2 \times 10^4)$

63. $(1.2 \times 10^{-3})(3 \times 10^5)$

64. $(4.2 \times 10^{-2})(1.83 \times 10^{-4})$

65. $(8 \times 10^{-2}) \div (2 \times 10^3)$

66. $(4.8 \times 10^4) \div (1.6 \times 10^{-3})$

67. $\dfrac{(5.44 \times 10^8)(2.1 \times 10^{-3})}{(3.4 \times 10^6)}$

68. $\dfrac{(1.4752 \times 10^{-2})(5.7 \times 10^4)}{(4.61 \times 10^{-3})}$

69. $\dfrac{(9.6368 \times 10^3)(4.15 \times 10^{-6})}{(1.52 \times 10^4)}$

70. $\dfrac{(3.5445 \times 10^{-3})(2.8 \times 10^{-5})}{(8.34 \times 10^6)}$

Rewrite each number using scientific notation before performing the operation. Leave your answer in scientific notation form.

71. $67,300,000 \times 1,200$

72. $83,600 \times 4,200,000$

73. $6,800,000 \times 2,300,000$

74. $1,750,000 \times 3,400,000$

75. 0.00016×0.0025

76. 0.000325×0.000008

Applying What You've Learned

In Exercises 77–94, use scientific notation to perform all calculations. Give your answer in scientific notation. Use the fact that one million is $1,000,000 = 10^6$, one billion is $1,000,000,000 = 10^9$, and one trillion is $1,000,000,000,000 = 10^{12}$.

---

*Before doing these exercises, you may find it useful to review the note *How to Succeed at Mathematics* on page xix.

Use the following graphs to solve Exercises 77–80. (Note: The 2015 and 2025 numbers are projections.)

U.S. Population Growth

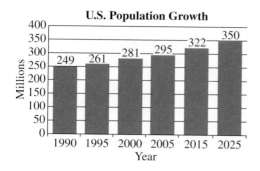

U.S. National Debt

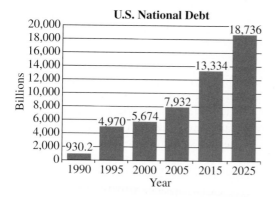

77. In 1990, the U.S. population was 249 million and the national debt was $930.2 billion. Calculate the amount of the national debt owed by each person in the United States.

78. Do the same for the years a) 2005 and b) 2025.

79. Find the increase expected in the national debt from 1995 to 2025.

80. In 2005, the world's population was 6.45 billion. How many times larger was the world's population than the U.S. population?

81. Comparing weights. How many times heavier is a 130-pound person than a mosquito? Assume that there are 454 grams to a pound and that a mosquito weighs 1×10^{-3} gram.

a. First find the number of grams that the person weighs.

b. Divide the number that you found in part a) by 1×10^{-3}.

c. Interpret your answer in part b).

82. Land per person in the United States. By 2005, the U.S. population had grown to approximately 296 million, and the land area of the United States was about 3.5×10^6 square miles. How many people per square mile were there in the United States in 2005?

83. In 2007, the U.S. population was 301 million and the country's defense budget was $439.3 billion. How much was spent on defense for each person in the United States?

84. Convert 1 million seconds to days. (*Hint:* There are $60 \times 60 \times 24$ seconds in a day.)

85. Convert 1 billion seconds to years.

86. On July 4, 2006, how many seconds old was the United States? (Ignore leap years.)

87. Land per person in the United States. In 2008, the population of the United States was approximately 304 million and the land area of the United States was roughly 3.5 million square miles. How many people per square mile were there in the United States in 2008?

88. Health care spending. By 2016, it is estimated that health care spending will be $4.1 trillion, or $12,800 per resident. Use this information to calculate the estimated U.S. population in 2016.

89. Space flight.* The Sun is about 93,000,000 miles away. If a rocket ship travels from Earth to the Sun at approximately 25,000 miles per hour, how many hours will the flight take?

90. Space flight. In 1977, scientists sent the spaceship Voyager II to Neptune, which is about 2.8 billion miles away. If the spacecraft averaged about 25,000 miles per hour, how long did the journey take?

91. Comparing astronomical distances. Earth is 93 million miles from the Sun and Pluto is 3.6 billion miles from the Sun. How many more times is Pluto's distance from the Sun than Earth's?

---

*In dealing with astronomical distances, we have simplified these problems in the sense that we are assuming the distances are straight-line distances. In reality, a spaceship would not fly in a straight line through space. However, to take these factors into account would complicate the problem.

92. Comparing disk drives. I wrote the first edition of this text on a Centris 650 computer that had a 230-megabyte hard disk. I wrote this edition on a Pentium IV that has an 80-gigabyte hard disk. How many times more disk space do I have now than I had on my Centris?

93. Frames of an animated movie. Computer-generated animations require tremendous amounts of computing power. A complex scene can require thousands of processors to spend hours of computer time just to produce $\frac{1}{24}$ of a second of an animated movie. Suppose that a complex frame requires 8.424×10^{11} calculations. If a bank of 10,000 processors are each doing 2.6 million calculations per second, how long will it take to generate that frame?

94. Distance light travels. Light travels at 186,000 miles per second. How far will light travel in one day?

Communicating Mathematics

95. Give an example that would help you remember what to do in simplifying $x^8 x^{-13}$.

96. Give an example that would help you remember what to do in simplifying $(x^{-11})^{-2}$.

97. Give an example that would help you remember what to do in simplifying $\dfrac{x^{-6}}{x^{-8}}$.

98. Explain why you must rewrite the following numbers if they are to be in scientific notation. Then, rewrite them properly in scientific notation.

   a. 125.436×10^3 **b.** 0.537×10^8

99. Explain the difference in evaluating -2^4 and $(-2)^4$. How does that affect your answer when calculating these expressions?

100. Provide a counterexample to show that $(x + y)^2 = x^2 + y^2$ is not true. How should you evaluate the left side of the equation versus the right side of the equation?

Using Technology to Investigate Mathematics

101. Search the Internet to locate some very large or very small quantities, such as the distance between astronomical bodies, the national debt, various populations, the size of an electron or a virus, and so on. Use scientific notation to express the quantities.

102. Search the Internet for data and make up problems similar to Exercises 77–94 that require the use of scientific notation. It is particularly interesting to see how changes in certain populations, costs, and so on change over a period of time.

For Extra Credit

103. Does $(x^m)^n = x^{(m^n)}$? Consider simple examples.

104. Look at simple examples to complete the following rule of exponents. $\left(\dfrac{a}{b}\right)^n =$

105. Use almanacs, encyclopedias, science textbooks, or other sources to locate examples, like the ones we gave you in this section, that involve very large or very small numbers. Make up a question similar to Exercises 77–94 and answer that question. For example, if a dollar bill is 0.003 inch thick, how high would a stack of one-dollar bills be that would represent the U.S. federal budget in 2010?

106. You are allowed to enter only three digits in your calculator but can use any other combination of keystrokes to form numbers. For example, you might enter $4*(5^6)$ or $(5^6)^8$. On my calculator, $4^{(4^4)}$ causes an overflow. Using the previous restrictions, find the largest number that your calculator will display without causing an overflow.

Looking Deeper

6.6 Sequences

Objectives

1. Understand arithmetic sequences.
2. Understand the difference between geometric and arithmetic sequences.
3. Recognize applications of the Fibonacci sequence.

Congratulations. When you signed your contract for your new job, you agreed to a monthly salary of $2,750 per month; however, today your supervisor gave you a pleasant surprise. During your first year, you will receive a salary increase of $50 each month. Because your salary will be different every month, you wonder what your total income will be for the next year. You could get out your calculator and add the amounts

$$2750, 2800, 2850, 2900, \ldots ,$$

but, as you will soon see, there is a faster way to do this.

In spite of your good news, you don't seem to be feeling so well. In fact, several days ago, you picked up a single virus that has been doubling ever since. The following list of numbers describes the growth of the virus in your body:

$$1, 2, 4, 8, 16, 32, 64, \dots.$$

You have just had a good experience and a bad experience with lists of numbers called sequences. We will study the patterns found in several different types of sequences in this section.

> **DEFINITION** A **sequence** is a list of numbers that follows some rule or pattern. The numbers in the list are called the **terms** of the sequence. We often name the terms $a_1, a_2, a_3, a_4, a_5, \dots$. We read this list "$a$ sub one, a sub two, a sub three, and so on."

In the sequence of monthly salaries, we get each successive term by adding 50 to the previous number. In this sequence

$$a_1 = 2{,}750,$$
$$a_2 = a_1 + 50 = 2{,}750 + 50 = 2{,}800,$$
$$a_3 = a_2 + 50 = 2{,}800 + 50 = 2{,}850, \text{ and so forth.}$$

In the sequence describing the virus growth, we get successive terms by doubling the previous terms. Here,

$$a_1 = 1,$$
$$a_2 = 2 \cdot a_1 = 2 \cdot 1 = 2,$$
$$a_3 = 2 \cdot a_2 = 2 \cdot 2 = 4,$$
$$a_4 = 2 \cdot a_3 = 2 \cdot 4 = 8, \text{ and so on.}$$

 **KEY POINT**

Terms in an arithmetic sequence differ by a fixed constant.

Arithmetic Sequences

The salary sequence is an example of an arithmetic sequence.

> **DEFINITION** An **arithmetic sequence** is a sequence in which each term after the first term differs from the preceding term by a fixed constant amount called the **common difference** of the sequence.

EXAMPLE 1 *Writing Terms of Arithmetic Sequences*

Find the common difference for the following arithmetic sequences. Then write the next three terms of the sequence.

a) 3, 7, 11, 15, . . .   b) 5, 3, 1, −1, . . .

SOLUTION:

a) We can find the common difference of an arithmetic sequence by subtracting the first term from the second term, so the common difference is $7 - 3 = 4$. The next three terms of the sequence are $15 + 4 = 19$, $19 + 4 = 23$, and $23 + 4 = 27$.

b) The common difference is $3 - 5 = -2$. Therefore, the next three terms of the sequence are −3, −5, and −7. ✳

In order to get more insight as to how arithmetic sequences behave, we will rewrite sequence a) in Example 1 as follows:

1st	2nd	3rd	4th		nth
3,	7,	11,	15, . . .		
$3 + 0 \cdot 4$,	$3 + 1 \cdot 4$,	$3 + 2 \cdot 4$,	$3 + 3 \cdot 4$, . . .		$3 + (n-1) \cdot 4$

From this, we see that in the first term we have added 4 *zero* times; in the second term, we have added 4 *one* time; in the third term, we have added 4 *two* times; and so on. In general, in each term we have added one less 4 than the number of the term. From this pattern, you can see that the sixteenth term would have fifteen 4s added, or would equal $3 + 15 \cdot 4 = 63$. This leads us to the following formula for the *n*th term of an arithmetic sequence:

> **FORMULA 1 The *n*th Term of an Arithmetic Sequence.** The *n*th term of an arithmetic sequence with first term a_1 and common difference *d* is
>
> $$a_n = a_1 + (n - 1)d.$$

EXAMPLE 2 *Finding the nth Term of Arithmetic Sequences*

a) Find the 12th term of the arithmetic sequence whose first term is 2,750 and whose common difference is 50.

b) Find the 20th term of the arithmetic sequence whose first term is 5 and whose common difference is -2.

SOLUTION:

a) Here, $n = 12$, $a_1 = 2{,}750$, and $d = 50$. Using Formula 1, we have

$$a_{12} = a_1 + (n - 1)d = 2{,}750 + (12 - 1)50 = 2{,}750 + 11 \cdot 50 = 3{,}300.$$

b) Now, $n = 20$, $a_1 = 5$, and $d = -2$. Using Formula 1, we have

$$a_{20} = a_1 + (n - 1)d = 5 + (20 - 1)(-2)$$
$$= 5 + 19(-2) = 5 - 38 = -33. \quad ✳ \quad \textbf{27}$$

Quiz Yourself **27**

Find the 16th term in the arithmetic sequence 5, 8, 11, 14, ...

As in the salary example, we sometimes want to add the first *n* terms in an arithmetic sequence. If we look carefully at a concrete example, we will see a pattern that leads us to a general formula for finding such sums.

Suppose that we wish to find the sum of the first six terms of the arithmetic sequence 2, 5, 8, 11, 14, 17, ... Rather than finding this sum directly, we will look at it in a slightly different way. Consider the sum $2 + 5 + 8 + 11 + 14 + 17$ and the same sum written in reverse order as $17 + 14 + 11 + 8 + 5 + 2$. If we place these sums under one another and add, we get

$$
\begin{array}{r}
2 + \ 5 + \ 8 + 11 + 14 + 17 \\
+\ 17 + 14 + 11 + \ 8 + \ 5 + \ 2. \\
\hline
19 + 19 + 19 + 19 + 19 + 19
\end{array}
$$

What you are seeing is that we got the sum, 19, which is $a_1 + a_6$, six times. Therefore, because we have added every term twice, $6(a_1 + a_6)$ is exactly twice the sum that we originally wanted. Therefore, $\dfrac{6(a_1 + a_6)}{2}$ is the sum we want. This example leads us to the following formula:

> **FORMULA 2 The Sum of the First *n* Terms of an Arithmetic Sequence.** The sum of the first *n* terms of an arithmetic sequence is given by
>
> $$\frac{n(a_1 + a_n)}{2}.$$

EXAMPLE 3 *Summing Terms in an Arithmetic Sequence*

Find the sum of the first 20 terms in the arithmetic sequence 3, 7, 11, 15, ...

SOLUTION: We will use Formula 2, $\dfrac{n(a_1 + a_n)}{2}$, but to do so, we first need to find a_{20}. We know that $a_1 = 3$ and $d = 4$, so using Formula 1 for finding the nth term of a sequence we get

$$a_{20} = a_1 + (20 - 1)d = 3 + 19 \cdot 4 = 3 + 76 = 79.$$

Now we can now apply Formula 2. We know that $a_1 = 3$, $a_{20} = 79$, and $n = 20$. Therefore, the sum is

$$\frac{20(3 + 79)}{2} = \frac{20(82)}{2} = 820.$$

Now try Exercises 17 to 24. ✳

We can now answer the question we posed at the beginning of this section.

EXAMPLE 4 *Finding the Yearly Total of an Arithmetic Sequence of Monthly Salaries*

Assume that you receive $2,750 as a salary this month, and for the next 11 months, you receive a monthly raise of $50. What is the total salary that you receive over the 12 months?

SOLUTION: The monthly salaries form an arithmetic sequence with the first term 2,750 and the monthly difference of $d = 50$. In Example 2, part a), we found that for this sequence, $a_{12} = 3,300$. So, using Formula 2, the sum of your salaries over 12 months is

$$\frac{n(a_1 + a_2)}{2} = \frac{12(2,750 + 3,300)}{2} = \frac{\overset{6}{\cancel{12}}(2,750 + 3,300)}{\cancel{2}} = 6 \cdot 6,050 = 36,300.$$

✳ **28**

Quiz Yourself **28**

Find the sum of the first 16 terms in the sequence 5, 8, 11, 14, ...

✎ **KEY POINT**

We multiply by a fixed constant to get new terms in a geometric sequence.

Geometric Sequences

You have seen that we generate each new term in an arithmetic sequence by adding the same fixed constant to the previous term. We generate each new term in a geometric sequence by multiplying the previous term by the same fixed constant. The virus sequence, 1, 2, 4, 8, 16, 32, . . . that we discussed earlier is an example of a geometric sequence because to get each new term, we multiply the previous term by 2.

> **DEFINITION** A **geometric sequence** is a sequence in which each term, after the first term, is a nonzero constant multiple of the preceding term. The constant by which we are multiplying is called the **common ratio**.

EXAMPLE 5 *Writing Terms in a Geometric Sequence*

Find the common ratio of each of the following geometric sequences. Then write the next three terms of the sequence.

a) 2, 6, 18, 54, . . .   b) 3, −6, 12, −24, . . .

SOLUTION:

a) We can find the common ratio of a geometric sequence by dividing the second term of the sequence by the first term. So, the common ratio of this sequence is $\frac{6}{2} = 3$. The next three terms of the sequence are

$3 \cdot 54 = 162$, $3 \cdot 162 = 486$, and $3 \cdot 486 = 1{,}458$.

b) The common ratio is $-\frac{6}{3} = -2$. The next three terms of the sequence are

$$(-2) \cdot (-24) = 48, \ (-2) \cdot 48 = -96, \text{ and } (-2) \cdot (-96) = 192. ✳$$

To understand how geometric sequences behave, we will rewrite sequence a) in Example 5 as follows:

$$\begin{array}{ccccc} 1^{\text{st}} & 2^{\text{nd}} & 3^{\text{rd}} & 4^{\text{th}} & n^{\text{th}} \\ 2, & 6, & 18, & 54, \ldots & \\ 2 \cdot 3^0, & 2 \cdot 3^1, & 2 \cdot 3^2, & 2 \cdot 3^3, \ldots & 2 \cdot 3^{(n-1)} \end{array}$$

From this we see that in the first term, we have multiplied 2 by 3^0; in the second term, we have multiplied 2 by 3^1; in the third term, we have multiplied 2 by 3^2; and so on. In general, in each term we have multiplied 2 by one less 3 than the number of the term. In general, in the nth term there are $n - 1$ factors that are 3. This gives us the following formula:

FORMULA 3 **The nth Term of a Geometric Sequence.** The nth term of a geometric sequence with common ratio r is

$$a_n = a_1 \cdot r^{n-1}.$$

We will use this formula in Example 6.

EXAMPLE 6 *Finding the nth Term of Geometric Sequences*

a) Find the eighth term of the geometric sequence whose first term is 5 and whose common ratio is 3.

b) Find the sixth term of the geometric sequence whose first term is 3 and whose common ratio is -2.

c) Find the number of viruses in your body after 18 hours if initially you have 1 and the virus doubles every hour.

SOLUTION:

a) Here $n = 8$, $a_1 = 5$, and $r = 3$. Using Formula 3, we have

$$a_8 = a_1 \cdot r^{8-1} = 5 \cdot 3^7 = 10{,}935.$$

b) Now, $n = 6$, $a_1 = 3$, and $r = -2$. Again using Formula 3, we have

$$a_6 = a_1 \cdot r^{6-1} = 3 \cdot (-2)^5 = 3 \cdot (-32) = -96.$$

c) In this case, $n = 18$, $a_1 = 1$, and $r = 2$, so

$$a_{18} = a_1 \cdot r^{18-1} = 1 \cdot (2)^{17} = 1 \cdot 131{,}072 = 131{,}072.$$

So, there are 131,072 copies of the virus in your body after 18 hours. ✳ **29**

Quiz Yourself **29**

Find the twelfth term in the sequence 2, 6, 18, 54, ...

We will use the following formula for finding the sum of the first n terms of a geometric sequence. We will use algebra to explain why this formula is valid in Exercise 50.

FORMULA 4 **The Sum of the First n Terms of a Geometric Sequence.** The sum of the first n terms of a geometric sequence is

$$\frac{a_1 \cdot (r^n - 1)}{r - 1}.$$

Example 7 illustrates Formula 4.

Math in Your Life

If it Sounds Too Good to Be True . . .

You have perhaps received a letter assuring you of riches that goes something like this:

> *This letter is not a joke. It really works! Follow these instructions carefully and do not break the chain. Send one dollar to the person who is first on the list, cross off that person's name and add your own name to the bottom of the list. Then send the new list to five people asking them to follow these same instructions. If you do not break the chain, after several weeks, you will receive over 1 million dollars in the mail.*

At first, this plan seems like it might work, provided that no one breaks the chain. In step 1, you send five letters, then each of these five people send five more letters for 25 or 5^2 letters. At the next stage, these 25 people send five letters each to get $125 = 5^3$ letters. If the list has 15 names, it would seem that by the time your name gets to the top of the list, you would be on easy street.

The problem, however, is that 5, 25, 125, 625, . . . is a geometric sequence with $a_1 = 5$ and $r = 5$. If the original list has 15 names, by the time your name reaches the top of the list and you start cashing in, $a_{15} = 5(5^{14}) = 5^{15}$. However, 5^{15} is roughly 30 billion which is about five times the size of the whole world's population! Even if no one breaks the chain, you will probably never get to see any money because we run out of people before your name ever gets to the top of the list.

Schemes, such as chain letters, that rely on geometric growth, or "pyramid," cannot work and are usually illegal.

EXAMPLE 7 *Finding the Sum of the First n Terms of a Geometric Sequence*

Find the sum of the first six terms of the geometric sequence whose first term is 4 and whose common ratio is 3.

SOLUTION: For this sequence, we have $n = 6$, $a_1 = 4$, and $r = 3$. Using Formula 4, we get

$$\frac{a_1 \cdot (r^n - 1)}{r - 1} = \frac{4 \cdot (3^6 - 1)}{3 - 1} = \frac{4 \cdot (729 - 1)}{3 - 1} = \frac{4 \cdot 728}{2} = \frac{2,912}{2} = 1,456.$$

Now try Exercises 25 to 30. ✳

Example 8 shows the dramatic difference between arithmetic and geometric growth.

EXAMPLE 8 *Comparing Arithmetic and Geometric Growth*

Suppose that you have won a lottery and can take your choice of either of the following two options:

a) You receive $50,000 this month, $60,000 next month, $70,000 the following month, $80,000 the month after that, and so on for 30 months.

b) You receive 1 cent this month, 2 cents next month, 4 cents the following month, 8 cents the month after that, and so on for 30 months.

Which option is the better deal? (Before you do the calculations, decide which option you would choose.)

SOLUTION: With option a), we are summing the first 30 terms of an arithmetic sequence in which $a_1 = 50,000$, $d = 10,000$, and $n = 30$. To use Formula 2, we need to know the 30th term, a_{30}. Using Formula 1, we get

$$a_{30} = a_1 + (n - 1)d = 50,000 + (30 - 1)10,000$$
$$= 50,000 + 290,000 = 340,000.$$

Now, we use Formula 2 to get the sum of the 30 months of payments as follows:

$$\frac{n(a_1 + a_n)}{2} = \frac{30(50,000 + 340,000)}{2} = \frac{30 \cdot 390,000}{2} = 5,850,000.$$

So, with option a) you will receive $5,850,000.

With option b), we are summing the first 30 terms of a geometric sequence in which $a_1 = 1$, $r = 2$, and $n = 30$. We will use Formula 4 to get

$$\frac{a_1 \cdot (r^n - 1)}{r - 1} = \frac{1 \cdot (2^{30} - 1)}{2 - 1} = \frac{1,073,741,823}{1} = 1,073,741,823.$$

The number 1,073,741,823 represents cents, so we convert this to dollars to get 10,737,418.23 dollars. Therefore, we see that the second option will give you almost $11 million versus the $5,850,000 of the first option. ✳ **30**

Example 8 shows how remarkably fast quantities can grow if we are using a geometric sequence versus an arithmetic sequence.

The Fibonacci Sequence

In 1202, Leonardo of Pisa, also known as Fibonacci, wrote a book, *Liber Abaci*, in which he introduced the Hindu-Arabic numeration system to Europe. His intention was to explain this new system in order to replace the more tedious Roman system of numeration for doing calculations. He included in his text the following problem that has given rise to perhaps the most famous sequence in the history of mathematics.

EXAMPLE 9   *The Fibonacci Sequence*

Assume that every pair of baby rabbits will mature during their second month and produce a pair of baby rabbits during their third month. If we place a single pair of adult rabbits in a pen, how many rabbits (both babies and adults) will there be in the pen in the eighth month? We are assuming that once adults start producing babies they do so every month from then on.

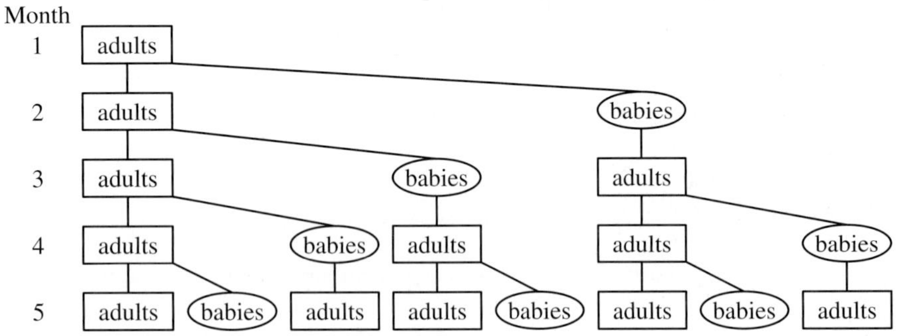

SOLUTION: Figure 6.25 will help you understand the problem.

If we count the pairs of rabbits each month, we get the following sequence: 1, 2, 3, 5, 8, ... The pattern that is beginning to emerge is that from the third term on, each term in the sequence is the sum of the previous two terms. Study Figure 6.25 carefully to see why this must be so. If we continue this pattern, there will be $5 + 8 = 13$ pairs in the sixth month, $8 + 13 = 21$ pairs in the seventh month, and $13 + 21 = 34$ pairs in the eighth month.

The pattern that we saw in Example 9 is essentially the Fibonacci sequence. ✳ **31**

FIGURE 6.25 The growth of rabbits leads to the Fibonacci sequence.

> **DEFINITION** The **Fibonacci sequence** is the sequence that begins
>
> 1, 1, 2, 3, 5, 8, 13, 21, 34, 55, 89, . . . ,
>
> where each term after the second is the sum of the previous two terms. We often label the terms in the Fibonacci sequence by $F_1, F_2, F_3, F_4, \ldots$ rather than $a_1, a_2, a_3, a_4, \ldots$

The Fibonacci sequence occurs in nature in many different ways. The seeds of some flowers, for example, daisies and sunflowers, are arranged in spirals going in two different directions. If you count the number of spirals going in one direction and then count

Quiz Yourself **30**

Find the sum of the first eight terms in the sequence 2, 6, 18, 54, ...

✏ **KEY POINT**

We add consecutive pairs of terms to generate terms in a Fibonacci sequence.

Quiz Yourself **31**

Write the next two terms in the Fibonacci sequence.

(a)   (b)

FIGURE 6.26 The Fibonacci sequence appears in (a) the spirals in a sunflower and (b) the patterns on a pineapple.

$\frac{1}{1} = 1$
$\frac{2}{1} = 2$
$\frac{3}{2} = 1.5$
$\frac{5}{3} = 1.66$
$\frac{8}{5} = 1.6$
$\frac{13}{8} = 1.625$
$\frac{21}{13} = 1.615$
$\frac{34}{21} = 1.619$
$\frac{55}{34} = 1.617$
$\frac{89}{55} = 1.618$
$\frac{144}{89} = 1.618$
$\frac{233}{144} = 1.618$

TABLE 6.4 Quotients obtained by dividing consecutive Fibonacci numbers.

the number of spirals going in the other direction, you will often find that the numbers you get are a pair of consecutive numbers in the Fibonacci sequence, such as 21 and 34, or 34 and 55 (see Figure 6.26(a)). The hexagonal "bumps" on the skin of a pineapple and the woody leaf-like structures on a pine cone also occur in two types of spirals (see Figure 6.26(b)). Again, if you count the spirals, you will get a pair of Fibonacci numbers. Later in this section, we will show you how the Fibonacci sequence describes the pattern in the shell of the chambered nautilus (see Figure 6.29(b)).

So much research has been done on the Fibonacci sequence that there is a Fibonacci Association and even a research journal called *The Fibonacci Quarterly* devoted to publishing research on its properties.

One of the remarkable patterns that was discovered about the Fibonacci sequence is that if you go through the sequence, dividing each term by the term preceding it, you get the numbers that we see in Table 6.4.* It seems as though these numbers are settling down and going toward some fixed number. And, in fact, they are. It was proved about 200 years ago that the number these quotients are approaching is $\frac{\sqrt{5}+1}{2}$, which we will call ϕ (phi), and approximate it by 1.618.

The number ϕ, which we discussed briefly in Section 6.4, is often called the **golden ratio** and was known to the ancient Greeks, although they obtained it in a different way. They believed that any rectangle whose one side was $\frac{\sqrt{5}+1}{2}$ times as long as its other side was perfectly proportioned, and they used this principle in art and architecture. Such a rectangle is called a **golden rectangle**. For example, they used golden rectangles in designing the Parthenon (Figure 6.27(a)). In modern times, architects used golden rectangles to

(a)   (b)

FIGURE 6.27 The architectures of (a) the Parthenon and (b) the United Nations building are based on the golden ratio.

*We used a spreadsheet to generate these numbers, and in many cases, we are approximating the true quotient, which has an infinite, repeating expansion.

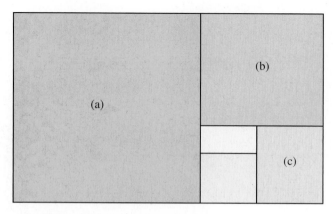

FIGURE 6.28 Cutting off squares from a golden rectangle gives other golden rectangles.

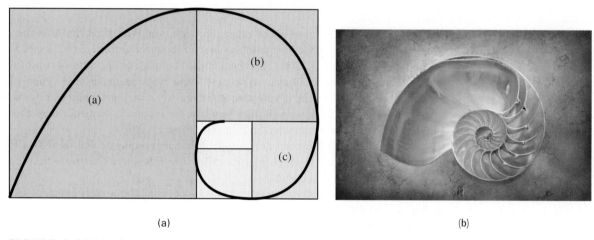

(a)

(b)

FIGURE 6.29 (a) Spiral generated from golden rectangles. (b) Chambered nautilus shell.

determine the rectangular shapes in the sides of the United Nations building (Figure 6.27(b)). The pleasing shape of the golden rectangle is also found in art and in sculpture from ancient to modern times.

An interesting property of a golden rectangle is that if you cut off square (a) on one end of the rectangle, as we show in Figure 6.28, the smaller rectangle that remains is again a golden rectangle. If you then cut off square (b) again, the part that remains is a golden rectangle. If you continue doing this, and connect the opposite diagonals of the squares that you are cutting off with a smooth curve, you generate a spiral that is much like the shape of a chambered nautilus shell (see Figure 6.29).

Exercises 6.6

Looking Back*

These exercises follow the general outline of the topics presented in this section and will give you a good overview of the material that you have just studied.

1. Explain how we used the example prior to Formula 1 to derive the formula for finding the nth term of an arithmetic sequence.

2. Explain how we used the example prior to Formula 3 to derive the formula for finding the nth term of a geometric sequence.

3. Make up an example similar to the one we gave prior to Formula 2 to illustrate how we sum the first n terms of an arithmetic sequence.

4. What type of sequence describes the rapid growth of letters in a chain letter?

Sharpening Your Skills

In Exercises 5–16, identify the sequence as either arithmetic or geometric. List the next two terms of each sequence.

5. $5, 8, 11, 14, \ldots$

6. $11, 7, 3, -1, \ldots$

7. $8, 24, 72, 216, \ldots$

8. $4, 12, 20, 28, \ldots$

9. $1, \dfrac{1}{2}, \dfrac{1}{4}, \dfrac{1}{8}, \ldots$

10. $0.1, 0.01, 0.001, 0.0001, \ldots$

11. $10, 5, 0, -5, \ldots$

12. $4, 17, 30, 43, \ldots$

13. $2, 4, 8, 16, \ldots$

14. $1, \dfrac{1}{3}, \dfrac{1}{9}, \dfrac{1}{27}, \ldots$

15. $1.5, 2.0, 2.5, 3.0, \ldots$

16. $1, -1, 1, -1, \ldots$

For each arithmetic sequence: a) find the specified term a_n and b) find the sum of the terms from a_1 to a_n, inclusive.

17. $5, 8, 11, 14, \ldots$; find a_{11}.

18. $11, 17, 23, 29, \ldots$; find a_9.

19. $2, 8, 14, 20, \ldots$; find a_{15}.

20. $-6, -2, 2, 6, \ldots$; find a_{22}.

21. $1, 1.5, 2.0, 2.5, \ldots$; find a_{20}.

22. $3, 3.25, 3.5, 3.75, \ldots$; find a_{11}.

23. In Exercise 17, find the sum of the terms from a_{14} to a_{21}, inclusive.

24. In Exercise 18, find the sum of the terms from a_{21} to a_{37}, inclusive.

For each geometric sequence: a) find a_n and b) find the sum of the terms from a_1 to a_n.

25. $1, 3, 9, 27, \ldots$; find a_{11}.

26. $3, 6, 12, 24, \ldots$; find a_9.

27. $1, \dfrac{1}{2}, \dfrac{1}{4}, \dfrac{1}{8}, \ldots$; find a_7.

28. $2, -4, 8, -16, \ldots$; find a_{10}.

29. $2, 0.2, 0.02, 0.002, \ldots$; find a_6.

30. $5, 50, 500, 5{,}000, \ldots$; find a_9.

In Exercises 31–34, we give you two terms in the Fibonacci sequence, and you must find the specified term.

31. $F_{11} = 89$ and $F_{13} = 233$. Find F_{12}.

32. $F_{22} = 17{,}711$ and $F_{23} = 28{,}657$. Find F_{21}.

33. $F_{13} = 233$ and $F_{15} = 610$. Find F_{14}.

34. $F_{23} = 28{,}657$ and $F_{24} = 46{,}368$. Find F_{25}.

Applying What You've Learned

35. **Finding total salary.** Redo Example 4, but now assume that your starting salary is \$2,875 and you are getting an increase of \$35 per month. Find the amount that you will earn in a year.

36. **Displaying products.** There is a pyramid of cans against the wall of a supermarket. There are 11 cans on the first row, 10 on the second row, 9 on the third row, and so on. What is the total number of cans on the stack?

37. **Growth in a bank account.** You have \$1,200 in your ING direct account that is paying an interest rate of 3.5% (0.035) per year. How much money will you have in that account in 6 years?

38. **Chain letters.** Reconsider the discussion of chain letters in the Math in Your Life on page 285. Assume that you receive a chain letter and at each stage, the person who receives the letter must send the letter to eight people. If you send your letters and the chain is unbroken, at what stage will the number of letters sent at that stage exceed the population of the world, which in 2008 was about 6.7 billion people?

39. **Height of a bouncing ball.** A ball is dropped from a height of 8 feet. The ball always bounces $\frac{7}{8}$ of the distance from which it was dropped. What will be the height of the ball after the fifth bounce?

40. **Gambling strategy.** A surefire way to never lose when betting at a casino game is to use the following strategy: If you lose, then always double your bet and play again until you win. Assume that you have \$2,000. If you are gambling at a roulette wheel and you begin with a \$3 bet and continue to lose, at what stage will you be unable to continue doubling your bet?

You are given the population and the growth rate as of 2008 for the following countries. Assume that the growth rate remains the same from year to year. Determine the size of the population in 2020.

41. Brazil: population = 185 million; growth rate = 0.91%

42. China: population = 1,322 million; growth rate = 0.61%

---

*Before doing these exercises, you may find it useful to review the note How to Succeed at Mathematics on page xix.

Communicating Mathematics

43. Make up an example to verify Formula 1.

44. Make up an example to verify Formula 2.

45. Make up an example to verify Formula 3.

46. Explain, by referring to sequences, why chain letters are doomed to fail.

Using Technology to Investigate Mathematics

47. Search the Internet to find information about the golden ratio in art and architecture. Report on your findings.

48. There are many interesting sites on the Internet that deal with the Fibonacci sequence. Find several such sites, and write a brief report about the information you discover about this sequence.

For Extra Credit

49. Find a formula to add the terms from a_k to a_n in an arithmetic sequence. This pattern is exactly like the pattern for adding the first n terms. Look at examples and add the terms from a_k to a_n in two different ways, as we did in the discussion prior to Formula 2.

50. Derive Formula 4 by explaining the following steps.

   a. The first n terms of a geometric sequence are
   $$a + ar + ar^2 + ar^3 + ar^4 + \cdots + ar^{n-1}.$$

   b. Rewrite this as $a(1 + r + r^2 + r^3 + r^4 + \cdots + r^{n-1})$.

   c. But $1 + r + r^2 + r^3 + r^4 + \cdots + r^{n-1} = \frac{1 - r^n}{1 - r}$.

   d. Therefore,

   $$a_1 + a_1 r + a_1 r^2 + a_1 r^3 + a_1 r^4 + \cdots + a_1 r^{n-1}$$
   $$= a_1(1 + r + r^2 + r^3 + r^4 + \cdots + r^{n-1})$$
   $$= a_1 \frac{1 - r^n}{1 - r} = \frac{a_1(r^n - 1)}{r - 1}.$$

51. By looking at concrete examples, derive a formula for the sum of the first n Fibonacci numbers, $F_1 + F_2 + F_3 + \cdots + F_n$.

52. By looking at concrete examples, derive a formula for the sum of the squares for two consecutive Fibonacci numbers, F_n and F_{n+1}.

The Binet form of Fibonacci numbers uses the formula

$$F_n = \frac{\left(\frac{1 + \sqrt{5}}{2}\right)^n + \left(\frac{1 - \sqrt{5}}{2}\right)^n}{\sqrt{5}}.$$

For example, to calculate F_8, *substitute 8 for* n *in the previous formula. Using a calculator, due to roundoff errors, the author got 21.00736 for* F_8, *whose value is actually 21. Use this formula and a calculator to approximate the Fibonacci numbers in Exercises 53–56.*

53. F_5 **54.** F_7 **55.** F_9 **56.** F_{11}

CHAPTER SUMMARY*

SECTION	SUMMARY	EXAMPLE
SECTION 6.1	The numbers that we use for counting, 1, 2, 3, . . . , are called the **natural numbers**, or **counting numbers**. The study of the natural numbers is called **number theory**. If a and b are natural numbers, then we say a **divides** b and we write $a|b$, if there is a natural number q such that $b = q \cdot a$. We also say that a is a **divisor** of b, a is a **factor** of b, or b is a **multiple** of a.	Discussion, p. 228
	A natural number, greater than 1, which has only itself and 1 as factors, is called a **prime number**. A natural number greater than 1 that is not prime is called **composite**.	Definition, p. 228
	We can use the **Sieve of Eratosthenes** to find prime numbers.	Example 1, p. 229
	There are simple ways to test a number for **divisibility** by 2, 3, 4, 5, 8, 9, and 10. We can use divisibility tests and **factor trees** to find the **prime factorization** of numbers. The **fundamental theorem of arithmetic** tells us that, except for order, this factorization will be unique.	Table 6.1, p. 231 Example 3, p. 232 Example 4, p. 232
	The **greatest common divisor (GCD)** of two natural numbers is the largest natural number that divides both numbers. The **least common multiple (LCM)** of two natural numbers is the smallest natural number that is a multiple of both numbers. We use prime factorization to find the GCD and LCM of two numbers.	Example 5, p. 234 Example 6, p. 235
	We use the GCD and LCM to solve applied problems.	Example 7, p. 236
SECTION 6.2	The set of **whole numbers** is the set $\{0, 1, 2, 3, . . .\}$. The set of **integers** is the set $\{. . . , -3, -2, -1, 0, 1, 2, 3, . . .\}$.	Discussion, p. 239
	We explain **addition rules** for integers by considering movement on a number line.	Example 1, p. 240
	Subtraction is a subsidiary operation to addition. If a and b are integers, then we define $a - b$ as $a + (-b)$.	Discussion, p. 241
	We use a **money model** to explain integer multiplication. If a and b have the *same* sign, then $a \cdot b$ is *positive*. If a and b have *opposite* signs, then $a \cdot b$ is *negative*.	Discussion p. 242 Example 2, p. 242
	Division is a subsidiary operation to multiplication. If a, b, and c are integers, then $\dfrac{a}{b} = c$ means that $a = b \cdot c$.	Definition, p. 243
	If a and b have the *same* sign, then $\dfrac{a}{b}$ is *positive*. If a and b have *opposite* signs, then $\dfrac{a}{b}$ is *negative*.	Example 3, p. 243
SECTION 6.3	The set of **rational numbers**, **Q**, is the set of all numbers that can be written in the form $\dfrac{a}{b}$, where a and b are both integers and $b \neq 0$. The top number a is called the **numerator** and the bottom number b is called the **denominator**.	Discussion. p. 247
	Two rational numbers $\dfrac{a}{b}$ and $\dfrac{c}{d}$ are **equal** if and only if $a \cdot d = b \cdot c$. We can also show two rational numbers to be equal by **canceling common factors** from the numerator and the denominator, that is, $\dfrac{a \cdot c}{b \cdot c} = \dfrac{a}{b}$. When all common factors have been canceled, the rational number is in **lowest terms**.	Example 1, p. 248 Example 2, p. 249
	We **add** and **subtract** rational numbers as follows: $$\frac{a}{b} + \frac{c}{d} = \frac{a \cdot d + b \cdot c}{b \cdot d} \quad \text{and} \quad \frac{a}{b} - \frac{c}{d} = \frac{a \cdot d - b \cdot c}{b \cdot d}$$	Example 3, p. 250

*Before studying this chapter's material, it would be useful to reread the note *How to Succeed at Mathematics* on page xix.

We **multiply** and **divide** rational numbers as follows:		Example 4, p. 251

$$\frac{a}{b} \cdot \frac{c}{d} = \frac{a \cdot c}{b \cdot d} \quad \text{and} \quad \frac{\dfrac{a}{b}}{\dfrac{c}{d}} = \frac{a}{b} \cdot \frac{d}{c}$$

Example 5, p. 252

To convert an **improper fraction** $\dfrac{a}{b}$ to a **mixed number**, divide a by b to get the quotient q and remainder r. Then,

Example 7, p. 254

$$\frac{a}{b} = q\frac{r}{b}.$$

The mixed number $q\dfrac{r}{b}$ equals the improper fraction $\dfrac{q \cdot b + r}{b}$.

Example 8, p. 254

Rational numbers have **repeating decimal expansions**.

Example 9, p. 255

SECTION 6.4   A number that is not a rational number is called an **irrational number** and has a *nonrepeating* decimal expansion.

Discussion, p. 260

If $a \geq 0$ and $b \geq 0$, then

Discussion, p. 262

$$\sqrt{ab} = \sqrt{a}\sqrt{b} \quad \text{and} \quad \sqrt{\frac{a}{b}} = \frac{\sqrt{a}}{\sqrt{b}}.$$

Example 2, p. 262

The real numbers have the following properties. For any real numbers a, b, and c:

Discussion. p. 265

Closure: $a + b$ and $a \cdot b$ are real numbers.

Example 6, p. 265

Commutativity: $a + b = b + a$ and $a \cdot b = b \cdot a$.

Example 7, p. 267

Associativity: $(a + b) + c = a + (b + c)$ and $(a \cdot b) \cdot c = a \cdot (b \cdot c)$.

Identity element: $a + 0 = 0 + a$ and $a \cdot 1 = 1 \cdot a = a$.

Inverse property: $a + (-a) = (-a) + a = 0$ and if $a \neq 0$, then $a \cdot \dfrac{1}{a} = \dfrac{1}{a} \cdot a = 1$.

Example 8, p. 267

Distributivity: $a \cdot (b + c) = a \cdot b + a \cdot c$

Example 9, p. 268

SECTION 6.5   If a is any real number and n is a counting number, then

Definition, p. 271

$$a^n = \underbrace{a \cdot a \cdot a \cdot \cdots \cdot a}_{\text{product of } n \text{ } a\text{'s}}$$

Example 1, p. 271

If x is a real number and m and n are natural numbers, then

$$x^m x^n = x^{m+n}$$

Example 2, p. 272

$$(x^m)^n = x^{m \cdot n}$$

Example 3, p. 272

$$\frac{x^m}{x^n} = x^{m-n}, x \neq 0$$

Example 4, p. 273

$$x^0 = 1, \text{ and if } x \neq 0, \text{ then } x^{-n} = \frac{1}{x^n}$$

We write numbers in **scientific notation** in the form $a \times 10^n$, where $1 \leq a < 10$. We use the rules of exponents to calculate with numbers written in scientific notation.

Definition, p. 274
Examples 6 and 7, p. 276

We can use scientific notation to solve applied problems involving **very large** or **very small** numbers.

Examples 8 and 9, p. 277

SECTION 6.6   A **sequence** is a list of numbers, called **terms**, that follow some pattern. We often name the terms $a_1, a_2, a_3, a_4, \ldots$.

Definition, p. 281

An **arithmetic sequence** is a sequence in which any two consecutive terms differ by a fixed constant called the **common difference**.

Definition, p. 281
Example 1, p. 281

The **nth term** of an arithmetic sequence with first term a_1 and common difference d is $a_n = a_1 + (n - 1)d$.

Example 2, p. 282

The **sum of the first n terms** of an arithmetic sequence is $\dfrac{n(a_1 + a_n)}{2}$.

Examples 3 and 4, p. 283

A **geometric sequence** is a sequence in which each term (except the first) is a nonzero constant multiple of the preceding term. The constant that we multiply by is called the **common ratio**.

Definition, p. 283
Example 5, p. 283

The **nth term** of a geometric sequence with first term a_1 and common ratio r is $a_n = a_1 \cdot r^{(n-1)}$.

Example 6, p. 284

The **sum of the first n terms** of a geometric sequence is $\dfrac{a_1 \cdot (r^n - 1)}{(r - 1)}$.

Example 7, p. 285

The **Fibonacci sequence** is the sequence that begins

Definition, p. 286

$$1, 1, 2, 3, 5, 8, 13, 21, 34, 55, 89, \ldots,$$

where each term after the second is the *sum of the two previous* terms. We label the terms of a Fibonacci sequence by $F_1, F_2, F_3, F_4, \ldots$.

As n gets larger, the quotient $\dfrac{F_{n+1}}{F_n}$ approaches φ, the **golden ratio**, which is approximately 1.618.

Discussion, p. 287

CHAPTER REVIEW EXERCISES

Section 6.1

1. Use the Sieve of Eratosthenes to find all prime numbers between 70 and 90.

2. Estimate the square root of 180 by finding a natural number a such that $a < \sqrt{180} < a + 1$.

3. Are 191 and 441 prime or composite? If the number is composite, factor it into a product of primes.

4. Check 1,080,036 for divisibility by 3, 4, 5, 6, 8, 9, and 10.

5. Find the LCM and GCD of 1,584 and 1,320.

6. Explain how you use prime factorization to get the GCD and the LCM of two natural numbers.

Section 6.2

7. Is 2,772 divisible by 9? 4? 8?

8. How can you use a calculator to determine if 11 divides 2,585?

9. Find the following:

   a. $-5 + 14$ **b.** $-13 - (-24)$

   c. $(-12)(-6)$ **d.** $\dfrac{48}{-3}$

10. Use a money example to illustrate how to compute the product $(-4)(+3)$.

11. Use the definition of division in terms of multiplication to compute $\dfrac{-24}{-8}$.

12. If the high temperature in northern Alaska is 17 degrees Fahrenheit at the beginning of October, and at the beginning of November it is -3 degrees Fahrenheit, what is the difference in these two temperatures?

13. Use the definition of division in terms of multiplication to explain why we cannot divide 5 by 0.

14. Simplify $\left(\dfrac{-3+9}{-4-2}\right) \cdot \left(\dfrac{12-(-4)}{-7-1}\right)$.

15. If stock for Pineapple computers currently sells for $43 a share but has lost $6 for each of the past 4 days, what was the value of the stock 4 days ago?

Section 6.3

16. Are the numbers $\dfrac{28}{65}$ and $\dfrac{14}{35}$ equal? Explain your answer.

17. Perform the following computations:

   a. $\dfrac{4}{9} \cdot \left(\dfrac{3}{4} - \dfrac{1}{3}\right)$ **b.** $\dfrac{3}{7} \div \left(\dfrac{2}{3} + \dfrac{3}{14}\right)$

18. Convert $\dfrac{53}{8}$ to a mixed number.

19. Find $\left(3\dfrac{1}{2}\right)\left(4\dfrac{1}{7}\right)$.

20. Write the following numbers as quotients of integers:

   a. 0.375   **b.** $0.\overline{63}$

21. You have purchased a $1\dfrac{1}{2}$-pound, Vanilla Bean cheesecake from the Cheesecake Factory. If the cake is to be divided among 12 people, how much will each person's piece weigh?

22. If a recipe for a hot chili sauce that serves 16 requires $2\dfrac{1}{2}$ cups of hot peppers and $3\dfrac{1}{4}$ cups of tomatoes, how much peppers and tomatoes would be required to make an amount of sauce to serve 24 people?

23. Explain by giving a specific example to show why we invert and multiply when dividing rational numbers.

Section 6.4

24. How is the decimal expansion of an irrational number different from that of a rational number?

25. Simplify the following:

   a. $\sqrt{108}$

   b. $\dfrac{\sqrt{15}}{\sqrt{5}}$ (Rationalize the denominator, if necessary.)

26. Does every real number have a multiplicative inverse? Explain.

27. Give an example to show that division of rational numbers is not associative.

Section 6.5

28. Evaluate each expression:

   a. $3^6 \cdot 3^{-2}$ **b.** $(2^4)^{-2}$ **c.** $\dfrac{6^8}{6^5}$ **d.** $\dfrac{8^{-6}}{8^{-4}}$

29. Explain the difference between evaluating -2^4 and $(-2)^4$.

30. Write 0.000456 and 1,230,000 in scientific notation.

31. Rewrite 1.325×10^6 and 8.63×10^{-5} in standard notation.

32. Simplify $\dfrac{(3.6 \times 10^3)(2.8 \times 10^{-5})}{(4.2 \times 10^4)}$ and write the answer in scientific notation.

33. In 2008, the world's population was 6.6 billion and Mexico's population was 109 million. How many times larger was the world's population than Mexico's population? Express your answer in scientific notation.

34. Identify the following sequences as arithmetic or geometric and give the next two terms of the sequence:

   a. $6, -12, 24, -48, 96, \ldots$ **b.** $11, 16, 21, 26, 31, \ldots$

35. Consider the arithmetic sequence $4, 7, 10, 13, 16, \ldots$.

   a. What is the 30th term of this sequence?

   b. What is the sum of the first 30 terms of this sequence?

36. Consider the geometric sequence $4, 12, 36, 108, \ldots$.

   a. What is the 10th term of this sequence?

   b. What is the sum of the first 10 terms of this sequence?

37. If $F_{12} = 144$ and $F_{13} = 233$ are two terms in the Fibonacci sequence, what are F_{10} and F_{15}?

CHAPTER TEST

1. Use the Sieve of Eratosthenes to find all prime numbers between 100 and 120.

2. Estimate the square root of 200 by finding a natural number a such that $a < \sqrt{200} < a + 1$.

3. Are the numbers 241 and 539 prime or composite? If the number is composite, factor it into a product of primes.

4. Check 2,542,128 for divisibility by 3, 4, 5, 6, 8, and 9.

5. Find the GCD and LCM of 1,716 and 936.

6. If $a = 2^3 3^9 5^3 7^2$ and $b = 2^8 3^6 5^2 7^9$, what is the GCD of a and b? The LCM?

7. Find the following:

   a. $-15 - (-9)$ **b.** $-8 + 11$

   c. $(-18)(+3)$ **d.** $\dfrac{-56}{-7}$.

8. Identify the following sequences as arithmetic or geometric and give the next two terms of the sequence:

   a. $12, 15, 18, 21, 24, \ldots$ **b.** $6, 18, 54, 162, \ldots$

9. Use a money example to illustrate how to compute the product $(+7)(-5)$.

10. Use the definition of division in terms of multiplication to compute $\dfrac{-21}{3}$.

11. If the average daily temperature in Alaska is 50 degrees Fahrenheit for July and -19 degrees Fahrenheit for December, what is the difference in these two temperatures?

12. Use the definition of division in terms of multiplication to explain why we cannot divide 8 by 0.

13. If $F_{16} = 987$ and $F_{17} = 1,597$ are two terms in the Fibonacci sequence, what are F_{14} and F_{20}?

14. Are the numbers $\dfrac{198}{213}$ and $\dfrac{68}{72}$ equal? Explain your answer.

15. Perform the following computations:

   a. $\dfrac{3}{5} \cdot \left(\dfrac{7}{9} - \dfrac{3}{4} \right)$ **b.** $\dfrac{4}{5} \div \left(\dfrac{2}{5} - \dfrac{1}{4} \right)$

16. Convert $\dfrac{47}{3}$ to a mixed number.

17. Find $4\dfrac{1}{2} \div 3\dfrac{3}{8}$

18. Write the following numbers as a quotient of integers:

   a. 0.573 **b.** $0.\overline{57}$

19. Consider the arithmetic sequence $11, 20, 29, 38, 47, \ldots$.

   a. What is the 20th term of this sequence?

   b. What is the sum of the first 20 terms of this sequence?

20. If a digital photo that measures $4\dfrac{1}{2}$ inches by $6\dfrac{1}{3}$ inches is being increased to make a poster by enlarging it $5\dfrac{1}{4}$ times, what will be the dimensions of the poster?

21. Explain by giving a specific example to show why we invert and multiply when dividing rational numbers.

22. What is the difference between the decimal expansions of a rational number and an irrational number?

23. Simplify the following:

   a. $\sqrt{180}$ **b.** $\dfrac{3}{\sqrt{15}}$

24. Give an example to show that the division of rational numbers is not associative.

25. Evaluate each expression:

   a. $2^7 \cdot 2^{-4}$ **b.** $(3^2)^{-2}$ **c.** $\dfrac{5^7}{5^4}$ **d.** $\dfrac{3^{-2}}{3^{-5}}$

26. Consider the geometric sequence $-2, 6, -18, 54, \ldots$.

   a. What is the 10th term of this sequence?

   b. What is the sum of the first 10 terms of this sequence?

27. Explain the difference between evaluating $(-3)^2$ and -3^2.

28. Write 15,460,000 and 0.00000000623 in scientific notation.

29. In 2008, the population of China was approximately 1,322,000,000, whereas the population of the United States was roughly 304,000,000. How many times larger was China's population than the population of the United States?

GROUP EXERCISES

1. Consider Exercise 104 in Section 6.4 that discusses how changing the placement of commas can change the meaning of sentences. Make up several of your own that are similar to the examples given there.

2. The following table defines an abstract operation called *. This operation has no real meaning, but behaves as follows: If you choose any row, say the A row and any column, say the B column, then the result of the operation is found in the intersection of that row and column. So we see in the table that A*B = D. Similarly, C*P = A. Use this table to answer the following questions:

*	D	I	A	C	B	P
D	D	I	A	C	B	P
I	I	D	P	B	C	A
A	A	P	B	I	D	I
C	C	B	I	D	P	A
B	B	C	D	P	A	C
P	P	A	I	B	C	D

a. What is C*I? D*B?

b. Is A*B = B*A? Is C*D = D*C?

c. Is * commutative? (*Hint:* Recall the Always Principle.)

d. What is the identity element for *?

e. What is B's inverse?

3. The Fibonacci sequence is related to a number of different concepts in mathematics. Do research on the Internet to find connections between the Fibonacci sequence and each of the following and give an example illustrating the connection:

a. Pascal's triangle   **b.** Lucas numbers

Algebraic Models

How Do We Approximate Reality?

W hat is a model? To an automotive engineer in Detroit, it may be a series of computer graphic images describing the effect of wind resistance on the gas mileage of a sleek, futuristic car. To a parent in Florida, it could be a spreadsheet reflecting the effects of increasing tuition costs on the family budget. A meteorologist in the Midwest might consider a model to be a set of equations describing the conditions on a steamy summer evening that can lead to a devastating tornado. In each case, the model builder represents reality by stripping away unimportant details in order to understand the variables in a situation and predict future events.

In this chapter, we will use equations to represent reality just as the engineer, the parent, and meteorologist do. You will see that these equations can enable you to solve real-world problems, although less complex than the ones mentioned here. In Sections 7.1 and 7.2, we build simple models by using linear equations. Although data rarely conforms exactly to a linear equation, you will see that linear equations are often "good enough" to describe actual data. In Section 7.3, we will use quadratic equations to model situations where the graph of the data is curved in a way that makes linear models inappropriate.

(continued)

The discussion of exponential and logistic models in Section 7.4 will show you how to represent the explosive growth typical in credit card debt, populations, and the spread of disease.

By the end of this chapter, you will have studied a variety of algebraic equations and will have learned when they are appropriate models and when they are not. ●

7.1 Linear Equations*

Objectives

1. Solve linear equations.
2. Use intercepts to graph linear equations.
3. Apply the slope–intercept form of a line in solving problems.

Do you ever wish that you could predict the future? Mathematicians and scientists try to do just that all the time. In fact, you probably read about their predictions regarding global warming, worldwide famine, spread of AIDS in Africa, and so on. In this chapter, you will get a glimpse of the way we build mathematical models, and we hope that you also become aware of some of the limitations of these models. You will see in this section that we can use relatively simple equations to give good approximations to real situations.

 **KEY POINT**

Linear equations have a constant rate of change between the variables.

Solving Linear Equations

In spite of its simplicity, a linear equation often provides us with a reasonably good model for a complex set of data.

> **DEFINITION** A **linear equation**† in two variables is an equation that can be written in the form
>
> $$Ax + By = C,$$
>
> where A, B, and C are real numbers and A and B are not both zero. When a linear equation is written in this form, it is in **standard form**.

The equations $-3x + y = 6$ and $5p = 6q - 4$ are examples of linear equations in two variables. In a linear equation in two variables, there is a constant rate of change between the two variables. To understand what we mean by this, let's rewrite the first equation in the form $y = 6 + 3x$. Notice that if we increase x by a certain amount, then y increases by three times as much. For example, if we increase x from 10 to 15, which is an increase of 5, then y increases from 36 to 51, which is an increase of 15. If we increase x from 1,000 to 1,007, then y increases from 3,006 to 3,027, which is an increase of 21.

A **solution** for an equation is a number or numbers such that if we substitute them for the variables in the equation, the resulting statement is true. Solutions of linear equations are ordered pairs of numbers. For example, the **ordered pair** (3, 15)** is a solution for the equation $-3x + y = 6$ because if we substitute 3 for x and 15 for y, we get the true

---

*For a review of basic algebra, see Appendix A.
†Linear equations contain variables to the first power only.
**In the pair (3, 15), the number 3 is called the *first coordinate* and the number 15 is called the *second coordinate*. These coordinates are also called the x-coordinate and y-coordinate, respectively.

statement $-3(3) + 15 = 6$. Two equations are **equivalent** if they have the same solutions. In solving an equation, we often use the following rules to rewrite it in a simpler, equivalent form. These rules apply to all types of equations.

REWRITING EQUATIONS IN AN EQUIVALENT FORM

1. You can add or subtract the same expression from both sides of an equation to get an equivalent equation.
2. You can multiply or divide both sides of an equation by the same *nonzero* expression to get an equivalent equation.

We will apply some of these rules in Example 1 to solve for one variable in terms of another—a technique that we often use in solving applied problems.

EXAMPLE 1   *Rewriting Equations in Equivalent Forms*

a) Solve the equation $6x + 4y = 12$ for y.

b) The equation $F = \dfrac{9}{5}C + 32$ converts the temperature in degrees Celsius, C, to the temperature in degrees Fahrenheit, F. Solve this equation for C.

SOLUTION:

a) "Solve the equation for y" means that we want the equation in the form $y = \ldots$, so we perform the following steps:

$$6x + 4y = 12 \qquad \text{Original equation}$$
$$4y = 12 - 6x \qquad \text{Subtract } 6x \text{ from both sides.}$$
$$y = \frac{12 - 6x}{4} \qquad \text{Divide both sides by 4.}$$
$$y = \frac{12}{4} - \frac{6}{4}x \qquad \text{Rewrite using distributivity.}$$
$$y = 3 - \frac{3}{2}x. \qquad \text{Simplify.}$$

b) We can solve for C by rewriting this equation in the following series of steps:

$$F = \frac{9}{5}C + 32 \qquad \text{Original equation}$$
$$F - 32 = \frac{9}{5}C \qquad \text{Subtract 32 from both sides.}$$
$$\frac{5}{9}(F - 32) = \frac{5}{9} \cdot \frac{9}{5}C \qquad \text{Multiply both sides by } \frac{5}{9}.$$
$$\frac{5}{9}(F - 32) = C \qquad \text{Simplify.}$$

Quiz Yourself ❶ *

a) Solve the equation
$6x + 4y = 12$ for x.

b) Solve the equation
$P = 2L + 2W$ for W.

We can now use this equation to convert degrees Fahrenheit to degrees Celsius. For example, a nice, warm 95-degree Fahrenheit day at the beach converts to a chilly sounding $C = \dfrac{5}{9}(95 - 32) = \dfrac{5}{9}(63) = 35$ degrees Celsius.

Now try Exercises 19 to 30. ✸ ❶

---

*Quiz yourself answers begin on page 778.

HISTORICAL HIGHLIGHT ❀ ❀ ❀

René Descartes and Analytical Geometry

René Descartes was born in France in 1596. Because of his poor health, his early teachers allowed him the luxury of remaining in bed as late as he pleased. He attributed his later success to this habit because it gave him time to think and reflect on philosophy and mathematics. In his famous work *Discourse on Method*, Descartes developed analytical geometry, in which we represent points, lines, circles, and planes as numbers and equations.* The advantage that analytical geometry has over traditional geometry is that we can use algebraic methods such as solving equations to solve geometric problems.

Descartes also contributed to many other areas of science. His work in biology was so important that some have called him the "father of modern biology." He made significant discoveries in several areas of physics, and he is credited with being the first person to satisfactorily explain the rainbow.

Unfortunately, this genius met with an untimely end while tutoring Queen Christine of Sweden, who insisted on rising at 5 o'clock to study philosophy with him. The early hour and the severe Swedish winter were too much for Descartes—he caught pneumonia and died on February 11, 1650.

 Some Good Advice

When solving equations, you may have been told that "when a quantity is taken from one side of an equation to the other, plus signs change to minus signs, and vice versa." It is unwise to rely on memory devices such as this. Our advice for working with equations can be traced back to the early Greeks, who had rules such as "equals added to equals give equals." Remember when rewriting equations, you are *adding*, *subtracting*, *multiplying*, or *dividing* both sides of the equation by the same quantity.

If you were to plot several solutions for the linear equation $-3x + y = 6$, such as $(3, 15)$, $(-2, 0)$, $(1, 9)$, and $(2, 12)$, you would find that they all seem to lie on a line, as we see in the graph below.

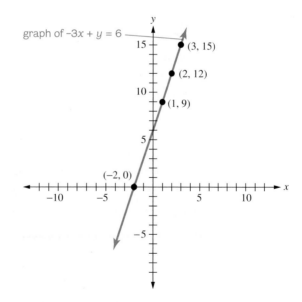

What you see here is no coincidence; in general, we have the following result.

> **THE GRAPH OF A LINEAR EQUATION**   The graph of a linear equation is always a straight line.

---

*As he lay in bed watching a fly walk on the ceiling, Descartes realized that he could describe the fly's position by the pair of numbers telling its distance from two perpendicular walls.

KEY POINT

We use intercepts to graph a linear equation.

Intercepts

The graph of a linear equation in two variables is a straight line, so to graph a linear equation, all you have to do is plot two points and then draw the line through those points. The graph of a linear equation gives us a visual way of understanding the relationship between the two variables in a model. Often, the two simplest points that you can use to graph a linear equation, are called the *x*- and *y*-intercepts.

> **DEFINITIONS** The **x-intercept** of the graph of a linear equation is the point where the graph crosses the *x*-axis. The **y-intercept** is the point where the graph crosses the *y*-axis. (See Figure 7.1.)

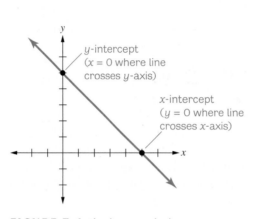

FIGURE 7.1 Plotting *x*- and *y*-intercepts.

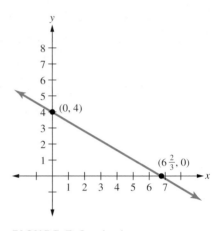

FIGURE 7.2 Using intercepts to graph $3x + 5y = 20$.

 Some Good Advice

You may mistakenly try to calculate the *x*-intercept by setting $x = 0$. Keep in mind that the term "*x*-intercept" means the point where the line "intercepts" or crosses the *x*-axis, so the *y*-coordinate of the intersection point is 0.

In Example 2, we use simple algebra to find intercepts.

EXAMPLE 2   *Using Intercepts to Graph a Linear Equation*

Find the intercepts and graph the equation $3x + 5y = 20$.

SOLUTION: To find the *x*-intercept, we set $y = 0$. This gives us $3x + 5(0) = 20$, or $3x = 20$. Thus, $x = 6\frac{2}{3}$, so the *x*-intercept of the line is the point $\left(6\frac{2}{3}, 0\right)$.

Setting $x = 0$, we get $3(0) + 5y = 20$. This simplifies to $5y = 20$, which means that the *y*-intercept is the point $(0, 4)$. We graph this line in Figure 7.2.

Now try Exercises 31 to 42. ❉ **②**

 *Quiz Yourself* **②**

Find the intercepts and graph the equation $4x + 3y = 18$.

EXAMPLE 3   *Interpreting Intercepts in a Financial Situation*

Suppose you have saved \$2,205 to use for living expenses next semester that you expect to be \$315 per month. Express this information in an equation using the variables *r* (for remaining cash) and *m* (for the number of months you are spending the money). Interpret the meaning of the intercepts for the graph of this equation.

SOLUTION: You are starting with $2,205, and each month you will reduce this amount by $315. Therefore, the equation

original amount ⌐   ⌐ monthly expense

$$r = 2{,}205 - 315m$$

describes your financial state at the end of m months. The r-intercept corresponds to the time when $m = 0$; that is,

no months have elapsed

$$r = 2{,}205 - 315(0) = 2{,}205.$$

At this point, you still have all your money.

To find the m intercept, we set $r = 0$. This is the time at which you run out of savings. To find this time, we will solve the equation

⌐ no remaining cash

$$0 = 2{,}205 - 315m.$$

Adding $315m$ to each side of the equation, we get $315m = 2{,}205$, and then, dividing both sides by 315, we obtain $m = 7$. Therefore, you can expect to deplete your savings at the end of 7 months. ✳

When using linear equations as models, we often want to know the steepness of an equation's graph. An economist may want to know how rapidly inflation is affecting the cost of a college education or, perhaps during a drought, government officials would want to know how rapidly a town's water supply is being depleted. We measure the steepness of a line by its slope.

DEFINITION If (x_1, y_1) and (x_2, y_2) are two points on a line and $x_1 \neq x_2$, then the **slope** of the line is defined as

$$m = \frac{\text{rise}}{\text{run}} = \frac{\text{change in } y}{\text{change in } x} = \frac{y_2 - y_1}{x_2 - x_1}.$$

HIGHLIGHT 🕸 🕸 🕸

Using Technology to Draw Graphs

Many equations are much too complicated for you to get an accurate graph by merely plotting a few points. Fortunately, technology such as graphing calculators can make your job easier. We used a graphing calculator to get the graph of $y = \dfrac{8x}{\sqrt{x^4 + 10}}$ shown in Figure 7.3.

If you had been graphing this equation by hand, you might have accidentally chosen to plot the three points that we have selected in Figure 7.3 and been completely fooled as to what the graph looked like. You might ask then, "How many points should I plot?" That is not an easy question to answer and the truth is that mathematicians who work with such equations do not plot points, but use

areas of mathematics beyond algebra to understand their complexity.

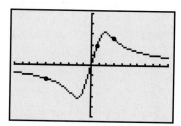

FIGURE 7.3 Graph of a complicated equation drawn with a graphing calculator.

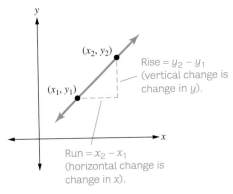

FIGURE 7.4 The rise and run determine a line's slope.

You can remember the meaning of the words *rise* and *run* by thinking about their use in everyday language. We watch the Sun rise and see a cake or a hot air balloon rising. In each case something is going up, so rise corresponds to a vertical change. For the word *run*, think of a person running across a field or a train running down a track. In these cases, *run* means a horizontal change. Figure 7.4 illustrates the relationship between slope, rise, and run.

EXAMPLE 4 *Finding the Slope of a Line*

Calculate the slope of the line containing the points (1, 5) and (6, 45).

SOLUTION: The slope of the line is

$$\text{slope} = \frac{\text{rise}}{\text{run}} = \frac{\text{change in } y}{\text{change in } x} = \frac{45 - 5}{6 - 1} = \frac{40}{5} = 8.$$

Now try Exercises 43 to 50. ✻

Quiz Yourself **3**

Find the slope of the line through the points (7, 5) and (11, 10).

꩜ Some Good Advice

In the definition of slope, if the run is equal to 1, then the slope is simply $y_2 - y_1$. This means that we can interpret the slope of a line as the amount of change in y corresponding to a change of 1 in x.

By estimating the rise and the run (including whether it is positive or negative), we can get a good idea of the slope of the line without doing any calculations. The line in Figure 7.5(a) has a positive slope because both rise and run are positive. The line in Figure 7.5(b) has negative slope because the run is positive but the rise is negative.

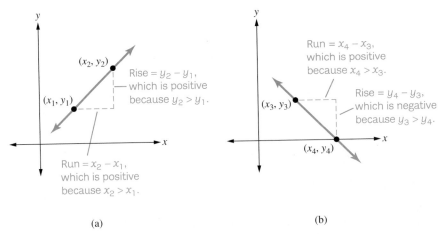

(a)　　　　　　　　　(b)

FIGURE 7.5 (a) A line with a positive slope rises from left to right. (b) A line with a negative slope falls from left to right.

Notice that if a line is horizontal, then its rise is zero, so $\text{slope} = \frac{\text{rise}}{\text{run}} = \frac{0}{\text{run}}$, which is equal to zero. Similarly, if a line is vertical, then $\text{slope} = \frac{\text{rise}}{\text{run}} = \frac{\text{rise}}{0}$, which is undefined because we cannot divide by zero.

 *KEY POINT*

The slope–intercept form of a linear equation tells us geometric information about its graph.

The Slope–Intercept Form of a Line

> **DEFINITION** A linear equation is in **slope–intercept form** if it is written in the form $y = mx + b$. The number m is the slope of the line that is the graph of the equation and $(0, b)$ is the y-intercept.

The equation $y = 2x + 3$ is in slope–intercept form. The slope is 2, and the y-intercept is $(0, 3)$. Example 5 shows that we can recognize useful geometric information about a line if we write its equation in slope–intercept form.

EXAMPLE 5 *Graphing a Linear Equation in Slope–Intercept Form*

Graph the linear equation $y = 3x + 4$.

SOLUTION: We can read off the y-intercept of the graph immediately; because

$$y = 3x + 4,$$
$$\underset{\text{\_\_ } y\text{-intercept}}{}$$

we see that the y-intercept is the point $(0, 4)$. We can easily find another point to graph the line if we let $x = 2$.* This give us

$$y = 3(\overset{x}{2}) + 4 = 6 + 4 = 10,$$

so $(2, 10)$ is a second point on the graph. We then plot the points $(0, 4)$ and $(2, 10)$ to draw the graph as in Figure 7.6.

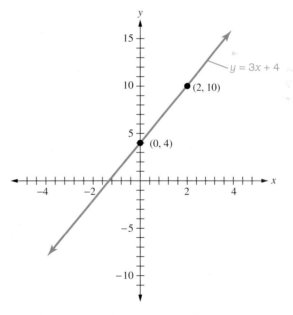

FIGURE 7.6 Graph of $y = 3x + 4$. ✳

Writing a linear equation in slope–intercept form often helps you to recognize some useful geometric information about the line.

---

*We can choose any value for x that we want; however, you will get a better-looking graph if you separate the points a little bit.

EXAMPLE 6 *Comparing Jet-Ski Rental Plans*

A store that rents jet skis has two rental plans. In plan A, the customer pays a base fee of $35.00 plus $12.50 per hour. In plan B, the customer pays a base fee of $22.50 plus $15 per hour. Model each of these plans by a linear equation in slope–intercept form. Graph the equations and estimate at what point rental plan A is preferable to plan B.

SOLUTION: The cost of renting a jet ski for x hours using plan A is the base fee of $35 plus x times $12.50. If we let y represent the total rental cost of a jet ski, we can model plan A by the equation

hourly rate = slope ⟍   ⟋ base fee = y-intercept

$$y = (12.50)x + 35. \qquad \text{(Plan A)}$$

Similarly, we model plan B by

hourly rate = slope ⟍   ⟋ base fee = y-intercept

$$y = (15)x + 22.50. \qquad \text{(Plan B)}$$

From these equations, we see that the slope for the graph of plan A is 12.50, which is less than 15, which is the slope of B's graph. So, even though B's cost is initially cheaper, if you rent the jet ski for enough hours, plan B will eventually be more expensive.

In A's equation, we see that the y-intercept is $(0, 35)$. To find another point on this graph, we can let $x = 1$, so

$$y = 12.50 \cdot \overset{\displaystyle x}{(1)} + 35 = 47.50.$$

Thus, $(1, 47.50)$ is a second point on A's graph.

For plan B's graph, we see that $(0, 22.50)$ is the y-intercept and, again letting $x = 1$, we get

$$y = 15 \cdot \overset{\displaystyle x}{(1)} + 22.50 = 37.50.$$

So, $(1, 37.50)$ is a second point on B's graph.

In Figure 7.7, we use the points $(0, 35)$ and $(1, 47.50)$ to graph A's equation and $(0, 22.50)$ and $(1, 37.50)$ to graph B's.

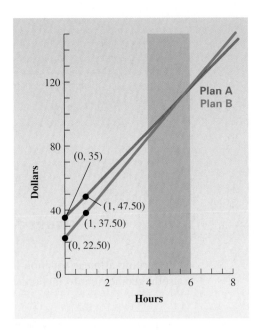

FIGURE 7.7 Graph of plan A versus plan B

We can see from Figure 7.7, that until 4 hours, B's graph lies below A's, so plan B is cheaper, but by the time we reach 6 hours, B has become more expensive. So, it appears from the graphs that around 5 hours, B becomes the more expensive plan.

Now try Exercises 85 and 86. ❋

We also can use algebra to solve Example 6. The point of intersection of the two lines in Figure 7.7 tells us where the two costs will be equal. This gives us the equation

$$12.50x + 35 = 15x + 22.50.$$

If you were to solve this equation, you would find that $x = 5$ is a solution.

The fact that we got the same solution both algebraically and geometrically demonstrates Descartes's reasoning in developing analytical geometry—what we can represent geometrically we can represent algebraically, and vice versa.

It is important when selecting a model, that the model is appropriate for the situation you are modeling. When you write a linear equation in slope–intercept form as $y = mx + b$, realize that an increase of 1 for x causes an increase of m for y, an increase of 2 for x causes an increase of $2m$ for y, an increase of 5 for x causes an increase of $5m$ for y, and so on. The point is that a change in x causes an increase of m times that change for y.

If the situation you are modeling does not satisfy this property, then a linear equation is not a good choice for a model. In subsequent sections of this chapter, we discuss several real-life situations that do not satisfy this property; in those cases, we use nonlinear equations to model them.

Exercises 7.1

Looking Back*

These exercises follow the general outline of the topics presented in this section and will give you a good overview of the material that you have just studied.

1. In solving equations, we recommended that you do not focus on memory devices such as "moving quantities to the other side of the equation and changing signs." What did we recommend instead?

2. How does the drawing of the balloon and train before Example 4 help you to remember the definition of slope?

3. We said that in a linear equation, the variables are related by a constant rate of change. What were the constant rates of change in Example 3? In Example 6?

4. In Example 6, why did the graph of plan B's pricing eventually rise above the graph of plan A's pricing?

5. What is analytical geometry?

6. What is the danger in simply plotting some points to get the graph of a complex equation?

Sharpening Your Skills

Solve each equation by using the rules for rewriting equations. It is easier to solve an equation which has fractions or decimals if you multiply each side of the equation by a suitable constant to make the coefficients easier to work with.

7. $3x + 4 = 5x - 6$

8. $4 - 2x = 9x + 13$

9. $4 - 2y = 8 + 3y$

10. $5y - 6 = 14y + 12$

11. $\frac{1}{2}x + 4 = \frac{3}{4}x - 6$

12. $\frac{1}{2}x - 6 = \frac{1}{5}x + 3$

13. $\frac{1}{3}y + 4 = \frac{1}{4}y + 3$

14. $\frac{5}{6}y + 1 = \frac{1}{3}y - 2$

15. $0.2x + 6 = 3x - 0.4$

16. $0.25x + 0.35 = 0.2x - 4$

17. $0.3y + 2 = 0.5y - 3$

18. $0.4y - 0.2 = 0.6y + 3$

Solve each equation for the stated variable.

19. $P = 2l + 2w$; solve for w.

20. $m = \dfrac{a + b}{2}$; solve for a.

21. $z = \dfrac{x - \mu}{\sigma}$;† solve for μ.

22. $z = \dfrac{x - \mu}{\sigma}$; solve for σ.

23. $A = P(1 + rt)$; solve for r.

24. $A = \dfrac{1}{2}h(b + B)$; solve for b.

---

*Before doing these exercises, you may find it useful to review the note *How to Succeed at Mathematics* on page xix.

†μ and σ are the lowercase Greek letters mu and sigma.

25. $2x + 3y = 6$, solve for x. **26.** $4x - 5y = 3$, solve for y.

27. $V = lwh$, solve for l. **28.** $A = \frac{1}{2}hb$, solve for b.

29. $S = 2\pi rh + 2\pi r^2$, solve for h.

30. $A = 2lw + 2lh + 2hw$, solve for w.

Graph each equation by first finding the x- and y-intercepts.

31. $3x + 2y = 12$ **32.** $x - 5y = 10$

33. $4x - 3y = 16$ **34.** $5x + 4y = -20$

35. $\frac{1}{3}x + \frac{1}{2}y = 3$ **36.** $x - \frac{1}{5}y = 2$

37. $\frac{1}{6}x - 2y = \frac{3}{4}$ **38.** $x - \frac{1}{4}y = 2$

39. $0.2x = 4y + 1.6$ **40.** $0.3y = 1.2x + 0.6$

41. $0.4x - 0.3y = 1.2$ **42.** $0.2x + 0.5y = 2$

Find the slope of the line passing through the two given points.

43. $(2, 5)$ and $(6, 8)$ **44.** $(4, 1)$ and $(7, 3)$

45. $(3, 6)$ and $(8, 2)$ **46.** $(9, 1)$ and $(6, 4)$

47. $(3, -4)$ and $(5, 1)$ **48.** $(8, -5)$ and $(9, 2)$

49. $(6, 5)$ and $(6, 8)$ **50.** $(2, 3)$ and $(7, 3)$

In Exercises 51–54, list all the lines in the given figure that satisfy the stated condition.

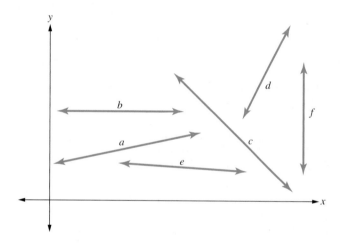

51. Slope is positive. **52.** Slope is negative.

53. Slope does not exist. **54.** Slope is zero.

55. Explain why the slope of a horizontal line is 0.

56. Explain why a vertical line has no slope.

In Exercises 57–60, state the y-intercept and slope of the graph of each equation.

57. $y = 4x - 3$ **58.** $y = 3x + 5$

59. $y = -5x - 3$ **60.** $y = -2x + 5$

Applying What You've Learned

To solve Exercises 61–64, first write an equation that describes each situation. Use meaningful names for the variables.

61. Computing health club charges. A health club charges a yearly membership fee of $95, and members must pay $2.50 per hour to use its facilities. How many hours did Jillian use the club last year if her bill was $515?

62. Computing movie club charges. Netflix is now selling its used DVDs at $12 apiece. There is a $5 membership fee to join. If Roeper had a bill of $233 dollars, how many DVDs did he buy?

63. Computing charges for word processing. Thiep does word processing and charges $16 for documents up to 10 pages and $1.30 per page extra for pages beyond 10. If he billed a customer $44.60, how long was the customer's document?

64. Computing overtime. Darryl works at a warehouse where he is paid $9 per hour up to 40 hours. If he works over 40 hours, he is paid 1.5 times his usual wage. If he earned $468 last week, how many hours did he work?

According to the U.S. Census, there were 64 people per square mile (this is called the population density) in the United States in 1980. By 2000, the number of people per square mile had grown to 80. Use this information to develop a linear equation in slope–intercept form to solve Exercises 65–68. In developing your equation, think of 1980 as year zero.

65. Write the equation that you are using to model this information.

66. Estimate the population density in 1960.

67. What do you expect the population density to be in 2015?

68. When will the population density be 100?

69. College expenses. According to *U.S. News and World Report*, the average cost of attending an in-state public college in 2007 was $11,900 per year and college expenses were increasing at a rate of 4.1% of the 2007 cost per year ($4.1\% = 0.041$). Write a linear equation* in slope–intercept form to predict what costs will be at these institutions in 2015.

70. College expenses. Use the information in Exercise 69 to estimate the average cost of attending an in-state public college in 1990.

In Exercises 71 and 72, write a linear equation to describe each situation before answering the question. Use good names for the variables.

71. Computing net pay. Dani earns a gross pay from which the following deductions are subtracted:

$10 per capita tax

6.75% Social Security tax

14% federal income tax

2.1% state tax

$17 unemployment tax

What net pay corresponds to her gross pay of $400?

72. Computing the price of a car. Christian intends to buy a used car. In addition to the base cost of the car, he must pay the following:

$35 title and tag fee

6% sales tax

1.2% uninsured motorist fund

$45 automobile club fee

What is his total cost for a car that has a base price of $6,800?

---

*You will see a more sophisticated and more accurate way to model this information in Section 7.4.

In Exercises 73–78, write an equation that models each situation.

73. **Calculating earnings.** Juno has two part-time jobs. Her job in a diner pays $5.60 per hour and her work in a pet store pays $7.35 per hour. She earns $133 in a given week.

74. **Determining rental payments.** Jared furnished his new apartment by renting a living room set and a TV from a local "rent to own" store. He has l more payments on the living room set of $22 each, and t payments remaining on the TV of $13 each. The total he still owes is $341.

75. **Investing in stock.** Tamyra has invested $14,500 in the stock market. Some of the stock is Time-Warner that costs $35 a share and the rest is in CDW that costs $55 a share.

76. **Grading.** RJ's psychology professor is grading the course based on a contract system. A project is worth 40 points, and a report on a journal article is worth 10 points. RJ earned 250 points toward his final grade.

77. **Manufacturing furniture.** Aidan owns a small woodworking company. He knows that it takes 9 board feet to construct an end table and 15 to construct a coffee table. He has 342 board feet available.

78. **Advertising a business.** Bobby Flay is deciding how to advertise a new gourmet restaurant that he is opening. An online ad costs $300 and a TV ad costs $700. He has $7,500 to spend on the ads.

79. **Comparing commuting costs.** You travel a toll road that requires an 85-cent token. If you buy a special car-pool sticker for $9.00, then tokens cost only 70 cents and you can use the express lane. At what point is the car-pool plan cheaper?

80. **Comparing food plans.** College students can purchase points that can be used in the food service areas instead of cash. If you initially pay the basic food service fee of $35, then points can be purchased for 23 cents each; otherwise, points cost 30 cents each. How many points must you use for it to be cheaper to pay the basic food service fee?

Communicating Mathematics

81. When a line crosses the x-axis, which coordinate of the intercept is zero?

82. We said that slope is "rise over run." What do we mean by "rise" and "run?"

83. What are usually the two easiest points to plot when graphing a line?

84. In a linear equation of the form $y = mx + b$, what do the m and the b represent?

We can describe each situation in Exercises 85 and 86 by a pair of linear equations as we did in Example 6. Use the algebraic method that we described after Example 6 to determine when both deals are the same.

85. Chuck can work at the Buy More for a base pay of $225 per week plus a $45 commission for each computer system that he sells. At Circuit Town, his base pay would be $400 with a $20 commission for each computer system that he sells. Interpret what the solution to this system tells you. How should Chuck decide which position to take?

86. Cassandra is comparing two cell phone calling plans. BT&T charges $12.75 per month and 7 cents per minute. Cingleton charges $14.15 per month and 5 cents per minute. Interpret what the solution to this system tells you. How should Cassandra decide which plan to take?

87. Explain when it is appropriate to use a linear equation as a model. Give an example of a situation we have not covered in this section that you believe fits a linear equation as a model.

88. Give an example for which you feel it would not be appropriate to use a linear equation as a model. Explain why you feel this way.

Using Technology to Investigate Mathematics

89. See your instructor for tutorials on how to use a graphing calculator to graph and solve equations. Use this information to duplicate some of the computations in this section.

90. Search the Internet for interactive programs that you can use to graph and solve equations. Use this information to duplicate some of the computations in this section.

For Extra Credit

91. **Comparing investments.** You want to invest $5,000 that a wealthy relative has given to you. There is a certificate of deposit that pays 3.8% interest. You are in the 14% federal income tax bracket, however, and will have to pay federal income tax on interest earned by the CD. There are several bonds that you could also purchase that are exempt from federal income tax; however, they pay less interest. What interest do you need to earn on these bonds to equal the return that you would get from the CD?

92. **Comparing investments.** Redo Exercise 91, except now you are considering investing in bonds that are exempt from both the federal tax of 14% and the 2.1% state tax.

The table shows the average number of currency units per dollar in January 2008 according to the Federal Reserve Bank.

Country	One Dollar Equals
Japan	105 yen
France	0.68 euro
India	39 rupees
Colombia	2,380 pesos
Mexico	11 pesos

Use this information to make the conversions in Exercises 93–96.

93. Converting currency. a. dollars to yen   **b.** yen to dollars

94. Converting currency. a. dollars to rupees   **b.** rupees to dollars

95. Converting currency. euros to Colombian pesos

96. Converting currency. yen to Mexican pesos

97. Wheelchair accessibility. To comply with the Americans with Disabilities Act, the town library must be made wheelchair accessible. Consider the following diagram of the entrance, which has enough room to construct a ramp as shown. Presently, there are three steps, each 6 inches high and 6 inches wide, leading up to this entrance. How far should point *P* be from point *B* (at the base of the first step) for the ramp to have a slope of 8% (that is, 0.08)?

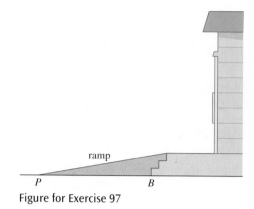

Figure for Exercise 97

98. Wheelchair accessibility. Repeat Exercise 97, but now assume that there are six steps instead of three.

7.2 Modeling with Linear Equations

Objectives

1. Build a linear model using a point and the slope.
2. Use two points to build a linear model.
3. Understand how to use the line of best fit to model real data.

If you were to search the Internet to find examples of linear equations being used to model real data (see Exercise 41), you might be astounded by the number of applications of this idea that you would find. Realize, however, that a model is simply that—a model. You have no doubt canceled a trip to the beach or the ballpark because the meteorologist's prediction (using a model of the weather) was off-target. Parents who invest in college savings accounts and others who retire early, based on a slick-talking salesman's financial model, often have to make an adjustment years later when the model fails to accurately predict their future.

Similar to the simplifications we do when building physical models, we also simplify mathematical models. The engineer we described at the start of the chapter was interested only in improving the aerodynamics of the car; therefore, she could ignore features such as the sound system and seats that did not affect air resistance. Similarly we often decide to omit certain features of a mathematical model in order to make it clearer.

We have seen in Section 7.1 that we can describe the same linear equation in several different ways. We will now discuss some other ways we can specify information that determines a linear equation as a model.

Method of Specifying Equation	Information Provided
Write the equation in standard form.	$3x + 2y = 6$
State the *x*- and *y*-intercepts of the graph of the equation.	*x*-intercept is $(2, 0)$. *y*-intercept is $(0, 3)$.
Specify the slope and *y*-intercept of the graph of the equation.	Slope is $-\frac{3}{2}$ and *y*-intercept is 3. Slope–intercept form of the equation is $y = -\frac{3}{2}x + 3$.

KEY POINT

A point and the slope determine a line.

Building a Model with a Point and the Slope

We can find a linear equation if we know the slope of its graph and a point on its graph.

EXAMPLE 1   *Using the Slope and a Point to Determine a Linear Equation*

Find a linear equation of the line with slope 3 passing through the point (4, 5).

SOLUTION: We will assume that we can write the equation of the line in slope–intercept form, $y = mx + b$. The slope of the graph of the equation is $m = 3$, so we can rewrite this equation as

$$y = \overset{\overbrace{}^{m}}{3}x + b.$$

We need to find b. Because (4, 5) lies on the line, we substitute 4 for x and 5 for y, to get

$$\overset{y}{5} = 3(\overset{x}{4}) + b.$$

Subtracting 12 from both sides gives us $-7 = b$. The equation we want is therefore

$$y = 3x - 7.$$

Now try Exercises 5 to 12. ✳ **4**

Quiz Yourself **4**

Find an equation of the line passing through (2, −3) having slope −4. Write your answer in slope–intercept form.

EXAMPLE 2   *Building a Model Using a Point and the Slope*

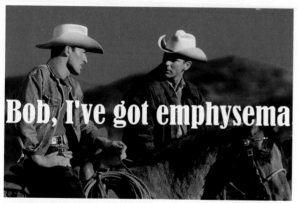

As part of a settlement of lawsuits against tobacco companies, the U.S. government required cigarette manufacturers to pay for antismoking education programs. Assume that, as a result of such programs, the smoking rate among 18- to 25-year-olds is dropping at a rate of 0.6% per year and that after 3 years the smoking rate is 38.8%. If we model this reduction by a linear equation, what would we predict the smoking rate among 18- to 25-year-olds to be in 10 more years?

SOLUTION: We will model this decrease in the smoking rate by a linear equation in slope–intercept form written as

$$r = mt + b,$$

where t represents the time in years from the beginning of the campaign, and r represents the rate of smoking in this age group. To determine the relationship between the time t and the smoking rate r, we must first find m and b. (*Note:* In this example, t plays the role of x, and r plays the role of y.)

The change in the smoking rate per year is −0.6% per year. This means that the slope of the graph is $m = -0.6$. Substituting this value for m, we get

$$r = \overset{\overbrace{}^{\text{slope}}}{-0.6}t + b.$$

We now use our rules for rewriting equations to find b.

$$38.8 = -0.6(3) + b \qquad \text{Original equation}$$
$$38.8 = -1.8 + b \qquad \text{Simplify product.}$$
$$38.8 + 1.8 = b \qquad \text{Add 1.8 to both sides.}$$
$$40.6 = b \qquad \text{Simplify.}$$

So our equation describing the reduction in smoking is therefore

$$r = -0.6t + 40.6.$$

We now use this equation to predict the smoking rate 10 years from now when $t = 13$. Substituting the value $t = 13$, we get

$$r = (-0.6)13 + 40.6 = -7.8 + 40.6 = 32.8.$$

Therefore, if the decline in smoking continues at the same rate, in 10 more years the smoking rate among 18- to 25-year-olds will be 32.8%.

Now try Exercises 21 to 26. ✳

Although the model we developed in Example 2 was simple to construct and to use, you may feel that it has several drawbacks that outweigh its simplicity. First, you could argue that a linear equation is not appropriate to describe accurately the relationship between the time and the smoking rate because a change in time does not produce a corresponding change in the smoking rate. Or, you could argue that we have not considered factors other than the advertising campaign that affect the smoking rate.

It also may seem unreasonable to use the model over such a long period of time. As a result of such concerns, a researcher studying this problem might develop a more complex model than a linear equation or might modify the model every few years to make it more accurate.

✏️ *KEY POINT*

We can use two points to find a linear equation.

Building a Model with Two Points

In modeling data, we often use two data points to write a linear equation. Example 3 shows how to find a linear equation if we know two points on its graph.

EXAMPLE 3 *Using Two Points to Find a Linear Equation*

Find an equation of the line passing through the points (4, 2) and (8, 5).

SOLUTION: We will write the equation in slope–intercept form as $y = mx + b$, so we need to find m and b. Using points (4, 2) and (8, 5), we find the slope:

$$m = \frac{\text{rise}}{\text{run}} = \frac{5 - 2}{8 - 4} = \frac{3}{4}.$$

Substituting for m, we get

$$y = \overset{m}{\frac{3}{4}}x + b.$$

Because (4, 2) lies on the graph,* we substitute 4 for x and 2 for y to get

$$\overset{y}{2} = \frac{3}{4}\overset{x}{(4)} + b = 3 + b.$$

Subtracting 3 from both sides, we find that $b = -1$. The equation we want is therefore

$$y = \frac{3}{4}x - 1.$$

Now try Exercises 13 to 20. ✳ **5**

Quiz Yourself **5**

Find an equation of the line passing through (−3, 5) and (6, 0). Write the equation in slope–intercept form.

🧩 **PROBLEM SOLVING**

Convert a New Problem into an Older One

In Example 3, we used the *Convert a New Problem into an Older One Strategy* that we discussed in Section 1.1. We did not approach Example 3 as a brand-new problem. Rather, we recognized that once we had the slope, we could solve the problem as we did in Example 1. A good problem solver often modifies a technique that is used in one situation and applies it in another.

---

*We could have also used (8, 5); however, we generally use the point with the simpler coordinates.

EXAMPLE 4   *Finding a Linear Equation for a Model Based on Two Data Points*

Sheena sells gourmet pastries on the Internet. In the fourth month of operation, she sold 480 dozen pastries, and in the seventh month she sold 792 dozen. Assume that we can model the increase in her business by a linear equation. She estimates that with her current equipment, she can bake a maximum of 1,500 dozen pastries per month. If she wants her business to keep growing, during what month will she exceed her current capacity to produce baked goods?

SOLUTION: We will model Sheena's situation by a linear equation of the form

$$d = mt + b,$$

where t is the time in months that her business has been in operation and d is the number of dozens of pastries she can sell. (In this problem, t plays the role of x, and d plays the role of y.)

We will use the points (4, 480) and (7, 792) to find the slope of the graph of the equation as follows:

$$m = \frac{\text{rise}}{\text{run}} = \frac{792 - 480}{7 - 4} = \frac{312}{3} = 104.$$

Substituting for m in the original equation, we get

$$d = 104t + b. \quad \text{—slope}$$

Now we can use either (4, 480) or (7, 792) to find b. As a rule of thumb, whenever possible, we use smaller numbers rather than larger ones, so we will use (4, 480). Substituting 4 for t, gives us the following equation, which we then solve for b:

$$480 = 104(4) + b \qquad \text{Original equation}$$
$$480 = 416 + b \qquad \text{Simplify.}$$
$$480 - 416 = b \qquad \text{Subtract 416 from both sides.}$$
$$64 = b. \qquad \text{Simplify.}$$

Thus, a linear equation describing the growth in Sheena's sales is

$$d = 104t + 64.$$

We now want to find when Sheena will be able to sell 1,500 dozen pastries, so we set $d = 1,500$ and solve the equation $1,500 = 104t + 64$ in the usual way.

$$1,500 = 104t + 64$$
$$1,436 = 104t \qquad \text{Subtract 64.}$$
$$\frac{1,436}{104} = t \qquad \text{Divide by 104.}$$

Therefore, $t = \frac{1,436}{104} \approx 13.8$, which means that she will exceed her capacity to produce pastries during the fourteenth month.

Now try Exercises 27 to 32. ✳

✎ **KEY POINT**

The line of best fit gives the best linear approximation of data.

The Line of Best Fit

The next example illustrates model building with real data. Real data are often messier to work with than the example data in problems because the data points usually do not fall on a straight line. Instead, they may have a somewhat linear pattern and we must find a line that best fits the data.

EXAMPLE 5   *Modeling the Increase of Music Sales over the Internet*

Over the past several years, an increasing percentage of consumers are using the Internet to purchase music. The percentages of music purchases in 2003, 2004, and 2005 were 5.0, 5.9, and 8.2, respectively.* We have plotted the points $A = (2003, 5.0)$, $B = (2004, 5.9)$, and $C = (2005, 8.2)$ in Figure 7.8.

Although these points do not lie exactly on a straight line, we still may want to model these data with a linear equation.

a) Model these data with an equation of the line passing through points A and C.

b) Use this model to predict the percentage of music sales over the Internet in 2006.

c) The actual percentage of music sales over the Internet in 2006 was 9.1%. Evaluate the accuracy of this model.

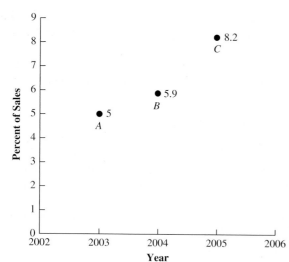

FIGURE 7.8 Internet music sales.

SOLUTION:

a) Using the technique of Example 4, (calling year 2003 year 0, etc.) we find that the equation of the line through points $(0, 5)$ and $(2, 8.2)$ to be $p = 1.6t + 5$ where t is the time from 2003 in years and p is the percent of music sales made over the Internet.

b) This equation predicts that in 2006 (year 3), the percent of Internet music sales will be

$$p = 1.6(3) + 5 = 4.8 + 5 = 9.8.$$

c) According to the Recording Industry Association of America, the actual percentage of Internet music sales was 9.1%. As shown in Figure 7.9, our prediction seems a little high, so maybe we should not trust this model.

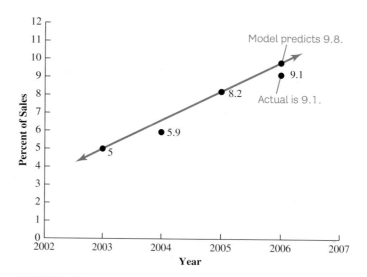

FIGURE 7.9 Graph of linear equation modeling sales.

Now try Exercises 33 to 34. ❁

*Data are from the Recording Industry Association of America.

HIGHLIGHT ✸ ✸ ✸

Finding the Line of Best Fit with a Graphing Calculator*

The computations necessary to find the line of best fit for a set of data points are so tedious that researchers rarely do them with pencil-and-paper calculations. Many calculators have a built-in capability to find the coefficients for the line of best fit, which we will now illustrate.

We will find the line of best fit for the data points (0, 5), (1, 5.9), and (2, 8.2). Calculator screen 1 shows that we have entered the x-coordinates of the data points in a list called

L1 and the y-coordinates in a list called L2. Screen 2 shows a menu of choices for statistical computations, and we have chosen LinReg($ax + b$) for linear regression.

After we select the command to perform linear regression, screen 3 shows that we told the calculator to use lists L1 and L2 to get the coordinates of the data points. Screen 4 indicates the slope a and y-intercept b for the equation $y = ax + b$ for the line of best fit.

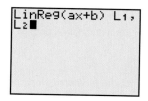	

Screen 1   Screen 2   Screen 3   Screen 4

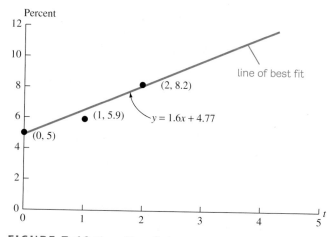

FIGURE 7.10 Line of best fit for Internet music sales data.

Keep in mind that there is nothing wrong with the mathematical computations in Example 5. However, we made a poor decision in choosing the line through points A and C to model the data. It may very well be the case that no linear equation will model the data accurately.

In Example 5, we imposed a strong condition by requiring that the line modeling Internet music sales pass *through* two of the data points. There is a technique in statistics called **linear regression** that we can use to find an equation of a line to approximate data. Because a set of real data points rarely lies on a straight line, when we approximate data with a linear equation, the line will pass above some data points and below others.

Linear regression finds a line that is called the **line of best fit**, which is frequently used to model data.

We used a graphing calculator (see Highlight box) to find that the line of best fit for the original three data values in Example 5 is

$$y = 1.6x + 4.77.$$

In Figure 7.10, we graphed this equation and the three data values from Example 5. Notice that the line is below data points A and C and above B.

This means that if we insist on using a linear equation to model the Internet sales data, this is the best line that we can find. The question still remains, however: "Should we be using a linear equation to model the data?" In Section 7.3, we will consider some alternative modeling methods that might produce better results.

---

*See your instructor for a tutorial on using a graphing calculator to find the line of best fit.

Exercises 7.2

Looking Back*

These exercises follow the general outline of the topics presented in this section and will give you a good overview of the material that you have just studied.

1. What are three ways that we can describe an equation?

2. In finding an equation of the line in Example 3, what did we do to convert the problem into one like in Example 1?

3. How is the method we used in Example 4 like the method we used in Example 3?

4. What was the point of our calculator computations on page 313?

Sharpening Your Skills

Find a linear equation whose graph has the indicated slope and passes through the given point.

5. slope 3, point (2, 1)   **6.** slope 4, point (5, 2)

7. slope 4, point (−2, 6)   **8.** slope 3, point (−5, 3)

9. slope −2, point (4, 3)   **10.** slope −3, point (6, 4)

11. slope −5, point (−6, −1)

12. slope −8, point (−4, −9)

Find a linear equation whose graph passes through the given points. Write the equation in slope–intercept form.

13. (2, 3) and (5, 9)   **14.** (3, 5) and (5, 17)

15. (17, 12) and (9, 10)   **16.** (14, 10) and (5, 7)

17. (11, −4) and (−8, 2)

18. (2, −6) and (−14, 9)

19. (−6, −8) and (−4, −1)

20. (−2, −9) and (−1, −4)

Applying What You've Learned

In Exercises 21–30, it is useful to think of slope as representing the average rate of change of one variable corresponding to a change in the other variable. To keep the numbers you work with small, represent the first year for which you have data as year 0. For example, in Exercise 21, the year 2007 is year 0. In Exercise 28, the year 2002 is year 0.

21. Life expectancy. In 2007, the life expectancy for a female born in the United States was 80.97 years and was increasing at a rate of 0.3 year per year. Assume that this rate of increase remained constant.

   a. Model this situation by a linear equation.

   b. Use the equation in part a) to estimate the life expectancy of a female born in the United States in 2020.

22. Life expectancy. In 2007, the life expectancy for a male born in the United States was 75.15 years and was increasing at a rate of 0.4 year per year. Assume that this rate of increase remained constant.

   a. Model this situation by a linear equation.

   b. Use the equation in part a) to estimate the life expectancy of a male born in the United States in 2020.

23. College enrollment. In 2005, there were 7.4 million males enrolled in degree-granting institutions of higher education and that number was increasing at a rate of 133,000 per year.

   a. Model this information with a linear equation.

   b. Use this model to project how many males will be in degree-granting institutions of higher education in 2020.

   c. Use this model to determine in what year we expect to have 10 million males enrolled in degree-granting institutions of higher education.

24. College enrollment. In 2005, there were 9.9 million females enrolled in degree-granting institutions of higher education and that number was increasing at a rate of 300,000 per year.

   a. Model this information with a linear equation.

   b. Use this model to project how many females will be in degree-granting institutions of higher education in 2020.

   c. Use this model to determine in what year we expect to have 15 million females enrolled in degree-granting institutions of higher education.

25. College expenses. According to the U.S. Department of Education, in 2005, college expenses at a public four-year institution averaged $11,400 and were increasing at a rate of $760 per year.

   a. Model this information with a linear equation.

   b. Use your model to predict the cost of attending a four-year college in 2015.

26. College expenses. According to the U.S. Department of Education, in 2005, college expenses at a public two-year institution averaged $6,300 and were increasing at a rate of $310 per year.

   a. Model this information with a linear equation.

   b. Use your model to predict the cost of attending a two-year college in 2015.

27. Crime statistics. The graph shows the decrease in violent crimes in the United States from 2002 to 2004, according to the U.S. Bureau of the Census. Assume that this rate of decrease continues for the next several years and can be modeled by a linear equation of the form $v = mt + b$.

   a. Use the given data to find the slope of the graph of this equation.

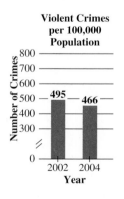

Violent Crimes per 100,000 Population

---

b. Use the slope from part a) and either of the data points to find a linear equation that models this decrease in violent crime. (2002 = year 0.)

c. Use the equation you found in part b) to predict the number of violent crimes per 100,000 U.S. inhabitants in 2015.

28. Poverty data. The graph shows the increase in the percentage of Americans living below the poverty level from 2002 to 2005, according to the U.S. Bureau of the Census. Assume that the rate of increase continues for the next several years and can be modeled by a linear equation of the form $p = mt + b$.

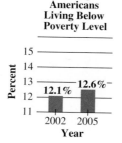

a. Use the given data to find the slope of the graph of this equation.

b. Use the slope from part a) and the data from 2002 to write an equation that describes this increase in poverty. (2002 = year 0.)

c. Use the equation you found in part b) to predict the percentage of Americans living below the poverty level in 2015.

29. Employment data. According to the Bureau of Labor estimates, the demand for database administrators increased from 87,000 in 1998 to 155,000 in 2008. Model these data with a linear equation, and estimate the demand for database administrators in 2015.

30. Employment data. According to the Bureau of Labor estimates, the demand for desktop publishing specialists increased from 26,000 in 1998 to 44,000 in 2008. Model these data with a linear equation, and estimate the demand for desktop publishing specialists in 2015.

31. Sleep data. According to the National Sleep Foundation, in 2004, Americans averaged 6.9 hours of sleep each weeknight. In 1900, they averaged about 8 hours and 30 minutes of sleep per weeknight. Model these data by a linear equation, and use it to predict the year when Americans will not be sleeping at all on a weeknight.

32. Satellite dish data. According to the Consumer Electronics Association, the number of direct satellite dish receivers rose from 4.3 million in 2000 to 16.8 million in 2004. Model these data with a linear equation, and estimate when there will be 40 million of these receivers.

33. U.S. labor force data. The graph shows the increase in the civilian labor force from 2003 to 2006, according to the U.S. Department of Labor. Assume that the rate of increase continues for the next several years and can be modeled by a linear equation of the form $l = mt + b$.

a. What is the average yearly increase from 2003 to 2006?

b. Use the result from part a) as the slope and the point that represents the 2003 data to write an equation that describes this increase in the labor force.

c. Use the equation you found in part b) to estimate in what year the size of the labor force will reach 200 million.

34. Travel data. The graph shows the amount that was spent on foreign travel by U.S. residents from 2002 to 2005.

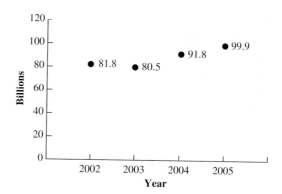

Model these data with a linear equation as follows:

a. What is the average yearly increase from 2002 to 2005?

b. Use the result from part a) as the slope and the point that represents the 2002 data as a point to write an equation describing the change in foreign travel expenditures from 2002 to 2005. As usual, treat the year 2002 as year 0.

c. Use the equation that you found in part b) to predict when expenditure on foreign travel by U.S. citizens will reach 120 billion.

Communicating Mathematics

35. When we used the phrase "rate of change" as in Example 2, what information were we telling you?

36. When we used a line to model the situation in Example 2, what were we assuming about the relationship between time and the rate of smoking?

37. Discuss the appropriateness of using a linear equation as a model for the music sales data in Example 5.

38. Discuss the appropriateness of using a linear equation as a model for the U.S. civilian labor force data in Exercise 33.

Using Technology to Investigate Mathematics

39. There are a number of sites on the Internet that have interactive programs illustrating the line of best fit. (You may want to search by the term "linear regression.") Find an interesting site, and report on your findings.

40. Find some interesting applications of linear regression in an area that interests you. When researching this question, I found sites applying the line of best fit to sociology, political science, and even music. Write a brief report on your findings.

For Extra Credit

41. **Line of best fit for labor force data.** We can use the four data points in Exercise 33 to find an equation of the line of

	2003 (Year 0)	2004 (Year 1)	2005 (Year 2)	2006 (Year 3)
Actual data	146	147	149.2	151.3
Values predicted by your model				151.3
Values predicted by line of best fit				

best fit to be $y = 1.81t + 145.66$. Complete the table, which compares the actual data values, your estimate of the data values using the equation you found in Exercise 33(b), and the data values that would be predicted by using the equation of the line of best fit.

42. **Line of best fit for travel data.** Repeat Exercise 41 for the data given in Exercise 34. The line of best fit for these data points is given by the equation $y = 6.56t + 78.66$. Complete the following table.

	2002 (Year 0)	2003 (Year 1)	2004 (Year 2)	2005 (Year 3)
Actual data	81.8	80.5	91.8	99.9
Values predicted by your model			93.86	
Values predicted by line of best fit			91.78	

43. Ask your instructor for a tutorial on using a graphing calculator to find the line of best fit, and then use your calculator to verify that the lines we gave in Exercises 41 and 42 are correct.

7.3 Modeling with Quadratic Equations

Objectives

1. Use the quadratic formula to solve equations.
2. Graph a quadratic equation.
3. Use a quadratic equation to model data.

According to a model that I made using real data, I found that in February 2008, *Star Trek* would be a more popular TV show than *Survivor*, *Lost*, and the Super Bowl between the New York Giants and the New England Patriots. For my model, I used actual Nielsen TV ratings for *Star Trek* from 1987 to 1990 to calculate the line of best fit. My mathematics is correct, but obviously, something is wrong—I used the wrong kind of model. Our experience tells us that the popularity of a TV show does not grow continually in a straight line. Some shows start out, grow in popularity, and eventually fade as newer shows take over the market. So, if we want to model this phenomenon, we cannot use a linear equation. In this section, we will explain how to use quadratic equations as models.

A **quadratic equation** is an equation of degree two. Although there are different types of quadratic equations, in this section we will restrict our discussion to quadratic equations of the form $y = ax^2 + bx + c$, where a, b, and c are real numbers and $a \neq 0$. The equations

$y = 2x^2 - 3x + 5$ and $y = \frac{3}{2}x^2 + 6x - 17$ are examples of the type of quadratic equations we will be using as models.

As with linear equations in two variables, solutions to quadratic equations will be ordered pairs of numbers.

EXAMPLE 1 *Verifying Solutions for Quadratic Equations in Two Variables*

Determine whether each ordered pair is a solution for the quadratic equation $y = 2x^2 - 3x + 5$.

a) (4, 25) b) (1, 2)

SOLUTION:

a) Substituting 4 for x and 25 for y gives us the true statement $25 = 2(4)^2 - 3(4) + 5$. This tells us that (4, 25) is a solution.

b) When we substitute 1 for x and 2 for y, we get the false statement $2 = 2(1)^2 - 3(1) + 5$. So, (1, 2) is not a solution. ✳

 **KEY POINT**

The graph of $y = ax^2 + bx + c$ is a parabola.

If we plot all the solutions to a quadratic equation, we get a graph that is a geometric figure called a **parabola**. We show two parabolas in Figure 7.11. If a is positive, the graph opens up as in Figure 7.11(a). If a is negative, the parabola opens down as in Figure 7.11(b). If a parabola is opening up, the lowest point on the parabola is called the **vertex** of the parabola. Similarly, for a parabola opening down, the highest point is called the vertex.

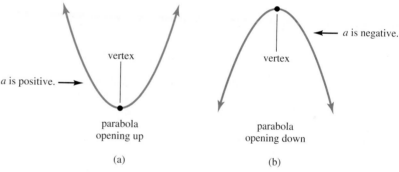

FIGURE 7.11 The graph of $y = ax^2 + bx + c$ is a parabola.

We will frequently use the following fact about quadratic equations without proof.

DEFINITION The **vertex** of the graph of the quadratic equation $y = ax^2 + bx + c$ occurs when $x = \dfrac{-b}{2a}$.

EXAMPLE 2 *Finding the Vertex of a Parabola*

Find the vertex of the graph of $y = 2x^2 - 4x + 5$.

SOLUTION: In this equation, $a = 2$, $b = -4$, and $c = 5$. The x-coordinate of the vertex is

$$x = \frac{-b}{2a} = \frac{-(-4)}{2(2)} = \frac{4}{4} = 1.$$

Substituting this value for x, we get the y-coordinate of the vertex, $y = 2(1)^2 - 4(1) + 5 = 2 - 4 + 5 = 3$. The vertex of the parabola, which is the graph of $y = 2x^2 - 4x + 5$, is therefore the point (1, 3). ✳

We use the quadratic formula to find the intercepts of a parabola.

The Quadratic Formula

In order to graph quadratic equations, we need another tool. We find x- and y-intercepts for the graphs of quadratic equations exactly as we did for linear equations. Recall that to find the x-intercept of a linear equation, we set y equal to 0 and then solve the resulting equation. Doing this for a quadratic equation, we get an equation of the form $ax^2 + bx + c = 0$, which is a quadratic equation in a single variable. Such an equation may or may not have solutions. If it has solutions, they can always be found using one of the most famous formulas in algebra, called the **quadratic formula**.

THE QUADRATIC FORMULA   The solutions of the quadratic equation $ax^2 + bx + c = 0$ will be

$$x = \frac{-b + \sqrt{b^2 - 4ac}}{2a} \quad \text{and} \quad x = \frac{-b - \sqrt{b^2 - 4ac}}{2a}.$$

We often combine these two formulas into the single formula

$$x = \frac{-b \pm \sqrt{b^2 - 4ac}}{2a}.$$

EXAMPLE 3   *Using the Quadratic Formula to Solve an Equation*

Use the quadratic formula to solve the equation $x^2 + 5x - 84 = 0$.

SOLUTION: In this equation, $a = 1$, $b = 5$, and $c = -84$. We substitute these values into the quadratic formula to get the solutions:

$$\begin{aligned} x &= \frac{-5 \pm \sqrt{5^2 - 4(1)(-84)}}{(2)(1)} \\ &= \frac{-5 \pm \sqrt{25 + 336}}{2} \\ &= \frac{-5 \pm \sqrt{361}}{2} = \frac{-5 \pm 19}{2}. \end{aligned}$$

The solutions are therefore

$$x = \frac{-5 + 19}{2} = 7 \quad \text{and} \quad x = \frac{-5 - 19}{2} = -12.$$

Now try Exercises 5 to 12. ✳ **6**

Quiz Yourself **6**

Use the quadratic formula to solve $x^2 + 7x - 44 = 0$.

Some Good Advice

In solving a quadratic equation in a single variable, we must calculate the quantity $b^2 - 4ac$, which is called the **discriminant** of the equation. There are three cases that can arise:

- $b^2 - 4ac$ is greater than 0—in this case, there are two distinct solutions to the equation.

- $b^2 - 4ac$ equals 0—in this case, there is only one solution to the equation.

- $b^2 - 4ac$ is less than 0—in this case, there are no real number solutions because we cannot take the square root of a negative number and get a real number.

Graphing Quadratic Equations

EXAMPLE 4 *Graphing a Quadratic Equation*

Graph the quadratic equation $y = x^2 - 4x - 12$, which is a parabola, by doing the following:

a) Determine if the parabola is opening up or down.

b) Find the vertex of the parabola.

c) Find the x- and y-intercepts of the graph.

d) Draw the graph.

SOLUTION:

a) Because the coefficient for x^2 is 1, the parabola is opening up.

b) For this equation, $a = 1$, $b = -4$, and $c = -12$, so the x coordinate of the vertex is

$$x = \frac{-b}{2a} = \frac{-(-4)}{2 \cdot 1} = \frac{4}{2} = 2.$$

The y-coordinate of the vertex is therefore

Substitute $x = 2$.

$$y = 2^2 - 4 \cdot 2 - 12 = 4 - 8 - 12 = -16.$$

This means that the vertex of the parabola is the point $(2, -16)$.

c) To find the x-intercepts of the graph, we set $y = 0$ and get the equation

$$0 = x^2 - 4x - 12.$$

Using the quadratic formula, we obtain

$$x = \frac{-(-4) \pm \sqrt{(-4)^2 - 4(1)(-12)}}{2 \cdot 1} = \frac{4 \pm \sqrt{16 + 48}}{2} = \frac{4 \pm \sqrt{64}}{2} = \frac{4 \pm 8}{2},$$

so $x = 6$ or $x = -2$. The x-intercepts are therefore $(-2, 0)$ and $(6, 0)$.

To find the y-intercept, we set $x = 0$ to get $y = 0^2 - 4 \cdot 0 - 12 = -12$. The y-intercept is the point $(0, -12)$.

d) With this information, it is now easy to draw a reasonable graph, as shown in Figure 7.12(a). Notice how the parabola is symmetric with respect to the vertical line through the vertex, which is called its **axis of symmetry**. Figure 7.12(b) shows the same graph produced on a graphing calculator.

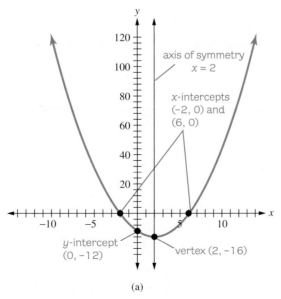

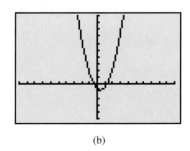

(a)   (b)

Quiz Yourself

Graph the equation
$y = x^2 + x - 12$.

FIGURE 7.12 (a) Graph of $y = x^2 - 4x - 12$. (b) Same graph produced by a graphing calculator.

Now try Exercises 13 to 22. ❊ **7**

❀ ❀ ❀ HISTORICAL HIGHLIGHT

Solving Equations

Although the ancient Babylonian, Greek, and Hindu mathematicians had algebraic methods to solve quadratic equations, it was the work of a ninth-century Arabic mathematician named Mohammed ibn-Musal al-Khowarizmi who made the term *algebra* a household word. In his book *Al-jabr wa'l muqabalah*, he introduced Europeans to the methods of earlier mathematicians for solving quadratic equations. Some believe his title refers to balancing and completing algebraic expressions, which are processes we commonly use in algebra.

Having mastered the quadratic, it was natural to next ask if there was a method to solve *cubic equations*—those that have an x^3 term in them. In 1545, the Italian Girolamo Cardano published his famous book *Ars Magna* in which he showed not only how to solve cubics but also *quartic equations* (equations with an x^4 term). At this point, it seemed mathematicians would invent methods to solve equations having any power of x. This turned out not to be the case.

In the nineteenth century, Niels Henrik Abel from Norway and Evariste Galois of France proved the remarkable result that there is no mechanical formula similar to the quadratic formula for solving *quintic equations* (equations with an x^5 term).

Modeling with Quadratic Equations

We will now apply our knowledge of quadratic equations to model building. After the release of a new book, movie, or DVD, there are four periods in its life cycle.

- Stage 1: Sales increase rapidly. With a new DVD, for example, the number of purchasers grows rapidly soon after the DVD is introduced.

- Stage 2: The sales are still growing, but the increase from week to week is not as great as in the early phase.

- Stage 3: The product is still selling, but now each week's sales are a little lower than the week before.

- Stage 4: The market is saturated, and sales are now dropping rapidly.

We can represent this situation by a parabola opening downward, as in Figure 7.13.

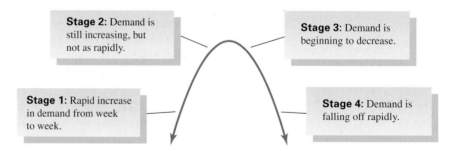

Stage 2: Demand is still increasing, but not as rapidly.

Stage 3: Demand is beginning to decrease.

Stage 1: Rapid increase in demand from week to week.

Stage 4: Demand is falling off rapidly.

FIGURE 7.13 A parabola models the product life cycle for a new product.

We use this pattern to model sales of a particular DVD in Example 5.

EXAMPLE 5 *Modeling DVD Sales with a Quadratic Equation*

Assume that a producer knows that the demand for a live concert by a popular artist on DVD can be modeled by the equation $S = -13n^2 + 169n$. Here, n is the number of weeks since the introduction of the DVD, and S is the dollar value in thousands of the DVDs sold during week n.

a) When do we expect the sales for the DVD to peak?

b) After sales have peaked, when does this model predict that the sales will sink below $100,000 per week?

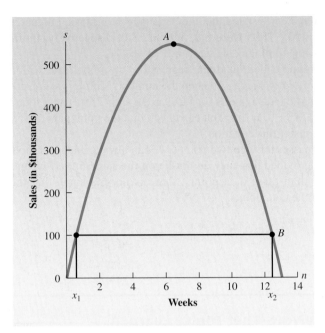

FIGURE 7.14 Parabola modeling DVD sales.

SOLUTION:

a) The equation is quadratic with a negative n^2 coefficient, so its graph is a parabola opening downward, as shown in Figure 7.14.

Point A, the vertex of the parabola, shows where the weekly sales are the greatest. Because $S = -13n^2 + 169n$, $a = -13$, $b = 169$, and $c = 0$. Therefore, the first coordinate of the vertex is

$$n = \frac{-b}{2a} = \frac{-169}{2(-13)} = \frac{-169}{-26} = 6.5.$$

Therefore, sales should peak during the seventh week.

Substituting 6.5 for n, we find the greatest dollar amount for weekly sales to be

Substitute n. ⌐————————————⌐

$$S = -13(6.5)^2 + 169(6.5) = -13(42.25) + 169(6.5)$$
$$= -549.25 + 1,098.5 = \$549.25.$$

Therefore, the highest weekly sales should be about \$549,250.

b) Point B shows where the sales, after peaking, drop to \$100,000 per week. From Figure 7.14, it appears that there are two values for x that will give sales of \$100,000. The first, x_1, appears to be between 0 and 2, and the other, x_2, is slightly past 12. To find these values exactly, we will solve the quadratic equation

$$-13n^2 + 169n = 100.$$

Subtracting 100 from both sides of this equation we get

$$-13n^2 + 169n - 100 = 0.$$

Using the quadratic formula, we find

$$n = \frac{-169 \pm \sqrt{169^2 - 4(-13)(-100)}}{2(-13)}$$
$$= \frac{-169 \pm \sqrt{28,561 - 5,200}}{-26} = \frac{-169 \pm \sqrt{23,361}}{-26}.$$

Thus, $n = 0.62$ and $n = 12.38$. Because we are interested in the point after sales have peaked, we ignore $n = 0.62$ and use $n = 12.38$ weeks as our solution. According to this model, by the 13th week, sales will have dropped below \$100,000. ❋

Math in Your Life

Who's Watching You Watch TV*

As you recall from our *Star Trek* example at the beginning of this section, although we have correct data, using an inappropriate model may result in an unreliable prediction. The opposite can happen. The model might be valid, but the data are wrong.

In the TV ratings wars, there is a fierce competition that influences millions of advertising dollars for networks. However, because of the intermixing of technologies—radio is on the Web, TV is on cell phones, and the Web is on TV—it is becoming more

difficult for ratings companies to gather reliable data. One company, Arbitron, has an innovative way to find out what you watch. They ask people to wear a device called the portable people meter, or PPM, which is about the size of a pager, to record exactly how much radio and TV programming you are exposed to each day. The PPM records everything about your viewing—including your flipping habits—and then transmits this information to be analyzed while you sleep.

*This note is based on the article "Our Ratings, Ourselves" in the *New York Times* by Jon Gertner, April 10, 2005.

Recall in Example 5 in Section 7.2 we stated that the percentages of music purchases made over the Internet in 2003, 2004, 2005 were 5.0, 5.9, and 8.2, respectively. In the Highlight, we found that the best line to fit the data was $y = 1.6x + 4.77$.

Similarly, we can use a technique called quadratic regression* to model data with a quadratic equation called the *parabola of best fit*. To keep the numbers simple, we considered year 2003 to be year 0, entered the data points (0, 5), (1, 5.9), and (2, 8.2), and found the parabola of best fit to be $y = 0.7x^2 + 0.2x + 5$. You can check that the point pairs (0, 5), (1, 5.9), and (2, 8.2) are all solutions of this equation.

Just as we can find a line that passes through two points, we can always find a parabola that passes through three points, provided that they do not lie on the same line. If we use quadratic regression with more than three data points, there is no guarantee that the parabola will pass through any of the data points.

Exercises 7.3

Looking Back[†]

These exercises follow the general outline of the topics presented in this section and will give you a good overview of the material that you have just studied.

1. Why did we use the quadratic formula in Example 3?

2. In graphing a quadratic equation, how do you find the vertex of the parabola?

3. What was our point in discussing the modeling of the *Star Trek* data at the beginning of the chapter?

4. What did Evariste Galois prove?

Sharpening Your Skills

Solve each quadratic equation.

5. $x^2 - 10x + 16 = 0$
6. $x^2 - 7x + 12 = 0$
7. $2x^2 - 5x + 3 = 0$
8. $6x^2 - 11x + 4 = 0$
9. $3x^2 + 7x - 6 = 0$
10. $2x^2 + x - 3 = 0$
11. $5x^2 - 17x - 12 = 0$
12. $4x^2 + 12x + 9 = 0$

Answer the following questions for each quadratic equation. Then draw the graph.

Is the equation's graph opening up or down?
What is the vertex of the graph?
What are the x-intercepts?
What is the y-intercept?

13. $y = -x^2 + 6x - 8$
14. $y = x^2 - 4x - 5$
15. $y = -4x^2 + 8x + 5$
16. $y = x^2 + 5x + 4$
17. $y = 4x^2 - 4x - 2$
18. $y = -2x^2 - 10x - 8$
19. $y = 3x^2 + 7x - 6$
20. $y = 2x^2 + x - 3$
21. $y = -x^2 + 7x - 12$
22. $y = -4x^2 - 12x - 9$

Applying What You've Learned

23. **Movie attendance.** Dreamworks movie studio tries to release a "blockbuster" movie each summer. Assume that these statistics describe ticket sales for such a movie:

    Week 2   5 million tickets sold

    Week 4   7 million tickets sold

    Week 6   8 million tickets sold

    We did quadratic regression to find that $A = -0.125x^2 + 1.75x + 2$ is an equation of the parabola of best fit for these data.

    a. Show that the graph of this equation passes through the three data points regarding attendance.

    b. According to this model, what is the maximum value that A can attain?

24. **Business profits.** New business owners typically experience a loss for a certain amount of time before the losses bottom out. Eventually the businesses show a profit. This pattern of profit and loss suggests a situation that could be modeled by a quadratic equation whose graph is a parabola opening  up. Assume that you have begun a business photographing weddings and that over the first three months of business you sustain losses of $200, $130, and $70. (We can rephrase this by saying your cumulative profit at the end of month 1 was −$200, at the end of month 2 was −$330, and at the end of month 3 was −$400). Using quadratic regression, we can

---

*See your instructor for a tutorial that shows how to perform quadratic regression on the TI-83 graphing calculator.
[†]Before doing these exercises, you may find it useful to review the note *How to Succeed at Mathematics* on page xix.

show that $P = 30x^2 - 220x - 10$ is the best quadratic equation that fits these data.

a. Verify that this equation fits the given data.

b. Using this model, at the end of what month do you expect to sustain your greatest cumulative loss? What is this loss?

c. At the end of what month do you expect to show your first cumulative profit? (By that, we mean that your total profits from the time the business started are greater than your total losses.)

d. What will your cumulative profit be by the end of month 10?

e. How much do you expect to earn during month 10?

25. DVD sales. Assume that the sales for the DVD in Example 5 is modeled by the equation $S = -4n^2 + 16n - 12$, where S is the number of millions of dollars of sales in week n.

a. When do we expect the sales for the DVD to peak?

b. According to this model, what is the largest value for S?

c. When do we expect the sales to drop to zero?

26. Presidential popularity ratings. After a U.S. president makes a politically unpopular decision, his approval rating usually drops. After some time, the approval rating rises again. We will model this drop in approval, bottoming out, and then rise in approval by a quadratic equation. Assume that before the president signs an unpopular tax bill, the approval rating is at 48%. One week after the president signs the bill, the rating is 41%, at two weeks it is 39%, and at three weeks it is 42%. Using quadratic regression we can show that $A = 2.5x^2 - 9.5x + 48$ is the best quadratic equation that fits the data.

How many weeks after the signing of the tax bill will the approval rating be back to what it was before the signing of the tax bill?

We model many physical relationships using quadratic equations.

27. Falling bodies. A plane is dropping emergency food supplies to relieve famine in a less-developed country. Crates are dropped, and the height of the crate above the ground at time t is given by the equation $H = 160 - 16t^2$.

a. Graph this equation.

b. Are there any values for t that it would not make sense to use for this equation because of the physical characteristics of a falling crate?

c. Find the time at which the crate will strike the ground.

28. Rocket flight. Assume that if a model rocket is fired directly upward (with a certain velocity), its distance above the ground at time t is given by the equation $d = 100t - 16t^2$.

a. Graph this equation.

b. Are there any values for t that it would not make sense to use for this equation because of the physical characteristics of a rocket?

c. Find the time at which the rocket reaches its highest point.

d. Find the time at which the rocket returns to the ground.

29. Prisoner population. Between 1970 and 2000 the number of federal and state prisoners grew at a rate that was 16 times as great as the rate of growth of the U.S. population.* The given graph shows the number of prisoners for various years.

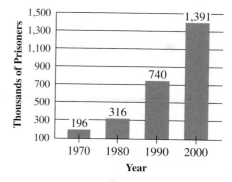

Using only the three data points (0, 196), (10, 316), and (20, 740), the equation of the parabola of best fit passing through those points is $y = 1.52x^2 - 3.2x + 196$.

a. Verify that the three ordered pairs given are solutions for this equation.

b. Use this equation to estimate the number of prisoners in 2000. Compare it with the actual number.

30. Use the model in Exercise 29 to predict the number of prisoners in 2015.

Communicating Mathematics

31. What is the discriminant of the equation $ax^2 + bx + c = 0$? What does this expression tell us about the solutions of the equation?

32. What is the purpose of quadratic regression?

33. Returning to Exercise 27, explain why it is reasonable for the graph representing the falling crate to be shaped as it was. How does this conform to your intuition? Would a linear equation have been an acceptable model?

---

*If the U.S. population had grown at the same rate as its prisoner population, the United States would now be the most populous country in the world with almost 1.5 billion people. (James M. Henslin, *Sociology*, 8th ed., Boston: Allyn & Bacon, 2007.)

34. Returning to Exercise 28, explain why it is reasonable for the graph representing the rocket flight to be shaped as it was. How does this conform to your intuition? Would a linear equation have been an acceptable model?

35. Running a race. The equation $d = 0.15t^2 + 8t$ describes the distance Dom Bowden who is running a 100-yard dash has traveled in t seconds.

   a. Graph this equation.

   b. Are there any values for t that it would not make sense to use in this equation due to the physical characteristics of this problem?

   c. From the graph, determine where the runner is running more slowly and where the runner is running faster.

   d. How long does it take for him to finish the race?

36. Running a race. Would it make sense to use the model in Exercise 35 for Bowden running a mile race? Explain.

Using Technology to Investigate Mathematics

37. Ask your instructor for a tutorial that will show you how to use the TI graphing calculators to calculate the parabola of best fit. Use the calculator to verify some of the examples in this section.

38. Search for an applet on the Internet that will allow you to do quadratic regression interactively. Use the applet to verify some of the results of this section.

For Extra Credit

39. The accompanying figure shows a time-lapse photo of a bouncing golf ball. The photographer captured all the images of the golf ball by using a strobe light that flashed once every 0.03 second. It is known from basic physics that the model for the path of a bouncing ball is a quadratic equation. Discuss how you would obtain the numerical data that then could be used to do quadratic regression to obtain the equation.

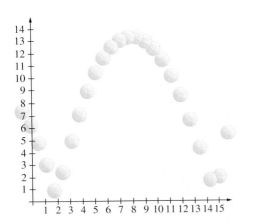

40. This exercise is based on Exercises 33 and 41 in Section 7.2. In Exercise 41 from Section 7.2, you were given the line of best fit for four data points. The equation was $y = 1.8t + 145.66$. We can use those same four data points to find the parabola of best fit whose equation is $y = 0.275t + 0.985t + 145.935$.

   a. Add a line to the table that you began in Exercise 35 from Section 7.2 as follows:

	2003 (Year 0)	2004 (Year 1)	2005 (Year 2)	2006 (Year 3)
Actual data	146	147	149.2	151.3
Values predicted by your model	146	147.77	149.54	151.3
Values predicted by line of best fit	145.66	147.47	149.54	151.3
Values predicted by parabola of best fit				

   b. Does the line of best fit or the parabola of best fit seem to model the data better? Explain your answer.

41. The following graph shows the way Medicare expenditures have risen in recent years. (*Source: Statistical Abstract, 2005*)

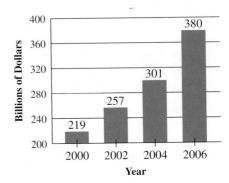

The equation $y = 2.5625x^2 + 10.975x + 220.45$ describes the parabola of best fit for these data. (We are treating year 2000 as year 0 to simplify calculations.)

   a. Compare the predicted expenses with the actual expenses for 2004 and 2006.

   b. Use the quadratic formula to estimate in what year Medicare expenditures will reach 1 trillion (1,000 billion) dollars.

42. Use the data in Exercise 29 and the quadratic formula to predict the year in which there will be 2 million prisoners in U.S. and state prisons.

Exponential Equations and Growth

7.4

Objectives

1. Understand the difference between linear, quadratic, and exponential growth.
2. Use exponential equations to model growth.
3. Solve exponential equations using the log function.
4. Use logistic models to describe growth.

It seems that every few years we are bombarded by the media with a new crisis—the Avian flu, the eventual collapse of the Social Security system, global warming, famine in developing countries, the mortgage crisis, the spread of AIDS in Africa—the list seems to go on forever. In this section, we will introduce you to a new type of equation that can be used to model many of these situations. You can begin to understand the relationships that we model in this section by recalling the way money grows in a bank account.*

 **KEY POINT**

Money compounded in a bank account grows exponentially.

Exponential Growth

Suppose that you deposit $1,000 (called the *principal*) in an account paying 8% interest per year. To keep our calculations simple, we will assume that you deposit the money at the beginning of the year and the interest is paid all at once at the end of the year.

At the end of the first year, you have the original $1,000 plus the interest it has earned in 1 year. That is, the amount at end of the first year equals

$$\$1,000 + \underbrace{8\% \times \$1,000}_{\text{Interest}} = \$1,000 + \underbrace{0.08 \times \$1,000}_{\text{Interest}}$$

$$= \$1,000 + \underbrace{\$80}_{\text{Interest}} = \$1,080. \quad \text{——amount at end of first year}$$

So, at the beginning of the second year, you have $1,080 earning 8% interest. Therefore, the amount in your account at the end of the second year is

$$\$1,080 + \underbrace{8\% \times \$1,080}_{\text{Interest}} = \$1,080 + \underbrace{0.08 \times \$1,080}_{\text{Interest}}$$

$$= \$1,080 + \underbrace{\$86.40}_{\text{Interest}} = \$1,166.40. \quad \text{——amount at end of second year}$$

To compute the amount for successive years, we perform similar calculations. That is, we leave the interest in the account and then calculate the new interest based on the total of the principal and interest previously earned. The process of earning interest on interest is called **compounding**.

Table 7.1 shows how compounding works and shows the pattern in the increased value of the account.

Year	Beginning Balance	+	Interest for Current Year	=	Balance at End of Year
1	1,000	+	$0.08 \times 1{,}000$	=	$1{,}000(1.08) = 1{,}080$
2	1,080	+	$0.08 \times 1{,}080$	=	$1{,}080(1.08) = 1{,}000(1.08)(1.08)$
3	1,166.40	+	$0.08 \times 1{,}166.40$	=	$1{,}166.40(1.08) = 1{,}000(1.08)(1.08)(1.08)$

TABLE 7.1 Calculating compound interest.

*We will cover compound interest in much greater depth in Chapter 9.

Notice in Table 7.1 that the amount in your account

at the end of the *first* year is: (original deposit) × $(1.08)^1$,

after the *second* year is: (original deposit) × $(1.08)^2$,

and, after the *third* year is: (original deposit) × $(1.08)^3$.

In general, we have the following formula:

> **THE COMPOUND INTEREST FORMULA** If we invest the amount P, called the *principal*, in an account earning a yearly interest rate r and we compound the interest for n years, then the amount in the account, A, is
>
> $$A = P(1 + r)^n. *$$

We apply this formula in Example 1.

EXAMPLE 1   *Using the Compound Interest Formula*

Suppose that you deposit $1,000 in an account that is compounded annually at a rate of 8%. Find the amount in this account after 30 years.

SOLUTION: The principal P is $1,000, the number of years n is 30, and the rate r is 0.08. Thus, the amount in the account at the end of 30 years will be

$$P(1 + r)^n = 1,000(1 + 0.08)^{30} = 1,000(1.08)^{30} = \$10,062.66.$$

Now try Exercises 11 to 18. ❀ **8**

In Example 1, we compounded the interest once a year. Many financial institutions compound interest more frequently. To learn how to do this, see Section 9.2.

Let's plot the amount in the account described in Example 1 over a period of years to see better how the value of the account is growing. In Table 7.2, we first calculate the amount in this account for each of the first 5 years and plot these values in Figure 7.15, which shows the relationship between time and the amount in the account.

Quiz Yourself **8**

Assume that you have invested $1,000 into an account paying 1.25% interest that is compounded yearly. What will be the value of this account in 30 years?

Year	Balance at End of Year
1	$1,000(1.08)^1 = 1,080$
2	$1,000(1.08)^2 = 1,166.40$
3	$1,000(1.08)^3 = 1,259.71$†
4	$1,000(1.08)^4 = 1,360.48$
5	$1,000(1.08)^5 = 1,469.32$

TABLE 7.2 Growth of an account over 5 years.

❀ ❀ ❀ **HIGHLIGHT** ———————————

Between the Numbers—Did Anything Bother You in the Last Example?

Over the course of our marriage, my wife and I have often been asked by eager young investment counselors for a few minutes of our time. After introducing their investment plan, they begin talking about the details with something like this:

> *Assuming a rate of return of 8%, in 30 years,*** your investment will be worth . . . let me punch it in the computer here . . . ah, there it is . . .*

and up pops some fantastic amount on the screen.

The operative words here are "***assuming a rate of return of 8%.***" When I was rewriting this chapter, the stock market had fallen over 1,000 points in less than a month and many investments were paying only a fraction of 1% interest. People were losing their homes due to increased interest rates on adjustable rate mortgages and still others were drowning in insurmountable credit card debt.

When the exponential model is working for you, it can be a wonderful thing—when it is working against you, it can be a financial tragedy.

---

*To raise a quantity to a power on your calculator, you must use the y^x key (or the $^\wedge$ key). For example, to compute $(1 + 0.08)^{12}$, you first enter 1.08, then press the y^x key (or the $^\wedge$ key), then press 12, and then the = key or Enter.

†In calculating the amount in the account, we will round the answer *down* to the nearest cent. For example, if the amount in the account is $11,255.47764, we will express the answer as $11,255.47.

**Notice the difference between the amount that you would earn in Example 1 with an interest rate of 8% versus what you earn in Quiz Yourself 8 with an interest rate of 1.25%.

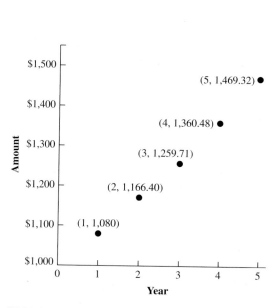

FIGURE 7.15 The balance in a bank account over 5 years.

FIGURE 7.16 Graph comparing the balance in a bank account and the line and parabola of best fit.

Because the points in Figure 7.15 appear to be close to lying on a straight line, you might want to use the line of best fit to approximate these data. However, if you look very closely, you will see a slight upward curving, so you may decide instead to find the parabola of best fit.

We used a graphing calculator to find the linear equation that best models the data to be

$$A = 97.272n + 975.366,$$

and the best quadratic model is given by

$$A = 3.74n^2 + 74.84n + 1001.54,$$

where A is the amount in the account after n years.

We graphed the line and parabola of best fit in Figure 7.16. In addition, we graphed the original five data points and also the points $(30, 10{,}062.65)$, $(35, 14{,}785.34)$, and $(40, 21{,}724.52)$, which correspond to the value of the account after 30, 35, and 40 years, respectively.

As time increases, the points representing the amounts in the bank account are farther and farther from both the line and parabola of best fit. The equation describing these data is neither linear nor quadratic.

The type of equation we used to calculate compound interest is also useful for modeling other important phenomena, so we will give it a name.

> **DEFINITION** An **exponential equation** is an equation of the form
> $$y = a \cdot b^x.$$

The compound interest formula, $A = P(1 + r)^n$, is one example of an exponential equation.

 Some Good Advice

We use an exponential equation to model situations in which *the rate of change of a quantity is proportional to its size.* For example, the more money you have in a bank account, the more interest you earn; the larger the population of a country, the more children will be born to increase the population.

KEY POINT

An exponential equation is a simple model of population growth.

Exponential Models

We can use an exponential equation to model population growth.* To do this, we use an initial population instead of an initial bank balance, and a yearly growth rate instead of an interest rate.

EXAMPLE 2 *Modeling Population Growth with an Exponential Function*

According to the U.S. Bureau of the Census, in 2008 the U.S. population was approximately 301 million, with an annual growth rate of 0.894%. If this growth rate continues at that rate until 2053, what will the population of the United States be then?

SOLUTION: We will use an exponential equation with

$$P = 301 \text{ million}, \qquad r = 0.00894, \qquad \text{and} \qquad n = 2053 - 2008 = 45.$$

Therefore,

$$P(1 + r)^n = 301(1 + 0.00894)^{45} = 301(1.00894)^{45} = 449.3 \text{ million}.$$

According to this exponential model, we expect 449.3 million people to be living in the United States in 2053.

You may think it strange that I chose the year 2053, but as I was rewriting this example, I compared it with a similar example from the previous edition of this text, using data from 2000 with a population of 281 million and an annual growth rate of 1.2%. In that example, the population projection for 45 years was over 480 million! The point is that even with a larger base population in 2008, the seemingly slight difference in the growth rate (1.2 − 0.894 = 0.306%) yielded a projection of roughly 30 million fewer people in our projection over 45 years.

Now try Exercises 19 to 26. ❋ **9**

Quiz Yourself **9**

In 2008, Mexico had a population of roughly 108 million and a yearly growth rate of 1.153%. Assuming that this growth rate continues for the next 10 years, what will Mexico's population be at that time?

When using a mathematical model as in Example 2, it is always important to consider the appropriateness of the model. For example, you should ask whether it is reasonable to assume that the growth rate in the United States will stay at 0.894% for 45 years.

Often in the media, researchers make dire predictions based on mathematical models regarding such things as population growth, global warming, and the bankruptcy of the Social Security system. These predictions may or may not be valid, depending on the assumptions made by the researcher. As an educated person, you should have an understanding of what a mathematical model can tell us and what it cannot. If the assumptions underlying a model are not sound, then its predictions are worthless.

In predicting future bank balances, population, or similar quantities that are growing exponentially, we often want to know, "How long will it take for this quantity to double?"

Before we can answer this question, we need another tool. Many calculators have a key labeled either "log" or "log x," which stands for the *common logarithmic function.*† Pressing this key has the effect of reversing the operation of raising 10 to a power. For example, suppose that you compute $10^5 = 100,000$ on your calculator. If you next press the "log" key, the display will show "5." If you enter 1,000, which is 10 raised to the third power, and press the "log" key, the display will show "3." Practice finding the log of powers of 10, such as 100 and 1,000,000.

The log function has an important property that will help us solve equations of the form $a = b^x$ for x. For example, we may want to find a value of x such that $5 = 3^x$. Although it may seem strange to you, it is possible to find a number, although not an integer, such that

KEY POINT

We use the log function to find the time it takes a quantity that is growing exponentially to double.

---

*We will discuss a more complex model of population growth later in this section.
†We will discuss the log function in greater detail in Section 9.2.

if we raise 3 to that power, we will get 5. In order to do that, we need the following property of the log function:

EXPONENT PROPERTY OF THE LOG FUNCTION

$$\log y^x = x \log y$$

To understand this property, use your calculator to verify the following:

$$\log 3^5 = 5 \log 3 \quad \text{and} \quad \log 8^2 = 2 \log 8$$

EXAMPLE 3   *Using the Log Function to Solve an Equation*

Solve the equation $5 = 3^x$.

SOLUTION: We take the log of both sides of the equation to get

$$\log 5 = \log 3^x.$$

We then use the exponent property of the log function to rewrite the equation as

$$\log 5 = x \log 3.$$

If we divide both sides of the equation by $\log 3$, we get

$$\frac{\log 5}{\log 3} = x.$$

We now find $x = \dfrac{\log 5}{\log 3} \approx \dfrac{0.69897}{0.47712} \approx 1.46.$

If you use your calculator, you will find that $3^{1.46}$ is approximately 5. (Because of the way we rounded numbers, you will not get exactly 5 as a result.)

Now try Exercises 27 to 34. ❋

We now can find how long it takes for a quantity to double.

EXAMPLE 4   *Doubling a Population*

In 2007, Cambodia's population was 14 million, with an annual growth rate of roughly 1.73%. Assuming that the growth rate remains the same, in what year will Cambodia's population double?

SOLUTION: We want to know when the population will be 28 million. We will use the growth model $A = P(1 + r)^n$, where $P = 14$, $r = 0.0173$, and A is the future population of 28 (million). So we want to solve the equation

$$\overset{A}{}\overset{P}{}\overset{r}{}$$
$$28 = 14(1 + 0.0173)^n$$

for n.

First, we divide both sides of the equation by 14 to get

$$2 = (1 + 0.0173)^n.$$

Next, we take the log of both sides, giving us

$$\log 2 = \log(1.0173)^n.$$

Then, using the exponent property for the log, we get

$$\log 2 = n \log(1.0173).$$

Dividing both sides of the equation by $\log 1.0173$, we find that

$$n = \frac{\log 2}{\log 1.0173} \approx 40.4.$$

Quiz Yourself ❿

Redo Example 4 but now assume that the growth rate is 2.1%.

So, we expect the population of Cambodia to double by 2048.

Now try Exercises 45 to 48. ❊ ❿

We will now show you how to build exponential models using two data points. Recall that an exponential model has the form $y = a \cdot b^x$.

EXAMPLE 5 *Building an Exponential Model for Inflation*

Just as money grows exponentially in a bank account, the price of a product increases over the years exponentially due to inflation. Suppose that in 2003, a pair of Ugg® boots costs $130 and in 2008, due to inflation, the same pair costs $150.

a) Develop an exponential model to describe the inflation in this price.

b) Use the model in part a) to estimate the cost of this pair of boots in 2014.

SOLUTION:

a) We will assume the model that we want to find has the form $y = a \cdot b^x$, so we must find a and b. We will call 2003 year 0 and year 2008 year 5.

First we will find a. If we let $x = 0$ and $y = 130$, then our model becomes $130 = a \cdot b^0$. Recall that any number except 0 raised to the 0 power is equal to 1, so $b^0 = 1$. Therefore, $130 = a \cdot b^0 = a \cdot 1 = a$. We can now write our model as $y = 130 \cdot b^x$.

Next, we have to find b. We know that when $x = 5$, $y = 150$. Substituting these values for x and y in our model gives us the following equation, which we solve for b:

Quiz Yourself ⓫

Assume that the price of a leather coat was $180 in 2006 and the same coat cost $210 in 2010.

a) Find an exponential equation to model this inflation.

b) Use your model in part a) to estimate the cost of the coat in 2015.

$$150 = 130 \cdot b^5 \qquad \text{Substitute 130 for a.}$$

$$b^5 = \left(\frac{150}{130}\right) \approx 1.154 \qquad \text{Divide both sides by 130.}$$

$$b = \sqrt[5]{1.154} \approx 1.03 \qquad \text{Take the fifth root of both sides.}$$

Our model is now $y = 130 \cdot (1.03)^x$.*

b) We want to find the value for y when $x = 2014 - 2003 = 11$. So, the estimated cost of the boots will be $130 \cdot (1.03)^{11} = \$179.95$.

Now try Exercises 57 to 62. ❊ ⓫

✎ **KEY POINT**

Limited resources may cause a population to grow according to a logistic model.

Logistic Models

In the 1980s, many were alarmed by the rapid growth of AIDS. Figure 7.17 contains data from a report by the Global Aids Policy Commission showing the increase of AIDS cases in North America from 1985 through 1991.

Looking at the graph in Figure 7.17, it might appear that the growth of AIDS was following an exponential pattern. However, notice that in Figure 7.17, unlike our earlier bank account example, *the growth rate is not the same from year to year.*

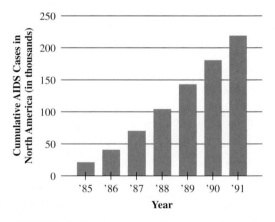

FIGURE 7.17 The growth of AIDS cases in North America, 1985–1991.

---

*You can take the fifth root of 1.154 by using the y^x key or calculating $y \wedge x$, with $y = 1.154$ and $x = \frac{1}{5}$.

Years	Cumulative AIDS Cases in North America (in thousands)	Percent Increase from Previous Year
1985	22	
1986	41	86
1987	70	71
1988	104	49
1989	143	38
1990	180	26
1991	219	22

Note decrease in rate of growth from year to year.

KEY POINT

A logistic model takes into account limits on population growth.

Thus, an exponential model is not appropriate to model the data, so we need to introduce another type of model.

In studying the growth of populations, demographers often use a logistic model. A *logistic model* takes into account the fact that as populations grow, there are limits in terms of space, food, and so on that prevent the population from following a true exponential growth pattern.

Recall that in computing compound interest, we used equations of the type

balance at end of year = (1 + rate)(balance at end of previous year).

Translating this into population growth, we would write

population at end of year = (1 + growth rate)(population at end of previous year).

In both cases, we assumed that the money in the bank account and the population would grow at the same rate from year to year. However, with restrictions on the amount of food, water, space, and so on, any environment can sustain only a limited population. As the population grows, the percentage of this capacity that has been used up influences the rate of growth. If a population has a growth rate of 3%, in the beginning it is reasonable to multiply the existing size of the population by (1 + 0.03) to estimate the population the next year. However, as the capacity to sustain further population diminishes, we must reduce the growth rate accordingly.

In a logistic model, it is common to represent the maximum capacity that an environment can support by 1, or 100%. For our model, we will define a quantity P_n, read "P sub n," that represents the percentage of the maximum capacity that has been attained by the population in year n.

For example, to say $P_5 = 0.40$ means that at the end of year 5, the population we are studying has attained 40% of its maximum capacity. If this same population had an original growth rate of 0.03, to calculate P_6 we would reduce this growth rate by the factor $1 - P_5 = 0.60$ to get a growth rate of 0.018, as shown in the following diagram.

original growth rate ⌐

⌐40% of the capacity to sustain a population has been used up.

$$0.03 \times (1 - P_5) = 0.03 \times 0.6 = 0.018$$

Only 60% of the capacity to sustain the population remains.

During year 6, we have only 60% of the original growth rate.

The idea is that because 40% of the capacity for growth already has been used up by the population, future growth rate can only be 60% of what it was originally.

We are now ready to give a precise definition of a logistic growth model.

DEFINITIONS Logistic Growth Model

Assume that a population is growing originally at rate r. We let P_n denote the percentage of the maximum capacity that the population has attained in year n. Moreover, P_n satisfies the following equation:

$$P_{n+1} = [1 + r(1 - P_n)]P_n.$$

This collection of equations for $n = 0, 1, 2, \ldots$ is called a **logistic model**.

To calculate P_{n+1}, we reduce the growth rate by multiplying it by $1 - P_n$. We will refer to this quantity as the *rate reduction factor*. So the logistic growth equation can be written as

$$P_{n+1} = [1 + r(\text{rate reduction factor})]P_n.$$

It is useful to recompute the rate reduction factor each time we calculate a new value for P_{n+1}. We illustrate how to use the logistic growth model in Example 6.

EXAMPLE 6 *Using the Logistic Growth Model to Predict Population Growth*

Assume that a population is growing initially at a rate of 3% per year. Also assume that at the end of the fifth year, the population is at 40% of its maximum size. At what percentage of its maximum size will the population be 1 year later?

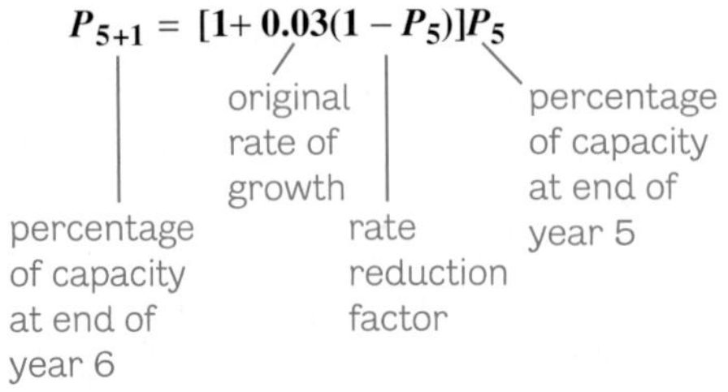

SOLUTION: We are told that $P_5 = 0.40$, $r = 0.03$, and we want to find P_6. Figure 7.18 shows how to use the logistic growth model.

At the end of year 5, the rate reduction factor will be $1 - 0.4 = 0.6$; thus,

$$P_{5+1} = [1 + 0.03(1 - P_5)]P_5$$

percentage of capacity at end of year 6

original rate of growth

rate reduction factor

percentage of capacity at end of year 5

FIGURE 7.18 Using P_5 to calculate P_6.

$$P_6 = [1 + (0.03)(1 - P_5)]P_5$$
$$= [1 + (0.03)(0.6)](0.4)$$
$$= (1 + 0.018)(0.4)$$
$$= (1.018)(0.4)$$
$$= 0.4072.$$

So we see that at the end of the sixth year, the population is at 0.4072, or 40.72% of its maximum capacity.

Now try Exercises 35 to 38. ❋

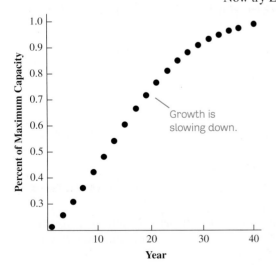

FIGURE 7.19 Graph illustrating logistic growth.

Growth is slowing down.

Notice that in Example 6, we did not compute P_6 by substituting 6 into some equation, as we did in previous models. Instead, we described P_6 in terms of P_5. If we had not been given P_5 explicitly, then we would have to calculate it by finding P_4. But to find P_4 would require knowing P_3, and so on. If we keep going, eventually we would need to know P_0. In our population examples, we will always assume that P_0 is the percentage of the capacity we have before the first year begins. An equation that is defined so that to find a given value requires knowing previous values is called a **recursive equation**.

In Figure 7.19, we have plotted the population growth for a hypothetical population over a period of 40 years based on a logistic growth model. It is easy to see that at first the population grows rapidly in a way similar to exponential growth, but as we reach year 15, because the population is at about 60% of its maximum capacity, the growth slows considerably.

EXAMPLE 7   *Using the Logistic Model to Predict Future Population*

Biologists have chosen a nearly extinct type of lemur to populate a small island. They have introduced a number of lemurs that constitute 20% of the population that the island is capable of supporting. If the growth rate of the lemur population is 10%, find what percentage of the maximum lemur population will be on the island in 3 years.

SOLUTION: To find the percentage of the maximum population at the end of 3 years, we will find P_1, P_2, and finally P_3.

P_0 is the initial percentage of the maximum population of lemurs, which is 0.20. We will round off calculations to three decimal places to keep them from becoming too unwieldy. The first rate reduction factor is $1 - P_0 = 1 - 0.20 = 0.80$.

$$P_1 = [1 + (0.10)(\text{rate reduction factor})]P_0$$
$$= [1 + (0.10)(0.8)](0.20)$$
$$= (1 + 0.08)(0.20) = (1.08)(0.20) = 0.216$$

The rate reduction factor is now $1 - P_1 = 1 - 0.216 = 0.784$.

$$P_2 = [1 + (0.10)(0.784)](0.216)$$
$$= (1 + 0.0784)(0.216) = (1.0784)(0.216) = 0.233$$

The new rate reduction factor is $1 - P_2 = 1 - 0.233 = 0.767$.

$$P_3 = [1 + (0.10)(0.767)](0.233)$$
$$= (1 + 0.0767)(0.233) = (1.0767)(0.233) = 0.251$$

At the end of 3 years, the island will have slightly more than 25% of its maximum lemur population.

Now try Exercises 39 to 44. ❋ ⑫

Quiz Yourself ⑫

Assume that a population is growing initially at a rate of 4% per year. Also assume that at the end of the third year, the population is at 60% of its maximum size. At what percentage of its maximum size will the population be 1 year later?

Exercises [7.4]

Looking Back*

These exercises follow the general outline of the topics presented in this section and will give you a good overview of the material that you have just studied.

1. What are the meanings of A, P, r, and n in the compound interest formula?

2. What was our purpose in Figure 7.16?

3. When we use an exponential equation to model the growth of a bank account, we talk about the initial deposit and the interest rate. What are the corresponding terms when we model the growth of a population?

4. In Example 6, we multiplied the initial growth rate by the quantity $1 - P_5$. What was P_5? What was the meaning of $1 - P_5$? Why did we multiply by this quantity?

5. Why did we not use a linear or quadratic equation to model the growth of money in a bank account?

6. What was the point of our "Between the Numbers" highlight?

Sharpening Your Skills

In Exercises 7–10, you are given the principal in a bank account at the beginning of a year and a rate of interest that is compounded annually. Calculate the amount in the account at the end of the year.

7. $1,000; 5%

8. $3,000; 6%

9. $4,000; 2.5%

10. $3,000; 6.5%

In Exercises 11–14, you are given the principal in a bank account, a yearly interest rate, and the time the money is in the account. Assuming that no withdrawals are made, use the compound interest

*Before doing these exercises, you may find it useful to review the note *How to Succeed at Mathematics* on page xix.

formula to compute the amount in the account after the specified time period. Assume compounding is done annually.

11. $5,000; 5%; 5 years

12. $7,500; 7%; 6 years

13. $4,000; 8%; 2 years

14. $8,000; 4%; 3 years

In Exercises 15–18, you are given an initial deposit in a bank account and the amount in the account after a certain number of years. Assume that compounding is done annually, and that no withdrawals are taken. Find the annual interest rate on the investment.

15. Initial deposit $15,000; amount after 12 years is $22,000

16. Initial deposit $20,000; amount after 20 years is $50,000

17. Initial deposit $10,000; amount after 8 years is $13,000

18. Initial deposit $12,000; amount after 10 years is $18,000

In Exercises 19–22, you are given the population and the growth rate as of 2007 for each country. Assume that the growth rate remains the same from year to year. Use an exponential function to model population growth and determine the size of the population in the specified year.

19. India: population $= 1,130$ million; growth rate $= 1.6\%$; year, 2020

20. Nigeria: population $= 124$ million; growth rate $= 3.2\%$; year, 2020

21. Cambodia: population $= 14$ million; growth rate $= 1.73\%$; year, 2020

22. Brazil: population $= 192$ million; growth rate $= 1.0\%$; year, 2020

In Exercises 23–26 you are given the population of a country for 2001 and 2007. Find the annual growth rate over that time period.

23. China: 2001—1,285 million; 2007—1,322 million

24. Brazil: 2001—173 million; 2007—192 million

25. Nigeria: 2001—117 million; 2007—124 million

26. India: 2001—1,025 million; 2007—1,130 million

In Exercises 27–34, solve each equation for x.

27. $5^x = 20$ **28.** $2^x = 15$

29. $3^x = 10$ **30.** $10^x = 3$

31. $10^x = 3.2$ **32.** $8^x = 4.65$

33. $(3.4)^x = 6.85$ **34.** $(15.7)^x = 155.5$

In Exercises 35–38, assume that a population is growing initially at the specified rate. You are given a value for P_n and are to use the logistic growth model to compute the value of P_{n+1}. Explain what your calculations tell you.

35. rate $= 3\%$; $P_4 = 0.36$

36. rate $= 5\%$; $P_7 = 0.48$

37. rate $= 4.5\%$; $P_8 = 0.72$

38. rate $= 5.5\%$; $P_3 = 0.51$

In Exercises 39–44, redo the calculations of Example 7 regarding the growth of the lemur population. Use the specified initial growth

rate and the given value for P_0 to find P_3. Your answers may vary from ours due to roundoff error.

39. rate $= 8\%$; $P_0 = 0.30$

40. rate $= 12\%$; $P_0 = 0.40$

41. rate $= 10\%$; $P_0 = 0.25$

42. rate $= 15\%$; $P_0 = 0.35$

43. rate $= 4.5\%$; $P_0 = 0.60$

44. rate $= 4.25\%$; $P_0 = 0.20$

Applying What You've Learned

45. **Compound interest.** If your parents have $10,000 in a college savings account that is paying an interest rate of 5%, which is being compounded annually, how many years will it take to double if the interest rate stays the same?

46. **Compound interest.** For the account in Exercise 45, how many years will it take to triple?

47. **Population growth.** In 2008, the population of the United States was 301 million and the growth rate was 0.9%. If the growth rate remains the same, in what year will the population be double what it was in 2008?

48. **Population growth.** In 2007, the population of Brazil was 192 million and the growth rate was 1.0%. If the growth rate remains the same, in what year will the population be double what it was in 2007?

If the rate of growth is negative, then we refer to exponential decay instead of exponential growth. Exercises 49–52 involve exponential decay.

49. **Population reduction.** Large, highly populated countries are often concerned about reducing the size of their population. Assume that in the year 2007 China had a population of 1,322 million. If China could maintain a growth rate of -0.5%, what would the population of China be in 2020?

50. **Population reduction.** Repeat Exercise 49 for India with a population of 1,092 million in 2005 and a growth rate of -0.6%.

51. **Radioactive decay.** Assume that a radioactive material decays with an annual growth rate of -0.35%. How many years will it take 100 pounds of the material to decay to 50 pounds?

52. **Radioactive decay.** Repeat Exercise 51, but now assume that the rate of growth is -0.14%.

When a drug such as a pain killer or an antibiotic is introduced into the body, the kidneys work to eliminate the drug. We will assume that after an hour, 15% of any drug in the body is eliminated. After another hour, 15% of the remaining drug is eliminated, and so on.

53. **Modeling drug concentration.** If you receive an injection of 500 mg of Novocain, how much of the drug will remain in your body after 3 hours?

54. **Modeling drug concentration.** If you take a 500-mg dose of erythromycin, how much of the drug will be in your body after 4 hours?

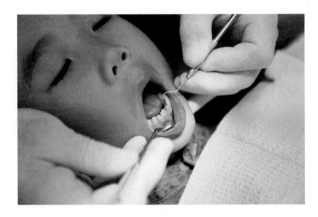

55. Modeling drug concentration. If a patient experiences some numbness as long as 150 mg of Novocain remain in the body, after how many complete hours should a patient stop feeling numbness if initially injected with 500 mg of Novocain?

56. Modeling drug concentration. If your physician wants to keep the level of erythromycin in your body above 200 mg and you are given an initial dose of 500 mg, at what time (to the nearest hour) should you take your next dose?

57. Modeling inflation. A pair of women's Saucony running shoes costs $120 in 2004 and the same model shoes cost $135 in 2008 due to inflation.

   a. Develop an exponential model to describe the rate of inflation over this time period.

   b. If inflation continues at the same rate, use your model in part a) to estimate the price of the shoes in 2014.

58. Modeling inflation. Repeat Exercise 57 for a Nikon digital camcorder that cost $320 in 2003 and $385 in 2008.

59. Modeling inflation. In the 1990s, Albania suffered from a period of runaway inflation. The inflation was so bad that if it were to continue, a fast-food meal that cost $4.65 would cost $1,712 10 years later. Develop an exponential model to determine the yearly rate of inflation over this time period.

60. Modeling inflation. In the 1990s, Hungary also suffered from severe inflation. If that inflation continued at the same rate, a pair of athletic shoes that cost $80 would cost $945 10 years later. Develop an exponential model to determine the yearly rate of inflation over this time period.

61. Modeling inflation. If $1,000 grows to $1,302 over 6 years, determine the interest rate during that period.

62. Interestrate reduction. If a $10,000 investment decreases to $8,850 over a four-year period, determine the (negative) interest rate over that time.

Communicating Mathematics

63. What does the log function do?

64. In Example 4, why did we have to use the log function?

65. When is an exponential model appropriate? Give examples.

66. How does a logistic model differ from an exponential model?

Using Technology to Investigate Mathematics

67. Search for applets on the Internet that will allow you to do exponential and logistic regression interactively. Use these applets to model some data of your choice, and report on your findings.

68. A number of sites on the Internet discuss exponential and logistic growth and environmental issues. For example, you can search for "exponential growth and environment," or, if you wish to be more specific, search for "exponential growth and greenhouse gases," or "exponential growth and population." Find several interesting sites and report on your findings.

For Extra Credit

Use the logistic growth model to solve Exercises 69 and 70. You may use a trial-and-error method to find your answer.

69. Assume that Carl dammed a stream on his land and intends to use the pond that has formed for bass fishing. He stocks the pond with 200 adult bass and believes that at maximum capacity, the pond could support 800 adult bass. Because he wants the bass population to grow, he will not fish from the pond until it has reached a population of 300 fish. Assuming a yearly growth rate of 18%, during which year will he be able to begin fishing?

70. Suppose that the pond in Exercise 69 can support 1,000 fish, the growth rate is 25%, and Carl wants to have 300 fish in the pond by the beginning of the fourth year. Assume that he must purchase fish for stocking in multiples of 50. What is the smallest amount that he can purchase to achieve his goal?

7.5 Proportions and Variation

Objectives

1. Solve problems using ratios and proportions.
2. Understand the capture–recapture method for estimating populations.
3. Use direct and inverse variation to solve problems.

Do you know how many tigers there are in the world? What about Florida manatees? Or wild yaks? How would you go about counting such animals? Is it possible to search all the oceans, bays, and rivers of the world and count the number of leatherback turtles? Yet, biologists and ecologists need to do this. In Example 2, we will show you a surprisingly simple method to estimate such populations, which is based on the notions of ratio and proportion that we will introduce next.

Ratio and Proportion

We encounter the notion of ratio in many situations. For example, if you drive 86.4 miles and use 2.7 gallons of gasoline, then the quotient $\frac{86.4}{2.7} = 32$ miles per gallon is an example of a ratio.

> **DEFINITIONS** A **ratio** is a quotient of two numbers. We write the ratio of the numbers a to b as $a:b$ or $\frac{a}{b}$. A **proportion** is a statement that two ratios are equal.

The statement $\frac{3}{4} = \frac{6}{8}$ is an example of a proportion. Notice in the proportion $\frac{3}{4} = \frac{6}{8}$ if you cross multiply, $\frac{3}{4} \diagup\!\!\!\!\diagdown \frac{6}{8}$, you get the products $3 \cdot 8 = 4 \cdot 6$. This illustrates a general principle that you will use in working with proportions.

> **CROSS-MULTIPLICATION PRINCIPLE** If $\frac{a}{b} = \frac{c}{d}$, then $a \cdot d = b \cdot c$. The quantities $a \cdot d$ and $b \cdot c$ are called **cross products.**

Sometimes one of the quantities in a proportion is unknown and we have to solve for it, as we see in Example 1.

Recently I had the opportunity to visit for several weeks in Colombia, South America. Colombia's currency is called the *peso*, and as we traveled around the country, I frequently had to make the type of conversion that we see in our first example.

EXAMPLE 1 *Estimating Prices in a Foreign Country*

Suppose you are shopping for a mummified piranha from the Amazon River as a gift for your cousin and see a price tag of 20,000 pesos in a gift shop. If earlier that morning, you exchanged 120 American dollars for 237,960 pesos, what is the cost in dollars for this piranha?

SOLUTION: We can solve this problem easily by setting up the following proportion that compares the ratio of dollars to pesos this morning with the ratio of dollars to pesos in the gift shop:

$$\text{dollars} \longrightarrow \frac{120}{237{,}960} = \frac{x}{20{,}000}. \longleftarrow \text{dollars}$$
$$\text{pesos} \longrightarrow \qquad\qquad\qquad\qquad \longleftarrow \text{pesos}$$

If we now cross multiply, we get the equation

$$(120)(20{,}000) = 237{,}960 \cdot x, \text{ or } 2{,}400{,}000 = 237{,}960 \cdot x.$$

Dividing both sides of this equation by 237,960, we get

$$x = \frac{2{,}400{,}000}{237{,}960} \approx 10.09.$$

So the piranha will cost slightly over 10 American dollars.*

Now try Exercises 5 to 12. ✳ ⓭

Some Good Advice

In Example 1, we set up the proportion

$$\frac{\text{dollars this morning}}{\text{pesos this morning}} = \frac{\text{dollars in the gift shop}}{\text{pesos in the gift shop}}.$$

It would not have mattered if we had started with

$$\frac{\text{pesos}}{\text{dollars}} = \frac{\text{pesos}}{\text{dollars}},$$

because we would have obtained the same answer.

However, it is important to be consistent. Whatever comparison you make on the left-hand side of the equation, you must be sure to make the comparison in the same way on the right-hand side.

The Capture–Recapture Method

As we alluded to in the introduction of this section, an interesting application of proportions occurs in estimating the size of wildlife populations. If, for example, we wanted to survey the population of leatherback turtles, we could capture a number of turtles, tag

Math in Your Life

What's a Country to Do?†

The U.S. Constitution mandates that a census be conducted every 10 years to count each and every person residing in the United States on the night of April 1 of the census year. From 1790, with marshals riding from town to town on horseback, to the present, we have tried our best to do what is increasingly an impossible task. What's a country to do?

Throughout the 1990s, there have been political debates, lawsuits, and court decisions as to whether or not the capture–recapture method could be used to estimate certain portions of the U.S. population (such as the homeless) rather than doing an actual count as the Constitution

specifies. In 1996, a blue-ribbon panel of the American Statistical Association found sampling to be a scientifically sound method for conducting the census. However, in 1997, Congress tried to prevent the U.S. Bureau of the Census from using any sampling method. Eventually the U.S. Supreme Court made a compromise ruling which said that although sampling could be used to count portions of the population, it could not be used in apportioning representatives to the House of Representatives.

Currently, statisticians and the U.S. Bureau of the Census are working to improve methods for counting the population before the next census in 2010.

*Actually, when I really had to do this type of conversion I remembered that there were 1,983 pesos in 1 dollar, so I estimated that there were roughly 2,000 pesos to a dollar. For a 20,000-peso purchase, I would knock off three zeros and divide by 2 to get a reasonable estimate of the cost in dollars. Doing this with the piranha would give me an estimate of $10.
†This note is based on the article "The Census Count: Who Counts? How Do We Count? When Do We Count?" by Lynne Billard (past president of the American Statistical Association) in *PS: Political Science and Politics*, December, 2000.

Quiz Yourself ⓭

If you can type 30 pages of your term paper in 2.5 hours, how long will it take you to type the entire paper, which is 54 pages long?

them, and release them back to the wild. At a later time, after the tagged turtles have had sufficient time to mix thoroughly with the population, we would capture a new sample of turtles. The proportion of tagged turtles in the new sample allows us to estimate the total turtle population. We illustrate this method in Example 2.

EXAMPLE 2   *The Capture–Recapture Method for Estimating Population Size*

Marine biologists around the world capture 832 female breeding leatherback turtles, tag them, and release them back into the wild. Several months later, the biologists take a second sample of 900 female breeding leatherback turtles, of which 7 have been tagged. Use these data to estimate the population of female leatherback breeding turtles.

SOLUTION: We assume that the ratio of all tagged turtles to the total number of all turtles in the population is equal to the ratio of tagged turtles in the second sample to the total number in the second sample. In other words, we can say that

$$\frac{\text{number of tagged turtles in population}}{\text{number of turtles in population}} = \frac{\text{number of tagged turtles in sample}}{\text{number of turtles in sample}}.$$

Let n be the number of turtles in the population. Also, there are 832 tagged turtles in the population and 7 tagged turtles in the sample of 900 turtles. Thus, the previous equation becomes

$$\begin{array}{r}\text{tagged in population} \longrightarrow \\ \text{number in population} \longrightarrow\end{array} \frac{832}{n} = \frac{7}{900}. \begin{array}{l}\longleftarrow \text{tagged in sample} \\ \longleftarrow \text{number in sample}\end{array}$$

If we cross multiply, we get $7n = 832 \cdot 900 = 748{,}800$. Dividing both sides by 7 gives us that $n \approx 106{,}971$. Thus, from this survey, the number of female leatherback breeding turtles is almost 107,000.

Now try Exercises 35 to 38. ❋ **14**

Quiz Yourself **14**

What would the number of turtles in the population have been in Example 2 if the second sample contained 1,000 turtles of which 8 had been tagged?

✎ **KEY POINT**

Direct variation relates quantities that increase or decrease in the same way.

Variation

Do you have a text messaging plan? If you do, then you are familiar with the mathematical notion of direct variation. For example, if your plan charges you 5 cents per text message, then the equation $c = 0.5t$ models the relationship between the number of text messages, t, and the cost, c. This is an example of direct variation; as the number of text messages increases or decreases, your cost varies in exactly the same way.

In another type of variation, called *inverse variation*, as one quantity increases, the other quantity decreases. The number of gallons of heating oil used per month and the outside average monthly temperature might be related in this way because the higher the temperature, the fewer gallons of heating oil you would use, and vice versa. We will now make these notions more precise.

> **DEFINITIONS** We say that y **varies directly** as x, or that y is **directly proportional** to x, if $y = kx$, where k is a nonzero constant. The constant k is called **the constant of variation** or **the constant of proportionality**.

EXAMPLE 3   *Solving a Direct Variation Problem*

Suppose that y varies directly as x and that $y = 50$ when $x = 15$. Find y when $x = 24$.

SOLUTION: Often, the first step in solving a variation problem is to use the information that you are given about x and y to calculate the constant of variation. Because we are told that y varies directly as x, we begin with the equation $y = kx$ and substitute 50 for y and 15 for x to get

$$50 = k(15).$$

Next, divide both sides of this equation by 15 to get

$$k = \frac{50}{15} = \frac{10}{3}.$$

Quiz Yourself 15

Suppose that p varies directly as q and that $p = 154$ when $q = 22$. Find p when $q = 19$.

The second step is to use our knowledge that $k = \frac{10}{3}$ to rewrite the equation $y = kx$ as $y = \frac{10}{3}x$. Then, substituting 24 for x gives us $y = \frac{10}{3} \cdot 24 = \frac{240}{3} = 80$. ❋ 15

Sometimes in dealing with direct variation, there may be more than two related quantities. This more complex relationship is called *joint variation*. However, as you will see in Example 4, the two-step solution process that we used in Example 3 is essentially the same.

EXAMPLE 4   *Calculating the Strength of a Beam*

A beam in a viewing booth for officials at the Daytona 500 auto race is being replaced. The current beam is 3.5 inches wide and 6 inches deep and will support a weight of 1,200 pounds. If a replacement beam with the same length is 3 inches wide and 7 inches deep, how much weight can it support? Use the fact that the strength of a beam is directly proportional to its width and the square of its depth.

SOLUTION:

Step 1: We begin with the equation $s = kwd^2$ and substitute the values given to us for w, d, and s to find the constant of variation k. That is,

$$1{,}200 = k(3.5)(6)^2 \text{ , or } 1{,}200 = k \cdot 126.$$

Dividing both sides of this equation by 126, we get

$$k = \frac{1{,}200}{126} = \frac{200}{21}.$$

Step 2: We can now use this value for k to rewrite the equation $s = kwd^2$ as $s = \frac{200}{21}wd^2$. If we substitute the values 3 for w and 7 for d, we find that

$$s = \frac{200}{21} wd^2 = \frac{200}{21} \cdot 3 \cdot (7^2) = 1{,}400 \text{ pounds.}$$

 KEY POINT

Inverse variation relates quantities that increase or decrease in opposite ways.

If we say that two quantities vary inversely, this means that as one increases, the other decreases.

DEFINITIONS We say that y **varies inversely** as x, or that y is **inversely proportional** to x, if $y = \frac{k}{x}$, where k is a nonzero constant.

A good example of inverse variation is the relationship between speed and time to travel a certain distance—the faster you go, the shorter the time it takes to travel a certain distance.

EXAMPLE 5   *Time Saved by Speeding*

Suppose that the speed limit is 65 miles per hour and a person (not you, of course—your friend) likes to drive just a little over the speed limit. If your friend takes a trip and it takes him $1\frac{1}{2}$ hours going at 65 miles per hour, how much time will he save by going 70 miles per hour? Use the fact that time is inversely proportional to speed.

Quiz Yourself 16

Repeat Example 5 using a speed of 75 miles per hour.

SOLUTION: We will use the equation, $t = \frac{k}{s}$, where t is time and s is speed. We will think of $1\frac{1}{2}$ hours as 90 minutes. So the previous equation becomes $90 = \frac{k}{65}$. Multiplying both sides of this equation by 65 gives us $90 \cdot 65 = 1 \cdot k = k$. Multiplying this out, we find that $k = 90 \cdot 65 = 5,850$.

Next, we rewrite our time–speed equation as $t = \frac{5,850}{s}$. Substituting the value 70 for t gives us $t = \frac{5,850}{70} \approx 83.6$ minutes, or 1 hour and 23.6 minutes. So all your friend has saved by speeding for over an hour is $90 - 83.6 = 6.4$ minutes. ✳ 16

It is possible to have a combination of both direct variation and inverse variation present in a relationship among several quantities. We will refine the model of the strength of a beam in Example 4 to illustrate this *combined variation.*

EXAMPLE 6 *Calculating the Strength of a Beam*

As we stated in Example 4, the strength of a beam is directly proportional to its width and the square of its depth. However, as you probably know from practical experience, the longer a beam is, the less weight it will support. That is, the strength of a beam varies inversely as its length. Suppose a wooden beam that is used to support lighting for an outdoor Shakespearean theater is 10 feet long, 3 inches wide, and 4 inches deep and will support a load of 600 pounds.

a) If the length of the beam is increased to 15 feet, how many pounds will the beam support?

b) If we want the 15-foot beam to support the same weight of 600 pounds, how wide should the beam be to give us the same strength?

SOLUTION:

a) We will model this situation with the equation

$$s = k\frac{w \cdot d^2}{l}. \quad \begin{array}{l}\text{—— } s \text{ varies directly as } w \text{ and } d^2. \\ \text{—— } s \text{ varies inversely as } l.\end{array}$$

We find the constant of variation, k, by substituting for the variables as follows:

$$600 = k\frac{3 \cdot 4^2}{10}. \quad \text{We substituted 3 for } w, \text{ 4 for } d, \text{ 10 for } l, \text{ and 600 for } s.$$

Therefore,

$$6,000 = k(48). \quad \begin{array}{l}\text{We multiplied both sides by} \\ \text{10 and simplified } 3 \cdot 4^2.\end{array}$$

Dividing both sides by 48, we get

$$k = \frac{6,000}{48} = 125.$$

We can then rewrite the strength equation as

$$s = 125\frac{w \cdot d^2}{l}.$$

If we now keep the values 3 for w and 4 for d, but substitute 15 for l, we get

$$s = 125\frac{3 \cdot 4^2}{15} = \frac{125 \cdot 3 \cdot 16}{15} = 400.$$

Thus, the longer beam will support only 400 pounds.

b) To solve this problem, we will use the values 4 for d, 15 for l, and 600 for s, and we must solve for the value of w. Substituting, we get the equation

$$600 = 125 \, \frac{w \cdot 4^2}{15} = \frac{125 \cdot w \cdot 16}{15}$$

or

$$600 \cdot 15 = 125 \cdot 16 \cdot w.$$

Therefore, $9,000 = 2,000 \cdot w$, or $w = \frac{9,000}{2,000} = \frac{9}{2} = 4.5$. So, the longer beam should be 4.5 inches wide in order to support 600 pounds. ✻

Now try Exercises 41 to 44.

Exercises 7.5

Looking Back*

These exercises follow the general outline of the topics presented in this section and will give you a good overview of the material that you have just studied.

1. Explain the comparisons that we made in setting up the proportion in Example 1.

2. What was the basic proportion that we set up in Example 2 to illustrate the capture–recapture method?

3. What were the two steps that we did in Example 4 that we often do in solving variation problems?

4. What did the Supreme Court decide about using the capture–recapture method for conducting the census?

Sharpening Your Skills

Solve for x in the following proportions.

5. $24:x = 18:3$

6. $35:4 = x:2$

7. $\dfrac{50}{4} = \dfrac{x}{5}$

8. $\dfrac{x}{8} = \dfrac{14}{4}$

9. $x:12 = 3:2$

10. $8:10 = 32:x$

11. $\dfrac{30}{40} = \dfrac{27}{x}$

12. $\dfrac{150}{x} = \dfrac{60}{40}$

Solve each of the following variation problems by first writing the variation as an equation and finding the constant of variation. Then answer the question that we ask.

13. Assume that y varies directly as x. If $y = 37.5$ when $x = 7.5$, what is the value for y when $x = 13$?

14. Assume that y varies inversely as x. If $y = 10$ when $x = 4$, what is the value for y when $x = 6$?

15. Assume that r varies inversely as s. If $r = 12$ when $s = \frac{2}{3}$, what is the value for r when $s = 8$?

16. Assume that d varies directly as the square of t. If $d = 24$ when $t = 4$, what is the value for d when $t = 10$?

17. Assume that a varies directly as the square of b. If $a = 16$ when $b = 6$, what is the value for a when $b = 15$?

18. Assume that y varies jointly as x and z. If $y = 60$ when $x = 4$ and $z = 5$, what is the value for z when $x = 6$ and $y = 45$?

19. Assume that D varies inversely as C. If $D = \frac{3}{4}$ when $C = 2$, what is the value for D when $C = 24$?

20. Assume that A varies directly as the square of r. If $A = 314$ when $r = 10$, what is the value for A when $r = 6$?

21. Assume that r varies jointly as x and y. If $r = 12.5$ when $x = 2$ and $y = 5$, what is the value for r when $x = 8$ and $y = 2.5$?

22. Assume that m varies inversely as n. If $m = 6$ when $n = \frac{2}{3}$, what is the value for m when $n = 15$?

23. Assume that y varies jointly as w and x^2. If $y = 504$ when $w = 4$ and $x = 6$, what is the value of x when $w = 10$ and $y = 6,860$?

24. Assume that r varies jointly as s and t^2. If $r = 5,600$ when $s = 14$ and $t = 8$, what is the value of s when $t = 22$ and $r = 18,150$?

25. Assume that y varies directly as w and inversely as x. If $y = 4$ when $x = 10$ and $w = 6$, what is the value of y when $x = 15$ and $w = 3$?

26. Assume that p varies directly as q and inversely as r. If $p = 6$ when $q = 8$ and $r = 5$, what is the value of r when $p = 6$ and $q = 4$?

27. Assume that y varies jointly as x^2 and w and inversely as z. If $y = 15$ when $x = 2.5$, $w = 8$, and $z = 20$, what is the value of y when $x = 8$, $w = 7$, and $z = 14$?

28. Assume that d varies jointly as a^2 and b and inversely as c. If $d = 288$ when $a = 6$, $b = 10$, and $c = 4$, what is the value of d when $a = 20$, $b = 11$, and $c = 4$?

Applying What You've Learned

In Exercises 29–34, set up a proportion to solve the given problem.

29. **Calculating drug dosage.** The dosage of a particular drug is proportional to the patient's body weight. If the dosage for a 150-pound woman is 6 milligrams, what would the dosage be for her daughter Maria who weighs 65 pounds?

30. **Calculating a speed limit.** While volunteering in Africa, Brad and Angelina rented a car whose speedometer is calibrated in both miles and kilometers per hour. The marking on the speedometer

---

*Before doing these exercises, you may find it useful to review the note *How to Succeed at Mathematics* on page xix.

shows that 30 miles per hour corresponds to 48 kilometers per hour. If the speed limit in Africa is 100 kilometers per hour, what is the speed limit in miles per hour?

31. **Estimating a distance.** While driving from Kenya to Tanzania, Brad sees a road sign that says "Kilimanjaro 56 kilometers." How many miles does he still have to drive? (See Exercise 30.)

32. **Buying fertilizer.** Garth's front lawn is a rectangle measuring 120 feet by 40 feet. If a 25-pound bag of "Weed 'N Feed" will treat 2,000 square feet, how many bags must Garth buy to treat his lawn? (Assume that he must buy whole bags.)

33. **Mowing grass.** Caroline has a part-time job on campus mowing grass. If it takes her $1\frac{1}{2}$ hours to mow a 60,000-square-foot lawn, how long will it take her to mow the rectangular lawn in front of the student center that is 200 feet wide and 650 feet long?

34. **Cost of carpet.** Jose is a hotel manager who paid $864 to have a carpet installed in a conference room that measures 18 feet by 27 feet. How much would he have to pay to have the same carpet installed in a room that measures 24 by 33 feet?

Use the capture–recapture method in Exercises 35–38. Use the word equation stated in Example 2.

35. **Estimating a wildlife population.** Biologists capture, tag, and release 400 bald eagles. Several months later, of 240 bald eagles that are captured, 8 are tagged. Estimate the population of bald eagles.

36. **Estimating a wildlife population.** Marine biologists capture, tag, and release 100 Florida manatees. Several months later, of 90 captured manatees, 5 have tags. Estimate the population of Florida manatees.

37. **Estimating a wildlife population.** It is estimated that there are 1,000 grizzly bears living in a certain region. Assume that biologists capture, tag, and release 55 bears. Several months later, they capture a sample of 95 bears. How many would you expect to find tagged?

38. **Estimating a wildlife population.** A lake contains 1,530 largemouth bass. Employees of the state fish commission catch, tag, and release 60 of the bass. If two months later, they recapture 106 largemouth bass, how many would you expect to find tagged?

Exercises 39 and 40 are based on Example 5.

39. **Speeding.** If the speed limit is 60 miles per hour and your trip takes 2 hours, how much time do you save by traveling at 65 miles per hour?

40. **Speeding.** If the speed limit is 65 miles per hour and your trip takes 2 hours, how much time do you save by traveling at 75 miles per hour?

Exercises 41–44 are based on the strength of a beam model in Example 6.

41. **Strength of a beam.** If a beam that is 6 inches wide, 8 inches deep, and 4 feet long can support a weight of 672 pounds, how much weight could the same type of beam that is 4 inches wide, 6 inches deep, and 8 feet long support?

42. **Strength of a beam.** If a beam that is 3 inches wide, 10 inches deep, and 10 feet long can support a weight of 480 pounds,

how much weight could the same type of beam that is 4 inches wide, 6 inches deep, and 8 feet long support?

43. **Strength of a beam.** A beam that is 4 inches wide, 8 inches deep, and 12 feet long can support a weight of 1,280 pounds. If the same type of beam that is 3 inches wide and 6 inches deep can support a weight of 540 pounds, how long is it?

44. **Strength of a beam.** A beam that is 5 inches wide, 10 inches deep, and 16 feet long can support a weight of 875 pounds. If the same type of beam with the same length is 4 inches wide and can support a weight of 343 pounds, how deep is it?

45. **Estimating water usage.** Assume that the amount of water that Mario uses for irrigation at his vineyard is inversely proportional to the amount of rainfall. If he uses 30,000 gallons during a month in which there is 3 inches of rain, how much water would he use in a month that has 5 inches of rain?

46. **Determining the maximum strength of a spring.** Hooke's law states that the length a spring can be stretched is directly proportional to the force applied to the spring. However, if too much force is applied to the spring, the spring can be stretched to that point at which it can no longer return to its original shape. If a force of 8 pounds stretches a spring 6 inches and stretching the spring beyond 10 inches will ruin the spring, what is the maximum force that can be applied to the spring without damaging it?

47. **Finding the distance a skydiver falls.** The distance that a body falls varies directly as the square of the time that it is falling. If Eileen jumps from a plane and falls 144 feet in the first 3 seconds, how many feet will she fall in 5 seconds?

48. **Adjusting a photographer's lighting.** The illumination from a light source is inversely proportional to the square of the distance from the light source. Ansel is taking a portrait with his light source set 4 feet from his subject and finds that the illumination is twice as bright as it should be. If he wants to reduce the illumination to one-half of what it is now, at what distance should he place his light?

49. **Calculating the amount of heating oil needed.** Assume that the amount of heating oil used during a given month is inversely proportional to the outside temperature. If Andrea's pet store uses 504 gallons of oil in a month that has an average temperature of 42°F, how much oil would she expect to use in a month (of the same length) with an average daily temperature of 36°F?

50. **Determining waterpark attendance.** In the summer, the monthly attendance at Six Flag's waterpark varies directly as the temperature and inversely as the number of days of rain during the month. If during a given month the average daily temperature is 88°F and it rains 8 days, the total attendance for the month is 3,200. What attendance should we expect for a month (of the same length) if it rains 12 days and the average daily temperature is 92°F?

51. **Finding gas pressure.** If the temperature is held constant, the pressure of the gas in a container varies inversely as the volume of the container. Assume that the pressure in a large piston filled with gas is 4 pounds per square inch when the volume of

the gas is 120 cubic inches. If the volume of the gas in the piston is reduced to 75 cubic inches, what is the pressure of the gas?

52. **Finding gas pressure.** Redo Exercise 51. Assume now that the pressure in a large piston filled with gas is 12 pounds per square inch when the volume of the gas is 6.5 cubic inches. If the pressure of the gas in the piston is increased to 48 pounds per square inch, what is the volume of the gas?

Communicating Mathematics

53. If y varies directly as x and x increases, what does y do?

54. If y varies inversely as x and x increases, what does y do?

55. If y varies inversely as x, does x vary inversely or directly as y?

56. Assume that last semester at Urbanopolis City College, out of a student body of 5,250 there were 1,470 students who made the dean's list. If this semester the percentage who made the dean's list is the same and 1,554 made the dean's list, what is the current size of the student body at UCC? In solving this problem, Justin set up the equation

$$\frac{5{,}250}{1{,}470} = \frac{1{,}554}{x}.$$

What answer did Justin get and why is it obviously incorrect? Explain what is wrong with Justin's approach.

57. Suppose that in solving a problem, you encounter the ratio $m{:}n$. What would the ratio $n{:}(m + n)$ represent? Explain your answer.

58. What is $(m{:}n) \times (n{:}m)$? Explain. Give an example.

Using Technology to Investigate Mathematics

59. From your instructor, obtain a TI-83 program that implements the capture–recapture method. Use this program to reproduce some of the computations and solve some of the exercises in this section.

60. There are many sites on the Internet that discuss the capture–recapture method. You might search for the history of the method or some interesting applications. Find such a site and report on your findings.

For Extra Credit

61. If the ratio of PCs to Macintoshes on East Central State's campus is 7:2, what is the ratio of PCs to all computers on campus? (Assume that there are no other computers except PCs and Macs.) Explain your thinking. You may want to give an example to support your conclusion.

As you saw in Example 6, the strength of a beam is directly proportional to its width and the square of its depth and inversely proportional to its length. An equation describing this combined variation is

$$s = k\,\frac{w \cdot d^2}{l}.$$

Use this information to explain how the strength of a beam would change in each of the following situations. It may help if you make up concrete examples.

62. The width of the beam is doubled.

63. Both the width and the length of the beam are doubled.

64. Both the length and the depth of the beam are tripled.

65. The length, width, and depth of the beam are all doubled.

7.6 Functions

Objectives

1. Understand and use function notation.
2. Represent functions as sets of ordered pairs and graphically.
3. Use functions as models.

Have you noticed the rising gas prices? Let's suppose that you've been keeping a record of the relationship between the amount of your driving and its effect on your budget. Assume that your car currently gets 30 miles per gallon and the price of gasoline is $3.45 per gallon. We could model your cost of driving as follows

$$\text{cost} = \frac{\text{number of miles driven}}{\text{miles per gallon}} \times \text{cost per gallon}.$$

If we think of the cost per gallon, and the miles per gallon as not changing, we could rewrite this equation as

$$c = \frac{n}{30} \times 3.45 = n \times \frac{3.45}{30} = 0.115n,$$

where c represents the cost of driving and n is the number of miles driven. This tells you that your cost of driving is 11.5 cents per mile. As you took various trips, you might calculate your cost as in the following table:

Miles Driven	100	150	80	120	200
Cost	$0.115(100) = 11.5$	$0.115(150) = 17.25$	9.20	13.80	23

This table shows a *relationship* between the numbers of miles you drive and the cost to you. This idea of such relationships occurs frequently in mathematics.

KEY POINT

A function is a relationship in which each input has exactly one output associated with it.

Functions

So far, you've seen nothing new. The equation $c = 0.115n$ is a relatively simple linear equation. However, it is sometimes useful to think of this relationship between miles driven and your cost in a slightly different way. We will say that the previous table represents a **function** f, where f computes the cost of driving a certain number of miles. We will use the notation $c = f(n)$, and read this notation as "c equals f of n." This is called **function notation**, and it is important to realize that when we write $f(n)$ we are not multiplying f and n. The notation $f(n)$ simply tells us what number to associate with n. Another way to express that $c = f(n)$ is to say, "c is a function of n."

You can think of f as being a small computer that calculates your driving cost for you. If you give f the input of 100, then f gives back to you the output 11.5. If you give f the input 120, it gives back the output 13.80.

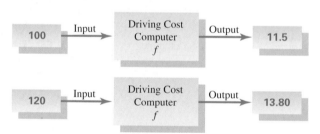

To say these facts using function terminology, we would say, "f of 100 equals 11.5," and "f of 120 equals 13.80." The following table will help you remember how to read this notation:

Function Notation	You Read As
$f(\text{Input}) = \text{Output}$	"f of Input = Output"
$f(100) = 11.5$	"f of $100 = 11.5$"
$f(120) = 13.8$	"f of $120 = 13.8$"

Quiz Yourself 🔢 **17**

How would you read the notation $f(80) = 9.20$?

Now try Exercises 5 to 8.   **17**

In the equation $c = f(n)$, the variable c depends on the variable n, so we call c the **dependent variable**. Because the variable n does not depend on anything else, n is called the **independent variable**. The following diagram summarizes what we have said so far:

The set of inputs that we can use with a function is called the **domain** of the function, and the set of associated outputs is called the **range**. In order for a relationship to be a function, *each input must have exactly one output* associated with it. We will now summarize what we have said so far.

> **DEFINITION** A **function** is a relationship that associates a unique range element with each element in its domain.

We can think of each of the following relationships as a function because in each case if we give one object (a domain element) to the function, it will return to us one object (a range element).

Input	Function	Output
Cost of a music DVD	Calculates tax and adds to purchase price	Final total cost of DVD
College basketball team that played in NCAA tournament	Finds ranking	Final ranking for the season
Number of courses you are taking this semester	Adds tuition plus other fees	Total bill you must pay this semester

EXAMPLE 1 *A Function Computing the Total Cost of a New Car*

Assume that when you purchase a new car, you must pay a 2.1% sales tax, a 1.3% clean-air emissions tax, a $470 dealer preparation fee, and a $165 state licensing fee. Assume that your car is on sale with a 5% discount on the price of the car. Assume taxes are paid on the base price of the car.

a) Write a function f that takes as input p, which is the base price of your car, and outputs t, the total price that you must pay for the car after taxes and fees.

b) Use your function to find the total cost of a new car that costs $23,000.

SOLUTION:

a) We will begin by writing a word equation that describes the function before we translate it into symbols.

$$\text{total cost} = (95\%)\text{ base price} + \text{sales tax} + \text{clean air tax} + \text{dealer preparation fee} + \text{state licensing fee}$$

If the base price is p and the total cost is t, the previous equation becomes

$$t = (0.95)p + (0.021)p + (0.013)p + 470 + 165.$$

price after 5% discount —⌡
2.1% sales tax —⌡
1.3% clean air tax —⌡
⌐— state license fee
⌐— dealer preparation fee

By combining like terms, we can simplify this to

$$t = (0.95 + 0.021 + 0.013)p + 635 = (0.984)p + 635.$$

We now write this equation as a function f:

$$t = f(p) = (0.984)p + 635.$$

This equation tells us that if the function f is given the input p, it will give us the corresponding value for t. It calculates the value for t by using the expression $(0.984)p + 635$.

b) The base price, p, of the car is 23,000, so

$$t = f(p) = f(23{,}000) = (0.984)(23{,}000) + 635 = 22{,}632 + 635 = 23{,}267.$$

"f of p" ⌐   ⌐ "f of 23,000"

Now try Exercises 9 to 16. ✳

🖉 **KEY POINT**

We can represent a function by a set of ordered pairs.

Representing Functions

Sometimes we like to think of a function as a set of ordered pairs. For example in our cost-of-driving example at the beginning of this section, we can think of the table of information as being the set of pairs

$$\{(100, 11.5), (150, 17.25), (80, 9.2), (120, 13.8), (200, 23)\}.$$

Here, 100, 150, and 80 are some of the domain elements of the function f, and 11.5, 17.25, and 9.2 are some of its range elements. This set represents some of the pairs that describe the function, but they do not represent the entire function. To describe the function f completely, we might say that $c = f(n) = 0.115n$, where $n = 0, 1, 2, 3, \ldots$. Now we are saying that the domain of f could be any nonnegative integer.

EXAMPLE 2   *Functions as Ordered Pairs*

Which of the following sets of ordered pairs represent functions?

a) $f = \{(1, 9), (2, 11), (3, 14), (4, 17), (5, 29)\}$
b) $g = \{(1, 2), (3, 4), (5, 6), (3, 8), (7, 9)\}$
c) $h = \{(1, 1), (2, 1), (3, 1), (4, 1)\}$

SOLUTION:

a) This is a function because for each domain element, we have only one range element corresponding to it.

b) This is not a function because the domain element 3 has both 4 and 8 associated with it.

c) This may surprise you, but it is a function. The domain element 1 has only one range element associated with it, the domain element 2 has only one range element associated with it, and so on. In a function, we are permitted to use the same range element with different domain elements, but we cannot use the same domain element with different range elements.

Now try Exercises 17 to 28. ✳ **18**

Quiz Yourself **18**

Which of the following sets of ordered pairs represent functions?

a) $f = \{(1, 7), (2, 4), (3, 5), (1, 8), (3, 6)\}$

b) $g = \{(2, 9), (5, 4), (3, 7), (4, 10), (8, 4)\}$

A set of ordered pairs such as b) in Example 2 is called a **relation**. A relation can have several range elements associated with one domain element, whereas a function cannot.

EXAMPLE 3   *Relations and Functions*

Explain why each of the following relations is *not* a function by giving one first element that is paired with at least two second elements.

a) $\{(x, y): x$ is a former U.S. president and y is his vice president$\}$*
b) $\{(x, y): y$ is a Major League Baseball team in city $x\}$
c) $\{(x, y): x$ won an Academy Award for movie $y\}$

SOLUTION:

a) Here we are looking for a U.S. president (a domain element) who had two vice presidents. One example would be Richard Nixon, whose vice presidents were Spiro Agnew and Gerald Ford.

---

*Recall that we read this set notation as "the set of all ordered pairs (x, y) such that x is a former U.S. president and y is his vice president."

b) The pairs (New York City, Yankees) and (New York City, Mets) are in this relation, so we have one domain element, New York City, associated with two different range elements, the Yankees and the Mets, so this is not a function.

c) We need an actor or actress who won Academy Awards for two different movies. Denzel Washington won an Oscar in 1989 for *Glory* and in 2002 for *Training Day*. If you want to go back a little further in movie history, Katherine Hepburn won four Oscars.

Now try Exercises 51 to 58. ❋

EXAMPLE 4 *Defining Functions Algebraically*

Determine which of the following define *y* as a function of *x*. If *y* is a function of *x*, call that function *f* and find $f(x)$ for the specified values of *x*. If the relation is not a function, give two ordered pairs to show that the function definition fails.

a) $y = 3x + 2$, $x = 5, 9, 0$
b) $y^2 = x^2$, $x = 4, 2, 0$
c) $y = \sqrt{x - 1}$, $x = 5, 0, 1$
d) $y \geq x + 5$, $x = 3, 2, 6$

SOLUTION:

a) This equation does define a function. If we substitute 5, 9, and 0 for *x*, we get

$$f(5) = 3 \cdot 5 + 2 = 17, \qquad f(9) = 3 \cdot 9 + 2 = 29, \qquad f(0) = 3 \cdot 0 + 2 = 2.$$

b) This equation does not define a function because both (4, −4) and (4, 4) are in this relation.

c) This equation does define a function. Substituting 5 for *x*, we get $f(5) = \sqrt{5 - 1} = \sqrt{4} = 2$. If we try to substitute 0 for *x*, we get $f(0) = \sqrt{0 - 1} = \sqrt{-1}$, which does not exist as a real number, so *0 is not in the domain of f*. If we substitute 1 for *x*, we get $f(1) = \sqrt{1 - 1} = \sqrt{0} = 0$.

d) This equation does not define a function. If $x = 3$, we can associate any number greater than or equal to 8, such as 9 or 10, with *x* because $9 \geq 3 + 5$ and $10 \geq 3 + 5$.

Now try Exercises 35 to 42. ❋

Sometimes we represent functions by diagrams such as those in Figure 7.20, which we will call *arrow diagrams*. Figure 7.20(a) represents a function because we are showing that each element in the domain {1, 2, 3, 4, 5} has only one range element associated with it. In Figure 7.20(b), the domain element 2 has both b and c associated with it, so this diagram does not represent a function.

Now try Exercises 29 to 34.

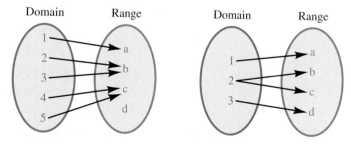

FIGURE 7.20 (a) This is a function. (b) This is not a function.

KEY POINT

We can think of a function graphically.

We can specify a function by graphing the pairs that make up the function as we do in the next example.

EXAMPLE 5   *Defining a Function Graphically*

What function *f* is given by the five points in Figure 7.21?

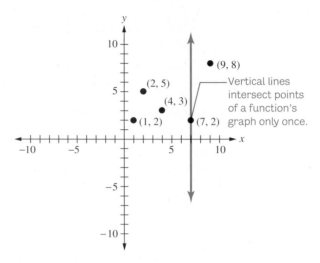

FIGURE 7.21 Graph of function *f*.

Quiz Yourself ⑲

In the graph in Figure 7.21, what is *f*(9)?

SOLUTION: The function *f* is the set of ordered pairs {(1, 2), (2, 5), (4, 3), (7, 2), (9, 8)}. Its domain is the set {1, 2, 4, 7, 9}, and its range is {2, 3, 5, 8}. ✳ ⑲

Notice in Figure 7.21 in Example 5, if you were to draw a vertical line through any point in the graph, it intersects points on the graph of the function only once. However, consider the graph in Figure 7.22 of the relation *g* = {(1, 2), (3, 4), (5, 6), (3, 8), (7, 9)} that you saw was not a function in Example 2.

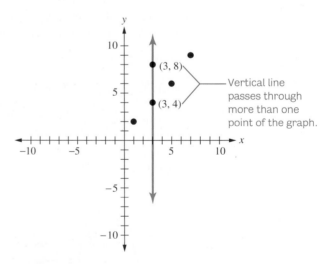

FIGURE 7.22 Not a function because a vertical line passes through more than one point on the graph.

The following test is a quick visual way for determining if a relation is a function.

> **VERTICAL LINE TEST**   If a vertical line passes through more than one point of the graph of a relation, then the relation is not a function.

EXAMPLE 6 *Using the Vertical Line Test*

Apply the vertical line test to decide which of the following graphs represent a function:

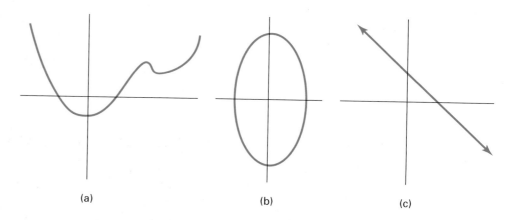

(a)   (b)   (c)

Quiz Yourself **20**

Does the relation {(2, 5), (3, 5), (4, 2), (2, 4), (1, 8)} pass the vertical line test?

SOLUTION: Graphs (a) and (c) are functions because if you draw any vertical line, it will only intersect the graphs once. Graph (b) is not a function because there are many places where a vertical line would intersect the graph twice.

Now try Exercises 43 to 46. ✳ **20**

✎ **KEY POINT**

Many models we studied earlier can be written as functions.

Modeling with Functions

By now you have probably realized that throughout this chapter you have been dealing with functions all along; however, we have not been using such a formal approach. Let's reconsider the linear equation $r = -0.6t + 40.6$, which we discussed in Example 2 in Section 7.2, that described the reduction in smoking over a period of years. We can now think of this relationship between time and the rate of smoking to be the following function f:

$$r = f(t) = -0.6t + 40.6.$$

To be complete, we really should specify the domain of this function, so let's say that we want this function to describe the reduction in smoking over a 15-year period, so the domain would then be the set {0, 1, 2, 3, . . . , 15}. Recall that in this example we wanted to know the smoking rate in year 13. Now, using the language of functions, we would express this as

$$f(13) = -0.6(13) + 40.6 = -7.8 + 40.6 = 32.8.$$

Notice that the computations we are doing are the same as before; however, we are now using a slightly more formal language.

A function that is defined by an equation of the form $f(x) = ax + b$, similar to the function f defined in our previous example, is called a **linear function**. Similarly, a function defined by an equation of the form $f(x) = ax^2 + bx + c$ is called a **quadratic function**. And, as you might guess, a function defined by the equation $f(x) = a \cdot b^x$ is called an **exponential function**.

Keep in mind that throughout this chapter you have really been working with functions, although we have not been using function notation. For another example, when we modeled the growth of the U.S. population in Example 2 of Section 7.4, instead of using the expression $301(1 + 0.00894)^n$ to represent the population in year n, we could have said that the population of the United States is modeled by the function defined by the equation $f(n) = 301(1 + 0.00894)^n$. Then if we wanted to know the population in 2053, which we said was year 45, we would write

$$P(1 + r)^n = 301(1 + 0.00894)^{45} = 301(1.00894)^{45} = 449.3 \text{ (million)}.$$

If we want to specify the domain of this function, we might indicate that $n = 0, 1, 2, 3, ...$

You will see more such applications of functions in Exercises 59 to 66.

Exercises 7.6

Looking Back*

These exercises follow the general outline of the topics presented in this section and will give you a good overview of the material that you have just studied.

1. In the equation $c = f(n)$ that we introduced at the beginning of this section, what is f? What does $f(n)$ mean? What is c?

2. What do you think is a common mistake that students make in reading the notation "$f(n)$?"

3. In Example 1, we defined the function by the equation $f(p) = (0.984)p + 635$. What is p? What is $f(p)$?

4. What pairs in the relation g in Example 2 cause g to fail the vertical line test?

Sharpening Your Skills

How would you read each of the following equations?

5. $y = f(x)$

6. $w = g(t)$

7. $g(x) = 8$

8. $f(3) = 14$

Evaluate the function for the given domain value.

9. $f(x) = 3x + 4, x = 2$

10. $g(t) = 4t^2 - 3, t = 3$

11. $h(a) = 3a^2 - 2a + 5, a = 5$

12. $f(x) = 2x^2 - 3x + 5, x = 3$

13. $g(t) = \sqrt{t - 4}, t = 40$

14. $h(a) = \sqrt{a - 5}, a = 30$

15. $f(x) = \sqrt{x - 9}, x = 5$

16. $g(t) = 5t - 2, t = 3$

Use the given table, which defines a function f, to answer the following questions. (Note: Some answers may not exist.)

x	y
3	5
4	-2
6	8
9	-14
7	21
5	8

17. What is the domain of f?

18. What is the range of f?

19. What is $f(6)$?

20. Find an x such that $f(x) = 21$.

21. Find an x such that $f(x) = -14$.

22. What is $f(7)$?

23. What is $f(8)$?

24. What is $f(9)$?

25. Find an x such that $f(x) = 7$.

26. Find an x such that $f(x) = 8$.

27. What is $f(4)$?

28. What is $f(21)$?

Draw an arrow diagram for each of the following relations. Which relations are functions?

29. $\{(2, 5), (3, 7), (3, 6), (7, 9), (1, 12), (2, 4)\}$

30. $\{(2, 5), (3, 9), (1, 4), (5, 6), (7, 9), (4, 1)\}$

31. $\{(3, 4), (4, 8), (2, 9), (1, 5), (6, 9), (5, 4)\}$

32. $\{(2, 5), (6, 1), (3, 2), (5, 5), (2, 8), (1, 9)\}$

33. $\{(7, 8), (3, 8), (4, 8), (5, 8), (6, 8)\}$

34. $\{(1, 2), (3, 1), (4, 1), (5, 1), (6, 1)\}$

Determine which of the following define y as a function of x. If y is a function of x, call that function f and find f(x) for the specified values of x. If the relation is not a function, give two ordered pairs to show that the function definition fails.

35. $y = 5x + 3, \quad x = 2, 0, -4$

36. $y \leq x + 8, \quad x = 4, 1, 7$

37. $y^2 = x^2, \quad x = 3, -1, 0$

38. $y = \sqrt{x - 5}, \quad x = 9, 0, 30$

39. $y = \sqrt{x + 3}, \quad x = 0, 6, -7$

40. $y \geq x - 4, \quad x = 3, 11, 0$

41. $y^2 = x^2 + 1, \quad x = 2, -1, 0$

42. $y = 3x - 5, \quad x = 1, 4, 7$

Which of the following graphs represent functions?

43.

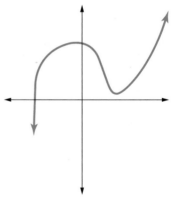

44.

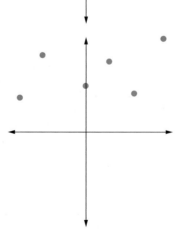

---

*Before doing these exercises, you may find it useful to review the note *How to Succeed at Mathematics* on page xix.

45.

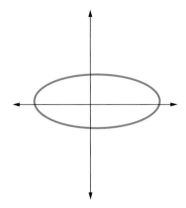

46.

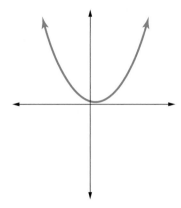

Rewrite each of the following equations to make the fact that y *is a function of* x *clearer. For example, you can rewrite* $y - x^2 = 2x + 3$ *as* $y = x^2 + 2x + 3$ *to say* $f(x) = x^2 + 2x + 3$.

47. $2x^2 = 5x - 3 + y$

48. $3x^2 = 2x - y$

49. $-3x^2 = -5x - y$

50. $-4x^2 = 2 - 3x + y$

Applying What You've Learned

Each line in the following table describes x *and some* y *that is related to* x. *Which of these relationships are functions?*

	x	**y**
51.	Any year since 1776	The population of the United States in that year
52.	A football team in the NFL	The number of victories the team had in 2009
53.	A movie	A star in the movie
54.	A country	The literacy rate of that country
55.	A person	A member of that person's immediate family
56.	A final grade in your math class	A person who received that grade
57.	A TV program	The night of the week that program airs
58.	A height in inches	A player in the WNBA who is that height

In Exercises 59–66, each equation was used to model a situation earlier in this chapter. Rewrite the function using function notation to define a function f, *and find the given quantity. We will use descriptive function names; for example, in Exercise 59, we will call the function* d *rather than* f.

59. $d = 104t + 64$; $d(14)$

60. $n = \dfrac{68}{2}t + 779$; $n(4)$

61. $y = 1.25t + 131.95$; $f(3)$

62. $H = 160 - 16t^2$; $H(5)$

63. $S = -13n^2 + 169n$; $S(6.5)$

64. $A = 1{,}000(1.08)^n$; $A(15)$

65. $A = 97n + 975$; $A(5)$

66. $A = 12(1.02)^n$; $A(35)$

Communicating Mathematics

67. Consider the function terminology: domain element, range element, dependent variable, and independent variable. Which of these terms mean the same thing?

68. What in a set of ordered pairs would cause that relation not to be a function?

69. Explain why if a relation fails the vertical line test, then that relation cannot be a function.

70. What is the difference between a relation and a function?

Using Technology to Investigate Mathematics

71. Investigate how to define functions on a calculator such as the TI-83, and then use it to duplicate some of the function computations in this section.

72. Investigate how to define functions using a spreadsheet such as Excel, and then use it to duplicate some of the function computations in this section.

73. Search the Internet for interactive applets that do function computations, and then use them to duplicate some of the function computations in this section.

For Extra Credit

If we think of functions as mathematical objects, we can perform operations on them similar to the way we operate on numbers. For example, if we have two functions f and g, we can define the sum of these functions, which we will call f + g, by saying that $(f + g)(x) = f(x) + g(x)$. This means that the name of the sum function is f + g, and the way we find the range element for a domain element x is to find the range elements for f and g separately and then add them together. For example, if $f(x) = 2x + 1$ and $g(x) = x^2 + 3$, then $(f + g)(x) = f(x) + g(x) = (2x + 1) + (x^2 + 3) = x^2 + 2x + 4$. We define the difference of two functions, called f − g, the product function f · g, and the quotient function f/g in a similar way. Of course, in the quotient function, we cannot use elements in the domain of f/g that would make the denominator equal to zero.

74. If $f(x) = 2x + 1$ and $g(x) = x^2 + 3$, what is $(f + g)(x)$ for $x = 0, 3, 5$?

75. Define what we mean by $(f − g)(x)$, and find $(f − g)(x)$ for $x = 1, 4, 6$.

76. Define what we mean by $(f · g)(x)$, and find $(f · g)(x)$ for $x = 0, 2, 4$.

77. Define what we mean by $(f/g)(x)$, and find $(f/g)(x)$ for $x = 0, 3, 5$.

78. If $h(x) = x − 5$, what is the meaning of $(h · (f + g))(x)$? Is $(h · (f + g))(x)$ the same as $(h · f + h · g))(x)$? Make examples to support your answer.

Looking Deeper

7.7 Dynamical Systems

Objectives

1. Understand dynamical systems as models.

2. Identify stable and unstable dynamical systems.

Why are weather predictions sometimes so unreliable? With the vast technology we have to gather and analyze data, it would seem possible to build mathematical models that are more reliable than they are at present. The problem in modeling weather is not only that it is much more complex than the situations we discussed in this chapter but also that many models of systems are unstable. This means that small changes in what is put into the model may result in very large changes in what comes out of the model.

We can use a coin to illustrate an unstable state. A coin can be in one of three resting states: head, tail, or on edge. If the coin is very close to being in the head state (perhaps the coin is held with one edge touching a table and the opposite side lifted $\frac{1}{10}$ of an inch), when the coin is released, it will go to the head state. There are obviously many states close to the head state such that if the coin is in one of these states the coin approaches the head state.

Contrast the head state with the edge state. If we place the coin almost on edge and release it, the coin will not move toward the edge state, but rather it will approach either the head state or the tail state. If a system is close to a *stable state*, then the system will move toward that state. In contrast, if a system is close to an *unstable state*, the system's behavior may be very unpredictable.

Dynamical Systems

The systems we study in this section are similar to the bank account and logistic growth models you studied in Section 7.4. In the banking example, knowing the amount in the account at time n allowed us to compute the amount at time $n + 1$. Similarly, in the logistic growth model, knowing the lemur population at time n enabled us to predict the population at time $n + 1$.

In both of these cases, we could think of the model as giving us a sequence of numbers where we can calculate each number in the sequence provided that we know the previous one. For the banking example, we started with a deposit of $1,000 and then for each successive year, the amount in the account was 1.08 times the amount in the account the previous year, as you can see below.

$$A_0 = 1{,}000, \quad A_1 = 1{,}080, \quad A_2 = 1{,}166.40, \quad A_3 = 1{,}259.71, \ldots, A_{n+1}, \ldots$$

| initial amount | $1.08 \cdot A_0$ | $1.08 \cdot A_1$ | $1.08 \cdot A_2$ | $1.08 \cdot A_n$ |

For the growth of the lemurs, we calculated the percentage of maximum population attained using the percentage attained the previous year, as follows:

$$P_0 = 0.20, \quad P_1 = 0.216, \quad P_2 = 0.233, \quad P_3 = 0.251, \ldots, P_{n+1}, \ldots$$

| initial percentage | depends on P_0 | depends on P_1 | depends on P_2 | depends on P_n |

In both examples, the amount that we calculated depended upon the amount present during the previous time period.

We can describe both the banking example and the logistic growth example by a mathematical model called a *dynamical system*.

KEY POINT

Each state of a dynamical system is determined by its previous state.

> **DEFINITION** A **dynamical system*** is a sequence of numbers A_0, A_1, A_2, A_3, and so on such that for each value of n,
>
> $$A_{n+1} = \text{an expression involving } A_n.$$

We now look at Example 1 from Section 7.4 as a dynamical system.

EXAMPLE 1 *A Bank Account as a Dynamical System*

The amount of $1,000 is placed in an account that is paying 8% interest compounded yearly. Model this situation by a dynamical system.

SOLUTION: From our earlier discussion, we see that the amount present in the account at the end of each year is 1.08 times as great as the amount present the previous year. We can put this a little more compactly by saying that

$$A_{n+1} = 1.08 \cdot A_n, \quad \text{for years } n = 0, 1, 2, 3, \ldots, \text{ where } A_0 = 1{,}000.$$

Notice that this dynamical system is really an infinite number of equations. ❀

Like the changes in a bank account and changes in the lemur population, if we had enough data and sufficiently complicated equations to describe weather changes, we could view the present weather as an initial state W_0. Then, W_1 would be the weather 1 hour later, W_2 the weather 2 hours later, and so on. If our model were accurate (and stable), then $W_{8,760}$ would be the state of the weather in 8,760 hours, or 1 year from now.

Just as equations can stand alone as mathematical objects without being explained in real-life terms, so can dynamical systems. It is common to identify a dynamical system with a rule that defines it. Here are two examples.

$$(1) \quad A_{n+1} = 5A_n + 3$$
$$(2) \quad A_{n+1} = 2A_n^2 + A_n - 5$$

System (1) is an example of a *linear system*, and system (2) is called *nonlinear*. Can you see why these names would be appropriate? Notice that a dynamical system is really an

---

*There are many different categories of dynamical systems, depending on what properties the system satisfies. For example, systems such that A_{n+1} depends only on A_n are sometimes called *first-order systems*. A *second-order system* would be one in which A_{n+1} depends on both A_n and A_{n-1}. We will not deal with such systems for now, and in order to make the discussion clearer, we have slightly simplified the terminology.

infinite number of equations. For example, system (1) corresponds to the collection of equations

$$A_1 = 5A_0 + 3,$$
$$A_2 = 5A_1 + 3,$$
$$A_3 = 5A_2 + 3,$$

KEY POINT

Dynamical systems model many phenomena.

and so on.

If you have ever been prescribed antibiotics by a physician, you have had personal experience with a dynamical system.

EXAMPLE 2 *Modeling Antibiotics in the Blood by a Dynamical System*

Suppose that your doctor has prescribed a 500-mg dose of an antibiotic to be taken three times a day. Assume that the drug enters your bloodstream instantly and that over an 8-hour period, your body eliminates 60% of the drug. Model the amount of drug in your blood by a dynamical system. Use this model to compute the amount of the drug in your system after three doses.

SOLUTION: Because you are taking the drug every 8 hours, each unit of time in our model corresponds to an 8-hour time interval. Let us represent the amount of drug present at time interval n by D_n. At $n = 5$, for example, we see that your body has eliminated 60% of the amount of drug that was present 8 hours earlier, so we see that before you take your next dose, your blood contains

We are beginning the fifth interval.

60% of the drug present at the beginning of the fourth time interval has been eliminated.

$$D_5 = 0.40(D_4).$$

amount present at beginning of fourth time interval

However, you now take a dose of the drug that puts 500 mg of the drug into your body. So we get

new dose

$$D_5 = 0.40(D_4) + 500$$

drug that was remaining in your blood

What is true for 5 is true for any time, so we see that

$$D_{n+1} = 0.40(D_n) + 500.$$

The only question that we still must answer is, what should be the value of D_0? If we want our first dose to correspond to time 1, the second to time 2, and so on, then we could say that at time 0, which is 8 hours before we took our first dose, there is no drug in the bloodstream. Therefore,

$$D_{n+1} = 0.40(D_n) + 500, \qquad n = 0, 1, 2, \ldots,$$

with the initial value $D_0 = 0$, describes this dynamical system.

We now calculate D_3, but to do this, we need to first compute D_0, D_1, and D_2.

$$D_0 = 0,$$
$$D_1 = 0.40(D_0) + 500 = 0.40(0) + 500 = 500,$$

Redo Example 2, but now assume that the body eliminates 80% of the drug between doses and that each dose is 400 mg.

KEY POINT

Once a dynamical system reaches an equilibrium value, it stays at that value.

n	D_n (rounded to one decimal place)
1	500
2	700
3	780
4	812
5	824.8
6	829.9
7	832.0
8	832.8
9	833.1
10	833.2
11	833.3
12	833.3

$$D_2 = 0.40(D_1) + 500 = 0.40(500) + 500 = 700,$$
$$D_3 = 0.40(D_2) + 500 = 0.40(700) + 500 = 780.$$

Now try Exercises 5 to 10. ❄ ㉑

Equilibrium Values and Stability

In Example 2, the amount of drug in the bloodstream is increasing, and we might wonder what are the effects of taking the drug for a long period of time. Is there the possibility that we could become poisoned by the amount of drug in the bloodstream? To look at this question more closely, we generate a table of values for D_n.

We see that as n gets larger, D_n appears to be getting closer to 833.3. In fact, as you will see shortly, if D_n ever equals $\frac{5,000}{6} = 833.3333\ldots$, then from that point on, all subsequent values of D_n will also equal $833.333\ldots$.

The value $\frac{5,000}{6}$ is an example of an equilibrium value for the dynamical system in Example 2.

> **DEFINITION** An **equilibrium value** for a dynamical system is a number a such that if $A_n = a$, then A_{n+1} will also equal a.

Equilibrium values for a system are important because they frequently tell us about the long-term behavior of the system. It is easy to find equilibrium values for some systems. For example, consider the system $A_{n+1} = 5A_n + 3$. To say that both $A_n = a$ and $A_{n+1} = a$, by substitution, means that $a = 5a + 3$. When we solve this for a, we get $a = -\frac{3}{4}$ as an equilibrium value for this system.

> **FINDING EQUILIBRIUM VALUES FOR DYNAMICAL SYSTEMS** To find an equilibrium value a for a dynamical system, we do the following:
>
> 1. Write the equation $A_{n+1} =$ an expression involving A_n.
> 2. Substitute a for both A_n and A_{n+1} in the equation in step 1.
> 3. Solve the equation in step 2 for a.

KEY POINT

An equilibrium value may be unstable.

An equilibrium value for a system may or may not be stable. You saw in the drug example that the numbers seemed to get closer and closer to $833.333\ldots$ and, as you will see in a moment, this is indeed what is happening. In Example 3, we look at another, similar dynamical system, but you will see very different behavior.

EXAMPLE 3　*A Nonstable Equilibrium Value for a Dynamical System*

a) Find an equilibrium value for the dynamical system $A_{n+1} = 4A_n - 5$.

b) Is this equilibrium value stable?

SOLUTION:

a) Substituting a for A_n and A_{n+1} in the equation $A_{n+1} = 4A_n - 5$, we get $a = 4a - 5$. Solving this equation, we find that $a = \frac{5}{3} \approx 1.7$.

b) We now generate a table of values for A using two values for A_0 that are extremely close to $\frac{5}{3}$ (which equals $1.666666\ldots$). The first value for A_0 is 1.66, which is slightly less than $\frac{5}{3}$; the second value for A_0 is 1.67, which is just a little bit greater than $\frac{5}{3}$.

n	A_n	A_n	A_n
0	$\frac{5}{3}$	1.66	1.67
1	$\frac{5}{3}$	1.64	1.68
2	$\frac{5}{3}$	1.56	1.72
3	$\frac{5}{3}$	1.24	1.88
4	$\frac{5}{3}$	$-.04$	2.52
5	$\frac{5}{3}$	-5.16	5.08
6	$\frac{5}{3}$	-25.64	15.32
7	$\frac{5}{3}$	-107.56	56.28
8	$\frac{5}{3}$	-435.24	220.12
9	$\frac{5}{3}$	$-1,745.96$	875.48
10	$\frac{5}{3}$	$-6,988.84$	3,496.92

Quiz Yourself 22

Find equilibrium values for the given dynamical systems and determine whether these equilibrium values are stable.

a) $A_{n+1} = 3A_n - 1$

b) $B_{n+1} = 0.5B_n - 3$

Notice that although we take values very close to $\frac{5}{3}$ as initial values A_0 for this system, we see that by the time we compute A_{10}, the values we are getting are very far away from the initial value. Thus, the equilibrium value $\frac{5}{3}$ is *not stable*. Much like the coin standing on edge, if we begin even a little bit away from $\frac{5}{3}$, we wind up with values that are quite far away from $\frac{5}{3}$.

Now try Exercises 11 to 14. ✸ 22

There is a theorem, which we will state without proof, that tells us when an equilibrium value is stable and when it is not.

STABILITY OF EQUILIBRIUM VALUES FOR DYNAMICAL SYSTEMS
Suppose that a is an equilibrium value for the system

$$A_{n+1} = mA_n + b.$$

1. If $-1 < m < 1$, then the equilibrium value a is stable.
2. If $m < -1$ or $m > 1$, then a is unstable.
3. If $m = -1$, then the values for A_n will oscillate between two values.

Dynamical systems have many applications. An interesting note about one of the causes of war was given by Lewis F. Richardson in his book *Arms and Insecurity: A Mathematical Study of the Causes and Origins of War.** Example 4 is based on Richardson's work.

EXAMPLE 4   *Modeling an Arms Race with a Dynamical System*

In the early part of the twentieth century, France-Russia and Germany-Austria-Hungary were opposing alliances. According to Richardson, the dynamical system[†]

$$D_{n+1} = \frac{5}{3}D_n - \frac{380}{3}, \qquad \text{where } D_0 = 199,$$

---

*L. F. Richardson, *Arms and Insecurity: A Mathematical Study of the Causes and Origins of War* (Boxwood Press: Pacific Grove, CA, 1960).
[†]For a derivation of this system, see James T. Sandefur, *Discrete Dynamical Systems: Theory and Applications* (Oxford University Press: New York, 1990), p. 76.

Year = n	Amount Spent on Defense = D_n
0	199
1	205
2	215
3	231.7
4	259.5
5	305.7
6	382.9
7	511.5
8	725.8

TABLE 7.3 Growth in defense spending, according to Richardson's model of the European arms race.

models total annual defense spending between the two alliances beginning in 1909 (year 0 in our model).

a) Find an equilibrium value for this system.

b) Is this equilibrium value stable?

c) What does this model predict will happen as n increases?

SOLUTION:

a) We solve the equation $a = \frac{5}{3}a - \frac{380}{3}$ to find an equilibrium value. Multiplying this equation by 3, we get $3a = 5a - 380$. This simplifies to $380 = 2a$, so the equilibrium value is 190.

b) Because the coefficient of D_n is $\frac{5}{3}$, which is greater than 1, the equilibrium value 190 is unstable.

c) The equilibrium value of 190 is not stable. Therefore, if we begin with the value $D_0 = 199$, as n increases, the values for D_n might be quite far away from 190. We calculate some values for D_n in Table 7.3.

You can see in Table 7.3 that as n increases, D_n (the amount spent on defense) also increases rapidly. Because it is impossible for both sides to increase defense expenditures indefinitely, eventually one or the other will be in a position where it feels threatened and will declare war. ❦

Exercises 7.7

Looking Back*

These exercises follow the general outline of the topics presented in this section and will give you a good overview of the material that you have just studied.

1. How are Examples 1 and 2 similar?

2. What is an equilibrium value for a dynamical system? What was an equilibrium value in Example 2?

3. Find an example of a stable equilibrium value and a nonstable equilibrium value in this section.

4. How did we know that the dynamical system in Example 4 was unstable?

Sharpening Your Skills

For each dynamical system, calculate the value of A_n.

5. $A_{n+1} = 2A_n - 1$, $A_0 = 3$; find A_1 and A_2.

6. $A_{n+1} = 3A_n + 2$, $A_0 = 1$; find A_1 and A_2.

7. $A_{n+1} = -3A_n + 4$, $A_0 = -2$; find A_3.

8. $A_{n+1} = 2.5A_n - 3$, $A_0 = 2$; find A_3.

9. $A_{n+1} = 1.8A_n - 2$, $A_0 = 4$; find A_4.

10. $A_{n+1} = -0.8A_n - 2$, $A_0 = 1.5$; find A_4.

Find the equilibrium value for each dynamical system. Comment on whether the value you find is stable or unstable.

11. $A_{n+1} = 2A_n + 3$ 12. $A_{n+1} = 4A_n - 5$

13. $B_{n+1} = 0.25B_n + 4$ 14. $B_{n+1} = 0.10B_n - 2$

Applying What You've Learned

Model each situation with a dynamical system. Use your model to answer the question.

15. **Compound interest.** You deposit $1,000 in a bank account paying 5% yearly interest that is compounded annually. How much will be in your account at the end of 2 years?

16. **Compound interest.** You deposit $1,500 in a bank account paying 6% yearly interest that is compounded annually. How much will be in your account at the end of 3 years?

17. **Wildlife growth.** An island is populated with lemurs that have a growth rate of 8%. The island is initially populated with 30% of the island's capacity to sustain these lemurs. What percentage of the lemur population's maximum capacity will be attained at the end of 2 years?

18. **Wildlife growth.** Repeat Exercise 17, but now assume the growth rate is 12% and the island is initially populated with 20% of its maximum capacity.

19. **Antibiotic level.** You take a 250-mg dose of an antibiotic every 4 hours. Your body eliminates 40% of the drug in a 4-hour period. How much antibiotic will be in your bloodstream after three doses?

20. **Antibiotic level.** You take a 1,000-mg dose of an antibiotic every 12 hours. Your body eliminates 75% of the drug in a 12-hour period. How much antibiotic will be in your bloodstream after three doses?

---

*Before doing these exercises, you may find it useful to review the note *How to Succeed at Mathematics* on page xix.

Communicating Mathematics

21. What do we mean when we say that a dynamical system is really an infinite number of equations?

22. What did we mean when we said that the equilibrium value in Example 3 was not stable?

23. If a is an equilibrium value for the dynamical system $A_{n+1} = mA_n + b$, how do you tell whether a is stable?

24. What did the fact that the dynamical system in Example 4 was unstable predict?

Using Technology to Investigate Mathematics

25. Ask your instructor for a tutorial for working with dynamical systems on a graphing calculator, and then use your calculator to reproduce some of the calculations in this section.

26. In this section, we have barely scratched the surface of dealing with dynamical systems. Do a search on the Internet and download some applets that illustrate dynamical systems. Write a report describing what you have found.

27. Search the Internet for applications of dynamical systems. Write a report describing what you have found.

For Extra Credit

Living plants and animals all contain the chemical element carbon. A certain percentage of that carbon is radioactive, and scientists believe that the percentage has remained constant for thousands of years. Radioactive carbon decays, so that when an animal dies, a tiny bit of the radioactive carbon is lost each year. It is known that the amount of radioactive carbon that remains in a fossil at the end of a year is approximately 0.99988 of the amount that was present at the beginning. Thus, the following dynamical system describes radioactive carbon decay in a fossil:

$$C_{n+1} = 0.99988 \cdot C_n \qquad \text{for } n = 0, 1, 2, 3, \ldots.$$

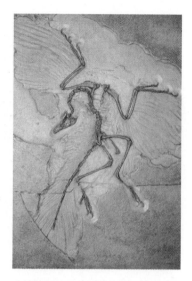

This system behaves exactly like the compound interest situation (except the amount of radioactive carbon is decreasing), so it is easy to see that after k years, the amount of radioactive carbon in the fossil will be $C_k = 0.99988^k \cdot C_0$. We will always assume that the amount of radioactive carbon at time 0 will be 1; that is, $C_0 = 1$. Use this model to answer Exercises 28 and 29. You can solve these equations using the log function, as you did in Section 7.4.

28. Carbon dating. A fossilized bone is found that contains 90% of the original radioactive carbon that was present. To the nearest 100 years, how old is the bone?

29. Carbon dating. A leaf of a fossilized plant is found that contains 60% of the original radioactive carbon that was present. To the nearest 100 years, how old is the plant?

Assume that the logistic equation $P_{n+1} = [1 + r(1 - P_n)]P_n$ models the growth of wild turkeys on a large parcel of state game land. If we want to allow hunting on this land, we adjust this equation by subtracting some number from P_{n+1} to account for the turkeys killed by hunting. In the following questions, suppose that the game land is capable of supporting a maximum of 1,000 turkeys.

30. Managing wildlife. Assume that currently there are 500 turkeys and we want to allow 80 to be harvested per year. The growth rate of the turkeys is 10% per year. Modify the growth equation accordingly, and use it to predict the turkey population at the end of 3 years.

31. Managing wildlife. Assume that currently there are 750 turkeys and we want to allow 100 to be harvested per year. The growth rate of the turkeys is 10% per year. Modify the growth equation accordingly, and use it to predict the turkey population at the end of 3 years.

CHAPTER SUMMARY

SECTION	SUMMARY	EXAMPLE
SECTION 7.1	A **linear equation** in two variables, written in **standard form**, has the form $Ax + By = C$. A **solution** for an equation is a number (or numbers) such that if we substitute them for the variables in the equation, the resulting statement is true. Two equations are **equivalent** if they have the same solutions. In **solving equations**, we can add or subtract the same expression from both sides of the equation. We can also multiply or divide both sides of the equation by the same *nonzero* expression.	Definition, p. 297 Discussion, p. 298 Example 1, p. 298
	The **x-intercept** (where $y = 0$) of the graph of a linear equation is the point where the graph crosses the *x*-axis. The **y-intercept** (where $x = 0$) of the graph of a linear equation is the point where the graph crosses the *y*-axis. We can use the *x*- and *y*-intercepts to graph a linear equation.	Example 2, p. 300
	If (x_1, y_1) and (x_2, y_2) are two points on a line and $x_1 \neq x_2$, then the **slope** of the line, m, is defined as $m = \dfrac{\text{rise}}{\text{run}} = \dfrac{y_2 - y_1}{x_2 - x_1}$. A linear equation is in **slope–intercept** form if it is written in the form $y = mx + b$, where m is the slope and b is the *y*-intercept of the line.	Example 4, p. 302 Definition, p. 303 Example 5, p. 303
SECTION 7.2	We can find a linear equation if we know the slope of its graph and one point on its graph.	Examples 1 and 2, p. 309
	We can use two points to find a linear equation by first using the points to find the slope and then applying an earlier technique.	Examples 3 and 4, pp. 310, 311
	We can use a graphing calculator to find the **line of best fit** for a set of data points.	Discussion, p. 313 Highlight, p. 313
SECTION 7.3	A **quadratic equation** is an equation in the form $y = ax^2 + bx + c$, where a, b, and c are real numbers and $a \neq 0$. A **solution** of a quadratic equation is a pair of numbers that satisfies the equation.	Discussion, p. 316 Example 1, p. 317
	The graph of a quadratic equation is a **parabola**. The lowest or highest point on a parabola is called its **vertex**. The vertex of the graph of $y = ax^2 + bx + c$ occurs when $x = \dfrac{-b}{2a}$. The **solution(s)** of the quadratic equation $ax^2 + bx + c = 0$ are	Discussion, p. 317 Example 2, p. 317
	$$x = \frac{-b \pm \sqrt{b^2 - 4ac}}{2a}.$$	Example 3, p. 318
	In the quadratic equation $y = ax^2 + bx + c$, if $a > 0$, then the graph of the parabola opens up. If $a < 0$, then the graph of the parabola opens down. We **graph** a quadratic equation by 1. Determining if the parabola opens up or down, 2. Finding its vertex, 3. Finding the *x*- and *y*-intercepts of the graph.	Example 4, p. 319
	We can use a quadratic equation to **model** data.	Example 5, p. 320
SECTION 7.4	The **compound interest formula** says that money grows in a bank account according to the formula $A = P(1 + r)^n$, where P is the principal, r is the yearly interest rate, and n is the number of years.	Example 1, p. 326
	We can use an exponential equation to model **population growth**.	Example 2, p. 328
	The **log function** reverses the operation of raising 10 to a power. The log function has the property that $\log y^x = x \log y$	Examples 3 and 4, p. 329
	A **logistic model** takes into account limits of population growth. The equation $P_{n+1} = [1 + r(1 - P_n)]P_n$ describes logistic growth.	Definition, p. 331 Examples 6 and 7, pp. 332, 333

SECTION 7.5	A **ratio** is a quotient of two numbers. A **proportion** is a statement that two ratios are equal. The **cross-multiplication principle** says that if $\frac{a}{b} = \frac{c}{d}$, then $a \cdot d = b \cdot c$. The quantities $a \cdot d$ and $b \cdot c$ are called **cross products**.	Discussion, p. 336 Example 1, p. 336
	The **capture–recapture method** is used for estimating the size of populations.	Example 2, p. 338
	We say that y **varies directly** as x, or that y is directly proportional to x, if $y = kx$, where k is a nonzero constant, called the **constant of variation** or the **constant of proportionality**. Joint variation is direct variation in which there are several quantities.	Definitions, p. 338 Example 3, p. 338 Example 4, p. 339
SECTION 7.6	A **function** is a relationship, often given by an equation, in which each *input* has *exactly one output* associated with it. The input, called an **independent variable**, determines a unique output, called a **dependent variable**. The set of inputs to the function is called its **domain**, and the set of outputs is called its **range**.	Discussion, p. 344–345 Example 1, p. 345
	We can often represent a function as a set of **ordered pairs**. A **relation** is a set of ordered pairs; however, it may or may not be a function. The **vertical line test** says that if a vertical line passes through *more than one* point of a relation, then the relation is *not* a function.	Examples 2 and 3, p. 346 Examples 5 and 6, pp. 348, 349
	We can write functions to model many of the situations that we described earlier with equations.	Discussion, p. 349
SECTION 7.7	A **dynamical system** is a sequence of numbers $A_0, A_1, A_2, A_3,$ and so on such that for each value of n, $A_{n+1} =$ an expression involving A_n.	Discussion, p. 352 Examples 1 and 2, pp. 353–354
	An **equilibrium value** for a dynamical system is a number a such that if $A_n = a$, then A_{n+1} will also equal a. An equilibrium value may be **stable** or **unstable**.	Discussion, p. 355 Example 3, p. 355

CHAPTER REVIEW EXERCISES

Section 7.1

1. Solve the following equations:

    a. $\frac{2}{3}x + 2 = \frac{1}{6}x + 4$ **b.** $0.3x - 2 = 3.5x - 0.4$

2. Solve $A = P(1 + rt)$ for r.

3. Darryl works at a warehouse job where he is paid $5 per hour up to 40 hours. If he works over 40 hours, he is paid twice his usual wage. Assume that he always works at least 40 hours per week.

    a. Model this situation with a linear equation.

    b. How much will he earn if he works 46 hours in a given week?

4. Graph $3x + 5y = 20$ by plotting the intercepts and drawing a line through them.

5. Find the slope of the line passing through (2, 5) and (6, 8).

6. Match the line in the diagram with the given information regarding its slope. Assume that the scale on both axes is the same.

    Line 1: positive slope; less than 1

    Line 2: negative slope; between −1 and 0

    Line 3: slope does not exist

    Line 4: slope greater than 1

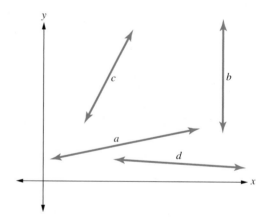

7. Nugyen is considering two satellite TV systems. Global Communications charges $240 for installation and $39 per month. World Communications charges $350 for installation and $28 per month. After how many months will World Communications be the better buy?

Section 7.2

8. Find an equation of the line with slope −4 that passes through (2, 5).

9. Find a linear equation whose graph passes through (3, 4) and (6, 9).

10. The amount spent by Americans on foreign travel increased from $81.8 billion in 2002 to $99.9 billion in 2005.

  a. Model this information with a linear equation.

  b. Use your model to estimate the amount that will be spent by Americans on foreign travel in 2015.

11. What do we mean by the line of best fit?

12. Explain when it is appropriate to use a linear function as a model.

Section 7.3

13. Solve the quadratic equation $2x^2 + 7x = 4$.

14. Answer the following questions for the graph of the equation

$$y = -2x^2 - 10x - 8.$$

  a. Is its graph opening up or down?

  b. What is the vertex of the graph?

  c. What are the x-intercepts?

  d. What is the y-intercept?

  e. Graph the equation.

15. What is quadratic regression?

16. Assume that the equation $A = -0.125x^2 + 2x + 1.125$ models attendance of a recently released movie for x weeks. During what week will the movie attain its highest attendance?

Section 7.4

17. If $10,000 is placed in an investment paying 4.8% yearly interest compounded annually, how much will be in the account after 5 years?

18. How long will it take for the account in Exercise 17 to double?

19. Assume that a population is growing initially at a 3% rate and that we are using a logistic growth model to describe the population growth. If $P_3 = 0.50$, what is P_4?

20. Assume that a room at a hotel in Disney World costs $220 a night in 2004 and that same room (due only to inflation) costs $280 a night in 2008. Use an exponential model to estimate the cost of that room in 2012.

21. What is the difference between an exponential model and a logistic model?

Section 7.5

22. Solve for x in the following proportions:

  a. $25:8 = x:2$ **b.** $\dfrac{30}{4} = \dfrac{x}{5}$

23. If it requires 3.5 gallons of sealer to coat a 840 square-foot driveway, how much sealer will be needed to coat a 1,500 square-foot driveway?

24. Marine biologists capture, tag, and release 180 penguins. Several months later, of 55 penguins that are captured, 12 are tagged. Estimate the population of penguins.

25. Assume that y varies inversely as x. If $y = 14$ when $x = 3$, what is the value of y when $x = 18$?

26. Assume that d varies jointly as a^2 and b, and inversely as c. If $d = 144$ when $a = 3$, $b = 7$, and $c = 14$, what is the value of d when $a = 8$, $b = 11$, and $c = 16$?

27. The strength of a beam is directly proportional to its width w and the square of its depth d, and inversely proportional to its length l. How would the strength of the beam change if the width and length are doubled?

Section 7.6

28. Evaluate the functions for the given domain value.

  a. $f(x) = 2x - 3$, $x = 5$ **b.** $g(t) = 2t^2 + 4$, $t = 6$

29. Use the set of ordered pairs $\{(2, 6), (3, 8), (1, 4), (5, -1), (7, 8), (4, -2)\}$, which defines a function f, to answer the following questions:

  a. What is the domain of f?

  b. What is the range of f?

  c. What is $f(5)$?

  d. Find x such that $f(x) = 8$.

30. Which of the following define y as a function of x?

  a. $y = 3x - 4$ **b.** $y \le x + 3$ **c.** $y^2 = x^2$ **d.** $y = \sqrt{x - 4}$

31. Which of the following graphs represent functions?

  a.

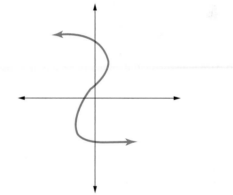

  b.

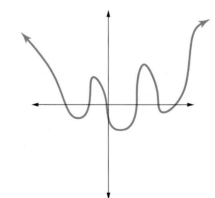

32. Which of the following define y as a function of x?

  a. The domain element x is any person in the United States. The range element y is that person's annual income.

  b. The domain element x is any person currently enrolled at your school. The range element y is the number of credits that person has this current semester.

Section 7.7

33. Calculate A_1, A_2, A_3 for the dynamical system $A_{n+1} = 4A_n + 3$, where $A_0 = 6$.

34. Find an equilibrium value for the dynamical system in Exercise 33. Is that value stable or unstable?

CHAPTER TEST

1. Solve the following equations:

  a. $\frac{3}{4}x + 5 = \frac{2}{3}x + 4$ **b.** $0.25x + 4 = 1.5x - 0.2$

2. Solve $X = a(1 + b)$ for b.

3. Find the slope of the line passing through the points $(3, 5)$ and $(8, 12)$.

4. Graph $5x - 4y = 10$ by plotting the intercepts and drawing a line through them.

5. Brandon subscribes to a cell phone plan that charges $30 per month for 1,200 minutes and $0.045 per minute for each minute over 1,200.

  a. Model Brandon's plan with a linear equation.

  b. If he talked on his cell phone 1,520 minutes, what was his bill?

6. Assume that a Wilson mini-autographed basketball cost $16 in 2002, and in 2008 (due only to inflation) that cost rose to $18.50. Use an exponential model to predict the cost of that same basketball in 2015.

7. Match the terms in the left column with those in the right column in the following table:

1. Positive slope	a. Line is falling from left to right.
2. No slope	b. Line is rising from left to right.
3. Negative slope	c. Line is horizontal.
4. Slope is zero.	d. Line is vertical.

8. Wilfredo is considering two DVD rental plans. Fliks 'R Us costs $40 to join and charges $1.50 for each DVD rented. DVD Mania costs $22 to join and charges $2 per rental. At what point is Fliks 'R Us a better deal?

9. When is it appropriate to use a linear equation as a model?

10. Find an equation of the line that passes through the points $(2, 5)$ and $(6, 8)$.

11. Model the following situation with a linear equation, but do not try to solve it. Shanaya has two part-time jobs on campus. Her job at the dining hall pays $6.30 per hour, and her work in the cognitive psychology lab pays $8.25 per hour. She earns $137.70 in a given week.

12. What is linear regression?

13. There were 36.5 million people living in poverty in the United States in 2008, up 1.8 million from 2002. Model this information

with a linear equation, and use this model to predict the number living in poverty in 2020.

14. Calculate A_1, A_2, A_3 for the dynamical system $A_{n+1} = 4A_n + 2$, where $A_0 = 5$.

15. Find an equilibrium value for the dynamical system in Exercise 14. Is that value stable or unstable?

16. Answer the following questions for the graph of the equation
$$y = -x^2 + 11x - 24.$$

  a. Is the graph opening up or down?

  b. What is the vertex of the graph?

  c. What are the x-intercepts?

  d. What is the y-intercept?

  e. Graph the equation.

17. If $5,000 is placed in an investment paying 2.4% yearly interest compounded annually, how much will be in the account after 8 years?

18. How long will it take the account in Exercise 17 to double?

19. What is the formal name for finding the parabola of best fit?

20. Assume that the equation $S = -1.2x^2 + 8x$ models the sales of a recently released music CD, where x is the number of weeks since the release. During what week will sales reach their highest point?

21. Use the set of ordered pairs $\{(1, 6), (2, 8), (8, 4), (5, -3), (7, 8), (4, -2)\}$, which define a function f, to answer the following questions:

  a. What is the domain of f?

  b. What is the range of f?

  c. What is $f(5)$?

  d. Find x such that $f(x) = 8$.

22. Evaluate the functions for the given domain value.

  a. $f(x) = 3x - 8, x = 5$ **b.** $g(t) = 5t^2 - 4, t = 6$

23. You are taking a 400-mg dose of a pain reliever every 12 hours. Your body eliminates 65% of the drug in 12 hours. How much of the drug will be in your system after four doses?

24. Assume that a population is growing initially at a 2% rate and that we are using a logistic model to describe the population growth. If $P_8 = 0.40$, what is P_9?

25. Why do we use the rate reduction factor in a logistic model?

26. Census takers are using the capture–recapture method to count the number of homeless people in a large city. They first identify

35. You are taking an 800-mg dose of an antibiotic every 8 hours. Your body eliminates 50% of the drug in 8 hours. How much of the drug will be in your system after three doses?

75 people as being homeless and several weeks later, in another survey of 120 homeless people, 5 are from the first group of homeless that was identified. How many homeless people are there in the city?

27. Which of the following define y as a function of x?

    a. The domain element x is any person in the United States. The range element y is that person's favorite ice cream flavor.

    b. The domain element x is any person currently enrolled at your school. The range element y is a course that person is taking this semester.

28. Solve for x in the following proportions.

    a. $40:6 = x:9$ **b.** $\dfrac{8}{3} = \dfrac{x}{10}$

29. If the pitcher Curt Shilling gets 4 hits in his first 35 at bats and continues at that rate, how many hits can we expect if he bats 90 times during the season?

30. The strength of a beam varies directly as the square of its depth d and inversely as its length l. How would the strength of a beam change if its depth and length were both doubled?

31. Assume that y varies inversely as x. If $y = 8$ when $x = \frac{2}{3}$, what is the value of y when $x = 4$?

32. Assume that s varies jointly as x and y and inversely as t. If $s = 16$ when $x = 2$, $y = 4$, and $t = 6$, what is the value of s when $x = 3$, $y = 5$, and $t = 9$?

33. Which of the following define y as a function of x?

    a. $y \geq x + 2$ **b.** $y = 5x - 4$ **c.** $y^2 = x^2$ **d.** $y = \sqrt{x + 5}$

34. Which of the following graphs represent functions?

    a.

    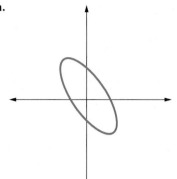

    b.

    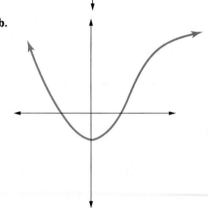

GROUP EXERCISES

1. Use an almanac or the Internet to gather data that you would like to model using the methods we have discussed in this chapter. Use a calculator and different models to see which best fits the data. One way of determining how well the model fits the data is to consider the correlation coefficient r (see Section 15.5). Your calculator should calculate this for you. The value for r will be between -1 and 1. The closer r is to 1, the better your model fits the data. It is particularly interesting if you use some of the data that you gathered to build the model and then see how well your model fits the data. For example, you could use data from the years 2002 to 2006 and then see how well your model predicts the existing data in 2008 and 2009. See who in your group can build the best model for the data.

2. Many graphing calculators have functions that allow you to find the solutions of equations. On some calculators this is called the *solve* command, or *solver*. You can also solve equations by using a function that shows the coordinates of points as you trace the graph of the equation. Find out how to do this by reading your calculator manual; perhaps your instructor can point you in the right direction. Use your calculator to solve some of the equations in this chapter. For, example, you could solve some of the quadratic equations in Section 7.3.

Modeling with Systems of Linear Equations and Inequalities

8

What's the Best Way to Do It?

*I*n 2005, Hurricane Katrina struck the Gulf Coast causing devastation on a scale that rarely has been seen in the United States.

It may seem strange to you to begin this chapter with that sentence, but the fact is when a crisis of this magnitude hits, there are a great many problems that need to be solved. Large amounts of materials, workers, and equipment need to be brought in and organized to accomplish recovery as efficiently as possible.

Researchers in an area of mathematics called *operations research* study problems such as this to understand how to best allocate resources to accomplish a large, complex task. One of the tools that operations researchers use is called *linear programming*, which has many practical applications such as scheduling airline flights, planning effective diets, and organizing industrial and military operations.

In Section 8.3, you will learn how to use linear programming to make the best choice from a large set of alternatives. Before we can do this, however, you will need to know how to solve systems of equations and systems of inequalities. In this

(continued)

chapter, you will learn how to solve a variety of real-life problems, including how to choose the best diet, make the right investment, and schedule work in a factory most efficiently. The topics that you will study in this chapter are certainly among the most widely used areas of applied mathematics. ●

8.1 Systems of Linear Equations

Objectives

1. Represent a system of linear equations graphically.
2. Solve a system of linear equations using the elimination method.
3. Use systems of linear equations as models.

So you are finally going to get in shape. Let's see, you'll start exercising and, of course, watch your diet. This seems like a good idea, . . . but why are we talking about this in a mathematics text? Actually, the diet problem is a very famous problem in mathematics, and when I did a search on the Internet, I got over 77,000 hits for "diet and linear programming."* The Web pages I found had a wide variety of applications—from feeding infants to feeding dairy cows.

The idea in the diet problem is to use a collection of linear equations to model the amounts of the various nutrients that you consume. There could be one equation modeling your calories, another for vitamin C intake, a third for calcium, and so on. In fact, in a real diet problem, there may be dozens of equations. More complex applications, such as modeling the weather or the U.S. economy, could require tens of thousands of linear equations.

A collection of linear equations that are related to each other, such as those we have mentioned in modeling your diet, the economy, or the weather, is called a **system of linear equations**. For the time being, we will consider very simple systems of equations that have only two equations and two variables (or unknowns).

KEY POINT

A system of linear equations represents a pair of lines.

Quiz Yourself ❶ †

Which ordered pair is a solution for the system

$$3x - 5y = 4$$
$$2x + 4y = 10?$$

a) $(2, 3)$

b) $(3, 1)$

c) $(-3, -1)$

Systems of Linear Equations

A **solution** of a pair of linear equations is an ordered pair of numbers that satisfies both equations. For example, if you substitute 5 for x and 3 for y in the system

$$2x + 4y = 22$$
$$x - 6y = -13,$$

you will get

$$\begin{array}{c} x = 5 \diagdown \qquad \diagup y = 3 \\ 2 \cdot 5 + 4 \cdot 3 = 22 \\ 5 - 6 \cdot 3 = -13. \end{array}$$

Because the ordered pair (5, 3) makes both equations true, it is a solution for the system. Although (7, 2) makes the first equation true in the system, it does not make the second equation true (verify). Therefore, it is not a solution for the system. ❶

Recall from Chapter 7 that the graph of a linear equation in two unknowns is a straight line. Therefore, if you graph a system of two linear equations in two unknowns, you will get a *pair* of lines. There are three possible situations, as we show in Figure 8.1 on the following page.

---

*We will discuss linear programming in Section 8.3 after we have developed some preliminary ideas.
†Quiz Yourself answers begin on page 778.

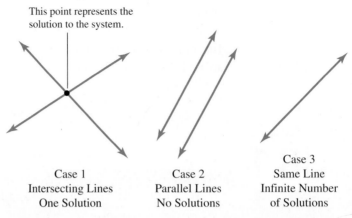

This point represents the
solution to the system.

Case 1	Case 2	Case 3
Intersecting Lines	Parallel Lines	Same Line
One Solution	No Solutions	Infinite Number
		of Solutions

FIGURE 8.1 A geometric representation of the three cases that
may arise in solving a system of linear equations.

Case 1: The lines intersect in a single point. The ordered pair that represents this point
is the *unique solution* for the system.

Case 2: The lines are distinct parallel lines and therefore don't intersect at all. Because
the lines have no common points, this means that the system has *no solutions.*

Case 3: The two lines are the same line. Because the lines have an infinite number
of points in common, this means that the system will have *an infinite number
of solutions.*

KEY POINT

We use the elimination
method to solve systems
of linear equations.

Solving Systems Using the Elimination Method

We will illustrate these three cases by using a technique called the **elimination method** to
solve systems of linear equations.* The basic strategy of the elimination method is to replace
the system of equations by a single equation that is easier to solve, as we see in Example 1.

EXAMPLE 1 *Solving a System That Has One Solution*

Solve the system

$$3x + 4y = 10$$
$$5x - 6y = 4. \qquad (1)$$

SOLUTION: To use the elimination method, we will multiply both equations in the system
by some constants so that one of the variables appears with *opposite coefficients* in the two
equations. If we then add these two equations, the variable with the opposite coefficients
will drop out, leaving us with a single equation having one unknown.

In system (1), if we multiply the top equation by 3 and the bottom equation by 2, we
get opposite coefficients for y. This gives us the system

$$9x + 12y = 30$$
$$10x - 12y = 8. \qquad (2)$$

Now, adding corresponding sides of both equations in (2) causes the y to drop out, giving us

$$9x + 12y = 30$$
$$\underline{+\,10x - 12y = 8}$$
$$19x + 0 = 38.$$

Dividing both sides of this equation by 19, we get

$$\frac{19x}{19} = \frac{38}{19}, \quad \text{or} \quad x = 2.$$ This tells you that there is
one solution to the system.

---

*We will explain another method for solving systems of equations, called the **substitution method**, in the
exercises.

Because we found a single value for x, we know that we are dealing with Case 1.

There will also be a unique value for *y*, which we will find next.

When you are using the elimination method, once you find one unknown, you have two choices. If the number you find is not easy to work with, you can go back to the original system, eliminate the other variable, and solve for the second unknown.

If the value you find is easy to use, as in this case, then substitute it in either equation of the original system to find the second unknown. Substituting 2 for *x* in the first equation, we get

Quiz Yourself **2**

Use the elimination method to solve the following system:

$$5x + 2y = -10$$
$$2x + 3y = 7$$

$$9(\overset{x=2}{2}) + 12y = 30.$$

Subtracting 18 from both sides, this simplifies to $12y = 12$, so $y = 1$. Thus, the solution for this system is (2, 1). You should verify that (2, 1) solves both equations. ❋ **2**

![Some Good Advice]

If you look at a situation from both a geometric and algebraic point of view, you often gain insight into how to solve a problem or check your solution. If the geometric and algebraic representations of a solution do not agree, then you must go back and find your error.

In Figure 8.2, we graph system (1) from Example 1 on a graphing calculator,* where you can see that the lines appear to intersect at the point (2, 1).

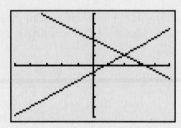

FIGURE 8.2 Graph of system (1).

EXAMPLE 2   *Trying to Solve a System That Has No Solutions*

Solve the system

$$-\frac{3}{2}x + y = \frac{5}{4}$$
$$3x - 2y = 1.$$

SOLUTION: It is a good idea to make your work a little easier in a situation such as this by multiplying both sides of the top equation by 4 to clear the fractions.

$$(\overset{2}{4})\left(-\frac{3}{2}\right)x + (4)y = (4)\frac{5}{4}$$

So our system becomes the simpler-looking system

$$-6x + 4y = 5$$
$$3x - 2y = 1.$$

If we multiply the bottom equation by 2 and add the equations to eliminate *x* from the system, we get a surprising result.

$$-6x + 4y = 5$$
$$\underline{+6x - 4y = 2}$$
$$0 + 0 = 7$$

This tells you that there is no solution to the system.

---

*For more on using a graphing calculator to solve systems of equations, see your instructor for a TI-83 tutorial.

We have assumed that there is a solution to this system. By doing this, we are now forced to accept that $0 + 0 = 7$. Our only way out of this predicament is to conclude that our assumption is wrong, so a solution for this system does not exist.*

In trying to eliminate one of the variables, we obtained a false statement. We conclude that there are no points common to both lines and therefore we are dealing with Case 2. ❄

A system that has no solutions, such as the one in Example 2, is said to be **inconsistent**.

EXAMPLE 3   *Solving a System Having Infinitely Many Solutions*

Solve the system

$$0.2x - 0.1y = 0.3$$
$$0.1x - 0.05y = 0.15. \tag{1}$$

SOLUTION: Again we will make it easy on ourselves by getting rid of the decimals. If we multiply the top equation by 10 and the bottom equation by 100, we can rewrite system (1) as

$$2x - y = 3$$
$$10x - 5y = 15. \tag{2}$$

If we now multiply the top equation by -5, and add the two equations, we can eliminate x from the system. Again we get a surprising result.

$$-10x + 5y = -15$$
$$\underline{+10x - 5y = 15}$$
$$0 + 0 = 0$$

This tells you that there are an infinite number of solutions to the system.

Quiz Yourself 3

a) Solve the system

$$5x - 3y = 3$$
$$-10x + 6y = 4.$$

b) What is your geometric interpretation of the result from part (a)?

You might be tempted to confuse this with Case 2. However, unlike Example 2, the statement $0 + 0 = 0$ is a *true* statement. In trying to solve for x, we obtained a statement that in no way restricts x. This means that we can use any x we want. If we substitute 5 for x in the top equation of system (2), we get $2 \cdot 5 - y = 3$. Solving for y gives us $y = 7$. Thus, $(5, 7)$ is one of the many pairs that are solutions for this system.

If in solving the system we obtain a true statement that does not involve either x or y, this tells us that there are an infinite number of ways to solve the system, so the two lines must be the same.

Now try Exercises 15 to 28. ❄ 3

A system that has an infinite number of solutions, such as the system in Example 3, is said to be **dependent**.

We summarize what you saw happening in Examples 1 to 3.

USING THE ELIMINATION METHOD TO SOLVE SYSTEMS OF LINEAR EQUATIONS

Case 1— We find a single value for both x and y. The pair (x, y) corresponds to the point of intersection of the lines represented by the equations in the system.

Case 2— We obtain an obviously false statement. We conclude that there are no solutions to this system, which represents a pair of distinct parallel lines.

Case 3— We obtain a statement that is always true and that does not contain either x or y. Any value can be used for x as the first coordinate of a solution of the system. There are an infinite number of solutions to the system, and the two equations of the system represent the same line.

---

*If you have studied the logic chapter, you can look at what we have done as: If there is a solution, then $0 + 0 = 7$. The contrapositive states: If $0 + 0 \neq 7$, then there is no solution.

KEY POINT

A system of equations models a set of relationships between two quantities.

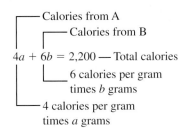

FIGURE 8.3 Equation describing calories consumed per day.

Modeling with Systems of Equations

Now that you know how to solve systems of equations, you can use them to model situations involving several relationships among a set of variables. The next several examples illustrate this.

EXAMPLE 4 *Modeling Nutritional Requirements*

Yuki is taking two types of food with her on her hike in the Grand Canyon. Each gram of food A provides 4 calories and 0.08 gram of protein, and each gram of food B provides 6 calories and 0.2 gram of protein. If she requires 2,200 calories and 50 grams of protein daily, how much of each should she consume to satisfy these requirements?

SOLUTION: We will assume that Yuki will eat a grams of food A and b grams of food B each day.

We will first write a linear equation to describe the calories that Yuki consumes each day. Because each gram of A provides 4 calories and each gram of B provides 6 calories, the number of calories provided by a grams of A and b grams of B is $4a + 6b$. The number of calories must equal 2,200, so we get the equation shown in Figure 8.3.

Similarly, the number of grams of protein in this diet is $0.08a + 0.2b$, and because we want to have 50 grams of protein daily, we get the equation

$$\underset{\text{protein from A}}{0.08a} + \underset{\text{protein from B}}{0.2b} = \underset{\text{total protein}}{50}.$$

Now, to find the solution to our problem, we solve the system

$$4a + 6b = 2{,}200 \quad\text{— calories}$$
$$0.08a + 0.2b = 50. \quad\text{—— protein}$$

First, we multiply the second equation by 100 to eliminate decimal coefficients, giving us

$$4a + 6b = 2{,}200$$
$$8a + 20b = 5{,}000.$$

Now multiply the top equation by -2 to get

$$-8a - 12b = -4{,}400$$
$$8a + 20b = 5{,}000.$$

Adding these equations eliminates the a to give us $8b = 600$, or $b = 75$. Because 75 is easy to work with, we substitute it for b in the top equation and solve for a.

Math in Your Life

Linear Equations and Your Nintendo® Wii™

In building models, linear equations are easy to work with so mathematicians often use them instead of more complex equations. Many times a good approximation can give us an acceptable answer to our problem.

For example, if you look carefully at a video game's screen, you may be able to see angular edges instead of the smooth face of a character. Curved surfaces take too long to compute, so game designers use tiny planar surfaces and sophisticated shading techniques instead of drawing the true curved surfaces.* Planes are the geometric representation of linear equations in three variables. So, the game designer is replacing nonlinear surfaces in three-dimensional space with linear approximations to speed up the computations. In other words, using linear equations for models allows you to play Super Smash Brothers on your Wii at real-time speed.

---

*As video game technology improves, it is harder to see these edges; however, be assured that in creating three-dimensional figures, the common technique is to approximate curved surfaces by flat polygons and then render them by using exotic shading methods. The same technique is used in creating an animated movie such as *Fraggle Rock*.

Making this substitution gives us

$$4a + 6(\overset{b}{75}) = 2{,}200, \quad \text{or}$$

$$4a + 450 = 2{,}200.$$

Subtracting 450 from both sides, we get $4a = 2{,}200 - 450 = 1{,}750$. Dividing by 4 yields $a = \frac{1{,}750}{4} = 437.5$. This tells us that if Yuki consumes 437.5 grams of food A and 75 grams of food B each day, she meets the dietary requirements for calories and protein. ❈

KEY POINT

A system of equations describes the relationship between supply and demand.

The **law of demand** in economics states that as the price of an item increases, the consumer is less willing to purchase it. According to the **law of supply**, as the price of an item increases, the producer is willing to produce more of the item. If the price is low, consumer demand increases, but because producers are not willing to produce much, there will be a shortage of the product. On the other hand, if the price is high, producers will produce more of the product, but because consumers are not willing to pay the high prices, there is a surplus of the item. Eventually the market for the item adjusts so that there will be a price at which the quantity demanded and the quantity supplied are equal. This point is called an **equilibrium point**. We illustrate this notion of equilibrium in Example 5.

EXAMPLE 5   *The Supply and Demand for Hand-Crafted Jewelry*

Malik sells hand-crafted turquoise, Native American jewelry at regional craft shows. At a price of $20 per necklace, he is willing to buy 30 necklaces from his suppliers. However, at this price, his suppliers will only provide him with 9 necklaces. However, if he will pay $60 per necklace, he can only sell 15 necklaces per show. At this higher price, his suppliers will provide him with 29 necklaces. Assuming that the equations relating price to demand and to supply are both linear, what should the price per necklace be for supply to equal demand?

SOLUTION: We summarize the information we have in Table 8.1.

Supply			Demand		
	Price	Necklaces Supplied		Price	Necklaces Demanded
Lower price—less supplied —	20	9		20	30	— Lower price—more demand
Higher price—more supplied —	60	29		60	15	— Higher price—less demand

TABLE 8.1 The supply and demand for turquoise necklaces.

Because the supply equation is linear, its graph is a line that passes through the points (20, 9) and (60, 29). The slope of this line is

$$\text{slope} = \frac{\text{rise}}{\text{run}} = \frac{29 - 9}{60 - 20} = \frac{20}{40} = \frac{1}{2}.$$

If we write the supply equation in slope–intercept form* as $y = mx + b$, we can substitute $\frac{1}{2}$ for m to rewrite this equation as

$$y = \frac{1}{2}x + b.$$

_____

*If you are rusty on this, review Section 7.1 and also see Appendix A.

Because (20, 9) lies on this line, we can substitute 20 for x and 9 for y, to get the equation

$$9 = \frac{1}{2}(20) + b, \quad \text{or} \quad 9 = 10 + b.$$

Subtracting 10 from both sides, we get $b = -1$. The supply equation is therefore

$$y = \frac{1}{2}x - 1.$$

Multiplying by 2, we get $2y = x - 2$. We rewrite this in standard form as

$$x - 2y = 2.$$

The points (20, 30) and (60, 15) lie on the graph of the demand equation, so the slope of this line is $\frac{15 - 30}{60 - 20} = \frac{-3}{8}$. Therefore, we can write the demand equation in slope–intercept form as

$$y = \frac{-3}{8}x + b.$$

The point (20, 30) is on this line, so we substitute 20 for x and 30 for y to get

$$30 = \frac{-3}{8} \cdot 20 + b, \quad \text{or} \quad 30 = -\frac{15}{2} + b.$$

Adding $\frac{15}{2}$ to both sides, we find that $b = 30 + \frac{15}{2} = \frac{60}{2} + \frac{15}{2} = \frac{75}{2}$. The equation representing demand for necklaces is therefore

$$y = \frac{-3}{8}x + \frac{75}{2}.$$

Multiplying by 8 we get $8y = -3x + 300$. We rewrite this equation in standard form as

$$3x + 8y = 300.$$

Therefore, the following system describes this supply and demand situation:

$$x - 2y = 2 \quad \text{(supply)}$$
$$3x + 8y = 300. \quad \text{(demand)}$$

We will now solve this system to find an equilibrium point. Multiplying the top equation by 4 gives us the system

$$4x - 8y = 8$$
$$3x + 8y = 300.$$

Adding these equations results in the equation $7x = 308$. So, $x = \frac{308}{7} = 44$.

We can now substitute 44 for x in either equation to find y. Let's use the bottom equation $3x + 8y = 300$, so

$$3(44) + 8y = 300, \text{ or } 132 + 8y = 300.$$

Subtracting 132 from both sides, we get $8y = 300 - 132 = 168$. So, $y = \frac{168}{8} = 21$.

This means that at a price of \$44 per necklace, Malik will be willing to buy 21 necklaces from the supplier and the supplier will be willing to sell 21 necklaces at this price.

Now try Exercises 49 to 52. ❋

Following the spirit of Descartes's analytic geometry, it is helpful to interpret the solution to Example 5 geometrically. We graph both the supply and demand equations in Figure 8.4. The region to the left of the equilibrium point shows where demand exceeds supply. The region to the right of the equilibrium point shows where supply exceeds demand.

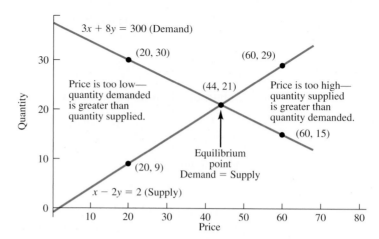

FIGURE 8.4 Supply and demand graphs intersect at the equilibrium point.

EXAMPLE 6 *Solving a System of Equations in a Manufacturing Situation*

Aidan's Custom Woodworking Company builds cherry furniture. This week, they will manufacture only end tables and coffee tables. An end table requires 6 board feet and a coffee table requires 8 board feet. It takes 2 hours of labor to make an end table and 4 hours of labor to make a coffee table. The company has 1,200 board feet of cherry wood available and also 480 hours of labor. Assume that we want to use all the labor and wood. Describe the conditions on wood and labor as a system of two linear equations in two unknowns. Solve this system and interpret your answer.

SOLUTION: It is helpful to organize this information as in Table 8.2, which we will call a **resource table**. We will represent the number of end tables by e and the number of coffee tables by c.

Resources	Needed for an End Table (e)	Needed for a Coffee Table (c)	Available
Wood	6 board feet	8 board feet	1,200 board feet
Labor	2 hours	4 hours	480 hours

TABLE 8.2 Resource table for furniture manufacturing problem.

$6e + 8c = 1,200$ — total wood available

wood needed for c coffee tables

wood needed for e end tables

FIGURE 8.5 Equation describing use of wood.

First let's consider the condition on the wood needed to manufacture the tables. Because we use 6 board feet for each end table and 8 board feet for each coffee table and have 1,200 board feet available, we get the equation as seen in Figure 8.5.

Let's now look at the restrictions on labor. We use 2 hours for each end table and 4 for each coffee table. Therefore, the equation $2e + 4c = 480$ describes how to use the 480 available hours to construct e end tables and c coffee tables. We will now solve the system:

$$6e + 8c = 1,200 \quad \text{(wood)}$$
$$2e + 4c = 480. \quad \text{(labor)}$$

We multiply the second equation by -2 to get

$$6e + 8c = 1,200$$
$$-4e - 8c = -960.$$

When we add these two equations, we get the equation $2e = 240$, whose solution is $e = 120$. If we substitute 120 for e in the second equation of the original system, we get $2(120) + 4c = 480$, which means that $c = 60$. Therefore, if the company manufacturers 120 end tables and 60 coffee tables, all 1,200 board feet of wood and 480 hours of labor will be used up.

Now try Exercises 37 to 46. ✳

Exercises 8.1

Looking Back*

These exercises follow the general outline of the topics presented in this section and will give you a good overview of the material that you have just studied.

1. What are the three cases that can arise if we are dealing with a pair of lines? What do these three cases have to do with Examples 1, 2, and 3?

2. Which example illustrated an inconsistent system of equations?

3. What is the benefit of simplifying coefficients in a system before solving it as we did in Examples 2 and 3?

4. What did the two equations in Example 4 model? ☐

5. How do game designers speed up the graphics in video games?

6. What was the point of Figure 8.2?

Sharpening Your Skills

The solution for each system is a pair of integers. Determine the solution by graphing each system, and check your answer by substituting values in both equations.

7. $\begin{aligned} 2x + 5y &= 18 \\ -3x + 4y &= -4 \end{aligned}$

8. $\begin{aligned} 3x + 4y &= 1 \\ 2x + 6y &= -6 \end{aligned}$

9. $\begin{aligned} 3x + 2y &= 9 \\ -x + 2y &= -3 \end{aligned}$

10. $\begin{aligned} 5x - 2y &= -2 \\ 2x + 3y &= -16 \end{aligned}$

11. $\begin{aligned} 2x - 5y &= 18 \\ 3x + y &= 10 \end{aligned}$

12. $\begin{aligned} -x + 2y &= -8 \\ -2x + 3y &= -11 \end{aligned}$

13. $\begin{aligned} -2x + 3y &= -5 \\ -x + y &= -2 \end{aligned}$

14. $\begin{aligned} 5x + 3y &= -11 \\ x - y &= -7 \end{aligned}$

Use the elimination method to solve the following systems of linear equations:

15. $\begin{aligned} -3x + 2y &= 5 \\ 6x + 4y &= 22 \end{aligned}$

16. $\begin{aligned} 4x - y &= 13 \\ 5x + 8y &= -30 \end{aligned}$

17. $\begin{aligned} 3x - 2y &= -14 \\ -4x + y &= 22 \end{aligned}$

18. $\begin{aligned} -5x + 4y &= -44 \\ 2x + y &= 15 \end{aligned}$

19. $\begin{aligned} 8x - 2y &= -2 \\ -4x + y &= 3 \end{aligned}$

20. $\begin{aligned} 6x + 9y &= -4 \\ 2x + 3y &= -2 \end{aligned}$

21. $\begin{aligned} x - y &= -2 \\ -4x + 4y &= 8 \end{aligned}$

22. $\begin{aligned} 10x - 4y &= -4 \\ -5x + 2y &= 2 \end{aligned}$

23. $\begin{aligned} 6x - 9y &= 8 \\ -4x + 6y &= 10 \end{aligned}$

24. $\begin{aligned} 12x - 4y &= 2 \\ -9x + 3y &= 11 \end{aligned}$

25. $\begin{aligned} 12x - 8y &= -1 \\ -2x + 5y &= 2 \end{aligned}$

26. $\begin{aligned} 12x - 27y &= -1 \\ 5x + 2y &= 4 \end{aligned}$

27. $\begin{aligned} 3x - 9y &= -3 \\ -4x + 2y &= 0 \end{aligned}$

28. $\begin{aligned} 4x - 5y &= -4 \\ -8x + 15y &= 13 \end{aligned}$

The **substitution method** is another method for solving systems of linear equations and involves the following steps:

Step 1. Solve either equation in the system for x or y. (You have four choices here: You can solve equation 1 for either x or y or you can solve equation 2 for either x or y. You make your decision as to which to do based on what will give you the simplest expression to work with.)

Step 2. Substitute the expression you found in step 1 for the variable in the other equation. This gives you a linear equation in a single variable.

Step 3. Solve the equation in step 2 for the variable.

Step 4. Substitute the value for the variable that you found in step 3 into the equation that you chose in step 1. This gives you a linear equation in the other variable that you can solve easily.

We will use this method to solve the system

$$3x - 4y = 13$$
$$x - 2y = 5$$

Step 1. In considering our four choices, it is clear that our best choice is to solve x − 2y = 5 for x in equation 2. This gives us x = 2y + 5.

Step 2. We now substitute 2y + 5 for x in equation 1. This gives us the equation

$$3(2y + 5) - 4y = 13.$$

Step 3. We can rewrite this equation as

$$6y + 15 - 4y = 13,$$

and again as

$$2y = -2.$$

Thus, y = −1.

---

*Before doing these exercises, you may find it useful to review the note *How to Succeed at Mathematics* on page xix.

Step 4. *Now we substitute* -1 *for y in equation 1 to get*

$$3x - 4(-1) = 13.$$

Thus, $3x + 4 = 13$, *or* $x = 3$. *The solution to the system is therefore* $(3, -1)$.

Use the substitution method to solve the systems in Exercises 29–36. As with the elimination method, you may discover some systems to be either inconsistent or dependent.

29. $x - 2y = -7$
$3x + 5y = 12$

30. $x + 3y = 11$
$-x + 4y = 10$

31. $-2x + y = 6$
$-2x + 3y = 14$

32. $5x + 4y = 9$
$3x + y = 11$

33. $x + y = 6$
$2x = 8 - 2y$

34. $x + y = 3$
$x - y = -7$

35. $x - y = 4$
$3y = 3x - 12$

36. $2x + 6y = -36$
$2y - x = -7$

Applying What You've Learned

Use the elimination method to solve each system of two linear equations in two unknowns.

37. Sports. In 2007, the Boston Red Sox and the Cleveland Indians were tied for the best record in baseball. In a 162-game season, they both won 30 more games than they lost. What were their records?

38. Sports. The U.S Olympic basketball team won a basketball game by 13 points over Germany. If the two teams together scored a total of 109 points, what was the final score?

39. Dietary requirements. An average bagel contains 30 mg of calcium and 2 mg of iron.* One ounce of cream cheese contains 25 mg of calcium and 0.4 mg of iron. If Leyla wants to eat a combination of bagels and cream cheese that contains exactly 245 mg of calcium and 10 mg of iron, how much of each should she eat?

40. Dietary requirements. A slice of cheese pizza contains 40 g of carbohydrates and 220 mg of calcium. A 12-oz cola contains 40 g of carbohydrates and 15 mg of calcium. If Karl eats several slices of cheese pizza and drinks cola, how much of each must Karl eat to get a nutritional benefit of exactly 200 g of carbohydrates and 690 mg of calcium?

41. Comparing satellite systems. Maneka is considering two satellite TV systems. WorldCom charges $179 dollars for installation and $17.50 per month. Satellite, Inc. charges $135 for installation and $21.50 per month. Interpret what the solution to this system tells you. How should Maneka decide which system to take?

42. Comparing job offers. Betty can work as freelance editor for *Mode* magazine for a base salary of $20 per hour plus 25 cents per page. At *Blush* magazine, she can earn $22.10 per hour plus 18 cents per page. Interpret what the solution to this system tells you. How should Betty decide which position to take?

43. Sports. In 2007, the WNBA San Antonio Silver Stars played 34 games. Half their number of wins was 3 more than half their number of losses. What was their record?

44. Sports. In 2007, the WNBA Los Angeles Sparks played 34 games. They had 4 more wins than twice their number of losses. What was their record?

45. Airport traffic. According to the Airports Association Council International, in 2007, the two busiest airports in the world were Atlanta's Hartsfield International and Chicago's O'Hare International. The two airports together handled 162 million passengers, and Hartsfield handled 8 million more than O'Hare. How many passengers went through each airport?

46. Tourist travel. According to the World Tourism Organization, the two European countries that had the most tourists in 2007 were France and Spain. The two countries together had 138 million visitors, and France had 20 million more visitors than Spain. How many tourists visited each country?

The given graph summarizes data regarding production of motor vehicles from 2000 to 2005 in the United States, Japan, and Western Europe. Model this information with systems of linear equations to solve Exercises 47 and 48.

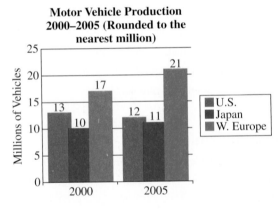

Motor Vehicle Production 2000–2005 (Rounded to the nearest million)

47. Motor vehicle production. If these trends continue, in what year will Japan's production exceed that of the United States?

48. Motor vehicle production. If these trends continue, in what year will Western Europe's production be twice that of the United States?

Use the elimination method in Exercises 49–52 to find an equilibrium point for the described supply–demand situation. You may assume that the supply and demand equations are both linear.

49. Supply and demand—tutoring. A tutoring service provides tutoring for a fee. When the tutoring service charges $8/hour, there is a demand for 30 tutors per week. When the hourly rate rises to $15/hour, the demand drops to 9 tutors per week. On the other hand, at $8/hour, the service is able to supply only 9 tutors per week, whereas at the rate of $15/hour, it is able to provide 37 tutors per week. Find the price per hour at

---

*The nutritional content of the foods mentioned in these exercises is based on the *Home and Garden Bulletin No. 72*, U.S. Department of Agriculture.

which the number of tutors demanded and the number of tutors supplied will be equal.

50. Supply and demand—housing. In doing a low-income housing survey, the local chamber of commerce found that if an apartment rents for $400 per month, there were only 275 available; however, when the rental price rose to $550 per month, the supply increased to 350. In contrast, at a rental price of $400 per month, there was a demand for 450 units, but when the rental price increased to $550 per month, the demand dropped to only 200 units. Find the monthly rental price at which the number of apartments demanded and the number of apartments supplied will be equal.

51. Supply and demand—used books. A bookstore buys and sells used textbooks. For a current sociology text, it has found that if it pays an average price of $30 for each used book, it will be able to buy 60 texts for resale and will have 95 customers who are willing to buy the texts after markup. If it pays $42 per used text, it will be able to buy 120 texts but will be able to sell only 50 of them. What price should it offer for the used texts in order to sell all that it buys?

52. Supply and demand—selling bagels. Dunkin' Donuts makes fresh bagels each morning. The proprietors will make 4 dozen bagels if the price is $7/dozen and the demand will be 10 dozen. At a price of $16/dozen, they will make 16 dozen but will be able to sell only 4 dozen. How should the bagels be priced so that all bagels made will be sold?

Communicating Mathematics

53. When you are solving a system of equations, how do you recognize that the system has exactly one solution?

54. When you are solving a system of equations, how do you recognize that the system has no solutions?

55. When you are solving a system of equations, how do you recognize that the system has an infinite number of solutions?

56. Does an equation representing demand, as in Example 5, have positive or negative slope? Why is this so? What about the equation representing supply?

57. How could you make up a word problem that has no solution? (*Hint:* Think about this geometrically and determine which case you are dealing with.) Take any one of the preceding word problems and change one of the conditions so that the problem has no solution.

58. Make up a system of two equations in two unknowns which has the solution (2, 3). Explain how you made this example.

Using Technology to Investigate Mathematics

59. Ask your instructor for a tutorial to show you how to graph and solve systems of equations on a graphing calculator. Use your calculator to reproduce some of the examples in this section.*

60. Search the Internet for interactive programs that allow you to graph and solve systems of linear equations. Use these programs to duplicate some of the examples in this section.

For Extra Credit

We can use the elimination method to solve systems that have more than two equations and two unknowns. We outline this technique in Exercises 61 and 62.

61. Reduce this system to a system with two equations and two unknowns by completing the following steps:

$$2x + 3y + z = 10$$
$$x - y + 2z = 7$$
$$3x + y - z = 4$$

a. Consider the first and second equations:

$$2x + 3y + z = 10$$
$$x - y + 2z = 7.$$

Multiply the first equation by -2 and then add this new equation to the second equation to get an equation containing only the variables x and y.

b. Consider the first and third equations:

$$2x + 3y + z = 10$$
$$3x + y - z = 4.$$

If you add these equations together, you will get another equation that contains only the variables x and y.

62. a. Now solve the system of equations obtained in Exercise 61. Because this system has two linear equations with two unknowns, we can solve for x and y as we did in this section.

b. Choose any of the three original equations and substitute the values for x and y that you found in part a). You are now able to solve for z.

Use the method that we outlined in Exercises 61 and 62 to solve the following systems:

63.
$$x - 3y + 2z = 6$$
$$x + y - 2z = 2$$
$$3x + 2y + z = 13$$

64.
$$2x - 4y + 3z = 7$$
$$-2x + y + 2z = 5$$
$$x + 2y - z = 0$$

65.
$$x - y - z = 2$$
$$4x + y + 2z = 24$$
$$x + y - z = 6$$

66.
$$x + 4y + 3z = 9$$
$$-x + 2y + z = -3$$
$$2x - 2y - 2z = 4$$

8.2 Systems of Linear Inequalities

Objectives

1. Solve a single linear inequality in two variables graphically.
2. Solve a system of linear inequalities in two variables.
3. Use systems of linear inequalities as models.

In life you often deal with quantities that are not equal. Your parents may expect you to earn *at least* $2,000 toward next year's tuition. Your doctor may want you to keep your weight *below* 160 pounds. Your math instructor may tell you that if you cut class *more than* three times, you will suffer a penalty. If you are concerned with your diet, you may want to be sure to consume *at least* 1,500 mg of calcium daily, but *no more than* 400 international units of vitamin D because too much vitamin D can be toxic.

Because we cannot model these situations with linear equations, we need to develop a mathematical theory of inequalities. The techniques we use for working with inequalities are similar to those that you have already learned for working with equations.

 **KEY POINT**

The solution to a linear inequality in two variables is a half-plane.

Solving Linear Inequalities

Like equations, inequalities can be linear or nonlinear. In this section, we will be concerned with linear inequalities in two variables.

> **DEFINITION** A **linear inequality in two variables** is a statement we can write in one of the following forms:
>
> $$ax + by \geq c \text{ (read } ax + by \text{ is greater than or equal to } c\text{)},$$
> $$ax + by > c \text{ (read } ax + by \text{ is greater than } c\text{)},$$
> $$ax + by \leq c \text{ (read } ax + by \text{ is less than or equal to } c\text{)},$$
> $$ax + by < c \text{ (read } ax + by \text{ is less than } c\text{)},$$
>
> where a, b, and c are real numbers with both a and b not equal to zero.

The following are examples of linear inequalities:*

$$2x + 3y \geq 6,$$
$$4x - 5y > 12,$$
$$y \leq 8,$$
$$x < 5.$$

As was the case with linear equations in two variables, solutions of linear inequalities are ordered pairs of numbers.

EXAMPLE 1   *Solutions to Inequalities*

Which of the following ordered pairs are solutions of the inequality $2x + 3y \leq 6$?

a) $(2, 1)$ b) $(0, 2)$ c) $(-4, 2)$

SOLUTION: To determine if these ordered pairs are solutions, we substitute the first coordinate of the pair for x and the second coordinate for y. If we get a true statement, the pair is a solution of the inequality; if the statement is false, we do not have a solution.

---

*In this section, we will be dealing only with linear inequalities in two variables; therefore, we will frequently omit the reference to the number of variables in the inequality.

Quiz Yourself ④

Which of the following ordered pairs are solutions of the inequality $3x - 4y \geq 5$?

a) $(4, 1)$ b) $(0, 2)$ c) $(1, 1)$

a) $2 \cdot 2 + 3 \cdot 1 \leq 6$ is false; therefore $(2, 1)$ is not a solution.

b) $2 \cdot 0 + 3 \cdot 2 \leq 6$ is true, so $(0, 2)$ is a solution.

c) $2 \cdot (-4) + 3 \cdot 2 \leq 6$ is true, so $(-4, 2)$ is a solution.

Now try Exercises 5 to 8. ✾ ④

EXAMPLE 2   *Solving a Linear Inequality Graphically*

Solve the inequality $3x + 4y \geq 11$.

SOLUTION: We can rephrase this inequality as the pair of statements

$$3x + 4y > 11 \quad \text{or} \quad 3x + 4y = 11.$$

This means that the solutions to the original inequality include the solutions to the linear equation $3x + 4y = 11$. We can visualize *some* of the solutions to the original inequality by graphing this line, as in Figure 8.6(a).

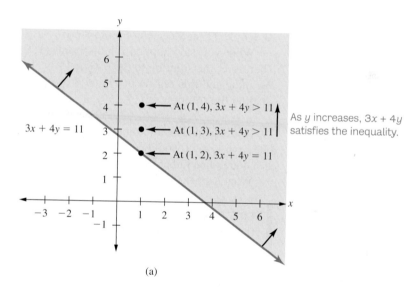

(a)

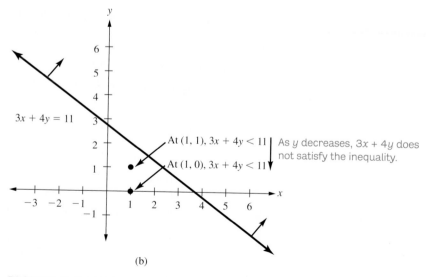

(b)

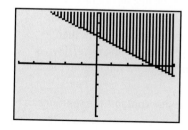

Graph of $3x + 4y \geq 11$ drawn with a graphing calculator.

FIGURE 8.6 (a) Points on or above the line $3x + 4y = 11$ are solutions. (b) Those below are not.

We now want to convince you that all points lying above the line are also part of the solution set of $3x + 4y \geq 11$. Let's start by plotting the point $(1, 2)$, which satisfies the equation $3x + 4y = 11$ and is therefore on the line. As we see in Table 8.3, if we increase y from 2 to 3 and then to 4, we see that the expression $3x + 4y$ gets larger. Because all the points above $(1, 2)$ make $3x + 4y$ greater than 11, all these points are solutions to the inequality $3x + 4y \geq 11$.

Point	$3x + 4y$	Comment
$(1, 4)$	$3 \cdot 1 + 4 \cdot 4 = 19$	Expression is greater than 11
$(1, 3)$	$3 \cdot 1 + 4 \cdot 3 = 15$	Expression is greater than 11
$(1, 2)$	$3 \cdot 1 + 4 \cdot 2 = 11$	Expression equals 11
$(1, 1)$	$3 \cdot 1 + 4 \cdot 1 = 7$	Expression is less than 11
$(1, 0)$	$3 \cdot 1 + 4 \cdot 0 = 3$	Expression is less than 11

TABLE 8.3 Points above the line satisfy the inequality, but points below the line do not.

Similarly, points such as $(1, 1)$, $(1, 0)$, and $(1, -1)$, which are below $(1, 2)$, make the expression $3x + 4y$ smaller than 11 and therefore are not solutions to the inequality $3x + 4y \geq 11$, as we see in Figure 8.6(b).

What you saw happen for the point $(1, 2)$ is true for any point lying on the line. Therefore, the solution to $3x + 4y \geq 11$ is all points *on* or *above* the line $3x + 4y = 11$. We indicated this region in Figure 8.6(a) by placing small arrows on the line and also shading the solution set. ❈

If the inequality in Example 2 were a strict inequality, namely $3x + 4y > 11$, then the line $3x + 4y = 11$ would not be part of the solution set. In this case, we would draw the line as a dotted line rather than a solid line.

What happened in Example 2 is true for any linear inequality in two variables.

THE SOLUTION SET FOR A LINEAR INEQUALITY IN TWO VARIABLES
The solution set for a linear inequality in two variables is always a half-plane with either a solid or dotted border.

✎ **KEY POINT**

The one-point test determines the half-plane in solving a linear inequality.

The method that we used in Example 2 always works; however, there is a simpler way to solve linear inequalities. The following method is a quick and reliable way for you to determine the correct half-plane when solving a linear inequality.

SOLVING LINEAR INEQUALITIES USING THE ONE-POINT TEST
To solve a linear inequality in two variables, follow these steps:

1. Change the inequality to an equation and graph the line. If the inequality contains $\leq$ or $\geq$, draw a solid line. If the inequality contains $<$ or $>$, draw a dotted line.

2. Choose a point that is clearly either above or below the line you drew in step 1. The point $(0, 0)$ is usually the best point to use unless the line passes through or is very close to $(0, 0)$.

3. If the coordinates of the point you chose in step 2 satisfy the inequality, then the side of the line containing that point contains solutions for the inequality. Otherwise, the opposite side of the line contains solutions. We will refer to this test as **the one-point test**.

4. Use arrows or shading to indicate which side of the line contains the solutions to the inequality.

We will illustrate this method in Example 3.

EXAMPLE 3   *Using the One-Point Test to Solve an Inequality*

Use the steps we listed previously to solve $4x - 3y \geq 9$.

SOLUTION:

Step 1: Graph the line $4x - 3y = 9$. Because the inequality is $\geq$, rather than the $>$, we will graph $4x - 3y = 9$ with a solid line in Figure 8.7.

Step 2: Use the one-point test to determine which side of the line contains solutions to the inequality. If we test $(0, 0)$, we see that $4(0) - 3(0) \geq 9$ is not true. Therefore, the $(0, 0)$ side of the line does not contain solutions. We choose the other side instead. The solution consists of all points on or below the line, as we show in Figure 8.7.

Now try Exercises 9 to 16. ✳ ⑤

Quiz Yourself ⑤

Solve $2x - 3y > 6$.

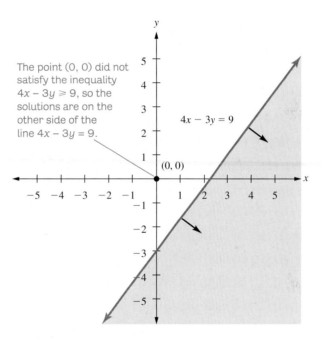

The point $(0, 0)$ did not satisfy the inequality $4x - 3y \geq 9$, so the solutions are on the other side of the line $4x - 3y = 9$.

$4x - 3y = 9$

$(0, 0)$

FIGURE 8.7 Using the one-point test to graph $4x - 3y \geq 9$.

 **Some Good Advice**

In working with inequalities, do not look for shortcuts to obtain the graphs more quickly. You cannot assume that if the symbols $\leq$ or $<$ appear in the inequality that the solution to the inequality occurs below the line. Also, don't assume that the symbols $\geq$ or $>$ mean that you should select the upper half-plane. In order to choose the proper half-plane, you should always use the one-point test.

✎ **KEY POINT**

The solution of a system of inequalities is the intersection of the solution sets for the individual inequalities.

Solving Systems of Inequalities

Just as we can have systems of equations, we can construct systems of two or more inequalities. In solving a system of inequalities, we solve the inequalities separately and then find the intersection of the individual solution sets.

✸ ✸ ✸ HIGHLIGHT

Higher Dimensions, Graphic Novels, and Mathematics

Have you ever wondered, "What is the fourth dimension?" To the graphic novel artist, the fourth dimension may be a bizarre world, existing parallel to our three-dimensional world and occupied by grotesque and sinister beings.

The mathematician, however, views higher dimensions less dramatically as sets of ordered arrangements of numbers. A point in the fourth dimension is simply an ordered 4-tuple of numbers such as $(2, -5, 6, 9)$. Scientists frequently use these higher dimensional spaces to solve real problems. In one application, a paper in an area called *wavelet theory* used mathematics done in a 64,000-dimensional space. Imagine how a cartoonist would draw Superman if he were to accidentally fly through the portal to that space!

To work in a higher dimensional space, you do not have to be able to visualize it. However, you would have to learn techniques that rely less on geometry and more on other abstract mathematical concepts.

EXAMPLE 4 *Solving a System of Inequalities*

Solve the system

$$2x - 3y < -6$$
$$x + y \leq 7.$$

SOLUTION: We first graph $2x - 3y = -6$ using a dotted line and $x + y = 7$ using a solid line, as in Figure 8.8.

Testing $(0, 0)$ with the first inequality, we get $2 \cdot 0 - 3 \cdot 0 = 0$, which is greater than -6, so $(0, 0)$ is not a solution. This means that the region above the line is the desired half-plane to solve $2x - 3y < -6$.

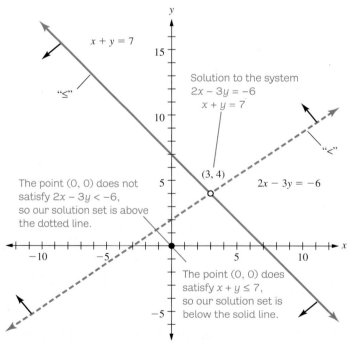

FIGURE 8.8 The dotted line contains points that are not part of the solution of $2x - 3y < -6$.

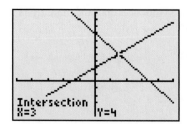

TI-83 screen showing results of using the intersect command.

When we test $(0, 0)$ with the second inequality, we get $0 + 0 = 0$, which is less than or equal to 7. Thus, the $(0, 0)$ side of the line $x + y = 7$ is the half-plane we want.

We have indicated these regions with arrows in Figure 8.9 and have also shaded the intersection of the two half-planes to show the solution to the system of inequalities.

Notice that the point $(3, 4)$ is a solution to the system of linear equations $2x - 3y = -6$ and $x + y = 7$. We found this solution using the elimination method from Section 8.1. The dot is open rather than solid because $(3, 4)$ lies on the line $2x - 3y = -6$, which is not part of the solution of $2x - 3y < -6$. If both boundary lines had been drawn solid, then we would also have drawn $(3, 4)$ as a solid dot. A point, such as $(3, 4)$, which is the intersection of two boundary lines of a solution set, is called a **corner point**.

Now try Exercises 17 to 26. ❁ **6**

Quiz Yourself 6

Solve the system $\begin{array}{l} 3x - 4y \leq 8 \\ x + 2y \geq 6 \end{array}$.

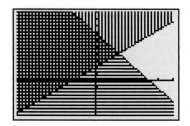

Graphing calculator graph of the solution set of $\begin{array}{l} 2x - 3y < -6 \\ x + y \leq 7 \end{array}$.

KEY POINT

We use a system of linear inequalities in two variables to represent a set of conditions on two quantities.

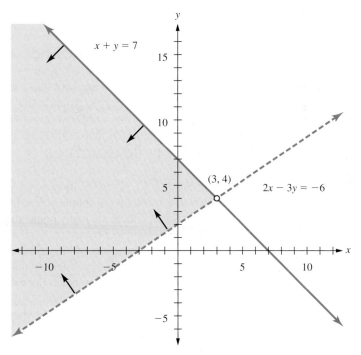

FIGURE 8.9 Graph of the solution of $\begin{array}{l} 2x - 3y < -6 \\ x + y \leq 7 \end{array}$.

Modeling with Systems of Inequalities

We will now consider a situation involving several conditions imposed on a diet. Example 5 shows how to model these conditions with a system of linear inequalities.

EXAMPLE 5 *Using Inequalities to Represent Nutritional Requirements*

Khalid is a marathon runner who is interested in the amount of protein and calcium in his diet. Two of his favorite foods are fried shrimp and broccoli. A serving of fried shrimp contains approximately 15 g of protein and 60 mg of calcium. A spear of broccoli contains 5 g of protein and 80 mg of calcium. Assume that, as part of his diet, he wants to get at least 60 g of protein and 600 mg of calcium from fried shrimp and broccoli. Express this pair of conditions as a system of inequalities and graph its solution set.

SOLUTION: Assume that Khalid eats s servings of shrimp and b spears of broccoli. Because each serving of shrimp has 15 g of protein, if Khalid eats s servings of shrimp, he will consume $15s$ g of protein. Each spear of broccoli has 5 g of protein, so eating b spears of broccoli provides 5*b* g of protein. He wants the amount of protein to be at least 60 g, which we can model with the inequality

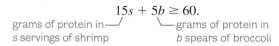

$$15s + 5b \geq 60.$$

grams of protein in—/ ⎞ ⎞—grams of protein in
s servings of shrimp   *b* spears of broccoli

Also, *s* servings of shrimp provides 60*s* mg of calcium and *b* spears of broccoli have 80*b* mg of calcium. Because Khalid wants the amount of calcium to be at least 600 mg, we get the inequality

$$60s + 80b \geq 600$$

milligrams of calcium—/ ⎞ ⎞—milligrams of calcium
in *s* servings of shrimp   in *b* spears of broccoli

$$15s + 5b \geq 60$$
$$60s + 80b \geq 600$$

We solve this system as we did in Example 4 and show the solution in Figure 8.10. You should verify this.

In Figure 8.10, it does not make sense to shade the part of the solution set below the horizontal axis because points in that region have a negative second coordinate and correspond to eating a negative number of broccoli spears. Similarly, we do not shade points to the left of the vertical axis.

Now try Exercises 27 to 38. ✳

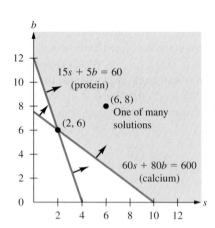

FIGURE 8.10 Shaded region represents the number of servings of shrimp and broccoli needed to satisfy the protein and calcium conditions.

In Section 8.1, when we used a system of linear equations in two variables as a model, a solution was usually* a unique ordered pair of numbers. The solution in Example 5 is very different because there are many points that satisfy the nutritional requirements stated in this problem. For example, because (6, 8) is in the solution set, this means that eating six servings of shrimp and eight spears of broccoli would provide at least 60 g of protein and 600 mg of calcium. Often, among all possible solutions, we want to find a solution that is the best with respect to some other requirement. For example, in Figure 8.10, we may want a combination of shrimp and broccoli that minimizes the amount of fat in the diet. We will discuss these kind of considerations in Section 8.3 when we introduce the topic linear programming.

Exercises 〔8.2〕

Looking Back†

These exercises follow the general outline of the topics presented in this section and will give you a good overview of the material that you have just studied.

1. What did Example 2 tell us about the solution to a linear inequality in two variables?

2. In Example 4, why did we use a dotted line to graph $2x - 3y = -6$ and a solid line to graph $x + y = 7$?

3. How many solutions did we have to the system of inequalities in Example 5? Why did this happen?

4. Sometimes when students see a linear inequality with a > or ≥ in it, they automatically assume that the solution set for the inequality is above the line. What does the "Some Good Advice" box say about doing that?

Sharpening Your Skills

Determine which points satisfy the given inequality.

5. $3x + 4y \geq 2$
   a. (3, 5) **b.** (1, −2) **c.** (0, 0) **d.** (−4, 6)

6. $2x - 4y \geq 5$
   a. (5, 2) **b.** (6, 0) **c.** (0, −3) **d.** (−2, 3)

7. $5y - 3x < 2$
   a. (2, 3) **b.** (1, 1) **c.** (0, 0) **d.** (−3, −2)

8. $8x - y < 2$
   a. (4, 6) **b.** (−1, −2) **c.** (0, 0) **d.** (−4, 3)

Use the one-point test to solve the inequality. Shade the solution.

9. $3x + 4y \geq 12$ 10. $2x - 4y \leq 6$

11. $2x - 4y < 12$ 12. $5x + 4y > 10$

---

*Cases 2 and 3 that we considered in solving linear systems in Section 8.1 are really extreme cases. Most times, when you are solving a system of equations, you are dealing with Case 1, in which there is one, unique solution.

†Before doing these exercises, you may find it useful to review the note *How to Succeed at Mathematics* on page xix.

13. $x \geq 3y - 9$ **14.** $4y \leq 10 - 2x$

15. $4x - 8 < 2y$ **16.** $2x - 6 < 3y$

Solve each system. Indicate all corner points and shade the solution set.

17. $\begin{array}{l} 2x - 3y \leq -5 \\ x - 2y \leq -8 \end{array}$ **18.** $\begin{array}{l} 2x - 5y \leq 23 \\ 3x + y \leq 9 \end{array}$

19. $\begin{array}{l} 2x - y \geq 3 \\ x - y \leq -1 \end{array}$ **20.** $\begin{array}{l} 3y \geq 13 + x \\ 9y \leq 23 - x \end{array}$

21. $\begin{array}{l} -4x + 3y \leq 23 \\ 3x > 19 - 5y \end{array}$ **22.** $\begin{array}{l} 5x - 6y \geq 21 \\ 3y - 6 < 8x \end{array}$

23. $\begin{array}{l} 3x + 5y \leq 32 \\ y \geq 4 \end{array}$ **24.** $\begin{array}{l} 5x - 2y \geq 3 \\ x \leq 3 \end{array}$

25. $\begin{array}{l} 2x + 5y > 26 \\ x < 8 \end{array}$ **26.** $\begin{array}{l} -3x + 2y < 12 \\ y \geq 6 \end{array}$

Applying What You've Learned

Use a system of inequalities to represent the situations described in Exercises 27–38, then solve the system using the methods developed in this section. In drawing the graph of your solution set, you may want to take into consideration that certain quantities cannot be negative and therefore you may want to restrict the solution set appropriately as we did in Example 5.

27. Manufacturing. Scott owns a manufacturing company that produces two models of entertainment centers. The Athens requires 4 feet of fancy molding and on average takes 4 hours to manufacture. The Barcelona needs 15 feet of molding and 3 hours to manufacture. In a given week, there are 120 hours of labor available, and the company has 360 feet of molding to use for the centers.

28. Investing. Gina is considering investing no more than $3,000 total in a pharmaceutical company and Facebook stock. She intends to invest at least three times as much money in the pharmaceutical stock as she puts into Facebook stock.

29. Small business. Tami Taylor sells decorated hats and T-shirts at a Dillon Panther's football game. It takes 30 minutes to decorate a hat and 20 minutes to decorate a T-shirt. She has a total of 600 minutes to spend decorating the items.

30. Small business. Jaleel has a part-time business preparing slides for small businesses. It takes him 10 minutes to produce a plain-text slide and 18 minutes to produce a slide with graphics. He intends to spend 300 minutes per week at his business.

31. Investing. Ari Gold is planning to invest no more than $18,000 in a bond fund and a mutual fund. He plans to invest at least twice as much in the bond fund as in the mutual fund.

32. Diet. As part of her preparation for the World Cup in women's soccer, Cassandra is taking several nutritional supplements. Quantum contains 4 mg of niacin and 80 mg of calcium. NutraPlus contains 2 mg of niacin and 220 mg of calcium. She wants to supplement her diet by taking at least 12 mg of niacin and 960 mg of calcium.

33. Diet. Raphael is training for the Mr. Universe competition and is supplementing his diet with PowerUp and StressTabs. PowerUp contains 30 mg of niacin and 200 mg of vitamin C. StressTabs has 40 mg of niacin and 400 mg of vitamin C. Raphael wants to take at least 180 mg of niacin and 1,600 mg of vitamin C.

34. Small business. Joleen makes redware plates and cups to commemorate special occasions. It takes her 5 minutes to make a plate and 11 minutes to make a cup. A plate requires 1.5 pounds of red clay, and a cup requires 1.1 pounds of clay. She has on hand 33 pounds of clay and plans to work 3 hours and 40 minutes today.

35. Diet. Jared is dieting by eating sandwiches from Subway®. A small ham sandwich has 300 calories and one slice of American cheese has 60 calories. Jared wants to eat ham sandwiches with cheese and consume no more than 900 calories. He will eat at least 2 sandwiches.

36. Small business. Grace Adler owns an interior decorating business. She wants to have wooden furniture refinished. It requires 3 hours and $4 worth of materials to refinish a table and 1 hour and $1 worth of materials to refinish a chair. There are 27 hours per week available for labor and $32 per week for materials.

37. Advertising. A newly opened Espresso Bar is deciding on advertising to generate business. The owners have $2,400 total to spend for ads in the local newspaper and on the local radio station. A radio ad costs $100, and a newspaper ad costs $300. They want to have at least three times as many radio ads as they have newspaper ads.

38. Diet. Caitlin is deciding how to feed her new dog. Premium dog food provides 180 calories and 200 mg of calcium per cup, whereas the cheaper dog food provides only 150 calories and 50 mg of calcium per cup. She wants to provide her dog with at least 1,500 calories and 1,200 mg of calcium.

Communicating Mathematics

39. Why do we use the one-point test?

40. What point is usually the easiest point to use in applying the one-point test?

41. In some problems involving systems of linear inequalities, we may decide not to shade the region that lies below the *x*-axis or to the left of the *y*-axis. Why might we do this?

42. If two lines are boundaries for a solution set of a pair of linear inequalities, and one line is drawn solid and the other is drawn dotted, how should you represent the intersection point of the two lines?

In Exercises 43–46, each system of inequalities has no solutions. Solve each system graphically and then explain why the system has no solution.

43. $\begin{aligned} 2x + 3y &\geq 18 \\ 4x + 6y &\leq 16 \end{aligned}$

44. $\begin{aligned} 5x - 2y &< 10 \\ -5x + 2y &\leq -20 \end{aligned}$

45. $\begin{aligned} -2x + 5y &\geq 20 \\ 4x - 10y &> -10 \end{aligned}$

46. $\begin{aligned} -3x + 2y &\leq 12 \\ -9x + 6y &\geq 60 \end{aligned}$

Using Technology to Investigate Mathematics

47. The TI-83 screens shown in this section show how a graphing calculator can be used to solve systems of equations and inequalities. Ask your instructor for a tutorial that will show you how to use a graphing calculator and then use it to reproduce several of the examples in this section.*

48. Search the Internet to find interactive programs that graph and solve systems of inequalities. Use these programs to reproduce several of the calculations that we did in this section.

For Extra Credit

We can have more than two inequalities in a system. Solve the following systems and find all corner points:

49. $\begin{aligned} 2x + 3y &\leq 25 \\ 5y &\geq 20 + x \\ y &\leq x + 5 \\ x \geq 0,\, y &\geq 0 \end{aligned}$

50. $\begin{aligned} 3x + 2y &\leq 22 \\ x + y &\leq 8 \\ x + 4y &\leq 24 \\ x \geq 0,\, y &\geq 0 \end{aligned}$

51. $\begin{aligned} x - 2y &\leq 8 \\ 2x + y &\geq 19 \\ y &\leq x + 5 \\ x \geq 0,\, y &\geq 0 \end{aligned}$

52. $\begin{aligned} 6y &\leq 14 + 5x \\ 2y &\leq 23 - 2x \\ 13y &\geq 56 - 2x \end{aligned}$

In Exercises 53–56, use the graph to write a system of inequalities whose solution is the indicated region. Ignore the axes in describing these regions.

53. Region 1

54. Region 2

55. Region 3

56. Region 4

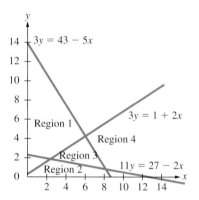

Graph showing lines $3y = 43 - 5x$, $3y = 1 + 2x$, $11y = 27 - 2x$, with Region 1, Region 2, Region 3, and Region 4 labeled.

Looking Deeper

8.3 Linear Programming

Objectives

1. Solve linear programming problems.
2. Use linear programming to optimize resources.

Do you remember Khalid, the marathon runner in Example 5 of Section 8.2, who was concerned about the amount of protein and calcium in his diet? We found that there were an infinite number of acceptable solutions to his diet problem, so you might ask, "Does it matter which one we pick?" That's a very good question, because Khalid probably has other concerns about his diet that we didn't mention. For example, he may want to keep the amount of sodium in his diet to a minimum, or he might want to keep his cost low. When you look at the problem this way, you can see that not all solutions are equally desirable.

As another example, consider Erin who makes antique copper chandeliers and sconces. In a given week, she may have on hand a certain amount of copper to use and a limited amount of time to spend in crafting the items. These two restrictions force her to make decisions as to how she will allocate her resources of copper and labor. There is always a trade-off. If she makes more chandeliers, then she has less material for making sconces. If Erin makes more sconces, then she has less time to devote to making chandeliers.

The question she would probably ask in deciding how to allocate her resources would be, "How many of each should I make to earn the biggest profit?" To answer this question, we would first model the restrictions on materials and labor by linear inequalities. Then we would represent her profit by an algebraic expression and consider how to make this expression take on its largest value. A problem such as this is an example of what we call a **linear programming problem**.

Linear Programming Problems

There is a great similarity between all linear programming problems.

 *KEY POINT*

All linear programming problems have the same three components.

COMPONENTS OF A LINEAR PROGRAMMING PROBLEM

1. The linear programming problems we study in this section will all have *two variables*, say *x* and *y*.*

2. We will represent *a set of conditions* on *x* and *y* by a system of linear inequalities. These conditions are called **(linear) constraints** and we will call the solution set for this system the **set of feasible solutions**.

3. There will be *some quantity that we wish to minimize or maximize*. We represent this quantity by a linear expression of the form $Ax + By$, which we call the **objective function**.

With these definitions, we can state the linear programming problem as

Minimize (or maximize) a linear objective function in two variables subject to a set of linear constraints.

KEY POINT

We find the solutions to a linear programming problem at corner points.

The intersection of two boundary lines for a set of feasible solutions is called a **corner point**. There is a well-known theorem in linear programming which says that we always find the solutions to linear programming problems at the corner points of the set of feasible solutions. To keep our discussion of linear programming simple, we will only consider sets of feasible solutions that are bounded on all sides by lines.

THEOREM: Linear Programming Theorem
In a linear programming problem, the objective function attains its maximum and minimum values at corner points of the set of feasible solutions.

The linear programming theorem gives us a straightforward way to solve linear programming problems.

METHOD FOR SOLVING LINEAR PROGRAMMING PROBLEMS

1. Graph the set of feasible solutions.

2. Find all corner points.

3. Evaluate the objective function for each corner point you find in step 2.

4. The largest value you find in step 3 is the maximum value for the objective function and the smallest value you find is the minimum.

We illustrate this method in Example 1.

---

*Linear programming problems can have many variables, but we will not consider such problems in this text.

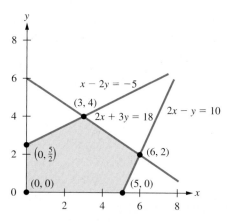

FIGURE 8.11 The set of feasible solutions for the system $x - 2y \geq -5$,
$\qquad\qquad 2x + 3y \leq 18$,
$\qquad\qquad 2x - y \leq 10$,
$\qquad\qquad x \geq 0, y \geq 0$.

EXAMPLE 1   *Solving a Linear Programming Problem by Using Corner Points*

Maximize the quantity $P = 2x - 5y$ subject to the constraints

$$x - 2y \geq -5,$$
$$2x + 3y \leq 18,$$
$$2x - y \leq 10,$$
$$x \geq 0, y \geq 0.$$

SOLUTION: We first graph the set of feasible solutions in Figure 8.11.

Next, we find the coordinates of each of the five corner points in Figure 8.11. We do this by using the elimination method to solve the system of linear equations that represents the pair of lines passing through each corner point. For example, the intersection of the lines $2x + 3y = 18$ and $2x - y = 10$ is the point $(6, 2)$ (see Table 8.4).

System	Solution	Value of $P = 2x - 5y$
$x = 0$ $x - 2y = -5$	$\left(0, \frac{5}{2}\right)$	$2 \cdot 0 - 5 \cdot \left(\frac{5}{2}\right) = -\frac{25}{2}$
$x - 2y = -5$ $2x + 3y = 18$	$(3, 4)$	$2 \cdot 3 - 5 \cdot 4 = -14$
$2x + 3y = 18$ $2x - y = 10$	$(6, 2)$	$2 \cdot 6 - 5 \cdot 2 = 2$
$2x - y = 10$ $y = 0$	$(5, 0)$	$2 \cdot 5 - 5 \cdot 0 = 10$ (maximum value for P)
$x = 0$ $y = 0$	$(0, 0)$	$2 \cdot 0 - 5 \cdot 0 = 0$

TABLE 8.4 Evaluating the objective function at corner points.

We finish the linear programming problem by evaluating the objective function $P = 2x - 5y$ for each of these corner points. From Table 8.4 we see that the maximum value of P is 10 and occurs at the corner point $(5, 0)$.

Now try Exercises 9 to 18. ✹

In many linear programming problems, there is only one point where the maximum or minimum of the objective function occurs. However, there are three other special situations that can arise in solving linear programming problems.

1. The set of feasible solutions can be empty.
2. There may be more than one point where the objective achieves its maximum and minimum values.
3. If the set of feasible solutions is unbounded, there may not be a maximum or minimum for the objective function.

We will examine these situations in the exercises.

KEY POINT

Linear programming problems help us find the "best" way to allocate resources.

Allocating Resources

In Example 2, we return to the question of how Erin should decide how many sconces and chandeliers to make.

HISTORICAL HIGHLIGHT ✺ ✺ ✺

George Dantzig—The Father of Linear Programming

In 1947, George Dantzig invented linear programming and a method for solving linear programming problems called the *simplex method*. Dantzig credits his father, Tobias Dantzig, with developing his mathematical ability. Throughout high school, his father challenged him with geometry problems, and by solving more than 10,000 of these problems, young George developed the skills that later would make him a world-class mathematician.

During World War II, Dantzig worked for the Air Force on ways to improve the deployment of troops and supplies. He felt that in order to decide on the best course of action in a situation, it was necessary to consider all possibilities, which can be an overwhelming task.

For example, suppose that you wanted to consider all the different ways you could assign 27 people to 27 different jobs. A high-speed computer that could examine 1 billion different assignments per second would take more than 300 billion years to complete this task! Yet Dantzig, using his remarkable theory of linear programming, was able to solve much larger problems in hours.

Many believed that Dantzig should have won the Nobel Prize for economics, which was awarded for economic theories based on his work. However, because his work was in mathematics, rather than economics, he did not share the prize.*

EXAMPLE 2   *Using Linear Programming to Maximize Profit*

Erin makes antique copper sconces and chandeliers. She needs 2 feet of copper to make a sconce and 16 feet of copper to make a chandelier. It takes her 1 hour to make a sconce and 3 hours to make a chandelier. She has 160 feet of copper available and plans to work 40 hours this week. Her profit on a sconce is $20, and she makes a profit of $75 on a chandelier. How many sconces and chandeliers should Erin make to achieve the greatest profit?

SOLUTION: Let us identify the components of this linear programming problem.

Variables—Let s be the number of sconces and c be the number of chandeliers.

Constraints—(1) The number of feet of copper used in making the sconces + the number of feet of copper used in making the chandeliers must be less than or equal to 160 feet.

(2) The amount of labor for making the sconces + the amount of labor for making the chandeliers must be less than or equal to 40 hours.

Objective Function—The artist wants to maximize her profit. She makes $20 per sconce and $75 per chandelier.

The information that you are given in a linear programming problem such as this usually falls into a pattern. You can understand it better if you organize it in a **constraints–objective table** as in Table 8.5.

	Sconce s	Chandelier c	Constraints
Copper	2 feet	16 feet	160 feet available
Labor	1 hour	3 hours	40 hours available
Objective: Profit	20s	75c	

TABLE 8.5 Constraints–objective table for sconce–chandelier linear programming problem.

We can write the constraints in Table 8.5 algebraically as

$$2s + 16c \leq 160 \quad \text{(copper)}$$
$$1s + 3c \leq 40 \quad \text{(labor)}$$
$$s \geq 0, c \geq 0.$$

---

*You might find it interesting to research why there is no Nobel Prize in mathematics.

The objective function has the form

$$P = 20s + 75c.$$

We graph the set of feasible solutions in Figure 8.12 and determine its corner points in Table 8.6.

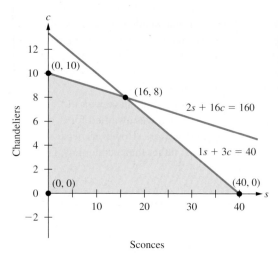

FIGURE 8.12 Set of feasible solutions for the sconce–chandelier problem.

System	Solution	Value of $P = 20s + 75c$
$s = 0, c = 0$	$(0, 0)$	$20 \cdot 0 + 75 \cdot 0 = 0$
$s = 0$ $2s + 16c = 160$	$(0, 10)$	$20 \cdot 0 + 75 \cdot 10 = 750$
$2s + 16c = 160$ $1s + 3c = 40$	$(16, 8)$	$20 \cdot 16 + 75 \cdot 8 = 920$
$1s + 3c = 40$ $c = 0$	$(40, 0)$	$20 \cdot 40 + 75 \cdot 0 = 800$

TABLE 8.6 Maximum profit occurs at $(16, 8)$.

We see that the maximum profit occurs at the point $(16, 8)$, which means Erin should make 16 sconces and 8 chandeliers. If she does this, she will make $920.

Now try Exercises 19 to 26. ❋

❋ ❋ ❋ HIGHLIGHT

Using Technology to Solve Linear Programming Problems

Throughout this chapter you have seen that to solve a linear programming problem by hand, you have to—graph lines—solve systems of linear equations to find corner points—solve a system of linear inequalities—evaluate the objective function at each corner point of the solution set.

Although each of these tasks in itself can be done in a reasonable amount of time, to do all of these tasks can be quite time-consuming. Throughout this chapter we have displayed various TI-83 screens showing how to do these tasks more quickly. Although a calculator can save you work, it is important for you to understand the problem well enough so that you can use the technology properly. And, it is equally important that you evaluate the answers you get as to whether or not they seem reasonable.

Exercises $\boxed{8.3}$

Looking Back*

These exercises follow the general outline of the topics presented in this section and will give you a good overview of the material that you have just studied.

1. In Example 1 and also in several examples in Section 8.2, we solved systems of inequalities. What are we doing differently in Example 1 that we didn't do in Section 8.2?

2. Why were the corner points of the set of feasible solutions important in Example 1?

3. What were the two constraints in Example 2? What was the purpose of the objective function?

4. Name one of George Dantzig's contributions to mathematics.

Sharpening Your Skills

Find all the corner points for each of the following sets of feasible solutions

5.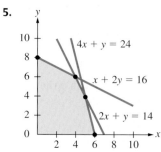

---

*Before doing these exercises, you may find it useful to review the note *How to Succeed at Mathematics* on page xix.

6.

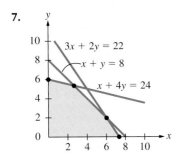

7.

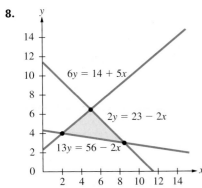

8.

Solve the given linear programming problems. Find the requested value and state at which corner point it occurs.

9. Maximize $P = 3x + 4y$ subject to the constraints

$$x + 2y \leq 16,$$
$$x + y \leq 10,$$
$$2x + y \leq 16,$$
$$x \geq 0, y \geq 0.$$

10. Maximize $P = x + y$ subject to the constraints

$$x - 3y \geq -9,$$
$$x + 2y \leq 11,$$
$$3x + y \leq 18,$$
$$x \geq 0, y \geq 0.$$

11. Minimize $P = 5x - y$ subject to the constraints

$$x - 6y \geq -18,$$
$$3x + 2y \leq 26,$$
$$x - 3y \leq 5,$$
$$x \geq 0, y \geq 0.$$

12. Minimize $P = x + 3y$ subject to the constraints

$$x + 5y \leq 40,$$
$$3x + 2y \leq 29,$$
$$4x + y \leq 32,$$
$$x \geq 0, y \geq 0.$$

13. Minimize $P = 20x - 12y$ subject to the constraints

$$x - y \geq -1,$$
$$y \leq 5,$$
$$y \geq 3,$$
$$2x + y \leq 21.$$

14. Maximize $P = 10x - 6y$ subject to the constraints

$$x - y \geq -2,$$
$$x \leq 6,$$
$$x \geq 2,$$
$$x + 4y \geq 14.$$

15. Maximize $P = 5x + 12y$ subject to the constraints

$$x - 2y \geq -2,$$
$$x + y \leq 10,$$
$$x + 4y \leq 16,$$
$$x + y \leq 4.$$

16. Maximize $P = x + y$ subject to the constraints

$$x - y \geq 2,$$
$$3x - y \geq 12,$$
$$3x + y \leq 30,$$
$$x \geq 0 \text{ and } y \geq 0.$$

17. Minimize $P = 4x + 2y$ subject to the constraints

$$6x - 7y \geq -6,$$
$$3x + 4y \leq 42,$$
$$6x - 22y \leq -51.$$

18. Minimize $P = 3x + 6y$ subject to the constraints

$$x + y \geq 13,$$
$$3x + 5y \leq 61,$$
$$2x + y \leq 22.$$

Applying What You've Learned

For each of the following linear programming problems, construct a constraints–objective table and then solve the problem by evaluating the objective function at the corner points of the set of feasible solutions as we did in Example 2.

19. Manufacturing. Scott owns a manufacturing company that produces two models of entertainment centers. The Athens requires 4 feet of fancy molding and, on average, takes 4 hours to manufacture. The Barcelona needs 15 feet of molding and 3 hours to manufacture. In a given week, there are 120 hours of labor available, and the company has 360 feet of molding to use for the centers. The company makes a profit of $9 on the Athens and $12 on the Barcelona. How many of each model should the company manufacture to maximize its profit?

20. Investing. Gina is considering investing no more than $3,000 total in a pharmaceutical company and Facebook stock. She intends to invest at least three times as much money in the pharmaceutical stock as she puts into Facebook stock. The pharmaceutical stock is currently paying a return of 7.5%, and Facebook stock is paying 12.5%. How should she divide her investment in order to maximize her return?

21. Small business. Nicole sells decorated hats and T-shirts at an arts festival. It takes 10 minutes to decorate a hat and 9 minutes to decorate a T-shirt. She makes a profit of $3 on each T-shirt and $4 on each hat and has a total of 140 minutes to spend decorating the items. She intends to decorate at least twice as many T-shirts as hats. How many T-shirts and hats should Nicole decorate to obtain the largest profit?

22. **Small business.** Jaleel has a part-time business preparing slides for small businesses. It takes him 10 minutes to produce a plain-text slide and 15 minutes to produce a slide with graphics. He intends to spend 350 minutes per week at his business. He also intends to make at least 5 more plain-text slides than graphics slides. If he makes a profit of $1 on each text slide and $2.30 on each graphics slide, what combination of slides will earn the most profit for him?

23. **Diet.** Raphael is training for the Mr. Universe competition and is supplementing his diet with PowerUp and StressTabs. PowerUp contains 30 mg of niacin and 200 mg of vitamin C. StressTabs have 40 mg of niacin and 400 mg of vitamin C. Raphael wants to take at least 180 mg of niacin and 1,600 mg of vitamin C. He wants to take no more than eight tablets. If PowerUp costs 9 cents per tablet and StressTabs costs 13 cents per tablet, what combination of tablets will be the cheapest for him to take and meet his nutritional requirements?

24. **Small business.** Joleen makes redware plates and cups to commemorate special occasions. It takes her 10 minutes to make a plate and 20 minutes to make a cup. A plate requires 1 pound of red clay, and a cup requires $\frac{1}{2}$ pound of clay. She has on hand 20 pounds of clay and plans to work 8 hours today. She expects to make no more than 28 pieces. Her profit on a plate is $12 and the profit on a cup is $8. How many plates and cups should she make today to maximize her profit?

25. **Small business.** The Reptile Farm has 400 square feet in which to house a collection of new lizards and frogs. A lizard requires 2 square feet of living space and costs $6 per month to feed. A frog also requires 2 square feet of living space but costs only $1 per month to feed. The farm has budgeted $600 per month for food. The farm makes a profit of $17 on each lizard and $6 on each frog. What combination of frogs and lizards will give the most profit?

26. **Real estate development.** Assume that on Donald Trump's *Celebrity Apprentice*, boxer Lennox Lewis and *Sopranos* star

Vincent Pastore have been assigned to work on a team that is designing an apartment building consisting of two- and three-bedroom apartments. For the project to be profitable, it must have at least 26,000 square feet devoted to the apartments, but no more than 30,000 square feet. Each two-bedroom apartment will contain 600 square feet of floor space and each three-bedroom apartment will contain 700 square feet. There will be at least as many three-bedroom apartments as there are two-bedroom apartments. The school board is concerned about the impact of this building on the school system. They believe that for every two-bedroom apartment there will be an increase of one child in the school system and for every three-bedroom apartment the increase will be two children. How many of each type of apartment should be constructed to minimize the increase in school children?

Communicating Mathematics

27. What are the three components of a linear programming problem?

28. How do we find the corner points for the set of feasible solutions in a linear programming problem?

29. Describe briefly the method for solving a linear programming problem.

30. What is the purpose of having constraints in a linear programming problem?

Using Technology to Investigate Mathematics

31. As we mentioned in the Highlight on page 388, your instructor can provide you with a graphing calculator tutorial to show you how to use a calculator to solve linear programming problems. Use your calculator to verify some of the examples in this section.

32. Search the Internet for "linear programming and ????," where ???? means some other topic. For example you could search for linear programming and some topic of interest to you in your major. I got the motivation for the chapter opener by searching for "linear programming and Hurricane Katrina."

For Extra Credit

33. The following steps describe how we would create a linear programming problem to guarantee that it has an integer solution:
    a. Select the corner points of the set of feasible solutions.
    b. Find equations of the lines passing through consecutive pairs of these points.
    c. State inequalities corresponding to the lines found in b).
    d. Experiment by choosing different objective functions until you find one that has its maximal value at (3, 4).

    Follow this procedure to construct a linear programming word problem so that the maximal value of the objective function occurs at the point (3, 4). Write a paragraph explaining how you constructed your example.

34. The following linear programming problem has no solutions. Explain why this is the case. Maximize $A = 5x - 3y$ subject to the restrictions

$$x + 2y \leq 8,$$
$$x + y \leq 6,$$
$$x + y \geq 10,$$
$$x \geq 0, y \geq 0.$$

35. The following linear programming problem has several solutions. Pick several points along the line segment joining those solutions and see what happens when you substitute their coordinates into the objective function. Describe the results of your investigations. Maximize $A = 4y + 2x$ subject to the restrictions

$$x + 2y \leq 8$$
$$x + y \leq 6$$
$$x \geq 0, y \geq 0.$$

36. The objective function for the following linear programming has no maximum. Explain why this is the case. (*Hint:* Think carefully about the set of feasible solutions.) Maximize $B = 3x + 5y$ subject to the restrictions

$$5x + 4y \geq 40,$$
$$3x + 4y \geq 32,$$
$$x \geq 0, y \geq 0.$$

CHAPTER SUMMARY

SECTION	SUMMARY	EXAMPLE
SECTION 8.1	A collection of linear equations is called a **system of linear equations**. A **solution** of the system is an ordered pair of numbers that makes all equations in the system true. We represent a system of linear equations as a **pair of lines**.	Discussion, pp. 365, 366
	In using the **elimination method** to solve a pair of linear equations, we replace the original system with an equivalent system in which the coefficients of one of the unknowns are *opposites* of each other. Adding the two equations results in a single equation in one unknown, which is easy to solve.	Discussion, p. 366
	A system of linear equations is one of three types:	
	Case 1—The *unique solution* to the system corresponds to the coordinates of the point of intersection of the lines represented by the system.	Example 1, p. 366
	Case 2—There are *no solutions* to the system that represents a pair of parallel lines.	Example 2, p. 367
	Case 3—There are an *infinite number of solutions* to the system and the two equations in the system represent the same line.	Example 3, p. 368
	We can use a system of linear equations to **model** a set of relationships between two quantities.	Example 4, p. 369
	The **law of demand** states that as the cost of an item increases, the consumer is less willing to purchase it. The **law of supply** states that as the price of an item increases, the producer is more willing to produce more of it. An **equilibrium point** is a point at which supply and demand are equal.	Example 5, p. 370
SECTION 8.2	A **linear equation in two variables** is a statement that can be written in one of the following forms: $$ax + by \geq c, \qquad ax + by > c, \qquad ax + by \leq c, \qquad ax + by < c,$$ where a, b, and c are real numbers with $a \neq 0$ and $b \neq 0$.	Discussion, p. 376
	A **solution** of a linear inequality in two variables is an ordered pair of numbers that, when substituted into the inequality, makes a true statement. The set of all solutions of a linear inequality is a **half-plane**.	Example 1, p. 376 Example 2, p. 377
	To **solve** a linear inequality in two variables:	Example 3, p. 379
	1. We change the inequality to an equation and graph the line.	
	2. If the inequality contains $\leq$ or $\geq$, we draw a solid line; otherwise, we draw a dotted line.	
	3. We use the **one-point test** to determine which half-plane is the solution set.	
	The **solution of a system of inequalities** is the intersection of the solution sets for the individual inequalities. The intersection of two boundary lines of a solution set is called a **corner point**.	Example 4, p. 380
	We can use a system of inequalities to **model** a set of relationships among two variables.	Example 5, p. 381
SECTION 8.3	All **linear programming problems** have the same components:	Discussion, p. 385
	1. Two **variables**—say, x and y.	
	2. A set of linear inequalities called **linear constraints**. The solution of this system is called the **set of feasible solutions**.	
	3. A linear expression of the form $Ax + By$, called the **objective function**, which we wish to minimize or maximize.	
	To **solve a linear programming problem**, we do the following:	Example 1, p. 386
	1. Graph the set of feasible solutions.	
	2. Find all corner points.	
	3. Evaluate the objective function at each corner point to find its maximum and minimum values.	
	We can use linear programming to **allocate resources** effectively.	Example 2, p. 387

CHAPTER REVIEW EXERCISES

Section 8.1

In Exercises 1–3, a) solve the system and b) explain the geometric significance of your solution.

1. $4x - 3y = 27$
   $x + 2y = -7$

2. $\dfrac{3}{4}x - \dfrac{1}{2}y = \dfrac{5}{4}$
   $-x + \dfrac{2}{3}y = -\dfrac{5}{3}$

3. $-4x + 5y = 3$
   $12x = 15y - 10$

4. The U.S. Olympic basketball team won a basketball game by 17 points. If 123 total points were scored, what was the final score?

5. Last month, Carmen had five Big Macs and three McChicken sandwiches for a total of 3,780 calories. If instead Carmen had eaten three Big Macs and five McChicken sandwiches, the total calories would have been 3,420. How many calories are there in each sandwich?

6. West Central State College is making holiday wreaths to raise money for a homeless shelter. From past sales experience, it was learned that if the wreaths are priced at $12 a piece, it will sell 32, but if the price is raised to $24, it will sell only 24. On the other hand, at a price of $12, the college is willing to make only 26 wreaths, but at a price of $24, it will make 30. What price should the college set so that its supply will equal the demand for the wreaths?

Section 8.2

7. Use the one-point test to solve $6x + 5y > 20$. Shade the solution.

8. Solve the system $\begin{array}{l} 2x + 5y \geq 24 \\ 2y \leq 2x + 18 \end{array}$. Find all corner points of the solution set.

9. The art club is selling custom-decorated T-shirts and hats to raise money for a homeless shelter. There is a profit of $4 on each T-shirt and $2 on each hat. The number of hats sold is at least 15 more than the number of T-shirts sold and the total profit is no more than $180. Use a system of inequalities to model this situation. Then solve the system using the methods developed in this chapter. In drawing your solution set, take into consideration that certain quantities cannot be negative and restrict the solution set appropriately.

Section 8.3

10. Solve the linear programming problem. Maximize $2x + 5y$, subject to the following constraints:
    $x \geq 0$, $\quad y \geq 0$, $\quad 3x + 4y \leq 20$, $\quad x + 2y \leq 8$.

CHAPTER TEST

In Exercises 1–3, a) solve the system and b) explain the geometric significance of your solution.

1. $6x = 2y - 4$
   $8y - 24x = 10$

2. $-3y + 4x = 6$
   $8x - 12 = 6y$

3. $3x + 4y = -6$
   $2x - 3y = 13$

4. At Dale Earnhart's car dealership, a shipment of new Hummers has just arrived. There were 3 H2s and 6 H3s and Dale's cost was $399 thousand. Last week in a similar shipment, Dale received 2 H2s and 9 H3s for a cost of $466 thousand. What is Dale's cost for the H2 and the H3 Hummers?

5. Shandra works in a fast-food restaurant for $5.65 per hour and is paid $8.50 per hour to give swimming lessons. Last week she spent 5 more hours giving swimming lessons than she worked in the restaurant. If she earned $212.30, how many hours did she work in the restaurant?

6. Krispy Kreme has found that if it prices a box of six deluxe donuts at $3, it will sell 27 boxes; if it raises the price to $8, it will sell only 17 boxes. On the other hand, Krispy Kreme will make only 17 boxes if the price is $3, but will make 32 boxes if the price is $8. How many boxes should it make if it wants the supply to equal the demand?

7. Use the one-point test to solve $4x - 3y \geq 8$. Shade the solution.

8. Solve the system $\begin{array}{l} y \leq -x + 14 \\ 2x - 5y \leq -14 \\ 5x - 2y \geq 7 \end{array}$. Find all corner points of the solution set.

9. Mickal sells decorative inlaid wood boxes as souvenirs to visitors to the Grand Canyon. A small box requires 2 square feet of wood and the large box requires 6 square feet of wood. It takes 1 hour to make a small box and 2 hours to make a large box. He has 90 square feet of wood and plans to work no more than 40 hours next week. Model this situation with a system of inequalities, and then solve this system.

10. Solve the linear programming problem. Maximize $x + 4y$, subject to the following constraints:
    $x \geq 0$, $\quad y \geq 0$, $\quad x + 3y \leq 11$, $\quad -x + 4y \leq 10$.

GROUP EXERCISES

1. In making up exercises, mathematics instructors often begin by selecting an answer and then writing a question to give that answer. Think about how you would use this approach to create a supply and demand problem. For example, you could get the supply equation by finding a line through two points. Pick a real situation, and write a complete problem using your method.

2. Similar to Exercise 1, make up a realistic linear programming problem and solve it. Perhaps you are selling hand-decorated shirts and hats as we did in one of the examples. Find actual prices, estimate times required to create hats and shirts, etc. Think about a reasonable set of constraints. You might think of more constraints than we gave in the exercises in Section 8.3. Then formulate and solve your problem.

Consumer Mathematics

The Mathematics of Everyday Life

9

As I was rewriting this chapter, there was a swirl of excitement going on in the world of personal finance. Credit card debt was at a record high and major lenders such as Citibank, American Express, and Capital One were setting aside billions of dollars to cover a flood of defaults by those who were unable to pay off their cards. In addition to credit card debt, foreclosures were increasing at an alarming rate and millions of families were in danger of losing their homes. On top of that, many students graduating from college were facing a future with overwhelming student loan payments.

If you research student loans, credit card regulations, mortgages, and various other forms of personal finance on the Internet, you would be amazed at the financial dangers that can befall an unwary consumer.

On the other hand, by understanding and applying the financial principles that you will learn in this chapter, you can avoid the financial pitfalls mentioned above and use your knowledge of the mathematics of how money works to your advantage.

9.1 Percent Change and Taxes

Objectives

1. Understand how to calculate with percent.
2. Use percents to represent change.
3. Apply the percent equation to solve applied problems.
4. Use percent in calculating income taxes.

Throughout this chapter, we will be discussing various aspects of your future financial life—student loans, credit card borrowing, investments, and mortgages. To understand these ideas and also much of the other information that you encounter daily, you must be comfortable with the notion of percent.

✏️ **KEY POINT**

Percent means "per hundred."

Percent

The word *percent* is derived from the Latin "per centum," which means "per hundred." Therefore, 17% means "seventeen per hundred." We can write 17% as $\frac{17}{100}$ or in decimal form as 0.17. In this chapter, we will usually write percents in decimal form.

EXAMPLE 1 *Writing Percents as Decimals*

a) Write each of the following percents in decimal form:

   36%   19.32%

b) Write each of the following decimals as percents:

   0.29   0.354

SOLUTION:

a) Think of 36% as the decimal thirty-six hundredths, or 0.36, as we show in Figure 9.1(a). To write 19.32%, recall the problem-solving advice from Section 1.1 that it is often helpful to solve a simpler problem instead. If we had asked you to write 19% as a decimal, you would write it as 0.19. Now, once you have positioned the 1 and the 9 properly in the decimal, write the 3 and the 2 immediately to their right, as we show in Figure 9.1(b). Thus, 19.32% is equal to 0.1932.

Quiz Yourself ❶ *

a) Write 17.45% as a decimal.
b) Write 0.05% as a decimal.
c) Write 2.45 as a percent.
d) Write 0.025 as a percent.

┌─ At first, ignore 32.

36 percent	19. 32 percent
36 hundredths	19. 32 hundredths
.36 (decimal form of 36 hundredths)	.19 32 (decimal form of 19 hundredths)
	.19 32

└ hundredths   └ hundredths

(a)   (b)

FIGURE 9.1 The key to converting percents to decimals is what you write in the hundredths place.

b) To translate decimals to percents, we first examine what numbers are in the tenths and hundredths place. To rewrite 0.29, we see that we have 29 hundredths, so 0.29 equals 29%. To rewrite 0.354, recognize that 0.35 would be 35%, so 0.354 is 35.4%.

Now try Exercises 5 to 20. ✳ ❶

PROBLEM SOLVING
The Analogies Principle

You may have been told that when converting a percent to a decimal, you move the decimal point two places to the left and that in converting a decimal to a percent, you move the decimal point two places to the right. It is all right to use such memory devices, provided that you understand where they come from.

Always remember that percent means hundredths. If you understand how to rewrite 0.29 as 29%, then you also know how to rewrite 1.29 as a percent. Similarly, if you know that 19% equals 0.19, then you also know how to rewrite 19.32% as a decimal.

If you want, you can use the following rules:

To convert from a percent to a decimal, drop the percent sign and divide by 100.

To convert from a decimal to a percent, multiply by 100 and add a percent sign.

We often have to convert a fraction to a percent. Example 2 shows how to do this.

EXAMPLE 2   *Converting a Fraction to a Percent*

Write $\frac{3}{8}$ as a percent.

SOLUTION: Because we already know how to rewrite a decimal as a percent, we first convert $\frac{3}{8}$ to a decimal and then rewrite the decimal as a percent.

If we divide the denominator into the numerator, we get $\frac{3}{8} = 0.375$. Next, we see that 0.375 is equal to 37.5%. Thus, $\frac{3}{8} = 37.5$ percent.

Now try Exercises 21 to 28. ✻

Percent of Change

Example 3 shows how you can use the simple notion of percent to compare changes in data over long periods of time.

EXAMPLE 3   *Changes in Defense Spending as a Percent of the Federal Budget over Time*

According to the Office of Management and Budget, in 1970 the U.S. government spent $82 billion for defense at a time when the federal budget was $196 billion. Thirty-seven years later, in 2007, spending for defense was $495 billion and the budget was $2,472 billion. What percent of the federal budget was spent for defense in 1970? In 2007?

SOLUTION: In 1970, $82 billion out of $196 billion was spent for defense. We can write this as the fraction $\frac{82}{196} \approx 0.418 = 41.8\%$. In 2007, this fraction was $\frac{495}{2,472} \approx 0.2002 \approx 20\%$.

Thus, you see, as a percentage of the federal budget, defense spending was considerably less in 2007 than it was in 1970. ✻

The media often uses percentages to explain the change in some quantity. For example, you may hear that, on a particularly bleak day on Wall Street, the stock market is down 1.2%. Or on a good day, the evening news tells us that consumer confidence in the economy is up 13.5% over last month. To make such statements, we have to understand several quantities.

The percent of change is always in relationship to a previous, or **base amount**. We then compare a **new amount** with the base amount as follows:

$$\text{percent of change} = \frac{\text{new amount} - \text{base amount}}{\text{base amount}}.\,^*$$

We illustrate this idea in Example 4.

EXAMPLE 4   *Finding the Percent of a Tuition Increase*

This year the tuition at Good Old State was $7,965, and for next year, the board of trustees has decided to raise the tuition to $8,435. What is the percent of increase in tuition?

SOLUTION: In this example, the base amount is $7,965 and the new amount is $8,435. To find the percent of tuition increase, we calculate

$$\text{percent of change} = \frac{\text{new amount} - \text{base amount}}{\text{base amount}}$$

$$= \frac{8,435 - 7,965}{7,965} = \frac{470}{7,965} \approx 0.059 = 5.9\%.$$

Thus, the tuition will increase almost 6% from this year to the next. ✳ **2**

Merchants often use percents to describe the deals that they are giving to the public when they have a sale. The increase that a merchant adds to his base price is called **markup**.

EXAMPLE 5   *Investigating a Sale Price on a New Car*

Monte's Autorama is having an end-of-year clearance in which the TV ads proclaim that all cars are sold at 5% markup over the dealer's cost. Monte has a new Exfinity for sale for $18,970. On the Internet, you find out that this particular model has a dealer cost of $17,500. Is Monte being honest in his advertising?

SOLUTION: To find the percent markup on this particular Exfinity, you can calculate the percent of markup, which is the same as the percent of change in the base price (dealer's cost) of the car.

We calculate as in Example 4:

$$\text{percent of markup} = \frac{\overset{\text{new amount}}{\overbrace{\text{selling price}}} - \overset{\text{base amount}}{\overbrace{\text{dealer cost}}}}{\underset{\text{base amount}}{\underbrace{\text{dealer cost}}}}$$

$$= \frac{18,970 - 17,500}{17,500} = \frac{1,470}{17,500} = 0.084 = 8.4\%.$$

Thus, Monte is not being quite truthful here because his markup on this model is 8.4%, which is well above the amount he stated in his TV ads.

Now try Exercises 57 and 58. ✳

Quiz Yourself **2**

The population of Florida increased from 12.9 million in 1990 to 16 million in 2000. What was the percent of increase of Florida's population over these 10 years?

✏ **KEY POINT**

Many percent problems are based on the same equation.

The Percent Equation

We will conclude this section with several examples of percent problems; however, it is important for you to recognize that these problems are all variations of the same equation. In each case, we will be taking some *percent* of a *base* quantity and setting it equal to an *amount*. We can write this as the equation

$$\text{percent} \times \text{base} = \text{amount}.$$

We will call this equation, *the percent equation*.

---

*If the new amount is less than the base amount, then the percent of change will be negative.

You saw this pattern in Example 5, where the percent was 8.4% = 0.084, the base (dealer's price) was $17,500, and the amount (markup) was $1,470. Notice that 0.084 × 17,500 = 1,470. In the remaining examples, we will give you two of the three quantities, percent, base, and amount, and ask you to find the third.

EXAMPLE 6   *Using the Percent Equation*

a) What is 35% of 140?

b) 63 is 18% of what number?

c) 288 is what percent of 640?

SOLUTION: We will illustrate the percent equation graphically to solve each problem.

a) The base is 140 and the percent is 35% = 0.35.

$$\text{percent} \times \text{base} = \text{amount}$$
$$\underset{0.35}{\swarrow} \qquad \underset{140}{\searrow}$$

So the amount is 0.35 × 140 = 49.

b) Using the percent equation again, we get

$$\text{percent} \times \text{base} = \text{amount.}$$
$$\underset{0.18}{\swarrow} \qquad \qquad \underset{63}{\searrow}$$

Therefore, we have 0.18 × base = 63, or, dividing both sides of this equation by 0.18, we get

$$\text{base} = \frac{63}{0.18} = 350.$$

c) Here the base is 640 and the amount is 288.

$$\text{percent} \times \text{base} = \text{amount.}$$
$$\underset{640}{\swarrow} \qquad \underset{288}{\searrow}$$

Thus, we have percent × 640 = 288. Dividing both sides of this equation by 640, we get

$$\text{percent} = \frac{288}{640} = 0.45 = 45\%.$$

Now try Exercises 29 to 38. ❋ ③

Quiz Yourself ③

a) What is 15% of 60?

b) 18 is 24% of what number?

c) 96 is what percent of 320?

EXAMPLE 7   *Calculating Sports Statistics*

In the 2006–2007 season, the Detroit Pistons, of the National Basketball Association, had a record of 53 wins and 29 losses. What percent of their games did they win?

SOLUTION: Again, you can use the percent equation to solve this problem; however, you have to be careful. The base is not 53, but rather the total number of games played, which is 53 + 29 = 82. The amount is the number of victories, 53. So we have

$$\text{percent} \times \text{base} = \text{amount.}$$
$$\underset{82}{\swarrow} \qquad \underset{53}{\searrow}$$

Dividing both sides of the equation—percent × 82 = 53—by 82 gives us

$$\text{percent} = \frac{53}{82} \approx 0.646 = 64.6\%. \ ❋$$

✻ ✻ ✻ HIGHLIGHT

Between the Numbers—Pay Careful Attention to What They *Don't* Tell You

When negotiating, people often want you to focus on their numbers and distract you from noticing other information that might be useful in making your decision. For example, consider the following situation that occurred several years ago when our faculty union was negotiating with the Commonwealth of Pennsylvania for a new contract.

The state negotiator recommended that we "back-load" the contract over 3 years by accepting raises of 0%, 2%, and 3% over 3 years instead of raises of 3%, 2%, and 0%. He stated that at the end of three years, the percentage of increase would be the same in either case, because

$1 \cdot (1.02)(1.03) = (1.03)(1.02) \cdot 1 = 1.0506$, giving a 5.06% increase. If we focus just on the percent of increase by the third year, the negotiator was telling the truth. So you might ask, "What's the difference?"

In the spirit of the Three-Way Principle, let's look at what effect both types of raises will have on a salary of $100 in Table 9.1.

With front-loading, we earn $13.12 - $7.06 = $6.06 more money on our initial $100 than with back-loading. Over three years, a person earning $50,000 (which is 500 times as large as $100) would earn 500($6.06) = $3,030 more with front-loading versus back-loading.

	Back-Loading	**Front-Loading**
Original Amount	$100	$100
Amount in 1st year	$100 + 0% ($100) = $100	$100 + 3% (100) = $103
Amount in 2nd year	$100 + 2% (100) = $102	$103 + 2% (103) = $105.06
Amount in 3rd year	$102 + 3% (102) = $105.06	$105.06 + 0% (105.06) = $105.06
	In 3 years, we gain 2 + 5.06 = $7.06 more than if we had no raise at all.	In 3 years, we gain 3 + 5.06 + 5.06 = $13.12 more than if we had no raise at all.

TABLE 9.1 Comparing back-loading versus front-loading.

EXAMPLE 8   *Increase in Student Loan Debt*

According to the American Association of State Colleges and Universities, in 2006 the average borrower who graduated from a public college owed $17,250 from student loans. This amount was up 115.625% from 1996. Find the average amount of student loan debt that graduates from these schools owed in 1996.

SOLUTION: In this case, the base is unknown. In deciding what percent to use in the percent equation, you have to be careful. The 115.625% is only the increase. The amount $17,250 represents 100% of the debt owed in 1996 plus the 115.625% increase. Therefore, the percent that we will use in the percent equation is

$$100\% + 115.625\% = 215.625\% = 2.15625.$$

Substituting in the percent equation, we get

$$\underset{\text{percent}}{2.15625} \times \text{base} = \underset{\text{amount}}{17{,}250}.$$

Dividing both sides of this equation by 2.15625, gives us

$$\text{base} = \frac{17{,}250}{2.15625} = 8{,}000.$$

So, the average student loan debt in 1996 was $8,000. ✻

Taxes

Calculating various kinds of taxes relies heavily on using percents properly.

EXAMPLE 9   *Calculating Your Income Tax*

Table 9.2 is taken from the instructions for filling out Form 1040 to compute the federal income tax for a person whose marriage status is single.

	If your taxable income is Over—	But not Over—	The tax is	Of the amount Over—
Line 1	$0	7,550	. . . . 10%	0$
Line 2	7,550	30,650	$755.00 + 15%	7,550
Line 3	30,650	74,200	$4,220.00 + 25%	30,650
Line 4	74,200	154,800	$15,107.50 + 28%	74,200
Line 5	154,800	336,550	$37,675.50 + 33%	154,800
Line 6	336,550	. . . .	$97,653.00 + 38%	336,550

TABLE 9.2 Federal income taxes due for a single person.

a) If Jaye is unmarried and has a taxable income* of $41,458, what is the amount of federal income tax she owes?

b) How did the IRS arrive at the $4,220 amount in column 3 of line 3?

SOLUTION:

a) In calculating this tax, you first must identify the line of the table that is relevant to Jaye's situation. Because her income is above $30,650 and below $74,200, we will use line 3 (highlighted) from Table 9.2.

 So Jaye must pay $4,220 + 25% of the amount of taxable income over $30,650. Therefore, her tax is

$$4,220 + (0.25)(41,458 - 30,650) = 4,220 + (0.25)(10,808) = 4,220 + 2,702 = \$6,922.$$

with labels: 25% and amount over $30,650.

Quiz Yourself 4

Use Table 9.2 to calculate Aliyah's federal income tax that is due if her taxable income last year was $85,500.

b) The table is treating Jaye's income as being divided into two parts. Up to $30,650, she is being taxed according to the instructions on line 2 of the table. The tax on $30,650 would be $755 + 15% of the amount of taxable income over $7,550. So her tax is

$$755 + (0.15)(30,650 - 7,550) = 755 + (0.15)(23,100) = 755 + 3,465 = \$4,220.$$

Now try Exercises 67 to 70. ✳ 4

---

*It is too complicated to get into detail here, but in essence, after you have totaled your wages, tips, interest earned, etc., you then reduce this total by making various kinds of adjustments such as exemptions, deductions for charitable contributions, work-related expenses, and other deductions to calculate what is called your *taxable income*.

Exercises [9.1]

Looking Back*

These exercises follow the general outline of the topics presented in this section and will give you a good overview of the material that you have just studied.

1. In Example 1, what was the key in writing 19.32% as a decimal?

2. In converting a fraction to a percent as we did in Example 2, what did we do first?

3. How did we find the percent of markup in Example 5?

4. Why is it not a good idea to make conversions between decimals and percents by thinking only about moving the decimal point one way or the other?

Sharpening Your Skills

Convert each of the following percents to decimals.

5. 78%   6. 65%   7. 8%   8. 3%

9. 27.35%   10. 83.75%   11. 0.35%   12. 0.08%

Write each of the following decimals as percents.

13. 0.43   14. 0.95   15. 0.365   16. 0.875

17. 1.45   18. 2.25   19. 0.002   20. 0.0035

Convert each of the following fractions to percents.

21. $\frac{3}{4}$ 22. $\frac{7}{8}$ 23. $\frac{5}{16}$ 24. $\frac{9}{25}$

25. $\frac{5}{2}$ 26. $\frac{11}{8}$ 27. $\frac{4}{250}$ 28. $\frac{3}{500}$

29. 12 is what percent of 80?   30. What is 24% of 125?

31. 77 is 22% of what number?

32. 33.6 is what percent of 96?

33. What is 12.25% of 160?

34. 47.74 is 38.5% of what number?

35. 8.4 is what percent of 48?   36. What is 23% of 140?

37. 29.76 is 23.25% of what number?

38. 149.5 is what percent of 130?

Applying What You've Learned

39. **Cookie sales.** In a recent year, the top-selling cookie in America was Nabisco's Chips Ahoy with sales of $294.6 million. The total cookie sales for that year were $3,124 million. What percent of the total cookie sales was due to Chips Ahoy? (*Source:* Information Resources, Inc.)

40. **Pizza sales.** In a recent year, DiGiorno sold $478.3 million worth of frozen pizzas. If total frozen pizza sales were $2,844.8 million, what percent of frozen pizza sales was due to DiGiorno? (*Source:* Information Resources, Inc.)

41. **Price of new homes.** According to the U.S. Bureau of the Census, from 2005 to 2006, the average price of a new home in the Northeast increased by 7.88% to $428,000. What was the average price of a new home in the Northeast in 2005? Round your answer to the nearest thousand.

42. **Music stations.** The number of country music stations decreased by 6.6% from 2004 to 2006. If there were 2,045 country music stations in 2006, how many were there in 2004? (*Source:* M Street Corporation.)

According to USA Today, the number of visitors (in millions) to the top five Web sites in 2006 is given in the following graph. Use this information to solve Exercises 43–46.

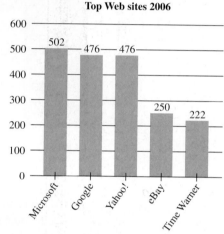

43. **Comparing Web site visitors.** What percent of the total visitors to these five sites visited Microsoft?

44. **Comparing Web site visitors.** What percent of the visitors to these sites was due to Google and eBay combined?

45. **Comparing Web site visitors.** What percent greater was the number of visitors to Yahoo! than Time Warner?

46. **Comparing Web site visitors.** What percent smaller was the number of visitors to Time Warner than eBay?

The graph below shows the sales (in thousands) of the five top-selling cars in 2006. Use this information to solve Exercises 47–50. (Source: The New York Times 2007 Almanac.)

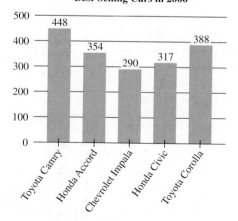

---

*Before doing these exercises, you may find it useful to review the note *How to Succeed at Mathematics* on page xix.

47. **Comparing car sales.** What percent of the sales of these five cars was due to the Toyota Camry?

48. **Comparing car sales.** What percent of the sales of these five cars was due to Hondas?

49. **Comparing car sales.** What percent greater was sales of the Camry than the Civic?

50. **Comparing car sales.** What percent smaller was the sales of the Impala than the Honda models?

51. **Autographs.** According to *The Useless Information Society*, only 6% of Beatles' autographs in circulation are estimated to be real. If there are 81 authentic autographs of the Beatles in circulation, then how many nonauthentic autographs are there in circulation?

52. **Happy Meal sales.** In the fourth quarter of 2007, McDonald's profits from Happy Meals were $508 million. If that accounted for 40% of the profits, how much profit came from selling other items?

53. **Population.** According to the U.S. Bureau of the Census, from 2000 to 2007 the population of the United States grew from 281 million to 301.6 million. What was the percent of increase?

54. **Population.** From 2000 to 2007 California's population grew from 33.9 million to 36.6 million. What was the percent of increase?

55. **Music downloads.** The Recording Industry Association of America found that downloaded album sales increased by 103.3% from 2005 to 2006. If there were 13.6 million albums downloaded in 2005, how many were downloaded in 2006?

56. **Music video sales.** The Recording Industry Association of America found that music video sales decreased by 31.7% from 2005 to 2006. If there were 33.8 million videos sold in 2005, how many were sold in 2006?

57. **Dealer markup on a car.**

   a. If a dealer buys a car from the manufacturer for $14,875 and then sells it for $16,065, what is his markup?

   b. In doing this problem, Angela got an incorrect answer of 7.4%. What mistake did she make?

58. **Dealer markup on a computer.**

   a. A computer retailer buys a multimedia computer for $1,850 and then sells it for $2,081. What is her markup on the computer?

   b. In doing this problem, Deepak got an incorrect answer of 11.1%. What mistake did Deepak make?

59. **Salary increase.** Marcy is working in a 2-year probationary period as a paralegal. She currently earns $28,000 a year and after her probationary period, her salary will increase by 35%. What will her yearly salary be at that time?

60. **Salary decrease.** Jed works for Metal Fabricators, Inc., which just lost a large contract and has asked employees to take a 12% pay cut. If Jed now makes $34,500 per year, what will his yearly pay be after the pay cut?

61. **Buying a grill.** Kirby is buying a new gas barbecue grill that has been reduced for an end-of-summer sale by 15% to $578. What was the original price of the grill?

62. **Increasing workload.** Gisela is an insurance claims adjuster. Last quarter she settled 124 claims and this quarter she settled 155 claims. What is the percent increase in the claims that she settled this quarter over last quarter?

63. **Markup on a boat.** Carlos is a boat dealer who bought a sailboat for $11,400 and then sold it for $12,711. What percent was his markup on the boat?

64. **Markup on an appliance.** Anna, who owns a small appliance store, bought a food processor for $524 and then sold it for $589.50. What percent was her markup on the food processor?

65. **Change in the stock market.** Due to a slump in the economy, Omarosa's mutual fund has dropped by 12% from last quarter to this quarter. If her fund is now worth $11,264, how much was her fund worth last quarter?

66. **Gas mileage.** Renaldo bought a new Honda Civic hybrid that gets 26.3% more miles per gallon than he got with his old Civic. If his new car gets 48 miles per gallon, what was the gas mileage on his old car?

For Exercises 67–70, use Table 9.2 in Example 9 to calculate the federal income tax due for the given taxable income.

67. $148,000

68. $28,750

69. $47,800

70. $440,000

Communicating Mathematics

Assume that you are studying with a friend for a quiz on percents. Your friend insists on blindly memorizing which way to "move the decimal point" in doing percent problems. Explain how you would help your friend understand how to do the conversions in Exercises 71–74 without relying solely on memorization.

71. You are converting 28.35% to a decimal.

72. You are converting 1.285 to a percent.

73. You are converting 0.0375 to a percent.

74. You are converting 1.375% to a decimal.

75. What is the meaning of *percent*?

76. What is the percent equation?

77. How could we have used the percent equation to solve Example 5?

78. Explain how the amount $15,107.50 was determined in line 4 of Table 9.2.

For Extra Credit

79. **Depreciation on a new car.** If a new car costs $18,000 and depreciates at a rate of 12% per year, what will be the value of the car in 4 years?

80. **Compounding raises.** If you are given a raise of 8% this year and a raise of 5% the following year, what single raise would give you the same yearly salary in the second year?

81. **Calculating prices.** If a merchant increases the price of a home entertainment system by x% and then later reduces the price by x%, is the price of the system the same as the original price? If not, what is the relationship between the two prices? Explain your answer by using appropriate examples.

82. Calculating prices. If a merchant reduces the price of a luxury speedboat by $x\%$ and then later increases the price by $x\%$, is the price of the boat the same as the original price? If not, what is the relationship between the two prices? Explain your answer by using appropriate examples.

83. Compounding percents. If you increase an amount by 10% and then again by 20%, is that the same as increasing the amount in one step by 30%? Explain your answer by using appropriate examples.

9.2 Interest

Objectives

1. Understand the simple interest formula.
2. Use the compound interest formula to find future value.
3. Solve the compound interest formula for different unknowns, such as the present value, length, and interest rate of a loan.

The most powerful force in the universe is. . . .

How would you finish this quote? The world-renowned physicist Albert Einstein said,

. . . compound interest.

Are you surprised that of all the forces that he might pick, Einstein chose this one? In this section, we will explain how interest can either work for you—or against you. As you will see, used properly, it can help you build a fortune; used improperly, it can lead you to financial ruin.

If you want to accumulate enough money to buy a newer car or go on a vacation, you could deposit money in a bank account. The bank will use your money to make loans to other customers and pay you interest for using your funds. However, if you borrow money from the bank, say to take a college course, then you will pay interest to the bank. In essence, **interest** is the money that one person (a borrower) pays to another (a lender) to use the lender's money. Savers earn interest; borrowers pay interest.

We will discuss simple and compound interest in this section, and discuss the cost of consumer loans in Section 9.3.

✎ **KEY POINT**

Simple interest is a straightforward way to compute interest.

Simple Interest*

The amount you deposit in a bank account is called the **principal**. The bank specifies an **interest rate** for that account as a percentage of your deposit. The rate is usually expressed as an annual rate. For example, a bank may offer an account that has an annual interest rate of 5%. To find the interest that you will earn in such an account, you also need to know how long the deposit will remain in the account. The time is usually stated in years. There is a simple formula that relates principal, interest earned, interest rate, and time. In words,

interest earned = principal × interest rate × time.

When we compute interest this way, it is called **simple interest**.

> **FORMULA FOR COMPUTING SIMPLE INTEREST** We calculate simple interest using the formula
>
> $$I = Prt,$$
>
> where I is the interest earned, P is the principal, r is the interest rate, and t is the time in years.

*If you want some practice with basic algebra, see Appendix A.

EXAMPLE 1 *Calculating Simple Interest*

If you deposit $500 in a bank account paying 6% annual interest, how much interest will the deposit earn in 4 years if the bank computes the interest using simple interest?

SOLUTION: In this example:

P is the principal, which is $500

r is the annual interest rate, which is 6% (written as 0.06)

t is the time, which is 4 (years)

Thus, the interest earned is

$$I = Prt = 500 \times 0.06 \times 4 = 120.$$

In 4 years, this account earns $120 in interest.

Now try Exercises 5 to 8. ❋

✏️ **KEY POINT**

Future value equals principal plus interest.

To find the amount that will be in your account at some time in the future, called the **future value** (or sometimes called the **future amount**) we add the principal and the interest earned. We will represent future value by A, so we can say

$$A = \text{principal} + \text{interest} = P + I.$$

If we replace I by Prt, we get the formula $A = P + Prt = P(1 + rt)$.

> **COMPUTING FUTURE VALUE USING SIMPLE INTEREST** To find the future value of an account that pays simple interest, use the formula
>
> $$A = P(1 + rt),$$
>
> where A is the future value, P is the principal, r is the annual interest rate, and t is the time in years.

EXAMPLE 2 *Computing Future Value Using Simple Interest*

Assume that you deposit $1,000 in a bank account paying 3% annual interest and leave the money there for 6 years. Use the simple interest formula to compute the future value of this account.

SOLUTION: We see that $P = 1,000$, $r = 0.03$, and $t = 6$. Therefore,

$$A = \overset{P}{1,000}(1 + \overset{r}{(0.03)}\overset{t}{(6)}) = 1,000(1 + 0.18) = 1,000(1.18) = 1,180.$$

Thus, your bank account will have $1,180 at the end of 6 years. ❋

In contrast to future value, the principal that you have to invest in an account now to have a specified amount in the account in the future is called the **present value** of the account. Notice that the formula for computing future value has four unknowns. If we want, we can use this formula for finding the present value of an account provided we know the future value, interest rate, and time.

EXAMPLE 3 *Finding the Present Value of an Account*

Assume that you plan to save $2,500 to take a white-water rafting trip in Costa Rica in 2 years. Your bank offers a certificate of deposit (CD) that pays 4% annual interest computed using simple interest. How much must you put in this CD now to have the necessary money in 2 years?

SOLUTION: We can use the formula

$$A = P(1 + rt).$$

We know that $A = 2,500$, $r = 4\% = 0.04$, and $t = 2$. Therefore,

$$2,500 = P(1 + (0.04)(2)).$$

We can rewrite this equation as

$$2,500 = P(1.08).$$

Dividing both sides of the equation by 1.08, we get

$$P = \frac{2,500}{1.08} \approx 2314.814815.$$

We will round this *up* to \$2,314.82 to guarantee that if you put this amount in the CD now, in 2 years you will have the \$2,500 you need for your white-water rafting trip.*

Now try Exercises 9 to 14. ❋ **5**

Redo Example 3, but now assume that you want to save \$2,400 in 4 years and the CD has an annual interest rate of 5%.

 Some Good Advice

In Example 3, we used the earlier formula for computing future value to find the present value rather than stating a new formula to solve this specific problem. You will find it easier to learn a few formulas well and use them, together with simple algebra, to solve new problems rather than trying to memorize separate formulas for every type of problem.

✏️ *KEY POINT*

Compounding pays interest on previously earned interest.

Compound Interest

It seems fair that if money in a bank account has earned interest, the bank should compute the interest due, add it to the principal, and then pay interest on this new, larger amount. This is in fact the way most bank accounts work. Interest that is paid on principal plus previously earned interest is called **compound interest**. If the interest is added yearly, we say that the interest is *compounded annually*. If the interest is added every three months, we say the interest is *compounded quarterly*. Interest also can be compounded monthly and daily.

EXAMPLE 4 *Calculating Compound Interest the Long Way*

Assume that you want to replace your sailboat with a larger one in 3 years. To save for a down payment for this purchase, you deposit \$2,000 for 3 years in a bank account that pays 10% annual interest,[†] compounded annually. How much will be in the account at the end of 3 years?

SOLUTION: We will perform the compound interest calculations one year at a time in the following table. In compounding the interest, we will use the future value from the previous year as the new principal at the beginning of the year. Notice that the quantity $(1 + rt) = (1 + 0.10 \times 1) = (1.10)$ remains the same throughout the computations.

Continue Example 4 to calculate the amount in your account at the end of the fourth year.

Year	Principal (Beginning of Year) P	Future Value (End of Year) $P(1 + rt) = P(1.10)$
1	\$2,000	\$2,000(1.10) = \$2,200
2	\$2,200	\$2,200(1.10) = \$2,420
3	\$2,420	\$2,420(1.10) = \$2,662

❋ **6**

---

*When calculating a deposit to accumulate a future amount, we will always round up to the next cent.
†An interest rate of 10% would be extraordinarily high. However, we will often choose rates in examples and exercises to keep the computations simple.

PROBLEM SOLVING
Verify Your Answer

You should always check answers to see whether they are reasonable. In Example 4, if we had used simple interest to find the future value, we would have obtained $A = 2,000 (1 + (0.10)(3)) = 2,000(1.30) = 2,600$. The interest we found in Example 4 is a *little* larger because as the interest is added to the principal each year, the bank is paying interest on an increasingly larger principal.

If we were to continue the process that we used in Example 4 for a longer period of time, say for 30 years, it would be very tedious. In Figure 9.2 we look at the same computations in a different way, keeping in mind that the amount in the account at the end of each year is 1.10 times the amount in the account at the beginning of the year.

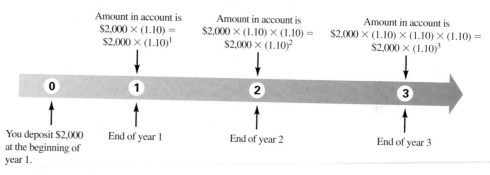

FIGURE 9.2 10% interest being compounded annually.

If we were to continue the pattern shown in Figure 9.2 to compute the future value of the account at the end of 30 years, we would see that

$$A = 2,000(1.10)^{30} \approx 2,000(17.44940227) \approx 34,898.80. *$$

This large amount shows how your money can grow if it is compounded over a long period of time.

In general, if we deposit a principal P in an account paying an annual interest rate r for t years, then the future value of the account is given by the formula

money you will
have in the future ⟶ money you have now

$$A = P(1 + r)^t.$$

In the example that we just calculated, $P = 2,000$, $r = 0.10$, and $t = 30$. It is important to understand that this formula for calculating compound interest only works for the case when r is the *annual interest rate* and t is time being measured in *years*. Do not bother to learn this formula because in just a moment we will give you a similar compounding formula that works for more general situations. **7**

Solving for Unknowns in the Compound Interest Formula

All banks and most other financial institutions compound interest more frequently than once a year. For example, many banks send savings account customers a monthly statement showing the balance in their accounts. So far in our discussion of compounding, we have used a yearly interest rate. If compounding takes place more frequently, then the interest rate must be adjusted accordingly. For example, a yearly interest rate of $12\% = 0.12$

Quiz Yourself

Calculate the future value of an account containing $3,000 for which the annual interest rate is 4% compounded annually for 10 years.

✎ **KEY POINT**

Knowing the principal, the periodic interest rate, and the number of compounding periods, it is easy to determine future value.

---

*To ensure greater accuracy, we often show calculations with eight decimal places. If your calculations do not agree with ours, it may be due to the difference in the way we are rounding our calculations.

corresponds to a monthly interest rate of $\frac{12\%}{12} = \frac{0.12}{12} = 0.01 = 1\%$. If the interest is being compounded quarterly, the quarterly interest rate would then be $\frac{12\%}{4} = \frac{0.12}{4} = 0.03 = 3\%$.

In order to handle situations such as these, we will modify the formula $A = P(1 + r)^t$ slightly.

> **THE COMPOUND INTEREST FORMULA** Assume that an account with principal P is paying an annual interest rate r and compounding is being done m times per year. If the money remains in the account for n time periods, then the future value, A, of the account is given by the formula
>
> $$A = P\left(1 + \frac{r}{m}\right)^n.$$
>
> Notice that in this formula, we have replaced r by $\frac{r}{m}$, which is the annual rate divided by the number of compounding periods per year, and t by n, which is the number of compounding periods.

You can use the compound interest formula for computing compound interest to compare investments.

EXAMPLE 5 *Understanding How "No Payments Until . . ." Works*

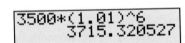

You have seen a home fitness center on sale for $3,500 and what really makes the deal attractive is that there is no money down and no payments due for 6 months. Realize that although you do not have to make any payments, the dealer is not loaning you the money for 6 months for nothing. You have borrowed $3,500 and, in 6 months, your payments will be based upon that fact. Assuming that your dealer is charging an annual interest rate of 12%, compounded monthly, what interest will accumulate on your purchase over the next 6 months?

SOLUTION: To determine the interest that has accumulated, we will find the future value of your "loan" (assuming that you make no payments) and subtract $3,500 from that. We will use the formula for calculating future value with $P = 3,500$, $r = 0.12$, $m = 12$, and $n = 6$. Therefore,

monthly interest rate ⌐ ⌐number of months

$$A = P\left(1 + \frac{r}{m}\right)^n = 3,500\left(1 + \frac{0.12}{12}\right)^6 = 3,500(1.01)^6 = 3,715.33.$$

So the accumulated interest is $3,715.33 - $3,500 = $215.33.

Now try Exercises 19 to 26. ❋ ⑧

Quiz Yourself ⑧

Sarah deposits $1,000 in a CD paying 6% annual interest for 2 years. What is the future value of her account if the interest is compounded quarterly?

❋ ❋ ❋ HIGHLIGHT

Between the Numbers—It Doesn't Hurt to Ask

In Example 5, you might ask yourself if you would be better off borrowing the $3,500 from another source that has a lower interest rate and paying for the fitness center outright.

If you have the money, sometimes a dealer might give you a better price if you offer to pay for an item with cash. The trick, of course, is to be able to put money aside so that when you want to make a deal, you are not at the mercy of someone else's money.

— HIGHLIGHT ✺ ✺ ✺

Doing Financial Calculations with a Calculator*

When doing financial computations, often technology can speed up your work. We will use a calculator to reproduce the solution to Example 6.

On my calculator, if we press the [2nd] [Finance] keys, Screen 1 comes up. The letters TVM stand for "Time Value of Money." Then by choosing option 1, we get Screen 2. Now we can enter the values 18 for N, the number of years; 4.8 for I%,

the annual interest rate; 60,000 for FV, the future value; and 4 for C/Y, the number of compounding periods per year. Next we position the cursor over PV (present value) and press the keys [Alpha] [Solve]. The amount −25418.75939 for present value means that we must deposit $25,418.76 now to have the desired $60,000 in 18 years (Screen 3).

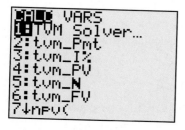

Screen 1

Screen 2

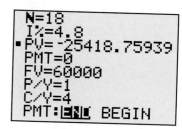

Screen 3

Example 6 illustrates a different way to use the compound interest formula.

EXAMPLE 6 *Finding the Present Value for a College Tuition Account*

Upon the birth of a child, a parent wants to make a deposit into a tax-free account to use later for the child's college education. Assume that the account has an annual interest rate of 4.8% and that the compounding is done quarterly. How much must the parent deposit now so that the child will have $60,000 at age 18?

SOLUTION: We will use the compound interest formula $A = P\left(1 + \frac{r}{m}\right)^n$. Because we know $A = 60{,}000$, $r = 0.048$, $n = 72$, and $m = 4$, we can find the present value by solving the equation

$$60{,}000 = P\left(1 + \frac{0.048}{4}\right)^{72} = P(1 + 0.012)^{72}$$

for P. Therefore,

$$P = \frac{60{,}000}{(1.012)^{72}} = \frac{60{,}000}{2.360461386} \approx 25{,}418.76.$$

A deposit slightly over $25,400 now will guarantee $60,000 for college in 18 years.

Now try Exercises 33 and 34. ✳

Although $60,000 may seem like a lot of money, realize that *inflation*, the increase in the price of goods and services, will also cause the cost of a college education to increase. We will consider the effects of inflation in the exercises.

So far we have used the formula $A = P\left(1 + \frac{r}{m}\right)^n$ to find A and P. Sometimes we want to find r or n. To do this, we need to introduce some new techniques.

If you want to solve for n in the formula $A = P\left(1 + \frac{r}{m}\right)^n$, you need to be able to solve an equation of the form $a^x = b$, where a and b are fixed numbers. A property of logarithmic functions enables you to solve such equations. Many calculators have a key labeled either "log" or "log *x*," which stands for the **common logarithmic function**. Pressing this key

✏ **KEY POINT**

We use the log function to solve for *n* in the formula

$$A = P\left(1 + \frac{r}{m}\right)^n.$$

---

*For this example, I am using a TI-83 calculator, but many other calculators have similar features for doing financial calculations. On the TI-83 plus and TI-84, press the [APPS] key and then choose option 1 to get screen 1.

reverses the operation of raising 10 to a power. For example, suppose that you compute $10^5 = 100{,}000$ on your calculator. If you next press the log key, the display will show 5. If you enter 1,000, which is 10 raised to the third power, and press the log key, the display will show 3. Practice finding the log of powers of 10 such as 100 and 1,000,000. If you enter 23 and then press the log key, the display will show 1.361727836. The interpretation of this result is that $10^{1.361727836} = 23$.* The log function has an important property that will help us solve equations of the form $a^x = b$.

EXPONENT PROPERTY OF THE LOG FUNCTION

$$\log y^x = x \log y$$

To understand this property, you should use your calculator to verify the following:

$$\log 4^5 = 5 \log 4$$
$$\log 6^3 = 3 \log 6$$

Example 7 illustrates how to use the exponent property to solve equations.

EXAMPLE 7 *Solving an Equation Using the Exponent Property of the Log Function*

Solve $3^x = 20$.

SOLUTION: We illustrate the steps required to solve this equation.

Step 1	Take the log of both sides of the equation.	$\log 3^x = \log 20$
Step 2	Use the exponent property of the log function.	$x \log 3 = \log 20$
Step 3	Divide both sides by log 3.	$x = \dfrac{\log 20}{\log 3}$
Step 4	Use a calculator to evaluate the right side of the equation (your calculator may give a slightly different answer).	$x = 2.726833028$

Quiz Yourself **9**

Solve $6^x = 15$.

Now try Exercises 35 to 42. ❋ **9**

In Example 8, we use the exponent property of the log function to find the time it takes an investment to grow to a certain amount.

EXAMPLE 8 *Saving for Equipment for a Business*

Mara wants to buy lighting equipment from her cousin to start a dance studio. He will sell his equipment for $2,800. She presently has $2,500 and found an investment that will pay her 9% annual interest, compounded monthly. In how many months will Mara be able to pay her cousin for the equipment?

SOLUTION: We know that the future value that Mara must pay her cousin is $A = 2{,}800$. She presently has $2,500 and the monthly interest rate is $\frac{r}{m} = \frac{0.09}{12} = 0.0075$. We must solve the compound interest formula $A = P\left(1 + \frac{r}{m}\right)^n$ for n, which represents the number of months of the compounding. Substituting for A, P, and $\frac{r}{m}$, we get the equation

$$2{,}800 = 2{,}500\left(1 + \frac{0.09}{12}\right)^n.$$

---

*We will not discuss what it means to raise 10 to a power such as 1.361727836.

We solve this equation by the following steps:

$$1.12 = (1.0075)^n$$ Divide both sides of the equation by 2,500 and simplify.

$$\log(1.12) = \log(1.0075)^n$$ Take the log of both sides.

$$\log(1.12) = n\log(1.0075)$$ Use the exponent property of the log function.

Quiz Yourself 10

Do Example 8 again, but now assume that the interest rate is 6%.

Solving for n, we get the equation

$$n = \frac{\log(1.12)}{\log(1.0075)} \approx 15.17.$$

This means that Mara will have the money she needs by the end of the 16th month. ✹ 10

The last situation that we will consider is how to solve the compound interest equation $A = P\left(1 + \frac{r}{m}\right)^n$ for r. To do this, we have to be able to solve an equation of the form $x^a = b$, where a and b are fixed numbers. We show how to solve such an equation in Example 9.

EXAMPLE 9   *Negotiating a Basketball Contract*

Kobe is negotiating a new basketball contract with the Lakers and expects to retire after playing one more year. In order to reduce his current taxes, his agent has agreed to defer a bonus of $1.4 million to be paid as $1.68 million in 2 years. If the Lakers invest the $1.4 million now, what rate of investment would they need to have $1.68 million to pay Kobe in 2 years? Assume that you want to find an annual interest rate that is compounded monthly.

SOLUTION: To solve this compound interest problem, we again use the formula $A = P\left(1 + \frac{r}{m}\right)^n$. We know that $A = 1.68$, $P = 1.4$, $m = 12$, and $n = 24$.

Substituting for A, P, m, and n, we get the equation

$$1.68 = 1.4\left(1 + \frac{r}{12}\right)^{24}.$$

Dividing both sides of the equation by 1.4 gives us $1.2 = \left(1 + \frac{r}{12}\right)^{24}$. We can get rid of the exponent 24 if we raise both sides of the equation to the $\frac{1}{24}$ power. This gives us the equation

$$(1.2)^{1/24} = \left(\left(1 + \frac{r}{12}\right)^{24}\right)^{1/24} = 1 + \frac{r}{12}.^*$$

Subtracting 1 from both sides of the equation, we get

$$\frac{r}{12} = (1.2)^{1/24} - 1 = 1.00762566 - 1 = 0.00762566.$$

Now, multiplying this equation by 12, we find the annual interest rate, r, to be $12(0.00762566) \approx 0.0915$. Thus, the Lakers need to find an investment that pays an annual interest rate of about 9.15% compounded monthly.

Now try Exercises 43 to 46. ✹

 Some Good Advice

Be careful to distinguish between the situations in Examples 8 and 9. In Example 8, we used the log function to solve an equation of the form $a^x = b$. In Example 9, we solved an equation of the form $x^a = b$ by raising both sides of the equation to the $\frac{1}{a}$ power.

---

*In algebra, $(a^x)^y = a^{xy}$. That is why $\left(\left(1 + \frac{r}{12}\right)^{24}\right)^{1/24} = \left(1 + \frac{r}{12}\right)^{(24)(1/24)} = \left(1 + \frac{r}{12}\right)^1 = 1 + \frac{r}{12}$.

Exercises 9.2

Looking Back*

These exercises follow the general outline of the topics presented in this section and will give you a good overview of the material that you have just studied.

1. How did we find the present value in Example 3?

2. Why did we divide the yearly interest rate of 0.12 by 12 in Example 5?

3. What property of the log function did we use to solve the equation $3^x = 20$ in Example 7?

4. What was our recommendation in the "Between the Numbers" Highlight following Example 5?

Sharpening Your Skills

In Exercises 5–8, use the simple interest formula $I = Prt$ and elementary algebra to find the missing quantities in the table below.

	I	P	r	t
5.		$1,000	8%	3 years
6.	$196		7%	2 years
7.	$700	$3,500		4 years
8.	$1,920	$8,000	6%	

In Exercises 9–14, use the future value formula $A = P(1 + rt)$ and elementary algebra to find the missing quantities in the table below.

	A	P	r	t
9.		$2,500	8%	3 years
10.		$1,600	4%	5 years
11.	$1,770		6%	3 years
12.	$2,332		3%	2 years
13.	$1,400	$1,250		2 years
14.	$966	$840	5%	

In Exercises 15–18, you are given an annual interest rate and the compounding period. Find the interest rate per compounding period.

15. 18%; monthly

16. 8%; quarterly

17. 12%; daily†

18. 10%; daily

In Exercises 19–26, you are given the principal, the annual interest rate, and the compounding period. Use the formula for computing future value using compound interest to determine the value of the account at the end of the specified time period.

19. $5,000, 5%, yearly; 5 years

20. $7,500, 7%, yearly; 6 years

21. $4,000, 8%, quarterly; 2 years

22. $8,000, 4%, quarterly; 3 years

23. $20,000, 8%, monthly; 2 years

24. $10,000, 6%, monthly; 5 years

25. $4,000, 10%, daily; 2 years

26. $6,000, 4%, daily; 3 years

Savings institutions often state two rates in their advertising. One is the nominal yield, which you can think of as an annual simple interest rate. The other is called the effective annual yield, which is the actual interest rate that the account earns due to the compounding. If $1,000 is in an account that pays a nominal yield of 9% and if the compounding is done monthly, then after 1 year, the account would contain $1,093.80, which corresponds to a simple interest rate of 9.38%. We would say that this account has an effective annual yield of 9.38%. In Exercises 27–30, find the effective annual yield for each account.

27. nominal yield, 7.5%; compounded monthly

28. nominal yield, 10%; compounded twice a year

29. nominal yield, 6%; compounded quarterly

30. nominal yield, 8%; compounded daily

In Exercises 31 and 32, you are given an annual interest rate and the compounding period for two investments. Decide which is the better investment.

31. 5% compounded yearly; 4.95% compounded quarterly

32. 4.75% compounded monthly; 4.70% compounded daily

In Exercises 33 and 34, Ann and Tom want to establish a fund for their grandson's college education. What lump sum must they deposit in each account in order to have $30,000 in the fund at the end of 15 years?

33. **Saving for college.** 6% annual interest rate, compounded quarterly

34. **Saving for college.** 7.5% annual interest rate, compounded monthly

In Exercises 35–42, solve each equation.

35. $3^x = 10$

36. $2^x = 12$

37. $(1.05)^x = 2$

38. $(1.15)^x = 3$

39. $x^3 = 10$

40. $x^2 = 10$

41. $x^4 = 10$

42. $x^4 = 25$

In Exercises 43–46, use the compound interest formula $A = P(1 + r)^t$ and the given information to solve for either t or r. (We are assuming that $n = 1$.)

43. $A = \$2,500, P = \$2,000, t = 5$

44. $A = \$400, P = \$20, t = 35$

45. $A = \$1,500, P = \$1,000, r = 4\%$

46. $A = \$2,500, P = \$1,000, r = 6\%$

---

*Before doing these exercises, you may find it useful to review the note *How to Succeed at Mathematics* on page xix.

†We will assume there are 365 days in a year.

Applying What You've Learned

47. Buying an entertainment system. You have purchased a home entertainment system for $3,600 and have agreed to pay off the system in 36 monthly payments of $136 each.

   a. What will be the total sum of your payments?

   b. What will be the total amount of interest that you have paid?

48. Buying a car. You have purchased a used car for $6,000 and have agreed to pay off the car in 24 monthly payments of $325 each.

   a. What will be the total sum of your payments?

   b. What will be the total amount of interest that you have paid?

Often, through government-supported programs, students may obtain "bargain" interest rates such as 6% or 8% to attend college. Frequently, payments are not due and interest does not accumulate until you stop attending college. In Exercises 49 and 50, calculate the amount of interest due 1 month after you must begin payments.

49. Borrowing for college. You have borrowed $10,000 at an annual interest rate of 8%.

50. Borrowing for college. You have borrowed $15,000 at an annual interest rate of 6%.

In Exercises 51–54, we will assume that the lender is using simple interest to compute the interest on the loan.

51. Borrowing for a trip. You plan to take a trip to the Grand Canyon in 2 years. You want to buy a certificate of deposit for $1,200 that you will cash in for your trip. What annual interest rate must you obtain on the certificate if you need $1,500 for your trip?

52. Paying interest on late taxes. Jonathan wants to defer payment of his $4,500 tax bill for 4 months. If he must pay an annual interest rate of 15% for doing this, what will his total payment be?

53. Borrowing from a pawn shop. Sanjay has borrowed $400 on his father's watch from the Main Street Pawn Shop. He has agreed to pay off the loan with $425 one month later. What is the annual interest rate that he is being charged?

54. Borrowing from a bail bondsman. If a person accused of a crime does not have sufficient resources, he may have a bail bondsman post bail to be released until a

trial is held. Assume that a bondsman charges a $50 fee plus 8% of the amount of the bail. If a bondsman posts $20,000 for a trial that takes place in 2 months, what is the interest rate being charged by the bondsman? (Treat the $50 fee plus the 8% as interest on a $20,000 loan for two months.)

The computations for dealing with **inflation** *are the same as for determining future value. If an item sells for $100 today and there is an annual inflation rate of 4% for 10 years, the item would then cost* $100(1.04)^{10} = \$148.02$. *The Bureau of Labor Statistics maintains an index called the* **consumer price index** *(CPI), which is a measure of inflation. The accompanying table shows the CPI for several recent years. The CPI of 207.3 for 2007 means that the price of certain basic items such as clothing, food, energy, automobiles, etc. that would have cost $100 in 1982 to 1984, which are the base years for the index, would now cost $207.30.*

Year	2002	2003	2004	2005	2006	2007
CPI	179.9	184.0	188.9	195.3	201.6	207.3
Percent Increase		2.3	2.7	3.4	3.2	2.8

In Exercises 55–58, you are given a year and the price of an item. Use the percent increase in the CPI as the rate of inflation for the next 10 years to calculate the price of that item 10 years later.

55. Inflation. 2004, fast-food meal, $4.65

56. Inflation. 2006, automobile, $17,650

57. Inflation. 2007, gallon of gasoline, $3.25

58. Inflation. 2005, athletic shoes, $96

59. Inflation. From 1992 to 1995, Albania experienced a yearly inflation rate of 226%. Determine the price of the fast-food meal in Exercise 55 after 5 years at a 226% inflation rate.

60. Inflation. The inflation rate in Hungary during the mid-1990s was about 28%. Determine the price of the athletic shoes in Exercise 58 after 10 years at a 28% inflation rate.

61. Comparing investments. Jocelyn purchased 100 shares of Jet Blue stock for $23.75 per share. Eight months later she sold the stock at $24.50 per share.

   a. What annual rate, calculated using simple interest, did she earn on this transaction?

   b. What annual rate would she have to earn in a savings account compounded monthly to earn the same money on her investment?

62. Comparing investments. Dominick purchased a bond for $2,400 to preserve a wildlife sanctuary and 10 months later he sold it for $2,580.

   a. What annual rate, calculated using simple interest, did he earn on this transaction?

   b. What annual rate would he have to earn in a savings account compounded monthly to earn the same money on his investment?

63. Investment earnings. Emily purchased a bond valued at $20,000 for highway construction for $9,420. If the bond pays 7.5% annual interest compounded monthly, how long must she hold it until it reaches its full face value?

64. Investment earnings. Lucas purchased a bond with a face value of $10,000 for $4,200 to build a new sports stadium. If the bond pays 6.5% annual interest compounded monthly, how long must he hold it until it reaches its full face value?

Communicating Mathematics

65. What formula do we use to compute simple interest?

66. What is the difference between simple interest and compound interest?

67. What is the meaning of each variable in the compound interest formula $A = P\left(1 + \frac{r}{m}\right)^n$?

68. Explain the relationship between the formulas $A = P(1 + r)^t$ and $A = P\left(1 + \frac{r}{m}\right)^n$.

69. Under what circumstances will $A = P(1 + r)^t$ and $A = P\left(1 + \frac{r}{m}\right)^n$ give you the same answers to a compound interest problem?

70. Explain the difference in the techniques that you have to use to solve a problem like Example 8 versus a problem like Example 9.

Using Technology to Investigate Mathematics

71. Get a tutorial from your instructor that explains in more detail how to use a calculator to solve finance problems. Use your calculator to reproduce some of the examples in this section. Your instructor also has Excel spreadsheets available for doing financial computations; use them to reproduce some of the computations in this section.*

72. There are many good interactive financial calculators available on the Internet. Find several and use them to verify some of the computations that we did in this section.

For Extra Credit

Some banks advertise that money in their accounts is compounded continuously. To get an understanding of what this means, apply the compound interest formula using a very large number of compounding periods per year. In Exercises 73 and 74, divide the year into 100,000 compounding periods per year. Apply the compound interest formula for finding future value to approximate what the effective annual yield would be if the compounding were done continuously for the stated nominal yield.

73. nominal yield, 10%

74. nominal yield, 12%

If the principal P is invested in an account that pays an annual interest rate of r% and the compounding is done continuously, then the future value, A, that will be in the account after t years is given by the formula

$$A = Pe^{rt}.$$

The number e is approximately 2.718281828.

75. Use the formula for continuous compounding to find the effective annual yield if the compounding in Exercise 73 is done continuously.

76. Use the formula for continuous compounding to find the effective annual yield if the compounding in Exercise 74 is done continuously.

9.3 Consumer Loans

Objectives

1. Determine payments for an add-on loan.
2. Compute finance charges on a credit card using the unpaid balance method.
3. Use the average daily balance method to compute credit card charges.
4. Compare credit card finance charge methods.

Debt is a bottomless sea.   —Thomas Carlyle

In the spring of 2008, publications such as *Fortune* and *Money* magazines featured articles highlighting a record U.S. credit card debt of $915 billion, and financial institutions such as Citigroup, American Express, and Bank of America were, to quote *Fortune* magazine, "strapping on their Kevlar vests," anticipating an impending financial explosion.

From ancient times to the present, the Bible, William Shakespeare, Benjamin Franklin, and others have warned about the dangers of unbridled credit. In this section, we will explain the mathematics of credit cards and show you how you can use credit wisely, to avoid drowning in Carlyle's bottomless sea.

Imagine that you have just signed the lease for your first apartment and now all you have to do is furnish it. If you buy living room furniture for $1,100, which you pay for in payments, you are taking out an installment loan. Loans having a fixed number of payments are called *closed-ended credit* agreements (or *installment loans*). Each payment is called an *installment*. The size of your payments is determined by the amount of your purchase and also by the interest rate that the seller is charging. The interest charged on a loan is often called a *finance charge*.

The Add-On Interest Method

KEY POINT

The add-on interest method is a simple way to compute payments on an installment loan.

We use the simple interest formula from Section 9.2 to calculate the finance charge for an installment loan. To determine the payments for an installment loan, we add the simple interest due on the loan to the loan amount and then divide this sum by the number of monthly payments.

> **FORMULA FOR DETERMINING THE MONTHLY PAYMENT OF AN INSTALLMENT LOAN**
>
> $$\text{monthly payment} = \frac{P+I}{n},$$
>
> where P is the amount of the loan, I is the amount of interest due on the loan, and n is the number of monthly payments.

This method is sometimes called the **add-on interest method** because we are adding on the interest due on the loan before determining the payments.

EXAMPLE 1 *Determining Payments for an Add-On Interest Loan*

A new pair of Bose speakers for your home theater system costs $720. If you take out an add-on loan for 2 years at an annual interest rate of 18%, what will be your monthly payments?

SOLUTION: We first use the simple interest formula to calculate the interest:

$$I = Prt = 720(0.18)2 = 259.20.$$

Next, we add the interest to the purchase price:

$$720 + 259.20 = 979.20.$$

To find the monthly payments, we divide this amount by 24:

$$\frac{979.20}{24} = 40.80.$$

Suppose that you take an installment loan for $360 for 1 year at an annual interest rate of 21%. What are your monthly payments?

Monthly payments are therefore $40.80.

Now try Exercises 5 to 12. ❋ ⓫

In Example 1, the annual interest rate of 18% is quite misleading. If we think about it, the purchase price was $720, so it would be fair to say that you are paying off $720/24 = $30 of the loan amount each month; the other $10.80 is interest. When you reach the last month, although you only owe $30 on the purchase, you are still paying $10.80 in interest. Simple arithmetic shows that 10.80/30 = 0.36. So in a certain sense, the interest rate for the last month of the loan is actually 36%. Because you are paying 36% interest for one month, this is equivalent to an annual interest rate of 12 × 36% = 432%. What we want to point out here is that although simple interest is easy to compute, as you pay off the loan amount, the actual interest you are paying on the outstanding balance is higher than the stated interest rate.

When you use your credit card to pay for gas at a gas station, you are using *open-ended credit*. With open-ended credit, the calculation of finance charges can be more complicated than with closed-ended credit. Although you may be making monthly payments on your loan, you may also be increasing the loan by making further purchases.

There are several ways that credit card companies compute finance charges. We will look at two methods and compare them at the end of this section. You will see that if you understand the method being used to compute your finance charges, you can use this information to reduce the cost of borrowing money.

The Unpaid Balance Method

The first method that we will discuss for computing finance charges is called the **unpaid balance method**. With this method, the interest is based on the previous month's balance.

✎ *KEY POINT*

The unpaid balance method computes finance charges on the balance at the end of the previous month.

> **THE UNPAID BALANCE METHOD FOR COMPUTING THE FINANCE CHARGE ON A CREDIT CARD LOAN** This method also uses the simple interest formula $I = Prt$; however,
>
> P = previous month's balance + finance charge + purchases made − returns − payments.
>
> The variable r is the annual interest rate, and $t = \frac{1}{12}$.

EXAMPLE 2 *Using the Unpaid Balance Method for Finding Finance Charges*

Assume that the annual interest rate on your credit card is 18% and your unpaid balance at the beginning of last month was $600. Since then, you purchased ski boots for $130 and sent in a payment of $170.

a) Using the unpaid balance method, what is your credit card bill this month?

b) What is your finance charge next month?

SOLUTION:

a) We will list the items that we need to know to compute this month's balance.

Previous month's balance: $600
Finance charge on last month's balance; $600 \times 0.18 \times \left(\frac{1}{12}\right) = \$9$

annual interest rate ⤴   Time is 1 month.

Purchases made: $130
Returns: $0
Payment: $170

Therefore, you owe

Previous month's balance + finance charge + purchases made − returns − payments
$$= 600 + 9 + 130 - 0 - 170 = \$569.$$

b) The finance charge for next month will be $\$569 \times 0.18 \times \left(\frac{1}{12}\right) = \$8.54$.

Now try Exercises 21 to 26. ❋ **12**

Quiz Yourself **12**

Assume that the annual interest rate on your credit card is 21%. Your outstanding balance last month was $300. Since then, you have charged a purchase for $84 and made a payment of $100. What is the outstanding balance on your card at the end of this month? What is next month's finance charge on this balance?

Note that you can use the unpaid balance method to your advantage by making a large purchase early in the billing period and then paying it off just before the billing date. This is not to the credit card company's advantage because you can use the credit card company's money for free for almost a whole month.

It is easy to use credit cards for purchases and so tempting to pay only the minimum payment that appears on your credit card bill that you may find your debt increasing even though you are sending in a payment every month. Example 3 illustrates how credit card debt can get out of hand.

EXAMPLE 3   *Paying Off a Credit Card Debt*

Assume that you want to pay off your credit card debt of $6,589 by making the minimum payment of $100 a month. What will your balance be at the end of 1 month? Assume that the annual interest rate on your card is 18% and that the credit card company is using the unpaid balance method to compute your finance charges.

SOLUTION: When you send in your $100, it is credited to your account, so the next month you have an unpaid balance of $6,589 - 100 = \$6,489$. The annual interest rate is 18%, so your monthly interest rate is $\frac{18}{12} = 1.5\%$.

Therefore, at the end of the month you still owe

$$\underset{\text{balance}}{6,489} + \underset{\text{interest}}{(0.015)(6,489)} = 6,489 + 97.34 = \$6,586.34.$$

Thus, your $100 payment has reduced your debt by only $6,589 - 6,586.34 = \$2.66$. ❋

Example 3 illustrates how difficult it is to pay off a large credit card bill. The best practice is to pay off as much of your outstanding balance as you can to avoid paying a large amount of interest. Although it is possible to borrow cash from one company to pay off debts at another, doing this is not reducing the size of your debt. Instead, the debt may still grow at a rate of 18% or more per year.

When you use credit, always look at the annual interest rate. In some cases, if you read the fine print in the agreement, you will find rates as high as 24% or 25%. Also, be very careful when taking a cash advance on your credit card because the interest rate is often much higher than the rate you are charged for making purchases on the card.

The Average Daily Balance Method

KEY POINT

The average daily balance method computes finance charges based on the balance in the account for each day of the month.

A more complicated method for determining the finance charge on a credit card is called the **average daily balance method**, which is one of the most common methods used by credit card companies. With this method, the balance is the average of all daily balances for the previous month.

> **THE AVERAGE DAILY BALANCE METHOD FOR COMPUTING THE FINANCE CHARGE ON A CREDIT CARD LOAN**
>
> 1. Add the outstanding balance for your account for each day of the month.
> 2. Divide the total in step 1 by the number of days in the month to find the average daily balance.
> 3. To find the finance charge, use the formula $I = Prt$, where P is the average daily balance found in step 2, r is the annual interest rate, and t is the number of days in the month divided by 365.

EXAMPLE 4   *Using the Average Daily Balance Method for Finding Finance Charges*

Suppose that you begin the month of September (which has 30 days) with a credit card balance of $240. Assume that your card has an annual interest rate of 18% and that during September the following adjustments are made on your account:

September 11: A payment of $60 is credited to your account.

September 18: You charge $24 for iTune downloads.

September 23: You charge $12 for gasoline.

Use the average daily balance method to compute the finance charge that will appear on your October credit card statement.

SOLUTION: To answer this question, we must first find the average daily balance for September. The easiest way to calculate the balance is to keep a day-by-day record of what you owe the credit card company for each day in September, as we do in Table 9.3.

Day	Balance	Number of Days × Balance
1, 2, 3, 4, 5, 6, 7, 8, 9, 10	$240	$10 \times 240 = 2{,}400$
11, 12, 13, 14, 15, 16, 17	$180	$7 \times 180 = 1{,}260$
18, 19, 20, 21, 22	$204	$5 \times 204 = 1{,}020$
23, 24, 25, 26, 27, 28, 29, 30	$216	$8 \times 216 = 1{,}728$

TABLE 9.3 Daily balances for September.

Quiz Yourself ⓭

Recalculate the average daily balance in Example 4, except now assume you bought the iTunes downloads on September 3 instead of September 18. Make a table similar to Table 9.3.

The average daily balance is therefore

$$\frac{(10 \times 240) + (7 \times 180) + (5 \times 204) + (8 \times 216)}{30}$$

$$= \frac{2{,}400 + 1{,}260 + 1{,}020 + 1{,}728}{30} = \frac{6{,}408}{30} = 213.6.$$

We next apply the simple interest formula, where $P = \$213.60$, $r = 0.18$, and $t = \frac{30}{365}$.* Thus, $I = Prt = 213.6(0.18)\left(\frac{30}{365}\right) = 3.16$. Your finance charge on the October statement will be $3.16.

Now try Exercises 27 to 30. ✳ ⓭

As you will see in Example 5, the amount of finance charges you pay on a loan will vary depending on the method used to compute the charges.

Math in Your Life

Will You Be Part of "Generation Broke?"†

You've heard of "Generation X" and "Generation Y" but have you ever heard of "Generation Broke?" Article after article, from *Time Magazine* to *USA Today*, warns of impending financial doom for young adults. Depending on which study you read, the average graduating college senior now owes between $5,000 and $20,000 in student loans and another $3,000 to $4,000 in credit card debt. According to a public policy group in New York called Demos, ". . . young adults are doing everything society tells them to do . . . [but] they can't get ahead because of the debt they went into to get the degree and get the good job."

Demos sees two trends fueling this rise in debt among younger Americans: increasing college costs and aggressive credit card marketing on college campuses. In a report in the *Christian Science Monitor*, one student laments not having had a class in personal financial management, "If I had that type of class, I wouldn't have gotten into such a credit mess."

---

*$t = \frac{30}{365}$ because we are using the credit card for 30 days out of a 365-day year.
†To read the full report, "Generation Broke: The Growth of Debt Among Younger Americans," refer to www.demos-usa.org/pub295.cfm.

—— HISTORICAL HIGHLIGHT ✸ ✸ ✸

Credit and Interest*

Credit cards were not widely used in the United States until the 1950s when cards such as Diners Club, Carte Blanche, and American Express made the use of plastic money more popular. Today, Americans charge about $1 trillion per year on their cards.

Credit is not a modern idea. Surprisingly, there are ancient Sumerian documents dating back to about 3000 BC that show the regular use of credit in borrowing grain and metal. Interest on these loans was often in the 20% to 30% range—similar to the 18% to 21% charged on many of today's credit cards. As the use of credit increased, so did its

misuse. Many societies wrote laws to prevent its abuse—particularly the charging of unfairly high interest rates, which is called *usury*.

Credit and interest can appear in many diverse forms. The Ifugao tribe of the Philippines charges 100% on a loan. If rice is borrowed, then at the next harvest, the loan must be paid in double. In Vancouver, Canada, the Kwakiutl have a system of credit based on blankets. The rules of interest state that if five blankets are borrowed, in 6 months they become seven. In Northern Siberia, loans are made in reindeer, usually at a 100% interest rate.

Comparing Financing Methods

EXAMPLE 5 *Comparing Methods for Finding Finance Charges*

Suppose that you begin the month of May (which has 31 days) with a credit card balance of $500. The annual interest rate is 21%. On May 11, you use your credit card to pay for a $400 car repair, and on May 29, you make a payment of $500. Calculate the finance charge that will appear on the statement for next month using the two methods we have discussed.

SOLUTION:

Method	P	r	t	Finance Charge $= I = Prt$
Unpaid balance	last month's balance + finance charge − payment + charge for car repair $= 500 + 8.75 + 400 - 500$ $= 408.75$	21%	$\dfrac{1}{12}$	$(408.75)(0.21)\left(\dfrac{1}{12}\right) = \$7.15$
Average daily balance	$\dfrac{10 \times 500 + 18 \times 900 + 3 \times 400}{31}$ $= \dfrac{22{,}400}{31}$ $= 722.58$	21%	$\dfrac{31}{365}$	$(722.58)(0.21)\left(\dfrac{31}{365}\right) = \$12.89$

With the unpaid balance method, the finance charge is $7.15; with the average daily balance method, the finance charge is $12.89. ✸

Example 5 shows that the exact same charges on two different credit cards having the same annual interest rate can result in very different finance charges. If you understand the method your credit card company is using, you may be able to schedule your purchases and payments to minimize your finance charges.

In deciding how to use credit, you must consider many other issues that we have not discussed in this section. Some credit card companies charge an annual fee; others return part of your interest payments. For some credit cards, there is a grace period. If you reduce your balance to zero during this grace period, then you pay no finance

*This Historical Highlight is based on S. Homer and R. Sylla, *A History of Interest Rates*, 3rd ed. (Piscataway, NJ: Rutgers University Press, 1991), pp. 21–30.

charges. A credit card may have a low introductory rate that changes to a much higher rate at a later time.

One common enticement is that you can make a purchase and pay no interest payments until several months later. With such deals, however, you must be careful. Often, if you do not pay off the purchase completely at the end of the interest-free period, then all the interest that would have accumulated is added to your balance. It is difficult to understand all the pros and cons of the many different types of credit contracts. However, if you read credit agreements carefully and remember the principles that you learned in this section, you will be an intelligent consumer who will be able to use credit wisely.

Exercises 9.3

Looking Back*

These exercises follow the general outline of the topics presented in this section and will give you a good overview of the material that you have just studied.

1. Describe briefly how we calculated the payments for the add-on interest loan in Example 1.

2. In Example 2, we computed the finance charge from the previous month using the expression *Prt*. What did each variable represent?

3. What does Table 9.3 show in Example 4?

4. In reference to the Math in Your Life on "Generation Broke," what do you think are some of the causes of this phenomenon?

Sharpening Your Skills

In Exercises 5–8, compute the monthly payments for each add-on interest loan. The amount of the loan, the annual interest rate, and the term of the loan are given.

5. $900; 12%; 2 years

6. $840; 10%; 3 years

7. $1,360; 8%; 4 years

8. $1,710; 9%; 3 years

9. **Paying off a computer.** Luis took out an add-on interest loan for $1,280 to buy a new laptop computer. The loan will be paid back in 2 years and the annual interest rate is 9.5%. How much interest will he pay? What are his monthly payments?

10. **Paying off furniture.** Mandy bought furniture costing $1,460 for her new apartment. To pay for it, her bank gave her a 5-year add-on interest loan at an annual interest rate of 10.4%. How much interest will she pay? What are her monthly payments?

11. **Financing equipment.** Angela's bank gave her a 4-year add-on interest loan for $6,480 to pay for new equipment for her antiques restoration business. The annual interest rate is 11.65%. How much interest will she pay? What are her monthly payments?

12. **Paying for a sculpture.** Mikeal purchased an antique sculpture from a gallery for $1,320. The gallery offered him a 3-year add-on loan at an annual rate of 9.75%. How much interest will he pay? What are his monthly payments?

In Exercises 13–16, use the add-on method for determining interest on the loan. Determine the annual interest rate during the last month of the loan.

13. $900; 12%; 2 years

14. $840; 10%; 3 years

15. $1,360; 8%; 4 years

16. $1,710; 9%; 4 years

Applying What You've Learned

17. **Financing a boat.** Ben is buying a new boat for $11,000. The dealer is charging him an annual interest rate of 9.2% and is using the add-on method to compute his monthly payments.

    a. If Ben pays off the boat in 48 months, what are his monthly payments?

    b. If he makes a down payment of $2,000, how much will this reduce his monthly payments?

    c. If he wants to have monthly payments of $200, how large should his down payment be?

18. **Financing a swimming pool.** Mr. Phelps is buying a new swimming pool for $14,000. The dealer is charging him an annual interest rate of 8.5% and is using the add-on method to compute his monthly payments.

    a. If Mr. Phelps pays off the pool in 48 months, what are his monthly payments?

    b. If he makes a down payment of $3,000, how much will this reduce his monthly payments?

    c. If he wants to have monthly payments of $250, how large should his down payment be?

19. **Financing rare coins.** Anna is buying $15,000 worth of rare coins as an investment. The dealer is charging her an annual interest rate of 9.6% and is using the add-on method to compute her monthly payments.

---

*Before doing these exercises, you may find it useful to review the note *How to Succeed at Mathematics* on page xix.

a. If Anna pays off the coins in 36 months, what are her monthly payments?

b. If she makes a down payment of $3,000, how much will this reduce her monthly payments?

c. If she wishes to have monthly payments of $300, how large should her down payment be?

20. **Financing a musical instrument.** Walt is buying a music synthesizer for his rock band for $6,500. The music store is charging him an annual interest rate of 8.5% and is using the add-on method to compute his monthly payments.

a. If Walt pays off the synthesizer in 24 months, what are his monthly payments?

b. If he makes a down payment of $1,500, how much will this reduce his monthly payments?

c. If he wants to have monthly payments of $150, how large should his down payment be?

In Exercises 21–26, use the unpaid balance method to find the finance charge on the credit card account. Last month's balance, the payment, the annual interest rate, and any other transactions are given.

21. **Computing a finance charge.** Last month's balance, $475; payment, $225; interest rate, 18%; bought ski jacket, $180; returned camera, $145

22. **Computing a finance charge.** Last month's balance, $510; payment, $360; interest rate, 21%; bought exercise equipment, $470; bought fish tank, $85

23. **Computing a finance charge.** Last month's balance, $640; payment: $320; interest rate, 16.5%; bought dog, $140; bought pet supplies, $35; paid veterinarian bill, $75

24. **Computing a finance charge.** Last month's balance, $340; payment, $180; interest rate, 17.5%; bought coat, $210; bought hat, $28; returned boots, $130

25. **Computing a finance charge.** Last month's balance, $460; payment, $300; interest rate, 18.8%; bought plane ticket, $140; bought luggage, $135; paid hotel bill, $175

26. **Computing a finance charge.** Last month's balance, $700; payment, $480; interest rate, 21%; bought ring, $210; bought theater tickets, $142; returned vase, $128

In Exercises 27–30, use the average daily balance method to compute the finance charge on the credit card account for the previous month. The starting balance and transactions on the account for the month are given. Assume an annual interest rate of 21% in each case.

27. **Computing a finance charge.** Month: August (31 days); previous month's balance: $280

Date	Transaction
August 5	Made payment of $75
August 15	Charged $135 for hiking boots
August 21	Charged $16 for gasoline
August 24	Charged $26 for restaurant meal

28. **Computing a finance charge.** Month: October (31 days); previous month's balance: $190

Date	Transaction
October 9	Charged $35 for a book
October 11	Charged $20 for gasoline
October 20	Made payment of $110
October 26	Charged $13 for lunch

29. **Computing a finance charge.** Month: April (30 days); previous month's balance: $240

Date	Transaction
April 3	Charged $135 for a coat
April 13	Made payment of $150
April 23	Charged $30 for DVDs
April 28	Charged $28 for groceries

30. **Computing a finance charge.** Month: June (30 days); previous month's balance: $350

Date	Transaction
June 9	Made payment of $200
June 15	Charged $15 for gasoline
June 20	Charged $180 for skis
June 26	Made payment of $130

In Exercises 31–34, redo the specified exercise using the unpaid balance method to calculate the finance charges.

31. Exercise 27

32. Exercise 28

33. Exercise 29

34. Exercise 30

35. **Comparing financing methods.** Mayesha purchased a large-screen TV for $1,000 and can pay it off in 10 months with an add-on interest loan at an annual rate of 10.5%, or she can use her credit card that has an annual rate of 18%. If she uses her credit card, she will pay $100 per month (beginning next month) plus the finance charges for the month. Assume that Mayesha's credit card company is using the unpaid balance method to compute her finance charges and that she is making no other transactions on her credit card. Which option will have the smaller total finance charges on her loan?

36. **Comparing financing methods.** Repeat Exercise 35, but now assume that Mayesha purchased an entertainment center for $2,000, the rate for the add-on loan is 9.6%, and she is paying off the loan in 20 months.

37. Accumulated interest. A Home Depot advertises 0% financing for 3 months for purchases made before the new year. The fine print in the advertisement states that if the purchase is not paid off within 3 months, the purchaser must pay interest that has accumulated at a monthly rate of 1.75%. Assume that you buy a refrigerator for $1,150 and make no payments during the next 3 months. How much interest has accumulated on your purchase during this time?

38. Accumulated interest. Repeat Exercise 37, but now assume that the purchase is for $1,450 and the annual interest rate is 24%.

Communicating Mathematics

39. Why is an add-on interest loan called by that name?

40. What was the point of our discussion in the paragraph following Example 1?

41. How can you use the unpaid balance method to your advantage when borrowing?

42. How would you explain informally the difference between computing finance charges with the unpaid balance method and the average daily method to a classmate?

Using Technology to Investigate Mathematics

43. Your instructor has an Excel spreadsheet that will carry out the computations for the average daily balance method. Use this spreadsheet to duplicate the computations in Example 4 and also to solve some of the exercises.*

44. Use the Web to research the terms of some actual credit cards. See if you can determine how the finance charges are computed with those cards. What other terms are in those agreements that we have not discussed in this section?

For Extra Credit

45. In Example 5, we found that the average daily balance method gave the highest finance charges. Explain why this happened.

46. Make up transactions on a hypothetical credit card so that the unpaid balance method will give you lower finance charges than the average daily balance method does. Assume an annual interest rate of 18%.

47. Make up transactions on a hypothetical credit card so that the average daily balance method will give you lower finance charges than the unpaid balance method does. Assume an annual interest rate of 21%.

48. In our discussions about credit, we have ignored the fact that whatever money is not used to pay off a loan can be invested. Assume that you can earn 5% on any money that you do not use for paying off a loan. However, any money you earn as interest is subject to federal, state, and local taxes. Assume that these taxes total 20%. Discuss how that might affect your decision to pay off your credit card debt.

9.4 Annuities

Objectives

1. Calculate the future value of an ordinary annuity.
2. Perform calculations regarding sinking funds.

Somewhere over the rainbow . . . skies are blue, . . . and the dreams that you dare to dream . . . really do come true. —Lyman Frank Baum
(Author of *The Wizard of Oz*)

What are your financial dreams? Do you dream about owning a beautiful house? Visiting some exotic far-away place? Providing a college education for your children? Retiring happily in comfortable surroundings? If you have any long-term plans such as these, to achieve your dreams you will need to have a large sum of money in the future.

In this section, we will discuss how you can do just that by making a series of regular payments over many years to accumulate the money that you will need. This type of investment is called an *annuity*.

 *KEY POINT*

We make regular payments into an annuity.

Annuities

An **annuity** is an interest-bearing account into which we make a series of payments of the same size. If one payment is made at the *end* of every compounding period, the annuity is called an **ordinary annuity**. The **future value of an annuity** is the amount in the account, including interest, after making all payments.

To illustrate the future value of an annuity, suppose that in January you begin making payments of $100 at the end of each month into an account paying 12% yearly interest compounded monthly. How much money will be in this account for a summer vacation beginning on July 1?

This problem is different from those in Section 9.2. In the earlier problems, we deposited a lump sum that earned a stated interest rate for an entire period. In this problem, we are depositing a *series* of payments, and each payment earns interest for a different number of periods.

The January deposit earns interest for February, March, April, May, and June. Using the formula for compound interest, this deposit will grow to

$$100(1 + 0.01)^5 = \$105.10.$$

However, the May deposit earns interest for only 1 month and therefore grows to only

$$100(1 + 0.01)^1 = \$101.$$

The June deposit earns no interest at all. We illustrate this pattern with the timeline in Figure 9.3.

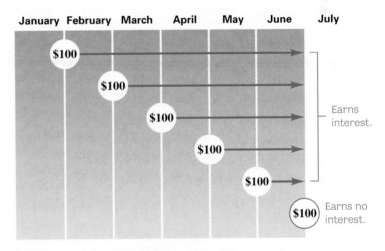

FIGURE 9.3 Timeline for ordinary annuity with deposits at end of January, . . . , June.

If we compute how much each deposit contributes to the account and sum these amounts, we will have the value of the annuity on July 1.

January	$100(1.01)^5 = \$105.10$
February	$100(1.01)^4 = \$104.06$
March	$100(1.01)^3 = \$103.03$
April	$100(1.01)^2 = \$102.01$
May	$100(1.01)^1 = \$101.00$
June	$100(1.01)^0 = \$100.00$
	Total = $615.20

We can express the value of this annuity as

$$100(1.01)^5 + 100(1.01)^4 + 100(1.01)^3 + 100(1.01)^2 + 100(1.01)^1 + 100.$$

By factoring out the 100, we can write the value of the annuity in the form

$$100[(1.01)^5 + (1.01)^4 + (1.01)^3 + (1.01)^2 + (1.01)^1 + 1]. \tag{1}$$

Notice that although you deposited $100 at the end of each month for *six* months, the first deposit earned interest for *five* months, the second deposit earned interest for *four* months, and so on.

Following this pattern, it is clear that if you had deposited the $100 in a vacation savings account at the end of each month for 10 months, the value of the annuity would have been

$$100[(1.01)^9 + (1.01)^8 + \cdots + (1.01)^2 + (1.01)^1 + 1]. \tag{2}$$

Notice that in equation (1) we have an expression of the form $x^5 + x^4 + x^3 + x^2 + x^1 + 1$, and in equation (2) we have an expression of the form $x^9 + x^8 + \cdots + x^2 + x^1 + 1$. Fortunately, there is a way to write these lengthy expressions in a simpler form.

EXAMPLE 1   *Simplifying Annuity Computations*

Show that $x^5 + x^4 + x^3 + x^2 + x^1 + 1 = \dfrac{x^6 - 1}{x - 1}$.

SOLUTION: To show this relationship, we multiply the polynomials $x^5 + x^4 + x^3 + x^2 + x^1 + 1$ and $x - 1$ in the usual way.

$$
\begin{array}{r}
x^5 + x^4 + x^3 + x^2 + x^1 + 1 \\
\times\, x - 1 \\
\hline
x^6 + x^5 + x^4 + x^3 + x^2 + x^1 \\
-\, x^5 - x^4 - x^3 - x^2 - x^1 - 1 \\
\hline
x^6 \hspace{4.5cm} -\, 1
\end{array}
$$

> *Quiz Yourself* ⑭
>
> Write $x^3 + x^2 + x^1 + 1$ as a quotient of two polynomials. (Do computations that are similar to those we did in Example 1.)

This result shows that $(x^5 + x^4 + x^3 + x^2 + x^1 + 1)(x - 1) = x^6 - 1$. Dividing this equation by $x - 1$, we obtain our desired relationship.

Now try Exercises 3 to 4. ❋ ⑭

Following the pattern of Example 1, we can prove that

$$x^n + x^{n-1} + x^{n-2} + \cdots + x^2 + x^1 + 1 = \frac{x^{n+1} - 1}{x - 1}. \tag{3}$$

Returning to our vacation savings account example, we can think of 1.01 as x in equation (1). Then we can use equation (3) to simplify our calculations. Because

$$(1.01)^5 + (1.01)^4 + (1.01)^3 + (1.01)^2 + (1.01)^1 + 1$$
$$= \frac{(1.01)^6 - 1}{1.01 - 1} = \frac{1.061520151 - 1}{1.01 - 1} = \frac{0.061520151}{0.01} \approx 6.1520,$$

we can write

$$100[(1.01)^5 + (1.01)^4 + (1.01)^3 + (1.01)^2 + (1.01)^1 + 1] \approx 100(6.1520) = \$615.20.$$

This is the same amount that we found earlier.

Doing similar computations, we find that the 10-month vacation savings account annuity has a value of

$$100[(1.01)^9 + (1.01)^8 + \cdots + (1.01)^2 + (1.01)^1 + 1]$$
$$= 100\left[\frac{(1.01)^{10} - 1}{1.01 - 1}\right] \approx 100(10.4622) = \$1{,}046.22.$$

 **KEY POINT**

The future value of an annuity depends on the size of the payment, the interest rate, and the number of payments.

We can generalize the patterns that we have just seen in a formula for finding the future value of an annuity.

> **FORMULA FOR FINDING THE FUTURE VALUE OF AN ORDINARY ANNUITY** Assume that we are making n regular payments, R, into an ordinary annuity. The interest is being compounded m times a year and deposits are made at the end of each compounding period. The future value (or amount), A, of this annuity at the end of the n periods is given by the equation
>
> $$A = R\frac{\left(1 + \dfrac{r}{m}\right)^n - 1}{\dfrac{r}{m}}.$$

To calculate this expression, you should do the following steps:

1st: Find $\frac{r}{m}$ and add 1.

2nd: Raise $1 + \frac{r}{m}$ to the n power and then subtract 1.

3rd: Divide the amount that you found in step 2 by $\frac{r}{m}$.

4th: Multiply the quantity that you found in step 3 by R.

EXAMPLE 2   *Finding the Future Value of an Ordinary Annuity*

Assume that we make a payment of $50 at the end of each month into an account paying a 6% annual interest rate, compounded monthly. How much will be in that account after 3 years?

SOLUTION: This account is an ordinary annuity. The payment R is 50, the monthly rate $\frac{r}{m}$ is $\frac{6\%}{12} = \frac{0.06}{12} = 0.005$, and the number of payments n is $3 \times 12 = 36$. Using the formula for finding the future value of an ordinary annuity, we get

$$A = 50\left[\frac{(1.005)^{36} - 1}{0.005}\right] = 50\left(\frac{1.19668053 - 1}{0.005}\right)$$

$$= 50\left(\frac{0.19668053}{0.005}\right)$$

$$= 50(39.336105) \approx \$1{,}966.81.^*$$

Now try Exercises 7 to 16. ❋ **15**

If you make regular payments into an annuity for many years, the value of the annuity can become enormous due to the compounding of interest. Figure 9.4 shows that if the annual interest rate is 6.6%, then in roughly 19 years, the amount of interest that the annuity has earned exceeds the amount of the deposits. The fact that the interest curve is rising so rapidly indicates that the future value of your account is also growing rapidly.

Sinking Funds

You may want to save regularly to have a fixed amount available in the future. For example, you may want to save $1,800 to travel to Jamaica in 2 years. The question is, how much should you save each month to accomplish this? The account that you establish for

---

Sidebar (left margin):

```
N=36
I%=6
PV=0
PMT=50
•FV=∎1966.805248
P/Y=12
C/Y=12
PMT:END BEGIN
```

TI-83 screen verifies calculations in Example 2.

Quiz Yourself **15**

Redo Example 2, except now assume that you are depositing $75 per month for 2 years.

 **KEY POINT**

With a sinking fund, we make payments to save a specified amount.

---

*When using your calculator, you should hold off on rounding your answers as long as possible. If you round off too soon, your answers will differ slightly from the answers in this text.

❈ ❈ ❈ HIGHLIGHT

Between the Numbers—Whom Do You "Trust"?

The "trust" that we are referring to is the Social Security trust fund. When you first started working, you may have been dismayed to see a deduction from your first paycheck labeled FICA, which is an acronym for the Federal Insurance Contributions Act. This law mandates that workers must contribute a certain amount of their wages to the Social Security trust fund.

When you contribute to Social Security, you are not actually saving your money for *your retirement*, but your taxes are paying for *someone else's retirement*. The idea is that when you retire, younger workers will then pay for your retirement. However, some see a huge problem with this. Right now, roughly 50 million Americans receive Social Security benefits with contributions supported by approximately 200 million U.S. workers. If we think of this as a ratio, every person on Social Security is supported by 4 workers. In 1950, the ratio was 16 workers for every Social Security beneficiary. It is projected that by 2030, the ratio will be 2 workers for every beneficiary and the fund will be in trouble. So what can you do to protect your retirement?

The government has been encouraging people to make plans for their retirement by establishing tax-deferred annuities to guarantee that when they retire they will have money to supplement Social Security benefits. *Tax deferred* means that the money you set aside in the annuity is not taxed now but at a later date when you start withdrawing from the annuity. As you will see in the exercises, there can be a huge financial benefit to saving this way.

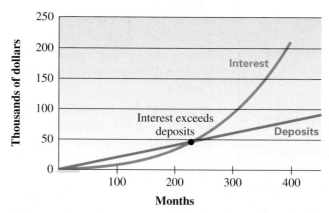

FIGURE 9.4 At an annual rate of 6.6%, the amount of interest earned in an ordinary annuity exceeds the amount of deposits in about 230 months.

your deposits is called a **sinking fund**. You could estimate the amount to save each month by simply dividing 1,800 by 24 months to get $\frac{1,800}{24} = \$75$ per month. Because your estimate ignores the interest that your deposits will generate, the actual amount you would need to put aside each month is somewhat less. Knowing exactly how much you need to save each month could be important if you were on a tight budget.

Because a sinking fund is a special type of annuity, it is not necessary to find a new formula to answer this question. We can use the formula for calculating the future value of an ordinary annuity that we have stated earlier. In this case, we know the value of A and we want to find R.*

EXAMPLE 3   *Calculating Payments for a Sinking Fund*

Assume that you wish to save $1,800 in a sinking fund in 2 years. The account pays 6% compounded quarterly and you will also make payments quarterly. What should be your monthly payment?

---

*In making payments into a sinking fund, we will always round the payment *up* to the next cent.

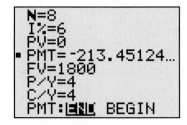

TI-83 screen verifies calculations in Example 3.

Quiz Yourself **16**

What payments must you make to a sinking fund that pays 9% yearly interest compounded monthly if you want to save $2,500 in 2 years?

 **KEY POINT**

We use the log function to find how long it takes for an annuity to accumulate a specified value.

SOLUTION: Recall the formula for finding the future value of an ordinary annuity:

$$A = R \frac{\left(1 + \frac{r}{m}\right)^n - 1}{\frac{r}{m}}. \tag{4}$$

We want the value A of the annuity to be $1,800, the monthly rate $\frac{r}{m}$ is $\frac{6\%}{4} = \frac{0.06}{4} = 0.015$, and the number of payments n is $2 \times 4 = 8$. Substituting these values in equation (4), we get

$$1,800 = R \frac{(1 + 0.015)^8 - 1}{0.015} = R(8.432839106).$$

Dividing both sides of the equation by 8.432839106, we get the monthly payment

$$R = \frac{1,800}{8.43289106} \approx \$213.46.$$

Now try Exercises 17 to 20. ❋ **16**

 *Some Good Advice*

You may be tempted to memorize a new formula to solve sinking fund problems. This is not necessary, because once you have learned to solve annuity problems, you can use the same formula (and a little bit of algebra) to solve sinking fund problems.

Sometimes in working with annuities, we want to know how long it will take to save a certain amount. That is, in the annuity formula we want to find n. This problem is a little more complicated than those we have solved so far. To solve such problems, we will use the exponent property for the log function, which we introduced in Section 9.3. We will show you how to use this property in Example 4.

EXAMPLE 4 *Finding the Time Required to Accumulate $1,000,000*

Suppose you have decided to retire as soon as you have saved $1,000,000. Your plan is to put $200 each month into an ordinary annuity that pays an annual interest rate of 8%. In how many years will you be able to retire?

SOLUTION: We can use the future value formula for an ordinary annuity to solve this problem:

$$A = R \frac{\left(1 + \frac{r}{m}\right)^n - 1}{\frac{r}{m}}$$

Here, A is the future value of $1,000,000, $\frac{r}{m}$ is the monthly interest rate of $\frac{0.08}{12} \approx 0.00667$, and R is 200. We must find n, the number of months for which you will be making deposits. Therefore, we must solve for n in the following equation:

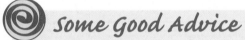

$$1,000,000 = 200 \left[\frac{(1 + 0.00667)^n - 1}{0.00667} \right]$$

We begin by multiplying both sides of the equation by 0.00667:

$$6,670 = 200[(1 + 0.00667)^n - 1].$$

Next, we divide both sides of the equation by 200 and then add 1 to both sides of the equation:

$$34.35 = (1.00667)^n.$$

Then we take the log of both sides:

$$\log 34.35 = \log(1.00667)^n.$$

Now we use the exponent property of the log function to simplify the equation:

$$\log 34.35 = n \log 1.00667.$$

We divide by log 1.00667 and use a calculator to find n:

$$n = \frac{\log 34.35}{\log 1.00667} = 531.991532 \approx 532.$$

This tells how many months it will take you to save $1,000,000. Dividing 532 by 12 gives us $\frac{532}{12} = 44.33$ years until your retirement.

Now try Exercises 29 to 34. ✳ **17**

Quiz Yourself **17**

If you put $150 per month into an ordinary annuity that pays an annual interest rate of 9%, how long will it take for the annuity to have a value of $100,000?

Exercises 9.4

Looking Back*

These exercises follow the general outline of the topics presented in this section and will give you a good overview of the material that you have just studied.

1. To calculate the future value of the annuity in Example 2, we used the expression

$$R\frac{\left(1 + \frac{r}{m}\right)^n - 1}{\frac{r}{m}}.$$ What is the meaning of $\frac{r}{m}$? What is the meaning of m? What is R?

2. What implications do you see of the situation presented in the Highlight regarding the Social Security trust fund?

Sharpening Your Skills

In Exercises 3 and 4, simplify each algebraic expression, as in Example 1.

3. $x^7 + x^6 + \cdots + x^2 + x^1 + 1$

4. $x^8 + x^7 + \cdots + x^2 + x^1 + 1$

Exercises 5 and 6 are based on the vacation account example at the beginning of this section. Assume that, beginning in January, you make payments at the end of each month into an account paying the specified yearly interest. Interest is compounded monthly. How much will you have available for your vacation by the specified date?

5. Monthly payment, $100; yearly interest rate, 6%; August 1

6. Monthly payment, $200; yearly interest rate, 3%; May 1

In Exercises 7 and −16, find the value of each ordinary annuity at the end of the indicated time period. The payment R, frequency of deposits m (which is the same as the frequency of compounding), annual interest rate r, and the time t are given.

7. Amount, $200; monthly; 3%; 8 years

8. Amount, $450; monthly; 2.4%; 10 years

9. Amount, $400; monthly; 9%; 4 years

10. Amount, $350; monthly; 10%; 10 years

11. Amount, $600; monthly; 9.5%; 8 years

12. Amount, $500; monthly; 7.5%; 12 years

13. Amount, $500; quarterly; 8%; 5 years

14. Amount, $750; quarterly; 9%; 3 years

15. Amount, $280; quarterly; 3.6%; 6 years

16. Amount, $250; quarterly; 4.8%; 18 years

In Exercises 17–20, find the monthly payment R needed to have a sinking fund accumulate the future value A. The yearly interest rate r and the time t is given. Interest is compounded monthly. Round your answer up to the next cent.

17. $A = \$2,000$; $r = 6\%$; $t = 1$

18. $A = \$10,000$; $r = 12\%$; $t = 5$

19. $A = \$5,000$; $r = 7.5\%$; $t = 2$

20. $A = \$8,000$; $r = 4.5\%$; $t = 3$

Solve each equation for x.

21. $3^x = 20$

22. $5^x = 15$

23. $8^x = 10$

24. $20^x = 100$

25. $\dfrac{8^x - 5}{6} = 20$

26. $\dfrac{4^x + 6}{3} = 10$

27. $\dfrac{8^x + 2}{5} = 12$

28. $\dfrac{5^x - 8}{10} = 14$

In Exercises 29–34, use the formula for finding the future value of an ordinary annuity,

$$A = R\frac{\left(1 + \frac{r}{m}\right)^n - 1}{\frac{r}{m}},$$

---

*Before doing these exercises, you may find it useful to review the note *How to Succeed at Mathematics* on page xix.

to solve for n. *You are given A, R, and r. Assume that payments are made monthly and that the interest rate is an annual rate.*

29. $A = \$10{,}000$; $R = 200$; $r = 9\%$
30. $A = \$12{,}000$; $R = 400$; $r = 8\%$
31. $A = \$5{,}000$; $R = 150$; $r = 6\%$
32. $A = \$8{,}000$; $R = 400$; $r = 5\%$
33. $A = \$6{,}000$; $R = 250$; $r = 7.5\%$
34. $A = \$7{,}500$; $R = 100$; $r = 8.5\%$

Applying What You've Learned

In Exercises 35–40, assume that the compounding is being done monthly.

35. **Saving for a scooter.** Matt is saving to buy a new Vespa. If he deposits $75 at the end of each month in an account that pays an annual interest rate of 6.5%, how much will he have saved in 30 months?

36. **Saving for a trip.** Angelina wants to save for an African safari. She is putting $200 each month in an ordinary annuity that pays an annual interest rate of 9%. If she makes payments for 2 years, how much will she have saved for her trip?

37. **Saving for a car.** Kristy Joe deposits $150 each month in an ordinary annuity to save for a new car. If the annuity pays a monthly interest rate of 0.85%, how much will she be able to save in 3 years?

38. **Saving for retirement.** Cohutta is saving for his retirement in 10 years by putting $500 each month into an ordinary annuity. If the annuity has an annual interest rate of 9.35%, how much will he have when he retires?

39. **Saving for a vacation home.** Wendy has set up an ordinary annuity to save for a retirement home in Florida in 15 years. If her monthly payments are $400 and the annuity has an annual interest rate of 6.5%, what will be the value of the annuity when she retires?

40. **Saving for retirement.** Thiep has set up an ordinary annuity to save for his retirement in 20 years. If his monthly payments are $350 and the annuity has an annual interest rate of 7.5%, what will be the value of the annuity when he retires?

41. **Saving for a condominium.** Kanye wants to save $14,000 in 8 years by making monthly payments into an ordinary annuity for a down payment on a condominium at the shore. If the annuity pays 0.7% monthly interest, what will his monthly payment be?

42. **Saving to start a business.** Victory is making monthly payments into an annuity that pays 0.8% monthly interest to save enough for a down payment to start her own business. If she wants to save $10,000 in 5 years, what should her monthly payments be?

43. **Saving for consumer goods.** Sandra Lee is making monthly payments into an annuity. She wants to have $600 in the fund to buy a new convection range in 6 months, and the account pays 8.2% annual interest. What are her monthly payments to the account?

44. **Saving for consumer goods.** Lennox is making monthly payments into an annuity. He wishes to have $1,150 in 10 months to buy exercise equipment. His account pays 9% annual interest. What are his monthly payments to the account?

Tax-deferred annuities work like this: If, for example, you plan to set aside $400 per month for your retirement in 30 years in a tax-deferred plan, the $400 is not taxed now, so all of the $400 is invested each month. In a nondeferred plan, the $400 is first taxed and then the remainder is invested. So, if your tax bracket is 25%, after you pay taxes, you would have only 75% of the $400 to invest each month. However, in the tax-deferred plan, all of your money is taxed when you withdraw the money. In the nondeferred plan, only the interest that you have earned is taxed.

In Exercises 45–50, we give the amount you are setting aside each month, your current tax rate, the number of years you will contribute to the annuity, and your tax rate when you begin withdrawing from the annuity. Answer the following questions for each situation:

a) *Find the value of the tax-deferred and the nondeferred accounts.*

b) *Calculate the interest that was earned in both accounts. This will be the value of the account minus the payments you made.*

c) *If you withdraw all money from each account and pay the relevant taxes, which account is better and by how much?*

	Monthly Payment	Number of Years	Annual Interest Rate	Current Tax Rate	Future Tax Rate
45.	$300	30	6%	25%	18%
46.	$400	25	4.5%	25%	15%
47.	$400	20	4%	30%	30%
48.	$600	30	4.6%	25%	25%
49.	$500	35	3.4%	25%	30%
50.	$500	30	4.8%	18%	25%

In Exercises 51–54, assume that monthly deposits are being placed in an ordinary annuity and interest is compounded monthly.

51. **Saving for a fire truck.** The Reliance Volunteer Fire Company wants to take advantage of a state program to save money to purchase a new fire truck. The truck will cost $400,000, and members of the finance committee estimate that with community and state contributions, they can save $5,000 per month in an account paying 10.8% annual interest. How long will it take to save for the truck?

52. **Saving for new equipment.** BioCon, a bioengineering company, must replace its water treatment equipment within 2 years. The

new equipment will cost $80,000, and the company will transfer $3,800 per month into a special account that pays 9.2% interest. How long will it take to save for this new equipment?

53. **Saving for a condominium.** Kirsten wants to save $30,000 for a down payment on a condominium at a ski resort. She feels that she can save $550 per month in an account that has a 7.8% annual interest rate. How long will it take to acquire her down payment?

54. **Saving for a business.** Leo needs to save $25,000 for a down payment to start a photo restoration business. He intends to save $300 per month in an account that pays an annual interest rate of 6%. How long will it take for him to save for the down payment?

Communicating Mathematics

55. What property of the log function would you use to solve an equation of the form $y = a^x$ for x?

56. What is similar about the problem of finding the future value of an annuity and finding the payments to make into a sinking fund? How are the two problems related?

Using Technology to Investigate Mathematics

57. See your instructor for tutorials for the TI-83 to do financial calculations. Use your calculator to reproduce some of the examples in this section.*

58. You can find many interactive annuity calculators on the Internet. Some implement the calculations that you have learned in this section and others implement other types of annuity problems. Find some interesting annuity calculators, experiment with them, and report on your findings.

For Extra Credit

59. **Saving for retirement.** Reconsider Exercise 38. Suppose that Cohutta had started his annuity 10 years earlier. What would his monthly payments have been to accumulate the same future value in his annuity as you found in Exercise 38?

60. **Saving for retirement.** Reconsider Exercise 40. Suppose that Thiep had started his annuity 10 years earlier. What would his monthly payments have been to accumulate the same future value in his annuity as you found in Exercise 40?

61. **Saving for retirement.** Carlos began to save for his retirement at age 25, and for 10 years he put $200 per month into an ordinary annuity at an annual interest rate of 6%. After the 10 years, he could no longer make payments, so he placed the value of the annuity into another account that paid 6% annual interest compounded monthly. He left the money in this account for 30 years until he was ready to retire. How much did he have for retirement?

62. **Saving for retirement.** If Carlos had waited until age 45 to think about retirement and then decided to put money into an ordinary annuity for 20 years, what would his monthly payments have to be to accumulate the same amount for retirement as you found in Exercise 61? (We assume that the interest rate for the annuity is the same.)

63. **Saving for retirement.** Examine your solutions to Exercises 61 and 62. What do you notice? Explain why this is so.

64. **Comparing annuities.** The difference between an *annuity due* and an ordinary annuity is that with an annuity due, the payment is made at the *beginning* of the month rather than at the end of the month. This means that each payment generates one more month of interest than with an ordinary annuity.

    a. How does this change the formula for finding the future value of an annuity?

    b. Use this formula to find the value of the annuity in Example 2, assuming that the annuity is an annuity due.

9.5 Amortization

Objectives

1. Calculate the payment to pay off an amortized loan.
2. Construct an amortization schedule.
3. Find the present value of an annuity.
4. Calculate the unpaid balance on a loan.

Congratulations! You just bought a new home—it's lovely—and in a good neighborhood. Only 360 more payments and it's all yours. When you make such a large purchase, you usually have to take out a loan that you repay in monthly payments. The process of paying off a loan (plus interest) by making a series of regular, equal payments is called **amortization**, and such a loan is called an **amortized loan**.

If you were to make such a purchase, one of the first questions you might ask is, "What are my monthly payments?" Of course, the lender can answer this question, but you may find it interesting to learn the mathematics involved with paying off a mortgage so that you can answer that question yourself.

KEY POINT

Paying off a loan with regular payments is called amortization.

Amortization

Assume that you have purchased a new car and after your down payment, you borrowed $10,000 from a bank to pay for the car. Also assume that you have agreed to pay off this loan by making equal monthly payments for 4 years. Let's look at this transaction from two points of view:

Banker's point of view: Instead of thinking about your payments, the banker might think of this transaction as a future value problem in which she is making a $10,000 loan to you now and compounding the interest monthly for 4 years. At the end of 4 years, she expects to be paid the full amount due. Recall from Section 9.2 that this future value is

$$A = P\left(1 + \frac{r}{m}\right)^n.$$

Your point of view: For the time being, you could also ignore the question of monthly payments and choose to pay the banker in full with one payment at the end of 4 years. In order to have this money available, you could make monthly payments into a sinking fund to have the amount A available in 4 years. As you saw in Section 9.4, the formula for doing this is

$$A = R\frac{\left(1 + \frac{r}{m}\right)^n - 1}{\frac{r}{m}}.$$

Thus, to find your monthly payment, we will set the amount the banker expects to receive equal to the amount that you will save in the sinking fund and then solve for R.

FORMULA FOR FINDING PAYMENTS ON AN AMORTIZED LOAN
Assume that you borrow an amount P, which you will repay by taking out an amortized loan. You will make m periodic payments per year for n total payments and the annual interest rate is r. Then, you can find your payment by solving for R in the equation

$$P\left(1 + \frac{r}{m}\right)^n = R\left(\frac{\left(1 + \frac{r}{m}\right)^n - 1}{\frac{r}{m}}\right).^*$$

Do not let this equation intimidate you. You have done the calculation on the left side many times in Section 9.2 and the computation on the right side in Section 9.4. Once you find these two numbers, you do a simple division to solve for R, as you will see in Example 1.

EXAMPLE 1 *Determining the Payments on an Amortized Loan*

Assume that you have taken out an amortized loan for $10,000 to buy a new car. The yearly interest rate is 18% and you have agreed to pay off the loan in 4 years. What is your monthly payment?

---

*Certainly we could do the necessary algebra to solve this equation for R. Then we could use this new formula for solving problems to find the monthly payments for amortized loans. We chose not to do this because our philosophy is to minimize the number of formulas that you have to memorize to solve the problems in this chapter. We will round payments on a loan *up* to the next cent.

TI-83 calculator confirms our computations in Example 1.

Quiz Yourself **18**

What would your payments be in Example 1 if you agree to pay off the loan in 5 years?

SOLUTION: We will use the preceding equation. The values of the variables in this equation are

$$P = 10,000$$
$$n = 12 \times 4 = 48$$
$$\frac{r}{m} = \frac{18\%}{12} = 0.015$$

We must solve for R in the equation

$$10,000(1 + 0.015)^{48} = R\left[\frac{(1 + 0.015)^{48} - 1}{0.015}\right].$$

monthly interest rate ⟍

amount of loan ⟋

⟍ number of payments

If we calculate the numerical expressions on both sides of this equation as we did in Sections 9.2 and 9.4, we get

$$20,434.78289 = R(69.56521929).$$

Therefore, your monthly payment is

$$R = \frac{20,434.78289}{69.56521929} \approx \$293.75.$$

Now try Exercises 3 to 10. ✳ **18**

Amortization Schedules

Payments that a borrower makes on an amortized loan partly pay off the principal and partly pay interest on the outstanding principal. As the principal is reduced, each successive payment pays more toward principal and less toward interest. A list showing payment-by-payment how much is going to principal and interest is called an **amortization schedule**. We illustrate such a schedule in Example 2.

EXAMPLE 2 *Constructing an Amortization Schedule*

To expand your business selling collectibles on the Internet, you need a loan of $5,000. Your banker loans you the money at a 12% annual interest rate, which you agree to pay back in three equal monthly installments of $1,700.12.* Construct an amortization schedule for this loan.

SOLUTION: At the end of the first month, you have borrowed $5,000 for 1 month at a 1% monthly interest rate. So using the simple interest formula, the interest that you owe the bank is

$$\overset{P}{\$5,000} \times \overset{r}{0.01} \times \overset{t}{1} = \overset{I}{\$50}.$$

Payment Number	Amount of Payment	Interest Payment	Applied to Principal	Balance
				$5,000.00
1	$1,700.12	$50.00	$1,650.12	$3,349.88
2	$1,700.12	$33.50	$1,666.62	$1,683.26
3	$1,700.12	$16.83	$1,683.29	−$0.03

TABLE 9.4 An amortization schedule.

Your payment is $1,700.12; therefore, $50 pays the interest, and the rest, $1,700.12 − $50 = $1,650.12, is applied to the principal.

For the second month, you are now borrowing $5,000 − $1,650.12 = $3,349.88 at 1% monthly interest. We complete the computations for the payments on this loan in Table 9.4.

*We used the method from Example 1 to calculate the exact payment to be $1,700.110557. Because we increase this ever so slightly to $1,700.12, after the third payment we have overpaid by $0.03.

As expected, we ended with a negative balance because the payment of $1,700.12 is a fraction of a cent larger than it needs to be. In an actual banking situation, the bank would adjust the final payment so that the final balance is exactly $0.00. ❋

Example 3 illustrates how discouraging it can be when you make your first payment on a mortgage for a house and realize how little of your payment goes toward paying the principal.

EXAMPLE 3   *Constructing an Amortization Schedule*

Assume that you have saved money for a down payment on your dream house, but you still need to borrow $120,000 from your bank to complete the deal. The bank offers you a 30-year mortgage at an annual rate of 7%. The monthly payment is $798.37. Construct an amortization schedule for the first three payments on this loan.

SOLUTION: We compute Table 9.5* as we did Table 9.4 in Example 2.

Payment Number	Amount of Payment	Interest Payment	Applied to Principal	Balance
				$120,000.00
1	$798.37	$700.00	$98.37	$119,901.63
2	$798.37	$699.43	$98.94	$119,802.69
3	$798.37	$698.85	$99.52	$119,703.17

TABLE 9.5 Making an amortization schedule for a lengthy mortgage.

Quiz Yourself 19

Compute the fourth line of Table 9.5.

You see that for such a lengthy amortized loan, the early payments are mostly interest. Fortunately, because the debt is being reduced, each month a little more of the payment goes toward principal and a little less toward interest.

Now try Exercises 11 to 14. ❋ 19

HIGHLIGHT ❋ ❋ ❋

Between the Numbers—Can They Really Do That to You?

How would you feel if you took out a $200,000 mortgage for a house, faithfully made all of your payments on time, and at the end of 1 year owed $201,118? Incredibly, this can actually happen if you have an adjustable rate mortgage, or ARM. Some ARMs allow you to make payments that *do not even cover the interest* on the loan, so the amount you owe increases even though you make your payments on time.

ARMs can have other very serious problems for the consumer. With an ARM, it is possible to start with a low interest rate, say 4%, and with yearly increases after several years *your interest rate could be much higher.* Mortgage lenders use an index, often tied to government securities, to determine how much to increase your interest rate. There are many different types of ARMs—some limit the rate increase from year to year, and others limit the maximum rate that can be charged. However, even with these limits, your monthly payments in an ARM could increase from $900 to $1,400 over a 3-year period, causing you great financial distress.

The *Consumer Handbook on Adjustable Rate Mortgages*, available from the Federal Reserve Board, is an excellent guide to ARMs and contains numerous examples, cautions, and a worksheet to help you make sensible decisions regarding mortgages.

*If you verify these computations by hand, your answers may differ slightly from ours due to a difference in the way we are rounding off our intermediate calculations.

HIGHLIGHT ❀ ❀ ❀

Using a Spreadsheet to Make an Amortization Schedule

A spreadsheet can create an amortization schedule in the blink of an eye. The following is a spreadsheet that calculates the schedule for an amortized loan for $10,000 with 60 monthly payments of $202.77. We first show the spreadsheet displaying the formulas in the cells of the spreadsheet.

	A	B	C	D	E	F
1	End of Month	Payment	Interest	Principal	Balance	
2	0	$202.77			$10,000	
3	1	$202.77	E2*0.08/12	B3 – C3	E2 – D3	
4	2	$202.77	E3*0.08/12	B4 – C4	E3 – D4	
5	3	$202.77	E4*0.08/12	B5 – C5	E4 – D5	
6	4	$202.77	E5*0.08/12	B6 – C6	E5 – D6	
7	5	$202.77	·	·	·	
8	6	$202.77	·	·	·	
9	7	$202.77	·	·	·	

Here is the same spreadsheet when the formulas in the spreadsheet are evaluated.

	A	B	C	D	E	F
1	End of Month	Payment	Interest	Principal	Balance	
2	0	$202.77			$10,000.00	
3	1	$202.77	$66.67	$136.10	$9,863.90	
4	2	$202.77	$65.76	$137.01	$9,726.89	
5	3	$202.77	$64.85	$137.92	$9,588.96	
6	4	$202.77	$63.93	$138.84	$9,450.12	
7	5	$202.77	·	·		·
8	6	$202.77	·	·		·
9	7	$202.77	·	·		·

To generate a new schedule for a mortgage, all we have to do is change the formulas in several cells and the entire spreadsheet will be recalculated.

 **KEY POINT**

We use the formula for finding the size of monthly payments to determine the present value of an annuity.

Finding the Present Value of an Annuity

When buying a car, your budget determines the size of the monthly payments you can afford, and that determines how much you can pay for the car you buy. Assume that you can afford car payments of $200 per month for 4 years and your bank will grant you a car loan at an annual rate of 12%. We can think of this as a future value of an annuity problem where R is 200, $\frac{r}{m}$ is 1%, and n is 48 months. We know from Section 9.4 that the future value of this annuity is

$$A = 200\left[\frac{(1+0.01)^{48} - 1}{0.01}\right] = \$12,244.52.$$

This result does not mean that now you can afford a $12,000 car! This amount is the *future value* of your annuity, not what that amount of money would be worth in the *present*.

> **DEFINITION** If we know the monthly payment, the interest rate, and the number of payments, then the amount we can borrow is called the **present value of the annuity**.

We can find the present value of an annuity by setting the expression for the future value of an account using compound interest equal to the expression for finding the future value of an annuity and solving for the present value P.

> **FINDING THE PRESENT VALUE OF AN ANNUITY** Assume that you are making m periodic payments per year for n total payments into an annuity that pays an annual interest rate of r. Also assume that each of your payments is R. Then to find the present value of your annuity, solve for P in the equation
>
> $$P\left(1+\frac{r}{m}\right)^n = R\left(\frac{\left(1+\frac{r}{m}\right)^n - 1}{\frac{r}{m}}\right).$$

Again, you have done the computations on both sides of this equation many times in Sections 9.2 and 9.4.

EXAMPLE 4   *Determining the Price You Can Afford for a Car*

If you can afford to spend $200 each month on car payments and the bank offers you a 4-year car loan with an annual rate of 12%, what is the present value of this annuity?

SOLUTION: To solve this problem, we can use the formula for finding payments on an amortized loan:

$$P\left(1+\frac{r}{m}\right)^n = R\left(\frac{\left(1+\frac{r}{m}\right)^n - 1}{\frac{r}{m}}\right). \tag{1}$$

We know $R = 200$, $\frac{r}{m} = 1\% = 0.01$, and $n = 48$ months. If we substitute these for the variables in equation (1), we get

$$P(1 + 0.01)^{48} = 200\left[\frac{(1+0.01)^{48} - 1}{0.01}\right]. \tag{2}$$

Calculating the numerical expressions on both sides of equation (2) gives us

$$P(1.612226078) = 12{,}244.52155.$$

Now dividing both sides of this equation by 1.612226078, we find

$$P = \left[\frac{12.244.52155}{1.612226078}\right] \approx \$7.594.79.$$

You may find this answer surprising, but the mathematics of this problem are clear. If you can only afford payments of $200 per month, then you can only afford to finance a car loan for about $7,600!

Now try Exercises 25 to 30. ❋ **20**

N=48
I%=12
• PV=-7594.791899
PMT=200
FV=0
P/Y=12
C/Y=12
PMT:**END** BEGIN

TI-83 screen confirms our computations in Example 4.

Quiz Yourself **20**

Redo Example 4, but now assume that you can afford payments of $250 per month.

✎ **KEY POINT**

To refinance a loan, we must know the unpaid balance on the loan.

Finding the Unpaid Balance of a Loan

During times when interest rates are high, people are forced to borrow money at these high rates if they want to buy a car or a house on credit. If interest rates decline, then it is wise to consider paying off the remaining debt on the first loan by taking out a second loan at a lower interest rate. This procedure is called **refinancing** the loan. To understand refinancing, we must be able to compute how much debt remains on a loan after a certain number of payments have been made.

EXAMPLE 5   *Finding the Unpaid Balance on a Loan*

a) Assume that you take out a 30-year mortgage for $100,000 at an annual interest rate of 9%. If, after 10 years, interest rates drop and you want to refinance, how much remains to be paid on your mortgage?

b) If you can refinance your mortgage for the remaining 20 years at an annual interest rate of 7.2%, what will your monthly payments be?

c) How much will you save in interest in 20 years by paying the lower rate?

SOLUTION:

a) Doing the same kind of calculations as we did in Example 1, we find that the monthly payment is $804.63.

   Your monthly payment was based on the assumption that you would be paying the loan for 30 years. Therefore, after 10 years, you will not have accumulated enough in your annuity to pay off the banker. This means that the amount you owe, $P\left(1+\frac{r}{m}\right)^n$, must be larger than the amount that you have accumulated in your sinking fund,

$$R\left(\frac{\left(1+\dfrac{r}{m}\right)^n - 1}{\dfrac{r}{m}}\right).$$

The unpaid balance U on the loan is therefore

$$\overset{\text{what you owe}}{U = P\left(1+\frac{r}{m}\right)^n} - R\left(\frac{\overset{\text{what you have accumulated}}{\left(1+\dfrac{r}{m}\right)^n - 1}}{\dfrac{r}{m}}\right).^* \tag{3}$$

It is important to recognize in equation (3) that only 10 years have elapsed, so $n = 12 \times 10 = 120$, not 360.

   We can now substitute the values $P = \$100,000$, $\frac{r}{m} = 0.09/12 = 0.0075$, $n = 120$, and $R = \$804.63$ in equation (3).

$$U = 100,000\underset{\text{10 years}}{(1 + 0.0075)^{120}} - 804.63\left[\frac{(1 + 0.0075)^{\overset{\text{10 years}}{120}} - 1}{0.0075}\right] = \$89,428.32.$$

Therefore, you still owe $89,428.32 on this mortgage.

b) Because you have a balance of $89,428.32 on your mortgage, in effect you are now taking out a new mortgage for this amount at an annual interest rate of 7.2% for the remaining 20 years. In this case, $P = \$89,428.32$, $n = 12 \times 20 = 240$, and $\frac{r}{m} = 0.072/12 = 0.006$.

   We now solve for the monthly payment R in the equation

$$\underset{\substack{\text{amount borrowed}\\\text{for second loan}}}{89,428.32}(1 + \underset{\substack{\text{new monthly}\\\text{interest rate}}}{0.006})^{\overset{\text{20 years}}{240}} = R\left[\frac{(1 + 0.006)^{240} - 1}{0.006}\right]$$

---

*The unpaid balance appears in the balance column of an amortization table.

If we calculate the numerical expressions on both sides of this equation as we did in Sections 9.2 and 9.4, we get

$$375,829.1355 = R(533.7623389).$$

Therefore, your new monthly payment is

$$R = \frac{375,829.1355}{533.7623389} \approx \$704.12.$$

Therefore, by refinancing, you are able to reduce your monthly mortgage payment by $804.63 − $704.12 = $100.51 per month.

c) If you make monthly payments on the unpaid balance for 20 years at the old interest rate, your total payments will be 240 × $804.63 = $193,111.20. If you make monthly payments for 20 years at the new interest rate, you will pay 240 × $704.12 = $168,988.80. The difference

$$\$193,111.20 - \$168,988.80 = \$24,122.40*$$

is the amount that you save in interest over 20 years at the reduced rate.

Now try Exercises 31 to 36. ✳

Often, when refinancing a mortgage you must pay a refinancing fee. The refinancing fee is often calculated as a percentage of the balance remaining on the mortgage. Suppose in Example 5 that you had to pay a 2% refinancing fee. Two percent of $89,428.32 is $1,788.57. You would gain this fee back in 18 months with the reduced payments on the loan. In this case, it is clear that after 18 months, you would benefit from the refinancing.

Exercises 9.5

Looking Back[†]

These exercises follow the general outline of the topics presented in this section and will give you a good overview of the material that you have just studied.

1. What does the left side of the first equation in Example 1 represent? What does the right side represent?

2. After reading the Highlight on adjustable rate mortgages, name two possible dangers of ARMs.

Sharpening Your Skills

Solve the equation

$$P\left(1 + \frac{r}{m}\right)^n = R\left(\frac{\left(1 + \frac{r}{m}\right)^n - 1}{\frac{r}{m}}\right)$$

for R to find the monthly payment necessary to pay off the loan. You are given the loan amount, the annual interest rate, and the length of the loan.

3. Amount, $5,000; rate, 10%; time, 4 years

4. Amount, $6,000; rate, 8%; time, 3 years

5. Amount, $8,000; rate, 7.5%; time, 4 years

6. Amount, $10,000; rate, 8.4%; time, 4 years

7. Amount, $12,500; rate, 8.25%; time, 4 years

8. Amount, $10,500; rate, 9.75%; time, 4 years

9. Amount, $1,900; rate, 8%; time, 18 months

10. Amount, $1,050; rate, 6.5%; time, 15 months

In Exercises 11–14, complete the first three lines of an amortization schedule for each loan. Your answer should look like Table 9.4.

11. The loan described in Exercise 3

12. The loan described in Exercise 5

13. The loan described in Exercise 7

14. The loan described in Exercise 9

Applying What You've Learned

15. **Paying off a mortgage.** Assume that you have taken out a 30-year mortgage for $100,000 at an annual rate of 7%.

   a. Construct the first three lines of an amortization schedule for this mortgage.

   b. Assume that you have decided to pay an extra $100 per month to pay off the mortgage more quickly. Find the first three lines of your payment schedule under this assumption.

   c. What is the difference in interest that you will pay on the mortgage during the fourth month if you pay the extra $100 per month versus paying only the required payment?

---

*Note that you can find this amount more quickly by multiplying the difference in mortgage payments, $100.51, by 240 months.

[†]Before doing these exercises, you may find it useful to review the note *How to Succeed at Mathematics* on page xix.

16. Paying off a mortgage. Repeat Exercise 15, but assume that the mortgage is a 20-year mortgage for $80,000 and the annual rate is 8%.

In Exercises 17–20, a) the monthly payment for each amortized loan and b) the total interest paid on the loan. Assume that all interest rates are annual rates.

17. Paying off a boat. Wilfredo bought a new boat for $13,500. He paid $2,000 for the down payment and financed the rest for 4 years at an interest rate of 7.2%.

18. Paying off a car. Beatrice bought a new car for $14,800. She received $3,500 as a trade-in on her old car and took out a 4 year loan at 8.4% to pay the rest.

19. Paying off a consumer debt. Franklin's new skis cost $350. After his down payment of $75, he financed the remainder at 18% for 5 months.

20. Paying off a consumer debt. Richard's used motorcycle cost $3,500. He paid $1,100 down and financed the rest at 8.5% for 2 years.

*In Exercises 21–24, assume that all mortgages are 30-year, adjustable rate mortgages. In each situation, use the additional information given to calculate the monthly payment on the mortgage a) in year one and b) in year five.**

21. $P = \$200,000$; beginning interest rate, 4%; rate increases 2% per year

22. $P = \$180,000$; beginning interest rate, 3.5%; rate increases 7% by year five

23. $P = \$220,000$; beginning interest rate, 4.4%; rate increases 2%, then 2%, then 1%, then 1.8%

24. $P = \$160,000$; beginning interest rate, 3.6%; rate increases 2%, then 1.6%, then 1.8%, then 2%

In Exercises 25–30, find the present value of each annuity. Assume that all rates are annual rates.

25. The value of a lottery prize. Marcus has won a $1,000,000 state lottery. He can take his prize as either 20 yearly payments of $50,000 or a lump sum of $425,000. Which is the better option? Assume an interest rate of 10%.

26. The value of a lottery prize. Belinda has won a $3,400,000 lottery. She can take her prize as either 20 yearly payments of $170,000 or a lump sum of $1,500,000. Which is the better option? Assume an interest rate of 10%.

27. Present value of a car. If Addison can afford car payments of $350 per month for 4 years, what is the price of a car that she can afford now? Assume an interest rate of 10.8%.

28. Present value of a car. If Pete can afford car payments of $250 per month for 5 years, what is the price of a car that he can afford now? Assume an interest rate of 9.6%.

29. Planning for retirement. Shane has a retirement plan with an insurance company. He can choose to be paid either $350 per month for 20 years, or he can receive a lump sum of $40,000. Which is the better option? Assume an interest rate of 9%.

30. Planning for retirement. Nico has a retirement plan with an investment company. She can choose to be paid either $400 per month for 10 years, or she can receive a lump sum of $30,000. Which is the better option? Assume an interest rate of 9%.

In Exercises 31–36, find the unpaid balance on each loan.

31. Paying off a loan early. You have taken an amortized loan at 8.5% for 5 years to pay off your new car, which cost $12,000. After 3 years, you decide to pay off the loan.

32. Paying off a loan early. In order to pay for new scuba equipment, you took an amortized loan for $1,800 at 11%, which you agreed to repay in 3 years. After 18 payments, you decide to pay off the loan.

33. Paying off a mortgage early. The MacGuffs took out a 30-year mortgage for $120,000 on their vacation home at an annual interest rate of 7%. They decide to refinance the mortgage after 8 years.

34. Paying off a mortgage early. In order to modernize their restaurant, the Buccos took out a 25-year mortgage for $135,000 at an annual interest rate of 6%. They decide to refinance the mortgage after 10 years.

35. Paying off a loan early. Garrett took out an amortized loan for $8,000 for 2 years at 8% to finish culinary school. After 12 payments, he decided to pay off the loan.

36. Paying off a loan early. Sheila borrowed $14,000 to invest in her floral shop. She took out an amortized loan at 6% for 5 years. After making payments for 1 year, she decided to pay off the loan.

---

*To keep these exercises simple, during year five, use the same values for P and n as you used in year one. Technically, to get a more exact answer, you should take into account that by year five, some of the principal would have been paid off and also that only 26 years of payments remain. Ignoring these does not affect the spirit of the exercise because during the first few years, most of your mortgage payments are going towards interest.

In Exercises 37–42, you are given the amount of an amortized loan P, the annual interest rate r, the number of payments of the loan n, and the monthly payment R. After the specified number of months, the borrower decides to refinance at the new interest rate for the remaining length of the loan.

a) What is the new monthly payment?

b) How much does the borrower save on interest?

37. $P = \$10,000$; $r = 8\%$; $n = 48$; $R = \$244.13$; refinance after 24 months at an interest rate of 6.5%.

38. $P = \$20,000$; $r = 9\%$; $n = 48$; $R = \$497.71$; refinance after 12 months at an interest rate of 7%.

39. $P = \$100,000$; $r = 8\%$; $n = 240$; $R = \$836.45$; refinance after 60 months at an interest rate of 7%.

40. $P = \$100,000$; $r = 9.5\%$; $n = 360$; $R = \$840.86$; refinance after 120 months at an interest rate of 7.5%.

41. $P = \$40,000$; $r = 8\%$; $n = 48$; $R = \$976.52$; refinance after 24 months at an interest rate of 7.5%.

42. $P = \$50,000$; $r = 10\%$; $n = 60$; $R = \$1,062.36$; refinance after 24 months at an interest rate of 9.5%.

Communicating Mathematics

43. In the formula for finding the payments on an amortized loan, where have you seen the expressions on the left and right sides of the equation before?

44. What does Example 5 show you with regard to the benefit of refinancing a loan?

45. What do the words *amortize* and *mortality* have in common? In light of the Three-Way Principle, how does this help you understand the word *amortization*?

46. Why is the interest payment on the mortgage decreasing in Table 9.4?

Using Technology to Investigate Mathematics

47. Ask your instructor for tutorials and spreadsheets to perform the amortization computations that we did in this section. Duplicate some of the results that we obtained in our examples.*

48. Find mortgage calculators on the Internet and use them to reproduce some of the examples that we explained in this section. Report on your findings.

For Extra Credit

49. In deciding to refinance at a lower interest rate, how does it affect your payments if you decide to refinance early in the loan period versus later in the loan period? For example, for a 60-month loan, would the new payments be larger, smaller, or the same if you refinance after 12 months instead of 36 months? Make up numerical examples to answer this question. Explain your answer.

50. Some mortgage agreements allow the borrower to make payments that are larger than what is required. Because this extra money goes toward reducing principal, increasing your payments may allow you to pay off the mortgage many years earlier, thus saving a great amount of interest. Assume you take out a 30-year amortized loan at 8% for $100,000 and your monthly payments will be $733.77. Suppose that instead of making the specified payment, you increase it by $100 to $833.77. How much do you save on interest over the life of the loan if you make the larger payment?

Looking Deeper

9.6 Annual Percentage Rate

Objectives

1. Calculate the annual percentage rate from a table.
2. Estimate the annual percentage rate.

A fool and his money are soon parted.   —Dutch Proverb

Have you ever heard of that saying? It certainly applies when you are borrowing money. Because the mathematics of borrowing money is complicated, there are unscrupulous money lenders who will try to take advantage of you. That is why Congress passed a law requiring lenders to inform consumers of the true cost of borrowing money.

The Annual Percentage Rate

To illustrate the problem, assume that you agree to repay a loan for $3,000 (plus the interest) in three yearly payments using an add-on interest rate of 10%. What is your true interest rate?

It depends on how you look at this agreement. Using the add-on method described in Section 9.2, we compute the interest using the formula $I = Prt = (3,000)(0.10)(3) = \$900$. Thus the amount to be repaid in three equal installments is $3,000 + 900 = \$3,900$. Each payment is therefore $\frac{3,900}{3} = \$1,300$, of which $\$1,000$ is being paid on the principal and $\$300$ is interest. Therefore, for the first year of your loan, you have borrowed $\$3,000$ and paid $\$300$ in interest. Solving the equation $300 = (3,000)(r)(1)$, we see that your interest rate is actually 10%.

At the end of the second year, you make another payment of $\$1,300$, of which $\$1,000$ goes to reduce the principal and $\$300$ is interest. Thus, for the second year you have paid $\$300$ interest on a $\$2,000$ loan. Solving the equation $300 = (2,000)(r)(1)$ for r, we find $r = 0.15$. So, in reality, the interest rate on your loan for the second year is 15%.

But, it gets worse! At the end of the third year, you make the final payment of $\$1,300$. For this last year you have paid $\$300$ interest on the $\$1,000$ remaining on the balance of the loan. Solving the equation $300 = (1,000)(r)(1)$ for r, we find $r = 0.30$. Now the interest rate is 30%.

What is the true interest rate?

The "true" interest rate we are looking for is called the **annual percentage rate**, or APR, which we will denote by a. Looking at our previous calculations, we see that

interest for the first year + interest for the second year + interest for the third year = $900.

Using the interest formula $I = Prt$, we can rewrite this as

$$(3,000)(a)(1) + (2,000)(a)(1) + (1,000)(a)(1) = 900.$$

Collecting like terms, we get $(6,000)(a)(1) = 900$. Solving this equation gives us $a = 0.15$. Thus, the annual percentage rate is 15%. You can verify that if you borrow $\$3,000$ for 1 year at 15% and then $\$2,000$ for 1 year at 15% and then $\$1,000$ for 1 year at 15%, the total interest for the 3 years is $\$900$.

 ## Some Good Advice

In doing these calculations, it is easy to make a minor computational error that will throw the answer off by a large amount. To detect such errors, always ask yourself, does the answer seem reasonable?

Suppose that we borrowed $\$6,000$ at a simple add-on interest rate of 10% and agreed to repay it by making 60 monthly payments. In calculating the APR, we should remember that we are repaying $\$100$ per month, so we have really borrowed $\$6,000$ for 1 month, $\$5,900$ for 1 month, $\$5,800$ for 1 month, etc. Instead of having three terms on the left side of the equation as we did previously, we would have 60 terms. To avoid such lengthy computations, lenders use tables similar to Table 9.6 to determine the APR. Because most loans are repaid with monthly payments, for the remainder of this section on APR, to keep the discussion simple, we will only consider payment plans having monthly payments.

	APR						
	10%	**11%**	**12%**	**13%**	**14%**	**15%**	**16%**
Number of Payments	Finance Charge per $100						
6	$2.94	$3.23	$3.53	$3.83	$4.12	$4.42	$4.72
12	$5.50	$6.06	$6.62	$7.18	$7.74	$8.31	$8.88
24	$10.75	$11.86	$12.98	$14.10	$15.23	$16.37	$17.51
36	$16.16	$17.86	$19.57	$21.30	$23.04	$24.80	$26.57
48	$21.74	$24.06	$26.40	$28.77	$31.17	$33.59	$36.03

TABLE 9.6 Finding the annual percentage rate.*

*We have kept this table simple to emphasize how it is used. A real table would have more columns for the APR, such as 14.5% and 14.25%.

In order to use Table 9.6, you must first know the **finance charge** on a loan, which is the total amount the borrower pays to use the money. This amount may include interest and fees. Then you must find the finance charge per $100 of the amount financed. To find this, divide the finance charge by the amount financed and multiply by 100. For example, if you borrow $780 and pay a finance charge of $148.20, then the finance charge per $100 of the amount financed is

$$\frac{\text{finance charge}}{\text{amount borrowed}} \times 100 = \frac{148.20}{780} \times 100 = 0.19 \times 100 = \$19.$$

USING TABLE 9.6 TO FIND THE APR ON A LOAN

1. Find the finance charge on the loan if it is not already given to you.

2. Determine the finance charge per $100 on the loan.

3. Use the line of Table 9.6 that corresponds to the number of payments to find the number closest to the amount found in step 2.

4. The top of the column containing the number found in step 3 is the APR.

EXAMPLE 1   *Using the APR Table*

Hector has agreed to pay off a $3,500 loan by making 24 monthly payments. If the total finance charge on his loan is $460, what is the APR he is being charged?

SOLUTION: The finance charge per $100 financed is

$$\frac{\text{finance charge}}{\text{amount borrowed}} \times 100 = \frac{460}{3,500} \times 100 \approx 0.1314 \times 100 \approx \$13.14.$$

Because Hector is making 24 monthly payments, we use the row in Table 9.6 for 24 payments, as we show in Figure 9.5. Reading across this line, we find that the closest amount to $13.14 is $12.98. The top of this column shows the approximate APR for Hector's loan, which is 12%.

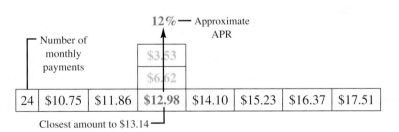

FIGURE 9.5 Using Table 9.6 to find the APR for Hector's loan.

Now try Exercises 5 to 12. ❈ **21**

We can find the APR if we know the number and size of payments on a loan.

EXAMPLE 2   *Finding the APR Using Table 9.6*

Jessica is considering buying a car costing $11,850. The terms of the sale require a down payment of $2,000 and the rest to be paid off by making 48 monthly payments of $250 each. What APR will she be paying on the car financing?

SOLUTION: The amount being financed is the purchase price minus the down payment, which is $11,850 - 2,000 = \$9,850$. Because her payments amount to $48 \times 250 = \$12,000$,

---

Quiz Yourself **21**

Assume that Jason will repay a loan for $11,250 by making 36 payments. Assume that the finance charge is $1,998.

a) What is the finance charge per $100 financed?

b) What is the APR?

this makes the finance charge equal to 12,000 − 9,850 = $2,150. The finance charge per $100 financed is therefore

$$\frac{\text{finance charge}}{\text{amount borrowed}} \times 100 = \frac{2,150}{9,850} \times 100 \approx 0.2183 \times 100 = \$21.83.$$

We now look at the row for 48 payments in Table 9.6. The number in this row that is closest to $21.83 is $21.74. Looking at the top of this column, we find that the APR for this car financing is about 10%.

Now try Exercises 17 to 24. ❄ **22**

Quiz Yourself **22**

In Example 2, assume that Jessica's payments are $265 instead of $250. What is the APR that she is being charged?

Because the Consumer Credit Protection Act was passed in 1968, it is not as common for lenders to offer add-on interest loans because they must reveal the APR. One way merchants can avoid having to state the APR is by offering the consumer the opportunity to rent rather than purchase the product outright. With a "rent-to-own" contract, because there is no loan, the merchant does not have to reveal the APR. If we consider these rental agreements as loans, we would often find that the APRs are outrageously high.

Knowing the APR enables consumers to determine the best deal when shopping for a loan.

EXAMPLE 3 *Computing the Cost of "Renting to Own"*

Jeff is considering renting a TV from a nearby rent-to-own store. He can rent the TV, which he saw priced at $479 in a local store, for $19.95 a month. If he rents the TV for 36 months, then the TV is his to keep. Analyze this rental agreement to determine whether Jeff is making a wise decision.

SOLUTION: There is really not much difference here between what Jeff is considering and purchasing the TV at another store with an agreement to make monthly payments of $19.95. Of course, with the rent-to-own agreement, Jeff can stop renting before 36 months. If Jeff were to rent the TV until he owned it, he would make 36 payments of $19.95, so he would pay a total of 36 × 19.95 = $718.20. His finance charge would be 718.20 − 479 = $239.20. The finance charge per $100 financed is therefore

$$\frac{\text{finance charge}}{\text{amount borrowed}} \times 100 = \frac{239.20}{479} \times 100 \approx 0.4994 \times 100 = \$49.94.$$

If we consider his rental as a 36-payment loan, we can try to use Table 9.6. Unfortunately, 49.94 is so large that we cannot find it in Table 9.6, which implies that the interest rate is quite high. Using other methods, which we explain in the Highlight on page 443, we find the APR to be about 28.5%. Jeff should carefully consider whether this rental is worth the cost. ❄

Math in Your Life

Short of Cash? We Can Help

If you are short of cash and were to search the Internet for "payday loans," you could find many sites that will offer you a very short-term loan to tide you over until your next payday. One site that I found will loan you $100 for 7 days provided you are willing to then pay back $125. If you take the time to check the page where the company states its annual percentage rates, you would find that you are going to be charged an APR of 1,303.57%.

Another way of looking at this loan, is to consider what would happen if you were not to repay this loan for 1 year. The loan P is $100, the weekly interest rate r is 0.25, and time t is 52 weeks. So, applying the compound interest formula, at the end of the year you would owe

$$P(1 + r)^t = 100(1.25)^{52} \approx \$10,947,644,25.$$

That's right, your $100 loan would have grown to a debt of almost $11 million!

HIGHLIGHT ❊ ❊ ❊

Using a Graphing Calculator to Find an APR

If we consider Jeff's rental in Example 3 as a loan with payments of $19.95 per month, then we can reason as we did in Section 9.5 when determining payments on a mortgage. Recall that we thought of the banker offering a loan where interest was computed monthly and we thought of the borrower as paying it off by making monthly payments into an annuity. The equation that we used was

$$P\left(1+\frac{r}{m}\right)^n = R\left(\frac{\left(1+\frac{r}{m}\right)^n - 1}{\frac{r}{m}}\right).$$

Knowing P, r, m, and n, it was easy to solve for R. The situation here is different. We know P, m, n, and R and we need to solve for r, which is quite difficult. However, using a calculator, such as the TI-83 we can solve for r as we show in the accompanying screen. Thus, the APR is roughly 28.5%.

```
N=36
■I%=28.52782389
 PV=479
 PMT=-19.95
 FV=0
 P/Y=12
 C/Y=12
 PMT:END BEGIN
```

Estimating the APR

As you saw in Example 3, it is difficult to calculate an APR without using technology; however, there is a formula that gives a good estimate of an APR for the special case of an add-on interest loan.

> **FORMULA TO APPROXIMATE THE ANNUAL PERCENTAGE RATE**
> We can approximate the annual percentage rate for an add-on interest loan by using the formula
> $$APR \approx \frac{2nr}{n+1},$$
> where r is the annual interest rate and n is the number of payments.

EXAMPLE 4 *Estimating an APR*

Minxia must borrow $4,000 to pay tuition for her last year in college. Her bank will give her an add-on interest loan at 7.7% for 3 years. Use the formula above to estimate Minxia's APR.

SOLUTION: In this problem, $n = 3 \times 12 = 36$ and $r = 7.7\% = 0.077$. So Minxia's annual percentage rate is

$$APR \approx \frac{2nr}{n+1} = \frac{2 \times 36 \times 0.077}{36+1} = \frac{5.544}{37} = 0.1499 = 14.99\%.$$

Now try Exercises 13 to 16. ❋ **23**

Quiz Yourself **23**

Use the formula to approximate the APR for an add-on interest loan for $5,500 at an annual interest rate of 9.6% that will be repaid in 48 months.

Exercises 9.6

Looking Back*

These exercises follow the general outline of the topics presented in this section and will give you a good overview of the material that you have just studied.

1. In our example on page 439, how did we argue that the interest rate for the last year of the loan was 30%?

2. In Example 2, how did we find the finance charge per $100 on Jessica's loan?

Sharpening Your Skills

In Exercises 3 and 4, use an approach similar to the discussion preceding Example 1 to find the APR of the loan. Realize that Table 9.6 does not apply to these situations because the payments are not monthly. You are given the amount of the loan, the number and type of payments, and the add-on interest rate.

3. Loan amount, $6,000; three yearly payments; rate = 8%

4. Loan amount, $8,000; four yearly payments; rate = 12%

*Before doing these exercises, you may find it useful to review the note *How to Succeed at Mathematics* on page xix.

Find the finance charge per $100 for each loan.

5. Loan, $1,800; finance charge, $270

6. Loan, $3,000; finance charge, $840

7. Loan, $2,000; finance charge, $260

8. Loan, $5,000; finance charge, $1,125

Use Table 9.6 to find the APR to the nearest whole percent. Assume that all interest rates are annual rates.

9. **Finding the APR on a loan.** Michael has agreed to pay off a $3,000 loan by making 24 monthly payments. The total finance charge on his loan is $420.

10. **Finding the APR on a loan.** Daisy has agreed to pay off a $4,500 loan by making 24 monthly payments. The total finance charge on her loan is $600.

11. **Finding the APR on a loan.** Luisa pays a finance charge of $165 on a 6-month, $4,000 loan.

12. **Finding the APR on a loan.** Wesley pays a finance charge of $310 on a 12-month, $5,000 loan.

In Exercises 13–16, estimate the annual percentage rate for the add-on loan using the given number of payments and annual interest rate. Use the formula on page 443.

13. $n = 36$; $r = 6.4\%$

14. $n = 48$; $r = 4.8\%$

15. $n = 42$; $r = 7\%$

16. $n = 30$; $r = 8\%$

Applying What You've Learned

Use Table 9.6 to find the APR to the nearest whole percent. Assume that all interest rates are annual rates.

17. **Finding the APR on a remodeling loan.** Thiep took out a $10,000 add-on loan to remodel his house and will repay it by making 24 payments of $485.

18. **Finding the APR on a car loan.** Diana took out an $8,000 add-on loan to buy a car and will repay it by making 36 payments of $270.

19. **Finding the APR on a consumer loan.** Pete took out a $4,500 add-on loan to buy a sound system for his band and will repay it by making 48 payments of $116.50.

20. **Finding the APR on a consumer loan.** Amanda took out a $1,500 add-on loan to buy a new computer and will repay it by making 24 payments of $71.25.

21. **Finding the APR on a travel loan.** Emily took out a 24-month, $2,000 add-on loan at an interest rate of 8% to go to China.

22. **Finding the APR on a vehicle loan.** John took out a 48-month, $26,000 add-on loan at an interest rate of 7.9% to pay off his truck.

23. **Finding the APR on a loan.** What is the APR of a 36-month add-on loan with an interest rate of 8.2%?

24. **Finding the APR on a loan.** What is the APR of a 48-month add-on loan with an interest rate of 8.75%?

In Exercises 25–28, decide which has the better APR to repay a $5,000 loan. Assume that you are making monthly payments.

25. **a.** An add-on interest loan at 8.4% for 3 years
    b. 24 payments of $230

26. **a.** An add-on interest loan at 7.2% for 2 years
    b. 36 payments of $165

27. **a.** An add-on interest loan at 8.9% for 4 years
    b. 36 payments of $165

28. **a.** An add-on interest loan at 8.4% for 1 year
    b. 24 payments of $240

In Exercises 29 and 30, think of the rent-to-own agreement as though it were an add-on loan. If the consumer rents until the item is paid for, find the finance charge per $100 financed. Although Table 9.6 does not contain enough columns to estimate the APR, guess as to what you think it might be.

29. **Evaluating a rent-to-own agreement.** Marcus rents a TV worth $375 for monthly payments of $18.75. After 2 years, he will own the TV.

30. **Evaluating a rent-to-own agreement.** Maria rents furniture worth $1,375 for monthly payments of $49. After 3 years, she will own the furniture.

Communicating Mathematics

31. How do you use Table 9.6 to find the APR on a loan?

32. In what way is a rent-to-own agreement different from an add-on interest loan?

Using Technology to Investigate Mathematics

33. Ask your instructor for a tutorial to show you how to find the APR with a graphing calculator. Use your calculator to reproduce the computations in Example 3 and in Exercises 29 and 30.＊

34. Search the Internet for payday loans or other types of loans such as car loans and mortgages. Try to find the APR on these loans and report on your findings.

For Extra Credit

35. We often see advertisements stating that we can consolidate our loans and have more manageable payments. The advertiser may do this for us by extending the length of our loans. Consider an add-on loan for $1,000 at an interest rate of 10% for 3 years and then for 4 years. How does this affect the APR? Is it more financially sound to pay off loans in a short period of time or a longer period of time? (Here we are only thinking about the APR—of course, there may be other considerations.)

36. Is the APR affected by the size of the loan?

CHAPTER SUMMARY*

SECTION	SUMMARY	EXAMPLE
SECTION 9.1	The word **percent** is derived from the Latin "per centum," which means "per hundred." Examining what numbers are in the tenths and hundredths place in a decimal will allow you to **convert a decimal to a percent** properly. Deciding which numbers to put in the tenths and hundredths place will help you to **convert a percent to a decimal** properly.	Discussion, p. 396 Examples 1–3, pp. 396–397
	Percent of change is given by	Examples 4 and 5, p. 398

$$\text{percent of change} = \frac{\text{new amount} - \text{base amount}}{\text{base amount}}.$$

	Many percent problems are based on the equation percent × base = amount. This equation is used to solve many applied problems.	Example 6, p. 399 Examples 7 and 8, pp. 399, 400
	Calculating with percents often occurs when **calculating taxes**.	Example 9, p. 401
SECTION 9.2	**Interest** is the money that a borrower pays to use a lender's money. The amount borrowed is called the **principal**. The **interest rate** is specified as a percentage of the principal. We use the formula $I = Prt$, where I is the interest, P is the principal, r is the interest rate, and t is the time. The equation $A = P(1 + rt)$ computes the **future value** of an account using simple interest. A is the future value (or amount), P is the principal, r is the interest rate, and t is the time. We find the **present value** of an account by solving the equation $A = P(1 + r\,t)$ for P.	Discussion, p. 404 Example 1, p. 405 Example 2, p. 405 Example 3, p. 405
	To find A, the **future value** of an account using **compound interest**, we use the formula $A = P\left(1 + \dfrac{r}{m}\right)^n$, where P is the **principal**, r is the annual **interest rate**, m is the number of **compounding periods** per year, and n is the total number of compounding periods.	Discussion, pp. 407, 408 Example 5, p. 408
	We find **present value** by solving the compound interest equation for P.	Example 6, p. 409
	The **common logarithmic function** reverses the operation of raising 10 to a power.	Example 7, p. 410
	To solve for n in the compound interest formula, take the log of both sides of the equation and use the exponent property of the log function. To solve for $\dfrac{r}{m}$, divide both sides of the equation $A = P\left(1 + \dfrac{r}{m}\right)^n$ by P, raise both sides of the resulting equation to the $\dfrac{1}{n}$ power, and then subtract 1 from both sides.	Example 8, p. 410 Example 9, p. 411
SECTION 9.3	When using the **add-on interest method** to compute the monthly payments on a loan, we use the following formula: payment $= \dfrac{P + I}{n}$, where P is the amount of the loan, I is the interest due on the loan, and n is the number of monthly payments.	Example 1, p. 415
	The **unpaid balance method** for computing finance charges uses the simple interest formula $I = Prt$, where P = previous month's balance + finance charge + purchases made – returns – payments. The variable r is the annual interest rate and $t = \frac{1}{12}$.	Example 2, p. 416
	Use the **average daily balance method** to compute a finance charge: 1. Add the outstanding balances for your account for each day of the previous month. 2. Divide this total by the number of days in the previous month. 3. Use the formula $I = Prt$, where P is the average daily balance found in step 2, r is the annual interest rate, and t the number of days in the previous month divided by 365.	Example 4, p. 417
	The exact same charges on two different credit cards may result in different finance charges.	Example 5, p. 419

---

*Before studying this chapter's material, it would be useful to reread the note *How to Succeed at Mathematics* on page xix.

SECTION 9.4

When we make a series of regular payments into an interest-bearing account, this account is called an **annuity**. If payments are made at the end of every compounding period, the annuity is called an **ordinary annuity**. The **interest** in an annuity is compounded with the same frequency as the payments. The sum of all the deposits plus all interest is called the **future value** of the account.

Discussion, p. 423

Assume that we are making a regular payment, R, at the end of each compounding period for an annuity that has an annual interest rate, r, which is being compounded m times per year. Then the value of the annuity after n compounding periods is

Example 2, p. 425

$$A = R\frac{\left(1 + \dfrac{r}{m}\right)^n - 1}{\dfrac{r}{m}}.$$

A **sinking fund** is an account into which we make regular payments for the purpose of saving some specified amount in the future. The interest is compounded with the same frequency as the payments. To find the regular **payments** that must be made into a sinking fund to save the amount A, solve for R in the following equation:

Discussion, p. 425

Example 3, p. 426

$$A = R\frac{\left(1 + \dfrac{r}{m}\right)^n - 1}{\dfrac{r}{m}}.$$

SECTION 9.5

The process of paying off a loan (plus interest) by making a series of regular equal payments is called **amortization**, and such a loan is called an **amortized loan**. We assume that P is the amount **borrowed**, r is the annual **interest rate**, m is the **number** of compounding periods per year, n is the total number of compounding periods, and R is the **payment** that is made regularly.

Discussion, p. 430

To find the regular **payment** due on an amortized loan, solve for R in the following equation:

Example 1, p. 431

$$P\left(1 + \frac{r}{m}\right)^n = R\frac{\left(1 + \dfrac{r}{m}\right)^n - 1}{\dfrac{r}{m}}.$$

A list showing payment by payment how much applies to principal and interest on an amortized loan is called an **amortization** schedule.

Examples 2 and 3, pp. 432, 433

To find the **present value** of your annuity, solve for P in the following equation:

Example 4, p. 435

$$P\left(1 + \frac{r}{m}\right)^n = R\frac{\left(1 + \dfrac{r}{m}\right)^n - 1}{\dfrac{r}{m}}.$$

The **unpaid balance**, U, on a loan after n payments is

Example 5, p. 436

$$U = P\left(1 + \frac{r}{m}\right)^n - R\frac{\left(1 + \dfrac{r}{m}\right)^n - 1}{\dfrac{r}{m}}.$$

SECTION 9.6	The **annual percentage rate**, or **APR**, is a standardized version of the "true" interest rate on a loan.	Discussion, pp. 439–440
	To use the APR table to calculate an annual percentage rate, we first find the **finance charge per $100** of the amount financed, which equals $\dfrac{\text{finance charge}}{\text{amount borrowed}} \times 100$. We can then use Table 9.6 to find the APR.	Example 1, p. 441 Example 2, p. 441
	We can **estimate** the annual percentage rate for an add-on loan using the formula $\text{APR} = \dfrac{2nr}{n+1}$, where n is the number of payments and r is the annual interest rate.	Example 4, p. 443

CHAPTER REVIEW EXERCISES

Section 9.1

1. Convert 0.1245 to a percent.

2. Convert 1.365 percent to a decimal.

3. Convert $\frac{11}{16}$ to a percent.

4. 2,890 is what percent of 3,400?

5. In 2007, M&M sales were $238.4 million, which was 13.2% of the total chocolate candy sales. What was the total amount spent on chocolate candy in 2007?

6. Use Table 9.2 to calculate the federal income tax that Maribel owes if her taxable income is $56,400.

Section 9.2

7. Find the future value of an account paying simple interest if $P = \$1,500$, $r = 9\%$, and $t = 2$ years.

8. You have agreed to pay off an $8,000 car loan with 24 monthly payments of $400 each. Use the simple interest formula to determine the interest rate that you are being charged.

9. Jacob wants to defer his income taxes of $11,400 for 6 months. If he must pay an 18% penalty (compounded monthly) to do this, what will his tax bill be?

10. Palma wants to establish a fund for her granddaughter's college education. What lump sum must she deposit in an account that pays an annual interest rate of 6%, compounded monthly, if she wants to have $10,000 in 10 years?

11. If you invest $1,000 in an account that pays an annual interest rate of 6.4%, compounded monthly, how long will it take for your money to double?

12. If $A = \$1,400$, $P = \$1,200$, and $t = 5$, solve $A = P(1 + r)^t$ for r.

Section 9.3

13. Bernie purchased a riding lawn mower for $1,320. The store offered him a 3-year add-on loan at an annual rate of 8.25%. How much interest will he pay? What are his monthly payments?

14. Use the unpaid balance method to calculate the finance charge on Joanna's credit account if last month's balance was $1,350, she made a payment of $375, she bought hiking boots for $120, and she returned a jacket for $140. Assume an annual interest rate of 21%.

15. Calculate the finance charges for the following credit card account for August (which has 31 days) using the average daily balance method. July's balance was $275 and the annual interest rate is 18%.

Date	Transaction
August 6	Made payment of $75
August 12	Charged $115 for clothes
August 19	Charged $20 for gasoline
August 24	Charged $16 for lunch

Section 9.4

16. Piers is saving for his retirement by putting $175 each month into an ordinary annuity. If the annuity pays an annual interest rate of 9.35%, how much will he save for his retirement in 10 years?

17. Find the monthly payment needed to have a sinking fund accumulate to $2,000 in 36 months if the annual interest rate is 6%.

18. Solve $\dfrac{3^x - 4}{2} = 10$ for x.

19. You are making monthly payments of $300 into an annuity that pays 9% annual interest. How long will it take to accumulate $10,000?

20. Assume that you are saving $350 a month in a retirement annuity that has an interest rate of 4.2%. Assume that your income taxes for the life of the annuity are 25% and that they drop to 18% when you retire. How much more do you earn in 30 years in a tax-deferred account than a nondeferred one?

Section 9.5

21. Find the monthly payment necessary to pay off a 4-year amortized loan of $5,000 if the annual interest rate is 10%.

22. Complete the first two lines of an amortization schedule for a 20-year, $100,000 mortgage if the annual interest rate is 8%.

23. Jesse has won a $1,000,000 state lottery. She can take her prize as either 20 yearly payments of $50,000 or a lump sum of $500,000. Which is the better option? Assume an interest rate of 8%.

24. You have taken an amortized loan at 7.5% for 5 years to pay off your new car, which cost $13,000. After 2 years, you decide to pay off the loan. What is your unpaid balance?

25. Assume that you borrow $180,000 in a 30-year adjustable rate mortgage with an initial interest rate of 4.5%. The interest rate increases over 4 years to 12.5%. What are your monthly payments in the first year? The fifth year?

Section 9.6

26. Ann took out a $1,800 add-on loan for a professional-quality color printer, which she will repay with 12 payments of $163. Use Table 9.6 to find her interest rate to the nearest percent.

27. Use the formula on page 443 to estimate the annual percentage rate on a loan that has an annual interest rate of 8% that will be repaid in 20 months.

CHAPTER TEST

1. Convert 0.3624 to a percent.

2. Convert 23.45 percent to a decimal.

3. Convert $\frac{7}{16}$ to a percent.

4. Use the unpaid balance method to calculate the finance charge on Marion's credit account if last month's balance was $950, she made a payment of $270, and she bought a ski parka for $217 and gloves for $23. Assume an annual interest rate of 24%.

5. Find the future value of an account paying simple interest if $P = \$3,400$, $r = 2.5\%$, and $t = 3$ years.

6. 994 is what percent of 2,840?

7. Best Buy is selling an MP3 player that normally sells for $169.99 for $149.99. What is the percent of the reduction on the price of this player?

8. Hiro took out a $1,600 add-on loan for a handmade, split-bamboo, Japanese fly-fishing rod that he will repay with 24 payments of $75. Use Table 9.6 to find his interest rate to the nearest percent.

9. You are paying off your large-screen projection TV that cost $3,000 with 24 monthly payments of $162.50. What is the annual simple interest rate that you are being charged?

10. If Danica deposits $4,000 in an account that has an interest rate of 3.6% that is compounded monthly, what is the value of the account in 4 years?

11. If you invest $1,000 in an account that pays an interest rate of 4.8% compounded monthly, how long will it take for your money to double?

12. Assume that you borrow $220,000 in a 30-year adjustable rate mortgage with an initial interest rate of 3.2%. The interest rate increases 1.5% per year for the next 4 years. What are your monthly payments in the first year? The fifth year?

13. José wants to establish a trust for his nephew who is now 3 years old. He will deposit a lump sum in an account with an annual interest rate of 4.2% compounded monthly. How much must he deposit now if he wants his nephew to have $15,000 when he turns 21?

14. To pay for scuba diving equipment costing $1,560, Carmen took out a 2-year add-on interest loan at an annual rate of 10.5%. How much interest will he pay? What are his monthly payments?

15. If $A = \$2,400$, $P = \$2,100$, and $t = 3$, solve $A = P(1 + r)^t$ for r.

16. From 2005 to 2006, the price of a new home rose from $250 thousand to $257 thousand. What was the percent of increase?

17. Calculate the finance charges for the following credit card account for April (which has 30 days) using the average daily balance method. March's balance was $425, and the annual interest rate is 21%.

Date	Transaction
April 4	Made payment of $85
April 10	Charged $25 for gasoline
April 15	Charged $15 for lunch
April 25	Charged $80 for concert tickets

18. Grace is saving for her son's college education by putting $200 each month into an ordinary annuity. If the annuity pays an annual interest rate of 5.15%, how much will she have saved in 8 years?

19. Solve $\dfrac{5^x - 4}{3} = 10$ for x.

20. Use the formula on page 443 to estimate the annual percentage rate on a loan that has an annual interest rate of 8% that will be repaid in 20 months.

21. Find the monthly payment needed to have a sinking fund accumulate to $1,800 in 36 months if the annual interest rate is 4%.

22. Find the monthly payment necessary to pay off an 8-year amortized loan of $20,000 if the annual interest rate is 9%.

23. Assume that you are making monthly payments of $450 into an annuity that pays 3.75% annual interest. How long will it take to accumulate $9,000?

24. Use Table 9.2 to calculate the federal income tax that Jakob owes if his taxable income is $48,600.

25. Mike has won the $1,000,000 prize on *Deal or No Deal*. He can take his prize as a lump sum of $750,000 or as a regular annuity with 10 yearly payments of $100,000. Which is the better option? Assume an annual interest rate of 4%.

26. Assume that you have taken out a 20-year amortized loan of $140,000 at an annual interest rate of 7.5%.

a. What is the monthly payment on your loan?

b. Write the first two lines of the amortization table for this loan.

27. Estelle has taken an amortized loan at 9.6% for 5 years to pay off her new car, which costs $16,500.

a. What is her monthly payment on the loan?

b. After 2 years, she decides to pay off the loan. What is her unpaid balance?

GROUP EXERCISES

1. Get federal income tax tables for filing single, married filing jointly, and married filing separately and investigate if there are any advantages for a married couple to file one way versus the other.

2. Go to the Web site www.federalreserve.gov/pubs/arms/arms_english.htm or some other Web site to learn more about adjustable rate mortgages. Also, try to find actual examples of current mortgage offers either online or by going to financial institutions. There are many varieties of ARMs other than the ones we have discussed. Make up new examples illustrating the dangers of ARMs.

Geometry

Ancient and Modern Mathematics Embrace

Geometry is all around us. If you stop to think about it for a moment, I believe you would agree with this statement because when you look around you see all kinds of geometric objects such as lines, rectangles, and spheres. But surprisingly, geometry is also inside of us, we wear it, we watch it at the movies and play with it on our Xboxes, Wiis, and PlayStations. And, if you were to fly from Florida to the Philippines, you would find it strange that geometry tells the pilot the best route goes through Alaska.* Also, geometry tells us that the paths that streams and rivers take in flowing from the mountains to the sea are geometrically similar to the paths that bacteria take while growing in a petri dish and the way airways branch inside our lungs.

(continued)

*See http://gc.kls2.com/ for an interactive site that will calculate flight paths such as this which contradict our intuition.

In this chapter, you will first study the classical geometry of points, lines, angles, and solids that the Greeks used as the basis for surveying, mapmaking, and architecture. But intertwined with that discussion, we will introduce you to a type of "curved" geometry that we use to navigate on a spherical Earth and that Albert Einstein used to describe our curved universe.

Finally in this chapter, we will show you a modern type of geometry where lines are neither straight nor curved, but rather have an infinite number of "wiggles." Scientists use this fractal geometry to describe the roughness of mountain peaks and shorelines, the ups and downs of the stock market, and the irregular beating of a healthy heart. ●

10.1 Lines, Angles, and Circles

Objectives

1. Understand the basic properties of geometric objects such as points, lines, and planes.
2. Work with the fundamental properties of angles.
3. Solve problems involving the relationships among angles, arcs, and circles.

Before we begin this section, let's take a moment to think about the future. I don't mean my future—or your future—or even a thousand years from now. Let's try to imagine 2,300 years into the future. What "important" ideas that you struggle to learn the night before a big exam in one of your courses will still be relevant then? It's hard to grasp the size of that question.

If you were to look backwards instead, you would see one idea that has shaped mathematics and science for over 2,000 years—the method of deductive reasoning that was put forth in the most famous textbook in the history of the world—*Elements* by Euclid of Alexandria. *Elements*, which was written around 300 BC, is an introduction to elementary mathematics—including number theory and geometry—that gave mathematicians a method of reasoning which we still use today.

KEY POINT

Two lines in the plane are either parallel or intersecting.

Points, Lines, and Planes

Euclid begins his discussion of geometry with intuitive descriptions of three *undefined* terms, saying that a **point** is "that which has no part," a **line** has "length but no breadth," and a **plane** has "length and breadth only."

To discuss some of the basic properties of lines and angles, we need to introduce some terminology and notation. We will label points with capital letters, such as A, B, and C, and lines with lowercase letters, such as l and m, or we may include subscripts, such as l_1 or l_2.

As we show in Figure 10.1, any point on a line divides the line into three parts—the point and two *half lines*. A **ray** is a half line with its endpoint included. In Figure 10.1(b), the open dot means that the point A is not included in the half line, whereas the solid dot in Figure 10.1(c) means that A is part of the ray. A piece of a line joining two points and including the points is called a **line segment**. ❶

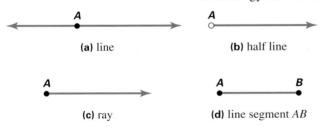

(a) line

(b) half line

(c) ray

(d) line segment AB

FIGURE 10.1 Some basic terminology regarding lines.

Although we do not try to define precisely what a plane is, you can think of it as being an infinite two-dimensional surface such as an infinitely large flat sheet of paper.

According to Euclid's geometry, lines either intersect or are parallel. **Parallel lines** are lines that lie on the same plane and have no points in common. In Figure 10.3, lines l_1 and l_2 are parallel. We express this as $l_1 \| l_2$. If two different lines lying on the same plane are not parallel, then they have a single point in common and are called **intersecting lines**. In Figure 10.3, lines l_3 and l_4 are intersecting lines.

Quiz Yourself ❶ *

Identify each object shown in Figure 10.2.

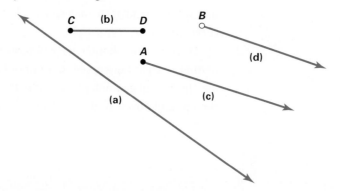

FIGURE 10.2 Line terminology.

parallel lines   intersecting lines

FIGURE 10.3 Parallel lines and intersecting lines.

Angles

Two rays having a common endpoint form an **angle**. In Figure 10.4, we form an angle by rotating ray *AB*, called the **initial side**, about point *B* to finish in the position corresponding to ray *BC*, called the **terminal side**. We use the symbol ∠ to denote an angle; therefore, we can call the angle in Figure 10.4 ∠*ABC*, or simply ∠*B*. Point *B* is called the **vertex** of the angle.

    We measure angles in units called *degrees.*† The symbol ° represents the word *degrees*. If you rotate the initial side of an angle one complete revolution about the vertex so that the terminal side coincides with the initial side, you will have formed a 360° angle, as we show in Figure 10.5(a). If you rotate the initial side only $\frac{1}{360}$ of the way around the vertex to reach the terminal side, that angle has size 1°. A 36° angle is shown in Figure 10.5(b).

✎ **KEY POINT**

We measure angles in degrees.

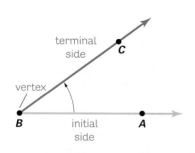

FIGURE 10.4 An angle formed by rotating a ray about point *B*.

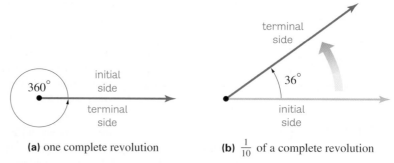

(a) one complete revolution   **(b)** $\frac{1}{10}$ of a complete revolution

FIGURE 10.5 We form a 36° angle by rotating the initial side $\frac{1}{10}$ of the way around the vertex.

*Quiz Yourself answers begin on page 778.
†There are other possible units that are used to measure angles, such as radians. We will not discuss these other measures in this book.

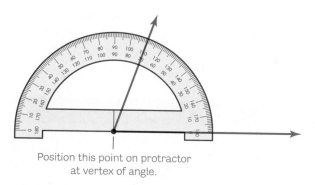

Position this point on protractor
at vertex of angle.

FIGURE 10.6 A protractor measuring a 70° angle.

Figure 10.6 shows a tool called a *protractor* measuring a 70° angle. We write the measure (in degrees) of ∠ABC as m∠ABC.

Certain types of angles occur so often that we give them special names (Figure 10.7). An angle whose measure is between 0° and 90° is called an **acute angle**. A **right angle** has a measure of 90°. We indicate a right angle by placing a square at the vertex of the angle, as shown in Figure 10.7. An **obtuse angle** has a measure between 90° and 180°, and a **straight angle** has a measure of 180°.

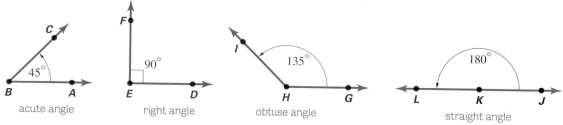

FIGURE 10.7 Some special types of angles.

Two intersecting lines form two pairs of angles called **vertical angles**. Figure 10.8 shows one pair of vertical angles, *ABC* and *EBD*.* Vertical angles have the following important property.

PROPERTY OF VERTICAL ANGLES Vertical angles have equal measures.

We call a pair of angles **complementary** if the sum of their measures is 90°. Two angles having an angle sum of 180° are called **supplementary** angles. In Figure 10.8, angles *PQR* and *RQS* are complementary, and angles *WXY* and *YXZ* are supplementary angles. Two lines that intersect forming right angles are called **perpendicular lines**. **2**

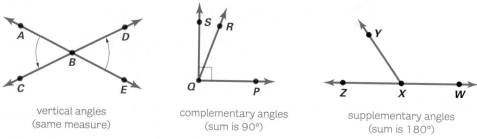

FIGURE 10.8 Some special pairs of angles.

Quiz Yourself **2**

Identify the angles shown in Figure 10.9.

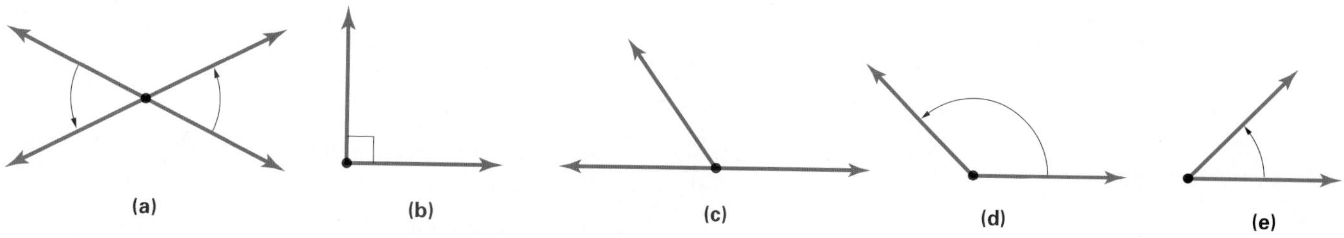

(a)   **(b)**   **(c)**   **(d)**   **(e)**

FIGURE 10.9

*DBA and CBE would be another pair of vertical angles.

KEY POINT

Parallel lines cut by a transversal form several pairs of equal angles.

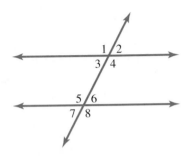

FIGURE 10.10 Special pairs of angles are formed when a transversal cuts parallel lines.

If we intersect a pair of parallel lines with a third line, called a **transversal**, we form eight angles, as shown in Figure 10.10. Certain pairs of these angles have special names and also special properties regarding their angle measures. We summarize some of these special properties in Table 10.1. **3**

Type of Angles	Examples	Property
Corresponding angles	Angles 1 and 5	Corresponding angles have equal measures.
Alternate interior angles	Angles 3 and 6	Alternate interior angles have equal measures.
Alternate exterior angles	Angles 1 and 8	Alternate exterior angles have equal measures.
Interior angles on the same side of the transversal	Angles 3 and 5	Interior angles on the same side of the transversal are supplementary angles (sum of measures is 180°).

TABLE 10.1 Properties of pairs of angles formed by cutting parallel lines with a transversal.

Quiz Yourself **3**

Fill in the blanks using angles other than those mentioned in Table 10.1 to make the following statements true:

a) Angles 2 and _____ are alternate exterior angles.

b) Angles 4 and _____ are corresponding angles.

c) Angles 4 and _____ are interior angles on the same side of the transversal.

d) Angles 5 and _____ are alternate interior angles.

PROBLEM SOLVING

The Analogies Principle

As we said in Section 1.1, it is easier to remember new mathematical terminology if you think about what the words mean in English. For example, when you use the term *alternate interior angles*, you are talking about "alternate" angles—one angle is on each side of the transversal—that are "interior"—inside, or between the parallel lines.

EXAMPLE 1 *Finding Measures of Angles*

In Figure 10.11, assume that lines *l* and *m* are parallel. Also assume that $m\angle A = 51°$ and $m\angle B = 76°$.

a) Find the measure of angle 9.   b) Find the measure of angle 2.

SOLUTION:

a) Angles 8 and *B* are equal because they are corresponding angles. Thus, $m\angle 8 = 76°$. Angles *A*, 8, and 9 form a straight angle, so

$$m\angle A + m\angle 8 + m\angle 9 = 180°.$$

Substituting the measures of angles *A* and 8 gives the equation

$$51° + 76° + m\angle 9 = 180°.$$

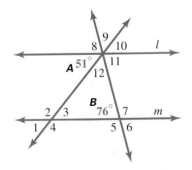

FIGURE 10.11 Finding measures of angles.

Quiz Yourself ④

Use Figure 10.11 to find a) the measure of angle 7 and b) the measure of angle 6.

KEY POINT

There is a correspondence between the measure of central angles and the length of arcs of circles.

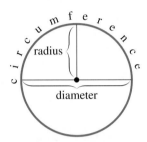

FIGURE 10.12 Radius (red), diameter (red), and circumference (blue) of a circle.

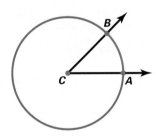

FIGURE 10.13 Angle *ACB* is a central angle.

Length of arc *AB* is proportional to measure of ∠*ACB*.

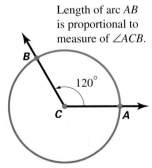

circumference = 12 meters

FIGURE 10.14 The measure of ∠*ACB* determines the length of arc *AB*.

Solving for $m\angle 9$, we get

$$m\angle 9 = 180° - 51° - 76° = 53°.$$

b) Because interior angles on the same side of the transversal are supplementary, we have $m\angle A + m\angle 2 = 180°$. Substituting $51°$ for $m\angle A$, we get $51° + m\angle 2 = 180°$. Subtracting $51°$ from both sides, we get $m\angle 2 = 180° - 51° = 129°$.

Now try Exercises 33 to 38. ✳ ④

Circles

Circles are common geometric objects that have many useful applications.

> **DEFINITIONS** A **circle** is the set of all points lying on a plane that are located at a fixed distance, called the **radius**,* from a given point called the **center**. A **diameter** of a circle is a line segment passing through the center with both endpoints lying on the circle. The **circumference** is the distance around the circle. (See Figure 10.12.)

An angle that has its vertex at the center of a circle is called a **central angle** (see Figure 10.13). In Figure 10.13, the length of the arc between *A* and *B* is proportional to the central angle *ACB*. By this we mean that we have the following ratio:

$$\frac{\text{the measure of angle } ACB}{360°} = \frac{\text{length of arc } AB}{\text{circumference of circle}}.$$

For example, if $m\angle ACB = 60°$, then the length of the arc from *A* to *B* is $\frac{60}{360} = \frac{1}{6}$ of the circumference of the circle.

EXAMPLE 2 *Using a Central Angle to Measure the Length of an Arc of a Circle*

Assume that a circle has a circumference of 12 meters. If central angle *ACB* has measure of 120°, then what is the length of the arc from *A* to *B*?

SOLUTION: It is a good problem-solving strategy to draw a diagram, as we do in Figure 10.14. We can now use the ratio

$$\frac{\text{length of arc } AB}{\underset{\underset{12\text{ meters}}{|}}{\text{circumference of circle}}} = \frac{\text{the measure of angle } ACB}{360°} \quad \overset{\overset{120°}{|}}{}$$

to solve the problem. This gives us the equation

$$\frac{\text{length of arc } AB}{12} = \frac{120}{360} = \frac{1}{3}.$$

Multiplying both sides by 12, we get that the length of arc $AB = \frac{1}{3} \cdot 12 = 4$ meters.

Now try Exercises 39 to 44. ✳ ⑤

Surprisingly, we can use the small amount of geometry we have developed so far to solve meaningful problems.

---

*We will use the word *radius* in two ways. (1) It is the *distance* between the center and any point on the circle. (2) It is a *line segment* joining the center to any point on the circle.

Quiz Yourself ⑤

Assume that a circle has a circumference of 40 inches. If a central angle has measure of 45°, what is the length of the arc of the circle determined by the sides of the angle?

EXAMPLE 3   *Finding the Circumference of Earth*

How can you use elementary geometry to estimate the circumference of Earth?*

SOLUTION: Consider Figure 10.15 below. Assume that lines *l* and *m* are parallel and cut by the transversal *t*. The point *C* is the center of the circle. Therefore, angles α and β are equal.

From our earlier discussion of central angles, we see that we have the following ratio:

$$\frac{\text{measure of angle } \beta}{360 \text{ degrees}} = \frac{\text{length of arc cut by sides of } \beta}{\text{total circumference of circle}}.$$

With this observation, it is easy to measure the circumference of Earth. Begin by placing a vertical pole in the ground and waiting until high noon when the rays of the Sun and the pole form an angle of 0°. Suppose at that very moment, you are talking to a friend who lives 1,000 miles away and who also has a similar vertical pole. Your friend tells you that the Sun's rays make an angle of 15° with his pole. If you redraw Figure 10.15 as in Figure 10.16, you will see how to find your answer.

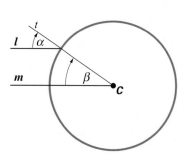

FIGURE 10.15 α = β because *l*‖*m* and α and β are corresponding angles.

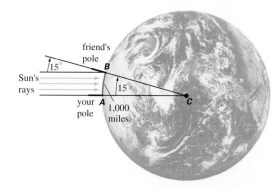

FIGURE 10.16 The measure of angle *ACB*, which is 15°, is the same fractional part of 360° as the length of the arc between *A* and *B* is of Earth's circumference.

Denote the circumference of Earth by *c* and set up the following ratio:

$$\frac{15°}{360°} = \frac{1,000}{c}. \tag{1}$$

Cross multiplying equation (1) results in the equation

$$15c = 360(1,000). \tag{2}$$

Dividing both sides of equation (2) by 15 and simplifying gives us *c* = 24,000. From this calculation, you can estimate the circumference of Earth to be 24,000 miles.

Now try Exercises 55 and 56. ❋

---

*Although Earth is not a perfect sphere, we will assume that it is in order to simplify our calculations.

HISTORICAL HIGHLIGHT 🌼 🌼 🌼

Non-Euclidean Geometry (Part 1)

Several years ago I had the opportunity to fly to France to visit my sister. As the jetliner climbed to an altitude of over 6 miles, a monitor displayed the path the flight was taking from Newark, New Jersey, to Paris. Surprisingly, instead of flying straight across the ocean, the plane went to the north towards Greenland and after several hours into the flight, it was clear that the flight path was curved. This might surprise you as well, because as we all know—the shortest distance between two points is a straight line—or is it? Although this statement *is true* for points on a plane, *it is not true* for other surfaces. On a sphere, the shortest distance between two points is an arc of a *great circle*.* A great circle, as shown in Figure 10.17, is a circle on a sphere that has the same center as the sphere. Therefore, in spherical geometry, we think of a "line" as a great circle.

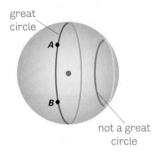

FIGURE 10.17 The distance between points *A* and *B* on a sphere is an arc of a great circle.

There can be no parallel lines in spherical geometry because any two great circles on a sphere intersect in two points. In Figure 10.18, the two great circles intersect at points *P* and *Q*.

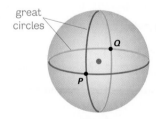

FIGURE 10.18 Two great circles intersect at points *P* and *Q*.

Notice that what we consider to be lines and what properties these lines have depends on the surface on which we are drawing the lines.

Euclid's fifth postulate† states that through a point not on a given line there is *exactly one* line that is parallel to the given line, as we show in Figure 10.19.

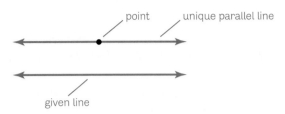

FIGURE 10.19 Euclid's fifth postulate.

However, for centuries mathematicians speculated that Euclid's fifth postulate could be proved from his other postulates. Unable to prove the fifth postulate, mathematicians decided to take another approach. They asked: "What if we assume that the fifth postulate is false? What would our geometry be like?"

There are two ways to deny the fifth postulate:

1) Assume that for a line and a point not on that line, there are no lines parallel to the given line.

2) Assume that for a line and a point not on that line, there are at least two lines parallel to the given line.

The first approach led to geometries such as spherical geometry that we mentioned previously. Around 1850, the great German mathematician Bernhard Riemann invented such a geometry.

In the early part of the nineteenth century, Karl Friedrich Gauss from Germany, Janos Bolyai from Hungary, and Nicolai Lobachevsky from Russia each independently developed a geometry using the second approach. We can visualize this type of geometry on a surface called a *pseudosphere*. To form a pseudosphere, we rotate a curve called a *tractrix* about a line, as shown in Figure 10.20.

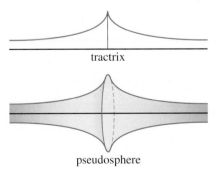

FIGURE 10.20 A pseudosphere is a model of a non-Euclidean geometry.

The surface looks somewhat like the bells of two trumpets joined together. We will discuss these non-Euclidean geometries further in Section 10.2.

---

*Arcs of circles that are not great circles will not give the shortest distance between two points on the surface of a sphere.
†This postulate was an assumption made by Euclid about geometry that could not be proved.

Exercises 10.1

Looking Back*

These exercises follow the general outline of the topics presented in this section and will give you a good overview of the material that you have just studied.

1. Use Figure 10.7 to name the type of angle described by each of the following angle measures:

   a. between 0° and 90°   **b.** 90°

   c. between 90° and 180°   **d.** 180°

2. Use Figure 10.10 to find an example of each of the following other than the examples mentioned in Table 10.1:

   a. corresponding angles

   b. alternate interior angles

3. In solving Example 2, what relationship were we using between the length of the arc *AB* and the central angle *ACB*?

4. What is the shortest distance between two points on a sphere?

Sharpening Your Skills

In Exercises 5–12, match each term with the numbered angles in the given figure. There may be several correct answers. We will state only one in the answer key. Lines l and m are parallel.

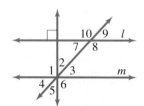

Figure for Exercises 5–12

5. vertical angles   6. complementary angles

7. alternate interior angles   8. right angle

9. obtuse angle   10. corresponding angles

11. supplementary angles   12. acute angle

In Exercises 13–22, determine whether each statement is true or false. Remember by the Always Principle that if a statement is true, it must always be true without exception. If you believe that a statement is false, you should try to find a counterexample.

13. Two lines lying on the same plane that do not intersect are parallel.

14. At least two of the angles formed by two intersecting lines are equal.

15. If two angles are complementary, then they must be equal.

16. If two angles (with measure greater than 0°) are complementary, then each must be an acute angle.

17. If two equal angles are supplementary, then each is a right angle.

18. An obtuse angle cannot be complementary to another angle.

19. An angle cannot be the complement of one angle and the supplement of another angle at the same time.

20. Four of the angles formed when parallel lines are cut by a transversal are equal.

21. The supplement of an acute angle must be an acute angle.

22. Alternate interior angles must be acute angles.

Use the given figure to answer Exercises 23–26. There may be several correct answers. We will state only one in the answer key.

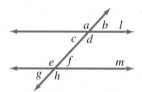

Figure for Exercises 23–26

23. Find a pair of obtuse, alternate interior angles.

24. Find a pair of acute, alternate exterior angles.

25. Find a pair of acute, corresponding angles.

26. Find a pair of obtuse, corresponding angles.

In Exercises 27–32, find the measure of a complementary angle and a supplementary angle for each angle.

27. 30°   **28.** 108°

29. 120°   **30.** 45°

31. 51.2°   **32.** 110.4°

In Exercises 33–38, find the measures of angles a, b, and c in each figure. In Exercises 35–38, lines l and m are parallel.

33.

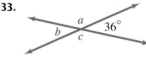

34.

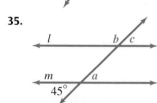

35.

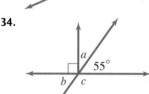

36.

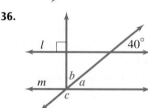

---

*Before doing these exercises, you may find it useful to review the note *How to Succeed at Mathematics* on page xix.*

37.

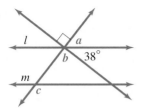

38.

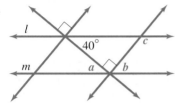

In Exercises 39–44, you are given two of the following three pieces of information: the circumference of the circle, the measure of the central angle ACB, and the length of arc AB. Find the third piece of information.

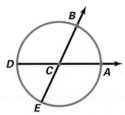

Figure for Exercises 39–44

39. circumference = 24 feet; $m\angle ACB = 90°$

40. circumference = 150 centimeters; $m\angle ACB = 72°$

41. circumference = 12 meters; length of arc $AB = 4$ meters

42. circumference = 240 inches; length of arc $AB = 40$ inches

43. $m\angle ACB = 30°$; length of arc $AB = 100$ millimeters

44. $m\angle ACB = 120°$; length of arc $AB = 9$ feet

Applying What You've Learned

Continuing the situation from Exercises 39–44, use the given information to answer Exercises 45–48.

45. circumference = 18 feet; $m\angle ACB = 60°$. Find the length of arc *BD*.

46. $m\angle ACB = 30°$; length of arc $AE = 10$ meters. Find the circumference.

47. circumference = 30 inches; length of arc $DE = 3$ inches. Find $m\angle BCD$.

48. circumference = 120 centimeters; length of arc $AB = 12$ centimeters. Find $m\angle DCE$.

In Exercises 49–52, solve for x. *Assume that lines* l *and* m *are parallel.*

49.

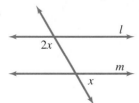

50.

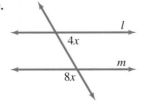

51.

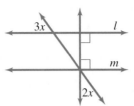

52.

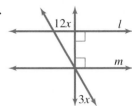

53. Two lines in a plane can intersect forming four angles (some may have the same measure). What is the greatest number of angles we can form using three lines?

54. Solve Exercise 53 for four lines.

55. In Example 3, your friend has erred in measuring the angle the vertical pole makes with the Sun's rays because the correct circumference of Earth is closer to 25,000 miles than 24,000 miles. If we use the circumference of 25,000 miles in Example 3, what should the angle measurement taken by your friend have been?

56. Reconsider Example 3. Assume that you do not know how far your friend is from you. At high noon, your time, your friend tells you that the Sun's rays make an 18° angle with the vertical pole he has in the ground. Assume that the true circumference of Earth is 25,000 miles. How far away is your friend?

Communicating Mathematics

57. What is the difference between supplementary and complementary angles?

58. When a pair of parallel lines is cut by a transversal, name three types of angles that are pairwise equal.

59. Explain how you remember the meaning of *alternate exterior angles*.

60. Explain how you remember the meaning of *interior angles on the same side of the transversal*.

In Exercises 61–64, use the fact that two intersecting lines form four angles.

61. Can all of the four angles be equal? Explain your answer.

62. Can all of the four angles be acute? Explain your answer.

63. What is the largest number of angles that can be obtuse? Explain your answer.

64. Could three of these angles be acute and one of them be obtuse? Explain your answer.

65. Can alternate interior angles be complementary? Explain.

66. Can alternate interior angles be supplementary? Explain.

Using Technology to Investigate Mathematics

67. You can find many interactive applets that illustrate the ideas of this section. For example, I found applets that illustrate Euclid's elements, tutorials that illustrate various properties of parallel lines and angles, and so forth. Search the Internet for some applets and duplicate some of the computations of this section. Report on your findings.

68. There are many Web sites that illustrate applications of geometry. You can search generally for "geometry and applications," or, you can be more specific by searching for combinations such as "geometry and art" or "geometry and medicine." Find some sites illustrating interesting applications of geometry and report on your findings.

For Extra Credit

69. Draw a diagram that has four lines and six points and in which each line passes through exactly three of the six points.

70. Draw a diagram that has 5 lines and 10 points and in which each line passes through exactly 4 of the 10 points.

71. Consider your solutions to Exercises 53 and 54. Without drawing a diagram, find the largest number of angles that can be formed by 10 lines. (*Hint:* Consider the number of intersections that can be formed by five lines, six lines, and so on.)

72. A boat lost at sea has a radio that transmits a distress signal. Anyone receiving the signal can determine the direction of the signal, but not the distance. Explain why if one ship receives the signal, it cannot notify a search plane of the exact location of the boat. If two different ships receive the signal, the exact location of the boat can now be determined. Explain how to find the boat's position with the information from both ships.

10.2 Polygons

Objectives

1. Understand the basic terminology and properties of polygons.
2. Solve problems involving angle relationships of polygons.
3. Use similar polygons to solve problems.
4. Be aware of some differences between Euclidean and non-Euclidean geometries.

Have you noticed the great number of geometric figures that you encounter each day? They are literally all around you. You see them in magazine ads, TV commercials, clothing design, art, architecture, product design, and religious symbols, as well as in many other places. In this section, we will build on our previous discussion of lines and angles to study a familiar class of geometric objects called *polygons*.

KEY POINT

Polygons are special types of plane figures made up of line segments.

Polygons

We begin with a few definitions (see Figures 10.21 and 10.22).

> **DEFINITIONS** A plane figure is **closed** if we can draw it without lifting the pencil and if the starting and ending points are the same. A plane figure is **simple** if we can draw it without lifting the pencil and in drawing it we never pass through the same point twice, with the possible exception of the starting and ending points.

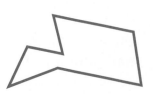

closed and simple   simple, not closed   closed, not simple   not simple, not closed

FIGURE 10.21 Closed and simple plane figures.

> **DEFINITIONS** A **polygon** is a simple, closed plane figure consisting only of line segments, called **edges**, such that no two consecutive edges lie on the same line. We call an endpoint of an edge a **vertex** (plural, *vertices*). A polygon is **regular** if all of its edges are the same length and all of its angles have the same measure.

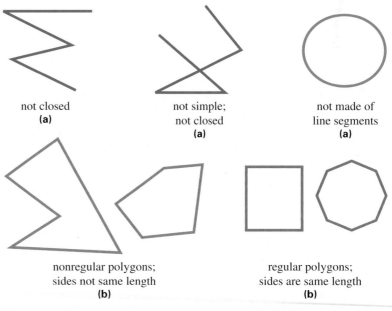

not closed
(a)

not simple;
not closed
(a)

not made of
line segments
(a)

nonregular polygons;
sides not same length
(b)

regular polygons;
sides are same length
(b)

FIGURE 10.22 (a) Nonpolygons. (b) Polygons.

Number of Sides	Name of Polygon
3	Triangle
4	Quadrilateral
5	Pentagon
6	Hexagon
7	Heptagon
8	Octagon
9	Nonagon
10	Decagon

TABLE 10.2 Names of polygons.

We classify polygons according to the number of their sides. Table 10.2 lists the names of polygons having up to 10 sides.

There is another property that a polygon may have regarding its shape.

> **DEFINITION** A polygon is **convex** if for any two points X and Y inside the polygon, the entire line segment XY also lies inside the polygon.

Figure 10.23 shows a convex and a nonconvex polygon.

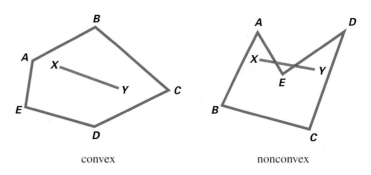

convex   nonconvex

FIGURE 10.23 A convex and a nonconvex polygon.

✺ ✺ ✺ HIGHLIGHT

Geometry in Everyday Things

If the floor to your new apartment is not perfectly even, you might consider buying a coffee table with three legs instead of four. Any three points always lie on a plane, so the three-legged table won't wobble and spill your drinks, but the four-legged table just might. It is for this same reason that surveyors and photographers often steady their equipment on three-legged tripods.

When carpenters build a new wall, they often nail a board diagonally across the beams, as shown in Figure 10.24. By introducing new lines that are not parallel to the existing lines in the structure, they add rigidity to the wall that keeps it from shifting as they lift it into place.

While driving in your car, you may want to locate the nearest Italian restaurant. By taking readings from different locations, your car's global positioning system (GPS) determines that your car lies on two nonparallel lines.

Because these lines intersect in a unique point, the system can find your exact location and direct you to the nearest restaurant.

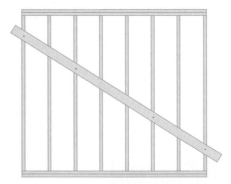

FIGURE 10.24 The diagonal board gives stability to the frame.

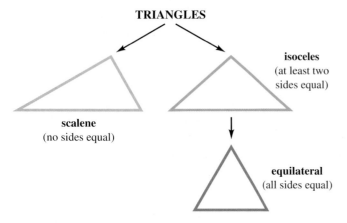

FIGURE 10.25 Classification of triangles.

Figures 10.25 and 10.26 show some special types of triangles and quadrilaterals. In these figures, an object following an arrow has all the properties of the objects preceding the arrow. For example, in Figure 10.25 we see that every equilateral triangle is also an isoceles triangle. Note that an isoceles triangle may not necessarily be an equilateral triangle.

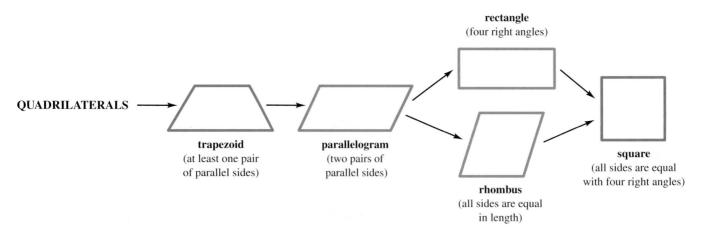

FIGURE 10.26 Classification of quadrilaterals.*

---

*Some define a trapezoid to be a quadrilateral that has *exactly one* pair of parallel sides. According to this alternate definition, a parallelogram would not be a trapezoid.

Polygons and Angles

We can use polygons to solve real-life problems such as designing a deck or building a house. However, to do this, we often need to know the sum of measures of the interior angles of the polygons. For example, imagine how difficult it would be to design a hexagon-shaped gazebo without knowing the measure of the interior angles of the floor.

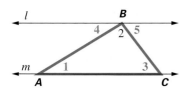

FIGURE 10.27 Lines *l* and *m* are parallel lines cut by transversals that form equal alternate interior angles.

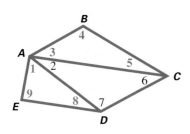

FIGURE 10.28 A pentagon divided into a collection of triangles.

Quiz Yourself **6**

Apply the method used in Example 2 to find the interior angle sum of quadrilateral *ABCD*.

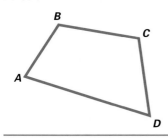

EXAMPLE 1   *The Angle Sum of a Triangle Is 180°*

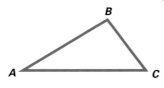

Find the sum of the measures of the interior angles in △*ABC*. (The notation △*ABC* is read "triangle *ABC*.")

SOLUTION: We begin by constructing a line *m* that contains line segment *AC* and a second line *l* through point *B* that is parallel to *m*, as shown in Figure 10.27.

Because angles 1 and 4 are alternate interior angles, they are equal. Similarly, angles 3 and 5 are equal. Therefore, the sum of the measures of angles 1, 2, and 3 equals the sum of the measures of angles 4, 2, and 5, which is 180°. ❋

Now that we know the sum of the angles of a triangle, we can use this information to find the interior angle sum of other polygons.

EXAMPLE 2   *The Angle Sum of a Pentagon*

Find the interior angle sum of the convex pentagon *ABCDE*.

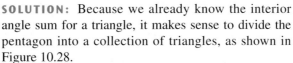

SOLUTION: Because we already know the interior angle sum for a triangle, it makes sense to divide the pentagon into a collection of triangles, as shown in Figure 10.28.

Notice in Figure 10.28 that ∠*A* is made up of three smaller angles 1, 2, and 3. Also, ∠*C* consists of angles 5 and 6 and ∠*D* is made up of angles 7 and 8.

The sum of the angles of the pentagon is

$$m\angle A + m\angle B + m\angle C + m\angle D + m\angle E.$$

If we replace $m\angle A$ by $m\angle 1 + m\angle 2 + m\angle 3$, $m\angle C$ by $m\angle 5 + m\angle 6$, and $m\angle D$ by $m\angle 7 + m\angle 8$, we see that the angle sum of the pentagon is

$$m\angle 1 + m\angle 2 + m\angle 3 + \cdots + m\angle 9.$$

This is the same as the angle sum of the three triangles △*AED*, △*ADC*, and △*ACB*, which equals $3 \times 180° = 540°$.

Now try Exercises 19 to 24. ❋ **6**

 PROBLEM SOLVING

Convert a New Problem to an Earlier One

As we mentioned in Section 1.1, you can often solve a new problem by relating it to one that you have seen before. We solved the problem in Example 2 by relating it to the earlier problem of finding the angle sum of a triangle.

KEY POINT

The number of sides of a polygon determines the sum of the measures of its interior angles.

Table 10.3 summarizes what you have seen so far about the angle sum of polygons.

Polygon	Number of Sides	Interior Angle Sum
Triangle	3	$180° = 1 \times 180°$
Quadrilateral	4	$360° = 2 \times 180°$
Pentagon	5	$540° = 3 \times 180°$

TABLE 10.3 Interior angle sum of polygons.

We can generalize this pattern to n-sided convex polygons.

> **ANGLE SUM OF A POLYGON** The sum of the measures of the interior angles of a convex polygon having n sides is $(n-2) \times 180°$.

If a polygon is regular, we can say a little more about the angles of the polygon. Because each angle of a regular polygon has the same measure, we find the following pattern:

each interior angle of a regular triangle (an equilateral triangle) has $\frac{180°}{3} = 60°$,

each interior angle of a regular quadrilateral (a square) has $\frac{360°}{4} = 90°$, and

each interior angle of a regular pentagon has $\frac{540°}{5} = 108°$.

We can also generalize this pattern.

Quiz Yourself ⑦

Find the measure of each interior angle of a regular octagon.

> **INTERIOR ANGLES OF A REGULAR POLYGON** Each interior angle of a regular polygon with n sides has measure $\frac{(n-2) \times 180°}{n}$. ⑦

Although many word-processing and computer illustration programs have graphics capabilities, you often need to understand basic geometry in order to draw a figure exactly the way you want it to look.

EXAMPLE 3 *Designing a Logo*

Imagine that *Dancing with the Stars* is taking its act on tour to your city and your advertising company has been hired to design a giant star-shaped billboard to advertise this event. Determine the angle measure of each point of the star.

SOLUTION: Consider the star in Figure 10.29.

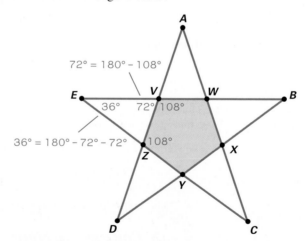

FIGURE 10.29 Constructing the five-point star.

The shaded pentagon *VWXYZ* is a regular pentagon, so each of its interior angles has measure $\frac{(5-2) \times 180°}{5} = 108°$. Because a straight angle equals 180°, $\angle EZV$ and EVZ each have measure $180° - 108° = 72°$. This means that angle E has measure $180° - 72° - 72° = 36°$. ❈

KEY POINT

Similar polygons have proportional sides and equal angles.

Similar Polygons

If you were hired to design a new open-air theater for your campus, you would first have to build a scale model of the project before a decision could be made to commit money for your project. As a graphic artist, you might create a small, preliminary version of an advertisement before enlarging it to billboard size. Similarly, fashion designers often make small sample garments that are later resized after they have sold the designs to stores. In each case, we are working with objects that have the same shape but different sizes. Likewise, in geometry, we often work with similar figures.

> **DEFINITION** Two polygons are **similar** if their corresponding sides are proportional and their corresponding angles are equal.

Polygons *A* and *B* in Figure 10.30 are similar. Sometimes it is useful to know if two triangles are similar. It can be proved in geometry that if one triangle has two angles equal to two angles in a second triangle, then the two triangles are similar. We will use this fact in Example 4.

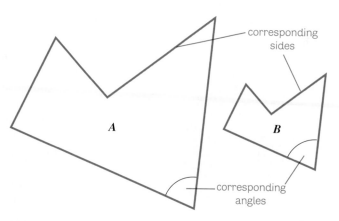

FIGURE 10.30 *A* and *B* are similar polygons.

EXAMPLE 4 *Using Similar Triangles to Build a Wilderness Bridge*

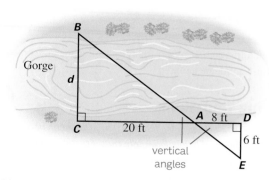

FIGURE 10.31 Building a bridge.

Yau-Man and James, racing against other teams in a wilderness survival contest, encounter a deep gorge whose bridge is missing. They plan to cross by cutting a tree to fall across the gorge to use as a temporary bridge. They have measured the distances in two right triangles, as shown in Figure 10.31. Use this information to find the distance, *d*, across the gorge.

SOLUTION: Angles *BAC* and *DAE* are vertical angles and therefore equal. Also, angles *D* and *C* are right angles, so they also are equal. Therefore, triangles *ACB* and *ADE* are similar, and so their corresponding sides are proportional. This gives us the ratio

$$\frac{\text{length of } BC}{\text{length of } ED} = \frac{\text{length of } AC}{\text{length of } AD}.$$

Substituting for these four lengths, we get $\frac{d}{6} = \frac{20}{8}$. Cross multiplying gives the equation $d \cdot 8 = 6 \cdot 20 = 120$. Dividing both sides of the equation by 8 gives us $d = \frac{120}{8} = 15$. Therefore, a 15-foot tree will do the job. ✤

❈ ❈ ❈ HISTORICAL HIGHLIGHT

Non-Euclidean Geometry (Part 2)

Because the fundamental axioms of non-Euclidean geometry are different from Euclid's axioms, it is not surprising that when we rephrase well-known Euclidean geometry theorems in non-Euclidean terms, they sound somewhat strange. The Euclidean theorem that the interior angle sum of a triangle is 180° in Riemannian geometry becomes

the interior angle sum of a triangle is greater than 180°.

If we consider the surface of a sphere, it is not hard to understand why this theorem should be true. Figure 10.32 shows a triangle whose interior angle sum is greater than 180°. Imagine beginning at the top of the sphere, the North Pole so to speak (call this point *A*), and draw an arc of a great circle to meet another great circle that is horizontal, comparable to the equator. Call this point *B*. Make a right angle, draw an arc along this "equator" to point *C*, and then make a right angle and draw an arc of a great circle from *C* back to *A*. The angles at *B* and *C* are each 90° so that when we add in the measure of angle *A*, we have an angle sum that exceeds 180°.

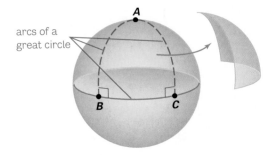

FIGURE 10.32 Triangle with angle sum greater than 180°.

On a pseudosphere, which has a shape like two bells of a trumpet, triangles look something like the curved triangle drawn in Figure 10.33. In this type of non-Euclidean geometry, triangles have an interior angle sum that is less than 180°.

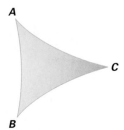

FIGURE 10.33
Triangle with an angle sum less than 180°.

At this point, you might be wondering which of the three geometries (Euclid's, Riemann's, Lobachevsky's) is the "correct" geometry. The answer is that there is no one "correct" geometry. Mathematicians have proved that all three types of geometries are perfectly consistent mathematical systems. Even more remarkable, they have also proved that for any one of these non-Euclidean geometries to be consistent, the other two types must also be consistent. So in this sense, no one of the geometries is better. As to which geometry we use, it all depends on what application we have in mind. For surveying land and building buildings, Euclidean geometry is appropriate; however, Einstein found that to understand the vastness of the universe, non-Euclidean geometry is a better tool.

Exercises 10.2

Looking Back*

These exercises follow the general outline of the topics presented in this section and will give you a good overview of the material that you have just studied.

1. What property did we use in Example 1 involving parallel lines cut by a transversal? What information did that property give us?

2. How did we arrive at the fact that the sum of the interior angles of the pentagon in Example 2 was 540°?

3. In Example 3, how did we determine that $m\angle EVZ = 72°$?

4. State one difference between triangles in Euclidean geometry and non-Euclidean geometry.

Sharpening Your Skills

In Exercises 5–10, determine whether each statement is true or false. Remember by the Always Principle that if a statement is true, it must always be true without exception.

5. A trapezoid is a parallelogram.

6. A rhombus is a parallelogram.

7. A regular polygon is convex.

---

*Before doing these exercises, you may find it useful to review the note *How to Succeed at Mathematics* on page xix.

8. If two polygons have corresponding angles equal, then the polygons are similar.

9. If two polygons have corresponding sides equal, then the polygons are similar.

10. A convex polygon can have an interior angle sum of 400°.

In Exercises 11–14, state whether each figure is a polygon. For those that are not polygons, state what part of the definition fails.

11.    **12.**

13.    **14.**

In Exercises 15–18, use the information that we give you to find the measures of angles A, B, *and* C *in* △ABC. *All measures are in degrees.*

15. $m\angle B = 2m\angle A$, $m\angle C = 3m\angle A$

16. $m\angle B = m\angle A$, $m\angle C = 5 + 3m\angle A$

17. $m\angle B = m\angle A + 10$, $m\angle C = 2m\angle A - 10$

18. $m\angle B = 3m\angle A$, $m\angle C = 2m\angle A + m\angle B$

19. If we divide a regular hexagon into triangles as we did in Example 2, how many triangles will there be? What will be the interior angle sum of the hexagon?

20. If we divide a regular octagon into triangles as we did in Example 2, how many triangles will there be? What will be the interior angle sum of the hexagon?

21. What is the measure of an interior angle of a regular 20-sided polygon?

22. What is the measure of an interior angle of a regular 12-sided polygon?

23. If each interior angle of a regular polygon measures 160°, how many sides does the polygon have?

24. If each interior angle of a regular polygon measures 135°, how many sides does the polygon have?

In Exercises 25 to 28, assume that each of the triangles in a pair are similar. Find length x.

25.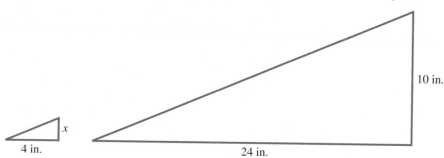

26.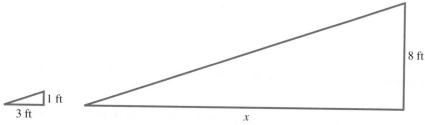

27.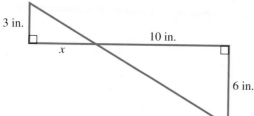

28.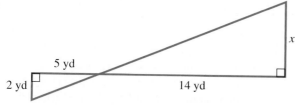

In Exercises 29–32, assume that in each pair the figures are similar. Given the lengths of sides and measures of angles in the left figure, what information do you know about the right figure?

29.

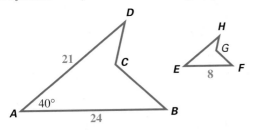

30.

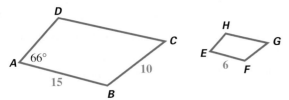

31.

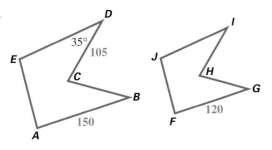

32.

Applying What You've Learned

33. The Russians erected the world's largest full-figure statue, *Motherland*, which is 270 feet tall, as a World War II memorial. If at a certain time of day, a 6-foot person casts a 10-foot shadow, how long will the statue's shadow be?

34. If Yao Ming, the 7-foot 6-inch center of the Houston Rockets casts a shadow of 18 feet 9 inches and Chauncy Billups, guard with the Detroit Pistons, standing beside him casts a shadow of 15 feet, $7\frac{1}{2}$ inches, how tall is Billups?

35. How wide is the river below at point *A*?

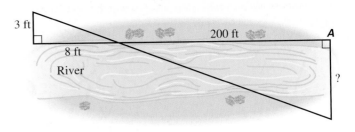

36. How wide is the river below at point *A*?

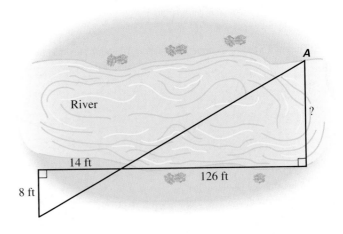

37. If triangles *ABC* and *ADE* are similar in this diagram, what is the length of the pond?

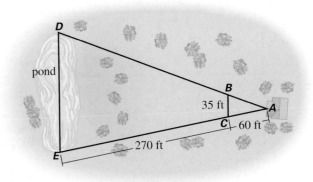

38. Jack Bauer wants to place a surveillance camera at location *C* to observe activity at location *X*. The camera cannot be seen directly from position *X* because of a large clump of bushes.

The camera will take pictures by aiming at a building, *M*, with shiny, mirror-like sides, as shown in the diagram. How far should the camera be positioned from point *I* for this setup to work?

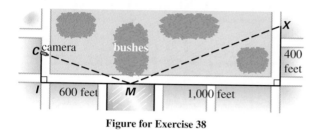

Figure for Exercise 38

Some Japanese furniture makers construct elegant wooden furniture without using any glue or mechanical fasteners. To assemble the furniture, the pieces of wood must be cut precisely so that they fit together tightly.

39. A furniture maker is constructing a chair using a horizontal beam with a cross section in the shape of a regular pentagon. The beam must fit into a notch cut into a support post so that an angle of the beam fits exactly in a notch cut in the post, as shown in the given diagram. What should the measure of the indicated angles be so that the beam fits exactly in the notch?

40. Repeat Exercise 39, but now assume that the horizontal beam has a cross section in the shape of a regular hexagon. The beam will rest in the support post as shown in the given diagram.

41. A gazebo has a floor in the shape of a regular hexagon with each side measuring 10 feet. The floor requires two supports, as shown, that measure roughly 17.4 feet. If we plan to build a slightly larger gazebo with 12-foot sides, how long must the support beams be?

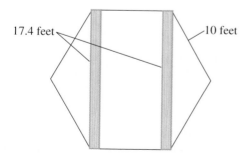

42. If the longest support beams in Exercise 41 that we can purchase are 15 feet, what would the length of the sides of the gazebo be?

Communicating Mathematics

43. What is the difference between an isosceles and an equilateral triangle?

44. What information do you know about the angles and the sides of similar polygons?

45. Can an isoceles triangle also be a scalene triangle? Explain.

46. Can an isoceles triangle also be an equilateral triangle? Explain.

47. What is the difference between a rhombus and a square?

48. What is the difference between a trapezoid and a parallelogram?

Using Technology to Investigate Mathematics

49. Search the Web for applications of geometry in different cultures. For example, you can find interesting sites if you search for "geometry and native American" or "geometry and Chinese." Find a site that interests you and report on your findings.

50. There are sites on the Internet devoted to non-Euclidean geometry. Some have applets that allow you to construct various geometric figures. Find such a site, run some of the applets, and report on your findings.

For Extra Credit

In Exercises 51–54, if it is possible to construct a triangle of the type described, then explain how you might draw one. If it is impossible to construct the triangle described, explain why it is impossible.

51. A scalene triangle with two acute angles

52. A right triangle that also has an obtuse angle

53. An equilateral triangle that has all obtuse angles

54. An equilateral triangle that has all acute angles

55. Make up a description of a triangle as we did in Exercises 51–54 that is impossible to construct.

56. Make up a description of a triangle as we did in Exercises 51–54 that is possible to construct.

57. What happens to the measure of the interior angles of a regular *n*-sided polygon as *n* gets larger and larger? Explain your answer.

58. The English surveyors Mason and Dixon surveyed the famous Mason–Dixon line between Maryland and Pennsylvania using a level made from wood in the shape of the letter A,* as shown in the given diagram. A string with a weight hangs from the tip of the level. Explain

*J. E. Thompson, *Geometry for the Practical Worker* (New York: Van Nostrand Reinhold, 1982), p. 82.

how you would build such a level and also explain how it would work. Be sure to discuss lengths of line segments and measures of angles.

59. Generalize Exercises 39 and 40. If the cross beam has a cross section in the shape of a regular polygon with *n* sides, what would be the measure of the indicated angles for the support beam? Assume that one vertex of the end of the cross beam points directly down, as was the case in Exercises 39 and 40.

60. In building scaffolding, often the scaffolding has many triangles, as shown in figure (a). What advantage do you see in using scaffolding of type (a) versus type (b)? (*Hint:* There is a fundamental property that triangles have that rectangles do not have.)

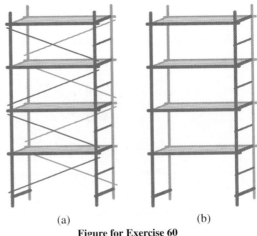

(a)   (b)

Figure for Exercise 60

10.3 Perimeter and Area

Objectives

1. Calculate the perimeter and area of geometric objects.
2. Understand how area formulas for geometric figures are related.
3. Use the Pythagorean theorem to solve problems involving right triangles.
4. Calculate the circumference and area of circles.

It's good that your vacation is coming up because you sure have a lot of jobs to get done around the house. Your new puppy Max is a lot of fun, but he just won't stay in the yard, so you'll have to think about getting a fence. Your dining room could use some touching up. Some fresh paint and decorative molding would certainly brighten up the room. Also, the deck is looking a little shabby and could use a new coat of sealer. If you have time, it would also be good to look into what size solar panels to have installed to take the bite out of your heating bill. What should you do first?

Perimeter and Area

Well, the first thing that you will need to do to get these jobs done is to recall some basic facts about perimeter and area.

> **DEFINITIONS** The **perimeter** of a polygon is the sum of the lengths of the sides of the polygon. The **area** is a measure of the amount of surface that the polygon covers.

We measure the perimeter of a figure with the same unit of measurement as we measure the sides, and we measure the area of the figure in square units. For example, as we see in Figure 10.34, a rectangle with length 6 feet and width 4 feet, has a perimeter of $6 + 4 + 6 + 4 = 20$ feet, and the area is $6 \times 4 = 24$ square feet.

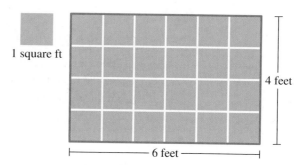

FIGURE 10.34 The perimeter of the rectangle is 20 feet; the area is 24 square feet.

The following are general formulas for calculating the perimeter and area of rectangles.

> **PERIMETER AND AREA OF A RECTANGLE** If a rectangle has length l and width w, then the perimeter of the rectangle is $P = 2l + 2w$ and the area is $A = l \cdot w$.

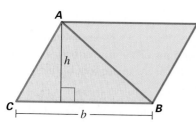

KEY POINT

We can use the formula for the area of a rectangle to derive formulas for the area of other polygons.

Deriving Area Formulas

Knowing how to do a calculation for one type of geometric figure often leads us to a method for doing that same computation for another type of figure. For example, once we know the formula for the area of a rectangle, we can easily derive the formula for the area of a parallelogram.

Consider the parallelogram in Figure 10.35(a), which has height h and base b. If we cut off the blue triangle on the left, slide it over, and attach it to the right side of the parallelogram, we get the rectangle shown in Figure 10.35(b). We know that the area of this rectangle is $h \cdot b$, so therefore the area of the parallelogram is also $h \cdot b$.

> **AREA OF A PARALLELOGRAM** The area of a parallelogram with height h and base b is $A = h \cdot b$.

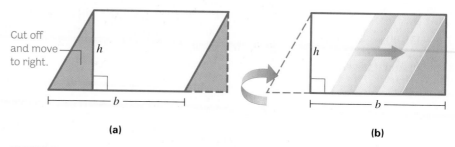

(a)   **(b)**

FIGURE 10.35 (a) Parallelogram. (b) The area of the parallelogram = the area of the rectangle = $h \cdot b$.

We next use the formula for the area of a parallelogram to derive the formula for the area of a triangle. In Figure 10.36, triangle ABC has height h and base b. It is easy to see that the area of $\triangle ABC$ is exactly one-half of the area of the parallelogram $AXBC$, which has area $h \cdot b$. Therefore, $\triangle ABC$ has area $\frac{1}{2}h \cdot b$.

FIGURE 10.36 The area of $\triangle ABC$ is one-half the area of the parallelogram, or $\frac{1}{2}h \cdot b$.

> **AREA OF A TRIANGLE** A triangle with height h and base b has area $A = \frac{1}{2}h \cdot b$.

Some Good Advice

If you practice deriving the area formulas as we have developed them, it will be easier for you to remember the details of the formulas. For example, remembering that a triangle is one-half of a parallelogram helps you recall that there is a factor of $\frac{1}{2}$ in the formula for the area of a triangle.

EXAMPLE 1   *The Area of a Playground*

a) Figure 10.37 shows a recreation area in the shape of a parallelogram. If a pound of grass seed covers 100 square yards, how much grass seed is needed to seed the entire area?

b) Suppose that we want to seed only the triangular area *ACD*. How much grass seed will be needed then?

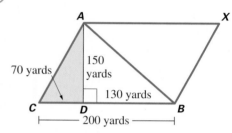

FIGURE 10.37 Recreation area shaped like a parallelogram.

SOLUTION:

a) The area of the parallelogram is $A = h \cdot b = 150 \cdot 200 = 30{,}000$ square yards. Dividing this area by 100, we see that 300 pounds of grass seed are needed for the entire recreation area.

b) Triangle *ACD* is a triangle with base 70 yards and height 150 yards. The area of this triangle is $\frac{1}{2}h \cdot b = \frac{1}{2} \cdot 150 \cdot 70 = 5{,}250$ square yards. Dividing this by 100 we get 52.5, which is the number of pounds of grass seed required for this area.

Now try Exercises 11 and 12. ❀ **8**

Sometimes we don't know the height of a triangle, but we do know the length of all three sides. In this case we can use Heron's formula to find the area of the triangle. Because this formula is not as intuitively clear as the formula we developed earlier, we will state it without proof.

Quiz Yourself **8**

Find the area of the given triangle.

(triangle with vertex A at top, base from C to B; height 12 marked inside, base labeled 21)

KEY POINT

Heron's formula uses the lengths of the sides of a triangle to find its area.

> **HERON'S FORMULA FOR THE AREA OF A TRIANGLE** Suppose that a triangle has sides *a*, *b*, and *c*. We define the quantity $s = \frac{1}{2}(a + b + c)$. Then the area of the triangle is
>
> $$A = \sqrt{s(s-a)(s-b)(s-c)}.$$

EXAMPLE 2   *Finding the Area of a Triangular Flower Bed Using Heron's Formula*

A gardener wants to fill a triangular flower bed outside the city museum with yellow tulips. Figure 10.38 shows the dimensions of the flower bed. The gardener estimates that he will need four bulbs for each square foot of the flower bed. How many dozen bulbs should he buy?

SOLUTION: Because we know the length of the sides of the flower bed, we can use Heron's formula. First we find

$$s = \frac{1}{2}(a + b + c) = \frac{1}{2}(18 + 12 + 24) = 27.$$

(triangular flower bed diagram with sides c = 24 feet, a = 18 feet, b = 12 feet, vertices A, B, C)

FIGURE 10.38 Triangular flower bed.

Then the area of the flower bed is

$$A = \sqrt{s(s-a)(s-b)(s-c)} = \sqrt{27(9)(15)(3)} = \sqrt{10{,}935} \approx 105 \text{ square feet.}$$

Therefore, the gardener needs about $105 \times 4 = 420 = 35$ dozen bulbs.

Now try Exercises 31 to 36. ❋

We can use the area formula for a triangle to develop area formulas for other figures such as the trapezoid shown in Figure 10.39 (a).

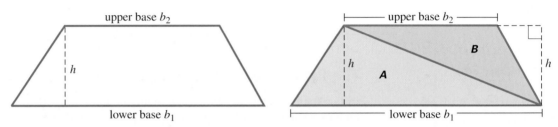

FIGURE 10.39 (a) A trapezoid. (b) The area of the trapezoid is the sum of the area of triangles A and B.

We will begin by dividing this trapezoid into two triangles, as we show in Figure 10.39 (b). Triangle A has base b_1 and height h, so the area of triangle A is $\frac{1}{2}b_1 \cdot h$. Similarly, triangle B has base b_2 and height h, so its area is $\frac{1}{2}b_2 \cdot h$. The area of the trapezoid is the area of A plus the area of B, which equals $\frac{1}{2}b_1 \cdot h + \frac{1}{2}b_2 \cdot h = \frac{1}{2}(b_1 + b_2) \times h$.

> **AREA OF A TRAPEZOID**   A trapezoid with lower base b_1, upper base b_2, and height h has area
>
> $$A = \frac{1}{2}(b_1 + b_2) \times h.$$

EXAMPLE 3   *Finding the Area of Trapezoids in the Base of a Statue*

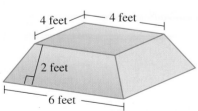

FIGURE 10.40 Base for statue formed by one square and four trapezoids.

A sculptor has created a statue for the lobby of the town's historical society building. Because the building is old and the town engineer wants to reduce the weight of the platform the statue will rest on, he recommends a hollow base instead of a solid one. The platform will be formed by joining four congruent trapezoids and one square, as shown in Figure 10.40. To determine which material will be best for the platform, the sculptor needs to know the surface area of the platform. Use the formulas we have developed so far to determine this surface area.

SOLUTION: We know that the area of the top of the platform is $4 \times 4 = 16$ square feet. Each side is a trapezoid with lower base 6 feet, upper base 4 feet, and height 2. Therefore, using the formula for the area of a trapezoid, the area of each face is

$$\frac{1}{2}(b_1 + b_2) \times h = \frac{1}{2}(6 + 4) \times 2 = 10 \text{ square feet.}$$

The total area of the four trapezoidal sides plus the top is $4 \times 10 + 16 = 56$ square feet.

Now try Exercises 9 and 10. ❋ **9**

Quiz Yourself **9**

Find the area of the given trapezoid.

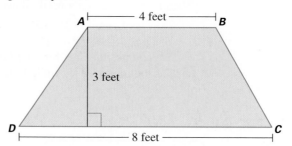

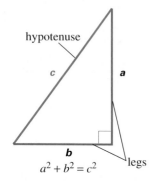

✏️ **KEY POINT**

We can use the Pythagorean theorem to find measurements involving right triangles.

FIGURE 10.42 The Pythagorean theorem.

$a^2 + b^2 = c^2$

The Pythagorean Theorem

In the sixth century BC, the Greek mathematician Pythagoras proved one of the most famous theorems in the history of mathematics. The Pythagorean theorem states that the sum of the squares of the lengths of the legs of a right triangle equals the square of the length of the hypotenuse.

> **THE PYTHAGOREAN THEOREM** If a and b are the legs of a right triangle and c is the hypotenuse (the side opposite the right angle, shown in red) then $a^2 + b^2 = c^2$ (see Figure 10.42).

Math in Your Life

That Geometry Looks Lovely on You*

A serious clothing designer needs to understand the basic notions of plane geometry that you are studying in this chapter. Two common methods of pattern making that designers use are the flat pattern method and draping.

A designer using the flat pattern method may begin with a generic pattern called a *bodice sloper*, as shown in Figure 10.41. This sloper is a rectangle from which various geometric shapes have been removed. By modifying this basic sloper, the designer can then create a blouse or a jacket. For women's clothes, there are also slopers for skirts, pants, and sleeves.

In draping, the designer drapes fabric cut into various shapes, such as rectangles, circles, and half circles, on a dress form or on a live model. A talented designer who creates fashions this way must have a good intuitive understanding of geometry.

So, you see that not only is geometry "all around us" but we wear it every day.

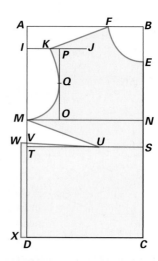

FIGURE 10.41 A bodice sloper.

---

*This example is based on the article "The Drafter, the Draper, the Flat Pattern," *Threads Magazine*, no. 11, June/July 1987, pp. 33–37.

If we know the lengths of any two sides of a right triangle, we can use the Pythagorean theorem to find the length of the third side.

EXAMPLE 4   *Using the Pythagorean Theorem*

Use the lengths of the two given sides to find the length of the third side in the triangles in Figure 10.43.

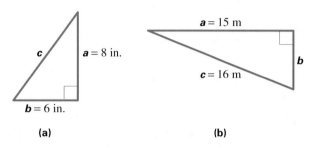

FIGURE 10.43 Using the Pythagorean theorem.

SOLUTION:

a) We are given $a = 8$ and $b = 6$, so we must find c. If we substitute the values for a and b in the Pythagorean theorem, we get

$$c^2 = 8^2 + 6^2 = 64 + 36 = 100.$$

Therefore, $c = \sqrt{100} = 10$ inches.

b) In this case, we are given that $a = 15$ and $c = 16$, and we must find b. Substituting these values in the Pythagorean theorem, we get

$$15^2 + b^2 = 16^2,$$

or

$$225 + b^2 = 256.$$

So,

$$b^2 = 256 - 225 = 31.$$

This means that

$$b = \sqrt{31} \approx 5.57 \text{ meters.}$$

Now try Exercises 37 to 42. ❋ **10**

Quiz Yourself **10**

Assume that the hypotenuse of a right triangle measures 20 inches and one leg measures 10 inches. What is the length of the other leg?

To solve a problem, it may be necessary to use the Pythagorean theorem several times, as we demonstrate in Example 5.

EXAMPLE 5   *Using the Pythagorean Theorem to Find the Height of a Pyramid*

The Great Pyramid, near Cairo, Egypt, is the tomb of the Egyptian pharaoh Khufu (also called Cheops by the Greeks) and was considered to be one of the seven wonders of the ancient world. An archaeologist wants to determine the height of this pyramid. This pyramid has a square base measuring 230 meters on each side, and she has found the distance from one corner of the base to the tip of the pyramid to be 219 meters. What is the height of the pyramid?

SOLUTION: If we imagine this pyramid to be hollow, we could drop a string with a weight from the tip of the pyramid, point T, to point M, which is the middle of the base. If we then drew a line segment from M to a corner of the base, calling this point C, we would

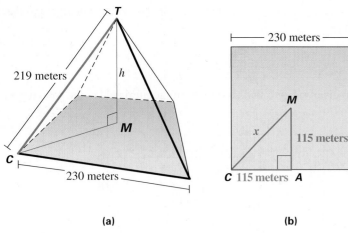

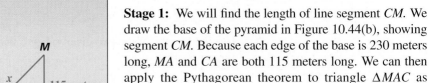

FIGURE 10.44 (a) The Great Pyramid. (b) The base of the pyramid; *M*, is the center of the base, *C* is the corner of the base, and *A* is the midpoint of a side of the base.

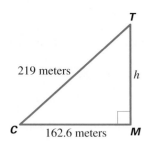

FIGURE 10.45

form a right triangle $\triangle TMC$, as we see in Figure 10.44(a). Our goal now is to find the length of line segment *TM*.

We will do the solution in two stages.

Stage 1: We will find the length of line segment *CM*. We draw the base of the pyramid in Figure 10.44(b), showing segment *CM*. Because each edge of the base is 230 meters long, *MA* and *CA* are both 115 meters long. We can then apply the Pythagorean theorem to triangle $\triangle MAC$ as follows:

$$\text{(length of } CM)^2 = \text{(length of } CA)^2 + \text{(length of } AM)^2,$$

or

$$x^2 = 115^2 + 115^2 = 13,225 + 13,225 = 26,450.$$

Thus, $x = \sqrt{26,450} = 115\sqrt{2} \approx 162.6$.

Stage 2: Now that we have found the length of *CM*, we can find the length of *TM* in the right triangle $\triangle TMC$ shown in Figure 10.45. Using the Pythagorean theorem again, we get

$$\text{(length of } TC)^2 = \text{(length of } CM)^2 + \text{(length of } TM)^2.$$

Substituting for the various lengths in $\triangle TMC$ gives the equation

$$219^2 = (115\sqrt{2})^2 + h^2,$$

or

$$h^2 = 219^2 - (115\sqrt{2})^2 = 47,961 - 26,450 = 21,511.$$

Taking the square root of both sides, we find that $h \approx 146.7$ meters.

Now try Exercises 57 and 58. ❋

✏️ **KEY POINT**

We use simple formulas to compute the circumference and area of circles.

Circles

As with polygons, we often need to calculate the circumference and area of circles. The derivation of these formulas is not as intuitive as those for the polygons we have been discussing, so we will state them without proof. The number π, which appears in these formulas, is the Greek letter pi. It stands for the measure of the circumference of a circle divided by the length of its diameter. We will use 3.14 to approximate pi.

HISTORICAL HIGHLIGHT ✺ ✺ ✺

Hypatia

Throughout the history of mathematics, fewer women than men are mentioned for their contributions because traditionally women were discouraged from studying mathematics. One notable exception is Hypatia, who was born in Greece in 370 AD. Her father, who was a professor of mathematics at the University of Alexandria, gave her a classical education, which included mathematics. She lectured on both philosophy and mathematics at Alexandria and wrote major papers on geometry, including the work of Euclid. She also

did work in philosophy and astronomy and is believed to have invented several astronomical devices.

Unfortunately, her work in science caused her problems with the Christian church, and moreover, as a Greek she was seen as a pagan. In 415, a mob attacked and brutally murdered her, which caused other scholars to flee Alexandria. This tragic event marked the end of the golden age of Greek mathematics and, some believe, the beginning of the Dark Ages in Europe.

> **CIRCUMFERENCE AND AREA OF A CIRCLE** A circle with radius r has circumference $C = 2\pi r$ and area $A = \pi r^2$.

These formulas tell us that a circle with radius 10 feet has a circumference of $C = 2\pi r = 2\pi(10) \approx 62.8$ feet and area $A = \pi r^2 = \pi(10)^2 \approx 314$ square feet.

EXAMPLE 6 *Laying Out a Basketball Court**

A man is designing a basketball court for his children. The court is in the form of a segment of a circle, as shown in Figure 10.46. There will be a fence at the rounded end of the court, and he will paint the court with a concrete sealer.

a) How much fencing is required?

b) What is the area of the surface of the court?

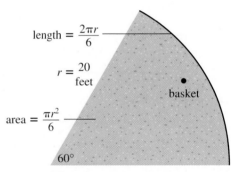

FIGURE 10.46 Basketball court.

SOLUTION: Since a circle contains 360°, the court is one-sixth of the interior of a circle with a radius of 20 feet.

a) The circumference of a circle with a 20-foot radius is $2\pi r = 2\pi(20) = 40\pi \approx 125.6$ feet. The fencing we need will be one-sixth of this, or approximately 20.93 feet. The man should buy about 21 feet of fencing.

b) The area of a circle with a 20-foot radius is $\pi(20)^2 = 400\pi \approx 1{,}256$ square feet. One-sixth of this is about 209.3 square feet.

Now try Exercises 59 and 60. ✸ **11**

Quiz Yourself **11**

Find the circumference and area of a circle that has radius 8.

Exercises (10.3)

Looking Back[†]

These exercises follow the general outline of the topics presented in this section and will give you a good overview of the material that you have just studied.

1. In Figure 10.35, what relationship were we showing between the area of a parallelogram and the area of a rectangle?

2. What was the point that we made in Figure 10.36?

3. How does Figure 10.39 help you remember the formula for the area of a trapezoid?

4. Who was Hypatia?

Sharpening Your Skills

In Exercises 5–16, find the area of each figure.**

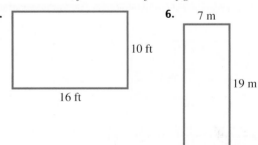

---

*Thanks to my friend Dr. Frank Siekman, of the KU music department, for asking me this question several years ago.

[†]Before doing these exercises, you may find it useful to review the note *How to Succeed at Mathematics* on page xix.

**Throughout this exercise set, approximate π by 3.14.

7.

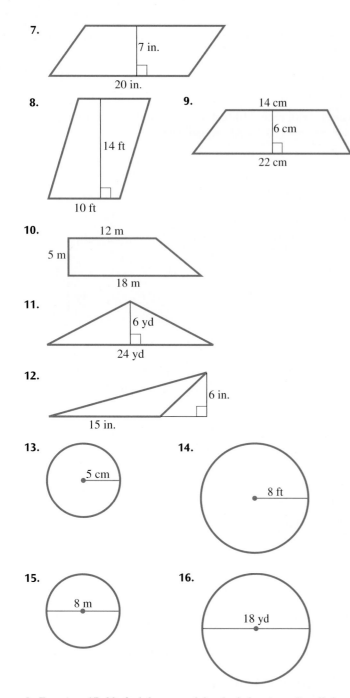

7 in.

20 in.

8.

14 ft

10 ft

9.

14 cm

6 cm

22 cm

10.

12 m

5 m

18 m

11.

6 yd

24 yd

12.

6 in.

15 in.

13.

5 cm

14.

8 ft

15.

8 m

16.

18 yd

In Exercises 17–22, find the area of the shaded regions. Recall that to solve a new problem, it is helpful to relate it to a problem that you have seen before.

17.

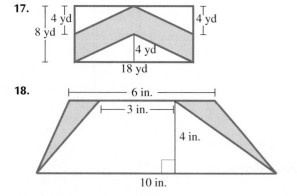

4 yd

8 yd

4 yd

4 yd

18 yd

18.

6 in.

3 in.

4 in.

10 in.

19.

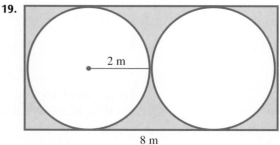

2 m

8 m

20.

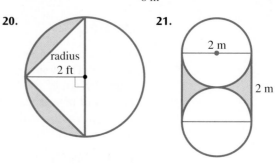

radius
2 ft

21.

2 m

2 m

22.

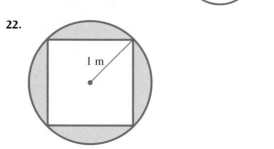

1 m

Use the following figure to answer Exercises 23 and 24. Assume that the area of the parallelogram ABCD is 60 square inches and the area of triangle BEC is 6 square inches.

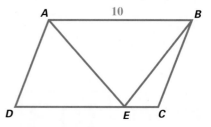

A   10   B

D   E   C

23. What is the area of triangle *ADE*?

24. What is the area of the trapezoid *ABCE*?

Use the following figure to answer Exercises 25 and 26. Assume that the area of triangle BAE is 30 square yards, the area of the trapezoid ABDF is 66 square yards, and the area of triangle BDC is twice the area of triangle AEF.

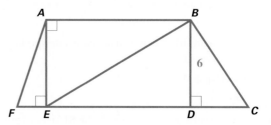

A   B

6

F   E   D   C

25. Find the area of triangle *BDC*.

26. Find the area of trapezoid *ABCF*.

A geoboard is a board with rows of nails spaced 1 inch apart in both the vertical and horizontal directions. In Exercises 27–30, we stretched a rubber band around some of the nails. Find the area of the enclosed figures.

27. 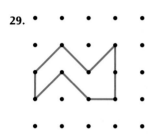 **28.**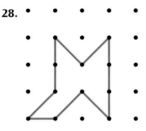

29. 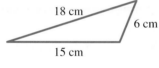 **30.**

In Exercises 31–34, use Heron's formula to find the area of each triangle.

31.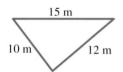

32.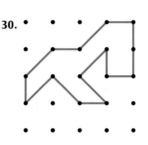

33.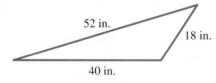

34.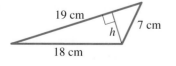

In Exercises 35 and 36, find the height h for each triangle.

35. 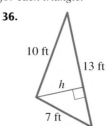 **36.**

In Exercises 37–40, find the length of side x for each triangle.

37. 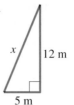 **38.**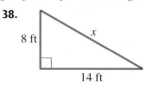

39.   **40.**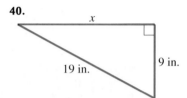

In Exercises 41 and 42, find the area of each triangle.

41.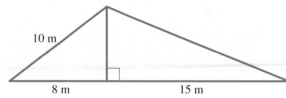

42.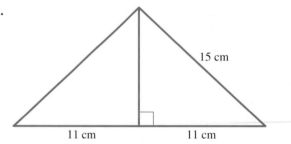

Applying What You've Learned

In Exercises 43–46, state whether perimeter or area would be the more appropriate quantity to measure.

43. You are covering an archeological plot with a tarpaulin.

44. Napoleon wants to buy fencing for a paddock to graze llamas.

45. You are putting a decorative stencil at the top of the walls in your bedroom.

46. You are surfacing your rooftop Japanese garden with slate tiles.

47. Finding distances on a baseball diamond. The bases on a baseball diamond are 90 feet apart. What is the distance from home plate to second base? (All angles in the diamond are right angles.)

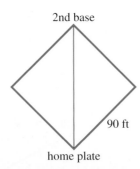

2nd base

90 ft

home plate

48. Finding the area of home plate. In baseball, home plate is shaped as shown ($m\angle FDE = 90°$).

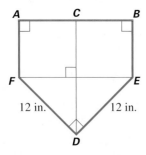

12 in.   12 in.

a. What is the length of line segment AB?

b. If we assume that line segment CD has the same length as segment AB, what is the area of home plate?

49. Find the length of line segment AB in the given figure.

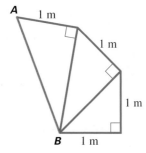

A   1 m

1 m

1 m

B   1 m

50. Continuing the pattern in Exercise 49, construct a line segment having length $\sqrt{6}$.

51. Making a stained glass window. A stained glass window in a museum is in the shape of a rectangle with a semicircle on top. The height of the rectangular part of the window is twice the base. If the base of the window measures 6 feet, what is the area of the window?

├── 6 ft ──┤

52. Making a stained glass window. Consider a window like the one in Exercise 51. Assume that the semicircular top has a radius of 2 feet. We want the rectangular part of the window to have the same area as the semicircular top. What should the dimensions of the rectangle be?

53. Comparing pizzas. At Cifaretto's Italian Ristorante, the medium pizza has a diameter of 12 inches and sells for $5.99. The large pizza has a diameter of 16 inches and sells for $8.99. Which pizza is the better buy? Explain.

54. Making a flower bed. A gardener has enough tiger lilies to fill a circular flower bed having an area of 50 square feet. Find the radius of the flower bed to the nearest foot.

55. Measuring a running track. A running track, 4 meters wide, has the dimensions shown in the following diagram. The ends of the track are semicircles with diameter 20 meters. What is the surface area of the track?

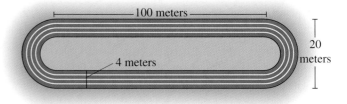

├──────── 100 meters ────────┤

20 meters

4 meters

56. Measuring a running track. In Exercise 55, if the 100-meter dimension is increased to 120 meters, the 20-meter dimension is increased to 40 meters, and the width of the track is increased to 6 meters, what is the surface area of the track?

Exercises 57 and 58 are based upon the information given about the Great Pyramid in Example 5.

57. What is the slant height of the pyramid—that is, the distance from its tip to the midpoint of one of its sides?

58. What is the area of one face of the pyramid?

59. Redo Example 6, but now assume that the sides of the basketball court are 18 feet long and the angle measures 72°.

60. Redo Example 6. However, now the tip of the court has been removed, as shown in the figure.

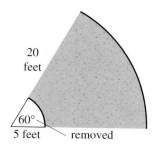

20
feet

60°
5 feet   removed

Communicating Mathematics

61. In this section, we derived formulas for the areas of the following figures. Arrange the terms in the order in which we did the derivation: triangle, trapezoid, rectangle, parallelogram.

62. The formula for the area of a trapezoid is $\frac{1}{2} \times (b_1 + b_2) \times h$. What is the meaning of each symbol in this formula?

63. In Example 5, why did we need to find the length of the line segment CM?

64. In Example 6, how did we determine the length of the fencing?

Using Technology to Investigate Mathematics

65. See your instructor for TI-83 programs to compute the areas of various geometric figures. Use them to verify some of the computations in this section.

66. There are many Web sites that have geometry calculators you can use to reproduce the calculations in this section. Run some of these programs and report on your findings. You might search on "triangles, circles, applets" or "geometry calculators" to find such pages.

For Extra Credit

67. A modern art museum has sides shaped like trapezoids with identical equilateral triangular windows with dimensions shown in the accompanying figure. What is the area of one side of the museum, excluding the area of the windows?

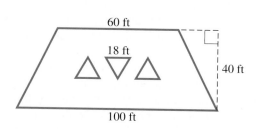

60 ft

18 ft

△▽△   40 ft

100 ft

68. One side of an air traffic control tower, with dimensions given, is shown in the accompanying figure. Assume that all polygons are trapezoids. What is the area of the side, excluding the two congruent windows?

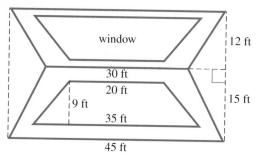

window   12 ft

30 ft
20 ft
9 ft   15 ft
35 ft

45 ft

69. **True or false?** If we double the radius of a circle, the area also doubles.

70. **True or false?** If we double the radius of a circle, the circumference also doubles.

71. In the figure below, W, X, Y, and Z are the midpoints of the line segments on which they lie. How does the area of $WXYZ$ compare with the area of rectangle $ABCD$? Explain how you got your answer.

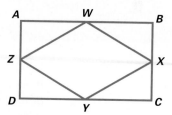

A   W   B

Z   X

D   Y   C

72. If W and Y were not the midpoints of the line segments on which they lie, would it change your answer in Exercise 71? Explain your answer.

73. You have 200 feet of fencing and want to enclose a rectangular area. What shape will enclose the most area? You might solve this problem by experimenting with different lengths and widths until you can make a reasonable guess.

74. Bernie is shaping round and square posts. In figure (a), he is cutting a circular post from a piece of wood with a square cross section. In figure (b), he is cutting a square post from a piece of wood with a circular cross section. In which situation is there the smaller *percentage* of waste? (*Hint:* Answer this question by making up numerical examples.)

(a)   (b)

75. *Extreme Makeover Home Edition* wants to create a housing development with lots clustered in circles and an open area in the center of the circle, as shown in the diagram. The side of a lot is 180 feet and the central common area is 11,304 square feet. Each lot in the cluster is the same size. What is the area of each lot? (We are using 3.14 for π)

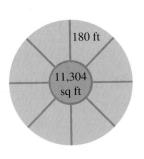

180 ft

11,304 sq ft

76. Ty Pennington in Exercise 75 wants to put a fence entirely around the lot, except for the curved section away from the common area. How much fencing is required?

In Exercises 77 and 78, express the shaded areas in terms of the radius r *of the circle.*

77.

78.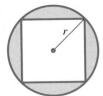

10.4 Volume and Surface Area

Objectives

1. Understand the idea of volume of basic three-dimensional objects.
2. Be able to apply the volume and surface area formulas for cylinders.
3. Understand the relationship between the volume and surface area formulas for cylinders, spheres, and cones.

In thinking about what to use for an opener for this section, I was reflecting on the many times that people have asked me to use mathematics to solve a real problem in their lives.* Several years ago, a colleague in the psychology department asked me to help her determine if her oil deliveryman was cheating her. In order to solve her problem, I needed to know the basic formulas for computing volumes that you will learn in this section.

We will now extend the ideas of two-dimensional Euclidean geometry to three dimensions. Instead of squares and rectangles, we will discuss boxlike figures called *parallelepipeds*; instead of circles we will talk about cylinders, spheres, and cones; and instead of measuring perimeters and areas, we will calculate the volumes and surface areas of three-dimensional geometric figures.

Many applications require a knowledge of three-dimensional geometry. When designing a skyscraper, an architect needs to know the volume of concrete to pour for a foundation. A manufacturer must be able to compute the amount of materials required to fabricate a cylindrical water tank. A state highway department engineer may want to know how much road salt is contained in a conical storage shed.

Volume

When we measure length in one dimension, we use units such as inches, feet, centimeters, and meters. In measuring area in two dimensions, we use square units such as square inches and square centimeters. We measure the **volume** of a three-dimensional figure using *cubic units*. Even though we will study three-dimensional geometric objects, *we will still measure surface area in square units.*

Figure 10.47 shows a cube whose edges are each 1 inch long, so the cube has a volume of 1 cubic inch. In determining the volume of a three-dimensional figure whose measurements are given in inches, we are really asking how many of these 1-inch cubes will fit inside the figure. Because it might be impossible to fit these cubes inside the figure exactly, we can

✏️ **KEY POINT**

The volume of a rectangular parallelepiped is the product of its length, width, and height.

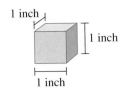

1 inch

1 inch

1 inch

FIGURE 10.47
A cube whose volume is 1 cubic inch.

---

*In Example 6 of Section 10.3, a friend asked me for help in laying out a strangely shaped basketball court for his son. The star problem in Example 3 of Section 10.2 was from a minister asking me to advise his youth group on how to construct a five-pointed star to place on their church steeple during advent.

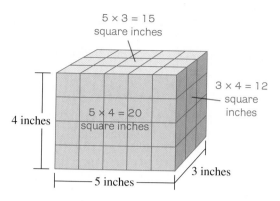

FIGURE 10.48 Rectangular solid sliced into 60 one-inch cubes.

imagine filling 1-cubic-inch containers with water and asking, "How many of these containers of water does it take to fill the figure exactly?"

It is easy to compute the volume of the box-shaped rectangular solid, called a *rectangular parallelepiped*,* shown in Figure 10.48. If we slice along the lines drawn at 1-inch intervals on the solid, we get four layers each containing 15 1-inch cubes. The volume of this solid is given by

$$\text{volume} = \text{length} \times \text{width} \times \text{height} = 5 \times 3 \times 4 = 60 \text{ cubic inches.}$$

It is also easy to find the surface area of the solid in Figure 10.48. Clearly the six sides of the solid have the following areas:

$$\text{areas of top and bottom} = \text{length} \times \text{width} = 5 \times 3 = 15,$$
$$\text{areas of front and back} = \text{length} \times \text{height} = 5 \times 4 = 20, \text{ and}$$
$$\text{areas of two sides} = \text{width} \times \text{height} = 3 \times 4 = 12.$$

The total surface area is therefore $(2 \times 15) + (2 \times 20) + (2 \times 12) = 94$ square inches. Figure 10.48 helps us remember these formulas for volume and surface area. **12**

Quiz Yourself **12**

Find the volume and surface area of a rectangular solid with length 8 centimeters, width 6 centimeters, and height 3 centimeters.

> **VOLUME AND SURFACE AREA OF A RECTANGULAR SOLID (RECTANGULAR PARALLELEPIPED)** If a rectangular solid has length *l*, width *w*, and height *h*, the volume of the solid is $V = lwh$ and the surface area of the solid is
>
> $$S = 2lw + 2lh + 2wh.$$

Another way to look at the way we calculated the area for the rectangular solid above is that we multiplied the area of the base, which is *lw*, times the height, *h*. We can apply this approach to other figures.

KEY POINT

For many solids, volume equals base area times height.

> **VOLUME EQUALS AREA OF BASE TIMES HEIGHT** If an object has a flat top and base and *sides perpendicular to the base*, as shown in Figure 10.49, then, if the area of the base is *A* and the height is *h*, the volume will be
>
> $$V = A \cdot h.$$

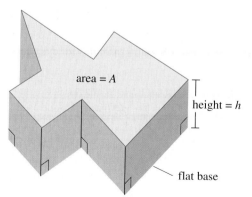

FIGURE 10.49 Volume = area of base times height.

EXAMPLE 1 *Comparing Volumes*

Figure 10.50 shows two blocks of cheese that are selling for the same price. Which block contains the greater volume?

SOLUTION: We calculate the volume of each block by multiplying the area of its base times its height. The area of the base of the trapezoidal block is

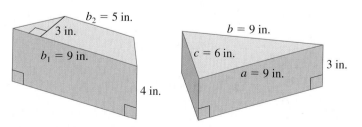

FIGURE 10.50 Which block has greater volume?

---

*A parallelepiped is a solid with six faces, each of which is a parallelogram. In this case, the parallelepiped is rectangular, which means the faces are rectangles.

$\frac{1}{2}(b_1 + b_2) \cdot h = \frac{1}{2}(9 + 5) \cdot 3 = 21$ square inches. Therefore, this block has a volume of $21 \times 4 = 84$ cubic inches.

We use Heron's formula to find the area of the base of the triangular block of cheese. We first must compute $s = \frac{1}{2} \times (9 + 9 + 6) = 12$. Then the area is

$$\sqrt{s(s-a)(s-b)(s-c)} = \sqrt{12(12-9)(12-9)(12-6)} = \sqrt{648} \approx 25.5.$$

Multiplying this area by the height of 3 gives us $25.5 \times 3 = 76.5$ cubic inches. Therefore, the trapezoidal block of cheese has slightly more volume and is the better deal.

Now try Exercises 13 to 18. ❋

✎ **KEY POINT**

Simple diagrams illustrate the formulas for finding the volume and surface area of a cylinder.

Cylinders

We can use this method of multiplying the base area times the height to find the volume of some common three-dimensional solids. A *right circular cylinder* is a solid that is shaped like a soup or tuna fish can (see Figure 10.51). We call these cylinders "right" because the sides are perpendicular to the base. The cylinder's base is a circle with area $A = \pi r^2$. * Thus,

$$\text{volume of cylinder} = \text{area of base} \times \text{height} = \pi r^2 \cdot h.$$

To find the surface area of the cylinder in Figure 10.51(a), imagine that we remove the top and bottom of the cylinder and then cut the side of the cylinder and open it up, as in Figure 10.51(b). If we flatten this curved surface, we get a rectangle with length $2\pi r$ and height h, shown in Figure 10.51(c). Therefore, the surface area of the side of the cylinder is $2\pi rh$. Adding to this the areas of the top and bottom of the cylinder, which are each πr^2, we get the total surface area of the cylinder.

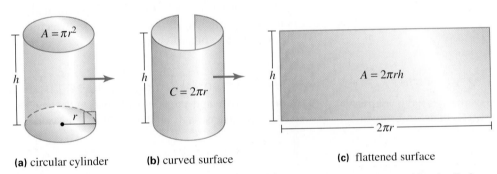

(a) circular cylinder   **(b)** curved surface   **(c)** flattened surface

FIGURE 10.51 (a) Right circular cylinder. (b) Remove top and bottom and cut side of cylinder. (c) Flatten side of cylinder.

The diagrams in Figure 10.51 will help you to remember the following formulas for cylinders.

VOLUME AND SURFACE AREA OF A RIGHT CIRCULAR CYLINDER
A right circular cylinder with radius r and height h has volume $V = \pi r^2 h$ and surface area $S = 2\pi rh + 2\pi r^2$.

We can use these formulas to compare purchases.

EXAMPLE 2 *Comparing Volumes of Commercial Products*

A Pearl art supply store sells paint solvent in a cylindrical can that has a diameter of 4 inches and a height of 6 inches. The giant economy size can of the same solvent has a diameter twice as large (the height of the can is the same) and costs three times as much as the smaller can.

a) Which can is the better deal?   b) Compare the surface areas of the two cans.

---

*Throughout this section, we again approximate π by 3.14.

SOLUTION:

a) The radius of the smaller can is 2, so its volume is

$$V = \pi r^2 h \approx 3.14 \cdot (2)^2 \cdot 6 = 75.4 \text{ cubic inches.}$$

The radius of the larger can is 4, so its volume is

$$V = \pi r^2 h \approx 3.14 \cdot (4)^2 \cdot 6 = 301.4 \text{ cubic inches.}$$

Because the large can contains *four* times as much solvent but costs only *three* times as much, the larger can is the better deal.

b) The surface area of the smaller can is

$$2\pi rh + 2\pi r^2 = 2\pi(2)(6) + 2\pi(2)^2 \approx 2(3.14) \cdot 2 \cdot 6 + 2(3.14)(2)^2 = 100.5 \text{ square inches.}$$

The surface area of the larger can is

$$2\pi rh + 2\pi r^2 = 2\pi(4)(6) + 2\pi(4)^2 \approx 2(3.14) \cdot 4 \cdot 6 + 2(3.14)(4)^2 = 251.2 \text{ square inches.}$$

Now try Exercises 29 to 32. ❋ **13**

Quiz Yourself 13

What is the volume and surface area of a right circular cylinder with a radius of 5 centimeters and height of 10 centimeters?

✏ **KEY POINT**

An efficient container has a small surface area-to-volume ratio.

You may have been surprised in Example 2 that although the larger can contains four times the amount of solvent contained in the smaller can, it takes only two and a half times as much material to make the larger can. Manufacturers are interested in such relationships so they can minimize the amount of materials they need to package a product and, by doing so, reduce their cost. What is the most efficient shape of a container to contain the largest amount of a product? For example, what is the most efficient shape of a soup can?

EXAMPLE 3 *The Most Efficient Shape of a Can*

What are the dimensions of a can that will contain 1 cubic foot of liquid and that will have the smallest amount of surface area?

SOLUTION: We must find the radius, r, and the height, h, of the can. Our intuition tells us that if the radius is small, as in Figure 10.52(a), then the height must be large. On the other hand, if the radius is large, as in Figure 10.52(b), then the height must be small. Our problem then is to find the ideal radius that gives the minimum surface area (Figure 10.52(c)).

Because the volume of the can is to be 1, we can set $\pi r^2 h = 1$. Dividing both sides of this equation by πr^2, we get $h = \frac{1}{\pi r^2}$. Thus, as our intuition tells us, the height depends on the choice of radius. Recall that the formula for finding the surface area of a cylinder is $S = 2\pi rh + 2\pi r^2$. We can substitute $\frac{1}{\pi r^2}$ for h in this equation to get

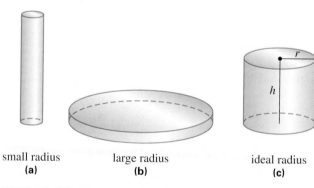

small radius   large radius   ideal radius
  (a)   **(b)**   **(c)**

FIGURE 10.52 A can with a small radius requires a large height, whereas a can with a large radius requires a small height.

$$S = 2\pi r \left(\frac{1}{\pi r^2} \right) + 2\pi r^2 = \frac{2\pi r}{\pi r^2} + 2\pi r^2. \qquad (1)$$

Canceling the π and an r from the quotient on the right side of equation (1), we get

$$S = \frac{2}{r} + 2\pi r^2. \qquad (2)$$

The neat thing about equation (2) is that now we have a formula for the surface area of a can (containing 1 cubic foot of liquid) that depends *only on the radius r*. In Table 10.4, we begin with a very small radius, 0.1, which we expect would give us a large surface area. Then as the size of the radius increases, the surface area decreases. Eventually, as the radius becomes too large, the can is becoming too flat, and as Table 10.4 shows, the surface area starts to increase again.

This TI-83 table confirms our calculations in Example 3.

Radius, r	Surface Area, $S = \frac{2}{r} + 2\pi r^2$	
0.1*	20.063	
0.2	10.251	decreasing
0.3	7.232	decreasing
0.4	6.005	decreasing
0.5	5.570	decreasing
0.6	5.594	increasing
0.7	5.934	increasing
0.8	6.519	increasing

Ideal radius is between 0.5 and 0.6.

TABLE 10.4 Surface area first decreases, then increases as r increases.

You can see in Table 10.4 that the ideal radius is somewhere between 0.5 and 0.6. To get a better estimate of the ideal radius, we could use a graphing calculator† to calculate the surface area for radii of sizes 0.51, 0.52, 0.53, and so on. If you were to do this, eventually you would find that $r \approx 0.5419$ gives us the smallest surface area for a can with volume 1. (Actually, if you were ever to take a calculus course, you would learn a much more powerful and faster way to find this value for r without doing so many tedious calculations.) ✳

KEY POINT

The formulas for the volume and surface area of cones and spheres are related to the formulas for cylinders.

Cones and Spheres

We will now work with formulas for the volume of two other solid figures: cones and spheres. Figure 10.53 shows a right circular cone with height h and base radius r. We call the cone *circular* because its base is a circle, and we use the adjective *right* because the line segment from its tip to the center of its base forms a right angle with the base.

> **VOLUME AND SURFACE AREA OF A RIGHT CIRCULAR CONE** A right circular cone with height h and base radius r has volume $V = \frac{1}{3}\pi r^2 h$ and surface area $S = \pi r \sqrt{r^2 + h^2}$.

FIGURE 10.53
Right circular cone.

In the formula for the surface area of a cone, if we want to include the area of the base we must add πr^2.

PROBLEM SOLVING

The Analogies Principle

We can use what we know about the volume of a cylinder to understand the formula for the volume of a cone. Figure 10.54 shows that a cone with height h and radius r easily fits inside a cylinder with height h and radius r. The cone does not appear to occupy even one-half of the volume of the cylinder, which is $V = \pi r^2 h$. Therefore, it is easy to remember that the volume of the cone is $V = \frac{1}{3}\pi r^2 h$. ✳✳

FIGURE 10.54
The cone has a volume $\frac{1}{3}$ that of the cylinder.

---

*Because the volume is expressed in cubic feet, the radius of the can is measured in feet, so 0.1 is one-tenth of a foot. The surface area is then measured in square feet, so 20.063 is slightly over 20 square feet.
†See your instructor for a tutorial that will show you how to use the Table command on a graphing calculator.
**Of course, we cannot tell just by looking at Figure 10.54 that the cone occupies exactly one-third of the cylinder; however, visualizing this diagram may help you remember the formula for the volume of a cone.

EXAMPLE 4 *Finding the Volume and Surface Area of a Cone*

What is the volume and surface area of a cone with a height of 10 meters and a base radius of 8 meters?

SOLUTION: The volume is $V = \frac{1}{3}\pi r^2 h = \frac{1}{3}\pi(8)^2(10) = \frac{640\pi}{3} \approx 669.87$ cubic meters. The surface area is

$$S = \pi r \sqrt{r^2 + h^2} = \pi(8)\sqrt{8^2 + 10^2} = 8\sqrt{164}\,\pi \approx 321.69 \text{ square meters.}$$

Now try Exercises 7 to 10. ✳ **14**

Quiz Yourself **14**

Find the volume and surface area of a right circular cone with a radius of 4 yards and a height of 5 yards.

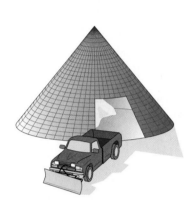

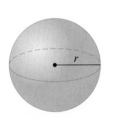

FIGURE 10.55
A sphere with radius *r*.

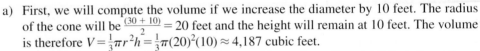

EXAMPLE 5 *The Volume of Conical Storage Buildings*

Many northern states stockpile salt to use for melting road ice. In one such state, the highway department builds sheds in the shape of right circular cones to store the salt. Current sheds have a diameter of 30 feet and a height of 10 feet. The department plans to build larger conical sheds to hold more salt. The larger shed will either have a 10-foot-longer diameter and the same height, or the same diameter and a 10-foot-greater height.

a) Calculate the volume for each of the proposed sheds.

b) Calculate the surface area for each of the proposed sheds.

c) Which design seems to be more economical?

SOLUTION:

a) First, we will compute the volume if we increase the diameter by 10 feet. The radius of the cone will be $\frac{(30 + 10)}{2} = 20$ feet and the height will remain at 10 feet. The volume is therefore $V = \frac{1}{3}\pi r^2 h = \frac{1}{3}\pi(20)^2(10) \approx 4{,}187$ cubic feet.

If we increase the height to 20 feet, and keep the radius at 15 feet, the volume will be

$$V = \frac{1}{3}\pi r^2 h = \frac{1}{3}\pi(15)^2(20) \approx 4{,}710 \text{ cubic feet.}$$

b) The surface area for the shed with radius 20 and height 10 is

$$S = \pi(r)\sqrt{r^2 + h^2} = \pi(20)\sqrt{20^2 + 10^2} \approx 1{,}404 \text{ square feet.}$$

A shed with radius 15 and height 20 has a surface area of

$$S = \pi(r)\sqrt{r^2 + h^2} = \pi(15)\sqrt{15^2 + 20^2} \approx 1{,}178 \text{ square feet.}$$

c) Increasing the height by 10 feet and keeping the diameter at 30 feet gives us more volume and smaller surface area than if we increase the diameter to 40 feet and keep the height at 10 feet. Therefore, increasing the height to 20 feet seems to be the better design. ✤

The last three-dimensional figure we consider is a sphere (Figure 10.55).

> **VOLUME AND SURFACE AREA OF A SPHERE** A sphere with radius *r* has volume $V = \frac{4}{3}\pi r^3$ and surface area $S = 4\pi r^2$.

area of base $= \pi r^2$

FIGURE 10.56 A sphere with radius *r* inside a cylinder with radius *r* and height 2*r*. The cylinder's volume is $\pi r^2 \times 2r = 2\pi r^3$, so the volume of the sphere is less.

Figure 10.56* will help you remember the formula for the volume of a sphere. A sphere with radius *r* fits exactly inside a cylinder having radius *r* and height 2*r*. The

---

*Again, we can't tell by just looking at Figure 10.56 that the volume of the sphere is exactly $\frac{4}{3}\pi r^3$, but the diagram may help you remember that the sphere's volume is somewhat less than $2\pi r^3$.

volume of this cylinder is $\pi r^2 \times 2r = 2\pi r^3$. Because the sphere does not completely fill the cylinder, you should remember that the volume of the sphere is *less than* $2\pi r^3$, or $\frac{4}{3}\pi r^3$.

Many containers, such as water tanks, are shaped like spheres.

EXAMPLE 6 *Measuring the Capacity of Water Tanks*

Metrodelphia is replacing existing spherical water tanks with larger spherical water tanks. The water commission insists that to provide for future expansion of the city, the new capacity of the tanks should be at least five times the capacity of the old water tanks. If the city purchases tanks that have a radius that is twice the radius of the old tanks, will these tanks satisfy the water commission?

SOLUTION: Let the radius of the old water tanks be r feet, so the volume of each tank will be $V = \frac{4}{3}\pi r^3$ cubic feet. The new tanks will each have a radius of $2r$ feet, so the volume of the new tanks will be

$$V = \frac{4}{3}\pi(2r)^3 = \frac{4}{3}\pi 8r^3 = 8\left(\frac{4}{3}\pi r^3\right) \text{ cubic feet.}$$

New radius is twice old radius.   original volume

The new volume is eight times the volume of the original tanks. Therefore, the new tanks will exceed the specifications of the water commission. ❋ **15**

Quiz Yourself **15**

Find the volume of a sphere with a radius of 6 centimeters.

Exercises 10.4

Looking Back*

These exercises follow the general outline of the topics presented in this section and will give you a good overview of the material that you have just studied.

1. How are the formulas for computing the volume the same in Figures 10.48 and 10.49?

2. In Example 2, why was the volume of the larger can of solvent four times the volume of the smaller can? Specifically, why did we get the four?

3. What was the point that we were making in Table 10.4?

4. Where did we use the Analogies Principle in this section?

Sharpening Your Skills

In Exercises 5–12, find a) the surface area and b) the volume of each figure. Approximate π by 3.14.

5.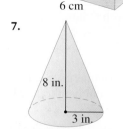
5 cm
6 cm   4 cm

6.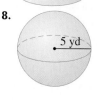
⊢3 ft⊣
6 ft

7.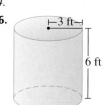
8 in.
3 in.

8.
5 yd

9.
⊢5 ft⊣
8 ft

10.
18 m
⊢5 m⊣

11.
20 cm

12.
15 in.
40 in.   10 in.

In Exercises 13–18, find the volume of each figure.

13.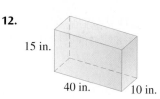
3 ft
area = 25 sq ft

14.
area = 40 sq ft
3 ft

15.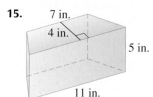
7 in.
4 in.
5 in.
11 in.

16.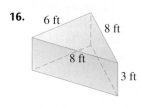
6 ft   8 ft
8 ft
3 ft

*Before doing these exercises, you may find it useful to review the note *How to Succeed at Mathematics* on page xix.

17.

18.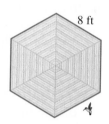

Applying What You've Learned

19. How many cubic inches are in a cubic foot?

20. How many cubic inches are in a cubic yard?

21. Punch bowl. A punch bowl is in the shape of a hemisphere (half of a sphere) with a radius of 9 inches. The cup part of a ladle is also in the shape of a hemisphere with a radius of 2 inches. If the bowl is full, how many full ladles of punch are there in the bowl?

22. Filling glasses. Assume that we are filling cylindrical glasses that have a diameter of 3 inches and a height of 4 inches from the punch bowl in Exercise 21. If the bowl is completely full, how many glasses can we fill?

23. Filling glasses. A cylindrical pitcher with radius 2.5 inches and 8 inches high is filled with a tropical drink. How many glasses shaped like an inverted cone with a height of 1.5 inches and radius 2 inches can be filled from this pitcher?

24. Filling juice containers. The Great American Orange Juice Factory is selling its Caribbean Orange Cocktail in collectible spherical containers. A standard half-gallon container of juice contains 115.5 cubic inches. How many of the spherical containers with a diameter of 3 inches, can be filled from a standard container?

25. Building a Japanese garden. Mariko needs about 16 wheelbarrows full of stone for her Japanese garden. Her wheelbarrow has vertical sides with the given shape and is $2\frac{1}{2}$ feet wide. How many cubic yards of stone should she order?

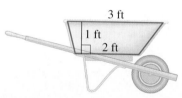

26. Juice barrel. A gallon contains 231 cubic inches. A cylindrical barrel full of juice concentrate has a diameter of 2 feet and is 3 feet high. How many gallons of concentrate are in the barrel?

27. Comparing cakes. One cake is rectangular with a length of 14 inches, a width of 9 inches, and a height of 3 inches. The other cake is a round cake with a radius of 5 inches and a height of 4 inches. Which cake has more volume?

28. Comparing ice cream scoops. Which has more volume: a single scoop of ice cream made with a 3-inch scoop, or two scoops of ice cream made with a $2\frac{1}{2}$-inch scoop? (Assume that both ice cream scoops form perfectly filled spheres of ice cream.)

In Exercises 29–32, we describe a geometric object. Compute the volume you obtain if a) you increase the radius of the object by 2 inches and keep the height the same and b) you increase the height of the object by 2 inches and keep the radius the same.

29. A cylinder with radius 10 inches and height 5 inches. Which causes the volume to increase more: increasing its radius or height? Explain.

30. A cylinder with radius 4 inches and height 8 inches. Which causes the volume to increase more: increasing its radius or height? Explain.

31. A cone with radius 6 inches and height 3 inches. Which causes the volume to increase more: increasing its radius or height? Explain.

32. A cone with radius 5 inches and height 12 inches. Which causes the volume to increase more: increasing its radius or height? Explain.

33. Building an aviary. Luis plans to build a hummingbird aviary with a concrete floor shaped like a regular hexagon. Each side will measure 8 feet, as shown in the diagram, and the floor will be 6 inches deep. How many cubic feet of concrete are needed? Assume that Luis cannot buy a fraction of a cubic foot. (*Hint:* Divide the hexagon into equilateral triangles and use Heron's formula to find the area of the top surface of the aviary.)

34. Constructing a patio. A homeowner is constructing a concrete patio that is 20 feet wide, 30 feet long, and 6 inches thick. Mixed concrete is sold by the cubic yard. How many cubic yards are needed for the project? Assume that the homeowner cannot buy a fraction of a cubic yard of mixed concrete. (*Hint:* You must convert cubic feet into cubic yards.)

35. Diameter of the moon. Earth has a diameter of approximately 7,920 miles and a volume that is roughly 49 times the volume of the moon. Find the diameter of the moon.

36. Diameter of Mars. The volume of Earth is roughly 6.7 times the volume of Mars. Find the diameter of Mars.

Communicating Mathematics

37. How do you "remember how to remember" the volume of a right circular cone?

38. How do you "remember how to remember" the volume of a sphere?

39. If we double the radius of a right circular cone, what effect does that have on the volume? Explain your answer.

40. If we double the radius of a sphere, what effect does that have on the volume? Explain your answer.

41. Assume that three tennis balls fit in a can with no room to spare. Which do you think is larger—the height of the can or the circumference of the can? Explain your answer.

42. A spherical tank has a certain volume. If another spherical tank has twice the volume of the first tank, how do the two radii compare? Explain why this is so.

Using Technology to Investigate Mathematics

43. Following the approach in Example 3, find the radius of a tin can that contains 2 cubic feet of liquid and has the smallest surface area.

44. Repeat Exercise 43, but now the can must contain 8 cubic feet.

45. See your instructor for TI-83 programs that will calculate the volumes and surface areas of the various objects you studied in this chapter. Use them to duplicate some of the computations that we did in this section.

46. There are many Web sites that have interactive geometry calculators that you can use to duplicate the computations that we did in this section. Run some of these programs and report on your findings.

For Extra Credit

47. **Cooling drinks.** Assume that ice cools a liquid proportionally to the amount of surface area of the ice that is in contact with the liquid. Determine whether a large ice cube or a number of smaller ice cubes whose total volume is equal to the larger cube will cool a drink faster. Give specific examples to support your claim.

48. **Comparing hamburgers.** Wendy's serves square hamburgers that are $4 \times 4 \times \frac{1}{4}$ inches. What should the radius be of a round hamburger that is also $\frac{1}{4}$ inch thick so that it has the same volume?

49. **Comparing hamburgers.** Burger King is serving a hamburger called the "tri-burger," which is shaped like an equilateral triangle and is $\frac{1}{4}$ inch thick. What should the length of its side be so that it has the same volume as the hamburgers in Exercise 48?

50. If we cut off the top of a cone by making a horizontal slice, we get a figure called a *frustum* of a cone, as shown in the accompanying figure. Suppose that we cut off the top half of a cone that has radius r and height h. What is the volume of the remaining frustum? (*Hint:* The radius of the top of the frustum is $\frac{r}{2}$.)

51. In Exercise 50, what is the surface area of the side of the frustum?

52. In Example 3, we determined the best shape for a cylinder that contains 1 cubic foot of liquid. Follow that example to determine the shape of a right circular cone with smallest surface area that contains 1 cubic foot. Estimate the radius to within one-tenth of a foot.

10.5 The Metric System and Dimensional Analysis

Objectives

1. Understand the basic units of measurement in the metric system.
2. Make conversions between metric measurements.
3. Use dimensional analysis to make conversions between different measurement systems to solve applied problems.

Would it surprise you if we were to tell you that the United States is an island? We don't mean to say that the United States is entirely surrounded by water, but we are surrounded by a world that uses a very different method for measuring length, volume, weight, and temperature. You are used to measuring length in inches, feet, yards, and miles, whereas the rest of the world uses centimeters, meters, and kilometers.

The Metric System

Unlike the United States, most of the world uses the **metric** system of measurement that is officially known as the **Systèm International d'Unités** or the **SI system**. The U.S. system of measurement is called the **U.S. customary system**.

In order to work in the metric system, you need to understand two things.

1. The basic units that we use to measure length, weight, and volume.
2. The way these different measures of length, volume, and weight relate to each other. As you will see, the relationship between different metric measures is much simpler than the relationship between different customary measures.

In the metric system, length measurement is based on the **meter**, which is slightly more than a yard. If you were to meet a tall man on a train in Argentina, he would be about 2 meters tall. Volume is based on the **liter**, which is a little more than a quart. Instead of buying a half-gallon of iced apple-pear drink in Poland, you would carry home 2 liters from the market. The basic unit of weight in the metric system is the **gram**, which is about the weight of a large paper clip. One thousand grams is called a **kilogram**, which is about 2.2 pounds. If you were visiting Thailand and wanted to make about 2 pounds of chicken breasts with herb-lemongrass crust, you would ask your meat cutter for a kilogram of chicken.

Although measurement in the metric system is based on meters, liters, and grams, we often use variations of these basic units to measure different quantities. For example, we might measure length in centimeters or kilometers, volume in milliliters, or weight in deci-grams or kilograms. Table 10.5 explains the meaning of prefixes such as milli-, deka-, kilo-, centi-, and so on.

KEY POINT

Units of measure in the metric system are based on powers of 10.

kilo-(k)	hecto-(h)	deka-(da)	base unit	deci-(d)	centi-(c)	milli-(m)
× 1,000	× 100	× 10		× 1/10 or × 0.1	× 1/100 or × 0.01	× 1/1,000 or × 0.001

TABLE 10.5 Some common metric prefixes.

For example, a kilometer is 1,000 meters, a centigram is $\frac{1}{100}$ of a gram or 0.01 gram, and a milliliter is $\frac{1}{1,000}$ of a liter, or 0.001 liter. Conversion of units in the metric system is quite simple. From Table 10.5, we see that 1,000 of something is always a "kilo" and $\frac{1}{100}$ of something is always a "centi." The same pattern of prefixes works the same for length, volume, and weight. In the U.S. customary system, this is certainly not the case. Three feet equal 1 yard, but 4 quarts equal 1 gallon. There are 16 ounces in a pound, but only 2 cups in a pint. There is no consistency or uniformity.

We have listed the abbreviations for the metric prefixes in Table 10.5. Also, we use m for meter, g for gram, and L for liter. For example, to represent kilograms, we could use k for kilo and g for grams, thus we could write 10 kilograms as 10 kg. Similarly, we could write 25 milliliters as 25 mL. ❋ **16**

Quiz Yourself **16**

a) 1 kiloliter is equal to how many liters?

b) 1 gram is equal to how many centigrams?

PROBLEM SOLVING

The Analogies Principle

It helps you to remember the meaning of some metric prefixes if you connect them with some common words. For example, "centi" reminds you of cent, which is one one-hundredth of a dollar. "Milli" might remind you of millennium, which is 1,000 years. "Deci" is found in the word *decimal* and a decimal system is based on powers of 10.

Metric Conversions

If you remember the meaning of the prefixes in Table 10.5, it is easy to convert from one unit to the other in the metric system. For example, in measuring length, a kilometer is 10 times as long as a hectometer and a meter is 100, or 10^2, times as long as a centimeter. A millimeter is $\frac{1}{1,000}$, or 10^{-3}, of a meter.

EXAMPLE 1 *Converting Units of Measurement in the Metric System*

Convert each of the following quantities to the unit of measurement that we specify.

a) 5 dekameters to centimeters   b) 2,300 milliliters to hectoliters

SOLUTION:

a) We can make this conversion as follows:

$$5 \text{ dekameters} = 5 \times (10 \text{ meters}) = 50 \text{ meters},$$

$$50 \text{ meters} = 50 \times (10 \text{ decimeters}) = 500 \text{ decimeters},$$

$$500 \text{ decimeters} = 500 \times (10 \text{ centimeters}) = 5,000 \text{ centimeters}.$$

Or, looking at Table 10.5 again, we see that every time we move one column to the right, we require 10 times as many objects. Note that because centimeters are much smaller than dekameters, we need many more centimeters than dekameters.

kilo-	hecto-	deka-	base unit	deci-	centi-	milli-
		1 of these	Equals 10 of these	Equals 100 of these	Equals 1,000 of these	

→

Moving three places to the right gives us 10^3 as many objects.

You can easily perform this calculation by moving the decimal point in 5.0 dekameters three places to the right to get 5,000 centimeters.

b) A quick way to solve this problem is to realize that because the prefix hecto- is five places to the left of milli- in Table 10.5, all we have to do to make the conversion is move the decimal point in 2,300 milliliters five places to the left to get

or 0.023 hectoliter.

It is good to check if this makes sense to you. Remember that milliliters are much smaller than hectoliters, so therefore it does not require many hectoliters to represent many milliliters.

Now try Exercises 15 to 20. ❈ **17**

Quiz Yourself 17

a) 5.63 kiloliters is equal to how many deciliters?

b) 4,850 milligrams is equal to how many hectograms?

![Some Good Advice icon] **Some Good Advice**

Although it is tempting to simply memorize how to move the decimal point in doing conversions such as those we did in Example 1, it is important to *understand* why you are moving the decimal point and in which direction. If you want to make a number larger, you are multiplying by powers of 10, which means that you move the decimal point to the right. To make the number smaller, you move the decimal point to the left. In checking your answer, remember that it takes many small things to make one large thing, and vice versa.

 KEY POINT

The meter is the basic unit of length in the metric system.

In 1790, the French Academy of Science defined the **meter** to be one ten-millionth of the distance from the North Pole to the equator—a distance of about 39.37 inches. Since that time, scientists have redefined the meter several times, and currently the meter is the distance that light travels in a vacuum in $\frac{1}{299,792,458}$ of a second. A good way to visualize a meter is that it is slightly longer than a yard stick. Figure 10.57 shows the relationship between millimeters $\left(\frac{1}{1,000} \text{ of a meter}\right)$, centimeters $\left(\frac{1}{100} \text{ of a meter}\right)$, and inches.

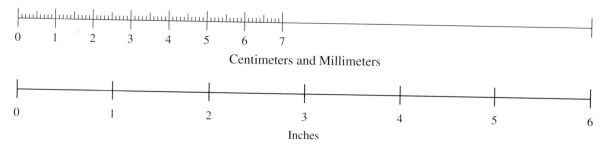

FIGURE 10.57 Comparing millimeters, centimeters, and inches.

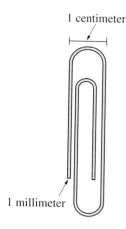

1 centimeter

1 millimeter

From Figure 10.57, you can see that a millimeter is about the thickness of a large paper clip, and the width of the paper clip is roughly one centimeter. One inch is about 2.5 centimeters, and a centimeter is roughly 0.4 inch. We often measure large distances in kilometers (1,000 meters). A kilometer is about 0.6 mile and a mile is approximately 1.6 kilometers.

EXAMPLE 2   *Estimating Metric Lengths*

Match each of the following items with one of these metric measurements: 2 m, 3 cm, 10 m, 16 km.

a) The length of a 30-foot yacht   b) The length of a 10-mile race

c) The length of your bed   d) The thickness of this book

SOLUTION:

a) Thirty feet equals 10 yards, which is about 10 m.

b) One mile is about 1.6 km, so a 10-mile race would be about 16 km long.

c) Your bed is probably a little more than 6 feet long, so it would be approximately 2 m long.

d) If you place the edge of your book on the diagram in Figure 10.57, you will see that the book is around 3 or 4 cm thick. ❀

🖉 KEY POINT

We use dimensional analysis to convert from one system to the other.

| 1 foot = 12 inches |
| 1 yard = 3 feet = 36 inches |
| 1 mile = 5,280 feet |

TABLE 10.6 Some relationships among units in the customary system.

Dimensional Analysis

Often we need to convert from customary units to metric units, or vice versa. Before we explain how to convert from one system to the other, let us examine how we convert one unit of length to another within our customary system. Recall the relationships among various units of length, which we give in Table 10.6.

To convert one quantity to another, we will use unit fractions. A *unit fraction* is a quotient such as $\frac{3\ feet}{1\ yard}$ or $\frac{1\ foot}{12\ inches}$ that has different units of measurement in the numerator and denominator and whose value is equal to 1. We will show you how to use unit fractions to make conversions in Example 3.

EXAMPLE 3   *Converting Yards to Inches*

Convert 5 yards to inches.

SOLUTION: We want to have an answer that is in inches instead of yards, so we will multiply by the unit fraction $\frac{36\ inches}{1\ yard}$.* Thus,

$$5\ \cancel{yards} \times \frac{36\ inches}{1\ \cancel{yard}} = 5 \times 36\ inches = 180\ inches. ❀$$

---

*Notice that because we wanted the final answer in terms of inches, we put "inches" in the numerator.

1 inch = 2.54 centimeters
1 foot = 30.48 centimeters
1 yard = 0.9144 meter
1 mile = 1.6 kilometers

TABLE 10.7 Some basic relationships between customary and metric units of length.

You can use dimensional analysis to make conversions between the customary and metric systems, but first you need to know some basic relationships* between units of length in the two systems, which we give in Table 10.7.

From Table 10.7, we see that there are a number of unit fractions that we can use to make conversions such as $\frac{1\ inch}{2.54\ centimeters}$ and $\frac{0.9144\ meters}{1\ yard}$.

EXAMPLE 4 *Converting Between the Metric and Customary Systems*

Make each of the following conversions:

a) 5 yards to centimeters   b) 0.3 kilometer to feet

SOLUTION:

a) To make this conversion, we need a unit fraction that will convert yards to meters and then another to convert meters to centimeters.[†] Multiplying by the unit fraction $\frac{0.9144\ meter}{1\ yard}$ will convert yards to meters, and because 1 meter = 100 *centimeters*, we can use the unit fraction $\frac{100\ centimeters}{1\ meter}$ to finish the conversion. Thus,

$$5\ yards = 5\ \underset{\text{converts yards to meters}}{\cancel{yards} \times \frac{0.9144\ \cancel{meter}}{1\ \cancel{yard}}} \times \underset{\text{converts meters to centimeters}}{\frac{100\ centimeters}{1\ \cancel{meter}}}$$
$$= 5 \times 0.9144 \times 100\ centimeters = 457.2\ centimeters.$$

b) We will solve this problem similarly to the way we solved part a). We need one unit fraction, $\frac{1\ mile}{1.6\ kilometers}$, to convert kilometers to miles and then another, $\frac{5,280\ feet}{1\ mile}$, to convert miles to feet. Therefore,

$$0.3\ kilometer = 0.3\ \underset{\text{converts kilometers to miles}}{\cancel{kilometer} \times \frac{1\ \cancel{mile}}{1.6\ \cancel{kilometers}}} \times \underset{\text{converts miles to feet}}{\frac{5,280\ feet}{1\ \cancel{mile}}}$$
$$= \frac{0.3 \times 5,280}{1.6} = 990\ feet.$$

Now try the Exercises 27 to 40 that deal with length. ❉ ⑱

Quiz Yourself ⑱

Convert 0.65 kilometers to feet.

✏️ **KEY POINT**

The liter is the basic unit of volume in the metric system.

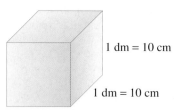

FIGURE 10.58 1 liter = 1 cubic decimeter = 1,000 cubic centimeters.

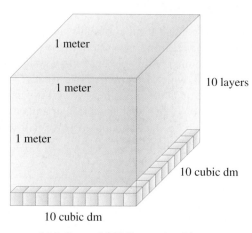

1 kiloliter = 1,000 liters = 1 cubic meter

FIGURE 10.59 Representing a cubic meter in terms of liters.

The **liter** is defined to be 1 cubic decimeter (see Figure 10.58). That is, a liter is the volume of a cube that measures 1 decimeter (or 10 centimeters) on each side. From Figure 10.58, you can see that a liter is 10 × 10 × 10 = 1,000 cubic centimeters.

As we mentioned earlier, a liter is slightly more than a quart. We can use the same prefixes (kilo-, hecto-, deka-, and so on) that we used in measuring length. We use a milliliter, $\frac{1}{1,000}$ of a liter, to measure small quantities, such as a dose of medicine. If you go to a clinic to get a flu shot, you might receive 2 milliliters of vaccine. We measure large volumes in kiloliters (1,000 liters). Figure 10.59 shows that a

---

*You can find more accurate approximations for these relationships in many science books and encyclopedias.
[†]Of course, we could convert meters to centimeters by moving the decimal point as we did in Example 1. Either approach will give you the same answer.

kiloliter is the same as a cubic meter. Each layer in Figure 10.59 contains $10 \times 10 = 100$ cubic decimeters, and because there are 10 layers, we have 1,000 cubic decimeters. In Canada, you might order 20 cubic meters of cement to build a patio floor. We also use cubic centimeters, cm^3 or cc, to measure dry materials.

EXAMPLE 5   *Estimating Metric Volumes*

Estimate the volume of each of the following items with one of these metric measurements: 3 mL (milliliters), 250 mL (milliliters), 100 L (liters), 500 m³ (cubic meters), 1 dm³ (cubic decimeter).*

a) The gas tank of a car   b) A large pile of gravel at a construction site

c) A glass of soda   d) A quart of orange juice

SOLUTION:

a) 100 liters is about 100 quarts, or 25 gallons, which is the size of the gas tank of a medium-size car.

b) 500 cubic meters is about 500 cubic yards, which is the size of a fairly good-size pile of gravel.

c) A glass of soda would be about $\frac{1}{4}$ of a quart. Because a quart is roughly equal to a liter and a liter is 1,000 milliliters, the glass of soda has a volume of about $\frac{1,000}{4} = 250$ milliliters.

d) A quart is roughly a liter and a liter is 1 cubic decimeter, so the volume of the orange juice is about 1 cubic decimeter. ✿

We make conversions among units of volume in the metric system in exactly the same way we made conversions among units of length.

EXAMPLE 6   *Estimating the Amount of Vaccinations*

The East Side Community Health Clinic has on hand four bottles that each contain 0.35 L of flu vaccine. Each vaccination requires 2 mL of vaccine. How many patients can be treated with the amount of vaccine that the clinic has on hand?

SOLUTION: The total amount of vaccine is $4 \times 0.35 = 1.4$ L. From Table 10.5, we see that to convert liters to millimeters, we can move the decimal point three places to the right. Therefore, the clinic has 1,400 mL of vaccine on hand, so it can give $\frac{1,400}{2} = 700$ vaccinations. ✿

You can use dimensional analysis to make conversions between units of volume in the metric and customary systems, but first you need to know some basic relationships, which we give in Table 10.8.

2 cups = 1 pint	1 cup = 0.2366 liter
2 pints = 1 quart	1 quart = 0.9464 liter
32 fluid ounces = 1 quart	1 cubic foot = 0.03 cubic meter
4 quarts = 1 gallon	1 cubic yard = 0.765 cubic meter

TABLE 10.8 Relationships between units of volume in the metric and customary systems.

You can now use these relationships to define unit fractions to make conversions from one system to the other as we did earlier in this section.

---

*We will denote a liter by L to avoid confusion between lowercase "el" and the numeral 1.

❋ ❋ ❋ HIGHLIGHT

Between the Numbers—But Officer—I Only Had . . .

In most states, you are considered legally drunk and unfit to drive if your blood alcohol content (BAC) is above 0.08. Now what exactly does that mean? The 0.08 stands for 0.08 gram per 100 milliliters of blood.

To put this in context, two beers contain about 30 grams of alcohol and an average-sized man has about 5 liters, or 5,000 milliliters of blood. So dividing 30 by 5,000, we get $\frac{30}{5,000} = 0.006$ gram of alcohol per milliliter in your blood. So in 100 milliliters, there would be $100 \times 0.006 = 0.6$ gram of alcohol. Actually, if this amount of alcohol were absorbed immediately, it would kill you—fatalities often occur with a BAC between 0.4 and 0.5. But, because the alcohol is absorbed over a period of time and the liver also works to eliminate the alcohol, we would not expect two beers to be fatal.

You can find BAC calculators on the Internet that will estimate your blood alcohol content based on your weight, what you are drinking, and so on. When I did this, I was thinking about the custom that some college students have of taking 21 drinks on their 21st birthday. When I tried to calculate my BAC if I had 21 drinks in 6 hours, the calculator responded by saying, "This is not a reasonable number of drinks, enter another number." So I entered 10 drinks for a 6 hour period and the estimated BAC was 0.22, or about three times the legal limit. If we were to double this to 20 drinks, it is not unreasonable to expect a BAC above 0.4, which is in the potentially fatal range.

EXAMPLE 7 *Calculating the Volume of a Car's Gas Tank*

Assume that an American car's gas tank has a capacity of 23 gallons. Supposing that you were driving this same car in France, how many liters of gasoline would you need to fill the tank?

SOLUTION: Looking at the shaded boxes in Table 10.8, we see that we can first use the unit fraction $\frac{4\ quarts}{1\ gallon}$ to convert gallons to quarts and then use the unit fraction $\frac{0.9464\ liter}{1\ quart}$ to convert quarts to liters. Thus,

$$\text{converts gallons to quarts} \qquad \text{converts quarts to liters}$$

$$23\ gallons = 23\ \cancel{gallons} \times \frac{4\ \cancel{quarts}}{1\ \cancel{gallon}} \times \frac{0.9464\ liter}{1\ \cancel{quart}}$$

$$= 23 \times 4 \times 0.9464 = 87.07\ liters.$$

So you would need to purchase 87.07 L of gasoline.

Now try Exercises 27 to 40 that deal with volume. ❋

KEY POINT

The gram is the basic unit of mass (weight) in the metric system.

The last type of measurement we will consider is mass, which, for simplicity's sake, you might think of as weight. Strictly speaking, mass and weight are not the same thing. The mass of an object depends on its molecular makeup and does not change. The weight of an object, however, depends on the gravitational pull on that object. Gravity is stronger on a large planet and weaker on a small planet. Because the moon's gravity is only $\frac{1}{6}$ of Earth's gravity, an elephant that weighs 2,400 pounds on Earth would weigh only 400 pounds on the moon. However, the mass of the elephant would be the same on both Earth and the moon.

The basic unit of mass in the metric system is the **gram**, which is defined to be the mass of 1 cubic centimeter (1 milliliter) of water at a certain specified temperature and pressure. A liter (roughly 1 quart) is 1,000 mL, so a liter of water has a mass of 1,000 g or 1 kg. It is common in the metric system to measure the mass of a large object in kilograms; 1 kilogram is approximately 2.2 pounds. So, if you were having a barbecue in Spain and wanted to broil about 8 or 9 quarter-pounders, you would go to the market and buy 1 kg of ground beef.

The pattern of prefixes, and abbreviations, and also the rules for making conversions are the same for mass as they were for length and volume.

EXAMPLE 8 *Estimating Mass in the Metric System*

Estimate the mass of each of the following items with one of these metric measurements: 1 g, 50 cg, 5 kg, 1,000 kg.

a) A decent-size steak   b) A large bag of sugar

c) A medium-size car   d) A large paper clip

SOLUTION:

a) The steak might weigh a pound or more. A kilogram is 2.2 pounds, so the steak would be about one-half of a kilogram, or 50 cg.

b) Typically, we buy sugar in 10-pound bags; 5 kg would be equal to $5 \times 2.2 = 11$ pounds, so the sugar would weigh a little less than 5 kg.

c) A car might weigh 2,200 pounds, which is 1,000 kg.

d) A paper clip is pretty light, so its weight might be about 1 g. ❊

To make mass conversions between the metric and customary systems, you can use the information in Table 10.9.

16 ounces = 1 pound	1 ounce = 28 grams
2,000 pounds = 1 ton	2.2 pounds = 1 kilogram
	1 metric tonne = 1,000 kilograms
	1.1 tons = 1 metric tonne

TABLE 10.9 Relationships between units of mass in the metric and customary systems.

Again, we can use dimensional analysis to convert units in one system to units in the other.

EXAMPLE 9 *Refurbishing a Church with Italian Marble*

A contractor who is refurbishing a church has ordered slabs of Italian marble that weigh 1.8 metric tonnes each. If he wants to rent a crane to lift the slabs of marble, how much, in pounds, should the cranes be able to lift?

SOLUTION: We will convert metric tonnes to tons and then tons to pounds. The unit fraction $\frac{1.1\ tons}{1\ metric\ tonne}$ converts metric tonnes to tons and the unit fraction $\frac{2,000\ pounds}{1\ ton}$ will convert tons to pounds. Therefore,

$$1.8\ metric\ tonnes = 1.8\ \overline{metric\ tonnes} \times \frac{1.1\ \overline{tons}}{1\ \overline{metric\ tonne}} \times \frac{2,000\ pounds}{1\ \overline{ton}}$$

$$= 1.8 \times 1.1 \times 2,000 = 3,960\ pounds$$

Converts metric tonnes to tons ╲   Converts tons to pounds

A crane that can lift 4,000 pounds would just barely do the job.

Now try Exercises 27 to 40 that deal with weight. ❊ **19**

We will use the equivalents between the metric and customary systems found in the table on the left to do the following exercises.

Quiz Yourself **19**

Convert 3.8 kg to ounces.

1 meter = 1.0936 yards
1 mile = 1.609 kilometers
1 pound = 454 grams
1 metric tonne = 1.1 tons
1 liter = 1.0567 quarts*

---

*Note that we are using this equivalence in the exercises, rather than using the equivalence 1 quart = 0.9464 liter that we stated earlier.

Exercises ⎡10.5⎤

Looking Back*

These exercises follow the general outline of the topics presented in this section and will give you a good overview of the material that you have just studied.

1. How did we use Table 10.5 in solving Example 1, part (a)?

2. How might you remember the meaning of the metric prefixes "centi-," "milli-," and "deci-?"

3. In Example 4, what unit fraction did we use to convert miles to kilometers?

4. What do you deduce from the Highlight regarding blood alcohol content?

Sharpening Your Skills

In Exercises 5–14, match the italicized words in the left column with the metric measurement in the right column that corresponds to it. Before calculating your final answer, it would be wise to first convert the given measurement to basic metric units such as meters, grams, or liters.

5. A *quarter-pound* hamburger   **a.** 0.02159 dam

6. A *gallon* of milk   **b.** 946.3 mL

7. A *15-foot* tall giraffe   **c.** 378.5 cl

8. *Five pounds* of potatoes   **d.** 1.77 dL

9. A *6-inch*-long ruler   **e.** 4,572 mm

10. A *quart* of motor oil   **f.** 0.0061 km

11. A book *eight and one-half inches* wide   **g.** 15.24 cm

12. A *12-ounce* gerbil   **h.** 1,135 dg

13. *Six ounces* of orange juice   **i.** 3.405 hg

14. A *20-foot*-long swimming pool   **j.** 2.27 kg

Make the indicated conversions in Exercises 15–20.

15. 2.4 kiloliters to deciliters

16. 240 centigrams to dekagrams

17. 28 decimeters to millimeters

18. 5.6 hectograms to centigrams

19. 3.5 dekaliters to deciliters

20. 7,600 centimeters to meters

Pick the most appropriate measurement for each of the following items. Explain your answer.

21. The volume of a small juice glass

    a. 125 mL   **b.** 500 mL   **c.** 2.5 L

22. The weight of books in your backpack

    a. 500 g   **b.** 80 hg   **c.** 6 kg

23. The length of your nose

    a. 4 dm   **b.** 3 mm   **c.** 5 cm

24. The volume of a bottle of wine

    a. 0.25 L   **b.** 0.2 kL   **c.** 750 mL

25. The height of a Boston terrier

    a. 100 mm   **b.** 0.06 dam   **c.** 0.6 hm

26. The weight of NFL lineman Orlando Pace

    a. 136 kg   **b.** 13,000 g   **c.** 20 dag

Use dimensional analysis[†] to make each of the following conversions. You may have to define several unit fractions to make the conversions. Also, because the constants that we give in the table are only approximations, depending on the method with which you do your computations, your answers may differ slightly from ours.

27. 18 meters to feet

28. 27 gallons to liters

29. 3 kilograms to ounces

30. 10,000 milliliters to quarts

31. 2.1 kiloliters to gallons

32. 47 pounds to kilograms

33. 10,000 deciliters to quarts

34. 507,820 milligrams to pounds

35. 176 centimeters to inches

36. 3 yards to millimeters

37. 45,000 kg to tons

38. 0.65 tonne to pounds

39. 2.6 feet to decimeters

40. 10 tons to tonnes

Applying What You've Learned

Rewrite each statement, replacing the metric measure by the corresponding customary measure.

41. Hold your ground. Don't give him 2.54 centimeters.

42. Tex was wearing a 9.5-liter hat.

43. It is first down and 9.14 meters to go.

44. 28 grams of prevention is worth 0.45 kilogram of cure.

45. **Volume of a tank.** A rectangular tank is 5 m wide, 8 m long, and 4 m deep.

    a. What is the volume of the tank in cubic meters?

    b. How many liters of water does the tank contain?

    c. What is the weight of the water in kilograms?

46. **Painting a wall.** If a rectangular wall of a gymnasium that measures 2.5 m by 30.8 m requires 8 L of paint, then how much paint would be needed to cover a wall that measures 3.2 m by 26.4 m?

---

*Before doing these exercises, you may find it useful to review the note *How to Succeed at Mathematics* on page xix.

[†]In some cases, you may find it quicker not to use dimensional analysis to get your answers.

47. Converting speed. In *The Amazing Race*, Rachel is driving a car that shows speed in both miles per hour and kilometers per hour. If she is traveling in Italy at 80 kilometers per hour, what is her speed in miles per hour?

48. Converting speed. TK is driving in Switzerland at a speed of 55 miles per hour. What is his speed in kilometers per hour?

49. Finding the volume of a swimming pool. Adrian's rectangular swimming pool is 20 ft wide, 40 ft long, and averages 6 ft in depth. How many kiloliters of water are needed to fill the pool?

50. Buying canned food. Marco purchased a large can of chili that has a diameter of 10 cm and a height of 14 cm. Determine the volume of the chili in the can.

    a. in liters.   **b.** in ounces.

51. Purchasing fruit in Europe. While visiting Poland, Anthony purchased some red plums that cost $2.75 per kilogram. What is the cost of the plums per pound?

52. Measuring a medication. Justin is taking the anti-inflammatory drug Niamoxin and the instructions say that the patient must receive 10 mg for each 15 kg of body weight. If Justin weighs 275 lb, what dosage should he receive?

53. Buying gasoline. If gasoline costs $2.18 per liter in Gambia (Africa), what is its cost in dollars per gallon?

54. Buying gasoline. Europeans pay very heavy taxes on gasoline. When I was writing these exercises, gasoline in Germany cost $8.00 per gallon. What would the cost be in dollars per liter?

55. Fencing a dog pen. Serina wants to fence in a rectangular exercise pen for her golden labrador retriever. The pen is 35 ft wide and 62 ft long. The fencing is sold in whole meters. How much fencing should she buy?

56. In Exercise 55, what is the area of Serina's pen in square meters?

57. Calculating gas mileage. If a car averages 30 miles per gallon, how many kilometers per liter would that be?

58. Calculating gas mileage. If a car averages 15 kilometers per liter, how many miles per gallon would that be?

59. Buying flooring. If oak flooring costs $8.00 a square foot, how much is that cost per square meter?

60. Buying flooring. If vinyl flooring costs $98.00 a square meter, how much is that per square foot?

In the metric system, temperatures are measured using the Celsius (also called centigrade) thermometer instead of Fahrenheit that we commonly use. On the Celsius scale, water freezes at 0 degrees and boils at 100 degrees. To convert a temperature using one thermometer to a temperature using the other, you can use the following equation:

$$F = \frac{9}{5}C + 32, *$$

where F is the Fahrenheit temperature and C is the Celsius temperature. Use this equation in Exercises 61–68 to convert each temperature to a corresponding measurement in the other system.

61. 149 degrees Fahrenheit   **62.** 95 degrees Fahrenheit

63. 60 degrees Celsius   **64.** 85 degrees Celsius

65. 20 degrees Celsius   **66.** 131 degrees Fahrenheit

67. 113 degrees Fahrenheit   **68.** 50 degrees Celsius

In Exercises 69–72, use the fact that a hectare (pronounced "HEKtaire") is a square that measures 100 meters on each side.

69. What is the relationship between hectares and square meters?

70. What is the relationship between hectares and square kilometers?

71. Thiep purchased a rectangular piece of land that measures 0.75 km by 1.2 km. How many hectares of land is that?

72. If Ivanka purchased a rectangular piece of land that contains 40 hectares and is 0.65 km long, how long is the other dimension?

Communicating Mathematics

73. What is the fundamental base for expressing measurements in a metric system?

74. In Table 10.5, an item in one column represents how many items in the column to its right?

75. If you were converting hectometers to decimeters, would you have more hectometers or more decimeters? Why?

76. What metric units do you use to measure mass? Volume? Length?

77. In converting milligrams to dekagrams, we would move the decimal point either four places to the right or to the left. Which is it? Explain how you would help a fellow classmate remember which to do.

78. If *a* kiloliters equals *b* dekaliters, which is larger, *a* or *b*? Explain how you arrived at your answer.

79. Explain the advantages that you see in using the metric system over the customary system. Be specific. Give concrete examples.

80. Why are conversions easy to do in the metric system? Give specific examples.

Using Technology to Investigate Mathematics

81. Find a blood alcohol content calculator on the Internet. Experiment with different scenarios and write a brief report on your findings.

82. Find an interactive Web site that converts measurements in the U.S. customary system into the metric system. Reproduce some of the conversions from this section. Write a report on your findings.

For Extra Credit

83. Research the history of the definition of an inch. Explain the results of your research.

---

*This equation converts degrees Celsius to degrees Fahrenheit. You can also use it (plus algebra) to convert Fahrenheit to Celsius, which we recommend that you do. If you really want another equation to convert Fahrenheit to Celsius, you can use the equation $C = \frac{5}{9}(F - 32)$. However, we strongly recommend that you memorize only one equation and use algebra to do the second conversion.

84. Research the history of the definition of a yard. Explain the results of your research.

85. Research the definitions of avoirdupois and troy weight. Use your research to explain why a pound of silver does not weigh the same as a pound of steak.

86. In 1870, Jules Verne wrote his famous novel *20,000 Leagues Under the Sea.* What is a league? If the submarine in Jules Verne's book traveled a distance of 20,000 leagues under water, how far did it travel? Give your answer in feet.

10.6 Geometric Symmetry and Tessellations

Objectives

1. Understand the relationship between symmetry and rigid motions.
2. Recognize the symmetries of an object.
3. Determine when it is possible to tessellate a plane with polygons.

At what age do you think a child becomes aware of beauty? Two years? Four years? Wouldn't you think that a child has to live in a society for a while before *learning* what is considered to be beautiful?

Surprisingly, this is not so. Psychologist Judith Langlois of the University of Texas, Austin, has found that babies seem to have a "built-in" appreciation for beauty that agrees with adults' concept of beauty. In her experiments, she found that children as young as 3 months will stare longer at an "attractive" female face, than at one who is not as "attractive." She repeated her experiments with caucasian females and males, Afro-American females, and even faces of other babies, and always got the same results.*

So, just what *are* the babies recognizing? It's the topic of this section—symmetry.

Rigid Motions

Although you no doubt have an intuitive idea of what we mean by symmetry, it is not an easy concept to describe. In mathematics, we begin with an intuitive concept, such as symmetry, and define it precisely so that we can measure it and calculate with it. For example, in Figure 10.60, it is easy to see that the star has more symmetry than the arrowhead. But exactly what do we mean when we say that?

To understand symmetry, imagine that the arrowhead is made of a very stiff wire and is resting in shallow grooves cut into a piece of wood. Now suppose that you close your eyes and a friend picks up the arrowhead, flips it over, and returns it to its resting place. Figure 10.61(a)

arrowhead

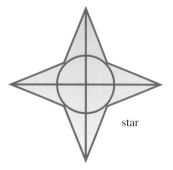

star

A   B   B   A
(a)   **(b)**

FIGURE 10.60 Objects possessing different amounts of symmetry.

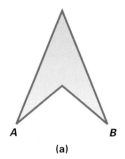

FIGURE 10.61 The arrowhead has been flipped over and returned to its resting place.

*For a further discussion of this topic and related links, see www.symonics.com/sci_balancing.html.

shows the arrowhead before the flip, and Figure 10.61(b) shows the arrowhead after the flip. Although many points on the arrowhead have been moved, the overall appearance of the arrowhead remains unchanged. When you open your eyes, you would not know whether the arrowhead was flipped. Other than this flip, there is really nothing else that we can do to the arrowhead that will move individual points yet keep the overall appearance of the object the same.

On the other hand, there are many different ways to move the star that will change the position of individual points but leave the overall appearance of the star the same. Figure 10.62 shows two possibilities. In Figure 10.62(b), we have rotated the star in a counterclockwise direction so that A moves to the bottom of the star, B moves to what was A's position, C moves to what was B's position, and so on. In Figure 10.62(c), we have flipped the original star over, interchanging vertices A and C.

Figures 10.61 and 10.62 show examples of rigid motions.

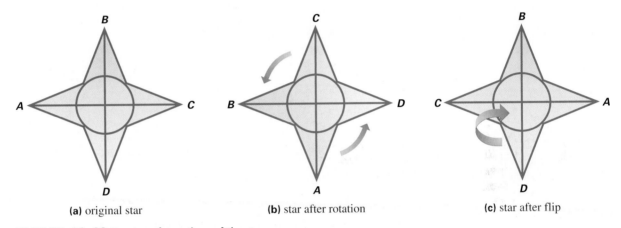

| (a) original star | (b) star after rotation | (c) star after flip |

FIGURE 10.62 Two transformations of the star.

> **DEFINITION** A **rigid motion** is the action of taking a geometric object in the plane and moving it in some fashion to another position in the plane without changing its shape or size.

When we consider a rigid motion, we are interested only in the beginning and ending position of the object. For example, suppose that in flipping the arrowhead, your friend drops it and it rolls across the floor. If your friend retrieves it and places it back in the grooves in the wood, most of what has happened to the arrowhead is irrelevant. All that matters mathematically is that from the beginning to the end of the process, the arrowhead has been flipped over.

You might think that there are many different ways that we could perform rigid motions; however, this is not the case.

KEY POINT

There are essentially only four possible rigid motions.

> Every rigid motion is essentially a reflection, a translation, a glide reflection, or a rotation.

By the word *essentially*, we mean that if we consider only the beginning and ending positions of the object, and ignore the intermediate motions, the rigid motion can be accomplished in only one of four ways.

We first discuss reflections.

> **DEFINITION** A **reflection** is a rigid motion in which we move an object so that the ending position is a mirror image of the object in its starting position.

In a reflection, there is a line, called the *axis of reflection*, that acts as a mirror to transform the figure from its original position to its final position. We say that the original figure has been *reflected about the axis of reflection* to produce the final figure. We show two reflections in Figure 10.63.

As in Figure 10.63, we label the vertices of the original polygons with A, B, C, and so on, and the vertices of the reflected polygons with A', B', C', and so on.

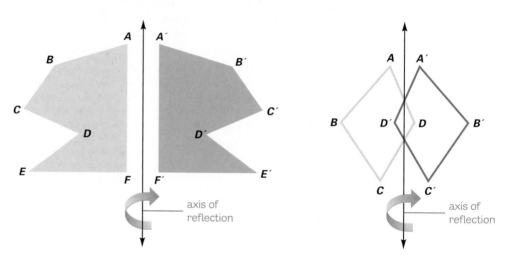

FIGURE 10.63 Original figures *ABCDEF* and *ABCD*; reflected figures *A'B'C'D'E'F'* and *A'B'C'D'*.

Quiz Yourself **20**

Reflect the quadrilateral *ABCD* about line *l*.

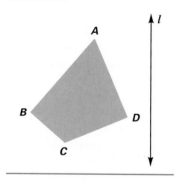

EXAMPLE 1 *Reflecting a Geometric Object*

Reflect polygon *ABCDE* about the axis of reflection *l* in Figure 10.64.

SOLUTION: In order to reflect the polygon about line *l*, we must reflect each vertex A, B, C, D, and E about *l*.

1st: Reflect A about *l* by drawing a line segment from A to A' that is perpendicular* to *l* and also so that the distance from A to *l* (in blue) is the same as the distance from A' to *l* (in red).

2nd: Draw segments *BB'*, *CC'*, *DD'*, and *EE'* in a similar way as shown in Figure 10.65.

3rd: Draw the reflected polygon by connecting vertices A', B', C', D' and E' as shown in Figure 10.66.

Now try Exercises 5 to 12. ❃ **20**

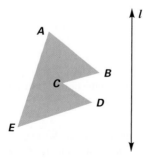

FIGURE 10.64 Polygon *ABCDE* is to be reflected about line *l*.

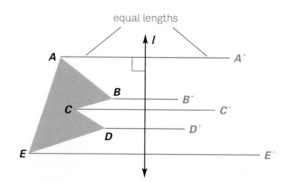

FIGURE 10.65 Reflecting points A, B, C, D, and E about *l*.

*Although we have not explained how to construct a line that is perpendicular to another line, you can draw a perpendicular freehand to get an idea of what the reflection looks like.

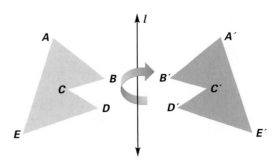

FIGURE 10.66 Reflecting polygon *ABCDE* about line *l*.

> **DEFINITION** A **translation** is a rigid motion in which we move a geometric object by sliding it along a line segment in the plane. The direction and length of the line segment completely determine the translation. We represent the distance and direction of a translation by a line segment with an arrow on it, called the **translation vector**.

Figure 10.67 shows a translation.

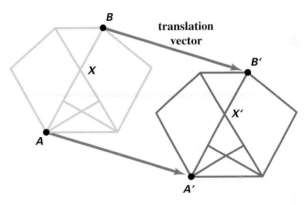

FIGURE 10.67 A translation vector determines a translation of an object.

Now try Exercises 15 to 18.

> **DEFINITION** A **glide reflection** is a rigid motion formed by performing a translation (the glide) followed by a reflection.

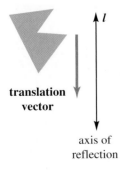

translation vector

axis of reflection

FIGURE 10.68
Forming a glide reflection of an object.

EXAMPLE 2 *Producing a Glide Reflection of a Geometric Object*

Use the translation vector and the axis of reflection to produce a glide reflection of the object shown in Figure 10.68.

SOLUTION: We will accomplish this glide reflection as follows:

1st: Place a copy of the translation vector at some point, say *Y*, on the object (Figure 10.69(a)).

2nd: Slide the object along the translation vector so that the point *Y* coincides with the tip of the translation vector (Figure 10.69(b)).

3rd: Reflect the object about the axis of reflection to get the final object Figure 10.69(c)).

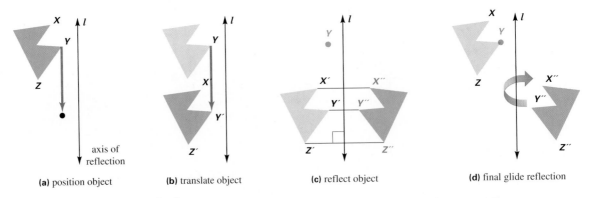

(a) position object   **(b)** translate object   **(c)** reflect object   **(d)** final glide reflection

FIGURE 10.69 A glide reflection.

Quiz Yourself **21**

Perform a glide reflection by first translating the triangle using the given translation vector, then reflecting the translated triangle about the axis of reflection.

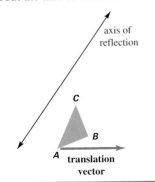

Figure 10.69(d) shows the final effect the glide reflection has in moving the original polygon to the polygon with vertices *X″*, *Y″*, and *Z″*.

Now try Exercises 19 and 20. ✻ **21**

PROBLEM SOLVING
The Order Principle

The Order Principle in Section 1.1 tells you to be careful about the order in which you perform a glide reflection. Performing a translation and then a reflection is not the same as if we reflect first and then translate.

The last rigid motion we will study is a rotation.

DEFINITION We perform a **rotation** by first selecting a point, called the *center of the rotation*, and then, while holding this point fixed, we rotate the plane about this point through an angle called the *angle of rotation*.

A good way to think of a rotation is to imagine that the plane is a piece of paper. If you stick a pin in the plane at the center of rotation and then rotate the plane, the plane will turn about the pin, and all points in the plane will move except the point where the pin is placed. We show a rotation in Figure 10.70.

Now try Exercises 21 to 24.

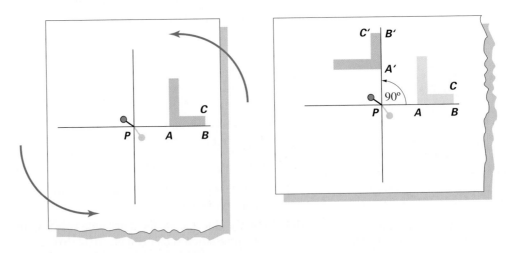

FIGURE 10.70 A rotation of 90° about point *P*.

 *KEY POINT*

A symmetry is a rigid motion that leaves the overall appearance of an object unchanged.

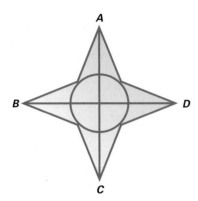

FIGURE 10.71 The star has many symmetries.

Symmetries

We can now define the concept of symmetry using the idea of a rigid motion.

> **DEFINITION** A **symmetry** of a geometric object is a rigid motion such that the beginning position and the ending position of the object by the motion are exactly the same.

Looking back at Figure 10.62, we can say that the rotation and flip are two examples of symmetries of the star. The rotation is of course a rotation as we defined earlier, whereas the flip is a reflection.

EXAMPLE 3   *Symmetries of a Star*

Find two symmetries (other than those shown in Figure 10.62) of the star in Figure 10.71.

SOLUTION: There are clearly many ways to reflect and rotate the star to produce symmetries of the star. For example, in Figure 10.72, we reflect the star about the line *l*. We call *l* a *line of symmetry* for the star.

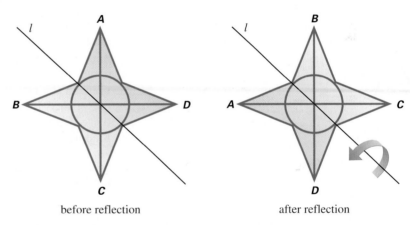

before reflection   after reflection

FIGURE 10.72 Reflection of the star about line *l*.

We can also rotate the star about its center, as shown in Figure 10.73.

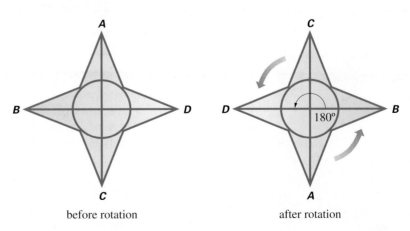

before rotation   after rotation

FIGURE 10.73 Rotation of the star about its center through an angle of 180°.

Now try Exercises 35 to 40. ✳

Math in Your Life

What Makes a Bug Beautiful?

Surprisingly, what makes one insect attracted to another insect is the same thing we mentioned earlier: symmetry. Randy Thornhill and Steven Gangestad, professors at the University of New Mexico, have found that female scorpion flies are attracted to males with symmetrical wings. In another experiment, scientists discovered that by clipping the tail feathers of male swallows, thereby making them less symmetrical, the birds were less attractive to females and less likely to mate.

Biologists believe that animals with a high degree of symmetry have greater genetic diversity, which enables them to withstand environmental stress better and makes them more resistant to parasites. Lower symmetry goes hand in hand with lower survival rates and fewer offspring.

Professor Thornhill worked with Professor Karl Grammer of the University of Vienna to investigate whether symmetry was a factor in human attractiveness. They devised an index to measure facial symmetry with regard to placement of eyes, cheekbones, nose, and several other factors. Their research found that there was indeed a high correlation between facial symmetry and perceived attractiveness.

On the downside, other research suggests that women with asymmetrical breasts have a higher rate of breast cancer, and a study of West Indian men indicates that less symmetrical men are more susceptible to disease then their more symmetrical brothers.

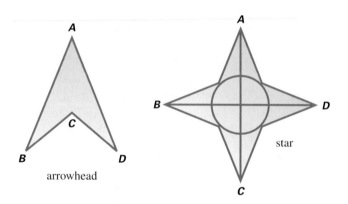

FIGURE 10.74 The star has more symmetries than the arrowhead.

 KEY POINT

Tessellations cover the plane with copies of polygons.

Let's return to our question about measuring the symmetry in the arrowhead and the star. In Figure 10.74, we see that the arrowhead has only two symmetries: a rotation about A through an angle of $0°$ (this symmetry leaves every point in the arrowhead unchanged) and a reflection about the vertical line through points A and C. The star also has these two symmetries as well as many more, as we saw in Example 3.

It can be shown that there are eight symmetries for the star. Using the language of set theory, we say that the set of symmetries of the arrowhead is a proper subset* of the set of symmetries of the star. It is in this sense that the star has greater symmetry than the arrowhead.

Tessellations

We can use rigid motions to place copies of a geometric figure at different positions in the plane. An interesting mathematical question is "Can we begin with a set of polygons and then completely cover the plane with copies of these polygons?"

> **DEFINITIONS** A **tessellation** (or *tiling*) of the plane is a pattern made up entirely of polygons that completely covers the plane. The pattern must have no holes or gaps, and polygons cannot overlap except at their edges. A **regular tessellation** consists of regular polygons of the same size and shape such that all vertices of the polygons touch other polygons only at their vertices.

A designer of wallpaper, a fabric pattern, or a company logo should know that equilateral triangles, squares, and regular hexagons tessellate the plane, as we see in Figure 10.75.

It is natural to ask, "Are there any other regular polygons that tessellate the plane?" To answer this question, recall that in Section 10.2 we stated that for a regular n-sided polygon,

*We discussed subsets in Section 2.2.

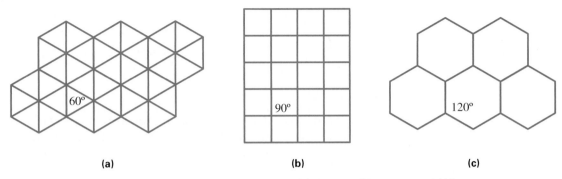

(a)   (b)   (c)

FIGURE 10.75 Regular tessellations of the plane using (a) triangles, (b) squares, and (c) hexagons.

each interior angle has measure $\frac{(n-2) \times 180°}{n}$. For example, the measure of each interior angle of a regular 12-sided polygon would be

$$\frac{(12-2) \times 180°}{12} = \frac{1{,}800°}{12} = 150°.$$

We will use this fact to determine which other regular polygons tessellate the plane.

EXAMPLE 4   *Tessellating the Plane with Regular Polygons*

Which regular polygons can tessellate the plane?

SOLUTION: To understand the problem, let us look at a part of a tessellation using hexagons (Figure 10.76).

Notice that around a vertex of any regular tessellation, we must have an angle sum of 360°, and we also must have three or more polygons of the same shape and size. In Figure 10.75(a), we see that each vertex is surrounded by six 60° angles; in Figure 10.75(b), each vertex is surrounded by four 90° angles.

Let us now consider if it is possible to have a regular tessellation of the plane using pentagons. Recall that interior angles of a regular pentagon measure

$$\frac{(5-2) \times 180°}{5} = \frac{540°}{5} = 108°.$$

If we have three pentagons surrounding the vertex of a tessellation, as shown in Figure 10.77, the sum of the angles around the vertex is $3 \times 108° = 324° < 360°$. We do not have enough pentagons to completely surround the vertex; however, if we include a fourth pentagon, the sum of the angles around the vertex exceeds 360°, which causes the pentagons to overlap. Thus, there cannot be a regular tessellation using pentagons.

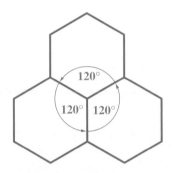

FIGURE 10.76 The angle sum around a vertex in a tessellation must add up to 360°.

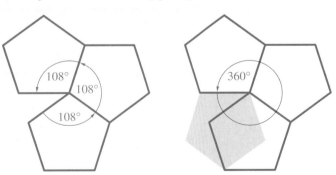

FIGURE 10.77 There is no regular tessellation of the plane using pentagons.

If a regular polygon has more than six sides, the measure of its interior angles exceeds 120°; therefore, we cannot have three such polygons surrounding a vertex of a tessellation.

Thus, we conclude that we can construct only three regular tessellations. They use equilateral triangles, squares, or regular hexagons. ✻

Although there are only three regular tessellations, many tessellations are not regular. Is it possible to construct a tessellation using two different types of polygons that have edges of the same length?

EXAMPLE 5   *Decorating with Nonregular Tessellations*

The Spanish architect Antoni Gaudi specializes in producing decorative mosaics using ceramic tiles. At present, he has a number of tiles shaped like equilateral triangles and squares. All tiles have sides of the same length. Is it possible to produce a tessellation using a combination that contains both types of tiles?

SOLUTION: We can solve this problem by considering the different possibilities systematically. We will investigate tessellations that have one square at a vertex, two squares at a vertex, and so on.

We will begin by trying to construct a tessellation using only one rectangle and the rest triangles; it would start out as shown in Figure 10.78.

Around the vertex 90° would be occupied by the square, and the remaining 270° would be occupied by triangles.

This gives us a problem similar to the one we had with the pentagons in Example 4. We need exactly 270 more degrees. Four triangles occupy only 240°, which is not enough, but five triangles occupy 300°, which is too much. Therefore, we *cannot* construct a tessellation with just one square and triangles as the remaining polygons.

Continuing this type of thinking, we see in Table 10.10 that the only combination of equilateral triangles and squares resulting in a total angle sum at each vertex of 360° occurs when we have two squares and three triangles. Such a configuration at a vertex is shown in Figure 10.79(a). We show a larger part of the tessellation in Figure 10.79(b).

Now try Exercises 45 to 48. ❈ **22**

FIGURE 10.78 One square leaves 270° to be used for triangles.

Number of Squares at a Vertex	Amount of 360° Used by the Squares	Amount of 360° Remaining for the Triangles	Does 60° Divide This Number?	Is This Configuration Possible?
1	90°	270°	No	No
2	180°	180°	Yes	Yes
3	270°	90°	No	No
4	360°	0°	Yes	No (there will be no triangles)

TABLE 10.10 Number of possible equilateral triangle–square combinations possible at the vertex of a tessellation.

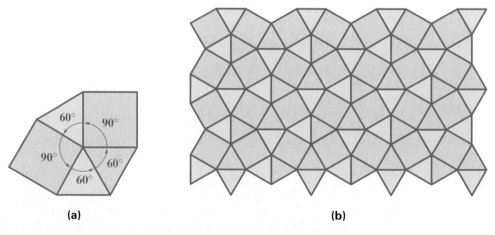

(a)   (b)

FIGURE 10.79 Two squares and three equilateral triangles around the vertex of a tessellation.

Quiz Yourself **22**

Explain why the following tessellation is possible by considering the angles at each vertex in the tessellation.

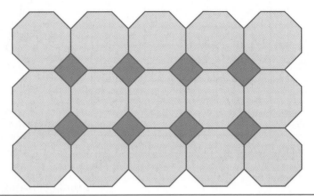

HIGHLIGHT ❈ ❈ ❈

Has All of Mathematics Been Discovered?

It may surprise you to know that mathematicians are still discovering new facts about the topics you studied in this section. You have seen that it is possible to tessellate the plane with certain types of regular polygons and nonregular polygons. For example, although we cannot tessellate the plane with regular pentagons, we can tessellate the plane with the following nonregular pentagon:

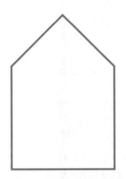

A simple question that a mathematician then asks is, "Exactly what types of pentagons can be used to tessellate the plane?" At one time, it was believed that there were only eight such pentagons. However, Marjorie Rice, a woman with only a high school background in mathematics, found a ninth pentagon that would tessellate the plane. By applying her new methods to this problem, she later went on to discover four more pentagonal tessellations. At this time, mathematicians do not know how many different types of convex pentagons will tile the plane. The example of Marjorie Rice is inspiring in that a person with little formal training in mathematics, but having interest and talent in mathematics, can make a significant contribution.

Exercises 10.6

Looking Back*

These exercises follow the general outline of the topics presented in this section and will give you a good overview of the material that you have just studied.

1. On page 506, we said, "It is in this sense that the star has greater symmetry than the arrowhead." What did we mean by this?

2. How did Figure 10.77 show that there is no regular tessellation of the plane using pentagons?

3. If a regular polygon has more than six sides, why can we not have a tessellation of the plane using that polygon?

4. In the Highlight "What Makes a Bug Beautiful," what seems to be the underlying basis for beauty?

Sharpening Your Skills

Use the figure on page 510 for Exercises 5–8. You may want to use graph paper to solve these exercises.

---

*Before doing these exercises, you may find it useful to review the note *How to Succeed at Mathematics* on page xix.

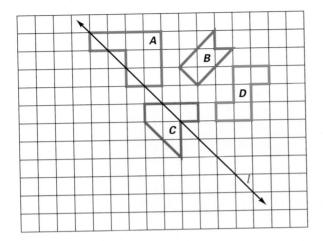

5. Reflect figure *A* about line *l*.

6. Reflect figure *B* about line *l*.

7. Reflect figure *C* about line *l*.

8. Reflect figure *D* about line *l*.

Use the following figure for Exercises 9 and 10.

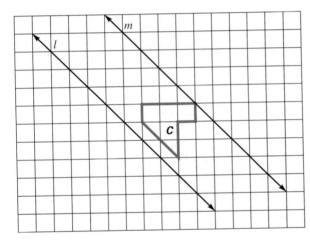

9. Reflect figure *C* about line *l*. Call the resulting figure *C'*. Then reflect figure *C'* about line *m*. Call the resulting figure *C''*.

10. Reflect figure *C* about line *m*. Call the resulting figure *C'*. Then reflect figure *C'* about line *l*. Call the resulting figure *C''*.

Use the following figure for Exercises 11 and 12.

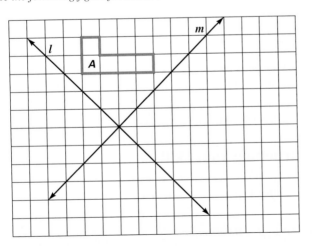

11. Reflect figure *A* about line *l*. Call the resulting figure *A'*. Then reflect figure *A'* about line *m*. Call the resulting figure *A''*.

12. Reflect figure *A* about line *m*. Call the resulting figure *A'*. Then reflect figure *A'* about line *l*. Call the resulting figure *A''*.

13. In Exercises 9 and 10, you were reflecting an object about one line and then another line that was parallel to the first line.

    a. Did the order in which you did the reflections make a difference? Explain.

    b. What is the effect of performing the two reflections on the object?

14. In Exercises 11 and 12, you were reflecting an object about one line and then another line that was perpendicular to the first line. Did the order in which you did the reflections make a difference? Explain.

Use the following figure for Exercises 15–18.

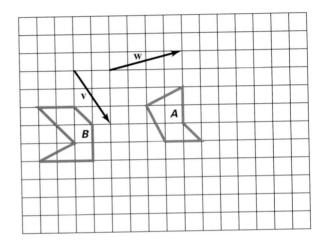

15. Translate figure *A* by vector **w**.

16. Translate figure *B* by vector **v**.

17. Translate figure *A* by vector **v**. Call the resulting figure *A'*. Then translate figure *A'* by vector **w**. Call the resulting figure *A''*.

18. If you reverse the order in which you do the translations in Exercise 17, does it make a difference in the resulting figure *A''*? Explain.

19. Perform the indicated glide reflection on figure *A*.

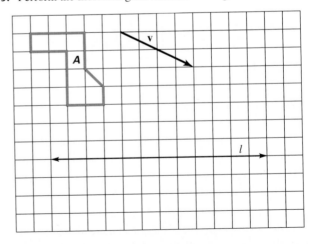

20. Perform the indicated glide reflection on figure *B*.

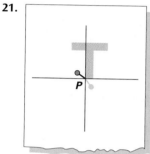

In Exercises 21–24, first rotate the object 45° about point P.
Then rotate the figure 90° about P. *Finally, rotate the object 180°
about* P.

21.

22.

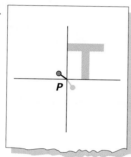

23.

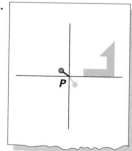

24.

25. What is the interior angle sum of a regular 12-sided polygon?
What is the measure of its interior angles?

26. What is the interior angle sum of a regular 15-sided polygon?
What is the measure of its interior angles?

27. Explain why we can tessellate the plane with a regular hexagon, but not with a regular pentagon.

In Exercises 28–30, tessellate the plane with the given figure.

28.

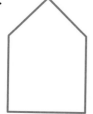

29.

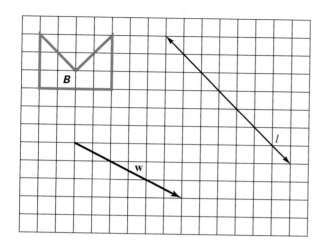

30.

Applying What You've Learned

*The following figure shows eight arrangements of tiles. Some of the
arrangements can be obtained by applying a rigid motion to other
arrangements. For example, we can obtain arrangement (e) by
reflecting arrangement (a) about a vertical axis. Use these arrangements to solve Exercises 31–34.*

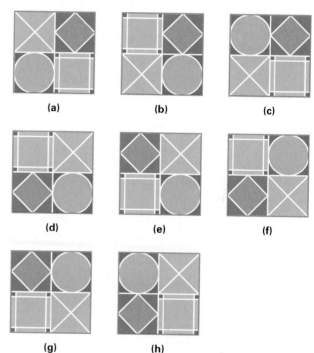

31. List all arrangements that we obtain by reflecting arrangement (b) about a line. (Don't forget reflections about diagonal
lines.)

32. List all arrangements that we obtain by rotating arrangement (g)
about its center.

33. List all arrangements that we obtain by applying a rigid motion
to arrangement (f).

34. Explain why it is impossible to find a rigid motion that we can
apply to arrangement (a) to obtain arrangement (b).

*In Exercises 35–40, list all the reflectional symmetries for each
object. Also find all rotational symmetries of the object using angles
between 1° and 359°.*

35.

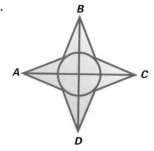

36.

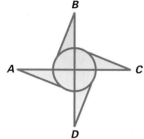

37.

(phone symbol)

38.

(hospital symbol)

39.

(Olympic symbol)

40.

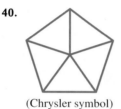

(Chrysler symbol)

Communicating Mathematics

41. What are the four types of rigid motions?

42. If we tessellate the plane with polygons, what is the angle sum around each vertex? Why does this imply that we can tessellate the plane with a hexagon but not a pentagon?

43. What is the difference between a symmetry of an object and a rigid motion?

44. In the Problem Solving box on page 504, we stated that in doing a glide-reflection you must do the glide first and then the reflection because if you reverse these rigid motions, you do not get the same result. Yet, in Example 2, it appears that if we had reflected first and then translated, we would get the same final figure. Is there a contradiction here?

In Exercises 45–48, explain why each tessellation is possible by considering the angles at each vertex in the tessellation, as we did in Quiz Yourself 22.

45.

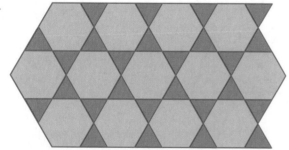

46.

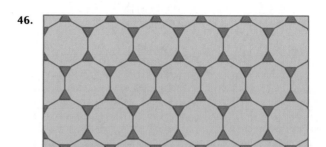

47.

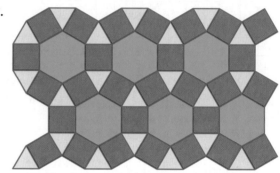

48.

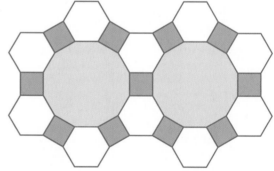

Using Technology to Investigate Mathematics

49. Search the Internet for the terms "mathematical symmetry and applets." You should find many interesting sites. Report on your findings.

50. In researching the Mathematics in Your Life feature on symmetry and beauty, I found many intriguing articles and far too much information to include in the feature. Do a similar search, and write a brief report on your findings.

51. There are many interesting and beautiful Internet sites devoted to tessellations. Some sites have movies describing how to tessellate the plane, and others allow you to experiment with creating your own tessellations. Find an interesting site, experiment with it, and report on your findings.

For Extra Credit

52. Use the given figure to explain why any convex quadrilateral will tessellate the plane.

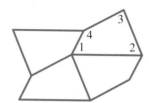

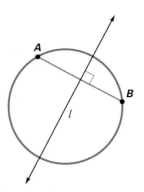

Figure for Exercises 54 and 55

53. Use a drawing similar to the one in Exercise 52 to show that the given quadrilateral tessellates the plane.

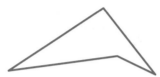

54.

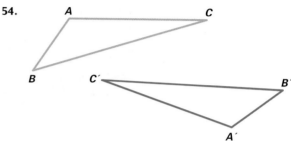

A perpendicular bisector of a chord* of a circle passes through the center of a circle. Because the line 1 divides the segment AB into two equal parts and is also perpendicular to the segment AB, we know that the center of the circle lies somewhere on 1. Use this information to estimate the center of rotation for the rotations shown in Exercises 54 and 55.

55.

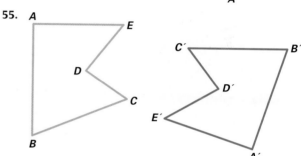

*A *chord* of a circle is a line segment with both endpoints on the circle.

Looking Deeper

10.7 Fractals

Objectives

1. Understand the self-similar nature of fractals.
2. Calculate the length and area of fractal objects.
3. Compute fractal dimension.

Everything in Nature can be viewed in terms of cones, cylinders, and spheres.*
—Paul Cezanne (nineteenth-century impressionist artist)

Clouds are not spheres, mountains are not cones, coastlines are not circles, and bark is not smooth, nor does lightning travel in a straight line.
—Benoit Mandelbrot (retired IBM research mathematician)

*Quotes are taken from Mandelbrot's Web site at Yale University, where he teaches a course in fractals.

So, who's right? If you look around you, it is easy to see that Mandelbrot seems to be describing nature better than Cezanne. Mountain peaks, the ocean shore, and the clouds do not have nice smooth edges as drawn in a child's storybook, but, rather, they have rough, jagged edges that we cannot explain using traditional Euclidean methods. In this section, you will study a relatively new and different type of geometry, called **fractal geometry**, that describes real-life objects and patterns more accurately than we can by using Euclidean geometry and that has many important real-life applications.

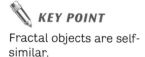

KEY POINT

Fractal objects are self-similar.

Fractals

To understand how fractal geometry differs from Euclidean geometry, imagine a photograph of the edge of a large cloud taken from a weather satellite hundreds of miles away. The edge would not be a smooth curve, as clouds are often drawn in children's books; rather, it would be extremely jagged, as shown in Figure 10.80(a). If we enlarged a small portion of this edge, we would see a jagged curve that looks something like Figure 10.80(b). If we further enlarged a tiny portion of this smaller piece, we would still see an edge containing the same type of jaggedness that was present in the original photograph. With a fractal object, no matter how much we magnify it, we still see patterns that are very similar to the patterns that were present in the original object. We say that an object with this property is *self-similar.*

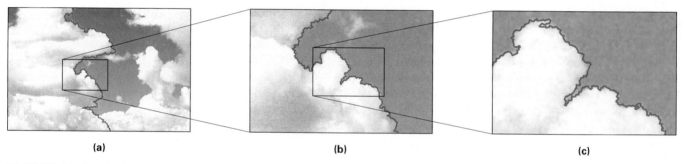

(a)   (b)   (c)

FIGURE 10.80 A cloud is self-similar.

Benoit Mandelbrot developed the theory of fractal geometry while working as a mathematician at IBM during the early 1960s. To get a better understanding of his geometry, we will construct a curve called the *Koch curve.*

EXAMPLE 1   *The Koch Curve Is a Fractal*

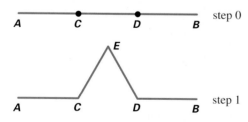

FIGURE 10.81 Beginning the Koch curve.

step 0   We begin the Koch curve by drawing a line segment* AB (step 0), which we divide into three equal parts. At segment CD, we construct an equilateral triangle △CED, and then remove segment CD, giving the object shown as step 1 in Figure 10.81.

step 1   To continue the construction of the curve, we divide each of the four line segments in step 1 into three parts and replace the middle segment by a triangular bump, as we did in going from step 0 to step 1. The resulting curve is shown in step 2 of Figure 10.82. If we repeat this process again for the 16 line segments in step 2, we get the curve shown in step 3 of Figure 10.82.

To finish the Koch curve, we must repeat indefinitely the process of subdividing each line segment and replacing it with a line segment having a triangular bump.

In Figure 10.83, we see why the Koch curve is a fractal. If we enlarge a small portion of the curve, we see that the enlargement has the exact same structure as the original

---

*To avoid cluttering fractal drawings, we will not show the endpoints of line segments.

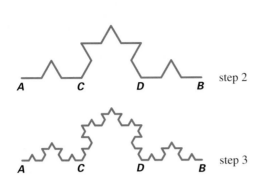

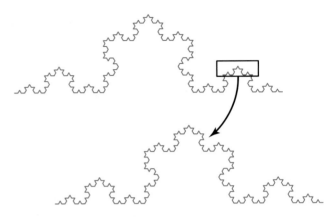

FIGURE 10.82 Steps 2 and 3 of the Koch curve.

FIGURE 10.83 Small portions of the Koch curve repeat the same pattern as the original curve.

Quiz Yourself 23

How many line segments would there be in step 4 of the Koch curve?

FIGURE 10.84 Fractal art.

Quiz Yourself 24

How many dark triangles would appear in step 5 of forming the Sierpinski gasket?

curve. No matter how many times we magnify this curve, we always see the same repeated pattern.

Now try Exercises 5 to 8. ✳ 23

Beautiful fractal art, such as in Figure 10.84, has the same self-similarity property as the Koch curve. If we were able to put the fractal in Figure 10.84 under a microscope, we would see the same beautiful patterns at every level of magnification.

We can begin with a two-dimensional object and, by applying some rule repeatedly, create a fractal, as we see in Example 2.

EXAMPLE 2   *The Sierpinski Gasket Is a Fractal*

We construct a fractal called the *Sierpinski gasket* by first constructing an equilateral triangle, as in Figure 10.85(a). We then divide this triangle into four smaller equilateral triangles and remove the middle triangle, as in Figure 10.85(b). Next, we apply this same rule to each of the three remaining triangles; that is, we divide each triangle into four smaller triangles and remove the center one. We show the results of this step in Figure 10.85(c).

As with the Koch curve, we must continue this process of removing the center of each solid equilateral triangle indefinitely to complete this fractal. Figure 10.86 shows the next step in forming the Sierpinski gasket.

We cannot draw the entire gasket, because to do this, we would have to draw smaller and smaller triangles that eventually become so small that their size would be finer than the resolution of any printing device.

Now try Exercises 27 and 28. ✳ 24

(a) step 0

(b) step 1

(c) step 2

step 3

FIGURE 10.85 The first two steps in forming the Sierpinski gasket.

FIGURE 10.86 Step 3 in forming the Sierpinski gasket.

Length and Area

As in Euclidean geometry, we are interested in the length, area, and volume of fractal objects. When we investigate the length of the Koch curve, we find a surprising result.

EXAMPLE 3   *The Length of the Koch Curve Is Infinite*

Find the length of the Koch curve.

SOLUTION: We began the Koch curve with a line segment that has length 1 (step 0). In step 1 we replaced this curve with a curve $\frac{4}{3}$ as long, as shown in Figure 10.87.

Figure 10.88 shows that in step 2, the curve consists of 16 small line segments with length $\frac{1}{9}$, so the length is now $\frac{16}{9}$.

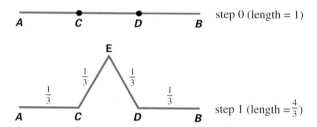

FIGURE 10.87 In step 1, the Koch curve is $\frac{4}{3}$ as long as the original line segment.

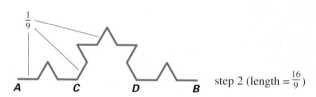

FIGURE 10.88 Each segment now has length $\frac{1}{9}$.

Quiz Yourself **25**

What is the length of the Koch curve at step 3 of the construction process?

At each successive step, the curve is $\frac{4}{3}$ as long as the curve in the previous step. This means that the curve's length keeps growing larger and larger as we construct further steps of the curve. At step 40, we would find that the length of the Koch curve is $\left(\frac{4}{3}\right)^{40}$, which is slightly more than 99,437 units long. At stage 100, the curve is over 3 trillion units long! Of course, we do not stop at stage 100. Because we must go through an infinite number of stages to construct the whole curve, the total length of the Koch curve is therefore infinite. ❋ **25**

EXAMPLE 4   *The Area of the Sierpinski Gasket Is Zero*

What is the area of the Sierpinski gasket?

SOLUTION: To make the computations easier to follow, let us assume that we begin the Sierpinski gasket with an equilateral triangle with area 1 (see Figure 10.89). In step 1 of the construction, we have removed $\frac{1}{4}$ of the area, so the gasket now consists of three triangles, each with area $\frac{1}{4}$. The area of the dark triangles at step 1 is therefore $\frac{3}{4}$.

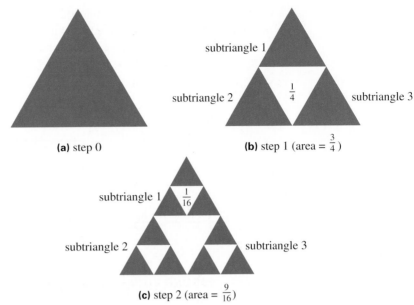

(a) step 0

(b) step 1 (area = $\frac{3}{4}$)

(c) step 2 (area = $\frac{9}{16}$)

FIGURE 10.89 Step 1 of the Sierpinski gasket has area $\frac{3}{4}$; step 2 has area $\frac{9}{16}$.

Now consider subtriangle 1. In step 2, we remove $\frac{1}{4}$ of its area, which is $\left(\frac{1}{4}\right)\left(\frac{1}{4}\right) = \frac{1}{16}$ of the original area, leaving three smaller triangles also with area $\frac{1}{16}$. The area that remains in subtriangle 1 is therefore $\frac{3}{16}$ of the original area. Likewise, $\frac{3}{16}$ of the original area remains in subtriangles 2 and 3 after we remove their centers. Therefore, in step 2 of constructing the gasket, the remaining area is

$$\frac{3}{16} + \frac{3}{16} + \frac{3}{16} = \frac{9}{16} = \left(\frac{3}{4}\right)\left(\frac{3}{4}\right).$$

Quiz Yourself **26**

What is the area of the
Sierpinski gasket at step 3
of the construction?

We see that at each successive step in constructing the gasket, we get an area that is $\frac{3}{4}$ the area of the previous step. Therefore, the area of the gasket keeps getting smaller and smaller with each successive step in the construction. For example, at step 20 the area is $\left(\frac{3}{4}\right)^{20} \approx 0.0032$ square unit. We conclude that the area of the gasket is 0, even though we have not removed all the points in the original triangle. ❈ **26**

Dimension

You may have a feeling that the Koch curve, because of all its "wiggling around," is somewhat thicker than the kinds of curves we draw in Euclidean geometry. Therefore, we might say that the Koch curve, in some sense, has a larger dimension than a smooth curve in Euclidean geometry. On the other hand, all the "wiggling" is not enough for the curve to fill entirely some region of the plane, which would make it a two-dimensional object. In order to make this concept of dimension more clear, consider the line segment, the square, and the cube in Figure 10.90.

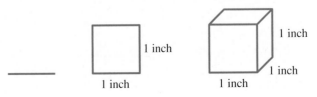

FIGURE 10.90 A unit of measurement in one, two, and
three dimensions.

Imagine that we place the line, the square, and the cube into a three-dimensional copying machine. This copying machine will increase or decrease the size of any object that we place into it. If we set the size of our copies to two times the original, the copies of the line, square, and cube would come out looking as in Figure 10.91.

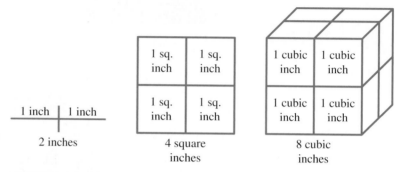

FIGURE 10.91 The line, square, and cube have been enlarged by a factor of 2.

❊ ❊ ❊ HIGHLIGHT

Applications of Fractals

It is surprising that the fractal geometry that produces such strange and beautiful images can also describe many important natural phenomena. Scientists use fractal geometry to explain the similarities and differences between the way a tree branch divides repeatedly to form finer subbranches and the way the bronchi of the lungs subdivide to form a tree of airways inside our lungs. This same sort of branching occurs in a reverse fashion as small tributaries join together to form streams and eventually a river.

Geographers classify the roughness of coastlines according to their fractal dimension. The South African coast is relatively smooth, with a fractal dimension close to 1, whereas the extreme irregularity of Norway's coast has a fractal dimension of 1.52.

Understanding the fractal pattern seen in a lightning bolt also helps explain how electrical insulators break down when subjected to high voltages. This same pattern occurs at a microscopic level when crystals form.

Economists studying the stock market have found that if they graph the market over hundreds of days, then over hundreds of hours, and finally over hundreds of 30-second intervals, the graphs look remarkably similar. Cardiologists have learned that the healthy heart beats according to a fractal rhythm rather than a steady, regular rhythm, as we used to think.

Using percolation theory, mathematicians apply fractal geometry to describe the way coffee percolates in a coffeepot and the way groundwater seeps into the soil. It is surprising that this same area of fractal mathematics also explains how a fire "percolates" through a forest, how an epidemic "percolates" through a population, and how galaxies "percolate" throughout the universe.

Figure 10.92 shows an interesting fractal similarity between a satellite view of the rivers of Norway and the growth of ice crystals on a window pane.

FIGURE 10.92 (a) Rivers of Norway. (b) Ice crystals.

A general way of looking at this is to say we used a scaling factor of s ($s = 2$ in this case). In the one-dimensional case, the copier returned a copy that had length equal to $s^1 = 2$ times the length of the original. In the two-dimensional case, the copier returned a copy containing an area equal to $s^2 = 4$ times the original area. For the cube, the copier returned a copy with $s^3 = 8$ times the volume of the original. It seems then that we can think of the dimension of an object as an exponent D that satisfies these types of equations. We now define the notion of dimension for fractals.

DEFINITION The **fractal dimension** of an object is a number D that satisfies the equation

$$n = s^D,$$

where s is a scaling factor and n is the amount by which the quantity we are measuring (length, area, volume) of the object changes when we apply the scaling factor to the object.

To understand the notion of fractal dimension, let's look again at the Koch curve.

EXAMPLE 5 *The Fractal Dimension of the Koch Curve*

What is the dimension of the Koch curve?

SOLUTION: We need to understand what it means to magnify the curve by a factor. Consider Figure 10.93(a), where we show a picture of the *entire* Koch curve but because the picture is so small, we cannot see much detail. We next enlarge Figure 10.93(a) by a factor of 3 and display this enlargement in Figure 10.93(b).

We see that when each of the 16 tiny line segments in Figure 10.93(a) is enlarged, it appears as a line segment with a "bump" on it, as shown in Figure 10.94. The line segment with the bump is $\frac{4}{3}$ times as long as the line segment before enlargement.

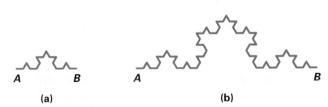

A B A B

(a)   (b)

FIGURE 10.93 The Koch curve enlarged by a factor of 3.

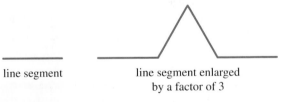

line segment   line segment enlarged
by a factor of 3

FIGURE 10.94 Each enlarged segment of the curve appears $\frac{4}{3}$ as long as the original.

What we are saying is that enlarging any portion of the curve by a scale factor of 3 appears to increase its length by 4. If we call the dimension of the curve D, this means that

$$3^D = 4. \tag{1}$$

We now solve for D. To do this, we need to use the log key on our calculator. The log function has the property that $\log a^x = x \log a$; we will use this property to solve equation (1).

Taking the log of both sides of equation (1) gives us

$$\log 3^D = \log 4. \tag{2}$$

Now using the property we just stated for log, we get

$$D \log 3 = \log 4. \tag{3}$$

Quiz Yourself **27**

Assume that for a fractal curve, enlarging the curve by a factor of 4 increases its length by a factor of 8. What is the fractal dimension of the curve?

Dividing both sides of equation (3) by log 3 and using a calculator to evaluate the result, we find that

$$D = \frac{\log 4}{\log 3}$$
$$\approx 1.26.$$

The fractal dimension of the Koch curve is therefore approximately 1.26.

Now try Exercises 13 and 14. ❋ **27**

The idea that the Koch curve has dimension 1.26 means that in a certain sense it is thicker, or fills space better, than a one-dimensional object such as a line segment. However, because this dimension is less than 2, the Koch curve does not fill space as well as a two-dimensional object such as a solid square.

Artists use fractal geometry in movies to create beautiful mountains, clouds, and other natural-looking objects. We will show you a simple example of how to create a tree using fractals.

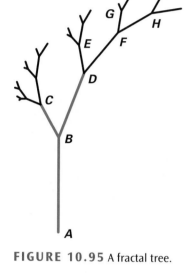

FIGURE 10.95 A fractal tree.

EXAMPLE 6   *Drawing a Fractal Tree*

Explain why the "tree" in Figure 10.95 is a fractal.

SOLUTION: The basic pattern in the tree is determined by the line segments joining points A, B, C, and D. At the end of segment AB is a branching represented by segments BC and BD. Smaller versions of this Y-shaped motif occur repeatedly throughout the tree. The fractal pattern is clear. To add finer branches to the tree, we choose a branch such as DF and replace it with a small Y-shaped figure.

Now try Exercises 29 and 30. ❋

Natural-looking scenes such as those shown in Figure 10.96 are generated using techniques that are similar to the method we described in Example 6.

FIGURE 10.96 Fractal landscape.

Exercises ❪10.7❫

Looking Back*

These exercises follow the general outline of the topics presented in this section and will give you a good overview of the material that you have just studied.

1. What was the point that we were making in the picture of the clouds in Figure 10.80?

2. In what sense are the Koch curve and the Sierpinski gasket fractals?

3. What equation did we set up to calculate the fractal dimension of the Koch curve? What did we use to solve for D?

4. State several applications that we mentioned in the Highlight on fractal applications.

Sharpening Your Skills

Exercises 5–8 pertain to the Koch curve in Example 1.

5. How many line segments will the curve have in step 5?

6. What is the length of the curve at step 5?

7. What is the length of the curve at step 10?

8. True or false: Doubling the number of steps doubles the length of the curve.

In Exercises 9 and 10, you are given steps 0 and 1 for constructing a fractal.

a) *Construct step 2 of the fractal.*

b) *Assume that the length of the line segment in step 0 is 1. Find the length of the curve in step 5.*

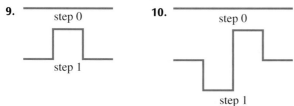

11. Solve $4^D = 6$.   12. Solve $4^D = 12$.

13. Find the fractal dimension of the curve described in Exercise 9.

14. Find the fractal dimension of the curve described in Exercise 10.

Applying What You've Learned

15. We can get a realistic "coastline" effect if we slightly vary the construction of the Koch curve. In the construction of the Koch curve in Example 1, whenever we added a bump to a line segment, we always added it above the curve. Now when we add a bump, we will use the following list of random numbers:

    87127 03570 73103 16946 81852 94819

    33108 72734 43411 31078

We will add bumps above the curve for even digits and below the curve for odd digits. The first random digit is even, so we will add the first bump above the curve; the second and third digits are odd, so we add the second and third bumps below the curve. The fourth digit is even, so we draw the fourth bump above the curve, and so on. We show steps 1 and 2 in the following diagram. Redraw steps 1 and 2 of this fractal; however, now use random digits beginning with 16946.

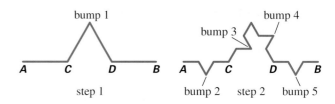

16. What is the dimension of the curve described in Exercise 15?

In Exercises 17 and 18, construct step 2 for each fractal.

17. The Sierpinski carpet

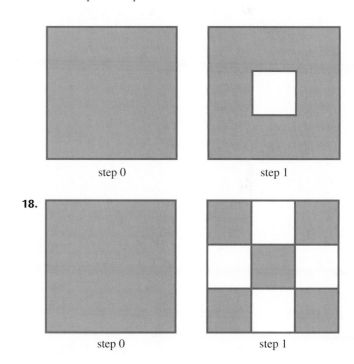

18.

Communicating Mathematics

19. What do we mean when we say that a fractal is self-similar?

20. How did we argue that the length of the Koch curve is infinite?

21. How did we argue that the area of the Sierpinski gasket is 0?

22. What is the intuitive meaning of fractal dimension?

---

*Before doing these exercises, you may find it useful to review the note *How to Succeed at Mathematics* on page xix.

Using Technology to Investigate Mathematics

23. If you search the Internet for "fractals and applets," you will find many beautiful sites devoted to fractals and fractal generation. Find several such sites, run the programs, and report on your findings.

24. There are many important real-life applications of fractals. You can do a general search on the Internet for "fractals and applications" or a more specific search, such as "fractals and medicine." Locate some interesting application sites and report on your findings.

For Extra Credit

25. Find a formula for the number of line segments at step n for the fractal in Exercise 9.

26. Find a formula for the number of line segments at step n for the fractal in Exercise 10.

27. Find a formula for the area of the Sierpinski gasket at step 10; at step n.

28. Find a formula for the area of the Sierpinski carpet in Exercise 17 at step 10; at step n.

29. Draw a fractal tree using the method of Example 6; however, vary the length and angles of the branches.

30. Draw a fractal tree using the method of Example 6; however, now use a motif with three branches.

CHAPTER SUMMARY*

SECTION	SUMMARY	EXAMPLE
SECTION 10.1	A **ray** is a half line with its endpoint included. A piece of a line joining two points and including the points is called a **line segment**. **Parallel lines** are lines that lie on the same plane and have no points in common. Two lines that have a single point in common are called **intersecting lines**.	Discussion, p. 451
	An **angle** is formed by two rays that have a common endpoint. An **acute** angle has a measure between 0° and 90°. An angle with a measure of 90° is called a **right** angle. An **obtuse** angle has a measure between 90° and 180°. A **straight** angle has a measure of 180°. Two intersecting lines form two pairs of **vertical angles** having equal measures. Two angles are **complementary** if the sum of their measures is 90°. Two angles that have an angle sum of 180° are called **supplementary** angles. Two lines that intersect forming right angles are called **perpendicular** lines. Parallel lines cut by a **transversal** form several pairs of **equal** angles.	Discussion, pp. 452–454

Example 1, p. 454 |
| | A **circle** is the set of all points lying on a plane that are located at a fixed distance, called the **radius**, from a given point, called the **center**. A **diameter** of a circle is a line segment passing through the center, with both endpoints lying on the circle. The **circumference** is the distance around the circle. | Discussion, p. 455
Example 2, p. 455
Example 3, p. 456 |
| **SECTION 10.2** | A plane figure is **closed** if we can draw it without lifting the pencil and the starting and ending points are the same. A plane figure is **simple** if we can draw it without lifting the pencil and in drawing it, we never pass through the same point twice, with the possible exception of the starting and ending points. A **polygon** is a simple, closed figure consisting only of line segments, called **edges**, such that no two consecutive edges lie on the same line. If all edges are the same length, the polygon is called **regular**. A polygon is **convex** if for any two points X and Y inside the polygon, the entire line segment XY also lies inside the polygon. | Definitions, p. 460

Definitions, p. 461

Definition, p. 461 |
	The **sum** of the measures of the interior angles of a convex polygon that has n sides is $(n-2) \times 180°$.	Examples 1 and 2, p. 463
	Two polygons are **similar** if their corresponding sides are proportional and their corresponding angles are equal.	Discussion, p. 465
	Non-Euclidean geometries have different properties than Euclidean geometry. The shortest distance between two points is not a straight line segment. Triangles do not have angle sums of 180°.	Highlight, p. 466
SECTION 10.3	If a rectangle has length l and width w, then the **perimeter** of the rectangle is $P = 2l + 2w$ and the **area** is $A = l \cdot w$.	Discussion, p. 471
	The **area of a parallelogram** with height h and base b is $A = h \cdot b$. The **area of a triangle** with height h and base b is $A = \frac{1}{2}h \cdot b$. **Heron's formula** states that for a triangle with sides of lengths a, b, and c, if we define the quantity $s = \frac{1}{2}(a + b + c)$, then the area of the triangle is $A = \sqrt{s(s-a)(s-b)(s-c)}$. A **trapezoid** with lower base b_1 and upper base b_2 and height h has **area** $A = \frac{1}{2}(b_1 + b_2) \times h$.	Discussion, p. 471

Example 1, p. 472
Example 2, p. 472
Example 3, p. 473 |
| | The **Pythagorean theorem** states that in a right triangle with legs of lengths a and b and hypotenuse (the side opposite the right angle) of length c, then $a^2 + b^2 = c^2$. | Example 4, p. 475
Example 5, p. 475 |
| | A circle with radius r has **circumference** $C = 2\pi r$ and **area** $A = \pi r^2$. | Example 6, p. 477 |
| **SECTION 10.4** | If a **rectangular solid** has length l, width w, and height h, then the **volume** of the solid is $V = lwh$ and the **surface area** of the solid is $S = 2lw + 2lh + 2wh$. If an object has a flat top and base and *sides perpendicular to the base*, and if the area of the base is A and the height is h, the volume will be $V = A \cdot h$. | Discussion, p. 483

Example 1, p. 483 |
| | A **right circular cylinder** with radius r and height h has **volume** $V = \pi r^2 h$ and **surface area** $S = 2\pi rh + 2\pi r^2$. | Example 2, p. 484 |
| | A **right circular cone** with base radius r and height h has **volume** $V = \frac{1}{3}\pi r^2 h$ and **surface area** $S = \pi r \sqrt{r^2 + h^2}$. A **sphere** with radius r has volume $V = \frac{4}{3}\pi r^3$ and **surface area** $S = 4\pi r^2$. | Examples 4 and 5, p. 487

Example 6, p. 488 |

*Before studying this chapter's material, it would be useful to reread the note *How to Succeed at Mathematics* on page xix.

SECTION 10.5	In the **metric system**, length measurement is based on the **meter**, which is slightly more than a yard. Volume is based on the **liter**, which is a little more than a quart. The basic unit of weight is the **gram**. A pound is 454 grams.
	Discussion, p. 491

The following table explains the **prefixes** that we use in the metric system: *Table 10.5, p. 491*

kilo- (k)	hecto- (h)	deka- (da)	base unit	deci- (d)	centi- (c)	milli- (m)
$\times 1{,}000$	$\times 100$	$\times 10$		$\times \dfrac{1}{10}$ or $\times 0.1$	$\times \dfrac{1}{100}$ or $\times 0.01$	$\times \dfrac{1}{1{,}000}$ or $\times 0.001$

We use the following **equivalents** between the metric and customary systems in doing **dimensional analysis**:   *Table, p. 497*

1 meter = 1.0936 yards
1 mile = 1.609 kilometers
1 pound = 454 grams
1 metric tonne = 1.1 tons
1 liter = 1.0567 quarts

SECTION 10.6	A **rigid** motion is the action of taking a geometric object in the plane and moving it in some fashion to some other place in the plane without changing its shape or size. Every rigid motion is essentially a reflection, a translation, a glide reflection, or a rotation. A **reflection** moves an object so that the ending position is a mirror image of the object in its starting position. A **translation** slides an object along a line segment, called the **translation vector**, in the plane. A **glide reflection** is formed by performing a translation (the glide) followed by a reflection. We perform a **rotation** by first selecting a point, called the **center of rotation**, and then while holding this point fixed, rotating the plane about this point through an angle called the **angle of rotation**.

Discussion, p. 500

Example 1, p. 502

Example 2, p. 503

Discussion, p. 504

A **symmetry** of a geometric object is a rigid motion such that the beginning position and the ending position of the object are exactly the same.   *Example 3, p. 505*

A **tessellation** (or tiling) of the plane is a pattern made up entirely of polygons that completely cover the plane with no gaps or overlapping polygons. The only regular polygons **that tessellate the plane** are triangles, squares, and hexagons.   *Discussion, p. 506*

Example 4, p. 507

SECTION 10.7	A **fractal** object is **self-similar** in the sense that if we magnify it, we see the same patterns that were present in the original object. The **Koch curve** and the **Sierpinski gasket** are examples of fractals.

Discussion, p. 514
Example 1, p. 514
Example 2, p. 515

The Koch curve has infinite **length**, and the **area** of the Sierpinski gasket is zero.   *Example 3, p. 516*
Example 4, p. 516

The **fractal dimension** of an object is a number D that satisfies the equation $n = s^D$, where s is a scaling factor and n is the amount by which the quantity we are measuring changes when we apply the scaling factor to the object.   *Example 5, p. 519*

CHAPTER REVIEW EXERCISES

Section 10.1

1. In the given figure:

   a. Find a pair of acute, alternate exterior angles.

   b. Find a pair of obtuse, alternate interior angles.

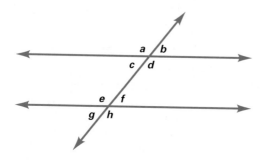

2. Find the measures of angles a, b, and c in the given diagram.

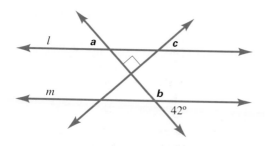

3. Assume that in the given diagram, the circumference is 24 inches and the length of arc DE is 3 inches. Find $m\angle BCD$.

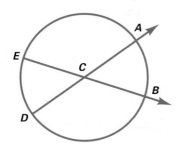

4. Solve for x in the given diagram.

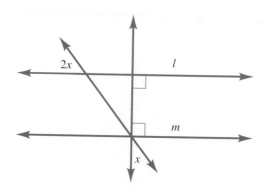

5. What is the shortest distance between two points on a sphere?

Section 10.2

6. What is the measure of an interior angle of a regular 18-sided polygon?

7. The given pair of figures are similar. What information do you know about the figure on the right?

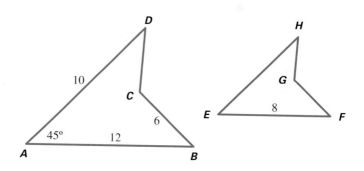

8. Solve for x in the given diagram.

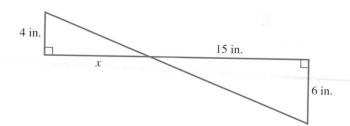

Section 10.3

9. Find the area of each figure.

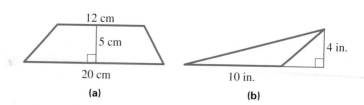

   (a)   **(b)**

10. Find the shaded area of each figure.

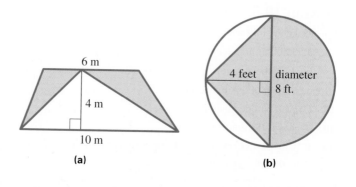

   (a)   **(b)**

11. a. Find the area of the triangle.

b. Find the height h of the triangle.

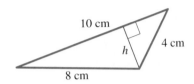

12. A running track, 5 meters wide, has the dimensions shown in the diagram. The ends of the track are semicircles with diameter 20 meters. What is the surface area of the track?

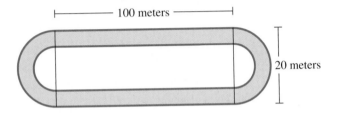

Section 10.4

13. Find the volume of each solid.

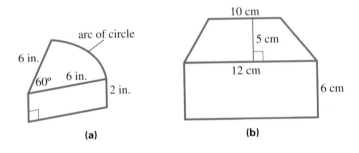

14. A punch bowl shaped like a hemisphere with a radius of 9 inches is full of punch. If we are filling cylindrical glasses that have a diameter of 3 inches and a height of 3 inches, how many glasses can we fill?

15. If we double the radius of a right circular cone, what effect does that have on the volume? Explain your answer.

Section 10.5

16. Make the following conversions:

a. 3,500 millimeters to meters

b. 4.315 hectograms to centigrams

c. 3.86 kiloliters to deciliters

17. Convert 514 decimeters to yards.

18. Convert 2.1 kiloliters to quarts.

19. If bamboo flooring costs $11.00 a square foot, how much is that cost per square meter?

Section 10.6

20. Perform the indicated glide reflection on figure B.

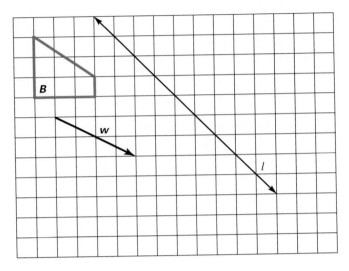

21. a. List all patterns that can be obtained by reflecting pattern (a) about a single line.

b. List all patterns that can be obtained by rotating pattern (a) about its center.

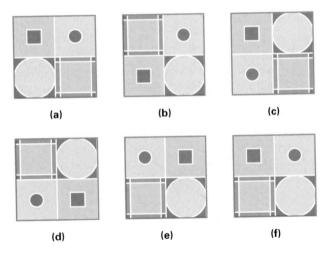

22. Find all reflectional symmetries and all rotational symmetries of the given object using angles between $1°$ and $359°$.

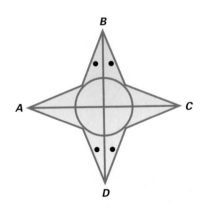

23. Tessellate the plane with the given quadrilateral.

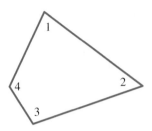

Section 10.7

24. How did we argue that the length of the Koch curve was infinite?

25. What was the area of the Sierpinski gasket?

26. You are given steps 0 and 1 for constructing a fractal. What is the length of the curve in step 8?

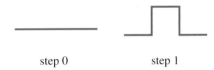

step 0   step 1

CHAPTER TEST

1. In the given figure, name each of the following pairs of angles. Assume $l \| m$.

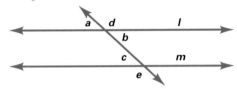

   a. a and b **b.** a and c **c.** d and e **d.** b and c

2. What is the sum of the measure of the interior angles of a regular 12-sided polygon?

3. Find the measure of angles a, b, and c in the given diagram. Assume $l \| m$.

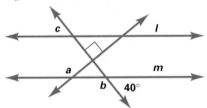

4. A spherical water tank that has a radius of 15 feet is being replaced with a cylindrical tank that also has a radius of 15 feet. How high must the new tank be to contain the same amount of water as the old tank?

5. Find the volume of each solid.

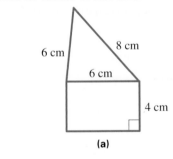

(a)

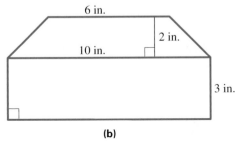

(b)

6. a. List all patterns that can be obtained by reflecting pattern (a) about a single line.

   b. List all patterns that can be obtained by rotating pattern (a) about its center.

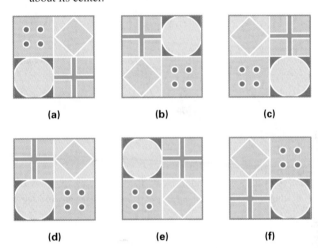

(a)   **(b)**   **(c)**

(d)   **(e)**   **(f)**

7. Assume that in the given diagram $m\angle BCD = 150°$ and the circumference is 36 inches. What is length of arc DE?

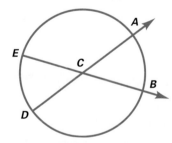

8. What is the length of the Koch curve after five steps?

9. Find the area of each figure.

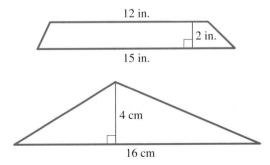

10. Find the shaded area of each figure.

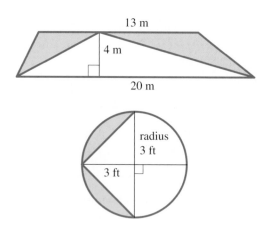

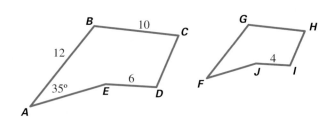

11. The given pair of figures are similar. What information do you know about the figure on the right?

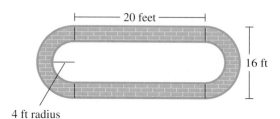

12. A pool is surrounded by a brick walkway as shown in the diagram. The pool is 3 feet deep and the walkway is 4 feet wide.

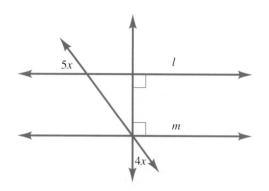

    a. Find the surface area of the pool.

    b. Find the volume of the pool.

    c. Find the area of the walkway.

13. Solve for x in the following diagram:

14. Reflect the given figure about the line $x = 1$, and then the line $y = 1$.

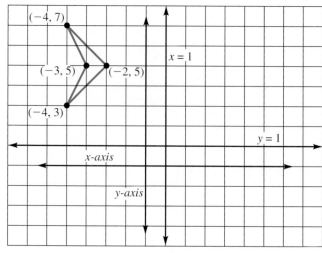

15. a. Find the area of the triangle.

    b. Find the height of the triangle.

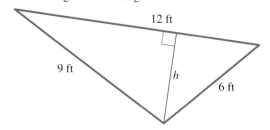

16. What was the area of the Sierpinski gasket?

17. Make the following conversions:

    a. 2,400 centimeters to meters

    b. 3.46 kilograms to milligrams

    c. 2.14 dekaliters to centiliters

18. If we double the radius of a sphere, what effect does that have on the surface area?

19. Convert 18 yards to decimeters.

20. Convert 2,614.35 quarts to kiloliters.

21. Find all reflectional symmetries and all rotational symmetries of the given figure using angles between 1° and 359°.

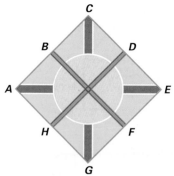

22. Tessellate the plane with the given quadrilateral.

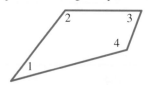

GROUP EXERCISES

1. **a.** Suppose that you have 24 inches of stiff wire to form a rectangle. Experiment by varying length and width to find the dimensions of the rectangle that contains the most area. (*Hint:* You can use the formula for the perimeter of a rectangle to express width in terms of length. Then, instead of doing hand calculations, you can use the table feature of a graphing calculator to calculate areas for you as you vary the length.)

   b. Do the same for triangles.

2. Collect a set of data and present it as a pie graph as we did in the exercise set for Section 1.3. Construct it by hand by finding the central angle of each "slice" of the pie. Then compare your graph with the same graph produced by a product such as Microsoft Word.

3. There are many sites on the Internet that discuss drawings similar to these two that were drawn by the Dutch artist Maurits Cornelis Escher. Escher formed these drawings by first creating a tessellation and then modifying it to create the image.

   a. Identify what tessellations Escher is using in Figures 1 and 2.

   b. Obtain other Escher drawings and analyze them as you did in part (a). A good place to start is www.mcescher.com.

   c. Research a Web site that explains how Escher made his drawings, and then try to make a simple one of your own.

Figure 1

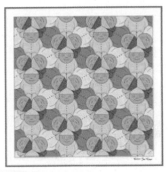

Figure 2

Apportionment
How Do We Measure Fairness?

11

The founding fathers were concerned that you and I would be fairly represented in the new government so they placed the requirements for state representatives at the very beginning of the U.S. Constitution. Article I, Section 2, states:

> Representatives and direct taxes shall be apportioned among the several states which may be included within the Union, according to their respective numbers.

You might think this is a straightforward task, but apportioning representatives is not as easy as it might seem. The problem is that one state might deserve exactly 13.465 representatives, whereas another deserves 11.702. Just like dividing 10 party favors among three small children, the question is, "How do we decide who gets the extra whole representative?"

(continued)

Daniel Webster believed that it was not possible to solve the apportionment problem perfectly. Speaking before the House of Representatives in 1832, he said,

*The Constitution . . . must be understood, not as enjoining an absolute relative equality. . . . That which cannot be done perfectly must be done in a manner as near perfection as can be.**

As you will see shortly, it turns out that Webster's intuition about apportionment was correct, and by the end of this chapter, you will have a better understanding of the mathematics of this sticky problem.

We begin this chapter by explaining how the method used to apportion early Congresses can lead to serious problems. Next, we will use inequalities to develop the current method for apportioning the U.S. House of Representatives. We then show you how to use apportionment methods to allocate resources other than political representatives. In Section 11.4, we will discuss several other apportionment methods suggested by Thomas Jefferson, John Quincy Adams, and Daniel Webster. Finally, in Section 11.5, we introduce you to an interesting method of fair division that can be used when settling an estate, dissolving a business, or selling stock holdings. ●

11.1 Understanding Apportionment

Objectives

1. Understand how the Hamilton method leads to the Alabama paradox.
2. Calculate the absolute and relative unfairness of an apportionment.

People often encourage you to use common sense when facing a problem, and this is often good advice; however, agreeing to a simple plan that seems to make sense now may lay a trap for you later. This is exactly what happened to our founding fathers when they approved a "commonsense" method, developed by Alexander Hamilton, to apportion the first U.S. Congress.

Years later, in 1881, Congress was surprised that, when using Hamilton's method, Alabama was entitled to 8 representatives in a House having 299 members but would receive only 7 representatives in a 300-member house. Under Hamilton's method, Alabama would receive fewer representatives in a larger House, *although no state had a change in population.* The strange situation, called the **Alabama paradox**, occurred again, following the 1890 census, when Arkansas lost a representative as the House increased from 359 to 360 members.

After the 1890 census, as a different apportionment plan caused the number of Maine's representatives to fluctuate, Representative Littlefield remarked:

. . . God help the state of Maine when mathematics reach for her and undertake to strike her down in this manner in connection with her representation on this floor. . . .

In order to develop an apportionment method that avoids an Alabama paradox, we must first understand why the problem occurs. The following example will help explain it.

Naxxon, Aroco, and Eurobile have formed a consortium to develop an oil drilling platform off the coast of Africa. The companies have agreed to form a nine-member board with executives from the three companies to oversee the project. Each company will have at least one representative on the board, and additional board members will be assigned in proportion to the number of stockholders in each company. Naxxon has 4,700 stockholders, Aroco has 3,700 stockholders, and Eurobile has 1,600 stockholders.

Because the total number of stockholders in the three companies is $4,700 + 3,700 + 1,600 = 10,000$, this means that Naxxon is entitled to

$$\frac{4,700}{10,000} = 0.47 = 47\%$$

---

*Daniel Webster, *The Works of Daniel Webster*, Vol. III, 16th ed. (Boston: Little Brown & Company, 1872).

of the board's nine members. Naxxon is therefore entitled to exactly 47% × 9 = 0.47 × 9 = 4.23 members. We show similar calculations for the other partners in Table 11.1.

Company	Percent of Stockholders	Board Members Deserved
Naxxon	$\frac{4,700}{10,000} = 0.47 = 47\%$	$0.47 \times 9 = 4.23$
Aroco	$\frac{3,700}{10,000} = 0.37 = 37\%$	$0.37 \times 9 = 3.33$
Euromobile	$\frac{1,600}{10,000} = 0.16 = 16\%$	$0.16 \times 9 = 1.44$
Total	100%	9.0

TABLE 11.1 Exact number of representatives allotted to each company.

A company cannot have a part of a board member. Therefore, because Naxxon is entitled to exactly 4.23 members, it will be given either 4 or 5. We will call 4 the *integer part* of 4.23 and 0.23 the *fractional part* of 4.23.

 **KEY POINT**

The Hamilton method uses fractional parts to apportion representatives.

The Hamilton Apportionment Method

If we give each company its integer part as its number of board members, then Naxxon will have four members, Aroco will have three, and Eurobile will have one. Thus there will be only eight members on the board. To have the required nine, we must now decide which company gets the additional member. It seems reasonable to assign the member to Eurobile, which has the highest fractional part—namely, 0.44. In fact, this is exactly how the **Hamilton apportionment method** would allocate the last board member.

❋ ❋ ❋ HIGHLIGHT

Using Technology to Perform Apportionment Computations

As you will see in this chapter, to perform apportionment you will have to do many repetitive computations. Technology such as graphing calculators and spreadsheets can be of great value in reducing the tedium of these calculations. For example, below are four screens showing how to use lists on a TI-83 calculator* to do the computations in Table 11.1. We begin by pressing ⎡STAT⎤ and then

selecting EDIT, as we show in Screen 1. After entering data in list L1, we see L1 in Screen 2.

In Screen 3, we divide all entries in list L1 by 10,000 and store these results in list L2, and then we multiply all entries in L2 by 9, storing the results in list L3. Screen 4 shows the contents of L1, L2, and L3 after performing these operations.

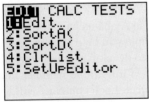

| Screen 1 | Screen 2 | Screen 3 | Screen 4 |

*See your instructor for a tutorial on how to work with lists on a TI-83 or TI-84 calculator.

HAMILTON APPORTIONMENT METHOD (APPLIED TO THE CONSORTIUM BOARD)

Follow these steps for each company.

1. Determine the exact number of board members due to the company by computing

    percent of stockholders $\times$ size of board.

2. Assign the integer part of the exact number of board members to each company.

If there are more members to be allocated, then go to step 3.

3. The first additional member goes to the company having the largest fractional part; the second additional member, if any, goes to the company with the second-largest fractional part. Continue in this manner until you have assigned all additional members.

To illustrate step 3, we add two columns to Table 11.1 to get Table 11.2.

Company	Percent of Stockholders	Step 1: Board Members Deserved	Step 2: Assign Integer Parts	Examine Fractional Parts	Step 3: Assign Additional Member
Naxxon	47	$0.47 \times 9 = 4.23$	4	0.23	4
Aroco	37	$0.37 \times 9 = 3.33$	3	0.33	3
Eurobile	16	$0.16 \times 9 = 1.44$	1	0.44	2
Total	100	9.0	8		9

TABLE 11.2 Using the Hamilton method to allocate nine board members.

We need to apportion one more.

The company with the largest fractional part gets the extra representative.

EXAMPLE 1   *Using the Hamilton Apportionment Method*

Suppose the consortium decides to increase the size of the board to 10 members. Use the Hamilton apportionment method to apportion the 10-member board.

SOLUTION: We show the steps in making this apportionment in Table 11.3.

Company	Percent of Stockholders	Step 1: Board Members Deserved	Step 2: Assign Integer Parts	Examine Fractional Parts	Step 3: Assign Additional Members
Naxxon	47	$0.47 \times 10 = 4.7$	4	0.7	5
Aroco	37	$0.37 \times 10 = 3.7$	3	0.7	4
Eurobile	16	$0.16 \times 10 = 1.6$	1	0.6	1
Total	100	10.0	8		10

TABLE 11.3 Using the Hamilton method to allocate 10 board members.

We need to apportion two more representatives.

The companies with the two largest fractional parts get the extra representatives.

Eurobile loses one representative.

The last column of Table 11.3 shows the assignment of members according to the Hamilton method. We assigned the first 8 members using the integer parts of the exact amounts deserved by each company. Then we gave the 2 additional members to Naxxon and Aroco because they have the highest fractional parts.

Quiz Yourself **1** *

Reapportion the oil consortium board assuming that there are now 12 members on the board.

Notice that increasing the size of the board to 10 members causes Eurobile to lose 1 board member.

Now try Exercises 3 to 8. ❄ **1**

The reason that the Alabama paradox occurred in Example 1 was that when the board increased to 10, *we reassigned the 9 board members that had been apportioned earlier.* This allowed for the possibility of some companies losing board members.

In order to prevent this problem, we need a method so that once we assign seats, they are never reassigned at a later date. In order to make assignments on a "once-and-done" basis, we need a clear measure of who is most deserving of the next representative at each stage in the apportionment process.

Measuring Fairness

To begin the development of such a measure, consider the representation of two hypothetical states A and B in the House of Representatives. Suppose state A has a population of 2,000,000 people and eight representatives and state B has 800,000 people and four representatives. Each of state A's representatives has an average of $\frac{2,000,000}{8} = 250,000$ constituents, whereas each representative from state B averages $\frac{800,000}{4} = 200,000$ constituents. Because a representative from state A serves more constituents than a representative from state B, it is fair to say that state A is more poorly represented in the House than state B. This leads to the following definitions.

KEY POINT

Average constituency measures the fairness of an apportionment.

> **DEFINITIONS** The **average constituency** of a state[†] is the quotient
>
> $$\frac{\text{population of the state}}{\text{number of representatives from the state}}.$$
>
> Comparing the representation of two states A and B, we say that state A is **more poorly represented** than state B if the average constituency of A is greater than the average constituency of B.

Math in Your Life

Just How Well Represented Are You?

It may surprise you that among the 165 countries that have a legislature, the United States is second only to India in the ratio of citizens to representatives in its legislature.** This was not always the case. In 1792, Congress passed an apportionment act[††] that set the size of the House at 103 representatives for a population of about 4 million, or a ratio of less than 40,000 people for each representative. This ratio has crept up until today we have 435 representatives for a population of about 301 million, or a ratio of 1 representative for every 692,000 people.

One way we might look at this is that you are only $\frac{40}{692}$ or $\frac{1}{17}$ as well represented by your congressman as were U.S. citizens more than 200 years ago. To put this in perspective, the United Kingdom, with a population that is about one-fifth that of the United States has a House of Commons with 659 members, or a ratio of about 1 representative for every 90,000 people. If the ratio of representatives to people were the same in the United States, we would have more than 3,000 representatives in our House of Representatives.

---

*Quiz Yourself answers begin on page 778.

[†]Although we will discuss apportionment in terms of assigning representatives to the U.S. House of Representatives, we can apply these ideas to many other situations.

**See James N. Danziger, *Understanding the Political World*, 7th ed. (New York: Pearson Longman, 2005), p. 142.

[††]Prior to signing this bill, George Washington, acting on the advice of Secretary of State Thomas Jefferson, had turned down another apportionment bill in what was the first presidential veto.

EXAMPLE 2 *Determining Which State Is More Poorly Represented*

According to U.S. Bureau of the Census estimates, in 2005, Wisconsin had a population of 5,479,000 and was allocated 8 representatives; Texas had a population of 21,487,000 and was allocated 32 representatives. Calculate the average constituency of each state, and determine which state was more poorly represented.

SOLUTION: The average constituency of Wisconsin was

$$\frac{\text{population of Wisconsin}}{\text{number of representatives allocated to Wisconsin}} = \frac{5,479,000}{8} = 684,875,$$

more people per representative

whereas the average constituency of Texas was

$$\frac{\text{population of Texas}}{\text{number of representatives allocated to Texas}} = \frac{21,487,000}{32} \approx 671,469.$$

fewer people per representative

Because Wisconsin has a higher average constituency, it is more poorly represented. ✤ **2**

It would be ideal, of course, to have an apportionment in which the average constituencies are the same for all states because it would be consistent with the "one man, one vote"* concept. However, it is usually not possible to achieve this ideal when making an actual apportionment. If we cannot have equality, then we should try assigning the representatives to make the average constituencies as equal as possible. One measure of how close we come to this goal is called *absolute unfairness.*

> **DEFINITIONS** Suppose that representatives are apportioned between two states A and B. We define the **absolute unfairness** of this apportionment as the difference between the larger average constituency and the smaller one. If state A has the larger average constituency, then the absolute unfairness is
>
> (average constituency of state A) − (average constituency of state B).
>
> If the two states have the same average constituencies, then we say that the two states are *equally well represented.*

Notice that in computing the absolute unfairness we subtract the smaller average constituency from the larger. Consequently, the absolute unfairness of an apportionment *cannot* be negative.

EXAMPLE 3 *Finding Absolute Unfairness*

Suppose the Weavers' Guild, with 1,542 members, has six delegates on the National Arts Commission and the Artists' Alliance, with 1,445 members, has five delegates. Calculate the absolute unfairness for this assignment of delegates.

SOLUTION: The average constituency of the Weavers' Guild is $\frac{1,542}{6} = 257$ and the average constituency of the Artists' Alliance is $\frac{1,445}{5} = 289$. The Artists' Alliance is more poorly represented than the Weavers' Guild; the absolute unfairness of this apportionment is $289 − 257 = 32$. ✤ **3**

*We have retained this terminology for historical reasons. See the Historical Highlight box in Section 11.4.

Quiz Yourself 2

a) If the 420-member electricians' union has three representatives on the United Labor Council, what is the average constituency of this group?

b) If the 440-member plumbers' union has four representatives on the council, are the electricians or the plumbers more poorly represented?

 **KEY POINT**

Absolute unfairness is the difference in average constituencies.

Quiz Yourself 3

Assume that state X has a population of 974,116 with four representatives and state Y has a population of 730,779 with three representatives. Compute the absolute unfairness for this apportionment.

KEY POINT

Relative unfairness considers the size of constituencies in calculating absolute unfairness.

Although absolute unfairness measures the imbalance of an apportionment between two states, it is inadequate to compare the unfairness of two apportionments. You may think that a larger absolute unfairness indicates a greater imbalance.

This conclusion is wrong!

When measuring the unfairness of an apportionment, it is important to take into consideration the size of the states involved. The following examples will help you understand why we need a different measure of unfairness.

In the 2008 Democratic Party primary in Texas, Hillary Clinton received 1,459,814 votes to Barack Obama's 1,358,785. Because Clinton defeated Obama by only 101,029 votes, newspapers proclaimed:

CLINTON DEFEATS OBAMA IN CLOSE RACE IN TEXAS PRIMARY

On the other hand, if Clinton and Obama were competing for a council seat in a small town and Clinton got 115 votes and Obama received 35, the local paper might read:

CLINTON BURIES OBAMA IN LANDSLIDE

Although the difference in the number of votes is important, the significance of that difference depends on the number of votes received. A difference of 101,000 votes is considered small when one candidate receives almost 1.5 million votes, whereas an 80-vote difference seems large if a candidate receives only 35 votes.

Similarly, the absolute unfairness of $684,875 - 671,469 = 13,406$ in Example 2 is *relatively* small due to the sizes of the average constituencies for Wisconsin and Texas. However, the absolute unfairness, 32, that we computed in Example 3 is *relatively* large when we compare it to the average constituencies of 289 and 257. From these examples, we see that we must consider the sizes of the average constituencies when measuring the unfairness of an apportionment.

> **DEFINITION** When apportioning the representatives for two states, the **relative unfairness** of the apportionment is defined as
>
> $$\frac{\text{the absolute unfairness of the apportionment}}{\text{the smaller average constituency of the two states}}.$$

EXAMPLE 4 *Determining Relative Unfairness*

Compute the relative unfairness for the apportionment of representatives to Wisconsin and Texas in Example 2 and the apportionment of delegates for the Weavers' Guild and the Artists' Alliance in Example 3.

SOLUTION: Recall that the absolute unfairness of the Wisconsin–Texas apportionment was 13,406. Because Texas had the smaller average constituency of 671,469, the relative unfairness of this apportionment is therefore $\frac{13,406}{671,469} \approx 0.02$.

The absolute unfairness of the apportionment in Example 3 is 32. Because the Weavers' Guild had the smaller average constituency, namely 257, the relative unfairness is $\frac{32}{257} \approx 0.125$.

The smaller relative unfairness for the apportionment for Wisconsin and Texas means that apportionment is fairer than the apportionment for the Weavers' Guild and the Artists' Alliance.

Now try Exercises 9 to 16. ❋ **4**

We now have the necessary background concepts on unfairness to develop an apportionment criterion and principle that avoids an Alabama paradox. We will show this in Section 11.2.

Quiz Yourself **4**

If state A has a population of 11,710 and five representatives and state B has a population of 16,457 and seven representatives, calculate the relative unfairness of the apportionment.

Exercises 11.1

Looking Back*

These exercises follow the general outline of the topics presented in this section and will give you a good overview of the material that you have just studied.

1. Why did the Alabama paradox occur in Example 1?

2. What did we mean in the Math in Your Life box when we said that you are now $\frac{1}{17}$ as well represented in the U.S. House of Representatives as citizens were in 1792?

Sharpening Your Skills

In Exercises 3–8, use the Hamilton apportionment method to make the assignment.

3. **Apportioning a water authority.** Assume that California, Arizona, and Nevada are cooperating to build a dam to provide water to communities currently lacking adequate water supplies. Seats on the 11-member Southwest Water Authority, which governs the project, are assigned according to the number of customers in each state who use the water from the project. There are 56,000 customers in California, 52,000 in Arizona, and 41,000 in Nevada. Assign the seats on this authority.

4. **Apportioning a negotiations committee.** The employees of the Six Flags theme park are negotiating a new contract. There are 213 performers, 273 food workers, and 178 maintenance workers. The nine-person negotiations committee has members in proportion to the number of employees in each of the three groups. Assign members to the negotiations committee.

5. **Assigning booths at an art show.** The Civic Arts Guild is having a show. There is room for 31 booths and the guild has decided that the booths will be assigned in proportion to the type of members in the guild. The guild has 87 painters, 46 sculptors, and 53 weavers. Assign the booths to the three groups.

6. **Allocating a resident council.** A campus has three freshman dormitories. Building A has 78 units, building B has 36 units, and building C has 51 units. A 12-person resident council will set rules governing the complex. Membership in the council is

to be proportional to the number of units in each building. Assign representatives to this council.

7. **Apportioning representatives.** In a recent census, Alabama's population in thousands was 4,041, Mississippi's was 2,573, and Louisiana's was 4,220. Allocate 19 members of the U.S. House of Representatives to these three states.

8. **Apportioning representatives.** In a recent census, Michigan's population in thousands was 9,295, Minnesota's was 4,375, and Wisconsin's was 4,892. Allocate 33 members of the U.S. House of Representatives to these three states.

9. If the American Nurses Association has 177,408 members and three representatives on the National Health Board, what is the association's average constituency?

10. If the International Brotherhood of Electrical Workers has 819,000 members and 13 representatives on the Building Trades Council, what is its average constituency?

11. Which state is more poorly represented: state A with a population of 27,600 and 16 representatives, or state B with a population of 23,100 and 14 representatives? What is the absolute unfairness of this apportionment? What is the relative unfairness of this apportionment?

12. Which state is more poorly represented: state C with a population of 85,800 and 11 representatives, or state D with a population of 86,880 and 12 representatives? What is the absolute unfairness of this apportionment? What is the relative unfairness of this apportionment?

13. Recall that on a 10-member board, Naxxon, with 4,700 stockholders, received five members and Aroco, with 3,700 stockholders, received four members. Calculate the absolute and relative unfairness of this apportionment.

14. Redo Exercise 13 for Aroco and Eurobile. Recall that Eurobile had 1,600 stockholders and one board member.

15. According to a recent census, Colorado had a population of approximately 2.2 million people, whereas Delaware had 0.6 million people. Colorado was allotted five seats in the House, and Delaware was allotted one. Calculate the absolute and relative unfairness of this apportionment.

16. According to a recent census, Iowa's population was approximately 2.8 million people while Utah had 1 million people. Iowa was allotted six seats in the House, and Utah was allotted two. Calculate the absolute and relative unfairness for this apportionment.

Applying What You've Learned

17. Suppose an electronics company has three divisions: (D)igital, (C)omputers, and (B)usiness products. Division D has 140 employees, C has 85, and B has 30. Assume that a 12-member quality improvement council has membership on

---

*Before doing these exercises, you may find it useful to review the note *How to Succeed at Mathematics* on page xix.

the council proportional to the number of employees in the three divisions.

  a. Apportion this council using the Hamilton method.

  b. Now increase the council's size to 13, and then to 14, and so on, redoing the apportionment each time until an Alabama paradox occurs.

  c. What is the first size above 12 when the paradox occurs, and what division loses a seat when the size of the council increases?

18. There are 47 local police, 13 federal agents, and 40 state police involved with drug enforcement in Metro City. A special nine-person task force will be formed to investigate a particular case. Officers will be assigned to this task force with membership proportional to the number of the three types of law enforcement officers.

  a. Apportion this task force using the Hamilton method.

  b. Now increase the task force's size to 10, and then to 11, and so on, redoing the apportionment each time until an Alabama paradox occurs.

  c. What is the first size above 10 when the paradox occurs, and what group loses an officer when the size of the task force increases?

19. Urbandelphia General Hospital has three emergency clinics throughout the city. Statistics have been gathered regarding the number of patients seen on night shift for each clinic.

Clinic	Center City	South Street	West Side
Number of Patients	213	65	300

  a. If there are 23 emergency physicians available, how should they be apportioned?

  b. Increase the number of physicians until an Alabama paradox occurs. At what increase does the paradox occur and what clinic loses a doctor?

20. Center City Community College has three branch campuses with student enrollment in media shown below. Currently the college has 12 computer labs, but with its new electronic media concentration, the administration wants to increase the number of labs.

Campus	Midtown	Northeast	East River
Number of Students	40	180	240

  a. How should the 12 labs be apportioned?

  b. Increase the number of labs until an Alabama paradox occurs. At what increase does the paradox occur and what branch loses a lab?

Communicating Mathematics

21. How can we avoid an Alabama paradox when apportioning representatives?

22. What was the point of our discussion about Clinton and Obama on page 536?

23. What is the difference between absolute and relative unfairness of an apportionment?

24. You have seen five examples of Alabama paradoxes in this section (Example 1 and Exercises 17–20). Can you describe a general pattern as to what "states" seem to be hurt by an Alabama paradox?

Using Technology to Investigate Mathematics

25. See your instructor for an Excel spreadsheet that will perform the computations to find an Alabama paradox. Use this spreadsheet to duplicate some of the calculations in this section and to solve Exercises 17 to 20.

26. Do a search on the Internet for "Hamilton apportionment applets." Use these programs to duplicate some of the calculations in this section and report on your findings.

27. Do a search on the Internet for apportionment. You will find sites devoted to the history of apportionment, the apportionment of the U.S. House of Representatives after the last census, and so on. You can also find data to compare the fairness of the latest apportionment between your state and others. Find a site that interests you and report on your findings.

For Extra Credit

28. Assume that on the oil consortium board, Naxxon currently receives five representatives and Aroco receives four. Suppose one additional representative can be given to either Naxxon or Aroco.

  a. Calculate the relative unfairness of the apportionment if the additional representative is given to Naxxon.

  b. Calculate the relative unfairness of the apportionment if the additional representative is given to Aroco.

  c. Based on your answers to parts (a) and (b), which company should get the additional representative? Why?

29. Assume that on the oil consortium board, Naxxon currently receives six representatives and Eurobile receives two. Suppose one additional representative can be given to either Naxxon or Eurobile.

  a. Calculate the relative unfairness of the apportionment if the additional representative is given to Naxxon.

  b. Calculate the relative unfairness of the apportionment if the additional representative is given to Eurobile.

  c. Based on your answers to parts (a) and (b), which company should get the additional representative? Why?

The Huntington–Hill Apportionment Principle

11.2

Objectives

1. Understand the apportionment criterion.
2. Compute Huntington–Hill numbers to apportion representatives.

You have seen that an Alabama paradox may occur if we reapportion seats that have already been assigned. Therefore, we will avoid this paradox if we apportion *only* the new seats when the representative body increases in size.

The Apportionment Criterion

We want a method that tells us at each stage in the apportionment process who gets the next seat. We use relative unfairness and the following criterion in making the decision.

> **APPORTIONMENT CRITERION** When assigning a representative among several parties, make the assignment so as to give the smallest relative unfairness.

EXAMPLE 1 *Using the Apportionment Criterion*

In a general election, state A has a population of 13,680 and five representatives, and state B has a population of 6,180 and two representatives. Use the apportionment criterion to determine which state is more deserving of one additional representative.

SOLUTION: We will show this solution in two stages. First, we will compute the relative unfairness of the apportionment if we give the additional representative to A. Then we will do the same computation if the representative is assigned to B.

Stage One: Assume that we give the representative to A instead of B. So A will have six representatives and B will have two. Their average constituencies are as follows:

$$\text{A's average constituency} = \frac{\text{A's population}}{\text{A's representatives}} = \frac{13,680}{6} = 2,280$$

└── smaller average constituency

$$\text{B's average constituency} = \frac{\text{B's population}}{\text{B's representatives}} = \frac{6,180}{2} = 3,090$$

Because state A has the smaller average constituency, the relative unfairness of this apportionment is

larger relative unfairness

$$\frac{\text{B's average constituency} - \text{A's average constituency}}{\text{A's average constituency}} = \frac{3,090 - 2,280}{2,280} = \frac{810}{2,280} \approx 0.355.$$

Stage Two: Assume that we give the representative to B instead of A. So A will have five representatives and B will have three. Their average constituencies are as follows:

$$\text{A's average constituency} = \frac{\text{A's population}}{\text{A's representatives}} = \frac{13,680}{5} = 2,736$$

Quiz Yourself ⑤

State A, with a population of 41,440, presently has seven representatives. State B, with a population of 25,200, has four representatives.

a) Determine the relative unfairness of the apportionment if we give an additional representative to state A.

b) Determine the relative unfairness of the apportionment if this representative is given instead to state B.

c) Use the apportionment criterion to decide which state should receive the additional representative.

$$\text{B's average constituency} = \frac{\text{B's population}}{\text{B's representatives}} = \frac{6{,}180}{3} = 2{,}060 \quad \overset{\text{smaller average}}{\underset{\text{constituency}}{}}$$

Now B has the smaller average constituency, so the relative unfairness of this apportionment is

$$\frac{\begin{array}{c}\text{A's average constituency} -\\ \text{B's average constituency}\end{array}}{\text{B's average constituency}} = \frac{2{,}736 - 2{,}060}{2{,}060} = \frac{676}{2{,}060} \approx 0.328. \quad \overset{\text{smaller relative}}{\underset{\text{unfairness}}{|}}$$

We see that we should give B a third representative before A receives a sixth one because that will result in smaller relative unfairness.

Now try Exercises 5 to 12. ❁ ⑤

Note that our solution in Example 1 violates the way the Hamilton method apportions eight representatives to states A and B. According to Hamilton's method, the exact number of representatives due state A is

$$\text{number of representatives} \times \frac{\text{population of A}}{\text{total population}} = 8 \times \frac{13{,}680}{13{,}680 + 6{,}180} = 8 \times \frac{13{,}680}{19{,}860} \approx 5.511,$$

whereas the exact number due state B is

$$\text{number of representatives} \times \frac{\text{population of B}}{\text{total population}} = 8 \times \frac{6{,}180}{13{,}680 + 6{,}180} = 8 \times \frac{6{,}180}{19{,}860} \approx 2.489.$$

Because A's fractional part (0.511) is larger than B's (0.489), Hamilton's method would assign the eighth representative to A.

Comparing the relative unfairness of two apportionments as we did in Example 1 is tedious. There is an easier way to determine which state is more deserving of an additional representative. To understand this method, let us return to the problem in Example 1 again.* We will represent the population of A by a and the population of B by b.

Stage One: We give the extra representative to A, so A has six representative and B still has two. Recall that in this stage, A had the smaller average constituency, now represented by $\frac{a}{6}$, and B's average constituency is $\frac{b}{2}$. So the relative unfairness of this assignment is

$$\frac{\begin{array}{c}\text{B's average constituency} - \text{A's average constituency}\end{array}}{\text{A's average constituency}} = \frac{\frac{b}{2} - \frac{a}{6}}{\frac{a}{6}} = \frac{\frac{b}{2}}{\frac{a}{6}} - \frac{\frac{a}{6}}{\frac{a}{6}} = \frac{\frac{b}{2}}{\frac{a}{6}} - 1 = \frac{6b}{2a} - 1. \quad (1)$$

$$\underset{\text{smaller average constituency}}{\nearrow} \qquad \overset{\text{relative unfairness if representative is given to A}}{\nearrow}$$

Stage Two: Now because we are assigning five representatives to A and three to B, the average constituency of A is $\frac{a}{5}$, and the average constituency of B is $\frac{b}{3}$. The relative unfairness with this assignment is

$$\frac{\begin{array}{c}\text{A's average constituency} - \text{B's average constituency}\end{array}}{\text{B's average constituency}} = \frac{\frac{a}{5} - \frac{b}{3}}{\frac{b}{3}} = \frac{\frac{a}{5}}{\frac{b}{3}} - \frac{\frac{b}{3}}{\frac{b}{3}} = \frac{\frac{a}{5}}{\frac{b}{3}} - 1 = \frac{3a}{5b} - 1. \quad (2)$$

$$\underset{\text{smaller average constituency}}{\nearrow} \qquad \overset{\text{relative unfairness if representative is given to B}}{\nearrow}$$

---

*This lengthy derivation is leading to the Huntington–Hill apportionment principle on page 541. You may want to come back to this derivation later.

In Example 1, we saw that state B should receive the additional representative because that resulted in a smaller relative unfairness. We can rewrite this as follows:

(relative unfairness if extra representative is given to B)

$<$ (relative unfairness if extra representative is given to A)

Using equations (1) and (2), we can express this algebraically as

$$\frac{a}{5}\cdot\frac{3}{b}-1<\frac{b}{2}\cdot\frac{6}{a}-1.$$

If we add 1 to both sides of this inequality, we get

$$\frac{a}{5}\cdot\frac{3}{b}<\frac{b}{2}\cdot\frac{6}{a}. \qquad (3)$$

Because a and b represent populations and are therefore positive, we see that $\frac{a}{6}\cdot\frac{b}{3}$ is also positive. Multiplying both sides of inequality (3) by this number we get

$$\left(\frac{a}{6}\cdot\frac{b}{3}\right)\frac{a}{5}\cdot\frac{3}{b}<\left(\frac{a}{6}\cdot\frac{b}{3}\right)\frac{b}{2}\cdot\frac{6}{a}.$$

Canceling common factors from the numerators and denominators of both sides of this inequality gives us the equivalent inequality

$$\frac{a^2}{5\cdot6}<\frac{b^2}{2\cdot3}. \qquad (4)$$

The point we are making is that the word inequality

(relative unfairness if extra representative is given to B)

$<$ (relative unfairness if extra representative is given to A)

is equivalent to the algebraic inequality*

$$\frac{a^2}{5\cdot6}<\frac{b^2}{2\cdot3}.$$

Therefore, to determine whether state A or state B deserves the additional representative, we could compute for each state a number of the form

$$\frac{\text{(population of the state)}^2}{\text{(number of representatives)}\cdot[\text{(number of representatives)}+1]}$$

and compare the two. The larger number indicates which state should receive the additional representative. These observations for states A and B lead us to the following principle.

The Huntington–Hill Method

> **THE HUNTINGTON–HILL APPORTIONMENT PRINCIPLE**   If states X and Y have already been allotted x and y representatives, respectively, then state X should be given an additional representative in preference to state Y provided that
>
> $$\frac{\text{(population of Y)}^2}{y\cdot(y+1)}<\frac{\text{(population of X)}^2}{x\cdot(x+1)}$$
>
> Otherwise, state Y should be given the additional representative. We will often refer to a number of the form $\dfrac{\text{(population of X)}^2}{x\cdot(x+1)}$ as a **Huntington–Hill number**.

---

*Although we began by comparing the relative unfairness of two apportionments, the quantities $\frac{a^2}{5\cdot6}$ and $\frac{b^2}{2\cdot3}$ are simply algebraic expressions. *They are not measures of relative unfairness.*

❋❋❋ HISTORICAL HIGHLIGHT

Is There a Perfect Apportionment Method?

The Huntington–Hill apportionment principle, developed by mathematicians Edward Huntington and Joseph Hill, was signed into law by Franklin D. Roosevelt in 1941 and is currently used to apportion the U.S. House of Representatives. Although this method avoids the Alabama paradox, it is not perfect.

You might wonder, "Does a perfect apportionment method exist?" To answer this question, consider two reasonable criteria that we expect an apportionment method to satisfy:

- The method should not be subject to the Alabama or other similar paradoxes.*

- An apportionment should satisfy the **quota rule**. That is, if the exact number of representatives due to

a state is 31.46, then the number apportioned must be either 31 or 32. The state cannot receive 30 or 33 representatives.

In 1980, Michael Balinski and H. Peyton Young proved a surprising theorem, called the **Balinski and Young's impossibility theorem**, which states:

> *There is no apportionment method that avoids all paradoxes and at the same time satisfies the quota rule.*

Because any apportionment method must be flawed, politics often plays as large a role as mathematics when Congress discusses an apportionment method.

Quiz Yourself ⑥

According to a recent census, Iowa had a population of approximately 2.9 million people and five representatives to the U.S. House of Representatives; and Nebraska had a population of 1.8 million people and three representatives. Use the Huntington–Hill apportionment principle to determine which state is most deserving of an additional representative.

EXAMPLE 2 *Using the Huntington–Hill Apportionment Principle*

There are 320 full-time nurses and 148 part-time nurses at Community General Hospital. The nursing supervisor has chosen 4 full-time nurses and 2 part-time nurses to serve on a committee to evaluate proposed nursing guidelines. Use the Huntington–Hill apportionment principle to decide whether the seventh nurse on the committee should be full-time or part-time.

SOLUTION: We compute the Huntington–Hill numbers for the full-time nurses and for the part-time nurses:

larger Huntington–Hill number

$$\frac{(\text{number of full-time nurses})^2}{(\text{current representation}) \cdot (\text{current representation} + 1)} = \frac{(320)^2}{4 \cdot 5} = 5{,}120$$

$$\frac{(\text{number of part-time nurses})^2}{(\text{current representation}) \cdot (\text{current representation} + 1)} = \frac{(148)^2}{2 \cdot 3} \approx 3{,}651$$

Comparing these numbers, we find that the next nurse selected for the committee should be a full-time nurse.

Now try Exercises 13 to 24. ❋ ⑥

Example 3 shows how we can use the Huntington–Hill apportionment principle to apportion representatives among more than two parties.

EXAMPLE 3 *Apportioning Representatives Among Three States*

Assume that the oil consortium board currently has two members from Naxxon, two from Aroco, and one from Eurobile. Use the Huntington–Hill apportionment principle to decide which company should receive the next member on the board.

---

*We will discuss other paradoxes in Section 11.4. Also, see M. L. Balinski and H. P. Young, *Fair Representation: Meeting the Ideal of One Man, One Vote* (New Haven, CT: Yale University Press, 1982).

SOLUTION: Recall that Naxxon had 4,700 stockholders, Aroco had 3,700, and Eurobile had 1,600. We compute the Huntington–Hill numbers for each company.

$$\text{Naxxon: } \frac{47^2}{2 \cdot 3} \approx 368 \qquad \text{Aroco: } \frac{37^2}{2 \cdot 3} \approx 228 \qquad \text{Eurobile: } \frac{16^2}{1 \cdot 2} = 128$$

Thus, Naxxon should get the next representative because it has the largest Huntington–Hill number. ✤

The Huntington–Hill apportionment principle tells us at any stage in an apportionment process which party most deserves the next seat. As we saw in Example 3, we can also use this principle when allocating representatives to more than two parties. Because this method meets the apportionment criterion we stated earlier, we will use it to apportion the entire oil consortium board in Section 11.3.

Now try Exercises 25 to 32.

Exercises 11.2

Looking Back*

These exercises follow the general outline of the topics presented in this section and will give you a good overview of the material that you have just studied.

1. In Example 1, why did we decide to give B a third representative before giving A its sixth one?

2. Examples 1 and 2 were illustrating essentially the same problem. What was the difference in our solutions?

3. How did we decide to assign the next board member in Example 3?

4. What did Balinski and Young prove regarding apportionment?

Sharpening Your Skills

5. The Musicians' Guild, with 908 members, has four representatives on the National Arts Advisory Board (NAAB); the Artists' Alliance, with 633 members, has three representatives. If either the musicians or the artists can be given one additional representative to the NAAB, determine which group should get it by using the following steps:

   a. Determine the relative unfairness of the apportionment if we give an additional representative to the musicians.

b. Determine the relative unfairness of the apportionment if this representative is given to the artists instead.

c. Use the apportionment criterion to decide which group deserves the additional representative more.

6. The 1,218-member carpenters' union has six representatives on the state labor council, and the 720-member plumbers' union has four representatives. If either the carpenters or the plumbers can have another representative on the council, which union is more deserving of this representative? Use the method outlined in Exercise 5.

7. The Metro City Transit System (MCTS) is made up of the red line and the blue line. The red line has nine cars and averages 405 passengers per run. The blue line has seven cars and averages 287 passengers per run. Use the method outlined in Exercise 5 to determine which line is more deserving of an additional car.

8. The Family Services Agency has two offices. The Grand Lakes office has seven caseworkers and a caseload of 595 clients. The Plains City office has 13 caseworkers and 819 clients. Use the method outlined in Exercise 5 to determine which office is more deserving of an additional caseworker.

In Exercises 9–12, we provide the populations of two states and their representation in the U.S. House of Representatives. Use the apportionment criterion to decide which state is more deserving of an additional representative in the House.

9. Alabama: population 4,461,130, 7 representatives

   New York: population 19,004,973, 29 representatives

10. Michigan: population 9,955,829, 15 representatives

    Pennsylvania: population 12,300,670, 19 representatives

11. Colorado: population 4,311,882, 7 representatives

    Delaware: population 785,068, 1 representative

---

*Before doing these exercises, you may find it useful to review the note *How to Succeed at Mathematics* on page xix.

12. Florida: population 16,028,890, 25 representatives

California: population 33,930,798, 53 representatives

In Exercises 13–20, calculate the Huntington–Hill number for each party. To keep the calculations manageable, round the populations to the nearest tenth of a million. For example, for Colorado use 4.3 million.

13. The Musicians' Guild in Exercise 5

14. The carpenters' union in Exercise 6

15. The blue line in Exercise 7

16. The Grand Lakes office in Exercise 8

17. Alabama (see Exercise 9)

18. Michigan (see Exercise 10)

19. Colorado (see Exercise 11)

20. Florida (see Exercise 12)

In Exercises 21–24, use this table of information regarding states' populations and representatives to calculate Huntington–Hill numbers to determine which state is most deserving of an additional representative.

State	Population	Representatives
Alaska	628,933	1
Georgia	9,206,975	13
Illinois	12,439,042	19
Indiana	6,090,782	9
Maine	1,277,731	2
New Hampshire	1,238,415	2
Tennessee	5,700,037	9
West Virginia	1,813,077	3
Wyoming	509,294	1

21. Indiana and Illinois

22. Maine and New Hampshire

23. Alaska, New Hampshire, and Wyoming

24. Tennessee, West Virginia, and Georgia

Applying What You've Learned

Exercises 25–28 refer to Exercises 5–8, respectively. In each case, we add one more party to the situation described earlier. Compute Huntington–Hill numbers to decide which of the three parties is most deserving of an additional object.

25. The Actors' Coalition has 420 members and two representatives on the NAAB.

26. The electricians' union has 681 members and three representatives on the labor council.

27. The MCTS adds a yellow line with three cars that average 156 passengers per run.

28. The new Family Services Agency office in Great Mountain has five caseworkers with 405 clients.

29. Reconsider Exercise 25. Suppose that instead of one member, we can add two members to the NAAB. How should these two members be allocated? (*Hint:* Do not assign both additional members at the same time. Use the Huntington–Hill apportionment principle to assign one member and then use the principle again to assign the second.)

30. Reconsider Exercise 27. Suppose that instead of one car, we can add two cars to the MCTS. How should these two cars be assigned?

31. Is the assignment of the 11 representatives to the NAAB in Exercise 29 consistent with the Hamilton apportionment method? Explain.

32. Is the assignment of the 21 cars to the MCTS in Exercise 30 consistent with the Hamilton apportionment method? Explain.

Communicating Mathematics

33. What is the apportionment criterion?

34. In the statement of the Huntington–Hill apportionment principle, we said that X should be assigned a representative in preference to Y if X's Huntington–Hill number is larger than Y's. What is the benefit if we make the assignment this way?

35. How do we use Huntington–Hill numbers in assigning representatives?

36. Why does using the Huntington–Hill method avoid an Alabama paradox?

For Extra Credit

37. Michigan has a population of approximately 9.3 million and 16 representatives, whereas Wisconsin has a population of approximately 4.9 million and 9 representatives. Suppose Wisconsin's population starts growing while Michigan's remains constant. Further, suppose with Wisconsin's growth, the 25 representatives between the two states would be redistributed using the Huntington–Hill method so that Wisconsin will have 10 seats in the House. Now, how large must Wisconsin's population become to take a representative away from Michigan? Explain how you arrived at your answer.

38. New Jersey has a population of 7.7 million and 13 representatives, whereas Washington has a population of 4.9 million and 9 representatives. Suppose Washington's population starts growing while New Jersey's remains constant. Further, suppose with Washington's growth, the 22 representatives between the two states would be redistributed using the Huntington–Hill method so that Washington will have 10 seats in the House. Now, how large must Washington's population become to take a representative away from New Jersey? Explain how you arrived at your answer.

39. Are the situations that we described in Exercises 37 and 38 examples of an Alabama paradox? Explain your answer.

11.3 Applications of the Apportionment Principle

Objectives

1. Understand how to use the Huntington–Hill method to do apportionment.
2. Use the Huntington–Hill method to allocate objects other than representatives.

Apportioning the Oil Consortium Board

We will now use the Huntington–Hill apportionment principle to apportion the entire nine-member oil consortium board. Recall that Naxxon has 47 hundred stockholders, Aroco has 37 hundred, and Eurobile has 16 hundred.

We will begin by giving one seat to each company, which is consistent with a provision in the U.S. Constitution that each state must have at least one representative. Next, we calculate a table of Huntington–Hill numbers to assign the remaining six seats. Recall that a Huntington–Hill number has the form

$$\frac{(\text{number of stockholders in the company})^2}{(\text{current representation}) \cdot (\text{current representation} + 1)}.$$

EXAMPLE 1 *Using Huntington–Hill Numbers to Allocate a Seat on the Council*

Table 11.4 shows the first line of a table of Huntington–Hill numbers where we are assuming that each company has one representative. Because Naxxon has the largest Huntington–Hill number in Table 11.4, it is given its second representative in preference to the other companies.

Naxxon	Aroco	Eurobile
$\frac{(47)^2}{1 \times 2} = 1{,}104.5$	$\frac{(37)^2}{1 \times 2} = 684.5$	$\frac{(16)^2}{1 \times 2} = 128$

Naxxon has largest Huntington–Hill number.

TABLE 11.4 Huntington–Hill numbers, assuming that each company has one representative.

Now that four representatives have been assigned, we will add lines to Table 11.4 giving us Table 11.5, which we will use to allocate the remaining representatives.

SOLUTION: We cross off the first entry under Naxxon because we have already used that Huntington–Hill number. We use the symbol ④ to indicate that Naxxon received the fourth representative.

Current Representation	Naxxon	Aroco	Eurobile
1	$\frac{(47)^2}{1 \times 2} = 1{,}104.5$ ④	$\frac{(37)^2}{1 \times 2} = 684.5$ ⑤	$\frac{(16)^2}{1 \times 2} = 128$
2	$\frac{(47)^2}{2 \times 3} = 368.2$	$\frac{(37)^2}{2 \times 3} = 228.2$	$\frac{(16)^2}{2 \times 3} = 42.7$
3	$\frac{(47)^2}{3 \times 4} = 184.1$	$\frac{(37)^2}{3 \times 4} = 114.1$	$\frac{(16)^2}{3 \times 4} = 21.3$
4	$\frac{(47)^2}{4 \times 5} = 110.5$	$\frac{(37)^2}{4 \times 5} = 68.5$	*
5	$\frac{(47)^2}{5 \times 6} = 73.6$	*	*

TABLE 11.5 More Huntington–Hill numbers.

Aroco deserves fifth representative.

Quiz Yourself 7

Use Table 11.5 to assign representative number 6 to the council.

Because 684.5 is the next-largest Huntington–Hill number, we should give representative number 5 to Aroco. ✳ 7

EXAMPLE 2   *Apportioning the Oil Consortium Council*

Complete the apportionment of the nine-member oil consortium council.

SOLUTION: We will use Table 11.6, which is a compact form of Table 11.5, to complete the assignment of representatives. Representatives 4 and 5 were assigned earlier and in Quiz Yourself 7 you should have found that the sixth representative goes to Naxxon, so we cross off those entries in Table 11.6. By examining Table 11.6, we assign the seventh, eighth, and ninth representatives to Aroco, Naxxon, and Eurobile, respectively, because they have the next three largest Huntington–Hill numbers (marked in red).

Current Representation	Naxxon	Aroco	Eurobile
1	1,104.5 ④	684.5 ⑤	128.0 ⑨
2	368.2 ⑥	228.2 ⑦	42.7
3	184.1 ⑧	114.1	21.3
4	110.5	68.5	*
5	73.6	*	*

Aroco has next largest H–H number.

TABLE 11.6 Huntington–Hill numbers to complete the apportionment of the nine-member oil consortium council.

Now try Exercises 3 to 10. ✳

Notice that if we increase the size of the council to 10, we see in Table 11.6 that 114.1 is the next-largest Huntington–Hill number. Therefore, the tenth representative would be assigned to Aroco—avoiding an Alabama paradox. Table 11.7 shows how to allocate a 10-member board.

Seat Number	Goes to	Number of Representatives Naxxon Has	Number of Representatives Aroco Has	Number of Representatives Eurobile Has
1	N	1	0	0
2	A	1	1	0
3	E	1	1	1
4	N	2	1	1
5	A	2	2	1
6	N	3	2	1
7	A	3	3	1
8	N	4	3	1
9	E	4	3	2
10	A	4	4	2

Alabama paradox no longer occurs.

TABLE 11.7 Apportionment of the oil consortium board.

HIGHLIGHT ❋ ❋ ❋

Using a Spreadsheet to Compute Huntington–Hill Numbers

Generating many Huntington–Hill numbers is a long and tedious process and is, of course, subject to error. A spreadsheet* can both avoid the tedium and ensure accuracy. The top spreadsheet shown here calculates the Huntington–Hill numbers listed in Table 11.6. (In fact, we used it to compute all the Huntington–Hill numbers in this section.) Spreadsheet 1 shows the formulas used to calculate the Huntington–Hill numbers, whereas Spreadsheet 2 shows the numbers after they have been computed.

If we were apportioning a large body such as the U.S. House of Representatives, a tool such as a spreadsheet helps keep the calculations manageable.

	A	B	C
1	47	37	16
2	(A1 * A1)/(1 * 2)	(B1 * B1)/(1 * 2)	(C1 * C1)/(1 * 2)
3	(A1 * A1)/(2 * 3)	(B1 * B1)/(2 * 3)	(C1 * C1)/(2 * 3)
4	(A1 * A1)/(3 * 4)	(B1 * B1)/(3 * 4)	(C1 * C1)/(3 * 4)
5	(A1 * A1)/(4 * 5)	(B1 * B1)/(4 * 5)	(C1 * C1)/(4 * 5)
6	(A1 * A1)/(5 * 6)	(B1 * B1)/(5 * 6)	(C1 * C1)/(5 * 6)
7	(A1 * A1)/(6 * 7)	(B1 * B1)/(6 * 7)	(C1 * C1)/(6 * 7)

SPREADSHEET 1 Formulas used to calculate Huntington–Hill numbers for apportioning the oil consortium board.

	A	B	C
1	47	37	16
2	1,104.5	684.5	128.0
3	368.2	228.2	42.7
4	184.1	114.1	21.3
5	110.5	68.5	12.8
6	73.6	45.6	8.5
7	52.6	32.6	6.1

SPREADSHEET 2 Huntington–Hill numbers for apportioning the oil consortium board.

Thus in a nine-member board, Naxxon receives four, Aroco receives three, and Eurobile two. We also see in Table 11.7 that for a 10-member board, Aroco should be given the additional representative, and the representation for the other companies remains the same. Now try Exercises 11 to 19.

The procedure we used in this example satisfies two important criteria. First, it prevents the Alabama paradox from occurring; second, using Table 11.5 to determine which company is to receive the next seat, the assignment can be made so that the relative unfairness between any two companies is minimal. Furthermore, it can be proved that the relative unfairness between any two states cannot be improved, even by transferring a representative from one state to another when the apportionment is done as we

*See your instructor for this spreadsheet.

described.* Although the Huntington–Hill method is not flawless because it can violate the quota rule, it is the method currently used to apportion Congress. It is still the subject of lively debate by mathematicians and politicians alike.

✏️ **KEY POINT**

We can use the Huntington–Hill method to allocate objects other than representatives.

Other Applications

If we view apportionment as a process by which we assign objects (namely, the representatives) to interested parties (namely, the states), then we can recognize other situations in which the same problem arises. In Example 3, we discuss a different situation that has some similarities to it.

Region 1	107 incidents
Region 2	65 incidents
Region 3	43 incidents
Total	215 incidents

TABLE 11.8 Number of police incidents.

EXAMPLE 3 *Using the Huntington–Hill Method to Assign Police Officers*

Suppose a small town has seven police officers. Furthermore, assume that the town is divided rather naturally into three regions. The police chief has gathered data regarding the number of incidents (crimes, traffic accidents, and so on) that have occurred in each region over the past several months, as listed in Table 11.8. If the chief decides to assign officers for duty in proportion to the number of incidents per region, how should he make these assignments?

SOLUTION: We can look at this as an apportionment problem in which the objects being apportioned are the officers and the parties receiving the objects are the regions. We do not want any region to lose a police officer if there are future increases in the size of the police force. Consequently, we will use the Huntington–Hill method to apportion the seven officers among the three regions.

We first compute the Huntington–Hill numbers in Table 11.9. Instead of using populations as we did earlier, we use the number of incidents per region to calculate these numbers. As usual, we start by giving one officer to each region. For example, $\frac{(107)^2}{1 \times 2} = 5{,}724.5$ is the Huntington–Hill number for region 1 if it already has one officer assigned.

If Given the Next Officer, the Number a Region Would Have Is:	Region 1	Region 2	Region 3
2	(a) 5,724.5	(b) 2,112.5	(e) 924.5
3	(c) 1,908.2	(f) 704.2	308.2
4	(d) 954.1	352.1	154.1
5	572.5	211.3	92.5

TABLE 11.9 Huntington–Hill numbers for assigning police officers.

Considering the entries in the table labeled (a) through (d), we see that after one officer has been assigned to each of the three regions, the remaining four officers should be assigned as follows:

a) Officer 4 goes to region 1.

b) Officer 5 goes to region 2.

c) Officer 6 goes to region 1.

d) Officer 7 goes to region 1.

If the town hires an eighth officer, looking at entries (e) and (f) we see that the next assignment should be made to region 3.

Now try Exercises 20 to 22. ❋

---

*E. V. Huntington, "The Apportionment of Representatives in Congress," *Transactions of the American Mathematical Society* 30 (1928), pp. 85–110.

Exercises 11.3

Looking Back*

These exercises follow the general outline of the topics presented in this section and will give you a good overview of the material that you have just studied.

1. How did we decide that Naxxon deserved its second representative before Aroco or Eurobile did?

2. Why did we choose to use the Huntington–Hill method rather than the Hamilton method in assigning police officers in Example 3?

Sharpening Your Skills

A labor council is being formed from the members of three unions. The electricians' union has 25 members, the plumbers have 18 members, and the carpenters have 31 members. The council will have seven representatives, with each union having at least one representative. Use the Huntington–Hill numbers listed in Table 11.10 to answer Exercises 3–6.

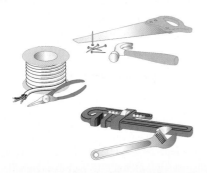

Current Representation	Electricians	Plumbers	Carpenters
1	312.5	162.0	480.5
2	104.2	54.0	160.2
3	52.1	27.0	80.1
4	31.3	*	48.1

TABLE 11.10 Huntington–Hill numbers for apportioning a labor council.

3. Knowing that each union currently has one representative, which entries in Table 11.10 do we use to assign the fourth seat on the council and which union gets that seat?

4. Assume that the carpenters have two representatives, whereas the electricians and plumbers each have one. Which entries in Table 11.10 do we use to assign the fifth seat on the council and which union gets that seat?

5. Suppose the current council has five seats, apportioned so that the electricians and carpenters have two seats each and the plumbers have one. Which entries in Table 11.10 do we use to assign the sixth seat on the council and which union gets that seat?

6. If the current council has six seats and each union has two representatives, which entries in Table 11.10 do we use to assign the seventh seat on the council and which union gets that seat?

We want to apportion 10 seats on a condominium association board to buildings Allen, Baker, and Carney. There are 23 units in Allen, 33 in Baker, and 43 in Carney. The board will have 10 seats, with each building having at least 1 seat. Use the Huntington–Hill numbers listed in Table 11.11 to answer Exercises 7–10.

Current Representation	Allen	Baker	Carney
1	264.5	544.5	924.5
2	88.2	181.5	308.2
3	44.1	90.8	154.1
4	26.5	54.5	92.5
5	17.6	36.3	61.6
6		25.9	44.0
7			33.0

TABLE 11.11 Huntington–Hill numbers for apportioning a condominium board.

7. Knowing that each building currently has one representative, which entries in Table 11.11 do we use to assign the fourth seat on the board and which building gets that seat?

8. Assume Carney has two representatives, whereas Allen and Baker each have one. Which entries in Table 11.11 do we use to assign the fifth seat on the board and which building gets that seat?

9. Suppose the current assignment of representatives to five seats on the council is Baker and Carney each with two and Allen with one. Which entries in Table 11.11 do we use to assign the sixth seat on the board and which building gets that seat?

10. Suppose the current assignment of representatives to six seats on the board is Carney with three, Baker with two, and Allen with one. Which entries in Table 11.11 do we use to assign the seventh seat on the board and which building gets that seat?

11. **Apportioning an arts board.** Ten seats on a state performing arts board are to be apportioned. In the state, there are 420 participating in theater, 435 in music, and 145 in dance. Use Table 11.12 on page 550, which contains the Huntington–Hill numbers, to apportion the 10 seats to this board. Begin by giving one representative to each area and then

assign the remaining 7 seats. List the order in which the representatives are apportioned.

Current Representation	(T)heater	(M)usic	(D)ance
1	88,200.0	94,612.5	10,512.5
2	29,400.0	31,537.5	3,504.2
3	14,700.0	15,768.8	1,752.1
4	8,820.0	9,461.3	*
5	5,880.0	6,307.5	*
6	4,200.0	4,505.4	*
7	3,150.0	3,379.0	*

TABLE 11.12 Huntington–Hill numbers for apportioning a performing arts board.

12. Allocating fellowships. Eight fellowships are to be apportioned among the schools of humanities, science, and business at a small university. The apportionment is based on the number of full-time graduate students in each school, which is 30 for humanities, 20 for science, and 40 for business. Use the Huntington–Hill numbers in Table 11.13 to carry out the apportionment. Begin by giving one fellowship to each school. List the order in which the fellowships are apportioned.

Current Representation	(H)umanities	(S)cience	(B)usiness
1	450.0	200.0	800.0
2	150.0	66.7	266.7
3	75.0	33.3	133.3
4	45.0	*	80.0
5	*	*	53.3

TABLE 11.13 Huntington–Hill numbers for allocating fellowships.

Applying What You've Learned

13. Apportioning a transit authority. The Triborough Transit Authority governs the three bridges connecting the New York City boroughs of Bronx, Manhattan, and Queens. Assume that representation on the 10-member authority is allocated in proportion with the population of each of the three boroughs. The Bronx has a population of 1.4 million, Manhattan has 1.6 million, and Queens has 2.2 million. Apportion the 10 authority seats

a. using the Hamilton method.

b. using the Huntington–Hill apportionment principle.

14. Assigning members to a labor council. The Unified Labor Council has 11 members representing the carpenters', electricians', plumbers', and painters' unions. The four trades are represented on the council in proportion to the size of their membership. If there are 84 carpenters, 48 electricians, 40 plumbers, and 28 painters, how many from each trade should be on the council

a. using the Hamilton method?

b. using the Huntington–Hill apportionment principle?

15. Apportioning representatives. Use the Huntington–Hill method to apportion 10 representatives among Utah, Idaho, and Oregon. The populations of the states are Utah, 2.4 million; Idaho, 1.5 million; and Oregon, 3.6 million. Begin by giving one representative to each state. List the order in which the representatives are apportioned.

16. Apportioning representatives. Use the Huntington–Hill method to apportion 11 representatives among Arizona, New Mexico, and Nevada. The populations of the states are Arizona, 5.2 million; New Mexico, 2.0 million; and Nevada, 2.1 million. Begin by giving one representative to each state. List the order in which the representatives are apportioned.

17. Apportioning representatives. Use the Huntington–Hill method to apportion 11 representatives among Arkansas, Kansas, and Nebraska. The populations of the states are Arkansas, 2.75 million; Kansas, 2.76 million; and Nebraska, 1.8 million. Begin by giving one representative to each state. List the order in which the representatives are apportioned.

18. Refer to your answer to Exercise 11 and Table 11.12. Suppose the arts board is increased to 12 representatives. How should the two new seats be distributed and how many representatives would each area have on the 12-member council?

19. Refer to your answer to Exercise 12 and Table 11.13. Suppose that three additional fellowships are to be awarded. How should the three new fellowships be distributed and which schools will get them?

20. Assigning medical personnel. A hospital administrator wants to assign seven emergency room medical teams to three community outreach centers. The number of patients treated last week at each center is listed in the given table. Begin by giving each center one team. Use the Huntington–Hill method to decide how the administrator should assign the medical teams to these three centers.

Center 1	98 patients
Center 2	34 patients
Center 3	57 patients
Total	189 patients

21. Assigning police officers. Consider the problem of apportioning police officers given in Example 3. Suppose we still want to apportion the seven officers among the three regions, but the number of incidents per region has changed. The number of related incidents for regions 1, 2, and 3 are listed in the table. How should the seven officers be assigned to the regions, given the new data?

Region 1	123 incidents
Region 2	44 incidents
Region 3	79 incidents
Total	246 incidents

22. **Scheduling fitness classes.** A health club instructor has a course load that allows her to teach six two-credit-hour classes. A preregistration survey indicates the following interests:

    56 want to take Pilates;

    29 want to take kick boxing;

    11 want to take yoga; and

    4 want to take spinning.

    Assume that the instructor will teach at least one class for each activity.

    a. Use the Huntington–Hill method to apportion the instructor's remaining two classes among the four activities.

    b. Does your solution in part (a) agree with your intuition? Explain.

Communicating Mathematics

23. What is the first step in using the Huntington–Hill apportionment method?

24. Why did we avoid an Alabama paradox this time when we assigned the tenth member to the oil consortium board?

Using Technology to Investigate Mathematics

25. See your instructor for an Excel spreadsheet that computes Huntington–Hill numbers and use it to duplicate some of the computations in this section. Also use it to solve some of the exercises.

26. You can find applets that illustrate the Huntington–Hill apportionment method. However, some of these use an approach based on divisor methods, which we explain in Section 11.4. Run one of these applets and, by adjusting the divisors appropriately, duplicate the apportionment assignments in this section. Report on your findings.

For Extra Credit

27. **Assigning police officers.** Consider again the problem of assigning officers as described in Example 3. As stated, it would not be appropriate to apply the Hamilton method because an Alabama paradox could arise if the number of officers was increased. Verify that this is indeed the case.

    a. Change the number of incidents in regions 1, 2, and 3.

    b. Use the Hamilton method to apportion officers. Increase the number of officers until a region has a decrease in the number of officers assigned to it.

    You may have to do step (a) several times until you get a reasonable example of an Alabama paradox.

28. **Assigning police officers.** Consider again the problem of assigning officers as described in Example 3. Table 11.9 contains the Huntington–Hill numbers used in solving this problem. Explain why the table is not adequate if the number of officers to be apportioned increased to nine—that is to say, why would we have to compute additional Huntington–Hill numbers to do the apportionment?

29. When constructing a table of Huntington–Hill numbers, it is not always necessary to continue the computation after a certain point. For example, consider the apportionment problem in Exercise 11 and Table 11.12. We did not compute the Current Representation entries 4 through 7 for dance. Explain why computing these entries is unnecessary. (*Hint:* Look at the last entries in the table under theater and music.)

30. Consider the Huntington–Hill numbers in Table 11.14, which we might use to apportion seats on a representative council. As usual, assume that we start the apportionment by giving each party one representative. Now, what would the smallest number of representatives on the council be so as to guarantee that B has three of them? Explain how to determine this number without actually carrying out an apportionment calculation.

Current Representation	A	B	C
1	32.00	60.50	112.50
2	10.67	20.17	37.50
3	5.33	10.08	18.75
4	3.20	6.05	11.25
5	2.13	4.03	7.50
6	1.52	2.88	5.36

TABLE 11.14 Huntington–Hill numbers for a representative council.

11.4 Other Paradoxes and Apportionment Methods

Objectives

1. Use standard divisors and quotients to do apportionment.
2. Understand how the population and new states paradoxes can occur with the Hamilton method.
3. Understand the similarities and differences among the Jefferson, Adams, and Webster apportionment methods.

We begin this section by showing you two paradoxes that can occur with the Hamilton method other than the Alabama paradox. Toward the end of this section, we will introduce methods devised by Thomas Jefferson, John Quincy Adams, and Daniel Webster that avoid these paradoxes—however, as Balinski and Young's impossibility theorem predicts, they all violate the quota rule.

Standard Divisors and Quotients

KEY POINT

We can explain apportionment using the standard divisor.

In order to understand the paradoxes and these other apportionment methods, we will have to look at apportionment in a slightly different way. Recall that to use the Hamilton method to compute the number of representatives* that a state deserved, we calculated the expression

$$\frac{\text{state's population}}{\text{total population}} \times \text{number of representatives being allocated}.$$

If we represent this expression by the algebraic expression $\frac{s}{t} \times n$, we can then rewrite it as

$$\frac{s}{t} \times n = s \times \frac{n}{t} = \frac{s}{\frac{t}{n}}.$$

where s is the state's population and $\frac{t}{n}$ is the total population divided by the number of representatives.

That is, in calculating the exact number of representatives that a state deserves, we could have first calculated

$$\frac{\text{total population}}{\text{number of representatives being allocated}}$$

and then divided this number into the state's population. Although this may not seem as intuitively clear as our original approach, this idea of first computing a divisor and then dividing into the state's population will be the central theme of our calculations for the rest of this section.

DEFINITIONS In allocating a group of representatives among several states, we define the **standard divisor** by

$$\text{standard divisor} = \frac{\text{total population}}{\text{number of representatives being allocated}},$$

---

*In this section, we will continue to use the language of states, populations, and representatives, even though you have seen in earlier sections that the apportionment principles we discuss apply to many other situations.

and a state's **standard quota** is defined to be

$$\text{standard quota} = \frac{\text{state's population}}{\text{standard divisor}}.$$

Intuitively, you can think of the standard divisor as follows:

$$\text{standard divisor} = \frac{\text{total population}}{\text{number of representatives}}$$

$$= \text{number each representative represents.}$$

We think of the standard quota as

$$\text{standard quota} = \frac{\text{state's population}}{\text{standard divisor}}$$

$$= \text{number of representatives state deserves.}$$

For example, if we want to allocate eight representatives among several states that have a total population of 4,000,000, then the standard divisor is

$$\frac{4,000,000}{8} = 500,000 = \text{number each representative represents.}$$

If a state has a population of 1,500,000, then its standard quota is

State has 1,500,000 people.

$$\text{standard quota} = \frac{\text{state's population}}{\text{standard divisor}} = \frac{1,500,000}{500,000} = 3.$$

Each representative serves 500,000 people.

So, the state deserves three representatives.

Example 1 shows that when we calculate the standard quotas, we get exactly the same results that we obtained in calculating the exact members deserved by Naxxon, Aroco, and Eurobile on the oil consortium board that we discussed in Section 11.1.

EXAMPLE 1 *Calculating Standard Quotas for the Oil Consortium Board*

Recall that we were apportioning a nine-member board to oversee the oil consortium formed by Naxxon, Aroco, and Eurobile. The members of the board would be apportioned according to the number of stockholders of each company. Naxxon had 4,700 stockholders, Aroco had 3,700, and Eurobile had 1,600.

a) Calculate the standard divisor for this apportionment.

b) Calculate the standard quota for Naxxon, Aroco, and Eurobile.

SOLUTION:

a) To find the standard divisor, we calculate

$$\text{standard divisor} = \frac{\text{total number of stockholders for Naxxon, Aroco, and Eurobile}}{\text{number of board members}}$$

$$= \frac{4,700 + 3,700 + 1,600}{9} = \frac{10,000}{9} \approx 1,111.11.$$

This means that each board member serves, on average, 1,111 stockholders.

b) To find the standard quota for each company, we divide the number of stockholders for the company by the standard divisor, as in Table 11.15.

Oil Company	Number of Stockholders	Standard Quota
Naxxon	4,700	$\frac{4,700}{1111.11} = 4.23$
Aroco	3,700	$\frac{3,700}{1111.11} = 3.33$
Eurobile	1,600	$\frac{1,600}{1111.11} = 1.44$

TABLE 11.15 Calculating the standard quotas for each oil company.

Notice that the standard quotas we found in Table 11.15 are the exact number of representatives that were due to each company as we calculated in Section 11.1. ✳

Some Good Advice

Keep in mind that the standard divisor is a single number that we calculate once and then use for the entire apportionment process. However, we must compute the standard quota individually for each state.

Recall that in using the Hamilton apportionment method, we always gave each state a number of representatives that was immediately below or immediately above its standard quota (the exact number of representatives that it was due). For example, if a state's standard quota was 4.375 representatives, we always gave it either 4 or 5 representatives. We will now make this notion more precise.

DEFINITIONS If we round the standard quota down, we call that number the **lower quota**; if we round the standard quota up, then we call that number the **upper quota**. If in making an apportionment, each state is allocated a number of representatives that is between its lower quota and upper quota, then we say the apportionment satisfies the **quota rule**.

EXAMPLE 2 *Apportioning Representatives to Three States*

Assume that we are apportioning eight representatives to three states: A, with population 5.4 million, B, with population 6.7 million, and C, with population 7.3 million. Find A's lower and upper quotas.

SOLUTION: To find A's upper and lower quotas, we must first find the standard divisor and then A's standard quota. The standard divisor for this apportionment is

$$\text{standard divisor} = \frac{\text{total population}}{\text{number of representatives}} = \frac{5.4 + 6.7 + 7.3}{8} = \frac{19.4}{8} = 2.425.$$

Dividing this into A's population, we get

$$\text{A's standard quota} = \frac{\text{A's population}}{\text{standard divisor}} = \frac{5.4}{2.425} \approx 2.227.$$

If we round 2.227 down, we get A's lower quota, which is 2. Rounding up, we get A's upper quota, which is 3. ✳

Quiz Yourself 8

Assume we are apportioning eight representatives among three states A, B, and C, which have populations of 3 million, 4 million, and 5 million, respectively.

a) Calculate the standard divisor for this apportionment.

b) Calculate each state's standard quota.

Quiz Yourself 9

In Example 2, find C's standard quota, lower quota, and upper quota.

> **HAMILTON'S APPORTIONMENT METHOD**
>
> a) Find the standard divisor for the apportionment (total population/total number of representatives).
>
> b) Find the standard quota (state's population/standard divisor) for each state and round it down to its lower quota. Assign that number of representatives to each state.
>
> c) If there are any representatives left over, assign them to states in order according to the size of the fractional parts of the states' standard quotas.

Although the terminology sounds different, the method we just described is exactly the way we apportioned the oil consortium board in Example 1 of Section 11.1.

Perhaps the Hamilton method would be the apportionment method used today if its only problem was the Alabama paradox. However, there are other paradoxes that we will discuss now.

✏️ **KEY POINT**

The Hamilton method can have the population paradox and the new states paradox.

More Apportionment Paradoxes

In the early 1900s, it was discovered that Hamilton's method had another serious flaw in that a state with a faster-growing population could lose a representative to a state whose population was growing more slowly.

> **DEFINITION** The **population paradox** occurs when state A's population is growing faster than state B's population, yet A loses a representative to state B. (We are assuming that the total number of representatives in the legislature is not changing.)

EXAMPLE 3 *The Population Paradox Can Occur with the Hamilton Method*

The graduate school at Great Eastern University used the Hamilton method to apportion 15 graduate assistantships among the colleges of education, liberal arts, and business based on their undergraduate enrollments, as shown in Table 11.16.

College	Standard Quota (Exact Number Deserved)	Lower Quota (Integer Part)	Fractional Part	Assign 2 Additional Assistantships
Education	$\frac{940}{267.33} = 3.52$	3	0.52 ⌐	4 ⌐
Liberal Arts	$\frac{1,470}{267.33} = 5.50$	5	0.50	5
Business	$\frac{1,600}{267.33} = 5.99$	5	0.99 ⌐	6 ⌐
Total		13		15

Divide enrollment of each college by the standard divisor.

Education and business have the largest fractional parts.

Colleges with largest fractional parts get extra assistantships.

TABLE 11.16 Apportioning the 15 graduate assistantships before the enrollment increase.

a) Use Hamilton's method to allocate the graduate assistantships to the three colleges.

b) Assume that after the allocation was made in part a) that education gains 30 students, liberal arts gains 46, and the business enrollment stays the same. Reapportion the graduate assistantships again using Hamilton's method.

c) Explain how this illustrates the population paradox.

SOLUTION:

a) The standard divisor for this apportionment is

$$\frac{\text{total number of undergraduate students}}{\text{number of assistantships}} = \frac{4{,}010}{15} \approx 267.33.$$

In Table 11.16, we compute the standard quota and the number of assistantships to be allocated to each college.

b) Table 11.17 shows how the apportionment would be done after the enrollment increases. The standard divisor is now

$$\frac{\text{new total number of undergraduate students}}{\text{number of assistantships}} = \frac{4{,}086}{15} \approx 272.4.$$

College	Number of Students	Standard Quota (Exact Number Deserved)	Lower Quota (Integer Part)	Fractional Part	Assign 2 Additional Assistantships
Education	$940 + 30 = 970$	$\frac{970}{272.4} = 3.56$	3	0.56	3
Liberal arts	$1{,}470 + 46 = 1{,}516$	$\frac{1{,}516}{272.4} = 5.57$	5	0.57	6
Business	1,600	$\frac{1{,}600}{272.4} = 5.87$	5	0.87	6
Total	4,086		13		15

Education loses one assistantship.

TABLE 11.17 Apportioning the 15 graduate assistantships after the enrollments increase.

c) Notice how the college of education lost an assistantship to the college of liberal arts. However, the percentage increase in the number of education students was

$$\frac{\text{increase in education students}}{\text{original number of education students}} = \frac{30}{940} = 0.0319 = 3.19\%.$$

But the increase in the number of liberal arts students was

$$\frac{\text{increase in liberal arts students}}{\text{original number of liberal arts students}} = \frac{46}{1{,}470} = 0.0313 = 3.13\%.$$

So, even though the college of liberal arts grew more slowly than the college of education, it was able to take one of education's assistantships away.

Now try Exercises 29 to 32. ❋

Admittedly, it was a close call in Example 3, and to be truthful it can be a lot of work to find such a counterexample (a spreadsheet* helps!), but the point is, Hamilton's method can allow the population paradox to occur.

*For a spreadsheet to do these calculations, see your instructor.

HISTORICAL HIGHLIGHT ❄ ❄ ❄

Apportionment and Political Inequalities

Until the 1960s, state legislatures had great power in how the representatives were apportioned within a state. Customarily, representation favored rural districts with small populations. For example, in the mid-1950s, Speaker of the House Sam Rayburn, of Bonham, Texas, presided over a district of 227,735 constituents, whereas Albert Thomas, of Houston, had 806,701 constituents. Several lawsuits were filed before

the Supreme Court, and the Court ruled that representation must be fairly distributed among the people within a state. This decision is sometimes called the "one man, one vote" rule. After these landmark decisions, representatives within a state were assigned more fairly. In 1973, again in Texas, the smallest district had 461,216 constituents, whereas the largest had 468,148.

When Oklahoma joined the union in 1907, the House of Representatives had to be reapportioned. Congress decided to increase the size of the House by five and give those five representatives to Oklahoma. However, when the House was reapportioned, New York was required to give one of its seats to Maine. This is called the *new states paradox*.

> **DEFINITION** The **new states paradox** occurs when a new state is added and its share of seats is added to the legislature, causing a change in the allocation of seats previously given to another state.

EXAMPLE 4   *The New States Paradox Can Occur with the Hamilton Method*

A small country, Namania, consists of three states A, B, and C with populations given in Table 11.18. Namania's legislature has 37 representatives that are to be apportioned to these states using the Hamilton method.

a) Apportion these representatives using the Hamilton method.

b) Assume that Namania annexes the country Darelia, whose population is 3,000 (thousands). Give Darelia its current share of representatives using the current standard divisor and add that number to the total number of representatives of Namania. Reapportion Namania again using the Hamilton method.

c) Explain how the new states paradox has occurred.

SOLUTION:

a) The standard divisor for this apportionment is the total population, 12,140 (thousand), divided by the number of representatives, 37. So the standard divisor is $\frac{12,140}{37} = 328.11$. We will use this to find each states' standard quota in Table 11.18.

State	Population (thousands)	Standard Quota (Exact Number Deserved)	Lower Quota (Integer Part)	Fractional Part	Assign Additional Representative
A	2,750	$\frac{2,750}{328.11} = 8.38$	8	0.38	8
B	6,040	$\frac{6,040}{328.11} = 18.41$	18	0.41	19
C	3,350	$\frac{3,350}{328.11} = 10.21$	10	0.21	10
Total	12,140		36		37

State with highest fractional part gets extra representative.

TABLE 11.18 Apportioning representatives to states A, B, and C before the annexation of Darelia.

b) According to the apportionment in part a), using the standard divisor of 328.11, Darelia's standard quota is $\frac{3,000}{328.11} = 9.14$. This means that Darelia deserves nine representatives under the current apportionment, so we will add nine representatives to Namania's legislature and reapportion it. Now, because Namania's population has increased and we have added nine representatives, we must recalculate the standard divisor.

$$\text{standard divisor} = \frac{\text{population}}{\text{number of representatives}} = \frac{12,140 + 3,000}{37 + 9} = \frac{15,140}{46} \approx 329.13$$

We will use this to compute the new standard quotas in Table 11.19.

State	Population (thousands)	Standard Quota (Exact Number Deserved)	Lower Quota (Integer Part)	Fractional Part	Assign Additional Representative	
A	2,750	$\frac{2,750}{329.13} = 8.36$	8	0.36	9	B loses one representative.
B	6,040	$\frac{6,040}{329.13} = 18.35$	18	0.35	18	
C	3,350	$\frac{3,350}{329.13} = 10.18$	10	0.18	10	
(D)arelia	3,000	$\frac{3,000}{328.13} = 9.11$	9	0.11	9	
Total	15,140		45		46	

TABLE 11.19 Apportioning representatives to states A, B, C, and D after the annexation of Darelia.

c) We saw in part b) that by adding Darelia, and increasing the size of the legislature according to the number of representatives that it deserved, B had to give one of its representatives to A. This illustrates the new states paradox.

Now try Exercises 33 to 36. ✺

For the remainder of this section, we will discuss some alternative apportionment methods proposed by Jefferson, Webster, and Adams. These three methods are similar in that each uses a divisor that is different from the standard divisor, which is called a **modified divisor**. (As you will see, we find this modified divisor by trial and error.) When we divide a state's population by the modified divisor, we get the state's **modified quota**.

Once we find the modified quotas for each state, the question is what do we do with them? Jefferson says round them down to get the required number of representatives, Adams says to round up, and Webster says to round in the usual way.

KEY POINT

Jefferson's method rounds quotas down.

Other Apportionment Methods

In Jefferson's method, we are going to round the modified quotas down, so we will need modified quotas which are *larger** than the standard quotas. This means that the modified divisor has to be *smaller* than the standard divisor to give us the larger quotas.

---
*Sometimes the standard divisor and quotas will work when using the Jefferson method.

> **JEFFERSON'S APPORTIONMENT METHOD**
>
> a) Use trial and error to find a modified divisor that is smaller than the standard divisor for the apportionment.
>
> b) Calculate the **modified quota** (state's population/modified divisor) for each state and round it down. Assign that number of representatives to each state. (Keep varying the modified divisor until the sum of these assignments is equal to the total number being apportioned.)

We will now reapportion the oil consortium board using Jefferson's method.

EXAMPLE 5 *Using Jefferson's Apportionment Method*

Recall in Example 1, we were apportioning a nine-member board to oversee the oil consortium formed by Naxxon, Aroco, and Eurobile. We used Hamilton's method to apportion the board according to the number of stockholders of each company. Naxxon had 4,700 stockholders, Aroco had 3,700, and Eurobile had 1,600. Now use Jefferson's method to apportion the board.

SOLUTION: In Example 1, we found that the standard divisor for this apportionment was $\frac{10,000}{9} \approx 1,111.11$, and the standard quotas for the three companies are listed in Table 11.15. When we used Hamilton's method and rounded the standard quotas down, the total was eight and we had to assign the ninth representative.

In applying Jefferson's method, we want to find modified quotas so that when we round these quotas down, the total is nine. This means that the modified quotas must be larger than the standard quotas. In order to get larger modified quotas, we must have a modified divisor that is smaller than the standard divisor of 1,111.11.

Let's try a modified divisor of 1,050 to see what happens. We show the calculations in Table 11.20.

Modified Divisor = 1,050			
	Naxxon	**Aroco**	**Eurobile**
Number of stockholders	4,700	3,700	1,600
Standard quota	4.23	3.33	1.44
Modified quota	$\frac{4,700}{1,050} = 4.48$	$\frac{3,700}{1,050} = 3.52$	$\frac{1,600}{1,050} = 1.52$
Round modified quota down	4	3	1
			Total = 8

Too small—we need a smaller modified divisor.

TABLE 11.20 Using the Jefferson method to apportion the oil consortium board using a modified divisor of 1,050.

Because the total number of board members assigned is too small, we need *larger* modified quotas. To do this, we must try *smaller* modified divisors. In making up this example, we used a spreadsheet and tried 1,000, and then 950, but neither of these modified divisors worked. Finally, when we used the modified divisor of 935, we found the results in Table 11.21. Of course, there are numbers other than 935 that would be acceptable modified divisors. We will ask you to investigate which other modified divisors would be acceptable in the exercises.

Quiz Yourself ⑩

Assume we are apportioning nine representatives among three groups A, B, and C, which have populations of 1,276, 1,427, and 2,697, respectively.

a) Calculate the standard divisor for this apportionment.

b) Use the modified divisor of 550 to calculate group C's modified quota.

c) Using Jefferson's method, how many representatives would C receive if we use the modified divisor 550?

KEY POINT

Adams's method rounds quotas up.

Modified Divisor = 935			
	Naxxon	**Aroco**	**Eurobile**
Number of stockholders	4,700	3,700	1,600
Standard quota	4.23	3.33	1.44
Modified quota	$\frac{4,700}{935} = 5.03$	$\frac{3,700}{935} = 3.96$	$\frac{1,600}{935} = 1.71$
Round modified quota down	5	3	1
			Total = 9

TABLE 11.21 Using the Jefferson method to apportion the oil consortium board using a modified divisor of 935. ✳ ⑩

We will next illustrate how to use Adams's method to apportion the oil consortium board.

In Adams's method, we are going to round the modified quotas up, so we will need modified quotas that are *smaller* than the standard quotas. This means that the modified divisor has to be *larger* than the standard divisor to give us the smaller quotas.

ADAMS'S APPORTIONMENT METHOD

a) Use trial and error to find a modified divisor that is larger than the standard divisor for the apportionment.

b) Calculate the modified quota (state's population/modified divisor) for each state and round it up. Assign that number of representatives to each state. (Keep varying the modified divisor until the sum of these assignments is equal to the total number being apportioned.)

We will now reapportion the oil consortium board using Adams's method.

EXAMPLE 6 *Using Adams's Apportionment Method*

Recall that Naxxon had 4,700 stockholders, Aroco had 3,700, and Eurobile had 1,600. Now use Adams's method to apportion the nine-member board.

SOLUTION: Because the standard divisor for this apportionment was $\frac{10,000}{9} \approx 1,111.11$, we need to try a larger modified divisor. We will first try a modified divisor of 1,200, as we show in Table 11.22.

Modified Divisor = 1,200			
	Naxxon	**Aroco**	**Eurobile**
Number of stockholders	4,700	3,700	1,600
Standard quota	4.23	3.33	1.44
Modified quota	$\frac{4,700}{1,200} = 3.92$	$\frac{3,700}{1,200} = 3.08$	$\frac{1,600}{1,200} = 1.33$
Round modified quota up	4	4	2
			Total = 10

Too large—we need a larger modified divisor.

TABLE 11.22 Using Adams's method to apportion the oil consortium board using a modified divisor of 1,200.

Because the total board members assigned is too large, we need *smaller* modified quotas. To do this, we must try *larger* modified divisors. When we try a modified divisor of 1,250 we get the results in Table 11.23.

Quiz Yourself 11

Assume we are apportioning 10 representatives among three groups A, B, and C, which have populations of 1,300, 950, and 2,550, respectively.

a) Calculate the standard divisor for this apportionment.

b) Use the modified divisor of 500 to calculate group A's modified quota.

c) Using Adams's method, how many representatives would A receive if we use the modified divisor 500?

Modified Divisor = 1,250			
	Naxxon	**Aroco**	**Eurobile**
Number of stockholders	4,700	3,700	1,600
Standard quota	4.23	3.33	1.44
Modified quota	$\frac{4,700}{1,250} = 3.76$	$\frac{3,700}{1,250} = 2.96$	$\frac{1,600}{1,250} = 1.28$
Round modified quota up	4	3	2
			Total = 9

TABLE 11.23 Using Adams's method to apportion the oil consortium board using a modified divisor of 1,250. ✳ **11**

As in the Jefferson method, there will be acceptable modified divisors other than 1,250 that we could have used in Example 6. We will pursue this question further in the exercises.

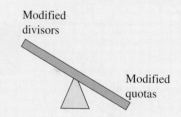

Some Good Advice

In order to know how to adjust the modified divisors when using the Jefferson and Adams methods, think of this diagram, where the modified divisors and quotas can be thought of as being on a teeter-totter. When one goes up, the other goes down.

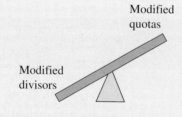

a) If you need more representatives, you need larger modified quotas, so you must use smaller modified divisors.

b) If you need fewer representatives, you need smaller modified quotas, so you must use larger modified divisors.

Finally, we will discuss Webster's apportionment method.

KEY POINT

Webster's method rounds quotas in the usual way.

In Webster's method, we round the modified quotas in the usual way. That is, if the quota has a decimal part of 0.5 or greater, we round up; otherwise, we round down. Again, as in the Jefferson and Adams method, we find the modified divisor by trial and error.

> **WEBSTER'S APPORTIONMENT METHOD**
>
> a) Use trial and error to find a modified divisor.
>
> b) Calculate the modified quota (state's population/modified divisor) for each state and round it in the usual way. Assign that number of representatives to each state. (Keep varying the modified divisor until the sum of these assignments is equal to the total number being apportioned.)

We will now reapportion the oil consortium board using Webster's method.

EXAMPLE 7 *Using Webster's Apportionment Method*

Naxxon has 4,700 stockholders, Aroco has 3,700, and Eurobile has 1,600. Use Webster's method to apportion the nine-member board.

SOLUTION: We need a starting place in using Webster's method, so as a first step, let's try a modified divisor of 1,111.11, which is the standard divisor (Table 11.24).

Modified Divisor = 1,111.11			
	Naxxon	**Aroco**	**Eurobile**
Number of stockholders	4,700	3,700	1,600
Standard quota	4.23	3.33	1.44
Modified quota	$\frac{4,700}{1,111.11} = 4.23$	$\frac{3,700}{1,111.11} = 3.33$	$\frac{1,600}{1,111.11} = 1.44$
Round modified quota in usual way	4	3	1
			Total = 8

Too small—we need a smaller modified divisor.

TABLE 11.24 Using the Webster method to apportion the oil consortium board using a modified divisor of 1,111.11.

Because the total board members assigned is too small, we need *larger* modified quotas. To do this, we must try *smaller* modified divisors. After trying several modified divisors, we tried the modified divisor of 1,060 to get the results in Table 11.25.

Quiz Yourself **12**

Assume we are apportioning eight representatives among three groups A, B, and C, which have populations of 2,700, 4,300, and 5,000, respectively.

a) Calculate the standard divisor for this apportionment.

b) Use the modified divisor of 1,540 to calculate group B's modified quota.

c) Using Webster's method, how many representatives would B receive if we use the modified divisor 1,540?

Modified Divisor = 1,060			
	Naxxon	**Aroco**	**Eurobile**
Number of stockholders	4,700	3,700	1,600
Standard quota	4.23	3.33	1.44
Modified quota	$\frac{4,700}{1,060} = 4.43$	$\frac{3,700}{1,060} = 3.49$	$\frac{1,600}{1,060} = 1.51$
Round modified quota in usual way	4	3	2
			Total = 9

TABLE 11.25 Using Webster's method to apportion the oil consortium board using a modified divisor of 1,060.

Now try Exercises 5 to 28. ❊ **12**

Again, there will be acceptable modified divisors other than 1,060. In the exercises, we will ask you to find the largest and the smallest integers that would be acceptable modified divisors for Example 7.

PROBLEM SOLVING

The Analogies Principle

In learning these apportionment methods, try to relate them to each other. In each case, first you must find a modified divisor and then round in one of three ways: Jefferson (down), Adams (up), or Webster (usual way).

The following diagram, although a little silly, will help you remember which way to round with each method. Think of the J in "Jefferson" as an arrow pointing down, and the A in "Adams" as the tip of an arrow pointing up. The W in "Webster" is pointing neither strictly up nor down.

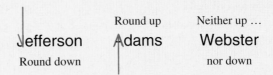

Don't bother memorizing how to select the modified divisor—you can always start with the standard divisor and remember that if your apportionment is too large, then you need smaller quotas and therefore you need a larger modified divisor. Likewise, if the apportionment is too small, you need a smaller modified divisor.

We summarize the four methods we have studied in Table 11.26. As you see from this table, the Jefferson, Adams, and Webster methods are not perfect because they all can fail to satisfy the quota rule. We will give examples of such violations in the exercises.

	Hamilton	**Jefferson**	**Adams**	**Webster**
Can have Alabama paradox	Yes	No	No	No
Can have population paradox	Yes	No	No	No
Can have new states paradox	Yes	No	No	No
Can violate quota rule	No	Yes	Yes	Yes

TABLE 11.26 Summary of four apportionment methods.

Exercises 11.4

Looking Back*

These exercises follow the general outline of the topics presented in this section and will give you a good overview of the material that you have just studied.

1. How do you calculate the standard divisor for an apportionment and the standard quota for a state?

2. In addition to the Alabama paradox, what are two paradoxes that can occur with the Hamilton method?

3. How did we generally find the modified divisor in using the Jefferson, Adams, and Webster methods?

4. Why were lawsuits filed in the Supreme Court against the state of Texas?

Sharpening Your Skills

In Exercises 5–8, we give you a total population, state A's population, and the total number of representatives that are to be apportioned. Find:

a) *The standard divisor for the apportionment and the standard quota for A.*

b) *The number of representatives that would be given to the state using Jefferson's method.*

c) *The number of representatives that would be given to the state using Adams's method.*

d) *The number of representatives that would be given to the state using Webster's method.*

---

*Before doing these exercises, you may find it useful to review the note *How to Succeed at Mathematics* on page xix.

5. Total population 512,000; A's population 84,160; number of representative's being apportioned 16

6. Total population 135,000; A's population 48,000; number of representative's being apportioned 9

7. Total population 102,000; A's population 19,975; number of representative's being apportioned 11

8. Total population 512,000; A's population 84,160; number of representative's being apportioned 15

Applying What You've Learned

9. **Assigning seats on a water authority.** Assume that California, Nevada, and Arizona are cooperating to build a dam to provide water to communities currently lacking adequate water supplies. Seats on the 11-member Southwest Water Authority, which governs the project, are assigned according to the number of customers in each state who use the water from the project. There are 56,000 customers in California, 52,000 in Arizona, and 41,000 in Nevada. Find the standard divisor and each state's standard quota.

10. Use the Jefferson method to assign the seats on the Southwest Water Authority.

11. Use the Adams method to assign the seats on the Southwest Water Authority.

12. Use the Webster method to assign the seats on the Southwest Water Authority.

13. **Choosing representatives on a negotiations committee.** The employees of Six Flags theme park are negotiating a new contract. There are 213 performers, 273 food workers, and 178 maintenance workers. The 20-person negotiations committee has members in proportion to the number of employees in each of the three groups. Find the standard divisor and each group's standard quota.

14. Use the Jefferson method to assign the members of the negotiations committee.

15. Use the Adams method to assign the members of the negotiations committee.

16. Use the Webster method to assign the members of the negotiations committee.

17. **Apportioning a labor council.** A labor council is being formed from the members of four unions. The electricians union has 25 members, the plumbers have 18 members, the painters have 29, and the carpenters have 31 members. The council will have 20 representatives. Find the standard divisor and each union's standard quota.

18. Use the Jefferson method to assign the representatives to the council.

19. Use the Adams method to assign the representatives to the council.

20. Use the Webster method to assign the representatives to the council.

21. **Awarding assistantships.** Nineteen fellowships are to be apportioned to students in a writing program at a major university. The apportionment is to be based on the number of full-time graduate students in each area of the writing program.

There are 30 fiction majors, 20 poetry majors, 17 majoring in technical writing, and 14 majoring in writing for the media. Find the standard divisor and each major's standard quota.

22. Use the Jefferson method to assign the fellowships.

23. Use the Adams method to assign the fellowships.

24. Use the Webster method to assign the fellowships.

25. **Scheduling fitness classes.** A health club instructor has a course load that allows her to teach six two-credit hour classes. A preregistration survey indicates the following interests:

    56 want to take Pilates;

    29 want to take kick boxing;

    11 want to take yoga; and

    4 want to take spinning.

    Find the standard divisor and the standard quotas for each interest area.

26. Use the Jefferson method to assign the number of classes to each area.

27. Use the Adams method to assign the number of classes to each area.

28. Use the Webster method to assign the number of classes to each area.

In Exercises 29–36, we use the Hamilton method to do the apportionment.

29. Ten representatives are apportioned among three states A, B, and C, which have the populations (in thousands) shown in Table 11.27. Ten years later, the population of A has increased by 20 thousand and the population of B has increased by 60 thousand, whereas C's population remained the same. The 10 representatives are reapportioned. Does this example illustrate the population paradox? Explain.

State	Population
A	570
B	2,557
C	6,873

TABLE 11.27

30. The Metrodelphia Regional Transportation Authority has three lines, the Arrow, the Blue Line, and the City division. The numbers of passengers that use each line per day are given in Table 11.28. Assume that 100 new cars are apportioned among the three lines according to the number of passengers using each line. One year later, the MRTA is able to purchase another 100 cars and again apportions them among the three divisions. We show this increase in ridership in Table 11.28. Does this example illustrate the population paradox? Explain.

Line	Passengers per Day (now)	Passengers per Day (one year later)
Arrow	23,530	23,930
Blue	5,550	5,650
City	70,920	71,100

TABLE 11.28

31. A group of three regional airports has hired a company to increase security. The contract calls for 150 security personnel to be apportioned to the airports according to the number of passengers using each airport each week. Two years later, the number of passengers per week has increased at the three airports, so the 150 security personnel are reallocated. We show these data in Table 11.29. Does this example illustrate the population paradox? Explain.

Airport	Passengers per Week (now)	Passengers per Week (one year later)
Allenport	120,920	121,420
Bakerstown	5,550	5,650
Columbia City	23,530	23,930

TABLE 11.29

32. The current enrollments at a university's three campuses are shown in Table 11.30. The New State University's administration has decided to upgrade 18 computer labs according to the enrollments at the three campuses. One year later, the enrollments have increased at each campus and again it is decided to upgrade 18 labs. Does this example illustrate the population paradox? Explain.

Campus	Students (now)	Students (one year later)
Altus	7,200	7,480
Brightsfield	20,500	21,140
Center City	52,300	52,900

TABLE 11.30

33. Two states A and B have formed a 100-member commission to explore ways in which they can work jointly to improve the business climate in their states. The members will be apportioned to the commission according to the populations of the two states. After seeing the success of the commission, a third state asks to join and appoint members to the commission. States A and B agree and will allow state C to add as many members to the commission as they would deserve now using the current standard divisor. The populations of the three states are given in Table 11.31. Does this example illustrate the new states paradox? Explain.

State	Population (thousands)
A	2,000
B	7,770
C	600

TABLE 11.31

34. Redo Exercise 33, but now use the populations in Table 11.32. Also assume that the original commission has 40 members.

State	Population (thousands)
A	12,550
B	4,450
C	8,250

TABLE 11.32

35. Redo Exercise 33, but now use the populations in Table 11.33. Also assume that the original commission has 56 members. Assume that A, B, and C are the original states and D is the new state.

State	Population (thousands)
A	8,700
B	4,300
C	3,200
D	6,500

TABLE 11.33

36. Redo Exercise 35, but now use the populations in Table 11.34. Also assume that the original commission has 70 members.

State	Population (thousands)
A	230
B	740
C	380
D	600

TABLE 11.34

Exercises 37–40 illustrate that the Jefferson and Adams apportionment methods can violate the quota rule. In each case, determine which method illustrates the violation, and explain how the quota rule is not satisfied.

37. We are apportioning 200 representatives among the states A, B, C, D, E, and F. The populations of these states are given in Table 11.35.

State	Population
A	700
B	1,500
C	820
D	4,530
E	2,200
F	550

TABLE 11.35

38. We are apportioning 100 representatives among the states A, B, C, D, E, and F. The populations of these states are given in Table 11.36.

State	Population
A	700
B	8,000
C	820
D	1,500
E	4,000
F	550

TABLE 11.36

39. We are apportioning 500 representatives among the states A, B, C, D, E, and F. The populations of these states are given in Table 11.37.

State	Population
A	1,400
B	2,000
C	1,500
D	9,000
E	4,300
F	500

TABLE 11.37

40. We are apportioning 400 representatives among the states A, B, C, D, E, and F. The populations of these states are given in Table 11.38.

State	Population
A	700
B	1,000
C	750
D	200
E	2,100
F	4,500

TABLE 11.38

Communicating Mathematics

41. What is the intuitive meaning of *standard divisor* and *standard quota*?

42. How do the Jefferson, Adams, and Webster methods use the modified quotas differently?

43. Explain why the modified divisor in the Jefferson method is usually smaller than the standard divisor.

44. Explain why the modified divisor in the Adams method is usually greater than the standard divisor.

45. In using the Adams method to apportion a legislature, Juanita got an assignment with too many representatives. What change must she make in her approach?

46. In using the Webster method to apportion a legislature, Joe got an assignment with too few representatives. What change must he make in his approach?

Using Technology to Investigate Mathematics

47. See your instructor for Excel spreadsheets that implement the various apportionment methods and illustrate the paradoxes we have discussed. Use these spreadsheets to duplicate several of the computations and solve some of the exercises in this section.

48. You can find applets on the Internet that illustrate the various apportionment methods we discussed in this section. For example, search for "applets and apportionment." Run several applets to duplicate some of the apportionments made in this section. Report on your findings.

For Extra Credit

49. In Example 5, we found that a modified divisor of 935 worked in using Jefferson's apportionment method. Use trial and error to find the smallest and the largest integers that we could have used in this example as modified divisors.

50. In Example 6, we found that a modified divisor of 1,250 worked in using Adams's apportionment method. Use trial and error to find the smallest and the largest integers that we could have used in this example as modified divisors.

51. In Example 7, we found that a modified divisor of 1,060 worked in using Webster's apportionment method. Use trial and error to find the smallest and the largest integers that we could have used in this example as modified divisors.

52. Make up several hypothetical countries each consisting of several large and several small states. Experiment with different apportionments to see if the Jefferson method seems to favor large or small states.

53. Repeat Exercise 52 for the Adams method.

54. Repeat Exercise 52 for the Webster method.

Looking Deeper

11.5 Fair Division

Objectives

1. Understand the difference between discrete and continuous division of items.
2. Use the method of sealed bids to make a fair division of items.

How often do you agree to a deal only to find out after it is done that you will receive more than you expected? You will now learn how to divide any group of items among several interested parties that does exactly that. This method works in dividing an inheritance among the heirs of an estate, awarding property to a couple getting divorced, or dissolving a business among partners.

Fair Division

There are various ways to solve these problems, which are called **fair division** problems. We will now explain one method for handling *discrete* fair division problems as opposed to *continuous* fair division problems.

> **DEFINITIONS** **Discrete fair division** means that the items being distributed cannot be subdivided into fractional parts. **Continuous fair division** means that the items being distributed can be subdivided into parts and the division can be done as finely as we want.

For example, we could not subdivide an antique car into parts to distribute it among heirs to an estate. On the other hand, we can subdivide a cake and distribute pieces of it among children at a birthday party. **13**

Method of Sealed Bids

We will now discuss the *method of sealed bids*, which is based on work done by the Polish mathematician Hugo Steinhaus.* We start by looking at the fair division of an estate consisting of one item being shared equally by several people. Our problem is to determine how each person, in his or her opinion, can get a fair share when the item cannot be physically divided. But first, we must explain what we mean by a person's *fair share* of an estate.

> **DEFINITION** A person's **fair share** of an estate is the person's estimate of the estate's value divided by the number of people sharing equally in the estate.

Wayne Malloy has died and left to three cousins—Dahlia, Cael, and DiDi—a rare painting that has been in the family for generations. The three cousins agree that the painting should not be sold, but each is entitled to an equal share of the bequest. Obviously, the painting cannot be divided, so they agree that one of them will get the painting and pay cash to the others.

Dahlia, Cael, and DiDi contact several appraisers to establish the painting's value. However, they receive such a wide range of estimates that it only causes confusion. Dahlia, who has taken a mathematics course in which she studied fair division, suggests that they proceed in the following way.

---

*Hugo Steinhaus, *Mathematical Snapshots*, 3rd ed. (New York: Oxford University Press, 1969), pp. 65–74.

Quiz Yourself **13**

Identify each situation as dealing with either discrete or continuous division of items.

a) Apportioning the seats in the U.S. House of Representatives.

b) Dividing a piece of land among three people.

c) Dividing a mixture of M&Ms, hard candy, and pieces of bubble gum among five children.

d) Sharing a can of soda among three people.

Each of them secretly writes down an estimate of the painting's value and places the bid in a sealed envelope. By doing this, the three of them will establish an opinion on the estate's value and, in turn, their fair share of the estate. For the process to work, each must promise to give an honest estimate and be content to receive a fair share. Next, they will open the bids and the person making the highest bid will get the painting.

Upon opening the envelopes, they find that Dahlia has valued the painting at $84,000, Cael feels the painting is worth $60,000, and DiDi thinks it is worth $72,000. As agreed, Dahlia gets the painting because she had the highest bid. But now she must pay the estate an amount of money. The amount is her estimate of the estate's value less her fair share of the estimate.

Because Dahlia had valued the estate at $84,000 and because the cousins are to share equally in the estate, Dahlia's estimate of her fair share is

$$\frac{1}{3} \times 84{,}000 = \$28{,}000.$$

Therefore, she pays the estate the difference:

value of painting ⌐ ⌐ Dahlia's fair share

$$\$84{,}000 - \$28{,}000 = \$56{,}000.\text{— Dahlia owes estate}$$

But should Cael and DiDi split that money and get $28,000 apiece? It would be unfair if they did, because they each estimated the estate at a lower value. Furthermore, they agreed to be content with receiving their fair share. Because Cael bid $60,000, he thinks that his fair share is

$$\frac{1}{3} \times 60{,}000 = \$20{,}000.$$

With DiDi's bid of $72,000, she perceives that her fair share of the estate is

$$\frac{1}{3} \times 72{,}000 = \$24{,}000.$$

Giving Cael $20,000 and DiDi $24,000 from the estate leaves a balance of

Dahlia paid to estate ⌐ ⌐ Cael and DiDi's fair shares

$$\$56{,}000 - \$20{,}000 - \$24{,}000 = \$12{,}000. \text{— remaining balance}$$

in the estate. Now what do we do with the balance? Because the three cousins have equal interests in the estate, we will divide this balance equally among them. Therefore, Cael and DiDi each get an additional $4,000, and the remaining $4,000 is returned to Dahlia.

Table 11.39 shows each person's estimate of the estate's value and his or her opinion of a fair share. Table 11.40 summarizes the accounting of cash based on our method of fair division. A plus sign indicates the person pays that amount to the estate, whereas a minus sign means the person receives that amount from the estate. Observe in Table 11.40 that the amount paid to the estate must equal the sum of the amounts received from the estate.

	Dahlia $\left(\frac{1}{3}\right)$	Cael $\left(\frac{1}{3}\right)$	DiDi $\left(\frac{1}{3}\right)$
Bid on painting	$84,000	$60,000	$72,000
Fair share of estate	$\dfrac{\$84{,}000}{3} = \$28{,}000$	$\dfrac{\$60{,}000}{3} = \$20{,}000$	$\dfrac{\$72{,}000}{3} = \$24{,}000$
Item obtained with high bid	Painting		

TABLE 11.39 Estimates of value and fair share.   Dahlia, Cael, and DiDi said they would be satisfied with these amounts.

	Dahlia $\left(\frac{1}{3}\right)$	Cael $\left(\frac{1}{3}\right)$	DiDi $\left(\frac{1}{3}\right)$
Pays to estate (+) or receives from estate (−)	$56,000 (+)	$20,000 (−)	$24,000 (−)
Division of estate balance ($12,000)	$4,000 (−)	$4,000 (−)	$4,000 (−)
Summary of cash	Pays $52,000	Receives $24,000	Receives $28,000

TABLE 11.40 Fair division accounting of estate cash.

It may appear at first that there is some unfairness in this division because DiDi received more money than Cael. However, keep in mind that each party agreed to be content with his or her estimate of a fair share when using this method. But what is interesting is that each party ends up with more than expected. First, Dahlia gets the painting and pays $4,000 less than her estimate of the others' fair share, namely, $52,000 versus $56,000. However, Cael and DiDi each received $4,000 more than their estimated fair share. This "win–win" outcome is due to Dahlia's higher assessment of the painting, which raised Cael and DiDi's share, whereas their smaller estimates lowered the amount that Dahlia had to pay to the estate. **14**

Quiz Yourself **14**

Rosa, Juan, Carlos, and Luis have inherited a house from their parents. Because they have equal interests in the house and want to keep it in the family, they will use the method of sealed bids to decide who gets the house. That person will pay cash to the others. The following table shows the four bids on the house. Complete the table by showing each person's estimate of his or her fair share of the estate and who gets the house.

	Rosa $\left(\frac{1}{4}\right)$	Juan $\left(\frac{1}{4}\right)$	Carlos $\left(\frac{1}{4}\right)$	Luis $\left(\frac{1}{4}\right)$
Bid on house	$250,000	$240,000	$260,000	$200,000
Fair share of estate				
Item obtained with highest bid				

Quiz Yourself **15**

We continue the fair division problem discussed in Quiz Yourself 14. Because Carlos gets the house, he must pay an amount of cash to the estate, whereas the others will receive cash. Construct a table to show an accounting of the estate's cash transactions.

EXAMPLE 1 *Dividing an Estate with Several Items*

Matt, Annie, Connor, and Wilber have inherited equal portions of an estate left by McNamara and Troy. The estate consists of a house, an expensive car, and a boat. They decide to use the method of sealed bids in dividing the estate, and each of the four submits a sealed bid on each item in the estate. Their bids, and therefore their individual estimates of a fair share of the estate, are listed in Table 11.41. Determine how the estate should be divided using the method of sealed bids. **15**

	Matt $\left(\frac{1}{4}\right)$	Annie $\left(\frac{1}{4}\right)$	Connor $\left(\frac{1}{4}\right)$	Wilber $\left(\frac{1}{4}\right)$
Bid on house	$165,000	$180,000	$160,000	$190,000
Bid on car	$55,000	$50,000	$35,000	$40,000
Bid on boat	$40,000	$36,000	$25,000	$30,000
Total value	$260,000	$266,000	$220,000	$260,000
Estimate of fair share of estate	$\frac{1}{4} \cdot \$260,000 = \$65,000$	$\frac{1}{4} \cdot \$266,000 = \$66,500$	$\frac{1}{4} \cdot \$220,000 = \$55,000$	$\frac{1}{4} \cdot \$260,000 = \$65,000$

TABLE 11.41 Estimates of value and fair share.

SOLUTION: We see from the last line of Table 11.41 that Matt and Wilber will be satisfied if they each receive a combination of items and cash worth at least $65,000, whereas Connor would be happy with $55,000, but Annie would not be satisfied with less than $66,500. Remember that no one knows the others' estimates of the estate's value until the bids are opened. When the bids are opened, Wilber gets the house because he has the high bid of $190,000. Because Matt bid the highest on both the car and the boat, he gets those items. Of course, Connor and Annie get none of the items but will receive cash from the estate. However, it is interesting to note that even though Annie did not submit the highest bid on any item, she had the largest estimate of the total value of the estate. Let us now do the final accounting of the cash transactions for the estate.

We start with those who received items from the estate. Wilber got the house, which is worth $190,000 to him. But he thinks that his fair share of the total estate is worth $65,000. Therefore, he received considerably more than his fair share, namely the difference

$$\overbrace{\$190,000}^{\text{value of house}} - \overbrace{\$65,000}^{\text{Wilber's fair share}} = \overbrace{\$125,000}^{\text{Wilber pays estate}}.$$

He must pay this difference to the estate. Matt received the car and the boat from the estate. Based on his bids for these items, he received a total value of

$$\$55,000 + \$40,000 = \$95,000.$$

However, this is more than what he feels is his fair share of the estate. Therefore, he must also pay the estate the difference between the value of items received and his fair share, namely,

$$\overbrace{\$95,000}^{\text{value of boat and car}} - \overbrace{\$65,000}^{\text{Matt's fair share}} = \overbrace{\$30,000}^{\text{Matt pays estate}}.$$

The combined cash paid to the estate by Wilber and Matt is

$$\$125,000 + \$30,000 = \$155,000.$$

Last, Connor and Annie receive their perceived fair share from these payments. The cash accounting thus far is tabulated in Table 11.42.

	Matt $\left(\frac{1}{4}\right)$	Annie $\left(\frac{1}{4}\right)$	Connor $\left(\frac{1}{4}\right)$	Wilber $\left(\frac{1}{4}\right)$
Item obtained with high bid	Car, boat			House
Pays to estate (+) or receives from estate (−)	$30,000 (+)	$66,500 (−)	$55,000 (−)	$125,000 (+)

TABLE 11.42 Fair division accounting of estate.

The last line of Table 11.42 shows that there is a cash balance left in the estate, namely,

$$\overbrace{\$30,000 + \$125,000}^{\text{Matt and Wilber paid to estate}} - \overbrace{\$66,500 - \$55,000}^{\text{Annie and Connor's fair shares}} = \overbrace{\$155,000 - \$121,000 = \$33,500}^{\text{balance}}.$$

Because the four heirs each have one-quarter interest in the estate, they each receive

$$\frac{1}{4} \cdot \$33,500 = \$8,375$$

of the balance. As with any fair division based on sealed bids, the four heirs get more than they expected. Both Matt and Wilber get items but pay less to the estate than they had expected. Although Annie and Connor do not receive items, they each get an amount of cash that is more than they think is their respective fair share. Table 11.43 summarizes the fair division of the estate. As before, the total paid to the estate equals the total received from the estate and the estate is now closed with a zero balance.

	Matt $\left(\frac{1}{4}\right)$	Annie $\left(\frac{1}{4}\right)$	Connor $\left(\frac{1}{4}\right)$	Wilber $\left(\frac{1}{4}\right)$
Item obtained with high bid	Car, boat			House
Pays to estate (+) or receives from estate (−)	$30,000 (+)	$66,500 (−)	$55,000 (−)	$125,000 (+)
Division of estate balance ($33,500)	$8,375 (−)	$8,375 (−)	$8,375 (−)	$8,375 (−)
Summary of cash	Pays $21,625	Receives $74,875	Receives $63,375	Pays $116,625

TABLE 11.43 Fair division accounting of estate. ✳

Let us summarize fair division through the method of sealed bids. We explain the process in terms of dividing an estate; however, the method can be applied to other, similar situations.

> **THE METHOD OF SEALED BIDS (EQUAL INTERESTS)** Each person having a stake in the fair division of the items makes a sealed bid on each item.
>
> 1. Each person's fair share is determined as follows:
>   - Add the person's bids on the various items.
>   - Divide this total by the number of people sharing in the estate.
>
>   Because different people make different bids, what is a fair share for one person may not be the same as a fair share to another.
>
> 2. Open the bids. The person bidding the highest amount on an item gets that item. If several people have the same highest bid, then the decision of who gets the item can be based on random selection.
>
> 3. If the total value of the items awarded to a person is more than the person's fair share, then the person pays the difference in cash to the estate. If the total value is less than the person's fair share, then the person receives the difference in cash from the estate. If a person does not receive any items, then the person gets his or her fair share in cash from the estate.
>
> 4. If, after completing step 3, there is a cash balance in the estate, then it is divided equally among the heirs.

The method of sealed bids will work successfully only when

- The parties involved make an honest bid on each item;
- The person making the highest bid on an item is willing to accept the item and pay cash to the estate if the bid is larger than the fair share; and
- Most important, each person makes a bid without knowledge of the others' bids.

We can also use this method when the parties have different percentage interests in the estate. For example, three heirs to an estate may have, respectively, 50%, 30%, and 20% interests as stated in a will. In this case, a person's fair share is his or her estimate of the estate's value times his or her percentage interest in the estate. The exercises have problems of this type.

Exercises 11.5

Looking Back*

These exercises follow the general outline of the topics presented in this section and will give you a good overview of the material that you have just studied.

1. What is the difference between discrete fair division and continuous fair division?

2. How did we calculate Dahlia's fair share of Wayne Malloy's painting?

3. In Example 1, why did Matt receive the boat and the car?

4. In Example 1, how did we get the balance of $33,500? What did we do with this balance?

Sharpening Your Skills

Identify each situation as dealing with either discrete or continuous division of items.

5. **a.** Dividing a collection of rare books among five people

   b. Dividing a chocolate bar among several children

   c. Apportioning the seats on a city council with representatives chosen from different neighborhoods of the city

6. **a.** Splitting candy collected at Halloween among three trick-or-treaters

   b. Dividing an estate consisting of a house and a boat among five heirs

   c. Dividing a piece of pie among four people

Use the method of sealed bids to complete the tables in Exercises 7–18.

7. **Dividing an inheritance.** Darnell and Joy want to determine who should inherit their rich aunt's antique car.

	Darnell $\left(\frac{1}{2}\right)$	Joy $\left(\frac{1}{2}\right)$
Bid on car	$110,000	$85,000
Fair share of estate	a.	b.
Item obtained with highest bid	c.	d.

8. **Dividing an inheritance.** Karl and Friedrich have inherited their father's copy of a rare book and want to decide who should receive it.

	Karl $\left(\frac{1}{2}\right)$	Friedrich $\left(\frac{1}{2}\right)$
Bid on book	$27,000	$35,000
Fair share of estate	a.	b.
Item obtained with highest bid	c.	d.

9. **Dissolving a partnership.** Dennis and Zadie have written a book together and now want to decide how to dissolve their partnership, with one of them obtaining sole rights to the copyright. Their current entitlement to the book is 40% belonging to Dennis and 60% belonging to Zadie.

	Dennis (40%)	Zadie (60%)
Bid on copyright	$20,000	$28,000
Fair share of copyright	a.	b.
Item obtained with highest bid	c.	d.

10. **Dissolving a partnership.** Matt and Tony own a small pizza stand at the shore, with Matt having 35% interest and Tony 65% interest. They want to dissolve their partnership, with one buying the other out.

	Matt (35%)	Tony (65%)
Bid on pizza stand	$120,000	$90,000
Fair share of pizza stand	a.	b.
Item obtained with highest bid	c.	d.

Applying What You've Learned

11. **Dividing an inheritance.** Three brothers—Ed, Al, and Jerry—want to decide who gets a valuable painting and statue that they have inherited from their grandfather.

	Ed $\left(\frac{1}{3}\right)$	Al $\left(\frac{1}{3}\right)$	Jerry $\left(\frac{1}{3}\right)$
Bid on painting	$13,000	$8,000	$9,000
Bid on statue	$14,000	$13,000	$15,000
Total value	$27,000	a.	b.
Fair share of estate	c.	d.	e.

12. **Dissolving a partnership.** Franz, Ida, Bill, and Monica are members of an investment club. Their portfolio consists of three stocks—a computer manufacturer, an oil company, and a pharmaceutical company. They want to dissolve their group and determine who gets the stocks and who gets cash.

---

*Before doing these exercises, you may find it useful to review the note *How to Succeed at Mathematics* on page xix.

	Franz $\left(\frac{1}{4}\right)$	Ida $\left(\frac{1}{4}\right)$	Bill $\left(\frac{1}{4}\right)$	Monica $\left(\frac{1}{4}\right)$
Bid on computer stock	$75,000	$80,000	$70,000	$90,000
Bid on oil stock	$35,000	$40,000	$45,000	$40,000
Bid on pharmaceutical stock	$40,000	$30,000	$25,000	$35,000
Total value	a.	b.	c.	d.
Fair share of portfolio	e.	f.	g.	h.

13. **Dividing an inheritance.** Consider your answer to Exercise 7. Complete the settlement of the estate by completing the fair division accounting of the estate cash as indicated in the table.

	Darnell $\left(\frac{1}{2}\right)$	Joy $\left(\frac{1}{2}\right)$
Pays to estate (+) or receives from estate (−)		
Division of estate balance ($)		
Summary of cash		

14. **Dividing an inheritance.** Consider your answer to Exercise 8. Complete the settlement of the estate by completing the fair division accounting of the estate cash as indicated in the table.

	Karl $\left(\frac{1}{2}\right)$	Friedrich $\left(\frac{1}{2}\right)$
Pays to estate (+) or receives from estate (−)		
Division of estate balance ($)		
Summary of cash		

15. **Dissolving a partnership.** Consider your answer to Exercise 9. Complete the settlement of the copyright by completing the fair division accounting of the cash as indicated in the table.

	Dennis (40%)	Zadie (60%)
Pays to pool (+) or receives from pool (−)		
Division of pool balance ($)		
Summary of cash		

16. **Dissolving a partnership.** Consider your answer to Exercise 10. Complete the settlement of the pizza stand by completing the fair division accounting of the cash as indicated in the table.

	Matt (35%)	Tony (65%)
Pays to pool (+) or receives from pool (−)		
Division of pool balance ($)		
Summary of cash		

17. **Dividing an inheritance.** Consider your answer to Exercise 11. Complete the settlement of the estate by completing the fair division accounting of the estate cash as indicated in the table.

	Ed $\left(\frac{1}{3}\right)$	Al $\left(\frac{1}{3}\right)$	Jerry $\left(\frac{1}{3}\right)$
Items obtained with highest bid			
Pays to estate (+) or receives from estate (−)			
Division of estate balance ($)			
Summary of cash			

18. **Dissolving a partnership.** Consider your answer to Exercise 12. Complete the settlement of the portfolio by completing the fair division accounting of the cash as indicated in the table.

	Franz $\left(\frac{1}{4}\right)$	Ida $\left(\frac{1}{4}\right)$	Bill $\left(\frac{1}{4}\right)$	Monica $\left(\frac{1}{4}\right)$
Stock obtained with highest bid				
Pays to pool (+) or receives from pool (−)				
Division of pool balance ($)				
Summary of cash				

In Exercises 19 and 20, use the method of sealed bids to determine how the items and cash are distributed.

19. **Dividing an inheritance.** Betty and Dennis want to divide a diamond ring, an antique desk, and a collection of rare books that were bequeathed to them by their dear aunt who passed away. Use the method of sealed bids to divide the items and determine a fair division of the estate. Betty's and Dennis's estimates of these items are given in the table.

	Betty	Dennis
Ring	$16,000	$18,000
Desk	$4,500	$5,000
Books	$4,000	$3,000

20. **Dissolving a partnership.** Matt, Dani, and Christian have a partnership in which they own a chain of four fast-food restaurants (call them A, B, C, and D). They have decided to dissolve their partnership but keep ownership of the restaurants individually among themselves. Determine a fair division by using the method of sealed bids. The table shows the partners' estimates of the values of the individual restaurants.

	Matt	Dani	Christian
A	$170,000	$180,000	$160,000
B	$145,000	$150,000	$155,000
C	$200,000	$190,000	$210,000
D	$70,000	$50,000	$45,000

Communicating Mathematics

21. Why are the fair shares of an estate usually different?

22. In dividing an estate, why can some parties receive no items?

23. In using the method of sealed bids, why can some parties receive more in cash than they expected?

24. What factors must be in place for the method of sealed bids to be successful?

Using Technology to Investigate Mathematics

25. Search the Internet for an applet that implements the method we described for solving fair division problems. Run the applet and report on your findings.

26. Search the Internet for applications of fair division. You might search for "fair division and applications" or "method of sealed bids and applications." Write a report on your findings.

For Extra Credit

In Exercises 27–30, we reconsider our opening example in which Dahlia, Cael, and DiDi are dividing Wayne Malloy's estate. In Exercises 31–34, we take another look at Example 1, in which Matt, Annie, Connor, and Wilber are dividing McNamara and Troy's estate. These questions are intended to be open ended, with no right or wrong answers. Just consider the implications of what might happen if the division is done when the rules of fairness are not being followed.

27. What can happen if Cael has prior knowledge of Dahlia's bid?

28. What can happen if Dahlia has prior knowledge of Cael's bid?

29. What can happen if Cael bids unfairly by bidding too low? Bidding too high?

30. What can happen if Dahlia bids unfairly by bidding too low? Bidding too high?

31. What can happen if Matt has prior knowledge of others' bids?

32. What can happen if Wilber has prior knowledge of others' bids?

33. What can happen if Annie bids unfairly by bidding too low? Bidding too high?

34. What can happen if Connor bids unfairly by bidding too low? Bidding too high?

CHAPTER SUMMARY*

You will learn the items listed in this chapter summary more thoroughly if you keep in mind the following advice:

1. Focus on "remembering how to remember" the ideas. What pictures, word analogies, and examples help you remember these ideas?

2. Practice writing each item without looking at the book.

3. Make up 3×5 flash cards to break your dependence on the text. Use these cards to give yourself practice tests.

SECTION	SUMMARY	EXAMPLE
SECTION 11.1	The **Alabama paradox** occurred in 1881 when Alabama lost one representative when the size of the legislature increased. The **Hamilton** method apportions representatives by giving each state the integer part of the exact number of representatives due to it. Additional representatives are allocated to states that have the largest fractional part of their exact representation.	Discussion, p. 531 Example 1, p. 533
	The **average constituency** of a state is defined as $$\frac{\text{population of the state}}{\text{number of representatives of the state}}.$$	Definitions, p. 534
	State A is **more poorly represented** than state B if A has a larger average constituency than B. If A is more poorly represented than B, the **absolute unfairness** is (average constituency of A) − (average constituency of B). The **relative unfairness** of an apportionment between two states is the absolute unfairness divided by the smaller average constituency.	Example 2, p. 535 Example 3, p. 535 Example 4, p. 536
Section 11.2	The **apportionment criterion** says that we assign a representative to a state to give the smallest relative unfairness.	Example 1, p. 539
	The **Huntington–Hill apportionment principle** says that if states X and Y have been allocated x and y representatives, respectively, then X should be given the next representative instead of Y provided $$\frac{(\text{population of Y})^2}{y \cdot (y+1)} < \frac{(\text{population of X})^2}{x \cdot (x+1)}.$$ These quotients are called **Huntington–Hill numbers**.	Example 2, p. 542
SECTION 11.3	In **apportioning the oil consortium** board, each time we allocated a representative to a company, we then calculated Huntington–Hill numbers to decide who gets the next representative.	Example 1, p. 545 Example 2, p. 546
	We can use the Huntington–Hill method in apportioning other objects where we do not want to reassign objects once they have already been assigned.	Example 3, p. 548
SECTION 11.4	In apportioning representatives, we defined the **standard divisor** by $$\text{standard divisor} = \frac{\text{total population}}{\text{number of representatives being allocated}},$$	Discussion, p. 552
	and a state's **standard quota** as $$\text{standard quota} = \frac{\text{state's population}}{\text{standard divisor}}.$$	Discussion, p. 553 Example 1, p. 553 Example 2, p. 554
	If we round the standard quota down, we call that number the **lower quota**; if we round the standard quota up, we call that number the **upper quota**. The **quota rule** requires that each state's allocation of representatives be between its lower and upper quotas. **Hamilton's method** is as follows:	
	a) Give each state its lower quota.	Discussion, p. 555
	b) Assign remaining representatives according to the size of the fractional parts of the states' standard quotas.	

---

*Before reviewing this chapter, you will find it useful to reread the note *How to Succeed at Mathematics* on page xix.

The **population paradox** occurs when A's population is growing faster than B's, yet A loses a representative to B. The **new states paradox** occurs when a new state is added and its share of seats is added to the legislature, causing a change in the allocation of seats previously given to another state.

Example 3, p. 555
Example 4, p. 557

Jefferson's method is as follows:

a) Find a modified divisor that is smaller than the standard divisor for the apportionment.

b) Calculate the modified quota for each state and round it down to determine the number of representatives assigned to the state.

Example 5, p. 559

Adams's method is as follows:

a) Find a modified divisor that is larger than the standard divisor for the apportionment.

b) Calculate the modified quota for each state and round it up to determine the number of representatives assigned.

Example 6, p. 560

Webster's method is as follows:

a) Use trial and error to find a modified divisor.

b) Calculate the modified quota for each state and round it in the usual way to determine the number of representatives assigned.

Example 7, p. 562

SECTION 11.5   **Discrete fair division** means that the items being distributed cannot be subdivided into fractional parts, as opposed to **continuous fair division**, where they can be subdivided.

Discussion, p. 567

A person's **fair share** of an estate is that person's estimate of the estate's value divided by the number of people sharing equally in the estate.

Discussion, pp. 567–568

In **the method of sealed bids (equal interests)**, each person makes a sealed bid on each item, then:

Example 1, p. 569

1. Divide a person's estimate of the value of the estate by the number of people sharing in the estate to determine that person's fair share.

2. The person making the highest bid on an item must take it.

3. If the total value of items awarded to a person is greater than his or her fair share, he or she must pay the estate; otherwise, that person receives cash from the estate.

4. If, after step 3, there remains a cash balance in the estate, then it is divided equally among the heirs.

CHAPTER REVIEW EXERCISES

Section 11.1

1. What is the Alabama paradox?

2. The governing board of the American History Roundtable Society has 11 members. The seats are apportioned among the four divisions of the society—Revolutionary War (560 members), Civil War (524), World War I (431), and World War II (485)—according to the number of members in each division. Use the Hamilton method to apportion seats on the board.

3. Explain why an Alabama paradox can occur when using the Hamilton apportionment method.

4. Suppose state A has a population of 935,000 and five representatives, whereas state B has a population of 2,343,000 and 11 representatives.

   a. Determine which state is more poorly represented, and calculate the absolute unfairness for this assignment of representatives.

   b. Determine the relative unfairness for this apportionment.

Section 11.2

5. Suppose that Florida has a population of approximately 16.03 million and has 25 representatives and Texas has a population of 21.49 million and has 32 representatives.

   a. Calculate the Huntington–Hill numbers for both Florida and Texas.

   b. According to the Huntington–Hill apportionment principle, which of these states is most deserving to receive an additional representative?

6. Explain why the Huntington–Hill apportionment principle will avoid an Alabama paradox.

Section 11.3

7. In the 2008 Summer Olympics, in addition to the host city of Beijing, several other cities—Qingdao, Tianjin, and Qinhuangdao—also hosted some events. You are to allocate 13 temporary train services to these cities. Each city is

guaranteed at least one train service and you are to use the Huntington–Hill method and the given table of Huntington–Hill numbers to allocate the train services.

Current Representation	(B)eijing	(Q)ingdao	(T)ianjin	Qin(H)uangdao
1	21,012.5	13,448.0	3,362.0	27,378.0
2	7,004.2	4,482.7	1,120.7	9,126.0
3	3,502.1	2,241.3	560.3	4,563.0
4	2,101.3	1,344.8	336.2	2,737.8
5	1,400.8	896.5	224.1	1,825.2
6	1,000.6	640.4		1,303.7

8. An instructor at the Physical Fitness Institute can teach eight classes. A preregistration survey indicates the following interests:

    66 want to take tae-bo;

    39 want to take karate;

    18 want to take weight training; and

    23 want to take tai chi.

    Assume that the instructor will teach at least one class for each area. Use the Huntington–Hill method to apportion the instructor's remaining four classes among the four areas.

Section 11.4

9. A health club instructor has a course load that allows her to teach eight classes. A preregistration survey indicates the following interests:

    8 want to take Pilates;

    64 want to take kick boxing;

    11 want to take yoga; and

    31 want to take spinning.

    Apportion her classes using the Jefferson method.

10. Apportion the classes in Exercise 9 using the Adams method.

11. Apportion the classes in Exercise 9 using the Webster method.

12. A group of three regional airports has hired a company to increase security. The contract calls for 150 security personnel to be apportioned to the airports according to the number of passengers using each airport each week. One year later, the number of passengers per week has increased at the three airports, so the 150 security personnel are reallocated. We show these data in the given table. (We are using the Hamilton method.) Does this example illustrate the population paradox? Explain.

Airport	Passengers per Week (now)	Passengers per Week (one year later)
A	80,500	87,700
B	6,800	7,410
C	93,300	101,300

13. What do we mean by the new states paradox?

14. Which apportionment method(s) do not satisfy the quota rule?

Section 11.5

15. Three cousins—Tito, Omarosa, and Piers—are arguing over how to divide an inheritance of a rare antique rifle and a jewel-encrusted sword that belonged to their grandfather. Use the method of sealed bids to complete the table below and solve their problem.

	Tito $\left(\frac{1}{3}\right)$	Omarosa $\left(\frac{1}{3}\right)$	Piers $\left(\frac{1}{3}\right)$
Bid on rifle	$12,000	$17,000	$10,000
Bid on sword	$15,000	$13,000	$11,000
Estimated total value of estate			
Fair share of estate			

16. Continuing with the situation in Exercise 15, complete the following table.

	Tito $\left(\frac{1}{3}\right)$	Omarosa $\left(\frac{1}{3}\right)$	Piers $\left(\frac{1}{3}\right)$
Items obtained with highest bid			
Pays to estate (+) or receives from estate (−)			
Division of estate balance ($)			
Summary of cash			

CHAPTER TEST

1. What is the Alabama paradox?

2. Suppose state C has a population of 1,640,000 and eight representatives and state D has a population of 1,863,000 and nine representatives. Determine the absolute and relative unfairness of this assignment of representatives.

3. The Metropolitan Community College Arts Council will consist of eight members. The seats are to be apportioned according to student participation in the areas of art (47 students), music (111 students), and theater (39 students). Use the Hamilton method to apportion the council.

4. Explain why an Alabama paradox can occur with the Hamilton method.

5. Suppose that Arizona has a population of 5.23 million and has eight representatives and Oregon with a population of 3.61 million has five representatives.

    a. Calculate the Huntington–Hill numbers for each state.

    b. According to the Huntington–Hill method, which state is most deserving of an additional representative?

6. Explain why the Huntington–Hill apportionment method avoids an Alabama paradox.

7. You are to allocate 14 shuttle buses to four areas in a theme park: Mystery Mountain, Jungle Village, Great Frontier, and City Sidewalks. Each area is to have at least one bus. Use the given table to apportion the remaining buses to each area using the Huntington–Hill method.

Current Number of Buses	Mystery Mountain	Jungle Village	Great Frontier	City Sidewalks
1	6,612.5	3,612.5	2,112.5	4,050.0
2	2,204.2	1,204.2	704.2	1,350.0
3	1,102.1	602.1	352.1	675.0
4	661.3	361.3	211.3	405.0
5	440.8	240.8	140.8	270.0

8. Your college's community outreach program has 11 volunteers for an afterschool tutoring program for elementary school students. Assume that any volunteer can tutor any subject. There are 16 requests for mathematics, 9 for reading, and 6 for study skills. Use the Huntington–Hill method to apportion the tutors to the three areas. Assume that each area will have at least one volunteer assigned.

9. Upscale Malls, Inc. has allocated 150 security personnel to its malls in Allenwood, Black Hills, and Colombia according to the weekly numbers of customers at each mall. Assume that Upscale builds a new mall in Devon, requiring new security personnel who are allocated to Devon according to the first apportionment. We show these data in the given table.

	Allenwood	Black Hills	Colombia	Devon
Number of customers per week	80,500	6,800	93,300	41,400

Use the Hamilton method to do the apportionment. Does this example illustrate the new states paradox? Explain.

10. What is the population paradox?

11. The Relaxation Institute can offer nine sections of stress management courses per week. A preregistration survey shows the following interest:

47 want to take massage;

32 want to take aromatherapy;

41 want to take yoga; and

21 want to take meditation.

Assume that the institute will offer at least one section in each area. Apportion the remaining sections using the Webster method.

12. Apportion the sections in Exercise 11 using the Jefferson method.

13. Apportion the sections in Exercise 11 using the Adams method.

14. Which apportionment method(s) that you have studied satisfy the quota rule?

15. Three brothers—Larry, Moe, and Curly—are dissolving their company Stooge Investments, Inc., which consists of two branches A and B. They want to keep the business in the family, so some of the brothers will buy the others out. Use the method of sealed bids to dissolve their partnership using the table below.

	Larry ($\frac{1}{3}$)	Moe ($\frac{1}{3}$)	Curly ($\frac{1}{3}$)
Estimate of A's worth	$13 million	$9 million	$10 million
Estimate of B's worth	$17 million	$18 million	$14 million

GROUP EXERCISES

1. a. Make up an example in which the Hamilton and the Huntington–Hill methods give you the same apportionment.

b. Make up an example in which the Hamilton and the Huntington–Hill methods give you different apportionments.

2. Make up new examples to illustrate the population paradox and new states paradox. Do this by trial and error by using spreadsheets or online applets that do apportionment.

3. Try to find an example where the Jefferson, Adams, and Webster methods all give different apportionments.

4. a. Make up an example of a method of sealed bids problem in which three people bid on three items and each person receives exactly one item.

b. Make up an example of a method of sealed bids problem in which three people bid on three items and one person receives all three items.

Voting 12
Using Mathematics to Make Choices

As a citizen of a free society, you have both a right and a duty to express your choice by voting in elections. In addition to selecting political leaders, you will vote on many other issues: Are you in favor of raising taxes to improve your state's clean air and water standards? As a school board member, do you support building a new high school or renovating the old one? Should we eliminate the death penalty? Do you favor using casino profits to give tax relief to those in your community with a fixed income? At work, whom do you support to lead your union or head your department? These are only a few of the political, environmental, and financial choices that you will make that affect the quality of your everyday life.

If there are only two choices, it is easy to determine the winner; the choice receiving the most votes wins. This voting system is called *majority rule*. However, when there are more than two choices, then selecting a winner is more complicated. We encounter voting situations with several alternatives when conducting primary elections, selecting the rookie of the year, and awarding a Grammy or an

Academy Award. What is the best method for conducting an election in these situations?

As you will see, the answer to that question is not simple. In this chapter, you will learn how to choose winners using a variety of voting systems. We will investigate the fairness of these systems and you will see that each system has curious quirks built into it that go against your intuition. For example, you will see a surprising example where, in order to win an election, a candidate asks, "Please vote against me so I can win."

Later in the chapter, we will also analyze weighted voting systems, in which not every voter has the same number of votes; you will also learn several ways to measure voters' power in such systems. ●

12.1 Voting Methods

Objectives

1. Use the plurality method to determine the winner of an election.
2. Understand the Borda count voting method.
3. Use the plurality-with-elimination method to determine the winner of an election.
4. Determine the winner of an election using the pairwise comparison method.

Are you ever troubled by the way our election system works? For example, does it seem fair that in the 2008 Republican primary in Florida, John McCain was awarded all 57 delegates to the Republican National Convention *even though 64% of the voters voted against him*? Mitt Romney, who garnered slightly over 30% of the votes, received none of the delegates. Would it not have been fairer to split the delegates among the candidates according to the number of votes they received? Or what about in 2000 when George W. Bush won the election even though Al Gore defeated him in the popular vote?*

Does it bother you (as it does me) that after a few small, early primaries for president, most candidates cannot raise enough money and have to withdraw from the race before most of us even get a chance to cast our vote? Certainly there has to be a voting system that would give us a better selection of candidates by the time our state's primary rolls around?

As you will see, voting can be more complicated than it first seems and the winner of an election depends not only on the votes cast but also on how we agree to use those votes. In this section, we will introduce you to several different voting systems, and in Section 12.2 we will investigate their weaknesses.

✎ **KEY POINT**

Using the plurality method, the person with the most votes wins an election.

The Plurality Method

The **plurality method** is the simplest way to determine the outcome of an election; the person earning the most votes is the winner.

> **DEFINITION The Plurality Method**
> Each person votes for his or her favorite candidate. The candidate receiving the most votes is declared the winner.

Many state and local elections use the plurality method because it is easy to determine the winner—all we need to do is tally the votes for each candidate.

---

*In U.S. presidential elections, each state is assigned a certain number of electoral votes. Without getting into details of exactly how the electoral college operates, it is fair to say that if a candidate wins a state, then that candidate is given the entire number of electoral votes allocated to that state. The winner of the presidential election is then the candidate who acquires the most electoral votes.

EXAMPLE 1 *Determining the Winner Using the Plurality Method*

Imagine that student employees on your campus have decided to organize a union in order to improve their salaries and working conditions. As a first step, a group of 33 students, calling themselves the Undergraduate Labor Council (ULC), have just had an election to choose their president. The results of this election are as follows:

Ann	10
Ben	9
Carim	11
Doreen	3

Using the plurality method, who is the winner of this election?

SOLUTION: Because Carim has the most votes, he is declared the winner. ❋

Note in Example 1 that using the plurality method, Carim won the election even though $\frac{22}{33} \approx 66.7\%$ of the ULC members voted against him.

 **KEY POINT**

Using the Borda count method, the person earning the most points wins an election.

The Borda Count Method

Although it was too late once the election had been conducted, Ann realized that even though she was not the first choice of most voters, she was the second choice of a great many voters. Unfortunately for Ann, the ballot did not allow voters to state their second, third, and fourth preferences. A voting method called the **Borda count method** permits a voter to "fine-tune" his or her vote in the sense that the voter can designate not only a first choice but also a second choice, a third choice, and so on. In the ULC election, we could have used the Borda method by specifying that on a voter's ballot the first choice would be given 4 points, the second choice 3 points, the third choice 2 points, and the fourth choice 1 point.

> **DEFINITION** **The Borda Count Method**
> If there are *k* candidates in an election, each voter ranks all candidates on the ballot. Then the first choice is given *k* points, the second choice is given *k* – 1 points, the third choice *k* – 2 points, and so on.* The candidate who receives the most total points wins the election.

To use the Borda count method in the ULC election in Example 1, voters must rank candidates on their ballots. For example, the given ballot would mean that a voter preferred Carim first, Ann second, Ben third, and Doreen fourth. Such a ballot is called a **preference ballot**.

1st	C	———— receives 4 points
2nd	A	———— receives 3 points
3rd	B	———— receives 2 points
4th	D	———— receives 1 point

With 33 voters, some preference ballots would be the same,† so in tallying the votes we group identical ballots together in a table called a **preference table**.

---

*In some variations of the Borda count method, different numbers of points are awarded for first place, second place, third place, and so on. For example, we may give five points to first place, three to second, and one to third. We investigate these variations in the exercises.

†When we discuss permutations in Section 12.4, you will see that there are only 24 possible different preference ballots for A, B, C, and D.

EXAMPLE 2   *Determining the Winner Using the Borda Count Method*

Assume that the ULC used the Borda count method to determine its president. Table 12.1 summarizes the preference ballots cast in the election. Who is the winner of the election?

Carim, Ann, Ben, and Doreen have the same number of first-place votes as before.

	Number of Ballots					
Preference	6	7	5	3	9	3
1st	C	A	C	A	B	D
2nd	A	C	D	D	A	A
3rd	B	B	B	B	D	C
4th	D	D	A	C	C	B

Ann has 18 second-place votes.

TABLE 12.1 Preference table for ULC election.

SOLUTION: The numbers at the top of the columns in Table 12.1 show how many voters submitted the particular preference ballot in that column. For example, the 6 means that six voters had preference ballots choosing Carim first, Ann second, Ben third, and Doreen fourth. Notice, as before, Carim has 11 first-place votes, Ann has 10, Ben 9, and Doreen 3.

We can use Table 12.1 to calculate each student's point total, awarding 4 points for each first-place vote, 3 points for each second-place vote, and so on. For example, Ann's total is

$$10 \times 4 + 18 \times 3 + 0 \times 2 + 5 \times 1 = 40 + 54 + 0 + 5 = 99$$

ten first-place votes times 4 points

eighteen second-place votes times 3 points

five fourth-place votes times 1 point

Table 12.2 summarizes the election results.

	Number of Points				
Candidates	**1st-Place Votes × 4 (Points)**	**2nd-Place Votes × 3 (Points)**	**3rd-Place Votes × 2 (Points)**	**4th-Place Votes × 1 (Points)**	**Total Points**
A	$10 \times 4 = 40$	$18 \times 3 = 54$	$0 \times 2 = 0$	$5 \times 1 = 5$	99
B	$9 \times 4 = 36$	$0 \times 3 = 0$	$21 \times 2 = 42$	$3 \times 1 = 3$	81
C	$11 \times 4 = 44$	$7 \times 3 = 21$	$3 \times 2 = 6$	$12 \times 1 = 12$	83
D	$3 \times 4 = 12$	$8 \times 3 = 24$	$9 \times 2 = 18$	$13 \times 1 = 13$	67
				Total	330

TABLE 12.2 Points earned using the Borda method for the ULC election.

When we examine the voting, we see that Ann now wins the election. ❋

 **PROBLEM SOLVING**

Verify Solutions

After completing a problem, double-check your answer for accuracy. In Example 2, we found that 330 points were awarded. This happened because each of the 33 voters was awarding 10 points.

HIGHLIGHT ✿ ✿ ✿

Between the Numbers—Would You Vote for This?

Year after year, after only a few small primaries, viable presidential candidates fall by the wayside before many states have had a chance to conduct their primary elections, leaving voters frustrated and feeling that their votes do not count. Some say that this problem arises from using the plurality voting method and suggest that we use another method called *approval voting* instead. In **approval voting**, voters may vote for as many candidates on the ballot as they want and the winner is the person who earns the most votes of approval. Supporters argue that approval voting tends to select the strongest candidate and reduces infighting among candidates.

In recent years another method, called **instant runoff voting** (IRV)* has been gaining support and in fact was supported by both Barack Obama and John McCain prior to the 2008 election. With IRV, voters rank their candidates and after the votes are counted, the weakest candidate is eliminated. However, if that happens to be your candidate, your vote is not wasted, because when your first choice is eliminated, your vote will then be counted towards your second choice.

Many polls use the Borda method to rank sports teams. For example, because there is no comprehensive playoff system to determine a national champion in college football, often there is controversy as to which team is number one. One of several polls used in determining this championship is conducted by the Associated Press. Each voter in this poll gives 25 points to his or her number-one choice, 24 to the second choice, and so on, to select the top 25 teams each week. **1**

Quiz Yourself **1** †

Use the preference table to determine the winner of this election using the Borda count method. How many points does the winning candidate receive?

	Number of Ballots			
Preference	**8**	**7**	**5**	**7**
1st	C	D	C	A
2nd	A	A	B	D
3rd	B	B	D	B
4th	D	C	A	C

Because the outcomes of the plurality method and the Borda count method are different, you may wonder which is the "correct" method for determining the ULC president. We will not answer this question. Rather, knowing that these two methods yield different results, which do you believe is more appropriate?

✏️ **KEY POINT**

The plurality-with-elimination method eliminates the weakest candidate before revoting.

The Plurality-with-Elimination Method

When we used the plurality method in the ULC election in Example 1, we obtained the following results.

Ann	10
Ben	9
Carim	11
Doreen	3

One could argue that because Doreen is so far out of the running, perhaps we should eliminate her and conduct a new election with just Ann, Ben, and Carim as candidates. If no candidate earns a majority (more than 50%) in this second election, then we could eliminate the person receiving the smallest number of votes and then conduct a third election. This voting method is called the **plurality-with-elimination method**.

---

*For entertaining explanations of how IRV works, go to www.fairvote.org/?page=1895, or www.fairvote.org/?page=1973.
†Quiz Yourself answers begin on page 778.

> **DEFINITION   The Plurality-with-Elimination Method**
> Each voter votes for one candidate. A candidate receiving a majority of votes is declared the winner. If no candidate receives a majority of votes, then the candidate (or candidates) with the fewest votes is dropped from the ballot and a new election is held.* This process continues until a candidate receives a majority of votes.

To avoid many rounds of voting, we can simplify the elimination process. If we assume that each voter has a ranking of all the candidates, we can construct a preference table as in the Borda count method. Of course, we will use this table differently than we do in the Borda method.

Throughout this section we will assume that once a voter has ranked the candidates, *this ranking does not change during subsequent rounds of voting.* That is, if a voter prefers Ann to Doreen and Doreen to Ben in round one, if Doreen is eliminated, then the voter will prefer Ann to Ben in round two.

EXAMPLE 3   *Determining the Winner Using the Plurality-with-Elimination Method*

Determine the winner of the ULC election using the plurality-with-elimination method.

SOLUTION: The preference table for this election is shown in Table 12.3.

Doreen has the fewest first-place votes, so she is eliminated.

Preference	Number of Ballots					
	6	7	5	3	9	3
1st	C	A	C	A	B	D
2nd	A	C	D	D	A	A
3rd	B	B	B	B	D	C
4th	D	D	A	C	C	B

TABLE 12.3 Preference table for the ULC election.

Remaining candidates move up.

From Table 12.3, we see that Doreen has only 3 first-place votes, so she is eliminated and a new election is held. Remember, *we are assuming that voters do not change their preferences from one round of voting to the next.* For example, because three voters had Doreen as their first choice, now that she is eliminated, those three voters now have Ann as their first choice. One way to think of this adjustment is to imagine that when we eliminate D from a column of the table, all candidates below D in that column move up one row. By eliminating Doreen, we get Table 12.4.

Preference	Number of Ballots					
	6	7	5	3	9	3
1st	C	A	C	A	B	A
2nd	A	C	B	B	A	C
3rd	B	B	A	C	C	B

TABLE 12.4 Preference table for the ULC election after Doreen is eliminated.

Combine identical ballots.

We combine identical columns in Table 12.4 to get Table 12.5.

---

*To reduce the number of rounds of voting, some variations of this method may drop more than one candidate from the second round of voting. For example, it may be specified that only the top two candidates are allowed to proceed to the second round of voting.

	Number of Ballots				
Preference	6	10	5	3	9
1st	C	A	C	A	B
2nd	A	C	B	B	A
3rd	B	B	A	C	C

Ben now has the fewest first-place votes, so he is eliminated.

TABLE 12.5 Preference table for the ULC election with three candidates.

We see now that Ann has 13 first-place votes, Ben has 9, and Carim has 11. Thus, Ben is eliminated and a third round of voting takes place. Table 12.6 summarizes the preferences at this point.

	Number of Ballots				
Preference	6	10	5	3	9
1st	C	A	C	A	A
2nd	A	C	A	C	C

Ann has 22 first-place votes; Carim has 11.

TABLE 12.6 Preference table for the ULC election after removing Ben.

In Table 12.6 we see that Ann has 22 first-place votes and Carim has 11, so Ann wins the ULC election. ❀ **2**

Quiz Yourself **2**

Use the preference table to determine the winner of this election using the plurality-with-elimination method.

	Number of Ballots				
Preference	8	9	5	4	2
1st	C	E	B	A	A
2nd	A	D	C	D	C
3rd	B	B	E	B	B
4th	E	C	A	C	E
5th	D	A	D	E	D

✏ **KEY POINT**

In the pairwise comparison method, the candidate who can beat the most other candidates "head-to-head" is the winner.

The Pairwise Comparison Method

It is natural to expect that the winner of an election should be capable of beating each of the other candidates "head-to-head." We show a variation of this in the next method, the **pairwise comparison method**.

> **DEFINITION　The Pairwise Comparison Method**
> Voters first rank all candidates. If A and B are a pair of candidates, we count how many voters prefer A to B and vice versa. Whichever candidate is preferred the most receives 1 point. If A and B are tied, then each receives $\frac{1}{2}$ point. Do this comparison, assigning points, for each pair of candidates. At the end, the candidate receiving the most points is the winner.

As with the plurality-with-elimination method, we assume that if a voter prefers A to B and B to C, then the voter will prefer A to C. In Example 4, we see that we can use a voting method to choose among alternatives as well as decide elections.

EXAMPLE 4 *Determining a Preference Using the Pairwise Comparison Method*

In order to decide which new item to add to its menu, the Chipotle chain did a market survey in which customers were asked to rank their preferences for (T)acos, (N)achos, and (B)urritos. The chain will use the pairwise comparison method to decide which item to add to its menu. The results of counting the ballots are shown in Table 12.7(a). Which item should the chain select?

Preference	Number of Ballots					
	2,108	**864**	**1,156**	**1,461**	**1,587**	**1,080**
1st	T	T	N	N	B	B
2nd	N	B	T	B	T	N
3rd	B	N	B	T	N	T

TABLE 12.7(a) Preference table for voting on menu items.

SOLUTION: We must compare a) T with N, b) T with B, and finally, c) N with B.

a) In comparing T with N, we will ignore all references to B in Table 12.7(b). You can see that 4,559 prefer T over N but only 3,697 prefer N over T, thus we award 1 point to option T.

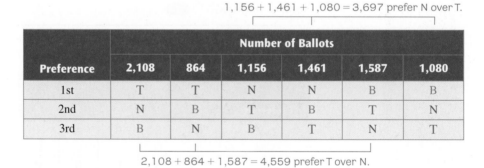

1,156 + 1,461 + 1,080 = 3,697 prefer N over T.

Preference	Number of Ballots					
	2,108	**864**	**1,156**	**1,461**	**1,587**	**1,080**
1st	T	T	N	N	B	B
2nd	N	B	T	B	T	N
3rd	B	N	B	T	N	T

2,108 + 864 + 1,587 = 4,559 prefer T over N.

TABLE 12.7(b) Comparing T with N.

b) In comparing T with B, we now ignore all references to N, as we see in Table 12.7(c). In this comparison, you see that 4,128 prefer T over B and also 4,128 prefer B over T. So T and B each receive $\frac{1}{2}$ point.

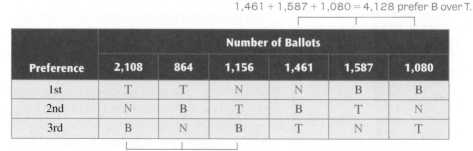

1,461 + 1,587 + 1,080 = 4,128 prefer B over T.

Preference	Number of Ballots					
	2,108	**864**	**1,156**	**1,461**	**1,587**	**1,080**
1st	T	T	N	N	B	B
2nd	N	B	T	B	T	N
3rd	B	N	B	T	N	T

2,108 + 864 + 1,156 = 4,128 prefer T over B.

TABLE 12.7(c) Comparing T with B.

c) Finally we compare N with B. Table 12.7(a) tells us that $2,108 + 1,156 + 1,461 = 4,725$ customers prefer N over B. Also, $864 + 1,587 + 1,080 = 3,531$ customers prefer B over N. We award 1 point to N.

Because T has $1\frac{1}{2}$ points, N has 1 point, and B has $\frac{1}{2}$ point, option T is declared the winner. ✳ ③

Quiz Yourself ③

Redo Example 4 using the following preference table to determine which item to add to Chipotle's menu. We are again using the pairwise comparison method. Also list how many points each item gets.

	Number of Ballots					
Preference	985	864	1,156	1,021	1,187	1,080
1st	T	T	N	N	B	B
2nd	N	B	T	B	T	N
3rd	B	N	B	T	N	T

Table 12.8 summarizes the voting methods that we have discussed in this section.

Method	How the Winning Candidate Is Determined
Plurality	The candidate receiving the most votes wins.
Borda count	Voters rank all candidates by assigning a set number of points to first choice, second choice, third choice, and so on; the candidate with the most points wins.
Plurality-with-elimination	Successive rounds of elections are held, with the candidate receiving the fewest votes being dropped from the ballot each time, until one candidate receives a majority of votes.
Pairwise comparison	Candidates are compared in pairs, with a point being assigned the voters' preference in each pair. (In the case of a tie, each candidate gets a half point.) After all candidates have been compared, the candidate receiving the most points wins.

TABLE 12.8 Summary of voting methods.

Which of these voting methods is the best? Unfortunately, each method we have introduced has serious flaws, as you will see in Section 12.2.

Exercises 12.1

Looking Back*

These exercises follow the general outline of the topics presented in this section and will give you a good overview of the material that you have just studied.

1. Why did Ann defeat Carim when we used the Borda count method in Example 2?

2. Why was Doreen the first to be eliminated in Example 3?

3. How were points awarded when comparing choices T and B in Example 4?

4. What are some of the claimed advantages of approval voting?

Sharpening Your Skills

5. Four candidates running for a vacant seat on the town council receive votes as follows: Edelson, 2,156; Borowski, 1,462; Callow, 986; Lopez, 428.

   a. Is there a candidate who earns a majority?

   b. Who wins the election using the plurality method?

*Before doing these exercises, you may find it useful to review the note *How to Succeed at Mathematics* on page xix.

6. Five candidates running for mayor receive votes as follows: Kinison, 415; Sanchez, 462; Lichman, 986; Devlin, 428; Saito, 381.

   a. Is there a candidate who earns a majority?

   b. Who wins the election using the plurality method?

The university administration has asked a group of student leaders to vote on the aspects of college life to target for improvement over the next year. The choices were (D)ining facilities, (A)thletic facilities, (C)ampus security, and the (S)tudent union building. The votes are summarized in the following preference table. Use this information to answer Exercises 7–10.

Preference	Number of Ballots					
	15	30	18	17	10	2
1st	A	D	A	C	C	S
2nd	C	C	S	D	S	A
3rd	S	S	D	S	A	C
4th	D	A	C	A	D	D

7. Voting on improving college life. What option is selected using the plurality method?

8. Voting on improving college life. What option is selected using the Borda count method?

9. Voting on improving college life. What option is selected using the plurality-with-elimination method?

10. Voting on improving college life. What option is selected using the pairwise comparison method?

The drama society members are voting for the type of play they will perform next season. The choices are a (D)rama, (C)omedy, (M)ystery, or (G)reek tragedy. The votes are summarized in the following preference table. Use this information to answer Exercises 11–14.

Preference	Number of Ballots					
	10	15	13	12	5	7
1st	C	D	C	M	M	G
2nd	M	M	G	D	G	C
3rd	G	G	D	G	C	M
4th	D	C	M	C	D	D

11. Selecting a play. What type of play is selected using the plurality method?

12. Selecting a play. What type of play is selected using the Borda count method?

13. Selecting a play. What type of play is selected using the plurality-with-elimination method?

14. Selecting a play. What type of play is selected using the pairwise comparison method?

Before a conference on "Trends in the Next Decade," a panel of participants voted on the topic for the keynote address. The choices were (T)echnology, (P)overty in Third World Countries, (E)conomy, (G)lobal Warming, and (F)oreign Policy. The votes are summarized in the following preference table. Use this information to answer Exercises 15–18.

Preference	Number of Ballots				
	15	7	13	5	2
1st	T	E	G	P	E
2nd	P	P	F	G	F
3rd	E	G	E	F	G
4th	F	F	P	E	P
5th	G	T	T	T	T

15. Choosing a speaker. What topic is selected using the plurality method?

16. Choosing a speaker. What topic is selected using the Borda count method?

17. Choosing a speaker. What topic is selected using the plurality-with-elimination method?

18. Choosing a speaker. What topic is selected using the pairwise comparison method?

A small employee-owned Internet company is voting on a merger with one of its competitors. The choices are (E)-Mall, (F)irestorm, (S)ecurenet, (T)echenium, and (L)ondram. The employees' votes are summarized in the following preference table. Use this information to answer Exercises 19–22.

Preference	Number of Ballots				
	15	11	9	10	2
1st	L	F	S	T	L
2nd	F	S	L	E	S
3rd	S	L	F	F	E
4th	E	E	T	L	F
5th	T	T	E	S	T

19. Deciding on a merger. What option is their first choice using the plurality method?

20. Deciding on a merger. What option is their first choice using the Borda count method?

21. Deciding on a merger. What option is their first choice using the plurality-with-elimination method?

22. Deciding on a merger. What option is their first choice using the pairwise comparison method?

Applying What You've Learned

Variations of the Borda count method are often used to determine winners of sports awards. In these methods, voters assign candidates a certain number of points for first place, fewer for second place, and so on. In Exercises 23 and 24, first determine the voting scheme being used and then complete the table.

23. **Awarding the Heisman Trophy.** The voting for the 2007 Heisman Trophy, which was awarded to Tim Tebow, is summarized in the following table. Find the missing entries in the table.

Player, Team	1st	2nd	3rd	Total
Tim Tebow, Florida	462	a. ___	113	1,957
Darren McFadden, Arkansas	b. ___	355	120	1,703
Colt Brennan, Hawaii	54	114	c. ___	632
Chase Daniel, Missouri	25	84	182	425

24. **Awarding the Cy Young Award.** The voting for the 2007 American League Cy Young Award, which was awarded to C. C. Sabathia, is summarized in the following table. Find the missing entries in the table.

Player, Team	1st	2nd	3rd	Total
C. C. Sabathia, Cleveland	19	a. ___	0	119
Josh Beckett, Boston	b. ___	14	4	86
John Lackey, Los Angeles Angels	1	5	c. ___	36
Fausto Carmona, Cleveland	0	1	4	7

We can use each voting method described in this section to rank candidates in addition to determining the winner. The plurality method and the Borda count method can rank candidates in order of the most first-place votes and the points earned, respectively. Use this information to answer Exercises 25 and 26.

25. Devise a method for ranking candidates using the plurality-with-elimination method.

26. Devise a method for ranking candidates using the pairwise comparison method.

In Exercises 27–30, refer to the preference table that was given prior to Exercise 7 regarding choices for improvement of college life. Use that table to rank the choices using the specified method.

27. The plurality method

28. The Borda count method

29. The plurality-with-elimination method

30. The pairwise comparison method

In Exercises 31–34, refer to the preference table that was given prior to Exercise 15 regarding choices for the conference keynote address. Use that table to rank the choices using the specified method.

31. The plurality method

32. The Borda count method

33. The plurality-with-elimination method

34. The pairwise comparison method

35. What is the total number of points that all candidates can earn in an election using the Borda count method if there are four candidates and 20 voters?

36. What is the total number of points that all candidates can earn in an election using the Borda count method if there are five candidates and 18 voters?

37. What is the maximum number of points that a candidate can earn in an election using the Borda count method if there are five candidates and 22 voters?

38. What is the minimum number of points that a candidate can earn in an election using the Borda count method if there are six candidates and 20 voters?

39. What is the total number of points that all candidates can earn in an election using the pairwise comparison method if there are four candidates?

40. What is the total number of points that all candidates can earn in an election using the pairwise comparison method if there are five candidates?

41. What is the maximum number of points that a candidate can earn in an election using the pairwise comparison method if there are five candidates?

42. What is the minimum number of points that a candidate can earn in an election using the pairwise comparison method if there are five candidates?

Communicating Mathematics

43. What is a benefit of the Borda count method over the plurality method?

44. When is the election decided using the plurality-with-elimination method?

45. What is the philosophy behind the pairwise comparison method?

In Exercises 46–48, assume that only two candidates are running in the election.

46. Explain why the winner using the Borda count method will have a majority of the votes.

47. Explain why the winner using the plurality-with-elimination method will have a majority of the votes.

48. Explain why the winner using the pairwise comparison method will have a majority of the votes.

Using Technology to Investigate Mathematics

49. See your instructor for Excel spreadsheets that will implement some of the voting methods that we discussed in this section.* Or, do searches on the Internet for applets that will implement the various voting methods that we discussed in this section. For example, if you search for "Borda Count Applets," you will find interactive programs that implement the Borda count method. Use the spreadsheets or applets to reproduce some of the examples in this section and to solve some of the exercises.

50. You may find it interesting to research some of the controversies and suggestions for improving the election process in the United States. You might search for "proportional voting," "electronic voting," "Internet voting," "approval voting," and so on. Write a report on your findings.

For Extra Credit

In approval voting, a person can vote for more than one candidate. In Exercises 51–54, we will assume that voters can vote for their top three choices. The candidate who receives the most votes of approval wins the election. Use the preference tables specified to conduct the election.

51. Use the preference table given before Exercise 7.

52. Use the preference table given before Exercise 11.

53. Use the preference table given before Exercise 15.

54. Use the preference table given before Exercise 19.

12.2 Defects in Voting Methods

Objectives

1. Understand the majority criterion and how it is not satisfied by the Borda count method.
2. See how the plurality method violates Condorcet's criterion.
3. Recognize that the pairwise comparison method can fail to satisfy the independence-of-irrelevant-alternatives criterion.
4. See how the plurality-with-elimination method fails the monotonicity criterion.

At this point, you may prefer one of the voting systems that we discussed in Section 12.1 over the others. Are you wondering which one of them is the best? You may be surprised by our answer at the end of this section.

We will now begin our discussion with some commonsense conditions that we want any election to have. An election should satisfy the following *four fairness criteria*:

- the majority criterion
- Condorcet's criterion
- the independence-of-irrelevant-alternatives criterion
- the monotonicity criterion

We will discuss which methods satisfy these criteria.

🖉 **KEY POINT**

According to the majority criterion, a candidate with more than half the first-place votes wins an election.

The Majority Criterion

The first criterion that we will consider is the majority criterion.

> **DEFINITION The Majority Criterion**
> If a majority* of the voters rank a candidate as their first choice, then that candidate should win the election.

---

*Majority means *more than* half.

It is clear that the plurality method and the pairwise comparison method satisfy the majority criterion. (Verify this.) However, this is not the case with the Borda count method, as we show in Example 1.

EXAMPLE 1   *The Borda Count Method Violates the Majority Criterion*

Motor Tendency Magazine has narrowed down its choice for car of the year to the following four cars: the high-end luxury (A)udi A8, the sporty (B)MW Z4, the classic (C)orvette Z06,

1st	C	C	D
2nd	D	D	A
3rd	A	B	B
4th	B	A	C

and the massive (D)odge Viper. A three-person editorial panel, using the Borda count method, will make the final selection. Preference ballots for the panel members are as shown at left. Determine the winner using the Borda count method, and show that this outcome violates the majority criterion.

SOLUTION: According to the Borda method, first-place votes are worth 4 points, second-place votes are worth 3 points, and so on. The following table summarizes the election results.

	Number of Points				
	1st-Place Votes (× 4 Points)	**2nd-Place Votes (× 3 Points)**	**3rd-Place Votes (× 2 Points)**	**4th-Place Votes (× 1 Point)**	**Total**
A	$0 \times 4 = 0$	$1 \times 3 = 3$	$1 \times 2 = 2$	$1 \times 1 = 1$	6
B	$0 \times 4 = 0$	$0 \times 3 = 0$	$2 \times 2 = 4$	$1 \times 1 = 1$	5
C	$2 \times 4 = 8$	$0 \times 3 = 0$	$0 \times 2 = 0$	$1 \times 1 = 1$	9
D	$1 \times 4 = 4$	$2 \times 3 = 6$	$0 \times 2 = 0$	$0 \times 1 = 0$	10

Although the Borda count winner is the Dodge Viper, the majority of first-place votes were cast for the Corvette Z06.

Now try Exercises 5 and 6. ❋ **④**

Quiz Yourself **④**

Use the Borda count method and the following preference table to determine whether the election satisfies the majority criterion.

	Number of Ballots			
Preference	**8**	**10**	**5**	**2**
1st	C	D	C	A
2nd	A	A	B	B
3rd	B	B	D	D
4th	D	C	A	C

✳ ✳ ✳ HISTORICAL HIGHLIGHT

The Marquis de Condorcet

Marie-Jean-Antoine-Nicolas Caritat, the Marquis de Condorcet, was an eighteenth-century French nobleman who was also a highly regarded philosopher and mathematician.

In 1785, Condorcet wrote a paper, *Essay on the Applications of Mathematics to the Theory of Decision Making*, in which he intended to prove that we could use mathematics to discover laws in the social sciences as precise as those in the physical sciences. He believed that scientists could employ mathematics to free society from its reliance on greedy capitalists and to develop insurance programs to protect the poor, and, once we discover appropriate moral laws, crime and war would disappear.

Although Condorcet's ideas may seem naive by twentieth-century standards, he was a forward thinker whose views are still relevant today. He believed in abolishing slavery, opposed capital punishment, advocated free speech, and supported early feminists lobbying for equal rights. As far as Condorcet was concerned, human rights were derived from people's ability to use reason in forming moral concepts.

Women, having these same qualities, must necessarily possess equal rights. Either no individual of the human species has any true rights, or all have the same. And he who votes against the rights of another, of whatever religion, color, or sex, has thereby abjured his own.*

 KEY POINT

Condorcet's criterion says that a candidate who defeats everyone else head-to-head is the winner.

Condorcet's Criterion

The next condition, which was proposed by the Marquis de Condorcet, is another desirable property of a voting method.

DEFINITION Condorcet's Criterion
If candidate X can defeat each of the other candidates in a head-to-head vote, then X is the winner of the election.

Clearly, a candidate with a majority of first-place votes can defeat every other candidate in a head-to-head election. However, Example 2 shows that the plurality method may violate Condorcet's criterion.

EXAMPLE 2   *The Plurality Method Violates Condorcet's Criterion*

A seven-member Olympic committee is voting using preference ballots to decide which one of (H)ungary, (I)ndia, and (T)aiwan will remain as a finalist country to host the 2016 Olympics. The partially completed preference ballots are as follows.

1st	H	H	H	I	I	T	T
2nd							
3rd							

Using the plurality voting method, Hungary is the winner. Complete the ballots so that in head-to-head voting India defeats both Hungary and Taiwan.

---

*This quote was taken from an article written by Condorcet in 1790, entitled "On Admission of Women to the Rights of Citizenship."

SOLUTION: Because Hungary defeats India in the first three ballots, we must force India to defeat Hungary on the remaining four ballots. We could do this as follows:

1st	H	H	H	I	I	T	T
2nd				H	H	I	I
3rd						H	H

At this point, India defeats Hungary by 4 to 3. We want India to also defeat Taiwan in head-to-head voting. Because Taiwan is currently leading India by 2 to 0, if we make Taiwan the last choice on the uncompleted ballots, we force India to defeat Taiwan. The final ballots are then as follows:

1st	H	H	H	I	I	T	T
2nd	I	I	I	H	H	I	I
3rd	T	T	T	T	T	H	H

We have shown that although Hungary wins the election using the plurality method, India defeats both Hungary and Taiwan in head-to-head competition, so Condorcet's criterion is not satisfied.

Now try Exercises 9, 10, 13, and 14.  ❋

You must be careful to understand what we are saying in Example 2 when we state that the plurality method *violates* Condorcet's criterion. We are not claiming that *every* vote using the plurality method violates Condorcet's criterion.* Example 2 simply provides a counterexample to the statement, "The plurality method satisfies Condorcet's criterion." ▪

Quiz Yourself ⑤

Complete the following ballots so that not only does India defeat Hungary and Taiwan in head-to-head voting but Taiwan also defeats Hungary in head-to-head voting.

1st	H	H	H	I	I	T	T
2nd	I	I	I	T	T	I	I
3rd	T	T	T	H	H	H	H

The Independence-of-Irrelevant-Alternatives Criterion

Suppose that after an election has been conducted using preference ballots we discover that one of the losing candidates should never have been listed on the ballot. Rather than have a new election, we could decide to strike each candidate's name from the ballot and do a recount. It is a desirable condition of voting that the winner using the original count should also be the winner using the modified ballots.

In Quiz Yourself 5, we see that although Hungary wins the election using the plurality method, most voters prefer both India and Taiwan over Hungary.

✎ KEY POINT

The independence-of-irrelevant-alternatives criterion says that removing losers from the ballot does not affect the outcome of the election.

DEFINITION   Independence-of-Irrelevant-Alternatives Criterion

If candidate X wins an election, some nonwinners are removed from the ballot, and a recount is done, then X still wins the election.

*Recall the Always Principle and the Counterexample Principle in Section 1.1.

EXAMPLE 3 *The Plurality Method Violates the Independence-of-Irrelevant-Alternatives Criterion*

The county board of supervisors is voting on methods to finance a new sports stadium. The options are levying a tax on (H)otel rooms, increasing the tax on (A)lcohol, and increasing the tax on (G)asoline. The board will use the plurality method to make its decision. The following preference table shows the results of the vote.

Number of Ballots			
Preference	8	6	6
1st	A	H	G
2nd	G	G	A
3rd	H	A	H

— A has most 1st-place votes.

Does this vote satisfy the independence-of-irrelevant-alternatives criterion?

SOLUTION: What we are really asking is whether the removal of one of the losing options changes the outcome of the election. If, due to lobbying pressure, the board removes the tax on hotel rooms as an option, we get the following:

Number of Ballots			
Preference	8	6	6
1st	A	G	G
2nd	G	A	A

— Now G has most 1st-place votes.

We see that the gasoline tax now wins by a vote of 12 to 8; thus, the plurality method does not satisfy the independence-of-irrelevant-alternatives criterion. ✷

Example 4 shows that the pairwise comparison method also violates the independence-of-irrelevant-alternatives criterion.

EXAMPLE 4 *The Pairwise Comparison Method Violates the Independence-of-Irrelevant-Alternatives Criterion*

The following table summarizes the preference ballots cast for candidates A, B, C, and D:

Number of Ballots				
Preference	8	4	5	1
1st	C	A	D	D
2nd	B	B	A	A
3rd	D	C	C	B
4th	A	D	B	C

Using the pairwise comparison method, who is the winner of this election? If any of the losing candidates are removed, does this change the results of the election?

SOLUTION: For each pair of candidates, we must determine the winner in a head-to-head vote. The following table shows the results of these comparisons:

	Vote Results	Points Earned
A vs. B	A wins 13 to 5.	A gets 1 point.
A vs. C	A wins 13 to 5.	A gets 1 point.
A vs. D	D wins 10 to 8.	D gets 1 point.
B vs. C	Tie—each has 9.	B and C get $\frac{1}{2}$ point.
B vs. D	B wins 13 to 5.	B gets 1 point.
C vs. D	C wins 13 to 5.	C gets 1 point.

Thus, A gets 2 points, B and C each get $1\frac{1}{2}$ points, and D gets 1 point. If we remove candidates B and C, the preference table looks like this:

Number of Ballots			
Preference	**8**	**4**	**5**	**1**
1st	A	D	D	D
2nd	D	A	A	A

We see that D defeats A by 10 votes to 8. Because the removal of some original losing candidates changes the outcome of the election, this method does not satisfy the independence-of-irrelevant-alternatives criterion.

Now try Exercises 11 and 12. ✻ ⑥

Quiz Yourself ⑥

Use the pairwise comparison method to determine the winner of the election summarized in the following preference table. Is the independence-of-irrelevant-alternatives criterion satisfied?

Number of Ballots			
Preference	**1**	**3**	**3**	**5**
1st	W	Z	Y	Z
2nd	X	W	X	W
3rd	Y	Y	Z	X
4th	Z	X	W	Y

The Monotonicity Criterion

Often before an election takes place, pollsters are hired to determine the preferences of the voters. If candidate X, the current front-runner, draws away support from one of his opponents, then it would seem likely that X has improved his chances of winning the election.* This idea motivates our final voting criterion, the monotonicity criterion.

KEY POINT

The monotonicity criterion states that if a candidate wins an election and then gains more support, then that candidate will win a reelection.

DEFINITION **The Monotonicity Criterion**

If X wins an election and in a reelection all voters who change their votes only change their votes to favor X, then X also wins the reelection.

If we are using the plurality method or the Borda count method, a candidate who wins an election and then gains more support will win any reelection.

Example 5 is the example that we described at the start of the chapter explaining why it may be an advantage for voters to vote against you in order for you to win an election.

---

*Of course, our intuition tells us that it depends on the voting method we are using.

❀ ❀ ❀ HIGHLIGHT

Using Spreadsheets* to Find Counterexamples

It is tedious to generate preference tables by hand to find flaws in a voting method. Instead, you can use spreadsheets to do the repetitive computations necessary to find a particular counterexample. We used the following spreadsheet to generate a preference table showing that the Borda count method fails Condorcet's criterion.

			=4*(E1+F1)+3*(B1+C1)+2*(G1)+1*(D1)				
C7	**A**	**B**	**C**	**D**	**E**	**F**	**G**
1		15	4	28	10	8	30
2	1st	C	D	C	A	A	B
3	2nd	A	A	B	D	B	C
4	3rd	B	B	D	B	C	A
5	4th	D	C	A	C	D	D
6							
7	BORDA	A pts	217				
8		B pts	286				
9		C pts	292	(C is Borda winner.)			
10		D pts	155				
11							
12		Total	950				
13		NVoters	95				
14							
15	CONDORCET						
16	A vs B	A=	37	B=	58		
17	A vs C	A=	22	C=	73		
18	A vs D	A=	63	D=	32		
19	B vs C	B=	52	C=	43		
20	B vs D	B=	81	D=	14		
21	C vs D	C=	81	D=	14		

Notice that although C is the Borda winner, B defeats every other candidate in a head-to-head contest. Thus, the Borda count method does not satisfy Condorcet's criterion.

We have highlighted cell C7, which contains a formula (shown at the top of the spreadsheet) for computing A's points using the Borda count method. By creating formulas such as this, we can simply change the numbers in row 1 to recalculate Borda point counts and tallies of who beats whom head-to-head. By adjusting the numbers in trial-and-error fashion, we quickly found the desired counterexample.

EXAMPLE 5 *The Plurality-with-Elimination Method Violates the Monotonicity Criterion*

Table 12.9 summarizes the preference ballots cast for the president of the International Students' Organization (ISO), in which Michael Chang, Rukevwe Kwami, and Anna Woytek are candidates. The plurality-with-elimination method is being used to determine the winner. The day before the election, three supporters of Woytek, who had preferred Chang over her, tell her that because they know she is going to win the election, they are going to throw their support

---

*See your instructor for a sample spreadsheet to do these calculations.

	Number of Ballots			
Preference	12	9	3	8
1st	W	K	C	C
2nd	C	W	W	K
3rd	K	C	K	W

K has fewest 1st-place votes and is eliminated. *(annotation pointing to 1st row)*

TABLE 12.9 Preference table for the ISO president election.

to her in the election tomorrow. Later that day, after talking with her uncle who is an expert in voting theory, Woytek calls the three new supporters and asks them to vote for Chang instead.

If the three voters indicated in the highlighted column in Table 12.9 change their votes so that Woytek is first, Chang is second, and Kwami is third, why should this cause Woytek concern?

SOLUTION: Let us first determine the winner of the election using the ballots in Table 12.9. We see that Woytek gets 12 first-place votes, Chang 11, and Kwami 9. Thus, Kwami is eliminated and a runoff election is held. Because voters do not change their preferences in the runoff, we can simply eliminate all references to Kwami on the original ballots and recount the votes. The ballots will look like this.

	Number of Ballots			
Preference	12	9	3	8
1st	W		C	C
2nd	C	W	W	
3rd		C		W

Woytek wins 21 to 11.

With Kwami removed from the ballots, we see that Woytek defeats Chang by a vote of 21 to 11.

Now let us see what would have happened if the three voters who preferred Chang to Woytek to Kwami had switched their votes. The ballots before the first elimination would have been as follows:

	Number of Ballots			
Preference	12	9	3	8
1st	W	K	W	C
2nd	C	W	C	K
3rd	K	C	K	W

Chang is eliminated.

These three voters have changed their votes. *(annotation pointing to column "3")*

Because Chang has the fewest first-place votes, he is eliminated and the ballots now look like this:

	Number of Ballots			
Preference	12	9	3	8
1st	W	K	W	
2nd		W		K
3rd	K		K	W

Kwami wins 17 to 15.

It is now clear why Woytek did not want the votes changed; in the runoff election between Woytek and Kwami, she loses 17 to 15. ❋

We have seen that each of the voting methods discussed in Section 12.1 violates an important voting criterion. Table 12.10 summarizes what we know about voting methods' defects at this point and indicates where we address this issue in the examples and exercises. In Table 12.10, a "Yes" entry means the criterion stated in the row is always satisfied by the method listed in the column. Otherwise, we reference a counterexample from this section showing that the method need not satisfy the stated criterion.

	Plurality	Borda Count	Plurality with Elimination	Pairwise Comparison
Majority	Yes	No—Example 1	Yes	Yes
Condorcet's	No—Example 2	No—Exercise 9	No—Example 2	Yes
Independence-of-irrelevant-alternatives	No—Example 3	No—Exercise 11	No—Exercise 24	No—Example 4
Monotonicity	Yes	Yes	No—Example 5	Yes

TABLE 12.10 Flaws in voting methods.

All the voting methods that you have studied so far have failed to satisfy at least one of the four fairness criteria, so you might be wondering, "Does any perfect voting method exist?" In 1951, an economist named Kenneth Arrow was studying decision making for a government think tank called the RAND Corporation. In his research, he discovered the following remarkable theorem that answers this question:

PROBLEM SOLVING
Arrow's Impossibility Theorem

In any election involving more than two candidates, there is no voting method that will satisfy all of the four fairness criteria.

Thus, any voting method that we decide to use must have flaws.

Exercises 12.2

Looking Back*

These exercises follow the general outline of the topics presented in this section and will give you a good overview of the material that you have just studied.

1. Which example in this section illustrated an undesirable property of the Borda count method?

2. What were the criteria we showed in this section that the plurality method failed to satisfy?

3. Which methods did we show failed the independence-of-irrelevant-alternatives criterion?

4. Why did we talk about spreadsheets in the Highlight on page 596?

Sharpening Your Skills

Some of these exercises have no fixed solution method. Often, constructing a preference table and adjusting the entries by trial and error will eventually lead to a solution.

5. In the preference table, A has the majority of first-place votes. Who wins the election if we use the Borda count method?

	Number of Ballots			
Preference	**12**	**15**	**9**	**13**
1st	A	B	C	A
2nd	B	C	B	D
3rd	C	A	D	B
4th	D	D	A	C

6. In the preference table on page 599, D has the majority of first-place votes. Who wins the election if we use the Borda count method?

---

*Before doing these exercises, you may find it useful to review the note *How to Succeed at Mathematics* on page xix.

Preference	Number of Ballots			
	4	10	3	2
1st	C	D	C	A
2nd	A	A	A	D
3rd	B	B	D	B
4th	D	C	B	C

Table for Exercise 6

Preference	Number of Ballots					
	15	4	8	10	8	2
1st	C	D	C	B	B	A
2nd	B	B	A	D	A	C
3rd	A	A	D	A	C	B
4th	D	C	B	C	D	D

7. Voters are choosing among three options. Make a preference table in which the Borda count winner violates the majority criterion.

8. Voters are choosing among five options. Make a preference table in which the Borda count winner violates the majority criterion.

9. **Determining the legal drinking age.** A state commission is voting on changing the legal drinking age. The options are A, lower the age to 18; B, lower the age to 19; C, lower the age to 20; and D, keep the age at 21. Use the preference table to determine the winner using the Borda count method. Show that Condorcet's criterion is not satisfied.

Preference	Number of Ballots				
	8	10	14	3	10
1st	C	D	C	A	B
2nd	A	A	B	D	C
3rd	B	B	D	B	A
4th	D	C	A	C	D

10. **Voting for the president of a club.** A chapter of the Sierra Club is voting for president. The candidates are (A)lvaro, (B)rown, (C)lark, and (D)ukevitch. Use the preference table to determine the winner using the Borda count method. Show that Condorcet's criterion is not satisfied.

Preference	Number of Ballots				
	4	23	8	3	12
1st	C	D	C	A	A
2nd	A	A	B	D	B
3rd	B	B	D	B	C
4th	D	C	A	C	D

11. **Choosing a location for a research facility.** Teach for America is considering (A)tlanta, (B)oston, (C)hicago, and (D)enver for a new training facility. A group of senior managers voted to determine where the facility will be located. Use the preference table to determine the city that was chosen using the Borda count method. Show that the independence-of-irrelevant-alternatives criterion is not satisfied.

12. **Locating a new factory.** The Land Mover Tractor Company is going to build a new factory in either (A)labama, (C)alifornia, (O)regon, or (T)exas. The vote of the board of directors is shown in the preference table. Determine the state that they chose using the Borda count method. Show that the independence-of-irrelevant-alternatives criterion is not satisfied.

Preference	Number of Ballots					
	9	2	4	5	4	1
1st	C	T	C	A	A	O
2nd	A	A	O	T	O	C
3rd	O	O	T	O	C	A
4th	T	C	A	C	T	T

13. **Reducing a budget.** Due to a budget problem, a citizens' committee is recommending to the school board ways to reduce expenses. The options are A, reduce sports programs; B, reduce expenditures on art and music programs; C, increase class size; and D, defer maintenance on buildings. Use the preference table on page 600 to determine the choice that the committee recommends using the plurality-with-elimination method. Show that Condorcet's criterion is not satisfied.

	Number of Ballots					
Preference	9	12	4	5	4	1
1st	C	D	C	A	A	B
2nd	A	A	B	D	B	C
3rd	B	B	D	B	C	A
4th	D	C	A	C	D	D

Table for Exercise 13

14. **Voting on an award for best restaurant.** A group of columnists is voting on the restaurant of the year. The choices are The (A)lamo, The (B)ar-B-Q, (C)hez Nous, and (D)anny's Place. Use the preference table to determine the winner using the plurality-with-elimination method. Show that Condorcet's criterion is not satisfied.

	Number of Ballots				
Preference	8	11	3	4	3
1st	B	D	B	C	C
2nd	C	C	A	D	A
3rd	A	A	D	A	B
4th	D	B	C	B	D

Use the following preference table for Exercises 15 and 16.

	13	10	5
1st	A	B	C
2nd	B	C	B
3rd	C	A	A

15. Who wins this election using the pairwise comparison method? Why does this election not violate the majority criterion?

16. **a.** Who wins the election using the plurality-with-elimination method?

    b. If the last five voters change their ballots to

C
A
B

    who now wins the election using the plurality-with-elimination method? Is this a violation of the monotonicity criterion? Explain.

Applying What You've Learned

17. Complete the preference table so that the Borda count winner violates Condorcet's criterion.

Preference	a.	b.
1st	B	A
2nd	A	C
3rd	C	B

18. Complete the preference table so that A is the Borda count winner, but when we remove C, then B is the Borda count winner, thus violating the independence-of-irrelevant-alternatives criterion.

Preference	a.	b.	c.
1st	A	B	C
2nd	C	A	B
3rd	B	C	A

19. Make a preference table, similar to the one given in Example 2, with nine voters to choose among three choices, in which the plurality method violates Condorcet's criterion.

20. Make a preference table similar to the one given in Example 3, but with at least four different types of ballots and three candidates, in which the plurality method violates the independence-of-irrelevant-alternatives criterion.

21. Complete the preference table so that the plurality-with-elimination method violates Condorcet's criterion.

	Number of Ballots				
Preference	20	—	8	8	12
1st	C	D	A	A	B
2nd	A	A	D	B	C
3rd	B	B	B	C	A
4th	D	C	C	D	D

22. Does the plurality method satisfy the majority criterion?

23. Does the plurality-with-elimination method satisfy the majority criterion?

24. Voters are choosing among four choices. Make a preference table in which the plurality-with-elimination method violates the independence-of-irrelevant-alternatives criterion.

Communicating Mathematics

25. What are four criteria that we would like an election to satisfy?

26. One of the voting methods we have been discussing satisfies three of the four criteria in Exercise 25. Which method is it and what criterion does it fail to satisfy?

27. What does Arrow's theorem tell us?

28. Explain why the pairwise comparison method satisfies the majority criterion.

29. Explain why the pairwise comparison method satisfies Condorcet's criterion.

30. Research how the electoral college system is used to elect the president of the United States. How does this procedure conflict with the majority criterion?

12.3 Weighted Voting Systems

Objectives

1. Understand the numerical representation of a weighted voting system.
2. Find the winning coalitions in a weighted voting system.
3. Compute the Banzhaf power index of a voter in a weighted voting system.

All animals are created equal, but some are more equal than others.

George Orwell wrote this famous line in his well-known book *Animal Farm.* The point was that the residents of Orwell's fictional animal farm did not all possess the same power. Some had more than others. The same is true for members of many other organizations. For example, not all members of the UN Security Council have the same voting power.

The present council consists of 5 permanent members (Great Britain, France, the United States, China, and Russia) and 10 nonpermanent members. According to its rules, the council cannot pass a resolution unless all the permanent members vote "yes," and, in addition, four of the nonpermanent members also vote "yes." There are other situations in which voting is done very differently from the methods we have discussed in this chapter. For example,

Using Technology to Investigate Mathematics

31. See your instructor for Excel spreadsheets that you can use to reproduce some of the examples in this section and to solve some of the exercises. Report on your findings.*

32. Search your library or the Internet on the topics of voting and social choice. You may also want to search for interactive applets. Briefly describe some aspect of voting that we have not discussed in this chapter.

For Extra Credit

33. Make a preference table, similar to the one given in Example 5, with at least five different types of ballots and four choices that shows that the plurality-with-elimination method violates the monotonicity criterion.

34. Make a preference table, similar to the one given in Example 4, in which voters use the pairwise comparison method to vote on five choices and the vote violates the independence-of-irrelevant-alternatives criterion.

35. Voters are choosing among five choices. Make a preference table in which the plurality-with-elimination method violates the monotonicity criterion.

36. Voters are choosing among four choices. Make a preference table in which the plurality-with-elimination method violates the monotonicity criterion.

37. A famous example called *Condorcet's paradox* is illustrated in the preference table.

Number of Ballots		
Preference	1	1	1
1st	A	B	C
2nd	B	C	A
3rd	C	A	B

Notice that two-thirds of the voters prefer A over B, two-thirds prefer B over C, and two-thirds prefer C over A. Make a preference table for five candidates A, B, C, D, and E such that 80% of the voters prefer A over B, 80% prefer B over C, 80% prefer C over D, 80% prefer D over E, and 80% prefer E over A.

38. In using the pairwise comparison method, as the number of candidates grows it is easy to see that the number of comparisons grows quite rapidly.

a. Complete the following table:

Candidates	Number of Comparisons
A, B	1
A, B, C	3
A, B, C, D	6
A, B, C, D, E	
A, B, C, D, E, F	

b. In Chapter 13, we will show that for k candidates there are $k(k - 1)/2$ comparisons necessary. How many comparisons are necessary for 10 candidates? For 20?

When a jury votes in a criminal trial, a vote of 11 to 1 is not enough to convict the accused. In corporate decision making, a large stockholder may possess 40 or 50 times the voting power of a smaller stockholder. In this section, we will develop a method to measure the power of voters in a system in which not everyone has the same strength.

Weighted Voting Systems

In order to understand the concept of a weighted voting system, consider the following situation. A corporation that owns the Phoenix Flames, a professional football team, has six stockholders, each of whom owns different amounts of stock. Let's say that Alicia Mendez and her son Ben each own 26% of the stock and Carl, Dante, Emily, and Felix each own 12%. (We represent these stockholders as A, B, C, D, E, and F.) Assume that each stockholder possesses as many votes as the percentage of stock owned. Voting in this corporation clearly does not reflect the "one person, one vote" principle. In fact, A and B possess a lot of power compared with C, D, E, and F. If a resolution requires a vote of more than 50% to pass, we see that A and B together can pass any resolution they want, whereas C, D, E, and F are considerably weaker in their ability to get their resolutions passed.

In this situation, we see a number of characteristics of weighted voting systems that are present in the examples we will study. First, there is a number of votes, 51, required for a resolution to pass. This number is called a *quota*. Second, there are voters, each of whom controls a number of votes. We will call this number the voter's *weight*. We will make these ideas more precise.

KEY POINT

We describe a weighted voting system by its quota and the weight of each voter.

DEFINITIONS A **weighted voting system** with n voters is described by a set of numbers that are listed in the following format:

[quota: weight of voter 1, weight of voter 2, . . . , weight of voter n]

The **quota** is the number of votes necessary in this system to get a resolution passed. The numbers that follow, called **weights**, are the amount of votes controlled by voter 1, voter 2, etc.

EXAMPLE 1   *Weighted Voting Systems*

Explain each of the following weighted voting systems:

a) [51 : 26, 26, 12, 12, 12, 12]   b) [4 : 1, 1, 1, 1, 1, 1, 1]

c) [14 : 15, 2, 3, 3, 5]   d) [10 : 4, 3, 2, 1]

e) [12 : 1, 2, 3, 1, 1, 2]   f) [12 : 1, 1, 1, 1, 1, 1, 1, 1]

g) [39 : 7, 7, 7, 7, 1, 1, 1, 1, 1, 1, 1, 1]

SOLUTION:

a) This is the stockholder situation that we described earlier. The following diagram explains how to interpret this system:

[51 : 26, 26, 12, 12, 12, 12]

Need 51 votes to pass a resolution.   A and B each have 26 votes.   C, D, E, and F have 12 votes each.

b) In this situation, there are seven voters having one vote each. Because the quota is four, a simple majority suffices to pass a resolution. This is an example of a "one person, one vote" situation.

[4 : 1, 1, 1, 1, 1, 1, 1]

Need 4 votes to pass a resolution.   A, B, C, D, E, F, G have 1 vote each.

c) The quota is 14; because voter 1 has 15 votes, he or she has total control. Because the other four voters have no power whatsoever in this system, we call voter 1 a **dictator**.

The quota is 14. ─┐   ┌─ The dictator is the only person
                  │   │   able to pass a resolution.
                  [14 : 15, 2, 3, 3, 5]

d) Notice that the sum of the votes is 10, which is also the quota. In this system, even though the first voter has greater weight than the others, in fact he or she has no more power, because the support of even the weakest voter is necessary to pass a resolution. A voter who can, by him- or herself, prevent a motion from passing has **veto power**.

The quota is 10. ─┐
                  [10 : 4, 3, 2, 1]
                      └────────┘
             Every voter is needed to
             pass every resolution—all
             have the same power.

e) This describes our jury system for trying criminal cases. Because the quota is 12, every voter must vote for a resolution for it to pass. Each voter has veto power.

The quota is 12. ─┐
                  [12 : 1, 1, 1, 1, 1, 1, 1, 1, 1, 1, 1, 1]
                       └──────────────────────────────┘
                       Every voter is needed to
                       pass every resolution—all
                       have the same power.

f) In this system, the sum of all the possible votes is less than the quota, so no resolutions can be passed.

The quota is 12. ─┐
                  [12 : 1, 2, 3, 1, 1, 2]
                       └──────────────┘
                   not enough votes to pass
                        any resolutions

g) This system describes the voting in the UN Security Council. Notice in the following diagram that the quota cannot be achieved unless all the first five voters vote for a resolution. In addition, 4 of the next 10 voters must also vote for it to pass a resolution.

                  [39 : 7, 7, 7, 7, 7, 1, 1, 1, 1, 1, 1, 1, 1, 1, 1]
need 39 votes to ─/   └───────────┘ └──────────────────────────┘
pass a resolution   Each of these voters   Four of these votes are
                     has veto power.   needed to pass a resolution.

Now try Exercises 5 to 16. ✳

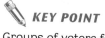

KEY POINT

Groups of voters form coalitions to pass resolutions.

Coalitions

The weighted voting system in Example 1(d) is interesting because even though it appears at first that voter 1 has more power than the other voters, this is not the case. The system [10 : 4, 3, 2, 1] behaves exactly the same as the system [4 : 1, 1, 1, 1]. Notice that the number of votes that a voter possesses is not the same thing as the power a voter has to pass resolutions. In order to describe the concept of power in a voting system, we need to introduce some more definitions. We first describe the subsets of voters that have the power to pass resolutions.

> **DEFINITIONS** Any set of voters who vote the same way is called a **coalition**. The sum of the weights of the voters in a coalition is called the **weight of the coalition**. If a coalition has a weight that is greater than or equal to the quota, then that coalition is called a **winning coalition**.

In a weighted voting system [4 : 1, 1, 1, 1, 1, 1, 1], any coalition of four or more voters is a winning coalition.

PROBLEM SOLVING

Be Systematic

Recall from Chapter 2 that a k-element set has 2^k subsets. Because one of these is the empty set, we see that a set of k voters can form $2^k - 1$ possible coalitions. For example, a set of five voters can form $2^5 - 1 = 32 - 1 = 31$ different coalitions. It is useful to list coalitions systematically. That is, consider all one-element sets, then all two-element sets, followed by the three-element sets, and so on.

EXAMPLE 2 *Finding Winning Coalitions*

A town has two large political parties, (R)epublican and (D)emocrat, and one small party, (I)ndependent. Membership on the town council is proportional to the size of the parties. We will assume that R has nine members on the council, D has eight, and I only three. Traditionally, each party votes as a single bloc, and resolutions are passed by a simple majority. List all possible coalitions and their weights, and identify the winning coalitions.

SOLUTION: A coalition is a nonempty subset of the set of all parties {R, D, I}. We list these subsets and their weights in the following table. Because there are 20 members on the council, any coalition with a weight of 11 or more is a winning coalition.

Coalition	Weight	
{R}	9	
{D}	8	
{I}	3	
{R, D}	17	Winning
{R, I}	12	Winning
{D, I}	11	Winning
{R, D, I}	20	Winning

Quiz Yourself ⑦

Consider the weighted voting system [5 : 3, 2, 2, 1].

a) How many voters are there?

b) What is the quota?

c) List all winning coalitions.

Now try Exercises 17 to 20. ✹

It is interesting that even though party I has less representation on the council than either R or D, it still appears in as many winning coalitions as does R or D.

In Example 2, it seems that all three parties have the same amount of voting power. We will make this idea of power more precise shortly. The key to understanding a voter's power is knowing how many coalitions need the voter to win.

DEFINITION A voter in a winning coalition is called **critical** if it is the case that if he or she were to leave the coalition, then the coalition would no longer be winning.

EXAMPLE 3   *Identifying Critical Voters in a Coalition*

Identify the critical voters in the winning coalitions in the town council in Example 2.

SOLUTION:

Coalition	Weight		Critical Voters
{R}	9		
{D}	8		
{I}	3		
{R, D}	17	Winning	R, D
{R, I}	12	Winning	R, I
{D, I}	11	Winning	D, I
{R, D, I}	20	Winning	none

Remove any of these voters and the coalition no longer wins.

Now try Exercises 21 to 24. ✳

The Banzhaf Power Index

✏️ **KEY POINT**

The Banzhaf power index measures voting power.

We now define how to measure voting power in a weighted voting system.

> **DEFINITION** In a weighted voting system, a voter's **Banzhaf power index*** is defined as
>
> $$\frac{\text{the number of times the voter is critical in winning coalitions}}{\text{the total number of times voters are critical in winning coalitions}}.$$

Quiz Yourself ⑧

Reconsider the weighted voting system [5 : 3, 2, 2, 1] from Quiz Yourself 7, which had winning coalitions {A, B}, {A, C}, {A, B, C}, {A, B, D}, {A, C, D}, {B, C, D}, and {A, B, C, D}.

a) Determine the critical voters in each winning coalition.

b) Compute the Banzhaf power index for voters in this system.

EXAMPLE 4   *Computing the Banzhaf Power Index*

In Example 3, we saw that R, D, and I each were critical voters twice. Thus, R's Banzhaf power index is

$$\frac{\text{the number of times R is critical in winning coalitions}}{\text{the total number of times voters are critical in winning coalitions}} = \frac{2}{6} = \frac{1}{3}.$$

Similarly, D and I each have a Banzhaf power index of $\frac{1}{3}$.

Now try Exercises 33 to 38. ✳ ⑧

EXAMPLE 5   *Calculating the Banzhaf Power Index Indirectly*

In the law firm of Krook, Cheatum, and Associates, there are two senior partners (Krook and Cheatum) and four associates (W, X, Y, and Z). To change any major policy of the firm, a vote must be taken in which Krook, Cheatum, and at least two associates agree to the change. Calculate the Banzhaf power index for each member of this firm.

SOLUTION: We can represent the members of the firm by {K, C, W, X, Y, Z}. At first it seems tempting to list all of the coalitions (subsets) of the firm and determine which are winning. However, this is a tedious job and is not necessary. Realize that every winning coalition includes {K, C}, so all we have to do is determine the subsets of {W, X, Y, Z} with two or more members and form the union of these with {K, C}.

2-Element Subsets of {W, X, Y, Z}	3-Element Subsets of {W, X, Y, Z}	4-Element Subsets of {W, X, Y, Z}
{W, X}, {W, Y}, {W, Z} {X, Y}, {X, Z}, {Y, Z}	{W, X, Y}, {W, X, Z}, {W, Y, Z}, {X, Y, Z}	{W, X, Y, Z}

*There are other slightly different ways of defining this index, depending on which author you read.

The winning coalitions of the firm and their critical members are therefore:

	Winning Coalitions	Critical Members
1	{K, C, W, X}	K, C, W, X
2	{K, C, W, Y}	K, C, W, Y
3	{K, C, W, Z}	K, C, W, Z
4	{K, C, X, Y}	K, C, X, Y
5	{K, C, X, Z}	K, C, X, Z
6	{K, C, Y, Z}	K, C, Y, Z
7	{K, C, W, X, Y}	K, C
8	{K, C, W, X, Z}	K, C
9	{K, C, W, Y, Z}	K, C
10	{K, C, X, Y, Z}	K, C
11	{K, C, W, X, Y, Z}	K, C

All voters are necessary to pass a resolution in these coalitions. *(brace alongside rows 1–6)*

Only K and C are critical in these coalitions. *(alongside rows 7–11)*

From this table we see that K and C are critical members 11 times, whereas W, X, Y, and Z are each critical members only 3 times. The Banzhaf power indices for the members of this firm are therefore:

Members	Banzhaf Power Index
K, C	$\dfrac{11}{11+11+3+3+3+3} = \dfrac{11}{34}$
W, X, Y, Z	$\dfrac{3}{11+11+3+3+3+3} = \dfrac{3}{34}$

If we add the Banzhaf power indices for the members of the firm as follows:

$$\underset{K}{\frac{11}{34}} + \underset{C}{\frac{11}{34}} + \underset{W}{\frac{3}{34}} + \underset{X}{\frac{3}{34}} + \underset{Y}{\frac{3}{34}} + \underset{Z}{\frac{3}{34}} = \frac{34}{34} = 1$$

we see that the total is 1. This is always the case when computing Banzhaf power indices in a weighted voting system.

Now try Exercises 39 to 42. ❋

Often the chair of a committee votes only in the case of breaking a tie. In fact, this is the way the U.S. Senate votes. The vice president of the United States presides over the 100-member U.S. Senate and votes only to break a tie. The mathematics to determine the Banzhaf power index for the vice president and the members of the Senate is too lengthy to include here. We will, however, analyze a much simpler example based on the same voting principle.

EXAMPLE 6   *Finding the Banzhaf Power Index of a "Tie Breaker"*

A five-person air safety review board is developing in-flight safety procedures to deal with skyjackings. The board is chaired by a federal aviation administrator (A) and consists of two senior pilots (S and T) and two flight attendants (F and G).

HISTORICAL HIGHLIGHT 🌀 🌀 🌀

Blocking Coalitions, Banzhaf, and the Electoral College

Political scientists use a variation of the method you have studied in this section to compute the Banzhaf power index of the states voting in the electoral college that elects the president of the United States.

A state in the electoral college has as many votes as the total of its senators and representatives. For example, New York presently has 29 representatives and 2 senators and therefore has 31 votes. We can think of the electoral college as a weighted voting system consisting of the 50 states plus the District of Columbia. California has the most votes, 55, and there are 8 states such as Delaware, having only 3.

Although there are more than 9 trillion coalitions to consider, a computer can calculate the Banzhaf indices for the electoral college fairly quickly. We list several of these indices for a recent year here.

State	Banzhaf Power Index (%)
California	11.44
Texas	6.20
New York	5.81
Florida	5.02
Pennsylvania	3.87
Ohio	3.68
New Jersey	2.75
Virginia	2.38
Arizona	1.83
Kentucky	1.46
New Mexico	0.91
Delaware	0.55

You can see in this table that California has almost three times the Banzhaf power of Pennsylvania and more than twenty times the power of Delaware. Keep in mind when you read these numbers that this is only one way that we can measure power in a weighted voting system. There are other methods and also much controversy about the way power is distributed in the electoral college. You can find many articles analyzing the electoral college on the Internet.

It was the intent when the board was established that the administrator have considerably less power than the pilots and flight attendants. Therefore, the administrator votes only in the case of a tie; otherwise, cases are decided by a simple majority. How much less power does the administrator have than the other members of the board?

SOLUTION: To answer this question, we compute the Banzhaf power index for each member of the board. We first list the winning coalitions and the critical members of these coalitions. Clearly if three or four of {S, T, F, G} vote together, they form a winning coalition in the table on page 608. We list these five coalitions first and then list the six ways ties can be formed, which then require the addition of the administrator to break them.

If we count all of the number of times members A, S, T, F, and G are critical members of some coalition, we get a total of 30. We see that any single board member (including the chair) is a critical member of exactly six coalitions. Therefore, each of the five board members has exactly the same Banzhaf power index, which is $\frac{6}{30}$. Even though it is not apparent at first, the chair of the board has exactly the same power as the other board members.

	Winning Coalitions	Critical Members
1	{S, T, F}	S, T, F
2	{S, T, G}	S, T, G
3	{S, F, G}	S, F, G
4	{T, F, G}	T, F, G
5	{S, T, F, G}	None
6	{A, S, T}	A, S, T
7	{A, S, F}	A, S, F
8	{A, S, G}	A, S, G
9	{A, T, F}	A, T, F
10	{A, T, G}	A, T, G
11	{A, F, G}	A, F, G

No tie { (rows 1–5)

Tie broken by administrator { (rows 6–11)

In Example 6, the administrator and each member of the board were critical in exactly the same number of circumstances, by being one of a bare majority in favor of a motion. Similarly, in the U.S. Senate, the vice president and each senator have exactly the same number of opportunities to be critical by being on the winning side of a 51-to-50 vote. Thus, the vice president has the same power as each senator. In Chapter 13, you will learn principles of counting that will enable you to calculate the number of cases that we have to consider to prove this.

Exercises 12.3

Looking Back*

These exercises follow the general outline of the topics presented in this section and will give you a good overview of the material that you have just studied.

1. What was the meaning of [51 : 26, 26, 12, 12, 12, 12] in Example 1?

2. How did we find the winning coalitions in Example 2?

3. In Example 5, why did K and C have larger Banzhaf power indices than W, X, Y, and Z?

4. Which voter has the most power in the voting system [10 : 4, 3, 2, 1]? Explain.

Sharpening Your Skills

In Exercises 5–16, the weights represent voters A, B, C, and so on, in that order. Identify a) the quota, b) the number and weights of the voters, c) dictators, and d) those having veto power.

5. [5 : 1, 1, 1, 1, 1]
6. [15 : 5, 4, 3, 2, 1]
7. [11 : 10, 3, 4, 5]
8. [6 : 6, 1, 2, 2]
9. [15 : 1, 2, 3, 4, 4]
10. [11 : 1, 2, 3, 4]
11. [12 : 1, 3, 5, 7]
12. [16 : 1, 5, 7, 9]
13. [25 : 4, 4, 6, 7, 9]
14. [21 : 3, 5, 6, 8, 9]
15. [51 : 20, 20, 20, 10, 10]
16. [67 : 15, 15, 15, 15, 10, 10]

In Exercises 17–20, write out all winning coalitions in each voting system. Do not approach this randomly. Be systematic by considering coalitions from smallest to largest.

17. [12 : 1, 3, 5, 7]
18. [16 : 1, 5, 7, 9]
19. [25 : 4, 4, 6, 7, 9]
20. [23 : 3, 5, 6, 8, 9]

21. Find the critical voters in the winning coalitions that you found in Exercise 17.

22. Find the critical voters in the winning coalitions that you found in Exercise 18.

23. Find the critical voters in the winning coalitions that you found in Exercise 19.

24. Find the critical voters in the winning coalitions that you found in Exercise 20.

Applying What You've Learned

25. **A theater guild.** The Theater Guild consists of (P)erformers, (T)echnicians, and (S)upport staff. Representation on the guild is proportional to the number in each group, and the

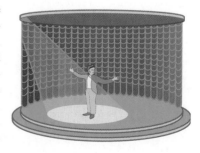

---

*Before doing these exercises, you may find it useful to review the note *How to Succeed at Mathematics* on page xix.

representatives of each group tend to vote in a bloc. A simple majority is required to pass a resolution. Assume there are five performers, four technicians, and two members of the support staff on the guild. List all the coalitions of {P, T, S} and their weights. Which are the winning coalitions?

26. A theater guild. Repeat Exercise 25, but now assume that there are eight performers, six technicians, and two members of the support staff on the guild.

27. A state committee. The college athletics procedures committee makes policy decisions affecting college athletics programs throughout the state. The committee consists of three (A)dministrators, four (C)oaches in athletics departments, three (T)eam captains, and two (N)onathlete students. Assume that each of these four groups votes in a bloc. To make a policy change, a vote of at least 8 is required. Find all the winning coalitions of {A, C, T, N}, and state their weights.

28. A state committee. Repeat Exercise 27, but now assume that there are four administrators, five coaches, four team captains, and three nonathletes. Also assume that a vote of 12 is required to make a policy change.

29. A theater guild. Determine the critical voters in the winning coalitions in Exercise 25.

30. A theater guild. Determine the critical voters in the winning coalitions in Exercise 26.

31. A state committee. Determine the critical voters in the winning coalitions in Exercise 27.

32. A state committee. Determine the critical voters in the winning coalitions in Exercise 28.

In Exercises 33–38, determine the Banzhaf power index for each voter in each weighted voting system.

33. The weighted voting system in Exercise 5

34. The weighted voting system in Exercise 7

35. The weighted voting system in Exercise 13

36. The weighted voting system in Exercise 15

37. The weighted voting system in Exercise 17

38. The weighted voting system in Exercise 19

In Example 5, we analyzed the voting power of the law partners in the firm of Krook, Cheatum, and Associates. Recall that to change a policy required the votes of Krook, Cheatum, and two associates. For each scenario described in Exercises 39–42, do the following:

a) Give an intuitive explanation as to how you think the voting power of Krook, Cheatum and each associate is changed.

b) Calculate the Banzhaf power index for each member of the firm to see how this calculation corresponds to your intuition.

39. A law firm. Another associate is added to the original firm, so the firm now has two senior partners and five associates.

40. A law firm. Another senior partner, named Fair, is added to the original firm, so the firm now has three senior partners and four associates. Assume that two senior partners and two associates must vote to change a policy.

41. A law firm. One of the original associates, named Howe, is promoted to senior partner. The firm now has three senior partners

and three associates. Assume that two senior partners and two associates must vote to change a policy.

42. A law firm. Krook resigns, leaving Cheatum as the only senior partner with four associates. Assume that Cheatum and two associates must vote to change a policy.

Communicating Mathematics

43. What does it mean to say a voter has veto power?

44. How do you compute the Banzhaf power index for a voter in a weighted voting system? What is the sum of the Banzhaf power indices in a weighted voting system?

45. The system [3 : 1, 1, 1, 1, 1] is an example of a "one person, one vote" situation.

   a. Calculate the Banzhaf power index for each person in this system.

   b. How does this conform with your intuition? Explain.

46. A 12-person jury corresponds to the weighted voting system

$$[12 : 1, 1, 1, 1, 1, 1, 1, 1, 1, 1, 1, 1].$$

   a. Calculate the Banzhaf power index for each person in this system.

   b. How does this conform with your intuition? Explain.

47. Consider the system [14 : 15, 2, 3, 3, 5] in which A is a dictator.

   a. Calculate the Banzhaf power index for each person in this system.

   b. How does this conform with your intuition? Explain.

48. Consider the system [12 : 1, 2, 3, 1, 1, 2] in which no resolution can be passed because the quota is too high.

   a. Explain why Banzhaf power indices cannot be calculated in this system.

   b. How does this conform with your intuition? Explain.

Using Technology to Investigate Mathematics

49. Find applets to compute the Banzhaf power index. Search for "Banzhaf applets" or "social choice applets" or "weighted voting applets."

50. Search for "electoral college applets," and report on your findings.

For Extra Credit

In Exercises 51 and 52, devise a voting system that behaves with specifications that are similar to the UN Security Council described in Example 1.

51. A committee has three standing members and six temporary members. The three standing members and two of the temporary members must vote for a resolution for it to be passed.

52. A committee has four standing members and eight temporary members. The four standing members and three of the temporary members must vote for a resolution for it to be passed.

It is interesting to compare the Banzhaf power indices for states in the electoral college with the percentage of electoral votes and with the states' percentage of the total U.S. population. Use a world almanac or similar source to gather the information needed to solve Exercises 53–55.

53. U.S. Congress. Make the suggested comparisons for California, New York, Florida, and Pennsylvania.

54. U.S. Congress. Make the suggested comparisons for Arkansas, New Mexico, Idaho, and Delaware.

55. In Exercises 53 and 54, what patterns do you notice in comparing the larger states with the smaller states? Can you explain this?

Looking Deeper

12.4 The Shapley-Shubik Index

Objectives

1. Determine all the permutations of a set.
2. Find the pivotal voters in a coalition.
3. Calculate the Shapley-Shubik index for a voter in a weighted voting system.

Because of your concern about rising tuition, you contacted your local state representative to ask for her help. She wrote back assuring you that she has always been a strong supporter of higher education and pointed out that she recently voted to increase appropriations for your school. Although she is being truthful, she neglected to tell you that initially she was neutral on this issue and only threw in her support when it was clear that the bill had enough votes to pass. In fact, her only motive for supporting the bill was to win votes in her district where many of her constituents attend college.

From this example you can see that it may be important to know the order in which members join a coalition to make it a winner. In the discussion of the Banzhaf power index, we only considered the members of a coalition, specifically the critical members. The method we study here, which is based on the work of the mathematician Lloyd Shapley and the economist Martin Shubik, focuses not only on the makeup of winning coalitions but also on the order in which winning coalitions are formed.

Permutations

To understand the Shapley-Shubik index, we must make a clear distinction between sets whose elements have an order to them versus sets in which the order is unimportant. Recall from Chapter 2, when we use the notation $\{A, B, C\}$, order is not important; if we want, we could write this set as $\{B, C, A\}$ or $\{C, A, B\}$ instead. If we want to emphasize that the order of the elements in a set is important, then we must use a different notation.

> **DEFINITION** An ordering of the elements of the set $\{x_1, x_2, x_3, \ldots, x_n\}$ in a straight line is called a **permutation*** of that set. We denote a permutation as follows:
>
> $$(x_{i_1}, x_{i_2}, x_{i_3}, \ldots, x_{i_n}),$$
>
> where x_{i_1} is the first element in the permutation, x_{i_2} is the second element in the permutation, and so on.

Using this definition we see that (A, B, C), (B, C, A), and (C, A, B) are all different permutations of the set $\{A, B, C\}$ because the elements are listed in different orders.

From now on, we will assume the following:

A permutation of voters specifies a coalition in which the voters were added one at a time.

---

*We will study permutations at much greater length in Chapter 13.

PROBLEM SOLVING
The Splitting-Hairs Principle

The Splitting-Hairs Principle in Section 1.1 stated that different notation usually means that you are dealing with different concepts. You can conclude that because {A, B, C} and (A, B, C) look different, they do not mean the same thing. Try to connect new notation with notation you have seen before. Recalling that the ordered pair of numbers (3, 4) does not mean the same thing as the ordered pair (4, 3) helps you understand the meaning of the notation (A, B, C).

It will be important to know how many different ways we can order the elements in a set. We will see how to generate different permutations of a set by looking at the problem graphically, as in the next example.

EXAMPLE 1   *Finding Permutations of a Set*

How many permutations are there of each set?

a) {A, B, C}   b) {A, B, C, D}

SOLUTION:

a) If we think about forming the orderings by deciding which is the first element of the ordering, then which is the second, and finally which is the third element, we can visualize the permutations as shown in Figure 12.1, which is called a *tree diagram*.

Following the six branches of this tree starting with "Begin," we generate the permutations (A, B, C), (A, C, B), (B, A, C), (B, C, A), (C, A, B), and (C, B, A). In this case, we found $3 \times 2 \times 1 = 6$ ways to order the three elements A, B, and C.

b) In this case, in Figure 12.2 we draw a tree beginning with four branches for our first choice, three branches at the next stage for our second choice, and so on.

From Figure 12.2, we see that there are $4 \times 3 \times 2 \times 1 = 24$ permutations of {A, B, C, D}. They are (A, B, C, D), (A, B, D, C), (A, C, B, D), . . . , (D, C, B, A).

Now try Exercises 5 to 8. ✳ **9**

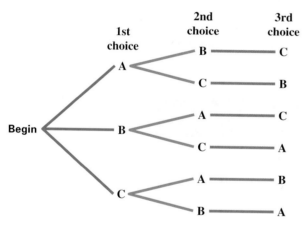

FIGURE 12.1 Tree diagram showing all permutations of {A, B, C}.

Quiz Yourself **9**

How many permutations are there of a five-element set?

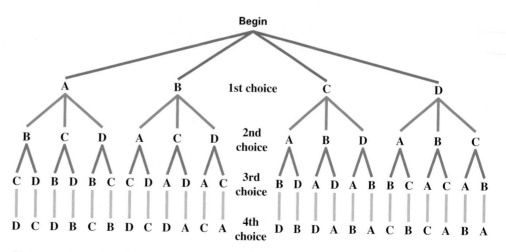

FIGURE 12.2 Tree diagram showing all permutations of {A, B, C, D}.

Example 1 leads us to state the following principle.

> **THE NUMBER OF PERMUTATIONS OF A SET** If we want to order a set with n elements, following the method in Example 1, there are n ways to choose the first element, followed by $n-1$ ways to choose the second element, followed by $n-2$ ways to choose the third element, and so on. Thus there are $n \times (n-1) \times (n-2) \times \cdots \times 2 \times 1$ permutations of these elements. This product is called n **factorial** and is written $n!$.

Pivotal Voters

In forming a coalition by adding voters one at a time, at some point we add a voter who makes the coalition a winning coalition. That voter is particularly important because he or she has changed a nonwinning coalition into a winning one.

> **DEFINITION** As we add voters to a coalition one at a time, the first player added who makes the coalition a winning coalition is called a **pivotal voter** for the coalition.

EXAMPLE 2   *Identifying the Pivotal Voter*

Consider the weighted voting system [5 : 3, 2, 2, 1] with voters A, B, C, and D, in that order. Recall that A has three votes, B and C have two votes each, and D has one vote. Find the pivotal voter in each coalition.

a) (A, B, C, D)   b) (A, D, C, B)   c) (D, A, B, C)   d) (D, C, A, B)

SOLUTION: In each coalition, we are looking for the first voter who makes the coalition a winning one by giving it a voting weight of 5 or more.

Quiz Yourself **10**

Consider the weighted voting system [5 : 2, 2, 1, 1]. Find the pivotal voter in each coalition:

a) (A, C, B, D)

b) (C, D, B, A)

	Sum Weights of Coalition Members Until We Reach Quota of 5	Pivotal Voter
a)	(A, B, C, D) 3 + 2	B
b)	(A, D, C, B) 3 + 1 + 2	C
c)	(D, A, B, C) 1 + 3 + 2	B
d)	(D, C, A, B) 1 + 2 + 3	A

B is the first member to make the coalition winning.

✳ **10**

The Shapely-Shubik Index

We are now in a position to define the Shapley-Shubik index for measuring voter power.

> **DEFINITION** In a weighted voting system, the **Shapley-Shubik index** for a voter is
> $$\frac{\text{the number of times the voter is pivotal in some permutation of the voters}}{\text{the total number of permutations of the voters}}.$$

EXAMPLE 3   *Finding the Shapley-Shubik Index*

Compute the Shapley-Shubik index for each voter in the weighted voting system [4 : 3, 2, 1].

SOLUTION: Consider the voters to be A, B, and C in that order. We first list all permutations of these voters and determine in each permutation which voter is pivotal.

Sum of Weights of Coalition Members Until We Reach Quota of 4	Pivotal Voter
(A, B, C) 3 + 2	B
(A, C, B) 3 + 1	C
(B, A, C) 2 + 3	A
(B, C, A) 2 + 1 + 3	A
(C, A, B) 1 + 3	A
(C, B, A) 1 + 2 + 3	A

A is pivotal four out of six times.

A's Shapley-Shubik index is

$$\frac{\text{the number of times A is pivotal in some permutation of the voters}}{\text{the total number of permutations of the voters}} = \frac{4}{6} = \frac{2}{3}.$$

Because B is pivotal one out of six times, B's index is $\frac{1}{6}$. Similarly, C's index is also $\frac{1}{6}$.

Now try Exercises 13 to 28. ✳

Note in Example 3 that even though B has twice the votes of C, both B and C have the same voting power according to the Shapley-Shubik index. In general, the number of votes a person has is not the same as that person's voting power as measured according to the Shapley-Shubik index.

 PROBLEM SOLVING

Verify Solutions

As we recommended in Section 1.1, it is always good to check to see if your answers are reasonable. Notice when computing the Shapley-Shubik index, the sum of the indices of all the voters is always 1.

As you might expect, the Banzhaf and Shapley-Shubik indices do not measure power the same way.

EXAMPLE 4   *The Banzhaf Power Index Differs from the Shapley-Shubik Index*

Let us compute the Banzhaf power index for the voters in the system [4 : 3, 2, 1] in Example 3.

SOLUTION: We summarize the computations in the following table:

Winning Coalition	Weight	Critical Voters
{A, B}	5	A, B
{A, C}	4	A, C
{A, B, C}	6	A

If we remove any voter, the coalition is no longer winning.

Because A is a critical voter three out of five times, her Banzhaf power index is $\frac{3}{5}$. Similarly, B and C have Banzhaf power indices of $\frac{1}{5}$ each. ✳

Notice that both the Shapley-Shubik index and the Banzhaf power index rate A as the most powerful member of the weighted voting system in Examples 3 and 4. However, also note that the power indices are not identical for both methods.

Often, the chair of a committee has more power than the other members of the committee. It is possible to measure exactly how much greater this power is using the Shapley-Shubik index.

EXAMPLE 5   *Comparing the Power of Committee Members*

Assume a campus blog has an editorial board consisting of the managing editor and four other members. In order to begin to develop a story, the managing editor and two other board members must agree that the story is newsworthy. Use the Shapley-Shubik index to compare the power of the managing editor with the other board members.

SOLUTION: Let us assume that the managing editor is A and the other board members are B, C, D, and E. We will solve this problem by first determining B's power. We can then deduce that C, D, and E also have this same power. From this information, we can easily determine the power that A has.

In forming permutations of {A, B, C, D, E}, we can think of filling slots in a list as follows:

$$\underline{\quad\quad}\ \ \underline{\quad\quad}\ \ \underline{\quad\quad}\ \ \underline{\quad\quad}\ \ \underline{\quad\quad}$$
$$\text{1st}\quad\ \text{2nd}\quad\ \text{3rd}\quad\ \text{4th}\quad\ \text{5th}$$

In order for B to be a pivotal voter, the managing editor A and one other board member must already have been selected for the 1st and 2nd slots and B must be in the 3rd slot. So, our slot diagram looks like this:

A is in slot 1 or 2.

$$\underline{\quad\quad}\ \ \underline{\quad\quad}\ \ \underline{\ \ \text{B}\ \ }\ \ \underline{\quad\quad}\ \ \underline{\quad\quad}$$
$$\text{1st}\quad\ \text{2nd}\quad\ \text{3rd}\quad\ \text{4th}\quad\ \text{5th}$$

We will next consider two cases depending on whether the editor A is in slot 1 or slot 2.

Case 1: Assume that A is in slot 1, so the diagram now looks like this:

$$\underline{\ \ \text{A}\ \ }\ \ \underline{\quad\quad}\ \ \underline{\ \ \text{B}\ \ }\ \ \underline{\quad\quad}\ \ \underline{\quad\quad}$$
$$\text{1st}\quad\ \text{2nd}\quad\ \text{3rd}\quad\ \text{4th}\quad\ \text{5th}$$

3 ways to fill this slot ⌐   2 ways to fill this slot   1 way to fill this slot

This means that we can fill the remaining three slots in $3! = 6$ ways.

Case 2: Assume that A is in slot 2. The diagram now looks like this:

$$\underline{\quad\quad}\ \ \underline{\ \ \text{A}\ \ }\ \ \underline{\ \ \text{B}\ \ }\ \ \underline{\quad\quad}\ \ \underline{\quad\quad}$$
$$\text{1st}\quad\ \text{2nd}\quad\ \text{3rd}\quad\ \text{4th}\quad\ \text{5th}$$

Again, the remaining three slots can be filled in six ways.

From Cases 1 and 2, we see that there are exactly 12 ways to form a permutation in which B is pivotal. Thus, B's Shapley-Shubik index is

$$\frac{12}{5!} = \frac{12}{120} = \frac{1}{10}.$$

By a similar analysis, we would find that C, D, and E also have a Shapley-Shubik index of $\frac{1}{10}$. Because B, C, D, and E each are pivotal in 12 permutations, it follows that A is pivotal in the other $120 - 12 - 12 - 12 - 12 = 72$ permutations. Thus, A's index is

$$\frac{72}{5!} = \frac{72}{120} = \frac{6}{10}.$$

Therefore, A is six times as powerful as the other editorial board members. ✳

Exercises 12.4

Looking Back*

These exercises follow the general outline of the topics presented in this section and will give you a good overview of the material that you have just studied.

1. If A, B, and C are voters, what is the difference in meaning between the notation {A, B, C} and (A, B, C)? Why is that important to us in this section?

2. How did we find the pivotal voter in the coalition (A, D, C, B) in Example 2(b)?

3. How did we find the Shapley-Shubik index of voter A in Example 3?

4. How many permutations are there in a coalition having *n* voters?

Sharpening Your Skills

In Exercises 5–8, use tree diagrams to find all the permutations of each set.

5. {X, Y, Z}

6. {P, Q, R}

7. {W, X, Y, Z}

8. {P, Q, R, S}

9. How many permutations are there of a six-element set?

10. How many permutations are there of an eight-element set?

11. How many permutations are there of a 12-element set?

12. How many permutations are there of a 13-element set?

13. For the weighted voting system [5 : 3, 2, 2], complete the following table, which is similar to the table given in Example 3:

Sum of Weights of Coalition Members	Pivotal Voter
(A, B, C)	
(A, C, B)	
(B, A, C)	
(B, C, A)	
(C, A, B)	
(C, B, A)	

14. For the weighted voting system [7 : 4, 3, 2], complete the following table, which is similar to the table given in Example 3:

Sum of Weights of Coalition Members	Pivotal Voter
(A, B, C)	
(A, C, B)	
(B, A, C)	
(B, C, A)	
(C, A, B)	
(C, B, A)	

In Exercises 15–20, determine the Shapley-Shubik index of each voter in each weighted voting system.

15. [6 : 3, 3, 2]

16. [8 : 4, 3, 3]

17. [8 : 3, 3, 2, 2]

18. [9 : 4, 3, 3, 1]

19. [4 : 2, 1, 1, 1]

20. [13 : 6, 6, 5, 3]

Applying What You've Learned

21. The system [3 : 1, 1, 1, 1, 1] is an example of a "one person, one vote" situation.

   a. What is the Shapley-Shubik index for each person in this system?

   b. Explain how you obtained your answer in part (a).

   c. How does your answer in part (a) conform with your intuition?

22. **Measuring power on a jury.** We can consider a 12-person jury as the weighted system [12 : 1, 1, 1, 1, 1, 1, 1, 1, 1, 1, 1, 1].

   a. What is the Shapley-Shubik index for each person in this system?

   b. Explain how you obtained your answer in part (a).

   c. How does your answer in part (a) conform with your intuition?

23. Consider the system [14 : 15, 2, 3, 3, 5], in which A is a dictator.

   a. What is the Shapley-Shubik index for each person in this system?

   b. Explain how you obtained your answer in part (a).

   c. How does your answer in part (a) conform with your intuition?

24. Consider the system [12 : 1, 2, 3, 1, 1, 2] in which no resolution can be passed because the quota is too high.

   a. What is the Shapley-Shubik index for each person in this system?

   b. Explain how you obtained your answer in part (a).

   c. How does your answer in part (a) conform with your intuition?

25. **Measuring power on a committee.** Committee {A, B, C, D} has chairperson A. For an item to be placed on the committee's agenda, the chairperson and at least one other committee member must agree to that item. What is the Shapley-Shubik index for each person in this system?

26. **Measuring power on a theater guild.** The Theater Guild consists of (P)erformers, (T)echnicians, and (S)upport staff. Representation on the guild is proportional to the number in each group, and the representatives of each group vote in a bloc. A simple majority is required to pass a resolution. There are five performers, four technicians, and two members of the support staff on the guild.

   a. List all permutations of {P, T, S}.

   b. Determine the pivotal voter in each permutation of {P, T, S}.

   c. Find the Shapley-Shubik index for each voter in this system.

---

*Before doing these exercises, you may find it useful to review the note *How to Succeed at Mathematics* on page xix.

27. **Measuring power on a state committee.** The college athletics procedures committee makes policy decisions affecting college athletics programs throughout the state. The committee consists of three (A)dministrators, four (C)oaches in athletics departments, three (T)eam captains, and two (N)onathlete students. Assume that each of these four groups votes in a bloc. To make a policy change, a vote of at least 8 is required.

    a. List all permutations of {A, C, T, N}.

    b. Determine the pivotal voter in each permutation of {A, C, T, N}.

    c. Find the Shapley-Shubik index for each voter in this system.

28. **Measuring power in a law firm.** In Example 5 of Section 12.3, we analyzed the voting power of the law partners in the firm of Krook, Cheatum, and Associates. Recall that to change a policy required the votes of Krook, Cheatum, and at least two of the associates: W, X, Y, and Z. Find the Shapley-Shubik index for each voter in this system.

Communicating Mathematics

29. What do we mean by a pivotal voter in a coalition?

30. The terms *critical voter* and *pivotal voter* sound somewhat similar. What is the difference in their meanings?

31. In Example 5 we said that in order to be pivotal, B must be in the third slot. Explain why B could not be pivotal if B were in slots 1 or 2.

32. In Example 5, explain why B could not be pivotal if B were in slots 4 or 5.

Using Technology to Investigate Mathematics

33. Search the Internet for an applet that implements the method we described for calculating the Shapley-Shubik index, and use it to duplicate some of the computations that we did in this section. Write a report on your findings.

34. The electoral college consists of the 50 U.S. states plus the District of Columbia. If we were to consider these 51 parties as voters in a weighted voting system and wanted to calculate the Shapley-Shubik index for each voter, you might want to consider all of the permutations of these 51 objects. There will be 51! permutations of these 51 objects. Compute this with your calculator. Your answer will be in scientific notation (see Section 6.5). If you had a computer that could list 1,000,000 permutations per second, how many years would it take to list all 51! permutations. (Assume a year has 365 days.)

CHAPTER SUMMARY*

You will learn the items listed in this chapter summary more thoroughly if you keep in mind the following advice:

1. Focus on "remembering how to remember" the ideas. What pictures, word analogies, and examples help you remember these ideas?

2. Practice writing each item without looking at the book.

3. Make up 3×5 flash cards to break your dependence on the text. Use these cards to give yourself practice tests.

SECTION	SUMMARY	EXAMPLE
SECTION 12.1	In the **plurality method**, each person votes for his or her favorite candidate and the candidate receiving the most votes is declared the winner.	Example 1, p. 581
	With the **Borda count method**, each voter ranks k candidates. The first choice is given k points, the second choice is given $k - 1$ points, the third is given $k - 2$ points, and so on. The candidate receiving the most points wins the election. A voter uses a **preference ballot** to rank the candidates. A **preference table** summarizes the preference ballots.	Discussion, p. 581 Example 2, p. 582
	Approval voting allows voters to vote for as many candidates on the ballot as they want. Candidates are not ranked and the candidate who receives the most votes of approval is the winner. In **instant runoff voting**, each voter ranks all candidates. If a voter's top candidate is eliminated, then that voter's vote is given to the second choice on the ballot.	Highlight, p. 583
	In the **plurality-with-elimination method**, each voter votes for one candidate. A candidate who receives a majority of votes is the winner. If no candidate receives a majority of votes, then the candidates with the fewest votes are dropped from the ballot and a new election is held. This process continues until a candidate receives a majority of votes.	Example 3, p. 584
	In the **pairwise comparison method**, voters rank all candidates. We then consider candidates two at a time. Whichever candidate, A or B, is preferred by most voters receives 1 point. If candidates A and B are tied, then each receives $\frac{1}{2}$ point. At the end, the candidate with the most points is the winner.	Example 4, p. 586
SECTION 12.2	The **majority criterion** states that if a majority of voters rank a candidate as their first choice, then that candidate wins the election.	Example 1, p. 591
	Condorcet's criterion says that if candidate X can defeat each of the other candidates in a head-to-head vote, then X is the winner of the election.	Example 2, p. 592
	The **independence-of-irrelevant-alternatives criterion** states that if candidate X wins an election, and some nonwinners are removed from the ballot, and then a recount is done, X stills wins the election.	Example 3, p. 594 Example 4, p. 594
	The **monotonicity criterion** says that if X wins an election and in a reelection all voters who change their votes only change their votes to favor X, then X wins the election.	Example 5, p. 596
	Arrow's impossibility theorem proves that there is no perfect voting method.	Statement, p. 598
SECTION 12.3	A **weighted voting system** with n voters is described by a set of numbers of the form [quota: weight of voter 1, weight of voter 2, . . . , weight of voter n]. The **quota** is the number of votes necessary to get a resolution passed. The **weights** are the amounts of votes controlled by voter 1, voter 2, and so on. A **dictator** is a voter such that resolutions can only be passed even if only the dictator votes for the resolution. A voter who can prevent a motion from passing has **veto power**.	Example 1, p. 602
	A set of voters who vote the same way is called a **coalition**. The sum of the weights of the voters in a coalition is called the **weight of the coalition**. A coalition that has weight greater than or equal to the quota is called a **winning coalition**. A voter in a winning coalition is called **critical** if he or she were to leave the coalition; then the coalition would no longer be winning.	Definition, p. 603 Example 2, p. 604 Example 3, p. 605

*Before studying this chapter's material, it would be useful to reread the note *How to Succeed at Mathematics* on page xix.

	In a weighted voting system, a voter's **Banzhaf power index** is defined as	Example 4, p. 605
	$$\frac{\text{the number of times the voter is critical in winning coalitions}}{\text{the total number of times voters are critical in winning coalitions}}.$$	Example 5, p. 605 Example 6, p. 606
	A **blocking coalition** is a set of voters with enough votes to defeat a resolution.	
SECTION 12.4	A **permutation** of the set $\{x_1, x_2, x_3, \dots, x_n\}$ is an ordering of the elements in a straight line. We denote a permutation as $\{x_{i_1}, x_{i_2}, x_{i_3}, \dots, x_{i_n}\}$, where x_{i_1} is the first element in the permutation, x_{i_2} is the second element in the permutation, and so on. A set with n elements has $n \times (n-1) \times (n-2) \times \cdots \times 2 \times 1 = n!$ permutations.	Example 1, p. 611 Discussion, p. 612
	As we add voters to a coalition one at a time, the first voter added who makes the coalition a winning coalition is called a **pivotal** voter.	Example 2, p. 612
	In a weighted voting system, the **Shapely-Shubik index** for a voter is	Example 3, p. 612
	$$\frac{\text{the number of times the voter is pivotal in some permutation of the voters}}{\text{the total number of permutations of the voters}}.$$	Example 5, p. 614

CHAPTER REVIEW EXERCISES

Section 12.1

1. Four candidates running for town council receive votes as follows: Myers, 2,156; Pulaski, 1,462; Harris, 986; Martinez, 428.

   a. Is there a candidate who earns a majority?

   b. Who wins the election using the plurality method?

2. Use the preference table to determine the winner of the election using the Borda count method.

	Number of Ballots					
Preference	8	5	7	4	3	6
1st	A	D	A	B	B	C
2nd	B	B	C	D	C	A
3rd	C	C	D	C	A	B
4th	D	A	B	A	D	D

3. Members of the chamber of commerce have been asked to vote on their preference for a topic for a speaker for their convention. The choices are (S)ocial justice, the (R)ole of government in a free society, (E)ducation in the future, and (G)lobal issues. Their preferences are summarized in the table. What option is their first choice using the plurality-with-elimination method?

	Number of Ballots				
Preference	1,531	1,102	906	442	375
1st	G	R	S	S	G
2nd	R	S	G	E	S
3rd	S	G	R	R	E
4th	E	E	E	G	R

4. Using the preference table, who wins the election using the pairwise comparison method?

	Number of Ballots			
Preference	8	4	5	6
1st	A	D	B	C
2nd	B	B	D	B
3rd	C	C	C	A
4th	D	A	A	D

Section 12.2

5. Consider the following three preference ballots:

Preference			
1st	R	R	D
2nd	D	D	P
3rd	P	Q	Q
4th	Q	P	R

Who is the winner of this election using the Borda count method? Does this election satisfy the majority criterion? Explain.

6. The Student Horror Organization of the College of Kokomo (SHOCK) is voting for their choice of a classic horror film to be featured as the theme of their spring horror festival. The finalists to be voted on by the members of SHOCK are *The (E)xorcist*, *(A)lien*, *The (N)ight of the Living Dead*, and *The (S)hining*. Use the preference table on page 619 to determine the winner using the Borda count method. Is Condorcet's criterion satisfied in this election?

	Number of Ballots				
Preference	**6**	**8**	**12**	**1**	**8**
1st	E	S	E	A	N
2nd	A	A	N	S	E
3rd	N	N	S	N	A
4th	S	E	A	E	S

Table for Exercise 6

7. Use the preference table to determine the winner using the Borda count method. Is the independence-of-irrelevant-alternatives criterion satisfied? Explain.

	Number of Ballots					
Preference	**11**	**3**	**6**	**9**	**4**	**3**
1st	C	D	C	B	B	A
2nd	B	B	A	D	A	C
3rd	A	A	D	A	C	B
4th	D	C	B	C	D	D

8. The Wetherholds have narrowed the location of their family reunion down to: (D)ollywood, (C)leveland (to visit the Rock and Roll Hall of Fame), or (B)ridgeville, Delaware, to see the Punkin Chunkin World Championship. Use the plurality-with-elimination method to determine the winner of the election using the preference table. Is the independence-of-irrelevant-alternatives criterion satisfied? Explain.

	Number of Ballots			
Preference	**10**	**7**	**2**	**4**
1st	D	B	C	C
2nd	C	D	D	B
3rd	B	C	B	D

Section 12.3

9. Find the quota, find the weights of the voters, determine whether there is a dictator, and find those having veto power in the weighted voting system [17 : 1, 5, 7, 8].

10. Write out all the winning coalitions in the voting system [11 : 2, 3, 5, 7].

11. Determine the Banzhaf power index for each voter in the weighted voting system [11 : 2, 3, 5, 7].

12. Determine the Banzhaf power index for each voter in each weighted voting system. Explain intuitively why you would expect the calculations to come out as they did.

    a. [10 : 1, 2, 3, 4]   **b.** [10 : 11, 1, 3, 3, 2]

13. Committee {A, B, C, D} has chairperson A. For a resolution to be passed, the chairperson and at least one other member of the committee must support the resolution. What is the Banzhaf power index for each person on this committee?

Section 12.4

14. How many permutations are there in a seven-element set?

15. We have completed the first line in the table for the weighted voting system [6 : 4, 3, 2]. Complete the rest of the table.

Sum of Weights of Coalition Members	**Pivotal Voter**
(A, B, C) 4 + 3	B
(A, C, B)	
(B, A, C)	
(B, C, A)	
(C, A, B)	
(C, B, A)	

16. Determine the Shapely-Shubik index of each voter in the voting system [6 : 4, 3, 2].

17. Consider the system [4 : 1, 1, 1, 1, 1, 1], which is an example of a "one person, one vote" situation. What is the Shapely-Shubik index for each person in this system?

CHAPTER TEST

1. Four candidates running for town council receive votes as follows: Molina, 2,543; Sobieski, 1,532; Wilson, 892; Gambone, 473.

    a. Is there a candidate who earns a majority?

    b. Who wins the election using the plurality method?

2. The Alliance of Women Scientists took a survey of its membership regarding issues they want addressed. The choices were (R)esearch funding, (E)quality in the workplace, (A)ttracting more women to science, and (Q)uality of life. Their preferences are summarized in the table. What option is their first choice using the plurality-with-elimination method?

	Number of Ballots				
Preference	**327**	**130**	**149**	**85**	**324**
1st	E	R	A	E	R
2nd	R	E	Q	R	E
3rd	A	A	E	Q	Q
4th	Q	Q	R	A	A

3. Use the preference table to determine the winner using the Borda count method. Is Condorcet's criterion satisfied in this election? Explain.

Preference	Number of Ballots				
	1,327	1,130	849	285	624
1st	A	B	C	A	B
2nd	B	A	D	B	A
3rd	C	C	A	D	D
4th	D	D	B	C	C

4. Find the quota, find the weights of the voters, determine if there is a dictator, and find those having veto power in the weighted system [15 : 5, 3, 1, 3, 4, 2].

5. How many permutations are there in a six-element set?

6. Use the given preference table to determine the winner of the election using the Borda count method.

Preference	Number of Ballots					
	3	5	8	2	6	5
1st	A	B	C	A	B	D
2nd	B	A	D	B	A	C
3rd	C	D	A	C	D	B
4th	D	C	B	D	C	A

7. Consider the following three preference ballots. Who is the winner of this election using the Borda method? Does this election satisfy the majority criterion? Explain.

Preference			
1st	A	A	B
2nd	B	B	D
3rd	C	C	C
4th	D	D	A

8. Consider the system [5 : 1, 1, 1, 1, 1, 1, 1, 1], which is an example of a "one person, one vote" situation. What is the Shapely-Shubik index for each person in this system?

9. Use the plurality-with-elimination method to determine the winner of the election. Is the independence-of-irrelevant-alternatives criterion satisfied? Explain.

Preference	Number of Ballots			
	35	71	36	14
1st	A	B	C	D
2nd	B	A	D	A
3rd	C	C	A	C
4th	D	D	B	B

10. Determine the Banzhaf power index for each voter in each voting system. Explain intuitively why you would expect the calculations to come out as they did.

   a. [15 : 2, 8, 3, 2]   **b.** [13 : 15, 2, 4, 1, 3]

11. We have completed the first line in the table for the weighted voting system [7 : 5, 3, 3]. Complete the rest of the table.

Sum of Weights of Coalition Members	Pivotal Voter
(A, B, C)	B
5 + 3	
(A, C, B)	
(B, A, C)	
(B, C, A)	
(C, A, B)	
(C, B, A)	

12. Use the given preference table to determine who wins the election using the pairwise comparison method.

Preference	Number of Ballots			
	23	47	83	21
1st	A	B	D	C
2nd	B	A	C	B
3rd	C	C	A	A
4th	D	D	B	D

13. Use the preference table to determine the winner using the Borda count method. Is the independence-of-irrelevant-alternatives criterion satisfied? Explain.

Preference	Number of Ballots					
	7	5	6	12	16	8
1st	A	B	C	A	B	D
2nd	B	A	D	D	A	C
3rd	C	D	A	B	C	B
4th	D	C	B	C	D	A

14. Write out all the winning coalitions in the voting system [15 : 3, 4, 6, 8].

15. Determine the Shapely-Shubik index of each voter in the voting system [7 : 4, 4, 2].

16. Committee {A, B, C, D} has co-chairpersons A and B. For a resolution to be passed, the two co-chairpersons and at least one other member of the committee must support the resolution. What is the Shapely-Shubik index for each person on this committee?

GROUP EXERCISES

1. Construct a preference table for an election with candidates A, B, C, and D such that using that table

   A wins using the plurality method,

   B wins using the Borda count method,

   C wins using the plurality-with-elimination method,

   D wins using the pairwise comparison method.

2. Choose a topic you think will be of interest to your class and have them use preference ballots to rank their preferences in choosing among four or five alternatives regarding that topic. Use the different voting methods we presented in this chapter to determine the winner. It would be good if, when you present the alternatives, there is not an obvious favorite, so that one alternative might win using one voting method but lose using another. Discuss why you think that one option wins using one method but loses using another.

3. **a.** Go to the Web site http://news.bbc.co.uk/2/hi/science/nature/3804841.stm for a description of the European Union as a weighted voting system.

   b. Next go to the Web site http://math.temple.edu/~cow/bpi.html to calculate the Banzhaf power index for various member countries of the European Union. For example, according to the Banzhaf power index how many more times powerful is Germany than Sweden?

*I*f someone promised that you would share in the winnings of a multi-million-dollar lottery—guaranteed—and all it would cost you would be $3,000 a share, would you buy into it? How many shares? Several years ago, I received just such an offer—but declined to invest.

A friend told me about an Australian-based syndicate that was raising $15 million to play small lotteries in which they could buy tickets to cover every possible combination of numbers. The plan was to then invest these winnings to buy tickets for more lucrative lotteries.

When offered this seeming "chance of a lifetime," my immediate question was, "How many tickets must the syndicate buy to guarantee that it had covered all possible ticket combinations?"

(continued)

My friend showed me a videotape in which the syndicate's founder explained that he had invented a new mathematical method for covering all the possibilities. He explained that, of course, he could not reveal his method but assured investors that he knew how to improve on the well-known counting principles (that you will learn in this chapter) to guarantee our success.

Because the videotape was not very convincing, I chose not to invest in the syndicate. Several months later, I was stunned to read the following headline:

Australian Syndicate Wins Virginia Lottery

Did I make a mistake? After learning the basic principles of counting in this chapter, you can evaluate my decision. ●

13.1 Introduction to Counting Methods

Objectives

1. Count elements in a set systematically.
2. Use tree diagrams to represent counting situations graphically.
3. Use counting techniques to solve applied problems.

As a young child, when you first learned how to count, you probably wanted to count for everyone. I recall driving with my young children, and they would ask, "Do you want to hear me count to 100?" About two minutes later they would finish with, ". . . , 99, 100." Suppose instead that they offered to count to a familiar number that we hear in the news every day—1 billion. At a rate of one number per second, they would finish almost 32 years later. Obviously, when counting a large number of objects, we need to stream-line the process.

As you study this chapter, you will begin to appreciate how easy it is to underestimate the number of ways that something can happen in a complex situation. Several years ago, a newspaper told the sad story of how a college student used his savings to buy lottery tickets in hopes of increasing his nest egg. Not realizing how many possible number combinations there were and how slim his chances were of selecting a winning combination, he quickly lost all his money.

In this section, you will begin to learn several techniques to count large sets of items systematically and effectively.

KEY POINT

We can count a set by listing its elements systematically.

Systematic Counting

One of the simplest ways to count a set is to list its elements.

EXAMPLE 1 *Counting Sets by Listing*

How many ways can we do each of the following?

a) Flip a coin.

b) Roll a single die (singular of *dice*).

c) Pick a card from a standard deck of cards (see Figure 13.1).

d) Choose a features editor from a five-person newspaper staff.

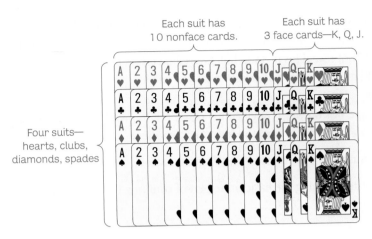

Each suit has 10 nonface cards.

Each suit has 3 face cards—K, Q, J.

Four suits— hearts, clubs, diamonds, spades

FIGURE 13.1 A standard deck of cards.

SOLUTION:

a) The coin can come up either heads or tails, so there are two ways to flip a coin.

b) The die has six faces numbered 1, 2, 3, 4, 5, and 6, so there are six ways the die can be rolled.

c) There are 52 different ways to choose a card from a standard deck (see Figure 13.1).

d) There are five ways to choose one of the five staff members to be editor. ✳

For more complex sets, it is useful to list the elements systematically, as we suggested in Section 1.1 on problem solving.

EXAMPLE 2 *Counting Animal Pairs*

An environmental group plans to develop a fund-raising campaign featuring two endangered species. The list of candidates includes the (C)heetah, the (O)tter, the black-footed (F)erret, and the Bengal (T)iger. One animal will appear in a series of TV commercials and a different animal will appear in online advertisements. In how many ways can we choose the two animals for the campaign?

SOLUTION: So as not to overlook any possibilities, we will list all pairs of animals systematically. We denote the animals by the letters C, O, F, and T. We begin by assuming that C is the animal selected for the TV commercials and then consider each of the other animals for the online campaign. That gives us

CO, CF, CT.

If we next list O as the animal used on TV with each of the other animals appearing online, we get

OC, OF, OT.

Continuing in this fashion, the complete list is

CO, CF, CT

OC, OF, OT

FC, FO, FT

TC, TO, TF.

Thus, there are 12 ways to select animals for the TV and online ads.

Now try Exercises 5 to 8. ✳ **1**

Quiz Yourself **1** *

A, B, C, D, E, F, and G are finalists in a downhill ski race. Medals will be awarded for first and second place. In how many different ways can we award these two medals? Do this either by listing all possible pairs or by reasoning abstractly, as we do in the discussion following the solution to Example 2.

Notice in Example 2 that if instead of 4 animals there were 10 animals under consideration, we could count without actually writing down the pairs. Clearly, if C were featured on TV, then there would be nine ways to complete the pair with an animal for the online ads. Similarly, if we used O on TV, there again would be nine ways to then choose an animal for the online campaign. It is easy to see that with each of the 10 choices for the TV animal, we could choose 9 animals to appear online. We would then have $10 \times 9 = 90$ ways to form the desired pairs.

🖉 **KEY POINT**

Tree diagrams help visualize counting situations that take place in stages.

Tree Diagrams

The Three-Way Principle in Section 1.1 recommended viewing situations graphically. It is good to keep this principle in mind when solving counting problems.

EXAMPLE 3 *A Tree Diagram Shows How Three Coins Are Flipped*

How many ways can three coins be flipped?

SOLUTION: To emphasize that the three coins are different, let's assume we are flipping a penny, a nickel, and a dime. A **tree diagram** is a handy way to illustrate the possibilities. First, in Figure 13.2(a) we draw a tree with two branches to show the possibility of a head or a tail for the penny. Next, in Figure 13.2(b), we see that if the penny shows either a head or a tail, the nickel also can show a head or a tail.

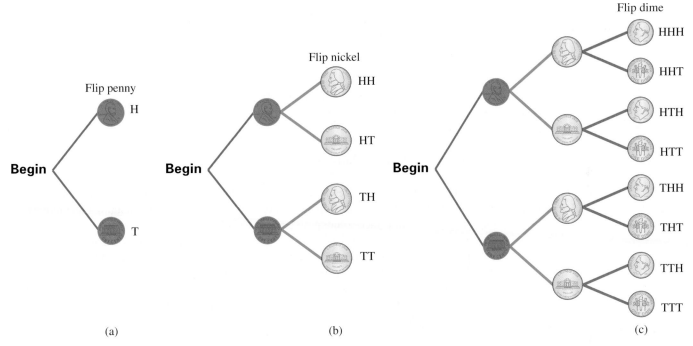

FIGURE 13.2 Tree diagrams representing all possible ways to flip (a) a penny; (b) a penny and then a nickel; (c) a penny, nickel, and a dime.

Quiz Yourself ❷

How many ways are there to flip four coins?

So far there are four ways to flip the two coins. Finally, with each of these four possibilities, the dime can show a head or a tail, as we see in Figure 13.2(c).

Moving through the tree from left to right, we can trace eight branches corresponding to the ways the three coins can be flipped:

HHH, HHT, HTH, HTT, THH, THT, TTH, TTT ✾ ❷

You will see the situation shown in Example 4 frequently in Chapter 14 when we discuss probability.

EXAMPLE 4 *Rolling Two Dice*

If we roll two dice, how many different pairs of numbers can appear on the upturned faces?

SOLUTION: To emphasize that the dice are different, we assume one is red and the other green. Clearly, a red 2 and a green 3 is not the same as a red 3 and a green 2. We will use ordered pairs of the form (red number, green number) to represent the pairs showing on the dice. For example, (4, 5) represents a red 4 and a green 5. Figure 13.3 on page 626 illustrates this situation.

The leftmost set of six branches in Figure 13.3 shows pairs corresponding to 1 on the red die and either a 1, 2, 3, 4, 5, or 6 on the green. These branches correspond to the pairs (1, 1), (1, 2), (1, 3), (1, 4), (1, 5), and (1, 6). Similarly, the second set of six branches

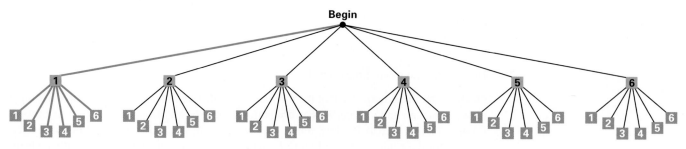

FIGURE 13.3 Tree diagram showing how many ways to roll two dice.

corresponds to the pairs (2, 1), (2, 2), (2, 3), (2, 4), (2, 5), and (2, 6). Continuing this pattern, we get the following 36 pairs:

$$(1, 1), \quad (1, 2), \quad (1, 3), \quad (1, 4), \quad (1, 5), \quad (1, 6),$$
$$(2, 1), \quad (2, 2), \quad (2, 3), \quad (2, 4), \quad (2, 5), \quad (2, 6),$$
$$(3, 1), \quad (3, 2), \quad (3, 3), \quad (3, 4), \quad (3, 5), \quad (3, 6),$$
$$(4, 1), \quad (4, 2), \quad (4, 3), \quad (4, 4), \quad (4, 5), \quad (4, 6),$$
$$(5, 1), \quad (5, 2), \quad (5, 3), \quad (5, 4), \quad (5, 5), \quad (5, 6),$$
$$(6, 1), \quad (6, 2), \quad (6, 3), \quad (6, 4), \quad (6, 5), \quad (6, 6)$$

Now try Exercises 11 to 14. ❊

To be able to solve a wide class of counting problems, including the lottery question we posed at the beginning of the chapter, we need to develop some more counting techniques—but first we will introduce some more terminology. In some counting problems, objects can be repeated; in others, they cannot. To open a combination lock, we can use the same number for each turn of the dial. For example, 23-23-23 could be a valid combination. In other situations, such as choosing the pairs of animals in Example 2, we cannot use the same animal twice. If objects are allowed to be used more than once in a counting problem, we will use the phrase *with repetition*. If we do not want objects to be used more than once, we will say *without repetition*.

 ## Some Good Advice

A good mathematician often takes a complex situation and breaks it into simpler components to solve a problem. We used this approach in Examples 2, 3, and 4. Frequently in counting problems it is useful to think of a situation not as occurring all at once but rather in distinct stages. Think of a first thing happening, then a second, then a third, and so on. If we count the possibilities at each stage, we then can combine this information to arrive at a final answer.

EXAMPLE 5 *Advertising Recording Stars*

Assume that as advertising director at a local TV station you have to fill three commercial spots and have decided to use promos for the latest albums by *American Idol* winners, (J)ordin Sparks, (T)aylor Hicks, and (C)arrie Underwood. In how many ways can you fill these spots (in some order) if

a) repetition is allowed?   b) repetition is not allowed?

SOLUTION:

a) If repetition is allowed, then you could choose Jordin Sparks, Taylor Hicks, and then Taylor Hicks again. We will abbreviate this ordering as JTT. Or, you could choose Jordin Sparks, Carrie Underwood, and then Jordin Sparks (JCJ). Figure 13.4 illustrates the 27 ways you could choose to fill the three spots.

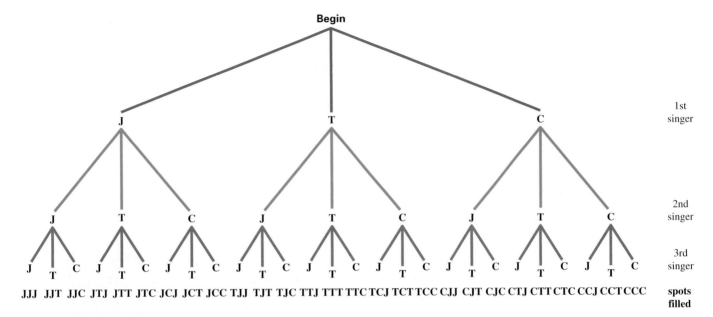

Begin

1st singer

2nd singer

3rd singer

JJJ JJT JJC JTJ JTT JTC JCJ JCT JCC TJJ TJT TJC TTJ TTT TTC TCJ TCT TCC CJJ CJT CJC CTJ CTT CTC CCJ CCT CCC   **spots filled**

FIGURE 13.4 Tree diagram representing all different ways to fill three commercial spots with repetition.

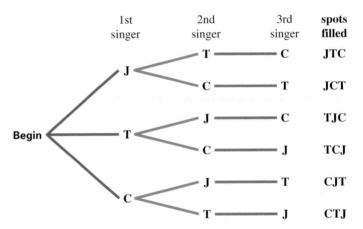

1st singer	2nd singer	3rd singer	spots filled

FIGURE 13.5 Tree diagram representing all different ways to fill three commercial spots without repetition.

KEY POINT

We can solve counting problems by imagining a tree instead of drawing it.

b) If repetition is not allowed, then you could not choose JTT or JCJ. Tracing the branches in Figure 13.5 shows that now there are only six possible ways that you can choose to fill the three spots.

Now try Exercises 15 to 18. ✾

Visualizing Trees

Drawing large tree diagrams is tedious. After you have drawn several trees, you should be able to "see" the tree in your mind without actually drawing it.

EXAMPLE 6 *Counting the Ways to Choose an Outfit*

Christian has designed a new collection for the final competition in *Project Runway*. He has created five tops, four pairs of pants, and three jackets. If we consider an outfit to be a top, pants, and jacket, how many different outfits can Christian's models wear without repeating the exact same outfit?

SOLUTION: We can imagine a tree diagram where the leftmost branches represent the choices of tops, the middle branches depict the choices of pants, and the rightmost branches show the choices for jackets.

Quiz Yourself ③

How many outfits could Christian have in Example 6 if he had four tops, two pairs of pants, and three jackets?

We begin the tree with five branches, one for each top. Then to each branch we attach four branches representing the pants. So far we would have a tree with 20 branches. Finally, to each of these branches we attach three more representing the jackets, to give us a total of 60 different branches. Therefore, Christian's models can wear 60 different outfits.

Now try Exercises 25 and 26. ✳ ③

Math in Your Life

Can We Use Mathematics to Defeat Terrorism?

"It all started when I saw the movie *A Beautiful Mind*," says Jonathan Farley, a visiting professor at the Massachusetts Institute of Technology. He was struck by the line, "Mathematicians won the war," referring to how the research of John Nash was applied successfully to implement Cold War military strategy. Farley believes that we can use mathematics to reveal the organization of terrorist cells and identify unknown terrorists. Farley's company, called Phoenix Mathematical Systems Modeling, offers advice on the mathematics of terrorism and other applications of mathematics to national defense.

Gary Nelson, a senior researcher at the Homeland Security Institute, a quasigovernment institute, agrees that new mathematical methods are necessary to help security agencies deal with enormous amounts of unorganized data. At the University of Southern California, Jafar Adibi is developing computer programs, using a process called *data mining*, to sift through large databases of

information. His programs apply mathematical techniques to search records regarding phone calls, places of worship, political affiliation, and family relations to find hidden links between known terrorists and their unknown confederates.

Exercises [13.1]

Looking Back*

These exercises follow the general outline of the topics presented in this section and will give you a good overview of the material that you have just studied.

1. How did we apply the problem-solving technique of being systematic to list the animal pairs in Example 2?

2. Why were there fewer branches in the tree diagram for the solution in Example 5(b) than there were for the solution of 5(a)?

3. Who is Jonathan Farley?

4. What is data mining?

Sharpening Your Skills

In Exercises 5–8, you are selecting American Idol winners from the set W = {(J)ordin Sparks, (T)aylor Hicks, (C)arrie Underwood, (F)antasia Barrino, (R)uben Studdard}.

5. List all the ways that you can select two *different* singers from W. The order in which you select the singers is not important. For example, JT is the same selection as TJ.

6. List all the ways you can select two singers from W. Repetition is allowed and order is not important. For example FF is allowed and CR is the same as RC.

7. List all the ways you can select two *different* singers from W. The order in which you select singers *is important*. For example, FR is not the same as RF.

8. List all the ways you can select two singers from W. Repetition is allowed and *order is important*. For example FF is allowed and CR is not the same as RC.

9. In problems such as Exercises 5–8, do you expect to get a larger answer if repetition is allowed or not allowed?

10. In problems such as Exercises 5–8, do you expect to get a larger answer if order is important or not important?

*Before doing these exercises, you may find it useful to review the note *How to Succeed at Mathematics* on page xix.

Draw a tree diagram that illustrates the different ways to flip a penny, nickel, dime, and quarter. Use this diagram to solve Exercises 11–14.

11. In how many ways can you get exactly one head?

12. In how many ways can you get no tails?

13. In how many ways can you get exactly two tails?

14. In how many ways can you get exactly three heads?

15. How many different two-digit numbers can you form using the digits 1, 2, 5, 7, 8, and 9 without repetition? For example, 55 is not allowed.

16. How many different two-digit numbers can you form using the digits 1, 2, 5, 7, 8, and 9 with repetition? For example, 55 is allowed.

17. How many different three-digit numbers can you form using the digits 1, 2, 5, 7, 8, and 9 without repetition?

18. How many different three-digit numbers can you form using the digits 1, 2, 5, 7, 8, and 9 with repetition?

In Exercises 19–24, assume you are rolling two dice: the first one is red, and the second one is green. Use a systematic listing to determine the number of ways you can roll each of the following. For example, a total of 3 can be rolled in two ways: (1, 2) and (2, 1).

19. Roll a total of 5.

20. Roll a total of 7.

21. Roll both numbers the same.

22. Roll a 3 on the red die.

23. Roll a total less than 6.

24. Roll a total greater than 9.

Recall in Example 6 that Christian has designed different tops, pants, and jackets to create outfits for Project Runway. *How many different outfits can his models wear if he has designed the following:*

25. Counting outfits. Six tops, five pants, four jackets

26. Counting outfits. Seven tops, six pants, three jackets

Applying What You've Learned

27. Assigning tasks. Angela's coworkers Pam, Phyllis, Jim, and Dwight have volunteered to help with the preparations for a party. How many ways can Angela assign someone to buy beverages, someone to arrange for food, and someone to send out invitations? Assume that no person does two jobs.

28. Making staff assignments. Suppose that the staff of a weekly newspaper, the *Southern California Sentinel*, consists of (A)drian, (B)rian, (C)armen, (D)avid, and (E)mily. The editor will choose a features editor and a sports editor from these five people. If the sports editor must be different from the features editor, in how many ways can the editor make this selection?

In Exercises 29 and 30, draw the tree diagram only if you must. Try to picture the tree mentally without actually putting it on paper.

29. If you drew a tree diagram showing how many ways five coins could be flipped, how many branches would it have?

30. If you drew a tree diagram showing how many ways six coins could be flipped, how many branches would it have?

31. The role-playing game *Dungeons and Dragons* uses a tetrahedral die that has four congruent triangular sides numbered 1, 2, 3, and 4. How many branches would a tree diagram have that shows the way two tetrahedral dice could be rolled?

32. *Dungeons and Dragons* also uses 12-sided dice. How many branches would a tree diagram have that shows the way two 12-sided dice could be rolled?

Use the given diagram to solve Exercises 33–36.

Squares such as ABCD *and* EFGH *are called* 1 × 1 *squares. A square such as* CXGY *is called a* 2 × 2 *square.*

33. How many 1 × 1 squares can be formed?

34. How many 4 × 4 squares can be formed?

35. How many 2 × 2 squares can be formed?

36. How many 3 × 3 squares can be formed?

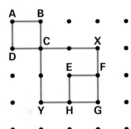

37. Answer Exercise 35 assuming the diagram has six rows with six dots in each row.

38. Answer Exercise 36 assuming the diagram has six rows with six dots in each row.

39. Counting license plates. An eyewitness to a crime said that the license plate of the getaway car began with the letters T, X, and L, but he could not remember the order. The rest of the plate had the numbers 8, 3, and 4, but again he did not recall the order. How many license plates must the police investigate to find this car?

40. Counting license plates. In a small state, the license plate for a car begins with two letters, which may be repeated, and ends with three digits, which also may be repeated. How many license plates are possible in that state?

In Exercises 41 and 42, you are taking a true or false quiz.

41. Assume that you are taking a five-question true or false quiz.

   a. How many different ways are there to answer all five questions?

   b. How many ways are there to get no questions wrong? One question? Two questions?

   c. If, not having studied, you simply guess at each answer, what are your chances of getting three or more correct?

42. Assume that you are taking a 10-question true or false quiz.

   a. How many different ways are there to answer all 10 questions?

   b. How many ways are there to get no questions wrong? One question? Two questions? Three questions?

   c. If, not having studied, you simply guess at each answer, what are your chances of getting at least seven correct?

Two couples, Beyoncé and Jay-Z and Brad and Angelina, have bought tickets for four adjacent seats to see the musical Spamalot. Use this information to answer Exercises 43–46. You may want to solve the problems by systematic listing, drawing a tree diagram, or by imagining a tree diagram. Also, drawing a diagram of the seats may help you.

43. In how many ways can the couples occupy their four seats?

44. In how many ways can the couples be seated if the women are to sit together?

45. In how many ways can the couples be seated if the women do not sit together?

46. In how many ways can the couples be seated if the couples are to sit together?

47. The base of a stack of oranges in a supermarket consists of five rows with five oranges in each row. The oranges in the next row will be placed as shown by the highlighted orange in the figure. How many oranges will there be in the stack?

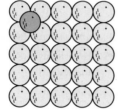

48. If the bottom row of oranges consisted of seven rows of seven oranges each, how many oranges would there be in the stack?

In Exercises 49–52, you are buying a triple-deck ice cream cone with vanilla, strawberry, and chocolate as possible flavors. The flavors can be repeated or not, and we will consider two cones to be different if the flavors are the same but occur in different order.

49. How many different cones are possible?

50. How many cones have only vanilla and strawberry?

51. How many possible cones have only two different flavored scoops?

52. How many possible cones have all different flavors?

Communicating Mathematics

53. In the Some Good Advice box on page 626, we advised you to take a complex problem and break it into smaller subproblems. How did we do that in Examples 3 and 4?

54. After we solved several problems by drawing tree diagrams, what did we encourage you to do?

Using Technology to Investigate Mathematics

55. Search the Internet for "mathematics and terrorism" or "mathematics and defense" to find sites dealing with this current area of mathematical research. You might also find it interesting to search for "Vanderbilt University and Jonathan Farley" to learn more about Dr. Farley and his research. Report on your findings.

56. Counting theory is part of a larger area of mathematics called *combinatorics*. Search the Internet for "combinatorics and applications." You will find an interesting collection of sites if you search for "real-life applications of nature-inspired combinatorial heuristics." Report on some of the applications that you find.

For Extra Credit

57. Building class schedules. Anthony is building his class schedule for next semester. He has decided to take four classes: mathematics, English, sociology, and music. He intends to schedule classes only on Monday, Wednesday, and Friday mornings. The possible classes that he can take are

Mathematics: MWF—9:00, 11:00, 12:00

English: MWF—9:00, 10:00, 12:00

Sociology: MWF—10:00, 11:00, 12:00

Music: MWF—9:00, 10:00, 11:00

Use a tree diagram to determine the possible schedules he can have and then list all those schedules.

58. Building appointment schedules. A pharmaceutical salesperson wants to schedule appointments with three doctors to introduce a new line of antibiotics. The times that the doctors have available are as follows:

Dr. House: Monday at 10:00, Wednesday at 11:00, Thursday at 11:00

Dr. Cuddy: Monday at 9:00, Monday at 10:00

Dr. Wilson: Monday at 10:00, Wednesday at 11:00, Thursday at 11:00

Use a tree diagram to determine all possible schedules that would allow the salesperson to meet with each doctor. List those schedules.

Assume that you are a contestant on a game show called Wheel of Destiny. *The final round of this game, called the* Gem Round, *is played as follows. There are four treasure chests, each*

containing a gem. Two of the gems are red, one is blue, and one is green. The game show host will ask you to open the chests one at a time in any order that you choose. If the fourth chest you open contains Your Final Gem, *then you win the grand prize.*

You must spin the Wheel of Destiny *to learn what your final gem will be. For example, when you spin the wheel, it may stop at "a green gem." This means that the round ends whenever you open a box containing the green gem.*

In Exercises 59–61, you are given the final gem. List the different orders in which you can select the colors red, blue, and green in order to win the grand prize. Drawing a tree will help you see the possibilities.

59. Enumerating game show possibilities. A green gem.

60. Enumerating game show possibilities. A blue gem.

61. Enumerating game show possibilities. A red gem.

62. If you were to draw tree diagrams to solve Exercises 59–61, how would they be different from the trees in Examples 3 and 4?

13.2 The Fundamental Counting Principle

Objectives

1. Understand the fundamental counting principle.
2. Use slot diagrams to organize information in counting problems.
3. Know how to solve counting problems with special conditions.

While carrying a demanding course load, working at a part-time job, and staying up late to hang out with your friends, you have been neglecting good health habits. With no time for exercise, snacking during late-night cram sessions, and substituting burgers and fries for Mom's home cooking, you have succumbed to the dreaded "Freshman Fifteen."* Maybe it's time to start an exercise program.

Assume that you have decided to work out your abs, arms, and cardiovascular system *in that order*, and your fitness center has six machines for the abs, four for arms, and eight for cardio. In how many different ways can you vary your workouts using different machines?

It is useful to relate a new problem such as this to problems that you have seen before. Whether flipping coins, rolling dice, or selecting clothes for a fashion show, you have seen the same basic pattern:

A first thing happens in *a* number of ways,
then a second thing happens in *b* number of ways,
then a third thing happens in *c* number of ways,
and so on.

To solve this problem, *imagine* a tree diagram as follows:

First, draw 6 branches representing your choices to work your abs.

Second, attach 4 branches to each of the 6 branches to represent your choice for the arm machines. This gives you a tree with 24 branches.

Third, attach 8 branches for the cardio machines to each of these 24 branches, which gives you a final tree with $6 \times 4 \times 8 = 192$ branches.

The Fundamental Counting Principle

KEY POINT

The fundamental counting principle solves problems without listing elements or drawing tree diagrams.

Continuing the above discussion, if you wanted to add a fourth or fifth type of exercise to your workout, we would add a fourth collection of branches and then a fifth collection of branches to the tree and so on. This thinking leads us to the following principle.

> **THE FUNDAMENTAL COUNTING PRINCIPLE (FCP)** If we want to perform a series of tasks and the first task can be done in *a* ways, the second can be done in *b* ways, the third can be done in *c* ways, and so on, then all the tasks can be done in $a \times b \times c \times \cdots$ ways.

We will revisit several situations from Section 13.1 to illustrate the power of the fundamental counting principle.

---

*Most associate the term "Freshman Fifteen" with the gain of 15 pounds by college freshmen during their first year at college.

EXAMPLE 1   *Using the FCP to Count Animal Pairs*

Recall that an environmental group plans to launch an advertising campaign featuring two endangered species from among the following: the (C)heetah, the (O)tter, the black-footed (F)erret, and the Bengal (T)iger. One animal will appear in a series of TV commercials and a different animal will be featured in online advertisements. In how many ways can this be done?

SOLUTION: We first have to choose the animal for the TV campaign, which can be done in four ways. The second task is to choose a *different* animal for the online ads. There are three ways to perform this task, so the total number of ways to choose both animals is

$$4 \times 3 = 12.$$

four ways to choose an ⌟   ⌞ . . . times three ways to choose a
animal for TV . . .   different animal for online ads

Now try Exercises 5 to 12. ❋

PROBLEM SOLVING
Draw Diagrams

If you get confused when using counting formulas as to when to multiply and when to add, you may find that a picture often helps you recall how we derived the formula. For example, by imagining a tree diagram, you will find the fundamental counting principle easier to remember.

EXAMPLE 2   *Using the FCP in Coin and Dice Problems*

a) How many ways can four coins be flipped?

b) How many ways can three dice (red, green, blue) be rolled?

SOLUTION:

a) The first task, flipping the first coin, can be done in two ways. The second, third, and fourth tasks also each consist of flipping a coin in one of two ways. Therefore, by the fundamental counting principle, the four coins can be flipped in

$$2 \times 2 \times 2 \times 2 = 16 \text{ ways.}$$

\ | | /
four tasks—each done in two ways

b) The first task is rolling the red die, which can be done in six ways. The second and third tasks also each can be done in six ways. Thus, the three dice can be rolled in $6 \times 6 \times 6 = 216$ ways.

Now try Exercises 13 to 16. ❋

EXAMPLE 3   *Counting Outfits Using the FCP*

Matthew is working as a summer intern for a TV station and wants to vary his outfit by wearing different combinations of coats, pants, shirts, and ties. If he has three sports coats, five pairs of pants, seven shirts, and four ties, how many different ways can he select an outfit consisting of a coat, pants, shirt, and tie?

SOLUTION: In selecting an outfit, we consider the following tasks:

Task	Number of Ways to Perform Task
Select coat	3
Select pants	5
Select shirt	7
Select tie	4

Quiz Yourself 4

Assume that you are shopping for a new car. There are two different models to choose from, there are eight different colors available, each model has three different interior packages, and each model comes with either the plain or sport exterior trim package. How many different cars do you have to choose from?

KEY POINT

Slot diagrams help organize information before applying the fundamental counting principle.

By the fundamental counting principle, Matthew has

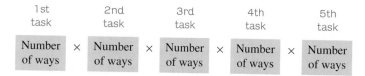

$$3 \times 5 \times 7 \times 4 = 420 \text{ outfits.}$$

number of coats . . . times number of pants . . . times number of shirts . . . times number of ties ✳ 4

Slot Diagrams

Sometimes special conditions affect the number of ways we can perform the various tasks. A useful technique for solving problems such as these is to draw a series of blank spaces, as shown in Figure 13.6, to keep track of the number of ways to do each task. We will call such a figure a **slot diagram**.

1st task		2nd task		3rd task		4th task		5th task
Number of ways	×	Number of ways	×	Number of ways	×	Number of ways	×	Number of ways

FIGURE 13.6 A slot diagram helps organize counting problems.

EXAMPLE 4 *Counting Keypad Patterns*

To open your locker at the fitness center, you must enter five digits in order from the set $0, 1, 2, \ldots, 9$. How many different keypad patterns are possible if

a) any digits can be used in any position and repetition of digits is allowed?

b) the digit 0 cannot be used as the first digit, but otherwise any digit can be used in any position and repetition is allowed?

c) any digits can be used in any position, but repetition is not allowed?

SOLUTION:

a) We can use any of the 10 digits for each of the five tasks, as shown in the slot diagram in Figure 13.7. Thus, we see that there are $10 \times 10 \times 10 \times 10 \times 10 = 100,000$ possible keypad patterns.

1st task		2nd task		3rd task		4th task		5th task
10	×	**10**	×	**10**	×	**10**	×	**10**
Use any digit		Use any digit		Use any digit		Use any digit		Use any digit

FIGURE 13.7 The slot diagram shows five tasks, each of which can be done in 10 ways.

b) In this situation, we have only nine possible ways to select the first digit. The slot diagram is shown in Figure 13.8.

1st task		2nd task		3rd task		4th task		5th task
9	×	**10**	×	**10**	×	**10**	×	**10**
Can't use 0		Use any digit		Use any digit		Use any digit		Use any digit

FIGURE 13.8 The first task can be done in 9 ways; the remaining tasks each can be done in 10 ways.

Therefore, there are $9 \times 10 \times 10 \times 10 \times 10 = 90,000$ possible keypad patterns.

c) We can use any of the 10 digits for the first number. However, because repetitions are not allowed, we have only nine digits for the second number. After two numbers have been chosen, there are only eight possibilities for the third number. Similarly, there are seven for the fourth number and six for the fifth number, as we see in Figure 13.9.

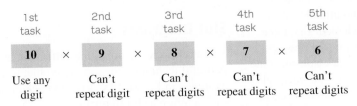

1st task		2nd task		3rd task		4th task		5th task
10	×	**9**	×	**8**	×	**7**	×	**6**
Use any digit		Can't repeat digit		Can't repeat digits		Can't repeat digits		Can't repeat digits

FIGURE 13.9 Because repetition of digits is not allowed, there are fewer possibilities for each keypad combination.

So we see that there are $10 \times 9 \times 8 \times 7 \times 6 = 30,240$ possible keypad patterns.

Now try Exercises 17 to 20. ❋

Now try Exercises 17 to 20.

KEY POINT

In solving counting problems, consider special conditions first.

Handling Special Conditions

We will now look at a counting situation requiring a slightly more complex analysis than we needed in the previous examples.

EXAMPLE 5 *Counting Seating Patterns with Special Conditions*

Professor Bartlett teaches an advanced cognitive psychology class of 10 students. She has a visually challenged student, Louise, who must sit in the front row next to her tutor, who is also a member of the class. If there are six chairs in the first row of her classroom, how many different ways can Professor Bartlett assign students to sit in the first row?

SOLUTION: In order to use the fundamental counting principle, we must identify the separate tasks in determining the seating.

We will first consider the special condition that Louise and her tutor must sit together.

Task 1: Decide which two seats Louise and her tutor will occupy.

Task 2: Decide where Louise and her tutor will sit in these two seats.

Task 3: Determine who sits in the remaining seats.

Task 1: As Figure 13.10 shows, there are five ways that Louise and her tutor can sit together.

Math in Your Life

Do You Want to Be a Millionaire?*

A survey sponsored by the Consumer Federation found that many Americans think that they have a better chance of acquiring money for retirement by playing the lottery than by systematically investing their money. Many do not understand that their chances of winning a lottery are remote and that by saving and investing regularly, they can accumulate a substantial sum if done over many years.

The survey found that younger and lower-income people were the most likely to believe that lotteries rather than saving were the path to wealth. An understanding of the counting principles that we present in this chapter, together with a knowledge of how investments grow (as we discussed in Chapter 9), should convince you that saving and investments rather than lotteries are a surer road to financial security.

*This highlight is based on an Associated Press article that appeared in the *New York Journal News.*

Seat 1		Seat 2	Seat 3	Seat 4	Seat 5	Seat 6
L	and	T	X	X	X	X
X		L	and	T	X	X	X
X		X	L	and	T	X	X
X		X	X	L	and	T	X
X		X	X	X	L	and	T

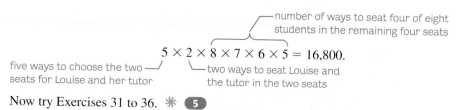

FIGURE 13.10 There are five ways to choose two seats for Louise and her tutor.

Task 2: Once we have determined which seats L and T occupy, there are two ways that L and T can sit within those seats—L sits either on the right or the left.

Task 3: After Louise and her tutor have been seated, we fill the remaining four seats from left to right. There are eight students left; thus, we have eight students for the first remaining seat, seven for the second seat, six for the third seat, and finally five for the last seat.

Therefore, the number of ways to seat the students in the first row is

Quiz Yourself **5**

Redo Example 5, but now assume that the class has 12 students and the first row has eight seats.

number of ways to seat four of eight students in the remaining four seats

$$5 \times 2 \times 8 \times 7 \times 6 \times 5 = 16{,}800.$$

five ways to choose the two seats for Louise and her tutor

two ways to seat Louise and the tutor in the two seats

Now try Exercises 31 to 36. ❋ **5**

We will use the fundamental counting principle in Section 13.3 to develop further counting tools that will enable us to answer the lottery question we posed at the beginning of the chapter.

Math in Your Life

How Can Counting Make You Rich?

Most games of chance are based on mathematical principles that slightly favor casinos, allowing them to earn billions of dollars from customers willing to be exploited by playing casino games. Blackjack, however, is one of the casino table games that can be legally beaten by a skilled player. By a method called *card counting*, a person, or team of individuals, can recognize when the remaining cards to be played will favor the player, rather than the house.

In the 1990s, a team of students in the Boston area, called the MIT blackjack team,* who were highly skilled in mathematics and computer science developed a devastatingly effective card-counting scheme that earned them

over $10 million. As a result of their stunning success, these players were banned from playing blackjack at many casinos.

In order to recognize card counters and prevent them from playing, casinos have embraced cutting-edge techniques such as data mining, facial recognition software, and iris scan technology. James X. Dempsey, from the Center for Democracy and Technology, says that Las Vegas is the incubator for a host of surveillance technologies that are now used by the government to identify terrorists and by malls and amusement parks to enhance security.†

---

*You might find it interesting to research the MIT blackjack team on the Internet, read about their exploits in the book *Bringing Down the House* by Ben Mezrich (Free Press, 2002), or see the 2008 movie *21*.
†See *From Casinos to Counterterrorism*, at www.washingtonpost.com, October 22, 2007.

Exercises 13.2

Looking Back*

These exercises follow the general outline of the topics presented in this section and will give you a good overview of the material that you have just studied.

1. How did we use the fundamental counting principle in Example 2?

2. Why are the slot diagrams in Example 4(a) and (c) different?

3. Why did the MIT blackjack team choose to play blackjack, rather than other casino games?

4. What technologies do casinos use to recognize card counters?

Sharpening Your Skills

5. **Assigning responsibilities.** The board of an Internet start-up company has seven members. If one person is to be in charge of marketing and another in charge of research, in how many ways can these two positions be filled?

6. **Assigning positions.** If there are 12 members on the staff of the *Sandpiper*, a newspaper covering shore news, in how many ways can we choose an entertainment editor and a sales manager?

7. **Assigning officers.** The Equestrian Club has eight members. If the club wants to select a president, vice president, and treasurer (all of whom must be different), in how many ways can this be done?

8. **Assigning officers.** If the Chamber of Commerce has 20 members, in how many ways can they elect a different president, vice president, and treasurer?

9. **Building a home theater system.** Elaine is building a home theater system consisting of a tuner, an optical disc player, speakers, and a high-definition TV. If she can select from four tuners, eight speakers, three optical disc players, and five TVs, in how many ways can she configure her system?

10. **Counting schedules.** Jorge is using his educational benefits from the Navy to enroll in four courses to prepare him for a career as an architect: design, science, math, and social science. If there are seven design courses, five science courses, four math courses, and six social science courses that fit his schedule, in how many ways can he select his courses?

11. **Counting meal possibilities.** The early bird special at TGI Friday's features an appetizer, soup or salad, entrée, and dessert. If there are five appetizers, six choices for soup or salad, 13 entrées, and four desserts, how many different meals are possible? (We assume that you make a selection from each category.)

12. **Counting meal possibilities.** What would your answer to Exercise 11 be if you are allowed to skip some categories? For example, you may choose to skip the appetizer and dessert.

In games such as Dungeons and Dragons, *in addition to the common 6-sided dice, dice having 4, 8, 12, or 20 sides are also used, as shown in the figure.*

In Exercises 13–16, determine the number of possibilities for each situation.

13. The 8-sided die is rolled twice.

14. The 12-sided die is rolled twice.

15. The 8-sided die is rolled and then the 12-sided die is rolled.

16. The 20-sided die is rolled and then the 6-sided die is rolled.

In Exercises 17–20, using the digits 0, 1, 2, . . . , 8, 9, determine how many of each type of four-digit number can be constructed.

17. Zero cannot be used for the first digit; digits may be repeated.

18. The number must begin and end with an odd digit; digits may not be repeated.

19. The number must be odd and greater than 5,000; digits may be repeated.

20. The number must be between 5,001 and 8,000; digits may not be repeated.

Applying What You've Learned

21. **Enumerating call letters.** Radio stations in the United States that are east of the Mississippi River have call letters consisting of a W followed by three letters (W can be used again). How many different call letters of this type are possible?

22. **Enumerating zip codes.** Postal zip codes currently consist of five digits. For example, the zip code for Honolulu, Hawaii, is 96820. If we used letters instead of digits, how many letters would be required to make as many zip codes as five digits?

---

*Before doing these exercises, you may find it useful to review the note *How to Succeed at Mathematics* on page xix.

23. Counting musical patterns. In the early part of the twentieth century, the Austrian composers Arnold Schoenberg, Alban Berg, and Anton Webern created a system of music called *atonal music*, in which a note cannot be repeated in a composition until all other notes are used. In atonal music, a melody, called a *tone row*, consists of a sequence of 12 different notes. How many different tone rows are possible in this system?

24. Counting license plates. In a certain state, license plates currently consist of two letters followed by three digits. How many such license plates are possible? If the state department of transportation decides to change the plates to have three letters followed by two digits, how many plates will now be possible?

A pin tumbler lock has a series of pins, each of which must be positioned at the proper height to allow the lock to open. The security of the lock is determined by the number of pins in the lock, as well as the number of different possible lengths for the pins. Inserting the proper key in the lock aligns the pins properly and allows the interior of the lock to be turned.

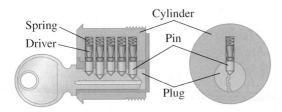

25. Counting keys. If a pin tumbler lock has five pins, each of three different lengths, how many different keys would you have to try in order to guarantee that you could open the lock?

26. Counting keys. If a pin tumbler lock has seven pins, each of four different lengths, how many different keys would you have to try to guarantee that you could open the lock?

27. Counting routes for an armored van. A Wells Fargo armored van must travel from the (D)iamond House, make a pickup at (J)avier's Jewelry, Inc, then stop at (E)mily's Emeralds, and finally go to the (B)ank. Use the following street map to answer the questions.

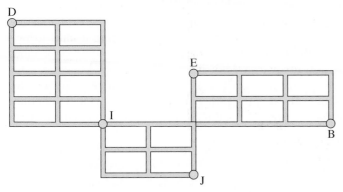

a. How many direct routes are there from D to I?

b. How many direct routes are there from I to J?

c. How many direct routes are there from J to E?

d. How many direct routes are there from E to B?

e. How many direct routes are there from D to B, passing through J and E?

28. Counting settings on an electronic device.

a. A computer interface for a Kawai digital studio piano has eight microswitches that can be set in either the "on" or "off" position. These switches must be set properly for the interface to work. In how many different ways can this group of switches be set?

b. If it takes 2 minutes to set the switches and test to see if the interface is working properly, what is the longest possible time that it would take to find the proper settings by trial and error?

29. Counting pizza combinations. Mamma's Pizza advertises a special in which you can choose a thin crust, thick crust, or cheese crust pizza with any combination of different toppings. The ad says that there are almost 200 different ways that you can order the pizza. What is the smallest number of toppings available? Explain your thinking by referring to the fundamental counting principle.

30. Counting sandwich combinations. The Underground Railway Sandwich Shop has a sandwich special. You can choose whole wheat, rye, or an onion roll, with mayo or mustard, and lettuce or tomato, and a choice of one meat. What is the minimum number of meats available if there are at least 300 possible sandwich combinations? In your counting, don't forget the possibility of "both" or "neither" in the mayo–mustard choice and the lettuce–tomato choice. Explain your thinking by referring to the fundamental counting principle.

Exercises 31 and 32 are alternative versions of the question posed in Example 5 about the seating of Louise and her tutor in Professor Bartlett's class.

31. Counting seating arrangements. Assume that there are 12 students in the class, the front row has eight chairs, and the tutor must be seated to the right of Louise.

32. Counting seating arrangements. Assume that there are 10 students in the class, the front row has six chairs, and Louise will sit between her tutor and her friend Jonathan. It does not matter on which side of Louise the tutor sits.

Assume that we wish to seat three men and three women—Alex, Bonnie, Carl, Daria, Edith, and Frank—in a row of six chairs. Answer Exercises 33–36 using the fundamental counting principle. To apply this principle, first identify the separate tasks involved in making the seating arrangements.

33. Counting seating arrangements. In how many ways can we seat these people with no restrictions?

34. Counting seating arrangements. In how many ways can we seat these people if men and women must alternate?

35. Counting seating arrangements. In how many ways can we seat these people if all the men must sit together and all the women must sit together?

36. Counting seating arrangements. In how many ways can we seat these people if all the women must sit together but no woman sits on an end seat?

Communicating Mathematics

37. Which example in this section illustrated the principle of considering special conditions first in a counting problem? What was the special condition?

38. We have advised you in Section 13.1 to think of a complex counting situation as occurring in stages. How did we use that technique in Example 5?

39. What does the fundamental counting principle have to do with the tree diagrams that we discussed earlier in this chapter?

40. Why would the number 29 be an unlikely answer for a problem that used the fundamental counting principle?

41. Explain in your own words why the fundamental counting principle works.

42. In Exercises 59–61 of Section 13.1, we described the *Gem Round* of the game show *Wheel of Destiny*. In this round, a contestant had to open a series of treasure chests containing a colored gem in a certain order to win the grand prize. There were two red gems, one blue gem, and one green gem. Explain why we cannot obtain the number of possible ways to select the gems in this round by using the fundamental counting principle.

43. Suppose that Yuki is selecting her schedule for next semester and has identified four English courses, three art courses, three psychology courses, and four history courses as possibilities to take. Assume that no two of the courses are offered at conflicting times.

   a. Can the fundamental counting principle be used in solving this problem? Explain.

   b. If any of the courses are offered at conflicting times, can you still use the fundamental counting principle? Explain.

44. Explain what must be present in a counting problem to make it possible to solve the problem using the fundamental counting principle.

For Extra Credit

In Exercises 45–48, create a counting problem whose answer is the given number.

45. 5×4 **46.** 7×8 **47.** $6 \times 5 \times 4$ **48.** $9 \times 8 \times 7$

13.3 Permutations and Combinations

Objectives

1. Calculate the number of permutations of *n* objects taken *r* at a time.
2. Use factorial notation to represent the number of permutations of a set of objects.
3. Calculate the number of combinations of *n* objects taken *r* at a time.
4. Apply the theory of permutations and combinations to solve counting problems.

A student, in a moment of frustration, once said to me, "When you first explain a new idea, I understand what you are doing. Then you keep on explaining it until I don't understand it at all!"* What I believe was frustrating this student is what you also may be dealing with in this chapter. At first we gave you simple, intuitive ways to count using systematic listing and drawing tree diagrams. Then, we went a little more deeply into the idea by introducing the fundamental counting principle. Now we are going to continue our line of thought by giving mathematical-sounding names and notation to the patterns that you have seen earlier.

For example, in Section 13.2, when we constructed keypad patterns in Example 4, you saw the product $10 \times 9 \times 8 \times 7 \times 6$. In Example 5, after seating Louise and her tutor, we used the pattern $8 \times 7 \times 6 \times 5$ to count the ways to fill the remaining four seats. We will now explore these patterns in more detail.

✎ KEY POINT

Permutations are arrangements of objects in a straight line.

Permutations

In both of these problems, we were selecting objects from a set and arranging them in order in a straight line. This notion occurs so often that we give it a name.

---

*I sincerely hope that this doesn't happen to you in this section.

> **DEFINITION** A **permutation** is an ordering of distinct objects in a straight line. If we select *r different* objects from a set of *n* objects and arrange them in a straight line, this is called a *permutation of n objects taken r at a time.* The number of permutations of *n* objects taken *r* at a time is denoted by *P*(*n, r*).*

The following diagram will help you to remember the meaning of the notation *P*(*n, r*).

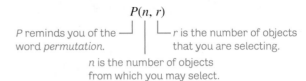

For example *P*(5, 3) indicates that you are counting permutations (straight-line arrangements) formed by selecting three different objects from a set of five available objects.

EXAMPLE 1   *Counting Permutations*

a) How many permutations are there of the letters *a, b, c,* and *d*? Write the answer using *P*(*n, r*) notation.

b) How many permutations are there of the letters *a, b, c, d, e, f,* and *g* if we take the letters three at a time? Write the answer using *P*(*n, r*) notation.

SOLUTION:

a) In this problem, we are arranging the letters *a, b, c,* and *d* in a straight line without repetition. For example, *abcd* would be one permutation, and *bacd* would be another. There are four letters that we can use for the first position, three for the second, and so on. Figure 13.11 is a slot diagram for this problem. From Figure 13.11, we see that there are $4 \times 3 \times 2 \times 1 = 24$ permutations of these four objects. We can write this number more succinctly as

$$P(4, 4) = 24.$$

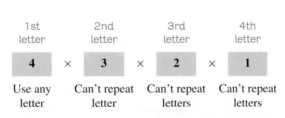

FIGURE 13.11 Slot diagram showing the number of ways to arrange four different objects in a straight line.

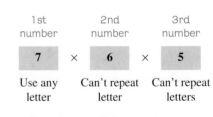

FIGURE 13.12 Slot diagram showing the number of ways to select three different objects from seven possible objects and arrange them in a straight line.

b) Figure 13.12 shows a slot diagram for this question. In this diagram, we see that because we have seven letters, we can use any one of the seven for the first position, one of the remaining six for the second position, and one of the remaining five for the third position.

We see that there are $7 \times 6 \times 5 = 210$ permutations of the seven objects taken three at a time. We write this as

$$P(7, 3) = 210.$$

✳ **6**

Quiz Yourself **6**

a) Explain in your own words the meaning of *P*(8, 3).

b) Find *P*(8, 3).

---

* $_nP_r$ is another common notation for *P*(*n, r*).

Some Good Advice

It is important to remember that an equation in mathematics is a symbolic form of an English sentence. In Example 1, when we write $P(7, 3) = 210$, we read this as "The number of permutations of seven objects taken three at a time is 210."

✏️ **KEY POINT**

We use factorial notation to express $P(n, r)$.

Factorial Notation

Suppose that we want to arrange 100 objects in a straight line. The number of ways to do this is $P(100, 100) = 100 \cdot 99 \cdot 98 \cdots \cdots 3 \cdot 2 \cdot 1$. Because such a product is very tedious to write out, we will introduce a notation to write such products more concisely.

> **DEFINITION** If n is a counting number, the symbol $n!$, called n *factorial*, stands for the product $n \cdot (n-1) \cdot (n-2) \cdot (n-3) \cdots \cdots 2 \cdot 1$. We define $0! = 1$.

EXAMPLE 2   *Using Factorial Notation*

Compute each of the following:

a) $6!$ b) $(8-3)!$ c) $\dfrac{9!}{5!}$ d) $\dfrac{8!}{5!3!}$

SOLUTION:

a) $6! = 6 \cdot 5 \cdot 4 \cdot 3 \cdot 2 \cdot 1 = 720$.

b) Remembering the Order Principle from Section 1.1, we work in parentheses first and do the subtraction *before* computing the factorial. Therefore,

$$(8-3)! = 5! = 5 \cdot 4 \cdot 3 \cdot 2 \cdot 1 = 120.$$

c) $\dfrac{9!}{5!} = \dfrac{9 \cdot 8 \cdot 7 \cdot 6 \cdot \cancel{5 \cdot 4 \cdot 3 \cdot 2 \cdot 1}}{\cancel{5 \cdot 4 \cdot 3 \cdot 2 \cdot 1}} = 9 \cdot 8 \cdot 7 \cdot 6 = 3{,}024.$

Cancel 5!.

Notice how canceling $5!$ from the numerator and denominator makes our computations simpler.

d) $\dfrac{8!}{5!3!} = \dfrac{8 \cdot 7 \cdot \cancel{6} \cdot \cancel{5 \cdot 4 \cdot 3 \cdot 2 \cdot 1}}{\cancel{5 \cdot 4 \cdot 3 \cdot 2 \cdot 1} \cdot \cancel{3 \cdot 2 \cdot 1}} = 8 \cdot 7 = 56.$

Cancel 5!.   Cancel 3!, which equals 6.

Now try Exercises 5 to 12. ❀

In Example 1(b), we first wrote $P(7, 3)$ in the form $7 \times 6 \times 5$. This product started as though we were going to compute $7!$; however, the product $4 \cdot 3 \cdot 2 \cdot 1$ is missing. Another way of saying this is that $P(7, 3) = \frac{7!}{4!} = \frac{7!}{(7-3)!}$. We state this as a general rule.

> **FORMULA FOR COMPUTING $P(n, r)$**
>
> $$P(n, r) = \frac{n!}{(n-r)!}$$

We will use this formula in the computations in Example 3.

EXAMPLE 3   *Counting Ways to Fill Positions at a Community Theater*

The 12-person community theater group produces a yearly musical. Club members select one person from the group to direct the play, a second to supervise the music, and a third to handle publicity, tickets, and other administrative details. In how many ways can the group fill these positions?

SOLUTION: This can be viewed as a permutation problem if we consider that we are selecting 3 people from 12 and then arranging those names in a straight line—director, music, administration. The answer to this question then is

$$P(12, 3) = \frac{12!}{(12 - 3)!} = \frac{12 \cdot 11 \cdot 10 \cdot \cancel{9 \cdot 8 \cdot 7 \cdot 6 \cdot 5 \cdot 4 \cdot 3 \cdot 2 \cdot 1}}{\cancel{9 \cdot 8 \cdot 7 \cdot 6 \cdot 5 \cdot 4 \cdot 3 \cdot 2 \cdot 1}} = 1{,}320$$

ways to select 3 people from the 12 available for the three positions.

Now try Exercises 17 to 20. ✳

 Some Good Advice

When talking with my students I often kid them by saying that to be a good mathematics student you have to be lazy—but in the right way. I don't mean that it is good to sleep through your math class or not do homework, but it is important to try to make your computations as easy as possible for yourself. For example, in Example 3 you had to compute the expression $P(12, 3)$. When doing this by hand, the weak student may compute this as

$$P(12, 3) = \frac{12!}{(12 - 3)!} = \frac{479{,}001{,}600}{362{,}880} = 1{,}320.$$

Working with such large numbers is tedious, time-consuming, and prone to error. However, if you simplify the computations as we did in Example 3, the computations go much more quickly. Some may argue that if you are using a calculator, then it doesn't make a difference as to how you do the computations. This is true provided that the numbers you are working with do not cause an overflow on your calculator. For example, try finding $\frac{500!}{498!}$ on your calculator without canceling before dividing.

 KEY POINT

In forming combinations, order is not important.

Directing	Music	Administration
A	B	C
A	C	B
B	A	C
B	C	A
C	A	B
C	B	A

TABLE 13.1 Six different assignments of responsibilities for the musical are now handled by one committee.

Combinations

In order to introduce a new idea, let's change the conditions in Example three slightly. Suppose that instead of selecting three people to carry out three different responsibilities, we choose a three-person committee that will work jointly to see that the job gets done. In Example 3, if A, B, and C were selected to be in charge of directing, music, and administration, respectively, that would be different than if B directed, C supervised the music, and A handled administrative details. However, with the new plan, it makes no difference whether we say A, B, C or B, C, A. From Table 13.1, we see that all of the six different assignments of A, B, and C under the old plan are equivalent to a single committee under the new plan.

What we are saying for A, B, and C applies to any three people in the theater group. Thus, the answer 1,320 we found in Example 3 is too large, so we must reduce it by a factor of 6 under the new plan. Therefore, the number of three-person committees we could form would be $\frac{1{,}320}{6} = 220$. Remember that the factor of 6 we divided out is really 3! (the number of ways we can arrange three people in a straight line). This means that the number of three-person committees we can select can be written in the form

$$\frac{1{,}320}{6} = \frac{P(12, 3)}{3!}.$$

The reason this number is smaller is that now we are concerned only with *choosing a set* of people to produce the play, but the *order* of the people chosen *is not important*.

✳ ✳ ✳ HISTORICAL HIGHLIGHT

The Origins of Permutations and Combinations

As is the case with so many ideas in mathematics, the origins of some of the fundamental counting concepts go back to ancient times. The notion of permutations can be found in a Hebrew work called the *Book of Creation*, which may have been written as early as AD 200.

Even more surprising, there is evidence that Chinese mathematicians tried to solve permutations and combinations problems in the fourth century BC. We can trace other early counting formulas back to Euclid in the third century BC and the Hindu mathematician Brahmagupta in the seventh century.

We can generalize what we just saw. If we are choosing r objects from a set of n objects and are not interested in the order of the objects, then to count the number of choices we must divide $P(n, r)$* by $r!$. We now state this formally.

> **FORMULA FOR COMPUTING $C(n, r)$** If we choose r objects from a set of n objects, we say that we are forming a **combination** of n objects taken r at a time. The notation $C(n, r)$ denotes the number of such combinations.[†] Also,
>
> $$C(n, r) = \frac{P(n, r)}{r!} = \frac{n!}{r! \cdot (n - r)!}.$$

 Some Good Advice

In working with permutations and combinations, we are choosing r different objects from a set of n objects. The big difference is whether the order of the objects is important. If the order of the objects matters, we are dealing with a permutation. If the order does not matter, then we are working with a combination.

Do not try to use the theory of permutations or combinations when that theory is not relevant. If a problem involves something other than simply choosing *different* objects (and maybe ordering them), then perhaps the fundamental counting principle may be more appropriate.

EXAMPLE 4   *Using the Combination Formula to Count Committees*

a) How many three-element sets can be chosen from a set of five objects?

b) How many four-person committees can be formed from a set of 10 people?

SOLUTION:

a) Because order is not important in considering the elements of a set, it is clear that this is a combination problem rather than a permutation problem. The number of ways to choose three elements from a set of five is

$$C(5, 3) = \frac{5!}{3! \cdot (5 - 3)!} = \frac{5 \cdot 4 \cdot \cancel{3} \cdot \cancel{2} \cdot \cancel{1}}{\cancel{3} \cdot \cancel{2} \cdot \cancel{1} \cdot 2 \cdot 1} = \frac{20}{2} = 10.$$

b) The number of different ways to choose four elements from a set of 10 is

$$C(10, 4) = \frac{10!}{4! \cdot (10 - 4)!} = \frac{10 \cdot \overset{3}{\cancel{9}} \cdot \cancel{8} \cdot 7 \cdot \cancel{6} \cdot \cancel{5} \cdot \cancel{4} \cdot \cancel{3} \cdot \cancel{2} \cdot \cancel{1}}{\cancel{4} \cdot \cancel{3} \cdot \cancel{2} \cdot \cancel{1} \cdot \cancel{6} \cdot \cancel{5} \cdot \cancel{4} \cdot \cancel{3} \cdot \cancel{2} \cdot \cancel{1}} = 210.$$

Now try Exercises 21 to 24. ✳ **7**

Quiz Yourself **7**

How many five-person committees can be formed from a set of 12 people?

---

*Recall that $P(n, r) = \frac{n!}{(n - r)!}$.

[†] $_nC_r$ is another common notation for $C(n, r)$.

We can now address the question about the lottery syndicate that we posed at the beginning of this chapter.

EXAMPLE 5 *How Many Tickets Are Necessary to Cover All Possibilities in a Lottery?*

Recall that a syndicate intended to raise $15 million to buy all the tickets for certain lotteries. The plan was that if the prize was larger than the amount spent on the tickets, then the syndicate would be guaranteed a profit. To play the Virginia lottery, the player buys a ticket for $1 containing a combination of six numbers from 1 to 44. Assuming that the syndicate has raised the $15 million, does it have enough money to buy enough tickets to be guaranteed a winner?

SOLUTION: Because we are choosing a combination of 6 numbers from the 44 possible, the number of different tickets possible is $C(44, 6) = 7,059,052$. Thus, the syndicate has more than enough money to buy enough tickets to be guaranteed a winner. ✳

In addition to wondering whether $15 million would cover all combinations in a lottery, there were some other practical concerns that caused me not to invest in the syndicate.

1) Suppose the syndicate did not raise all $15 million. What would be done with the money then? Would they play lotteries in which not all tickets can be covered? Doing this could lose all the money we invested.

2) What if there are multiple winners? In this case, the jackpot is shared and it is possible that the money won will not cover the cost of the tickets. This case would result in a loss to the syndicate.

3) Is it physically possible to actually purchase 7,059,052 tickets? Because there are $60 \times 60 \times 24 \times 7 = 604,800$ seconds in a week, it would take almost 12 weeks for one person buying one ticket every second to purchase enough tickets to cover all the combinations. In Virginia, the syndicate bought only about 5 million tickets, thus opening up the possibility that all the money could have been lost!

4) How much of the money raised will be used by the syndicate for its own administrative expenses?

5) Because the jackpot is to be shared with thousands of members of the syndicate and the prize money will be paid over a 20-year period, is the amount of return better than if the money had been placed in a high-interest-bearing account?

In light of these concerns, was I too cautious? What do you think? As a final note on this matter, since the Virginia lottery episode, some states have made it illegal to play a lottery by covering all ticket combinations.

As Example 6 shows, applications of counting appear in many gambling situations.*

EXAMPLE 6 *Counting Card Hands*

a) In the game of poker, five cards are drawn from a standard 52-card deck.† How many different poker hands are possible?

b) In the game of bridge, a hand consists of 13 cards drawn from a standard 52-card deck. How many different bridge hands are there?

---

*We cover counting as it applies to gambling in Section 13.4 at the end of this chapter.
†Figure 13.1 shows that a standard 52-card deck contains 13 hearts, 13 spades, 13 clubs, and 13 diamonds. Each suit consists of A, K, Q, J, 10, 9, . . . , 3, 2. See Table 13.2, page 652, for the possible poker hands.

✳ ✳ ✳ HIGHLIGHT

Counting and Technology

We rarely do the kind of computations that you saw in Example 6 with pencil and paper. Not only are these computations tedious but, more important, it is easy to make errors when doing pencil-and-paper calculations.

Provided that the numbers do not get too large, you can calculate the number of permutations and combinations on a graphing calculator such as the TI-83 or TI-84. To find $C(52, 5)$ you would

1st: Type 52 (but do not press ENTER).

2nd: Press the MATH key.

3rd: Move the cursor to the right to PRB and then down to nCr (see Screen 1).

4th: Press ENTER.

5th: Type 5 and press ENTER (see Screen 2).

Notice that we get the same answer as we found in Example 6 by hand.

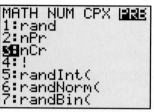

Screen 1

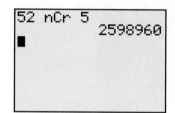

Screen 2

SOLUTION:

a) $C(52, 5) = \dfrac{52!}{5!47!} = \dfrac{\overset{13}{\cancel{52}} \cdot \overset{17}{\cancel{51}} \cdot \overset{10}{\cancel{50}} \cdot 49 \cdot \overset{24}{\cancel{48}}}{\cancel{8} \cdot \cancel{4} \cdot \cancel{3} \cdot \cancel{2} \cdot 1} = 2{,}598{,}960.$

b) $C(52, 13) = \dfrac{52!}{13!39!} = 635{,}013{,}559{,}600.$ ✳

Combining Counting Methods

In some situations, it is necessary to combine several counting methods to arrive at an answer. We see this technique in the next example.

EXAMPLE 7 *Combining Counting Methods*

The division of student services at your school is selecting two men and two women to attend a leadership conference in Honolulu, Hawaii. If 10 men and nine women are qualified for the conference, in how many different ways can management make its decision?

SOLUTION: It is tempting to say that because we are selecting four people from a possible 19, the answer is $C(19, 4)$. But this is wrong. If we think about it, certain choices are unacceptable. For example, we could not choose all men or three women and one man.

We can use the fundamental counting principle once we recognize that the group can be formed in two stages.

Stage 1: Select the two women from the nine available. We can do this in $C(9, 2) = \frac{9!}{2!7!} = 36$ ways.

Stage 2: Select two men from the 10 available. We can make this choice in $C(10, 2) = \frac{10!}{2!8!} = 45$ ways.

Thus, choosing the women and then choosing the men can be done in $36 \cdot 45 = 1{,}620$ ways. ✳

Example 8 shows another situation in which you have to use several counting techniques at the same time.

EXAMPLE 8 *Forming a Governing Committee*

Assume that you and 15 of your friends have formed a company called Net-Media, an Internet music and video provider. A committee consisting of a president, a vice president, and a three-member executive board will govern the company. In how many different ways can this committee be formed?

SOLUTION: We can form the committee in two stages:

a) Choose the president and vice president.

b) Select the remaining three executive members.

Thus, we can apply the fundamental counting principle.

Because order is important in stage a), we recognize that we can choose the president and vice president in $P(16, 2)$ ways.

Order is not important in choosing the remaining three committee members, so we can select the rest of the committee from the 14 remaining people in $C(14, 3)$ ways.

Thus, you see that we can do stage a) followed by stage b) in

$$P(16, 2) \times C(14, 3) = 87{,}360 \text{ ways.}$$

Choose president and vice ⎯⎯⎯⎯/   \⎯⎯ Choose 3 committee members
president from 16 people.   from remaining 14 people.   ✳

KEY POINT

The rows of Pascal's triangle count the subsets of a set.

Row

```
0                1
1              1   1
2            1   2   1
3          (1) (3)  3   1
4        1   4   6  (4) (1)
5      1   5  10  10  (5)  1
              ?
              ?
```

FIGURE 13.13 Pascal's triangle.

It is surprising how the same mathematical idea appears again and again in different cultures over a period of centuries. For example, Pascal's triangle in Figure 13.13 occurs in the writings of fourteenth-century Chinese mathematicians, which some historians believe are based on twelfth-century manuscripts. Pascal's triangle is also found in the work of eighteenth-century Japanese mathematicians.

In Pascal's triangle, we begin numbering the rows of the triangle with zero and also begin numbering the entries of each row with zero. For example, in Figure 13.13, the number 6 is the second entry in the fourth row. Each entry in this triangle after row 1 is the sum of the two numbers immediately above it, which are a little to the right and a little to the left.

Pascal's triangle has an interesting application in set theory. Suppose that we want to find all subsets of a particular set. Recall that the Be Systematic strategy from Section 1.1 encourages us to list possibilities systematically to solve problems. We can find all subsets of $\{1, 2, 3, 4\}$ by first listing sets of size 0, then sets of size 1, 2, 3, and 4, as in the following list:

Ø ⎯⎯ 1 zero-element set

$\{1\}, \{2\}, \{3\}, \{4\}$ ⎯⎯ 4 one-element sets

$\{1, 2\}, \{1, 3\}, \{1, 4\}, \{2, 3\}, \{2, 4\}, \{3, 4\}$ ⎯⎯ 6 two-element sets

$\{1, 2, 3\}, \{1, 2, 4\}, \{1, 3, 4\}, \{2, 3, 4\}$ ⎯⎯ 4 three-element sets

$\{1, 2, 3, 4\}$ ⎯⎯ 1 four-element set

We see that there is one subset of size 0, four of size 1, six of size 2, four of size 3, and one of size 4. Note that this pattern,

$$1 \quad 4 \quad 6 \quad 4 \quad 1,$$

occurs as the fourth row in Pascal's triangle.

Similarly, the pattern

$$1 \quad 5 \quad 10 \quad 10 \quad 5 \quad 1,$$

which is the fifth row of Pascal's triangle, counts the subsets of size 0, 1, 2, and so on of a five-element set. In fact, the following result is true. **8**

Quiz Yourself 8

a) What is the seventh row of Pascal's triangle?

b) How many subsets of size 3 are there in a seven-element set?

PASCAL'S TRIANGLE COUNTS THE SUBSETS OF A SET The nth row of Pascal's triangle counts the subsets of various sizes of an n-element set.

KEY POINT

The entries of Pascal's triangle are numbers of the form $C(n, r)$.

Let us look at the subsets of the set $S = \{1, 2, 3, 4\}$ again. The sets $\{1\}$, $\{2\}$, $\{3\}$, and $\{4\}$ are the four different ways we can choose a one-element set from S. The sets $\{1, 2\}$, $\{1, 3\}$, $\{1, 4\}$, $\{2, 3\}$, $\{2, 4\}$, and $\{3, 4\}$ are all the different ways that we can choose a two-element set from S. Because combinations are sets, this means that we can also use the entries in Pascal's triangle to count combinations.

> **ENTRIES OF PASCAL'S TRIANGLE AS $C(n, r)$** The rth entry of the nth row of Pascal's triangle is $C(n, r)$.

EXAMPLE 9 *Relating Entries in Pascal's Triangle to Combinations*

Interpret the numbers in the fourth row of Pascal's triangle as counting combinations.

SOLUTION: The fourth row of Pascal's triangle is

$$1 \quad 4 \quad 6 \quad 4 \quad 1.$$

Because the leftmost 1 is the zeroth entry in the fourth row, we can write it as $C(4, 0)$. That is to say, $C(4, 0) = 1$. The first entry in the fourth row is 4, which means that $C(4, 1) = 4$. Similarly, $C(4, 2) = 6$, $C(4, 3) = 4$, and $C(4, 4) = 1$. ❋ **9**

Quiz Yourself **9**

Write the numbers in the third row of Pascal's triangle using $C(n, r)$ notation.

For counting problems that are not too large, it is quicker to use entries in Pascal's triangle to count combinations rather than to use the formula for $C(n, r)$ that we stated earlier, as you will see in Example 10.

EXAMPLE 10 *Using Pascal's Triangle to Count Drug Combinations*

Frequently, doctors experiment with combinations of new drugs to combat hard-to-treat illnesses such as AIDS and hepatitis. Assume that a pharmaceutical company has developed five antibiotics and four immune system stimulators. In how many ways can we choose a treatment program consisting of three antibiotics and two immune system stimulators to treat a disease? Use Pascal's triangle to speed your computations.

SOLUTION: We can select the drugs in two stages: first select the antibiotics and then the immune system stimulators. Therefore, we can apply the fundamental counting principle, introduced in Section 13.2.

In selecting the antibiotics we are choosing three drugs from five, so this can be done in $C(5, 3)$ ways. Looking at the fifth row of Pascal's triangle, which is

$$1 \quad 5 \quad 10 \quad 10 \quad 5 \quad 1,$$
$$\diagup \quad \diagup \quad \diagup \quad \diagup$$
$$C(5, 0) \; C(5, 1) \; C(5, 2) \; C(5, 3)$$

we see that the third entry (remember, we start counting row entries with 0) is 10. Thus, $C(5, 3) = 10$.

Choosing two immune system stimulators from four can be done in $C(4, 2)$ ways. The fourth row in Pascal's triangle is

$$1 \quad 4 \quad 6 \quad 4 \quad 1.$$
$$\diagup \quad \diagup \quad \diagup$$
$$C(4, 0) \; C(4, 1) \; C(4, 2)$$

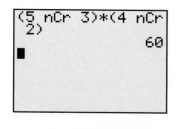

This means that $C(4, 2) = 6$. It follows that we can choose the antibiotics and then the immune system stimulators in $C(5, 3) \times C(4, 2) = 10 \times 6 = 60$ ways. The accompanying screen shows how to do the calculations with a graphing calculator. ❋

HISTORICAL HIGHLIGHT ✻ ✻ ✻

Blaise Pascal

While trying to answer questions about rolling dice and drawing cards, a group of seventeenth-century European mathematicians began to develop a theory of counting. The Frenchman Blaise Pascal, who was one of the central figures in this endeavor, wrote a paper on combinations in 1654.

Pascal was a child prodigy who became interested in Euclid's *Elements* at age 12. Within four years, he wrote a research paper of such quality that some of the leading mathematicians of the time refused to believe that a 16-year-old boy had written it.

Pascal temporarily abandoned mathematics to devote himself to philosophy and religion. However, while suffering from a toothache, he decided to take his mind off the pain by thinking about geometry. Surprisingly, the pain stopped. Pascal took this as a sign from heaven that he should return to mathematics. He resumed his research but soon became seriously ill with dyspepsia, a digestive disorder. Pascal spent the remaining years of his life in excruciating pain, doing little work until his death at age 39 in 1662.

Exercises 13.3

Looking Back*

These exercises follow the general outline of the topics presented in this section and will give you a good overview of the material that you have just studied.

1. In Example 1, you saw the patterns $4 \times 3 \times 2 \times 1$ and $7 \times 6 \times 5$. Why did we decrease the numbers in these products one at a time?

2. What is a quick way to compute the expression $\frac{9!}{5!}$ as we did in Example 2?

3. What point were we making in Table 13.1?

4. Why are we discussing Blaise Pascal in this section?

Sharpening Your Skills

In Exercises 5–16, calculate each value.

5. $4!$

6. $3!$

7. $(8 - 5)!$

8. $(10 - 5)!$

9. $\frac{10!}{7!}$

10. $\frac{11!}{9!}$

11. $\frac{10!}{7!3!}$

12. $\frac{11!}{9!2!}$

13. $P(6, 2)$

14. $P(5, 3)$

15. $C(10, 3)$

16. $C(10, 4)$

In Exercises 17–26, find the number of permutations.

17. Eight objects taken three at a time

18. Seven objects taken five at a time

19. Ten objects taken eight at a time

20. Nine objects taken six at a time

In Exercises 21–24, find the number of combinations.

21. Eight objects taken three at a time

22. Seven objects taken five at a time

23. Ten objects taken eight at a time

24. Nine objects taken six at a time

Explain the meaning of each symbol in Exercises 25 and 26.

25. $P(10, 3)$

26. $C(6, 2)$

27. Find the eighth row in Pascal's triangle.

28. Find the tenth row in Pascal's triangle.

Use the seventh row of Pascal's triangle to answer Exercises 29 and 30.

29. How many two-element subsets can we choose from a seven-element set?

30. How many four-element subsets can we choose from a seven-element set?

In Exercises 31–34, describe where each number would be found in Pascal's triangle.

31. $C(18, 2)$

32. $C(19, 5)$

33. $C(20, 6)$

34. $C(21, 0)$

Applying What You've Learned

In Exercises 35–46, specify the number of ways to perform the task described. Give your answers using P(n, r) or C(n, r) notation. The key in recognizing whether a problem involves permutations or combinations is deciding whether order is important.

35. **Quiz possibilities.** On a biology quiz, a student must match eight terms with their definitions. Assume that the same term cannot be used twice.

36. **Scheduling interviews.** Producers wish to interview six actors on six different days to replace the actor who plays the phantom in Broadway's longest running play, *The Phantom of the Opera*.

37. **Choosing tasks.** Gil Grissom on *CSI* wants to review the files on three unsolved cases from a list of 17. He does not care about the order in which he reads the files.

---

*Before doing these exercises, you may find it useful to review the note *How to Succeed at Mathematics* on page xix.

38. **Selecting roommates.** Annette has rented a summer house for next semester. She wants to select four roommates from a group of six friends.

39. **Baseball lineups.** Joe Torre has already chosen the first, second, fourth, and ninth batters in the batting lineup. He has selected five others to play and wants to write their names in the batting lineup.

40. **Race results.** There are seven boats that will finish the America's Cup yacht race.

41. **Journalism awards.** Ten magazines are competing for three identical awards for excellence in journalism. No magazine can receive more than one award.

42. **Journalism awards.** Redo Exercise 41, but now assume the awards are all different. No magazine can receive more than one award.

43. **Lock patterns.** In order to unlock a door, five different buttons must be pressed down on the panel shown. When the door opens, the depressed buttons pop back up. The order in which the buttons are pressed is not important.

44. **Lock patterns.** A bicycle lock has three rings with the letters *A* through *K* on each ring. To unlock the lock, a letter must be selected on each ring. Duplicate letters are not allowed, and the order in which the letters are selected on the rings does not matter.

45. **Choosing articles for a magazine.** The editorial staff of Oprah's *O Magazine* is reviewing 17 articles. They wish to select eight for the next issue of the magazine.

46. **Choosing articles for a magazine.** Redo Exercise 45, except that in addition to choosing the stories, the editorial staff must also decide in what order the stories will appear.

47. **Reading program.** Six players are to be selected from a 25-player Major League Baseball team to visit schools to support a summer reading program. In how many ways can this selection be made?

48. **Reading program.** If in Exercise 47 we not only want to select the players but also want to assign each player to visit one of six schools, how many ways can that assignment be done?

49. **Writing articles.** In Exercise 46, if it takes 1 minute to write a list of eight articles selected for the magazine, how many years

would it take to write all possible lists of 8 articles? Assume 60 minutes in an hour, 24 hours in a day, and 365 days in a year.

50. **Writing assignments.** In Exercise 48, if it takes 1 minute to write a list of six players with their assignments, how many years would it take to write all possible lists of assignments?

A typical bingo card is shown in the figure. The numbers 1–15 are found under the letter B, 16–30 under the letter I, 31–45 under the letter N, 46–60 under the letter G, and 61–75 under the letter O. The center space on the card is labeled "FREE."

51. **Bingo cards.** How many different columns are possible under the letter B?

52. **Bingo cards.** How many different columns are possible under the letter N?

53. **Bingo cards.** How many different bingo cards are possible? (*Hint:* Use the fundamental counting principle.)

54. **Bingo cards.** Why is the answer to Exercise 53 not *P*(75, 24)?

B	I	N	G	O
5	28	33	51	75
7	17	41	59	63
2	22	FREE	48	61
11	29	44	46	72
9	30	38	52	68

In Exercises 55–64, use the fundamental counting principle.

55. **Selecting astronauts.** NASA wants to appoint two men and three women to send back to the moon in 2018. The finalists for these positions consist of six men and eight women. In how many ways can NASA make this selection?

56. **Computer passwords.** A password for a computer consists of three different letters of the alphabet followed by four different digits from 0 to 9. How many different passwords are possible?

57. **Forming a public safety committee.** Nicetown is forming a committee to investigate ways to improve public safety in the town. The committee will consist of three representatives from the seven-member town council, two members of a five-person citizens advisory board, and three of the 11 police officers on the force. How many ways can that committee be formed?

58. **Choosing a team for a seminar.** HazMat, Inc., is sending a group of eight people to attend a seminar on toxic waste disposal. They will send two of eight engineers and three of nine crew supervisors and the remainder will be chosen from the five senior managers. In how many ways can these choices be made?

59. **Designing a fitness program.** To lose weight and shape up, Sgt. Walden is considering doing two types of exercises to

improve his cardiovascular fitness from running, bicycling, swimming, stair stepping, and Tae Bo. He is also going to take two nutritional supplements from AllFit, Energize, ProTime, and DynaBlend. In how many ways can he choose his cardiovascular exercises and nutritional supplements?

60. Counting assignment possibilities. Gina must write evaluation reports on three hospitals and two health clinics as part of her degree program in community health services. If there are six hospitals and five clinics in her vicinity, in how many ways can she complete her assignment?

61. Picking a team for a competition. The students in the 12-member advanced communications design class at Center City Community College are submitting a project to a national competition. They must select a four-member team to attend the competition. The team must have a team leader and a main presenter; the other two members have no particularly defined roles. In how many different ways can this team be formed?

62. Why is Exercise 61 neither a strictly permutations nor a strictly combinations problem?

63. Choosing an evaluation team. The academic computing committee at Sweet Valley College is in the process of evaluating different computer systems. The committee consists of five administrators, seven faculty, and four students. A five-person subcommittee is to be formed. The chair and vice chair of the committee must be administrators; the remainder of the committee will consist of faculty and students. In how many ways can this subcommittee be formed?

64. Why is Exercise 63 neither a strictly permutations nor a strictly combinations problem?

An armored car must pick up receipts at a shopping mall and at a jewelry store and deliver them to a bank. For security purposes, the driver will vary the route that the car takes each day. Use the map to answer Exercises 65 and 66.

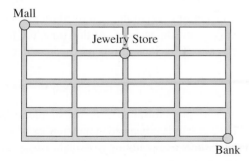

65. Armored car routes. How many different direct routes can the driver take from the shopping mall to the bank?

66. Armoured car routes. How many direct routes can the driver take from the mall to the bank if he must also stop at the jewelry store?

Communicating Mathematics

67. How do you distinguish a combination problem from a permutation problem?

68. In Example 7, we were choosing 4 people from a group of 19, yet our answer was not $C(19, 4)$. Explain this. How does the fundamental counting principle apply here?

69. Where would you find the number $C(n, r)$ in Pascal's triangle?

70. Recall that combinations are subsets of a given set and that $C(n, r)$ is the number of subsets of size r in an n-element set. Give an intuitive explanation as to why you would expect each of the following equations to be true.

a. $C(5, 0) = 1$ **b.** $C(8, 7) = 8$

71. Consider the question: How many four-digit numbers can be formed that are odd and greater than 5,000? Why is this not a permutation problem?

72. Consider the question: How many four-digit numbers can be formed using the digits 1, 3, 5, 7, and 9? Repetition of digits is not allowed. Why is this not a combination problem?

73. Whenever we used the formulas for computing $P(n, r)$ or $C(n, r)$, we saw that many factors from the numerator canceled with all the factors from the denominator so that the resulting number was never a fraction. Using the definitions of $P(n, r)$ and $C(n, r)$, explain why these numbers can never be fractions.

74. Explain why $C(n, r) = C(n, n - r)$. To understand this statement, recall the Three-Way Principle from Section 1.1, which states that making numerical examples is useful in gaining understanding of a situation.

75. We numbered the rows of Pascal's triangle beginning with 0 instead of 1. Explain why we did this.

76. We numbered the entries of each row in Pascal's triangle beginning with 0 instead of 1. Explain why we did this.

Using Technology to Investigate Mathematics

77. Learn to use either a graphing calculator or an Excel spreadsheet to find $P(n, r)$ and $C(n, r)$, and then solve some of the exercises in this section using those technologies.

78. Search the Internet for applets that compute $P(n, r)$ and $C(n, r)$ and use them to solve some of the exercises in this section.

For Extra Credit

79. Playing cards. How many five-card hands chosen from a standard deck contain two diamonds and three hearts?

80. Playing cards. How many five-card hands chosen from a standard deck contain two kings and three aces?

The following diagram shows a portion of Pascal's triangle found in a book called The Precious Mirror, *written in China in 1303. Use this diagram to answer Exercises 81–84.*

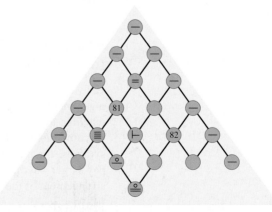

81. What symbol do you expect to find in the circle labeled 81?

82. What symbol do you expect to find in the circle labeled 82?

83. What number does the symbol represent?

84. What number does the symbol 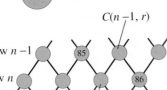 represent?

Consider rows n *and* n − 1 *of Pascal's triangle, shown in the diagram to the right. Use this diagram to answer Exercises 85–89.*

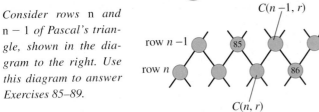

85. What expression will be in the circle labeled 85?

86. What expression will be in the circle labeled 86?

87. Complete the following equation: $\boxed{} + C(n-1, r) = \boxed{}$

88. Complete the following equation: $C(n, r) + \boxed{} = \boxed{}$

89. Notice that the arrangement of numbers in each row of Pascal's triangle is symmetric. See how the 4s are balanced in the fourth row, the 5s are balanced in the fifth row, and so on. Can you give a set theory explanation to account for this?

Looking Deeper
13.4 Counting and Gambling

Objectives

1. **Understand how to use the fundamental counting principle to analyze counting problems that occur in gambling.**
2. **Use the theory of permutations and combinations to compute the probability of drawing various card hands.**

A neon sign in a casino proclaims:

$100,000 jackpot !!

The jackpot was formed by putting one nickel of every dollar gambled on slot machines into the jackpot. Periodically, a lucky player wins the jackpot. At that time, the jackpot is emptied and must start building all over again. If we look at this jackpot another way, the neon sign should perhaps proclaim:

$2,000,000 gambled on slot machines since the last jackpot !!

Operators of gambling establishments earn huge fortunes using the simple mathematical principle that the number of ways they can win is greater than the number of ways that you can win. In this section, we will use the theory of counting to explain why people win (and lose) at games such as slot machines and poker.

Applying the Fundamental Counting Principle to Gambling

Early mechanical slot machines had three wheels with 20 positions on each wheel containing symbols such as oranges, bells, lemons, cherries, and bars. A gambler put a coin in the machine and pulled a lever, causing the wheels to spin. When the wheels stopped, the machine displayed a symbol from each wheel in a window. For certain arrangements of symbols, the player received a payoff. Arrangements of symbols that occur infrequently pay larger payoffs than those that occur more frequently. Figure 13.14 shows a typical arrangement of the symbols that might appear on each of the three wheels.

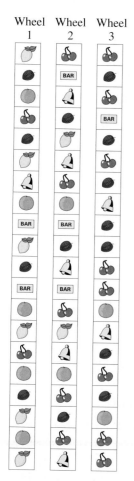

Wheel 1 Wheel 2 Wheel 3

FIGURE 13.14
Three wheels of a typical mechanical slot machine.

EXAMPLE 1 *Using the FCP to Analyze a Slot Machine*

1st wheel		2nd wheel		3rd wheel
20	×	**20**	×	**20**
Number of ways first wheel can stop		Number of ways second wheel can stop		Number of ways third wheel can stop

1st wheel		2nd wheel		3rd wheel
2	×	**15**	×	**12**
Number of ways first wheel can stop with cherries		Number of ways second wheel can stop with no cherries		Number of ways third wheel can stop with no cherries

a) In how many different ways can the three wheels of the slot machine depicted in Figure 13.14 stop?

b) One payoff is for cherries on wheel 1 and no cherries on the other wheels. In how many different ways can this happen?

SOLUTION:

a) It is useful to think of the first wheel stopping, then the second, and finally the third. The solution becomes apparent if we draw a slot diagram. Using the fundamental counting principle, we see that the total number of different ways the three wheels can stop is $20 \times 20 \times 20 = 8,000$.

b) As before, think of the wheels stopping in stages. Again, a slot diagram helps with the counting. By the fundamental counting principle, we see that the total number of ways we can get cherries only on the first wheel is $2 \times 15 \times 12 = 360$. ❀

EXAMPLE 2 *Counting the Number of Ways to Get Three Bars*

For the slot machine in Figure 13.14, the largest payoff is for three bars. In how many ways can three bars be obtained?

SOLUTION: There are two ways a bar can come up on the first wheel, three ways a bar can come up on the second wheel, and only one way a bar can come up on the third wheel. By the fundamental counting principle, a bar on the first, second, and third wheel can occur in $2 \times 3 \times 1 = 6$ ways.

Now try Exercises 3 to 8. ❀

From Examples 1 and 2 we would expect that the payoff for three bars to be much greater than the payoff for cherries on the first wheel only.

Counting and Poker

Counting principles also explain why one poker hand beats another (see Table 13.2 on page 652). In poker, a hand that is less likely to occur beats a hand that occurs more commonly. We first determine the total number of possible poker hands.

EXAMPLE 3 *Counting the Number of Poker Hands*

How many ways can a five-card poker hand be drawn from a standard 52-card deck?

SOLUTION: In drawing a five-card poker hand, we are choosing five cards from 52, which can be done in

$$C(52, 5) = \frac{52!}{5!47!} = \frac{52 \cdot 51 \cdot 50 \cdot 49 \cdot 48}{5 \cdot 4 \cdot 3 \cdot 2 \cdot 1} = 2,598,960 \text{ ways.} ❀$$

As we will show in Examples 4 and 6, the reason four of a kind beats a full house is that there are fewer ways to obtain four of a kind than there are to obtain a full house. In these examples, we will make heavy use of the fundamental counting principle.

Royal flush	10, J, Q, K, and A, all of the same suit
Straight flush	Five cards in sequence,* all of the same suit
Four of a kind	Four cards of one rank, plus another card
Full house	Three cards of one rank and two of another
Flush	All five cards are from the same suit
Straight	Any five cards in sequence
Three of a kind	Three cards of one rank and two other cards that are different from each other
Two pair	Two different pairs of cards and a fifth card that is different from the first four
Pair	Two cards of the same rank and three other cards, all of which are different from each other
Nothing	None of the above hands

TABLE 13.2 Poker hands in order of strength.

EXAMPLE 4   *Drawing Four of a Kind*

Determine the number of different ways we can obtain four of a kind when drawing five cards from a standard 52-card deck.

SOLUTION: We can construct a four-of-a-kind hand in two stages.

Step 1: Pick the rank of the card of which we will have four of a kind. For example, the four-of-a-kind cards could be aces, kings, threes, and so on. This decision can be made in 13 ways.

Step 2: Now that we have chosen four cards, we select the fifth card. For example, if we have selected four kings, then we could pick a jack for the fifth card. We can choose any one of the remaining 48 cards for the fifth card.

Therefore, by the fundamental counting principle, we can build a four-of-a-kind hand in $13 \times 48 = 624$ ways. ✷

Recall that a full house is a hand that has three cards of one rank and two of another rank. For example, three kings and two jacks make a full house. Before determining how many different ways we can draw a full house, we will attack a simpler question.

Quiz Yourself **10**

Determine the number of different ways to draw two cards of the same rank when drawing two cards from a standard 52-card deck.

EXAMPLE 5   *Drawing Three Cards of the Same Rank*

Determine the number of different ways to draw three cards of the same rank when drawing three cards from a standard 52-card deck.

SOLUTION: This problem can be divided into two stages. First, choose the rank of the three cards that will be the same. This decision can be made in 13 ways. Second, choose three cards from the four cards in the deck with this rank. This choice can be made in $C(4, 3) = 4$ ways. Using the fundamental counting principle, we see that the three cards of the same rank can be selected in $13 \times 4 = 52$ ways. ✷ **10**

EXAMPLE 6   *Drawing a Full House*

Determine the number of different ways we can obtain a full house when drawing five cards from a standard 52-card deck.

---

*In choosing a sequence of five cards, we allow the ace to be considered higher than the king and also as a one. For example, the sequence A, 2, 3, 4, 5 is allowable as well as the sequence 10, J, Q, K, A.

SOLUTION: We can think of this problem as consisting of two simpler subproblems.

Subproblem A: Determine the number of ways we can choose the three cards having the same rank. We solved this subproblem in Example 5, where we found that this selection can be made in 52 ways.

Subproblem B: Determine the number of ways we can choose the remaining two cards having the same rank. It is tempting to use the answer found in Quiz Yourself 10, but this would be a mistake. Realize that at this stage, we are selecting the last two cards *after* we have already decided on the three-of-a-kind cards.

Because we have already used one rank for the three-of-a-kind cards, we have only 12 ranks remaining. For example, if we have chosen three kings, then kings cannot be used again for the remaining two cards. Thus, the rank of the last two cards can be chosen in only 12 ways. After we have decided the rank of the last two cards, we must select two of the four cards with this rank. This can be done in $C(4, 2) = 6$ ways. Using the fundamental counting principle, we see that the two-of-a-kind cards can be selected in $12 \times 6 = 72$ ways.

Now we are ready to answer the original question. In constructing a full house, we can first choose the three-of-a-kind cards in 52 ways. Then we can choose the two-of-a-kind cards in 72 ways. Using the fundamental counting principle a final time, we see that a full house can be constructed in $52 \times 72 = 3,744$ ways. ❧

EXAMPLE 7 *Drawing Three of a Kind*

If we select five cards from a standard 52-card deck, in how many ways can we draw a hand with three of a kind?

SOLUTION: We first must count the number of ways that we can select three cards of the same rank. In Example 5, we found the answer to this question to be 52.

Next we must select the remaining two cards. The fourth card can be chosen among the 12 remaining ranks in 48 ways, and then the fifth card (which must be a different value than the first four) can be chosen in 44 ways. It would seem then that the fourth and fifth cards could be selected in 48×44 ways.

However, there is a slight flaw in our thinking! Suppose, for example, we began with three kings and then for our fourth and fifth cards chose a queen and an eight *in that order*. We could then rearrange the queen and eight in our hand and the hand would be unchanged. Thus, we recognize that in counting the ways to complete the hand with a fourth and fifth card, we have counted every possibility twice. Therefore, there are only $\frac{48 \times 44}{2} = 1,056$ ways to select the last two cards.

Since the first three cards (of the same rank) can be selected in 52 ways and the last two cards can be selected in 1,056 ways, a hand with three of a kind can be chosen in $52 \times 1,056 = 54,912$ ways.

Now try Exercises 9 to 14. ❧

Exercises 13.4

Looking Back*

These exercises follow the general outline of the topics presented in this section and will give you a good overview of the material that you have just studied.

1. How did we use the fundamental counting principle in analyzing the slot machine in Example 1?

---

*Before doing these exercises, you may find it useful to review the note *How to Succeed at Mathematics* on page xix.

2. How did we use the fundamental counting principle in analyzing how to obtain a full house in Example 6?

Applying What You've Learned

Exercises 3–8 are based on the slot machine shown in Figure 13.14.

3. Slot machines. In how many ways can we obtain cherries on only the first two wheels?

4. Slot machines. In how many ways can we obtain cherries on all three wheels?

5. **Slot machines.** In how many ways can we obtain oranges on all three wheels?

6. **Slot machines.** In how many ways can we obtain plums on all three wheels?

7. **Slot machines.** In how many ways can we obtain bells on all three wheels?

8. **Slot machines.** In how many ways can we obtain bars on all three wheels?

Exercises 9–14 deal with the poker hands described in Table 13.2. We assume that we are drawing five cards from a standard 52-card deck.

9. **Playing poker.** How many ways can we obtain a royal flush (10, J, Q, K, and A, all of the same suit)?

10. **Playing poker.** In constructing a straight flush, we first choose a suit and then choose a sequence of five cards within the suit.

    a. How many ways can we choose the suit?

    b. How many ways can we choose the sequence of five cards within the suit?

    c. How many ways can we construct a straight flush?

11. **Playing poker.** To construct a flush, we first select a suit and then choose five cards without regard to order from that suit.

    a. How many ways can we choose the suit?

    b. How many ways can we choose the five cards?

    c. How many ways can we construct a flush?

12. **Playing poker.** To construct a straight, we must choose five cards in sequence (not all of the same suit).

    a. How many ways can we choose a sequence of five cards?

    b. For each sequence in part (a) we must select a suit for the first card, a suit for the second card, and so on. In how many ways can we do this? (*Hint:* Draw a slot diagram for the five cards.)

    c. Multiply the results from parts (a) and (b).

    d. Subtract the number of royal flushes and straight flushes that are not royal, found in Exercises 9 and 10.

13. **Playing poker.** How many ways can we construct a five-card poker hand that contains two pairs?

14. **Playing poker.** How many ways can we construct a poker hand that contains only one pair and no other cards of any value?

Communicating Mathematics

15. Why is Example 5 not strictly a combinations problem?

16. Why are we using combinations rather than permutations in solving the examples in this section?

17. On the slot machine in Figure 13.14, would you expect the payoff for three cherries or three oranges to be larger? Explain.

18. On the slot machine in Figure 13.14, would you expect the payoff for three plums or three bells to be larger? Explain.

Using Technology to Investigate Mathematics

19. Search the Internet for "mathematics and gambling." Find a site that discusses counting and gambling. Report on your findings.

For Extra Credit

20. **Playing poker.** How many poker hands contain not even one pair? To solve this,

    a. Calculate the number of hands of each type in Table 13.2.

    b. Subtract the total of the hands found in part (a) from the total of all possible hands.

CHAPTER SUMMARY*

You will learn the items listed in this chapter summary more thoroughly if you keep in mind the following advice:

1. Focus on "remembering how to remember" the ideas. What pictures, word analogies, and examples help you remember these ideas?
2. Practice writing each item without looking at the book.
3. Make 3×5 flash cards to break your dependence on the text. Use these cards to give yourself practice tests.

SECTION	SUMMARY	EXAMPLE
SECTION 13.1	A small set can be counted by **listing** its elements systematically.	Example 1, p. 623 Example 2, p. 624
	Tree diagrams help visualize counting situations that happen in stages.	Example 3, p. 625 Example 4, p. 625 Example 5, p. 626
	It is often enough to **imagine** a tree without actually drawing it to solve a counting problem.	Example 6, p. 627
SECTION 13.2	The **fundamental counting principle** states that if we want to perform a series of tasks and the first task can be done in a ways, the second can be done in b ways, the third can be done in c ways, and so on, then all the tasks can be done in $a \times b \times c \times \cdots$ ways.	Examples 1, 2, and 3, p. 632
	Slot diagrams help organize information before applying the fundamental counting principle.	Example 4, p. 633
	Any **special conditions** in a counting problem should be considered first.	Example 5, p. 634
SECTION 13.3	A **permutation** is an ordering of objects in a straight line. We denote the **number of permutations of n objects taken r at a time** by the notation $P(n, r)$: $$P(n, r) = \frac{n!}{(n-r)!}$$	Example 1, p. 639 Examples 2 and 3, p. 640
	If we choose r objects from a set of n objects, we are forming a **combination of n objects taken r at a time**. The symbol $C(n, r)$ represents the number of such combinations.	Examples 4 and 5, pp. 642–643
	$$C(n, r) = \frac{n!}{r!\,(n-r)!}$$	Example 6, p. 643
	We can **combine counting methods** to solve applied counting problems.	Example 7, p. 644 Example 8, p. 645
	The nth line of **Pascal's triangle** counts the subsets of various sizes of a set having n elements. The rth entry of the nth row of Pascal's triangle is $C(n, r)$.	Example 9, p. 646 Example 10, p. 646
SECTION 13.4	We can use the fundamental counting principle to solve gambling problems.	Examples 1 and 2, p. 651
	We can use the theory of permutations and combinations to explain the ranking of poker hands.	Examples 3–6, pp. 651–652 Example 7, p. 653

---

*Before studying this chapter's material, it would be useful to reread the note *How to Succeed at Mathematics* on page xix.

CHAPTER REVIEW EXERCISES

Section 13.1

1. List all the ways you can select two different members from $S = \{P, Q, R, S\}$. The order in which you select the members is important. For example, PQ is not the same selection as QP.

2. The game *Dungeons and Dragons* uses a tetrahedral die that has four congruent triangular sides numbered 1, 2, 3, and 4. If you were to draw a tree diagram to show the number of ways that three tetrahedral dice could be rolled, how many branches would it have?

3. An outfit consists of a shirt, pants, tie, and jacket. If you have five shirts, four pairs of pants, six ties, and two jackets, how many different outfits can you make?

Section 13.2

4. The rock-climbing club has 14 members. If the club wants to select a president, vice president, and treasurer (all of whom must be different), in how many ways can this be done?

5. The early bird special at TGI Friday's consists of an appetizer, entrée, and dessert. If there are four appetizers, 12 entrées, and six desserts, how many different meals are possible? (Assume that you have to make a selection from each category.)

6. Hector is building a home theater system consisting of a tuner, an optical disc player, speakers, and a high-definition TV. If he can select from three tuners, six speakers, three optical disc players, and four TVs, in how many ways can he configure his system?

7. A campus bus must travel from a (D)ormitory to the (S)tudent center and then to the (F)itness center. How many direct routes are there from D to F, passing through S?

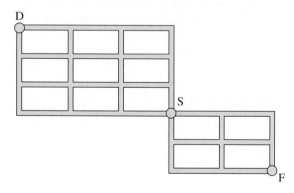

8. Assume that we want to seat Alex, Bonnie, Carl, Daria, Edith, and Frank in a row of six chairs. In how many ways can the seating be done if Alex and Bonnie must sit together?

Section 13.3

9. Explain the main difference between a permutation and a combination.

10. On a psychological test, a patient must match eight terms with their pictures. We are assuming that the same term cannot be used twice. In how many ways can this be done?

11. The research director of NASA must choose three experiments for the next space shuttle. If 17 experiments are proposed, in how many ways can this decision be made?

12. A password for a computer consists of three different letters of the alphabet followed by two different digits from 0 to 9. How many passwords are possible?

13. How is the formula for $P(n, r)$ related to the formula for $C(n, r)$?

14. Use the seventh row of Pascal's triangle to find the number of two-element subsets that can be chosen from a seven-element set.

15. Describe where $C(18, 2)$ would be found in Pascal's triangle.

Section 13.4

16. Using the slot machine in Figure 13.14, in how many ways can we obtain
    a. Plums on the first two wheels only?
    b. Plums on all three wheels?

17. In poker, a flush is a five-card hand all from the same suit. In how many different ways can we construct a flush? (*Hint:* Think of first selecting the suit and then selecting the five cards.)

CHAPTER TEST

1. List all the ways that you can select two different members from the set $S = \{A, B, C, D\}$. The order of the members is important. For example, AB is not the same as BA.

2. Some role-playing games use octahedral dice that have eight congruent triangular sides numbered 1, 2, . . . , 8. If you were to draw a tree diagram showing the number of ways that two octahedral dice could be rolled, how many branches would it have?

3. Allen, Burt, Consuella, Devaun, and Emily are attending a play. If Devaun and Emily want to sit together, in how many ways can we seat them in five seats?

4. Use the sixth row of Pascal's triangle to find the number of three-element subsets that can be chosen from a six-element set.

5. On the slot machine wheel in Figure 13.14, how many ways are there to get bells on all three wheels?

6. Senior communication design students must present three projects for their senior seminar from a list of 12 possibilities. In how many ways can they make their decision?

7. The Social Action Club has 15 members. The club wants to elect a president, vice president, secretary, and public relations

officer (all of whom must be different). In how many ways can this be done?

8. In poker, a straight flush is a sequence of consecutive cards all in the same suit. How many ways can you get a straight flush? (*Note:* AKQJ10 is one and 5432A is another.)

9. You are selecting your courses for next semester and you can choose one of three biology courses that will satisfy your science requirement. There are four courses that will satisfy your diversity requirement, and six courses that will fulfill your writing requirement. If you want to choose a biology course, a diversity course, and a writing course, in how many ways can you make your choice?

10. What is the main difference between a permutation and a combination?

11. Raphael is about to sign a sales agreement for his new car. He can choose one of 4 warranties, one of 3 exterior combos, one color from 11 available, and one of 6 interior packages. (Assume that he must choose a color and an interior but can

decide not to take any of the other options.) In how many ways can he make his decision?

12. Six men and six women are playing in a mixed doubles tennis tournament. In how many ways can we match all six men with all six women as teams?

13. Write an equation that relates the expressions $P(n, r)$, $C(n, r)$, and $r!$.

14. Where would you find $C(12, 5)$ in Pascal's triangle?

15. In choosing your new cell phone, you must choose one of three purchase agreements, one of nine exterior designs, and one of two calling plans and you have the option of buying extended service. In how many ways can you make your decision?

16. Viewers of the popular TV game show *Deal or No Deal* are encouraged to "play at home." Home contestants submit their entries with a viewer ID that consists of three different letters of the alphabet followed by four different digits from 0 to 9. How many different viewer IDs are possible?

GROUP EXERCISES

1. The gambling game keno, which originated in China over 2,000 years ago, is popular in many casinos. In keno there are balls numbered from 1 to 80. The player chooses from 1 to 20 numbers, or "spots," as they are called, and marks them on a card. The casino then randomly picks 20 balls, called the *lucky* balls, from the 80. The remaining 60 balls are called the *unlucky* balls. The more lucky numbers the player has among his spots, the more he wins. With this in mind, calculate the following:

   a. How many different ways can the casino choose the 20 lucky balls?

   b. If a player plays five spots, how many different ways can the player have all lucky numbers?

   c. If the player plays five spots, in how many different ways can the player have exactly three lucky numbers? (*Note:* Realize to have three lucky numbers, the player must have also two unlucky numbers. This is similar to our analysis of counting the number of ways to get a full house in poker in Section 13.4.)

   d. If the player plays 20 spots, in how many different ways can the player have 17 lucky numbers?

2. Research keno on the Internet and find other variations of the game than the one we described. For example, in some games a player can bet on how many of the 20 lucky numbers will be in the first half of the 80 numbers from which to choose. Also, you might find it interesting to research the odds and payoffs of playing different numbers of spots. Report on your findings.

Probability 14

What Are the Chances?

The 2008 presidential race . . . being hit by lightning . . . the Super Bowl . . . drilling for oil in the Arctic National Wildlife Refuge . . . car insurance . . . Las Vegas . . . . You might ask why we would start a chapter with these seemingly unrelated topics. The common thread that runs through all of these topics is the theme of this chapter—probability.

Whenever you hear a prediction as to who will win an election, the likelihood of finding oil, bets being placed on the Super Bowl, or deciding how much insurance you should carry on your car, you are dealing with uncertainty. But the good thing is there are well-known mathematical rules that enable us to make reasonable predictions out of uncertain situations. *(continued)*

By applying the rules of probability you will see how to determine if a flu drug is effective and to calculate how much profit an insurance company can expect to earn on its policies. You will find that playing roulette is a 10 times better gamble than playing the daily number and that the odds of being killed by lightning are thousands of times greater than the odds of winning a large state lottery. Later in the chapter, we will show you an example of how it is possible that a person who tests positive on a drug test has a surprisingly high likelihood of being innocent. ●

14.1 The Basics of Probability Theory

Objectives

1. Calculate probabilities by counting outcomes in a sample space.
2. Use counting formulas to compute probabilities.
3. Understand how probability theory is used in genetics.
4. Understand the relationship between probability and odds.

Have you ever been camping in the rain? If not, try to imagine it. You are sitting in a tent with the rain flaps down, with nothing to do—the weather report was wrong and you wish that you had picked another weekend for this experience. As we all know, predicting weather is not an exact science. The weather is an example of a random phenomenon. **Random phenomena** are occurrences that vary from day-to-day and case-to-case. In addition to the weather, rolling dice in Monopoly, drilling for oil, and driving your car are all examples of random phenomena.

Although we never know exactly how a random phenomenon will turn out, we can often calculate a number called a **probability** that it will occur in a certain way. We will now begin to introduce some basic probability terminology.

Sample Spaces and Events

KEY POINT

Knowing the sample space helps us compute probabilities.

Our first step in calculating the probability of a random phenomenon is to determine the sample space of an experiment.

> **DEFINITIONS** An **experiment** is any observation of a random phenomenon. The different possible results of the experiment are called **outcomes**. The set of all possible outcomes for an experiment is called a **sample space**.

If we observe the results of flipping a single coin, we have an example of an experiment. In the Oscar-winning film *No Country for Old Men*, Anton flips a coin to bet on a person's life. The possible outcomes are head and tail, so a sample space for the experiment would be the set {head, tail}.

EXAMPLE 1 *Finding Sample Spaces*

Determine a sample space for each experiment.

a) We select an iPhone from a production line and determine whether it is defective.

b) Three children are born to a family and we note the birth order with respect to gender.

c) We select one card from a standard 52-card deck,* and then without returning the card to the deck, we select a second card. We will assume the order in which we select the cards is important.

d) We roll two dice and observe the pair of numbers showing on the top faces.

---

*See Figure 13.1 in Chapter 13 for a picture of a standard 52-card deck.

SOLUTION: In each case, we find the sample space by collecting the outcomes of the experiment into a set.

a) This sample space is {defective, nondefective}.

b) In this experiment, we want to know not only how many boys and girls are born but also the birth order. For example, a boy followed by two girls is not the same as two girls followed by a boy. The tree diagram* in Figure 14.1 helps us find the sample space.

There are two ways that the first child can be born, followed by two ways for the second child, and finally two ways for the third. If we abbreviate "boy" by "b" and "girl" by "g," following the branches of the tree diagram gives us the following sample space:

{bbb, bbg, bgb, bgg, gbb, gbg, ggb, ggg}.

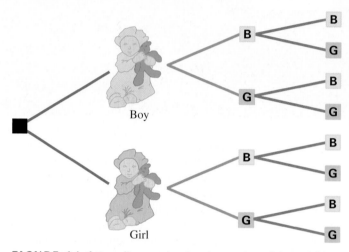

FIGURE 14.1 Tree diagram showing the genders of three children.

c) This sample space is too large to list; however, we can use the *fundamental counting principle* from Section 13.2 to count its members. Because we can choose the first card in 52 ways and the second card in 51 ways (we are not replacing the first card), we can select the two cards in $52 \times 51 = 2,652$ ways.

d) As we did in Example 4 of Section 13.1, we will think of the first die as being red and the second die as being green. As you can see, the pairs $(1, 3)$ and $(3, 1)$ are not the same. With this in mind, the sample space for this experiment consists of the following 36 pairs.

$$\begin{array}{cccccc}
\{(1, 1), & (1, 2), & (1, 3), & (1, 4), & (1, 5), & (1, 6), \\
(2, 1), & (2, 2), & (2, 3), & (2, 4), & (2, 5), & (2, 6), \\
(3, 1), & (3, 2), & (3, 3), & (3, 4), & (3, 5), & (3, 6), \\
(4, 1), & (4, 2), & (4, 3), & (4, 4), & (4, 5), & (4, 6), \\
(5, 1), & (5, 2), & (5, 3), & (5, 4), & (5, 5), & (5, 6), \\
(6, 1), & (6, 2), & (6, 3), & (6, 4), & (6, 5), & (6, 6)\}
\end{array}$$

 *KEY POINT*

An event is a subset of a sample space.

Imagine that you are playing Monopoly and are on the verge of going bankrupt. If you roll a total of seven, you will lose the game and you want to know the probability that this will happen. At this point, you are only concerned with the *set* of pairs $(1, 6)$, $(2, 5)$, $(3, 4)$, $(4, 3)$, $(5, 2)$, and $(6, 1)$. This focus on particular *subsets* of the sample space is a recurring theme in probability theory.

> **DEFINITION** In probability theory, an **event** is a subset of the sample space.

---

*We introduced tree diagrams in Chapter 13.

Keep in mind that *any* subset of the sample space is an event, including such extreme subsets as the empty set, one-element sets, and the whole sample space.

 Some Good Advice

Although we usually describe events verbally, you should remember that *an event is always a subset of the sample space.* You can use the verbal description to identify the set of outcomes that make up the event.

Example 2 illustrates some events from the sample spaces in Example 1.

EXAMPLE 2 *Describing Events as Subsets*

Write each event as a subset of the sample space.

a) A head occurs when we flip a single coin.
b) Two girls and one boy are born to a family.
c) A total of five occurs when we roll a pair of dice.

SOLUTION:

a) The set {head} is the event.
b) Noting that the boy can be the first, second, or third child, the event is {bgg, gbg, ggb}.
c) The following set shows how we can roll a total of five on two dice: {(1, 4), (2, 3), (3, 2), (4, 1)}.

Now try Exercises 7 to 14. ✳ **1**

We will use the notions of outcome, sample space, and event to compute probabilities. Intuitively, you may expect that when rolling a fair die, each number has the same chance, namely $\frac{1}{6}$, of showing. In predicting weather, a forecaster may state that there is a 30% chance of rain tomorrow. You may believe that you have a 50–50 chance (that is, a 50% chance) of getting a job offer in your field. In each of these examples, we have assigned a number between 0 and 1 to represent the likelihood that the outcome will occur. You can interpret probability intuitively as in the diagram below at left.

DEFINITIONS The **probability of an outcome** in a sample space is a number between 0 and 1 inclusive. The sum of the probabilities of all the outcomes in the sample space must be 1. The **probability of an event** E, written $P(E)$,[†] is defined as the sum of the probabilities of the outcomes that make up E.

One way to determine probabilities is to use *empirical information*. That is, we make observations and assign probabilities based on those observations.

EMPIRICAL ASSIGNMENT OF PROBABILITIES If E is an event and we perform an experiment several times, then we estimate the probability of E as follows:

$$P(E) = \frac{\text{the number of times } E \text{ occurs}}{\text{the number of times the experiment is performed}}.$$

This ratio is sometimes called the *relative frequency* of E.

Quiz Yourself **1** *

Write each event as a set of outcomes.

a) A total of six occurs when we roll two dice.

b) In a family with three children, there are more boys than girls.

 **KEY POINT**

The probability of an event is the sum of the probabilities of the outcomes in that event.

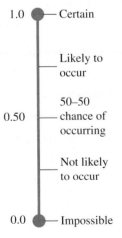

Intuitive meaning of probability.

Side Effects	Number of Times
None	72
Mild	25
Severe	3

TABLE 14.1 Summary of side effects of the flu vaccine.

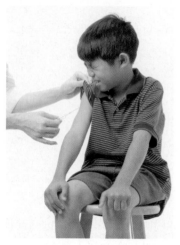

Quiz Yourself **2**

Use Table 14.1 to find the probability that a patient receiving the flu vaccine will experience no side effects.

We are only interested in males.

Quiz Yourself **3**

Using Table 14.2, if we select a widowed person, what is the probability that the person is a woman?

 **KEY POINT**

We can use counting formulas to compute probabilities.

EXAMPLE 3 *Using Empirical Information to Assign Probabilities*

A pharmaceutical company is testing a new flu vaccine. The experiment is to inject a patient with the vaccine and observe the occurrence of side effects. Assume that we perform this experiment 100 times and obtain the information in Table 14.1.

Based on Table 14.1, if a physician injects a patient with this vaccine, what is the probability that the patient will develop severe side effects?

SOLUTION: In this case, we base our probability assignment of the event that severe side effects occur on observations. We use the formula for the relative frequency of an event as follows:

$$P(\text{severe side effects}) = \frac{\text{the number of times severe side effects occurred}}{\text{the number of times the experiment was performed}}$$

$$= \frac{3}{100} = 0.03$$

Thus, from this empirical information, we can expect a 3% chance that there will be severe side effects from this vaccine.

Now try Exercises 29 to 32. ✳ **2**

EXAMPLE 4 *Investigating Marital Data*

Table 14.2* summarizes the marital status of men and women in the United States in 2006. All numbers represent number of thousands. If we randomly pick a male, what is the probability that he is divorced?

	Now Married (except separated)	Widowed	Divorced	Separated	Never Married
Males	60,955	2,908	10,818	2,210	39,435
Females	59,173	12,226	14,182	3,179	33,377

TABLE 14.2 Marital status of men and women (in thousands) in the United States in 2006.

SOLUTION: It is important to recognize that not all entries in Table 14.2 are relevant to the question that you were asked. We are only interested in males, so the line that we have highlighted in the table contains all the information that we need to solve the problem. Therefore, we consider our sample space to be the

$$60,955 + 2,908 + 10,818 + 2,210 + 39,435 = 116,326$$

males.

The event, call it D, is the set of 10,818 men who are divorced. Therefore, the probability that we would select a divorced male is

$$P(D) = \frac{n(D)}{n(S)} = \frac{10,818}{116,326} \approx 0.093.$$

Now try Exercises 41, 42, and 47 to 50. ✳ **3**

Counting and Probability

Another way we can determine probabilities is to use *theoretical information*, such as the counting formulas we discussed in Chapter 13.

---

*U.S. Bureau of the Census.

In order to understand the difference between using theoretical and empirical information, compare these two experiments:

Experiment one—Without looking, draw a ball from a box, note its color, and return the ball to the box. If you repeat this experiment 100 times and get 60 red balls and 40 blue balls, based on this *empirical information,* you would expect that the probability of getting a red ball on your next draw to be $\frac{60}{100} = 0.60$.

Experiment two—Draw a 5-card hand from a standard 52-card deck. As you will soon see, you can use *theoretical information,* namely, combination formulas from Chapter 13, to calculate the probability that *all* cards will be hearts.

EXAMPLE 5 *Using Counting Formulas to Calculate Probabilities*

Assign probabilities to the outcomes in the following sample spaces.

a) We flip three fair coins. What is the probability of each outcome in this sample space?

b) We draw a 5-card hand randomly from a standard 52-card deck. What is the probability that we draw one particular hand?

SOLUTION:

a) This sample space has eight outcomes, as we show in Figure 14.2. Because the coins are fair, we expect that heads and tails are equally likely to occur. Therefore, it is reasonable to assign a probability of $\frac{1}{8}$ to each outcome in this sample space.

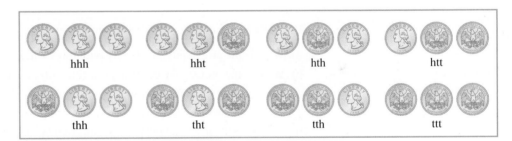

FIGURE 14.2 Eight theoretically possible outcomes for flipping three coins.

Quiz Yourself

If we were to flip four coins, how many outcomes would there be in the sample space?

b) In Chapter 13, we found that there are $C(52, 5) = 2,598,960$ different ways to choose 5 cards from a deck of 52. Because we are drawing the cards randomly, each hand has the same chance of being drawn. Therefore, the probability of drawing any one hand is $\frac{1}{2,598,960}$. ❋ **4**

In a sample space with equally likely outcomes, it is easy to calculate the probability of any event by using the following formula.

CALCULATING PROBABILITY WHEN OUTCOMES ARE EQUALLY LIKELY If *E* is an event in a sample space *S* with all *equally likely outcomes,* then the probability of *E* is given by the formula:

$$P(E) = \frac{n(E)}{n(S)}.\ *$$

---

*Recall that *n(E)* is the cardinal number of set *E* (see Section 2.1).

Math in Your Life*

Why Does It Always Happen to You?

When you wait at a toll booth, why does it seem that the line of cars next to you moves faster than your line? If you look into a sock drawer, do you notice how many unmatched socks there are? Why does the buttered side of a slice of bread almost always land face down if you drop the bread while making a sandwich? Do such annoyances affect only you? Or is there a mathematical explanation?

First let's consider the socks. Suppose you have 10 pairs of socks in a drawer and lose 1 sock, destroying a pair. Of the 19 remaining socks, there is only 1 unpaired sock. Therefore, if you lose a second sock, the probability of losing a paired sock is $\frac{18}{19}$. Now you have 2 unpaired socks and 16 paired socks. The probability that the third sock you lose will be part of a pair is still overwhelming. Continuing this line of thought, you see that it will be quite a while before probability theory predicts that you can expect to lose an unpaired sock.

The problem with the buttered bread is simple to explain. In fact you might conduct an experiment to simulate this situation with an object such as a computer mouse pad. If you slide the object off the edge of a table, as it begins to fall it rotates halfway around so that the top side is facing down. However, there is usually not enough time for the object to rotate back to the upward position before it hits the floor. You could conduct this experiment 100 times and determine the empirical probability that an object you slide off the table will land face down. So again, it's not bad luck—it's probability.

The slow-moving line at the toll booth is the easiest to explain. If we assume that delays in any line occur randomly, then one of the lines—yours, the one to the left, or the one to the right—will move fastest. Therefore, the probability that the fastest line will be yours is only $\frac{1}{3}$.

EXAMPLE 6 *Computing Probability of Events*

a) What is the probability in a family with three children that two of the children are girls?

b) What is the probability that a total of four shows when we roll two fair dice?

†c) If we draw a 5-card hand from a standard 52-card deck, what is the probability that all 5 cards are hearts?

†d) Harold, Kumar, Neil, and Vanessa are friends who belong to their college's 10-person international relations club. Two people from the club will be selected randomly to attend a conference at the United Nations building. What is the probability that two of these four friends will be selected?

SOLUTION: In each situation, we assume that the outcomes are equally likely.

a) We saw in Example 1 that there are eight outcomes in this sample space. We denote the event that two of the children are girls by the set $G = \{bgg, gbg, ggb\}$. Thus,

$$P(G) = \frac{n(G)}{n(S)} = \frac{3}{8}.$$

b) The sample space for rolling two dice has 36 ordered pairs of numbers. We will represent the event "rolling a four" by F. Then $F = \{(1, 3), (2, 2), (3, 1)\}$. Thus,

$$P(F) = \frac{n(F)}{n(S)} = \frac{3}{36} = \frac{1}{12}.$$

c) From Example 5, we know that there are $C(52, 5)$ ways to select a 5-card hand from a 52-card deck. If we want to draw only hearts, then we are selecting 5 hearts from the

---

*This Math in Your Life is based on Robert Matthews, "Murphy's Law or Coincidence." *Reader's Digest*, March 1998, pp. 25–30.
†Questions c) and d) require counting formulas from Chapter 13.

13 available, which can be done in $C(13, 5)$ ways. Thus, the probability of selecting all five cards to be hearts is

$$\frac{C(13, 5)}{C(52, 5)} = \frac{1{,}287}{2{,}598{,}960} \approx 0.000495.$$

Quiz Yourself 5

a) If we roll two fair dice, what is the probability of rolling a total of eight?

b) If we select 2 cards randomly from a standard 52-card deck, what is the probability that both are face cards?

d) The sample space, S, consists of all the ways we can select two people from the 10 members in the club. As you know from Chapter 13, we can choose 2 people from 10 in $C(10, 2) = \dfrac{10!}{8! \cdot 2!} = \dfrac{10 \cdot 9}{2} = 45$ ways. The event, call it E, consists of the ways we can choose two of the four friends. This can be done in $C(4, 2) = 6$ ways. Because all elements in S (choices of two people) are equally likely, the probability of E is

$$P(E) = \frac{n(E)}{n(S)} = \frac{C(4, 2)}{C(10, 2)} = \frac{6}{45} = \frac{2}{15}.$$

Now try Exercises 15 to 20. ❈ 5

 Some Good Advice

If the outcomes in a sample space are not equally likely, then you cannot use the formula $P(E) = \dfrac{n(E)}{n(S)}$. In that case, to find $P(E)$, you must add the probabilities of all the individual outcomes in E.

Suppose that we have a sample space with equally likely outcomes. An event, E, can contain none, some, or all of the outcomes in the sample space S, so we can say

$$0 \le n(E) \le n(S).$$

Dividing this inequality by the positive quantity $n(S)$, we get the inequality

$$\frac{0}{n(S)} \le \frac{n(E)}{n(S)} \le \frac{n(S)}{n(S)},$$

which simplifies to $0 \le \dfrac{n(E)}{n(S)} \le 1$. This gives us the first probability property listed below. The other properties are easy to see.

 **KEY POINT**

Probability theory helps explain genetic theory.

> **BASIC PROPERTIES OF PROBABILITY** Assume that S is a sample space for some experiment and E is an event in S.
>
> 1. $0 \le P(E) \le 1$ 2. $P(\varnothing) = 0$ 3. $P(S) = 1$

Probability and Genetics

In the nineteenth century, the Austrian monk Gregor Mendel noticed while cross-breeding plants that often a characteristic of the plants would disappear in the first-generation off-spring but reappear in the second generation. He theorized that the first-generation plants contained a hidden factor (which we now call a *gene*) that was somehow transmitted to the second generation to enable the characteristic to reappear.

To check his theory, he selected a characteristic such as seed color—some peas had yellow seeds and some had green. Then when he was sure that he had bred plants that would produce only yellow seeds or green seeds, he was ready to begin his experiment. Mendel believed that one of the colors was *dominant* and the other was *recessive*. Which turned out to be the case, because when yellow-seeded plants were crossed with green-seeded plants, the offspring had yellow seeds. When Mendel crossed these offspring for a second generation, he found that 6,022 plants had yellow seeds and 2,001 had green seeds,

which is almost exactly a ratio of 3 to 1. Because Mendel was skilled in mathematics as well as biology, he gave the following explanation for what he had observed.

We will represent the gene that produces the yellow seed by *Y* and the gene that produces the green seed by *g*. The uppercase *Y* indicates that yellow is dominant and the lowercase *g* indicates that green is recessive.*

Figure 14.3 shows the possible genetic makeup of the offspring from crossing a plant with pure yellow seeds and a plant with pure green seeds. Every one of these offspring has a *Yg* pair of genes. Because "yellow seed" is dominant over "green seed," every plant in the first generation will have yellow seeds.

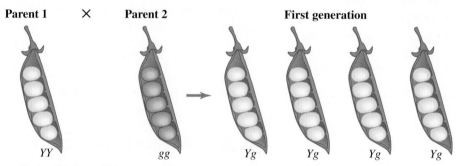

Each offspring will have one *Y* gene from parent 1 and one *g* gene from parent 2, and each will have yellow seeds.

FIGURE 14.3 The possible first-generation offspring we obtain by crossing pure yellow and pure green parents.

Figure 14.4 shows the possible genetic outcomes that can occur if we cross two first-generation pea plants. We summarize Figure 14.4 in Table 14.4, which is called a **Punnett square**.

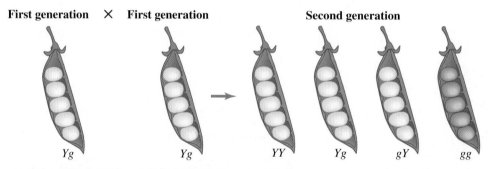

Each offspring will have either a *Y* gene or a *g* gene from each parent. Only the seeds of the offspring with the *gg* pair of genes will be green.

FIGURE 14.4 The possible second-generation offspring we obtain by crossing the first-generation offspring of pure yellow and pure green parents.

| | | **First-Generation Plant** | |
		Y	**g**
First-Generation	**Y**	YY	Yg
Plant	**g**	gY	gg

TABLE 14.3 The genetic possibilities of crossing two plants that each have one yellow-seed and one green-seed gene.

As you can see in Table 14.3, there are four things that can happen when we cross two first-generation plants. We can get a *YY*, *Yg*, *gY*, or *gg* type of plant. Of these four possibilities, only the *gg* will result in green seeds, which explains why Mendel saw the recessive characteristic return in roughly one fourth of the second-generation plants.

---

*Biology books customarily use slightly different notation to indicate genes. For example, to indicate that yellow is dominant over green, you may see the notation *Yy*. The capital *Y* represents the yellow dominant gene and the lowercase *y* indicates the recessive green gene.

EXAMPLE 7 *Using Probability to Explain Genetic Diseases*

Sickle-cell anemia is a serious inherited disease that is about 30 times more likely to occur in African American babies than in non–African American babies. A person with two sickle-cell genes will have the disease, but a person with only one sickle-cell gene will be a carrier of the disease.

If two parents who are carriers of sickle-cell anemia have a child, what is the probability of each of the following:

a) The child has sickle-cell anemia?

b) The child is a carrier?

c) The child is disease free?

SOLUTION: Table 14.4 shows the genetic possibilities when two people who are carriers of sickle-cell anemia have a child. We will denote the sickle-cell gene by *s* and the normal gene by *n*. We use lowercase letters to indicate that neither *s* nor *n* is dominant.

From Table 14.4, we see that there are four equally likely outcomes for the child.

		Second Parent	
		s	*n*
First Parent	*s*	*ss* has disease	*sn* carrier
	n	*ns* carrier	*nn* normal

TABLE 14.4 The genetic possibilities for children of two parents with sickle-cell trait.

- The child receives two sickle-cell genes and therefore has the disease.
- The child receives a sickle-cell gene from the first parent and a normal gene from the second parent and therefore is a carrier.
- The child receives a normal gene from the first parent and a sickle-cell gene from the second parent and therefore is a carrier.
- The child receives two normal genes and therefore is disease-free.

From this analysis, it is clear that

a) $P(\text{the child has sickle-cell anemia}) = \frac{1}{4}$.

b) $P(\text{the child is a carrier}) = \frac{1}{2}$.

c) $P(\text{the child is normal}) = \frac{1}{4}$.

Now try Exercises 33 to 40. ❁

KEY POINT

In computing odds remember "against" versus "for."

Odds

We often use the word *odds* to express the notion of probability. When we do this, we usually state the odds *against* something happening. For example, before the 2008 Kentucky Derby, Las Vegas oddsmakers had set the odds against Big Brown, the eventual winner of the derby, to be 4 to 1.

When you calculate odds against an event, it is helpful to think of what is against the event as compared with what is in favor of the event. For example, if you roll two dice and want to find the odds against rolling a seven, you think that there are 30 pairs that will give you a nonseven and six that will give you a seven. Therefore, the odds against rolling a total of seven are 30 to 6. We write this as 30:6. Just like reducing a fraction, we can write these odds as 5:1. Sometimes you might see odds written as a fraction, such as 30/6, but, for the most part, we will avoid using this notation.

> **DEFINITION** If the outcomes of a sample space are equally likely, then the **odds against an event** *E* are simply the number of outcomes that are against *E* compared with the number of outcomes in favor of *E* occurring. We would write these odds as *n*(*E*′):*n*(*E*), where *E*′ is the complement of event *E*.*

---

*We discussed the complement of a set in Section 2.3.

✸ ✸ ✸ HISTORICAL HIGHLIGHT

Early Probability Theory

We can trace probability theory back to prehistoric times. Archaeologists have found numerous small bones, called *astrogali*, that are believed to have been used in playing various games of chance. Similar bones were later ground into the shape of cubes and decorated in various ways. They eventually evolved into modern dice. Games of chance were played for thousands of years, but it was not until the sixteenth century that a serious attempt was made to study them using mathematics.

Among the first to study the mathematics of games of chance was the Italian Girolamo Cardano. In addition to his interest in probability, Cardano was an astrologer. This caused him frequent misfortune. In one instance, Cardano cast the horoscope of Jesus and for this blasphemy he was sent to prison. Shortly thereafter, he predicted the exact date of his own death and, when the day arrived, Cardano made his prediction come true by taking his own life.

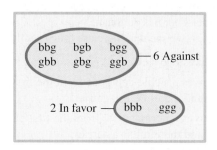

Recall that in Example 1, you saw there were eight ways for boys and girls to be born in order in a family. To find the odds against all children being of the same gender, you think of the six outcomes that are against this happening versus the two outcomes that are in favor, as in the accompanying diagram.

Therefore, the odds against all three children being of the same gender are 6:2, which we can reduce and rewrite as 3:1. We could also say that the *odds in favor* of all children being of the same gender are 1:3. It is important to understand that the odds in favor of an event are not the same as the probability of an event. In the example we are discussing, the probability of all children being the same gender is $\frac{2}{8} = \frac{1}{4}$.

PROBLEM SOLVING

The Analogies Principle

Making an analogy with a real-life situation helps you to remember the meaning of mathematical terminology. For example, when General Custer fought Chief Sitting Bull at the battle of Little Big Horn, Custer had about 650 soldiers and Sitting Bull had 2,500 braves. If we think of how many were for Custer and how many were against him, we might say that the odds against Custer winning were 2,500 to 650. Similarly, the odds against Sitting Bull winning were 650 to 2,500.

EXAMPLE 8   *Calculating Odds on a Roulette Wheel*

A common type of roulette wheel has 38 equal-size compartments. Thirty-six of the compartments are numbered 1 to 36 with half of them colored red and the other half black. The remaining 2 compartments are green and numbered 0 and 00. A small ball is placed on the spinning wheel and when the wheel stops, the ball rests in one of the compartments. What are the odds against the ball landing on red?

SOLUTION: This is an experiment with 38 equally likely outcomes. Because 18 of these are in favor of the event "the ball lands on red" and 20 are against the event, the odds against red are 20 to 18. We can write this as 20:18, which we may reduce to 10:9. ✸

Although we have defined odds in terms of counting, we also can think of odds in terms of probability. Notice in the next definition we are comparing "probability against" with "probability for."

> **PROBABILITY FORMULA FOR COMPUTING ODDS** If E' is the complement of the event E, then the odds against E are
>
> $$\frac{P(E')}{P(E)}.$$

You may have been surprised that we have not used the "$a{:}b$" notation in this definition of odds. However, we have a good reason for doing this. If the probability of an event E is 0.30, then it is awkward to say that the odds against E are 0.70 to 0.30. It is better to say

$$\text{odds against } E = \frac{\text{probability of } E'}{\text{probability of } E} = \frac{0.70}{0.30} = \frac{70}{30}.$$

We could then say that the odds against E are 70 to 30, or 7 to 3. We will use this approach in Example 9.

EXAMPLE 9   *Using the Probability Formula for Computing the Odds in a Football Game*

Suppose that the probability of Green Bay winning the Super Bowl is 0.35. What are the odds against Green Bay winning the Super Bowl?

SOLUTION: Recall from Section 1.1 on problem solving that a diagram often helps us remember what to do. Call the event "Green Bay wins the Super Bowl" G. We illustrate this in Figure 14.5.

Thus, the odds against Green Bay winning the Super Bowl are

$$\frac{P(G')}{P(G)} = \frac{0.65}{0.35} = \frac{0.65 \times 100}{0.35 \times 100} = \frac{65}{35} = \frac{13}{7}.$$

In this case, we would say the odds against Green Bay winning the Super Bowl are 13 to 7.

Now try Exercises 21 to 24. ✳

Now that we have introduced some of the basic terminology and properties of probability, in the next section we will learn rules for calculating probabilities.

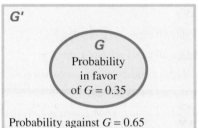

FIGURE 14.5 The odds against Green Bay winning are the probability of Green Bay losing versus the probability of Green Bay winning.

Exercises 14.1

Looking Back*

These exercises follow the general outline of the topics presented in this section and will give you a good overview of the material that you have just studied.

1. In Example 1(b), what method did we use to determine the sample space?

2. Although we usually describe events verbally, you should remember that an event is always what?

3. In Example 5(b), how did we find the number of objects in the sample space?

4. In Example 9, we knew that the probability of Green Bay winning the Super Bowl was 0.35. How did we find the odds against Green Bay winning?

5. In the Math in Your Life box, what was our explanation as to why the buttered side of a falling piece of bread lands face down? Why is it reasonable that your line at a toll booth would not be the fastest?

6. What was the point of our discussion about Custer and Sitting Bull?

Sharpening Your Skills

In Exercises 7–10, write each event as a set of outcomes. If the event is large, you may describe the event without writing it out.

7. When we roll two dice, the total showing is seven.

8. When we roll two dice, the total showing is five.

9. We flip three coins and obtain more (h)eads than (t)ails.

10. We select a red face card from a standard 52-card deck.

---

*Before doing these exercises, you may find it useful to review the note *How to Succeed at Mathematics* on page xix.

In Exercises 11–14, use the given spinner to write the event as a set of outcomes. Abbreviate "red" as "r," "blue" as "b," and "yellow" as "y."

11. Red appears exactly once when we spin the given spinner two times.

12. Yellow appears at least once when we spin the given spinner two times.

13. Blue appears exactly twice in three spins of the spinner.

14. Red does not appear in three spins of the spinner.

15. We are rolling two four-sided dice having the numbers 1, 2, 3, and 4 on their faces. Outcomes in the sample space are pairs such as (1, 3) and (4, 4).

   a. How many elements are in the sample space?

   b. Express the event "the total showing is even" as a set.

   c. What is the probability that the total showing is even?

   d. What is the probability that the total showing is greater than six?

16. You select three digital picture frames from a production line to determine if they are (d)efective or (n)ondefective. Outcomes in the sample space are represented by strings such as dnd and nnn.

   a. How many elements are in the sample space?

   b. Express the event "exactly one frame is defective" as a set.

   c. What is the probability that exactly one frame is defective?

   d. What is the probability that more frames are defective than nondefective?

17. Singers (C)arrie, (M)ariah, (K)eith, and (J)ustin are to perform in a talent contest and the order of their appearance will be chosen randomly. Outcomes in the sample space are represented by strings of letters such as CKMJ and JKCM.

   a. How many elements are in the sample space?

   b. Express the event "the women (C and M) do not perform consecutively" as a set.

   c. What is the probability that the women do not perform consecutively?

   d. What is the probability that Carrie will perform last?

18. We are flipping four coins. Outcomes in the sample space are represented by strings of Hs and Ts such as TTHT and HHTT.

   a. How many elements are in this sample space?

   b. Express the event "there are more heads than tails" as a set.

   c. What is the probability that there are more heads than tails?

   d. What is the probability that there are an equal number of heads and tails?

19. An experimenter testing for extrasensory perception has five cards with pictures of a (s)tar, a (c)ircle, (w)iggly lines, a (d)ollar sign, and a (h)eart. She selects two cards without replacement. Outcomes in the sample space are represented by pairs such as (s, d) and (h, c).

   a. How many elements are in this sample space?

   b. Express the event "a star appears on one of the cards" as a set.

   c. What is the probability that a star appears on one of the cards?

   d. What is the probability that a heart does not appear?

20. You have bought stock in Apple and Dell and each stock can either (i)ncrease in value, (d)ecrease in value, or (s)tay the same. Outcomes in the sample space are represented by pairs such as (i, d) and (d, d).

   a. How many elements are in this sample space?

   b. Express the event "Apple increases" as a set.

   c. What is the probability that Apple increases?

   d. What is the probability that Dell does not stay the same?

In Exercises 21–24, a) Find the probability of the given event. b) Find the odds against the given event.

21. A total of nine shows when we roll two fair dice.

22. A total of three shows when we roll two fair dice.

23. We draw a heart when we select 1 card randomly from a standard 52-card deck.

24. We draw a face card when we select 1 card randomly from a standard 52-card deck.

In Exercises 25–28, assume that we are drawing a 5-card hand from a standard 52-card deck.

25. What is the probability that all cards are diamonds?

26. What is the probability that all cards are face cards?

27. What is the probability that all cards are of the same suit?

28. What is the probability that all cards are red?

The residents of a small town and the surrounding area are divided over the proposed construction of a sprint car racetrack in the town. Use the following table to answer Exercises 29 and 30.

	Support Racetrack	Oppose Racetrack
Live in Town	1,512	2,268
Live in Surrounding Area	3,528	1,764

29. If a newspaper reporter randomly selects a person to interview from these people,

   a. what is the probability that the person supports the racetrack?

   b. what are the odds in favor of the person supporting the racetrack?

30. If a newspaper reporter randomly selects a person from town to interview,

   a. what is the probability that the person supports the racetrack?

   b. what are the odds in favor of the person supporting the racetrack?

Applying What You've Learned

31. In a given year, 2,048,861 males and 1,951,379 females were born in the United States. If a child is selected randomly from this group, what is the probability that it is a female?

32. Recently, the FBI reported that there were 7,160 hate crimes in the United States. Of these, 3,919 were based on race, 1,227 on religion, 1,017 on sexual orientation, 944 on ethnicity, and 53 on disability. (Assume that no crime was reported in two categories.) If you selected one of these crimes randomly, what is the probability that it would not be based on race?

33. The following table lists some of the empirical results that Mendel obtained in his experiments in cross-breeding pea plants.

Characteristics That Were Crossbred	First-Generation Plants	Second-Generation Plants
Tall vs. short	All tall	787 tall 277 short
Smooth seeds vs. wrinkled seeds	All smooth seeds	5,474 smooth 1,850 wrinkled

Assume that we are cross-breeding genetically pure tall plants with genetically pure short plants. Use this information to assign the probability that a second-generation plant will be short. How consistent is this with the theoretical results that Mendel derived?

34. Assume that we are cross-breeding genetically pure smooth-seed plants with genetically pure wrinkled-seed plants. Use the information provided in Exercise 33 to assign the probability that a second-generation plant will have smooth seeds. How consistent is this with the theoretical results that Mendel derived?

In Exercises 35 and 36, construct a Punnett square as we did in Table 14.4 to show the probabilities for the offspring.

35. One parent who has sickle-cell anemia and one parent who is a carrier have a child. Find the probability that the child is a carrier of sickle-cell anemia.

36. One parent who has sickle-cell anemia and one parent who is a carrier have a child. Find the probability that the child has sickle-cell anemia.

In cross-breeding snapdragons, Mendel found that flower color does not dominate, as happens with peas. For example, a snapdragon with one red and one white gene will have pink flowers. In Exercises 37 and 38, analyze the cross-breeding experiment as we did in the discussion prior to Example 7.

37. **a.** Construct a Punnett square showing the results of crossing a purebred white snapdragon with a purebred red one.

   b. What is the probability of getting red flowers in the first-generation plants? What is the probability of getting white? Of getting pink?

38. **a.** If we cross two pink snapdragons, draw a Punnett square that shows the results of crossing two of these first-generation plants.

   b. What is the probability of getting red flowers in the second-generation plants? What about white? Pink?

39. Cystic fibrosis is a serious inherited lung disorder that often causes death in victims during early childhood. Because the gene for this disease is recessive, two apparently healthy adults, called *carriers*, can have a child with the disease. We will denote the normal gene by N and the cystic fibrosis gene by c to indicate its recessive nature.

   a. Construct a Punnett square as we did in Example 7 to describe the genetic possibilities for a child whose two parents are carriers of cystic fibrosis.

   b. What is the probability that this child will have the disease?

40. From the Punnett square in Exercise 39, what is the probability that a child of two carriers will

   a. be normal?

   b. be a carrier?

41. Assume that the following table summarizes a survey involving the relationship between living arrangements and grade point average for a group of students.

	On Campus	At Home	Apartment	Totals
Below 2.5	98	40	44	182
2.5 to 3.5	64	25	20	109
Over 3.5	17	4	8	29
Totals	179	69	72	320

If we select a student randomly from this group, what is the probability that the student has a grade point average of at least 2.5?

42. Using the data in Exercise 41, if a student is selected randomly, what is the probability that the student lives off campus?

Use this replica of the Monopoly game board to answer Exercises 43–46.

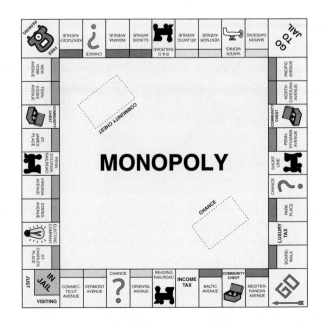

43. Assume that your game piece is on the Electric Company. If you land on either St. James Place, Tennessee Avenue, or New York Avenue, you will go bankrupt. What is the probability that you avoid these properties?

44. Assume that your game piece is on Pacific Avenue. If you land on either Park Place or Boardwalk, you will go bankrupt. What is the probability that you avoid these properties?

45. Your game piece is on Virginia Avenue. What is the probability that you will land on a railroad on your next move?

46. Your game piece is on Pennsylvania Avenue. What is the probability that you will have to pay a tax on your next move?

In Exercises 47–50 assume that we are randomly picking a person described in Table 14.2 from Example 4.

47. If we pick a divorced person, what is the probability that the person is a woman?

48. If we pick a woman, what is the probability that the person is married but not separated?

49. If we pick a never-married person, what is the probability that the person is a man?

50. If we pick a man, what is the probability that the person is a widower?

In horse racing, a trifecta is a race in which you must pick the first-, second-, and third-place winners in their proper order to win a payoff.

51. If eight horses are racing and you randomly select three as your bet in the trifecta, what is the probability that you will win? (Assume that all horses have the same chance to win.)

52. If 10 horses are racing and you randomly select 3 as your bet in the trifecta, what is the probability that you will win? (Assume that all horses have the same chance to win.)

53. If the odds against event *E* are 5 to 2, what is the probability of *E*?

54. If $P(E) = 0.45$, then what are the odds against *E*?

55. If the odds against the U.S. women's soccer team winning the World Cup are 7 to 5, what is the probability that they will win the World Cup?

56. If the odds against the U.S. men's volleyball team defeating China in the Summer 2008 Olympic Games is 9 to 3, what is the probability that the United States will defeat China?

57. Suppose the probability that the Yankees will win the World Series is 0.30.
   a. What are the odds in favor of the Yankees winning the World Series?
   b. What are the odds against the Yankees winning the World Series?

58. Suppose the probability that Rags to Riches will win the Triple Crown in horse racing is 0.15.
   a. What are the odds in favor of Rags to Riches winning the Triple Crown?
   b. What are the odds against Rags to Riches winning the Triple Crown?

59. In the New York Lotto, the player must correctly pick six numbers from the numbers 1 to 59. What are the odds against winning this lottery?

60. Go to the National Safety Council's Web site at www.nsc.org to find the odds against you being killed by lightning in your lifetime. How many more times likely is it that you will be killed by lightning than win the New York Lotto?

Communicating Mathematics

61. Suppose that we want to assign the probability of *E* empirically. Complete the following equation: $P(E) =$

62. What condition must we have in order to use the formula $P(E) = \dfrac{n(E)}{n(S)}$?

63. In the pea plant example, we found that all the first-generation plants had yellow seeds. Why was this the case?

64. If the odds against an event are *a* to *b*, what are the odds in favor of the event?

65. Explain the difference between an outcome and an event.

66. When rolling two dice, describe in words an event that is the empty set.

67. When rolling two dice, describe in words an event that is the whole sample space.

68. Can an event have just one single outcome in it? Give an example.

69. Explain the difference between the probability of an event and the odds *in favor* of the event.

70. You know that the probability of an event can never be greater than 1. Can the odds in favor of an event ever exceed 1? Explain.

Using Technology to Investigate Mathematics

71. See your instructor for a tutorial on using graphing calculators to demonstrate some of the ideas we have discussed in this section. Report on your findings.*

72. Search the Internet to find applets that illustrate some of the ideas we have discussed in this section. Report on your findings.

For Extra Credit

In Exercises 73 and 74, assume that to log in to a computer network you must enter a password. Assume that a hacker who is trying to break into the system randomly types one password every 10 seconds. If the hacker does not enter a valid password within 3 minutes, the system will not allow any further attempts to log in. What is the probability that the hacker will be successful in discovering a valid password for each type of password?

73. The password consists of two letters followed by three digits. Case does not matter for the letters. Thus, Ca154 and CA154 would be considered the same password.

74. The password consists of any sequence of two letters and three digits. Case does not matter for the letters. Thus, B12q5 and b12Q5 would be considered the same password.

75. Find some examples of advertising claims in the media. In what way are these claims probabilities?

76. a. Flip a coin 100 times. How do your empirical results compare with the theoretical probabilities for obtaining heads and tails?

b. Roll a pair of dice 100 times. How do your empirical results compare with the theoretical probabilities for rolling a total of two, three, four, and so on?

c. Toss an irregular object such as a thumbtack 1,000 times. After doing this, what probability would you assign to the thumbtack landing point up? What about point down? Does it matter what kind of thumbtack you use? Explain.

77. Investigate other genetic diseases, such as Tay-Sachs or Huntington's disease. Explain how the mathematics of the genetics of these diseases is similar or dissimilar to the examples we studied in this section.

78. Experiment with an object such as a mouse pad. Slide it off a table 100 times and record how often it lands face down. If the object slips off the table, what is the probability that the object will land face down?

79. Simulate the lost socks example discussed in the Math in Your Life box by doing the following. Take 20 three-by-five cards. Label two of them "pair one," two of them "pair two," two of them "pair three," and so on. Put the 20 cards into a box and draw cards randomly, without replacement, until you have drawn two cards that have the same label. Do this 30 times. On average, how many cards do you draw before you have drawn two cards with the same label?

Complements and Unions of Events

14.2

Objectives

1. Understand the relationship between the probability of an event and the probability of its complement.
2. Calculate the probability of the union of two events.
3. Use complement and union formulas to compute the probability of an event.

Suppose that you are attending a luncheon on your campus to develop a campaign to fight world hunger. Someone at your table might ask, "This room is pretty crowded. I wonder how many are here?" You look around and see that there are eight tables that seat 10 people each and notice that there are only six empty seats. So, without actually counting, you can quickly say that there are 74 people attending.

Similarly, you can often solve complex probability problems by applying a few simple, intuitive formulas. We will use some results from set theory in Chapter 2 to develop rules for computing the probability of the complement and the union of events.

KEY POINT

We can compute the probability of an event by finding the probability of its complement.

Complements of Events

You can frequently solve a mathematics problem by rephrasing it as an equivalent question that is easier to answer. For example, suppose that you want to calculate the probability of event *E*,* but find that *E* is too complicated to understand easily. If you remember that the total probability available in a sample space is 1, then it may be simpler to find the probability of *E*'s complement and subtract that number from 1.

---

*Throughout the rest of this chapter, unless we state otherwise, *E*, *F*, and so on will represent events in the sample space *S*.

> **COMPUTING THE PROBABILITY OF THE COMPLEMENT OF AN**
> **EVENT** If E is an event, then $P(E') = 1 - P(E)$.

Of course we can state this result differently as $P(E) = 1 - P(E')$ or $P(E) + P(E') = 1$. We illustrate this formula in Figure 14.6.

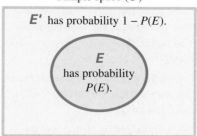

Sample space (**S**)

E' has probability $1 - P(E)$.

E
has probability
$P(E)$.

FIGURE 14.6 $P(E) + P(E') = 1$.

PROBLEM SOLVING
The Analogies Principle

The Analogies Principle in Section 1.1 says that you can learn a new area of mathematics more easily if you connect it with an area that you already know. By drawing Venn diagrams, you can visualize many of the rules of probability theory that will help you remember them and use them properly.

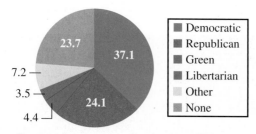

■	Democratic
■	Republican
■	Green
■	Libertarian
□	Other
■	None

Percent of voters according to party affiliation

EXAMPLE 1 Using the Complement Formula to Study Voter Affiliation

The accompanying graph shows how a group of first-time voters are classified according to their party affiliation. If we randomly select a person from this group, what is the probability that the person has a party affiliation?

SOLUTION: Let A be the event that the person we select has some party affiliation. Rather than compute the probability of this event, it is simpler to calculate the probability of A' which is the event that the person we select has no party affiliation. We illustrate this situation in Figure 14.7.

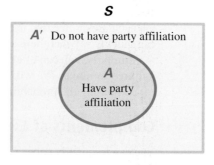

S

A' Do not have party affiliation

A
Have party
affiliation

FIGURE 14.7 Calculate $P(A)$ by first finding $P(A')$.

Quiz Yourself ❻

If a pair of dice are rolled, what is the probability of rolling a total that is less than 11? (Use the complement rule.)

Because 23.7% have no party affiliation, the probability of selecting such a person is 0.237. Thus, $P(A) = 1 - P(A') = 1 - 0.237 = 0.763$.

Now try Exercises 5 to 12. ✻ ❻

HIGHLIGHT ✺ ✺ ✺

Between the Numbers—What Are Your Chances?

Once, while we were studying the probability of lotteries and casino games, one of my students, upon realizing his slim chances of winning a lottery, blurted out, "I'm never going to gamble again!"

If you did Exercises 59 and 60 from Section 14.1, you saw that the odds against your winning the New York

Lotto are 45,057,474 to 1, which is much greater than being killed by lightning. To put in perspective what a long shot it is to win a big lottery, consider the odds on your dying from various events provided by the National Safety Council.

Event (Dying From . . .)	Odds Against This Event	Number of Times More Likely Than Winning The New York Lotto
Lightning	79,113 to 1	569
Drowning	9,641 to 1	4,674
Assault by Firearms	308 to 1	146,290
Motor Vehicle Accident	87 to 1	517,902

 KEY POINT

We often describe the union of events using the word *or*.

Unions of Events

We often combine objects in a mathematical system to obtain other objects in the system. For example, in number systems, we add and subtract pairs of numbers to get other numbers. In probability, we join events using the set operations union and intersection, which we discussed in Section 2.3. We will investigate the union of events in this section and the intersection of events in Section 14.3. We often describe the union of events using the word *or*.

Figure 14.8 shows the union of events E and F. In computing $P(E \cup F)$, it is a *common mistake* to simply add $P(E)$ and $P(F)$. Because some outcomes may be in both E and F, you will count their probabilities twice. To compute $P(E \cup F)$ properly, you must subtract $P(E \cap F)$.

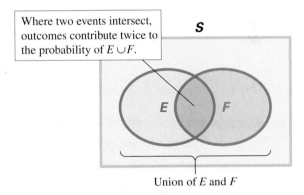

Where two events intersect, outcomes contribute twice to the probability of $E \cup F$.

Union of E and F

FIGURE 14.8 The union of two events.

> **RULE FOR COMPUTING THE PROBABILITY OF A UNION OF TWO EVENTS** If E and F are events, then
>
> $$P(E \cup F) = P(E) + P(F) - P(E \cap F). *$$
>
> If E and F have no outcomes in common, they are called *mutually exclusive events*. In this case, because $E \cap F = \varnothing$, the preceding formula simplifies to
>
> $$P(E \cup F) = P(E) + P(F).$$

*The counting version of this rule can be found in Section 2.3.

EXAMPLE 2 *Finding the Probability of the Union of Two Events*

If we select a single card from a standard 52-card deck, what is the probability that we draw either a heart or a face card?

SOLUTION: Let H be the event "draw a heart" and F be the event "draw a face card." We are looking for $P(H \cup F)$. There are thirteen hearts, twelve face cards, and three cards that are both hearts and face cards. Figure 14.9 helps us remember the formula to use.

S

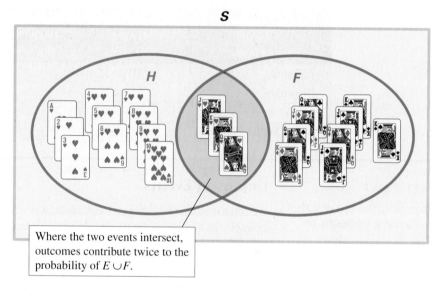

Where the two events intersect, outcomes contribute twice to the probability of $E \cup F$.

FIGURE 14.9 The union of the events "draw a heart" and "draw a face card."

Therefore,

$$P(H \cup F) = P(H) + P(F) - P(H \cap F) = \frac{13}{52} + \frac{12}{52} - \frac{3}{52} = \frac{22}{52} = \frac{11}{26}$$

probability of a heart — probability of a face card — probability of a heart that is a face card

Now try Exercises 13 to 16. ❋

 Some Good Advice

If you are given any three of the four quantities in the formula

$$P(E \cup F) = P(E) + P(F) - P(E \cap F),$$

you can use algebra to solve for the other. See Example 3.

EXAMPLE 3 *Using Algebra to Find a Missing Probability*

A magazine conducted a survey of readers age 18 to 25 regarding their health concerns. The editors will use this information to choose topics relevant to their readers. The survey found that 35% of the readers were concerned with improving their cardiovascular fitness and 55% wanted to lose weight. Also, the survey found that 70% are concerned with either improving their cardiovascular fitness or losing weight. If the editors randomly select one of those surveyed to profile in a feature article, what is the probability that the person is concerned with both improving cardiovascular fitness *and* losing weight?

SOLUTION: If we let C be the event "the person wants to improve cardiovascular fitness" and W be the event "the person wishes to lose weight," then the word "and" tells us that we need to find $P(C \cap W)$.

We are given that 35% want to improve cardiovascular fitness, so $P(C) = 0.35$. Similarly, $P(W) = 0.55$. The event "a person wants to improve cardiovascular fitness or lose weight," is the event $C \cup W$, and we are told that $P(C \cup W) = 0.70$. Figure 14.10 will now help you to see what to do.

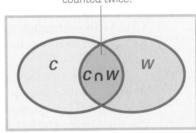

FIGURE 14.10 The union of events C and W.

From this diagram* you can see that

$$P(C \cup W) = P(C) + P(W) - P(C \cap W).$$
$$\quad 0.70 \qquad\ 0.35 \quad 0.55 \quad\ \text{unknown}$$

This gives us the equation

$$0.70 = 0.35 + 0.55 - P(C \cap W).$$

Rewriting this equation as

$$P(C \cap W) = 0.35 + 0.55 - 0.70,$$

we find that $P(C \cap W) = 0.20$.

This means that there is a 20% chance that the person chosen will be interested in both improving cardiovascular fitness and losing weight.

Now try Exercises 17 to 20. ✳ **7**

Quiz Yourself **7**

Suppose that A and B are events such that $P(A) = 0.35$, $P(A \cap B) = 0.15$, and $P(A \cup B) = 0.65$. Find $P(B)$.

 **KEY POINT**

We may use several formulas to calculate an event's probability.

Combining Complement and Union Formulas

In Example 4, we use both the complement formula and the union formula to compute an event's probability.

EXAMPLE 4 *Finding the Probability of the Complement of the Union of Two Events*

A survey of consumers comparing the amount of time they spend shopping on the Internet per month with their annual income produced the results in Table 14.5.[†]

Assume that these results are representative of all consumers. If we select a consumer randomly, what is the probability that the consumer neither shops on the Internet 10 or more hours per month nor has an annual income above $60,000?

---
*We will not always draw diagrams to illustrate the rule for calculating the probability of the union of two events; however, we do encourage you to do so to aid you in your computations.

[†]We will assume that all times are rounded to the nearest hour.

$n(T \cap A)$

Annual Income	10 + Hours (*T*)	3–9 Hours	0–2 Hours	Totals	
Above $60,000 (*A*)	192	176	128	496	— *n*(*A*)
$40,000–$60,000	160	208	144	512	
Below $40,000	128	192	272	592	
Totals	480	576	544	1,600	— *n*(*S*)

n(*T*)

TABLE 14.5 Survey results on Internet shopping.

SOLUTION: Although we could use the probability techniques that we introduced in Section 14.1 to answer this question, we will use instead the formulas for the complement and union of events.

We will let *T* be the event "the consumer selected spends 10 or more hours per month shopping on the Internet." Event *T* corresponds to the first column in Table 14.5. Also, let *A* be the event "the consumer selected has an annual income above $60,000," which is described by the first row in Table 14.5.

In Figure 14.11, you can see that the event "the consumer selected neither shops on the Internet 10 or more hours per month nor has an annual income above $60,000" is the region outside of $T \cup A$, which is the complement of $T \cup A$.

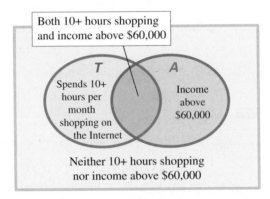

FIGURE 14.11 The event "neither *T* nor *A*" corresponds to the complement of $T \cup A$.

From Table 14.5, you see that the number of outcomes in *T* is

$$n(T) = 192 + 160 + 128 = 480.$$

The total number of outcomes in the sample space, *S*, is the total number of people surveyed, which is 1,600. Therefore,

$$P(T) = \frac{n(T)}{n(S)} = \frac{480}{1,600} = 0.30.$$

Similarly,

$$P(A) = \frac{n(A)}{n(S)} = \frac{496}{1,600} = 0.31.$$

Also, if you look at the intersection of the column labeled *T* and the row labeled *A*, you see that the number of people surveyed who shop on the Internet more than 10 hours per month and also earn above $60,000 is 192. Therefore,

$$P(T \cap A) = \frac{n(T \cap A)}{n(S)} = \frac{192}{1,600} = 0.12.$$

━━━━━━━━━━━━ **HISTORICAL HIGHLIGHT** ✺ ✺ ✺ ━━━

Modern Probability Theory

Modern probability theory began in the seventeenth century when Blaise Pascal and his friend Pierre de Fermat began to study the mathematical principles of gambling. They were answering several questions posed to Pascal by Antoine Gombauld, the Chevalier de Mere:

> "How many times should a single die be thrown before we could reasonably expect two sixes?"

> "How should prize money in a contest be fairly divided in the case that the contest, for some reason, cannot be completed?"

In 1812, Pierre-Simon de Laplace presented classical probability theory in his work *Theorie Analytique des Probabilities*. Laplace boldly affirmed that all knowledge could be obtained by using the principles he set forth.

Physicists now use probability theory to study radiation and atomic physics, and biologists apply it to genetics and mathematical learning theory. Probability also is a theoretical basis for statistics, which is used in scientific, industrial, and social research.

We can now calculate the probability of the complement of $T \cup A$ as follows:

$$P((T \cup A)') = 1 - P(T \cup A) = 1 - [P(T) + P(A) - P(T \cap A)]$$
$$= 1 - [0.30 + 0.31 - 0.12] = 1 - 0.49 = 0.51.$$

This means that if we select a consumer randomly, there is a 51% chance that the consumer neither spends 10 or more hours per month shopping on the Internet nor has a yearly income above $60,000.

Now try Exercises 21, 22, and 25 to 28. ✻

Exercises ⟮14.2⟯

Looking Back*

These exercises follow the general outline of the topics presented in this section and will give you a good overview of the material that you have just studied.

1. In Example 1, why did we compute the probability of A' rather than compute $P(A)$ directly?

2. Why did we subtract $\frac{3}{52}$ in calculating $P(H \cup F)$ in Example 2?

3. What was the point of the Between the Numbers Highlight?

4. How did Pascal and Fermat get interested in probability theory?

Sharpening Your Skills

In Exercises 5–12, use the complement formula to find the probability of each event.

5. If the probability that your DVD player breaks down before the extended warranty expires is 0.015, what is the probability that the player will not break down before the warranty expires?

6. If the probability that a vaccine you took will protect you from getting the flu is 0.965, what is the probability that you will get the flu?

7. If there is a 1 in 1,000 chance that you will pick the numbers correctly in tonight's lottery, what is the probability that you will not pick the numbers correctly?

8. If there is a 1 in 4 chance that it will rain for your Fourth of July barbecue, what is the probability that it won't rain?

9. If two dice are rolled, find the probability that neither die shows a five on it.

10. If two dice are rolled, find the probability that the total showing is less than 10.

11. If five coins are flipped, what is the probability of obtaining at least one head?

12. If five coins are flipped, what is the probability of obtaining at least one head and at least one tail?

13. If a single card is drawn from a standard 52-card deck, what is the probability that it is either a five or a red card?

14. If a single card is drawn from a standard 52-card deck, what is the probability that it is either a face card or a red card?

15. If a pair of dice is rolled, what is the probability that the total showing is either odd or greater than eight?

16. If a pair of dice is rolled, what is the probability that the total showing is either even or less than seven?

───────────

*Before doing these exercises, you may find it useful to review the note *How to Succeed at Mathematics* on page xix.

In Exercises 17–20, assume that A *and* B *are events.*

17. If $P(A \cup B) = 0.85$, $P(B) = 0.40$, and $P(A) = 0.55$, find $P(A \cap B)$.

18. If $P(A \cup B) = 0.75$, $P(B) = 0.45$, and $P(A) = 0.60$, find $P(A \cap B)$.

19. If $P(A \cup B) = 0.70$, $P(A) = 0.40$, and $P(A \cap B) = 0.25$, find $P(B)$.

20. If $P(A \cup B) = 0.60$, $P(B) = 0.45$, and $P(A \cap B) = 0.20$, find $P(A)$.

21. If we draw a card from a standard 52-card deck, what is the probability that the card is neither a heart nor a face card? (*Hint:* Draw a picture of this situation before trying to calculate the probability.)

22. If we draw a card from a standard 52-card deck, what is the probability that the card is neither red nor a queen? (*Hint:* Draw a picture of this situation before trying to calculate the probability.)

Applying What You've Learned

Use the following table from the U.S. Bureau of Labor Statistics, which shows the age distribution of those who earned less than the minimum wage in 2004, to answer Exercises 23 and 24.

Age	Working Below Minimum Wage (thousands)
16–19	329
20–24	420
25–34	320
35–44	175
45–54	125
55–64	61
65 and older	53

23. If we select a worker randomly from those surveyed, what is the probability that the person is younger than 55?

24. If we select a worker randomly from those surveyed, what is the probability that the person is older than 19?

Use the following table, that we presented in Example 4, relating the amount of time consumers spend shopping on the Internet per month with their annual income, to answer Exercises 25–28.

Annual Income	10+ Hours	3–9 Hours	0–2 Hours	Totals
Above $60,000	192	176	128	496
$40,000–$60,000	160	208	144	512
Below $40,000	128	192	272	592
Totals	480	576	544	1,600

25. What is the probability that a consumer we select randomly either spends 0–2 hours per month shopping on the Internet or has an annual income below $40,000?

26. What is the probability that a consumer we select randomly either spends 10 or more hours per month shopping on the Internet or has an annual income between $40,000 and $60,000?

27. What is the probability that a consumer we select randomly neither spends more than 2 hours per month shopping on the Internet nor has an annual income of $60,000 or less?

28. What is the probability that a consumer we select randomly neither spends more than 2 hours per month shopping on the Internet nor has an annual income below $40,000?

Probability of earning commissions. *Joanna earns both a salary and a monthly commission as a sales representative for an electronics store. The following table lists her estimates of the probabilities of earning various commissions next month. Use this table to calculate the probabilities in Exercises 29–32.*

Commission	Probability That This Will Happen
Less than $1,000	0.08
$1,000 to $1,249	0.11
$1,250 to $1,499	0.23
$1,500 to $1,749	0.30
$1,750 to $1,999	0.12
$2,000 to $2,249	0.05
$2,250 to $2,499	0.08
$2,500 or more	0.03

29. The probability that she will earn at least $1,000 in commissions

30. The probability that she will earn at least $1,500 in commissions

31. The probability that she will earn no more than $1,999 in commissions

32. The probability that she will earn less than $2,250 in commissions

A college administration has conducted a study of 200 randomly selected students to determine the relationship between satisfaction with academic advisement and academic success. They obtained the

following information: Of the 70 students on academic probation, 32 are not satisfied with advisement; however, only 20 of the students not on academic probation are dissatisfied with advisement. Use these data to answer Exercises 33–36. In each exercise, assume that we select a student at random.

33. What is the probability that the student is not on academic probation?

34. What is the probability that the student is satisfied with advisement?

35. What is the probability that the student is on academic probation and is satisfied with advisement?

36. What is the probability that the student is not on academic probation and is satisfied with advisement?

Communicating Mathematics

37. Why does $P(E) + P(E') = 1$?

38. What word in Example 4 suggested that we compute the probability of the *complement* of the union of T and A?

In Exercises 39–42, determine whether each statement is true or false for events A *and* B. *Explain your answer.*

39. $P(A) = P(A \cup B) - P(B)$

40. $P(A \cup B) - P(B) = P(A \cap B)$

41. $P(A) + P(B) - P(A \cup B) = P(A \cap B)$

42. $P(A \cup B) - P(B) = P(A)$

43. Many texts that discuss probability state that if events E and F are disjoint, then $P(E \cup F) = P(E) + P(F)$. Explain why it is really not necessary to state this formula.

44. If $P(E \cup F) = P(E) + P(F)$, what can you conclude about $P(E \cap F)$?

For Extra Credit

45. If events A, B, and C are as pictured in this Venn diagram, write a formula for $P(A \cup B \cup C)$. Explain your answer.

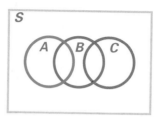

46. If events A, B, and C are as pictured in this Venn diagram, write a formula for $P(A \cup B \cup C)$. Explain your answer.

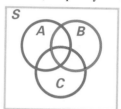

Use the given spinner to answer Exercises 47 to 50. Each green sector occupies 10% of the circle, each blue sector occupies 12%, each red sector occupies 9%, and the yellow sector occupies 4%. Assume that the spinner is spun once. What is the probability that

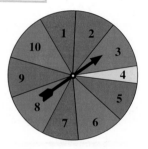

47. The spinner does not stop on yellow?

48. The spinner stops on an odd number or a green?

49. The spinner stops on neither an odd number nor a blue?

50. The spinner stops on neither green nor an even number?

14.3

Conditional Probability and Intersections of Events

Objectives

1. Understand how to compute conditional probability.
2. Calculate the probability of the intersection of two events.
3. Use probability trees to compute conditional probabilities.
4. Understand the difference between dependent and independent events.

You know how to compute the probability of complements and unions of events. We will now show you how to find the probability of intersections of events, but first you need to understand how the occurrence of one event can affect the probability of another event.

✏️ **KEY POINT**

Conditional probability takes into account that one event occurring may change the probability of a second event.

Conditional Probability

Suppose that you and your friend Marcus cannot agree as to which video to rent and you decide to settle the matter by rolling a pair of dice. You will each pick a number and then roll the dice. The person whose total showing on the dice comes up first gets to pick the video. With your knowledge of probability, you know that the number you should pick is 7 because it has the highest probability of appearing, namely $\frac{1}{6}$.

To illustrate the idea of conditional probability, let's now change this situation slightly. Assume that your friend Janelle will roll the dice before you and Marcus pick your numbers. You are not allowed to see the dice, but Janelle will tell you something about the dice and then you will choose your number before looking at the dice.

Suppose that Janelle tells you that the total showing is an even number. Would you still choose a 7? Of course not, because once you know the *condition* that the total is even, you now know that the probability of having a 7 is 0. A good way to think of this is that once you know that the total is even, you must exclude all pairs from the sample space such as (1, 4), (5, 6), and (4, 3) that give odd totals.

In a similar way, suppose that you draw a card from a standard 52-card deck, put that card in your pocket, and then draw a second card. What is the probability that the second card is a king? How you answer this question depends on knowing what card is in your pocket. If the card in your pocket is a king, then there are three kings remaining in the 51 cards that are left, so the probability is $\frac{3}{51}$. If the card in your pocket is not a king, then the probability of the second card being a king is $\frac{4}{51}$. Why? This discussion leads us to the formal definition of conditional probability.

DEFINITION When we compute the probability of event *F* assuming that the event *E* has already occurred, we call this the **conditional probability** of *F*, given *E*. We denote this probability as $P(F\,|\,E)$. We read $P(F\,|\,E)$ as "the probability of *F* given that *E* has occurred," or in a quicker way, "the probability of *F* given *E*."

Do not let this new notation intimidate you. The notation $P(F\,|\,E)$ simply means that you are going to compute a probability knowing that something else has already happened. For example, in our earlier discussion of Janelle, we said, "The probability of having a total of 7 knowing that the total is even is 0." We will restate this several times, each time increasing our use of symbols. So, we could say instead,

$$P(\text{having a total of 7 given that the total is even}) = 0;$$

or,

$$P(\text{having a total of 7} \mid \text{the total is even}) = 0.$$

If we now represent the event "total is 7" by *F* and "total is even" by *E*, then we could write our original statement as

┌─ Total is even.

$$P(F \mid E) = 0.$$

Total is 7. ─┘

Similarly, let's return to the example of drawing two cards, and let *A* represent the event that we draw a king on our first card and put it in our pocket and, let *B* represent the event that we draw a king on our second card. Then we could write, "The probability that we draw a second king given that the first card was a king is $\frac{3}{51}$," as

┌─ First card was a king.

$$P(B \mid A) = \frac{3}{51}.$$

Second card is a king. ─┘

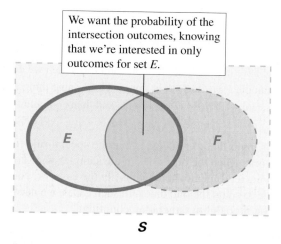

We want the probability of the intersection outcomes, knowing that we're interested in only outcomes for set E.

FIGURE 14.12 To compute the probability of F given E, we compare the outcomes in $E \cap F$ with the outcomes in E.

The Venn diagram in Figure 14.12 will help you remember how to compute conditional probability.

We drew E with a heavy line in Figure 14.12 to emphasize that when we assume that E has occurred, we can then think of the outcomes outside of E as being discarded from the discussion. In computing conditional probability, you will find it useful to consider the sample space to be E and the event as being $E \cap F$, rather than F. We will first state a special rule for computing conditional probability when the outcomes are equally likely (all have the same probability of occurring). We will state the more general conditional probability rule later.

> **SPECIAL RULE FOR COMPUTING $P(F|E)$ BY COUNTING**   If E and F are events in a sample space with equally likely outcomes, then $P(F|E) = \dfrac{n(E \cap F)}{n(E)}$.

EXAMPLE 1   *Computing Conditional Probability by Counting*

Assume that we roll two dice and the total showing is greater than nine. What is the probability that the total is odd?

SOLUTION: This sample space has 36 equally likely outcomes. We will let G be the event "we roll a total greater than nine" and let O be the event "the total is odd." Therefore,

$$G = \{(4, 6), (5, 5), (5, 6), (6, 4), (6, 5), (6, 6)\}.$$

The set O consists of all pairs that give an odd total. Figure 14.13 shows you how to use the special rule for computing conditional probability to find $P(O|G)$.

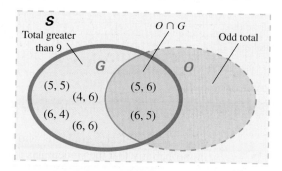

FIGURE 14.13 To compute $P(O|G)$, we compare the number of outcomes in $O \cap G$ with the number of outcomes in G.

Therefore,

$$P(O|G) = \frac{n(O \cap G)}{n(G)} = \frac{2}{6} = \frac{1}{3}.$$

Notice how the probability of rolling an odd total has changed from $\frac{1}{2}$ to $\frac{1}{3}$ when we know that the total showing is greater than nine.

Now try Exercises 5 to 22. ✳

PROBLEM SOLVING

The Order Principle

As we have emphasized often, the order in which we do things is important in mathematics. In Example 1, $P(O \mid G)$ does not mean the same thing as $P(G \mid O)$. To understand the difference, state what $P(G \mid O)$ represents and then compute $P(G \mid O)$.

It is important to remember that the special rule for computing conditional probability only works when the outcomes in the sample space are *equally likely*. Sometimes you may be solving a problem where that is not the case; or, you may have a situation where it is not possible to count the outcomes. In such cases, we need a rule for computing $P(F \mid E)$ that is based on probability rather than counting.

> **GENERAL RULE FOR COMPUTING $P(F \mid E)$** If E and F are events in a sample space, then $P(F \mid E) = \dfrac{P(E \cap F)}{P(E)}$.

We still can use Figure 14.12 to remember this rule; however, now instead of comparing the number of outcomes in $E \cap F$ with the number of outcomes in E, we compare the probability of $E \cap F$ with the probability of E.

EXAMPLE 2   *Using the General Rule for Computing Conditional Probability*

The state bureau of labor statistics conducted a survey of college graduates comparing starting salaries to majors. The survey results are listed in Table 14.6.

Major	$30,000 and Below	$30,001 to $35,000	$35,001 to $40,000	$40,001 to $45,000	Above $45,000	Totals (%)
Liberal arts	6*	10	9	1	1	27
Science	2	4	10	2	2	20
Social sciences	3	6	7	1	1	18
Health fields	1	1	8	3	1	14
Technology	0	2	7	8	4	21
Totals (%)	12	23	41	15	9	100

*These numbers are percentages.

TABLE 14.6 Survey comparing starting salaries to major in college.

If we select a graduate who was offered between $40,001 and $45,000, what is the probability that the student has a degree in the health fields?

SOLUTION: Each entry in Table 14.6 is the probability of an event. For example, the 8% that we highlighted is the probability of selecting a graduate in technology who earns between $40,001 and $45,000, inclusive. The 14% that we highlighted tells us the probability of selecting a graduate who majored in the health fields.

Let R be the event "the graduate received a starting salary between $40,001 and $45,000" and H be the event "the student has a degree in the health fields." *It is important in doing this problem that you identify clearly what you are given and what you must find.* We are given R and must find the probability of H, so we want to find $P(H \mid R)$, not $P(R \mid H)$.

Because we want the probability of H given R, we can, in effect, ignore all the outcomes that do not correspond to a starting salary of $40,001 to $45,000. We darken the columns we want to ignore in Table 14.7.

given

Major	$30,000 or Below	$30,001 to $35,000	$35,001 to $40,000	$40,001 to $45,000	Above $45,000	Totals (%)
Liberal arts	6	10	9	1	1	27
Science	2	4	10	2	2	20
Social sciences	3	6	7	1	1	18
Health fields	1	1	8	3	1	14
Technology	0	2	7	8	4	21
Totals (%)	12	23	41	15	9	100

$P(R)$ $P(H \cap R)$

TABLE 14.7 The columns we want to ignore are darkened.

In order to use the general rule for computing conditional probability, we first need to know $P(R)$ and $P(H \cap R)$, as you can see from Figure 14.14.

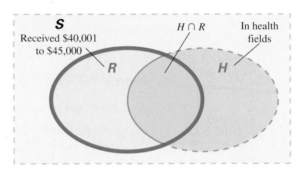

FIGURE 14.14 To compute $P(H \mid R)$, we compare $P(H \cap R)$ with $P(R)$.

Quiz Yourself ⑧

Assume that we select a graduate who received more than $45,000 as a starting salary. Use Table 14.7 to find the probability that the graduate has a degree in technology.

KEY POINT

We use conditional probability to find the probability of the intersection of two events.

Therefore,

$$P(H \mid R) = \frac{P(H \cap R)}{P(R)} = \frac{0.03}{0.15} = 0.20.$$

If a graduate receives an offer between $40,001 and $45,000, then the probability that the person is in the health fields is 0.20, or 20%.

Now try Exercises 49 to 52. ✻ ⑧

The Intersection of Events

We can now find a formula to compute the probability of the intersection of two events. The general rule for computing conditional probability states that

$$P(F \mid E) = \frac{P(E \cap F)}{P(E)}.$$

If we multiply both sides of this equation by the expression $P(E)$, we get

$$P(E) \cdot P(F \mid E) = P(E) \cdot \frac{P(E \cap F)}{P(E)}.$$

Canceling $P(E)$ from the numerator and denominator of the right side of this equation gives us the rule for computing the probability of the intersection of events E and F.

RULE FOR COMPUTING THE PROBABILITY OF THE INTERSECTION OF EVENTS If E and F are two events, then

$$P(E \cap F) = P(E) \cdot P(F \mid E).$$

This rule says that to find the probability of $P(E \cap F)$, you first find the probability of E and then multiply it by the probability of F, *assuming that E has occurred.*

EXAMPLE 3 *Estimating Your Grade in a Class*

Assume that for your literature final your professor has written questions on 10 assigned readings on cards and you are to randomly select two cards and write an essay on them. If you have read 8 of the 10 readings, what is your probability of getting two questions that you can answer? (We will assume that if you've done a reading, then you can answer the question about that reading; otherwise, you can't answer the question.)

SOLUTION: We can think of this event as the intersection of two events A and B where

A is "you can answer the first question;"

B is "you can answer the second question."

By the rule we just stated, you need to calculate

probability you can answer ⌐   ⌐ probability you can answer the second question,
the first question   given that you answered the first question

$$P(A \cap B) = P(A) \cdot P(B \mid A).$$

Because you have read 8 readings from 10 that were assigned, we see that $P(A) = \dfrac{8}{10}$. To find $P(B \mid A)$, think of this in words, as we want to find

the probability that you can answer the second question, given that you have answered the first question.

Quiz Yourself ❾

Suppose that we draw 2 cards without replacement from a standard 52-card deck. What is the probability that both cards are face cards?

There are now only seven questions that you can answer on the remaining nine cards so $P(B \mid A) = \dfrac{7}{9}$. Thus the probability that you get two questions on readings that you have done is

$$P(A \cap B) = P(A) \cdot P(B \mid A) = \frac{8}{10} \cdot \frac{7}{9} = \frac{56}{90} \approx 0.62.$$

Now try Exercises 27 to 40. ✳ ❾

 Some Good Advice

A common mistake that you can make when computing conditional probability is to use the formula $P(A \cap B) = P(A) \cdot P(B)$. That is, you may forget to take into account that event A has occurred. Notice that if we had used this incorrect formula in Example 3, the probability that you can answer the second question would have been $\dfrac{8}{10}$, not $\dfrac{7}{9}$. In effect, we would have computed the second probability as though the first question had been returned to the cards.

✎ **KEY POINT**

Trees help you visualize probability computations.

Probability Trees

Recall that the Three-Way Principle in Section 1.1 tells you that drawing a diagram is a good problem-solving technique. You will find that it is often very helpful to draw a probability tree to help you understand a conditional probability problem.

> **USING TREES TO CALCULATE PROBABILITIES** We can represent an experiment that happens in stages with a tree whose branches represent the outcomes of the experiment. We calculate the probability of an outcome by multiplying the probabilities found along the branch representing that outcome. We will call these trees **probability trees**.

In order to show you how to use trees to visualize conditional probabilities, we will look at the situation in Example 3 again.

EXAMPLE 4 *Using a Tree to Compute Probabilities*

Recall that you are taking an exam and will pick two questions. Eight of the 10 questions are on readings that you have done and 2 are on readings you have not done. What is the probability that you will get questions on two readings that you have not done? (We will assume that if you've done a reading, then you can answer the question about that reading; otherwise, you can't answer the question.)

SOLUTION: A will represent the event, "you can answer a question," and N^* will represent the event "you can't answer a question." The tree in Figure 14.15(a) shows that when you draw your first card you either can answer the question or not. Because 8 of the 10 cards have questions you can answer, $P(A) = \dfrac{8}{10}$, and because there are two questions you can't answer, $P(N) = \dfrac{2}{10}$.

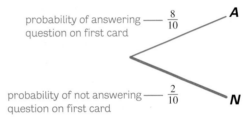

probability of answering ── $\frac{8}{10}$ **A**
question on first card

probability of not answering ── $\frac{2}{10}$ **N**
question on first card

FIGURE 14.15 (a) Tree showing the probabilities on the first draw.

We have indicated the two possibilities for the first draw, namely, A or N. Notice how we place the probabilities for A and N on the branches of the tree. In Figure 14.15(b), we add more branches to show the possibilities for drawing the second card.

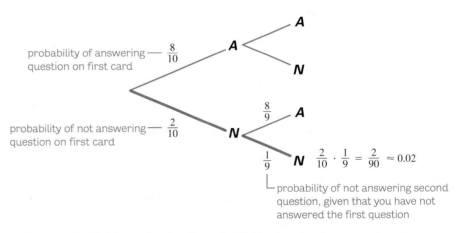

probability of answering ── $\frac{8}{10}$ **A**
question on first card

A
N

probability of not answering ── $\frac{2}{10}$ **N**
question on first card

$\frac{8}{9}$ **A**

$\frac{1}{9}$ **N** $\frac{2}{10} \cdot \frac{1}{9} = \frac{2}{90} \approx 0.02$

└ probability of not answering second
question, given that you have not
answered the first question

FIGURE 14.15 (b) Tree showing the probabilities when drawing two cards.

From Figure 14.15(b), you can see that the probability of not answering the first question is $P(N) = \dfrac{2}{10}$. If you did not answer the first question, then there is only one of the remaining nine cards with a question you can't answer. So the probability of not answering

*Recognize that the event N is just another name for A', the complement of A.

In Example 4, what is the probability that you can answer the first question, but not the second?

the second question is $\frac{1}{9}$. Thus, the probability that you can't answer the first question and also can't answer the second question is

$$P(N) \cdot P(N \mid N) = \frac{2}{10} \cdot \frac{1}{9} = \frac{2}{90} \approx 0.02.$$

probability of not ┘   └ probability of not answering second question,
answering first question   given that you did not answer first question   ✿ ⑩

The probability tree in Figure 14.15(b), describes four (non-equally-likely) outcomes: (A, A), (A, N), (N, A), (N, N). We can find the probability of each of these outcomes by multiplying the probabilities along the branch that corresponds to that outcome. In Example 3, we found the probability of (A, A) without drawing a tree. In Example 4, we found the probability of (N, N), by multiplying the probabilities along the bottom branch of the tree.

Dependent and Independent Events

✎ **KEY POINT**

Independent events have no effect on each other's probabilities.

As we have seen, knowing that a first event has occurred may affect the way we calculate the probability of a second event. Such was the case in Example 3 when we computed the probability of drawing a second card, given that a first card was drawn and not returned to the deck.

However, sometimes the occurrence of a first event has no effect whatsoever on the probability of the second event. This would be the case if you were drawing 2 cards with replacement from a standard 52-card deck. The probability of drawing a first king is $\frac{4}{52}$. If you return the king to the deck and draw again, the probability of the second king is also $\frac{4}{52}$. Therefore, drawing the first king would not affect the probability of drawing a second king. So you see that sometimes two events influence each other and other times they do not.

Math in Your Life

The Numbers Don't Lie—Or Do They?

- In an article in *Chance* magazine, Mary C. Meyer, a statistics professor at the University of Georgia, claims that a recent study shows the probability of your death in a automobile accident is slightly greater if your car has an airbag than if it doesn't.

- *The Ottawa Citizen* reports that Dennis Lindley, a professor of statistics at University College, London, has derived a formula based on probability that tells you the optimal age for you to settle down. His formula is $X + \frac{Y - X}{2.718}$, where X is the age at which you start looking for a spouse and Y is the age at which you expect to stop looking. According to Lindley's formula, if you start searching at age 16 and expect to stop looking at age 60, then

the best age for you to choose a mate is at age

$$16 + \frac{60 - 16}{2.718} \approx 32 \text{ years.}$$

- A third article from the Woods Hole Oceanographic Institute in Massachusetts states that commercial fishing is one of the least safe occupations. If you are a commercial fisherman, you are 16 times more likely to die on the job than a firefighter or a police officer.

Truly probability is all around us. Do you agree with the previous statements? Or do they go against your "common sense?" Do you feel that something is missing? As we have been emphasizing throughout this text, always remember that when someone builds a model, they decide what goes into the model and what is left out. Maybe the numbers don't lie—or, . . . maybe . . . sometimes they do.

> **DEFINITIONS** Events *E* and *F* are **independent** events if
> $$P(F \mid E) = P(F).$$
> If $P(F \mid E) \neq P(F)$, then *E* and *F* are **dependent**.

This definition says that if *E* and *F* are independent, then knowing that *E* has occurred does not influence the way we compute the probability of *F*.

PROBLEM SOLVING
The Analogies Principle

Recall that the Analogies Principle in Section 1.1 tells you that mathematical terminology, symbolism, and its equations are often based on real-life ideas. Although at times it may seem difficult, if you work hard to make the connection between the intuitive ideas and the mathematical formalism, you will be rewarded by your increased understanding of mathematics. If you understand how the everyday usage of the terms *independent* and *dependent* corresponds to their mathematical definitions, it will help you remember their meaning.

EXAMPLE 5   *Determining Whether Events Are Independent or Dependent*

Assume we roll a red and a green die. Are the events *F*, "a five shows on the red die," and *G*, "the total showing on the dice is greater than 10," independent or dependent?

SOLUTION: To answer this question, we must determine whether $P(G \mid F)$ and $P(G)$ are the same or different. There are three outcomes—(5, 6), (6, 5), and (6, 6)—that give a total greater than 10, so $P(G) = \dfrac{3}{36} = \dfrac{1}{12}$.

Now,

$$F = \{(5, 1), (5, 2), (5, 3), (5, 4), (5, 5), (5, 6)\}$$

and

$$G \cap F = \{(5, 6)\},$$

so

$$P(G \mid F) = \frac{P(G \cap F)}{P(F)} = \frac{1/36}{6/36} = \frac{1}{6}.$$

Because $P(G \mid F) \neq P(G)$, the events are dependent.

Now try Exercises 45 to 48. ❊ **11**

Quiz Yourself **11**

The situation is the same as in Example 5. Are the events *F*, "a five shows on the red die," and *O*, "an odd total shows on the dice," dependent or independent?

EXAMPLE 6   *Selecting a Dormitory Room*

Brianna is taking part in a lottery for a room in one of the new dormitories at her college. She is guaranteed a space, but she will have to draw a card randomly to determine exactly which room she will have. Each card has the name of one dormitory, *X*, *Y*, or *Z*, and also has a two-person room number or an apartment number. Thirty percent of the available spaces are in *X*, 50% in *Y*, and 20% of the spaces are in *Z*.

Half of the available spaces in X are in rooms, 40% of Y's spaces are in rooms, and 30% of the spaces in Z are in rooms.

a) Draw a probability tree to describe this situation.

b) Given that Brianna selects a card for dormitory Y, what is the probability that she will be assigned to an apartment?

c) What is the probability that Brianna will be assigned an apartment in one of the three dormitories?

SOLUTION:

a) In drawing the tree, we will think of Brianna's assignment happening in two stages. First, she is assigned a dormitory, and then she is assigned either a room or an apartment. We show the tree in Figure 14.16, which begins with three branches corresponding to the dormitories X, Y, and Z. Each of these three branches has two further branches representing the assignment of either a room or an apartment.

We wrote various probabilities along the tree's branches in Figure 14.16. If you look carefully at the lower branch, you see that the 0.20 is the probability that Brianna will be assigned to dormitory Z; thus, $P(Z) = 0.20$. The 0.70 is a conditional probability. Assuming the condition that she is assigned to dormitory Z, then there is a 0.70 chance that she will be assigned to an apartment. Symbolically, this means that $P(Apartment \mid Z) = 0.70$.

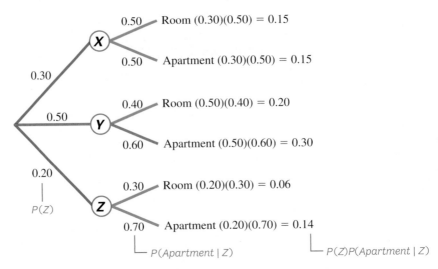

FIGURE 14.16 Probability tree for Brianna's room assignment.

The product $(0.20)(0.70) = 0.14$ is the probability that Brianna will be assigned to dormitory Z and also to an apartment.* We could have written this numeric equation more formally as

$$P(Z \cap Apartment) = P(Z) \cdot P(Apartment \mid Z),$$

which is the formula we gave you for computing the probability of the intersection of two events. In Quiz Yourself 12, we will ask you to interpret some more of these probabilities.

b) We can answer this question easily by looking at the branch that is highlighted in red in Figure 14.16. That branch shows that after Brianna has been assigned to dormitory Y, there is then a 0.60 chance that she will get an apartment. That is, $P(Apartment \mid Y) = 0.60$.

---

*It may seem that we have gotten away from thinking of events as subsets of the sample space, but we have not. The "Z" represents the set of cards that have dormitory Z written on them, and "*Apartment*" represents the set of all cards that have apartment written on them.

Quiz Yourself 12

a) Explain the meaning of the 0.30 on the top branch of the tree in Figure 14.16.

b) What is the probability that Brianna will be assigned a room, given that she is assigned to dormitory *Y*?

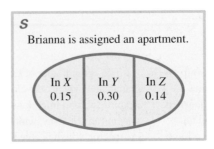

FIGURE 14.17 Probabilities that Brianna is assigned an apartment in dormitories *X*, *Y*, or *Z*.

c) Realize that if Brianna is assigned to an apartment, then she is in *exactly one* of dormitories *X*, *Y*, or *Z*. Figure 14.17 shows how to find the probability that she is assigned an apartment.

From Figure 14.17, we see that the event "Brianna is assigned an apartment" is the union of three mutually disjoint subevents with probabilities 0.15, 0.30, and 0.14. Therefore, the probability that Brianna is assigned an apartment is $0.15 + 0.30 + 0.14 = 0.59$.

Now try Exercises 57 to 60. ❋ 12

We will now return to the issue of drug testing that we mentioned in the chapter opener.

EXAMPLE 7 *Drug Testing**

Assume that you are working for a company that has a mandatory drug testing policy. It is estimated that 2% of the employees use a certain drug, and the company is giving a test that is 99% accurate in identifying users of this drug. What is the probability that if an employee is identified by this test as a drug user, the person is innocent?

SOLUTION: Let *D* be the event "the person is a drug user" and let *T* be the event "the person tests positive for the drug." We are asking, then, "if we are given that the person tests positive, what is the probability that the person does not use the drug?" Realize that the complement of *D*, namely *D'*, is the event "the person does not use the drug." So, we are asking for the conditional probability

$$P(D' \mid T).$$

Person does not use drug ⟍   ⟍ given that person tests positive.

The probability tree in Figure 14.18 will give you some insight into this problem.

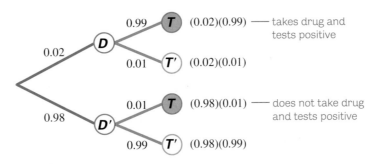

FIGURE 14.18 Tree showing probabilities for drug testing.

Recall that $P(D'|T) = \dfrac{P(D' \cap T)}{P(T)}$. Branches 1 and 3 in Figure 14.18 correspond to the drug test being positive. This means that $P(T) = (0.02)(0.99) + (0.98)(0.01)$. The event $D' \cap T$ corresponds to branch 3. Therefore, $P(D' \cap T) = (0.98)(0.01)$. From this, we see that the probability that an innocent person will test positive for the drug is

$$P(D'|T) = \frac{P(D' \cap T)}{P(T)} = \frac{(0.98)(0.01)}{(0.02)(0.99) + (0.98)(0.01)} = \frac{0.0098}{.0296} \approx 0.331.$$

In other words, if a person tests positive, there is a roughly $\frac{1}{3}$ chance that the person is innocent!

Now try Exercises 71 and 72. ❋

---

*This example is based on an article that can be found at www.intuitor.com/statistics/BadTestResults.html.

✸ ✸ ✸ HIGHLIGHT

Expert Systems

Whether playing chess, doing mathematics, making soup, or practicing law, what makes an expert different from the rest of us? Scientists studying this question have learned that organizing knowledge effectively makes a person an expert. Researchers in the area of artificial intelligence have developed complex computer programs called expert systems that rely heavily on conditional probability to draw conclusions from an organized body of information.

The "grandfather of expert systems" is an ingenious program called Mycin, developed at Stanford University, which diagnoses blood infections. Some other well-known expert systems are Hearsay, which understands spoken language; Prospector, which is able to predict the location of valuable mineral resources; and Internist, which can diagnose many internal diseases. With ongoing research in this area, expert systems continue to be a major application of probability theory.

Exercises 14.3

Looking Back*

These exercises follow the general outline of the topics presented in this section and will give you a good overview of the material that you have just studied.

1. In Example 1, what were we computing when we found $P(O \mid G)$?

2. We blanked out all columns in Table 14.7, except one. Why did we do this?

3. In Example 5, how did we determine that the events F and G were independent?

4. What is an expert system?

Sharpening Your Skills

In many of these exercises, you may find it helpful to draw a tree diagram before computing the probabilities.

In Exercises 5–8, assume that we are rolling two fair dice. First compute P(F) and then P(F | E). Explain why you would expect the probability of F to change as it did when we added the condition that E had occurred.

5. *E*—an odd total shows on the dice.

    F—the total is seven.

6. *E*—an even total shows on the dice.

    F—the total is four.

7. *E*—a three shows on at least one of the dice.

    F—the total is less than five.

8. *E*—a two shows on at least one of the dice.

    F—the total is greater than five.

In Exercises 9–12, we are drawing a single card from a standard 52-card deck. Find each probability.

9. *P*(heart | red)

10. *P*(king | face card)

11. *P*(seven | nonface card)

12. *P*(even-numbered card | nonface card)

You are to randomly pick one disk from a bag that contains the disks shown below. Find each of the following probabilities. For example, P(heart | yellow) means you are to find the probability of a heart being on the disk, given that the disk is yellow.

13. *P*(heart | yellow)

14. *P*(pink | smiley face)

15. *P*(yellow | heart)

16. *P*(heart | blue)

17. *P*(heart | pink)

18. *P*(smiley face | blue)

Playing a store's discount game. *To stimulate business, a department store has introduced a game called "Register Roulette." To play, a customer selects a first purchase and a second purchase to qualify for a discount. The customer then reaches into a container and randomly selects two game tokens. There are 50 green tokens that qualify a purchase for a 10% discount, 10 blue tokens that result in a 20% discount, and 2 red tokens that give a 50% discount.*

In Exercises 19–22, assume that you are randomly selecting two of these tokens, without replacement. Compute P(F | E).

19. *E*—you select a red token first.

    F—the second token is green.

20. *E*—you select a red token first.

    F—the second token is red.

21. *E*—you select a nonred token first.

    F—the second token is red.

22. *E*—you select a nonred token first.

    F—the second token is nonred.

According to U.S. government statistics, mononucleosis (mono) is four times more common among college students than the rest of the population. Blood tests for the disease are not 100% accurate. Assume that

---

*Before doing these exercises, you may find it useful to review the note *How to Succeed at Mathematics* on page xix.

the following table of data was obtained regarding students who came to your school's health center complaining of tiredness, a sore throat, and slight fever. Use these data to answer Exercises 23–26.

	Has Mono	Doesn't Have Mono	Totals
Positive Blood Test	72	4	76
Negative Blood Test	8	56	64
Totals	80	60	140

If a student is selected from this group, what is the probability of each of the following?

23. The student has mono, given that the test is positive.

24. The student does not have mono, given that the test is positive.

25. The test is positive, given that the student has mono.

26. The test is negative, given that the student does not have mono.

Probability and drawing cards. *In Exercises 27–32, assume that we draw 2 cards from a standard 52-card deck. Find the desired probabilities.*

a) *First assume that the cards are drawn without replacement.*

b) *Next assume that the cards are drawn with replacement.*

27. The probability that we draw two jacks

28. The probability that we draw two hearts

29. The probability that we draw a face card followed by a non-face card

30. The probability that we draw a heart followed by a spade

31. The probability that we draw a jack and a king

32. The probability that we draw a heart and a spade

33. We are drawing 2 cards without replacement from a standard 52-card deck. Find the probability that we draw at least one face card. (*Hint:* Consider the complement.)

34. We are drawing 2 cards with replacement from a standard 52-card deck. Find the probability that we draw at least one heart. (*Hint:* Consider the complement.)

35. We draw 3 cards without replacement from a standard 52-card deck. Find the probability of drawing three face cards.

36. We draw 3 cards without replacement from a standard 52-card deck. Find the probability of drawing three spades.

37. We draw 3 cards without replacement from a standard 52-card deck. Find the probability of drawing exactly two hearts.

38. We draw 3 cards without replacement from a standard 52-card deck. Find the probability of drawing exactly two kings.

39. We roll a pair of dice three times. Find the probability that a total of five is rolled each time.

40. We roll a pair of dice three times. Find the probability that a total of five is rolled exactly twice.

Assume that the Political Action Club on your campus has gathered the following information about the 2008 Democratic presidential election. Use this table to answer Exercises 41–44.

	Democrat	Republican	Independent	Totals
Preferred Obama	15	68	15	98
Preferred McCain	105	12	25	142
Totals	120	80	40	240

If a student is selected from this group, what is the probability of each of the following?

41. The student is a Democrat, given that he preferred McCain.

42. The student is an Independent, given that he preferred McCain.

43. The student preferred McCain, given that the student is an Independent.

44. The student preferred Obama, given that the student is a Republican.

Probability and rolling dice. *In Exercises 45–48, we are rolling a pair of fair dice. For each pair of events, determine whether E and F are dependent.*

45. *E*—the total showing is greater than nine.

   F—the total showing is even.

46. *E*—at least one 2 shows on the dice.

   F—the total showing is less than six.

47. *E*—a three shows on the first die.

   F—the total showing is even.

48. *E*—the total showing is odd.

   F—the total showing is greater than four.

Applying What You've Learned

Probability and starting salaries. *The following table from Example 2 relates starting salaries of college graduates and their majors. In Exercises 49–52, find* $P(F \mid E)$.

Major	$30,000 or Below	$30,001 to $35,000	$35,001 to $40,000	$40,001 to $45,000	Above $45,000	Totals (%)
Liberal arts	6*	10	9	1	1	27
Science	2	4	10	2	2	20
Social sciences	3	6	7	1	1	18
Health fields	1	1	8	3	1	14
Technology	0	2	7	8	4	21
Totals (%)	12	23	41	15	9	100

*These numbers are percentages.

49. *E*—a person majored in social science.

  F—a person received a starting salary between $30,001 and $35,000.

50. *E*—a person majored in science.

  F—a person received a starting salary between $35,001 and $40,000.

51. *E*—a person received a starting salary between $40,001 and $45,000.

  F—a person majored in technology.

52. *E*—a person received a starting salary between $30,001 and $35,000.

  F—a person majored in liberal arts.

Assume that a softball player has a .300 batting average—to keep this simple we'll assume that this means the player has 0.30 probability of getting a hit in each at bat. Use this information to answer Exercises 53–56. Assume that the player bats four times.

53. What is the probability that she gets a hit only in her first at bat?

54. What is the probability that she gets exactly one hit? (Realize that there are four ways to do this.)

55. What is the probability that she gets exactly two hits?

56. What is the probability that she gets at least one hit? (*Hint:* Think complement.)

Selecting a dormitory room. *Exercises 57–60 refer to the tree diagram that we drew in Figure 14.16 of Example 6.*

57. What is the meaning of the number 0.40 on the top middle branch of Figure 14.16?

58. What is $P(Room \cap Y)$?

59. What is $P(Apartment \mid X)$?

60. If Brianna gets an apartment, what is the probability that she is in dorm *Z*?

Testing a cold medication. *Imagine that you are taking part in a study to test a new cold medicine. Although you don't know exactly what drug you are taking, the probability that it is drug A is 10%, that it is drug B is 20%, and that it is drug C, 70%. From past clinical trials, the probabilities that these drugs will improve your condition are: A (30%), B (60%), and C (70%).*

61. Draw a tree to illustrate this drug trial situation.

62. What is the probability that you will improve given that you are taking drug B?

63. What is the probability that you will improve?

64. If you improve, what is the probability that you are taking drug B?

65. **Probability and exam questions.** Assume that either Professor Ansah or Professor Brunich has constructed the comprehensive exam that you must pass for graduation. Because each professor has extremely different views, it would be useful for you to know who has written the exam questions so that you can slant your answers accordingly. Assume that there is a 60% chance that Ansah wrote the exam. Ansah asks a question about international relations 30% of the time and Brunich asks a similar question 75% of the time. If there is a question on the exam regarding international relations, what is the probability that Ansah wrote the exam?

66. **Probability and exam questions.** Assume now that a third professor, Professor Ubaru, writes the exam 20% of the time, Brunich 30% of the time, and Ansah the rest. Ubaru asks a question about international relations 40% of the time, Brunich 35% of the time, and Ansah 25% of the time. If there is an international relations question on the exam, what is the probability that Brunich did not write the exam?

Product reliability. *You want to purchase a DVD drive for your laptop computer. Assume that 65% of the drives are made outside the United States. Of the U.S.-made drives, 4% are defective; of the foreign-made drives, 6% are defective. Determine each probability rounded to three decimal places.*

67. The probability that the drive you purchase is U.S. made and is not defective

68. The probability that the drive you purchase is foreign made and is defective

69. If your drive is defective, the probability that it is foreign made

70. If your drive is defective, the probability that it is made in the United States

Drug testing. *In Exercises 71 and 72, do computations similar to those in Example 7 using this revised information. Assume that 4% of the employees use the drug and that the test correctly identifies a drug user 98% of the time. Also assume that the test identifies a nonuser as a drug user 3% of the time.*

71. If an employee tests positive, what is the probability that the person is innocent?

72. If an employee tests negative, what is the probability that the person is a user?

Communicating Mathematics

73. If you know the conditional probability formula $P(F \mid E) = \dfrac{P(E \cap F)}{P(E)}$, how do you find the probability formula for $P(E \cap F)$?

74. We say that events *E* and *F* are independent if $P(F \mid E) = P(F)$. Give an intuitive explanation of what this equation is saying.

75. In what ways are the special and general rules for computing conditional events similar? How are they different?

76. If *A* and *B* are events, can $P(B \mid A) = P(B)$? When?

77. Explain how the formal definition of dependent and independent events corresponds to your intuitive understanding of these words in English.

78. The formula $P(E \cap F) = P(E) \cdot P(F)$ is, in general, not true. When can we use this formula instead of the correct formula, $P(E \cap F) = P(E) \cdot P(F \mid E)$?

Using Technology to Investigate Mathematics

79. Search the Internet for "expert systems." Report on an interesting site that you find.

80. Search the Internet for "amazing applications of probability and statistics." Report on an interesting site that you find.

For Extra Credit

The birthday problem. *A surprising result that appears in many elementary discussions on probability is called the birthday problem. The question simply stated is this: "If we poll a certain number of people, what is the probability that at least two of those people were born on the same day of the year?" For example, it may be that in the survey two people were both born on March 29.*

To solve this problem, we will use the formula for computing the probability of a complement of an event. It is clear that

P(*duplication of some birthdays*) = 1 − P(*no duplications*)

To illustrate this, assume that we have three people. To have no duplications, the second person must have a birthday that is different

from that of the first person, and the third person must have a birthday that is different from the first two. The probability that the second birthday is different from the first is $\frac{364}{365}$. The probability that the third person has a birthday different from the first two is $\frac{363}{365}$.

Therefore, for three people,

P(*duplication of birthdays*) = 1 − P(*no duplication of birthdays*)

$$= 1 - \left(\frac{364}{365}\right)\left(\frac{363}{365}\right) = 0.0082.$$

In Exercises 81–84, we will look at several other cases.

81. Assume that we have 10 people. Find the probability that at least 2 of the 10 people were born on the same day of the year.

82. Repeat Exercise 81 for 20 people.

83. Find the smallest number of people such that the probability of two of them being born on the same day of the year is greater than 0.50.

84. Conduct an experiment by surveying groups of various sizes* to see how your surveys conform to the predicted probabilities. If five people are working in a group, each person can survey four groups of 20, 30, 40, and so on to generate a reasonable amount of data.

---

*Instead of surveying people, you can write the numbers 1 to 365 on pieces of paper or cardboard and draw these from a box *with replacement*. Be sure to shake the container well before each draw. In this way you can quickly simulate interviews of groups of 20, 30, or 40 people, with the purpose of determining how close your experimental results conform to the predicted probabilities.

14.4 Expected Value

Objectives

1. Understand the meaning of expected value.
2. Calculate the expected value of lotteries and games of chance.
3. Use expected value to solve applied problems.

Life and Health Insurers' Profits Skyrocket 213% . . .*

How do insurance companies make so much money? When you buy car insurance, you are playing a sort of mathematical game with the insurance company. You are betting that you are going to have an accident—the insurance company is betting that you won't. Similarly, with health insurance, you are betting that you will be sick—the insurance company is betting that you will stay well. With life insurance, you are betting that, . . . well, . . . you get the idea.

Expected Value

Casinos also amass their vast profits by relying on this same mathematical theory—called **expected value**, which we will introduce to you in this section. Expected value uses probability to compare alternatives to help us make decisions.

Because of an increase in theft on campus, your school now offers personal property insurance that covers items such as laptops, iPods, cell phones, and even books. Although

---

*According to a report by Weiss Ratings, Inc., a provider of independent ratings of financial institutions.

the premium seems a little high, the insurance will fully replace any lost or stolen items. Our first example will help you get an idea of how probability can help you understand situations such as this.

EXAMPLE 1 *Evaluating an Insurance Policy*

Suppose that you want to insure a laptop computer, an iPhone, a trail bike, and your textbooks. Table 14.8 lists the values of these items and the probabilities that these items will be stolen over the next year.

a) Predict what the insurance company can expect to pay in claims on your policy.

b) Is $100 a fair premium for this policy?

Item	Value	Probability of Being Stolen	Expected Payout by Insurance Company
Laptop	$2,000	0.02	0.02($2,000) = $40
iPhone	$400	0.03	0.03($400) = $12
Trail bike	$600	0.01	0.01($600) = $6
Textbooks	$800	0.04	0.04($800) = $32

TABLE 14.8 Value of personal items and the probability of their being stolen.

SOLUTION:

a) From Table 14.8 the company has a 2% chance of having to pay you $2,000, or, another way to look at this is the company expects to lose on average 0.02($2,000) = $40 by insuring your computer. Similarly, the expected loss on insuring your iPhone is 0.03($400) = $12. To estimate, on average, what it would cost the company to insure all four items, we compute the following sum:

$$0.02(\$2,000) + 0.03(\$400) + 0.01(\$600) + 0.04(\$800) = \$90.$$

probability of computer being stolen | cost of computer | probability of iPhone being stolen | cost of iPhone | probability of books being stolen | cost of books | probability of bike being stolen | cost of bike

The $90 represents, *on average*, what the company can expect to pay out on a policy such as yours.

b) The $90 in part a) is telling us that if the insurance company were to write one million policies like this, it would expect to pay 1,000,000 × ($90) = $90,000,000 in claims. If the company is to make a profit, it must charge more than $90 as a premium, so it seems like a $100 premium is reasonable. ❋ **13**

The amount of $90 we found in Example 1 is called the *expected value* of the claims paid by the insurance company. We will now give the formal definition of this notion.

> **DEFINITION** Assume that an experiment has outcomes numbered 1 to n with probabilities $P_1, P_2, P_3, \ldots, P_n$. Assume that each outcome has a numerical value associated with it and these are labeled $V_1, V_2, V_3, \ldots, V_n$. The **expected value** of the experiment is
>
> $$(P_1 \cdot V_1) + (P_2 \cdot V_2) + (P_3 \cdot V_3) + \cdots + (P_n \cdot V_n).$$

In Example 1, the probabilities were $P_1 = 0.02$, $P_2 = 0.03$, $P_3 = 0.01$, and $P_4 = 0.04$. The values were $V_1 = 2,000$, $V_2 = 400$, $V_3 = 600$, and $V_4 = 800$.

Quiz Yourself **13**

In Example 1, if you were to drop coverage on your iPhone and add coverage on your saxophone that cost $1,400, what would the insurance company now expect to pay out in claims if the probability of the saxophone being stolen is 4% and the probability of your books being stolen is reduced to 3%?

Some Good Advice

Pay careful attention to what notation tells you to do in performing a calculation. In calculating expected value, you are told to *first* multiply the probability of each outcome by its value and *then* add these products together.

Expected Value of Games of Chance

EXAMPLE 2 *Computing Expected Value When Flipping Coins*

What is the number of heads we can expect when we flip four fair coins?

SOLUTION: Recall that there are 16 ways to flip four coins. We will consider the outcomes for this experiment to be the different numbers of heads that could arise. Of course, these outcomes are not equally likely, as we indicate in Table 14.9. If you don't see this at first, you could draw a tree to show the 16 possible ways that four coins can be flipped. You would find that 1 of the 16 branches corresponds to no heads, 4 of the 16 branches would represent flipping exactly one head, and 6 of the 16 branches would represent flipping exactly two heads, and so on.

We calculate the expected number of heads by first multiplying each outcome by its probability and then adding these products, as follows:

$$\left(\frac{1}{16}\cdot 0\right)+\left(\frac{4}{16}\cdot 1\right)+\left(\frac{6}{16}\cdot 2\right)+\left(\frac{4}{16}\cdot 3\right)+\left(\frac{1}{16}\cdot 4\right)=\frac{32}{16}=2$$

Thus we can expect to flip two heads when we flip four coins, which corresponds to our intuition. ✻

We can use the notion of expected value to predict the likelihood of winning (or more likely losing) at games of chance such as blackjack, roulette, and even lotteries.

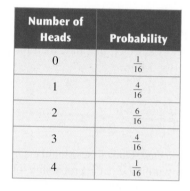

Number of Heads	Probability
0	$\frac{1}{16}$
1	$\frac{4}{16}$
2	$\frac{6}{16}$
3	$\frac{4}{16}$
4	$\frac{1}{16}$

TABLE 14.9 Probabilities of obtaining a number of heads when flipping four coins.

EXAMPLE 3 *The Expected Value of a Roulette Wheel*

Although there are many ways to bet on the 38 numbers of a roulette wheel,* one simple betting scheme is to place a bet, let's say $1, on a single number. In this case, the casino pays you $35 (you also keep your $1 bet) if your number comes up and otherwise you lose the $1. What is the expected value of this bet?

SOLUTION: We can think of this betting scheme as an experiment with two outcomes:

1. Your number comes up and the value to you is +$35.
2. Your number doesn't come up and the value to you is −$1.

Because there are 38 equally likely numbers that can occur, the probability of the first outcome is $\frac{1}{38}$ and the probability of the second is $\frac{37}{38}$. The expected value of this bet is therefore

$$\left(\frac{1}{38}\cdot 35\right)+\left(\frac{37}{38}\cdot(-1)\right)=\frac{35-37}{38}=\frac{-2}{38}=-\frac{1}{19}=-0.0526.$$

*See Example 8 in Section 14.1 for a description of a roulette wheel.

This amount means that, on the average, the casino expects you to lose slightly more than 5 cents for every dollar you bet.

Now try Exercises 3 to 8. ✳

The roulette wheel in Example 3 is an example of an unfair game.

> **DEFINITIONS** If a game has an expected value of 0, then the game is called **fair**. A game in which the expected value is not 0 is called an **unfair game**.

Although it would seem that you would not want to play an unfair game, in order for a casino or a state lottery to make a profit, the game has to be favored against the player.

EXAMPLE 4 *Determining the Fair Price of a Lottery Ticket*

Assume that it costs $1 to play a state's daily number. The player chooses a three-digit number between 000 and 999, inclusive, and if the number is selected that day, then the player wins $500 (this means the player's profit is $500 − $1 = $499.)

a) What is the expected value of this game?

b) What should the price of a ticket be in order to make this game fair?

Outcome	Value	Probability
You win	$499	$\dfrac{1}{1,000}$
You lose	−$1	$\dfrac{999}{1,000}$

TABLE 14.10 Values and probabilities associated with playing the daily number.

SOLUTION:

a) There are 1,000 possible numbers that can be selected. One of these numbers is in your favor and the other 999 are against your winning. So, the probability of you winning is $\dfrac{1}{1,000}$ and the probability of you losing is $\dfrac{999}{1000}$. We summarize the values for this game with their associated probabilities in Table 14.10. The expected value of this game is therefore

$$\left(\underbrace{\frac{1}{1,000}}_{\text{probability of winning}} \cdot \underbrace{499}_{\text{amount won}}\right) + \left(\underbrace{\frac{999}{1,000}}_{\text{probability of losing}} \cdot \underbrace{(-1)}_{\text{amount lost}}\right) = \frac{499 - 999}{1,000} = \frac{-500}{1,000} = -0.50.$$

This means that the player, on average, can expect to lose 50 cents per game. Notice that playing this lottery is 10 times as bad as playing a single number in roulette.

b) Let x be the price of a ticket for the lottery to be fair. Then if you win, your profit will be $500 - x$ and if you lose, your loss will be x. With this in mind, we will recalculate the expected value to get

$$\left(\frac{1}{1,000} \cdot (500 - x)\right) + \left(\frac{999}{1,000} \cdot (-x)\right) = \frac{(500 - x) - 999x}{1,000} = \frac{500 - 1,000x}{1,000}.$$

We want the game to be fair, so we will set this expected value equal to zero and solve for x, as follows:

$$\frac{500 - 1,000x}{1,000} = 0$$

$$\cancel{1,000} \cdot \frac{500 - 1,000x}{\cancel{1,000}} = 1,000 \cdot 0 = 0 \qquad \text{Multiply both sides of the equation by 1,000.}$$

$$500 - 1,000x = 0 \qquad \text{Cancel 1,000 and simplify.}$$

$$500 = 1,000x \qquad \text{Add 1,000x to both sides.}$$

$$\frac{500}{1,000} = \frac{\cancel{1,000}x}{\cancel{1,000}} = x \qquad \text{Divide both sides by 1,000.}$$

$$x = \frac{500}{1,000} = 0.50 \qquad \text{Simplify.}$$

HISTORICAL HIGHLIGHT ✳ ✳ ✳

The History of Lotteries

Lotteries have existed since ancient times. The Roman emperor Nero gave slaves or villas as door prizes to guests attending his banquets, and Augustus Caesar used public lotteries to raise funds to repair Rome.

The first public lottery paying money prizes began in Florence, Italy, in the early 1500s; when Italy became consolidated in 1870, this lottery evolved into the Italian National Lottery. In this lottery, five numbers are drawn from 1 to 90. A winner who guesses all five numbers is paid at a ratio of 1,000,000 to 1. The number of possible ways to choose these five numbers is $C(90, 5) = 43,949,268$. Thus, as with most lotteries, these odds make the lottery a very good bet for the state and a poor one for the ordinary citizen.

Lotteries also played an important role in the early history of the United States. In 1612, King James I used lotteries to finance the Virginia Company to send colonists to the New World. Benjamin Franklin obtained money to buy cannons to defend Philadelphia, and George Washington built roads through the Cumberland mountains by raising money through another lottery he conducted. In fact, in 1776, the Continental Congress used a lottery to raise $10 million to finance the American Revolution.

This means that 50 cents would be a fair price for a ticket to play this lottery. Of course, such a lottery would make no money for the state, which is why most states charge $1 to play the game.

Now try Exercises 9 to 12. ✳

Other Applications of Expected Value

Calculating expected value can help you decide what is the best strategy for answering questions on standardized tests such as the GMATs.

EXAMPLE 5 *Expected Value and Standardized Tests*

A student is taking a standardized test consisting of multiple-choice questions, each of which has five choices. The test taker earns 1 point for each correct answer; $\frac{1}{3}$ point is subtracted for each incorrect answer. Questions left blank neither receive nor lose points.

a) Find the expected value of randomly guessing an answer to a question. Interpret the meaning of this result for the student.

b) If you can eliminate one of the choices, is it wise to guess in this situation?

SOLUTION:

a) Because there are five choices, you have a probability of $\frac{1}{5}$ of guessing the correct result, and the value of this is +1 point. There is a $\frac{4}{5}$ probability of an incorrect guess, with an associated value of $-\frac{1}{3}$ point. The expected value is therefore

$$\left(\frac{1}{5} \cdot 1\right) + \left(\frac{4}{5} \cdot \left(-\frac{1}{3}\right)\right) = \frac{1}{5} + \frac{-4}{15} = \frac{3}{15} - \frac{4}{15} = -\frac{1}{15}.$$

Thus, you will be penalized for guessing and should not do so.

b) If you eliminate one of the choices and choose randomly from the remaining four choices, the probability of being correct is $\frac{1}{4}$ with a value of $+1$ point; the probability of being incorrect is $\frac{3}{4}$ with a value of $-\frac{1}{3}$. The expected value is now

$$\left(\frac{1}{4} \cdot 1\right) + \left(\frac{3}{4} \cdot \left(-\frac{1}{3}\right)\right) = \frac{1}{4} + \frac{-1}{4} = 0.$$

You now neither benefit nor are penalized by guessing.

Now try Exercises 19 to 22. ❄ **14**

Quiz Yourself **14**

Calculate the expected value as in Example 5(b), but now assume that the student can eliminate two of the choices. Interpret this result.

Businesses have to be careful when ordering inventory. If they order too much, they will be stuck with a surplus and might take a loss. On the other hand, if they do not order enough, then they will have to turn customers away, losing profits.

EXAMPLE 6   *Using Expected Value in Business*

Cher, the manager of the U2 Coffee Shoppe, is deciding on how many of Bono's Bagels to order for tomorrow. According to her records, for the past 10 days the demand has been as follows:

Demand for Bagels	40	30
Number of Days with These Sales	4	6

She buys bagels for $1.45 each and sells them for $1.85. Unsold bagels are discarded. Find her expected value for her profit or loss if she orders 40 bagels for tomorrow morning.

SOLUTION: We can describe the expected value in words as

$P(\text{demand is } 40) \times (\text{the profit or loss if demand is } 40)$
$\qquad\qquad + P(\text{demand is } 30) \times (\text{the profit or loss if demand is } 30).$

For 4 of the last 10 days the demand has been for 40 bagels, so $P(\text{demand is } 40) = \frac{4}{10} = 0.4$. Similarly, $P(\text{demand is } 30) = \frac{6}{10} = 0.6$.

We will now consider her potential profit or loss. If the demand is for 40 bagels, she will sell all of the bagels and make $40(\$1.85 - \$1.45) = 40(\$0.40) = \$16.00$ profit. If the demand is for 30 bagels, she will make $30(\$0.40) = \$12.00$ profit on the sold bagels, but will lose $10(\$1.45) = \$14.50$ on the 10 unsold bagels. So, she will lose $2.50. The following table summarizes our discussion so far if she orders 40 bagels.

Demand	Probability	Profit or Loss
40	0.4	$16.00
30	0.6	−$2.50

Quiz Yourself **15**

Redo Example 6 assuming that Cher orders 30 bagels.

Thus, the expected value in profit or loss for ordering 40 bagels is

$$(0.40)(16) + (0.60)(-2.50) = +6.40 + (-1.50) = 4.90.$$

So she can expect a profit of $4.90 if she orders 40 bagels. ❄ **15**

Exercises 14.4

Looking Back*

These exercises follow the general outline of the topics presented in this section and will give you a good overview of the material that you have just studied.

1. How did we arrive at the term 0.02($2,000) in the equation that we wrote for the expected value in Example 1?

2. What did Examples 3 and 4 show us about the expected values of playing roulette versus playing the daily number?

Sharpening Your Skills

In Exercises 3 and 4, you are playing a game in which a single die is rolled. Calculate your expected value for each game. Is the game fair? (Assume that there is no cost to play the game.)

3. If an odd number comes up, you win the number of dollars showing on the die. If an even number comes up, you lose the number of dollars showing on the die.

4. You are playing a game in which a single die is rolled. If a four or five comes up, you win $2; otherwise, you lose $1.

In Exercises 5 and 6, you pay $1 to play a game in which a pair of fair dice are rolled. Calculate your expected value for the game. (Remember to subtract the cost of playing the game from your winnings.) Calculate the price of the game to make the game fair.

5. If a six, seven, or eight comes up, you win $5; if a two or 12 comes up, you win $3; otherwise, you lose the dollar you paid to play the game.

6. If a total lower than five comes up, you win $5; if a total greater than nine comes up, you win $2; otherwise, you lose the dollar you paid to play the game.

In Exercises 7 and 8, a card is drawn from a standard 52-card deck. Calculate your expected value for each game. You pay $5 to play the game, which must be subtracted from your winnings. Calculate the price of the game to make the game fair.

7. If a heart is drawn, you win $10; otherwise, you lose your $5.

8. If a face card is drawn, you win $20; otherwise, you lose your $5.

In Exercises 9–12, first calculate the expected value of the lottery. Determine whether the lottery is a fair game. If the game is not fair, determine a price for playing the game that would make it fair.

9. The Daily Number lottery costs $1 to play. You must pick three digits in order from 0 to 9 and duplicates are allowed. If you win, the prize is $600.

10. The Big Four lottery costs $1 to play. You must pick four digits in order from 0 to 9 and duplicates are allowed. If you win, the prize is $2,000.

11. Five hundred chances are sold at $5 apiece for a raffle. There is a grand prize of $500, two second prizes of $250, and five third prizes of $100.

12. One thousand chances are sold at $2 apiece for a raffle. There is a grand prize of $300, two second prizes of $100, and five third prizes of $25.

13. Grace Adler is planning to buy a franchise from Home Deco to sell decorations for the home. The table below shows average weekly profits, rounded to the nearest hundred, for a number of the current franchises. If she were to buy a franchise, what would her expected weekly profit be?

Average Weekly Profit	Number Who Earned This
$100	4
$200	8
$300	13
$400	21
$500	3
$600	1

14. For the past several years, the Metrodelphia Fire Department has been keeping track of the number of fire hydrants that have been opened illegally daily during heat waves. These data (rounded to the nearest ten) are given in the table below. Use this information to calculate how many hydrants the department should expect to be opened per day during the upcoming heat wave.

Hydrants Opened	Days
20	13
30	11
40	15
50	11
60	9
70	1

Applying What You've Learned

In Exercises 15–18, we describe several ways to bet on a roulette wheel. Calculate the expected value of each bet. We show a portion of a layout for betting on roulette in the diagram on page 702. When we say that a bet pays "k to 1," we mean that if a player wins, the player wins k dollars as well as keeping his or her bet. When the player loses, he or she loses $1. Recall that there are 38 numbers on a roulette wheel.

15. A player can "bet on a line" by placing a chip at location A in the figure. By placing the chip at A, the player is betting on 1, 2, 3, 0, and 00. This bet pays 6 to 1.

*Before doing these exercises, you may find it useful to review the note *How to Succeed at Mathematics* on page xix.

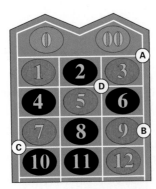

Figure for Exercises 15–18

16. A player can "bet on a square" by placing a chip at the intersection of two lines, as at location D. By placing a chip at D, the player is betting on 2, 3, 5, and 6. This bet pays 8 to 1.

17. A player can "bet on a street" by placing a chip on the table at location B. The player is now betting on 7, 8, and 9. This bet pays 8 to 1.

18. Another way to "bet on a line" is to place a chip at location C. By placing a chip at location C, the player is betting on 7, 8, 9, 10, 11, and 12. This bet pays 5 to 1.

In Exercises 19–22, a student is taking the GRE consisting of several multiple-choice questions. One point is awarded for each correct answer. Questions left blank neither receive nor lose points.

19. If there are four options for each question and the student is penalized $\frac{1}{4}$ point for each wrong answer, is it in the student's best interest to guess? Explain.

20. If there are three options for each question and the student is penalized $\frac{1}{3}$ point for each wrong answer, is it in the student's best interest to guess? Explain.

21. If there are five options for each question and the student is penalized $\frac{1}{2}$ point for each wrong answer, how many options must the student be able to rule out before the expected value of guessing is zero?

22. If there are four options for each question and the student is penalized $\frac{1}{2}$ point for each wrong answer, how many options must the student be able to rule out before the expected value of guessing is zero?

23. Assume that the probability of a 25-year-old male living to age 26, based on mortality tables, is 0.98. If a $1,000 one-year term life insurance policy on a 25-year-old male costs $27.50, what is its expected value?

24. Assume that the probability of a 22-year-old female living to age 23, based on mortality tables, is 0.995. If a $1,000 one-year term life insurance policy on a 22-year-old female costs $20.50, what is its expected value?

25. Your insurance company has a policy to insure personal property. Assume you have a laptop computer worth $2,200, and there is a 2% chance that the laptop will be lost or stolen during the next year. What would be a fair premium for the insurance? (We are assuming that the insurance company earns no profit.)

26. Assume that you have a used car worth $6,500 and you wish to insure it for full replacement value if it is stolen. If there is a 1% chance that the car will be stolen, what would be a fair premium for this insurance? (We are assuming that the insurance company earns no profit.)

27. A company estimates that it has a 60% chance of being successful in bidding on a $50,000 contract. If it costs $5,000 in consultant fees to prepare the bid, what is the expected gain or loss for the company if it decides to bid on this contract?

28. In Exercise 27, suppose that the company believes that it has a 40% chance to obtain a contract for $35,000. If it will cost $2,000 to prepare the bid, what is the expected gain or loss for the company if it decides to bid on this contract?

Communicating Mathematics

29. Explain in your own words the definition of expected value.

30. Explain in your own words how we determined the fair price of the game in Example 4. Why would the state not set this as the price to play the daily number?

Using Technology to Investigate Mathematics

31. Search the Internet for "the mathematics of lotteries." You might be surprised by the number of sites that you find selling software that can predict winning numbers. What is your reaction to these claims? Report on your findings.

32. Search the Internet for "expected value applets." Run some of these applets and report on your findings.

For Extra Credit

33. Nell's Bagels & Stuff, a local coffee shop, sells coffee, bagels, magazines, and newspapers. Nell has gathered information for the past 20 days regarding the demand for bagels. We list this information in the following table.

Demand for Bagels Sold	150	140	130	120
Number of Days with These Sales	3	6	5	6

Nell wants to use expected value to compute her best strategy for ordering bagels for the next week. She intends to order the same number each day and must order in multiples of 10; therefore, she will order either 120, 130, 140, or 150 bagels. She buys the bagels for 65 cents each and sells them for 90 cents.

a. What is Nell's expected daily profit if she orders 130 bagels per day? (*Hint:* First compute the profit Nell will earn if she can sell 150, 140, 130, and 120 bagels.)

b. What is Nell's expected daily profit if she orders 140 bagels per day?

34. Mike sells the *Town Crier*, a local paper, at his newsstand. Over the past 2 weeks, he has sold the following number of copies.

Number of Copies Sold	90	85	80	75
Number of Days with These Sales	2	3	4	1

Each copy of the paper costs him 40 cents, and he sells it for 60 cents. Assume that these data will be consistent in the future.

a. What is Mike's expected daily profit if he orders 80 copies per day? (*Hint:* First compute the profit Mike will earn if he can sell 90, 85, 80, and 75 papers.)

b. What is Mike's expected daily profit if he orders 85 copies per day?

35. a. Calculate the expected total if you roll a pair of standard dice.

b. Unusual dice, called *Sicherman dice*, are numbered as follows.

Red: 1, 2, 2, 3, 3, 4   Green: 1, 3, 4, 5, 6, 8

Calculate the expected total if you roll a pair of these dice.

36. Suppose we have two pairs of dice (these are called *Efron's dice*) numbered as follows.

Pair One: Red: 2, 2, 2, 2, 6, 6   Green: 5, 5, 6, 6, 6, 6

Pair Two: Red: 1, 1, 1, 5, 5, 5   Green: 4, 4, 4, 4, 12, 12

If you were to play a game in which the highest total wins when you roll, what is the better pair to play with?

37. In playing a lottery, a person might buy several chances in order to improve the likelihood (probability) of winning. Discuss whether this is the case. Does buying several chances change your expected value of the game? Explain.

38. A lottery in which you must choose 6 numbers correctly from 40 possible numbers is called a $\frac{40}{6}$ lottery. In general, an $\frac{m}{n}$ lottery is one in which you must correctly choose n numbers from m possible numbers. Investigate what kind of lotteries there are in your state. What is the probability of winning such a lottery?

Looking Deeper

14.5 Binomial Experiments

Objectives

1. Be able to compute binomial probabilities.
2. Use binomial probability to solve applied problems.

Have you ever taken a quiz for which you were not prepared? Perhaps it was a 10-question true–false quiz, or maybe a 5-question multiple-choice quiz. What would your chances of being successful on these types of tests be by purely guessing? Maybe you have bought a box of cereal containing a collectable figure. How many purchases must you make before you can reasonably expect to have the complete set of figures? In an apparently different situation, a pharmaceutical company makes a claim regarding the effectiveness of a new vaccine. How might we test the company's claim for its reliability?

 KEY POINT

Binomial trials have only two outcomes.

Binomial Probability

All these questions are related to an important class of experiments in probability theory that deserve special consideration. These experiments are similar in that they all have two outcomes—one of the outcomes is referred to as "success" and the other as "failure." Such experiments are called *binomial trials*.* We list several of these in Table 14.11. What we have decided to call a success or a failure is arbitrary. The point is that one outcome is considered a success and the other a failure.

---

*These are also often called Bernoulli trials, named after the brilliant seventeenth-century Swiss mathematician Jakob Bernoulli.

Experiment	Success	Failure
Flip one coin	Head	Tail
Roll two dice	Roll a total of seven	Roll a total other than a seven
Test a computer chip	Chip functions properly	Chip is defective
Inoculate a person with a flu vaccine	Person does not get the flu	Person gets the flu
Buy a box of cereal	Obtain a new collectable figure	Obtain a collectable figure that you already have
Guess at a multiple-choice question	Guess correct answer	Guess incorrectly

TABLE 14.11 Examples of binomial trials.

> **PROPERTIES OF A BINOMIAL EXPERIMENT** A sequence of binomial trials is called a **binomial experiment** and has the following properties.
>
> 1. The experiment is performed for a fixed number of trials.
> 2. The experiment has only two outcomes, "success" and "failure."
> 3. The probability of success is the same from trial to trial.
> 4. The trials are independent of each other.

EXAMPLE 1   *Experiments That Are Not Binomial Experiments*

Explain why each experiment is not a binomial experiment.

a) Roll a single die until a six comes up.
b) A student takes a course and earns either an A, B, C, D, or F.
c) Select four cards without replacement and observe whether a heart is drawn.

SOLUTION:

a) The number of trials is not fixed. We do not know how many times the die will have to be rolled before a six comes up.
b) There are more than two outcomes.
c) Because the cards are not being replaced, the probability of a heart will change from draw to draw.

Now try Exercises 3 to 8. ✺

✎ **KEY POINT**

The probabilities of binomial experiments have the same pattern.

In Example 2, we see a common pattern that occurs in computing binomial probabilities.

EXAMPLE 2   *A Binomial Experiment of Rolling a Pair of Dice Three Times*

What is the probability of rolling a total of seven exactly once if we roll a pair of dice three times?

SOLUTION: Because we can think of this experiment as occurring in three stages (roll dice first time, roll dice second time, roll dice third time), we can represent it by a tree diagram, as in Figure 14.19. We have placed the probability of each outcome of a stage of the experiment along the appropriate branch.

We have highlighted three branches in Figure 14.19 to show the three ways to roll exactly one seven in three rolls of the dice. If we consider rolling a seven a success and rolling a

1st roll   **2nd roll**   **3rd roll**

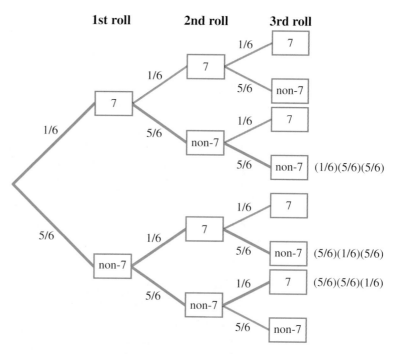

FIGURE 14.19 Rolling three dice can be thought of as a binomial experiment—in each roll we get a seven or a nonseven.

nonseven a failure, then we see that the probability of a success on the first roll and failure on rolls two and three is $\left(\frac{1}{6}\right)\left(\frac{5}{6}\right)\left(\frac{5}{6}\right)$. Similarly, we could get success on the second roll and failure on the other rolls with a probability of $\left(\frac{5}{6}\right)\left(\frac{1}{6}\right)\left(\frac{5}{6}\right)$. Finally, if the success occurs on the third roll, the probability is $\left(\frac{5}{6}\right)\left(\frac{5}{6}\right)\left(\frac{1}{6}\right)$. Thus, the probability of rolling exactly one seven on three rolls of the dice is $\left(\frac{1}{6}\right)\left(\frac{5}{6}\right)\left(\frac{5}{6}\right) + \left(\frac{5}{6}\right)\left(\frac{1}{6}\right)\left(\frac{5}{6}\right) + \left(\frac{5}{6}\right)\left(\frac{5}{6}\right)\left(\frac{1}{6}\right)$

$= \frac{75}{216} \approx 0.3472.$

Now try Exercises 9 and 10. ❋

In Example 2, we can write each one of the three probabilities in the form $\left(\frac{1}{6}\right)^1\left(\frac{5}{6}\right)^2$. We view this abstractly as

(probability of success)$^{\text{(number of successes)}}$ × (probability of failure)$^{\text{(number of failures)}}$

Suppose now that we want to compute the probability of obtaining exactly two sevens when a pair of dice is rolled five times. We will compute this probability in stages.

First, we have to choose which two of the five rolls correspond to sevens. From Chapter 13, recall that the number of ways to choose two objects from a set of five is

$C(5, 2) = \frac{5!}{2!3!} = 10.$

Second, by doing computations similar to Example 2, we see that each of these 10 outcomes has a probability of $\left(\frac{1}{6}\right)^2\left(\frac{5}{6}\right)^3$. Thus the probability of rolling exactly two sevens in five rolls of a pair of dice is probability of a seven ⌐   ⌐ probability of a nonseven

$$\underset{\substack{\text{number of ways}\\\text{to choose 2 of 5 rolls}}}{C(5, 2)}\left(\frac{1}{6}\right)^2\left(\frac{5}{6}\right)^3 = 10 \cdot \frac{1}{36} \cdot \frac{125}{216} = \frac{1,250}{7,776} \approx 0.1608.$$

We will now state this calculation in a general form.

FORMULA FOR COMPUTING BINOMIAL PROBABILITIES In a binomial experiment with n trials, if the probability of success in each trial is p, then the probability of exactly k successes is given by

$$C(n, k)(p)^k(1 - p)^{(n-k)}.$$

We will denote this probability by $B(n, k; p)$.

 Some Good Advice

If you connect new ideas with those you have seen before, it helps you understand mathematics as an integrated collection of concepts. Even though the binomial probability notation may seem complicated, the notation helps us remember its meaning. The B reminds us of the word *binomial*. We have seen the n, k pattern before in Chapter 13 in counting the number of combinations of n objects taken k at a time. Finally, the p reminds us of the word *probability*; in this case, the probability of a success in a binomial trial.

Application of Binomial Probability

Binomial probabilities arise in many real-life situations involving the testing of products.

EXAMPLE 3 *Binomial Probability and Drug Testing*

A pharmaceutical company claims that a new drug is effective 75% of the time. Assuming the company's claim is accurate, if we test the drug on 20 patients, what is the probability that exactly 15 people will benefit from the drug?

SOLUTION: This is a binomial experiment with 20 trials in which success is that the drug benefits the patient. According to the company, the probability of success is 0.75 and the probability of failure is 0.25. The probability of exactly 15 successes is therefore $B(20, 15; 0.75)$.

Using the binomial probability formula,* we get

$$B(20, 15; 0.75) = C(20, 15)(0.75)^{15}(1 - 0.75)^{(20-15)}$$
$$= 15{,}504(0.75)^{15}(0.25)^5 \approx 0.2023. \; ✳ \; \boxed{16}$$

We will investigate further how to test the pharmaceutical company's claim in the exercises.

 **16**

If we roll a single die eight times, what is the probability of getting exactly two threes?

EXAMPLE 4 *Binomial Probability and Guessing on Exams*

Maria has fallen behind in her anthropology class and her instructor just announced a pop quiz consisting of 10 true–false questions. If she must get 6 or more correct to pass the quiz, what is the probability of passing the quiz by guessing?

---

*Throughout this section, we will round these binomial probabilities to four decimal places. If you carry more decimal places, you will sometimes get answers that are slightly different from ours.

SOLUTION: To pass the quiz, Maria must get either 6, 7, 8, 9, or 10 correct. The probability of this is

probability of getting 6 correct
probability of getting 7 correct

$$B(10, 6; \tfrac{1}{2}) + B(10, 7; \tfrac{1}{2}) + B(10, 8; \tfrac{1}{2}) + B(10, 9; \tfrac{1}{2}) + B(10, 10; \tfrac{1}{2})$$
$$= 0.2051 + 0.1172 + 0.0439 + 0.0098 + 0.0010 = 0.377.$$

Thus, the probability that she will pass the quiz by randomly guessing is almost 40%.

Now try Exercises 15 to 24. ❋

We conclude this section with another interesting application involving binomial trials. Manufacturers often package sports cards in small groups so that a collector will have to buy numerous packages in order to obtain a complete set. A collector might ask, "If I know the number of items in the complete set, how many purchases should I expect to make to get a complete set?" Before we answer the collector's question, we need to know more about binomial trials.

Suppose that we are making a sequence of binomial trials where success has probability p and that we repeat the trial until we obtain a success. How many trials do we expect to perform before we get a success? If the probability of success is small, then our intuition tells us that we can expect to perform many trials before we get a success. On the other hand, if the probability of success is large, then we should not have to perform many trials before obtaining a success. In fact, the following can be proved (we will not prove it here):

> **THE NUMBER OF BINOMIAL TRIALS WE CAN EXPECT BEFORE A SUCCESS** If we repeat a binomial trial in which success has probability p, then the number of trials we can expect to perform before we get a success is $\dfrac{1}{p}$.

This is saying, for example, that if in a binomial trial the probability of success is $\dfrac{1}{100}$, then we should expect to perform the experiment 100 times before we get a success. We are now ready to answer a simplified version of the collector's question.

EXAMPLE 5   *How Many Packages Should You Expect to Buy to Obtain a Complete Set of Collectibles?**

Assume that a person is buying boxes of cereal each of which contains a figure of one of five characters in a current movie. The packages are sealed, so the buyer does not know which figure is included with the cereal. How many purchases can this collector expect to make before obtaining the entire set of five figures?

SOLUTION: We will call the total number of purchases to obtain a complete set T. Then,

T = the number of purchases to obtain a first new figure +
the number of purchases to obtain a second new figure +
the number of purchases to obtain a third new figure +
the number of purchases to obtain a fourth new figure +
the number of purchases to obtain a fifth new figure.

---

*For a more extensive discussion of the collector's problem, see Richard Isaac, The Pleasures of Probability (New York: Springer, 1995), Chapter 8.

The number of purchases required to obtain a first new figure is 1 because the very first purchase will give us a new figure.

At this point, we have one figure; before we open our next purchase, there are four possible new figures and one old figure. We can think of buying a box of cereal now as a binomial experiment in which the probability of getting a new figure (success) is $\frac{4}{5}$ and the probability of failure is $\frac{1}{5}$. The number of trials we can expect before we get a new figure is therefore $\frac{1}{4/5} = \frac{5}{4}$.

Once we have two different figures, we change our computations. Before we open the next box of cereal, there are possibly one of two old figures in the box or one of three new ones. Buying a box of cereal is now a binomial trial with probability of success $\frac{3}{5}$. Thus, we can expect to obtain the third new figure in $\frac{1}{3/5} = \frac{5}{3}$ trials.

Similarly, the expected number of purchases before we get the fourth figure is $\frac{5}{2}$ and the expected number of purchases before we get the fifth figure is $\frac{5}{1}$.

The total number of purchases T we can expect to make before we have a complete set is, therefore,

$$T = 1 + \frac{5}{4} + \frac{5}{3} + \frac{5}{2} + \frac{5}{1} = 11.42.$$

It is likely that the collector will obtain the complete set of five figures by the twelfth purchase.

Now try Exercises 27 and 28. ❋

Exercises $\boxed{14.5}$

Looking Back*

These exercises follow the general outline of the topics presented in this section and will give you a good overview of the material that you have just studied.

1. In Example 2, explain why rolling a pair of dice three times was an example of a binomial experiment.

2. Explain the meaning of each symbol in the expression $B(20, 15; 0.75)$ that we used in Example 3.

Sharpening Your Skills

In Exercises 3–8, determine whether each experiment is a binomial experiment. If not, explain what property of a binomial experiment fails.

3. Select 4 cards with replacement from a standard 52-card deck and observe whether a king is drawn.

4. Roll a pair of dice five times and observe whether doubles are obtained.

5. Three coins are flipped until three heads are obtained.

6. A person buys five chances on the Daily Number lottery.

7. A student takes a 10-question true–false quiz.

8. Twenty people are injected with a new cold vaccine.

9. If we roll a pair of dice three times, what is the probability of rolling a total of five exactly once?

10. If we roll a pair of dice three times, what is the probability of rolling a total of eight exactly twice?

In Exercises 11 and 12, explain the meaning of each expression and compute its value.

11. $B\left(5, 3; \frac{1}{4}\right)$

12. $B\left(6, 2; \frac{1}{2}\right)$

In Exercises 13 and 14, explain the error in using the notation to indicate a binomial probability.

13. $B\left(3, 5; \frac{1}{3}\right)$

14. $B(4, 3; 2)$

---

*Before doing these exercises, you may find it useful to review the note *How to Succeed at Mathematics* on page xix.

Applying What You've Learned

15. If you are given a 12-question true–false quiz, what is the probability that you will answer *exactly* 9 of the questions correctly by guessing?

16. If you are given an eight-question multiple-choice quiz, what is the probability that you will answer *exactly* six of the questions correctly by guessing? Assume that each question has four options.

17. If you are given a 12-question true–false quiz, what is the probability that you will answer *at least* 9 of the questions correctly by guessing?

18. If you are given a six-question multiple-choice quiz, what is the probability that you will answer *at least* four of the questions correctly by guessing? Assume that each question has five options.

19. Dr. Hahn has developed a new procedure that she believes can correct a life-threatening medical condition. If the success rate for this procedure is 80%, and the procedure is tried on 10 patients, what is the probability that at least 8 of them will show improvement?

20. You are planning a vacation at the beach and the weather forecast is that there is a 30% chance of rain for each of the next 3 days. What is the probability that you will have rain less than 2 days?

21. Chase Utley currently has a 0.250 batting average. If we ignore all other factors, what is the probability that if he comes to bat five times he gets exactly two hits in today's game?

22. Lisa Leslie, a WNBA basketball player, has a 50% foul shooting average. If she attempts eight foul shots today, what is the probability that she makes exactly four foul shots?

23. A hospital has an urgent need for three units of type A+ blood. Assume that approximately 30% of the population has this type of blood. If there are 12 people waiting to donate one unit of blood, what is the probability that the hospital will be able to meet its need? (*Hint:* Subtract the probability that fewer than three people have A+ blood from 1.)

24. A cellular phone communications network has redundancy built into it in the sense that if several of the components fail, the network may still be able to function properly. If a network has 15 components, each of which is 95% reliable, what is the probability that the network will fail if the network requires at least 12 of the components working in order to function properly? (*Hint:* Compute the probability that the network will not fail and subtract from 1.)

In Exercises 25 and 26, fill in the missing item to make the equation true. Give an intuitive reason why both sides of the equation must be equal.

25. $B\left(5, 3; \dfrac{1}{4}\right) = B(5, 2; ?)$

26. $B(10, 3; 0.4) = B(10, ?; 0.6)$

27. Assume that a child is buying a fast-food meal that also contains a (random) figure of one of six characters in a popular movie. How many purchases can she expect to make before she collects the entire set of six figures?

28. Assume that a child is buying packages of candy that also contain a (random) picture card of one of 10 players on the 2008 men's Olympic basketball team. How many purchases can he expect to make before he collects the entire set of 10 pictures?

Communicating Mathematics

29. What are the four properties of a binomial experiment?

30. If the probability of success in a binomial trial is $\frac{1}{5}$, how many trials can you expect to perform before you get a success?

Using Technology to Investigate Mathematics

31. Search the Internet for "binomial probability applets" or "binomial probability applications." Find a site that interests you and report on your findings.

32. Find a site on the Internet that allows you to download a crossword-puzzle-generating program. Make a crossword puzzle using key words of probability. Try to be clever in giving your clues.

For Extra Credit

33. Assume that during the winter, 50% of the population will come down with the common cold. Suppose a company claims that it has developed a cold vaccine to lower the infection rate to 25%. Ten people are vaccinated and we agree that we will accept the claim of the company if four or fewer catch a cold. What is the probability that we are accepting a worthless claim? (*Hint:* If the vaccine is worthless, the probability of catching a cold has not changed.)

34. Continuing the situation in Exercise 33, suppose we decide on a more strict requirement in order to accept the claim that the cold vaccine is effective. We agree that now we will accept the company's claim provided that 2 or fewer people of the 10 who were inoculated catch a cold. What is the probability that we reject a worthwhile vaccine?

CHAPTER SUMMARY*

You will learn the items listed in this chapter summary more thoroughly if you keep in mind the following advice:

1. Focus on "remembering how to remember" the ideas. What pictures, word analogies, and examples help you remember these ideas?
2. Practice writing each item without looking at the book.
3. Make 3×5 flash cards to break your dependence on the text. Use these cards to give yourself practice tests.

SECTION	SUMMARY	EXAMPLE
SECTION 14.1	An observation of a **random phenomenon** is called an **experiment**. The different possible observations of the experiment are called **outcomes**. The set of all possible outcomes is called the **sample space** of the experiment. An **event** is a subset of the sample space. The **probability of an outcome** in a sample space is a number between 0 and 1, inclusive. The **probability of an event** is the sum of the probabilities of the outcomes in the event. We can assign the probability of an event **empirically** as $$P(E) = \frac{\text{the number of times } E \text{ occurs}}{\text{the number of times the experiment is performed}}.$$ This ratio is called the **relative frequency** of E. If E is an event in a sample space S with all equally likely outcomes, then $$P(E) = \frac{n(E)}{n(S)}.$$ We can use **counting formulas** to compute probabilities. If E is an event in some sample space S, then 1. $0 \le P(E) \le 1$ 2. $P(\varnothing) = 0$ 3. $P(S) = 1$. Probability can help explain **genetic theory**. If the outcomes in the sample space are equally likely, the **odds against** event E are the number of outcomes against E occurring compared to the number of outcomes in favor of E occurring. That is, $\dfrac{n(E')}{n(E)}$. An alternate formula for computing odds using probability is $\dfrac{P(E')}{P(E)}$.	Definitions, p. 659 Example 1, p. 659 Example 2, p. 661 Examples 3 and 4, p. 662 Example 5, p. 663 Example 6, p. 664 Discussion, p. 665 Discussion, p. 666 Example 7, p. 667 Definition, p. 667 Examples 8 and 9, pp. 668, 669
SECTION 14.2	We can find the probability of event E by subtracting the probability of its **complement** from 1. That is, $P(E) = 1 - P(E')$. If E and F are events, we find the probability of their **union** as follows: $P(E \cup F) = P(E) + P(F) - P(E \cap F)$. If E and F and disjoint, this formula simplifies to $P(E \cup F) = P(E) + P(F)$. We sometimes combine the complement and union formulas to solve a problem.	Example 1, p. 674 Examples 2 and 3, p. 676 Example 4, p. 677
SECTION 14.3	If we compute the probability of an event F assuming that event E has already occurred, this is called the **conditional probability of F given E**. We denote this probability by $P(F \mid E)$. If E and F are events in a sample space with *equally likely outcomes*, then $P(F \mid E) = \dfrac{n(E \cap F)}{n(E)}$. In general, $P(F \mid E) = \dfrac{P(E \cap F)}{P(E)}$. If E and F are events, then we calculate the probability of their **intersection** by using the following formula: $$P(F \cap E) = P(E) \cdot P(F \mid E).$$	Definition, p. 682 Example 1, p. 683 Example 2, p. 684 Example 3, p. 686

---

*Before studying this chapter's material, it would be useful to reread the note *How to Succeed at Mathematics* on page xix.

	We use **probability trees** to visualize probability problems.	Example 4, p. 687 Example 6, p. 689 Example 7, p. 691
	Events E and F are **independent** if $P(F \mid E) = P(F)$. Otherwise, they are **dependent**.	Definition, p. 689 Example 5, p. 689
SECTION 14.4	If an experiment has n outcomes with probabilities $P_1, P_2, P_3, \ldots, P_n$, and $V_1, V_2, V_3, \ldots, V_n$, are values associated with the outcomes, the **expected value** of the experiment is $(P_1 \cdot V_1) + (P_2 \cdot V_2) + (P_3 \cdot V_3) + \cdots + (P_n \cdot V_n)$.	Example 1, p. 696 Definition, p. 696
	We use expected value to analyze games of chance.	Examples 2 and 3, p. 697 Example 4, p. 698
	Expected value has many applications, such as evaluating test-taking strategies and business decision making.	Example 5, p. 699 Example 6, p. 700
SECTION 14.5	A **binomial trial** is an experiment that has two outcomes, labeled "success" and "failure." A binomial is a sequence of binomial trials that has the following properties: 1. The experiment is performed for a fixed number of trials. 2. Each trial has only two outcomes. 3. The probability of success is the same from trial to trial. 4. The trials are independent of each other.	Definitions, p. 703 Examples 1 and 2, p. 704
	In a binomial experiment with n trials, if the probability of success in each trial is p, the probability of exactly k successes is given by $C(n, k)(p)^k (1 - p)^{(n-k)}$. We denote this probability by $B(n, k; p)$.	Discussion, p. 706
	Binomial probability can be used to solve wide-ranging applied problems.	Examples 3 and 4, p. 706 Example 5, p. 707

CHAPTER REVIEW EXERCISES

Section 14.1

1. Describe each event as a set of outcomes.

   a. When three coins are flipped, we obtain exactly two heads.

   b. When two dice are rolled, we obtain a total of eight.

2. If a single card is selected from a standard 52-card deck, what is the probability that a red face card is selected?

3. Explain the difference between empirical and theoretical probability. Give an example of each type of probability.

4. In cross-breeding pea plants, Mendel found that the characteristic "tall" dominated the characteristic "short." If we cross pure tall plants with pure short plants, what is the probability of tall plants in the second generation?

5. **a.** If the odds against the Dodgers winning the World Series are 17 to 2, what is the probability of them winning the Series?

   b. If the probability of rain tomorrow is 0.55, what are the odds against rain?

Section 14.2

6. **a.** State the formula for computing the probability of the complement of an event.

   b. Draw a diagram to explain this formula.

   c. In what sort of situations would you be likely to use this formula?

7. If a single card is drawn from a standard 52-card deck, what is the probability that we obtain either a face card or a red card? Draw a diagram to illustrate this situation.

8. On the given spinner, each green sector occupies 10% of the circle, each blue sector occupies 12%, each red sector occupies 9%, and the yellow sector occupies 4%. Assume that the spinner is spun once. What is the probability that

   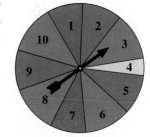

   a. The spinner does not stop on red?

   b. The spinner stops on an even number or blue?

Section 14.3

9. Explain in your own words what we mean by *conditional probability*.

10. If a pair of fair dice is rolled, what is the probability that we roll a total of five if we are given that the total is less than nine?

11. Assume that 2 cards are drawn without replacement from a standard 52-card deck.

    a. What is the probability that two hearts are drawn?

    b. What is the probability that a queen and then an ace is drawn?

12. A pair of fair dice are rolled. Are events E and F independent?

E—an odd total is obtained.

F—the total showing is less than six.

13. Assume that the incidence of the HIV virus in a particular population is 4% and that the test correctly identifies the virus 90% of the time. Assume that false positives occur 6% of the time. If a person tests positive for the virus, what is the probability that the person actually has the virus?

Section 14.4

14. Based on mortality tables, the probability of a 20-year-old male living to age 21 is 0.99. What is the expected value of a $1,000 one-year term life insurance on 20-year-old male from the insurance company's point of view? Assume that the yearly premium is $25.25.

15. A card is drawn from a standard 52-card deck. If a face card is drawn, you win $15; otherwise, you lose $4. Calculate the expected value for the game.

16. You are playing a game in which four fair coins are flipped. If all coins show the same (all heads or all tails), you win $5. Calculate the price to play the game that would make the game fair.

Section 14.5

17. Calculate $B\left(8, 3; \dfrac{1}{2}\right)$.

18. If you are guessing on a 10-question true–false quiz, what is the probability that you will get 8 correct?

CHAPTER TEST

1. Describe each event as a set of outcomes.

   a. When two dice are rolled, we obtain a total greater than nine.

   b. When four coins are flipped, we get more heads than tails.

2. If we select a single card from a standard 52-card deck, what is the probability that we select a black king?

3. a. If the odds against the Dolphins winning the Super Bowl are 28 to 3, what is the probability of them winning the Super Bowl?

   b. If the probability of rain tomorrow is 0.15, what are the odds against rain?

4. If we draw a single card from a standard 52-card deck, what is the probability that we obtain either a heart or a king?

5. What is the difference in the meaning between $P(B \mid A)$ and $P(A \mid B)$?

6. We are rolling a pair of fair dice. Are events E and F independent?

   E—both dice show the same number.

   F—the total is greater than eight.

7. It costs $1 to play a game in which two dice are rolled. If both dice show the same number, you win $5; otherwise, you lose. What is the expected value of this game?

8. Calculate $B\left(10, 2; \dfrac{1}{4}\right)$.

9. If you are guessing on a five-question multiple-choice quiz where each question has four possible answers, what is the probability that you will get three correct?

10. a. Complete the following equation $P(E) +$ _____ $= 1$.

   b. Draw a diagram to illustrate this formula.

11. In pea plants, purple flower color dominates white. With snapdragons, however, a pure red flowering plant crossed with pure white produces a pink flowering plant. If we begin by crossing pure red and white snapdragons, what is the probability of pink flowers in the second-generation plants?

12. Assume that 2% of the Brazilian population have dengue fever in 2008 and that a test correctly identifies the fever 95% of the time. Assume that a false positive occurs 5% of the time. If a person tests positive for the fever, what is the probability that the person actually has the fever?

13. It costs $2 to buy a raffle ticket. If there are 500 tickets sold, and there is one first prize of $250, three second prizes of $100, and five third prizes of $50, what is the expected value of this raffle?

14. If a pair of dice is rolled, what is the probability of rolling an even total given that the total is less than five?

15. Assume that 2 cards are drawn without replacement from a standard 52-card deck.

   a. What is the probability that two face cards are drawn?

   b. What is the probability that a king and an ace are drawn? (*Hint:* There are two cases.)

GROUP EXERCISES

1. **a.** Graphing calculators have random number generators built into them. For example, on the TI-83 and TI-84, if you press $\boxed{\text{MATH}}$, then move the cursor to the right to $\boxed{\text{PRB}}$, then press $\boxed{\text{ENTER}}$ twice, you will get a number like .5489861799. If you generate a list of such numbers, you can use the digits in the list to simulate flipping a coin. For example, if you think of an odd digit as heads and an even digit as tails, then the number above corresponds to HTTHTTHHHH. Generate 1,000 digits to see how close your simulated coin flipping probability of a head corresponds to the theoretical probability of $\frac{1}{2}$.

    b. Make up a similar simulation for flipping two coins and compare some simulated probabilities with the theoretical ones. For example, if you generate 1,000 pairs of numbers, how close do you get to the theoretical probability of $\frac{1}{4}$ for getting two heads on two coins?

    c. Do similar simulations for rolling one die and rolling two dice.

2. Look up some real statistics regarding seat belt usage and traffic fatalities. Then compute the conditional probabilities $P(\text{driver is killed} \mid \text{driver was wearing a seat belt})$ versus $P(\text{driver is killed} \mid \text{driver was not wearing a seat belt})$.

Descriptive Statistics

15

What a Data Set Tells Us

When the campaign for the Democratic nomination for president was in full swing in the spring of 2008, I would often log on to the Internet and be asked to take a poll as to whether I favored Obama or Clinton. After voting, I could then check state-by-state to see who the favorite candidate was. It was great fun! I would vote several times a week—sometimes saying that I was from Pennsylvania, other times from Arkansas, or Florida, or New York, or Texas. My point is that the data that was being reported as news was erroneous. As an educated citizen, it is important that you are an intelligent consumer of statistical information and are aware of how others may use statistics to manipulate you.

On the other hand, researchers gather and analyze statistical data to improve highway safety and increase the quality of the products you buy every day. Other scientists use statistical distributions to create computer simulations to study the traffic flow on the expressways that you take to school and to decrease the amount of time that you spend waiting in line to ride Space Mountain at Disney World. *(continued)*

In this chapter, you will learn how to organize data so that you can observe patterns and spot trends. You will also learn how to describe data numerically to understand relationships among the data better.

Toward the end of this chapter, you will see that it is no coincidence that your HD-DVD player breaks down just after its warranty expired. It almost seems as though the manufacturer knows the exact moment when a product will fail. In fact, manufacturers do need to know a product's life expectancy—at least statistically—to be able to assure consumers that the product will last for a reasonable length of time. After you study the normal distribution in Section 15.4, you will understand the mathematics behind writing an effective warranty. ●

15.1 Organizing and Visualizing Data*

Objectives

1. Understand the difference between a sample and a population.
2. Organize data in a frequency table.
3. Use a variety of methods to represent data visually.
4. Use stem-and-leaf displays to compare data.

In a survey of 100,000 women conducted by *Cosmopolitan Magazine*, it was found that over 70% of women who were married for more than 5 years had had an affair. These are truly shocking results, but before you swear off marriage completely, there are some questions that you should be asking, such as who conducted the study? How were the women selected for the study? Of the 100,000 women, how many responded to the questions in the study? Did those being surveyed have any particular biases?

It may reassure you to know that a larger survey of 200,000 women found that only 15% of the women reported that they had been unfaithful. These contradictory results make us wonder, which survey is correct? As you will learn shortly, perhaps we should trust neither survey.

Populations and Samples

In this chapter you will study **statistics**, an area of mathematics in which we are interested in gathering, organizing, analyzing, and making predictions from numerical information called **data**. In the two marriage surveys, it would have been ideal if the researchers had been able to contact each one of the millions of married woman in the United States. This set of all married women is called the **population**. Of course, this is impractical, so the researchers contacted only a subset of the population, called a **sample**. It is very important that the sample is typical of the population as a whole. In fact, in the two studies just mentioned, both samples were chosen very poorly, and we should trust neither survey.

We will describe a sample as **biased** if it does not accurately reflect the population as a whole with regard to the data that we are gathering. Bias often occurs if we use poor sampling techniques. There are many ways in which bias can creep into a sample. It could occur because of the way in which we decide how to choose the people to participate in the

---

*Many of the calculations that we do in this chapter can be done easily using computers and graphing calculators.

survey. This is called **selection bias**. As an example, if we were to do a phone survey in the middle of a weekday afternoon, we would probably get an overrepresentation of retirees and stay-at-home parents in our sample. Call-in surveys conducted by local news shows and radio stations are also prone to selection bias.

It might seem we would get better, more reliable information if we were to walk around a town and ask people "randomly" to take part in our survey. We put the word *randomly* in quotes because studies have found that selection bias can occur using this method since interviewers tend to choose people who are better dressed and who look cooperative, thus skewing the sample.

Another issue that can affect the reliability of a survey is the way we ask the questions, which is called **leading-question bias**. For example, in a Roper poll conducted for the American Jewish Committee on the Holocaust, people were asked, "Does it seem possible or does it seem impossible to you that the Nazi extermination of the Jews never happened?" The use of double negatives in this question caused confusion in the way people responded to the survey. When the question was worded this way, 22% of those surveyed said that it was possible that the Holocaust did not occur. A new survey was conducted in which the question was rephrased, "Does it seem possible to you that the Nazi extermination of the Jews never happened, or do you feel certain that it happened?" In the new survey, only 1% of those surveyed stated that it was possible that the Holocaust never occurred.

We have only touched on the notion of how important it is that statistical conclusions are based on reliable, nonbiased data. There are whole books written on sampling theory. For now, we just want you, as an educated consumer of technical information, to be aware that just because a study says something, as the song says, "It ain't necessarily so!"

We will now turn our focus on what to do once we have obtained reliable data.

Frequency Tables

When we gather information about a population, we often end up with a large collection of numbers. Unless we can organize the data in a meaningful way, it is nearly impossible to interpret these facts. For example, if you glance at the financial section of a newspaper, you will find several pages containing thousands of numerical facts about the daily performance of various stocks in the stock market. This amount of detail about the market's activity seems overwhelming; it is difficult to understand the general pattern of changes in stock prices from those lists of numbers. However, if you were to tune in to the evening news, the commentator might summarize this set of data by saying, "The Dow Jones lost 99.59 points today, closing at 12,251.7. Losers outnumbered winners by three to one." For any large amount of data to be comprehensible, we must organize it and present it so that we can see patterns, trends, and relationships.

> **DEFINITIONS** We refer to a collection of numerical information as **data** or a **distribution**. A set of data listed with their frequencies is called a **frequency distribution**.

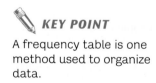 **KEY POINT**

A frequency table is one method used to organize data.

Sometimes we want to show the percent of the time that each item occurs in a frequency distribution. In this case, we call the distribution a **relative frequency distribution**. We often present a frequency distribution as a **frequency table**. In a frequency table, we list the values in one column and the frequencies of the values in another column, as we show in Example 1. We can also present a relative frequency distribution in table form.

HISTORICAL HIGHLIGHT ✻ ✻ ✻

Presidential Polls

In 1936, pollster George Gallup boldly declared that the *Literary Digest* would incorrectly predict Alfred Landon to defeat Franklin Roosevelt for reelection as president of the United States. Gallup's claim seemed far-fetched because he used a sample of only 50,000, whereas the *Digest* intended to survey 10 million.

When Roosevelt won, it was clear that Gallup's superior sampling methods were more important than having a large sample. Because the *Digest* used telephone directories, magazine subscription lists, and membership lists of clubs and organizations, it had sampled people in the higher economic classes. Thus, its survey suffered from extreme selection bias and greatly overrepresented Republicans.

After this fiasco, pollsters developed *quota sampling*, in which samples would reflect the makeup of the population. The idea was to have the same percentage of men, women, Catholics, Jews, blacks, whites, and so on in a sample as there were in the population. However, when pollsters used quota sampling to forecast the 1948 presidential election, disaster struck again. All the major polling organizations wrongly predicted that Thomas Dewey would defeat the incumbent Harry S. Truman.

One major reason that the 1948 polls failed was that the polling stopped too soon, missing a late trend toward Truman. Even though quotas were met, selection bias crept in because interviewers had too much freedom in choosing whom to interview within those categories.

Evaluation	Frequency
E	4
A	7
V	8
B	4
P	2
Total	25

TABLE 15.1 Frequency table summarizing viewer evaluations of a police drama.

EXAMPLE 1   *Using Tables to Summarize TV Program Evaluations*

CBS has asked 25 viewers to evaluate the latest episode of *CSI*. The possible evaluations are

(E)xcellent, (A)bove average, a(V)erage, (B)elow average, (P)oor.

After the show, the 25 evaluations were as follows:

A, V, V, B, P, E, A, E, V, V, A, E, P, B, V, V, A, A, A, E, B, V, A, B, V

Construct a frequency table and a relative frequency table for this list of evaluations.

SOLUTION: If we count the number of Es, As, and so on in the list, we get the results shown in Table 15.1.

By organizing the data in this table, we can see the distribution of favorable and unfavorable evaluations more quickly. Notice that the sum of the frequencies in Table 15.1 is 25, which is the number of viewers asked to evaluate the program.

We construct a relative frequency distribution for these data by dividing each frequency in Table 15.1 by 25. For example, because there are 4 Es, the relative frequency of the score E is $\frac{4}{25} = 0.16$. Table 15.2 shows the relative frequency distribution for the set of evaluations.

Evaluation	Relative Frequency
E	$\frac{4}{25} = 0.16$
A	$\frac{7}{25} = 0.28$
V	$\frac{8}{25} = 0.32$
B	$\frac{4}{25} = 0.16$
P	$\frac{2}{25} = 0.08$
Total	1.00

TABLE 15.2 Relative frequency table summarizing viewer evaluations of a police drama.

Quiz Yourself ❶ *

Construct a frequency table and a relative frequency table for the following distribution:

1, 2, 7, 2, 6, 5, 2, 7, 8, 8, 1, 3, 10, 7, 9, 1, 7, 3, 5, 2

The sum of the relative frequencies in Table 15.2 is 1; however, in other examples, the sum of the relative frequencies may not be exactly 1 due to rounding. ✻ ❶

If there are many different values in a data set, we may group the data values into classes to make the information more understandable. Although there is no hard-and-fast rule, generally using 8 to 12 classes will give a good presentation of the data. We illustrate how to group data in Example 2.

---

*Quiz Yourself answers begin on page 778.

EXAMPLE 2 *Grouping Data Values into Classes*

Suppose 40 health care workers take an AIDS awareness test and earn the following scores:

79, 62, 87, 84, 53, 76, 67, 73, 82, 68,
82, 79, 61, 51, 66, 77, 78, 66, 86, 70,
76, 64, 87, 82, 61, 59, 77, 88, 80, 58,
56, 64, 83, 71, 74, 79, 67, 79, 84, 68

Construct a frequency table and a relative frequency table for these data.

Range of Scores on AIDS Awareness Test	Frequency	Relative Frequency
50–54	2	0.05
55–59	3	$\frac{3}{40} = 0.075$
60–64	5	0.125
65–69	6	0.15
70–74	4	0.10
75–79	9	0.225
80–84	7	0.175
85–89	4	0.10
Total	40	1.00

TABLE 15.3 Frequency table and relative frequency table for scores on AIDS awareness test.

SOLUTION: Because there are so many different scores in this list, the frequency of each score will be very small; constructing a frequency table as we did in Example 1 would not give us any useful information.

We must decide how to group the scores before making a table. The smallest score is 51 and the largest is 88. The difference, $88 - 51 = 37$, suggests that if we take a range of 40 and divide it into equal parts, we might get a reasonable grouping of the data. We will group the data into classes, each containing five values. The first class contains numbers from 50 to 54, the second contains numbers from 55 to 59, and so on. Counting the numbers in each class gives us the frequencies in the second column of Table 15.3.

To find the relative frequencies, we divide each count in the first column by 40, which is the total number of scores. For example, in the row labeled 55–59, we divide 3 by 40 to get 0.075 in the third column. ✳

Table 15.3 helps us see patterns and trends in the data. For example, a large number of workers have test scores below 70, which might mean that these workers are not effective in treating AIDS patients and may require extra training.

 **KEY POINT**

We use bar graphs to represent frequency distributions graphically.

Representing Data Visually

The saying "A picture is worth a thousand words" certainly applies when working with large sets of data. By presenting data graphically, we can observe patterns more easily. A **bar graph** is one way to visualize a frequency distribution. In drawing a bar graph, we specify the classes on the horizontal axis and the frequencies on the vertical axis. If we are graphing a relative frequency distribution, then the heights of the bars correspond to the size of the relative frequencies, as we show in Example 3.

EXAMPLE 3 *Drawing a Bar Graph of the Viewer Evaluation Data*

a) Draw a bar graph of the frequency distribution of CBS viewers' responses summarized in Table 15.1 in Example 1.

b) Draw a bar graph of the relative frequency distribution of the CBS viewers' responses summarized in Table 15.2 in Example 1.

SOLUTION:

a) Because the largest frequency is 8, we labeled the vertical axis from 0 to 8. Next we drew five bars of heights 4, 7, 8, 4, and 2 to indicate the frequencies of the evaluations E, A, V, B, and P, as is shown in Figure 15.1(a).

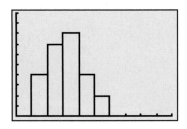

Graphing calculator representation of Figure 15.1(a).*

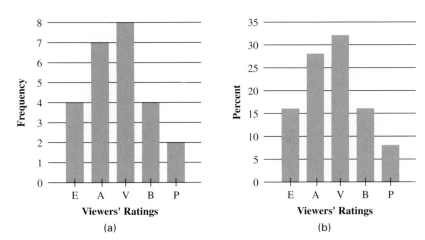

FIGURE 15.1 (a) Bar graph of frequency distribution of viewers' ratings.
(b) Bar graph of relative frequency distribution of viewers' ratings.

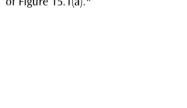

Quiz Yourself ②

Draw a bar graph representing the relative frequencies that you found in Quiz Yourself 1.

b) In Figure 15.1(b), we labeled the vertical axis from 0 to 35 because the largest relative frequency was 0.32, or 32%. Although both bar graphs have the same shape, it is usually better to draw a bar graph of relative frequency distributions when comparing two different data sets.

Now try Exercises 7 to 12. ✻ ②

If we are comparing two data sets of different sizes, graphing the relative frequencies, rather than the actual values in the data sets, allows us to compare the distributions. In this case, instead of drawing two separate bar graphs, we could show both distributions on a single graph, using, say, red for the bars in the first distribution and green for the bars in the second.

Until now, the data we have been organizing and graphing could not take on fractional values. By this we mean that in Example 1, a viewer could evaluate *CSI* as above average or excellent but could not give a rating between those two. Similarly, in Example 2, a score on the AIDS awareness test could be 78 or 79, but a score between these two numbers, such as 78.56, was not possible. A variable quantity that cannot take on arbitrary values is called *discrete*. Other quantities, called *continuous* variables, can take on arbitrary values. Weight is an example of a continuous variable. We may say that a person weighs 150 pounds; however, with a more accurate scale, we may find that the person actually weighs 150.3 or perhaps 150.314 pounds.

We use a special type of bar graph called a **histogram** to graph a frequency distribution when we are dealing with a continuous variable quantity. We also may use a histogram when the variable quantity is not continuous, but has a very large number of different possible values. Money is an example of such a quantity.

As with a bar graph, we specify classes for a histogram. With a histogram, however, we do not allow any spaces between the bars above each class. If a data value falls on the boundary between two data classes, then you must make it clear as to whether you are counting that value in the class to the right or to the left of the data value. We show how to draw a histogram for a frequency distribution in Example 4. As with bar graphs, we can also draw a histogram for a relative frequency distribution.

Pounds Lost	Frequency
0 to 10	14
10+ to 20	23
20+ to 30	17
30+ to 40	8
40+ to 50	3
Total	65

EXAMPLE 4　*Drawing a Histogram to Represent Weight-Loss Data*

The New You Clinic has the following data regarding the weight lost by its clients over the past 6 months. Draw a histogram for the relative frequency distribution for these data.

SOLUTION: We first must find the relative frequency distribution. Because there are 65 data values, we divide each frequency by 65 to obtain the corresponding relative frequency distribution in the third column of Table 15.4.

---

*Note that in this graphing calculator screen, there are no spaces between the bars as in the graph in Figure 15.1(a).

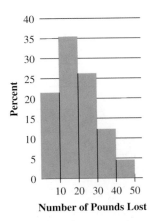

FIGURE 15.2 Histogram of weight loss at the New You Clinic.

Pounds Lost	Frequency	Relative Frequency
0 to 10	14	0.215
10+ to 20	23	0.354
20+ to 30	17	0.262
30+ to 40	8	0.123
40+ to 50	3	0.046
Total	65	1.00

TABLE 15.4 Frequency and relative frequency distributions of weight loss at the New You Clinic.

We now draw this histogram exactly like a bar graph, as shown in Figure 15.2, except that we do not allow spaces between the bars. Also, we label the endpoints of the class intervals on the horizontal axis.

Now try Exercises 13, 14, 17, and 18. ✳

When we look at the histogram in Figure 15.2, we can see that the majority of the clients lost between 10 and 30 pounds. There is really no strict rule as to how to construct a histogram. It is up to you to decide whether to group the data and how large each data class should be; however, it is customary to have data classes all of the same size.

EXAMPLE 5 *Determining Information from a Graph*

Figure 15.3 shows the number of Atlantic hurricanes* over a period of years. Use this bar graph to answer the following questions.

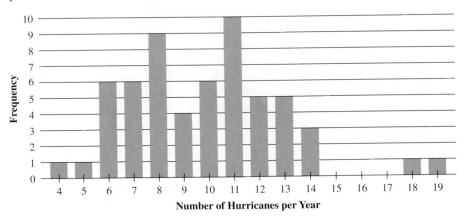

FIGURE 15.3 Number of hurricanes per year.

a) What was the smallest number of hurricanes in a year during this period? What was the largest?

b) What number of hurricanes per year occurred most frequently?

c) How many years were the hurricanes counted?

d) In what percentage of the years were there more than 10 hurricanes?

SOLUTION:

a) The smallest number of hurricanes in any year during this time period was 4. The largest number was 19.

b) The number of hurricanes per year that occurred most frequently corresponds to the tallest bar in Figure 15.3, which appears over the number 11. Therefore, 11 hurricanes occurred in 10 different years.

---

*These data are from the Colorado State Tropical Prediction Center.

c) To find the total number of years for which these data were gathered, we add the heights of all of the bars to get

$$1 + 1 + 6 + 6 + 9 + 4 + 6 + 10 + 5 + 5 + 3 + 1 + 1 = 58 \text{ years.}$$

d) First, we count the number of years in which there were more than 10 hurricanes. If we add the heights of the bars above these values, we get $10 + 5 + 5 + 3 + 1 + 1 = 25$. Because there are 58 years of data, we calculate $\frac{25}{58} = 0.431$, which is approximately 43%.

Now try Exercises 19 to 22. ❀

Stem-and-Leaf Displays

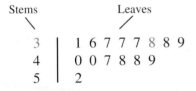 **KEY POINT**

A stem-and-leaf display is another way to display data.

A **stem-and-leaf display** is an effective way to present two sets of data "side by side" for analysis. This technique is used in an area called *exploratory data analysis*, developed by John Tukey, a mathematician who worked at Princeton University and Bell Labs.

Some sports fans believe that home run records have become meaningless in recent years because of the use of steroids by baseball players. In Example 6, we use a stem-and-leaf display to investigate whether there has been an increase in home run production in the National League recently.

Stems	Leaves
3	1 6 7 7 7 8 8 9
4	0 0 7 8 8 9
5	2

FIGURE 15.4 Stem-and-leaf display of home run data for 1975 to 1989.

EXAMPLE 6 *Using Stem-and-Leaf Home Run Records from Two Eras*

The following are the number of home runs hit by the home run champions in the National League for the years 1975 to 1989 and for 1993 to 2007.

a) 1975–1989: 38, 38, 52, 40, 48, 48, 31, 37, 40, 36, 37, 37, 49, 39, 47

b) 1993–2007: 46, 43, 40, 47, 49, 70, 65, 50, 73, 49, 47, 48, 51, 58, 50

Compare these home run records using a stem-and-leaf display.

4	0 3 6 7 7 8 9 9
5	0 0 1 8
6	5
7	0 3

FIGURE 15.5 Stem-and-leaf display of home run data for 1993 to 2007.

SOLUTION: We first examine the home run data for 1975 to 1989. In constructing a stem-and-leaf display, we view each number as having two parts. The left digit is considered the stem and the right digit the leaf. For example, 38 has a stem of 3 and a leaf of 8. The stems for the data in part a) are 3, 4, and 5. We first list the stems in numerical order and draw a vertical bar to their right. Next, we write the leaves corresponding to each stem to the right of its stem and vertical bar. We also list the leaves in increasing order away from the stem. Figure 15.4 shows the stem-and-leaf display for the data in part a). We show the stem-and-leaf display for the data in part b) in Figure 15.5.

HIGHLIGHT ❀ ❀ ❀

Using Technology to Graph Data*

In organizing a large amount of data, it is tedious to draw graphs by hand. Many software programs such as Microsoft Word and Excel and graphing calculators allow you to enter a frequency table and then will draw a graph for you. For example, you can use Microsoft Word to draw the following pie chart using the data from Example 1 regarding the TV program *CSI*.

*See your instructor for tutorials on using technology to graph data.

We can compare these data by placing these two displays side by side as we show in Figure 15.6. Some call this display a *back-to-back stem-and-leaf display*.

1975–1989		1993–2007
9 8 8 7 7 7 6 1	3	
9 8 8 7 0 0	4	0 3 6 7 7 8 9 9
2	5	0 0 1 8
	6	5
	7	0 3

FIGURE 15.6 Combined stem-and-leaf display of home run data.

Quiz Yourself ❸

Use a stem-and-leaf display to represent the following collection of scores:

92, 68, 77, 98, 88,
75, 82, 62, 84, 67,
62, 91, 82, 73, 66,
81, 63, 90, 83, 71

From Figure 15.6, we can clearly see the pattern that National League home run champions hit significantly more home runs from 1993 to 2007 than from 1975 to 1989.

Now try Exercises 15 and 16. ✳ ❸

Because the numbers in Example 6 were two-digit numbers, we used single digits for the stems. If we had to represent a number such as 325, we could use either a stem of 32 and a leaf of 5 or a stem of 3 and a leaf of 25, depending on which way presented the data more clearly.

Organizing and displaying a collection of data is generally not our final goal; we usually want a concise numerical description of a set of data. In Section 15.2, we will discuss how to analyze data once we have organized it.

✳ ✳ ✳ HISTORICAL HIGHLIGHT

Florence Nightingale*

It may surprise you to see a reference to the legendary nurse Florence Nightingale in a discussion of statistics. Although she is best known for her compassion toward the sick and her work in improving hospital sanitation, she was also trained in mathematics. As a young woman, she had the good fortune to study with James Sylvester, one of the most eminent British mathematicians of the nineteenth century.

While serving as a nurse at a military hospital during the Crimean War, she was disturbed by the high mortality rate of her patients. Using statistical methods that she invented, she persuaded her superiors to carry out hospital reforms. As a result of improved sanitation, deaths decreased, and some credit her with saving the British Army during the Crimean War.

In recognition for her revolutionary work in developing a statistical approach to medicine, Nightingale was elected to the Statistical Society of England. She also was an advisor on military health during the American Civil War, and in 1874, she became an honorary member of the American Statistical Association. The renowned statistician Karl Pearson described Florence Nightingale as a prophetess in the development of applied statistics.

*This note is based on the biography of Florence Nightingale by Cynthia Audain, located at the Biographies of Women Mathematicians Web site (www.agnesscott.edu/lriddle/women/women.htm) at Agnes Scott College, Atlanta, Georgia. Also, we used "Mathematical Education in the Life of Florence Nightingale," by Sally Lipsey, *The Newsletter of the Association for Women in Mathematics*, Vol. 23, No. 4 (1993), 11–12.

Exercises 15.1

Looking Back*

These exercises follow the general outline of the topics presented in this section and will give you a good overview of the material that you have just studied.

1. What are two kinds of bias that we mentioned that can affect the validity of a sample?

2. How did we find the relative frequencies in Table 15.2?

3. Why did we use a histogram in Example 4 rather than a bar graph?

4. What did we use to compare the home run data in Example 6?

5. Referring to the Historical Highlight on presidential polling on page 717, why were George Gallup's predictions more reliable than those of the *Literary Digest*?

6. How did Florence Nightingale use mathematics to improve health care?

Sharpening Your Skills

In Exercises 7 and 8, construct a frequency table, a relative frequency table, and a bar graph for the data given.

7. The number of hours of flight time for 20 amateur pilots last month:

    7, 8, 6, 5, 7, 10, 2, 7, 9, 5, 8, 8, 10, 9, 6, 5, 10, 7, 9, 8

8. The number of passengers per car on an Amtrak Acela train:

    38, 39, 38, 37, 40, 38, 38, 37, 40, 38, 38, 39, 38, 37, 40, 37, 40, 38, 38, 37, 40, 38, 39, 38, 37, 40, 38, 38, 39, 38

In Exercises 9 and 10, construct a bar graph for the data given in each frequency table.

9. This table contains the EPA mileage ratings for 38 domestic cars.

MPG	19	20	21	22	23	24	25	26	27	28	29	30
Frequency	2	5	3	2	1	8	1	2	3	2	7	2

10. This table contains the weight loss by 30 participants in the *Biggest Loser* show.

Weight Loss (Pounds)	0	1	2	3	4	5	6	7	8	9	10
Frequency	2	3	3	0	1	6	1	4	3	2	5

In Exercises 11 and 12, construct a bar graph for the relative frequency table for the data given.

11. The following are the ages of 60 people who have volunteered to work for the Katrina relief effort:

    21, 23, 27, 22, 23, 29, 28, 24, 24, 25, 27, 22, 26, 26, 23,
    23, 26, 28, 25, 24, 28, 27, 24, 23, 29, 28, 24, 22, 27, 26,
    22, 24, 26, 21, 24, 28, 24, 25, 22, 25, 27, 21, 23, 26, 23,
    23, 27, 27, 23, 21, 22, 27, 26, 23, 25, 29, 24, 27, 27, 26

12. The following are the ages of 40 ExecuCorps volunteers who are advising young entrepreneurs:

    51, 56, 57, 52, 53, 59, 58, 52, 54, 55,
    53, 56, 58, 55, 54, 58, 57, 52, 53, 59,
    52, 54, 53, 51, 54, 58, 54, 55, 52, 55,
    52, 57, 57, 53, 51, 52, 57, 56, 53, 55

In Exercises 13 and 14, group the data as indicated and construct a histogram.

13. The following are the heights (in inches) of the players in four Eastern Conference teams of the WNBA for the 2007 season:

    Chicago Sky: 69, 71, 74, 74, 78, 68, 68, 71, 75, 69,
                         76, 68, 69, 74, 65, 73

    NY Liberty: 72, 67, 72, 73, 75, 77, 73, 69, 67, 75,
                         76, 76, 73, 74, 69, 70

    Washington Mystics: 68, 72, 75, 68, 70, 69, 68, 74, 74, 69,
                         73, 76, 76, 80, 73, 78

    Atlanta Dream: 72, 77, 71, 80, 69, 73, 69, 75, 66, 68,
                         77, 74, 76, 73, 72, 75

    Use classes of width 2, starting at 64.5.
    (*Source*: www.wnba.com)

---

*Before doing these exercises, you may find it useful to review the note *How to Succeed at Mathematics* on page xix.

14. The following are the scores on a 100-point language aptitude test given to 60 people who are applying for the Peace Corps:

83, 71, 92, 87, 56, 64, 41, 95, 88, 91, 78, 73, 81, 79, 59,
73, 81, 93, 84, 66, 74, 51, 85, 78, 81, 98, 63, 91, 89, 64,
74, 61, 92, 77, 86, 79, 63, 91, 86, 91, 58, 83, 81, 77, 89,
83, 61, 83, 94, 76, 78, 61, 84, 88, 87, 68, 83, 71, 85, 64

Use classes of width 10, starting at 40.5.

In Exercises 15 and 16, represent the two sets of data on a single stem-and-leaf display.

15. A: 29, 32, 34, 43, 47, 43, 22, 38, 42, 39,
37, 33, 42, 18, 22, 39, 21, 26, 18, 43

B: 32, 38, 22, 39, 21, 26, 28, 16, 13, 20,
21, 29, 22, 24, 33, 47, 23, 22, 18, 33

16. X: 29, 42, 34, 44, 47, 43, 22, 38, 42, 59, 41, 16,
47, 43, 42, 18, 22, 49, 21, 26, 18, 45, 24, 40

Y: 32, 48, 22, 59, 21, 26, 28, 16, 14, 20, 17, 45,
21, 29, 22, 24, 34, 47, 23, 22, 18, 45, 21, 16

Applying What You've Learned

In Exercises 17 and 18, group the data as indicated and construct a histogram.

17. The following table contains the average price (in dollars) of 1 gallon of unleaded regular gasoline for 2005 to 2007. Use classes of width 20 cents, starting at $1.80. (*Source:* U.S. Bureau of Labor Statistics)

Year	Jan.	Feb.	Mar.	Apr.	May	Jun.	Jul.	Aug.	Sep.	Oct.	Nov.	Dec.
2005	1.823	1.918	2.065	2.283	2.216	2.176	2.316	2.506	2.927	2.785	2.343	2.186
2006	2.315	2.310	2.401	2.757	2.947	2.917	2.999	2.985	2.589	2.272	2.241	2.334
2007	2.274	2.285	2.592	2.860	3.130	3.052	2.961	2.782	2.789	2.793	3.069	3.020

18. The following data are the number of reports of mishandled baggage per 1,000 passengers for 10 U.S. airlines during 3 months of 2008. Use classes of width 1, starting at 3. (*Source:* U.S. Department of Transportation)

Month	AA	UA	NWA	Delta	Comair	US Air	SWA	Am. Eagle	Jet Blue	Cont.
Jan.	7.75	6.47	5.00	7.87	9.28	7.35	6.99	13.71	3.93	4.76
Feb.	6.85	5.44	4.68	6.90	8.45	6.96	5.63	12.81	3.27	4.60
Mar.	7.34	4.86	4.57	7.90	9.83	6.93	5.49	12.74	3.51	5.50

19. The manager at the local Starbucks counted the customers once each hour over a busy weekend. We summarize the results she obtained in the following bar graph; use it to answer the following questions.

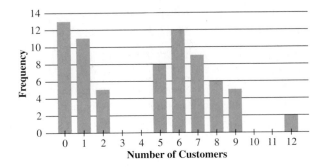

a. What was the smallest number of customers in the bar, and how often did it occur?

b. What was the largest number of customers in the bar, and how often did it occur?

c. What was the most frequently occurring nonzero customer count?

d. For how many hours were the customers counted?

e. For what fractional part of the total number of hours were there more than six customers in the bar?

20. Redo Exercise 19 using the following bar graph.

To estimate how much money to budget for snow removal, a town has kept records for the past 20 years of the number of days it has snowed each winter. The bar graph below summarizes this information. Use this graph to answer Exercises 21 and 22.

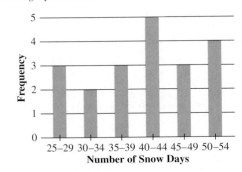

21. a. How many years had between 40 and 49 snow days?

   b. How many years had fewer than 30 snow days?

   c. How many years had at least 40 snow days?

   d. How many years had fewer than 50 snow days?

22. a. How many years had between 25 and 29 snow days?

   b. How many years had fewer than 40 snow days?

   c. How many years had at least 45 snow days?

   d. How many years had more than 34 snow days?

Comparing wage data. *The following bar graphs compare women's and men's hourly wages in a recent year. Use these graphs to answer Exercises 23–28. Because you are estimating your answers by looking at the graphs, your answers may not agree exactly with ours.*

**Percentage Earning
Less than $7.51 per Hour**

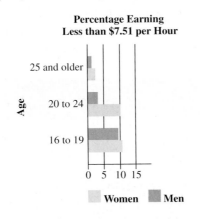

**Percentage Earning $7.51
to $9.99 per Hour**

**Percentage Earning at
Least $10.00 per Hour**

23. What percent of women ages 20 to 24 earned $7.50 or less per hour?

24. What percent of men age 25 and older earned between $7.51 and $9.99 per hour?

25. In what age and wage category do men seem to have the biggest advantage (in terms of a percentage difference) over women?

26. In what age and wage category do women seem to have the biggest advantage (in terms of a percentage difference) over men?

27. In what age and wage category do women and men seem to be most equal?

28. What general conclusions can you draw from these data?

29. Comparing training programs. Mary Kay gives sales training to its newly hired employees. To determine how effective the training is, the firm compared the monthly sales of a group that has completed the training with a group that has not. The following numbers indicate thousands of dollars of sales last month. Represent the two sets of data on a single stem-and-leaf display. Does the orientation program seem to be succeeding?

No Training: 19, 22, 34, 23, 27, 43, 42, 28,
                        32, 29, 41, 26, 28, 26, 43, 40

With Training: 29, 21, 39, 44, 41, 36, 37, 29,
                      43, 45, 28, 32, 28, 33, 36, 32

30. Comparing weight-loss programs. A hospital is testing two weight-loss programs to determine which program is more effective. The following data represent the amount of weight lost by comparable clients for a year in each program. Represent the two given sets of data on a single stem-and-leaf display. Which program seems to be more effective?

Program A: 19, 32, 27, 34, 33, 36, 47, 32, 25,
                 52, 29, 26, 37, 28, 26, 43, 31, 40

Program B: 29, 21, 39, 44, 41, 36, 37, 26, 26,
                 43, 45, 28, 32, 28, 33, 36, 53, 39

Communicating Mathematics

31. What is the difference between a population and a sample?

32. Is it better to use a frequency graph or a relative frequency graph to compare two different data sets?

33. What is the difference between a variable that is continuous versus a variable that is discrete?

34. What type of a graph do we use to graph a frequency distribution of a continuous variable quantity?

35. What do you see as an advantage in grouping data? A disadvantage?

36. How does grouping data affect the way a bar graph looks?

37. What do you think a bar graph (with grouped data) of the number of credits completed by members of your class would look like? Why?

38. What do you think a bar graph (with grouped data) of the day of the year that the members of your class were born (Jan. 1 = 1, Jan. 2 = 2, and so on) would look like? Why?

Using Technology to Investigate Mathematics

39. See your instructor for tutorials to use graphing calculators, spreadsheets, or other technology to demonstrate some of the ideas we have discussed in this section. Use the technology to reproduce some of the graphs in this section.

40. Search the Internet for "bar graph and applets" or "histogram and applets," or "stem-and-leaf display and applets." Find an interactive program that interests you, experiment with it, and report on your findings.

For Extra Credit

In Exercises 41 and 42, use the data to construct a histogram that conveys the information as well as you can. Explain how you decided on the class width and the first class. Mention any problems that the data presented to you in making an attractive histogram.

41. According to *Variety Magazine* (*World Almanac*, 2008), the all-time top grossing movies through 2007 were as follows:

Movie	Gross (millions $)
1. *Titanic* (1997)	601.0
2. *Star Wars Episode IV* (1977)	461.0
3. *Shrek 2* (2004)	437.0
4. *E.T.* (1982)	435.0
5. *Star Wars Episode I* (1999)	431.0
6. *Pirates of the Caribbean: Dead Man's Chest* (2006)	423.3
7. *Spider-Man* (2002)	404.0
8. *Star Wars Episode III— Revenge of the Sith* (2005)	380.3
9. *The Lord of the Rings: The Return of the King* (2003)	377.0
10. *Spider-Man 2* (2004)	373.4
11. *The Passion of the Christ* (2004)	370.3
12. *Jurassic Park* (1993)	357.1
13. *The Lord of the Rings: The Two Towers* (2002)	341.8
14. *Finding Nemo* (2003)	339.7
15. *Spider-Man 3* (2007)	336.5
16. *Forrest Gump* (1994)	329.7
17. *The Lion King* (1994)	328.5
18. *Shrek the Third* (2007)	321.0
19. *Harry Potter and the Sorcerer's Stone* (2001)	317.6
20. *The Lord of the Rings: The Fellowship of the Ring* (2001)	314.8
21. *Star Wars Episode II—Attack of the Clones* (2002)	310.7
22. *Star Wars Episode VI—Return of the Jedi* (1983)	309.2
23. *Transformers* (2007)	309.0
24. *Pirates of the Caribbean: At World's End* (2007)	308.3
25. *Independence Day* (1996)	306.2

42. According to Nielsen Media Research (*World Almanac*, 2008), the favorite prime-time shows in 2007 and their average audience were as follows:

Program	Avg. Audience (%)
1. *American Idol—Wednesday*	17.3
2. *American Idol—Tuesday*	16.8
3. *Dancing with the Stars*	13.3
4. *Dancing with the Stars—Monday*	12.7
5. *Dancing with the Stars— Results Show*	12.6
6. *CSI*	12.2
7. *Grey's Anatomy*	12.1
8. *Dancing with the Stars—Tuesday*	11.8
9. *House*	11.0
10. *Desperate Housewives* (tied with) *NBC Sunday Night Football*	10.8
12. *CSI: Miami*	10.7
13. *FOX NFL Sunday*	9.5
14. *Without a Trace*	9.4
15. *Deal or No Deal—Monday*	9.2
16. *Survivor: Cook Islands* (tied with) *Two and a Half Men*	9.1
18. *NCIS*	9.0
19. *Cold Case* (tied with) *CSI: NY*	8.9
21. *Criminal Minds*	8.8
22. *60 Minutes* (tied with) *Shark*	8.7
24. *Survivor: Fiji*	8.4
25. *Lost*	8.3

43. How might you present three sets of data in the same graph?

44. The following table is an example of a double-stem display. Without having this table explained to you, try to interpret its meaning and list the data items in the table.

(201–250)	2	34	45	49
(251–300)	2	57	68	77	82
(301–350)	3	23	45
(351–400)	3	62	73	78
(401–450)	4	12	34
(451–500)	4	82	89	93

45. You may find it interesting to search the Internet for "how to lie with statistics." In addition to many references to Darrell Huff's classic book *How to Lie With Statistics*, you will find other sites that explain how to use statistics to mislead your audience. Find a site that interests you and report on your findings.

15.2 Measures of Central Tendency

Objectives

1. Compute the mean, median, and mode of distributions.
2. Find the five-number summary of a distribution.
3. Apply measures of central tendency to compare data.

> *You look nice, but your voice is distinctly average, I'm afraid.* —Simon Cowell
>
> *The average adult laughs 15 times a day; the average child, more than 400 times.* —Martha Beck
>
> *In 1900 Americans on average lived for only 49 years and most working people died still on the job.* —William Greider
>
> *The average hacker lives at his parents' home and is still a student.* —SpringerLink.com

As you can see, we use the word *average* in many different ways. As practical applications, you might be concerned right now about your grade point average, and in the future, you may want to know the average starting salary in your chosen career or the average school taxes in the neighborhood where you buy a new home.

Statisticians are also interested in describing the average of a set of data and measure it in several different ways called **measures of central tendency**. The four measures of central tendency that we will discuss are the mean, median, mode, and quartiles. Except for the mode, each number will give you some sense of the center of a set of data.

 **KEY POINT**

The mean is the most common notion of average.

The Mean and the Median

When considering changing your cell phone company, you may want to know the average number of minutes that you have been calling over the past 6 months. To calculate this average, you would add the number of minutes for each of the 6 months and divide by 6. When we calculate an "average" this way, we are calculating the mean.

To calculate the mean and other numbers in statistics, we have to add lists of numbers. We use the Greek letter Σ (capital sigma) to indicate a sum.* If we have n data values, $x_1, x_2, x_3, \ldots, x_n$, then we will represent the sum of these data values by Σx. For example, we will write the sum of the data values 7, 2, 9, 4, 10 by $\Sigma x = 7 + 2 + 9 + 4 + 10$.

---

*Although we could avoid some of this formal notation in our discussion, it is good for you to become comfortable with it if you are to ever study statistics more deeply.

COMPUTING THE MEAN If a data set contains *n* data values, the **mean** $\bar{x}$ of the data set is

$$\bar{x} = \frac{\sum x}{n}.$$

We represent the mean of a *sample* of a population by $\bar{x}$ (read as "*x* bar"), and we will use the Greek letter μ (lowercase mu) to represent the mean of the whole *population*. Unless we state otherwise, we will assume that data sets in this chapter are samples rather than populations.

EXAMPLE 1 *Finding the Mean Number of Accidents*

Union leaders criticizing NAFTA's outsourcing of jobs to other countries claim that safety in factories in other countries is not as good as it is in the United States. As a result, The National Motor Corporation has been studying its safety record at its Mexican factory and found that the number of accidents over the past 5 years was 25, 23, 27, 22, and 26. Find the mean annual number of accidents for this 5-year period.

SOLUTION: To calculate the mean, we add the number of accidents and divide by 5, as follows:

$$\bar{x} = \frac{\overset{\text{Add data values.}}{\overbrace{\sum x}}}{\underset{\text{number of data values}}{\underbrace{n}}} = \frac{25 + 23 + 27 + 22 + 26}{5} = \frac{123}{5} = 24.6$$

Quiz Yourself ❹

Find each of the following for the data set 9, 12, 22, 6, 5, 15, 12, 25.
a) $\sum x$ b) n c) $\bar{x}$

This tells us that over the past 5 years, the factory in Mexico has averaged between 24 and 25 accidents per year. ✳ ❹

As you saw in Example 1, the five data values balance on either side of the mean, as shown in Figure 15.7. Notice that just like children on a seesaw, a few numbers further from the mean (the balancing point) will balance with more numbers that are close to the mean.

FIGURE 15.7 A set of data values balances about its mean.

We often use the mean to compare data to see trends. In Example 1, we can ask whether the 5-year mean of 24.6 accidents per year is good or bad. We do not know unless we compare it with some other data. For example, suppose that National Motor also has a Seattle, Washington, factory of roughly the same size and workload, but that factory had a mean of 31.8 accidents per year over the past 5 years. This difference in means shows that perhaps a problem exists at the Seattle factory instead, and management must take steps to correct it.

In Example 1, each data value occurred once. However, often in a data set, some values occur several times, in which case we use a frequency table to compute the mean.

EXAMPLE 2 *Computing the Mean of a Frequency Distribution of Water Temperatures*

The Environmental Protection Agency (EPA) suspects that hot water discharged from a nuclear power plant is responsible for a recent fish kill. To investigate this problem, the agency has recorded the water temperature at a point downstream from the plant for the last 30 days. We summarize this information in Table 15.5. What is the mean temperature for this distribution?

SOLUTION: We can simplify our calculations by remembering that repeated addition is the same as multiplication. For example, because 52 occurs four times in the distribution, we write $52 + 52 + 52 + 52$ as $52 \cdot 4$. The third column of Table 15.6 contains the products of the raw scores and their frequencies.

Temperature (°F), x	Frequency, f
52	4
53	6
54	3
55	8
56	4
57	5
Total	30

TABLE 15.5 Frequency table of water temperatures.

Temperature (°F), x	Frequency, f	Product, $x \cdot f$
52	4	$52 \cdot 4 = 208$
53	6	$53 \cdot 6 = 318$
54	3	$54 \cdot 3 = 162$
55	8	$55 \cdot 8 = 440$
56	4	$56 \cdot 4 = 224$
57	5	$57 \cdot 5 = 285$
Totals	$\Sigma f = 30$	$\Sigma(x \cdot f) = 1{,}637$

sum of frequencies ⎯⎯⎯⎯⎯⎯⎯⎯⎯⎯⎯⎯⎯⎯⎯⎯⎯⎯⎯   ⎯ sum of products

TABLE 15.6 The number of scores and the sum of the scores in the distribution of water temperatures.

The sum of the frequencies, Σf, is the same as n, the total number of scores in the distribution, and $\Sigma(x \cdot f)$ is the sum of all scores in the distribution. Therefore, the mean of the distribution is

$$\frac{\Sigma(x \cdot f)}{\Sigma f} = \frac{\text{sum of scores}}{\text{number of scores}} = \frac{1{,}637}{30} \approx 54.6°\text{F}.$$

With this information, a biologist may decide that the water is not too hot and look for another reason for the fish kill. ❋

As you saw in Example 2, when you compute the mean of a frequency distribution, *you must multiply the data values by their frequencies before adding them.*

COMPUTING THE MEAN OF A FREQUENCY DISTRIBUTION We use a frequency table to compute the mean of a data set as follows:

1. Write all products $x \cdot f$ of the scores times their frequencies in a new column of the table.

2. Represent the sum of the products you calculated in step 1 by $\Sigma(x \cdot f)$.

3. Denote the sum of the frequencies by Σf.

4. The mean is then $\dfrac{\Sigma(x \cdot f)}{\Sigma f}$.

Some Good Advice

A common mistake students sometimes make when computing the mean of a frequency distribution is to divide by the number of entries in the first column of the frequency table. This is the number of different data values, but it does not count repeated values. You must instead be sure to divide by the sum of the frequencies. **5**

Quiz Yourself **5**

A paramedic service kept track of the number of calls per day that it received over a 2-week period. Use the information in the following frequency table to find the mean number of daily calls over this period.

Number of Calls, x	4	5	6	7	8	9	10
Frequency, f	1	4	1	2	3	2	1

One of the drawbacks in using the mean to represent the average value in a data set is that one or two extreme scores can have a strong influence on the mean.

EXAMPLE 3   *The Effect of Extreme Scores on the Mean*

Table 15.7 lists the yearly earnings of some celebrities.

a)   What is the mean of the earnings of the celebrities on this list?

b)   Is this mean an accurate measure of the "average" earnings for these celebrities?

Rank	Celebrity	Earnings (2007) (millions of dollars)
1	Oprah Winfrey	260
2	Steven Spielberg	110
3	Simon Cowell	45
4	Celine Dion	45
5	David Letterman	40
6	David Beckham	33
7	Kobe Bryant	33
8	Donald Trump	32
9	U2	30
10	Angelina Jolie	20

TABLE 15.7 Earnings of some top celebrities.

Quiz Yourself **6**

To see the effect that Oprah Winfrey's and Steven Spielberg's earnings have on the mean in Example 3, recalculate the mean with these two extreme scores removed. (Note that the number of scores will now be 8, not 10.)

SOLUTION:

a)   Summing the salaries and dividing by 10 gives us

$$\frac{\sum x}{n} = \frac{260 + 110 + 45 + 45 + 40 + 33 + 33 + 32 + 30 + 20}{10} = \frac{648}{10} = \$64.8 \text{ million.}$$

b)   To answer our second question, notice that eight of the celebrities have earnings below the mean, whereas only two have earnings above the mean. Therefore, the mean in this example does not give an accurate sense of what is "average" in this set of data because it was unduly influenced by Oprah Winfrey's and Steven Spielberg's earnings. ❈ **6**

HIGHLIGHT ✺ ✺ ✺

Using a Graphing Calculator in Statistics*

Doing statistical calculations by hand when you have a large amount of data can be tedious and prone to errors. This is why people who do statistical computations in real applications use technology to make the work more manageable.

Graphing calculators such as the TI-83 and TI-84 have very powerful built-in statistical capabilities. To use these capabilities, we first store our data in lists. Figure 15.8(a) shows some of the data from Example 3 stored in list L1.

Then Figure 15.8(b) displays a menu of choices to process the data, and finally in Figure 15.8(c), we see the results of the calculations. In addition to other results, which are not relevant now, you can see the number of data, $n = 10$, the sum of the data, $\Sigma x = 648$, and the mean, $\bar{x} = 64.8$.

To calculate the mean of another set of data, we would simply change the data in L1 and, in a few seconds, redo the computations.

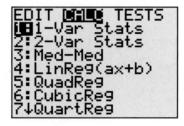 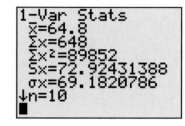

FIGURE 15.8 (a) Data.   (b) Menu of choices.   (c) Results of calculations.

Extreme scores in a data set, such as Steven Spielberg's and Oprah Winfrey's earnings, are called *outliers*, and you must decide what to do with them when you analyze data. In Example 3, you may believe that it is best to discard these two extreme values and calculate the mean based on the remaining eight earnings values. Or you may decide that it is better to describe the data using some other measure than the mean.

A measure that describes the middle of a data set is called the *median*.

 **KEY POINT**

The median tells us the middle of a set of data.

> **COMPUTING THE MEDIAN** If we arrange a set of numbers in increasing (or decreasing) order, the **median** is the middle value in the list of numbers. However, there are two cases to consider.
>
> 1. If there is an odd number of numbers, then the median is the number in the middle position.
>
> 2. If there is an even number of numbers, then the median is the average of the two middle numbers.

One reason that we often use the median to describe the middle of a set of data is that the median is not affected by a few outliers.

 Some Good Advice

A common error made when finding the median is forgetting to arrange the scores in increasing numerical order.

When Bill Clinton was campaigning for president, many political observers had the opinion that Clinton was too young and inexperienced to be president. How did Clinton's age compare with that of earlier presidents?

---

*See your instructor for a tutorial on using the TI calculators to do statistics.

T. Roosevelt	42
Taft	51
Wilson	56
Harding	55
Coolidge	51
Hoover	54
F. D. Roosevelt	51
Truman	60
Eisenhower	61
Kennedy	43
L. Johnson	55
Nixon	56
Ford	61
Carter	52
Reagan	69
G. H. W. Bush	64
Clinton	46

TABLE 15.8 Ages of U.S. presidents at inauguration.

EXAMPLE 4   *Finding the Median of Presidents' Ages*

Table 15.8 lists the ages at inauguration of the presidents who assumed office between 1901 and 1993. Find the median age for this distribution to see how Bill Clinton's age at inauguration compares with that of other presidents.

SOLUTION: The distribution of ages, when arranged in increasing order, is

Median is middle score.

$$42, 43, 46, 51, 51, 51, 52, 54, 55, 55, 56, 56, 60, 61, 61, 64, 69.$$

Because there are 17 scores in this distribution, the middle score is the ninth score, which is 55. Because Clinton's age when he took office is well below the median, we agree with the opinion that he was young when he took office. ❋

> ### PROBLEM SOLVING
> *The Analogies Principle*
>
> The Analogies Principle in Section 1.1 says that making associations with everyday words helps you understand the meaning of mathematical ideas. Recalling that the strip between a divided highway is called the *median* will help you remember that the median is the middle number in a distribution.

Often, consumer protection agencies test products to see whether they match the weight or volume stated on the package. In Example 5, we use the median to evaluate the accuracy of packaging information.

EXAMPLE 5   *Using a Frequency Table to Find the Median*

Acting on a tip from a dissatisfied customer, an agent of the consumer protection agency purchased 50 quarts of a particular brand of milk at various supermarkets to see whether they contained 32 ounces of milk. The results of this survey are reported in Table 15.9. What is the median for this distribution?

Median is between the 25th and 26th scores.

Number of Ounces, x	Frequency, f	
27	8	⎫
28	5	⎬ 25 scores here
29	12	⎭
30	16	⎫ 25 scores here
31	9	⎭
	$\Sigma f = 50$	

TABLE 15.9 Number of ounces of milk in 50 quart cartons.

SOLUTION: Notice that because the 50 scores in Table 15.9 are in increasing order, the two middle scores are in positions 25 and 26. Counting the frequencies, we see that 29 ounces is in position 25 and 30 ounces is in position 26. Therefore, the median for this distribution is $\frac{29 + 30}{2} = 29.5$. Because the median is far below the advertised 32 ounces, the variation is probably not due to randomness in filling the containers and perhaps the company packaging this milk should be fined.

Now try Exercises 5 to 12 and 17 to 20 (we discuss the mode later in this section). ❋ **7**

 *Quiz Yourself* **7**

Find the median for the following frequency distribution.

x	72	86	91	95	100
Frequency, f	2	4	3	7	3

KEY POINT

The five-number summary describes the position of data items in a data set.

The Five-Number Summary

> **DEFINITION The Five-Number Summary**
> The median divides a data set into two halves. The set of numbers below the median is called the **lower half** and the set of numbers above the median is called the **upper half**. The median of the lower half is called the **first quartile** and is indicated by Q_1; the **third quartile**, denoted by Q_3, is the median of the upper half. The **five-number summary** of a set of data values consists of the following:
>
> minimum value, Q_1, median, Q_3, maximum value

The five-number summary is an effective way to describe a set of data, as we see in Example 6.

EXAMPLE 6 *Finding the Five-Number Summary of Presidents' Ages*

Consider the list of ages of the presidents from Example 4:

$$42, 43, 46, 51, 51, 51, 52, 54, 55, 55, 56, 56, 60, 61, 61, 64, 69.$$

Find the following for this data set:

a) the lower and upper halves

b) the first and third quartiles

c) the five-number summary

SOLUTION:

a) It helps to consider the following diagram.

Median

$$\underbrace{42, 43, 46, 51, 51, 51, 52, 54,}_{\text{Lower half}} 55, \underbrace{55, 56, 56, 60, 61, 61, 64, 69}_{\text{Upper half}}$$

Because the median is in the ninth position in the list, the lower half is 42, 43, 46, 51, 51, 51, 52, 54 and the upper half is 55, 56, 56, 60, 61, 61, 64, 69.

b) The first quartile is the median of the lower half. In this diagram, you see that to find the median, we have to average the middle two scores of the lower half, so

$$Q_1 = \frac{51 + 51}{2} = 51.$$

The third quartile is

$$Q_3 = \frac{60 + 61}{2} = 60.5.$$

c) The five-number summary for this data set is 42, 51, 55, 60.5, 69.

Now try Exercises 21 to 24. ✳ **8**

Quiz Yourself **8**

a) Find the five-number summary of the entertainers' salaries listed in Example 3.

b) Represent the five-number summary in part a) by a box-and-whisker plot.

We represent the five-number summary by a graph called a **box-and-whisker plot**. The five-number summary for the president's ages in Example 6 is 42, 51, 55, 60.5, 69. We show the box-and-whisker plot for this summary in Figure 15.9.

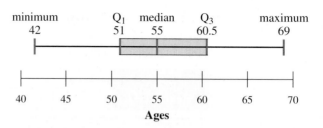

FIGURE 15.9 Box-and-whisker plot of ages of some U.S. presidents at inauguration.

The horizontal axis in Figure 15.9 represents the ages of the presidents and includes the smallest and largest ages. Next, we draw a box above the axis from the first quartile to the third quartile to show where the ages between Q_1 and Q_3 lie. We draw a vertical line through the box to show the median. Last, the lines extending to the left and right of the box—the "whiskers"—show the extreme values in the distribution. From this graph we can see quickly where the extreme scores, the median, and the middle 50% of the data lie.

We often want to know which data item occurs the most frequently in a data set. For example, a fashion designer might want to know what is the most common dress size, or an automobile manufacturer might be interested in the most common height of U.S. drivers. The mode is a measure that is easy to find to describe the most prevalent item in a data set.

> **KEY POINT**
>
> The mode is the most frequent score in a data set.

> **COMPUTING THE MODE** The **mode** of a set of data is the data item that occurs most frequently. If two items occur most frequently, then each is a mode. If more than two scores occur most frequently, we will say that there is no mode.*

EXAMPLE 7 *Finding the Mode of a Data Set*

Find the mode for each data set.

a) 5, 5, 68, 69, 70   b) 3, 3, 3, 2, 1, 4, 4, 9, 9, 9
c) 98, 99, 100, 101, 102   d) 2, 3, 4, 2, 3, 4, 5

SOLUTION:

a) The mode is 5 because that is the most frequently occurring item.

b) There are two modes, namely 3 and 9.

In parts c) and d) there is no mode because more than two items occur most often. ✳

Comparing Measures of Central Tendency

For a set of data, the mean, median, and mode are usually not the same. Therefore, when presenting a summary of data, you may want to emphasize one measure over another.

EXAMPLE 8 *Which Measure of Central Tendency Is the Best?*

Assume that you are negotiating the contract for your union at Magnum Industrial Corporation. To prepare for the next negotiation session, you have gathered annual wage data and found that three workers earn $30,000, five workers earn $32,000, three workers earn $44,000, and one worker earns $50,000. In your negotiations, which measure of central tendency should you emphasize?

SOLUTION: You want to make the wages seem as low as possible; therefore, you should choose the smallest measure of central tendency. We list the frequency distribution of wages in Table 15.10.

Salary (thousands $), x	30	32	44	50
Frequency, f	3	5	3	1

TABLE 15.10 Wage distribution at Magnum Industrial Corporation.

---
*Although some authors allow for three or more modes, we will not do so.

We see in Table 15.10 that the mode is $32,000. Because there are 12 salaries, we look at positions 6 and 7. Both contain $32,000; therefore, the median salary is also $32,000.

We use Table 15.10 to compute the mean as follows:

$$\bar{x} = \frac{\sum(x \cdot f)}{\sum f} = \frac{30 \cdot 3 + 32 \cdot 5 + 44 \cdot 3 + 50 \cdot 1}{12} = \frac{432}{12} = 36$$

The mean is therefore $36,000.

It is to the union's benefit to present the median (or the mode) of $32,000. No doubt management will claim that the mean of $36,000 is the average salary.

Now try Exercises 25 to 28. ❋

In describing data, which measure—mean, median, or mode—is most appropriate? This is not an easy question to answer. The three measures are often different, and it is really up to you to decide how you want to summarize the data.

If you are considering the mean, remember that one or two extreme values in a distribution have an undue influence on the mean. This is why, for example, in scoring ice skating at the Olympics the highest and lowest scores are discarded. If a data set has a few outliers, you may decide that it is best to discard them before computing the mean.

If the scores in a distribution are symmetrically balanced on either side of the mean, then the mean, median, and mode (assuming there is only one mode) will all be the same. However, some distributions are highly asymmetric, so there may be many small scores on one side of the mean and a few large scores on the other. In such a case, the mean is not the best measure of the center of the data. For this reason, when we read about the average family income or the average cost of a new home, the median is often used rather than the mean.

The mode emphasizes what scores occur most frequently. For example, if a shoe store sells mostly size 9s and size 11s, then it would not make sense in a report to say that the average size of the shoes sold was size 10.

In addition to describing the center of a data set, it is also important to understand how data values are spread out. In the next section, we will compute numbers to measure the spread of data that gives us a better understanding of a distribution.

Math in Your Life

You Can Observe a Lot by Just Watching (Yogi Berra)

Several years ago, I attended a town meeting held by a local state representative. As the presentation unfolded, I realized that the slide presentation was carefully crafted to appeal to the large number of senior citizens in the audience. Pointing to a graph (similar to the one

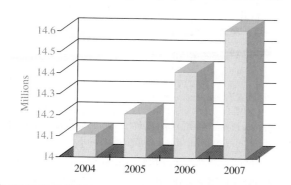

in this note*), he remarked that to keep taxes down, he was keeping an eye on the increases for higher education that had occurred over the past few years.

Does anything bother you about this graph? It should. Notice that this particular increase for higher education was $0.6 million over 3 years beginning with a base of $14 million. That amounts to an average increase of less than 1.5% per year. Also, the labels on the vertical axis are in light gray and begin with $14 million, which makes the data hard to see and causes the graph to appear to be rapidly increasing. How would you present these same data to make the case to your state legislature for greater appropriations for higher education? As an intelligent consumer of information, always be aware that people can mislead you by the way they present their data.

---

*To be honest, I did not write down the exact numbers from the presentation, but the data and the spirit of the talk were essentially as I describe it.

Exercises 15.2

Looking Back*

These exercises follow the general outline of the topics presented in this section and will give you a good overview of the material that you have just studied.

1. In computing the mean from Table 15.5, we divided the sum of the scores by $n = 30$. What is a common mistake that people make when doing this computation?

2. What was the significance of Steven Spielberg's and Oprah Winfrey's scores in Example 3?

3. What is a good way to remember the meaning of the median in Example 4?

4. What was the point of the Math in Your Life box on page 735?

Sharpening Your Skills

Find the mean, median, and mode for the following distributions.

5. 4, 6, 8, 3, 9, 11, 4, 7, 5

6. 8, 9, 4, 2, 10, 5, 5, 3, 3

7. 4, 6, 4, 6, 7, 9, 3, 9, 10, 11

8. 12, 11, 7, 9, 8, 6, 4, 5, 10, 1

9. 7, 3, 1, 5, 8, 6, 2, 5, 9, 4

10. 12, 4, 4, 8, 4, 7, 9, 8, 7, 7

11. 7, 8, 6, 5, 7, 10, 2, 7, 9, 5, 8, 8, 10, 9, 6, 5, 10, 7, 9, 8

12. 8, 8, 6, 6, 7, 10, 4, 7, 9, 5, 9, 8, 10, 10, 6, 5, 10, 8, 9, 8

In Exercises 13 to 16, use the given graph to find the mean, median, and mode of the distribution.

13.

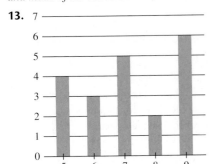

14.

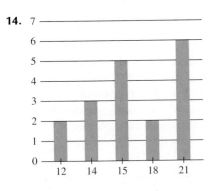

15.

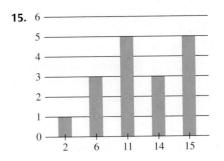

16.

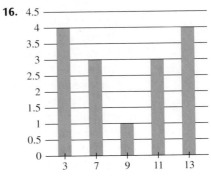

Find the mean, median, and mode for the following frequency distributions.

17.

x	f
3	2
4	5
6	3
7	1
8	4
10	3
11	2

18.

x	f
2	3
5	2
7	3
8	2
9	1
10	5
11	2

19.

x	f
5	1
8	3
11	6
12	2
14	5
15	4
18	3

20.

x	f
3	5
6	2
8	8
11	3
15	6
17	2
23	1

In Exercises 21–24, a) give the five-number summary for the distribution, and b) draw a box-and-whisker plot.

21. 11, 23, 25, 17, 26, 31, 45, 18, 41, 26, 31, 33, 48, 44, 53

22. 21, 24, 15, 45, 18, 31, 26, 41, 23, 18, 44, 27, 36, 21, 43

*Before doing these exercises, you may find it useful to review the note *How to Succeed at Mathematics* on page xix.

23. 31, 25, 41, 33, 28, 34, 37, 41, 33, 29, 49, 32, 38, 45, 30

24. 13, 24, 27, 45, 32, 29, 28, 39, 25, 21, 37, 36, 42, 34, 49

Applying What You've Learned

In Exercises 25–28, find the mean, median, and mode for each data set. In each case, explain which measure you believe best describes a typical value in the distribution.

25. Ozone pollution. Ozone is formed when sunlight acts on nitrogen oxides. The number of millions of short tons of nitrogen oxides emitted into the air each year from 1994 to 2006 are given in the accompanying table. (*Source:* U.S. Environmental Protection Agency)

Year	Millions of Tons of Nitrogen Oxides
1994	25
1995	25
1996	26
1997	26
1998	26
1999	25
2000	23
2001	22
2002	21
2003	21
2004	20
2005	19
2006	18

26. Baseball records. A minor league baseball team has had the following yearly numbers of wins over the past 20 years:

38, 38, 36, 27, 21, 26, 39, 37, 23, 27,
36, 26, 26, 37, 30, 28, 37, 27, 29, 27

27. Celebrity earnings. According to www.forbes.com, here are the earnings of a number of deceased celebrities for 2007.

Year	Earnings (millions $)
1. Elvis	54
2. John Lennon	44
3. Charles M. Schultz	35
4. George Harrison	22
5. Albert Einstein	18
6. Andy Warhol	15
7. Dr. Suess (Theodor Geisel)	13
8. Tupac Shakur	9
9. Marilyn Monroe	7
10. Steve McQueen	6

28. AIDS deaths. The following table shows the cumulative number of deaths due to AIDS from 1998 to 2005. For your data, use the number of AIDS deaths per year for the years 1998–2005.

Year	Cumulative Number of Deaths (thousands)
1998	432
1999	451
2000	468
2001	485
2002	502
2003	520
2004	538
2005	555

In Exercises 29 and 30, find the mean, median, and mode for each data set and explain which measure you believe best describes a typical value in the distribution.

29. Presidential vetoes. The following is a summary of the number of vetoes (including pocket vetoes) for U.S. presidents from Franklin D. Roosevelt to George W. Bush. (*Source: The World Almanac,* 2008)

President	Number of Vetoes
F. D. Roosevelt	635
Truman	250
Eisenhower	181
Kennedy	21
L. Johnson	30
Nixon	43
Ford	66
Carter	31
Reagan	78
G. H. W. Bush	44
Clinton	37
G. W. Bush	3

30. Ages of Supreme Court justices. At the start of the 2007–2008 term of the Supreme Court, the nine justices and their years of birth were Stevens (1920), Scalia (1936), Kennedy (1936), Souter (1939), Thomas, (1948), Ginsburg (1933), Breyer (1938), Roberts (1955), and Alito (1950). (*Source: The World Almanac,* 2008) For data, use the ages of the justices on December 31, 2007.

Many colleges assign numerical points to grades as follows: A − 4, B − 3, C − 2, D − 1, and F − 0. Then your grade point average (GPA) is computed by multiplying the number of credits for each course by its numerical grade, adding these products, and dividing by the number of credits. For example, if you get an A in a 3-credit history course and a D in a 2-credit personal fitness course, then your grade point average would be $\dfrac{3 \times 4 + 2 \times 1}{3 + 2} = \dfrac{14}{5} = 2.8$. In Exercises 31 and 32, use this method to compute the GPA for each semester grade report.

31.

Course	Credits	Grade
English	3	A
Speech	2	B
Mathematics	3	A
History	3	F
Physics	4	B

32.

Course	Credits	Grade
Philosophy	3	C
Health	2	A
Mathematics	3	B
Psychology	3	D
Music	1	B

33. Exam scores. Izzy had an 84 and an 86 on his first two tests and believed that he did well enough on his final exam to keep his B average. However, when he got his grade he received a D. He checked with his instructor and learned that the instructor had made a mistake by transposing the digits when he recorded the final exam grade. If Izzy's incorrect average (mean) for the class was 69, what was his correct final exam score?

34. Hours worked. The mean number of hours that Raphael worked over the past four weeks is 38.75. If he works 42 hours this week, what will the mean number of hours be that he will have worked over the 5-week period?

35. Mileage ratings. The EPA has determined the number of miles per gallon (MPG) for 58 foreign cars, as given in the following frequency table. Find the mean, median, and mode for these data.

MPG	19	20	21	22	23	24	25	26	27	28	29	30
Frequency	2	5	3	5	4	8	8	6	3	5	7	2

36. Mileage ratings. The Rolls-Royce and Jaguar XJ12 have ratings of 11 MPG. Suppose this new score with its frequency of 2 is included in the table in Exercise 35. What are the new mean, median, and mode of the distribution?

37. Exam scores. Assume that in your History of Film class you have earned test scores of 78, 82, 56, and 72, and only one test remains.

a. If you need a mean score of 70 to earn a C, then what must you obtain on the final test?

b. If you need a mean score of 80 to earn a B, what must you obtain on the final test?

38. Exam scores. Assume that in your Abnormal Psychology class you have earned test scores of 74, 81, 56, and 70, and only one test remains.

a. If you need a mean score of 70 to earn a C, then what must you obtain on the final test?

b. If you need a mean score of 80 to earn a B, what must you obtain on the final test?

Credit card companies often compute the average daily balance* on your account as follows: If you start on the first of a 31-day month with a balance due of $100, and then on the 5th you charge another $50 and on the 27th you charge another $20, they would say that on days 1, 2, 3, and 4 you owed $100; on days 5, 6, 7, . . . , 25, and 26 you owed $150; and on days 27, 28, 29, 30, and 31 you owed $170. Therefore, to calculate the average daily balance, they would compute

$$\frac{4 \cdot 100 + 22 \cdot 150 + 5 \cdot 170}{31} = \frac{4{,}550}{31} = \$146.77.$$

Use this method to answer Exercises 39 and 40.

39. Credit card balance. Assume that in a 31-day month you begin with a $50 balance due on your credit card, then charge an item for $75 on the 10th of the month and a $120 item on the 25th of the month. What is your average daily balance on your credit card for this month?

40. Credit card balance. Assume that in a 31-day month you begin with an $80 balance due on your credit card, then charge an item for $60 on the 5th of the month and a $100 item on the 20th of the month. What is your average daily balance on your credit card for this month?

41. Draw box-and-whisker plots for the distributions of the number of home runs hit for the two given time periods in Example 6 in Section 15.1.

42. In considering your answer to Exercise 41, which do you feel conveys the home run information better—the stem-and-leaf displays or the box-and-whisker plots?

A college placement office has made a comparative study of the starting salaries for graduates in various majors. Use the following box-and-whisker plots, which describe the starting salaries for business, education, engineering, and liberal arts, to answer Exercises 43–48.

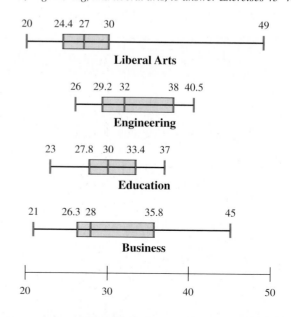

---

*The average daily balance method is covered thoroughly in Section 9.2.

43. Salaries for college graduates. Which major, in general, receives the worst salary offers?

44. Salaries for college graduates. If a student wants to receive $32,000 or more for a starting salary, which major offers the best chance of achieving that?

45. Salaries for college graduates. Does education or business have the higher median salary? Does this imply that salaries in education are generally higher than those in business? Discuss.

46. Salaries for college graduates. What is the significance of the long whisker in the liberal arts plot?

47. Salaries for college graduates. What is the significance of the short whisker in the engineering plot?

48. Salaries for college graduates. Try to describe specific situations that might lead a student to a misinterpretation of the information contained in these plots.

Communicating Mathematics

49. What is the difference between Σx and $\Sigma x \cdot f$?

50. What are the five numbers in the five-number summary?

51. What is a good way to represent the five-number summary graphically?

52. Give three separate real-life examples when the mean, median, or mode would be appropriate measures of the average in a distribution. Explain why you think that each measure would be appropriate in the given situation.

53. Sports. Statements similar to the following two statements are somewhat common in sports discussions:

"I read that the Golden Bears averaged 7 yards per carry in Saturday's game."

"I saw all of our team's basketball games and I would say that Gunner averages 21 points a game."

Which measure of central tendency do you think is being used in each statement?

54. Consumer costs. What measure of central tendency do you think is being used in each of the following statements?

"The average cost of a new house is $123,000."

"The typical new car is in the $14,000–$18,000 range."

Using Technology to Investigate Mathematics

55. See your instructor for tutorial to use graphing calculators, spreadsheets, or other technology to demonstrate some of the ideas we have discussed in this section. Use the technology to reproduce some of the graphs in this section.

56. Search the Internet for applets that will reproduce some of the computations and graphs that we have shown you in this section. Report on your findings.

For Extra Credit

57. The Super Bowl. Many have criticized the Super Bowl as uninteresting because the games have not been close contests. Use the methods that you have learned in this chapter to make an argument for or against this statement using as data the margins of victory for each Super Bowl since 1967.

58. Comparing sports heroes. If you are a Chicago Cubs fan, you might believe that Greg Maddux has been the most dominant pitcher in baseball over the last 15 years. A Boston Red Sox fan may disagree and think that Curt Schilling has been better. Use the methods of this chapter to compare these two pitchers. Use statistics from the last 15 years that you can find at www.baseball-reference.com. Write a brief report on your findings.

59. Suppose there are nine scores from 1 to 10 in a distribution. If possible, give an example of each of the following. If you think that it is not possible to give the requested distribution, try to explain why you believe this.

   a. a distribution in which the mean, median, and mode are each 5

   b. a distribution in which the median is 3 and the mean is 5

   c. a distribution in which the median is 5 and the mean is less than 5

60. Suppose there are nine scores from 1 to 10 in a distribution. If possible, give an example of each of the following. If you think that it is not possible to give the requested distribution, try to explain why you believe this.

   a. a distribution in which the mean is 5, the median is 4, and the mode is 2

   b. a distribution in which the mean is 3 and the median is 5

   c. a distribution in which the mean is 5 and the median is less than 5

If possible, give examples of data sets (sets of numbers) that have the following properties.

61. The mean, median, and mode are all the same.

62. The mean, median, and mode are all different.

63. The mean is larger than the median.

64. The median is larger than the mean.

65. Suppose that distributions X and Y have means $\bar{x}$ and $\bar{y}$, respectively. If we combine X and Y into a single distribution, is the mean of this new distribution $\bar{x} + \bar{y}$? Explain.

66. Suppose that distributions X and Y have medians a and b, respectively. If we combine X and Y into a single distribution, is the median of this new distribution $a + b$? Explain.

67. Give an example of two distributions that have the same mean, median, and mode. In one distribution, the data values should be tightly bunched together; in the other, the data values should be spread apart.

15.3 Measures of Dispersion

Objectives

1. Compute the range of a data set.
2. Understand how the standard deviation measures the spread of a distribution.
3. Use the coefficient of variation to compare the standard deviations of different distributions.

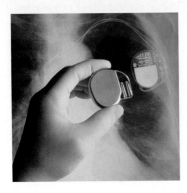

Let's imagine that you are a cardiac surgeon faced with deciding which of two heart pacemaker batteries you should choose. Because you must operate to replace a failing battery, it is important that you choose the one that will last the longest. Assume that battery A lasts for a mean time of 45,000 hours (slightly more than 5 years) and battery B has a mean time of 46,000 hours.

On the surface, it would seem that you should choose battery B. However, suppose that we also tell you that in testing the batteries, we found that all of A's times were within 500 hours of the mean, but B's times varied widely. In fact, some of B's times were as much as 2,500 hours below the mean. So B's times could be as low as $46,000 - 2,500 = 43,500$ hours, whereas A's times were never below 44,500. Based on this information, it appears that battery A is the better choice.

The point of our story is that although the mean and median tell you something about a distribution, they do not tell the whole story. For example, the following two distributions both have a mean and median of 25, but Y's values are much more spread out than X's:

$$X: 24, 25, 25, 25, 25, 26 \qquad Y: 1, 2, 3, 47, 48, 49$$

From these examples, it is clear that we need to develop some method for measuring the spread of a distribution.

✎ **KEY POINT**

The range is a crude measure of the spread of a data set.

The Range of a Data Set

It is clear from the discussion of pacemaker batteries that a numerical measure of the dispersion, or spread, of a data set can provide useful information. One simple way to describe the spread of a set of data is to subtract the smallest data value from the largest.

> **DEFINITION** The **range** of a data set is the difference between the largest and smallest data values in the set.

EXAMPLE 1   *Comparing Heights*

Find the range of the heights of the people listed in the accompanying table.

Person	Height	Height in Inches
Leonid Stadynk (World's Tallest Person)	8 feet, 6 inches	102 inches
LaDainian Tomlinson	5 feet, 10 inches	70 inches
Madge Bester (World's Shortest Person)	2 feet, 2 inches	26 inches
LeBron James	6 feet, 8 inches	80 inches

SOLUTION: The range of these data is

range = largest data value − smallest data value = $102 - 26 = 76$ inches = 6 feet, 4 inches. ❋

HISTORICAL HIGHLIGHT ❈ ❈ ❈

The Origins of Statistics

Ring a ring of roses,
A pocket full of posies,
*Asha Asha, We all fall down.**

It may have been King Henry VII's fear of the dreaded Black Plague that prompted him to begin publishing weekly Bills of Mortality in 1532. John Graunt, a merchant, noticed patterns in these reports regarding deaths due to accident, suicide, and disease and concluded that social phenomena do not occur randomly. He published his observations in a paper titled, *"Natural and Political Observations . . . Made upon the Bills of Mortality."* King Charles II, impressed by

Graunt's work, nominated him to the Royal Society of London even though he had few academic credentials.

Edmund Halley, the noted English astronomer, continued the work of Graunt in a paper titled *"An Estimate of the Degrees of the Mortality of Mankind, Drawn from Curious Tables of the Births and Funerals at the City of Breslaw; with an Attempt to Ascertain the Price of Annuities upon Lives."* By studying the mathematics of life expectancy, Halley helped lay the foundations for actuarial science, which insurance companies use today to determine their premiums.

The range is generally not useful to measure the spread of a distribution because it can be influenced by a single outlier. For example, the range of the distribution 2, 2, 2, 2, 3, 4, 100 is $100 - 2 = 98$.

🖊 KEY POINT

The standard deviation is a reliable measure of dispersion.

Data Value, x	Deviation from Mean, $x - \bar{x}$
16	−1
14	−3
12	−5
21	4
22	5
Total	0

TABLE 15.11 Deviations from the mean for data values in distribution *A*.

Standard Deviation

A better way to find the spread of data is to use the standard deviation, a measure based on calculating the distance of each data value from the mean.

> **DEFINITION** If x is a data value in a set whose mean is $\bar{x}$, then $x - \bar{x}$ is called x's *deviation from the mean.*

To illustrate the deviation from the mean, consider distribution *A*: 16, 14, 12, 21, 22, whose mean is 17. We list the deviation of each data value from the mean in Table 15.11.

It might seem reasonable to measure the spread in *A* by averaging the deviations from the mean in Table 15.11. However, this will not work. As you see in Figure 15.10, for scores above the mean that have positive deviations from the mean, there must be scores below the mean that have negative deviations from the mean. When we add the positive and negative deviations, they cancel each other, giving a total of zero.

To avoid this cancellation, we square each of these deviations, as shown in Table 15.12.

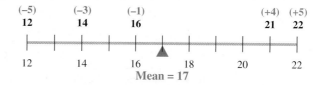

FIGURE 15.10 Values having positive and negative deviations from the mean must balance about the mean.

---

*Some see this rhyme as describing the rosy rash, the pockets full of medicinal herbs, the sneezing and eventual death associated with the Black Plague. If so, it owes its origin, like statistics, to that horrible time in European history.

Data Value, x	Deviation from Mean, $x - \bar{x}$	Deviation Squared, $(x - \bar{x})^2$
16	−1	1
14	−3	9
12	−5	25
21	4	16
22	5	25

TABLE 15.12 The squares of deviations from the mean for distribution *A*.

If we now average* these squared deviations, we get

$$\underbrace{\frac{1 + 9 + 25 + 16 + 25}{5 - 1}}_{n-1} = \frac{76}{4} = 19.$$

Clearly, this number is too large to represent the spread of scores from the mean. To compensate for the fact that we squared the deviations in doing our calculations, we take the square root of this number to get $\sqrt{19} \approx 4.36$. This number is a more reasonable measurement of the way scores vary from the mean. The quantity that we just calculated is called the *standard deviation*.

DEFINITION We denote the **standard deviation** of a *sample* of *n* data values by *s*,[†] which is defined as follows:

$$s = \sqrt{\frac{\sum(x - \bar{x})^2}{n - 1}}$$

We compute the standard deviation in four steps.

COMPUTING THE STANDARD DEVIATION To compute the standard deviation for a sample consisting of *n* data values, do the following:

1. Compute the mean of the data set; call it $\bar{x}$.

2. Find $(x - \bar{x})^2$ for each score x in the data set.

3. Add the squares found in step 2 and divide this sum by $n - 1$; that is, find

$$\frac{\sum(x - \bar{x})^2}{n - 1},$$

   which is called the **variance**.

4. Compute the square root of the number found in step 3.

EXAMPLE 2   *Calculating the Standard Deviation*

Dunder Mifflin has hired six interns. After 4 months, their work records show the following number of work days missed for each worker:

$$0, 2, 1, 4, 2, 3$$

Find the standard deviation of this data set.

---

*When doing this calculation for a *sample* (as opposed to a *population*), statisticians divide by $n - 1$ instead of n. There are technical reasons for doing this that are beyond the scope of this text. In doing this calculation for a population, we would divide by n rather than $n - 1$.

[†]We represent the standard deviation of a population by σ (instead of *s*), which we calculate using the formula

$$\sigma = \sqrt{\frac{\sum(x - \mu)^2}{n}}.$$

SOLUTION: The mean of this data set is $\dfrac{0+2+1+4+2+3}{6} = \dfrac{12}{6} = 2$. We calculate the squares of the deviations of the data values from the mean in the following table.

Number of Days Missed	Deviation from Mean	Square of Deviation from the Mean
0	−2	4
2	0	0
1	−1	1
4	2	4
2	0	0
3	1	1
		$\Sigma(x-2)^2 = 10$

There are six scores in the distribution, so the standard deviation is

$$s = \sqrt{\dfrac{\overbrace{\Sigma(x-2)^2}}{\underbrace{6-1}_{n-1}}} = \sqrt{\dfrac{10}{5}} \approx 1.41.$$

This relatively small standard deviation shows that the data values are not spread too far from the mean of 2.

Now try Exercises 5 to 14. ✳ **9**

Quiz Yourself **9**

Find the standard deviation of the following sample of data values:

3, 4, 5, 6, 4, 2, 0, 8, 4

In Example 3, we calculate the standard deviation of a distribution using a frequency table.

EXAMPLE 3 *Using a Frequency Table to Compute the Standard Deviation*

Anya is considering investing in WebSoft, a software company that specializes in Internet applications. She intends to compute the standard deviation of its recent price changes to measure how steady its stock price has been recently. The following are the closing prices for the stock for the past 20 trading sessions:

37, 39, 39, 40, 40, 38, 38, 39, 40, 41,
41, 39, 41, 42, 42, 44, 39, 40, 40, 41

What is the standard deviation for this data set?

SOLUTION: In Table 15.13, we see that the sum of the 20 closing prices is 800, so the mean is $\dfrac{800}{20} = 40$. We have added columns to the table for the deviations from the mean, the squares of these deviations, and so on.

Closing Price, x	Frequency, f	Product, $x \cdot f$	Deviation, $(x - 40)$	Deviation Squared, $(x - 40)^2$	Product, $(x - 40)^2 \cdot f$
37	1	37	−3	9	9
38	2	76	−2	4	8
39	5	195	−1	1	5
40	5	200	0	0	0
41	4	164	1	1	4
42	2	84	2	4	8
44	1	44	4	16	16
	$\Sigma f = 20$	$\Sigma(x \cdot f) = 800$			$\Sigma(x-40)^2 \cdot f = 50$

TABLE 15.13 Computations necessary to find the standard deviation of WebSoft closing prices.

When calculating the standard deviation, we must multiply the squared deviation of each price from the mean by its frequency. We list these products in the column labeled "Product, $(x - 40)^2 \cdot f$," and show the sum of these products at the bottom of the column.

We can now calculate the standard deviation. Remember that Σf is the number of data values that we have been representing by n. The standard deviation is therefore

$$s = \sqrt{\frac{\Sigma(x - 40)^2 \cdot f}{n - 1}} = \sqrt{\frac{50}{19}} \approx 1.62.$$

This relatively small standard deviation indicates that the closing prices for the Web-Soft stock have not been varying very much lately. If Anya is a cautious investor, she will find the stability of WebSoft's prices appealing.

Now try Exercises 15 to 20. ❀

We summarize the method we use for finding the standard deviation in the following formula.

> **FORMULA FOR COMPUTING THE SAMPLE STANDARD DEVIATION FOR A FREQUENCY DISTRIBUTION** We calculate the standard deviation, s, of a sample that is given as a frequency distribution as follows:
>
> $$s = \sqrt{\frac{\Sigma(x - \bar{x})^2 \cdot f}{n - 1}}$$
>
> where $\bar{x}$ is the mean of the distribution, f is the frequency of data value x, and $n = \Sigma f$, the number of data values in the distribution. **10**

Quiz Yourself **10**

The investor in Example 3 also gathered the closing stock prices for WebRanger, another Internet software company. We list these stock prices in Table 15.14. The mean closing price is 42. Find the standard deviation for this distribution.

Closing Price, x	36	37	38	39	40	41	42	43	44	45
Frequency, f	2	0	0	3	1	3	0	1	5	5

TABLE 15.14 WebRanger closing stock prices for 20 days.

Comparing the standard deviation of 1.62 for WebSoft with the standard deviation of 3.01 for WebRanger in Quiz Yourself 10, we see that WebRanger's stock is much more volatile than WebSoft's stock.

We have said that the standard deviation describes the spread of a distribution. In Figure 15.11, all three distributions have a mean and median of 5; however, as the spread of the distribution increases, so does the standard deviation.

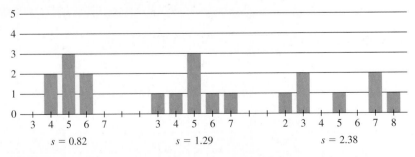

FIGURE 15.11 As the spread of the distribution increases, so does the standard deviation.

— HIGHLIGHT ✺ ✺ ✺

Using a Graphing Calculator to Find the Standard Deviation*

Because calculating the standard deviation can become tedious, we often use a calculator to save work, as we show in the accompanying TI calculator screen. We have stored the data from Example 3 in list L1, and then used the calculator to find the standard deviation, as we see in the screen below.

With such powerful technology available, it is very important that you think critically to set up problems correctly rather than just focusing on doing computations.

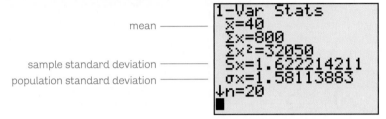

mean —————

sample standard deviation —————
population standard deviation —————

```
1-Var Stats
 x̄=40
 Σx=800
 Σx²=32050
 Sx=1.622214211
 σx=1.58113883
↓n=20
■
```

 *KEY POINT*

We use the coefficient of variation to compare the standard deviations of different data sets.

The Coefficient of Variation

In Example 3, we used the standard deviation with one set of data. If you use the standard deviation to compare two sets of data, the data must be similar and have comparable means. The following situation shows why.

Suppose we are comparing the body weights of two groups of people and find the standard deviation to be 3 pounds for the first group and 10 pounds for the second group. Can we state that there is more uniformity in the first group than the second? Before answering this question, here is more information. The first group consists of preschool children, whereas the second is a group of National Football League linemen. Clearly, the standard deviation of 3 is more significant, *relatively speaking*, for a group of preschoolers with a mean weight of 30 pounds than the standard deviation of 10 for a group of football players with a mean weight of 300 pounds. In order to use the standard deviation effectively to compare different sets of data, we must first make the numbers comparable. We do this by finding the coefficient of variation.

> **DEFINITION** For a set of data with mean $\bar{x}$ and standard deviation s, we define the **coefficient of variation**, denoted by CV, as
>
> $$CV = \frac{s}{\bar{x}} \cdot 100\%.$$

Note that the coefficient of variation compares the standard deviation to the mean and is expressed as a percentage. The coefficient of variation for the group of preschool children is

$$CV = \frac{3}{30} \cdot 100\% = 10\%.$$

In contrast, the coefficient of variation for the group of NFL football linemen is

$$CV = \frac{10}{300} \cdot 100\% = 3.3\%.$$

Because the coefficient of variation is larger for the group of preschoolers, we conclude that there is more variation, *relatively speaking*, in their weights than in the weights of the football players.

———————
*Ask your instructor for tutorials on using technology in statistics.

EXAMPLE 4 *Using the Coefficient of Variation to Compare Data*

Use the coefficient of variation to determine whether the women's 100-meter race (328 feet) or the men's marathon (26 miles) has had more consistent times over the five Olympics listed in Table 15.15.

	Women's 100 Meters	Men's Marathon
2004	10.93 sec	2 h, 10 m, 55 sec (7,855 sec)
2000	10.75 sec	2 h, 10 m, 11 sec (7,811 sec)
1996	10.94 sec	2 h, 12 m, 36 sec (7,956 sec)
1992	10.82 sec	2 h, 13 m, 23 sec (8,003 sec)
1988	10.54 sec	2 h, 10 m, 32 sec (7,832 sec)

TABLE 15.15 Olympic times in women's 100-meter race and men's marathon.

SOLUTION: Using a calculator, we found that the mean time for the 100-meter race is 10.796 and the standard deviation is approximately 0.163. For the marathon, the mean is 7,891.4 and the standard deviation is approximately 83.58.

Thus, the coefficient of variation for the women's 100-meter race is

$$\frac{0.163}{10.796} \cdot 100\% \approx 1.51\%.$$

For the men's marathon, the coefficient of variation is

$$\frac{83.5}{7,891.4} \cdot 100\% \approx 1.06\%.$$

Surprisingly, using the coefficient of variation as a measure, there is less variation in the times for the marathon than for the 100-meter race! ❊

Exercises 15.3

Looking Back*

These exercises follow the general outline of the topics presented in this section and will give you a good overview of the material that you have just studied.

1. What problem did we point out in Example 1 that can occur when we use the range to measure the spread of a distribution?

2. What do you get if you average the deviations from the mean without squaring them?

3. What is the purpose of the coefficient of variation?

4. What was our point in the technology Highlight?

Sharpening Your Skills

If you want, use a calculator or computer software to solve the following exercises. Some of your answers may differ slightly from ours due to differences in the way we round off our calculations.

Find the range, mean, and standard deviation for the following data sets.

5. 19, 18, 20, 19, 21, 20, 22, 18, 18, 17

6. 89, 72, 100, 87, 65, 98, 77, 92

7. 5, 7, 9, 4, 6, 8, 7, 10

8. 8, 4, 7, 6, 5, 5, 4, 9

9. 2, 12, 3, 11, 14, 5, 8, 9

10. 21, 3, 5, 11, 7, 1, 9, 7

11. 18, 3, 8, 7, 7, 9, 13, 7

12. 22, 18, 15, 21, 21, 15, 19, 13

13. 3, 3, 3, 3, 3, 3, 3

14. 4, 6, 4, 6, 4, 6, 4, 6

15. The following frequency table shows the number of fires per month in a city. Complete the table entries to find the mean and standard deviation for this distribution.

*Before doing these exercises, you may find it useful to review the note *How to Succeed at Mathematics* on page xix.

Number, x	Frequency, f	Product, $x \cdot f$	Deviation, $(x - \bar{x})$	Deviation², $(x - \bar{x})^2$	Product, $(x - \bar{x})^2 \cdot f$
2	1				
3	1				
4	0				
5	2				
6	3				
7	4				
8	4				
9	3				
10	2				
	$\Sigma f =$	$\Sigma x \cdot f =$			$\Sigma (x - \bar{x})^2 \cdot f =$

Table for Exercise 15

16. The following frequency table summarizes the number of job offers made to graduates of a network administrator certification program. Complete the table entries to find the mean and standard deviation for this distribution.

Number, x	Frequency, f	Product, $x \cdot f$	Deviation, $(x - \bar{x})$	Deviation², $(x - \bar{x})^2$	Product, $(x - \bar{x})^2 \cdot f$
4	2				
5	0				
6	3				
7	5				
8	4				
9	5				
10	1				
	$\Sigma f =$	$\Sigma x \cdot f =$			$\Sigma (x - \bar{x})^2 \cdot f =$

Find the mean and standard deviation for the following frequency distributions.

17.

x	f
2	2
3	4
4	2
5	0
6	4

18.

x	f
8	3
9	2
10	1
11	2
12	3

19.

x	f
3	1
4	1
5	0
6	2
7	5
8	3
9	3

20.

x	f
12	3
13	0
14	5
15	0
16	1
17	2
18	3

Use the following graphs to solve Exercises 21 and 22.

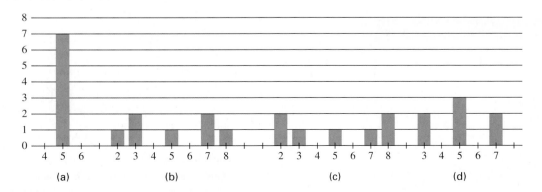

(a)   (b)   (c)   (d)

21. a. Which data set has the smallest standard deviation?

b. The largest?

22. Rank the data sets in order from smallest standard deviation to largest. You can do this just by looking at the graphs and without doing any calculations.

Applying What You've Learned

23. Summarizing test score data. The following are the scores of 20 people who took a paramedics licensing test. Find the mean and standard deviation for these data.

Score	72	73	78	84	86	93
Frequency	7	1	3	6	2	1

24. Summarizing age data. The following are the ages of 16 World War II veterans who are attending a reunion commemorating the Normandy invasion. Find the mean and the standard deviation for these data.

Age	82	83	84	86	93	95
Frequency	4	2	4	2	3	1

25. Summarizing tax data. The following table lists the state income tax for a person earning $50,000 per year for several states. Find the mean and sample standard deviation for these data. (*Source: The World Almanac*, 2008)

State	Income Tax (%)
Arizona	3.36
Colorado	4.63
Hawaii	8.25
Kansas	6.45
Massachusetts	5.3
New York	6.85
Pennsylvania	3.07
Virginia	5.75

26. Summarizing vital statistics. The following is the 2008 roster of the Houston Comets in the WNBA. Find the mean and the *population* standard deviation of these data.

Player	Weight (lbs)
Matee Ajavon	160
Latasha Byears	206
Tamecka Dixon	148
Sequoia Holmes	155
Shannon Johnson	152
Crystal Kelly	190
Sancho Lyttle	175
Mwadi Mabika	165
Hamchétou Maïga-Ba	160
Ashley Shields	155
Michelle Snow	158
Tina Thompson	178
Marcedes Walker	253
Erica White	135
Mistie Williams	184

Source: www.wnba.com

27. Summarizing customer data. The manager at the local Starbucks counted the customers once each hour over a busy weekend. We summarize the results she obtained in the bar graph below. Find the mean and standard deviation of these data.

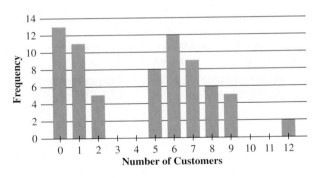

28. Find the mean and sample standard deviation for the distribution given by the following bar graph.

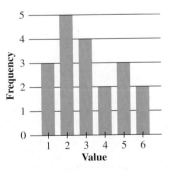

29. Family incomes. The following table gives the annual incomes for eight families, in thousands of dollars. Find the number of standard deviations family H's income is from the mean.

Family	A	B	C	D	E	F	G	H
Annual Income (in $thousands)	47	48	50	49	51	47	49	51

Table for Exercise 29

30. Family incomes. The following table gives the annual incomes for eight families, in thousands of dollars. Find the number of standard deviations family A's income is from the mean.

Family	A	B	C	D	E	F	G	H
Annual Income (in $thousands)	49	51	52	51	51	50	52	52

In Exercises 31 and 32, you are given a distribution of averages in a literature class. The professor in this class assigns grades as follows:*

A: *averages that are at least 1.5 standard deviations above the mean*

B: *averages that are between 0.5 standard deviation above the mean, and 1.5 standard deviations above the mean*

C: *averages that are between 0.5 standard deviation below the mean, and 0.5 standard deviations above the mean*

D: *averages that are between 1.5 standard deviations below the mean, and 0.5 standard deviation below the mean*

F: *averages that are more than 1.5 standard deviations below the mean*

31. Assigning grades. The averages in the class are 80, 76, 81, 84, 79, 80, 90, 75, 75, and 80. What grade does the person earning the 76 get?

32. Assigning grades. The grades in the class are 72, 71, 73, 70, 71, 79, 65, 73, 74, and 72. What grade does the person earning the 74 get?

In Exercises 33–36, we present information on the performance of various stocks during a given month. Stocks with greater coefficients of variation are considered more volatile.

33. Comparing stocks. Suppose for a given month that the mean daily closing price for Apple Computer common stock was 123.76 and the standard deviation was 12.3. For Dell stock, the mean daily closing price was 78.6 with a standard deviation of 7.2. Which stock was more volatile?

34. Comparing stocks. Suppose for a given month that the mean daily closing price for Netflix was 37.4 and the standard deviation was 4.1. For Blockbuster, the mean daily closing price was 18.6 with a standard deviation of 3.2. Which stock was more volatile?

35. Comparing stocks. The Dow Jones Industrial Average (DJIA) measures the stock prices of a large group of stocks. If, for a given week, the DJIA had a mean daily closing price of 11,261.12 with a standard deviation of 72.17 and WebMaster stock had a mean closing price of 37.6 with a standard deviation of 1.7, did WebMaster have more or less volatility than the DJIA during that week?

36. Comparing stocks. If, for a given week, the DJIA had a mean daily closing price of 10,834.8 with a standard deviation of 144.5 and WebRanger stock had a mean closing price of 123.6

with a standard deviation of 1.2, did WebRanger have more or less volatility than the DJIA during that week?

37. Price stability. In the following table, we list the monthly price (in dollars) of a pound of coffee and a gallon of unleaded gasoline for 2007. Find the *population* standard deviation of each set of data. (*Source:* Bureau of Labor Statistics)

2007	Coffee Prices $/lb	Gasoline Prices $/gal
Jan.	3.288	2.274
Feb.	3.456	2.285
Mar.	3.475	2.592
Apr.	3.437	2.860
May	3.308	3.130
Jun.	3.407	3.052
Jul.	3.529	2.961
Aug.	3.497	2.782
Sep.	3.537	2.789
Oct.	3.577	2.793
Nov.	3.607	3.069
Dec.	3.685	3.020

Source: Bureau of Labor Statistics, U.S. Department of Energy

38. Comparing price stability. Consider the table given in Exercise 37. Use the coefficient of variation to determine whether coffee prices or gasoline prices were more stable in 2007.

39. Comparing family incomes. Consider the tables given in Exercises 29 and 30, respectively. Suppose that the eight families are the same in both tables and the incomes are for two different years that are 2 years apart. Which of the families had the greatest improvement in annual income over the 2 year period with respect to the other families? Explain how you arrived at your conclusion.

40. Comparing family incomes. Suppose that the mean family income in the United States was $48,000 with a standard deviation of $1,000 in year X. Also, the mean family income was $51,000 with a standard deviation of $2,250 three years later, in year X + 3. If a family earned $50,000 during year X and $54,000 three years later, then in which of these 2 years did the family have a better annual income in relation to the rest of the population? Explain how you arrived at your answer.

Communicating Mathematics

41. What do we do differently when we calculate the standard deviation of a sample versus a population?

Try to answer each of the following as true or false without doing any computations. Explain your answers in words or give appropriate examples or counterexamples to support your answers.

42. The standard deviation of the set of numbers −2, 2, −2, 2, −2, 2, −2, 2 is zero.

43. If the standard deviation of a set of data is zero, then all the numbers in the data set are the same.

---

*If an average falls on the border of two grades, the professor will assign the higher grade.

44. The more numbers there are in a distribution, the larger the standard deviation.

Using Technology to Investigate Mathematics

45. Search the Internet for "statistical calculators." Duplicate some of the examples in Sections 15.2 and 15.3. Report on your findings.

46. See your instructor for tutorials on using a graphing calculator to do statistical calculations. Use a graphing calculator to redo some of the examples and exercises in this section.

47. Data on the Internet. Do a search on the Internet to locate data in an area that is of interest to you. For example, you can go to the *New York Times* Web site and find the number of weeks that the top 10 fiction books have been on the bestseller list. Or, you can locate statistics from the Web site of your favorite sports team. Other good sources of data are the Bureau of Labor Statistics and the U.S. Department of Transportation Web sites. After you have located a site, choose a data set and find its mean and standard deviation.

For Extra Credit

48. a. Pick a data set consisting of any five numbers. Compute the mean and the standard deviation of this data set. (Consider the data set to be a sample.)

  b. Add 20 to each number in the data set you created in part (a), and compute the mean and standard deviation for this new data set.

  c. Subtract 5 from each number in the data set you created in part (a), and compute the mean and standard deviation for this new data set.

d. What conclusion can you make about changes in the mean and the standard deviation when the same number is added to or subtracted from each score in a data set?

e. Use the conclusion reached in part (d) to simplify the calculation of the mean and standard deviation for the distribution 598, 597, 599, 596, 600, 601, 602, 603.

49. a. Pick any five numbers and compute the mean and standard deviation for this data set.

  b. Multiply each number in the data set you created in part (a) by 4, and compute the mean and standard deviation for this new data set.

  c. Multiply each number in the data set you created in part (a) by 9, and compute the mean and standard deviation for this new data set.

  d. What conclusion can you make about changes in the mean and standard deviation when each score in a data set is multiplied by the same number?

  e. The mean is 4 and the standard deviation is 2.2 for the data set 3, 4, 7, 1, 5. Consider your conclusion from part (d), and compute the mean and standard deviation for the data set 15, 20, 35, 5, 25.

50. Consider the two distributions given by the bar graphs in Exercises 27 and 28. Can you simply look at the graphs to decide which of the two distributions has the greater standard deviation? If so, describe what you look for in the graphs on which to base your decision.

51. Comparative cost of living. According to an Internet service that compares the cost of living in one location with the cost of living in another, a salary of $100,000 in Manhattan is comparable to $38,000 in Allentown, Pennsylvania. What statistical methods do you think this service uses in determining this information?

15.4 The Normal Distribution

Objectives

1. Understand the basic properties of the normal curve.
2. Relate the area under a normal curve to z-scores.
3. Make conversions between raw scores and z-scores.
4. Use the normal distribution to solve applied problems.

If the shoe fits, wear it. —Well-known proverb

In addition to shoes, what about the fit of movie seats, men's ties, and headroom in cars? How do manufacturers decide how wide the seats at your favorite movie theater should be? Or, how long must a tie be so that it is neither too long nor too short for most men? How do we design a car so that most people will not bump their heads on its roof? Or, how high should a computer table be so that it does not put an unnecessary strain on your wrists and arms? In the relatively new science of *ergonomics*, scientists gather data to answer questions such as these and to ensure most people fit comfortably into their environments. Much of the work of these scientists is based on a curve that you will study in this section called the **normal curve**, or **normal distribution**.*

✏️ *KEY POINT*

Many different types of data sets are normal distributions.

The Normal Distribution

The normal distribution is the most common distribution in statistics and describes many real-life data sets. The histogram shown in Figure 15.12 will begin to give you an idea of the shape of a normal distribution.

Distributions of such diverse data sets as SAT scores, heights of people, the number of miles before an automobile tire wears out, the size of hamburgers at a fast-food restaurant, and the number of hours of use before a certain brand of DVD player breaks down are all examples of normal distributions.

There are several patterns in Figure 15.12 that are common to all normal distributions. First, if we were to take a wire and attach it to the tops of the bars in the histogram and then smooth out the wire to make a curve, the curve would have a bell-shaped appearance, as shown in Figure 15.13. This is why a normal distribution is often called a *bell-shaped curve.*

Second, the mean, median, and mode of a normal distribution are the same. Third, the curve is symmetric with respect to the mean. This means that if you see some pattern in the graph on one side of the mean, then the mean acts as a mirror to give a reflection of that same pattern on the other side of the mean. Fourth, the area under a normal curve equals 1.

In Figure 15.14, we have marked the mean and two points on the normal curve called inflection points. An *inflection point* is a point on the curve where the curve changes from being curved upward to being curved downward, or vice versa. For a normal curve, inflection points are located 1 standard deviation from the mean. Also, because a normal curve is symmetric with respect to the mean, one-half, or 50%, of the area under the curve is located on each side of the mean.

In normal distributions, approximately 68% of the data values occur within 1 standard deviation of the mean, 95% of the data values lie within 2 standard deviations of the mean, and 99.7% of the data values lie within 3 standard deviations of the mean. We refer to these facts as the *68-95-99.7 rule*. Figure 15.14 illustrates the 68-95-99.7 rule. In discussing normal distributions, we usually assume that we are dealing with an *entire population* rather than a *sample*, so in Figure 15.14 we represent the mean by μ and the standard deviation by σ (rather than $\bar{x}$ and s).

We summarize the properties of a normal distribution.

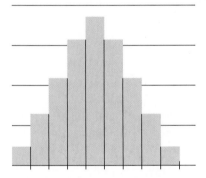

FIGURE 15.12 A normal distribution.

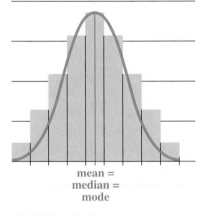

mean = median = mode

FIGURE 15.13 Smoothing a histogram of a normal distribution into a normal curve.

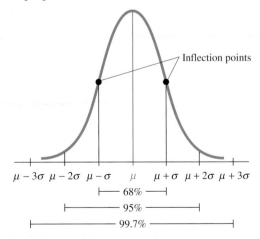

FIGURE 15.14 The 68-95-99.7 rule for a normal distribution.

---

*Recall that a *distribution* is just another name for a set of data.

> **PROPERTIES OF A NORMAL DISTRIBUTION**
>
> 1. A normal curve is bell shaped.
> 2. The highest point on the curve is at the mean of the distribution.
> 3. The mean, median, and mode of the distribution are the same.
> 4. The curve is symmetric with respect to its mean.
> 5. The total area under the curve is 1.
> 6. Roughly 68% of the data values are within 1 standard deviation from the mean, 95% of the data values are within 2 standard deviations from the mean, and 99.7% of the data values are within 3 standard deviations from the mean.*

We can use the 68-95-99.7 rule to estimate how many values we expect to fall within 1, 2, or 3 standard deviations of the mean of a normal distribution.

EXAMPLE 1   *The Normal Distribution and Intelligence Tests*

Suppose that the distribution of scores of 1,000 students who take a standardized intelligence test is a normal distribution. If the distribution's mean is 450 and its standard deviation is 25,

a) how many scores do we expect to fall between 425 and 475?

b) how many scores do we expect to fall above 500?

SOLUTION:

a) As you can see in Figure 15.15, the scores 425 and 475 are 1 standard deviation below and above the mean, respectively.

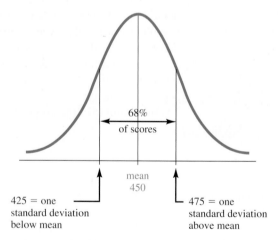

FIGURE 15.15 Sixty-eight percent of the scores lie within 1 standard deviation of the mean.

From the 68-95-99.7 rule, we know that 68%, or about 0.68 of the scores, lie within 1 standard deviation of the mean. Because we have 1,000 scores, we can expect that about $0.68 \times 1,000 = 680$ scores are in the range 425 to 475.

b) Recall in Figure 15.16, that 95% of the scores in a normal distribution fall between 2 standard deviations below the mean and 2 standard deviations above the mean.

---

*To keep this discussion simple, we are approximating these percentages. Shortly, we will use a table to slightly improve our estimate of the percentage of scores within 1 and 2 standard deviations from the mean.

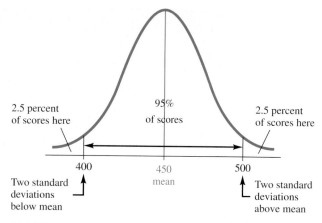

FIGURE 15.16 Five percent of the scores lie more than 2 standard deviations from the mean.

Quiz Yourself ⑪

Use the information given in Example 1 to determine the following.

a) How many scores do we expect to fall between 400 and 500?

b) How many scores do we expect to fall below 400?

KEY POINT

Areas under a normal curve represent percentages of values in the distribution.

This means that 5%, or 0.05, of the scores lie more than 2 standard deviations above or below the mean. Thus, we can expect to have $0.05 \div 2 = 0.025$ of the scores to be above 500. Again multiplying by 1,000, we can expect that $0.025 \times 1,000 = 25$ scores to be above 500.

Now try Exercises 5 to 16. ✳ ⑪

z-Scores

In Example 1, we estimated how many values were within 1 standard deviation of the mean. It is natural to ask whether we can predict how many values lie other distances from the mean. For example, knowing that the weights of women between ages 18 and 25 form a normal distribution, a clothing manufacturer may want to know what percentage of the female population is between 1.3 and 2.5 standard deviations above the mean weight for women. It is possible to find this by using Table 15.16, which is the table of areas under the standard normal curve. The standard normal distribution has a mean of 0 and a standard deviation of 1.

Math in Your Life

How Worried Are You About Your Next Test?*

People often argue about how much of your physical, intellectual, and psychological makeup is due to your heredity and how much is due to your environment. Scientists have identified genes that affect characteristics such as height, body fat, blood pressure, and IQ. Surprisingly, even human anxiety has been associated with genetic makeup.

It has been estimated that there are as many as 15 genes that can affect your anxiety level. If you have all these genes, then you have a greater tendency[†] to be anxious, and if you have none of them, then you tend to be less anxious. If we were to plot a histogram of the number of anxiety genes present in a large number of people, we would find very few people have almost none of these genes, many people have a moderate number of these genes, and very few people have almost all of them. If you were to plot the distribution of anxiety genes present in the population, it would look very much like the normal curves that you are studying in this section.

*This note is based on a series of PowerPoint slides by Dr. Lee Bardwell, professor of genetics at the University of California at Irvine.
[†]It is believed that 45% of the variance in human anxiety is due to genetic factors. The other 55% is due to other factors.

z	A	z	A	z	A	z	A	z	A	z	A
.00	.000	.56	.212	1.12	.369	1.68	.454	2.24	.488	2.80	.497
.01	.004	.57	.216	1.13	.371	1.69	.455	2.25	.488	2.81	.498
.02	.008	.58	.219	1.14	.373	1.70	.455	2.26	.488	2.82	.498
.03	.012	.59	.222	1.15	.375	1.71	.456	2.27	.488	2.83	.498
.04	.016	.60	.226	1.16	.377	1.72	.457	2.28	.489	2.84	.498
.05	.020	.61	.229	1.17	.379	1.73	.458	2.29	.489	2.85	.498
.06	.024	.62	.232	1.18	.381	1.74	.459	2.30	.489	2.86	.498
.07	.028	.63	.236	1.19	.383	1.75	.460	2.31	.490	2.87	.498
.08	.032	.64	.239	1.20	.385	1.76	.461	2.32	.490	2.88	.498
.09	.036	.65	.242	1.21	.387	1.77	.462	2.33	.490	2.89	.498
.10	.040	.66	.245	1.22	.389	1.78	.463	2.34	.490	2.90	.498
.11	.044	.67	.249	1.23	.391	1.79	.463	2.35	.491	2.91	.498
.12	.048	.68	.252	1.24	.393	1.80	.464	2.36	.491	2.92	.498
.13	.052	.69	.255	1.25	.394	1.81	.465	2.37	.491	2.93	.498
.14	.056	.70	.258	1.26	.396	1.82	.466	2.38	.491	2.94	.498
.15	.060	.71	.261	1.27	.398	1.83	.466	2.39	.492	2.95	.498
.16	.064	.72	.264	1.28	.400	1.84	.467	2.40	.492	2.96	.499
.17	.068	.73	.267	1.29	.402	1.85	.468	2.41	.492	2.97	.499
.18	.071	.74	.270	1.30	.403	1.86	.469	2.42	.492	2.98	.499
.19	.075	.75	.273	1.31	.405	1.87	.469	2.43	.493	2.99	.499
.20	.079	.76	.276	1.32	.407	1.88	.470	2.44	.493	3.00	.499
.21	.083	.77	.279	1.33	.408	1.89	.471	2.45	.493	3.01	.499
.22	.087	.78	.282	1.34	.410	1.90	.471	2.46	.493	3.02	.499
.23	.091	.79	.285	1.35	.412	1.91	.472	2.47	.493	3.03	.499
.24	.095	.80	.288	1.36	.413	1.92	.473	2.48	.493	3.04	.499
.25	.099	.81	.291	1.37	.415	1.93	.473	2.49	.494	3.05	.499
.26	.103	.82	.294	1.38	.416	1.94	.474	2.50	.494	3.06	.499
.27	.106	.83	.297	1.39	.418	1.95	.474	2.51	.494	3.07	.499
.28	.110	.84	.300	1.40	.419	1.96	.475	2.52	.494	3.08	.499
.29	.114	.85	.302	1.41	.421	1.97	.476	2.53	.494	3.09	.499
.30	.118	.86	.305	1.42	.422	1.98	.476	2.54	.495	3.10	.499
.31	.122	.87	.308	1.43	.424	1.99	.477	2.55	.495	3.11	.499
.32	.126	.88	.311	1.44	.425	2.00	.477	2.56	.495	3.12	.499
.33	.129	.89	.313	1.45	.427	2.01	.478	2.57	.495	3.13	.499
.34	.133	.90	.316	1.46	.428	2.02	.478	2.58	.495	3.14	.499
.35	.137	.91	.319	1.47	.429	2.03	.479	2.59	.495	3.15	.499
.36	.141	.92	.321	1.48	.431	2.04	.479	2.60	.495	3.16	.499
.37	.144	.93	.324	1.49	.432	2.05	.480	2.61	.496	3.17	.499
.38	.148	.94	.326	1.50	.433	2.06	.480	2.62	.496	3.18	.499
.39	.152	.95	.329	1.51	.435	2.07	.481	2.63	.496	3.19	.499
.40	.155	.96	.332	1.52	.436	2.08	.481	2.64	.496	3.20	.499
.41	.159	.97	.334	1.53	.437	2.09	.482	2.65	.496	3.21	.499
.42	.163	.98	.337	1.54	.438	2.10	.482	2.66	.496	3.22	.499
.43	.166	.99	.339	1.55	.439	2.11	.483	2.67	.496	3.23	.499
.44	.170	1.00	.341	1.56	.441	2.12	.483	2.68	.496	3.24	.499
.45	.174	1.01	.344	1.57	.442	2.13	.483	2.69	.496	3.25	.499
.46	.177	1.02	.346	1.58	.443	2.14	.484	2.70	.497	3.26	.499
.47	.181	1.03	.349	1.59	.444	2.15	.484	2.71	.497	3.27	.500
.48	.184	1.04	.351	1.60	.445	2.16	.485	2.72	.497	3.28	.500
.49	.188	1.05	.353	1.61	.446	2.17	.485	2.73	.497	3.29	.500
.50	.192	1.06	.355	1.62	.447	2.18	.485	2.74	.497	3.30	.500
.51	.195	1.07	.358	1.63	.449	2.19	.486	2.75	.497	3.31	.500
.52	.199	1.08	.360	1.64	.450	2.20	.486	2.76	.497	3.32	.500
.53	.202	1.09	.362	1.65	.451	2.21	.487	2.77	.497	3.33	.500
.54	.205	1.10	.364	1.66	.452	2.22	.487	2.78	.497		
.55	.209	1.11	.367	1.67	.453	2.23	.487	2.79	.497		

TABLE 15.16 Standard normal distribution.

Table 15.16 gives the area under this curve between the mean and a number called a *z*-score. A **z-score** represents the number of standard deviations a data value is from the mean. In Example 1, we found that for a normal distribution with a mean of 450 and a standard deviation of 25, the value 500 was 2 standard deviations above the mean. Another way of saying this is that the value 500 corresponds to a *z*-score of 2.

Notice that Table 15.16 gives areas only for positive *z*-scores—that is, for data that lie above the mean. We use the symmetry of a normal curve to find areas corresponding to negative *z*-scores.

Example 2 shows how to use Table 15.16. Because the total area under a normal curve is 1, we can interpret areas under the standard normal curve as percentages of the data values in the distribution. Also note that because the mean is 0 and the standard deviation for the standard normal curve is 1, the value of a *z*-score is also the same as the number of standard deviations that *z*-score is from the mean. After you have practiced working with the standard normal distribution, we will show you how to use the techniques that you have learned to solve problems involving real-life data.

 PROBLEM SOLVING

The Three-Way Principle

The Three-Way Principle in Section 1.1 suggests that you can often understand a problem graphically. This is certainly true in solving problems involving the normal curve. If you draw a good picture of the situation, it often becomes more clear to you as to what to do to set up and solve the problem.

EXAMPLE 2   *Finding Areas under the Standard Normal Curve*

Use Table 15.16 to find the percentage of the data (area under the curve) that lie in the following regions for a standard normal distribution:

a) between $z = 0$ and $z = 1.3$

b) between $z = 1.5$ and $z = 2.1$

c) between $z = 0$ and $z = -1.83$

SOLUTION:

a) The area under the curve between $z = 0$ and $z = 1.3$ is shown in Figure 15.17. We find this area by looking in Table 15.16 for the *z*-score 1.30. We see that A is 0.403 when $z = 1.30$. Thus, we expect 0.403, or 40.3%, of the data to fall between 0 and 1.3 standard deviations above the mean.

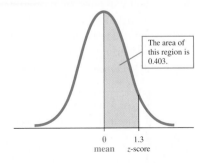

The area of this region is 0.403.

FIGURE 15.17 Area under the standard normal curve between $z = 0$ and $z = 1.3$.

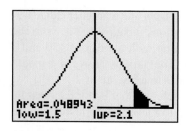

TI Screen showing area under standard normal curve between $z = 1.5$ and $z = 2.1$.

b) Figure 15.18 shows the area we want. Areas in Table 15.16 correspond to regions from $z = 0$ to the given z-score. Therefore, to find the area under the curve between $z = 1.5$ and $z = 2.1$, we must first find the area from $z = 0$ to $z = 2.1$ and then subtract the area from $z = 0$ to $z = 1.5$. From Table 15.16, we see that when $z = 2.1$, $A = 0.482$, and when $z = 1.5$, $A = 0.433$. To finish this problem, we can set up our calculations as follows:

$$\begin{array}{ll} \text{larger area from } z = 0 \text{ to } z = 2.1 & 0.482 \\ -\text{ smaller area from } z = 0 \text{ to } z = 1.5 & -0.433 \\ \hline & 0.049 \end{array}$$

— subtract areas, not z-scores

— area we want

This means that in the standard normal distribution, the area under the curve between $z = 1.5$ and $z = 2.1$ is 0.049, or 4.9%.

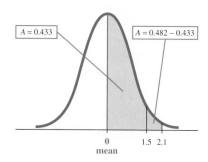

$A = 0.433$ $A = 0.482 - 0.433$

0
mean

1.5 2.1

FIGURE 15.18 Finding the area under the standard normal curve between $z = 1.5$ and $z = 2.1$.

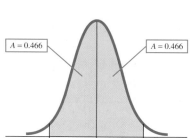

$A = 0.466$ $A = 0.466$

−1.83   0   1.83
mean

FIGURE 15.19 Finding the area under the standard normal curve between $z = -1.83$ and $z = 0$.

c) Due to the symmetry of the normal distribution, the area between $z = 0$ and $z = -1.83$ is the same as the area between $z = 0$ and $z = 1.83$ (Figure 15.19). From Table 15.16 we see that when $z = 1.83$, $A = 0.466$. Therefore, 46.6% of the data values lie between 0 and −1.83.

Now try Exercises 17 to 34. ❋ **12**

Quiz Yourself **12**

Use Table 15.16 to find the following areas under the standard normal curve:

a) between $z = 0$ and $z = 1.45$

b) between $z = 1.23$ and $z = 1.85$

c) between $z = 0$ and $z = -1.35$

KEY POINT

We convert values in a nonstandard normal distribution to z-scores.

 *Some Good Advice*

A common mistake in solving a problem such as Example 2(b) is to subtract 1.5 from 2.1 to get 0.6 and then wrongly use this for the z-score in Table 15.16. If you think about the graph of the standard normal curve, clearly the area between $z = 0$ and $z = 0.6$ is not the same as the area between $z = 1.5$ and $z = 2.1$.

Converting Raw Scores to z-Scores

A real-life normal distribution, such as the set of all weights of women between ages 18 and 25, may have a mean of 120 pounds and a standard deviation of 25 pounds. Such a distribution will have the properties that we stated earlier for a normal distribution, but because the distribution does not have a mean of 0 and a standard deviation of 1, we cannot use Table 15.16 directly, as we did in Example 2. We *can* use Table 15.16, however, if we first convert nonstandard values, called *raw scores*, to z-scores. The following formula shows the desired relationship between values in a nonstandard normal distribution and z-scores.

> **FORMULA FOR CONVERTING RAW SCORES TO Z-SCORES** Assume a normal distribution has a mean of μ and a standard deviation of σ. We use the equation
>
> $$z = \frac{x - \mu}{\sigma}$$
>
> to convert a value x in the nonstandard distribution to a z-score.

EXAMPLE 3 *Converting Raw Scores to z-Scores*

Suppose the mean of a normal distribution is 20 and its standard deviation is 3.

a) Find the *z*-score corresponding to the raw score 25.

b) Find the *z*-score corresponding to the raw score 16.

SOLUTION:

a) Figure 15.20 gives us a picture of this situation. We will use the conversion formula $z = \frac{x - \mu}{\sigma}$, with the raw score $x = 25$, the mean $\mu = 20$, and the standard deviation $\sigma = 3$. Making the substitutions, we have, $z = \frac{25 - 20}{3} = \frac{5}{3} = 1.67$. We can interpret this result as telling us that in this distribution, 25 is 1.67 standard deviations above the mean.

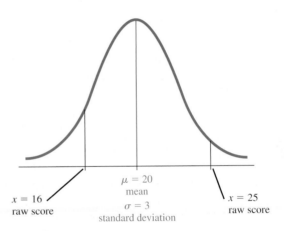

FIGURE 15.20 Normal distribution with mean of 20 and standard deviation of 3.

Quiz Yourself

Suppose a normal distribution has a mean of 50 and a standard deviation of 7. Convert each raw score to a z-score.

a) 59   b) 50   c) 38

b) We will use Figure 15.20 and the conversion formula again, but this time with $x = 16$, $\mu = 20$, and $\sigma = 3$. Therefore, the corresponding *z*-score is $z = \frac{16 - 20}{3} = \frac{-4}{3} = -1.33$. This tells us that 16 is 1.33 standard deviations below the mean.

Now try Exercises 43 to 48. ✳ **13**

The Three-Way Principle in Section 1.1 suggests that you can remember the formula for converting raw scores to *z*-scores by interpreting the formula geometrically. In Example 3, we had a normal distribution with a mean of 20 and standard deviation 3. In order to make the mean of 20 in the nonstandard distribution correspond to the mean of 0 in the standard normal distribution, you can think of "sliding" the distribution to the left 20 units, as we show in Figure 15.21(a). That is why we compute an expression of the form $x - 20$ in the numerator of the *z*-score formula. Because the standard deviation was 3, we divided $x - 20$ by 3 to "narrow" the distribution so that it would have a standard deviation of 1, as you can see in Figure 15.21(b), and therefore allow us to use Table 15.16.

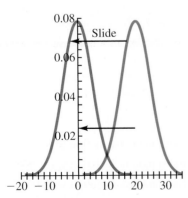

FIGURE 15.21 (a) First slide the distribution 20 units to the left to get a mean of 0.

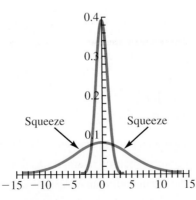

FIGURE 15.21 (b) Then squeeze the distribution to get a standard deviation of 1.

Applications

Normal distributions have many practical applications.

EXAMPLE 4 *Interpreting the Significance of an Exam Score*

Suppose that to qualify for a management training program offered by your employer you must score in the top 10% of those employees who take a standardized test. Assume that the distribution of scores is normal and you received a score of 72 on the test, which had a mean of 65 and a standard deviation of 4. What percentage of those who took this test had a score below yours?

SOLUTION: We will find the number of standard deviations your score is above the mean and then determine what percentage of scores fall below this number. First, we calculate the *z*-score, which corresponds to your 72:

$$z = \frac{72 - 65}{4} = \frac{7}{4} = 1.75$$

Now, looking in Table 15.16 for $z = 1.75$, we find that $A = 0.460$. Therefore, 46% of the scores fall between the mean and your score (Figure 15.22).

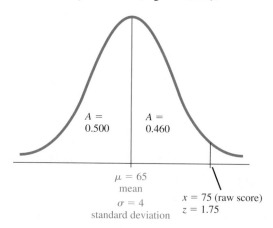

$A = 0.500$

$A = 0.460$

$\mu = 65$
mean

$\sigma = 4$
standard deviation

$x = 75$ (raw score)
$z = 1.75$

FIGURE 15.22 Normal distribution with mean of 65 and standard deviation of 4.

However, we must not forget the scores that fall below the mean. Because a normal curve is symmetric, another 50% of the scores fall below the mean. So, there are $50\% + 46\% = 96\%$ of the scores below yours. Congratulations! You qualify for the program. ✳

We can use the ideas we have been discussing about the normal curve to compare data.

EXAMPLE 5 *Using z-Scores to Compare Data*

Consider the following information:

Ty Cobb hit .420 in 1911.

Ted Williams hit .406 in 1941.

George Brett hit .390 in 1980.

In the 1910s, the mean batting average was .266 and the standard deviation was .0371.

In the 1940s, the mean batting average was .267 and the standard deviation was .0326.

In the 1970s, the mean batting average was .261 and the standard deviation was .0317.

Assume that the batting averages were normally distributed in each of these three decades. Use *z*-scores to determine which of the three batters was ranked the highest in relationship to his contemporaries.

SOLUTION: We will convert each of the three batting averages to a corresponding z-score for the distribution of batting averages for the decade in which the athlete played.

In the 1910s, Ty Cobb's average of .420 corresponded to a z-score of

$$\frac{.420 - .266}{.0371} = \frac{.154}{.0371} = 4.1509.$$

In the 1940s, Ted Williams's average of .406 corresponded to a z-score of

$$\frac{.406 - .267}{.0326} = \frac{.139}{.0326} = 4.2638.$$

In the 1970s, George Brett's average of .390 corresponded to a z-score of

$$\frac{.390 - .261}{.0317} = \frac{.129}{.0317} = 4.0694.$$

We see that in using the standard deviation to compare each hitter with his contemporaries, Ted Williams was ranked as the best hitter.

Now try Exercises 69 and 70. ❋

We conclude our discussion of normal distributions by explaining how manufacturers can use information regarding the life of their products as a basis for warranties.

EXAMPLE 6 *Using the Normal Distribution to Write Warranties*

To increase sales, a manufacturer plans to offer a warranty on a new model of the iPod Touch. In testing the iPod Touch, quality control engineers found that it has a mean time to failure of 3,000 hours with a standard deviation of 500 hours. Assume that the typical purchaser will use the device for 4 hours per day. If the manufacturer does not want more than 5% to be returned as defective within the warranty period, how long should the warranty period be to guarantee this?

SOLUTION: In solving this problem, we first work with the standard normal distribution and then convert the answer to fit our nonstandard normal distribution. Figure 15.23 gives a picture of the situation for the standard normal curve.

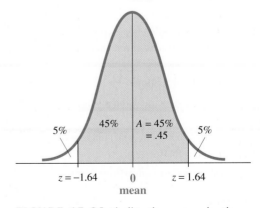

FIGURE 15.23 Finding the area under the standard normal curve for a negative z-score.

In Figure 15.23, we see that we need to find a z-score such that at least 95% of the entire area is *beyond* this point. Observe that this score is to the left of the mean and is negative. Therefore, we use symmetry and find the z-score such that 95% of the entire area is *below* this score. However, Table 15.16 gives the areas only for the upper half of the distribution. However, this is not a problem because we know that 50% of the entire area lies below the mean. Therefore, our problem reduces to finding a z-score greater than 0 such that 45% of the area lies between the mean and that z-score.

In Table 15.16, we see that if $A = 0.450$, the corresponding z-score is 1.64. This means that 95% of the area underneath the standard normal curve falls below $z = 1.64$. By symmetry, we also can say that 95% of the values lie above -1.64. We must now interpret this z-score in terms of original normal distribution of raw scores.

Recall the equation that relates values in a distribution to z-scores, namely,

$$z = \frac{x - \mu}{\sigma}. \tag{1}$$

Previously we used this equation to convert raw scores to z-scores. We now use the equation in answering the reverse question. Specifically, what is x when $\mu = 3{,}000$, $\sigma = 500$, and $z = -1.64$? Substituting these numbers in equation (1), we get

$$-1.64 = \frac{x - 3{,}000}{500}. \tag{2}$$

We solve for x in this equation by multiplying both sides by 500:

$$-1.64 \cdot (500) = \frac{x - 3{,}000}{500} \cdot (500)$$

Simplifying the equation, we get

$$-820 = x - 3{,}000.$$

Last, we add 3,000 to both sides to obtain $x = 2{,}180$ hours. We expect 95% of the breakdowns to occur beyond 2,180 hours.

Because we know that owners use the iPod Touch about 4 hours per day, we divide 2,180 by 4 to get $\frac{2{,}180}{4} = 545$ days. This is equivalent to $\frac{545}{31} = 17.58$ months if we use 31 days per month. Therefore, if the manufacturer wants to have no more than 5% of the breakdowns occur within the warranty period, the warranty should be for roughly 18 months. ❈ **14**

Quiz Yourself **14**

Redo Example 6, but now assume that the mean time to failure is 4,200 hours, with a standard deviation of 600, and the manufacturer wants no more than 2% to be returned as defective within the warranty period.

Exercises 15.4

Looking Back*

These exercises follow the general outline of the topics presented in this section and will give you a good overview of the material that you have just studied.

1. List six properties of the normal distribution.

2. What was the common error that we warned you against making in solving Example 2(b)?

3. What property of the normal distribution makes it unnecessary to have negative z-scores in Table 15.16?

4. State the equation that allows you to convert raw scores to z-scores?

Sharpening Your Skills

Assume that all distributions in this exercise set are normal. Assume that the distribution in Exercises 5–10 has a mean of 10

and a standard deviation of 2. Use the 68-95-99.7 rule to find the percentage of values in the distribution that we describe.

5. Between 10 and 12
6. Between 12 and 14
7. Above 14
8. Below 8
9. Above 12
10. Below 10

Assume that the distribution in Exercises 11–16 has a mean of 12 and a standard deviation of 3. Use the 68-95-99.7 rule to find the percentage of values in the distribution that we describe.

11. Below 9
12. Between 12 and 15
13. Above 6
14. Below 12
15. Between 15 and 18
16. Above 18

Use Table 15.16 to find the percentage of area under the standard normal curve that is specified.

17. Between $z = 0$ and $z = 1.23$
18. Between $z = 0$ and $z = 2.06$
19. Between $z = 0.89$ and $z = 1.43$

---

*Before doing these exercises, you may find it useful to review the note *How to Succeed at Mathematics* on page xix.*

20. Between $z = 0.76$ and $z = 1.52$

21. Between $z = 0$ and $z = -0.75$

22. Between $z = 0$ and $z = -1.35$

23. Between $z = 1.25$ and $z = 1.95$

24. Between $z = 0.37$ and $z = 1.23$

25. Between $z = -0.38$ and $z = -0.76$

26. Between $z = -1.55$ and $z = -2.13$

27. Above $z = 1.45$ 28. Below $z = -0.64$

29. Below $z = -1.40$ 30. Above $z = 0.78$

31. Below $z = 1.33$ 32. Above $z = -0.46$

33. Above $z = -0.84$ 34. Below $z = 1.22$

35. Find a z-score such that 10% of the area under the standard normal curve is above that score.

36. Find a z-score such that 20% of the area under the standard normal curve is above that score.

37. Find a z-score such that 12% of the area under the standard normal curve is below that score.

38. Find a z-score such that 24% of the area under the standard normal curve is below that score.

39. Find a z-score such that 60% of the area is below that score.

40. Find a z-score such that 90% of the area is below that score.

41. Find a z-score such that 75% of the area is above that score.

42. Find a z-score such that 60% of the area is above that score.

In Exercises 43–48, we give you a mean, a standard deviation, and a raw score. Find the corresponding z-score.

43. Mean 80, standard deviation 5, $x = 87$

44. Mean 100, standard deviation 15, $x = 117$

45. Mean 21, standard deviation 4, $x = 14$

46. Mean 52, standard deviation 7.5, $x = 61$

47. Mean 38, standard deviation 10.3, $x = 48$

48. Mean 8, standard deviation 2.4, $x = 6.2$

In Exercises 49–54, we give you a mean, a standard deviation, and a z-score. Find the corresponding raw score.

49. Mean 60, standard deviation 5, $z = 0.84$

50. Mean 20, standard deviation 6, $z = 1.32$

51. Mean 35, standard deviation 3, $z = -0.45$

52. Mean 62, standard deviation 7.5, $z = -1.40$

53. Mean 28, standard deviation 2.25, $z = 1.64$

54. Mean 8, standard deviation 3.5, $z = -1.25$

Applying What You've Learned

Due to random variations in the operation of an automatic coffee machine, not every cup is filled with the same amount of coffee. Assume that the mean amount of coffee dispensed is 8 ounces and the standard deviation is 0.5 ounce. Use Figure 15.14 to solve Exercises 55 and 56.

55. **Analyzing vending machines.**

   a. What percentage of the cups should have at least 8 ounces of coffee?

   b. What percentage of the cups should have less than 7.5 ounces of coffee?

56. **Analyzing vending machines.**

   a. What percentage of the cups should have at least 8.5 ounces of coffee?

   b. What percentage of the cups should have less than 8 ounces of coffee?

A machine fills bags of candy. Due to slight irregularities in the operation of the machine, not every bag gets exactly the same number of pieces. Assume that the number of pieces per bag has a mean of 200 with a standard deviation of 2. Use Figure 15.14 to solve Exercises 57 and 58.

57. **Analyzing vending machines.**

   a. How many of the bags do we expect to have between 200 and 202 pieces in them?

   b. How many of the bags do we expect to have at least 202 pieces in them?

58. **Analyzing vending machines.**

   a. How many of the bags do we expect to have between 198 and 200 pieces in them?

   b. How many of the bags do we expect to have at least 196 pieces in them?

59. **Product reliability.** Suppose that for a particular brand of LCD television set, the distribution of failures has a mean of 120 months with a standard deviation of 12 months. With respect to this distribution, 140 months corresponds to what z-score?

60. **Product reliability.** Redo Exercise 59 using 130 months instead of 140.

61. **Distribution of heights.** Assume that in the NBA the distribution of heights has a mean of 6 feet, 8 inches, and a standard deviation of 3 inches. If there are 324 players in the league, how many players do you expect to be over 7 feet tall?

62. **Distribution of heights.** Redo Exercise 61, but now we want to know how many players you would expect to be over 6 feet, 6 inches.

63. **Distribution of heart rates.** Assume that the distribution of 21-year-old women's heart rates at rest is 68 beats per minute with a standard deviation of 4 beats per minute. If 200 women are examined, how many would you expect to have a heart rate of less than 70?

64. **Distribution of heart rates.** Redo Exercise 63, but now we ask how many you would expect to have a heart rate less than 75.

In Exercises 65 and 66, we assume that the mean length of a human pregnancy is 268 days.

65. **Birth statistics.** If 95% of all human pregnancies last between 250 and 286 days, what is the standard deviation of the distribution of lengths of human pregnancies? What percentage of human pregnancies do we expect to last at least 275 days?

66. **Birth statistics.** Babies born before the 37th week (that is, before day 252 of a pregnancy) are considered premature. What percentage of births do we expect to be premature? (See Exercise 65.)

67. **Analyzing Internet use.** Comcast has analyzed the amount of time that its customers are online per session. It found that the distribution of connect times has a mean of 37 minutes and a standard deviation of 11 minutes. For this distribution, find the raw score that corresponds to a z-score of 1.5.

68. **Analyzing customer service.** The distribution of times that customers spend waiting in line in a supermarket has a mean of 3.6 minutes and a standard deviation of 1.2 minutes. For this distribution, find the raw score that corresponds to a z-score of −1.3.

In Exercises 69 and 70, use the information and methods of Example 5 to make the indicated comparisons.

69. **Comparing athletes.** In 1949, Jackie Robinson hit .342 for the Brooklyn Dodgers; in 1973, Rod Carew hit .350 for the Minnesota Twins. Using z-scores to compare each batter with his contemporaries, determine which batting average was more impressive.

70. **Comparing athletes.** In 1940, Joe DiMaggio hit .352 for the New York Yankees; in 1975, Bill Madlock hit .354 for the Pittsburgh Pirates. Using z-scores to compare each batter with his contemporaries, determine which batting average was more impressive.

71. **Writing a warranty.** A manufacturer plans to provide a warranty on Wii Fit. In testing, the manufacturer has found that the set of failure times for Wii Fit is normally distributed with a mean time of 2,000 hours and a standard deviation of 800 hours. Assume that the typical purchaser will use Wii Fit for 2 hours per day. If the manufacturer does not want more than 4% returned as defective within the warranty period, how long should the warranty period be to guarantee this? (Assume that there are 31 days in each month.)

72. **Writing a warranty.** Redo Exercise 71, but now assume that the mean of the distribution of failure times is 2,500 hours, the standard deviation is 600 hours, and the manufacturer wants no more than 2.5% returned during the warranty period.

73. **Analyzing investments.** A certain mutual fund over the last 15 years has a mean yearly return of 7.8% with a standard deviation of 1.3%.

    a. If you had invested in this fund over the past 15 years, in how many of these years would you expect to earn at least 9% on your investment?

    b. In how many years would you expect to earn less than 6%?

74. **Analyzing investments.** A certain bond fund over the last 12 years has a mean yearly return of 5.9% with a standard deviation of 1.8%.

    a. If you had invested in this fund over the past 12 years, in how many of these years would you expect to earn at least 8% on your investment?

    b. In how many years would you expect to earn less than 4%?

75. **Analyzing the SATs.** Assume that the math SAT scores are normally distributed with a mean of 500 and a standard deviation of 100. If you score 480 on this exam, what percentage of those taking the test scored below you?

76. **Analyzing the SATs.** Assume that the verbal SAT scores are normally distributed with a mean of 500 and a standard deviation of 100. If you score 520 on this exam, what percentage of those taking the test scored below you?

Communicating Mathematics

77. If a raw score corresponds to a z-score of 1.75, what does that tell you about that score in relationship to the mean of the distribution?

78. If a raw score corresponds to a z-score of −0.85, what does that tell you about that score in relationship to the mean of the distribution?

79. Explain why, if you want to compute the area under a normal curve between $z = 1.3$ and $z = 2.0$, it is not correct to subtract $2.0 - 1.3 = 0.7$ and then use the z-score of 0.7 to find the area.

80. Without looking at Table 15.16, can you determine whether the area under the standard normal curve between $z = 0.5$ and $z = 1.0$ is greater than or less than the area between $z = 1.5$ and $z = 2.0$? Explain.

81. Explain how you would estimate the mean and the standard deviation of the two normal distributions in the given figure.

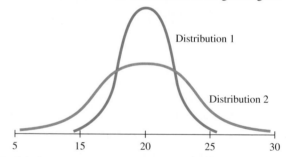

82. Explain how you would estimate the mean and the standard deviation of the two normal distributions in the given figure.

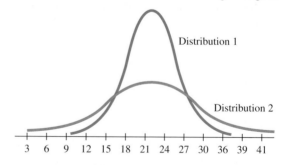

Using Technology to Investigate Mathematics

83. You can find many interactive programs that illustrate various properties of the normal distribution. Search the Internet for "normal distribution applets," run some, and report on your findings.

84. Search the Internet for some interesting applications of the normal distribution. While searching, I found many applications in medicine and even some in music. Report on your findings.

For Extra Credit

85. If you had a large collection of data, how might you go about determining whether the distribution was normal?

86. Discuss whether you feel that the method we used to compare Ty Cobb, Ted Williams, and George Brett would be a valid way to compare the great African American athletes who played in the Negro leagues with the great white athletes of the same era.

The nth percentile in a distribution is a number x such that n% of the values in that distribution fall below x. For example, 20% of the values in the distribution fall below the 20th percentile.

87. What z-score in the standard normal distribution is the 80th percentile?

88. What z-score in the standard normal distribution is the 30th percentile?

89. If a distribution has a mean of 40 and a standard deviation of 4, what is the 75th percentile?

90. If a distribution has a mean of 80 and a standard deviation of 6, what is the 35th percentile?

91. **Finding percentiles.** If you score a 75 on a commercial pilot's exam that has a mean score of 68 and a standard deviation of 4, to what percentile does your score correspond?

92. **Finding percentiles.** If you know that the mean salary for your profession is $53,000 with a standard deviation of $2,500, to what percentile does your salary of $57,000 correspond?

Looking Deeper
Linear Correlation

Objectives

1. Be able to construct a scatterplot to show the relationship between two variables.
2. Understand the properties of the linear correlation coefficient.
3. Use linear regression to find the line of best fit for a set of data points.

How can we tell whether two sets of data are related? For example, is there a relationship between

- the amount of dietary fat eaten and the number of pounds a person is overweight?
- time spent at tutoring sessions and grades?
- the amount of cell phone use and cancer?

In each case, we seek some connection between the first quantity and the second. We say there is a **correlation** between two variables if one of them is related to the other in some way—for example, "Is there a correlation between the amount of dietary fat and number of pounds a person is overweight?" Although there are many different types of correlation, later in this section we will discuss one particular type called *linear correlation*.

Scatterplots

In order to determine whether there is a correlation between two variables, we obtain pairs of data, called *data points*, relating the first variable to the second. For example, we might examine 100 patients and relate their dietary fat to the number of pounds they are overweight. To understand such data, we plot data points in a graph called a **scatterplot**.

EXAMPLE 1   *Constructing a Scatterplot*

An instructor wants to know whether there is a correlation between the number of times students attended tutoring sessions during the semester and their grades on a 50-point examination. The instructor has collected data for 10 students, as shown in Table 15.17. Represent these data points by a scatterplot and interpret the graph.

Number of Tutoring Sessions, x	18	6	16	14	0	4	0	10	12	18
Exam Grade, y	42	31	46	41	25	38	28	39	42	44

TABLE 15.17 Tutoring sessions attended and exam grades.

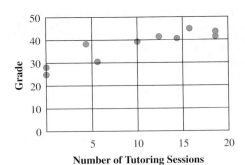

FIGURE 15.24 Scatterplot of data in Table 15.17.

SOLUTION: We plot the points (18, 42), (6, 31), (16, 46), and so on in Figure 15.24. The scatterplot shows that as the number of tutoring sessions increases, the grades also generally increase. Based on this scatterplot, the instructor might conclude that there is some relationship between the number of study sessions students attend and their grades. ❈ **15**

Quiz Yourself **15**

Draw a scatterplot of the data shown in Table 15.18 and interpret the graph.

x	5	16	14	0	10	20	3	2
y	39	20	29	41	30	15	42	44

TABLE 15.18 Paired data for variables *x* and *y*.

In Example 1, we must be careful not to conclude that there is a cause-and-effect relationship between the number of tutoring sessions attended and the students' grades. There are probably many factors that affect the students' grades, such as class attendance, motivation, and the amount of time spent doing homework. Statisticians often say, "Correlation does not imply causation." Based on the scatterplot, there seems to be some kind of relationship between the two variables; however, at this point we cannot say more than this.

Linear Correlation

We will now discuss one type of correlation called *linear correlation*. **Linear correlation** exists between two variables if, when we graph them in a scatterplot, the points in the graph tend to lie in a straight line. Although this definition may seem vague, there is a number called the *linear correlation coefficient* that allows us to compute to what degree the points of a scatterplot lie along a straight line. The derivation of this number is beyond the scope of this text, but we can explain its meaning and show you how to use it.

FORMULA FOR COMPUTING THE LINEAR CORRELATION COEFFICIENT If we have pairs of data for two variables *x* and *y*, the linear correlation coefficient* for these data, denoted by *r*, is given by the formula

$$r = \frac{n\sum xy - \left(\sum x\right)\left(\sum y\right)}{\sqrt{n\left(\sum x^2\right) - \left(\sum x\right)^2}\ \sqrt{n\left(\sum y^2\right) - \left(\sum y\right)^2}} \tag{1}$$

The number *n* is the number of pairs of data we have for *x* and *y*.

We will first show you how to compute the linear correlation coefficient and we will then explain how to interpret its meaning.

---

*In this section, we will assume that we are dealing with samples. In this case, the linear correlation coefficient is sometimes called the *sample* linear correlation coefficient.

EXAMPLE 2 *Computing the Linear Correlation Coefficient*

Compute the linear correlation coefficient for the data given in Table 15.17.

SOLUTION: We will first describe each expression in equation (1) and then compute these expressions in Table 15.19.

x	y	x^2	y^2	xy
18	42	324	1,764	756
6	31	36	961	186
16	46	256	2,116	736
14	41	196	1,681	574
0	25	0	625	0
4	38	16	1,444	152
0	28	0	784	0
10	39	100	1,521	390
12	42	144	1,764	504
18	44	324	1,936	792
$\Sigma x = 98$	$\Sigma y = 376$	$\Sigma x^2 = 1,396$	$\Sigma y^2 = 14,596$	$\Sigma xy = 4,090$

TABLE 15.19 Computations to calculate the correlation coefficient.

Σx is the sum of the first coordinates of the data pairs.

Σy is the sum of the second coordinates of the data pairs.

Σx^2 is the sum of the squares of the first coordinates of the data pairs.

Σy^2 is the sum of the squares of the second coordinates of the data pairs.

Σxy is the sum of the products of the first and second coordinates of the data pairs.

The linear correlation coefficient is therefore

$$r = \frac{n\sum xy - \left(\sum x\right)\left(\sum y\right)}{\sqrt{n\left(\sum x^2\right) - \left(\sum x\right)^2}\sqrt{n\left(\sum y^2\right) - \left(\sum y\right)^2}}$$

$$= \frac{10 \cdot 4{,}090 - (98)(376)}{\sqrt{10 \cdot (1{,}396) - 98^2}\sqrt{10 \cdot (14{,}596) - 376^2}}$$

$$= \frac{4{,}052}{\sqrt{4{,}356}\sqrt{4{,}584}} \approx 0.9068.$$

Now try Exercises 5 to 8. ✿ **16**

Quiz Yourself **16**

Compute the linear correlation coefficient for the data shown in Table 15.18.

We will now investigate the meaning of the correlation coefficient. There is a *positive correlation* between the variables x and y if whenever x increases or decreases, then y changes in the same way. We will say that there is a *negative correlation* between the variables x and y if whenever x increases or decreases, then y changes in the opposite way. The correlation coefficient, r, allows us to measure the linear correlation between two variables and has the following properties.

PROPERTIES OF THE LINEAR CORRELATION COEFFICIENT

1. r is a number between −1 and 1, inclusive. If $r = 1$ or −1, then the scatterplot lies on a straight line.

2. If r is positive, there is positive correlation between the variables. If r is negative, there is negative correlation between the variables.

(continued)

3. If *r* is close to 1, then there is a significant positive linear correlation between the variables and the scatterplot is close to lying along a straight line that rises from left to right.

4. If *r* is close to −1, then there is a significant negative linear correlation between the variables and the scatterplot is close to lying along a straight line that falls from left to right.

5. If *r* is close to 0, then there is little linear correlation between the variables.

We show the relationship between scatterplots and the linear correlation coefficient in Figure 15.25.

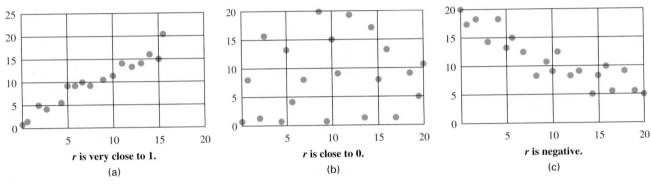

FIGURE 15.25 The value of *r* is related to the shape of the scatterplot.

n	α = .05	α = .01
4	.950	.999
5	.878	.959
6	.811	.917
7	.754	.875
8	.707	.834
9	.666	.798
10	.632	.765
11	.602	.735
12	.576	.708
13	.553	.684
14	.532	.661
15	.514	.641
16	.497	.623
17	.482	.606
18	.468	.590
19	.456	.575
20	.444	.561

TABLE 15.20 Critical values for the linear correlation coefficient.

If *r* is close to 1 or −1, then there is significant positive or negative linear correlation. But how close to 1 or −1 does *r* have to be for us to conclude that there is significant linear correlation? Table 15.20 contains a list of critical values that we can use to decide whether there is significant linear correlation between two variables. We use this table as follows:

1. Compute the linear correlation coefficient *r* for *n* pairs of data.

2. Go to line *n* in Table 15.20.

3. If the absolute value* of *r* exceeds the number in the column labeled α = .05 on line *n*, then there is less than a .05 (5%) chance that the variables do not have significant linear correlation. That is, we can be 95% confident that there is significant linear correlation between the variables.

4. If the absolute value of *r* exceeds the number in the column labeled α = .01 on line *n*, then there is less than a .01 (1%) chance that the variables do not have significant linear correlation. In other words, we can be 99% confident that there is significant linear correlation between the variables.

We will now use the linear correlation coefficient to determine how confident we are that there is a linear relationship between two variables.

EXAMPLE 3 *Determining Correlation between Car Weight and Mileage*

Let us suppose that you are considering buying a used car. After talking to several car dealers, you are convinced that there is a relationship between the weight of the car and gas mileage. To check this, you gather data† regarding gas mileage of cars with various weights, summarized in Table 15.21, to see whether there is a linear correlation between weight and gas mileage for these data.

---

*The absolute value of a nonnegative number is the number itself. The absolute value of a negative number is its opposite. For example, the absolute value of 5 is 5, and the absolute value of −13 is 13. The absolute value of 0 is 0.
†This set of data is based on actual mileage ratings.

Weight (hundreds of pounds)	29	28	31	24	25	30	24	28	32	26
City MPG	21	22	22	23	23	21	24	21	20	22

TABLE 15.21 Car weight versus gas mileage.

Find the correlation coefficient for these data and determine whether there is linear correlation at the 5% or 1% level.

SOLUTION: We will represent the weight by x and the gas mileage by y and by doing calculations similar to those in Example 2, we find that $\Sigma x = 277$, $\Sigma y = 219$, $\Sigma x^2 = 7,747$, $\Sigma y^2 = 4,809$, and $\Sigma xy = 6,040$. Substituting these values in the formula for the correlation coefficient, we get $r = -0.8507$. This number is close to -1, so we expect that there is significant negative linear correlation for this sample of data.

Because there are 10 pairs of data, we will use line 10 of Table 15.20 to determine how confident we can be that there is a significant linear correlation. The absolute value of r is 0.85, which exceeds .765 in line 10 of Table 15.20. Therefore, we can be 99% confident that there is significant negative linear correlation between the variables of car weight and gas mileage.

Now try Exercises 9 to 12. ❉

Although the computations in Example 3 show that we can be very sure that there is a significant linear correlation between the variables of car weight and gas mileage, remember that you cannot assume that there is a cause-and-effect relationship between these variables.

KEY POINT

We use linear regression to find the line of best fit.

The Line of Best Fit

In Example 2, you saw that there was a strong positive linear correlation between the number of times that students attended tutoring sessions and their scores on an examination. We will now show you how to find the line that best models the data that we

HIGHLIGHT ❉ ❉ ❉

Using Technology to Compute the Linear Correlation Coefficient*

When you are working with a large set of data, a statistical calculator will speed your work. The following screens are taken from a graphing calculator. Screen (a) shows the data for the car weights and gas mileage from Example 3.

Screen (b) shows the command (we chose 4) we enter to instruct the calculator to compute the linear correlation coefficient. Screen (c) shows the value of r, which is the same as we computed in Example 3.

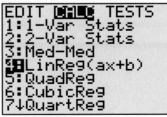

Data has been entered.
(a)

Select command.
(b)

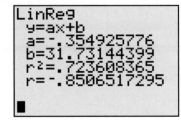

Calculator computes linear correlation coefficient.
(c)

*See your instructor for tutorials on using a graphing calculator to do statistical calculations.

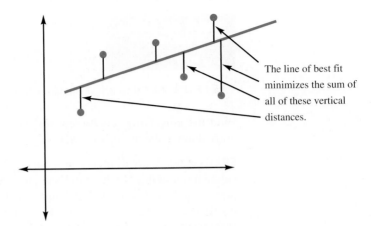

FIGURE 15.26 The line of best fit is the best linear approximation for a set of data points.

used in Example 2. This line is called **the line of best fit**.* Although the line of best fit usually does not pass through many of the data points, it is the line that minimizes the sum of all the vertical distances from the data points to the line, as we show in Figure 15.26.

Recall that (except for vertical lines) we can always write a linear equation in the form $y = mx + b$, where m is the slope of the line and b is the y-intercept.† Notice in the next definition that the calculations we do to find the slope and the y-intercept of the line of best fit are similar to the calculations that you did in finding the linear regression coefficient.

> **DEFINITION** The **line of best fit** for a set of data points of the form (x, y) is of the form $y = mx + b$, where
>
> $$m = \frac{n\sum xy - \left(\sum x\right)\left(\sum y\right)}{n\left(\sum x^2\right) - \left(\sum x\right)^2} \quad \text{and} \quad b = \frac{\sum y - m\left(\sum x\right)}{n}.$$

We will use this definition to find the line of best fit for the data points in Example 2.

EXAMPLE 4   *Finding the Line of Best Fit for a Set of Data Points*

Find the line of best fit for the set of data points we used in Example 2.

SOLUTION: Recall in Example 2 that $n = 10$, $\Sigma xy = 4{,}090$, $\Sigma x = 98$, $\Sigma y = 376$, and $\Sigma x^2 = 1{,}396$. Substituting these values in the formula for the slope of the line, we get

$$m = \frac{n\sum xy - \left(\sum x\right)\left(\sum y\right)}{n\left(\sum x^2\right) - \left(\sum x\right)^2} = \frac{10 \cdot 4{,}090 - 98 \cdot 376}{10 \cdot 1{,}396 - (98)^2}$$

$$= \frac{40{,}900 - 36{,}848}{13{,}960 - 9{,}604} = \frac{4{,}052}{4{,}356} \approx 0.9302.$$

---

*The line of best fit is also called the *regression line* or the *least squares line*.
†If you are a little rusty on this, see Section 7.1.

The y-intercept is given by

$$b = \frac{\sum y - m\left(\sum x\right)}{n} = \frac{376 - (0.9302)98}{10} = \frac{376 - 91.1596}{10} \approx \frac{284.84}{10} = 28.48.$$

Thus, the line of best fit for the data in Example 2 is

$$y = 0.93x + 28.48.$$

In Example 2, we found that the linear correlation coefficient was close to 1, which meant that it was reasonable for us to model the data by a line. The best line to do that is the line that we just found.

Now try Exercises 13 to 16. ❋

As you can see, doing calculations to find the regression coefficient and the line of best fit can be both lengthy and tedious. For that reason, most people who do these computations would use a graphing calculator or a computer program. The Highlight that we showed earlier to do the correlation computations in Example 3 regarding the gas mileage data also shows the slope and the y-intercept of the line of best fit. However, the TI-83 calls the slope a instead of m. As you can see from the Highlight, the line of best fit for the car mileage data is

$$y = -0.355x + 31.73.$$

Notice that the slope of the line is negative, which is consistent with the fact that we found negative linear correlation in Example 3.

HIGHLIGHT ❈ ❈ ❈

Health Scares

If you eat apples, drink coffee, or use a cell phone, you can breathe a little easier. According to the American Council on Science and Health, these are three of the greatest unfounded health scares of the last 50 years.

In 1989, Alar, a chemical used for ripening apples, was thought to cause childhood cancer. Later studies showed that Alar produced cancer only at a dosage of over 100,000 times what a child would consume by eating apples or drinking apple juice. Nonetheless, this scare cost the apple industry about $375 million.

In 1981, Harvard researchers found that drinking two cups of coffee a day doubled the risk of pancreatic cancer.

When other researchers were unable to reproduce these results, the council stated, "This brief scare illustrates the danger of putting too much credence in a single study without analyzing any possible biases or confounding factors."

More recently, several cancer patients claimed that electromagnetic fields from their cell phones caused their illness. The cell phone industry sponsored studies that determined there is no evidence that cell phones are a cause of cancer.*

These examples point out that you must look at reported studies very carefully to decide whether they are reliable.

---

*However, more recent studies show that there may be a correlation between the use of cell phones and certain types of cancer.

Exercises 15.5

Looking Back*

These exercises follow the general outline of the topics presented in this section and will give you a good overview of the material that you have just studied.

1. What did we conclude from the scatterplot in Figure 15.24?

2. In Example 2, we found that the linear correlation coefficient for the data points relating the number of study sessions to grades was 0.9068. What did that tell us?

Sharpening Your Skills

In Exercises 3 and 4, state what kind of correlation, if any, each scatterplot indicates.

3.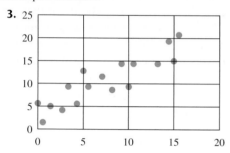

4.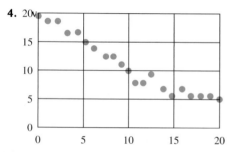

For Exercises 6–8, do the following.

a) *Plot the xy pairs in a scatterplot.*

b) *Estimate the linear correlation coefficient between the two data sets.*

c) *Calculate the linear correlation coefficient to see how accurate your guess was.*

5. (3, 5), (7, 8), (4, 6), (6, 7)

6. (4, 5), (6, 8), (5, 3), (5, 6)

7. (11, 5), (15, 8), (12, 3), (12, 6)

8. (3, 12), (7, 10), (4, 8), (6, 0)

In Exercises 9–12, use Table 15.20 and the linear correlation coefficient you found in the given exercise to determine whether we can be 95% or 99% confident that there is significant linear correlation between x and y.

9. Exercise 5
10. Exercise 6
11. Exercise 7
12. Exercise 8

In Exercises 13–16, find the line of best fit for the data in the specified exercise.

13. Exercise 5
14. Exercise 6
15. Exercise 7
16. Exercise 8

Applying What You've Learned

In Exercises 17–20, determine the linear correlation coefficient for the data and also determine whether we can be 95% or 99% confident that there is significant linear correlation between the two variables.

17. The following table lists the number of years of education past high school and the annual income of five people.

Years of Education Past High School	Annual Income (in $thousands)
0	23
1	22
2	27
2	28
5	35

18. The accompanying table lists the average SAT score and teachers' salaries for certain states. (*Source: Statistical Abstract of the United States*)

State	SAT Score	Teacher Salary (in $thousands)
Connecticut	901	44
Delaware	903	35
Kansas	1,040	30
Kentucky	994	29
Louisiana	993	26

19. As a car begins to accelerate, the gas mileage is poor. As the speed increases, the gas mileage continues to increase. The gas mileage increases for a while, then as the speed increases further, the mileage begins to decrease. The accompanying table of data illustrates this.

Speed in Miles per Hour	Mileage in Miles per Gallon
30	26
40	31
50	33
60	31
70	26

---

*Before doing these exercises, you may find it useful to review the note *How to Succeed at Mathematics* on page xix.

20. The accompanying table is based on data from the National Center for Health Statistics (www.cdc.gov/nchs).

Years of Education	Percentage Who Are Overweight
8	62.7
10	61.1
12	65.5
16	57.6
18	53.7

In Exercises 21–24, find the line of best fit for the data in the specified exercise.

21. Exercise 17   **22.** Exercise 18

23. Exercise 19   **24.** Exercise 20

Communicating Mathematics

25. In Example 3, we found the linear correlation coefficient r to be 0.85, which was greater than 0.765 in the line labeled 10 in Table 15.20. What did that tell us?

26. What is the purpose of finding the line of best fit?

27. a. Describe a situation in which you would expect to have a significant positive correlation between two data sets.

b. Describe a situation in which you would expect to have a significant negative correlation between two data sets.

c. Describe a situation in which you would expect to have no correlation between two data sets.

Using Technology to Investigate Mathematics

28. There are many well-done interactive applications that illustrate the topics of this section. Search the Internet for "linear regression applets" or "line of best fit applets." Experiment with a site that interests you, and report on your findings.

29. See your instructor for a tutorial that explains how to do statistical calculations on a graphing calculator. Use a graphing calculator to reproduce some of the computations in this section and report on your findings.

For Extra Credit

30. What is the linear correlation coefficient for a set of xy pairs if each y score is the double of the corresponding x score? We are looking at pairs of the sort (12, 24), (10, 20), (23, 46), and so on.

31. What is the linear correlation coefficient for a set of xy pairs if in each pair x and y are related by the equation $y = 3x + 5$? We are looking at pairs such as (4, 17), (10, 35), (21, 68), and so on.

CHAPTER SUMMARY*

You will learn the items listed in this chapter summary more thoroughly if you keep in mind the following advice:

1. Focus on "remembering how to remember" the ideas. What pictures, word analogies, and examples help you remember these ideas?
2. Practice writing each item without looking at the book.
3. Make 3×5 flash cards to break your dependence on the text. Use these cards to give yourself practice tests.

SECTION	SUMMARY	EXAMPLE
SECTION 15.1	We refer to a collection of numerical information as **data**. The entire set of data that we are studying is called the **population**. A subset of the population is called a **sample**. When selecting a sample for analysis, it is important that the sample is not **biased** and fairly represents the whole population.	Discussion, p. 715
	A set of data listed with their frequencies is called a **frequency distribution**. We often present a frequency distribution in a **frequency table**. A **relative frequency distribution** shows the percentage of time that each value occurs in a frequency distribution.	Example 1, p. 717 Example 2, p. 718
	In drawing a **bar graph**, we specify the classes on the horizontal axis and frequencies on the vertical axis. We draw bars (rectangles) above each class having a height equal to the frequency of the class. A **histogram** is a type of bar graph used to graph a frequency distribution when we are dealing with a continuous distribution.	Example 3, p. 718 Example 4, p. 719
	A **stem-and-leaf display** is an effective way to present two sets of data "side by side" for analysis.	Example 6, p. 721
SECTION 15.2	The **mean** of a data set is $\dfrac{\Sigma x}{n}$, where Σx is the sum of the data values and n is the number of scores. We represent the mean of a sample by $\bar{x}$ and the mean of a population by μ. We compute the mean of a frequency distribution using the formula $\dfrac{\Sigma(x \cdot f)}{\Sigma f}$, where $x \cdot f$ represent the product of a data value x and its frequency f, and Σf is the sum of the frequencies of the data values in the distribution. To find the **median**, first arrange the data in order. If there is an odd number of values, the median is the number in the middle. If there is an even number of values, the median is the mean of the two middle values. The **mode** is the most frequently occurring value. If two values are most frequent, then we say that there are two modes. We do not allow more than two modes.	Example 1, p. 728 Example 2, p. 729 Examples 4 and 5, p. 732 Example 7, p. 734
	If a data set is arranged in increasing order, the set of values below the median is called the **lower half**. The set of values above the median is called the **upper half**. The **first quartile**, Q_1, is the median of the lower half. The **third quartile**, Q_3, is the median of the upper half. The **five-number summary** of a data set is the minimum value, the first quartile, the median, the third quartile, and the maximum value. We draw a **box-and-whisker plot** as follows: First draw a horizontal axis with a scale that includes the extreme values of the data. Then draw a box from Q_1 to Q_3 with a vertical line in the box to indicate the median. Finally, draw whiskers (lines) from Q_3 to the largest score and from Q_1 to the smallest score.	Example 6, p. 733 Discussion, p. 733
	We can summarize data in various ways depending on which measure of central tendency we choose to use.	Example 8, p. 734
SECTION 15.3	The **range** is the difference between the largest and smallest values in a data set. The **standard deviation** of a sample, indicated by s, is defined as $$s = \sqrt{\frac{\sum(x - \bar{x})^2}{n - 1}},$$	Example 1, p. 740 Definition, p. 742 Example 2, p. 742

---

*Before studying this chapter's material, it would be useful to reread the note *How to Succeed at Mathematics* on page xix.

where $\bar{x}$ is the mean of the sample, and n is the number of data values. For a population, we use the formula

$$\sigma = \sqrt{\frac{\sum(x-\mu)^2}{n}}.$$

If f represents the frequency of data value x in the distribution, then the standard deviation is given by the formula

Example 3, p. 743

$$s = \sqrt{\frac{\sum(x-\bar{x})^2 \cdot f}{n-1}}.$$

The number $n = \Sigma f$ is the number of data values in the distribution.

The **coefficient of variation** expresses the standard deviation of a sample as a percentage of the mean and is calculated as follows:

$$CV = \frac{s}{\bar{x}} \cdot 100\%$$

Example 4, p. 746

SECTION 15.4	A **normal curve** has the following properties:	Discussion, p. 751

1. It is bell-shaped.
2. The highest point on the curve is at the mean of the distribution.
3. The mean, median, and mode of the distribution are the same.
4. The curve is symmetric with respect to its mean.
5. The total area under the curve is 1.
6. Roughly 68% of the data values are within 1 standard deviation from the mean, 95% of the values are within 2 standard deviations from the mean, and 99.7% of the values are within 3 standard deviations from the mean.

Example 1, p. 752

A **z-score** represents the number of standard deviations a data value is from the mean of a data set. We can use a table to relate positive z-scores to the amount of area under the standard normal curve between the mean and the given z-scores.

Example 2, p. 755

If a normal distribution has a mean of μ and a standard deviation of σ, the equation

$$z = \frac{x - \mu}{\sigma}$$ converts a data value to a z-score.

Example 3, p. 757

The normal distribution has many useful applications.

Example 4, p. 758
Example 5, p. 758
Example 6, p. 759

SECTION 15.5	A **scatterplot** is a graph of a set of data points.	Example 1, p. 763

The **linear correlation coefficient** measures to what degree points of a scatterplot lie along a straight line.

Example 2, p. 765
Example 3, p. 766

The **line of best fit** is the line that best models a set of data points.

Example 4, p. 768

CHAPTER REVIEW EXERCISES

Section 15.1

The following data set represents the number of automobile accidents that have occurred in a city each day over the past month.

8, 8, 6, 6, 7, 10, 4, 6, 5, 10, 8, 9, 8, 7, 8, 8,
6, 9, 5, 9, 8, 10, 8, 8, 6, 10, 6, 5, 10, 8, 9

1. Construct a frequency table for these data.

2. Construct a relative frequency table for these data.

3. Display the relative frequency table in Exercise 2 as a bar graph.

4. An operator of Six Flags' Superman roller coaster counted the number of people on the ride at 10-minute intervals for a certain period of time. The results he obtained are summarized in the accompanying bar graph. Use the graph to answer the following questions.

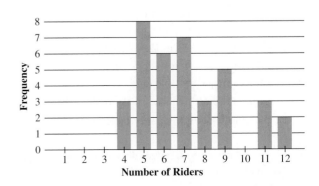

a. What was the smallest number of people on the ride and how often did it occur?

b. What was the most frequently occurring rider count?

c. For how many 10-minute intervals were the riders counted?

5. The lists represent the ages of actors (*M*) and actresses (*F*) at the time they won an academy award from 1980 to 2008.

> *F:* 31, 74, 33, 49, 38, 61, 21, 41, 26, 80, 42, 29, 33, 35, 45, 49, 39, 34, 26, 25, 33, 35, 35, 28, 30, 30, 29, 61, 32
>
> *M:* 37, 76, 39, 52, 45, 35, 61, 43, 51, 32, 42, 54, 52, 37, 38, 31, 45, 60, 46, 40, 36, 47, 29, 43, 37, 37, 38, 45, 50

a. Represent these two sets of data on a single stem-and-leaf plot.

b. Do you notice any patterns in comparing the two plots?

Section 15.2

6. The following is the number of regular-season wins by the Detroit Pistons over 12 seasons ending in 2008: 59, 53, 64, 54, 54, 50, 50, 32, 42, 29, 37, 54. Find the mean, median, and mode for this distribution.

7. Briefly describe what information is conveyed by the mean, median, and mode.

8. Calculate the five-number summary for the following distribution:

> 10, 8, 6, 12, 6, 3, 11, 7, 6, 17, 4, 9, 13, 20, 7

9. Construct a box-and-whisker plot for the distribution given in Exercise 8.

Section 15.3

10. Calculate the sample standard deviation for the following distribution:

> 4, 6, 7, 3, 5, 6, 4, 5

11. Explain what the standard deviation tells you about a distribution.

Section 15.4

12. State several basic properties of the normal distribution.

13. If a distribution has a mean of 80 and a standard deviation of 7, to what *z*-score does the raw score 85 correspond?

14. If a distribution has a mean of 60 and a standard deviation of 5, the *z*-score 1.35 corresponds to what raw score?

15. Assume that 1,000 scores are normally distributed with a mean of 72 and a standard deviation of 4. How many scores can we expect to be between 75 and 82?

16. Suppose you earned an 82 on a history exam that had a mean score of 78 and a standard deviation of 3, and you also earned an 84 on an anthropology exam that had a mean of 79 and a standard deviation of 4. On which exam did you do better in relationship to the rest of the class?

17. Redo Example 6 in Section 15.4 regarding setting a warranty on an iPod Touch. Now assume that the mean time to failure is 2,500 hours with a standard deviation of 500 and we want no more than 4% to be returned before the warranty expires.

Section 15.5

18. State what kind of correlation, if any, is indicated by the following scatterplots.

a.

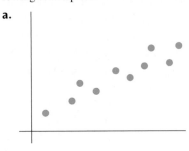

b.

19. a. Calculate the correlation coefficient for the following set of data points:

> (3, 5), (4, 9), (5, 10), (6, 13)

b. Find the line of best fit for these points.

CHAPTER TEST

The following data set represents the number of visits to the health center that have occurred at your school over the past month.

> 7, 7, 8, 8, 6, 11, 5, 7, 6, 12, 4, 9, 8, 6, 9, 8, 9, 7, 6, 7, 7, 8, 6, 10, 9, 11, 8, 8, 7, 9, 10, 6, 7, 5, 5, 6, 6, 7, 9, 5

1. Construct a frequency table for these data.

2. Construct a relative frequency table for these data.

3. Display the relative frequency table in Exercise 2 as a bar graph.

4. Explain what the standard deviation tells you about a distribution.

5. Calculate the standard deviation for the following distribution:

> 3, 4, 5, 6, 5, 6, 8, 9, 8

6. State several basic properties of the normal distribution.

7. We have listed the number of home runs that Babe Ruth hit during the seasons he played for the Yankees, and the number of home runs that Hank Aaron hit during his best 15 years. Represent the two sets of data on a single stem-and-leaf display.

> Ruth: 54, 59, 35, 41, 46, 25, 47, 60, 54, 46, 49, 46, 41, 34, 22
>
> Aaron: 44, 30, 39, 40, 34, 45, 44, 32, 44, 39, 44, 38, 47, 34, 40

8. The following is a list of wins by the Boston Red Sox from 1998 to 2007:

$$92, 94, 85, 82, 93, 95, 98, 95, 86, 96$$

a. Calculate the five-number summary for this distribution.

b. Construct a box-and-whisker plot for this distribution.

9. Suppose that you earned an 85 on a statistics exam that had a mean of 78 and a standard deviation of 4, and an 88 on a sociology exam that had a mean of 80 and a standard deviation of 5. On which exam did you do better in relationship to the rest of the class?

10. The webmaster at Match.com recorded the number of hits at 1-minute intervals for a certain period of time. The results she obtained are summarized in the accompanying bar graph. Use the graph to answer the following questions.

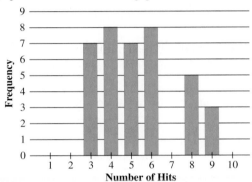

a. What was the smallest number of hits per minute, and how often did that occur?

b. What was the most frequent number of hits?

c. For how many 1-minute intervals were the number of hits counted?

11. State what kind of correlation, if any, is indicated by the following scatterplots:

a.

b.

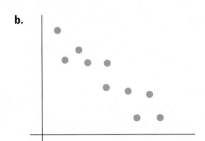

12. Briefly describe what information is conveyed by the mean, median, and mode.

13. Find the mean, median, and mode of the following distribution:

$$9, 8, 6, 11, 5, 9, 7, 6, 3, 10, 11, 9, 9, 6, 10, 7, 8, 9, 5, 9$$

14. If a distribution has a mean of 50 and a standard deviation of 6, the z-score 1.83 corresponds to what raw score?

15. If a distribution has a mean of 75 and a standard deviation of 5, to what z-score does the raw score 82 correspond?

16. Assume that 1,000 scores are normally distributed with a mean of 54 and a standard deviation of 5. How many scores can we expect to be above 58?

17. Redo Example 6 in Section 15.4 regarding setting a warranty on an iPod Touch. Now assume that the mean time to failure is 2,800 hours with a standard deviation of 400 and we want no more than 6% to be returned before the warranty expires.

18. a. Calculate the correlation coefficient for the following set of data points:

$$(3, 6), (4, 8), (5, 7), (6, 10)$$

b. Find the line of best fit for these points.

GROUP EXERCISES

1. Find some data that interests you and present it in a stem-and-leaf display similar to the way we presented the age data for men and women on winning academy awards in the Chapter Review Exercises. Write a brief report on some conclusions that you draw from your presentation.

2. Research other ways to present data than the ones that we have discussed in this chapter and use these methods to represent some of the data in this chapter.

3. Research some ways that people can mislead you when presenting data. An excellent place to start on this topic is the classic book *How to Lie with Statistics* by Darrell Huff.

4. Find some data of interest to you relating two quantities similar to our example regarding the weight of cars and gas mileage in Example 3 of Section 15.5. Find the correlation coefficient for these data points and also find the line of best fit. A professor in one of your nonmathematics classes might be able to suggest an interesting topic for this statistical analysis.

Basic Mathematics Review

Order of Operations

To do arithmetic and algebraic calculations correctly, you must understand the order in which we perform operations. Many people use the memory device, which we explain in the following diagram:

(P)lease (E)xcuse (M)y (D)ear (A)unt (S)ally

First, work inside parenthesis.

Second, evaluate expressions with exponents.

Third, perform all multiplications and divisions in order from left to right.

Last, perform all additions and subtractions in order from left to right.

For example, we would evaluate the expression $(3 + 5)^2 + 2 \cdot (7 + 4) + 9$ as follows:

$(3 + 5)^2 + 2 \cdot (7 + 4) + 9$	
$8^2 + 2 \cdot 11 + 9$	First, work inside parentheses.
$64 + 2 \cdot 11 + 9$	Second, evaluate 8^2.
$64 + 22 + 9$	Next, multiply 2 times 11.
95	Finally, add.

Solving Equations

To solve equations, first, if you can, simplify by collecting like terms on each side of the equation. Then you can add or subtract the same quantity from both sides of the equation. You can also multiply or divide both sides of the equation by the same quantity, provided you do not divide by zero. We could solve the equation $3x - 17 + 7x = 15 + 6x + 8$ as follows:

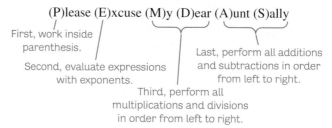

like terms   like terms

$3x - 17 + 7x = 15 + 6x + 8$	
$10x - 17 = 23 + 6x$	Collect like terms.
$10x - 17 + 17 = 23 + 6x + 17$	Add 17 to both sides.
$10x = 40 + 6x$	Simplify.
$10x - 6x = 40 + 6x + 6x$	Subtract $6x$ from both sides.
$4x = 40$	Simplify.
$x = 10$	Divide both sides by 4.

If an equation has decimal or fractional coefficients, you can make the equation easier to solve if you simplify the coefficients by multiplying both sides of the equation by a suitable quantity before solving.

For example, you could multiply both sides of the equation $2.4x + 4.12 = 1.3x + 5$ by 100 to eliminate decimals and then solve the resulting equation $240x + 412 = 130x + 500$.

To solve the equation $\frac{1}{3}x + \frac{1}{2} = x - \frac{5}{4}$, you could multiply both sides of the equation

by 12 to get the equivalent equation $(12)\frac{1}{3}x + (12)\frac{1}{2} = (12)x - (12)\frac{5}{4}$, which simplifies to

$4x + 6 = 12x - 15$.

Working with Fractions (Rational Numbers)

For a review of the properties and operations on rational numbers see the following examples in the text:

> Equality of Rational Numbers: Section 6.3, Example 1
>
> Reducing Rational Numbers: Section 6.3, Example 2
>
> Addition and Subtraction: Section 6.3, Example 3
>
> Multiplication: Section 6.3, Example 4
>
> Division: Section 6.3, Example 5
>
> Improper Fractions and Mixed Numbers: Section 6.3, Examples 7 and 8

Working with Decimals

To add or subtract decimals, line up the decimal points as in the following example:

$$\begin{array}{r} 27.36 \\ 142.405 \\ +\ 3.2 \\ \hline 172.965 \end{array}$$

$\llcorner$ Line up decimal points.

To multiply decimals, we add the decimal places to obtain the final answer as follows:

$$\begin{array}{r} 5.304 \\ \times 6.27 \\ \hline 33.25608 \end{array}$$

5.304 —— 3 decimal places

×6.27 —— 2 decimal places

33.25608 —— 3 + 2 decimal places

To divide decimals, move the decimal point in the divisor and dividend before dividing as in the following example:

$$1.27\overline{)93.472} \quad \to 73.6$$

Move decimal point two places to the right.

Working with Percents

For a review of the basic principles in working with percents see the following examples:

> Meaning of Percent: Introduction to Section 9.1
>
> Conversions between decimals and percents: Section 9.1, Example 1
>
> Conversions between fractions and percents: Section 9.1, Example 2
>
> Solving equations involving percents: Section 9.1, Example 6

Answers to Quiz Yourself Problems

Chapter 1

1.

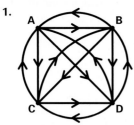

2. a. Students taking both swing and Latin

    b. Taking Latin but not swing

3. h and l

4.

Line 4:	No	No	Yes
Line 7:	No	Yes	Yes

5. a.   1   6   15   20   15   6   1
      1   7   21   35   35   21   7   1

    b. 1 and 100

6. 132   **7.** 4, 18, 28 tons   **8.** 1,024

9. Any numbers will work.

10. Almost any choice of numbers for a, b, and c will work. For example, let

$$a = 3, \ b = 4, \text{ and } c = 5. \text{ Then } \frac{a+b}{a+c} = \frac{3+4}{3+5} = \frac{7}{8}.$$

However, $\dfrac{b}{c} = \dfrac{4}{5}$.

11. a. We are first adding two numbers and then taking the square root of the sum.

    b. We are first taking the square roots of two numbers and then forming the sum of these square roots.

    c. To "first add and then take the square root" is not the same as to "first take the square roots and then add."

12. a. There is an extra line in the symbol on the right.

    b. The symbol on the left has a pair of braces around the $\varnothing$ symbol.

    c. The symbol on the left is rounded; the symbol on the right is pointed.

    d. The symbol on the right is the number 0; the symbol on the left is the number 0 with braces around it.

13. a. intersection of streets   **b.** happening at the same time

14. a. $31 + 7$ **b.** $5 + 41$ **15.** 479,001,600

16. a. 635,000 **b.** 724,000

17. $400,000,000 \times 0.15 = 60,000,000$

Chapter 2

1. a. {Monday, Tuesday, . . . , Sunday}

    b. $\{x : x \text{ is a natural number less than or equal to } 60\}$

2. a. well defined   **b.** not well defined

3. a. true **b.** true **c.** false   **4. a.** 10 **b.** 2 **c.** 50

5. a. true **b.** false   **6. a.** true **b.** false

7. $\varnothing$, $\{1\}$, $\{2\}$, $\{3\}$, $\{1, 2\}$, $\{1, 3\}$, $\{2, 3\}$, $\{1, 2, 3\}$

8. a. $2^5 = 32$ **b.** $2^{26} = 67{,}108{,}864$

9. a. r_1; volunteers who will neither cook nor serve food

    b. r_1 and r_4; the volunteers who will not cook

10. $M \cup B = \{m, a, t, h, e, i, c, s, b, u, y\}, M \cap B = \{a, t, e\}$

11. a. $C' = \{c, i, k, o, q\}$ **b.** $B - A = \{2, 7, 8\}; A - B = \{1, 5\}$

12. a. $\{1, 3, 4, 6, 8, 9, 10\}$ **b.** $\{3, 4, 6, 10\}$

    c. $\{1, 4, 8, 9, 10\}$ **d.** $\{1, 3, 4, 6, 8, 9, 10\}$

13. a. $A' \cap B' \cap C$ **b.** $(A \cap B \cap C') \cup (A' \cap B \cap C')$

14. 57

15. There are several correct answers; one would be

    Carreras   Domingo   Pavarotti
        $\updownarrow$ $\updownarrow$ $\updownarrow$
    Aguilera   Beyonce   Clarkson

16. There are many correct answers for this, one is

     1   2   3 $\cdots$ n $\cdots$
     $\updownarrow$ $\updownarrow$ $\updownarrow$ $\updownarrow$
     2   3   4 $\cdots$ $n+1$ $\cdots$

17. a. $6, \frac{5}{2}, \frac{4}{3}$ **b.** $\frac{2}{6}$

Chapter 3

1. a. simple **b.** compound **c.** compound

2. a. $\sim d \vee \sim i$ **b.** $\sim d \wedge i$

3. a. $\sim f \rightarrow \sim q$ **b.** $f \leftrightarrow q$

4. a. At least one professor is not friendly.

    b. No dogs bite.

5. a. true **b.** false **c.** false **d.** true

6.

			2	**1**	**5**	**3**	**4**	
p	*q*	(*p*	$\wedge$	~*q*)	$\vee$	(~*p*	$\vee$	*q*)
T	T	T	F	F	T	F	T	T
T	F	T	T	T	T	F	F	F
F	T	F	F	F	T	T	T	T
F	F	F	F	T	T	T	T	F

7.

		2	**1**		
p	*q*	~	(*p*	∨	*q*)
T	T	F	T	T	T
T	F	F	T	T	F
F	T	F	F	T	T
F	F	F	F	F	F

1	**3**	**2**
(~*p*)	∧	(~*q*)
F	F	F
F	F	T
T	F	F
T	T	T

The statements have identical truth tables and so are logically equivalent.

8. It is not the case that the car is more than five years old or has been driven over 50,000 miles.

9.

			2	**1**	**6**	**3**	**5**	**4**
p	*q*	(*p*	∧	~*q*)	→	(~*p*	∨	*q*)
T	T	T	F	F	T	F	T	T
T	F	T	T	T	F	F	F	F
F	T	F	F	F	T	T	T	T
F	F	F	F	T	T	T	T	F

10. Inverse: If the price of downloading videos does not increase, then people will not copy them illegally.

Contrapositive: If people will not copy them illegally, then the price of downloading videos does not increase.

11. a. If you travel on the Amazon, then you have updated your immunization.

b. If you increase your cardiovascular fitness, then you exercise three times a week.

12. The argument is invalid.

13. fallacy of the inverse; invalid

14.

invalid

15. valid

16. Answers will vary; however, here are some possibilities:

a. Some Bs are not As.

b. Some As are not Cs.

c. Some Bs are not Cs.

17. d. The truth value of "*Gone with the Wind* is not a great movie" is 0.27.

e. "You find mathematics uninteresting" has a truth value of 0.3.

18. a. 0.45 **b.** 0.82   **19. a.** 0.45 **b.** 0.55

20. a. 0.80 **b.** 0.55 **c.** 0.80

21. a. $(\sim i) \vee c = 0.95$ **b.** $(\sim i) \vee c = 0.75$

Chapter 4

1. The graph is connected. A, B, D, E, and F are even. C and G are odd. Edge CG is a bridge.

2. a. Can be traced; C, B, A, C, D, B, A, D.

b. Cannot be traced; it has four odd vertices.

3. a. CABCFEGDFIH

b. No. Because C and H are both odd vertices, we must begin at one and end at the other.

4. A, B, C, and D are the odd vertices in the graph. If we duplicate edge CD and also all edges along the indicated path from A to B, the new graph will be Eulerian.

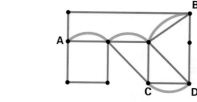

5.

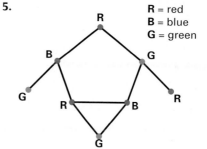

R = red
B = blue
G = green

6. a. BDACB **b.** ADBCA   **7.** 6! = 720

8. ABDCA has the minimal weight of 11. (Of course, the reverse circuit ACDBA also has the same weight.)

9. MAPNCM; $1,200

10. a. yes; ABC and AFBC **b.** no **c.** 4

11.

		To		
		A	**B**	**C**	**D**

		A	**B**	**C**	**D**
	A	0	1	0	0
From	**B**	0	0	0	0
	C	1	2	0	0
	D	1	2	1	0

12.

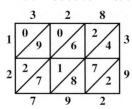

13. **a.** Begin, B, D, E, G

   b. Begin, B, D, E, G, End

   c. To begin on the tenth day of the project

   d. 15 days

Chapter 5

1. **a.** 21,346

   b. ⌒⌒ℓℓℓℓℓ𓎡99999∩∩∩||||||

2. **a.** 𓎡9∩∩∩∩∩∩∩| **b.** 𓎡𓎡999999||||

3. **a.** $1 + 8 + 16$ **b.** $43 + 344 + 688 = 1{,}075$

4. **a.** We cannot subtract L; we can only subtract I from V and X.

   b. 548   **c.** $9 \times 100 = 900; 5 \times 1{,}000 = 5{,}000$

5. **a.** $12 \times 60^2 + 23 \times 60 + 32 = 44{,}612$

   b. ▼▼ ▼▼▼▼▼ ⟨▼▼▼

6.

	3	2	8	
1	0/9	0/6	2/4	3
2	2/7	1/8	7/2	9
	7	9	2	

Product is 12,792.

7. $1, 2, 10_3, 11_3, 12_3, 20_3, 21_3, 22_3, 100_3, 101_3$

8. 142   **9.** 423_5 **10.** 11_5 and 10_5

11. **a.** $1{,}243_6$ **b.** 154_6 **12.** 13_5 and 31_5

13. **a.** $3{,}133_5$ **b.** $3{,}402_8$ **14.** **a.** 3   **b.** 6

15. **a.** false   **b.** true   **16.** **a.** 2   **b.** 7   **c.** 2

17. **a.** 5   **b.** 0, 3, 6   **18.** 33

Chapter 6

1. **a.** prime   **b.** composite   **c.** composite   **d.** prime

2. **a.** yes   **b.** yes   **c.** no   **d.** yes

   e. yes   **f.** no   **g.** yes   **h.** yes

3. **a.**

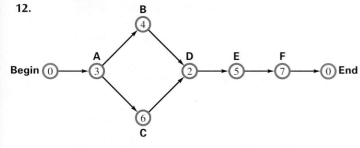

   b. $1{,}560 = 2 \cdot 2 \cdot 2 \cdot 3 \cdot 13 \cdot 5 = 2^3 \cdot 3 \cdot 5 \cdot 13$

4. 42   **5.** **a.** $+10$ **b.** -5 **c.** -10 **d.** $+22$

6. **a.** $(+5) - (+12) = (+5) + (-12) = -7$

   b. $(-3) + (+9) = +6$

7. **a.** $+24$ **b.** -56 **8.** -24

9. **a.** equal   **b.** not equal   **10.** $\dfrac{3}{4}$

11. $\dfrac{29}{24}$ and $\dfrac{17}{48}$ **12.** **a.** $\dfrac{14}{5}$ **b.** $\dfrac{10}{9}$

13. **a.** $\dfrac{5}{22}$ **b.** $\dfrac{12}{7}$ **14.** **a.** $5\dfrac{3}{4}$ **b.** $\dfrac{13}{5}$ **15.** $0.\overline{63}$

16. **a.** $\dfrac{2{,}548}{10{,}000} = \dfrac{637}{2{,}500}$ **b.** $\dfrac{6}{11}$ **17.** **a.** $3\sqrt{5}$ **b.** $\dfrac{\sqrt{5}}{4}$

18. $\dfrac{\sqrt{15}}{5}$ **19.** **a.** $11\sqrt{5}$ **b.** 0

20. **a.** $\sqrt{2} - \sqrt{2} = 0$ **b.** $(-2)(-3) = +6$

21. **a.** $(3 + 4) + 5 = 3 + (4 + 5)$

   b. $4 \cdot (-2) = (-2) \cdot 4$

22. $3^6 = 729$ **23.** **a.** 5^{12} **b.** $(-2)^8 = 256$

24. **a.** $\dfrac{1}{36}$ **b.** 64   **c.** 1

25. **a.** $3^8 \cdot 3^{-6} = 3^{8+(-6)} = 3^2 = 9$

   b. $(2^{-2})^{-1} = 2^{(-2)\cdot(-1)} = 2^2 = 4$

26. **a.** 5.372841×10^4 **b.** 0.00245   **27.** 50   **28.** 440

29. 354,294   **30.** 6,560   **31.** 144 and 233

Chapter 7

1. **a.** $x = 2 - \dfrac{2}{3}y$ **b.** $W = \dfrac{P - 2L}{2}$

2. x-intercept, $(4\tfrac{1}{2}, 0)$; y-intercept, $(0, 6)$

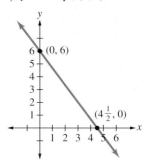

3. $\dfrac{5}{4}$

4. $y = -4x + 5$ **5.** $y = -\dfrac{5}{9}x + \dfrac{10}{3}$

6. $x = 4$ and $x = -11$

7.

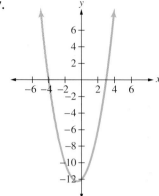

8. $1,451.61   **9.** slightly over 121.1 million

10. 2041   **11. a.** $y = 180 (1.04)^x$ **b.** $256.20

12. 0.6096, or 60.96%   **13.** 4.5 hours

14. 104,000   **15.** 133   **16.** 12 minutes

17. "f of 80 equals 6.40."

18. a. f is not a function.   **b.** g is a function.   **19.** 8

20. No. If we were to graph the pairs (2, 5) and (2, 4), we would see that a vertical line would pass through them both.

21. 496 mg   **22. a.** $\frac{1}{2}$; unstable   **b.** -6; stable

Chapter 8

1. (3, 1)   **2.** (-4, 5)

3. a. There are no solutions.
   b. The system represents a pair of distinct parallel lines.

4. a. yes   **b.** no   **c.** no

5.

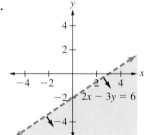

6.

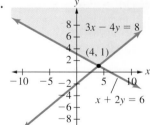

Chapter 9

1. a. 0.1745   **b.** 0.0005   **c.** 245%   **d.** 2.5%   **2.** 24%

3. a. 9   **b.** 75   **c.** 30%   **4.** $18,271.50

5. $2,000   **6.** $2,928.20   **7.** $4,440.73

8. $1,126.49   **9.** 1.511391594   **10.** 23 months

11. $36.30   **12.** $289.25; $5.06

13.

Day	Balance	Number of Days × Balance
1, 2	$240	$2 \times 240 = 480$
3, 4, 5, 6, 7, 8, 9, 10	$264	$8 \times 264 = 2,112$
11, 12, 13, 14, 15, 16, 17, 18, 19, 20, 21, 22	$204	$12 \times 204 = 2,448$
23, 24, 25, 26, 27, 28, 29, 30	$216	$8 \times 216 = 1,728$

average daily balance $= \dfrac{6,768}{30} = \$225.60$.

14. $\dfrac{x^4 - 1}{x - 1}$ **15.** $1,907.40   **16.** $95.47

17. 240 months, or 20 years

18. $253.94

19.

4	$798.37	$698.27	$100.10	$119,603.07

20. $9,493.49

21. a. $17.76   **b.** about 11%   **22.** 13%   **23.** 18.81%

Chapter 10

1. a. a line   **b.** line segment CD
   c. a ray   **d.** a half line

2. a. vertical angles   **b.** a right angle
   c. supplementary angles   **d.** an obtuse angle
   e. an acute angle

3. a. 7   **b.** 8   **c.** 6   **d.** 4

4. a. 104°   **b.** 76°   **5.** 5 in.

6. 360°   **7.** 135°

8. 126 square units   **9.** 18 sq ft

10. approximately 17.3 in.

11. circumference ≈ 50.24; area ≈ 200.96

12. volume $= 144$ cu cm; surface area $= 180$ sq cm

13. volume ≈ 785 cu cm; surface area ≈ 471 sq cm

14. volume ≈ 83.73 cu yd; surface area ≈ 80.42 sq yd

15. $\dfrac{4}{3}\pi(6)^3 \approx 904$ cu cm   **16. a.** 1,000   **b.** 100

17. a. 56,300   **b.** 0.04850

18. 2,145 ft   **19.** 133.76 oz

20.

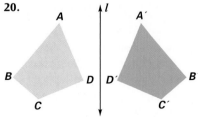

21.

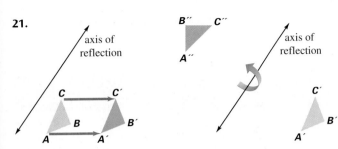

22. Each vertex is the vertex of two octagons and one square. The interior angle of each octagon is 135° and the interior angle of the square is 90°. The tessellation is possible because $135° + 135° + 90° = 360°$.

23. 256   **24.** 243   **25.** $\frac{64}{27}$ **26.** $\frac{27}{64}$ **27.** 1.5

Chapter 11

1. On a 12-member board, Naxxon gets 6 representatives, Aroco 4, and Eurobile 2.

2. a. The average constituency of the electricians is 140.
   b. The electricians are more poorly represented.

3. 64   **4.** $\frac{2{,}351 - 2{,}342}{2{,}342} = \frac{9}{2{,}342} \approx 0.004$

5. a. 0.216   **b.** 0.175
   c. State B should get the additional representative.

6. For Iowa, the Huntington–Hill number is $\frac{(2.9)^2}{5 \times 6} \approx 0.28$; for Nebraska, the Huntington–Hill number is $\frac{(1.8)^2}{3 \times 4} = 0.27$. Therefore, Iowa should get the additional representative.

7. The Huntington–Hill numbers for the three companies are as follows: Naxxon, 368.2; Aroco, 228.2; Eurobile, 128. Because 368.2 is the largest, Naxxon gets the sixth representative.

8. a. 1,500,000   **b.** A: 2   B: 2.67   C: 3.33

9. 3.01; 3; 4   **10. a.** 600   **b.** 4.9   **c.** 4

11. a. 480   **b.** 2.6   **c.** 3

12. a. 1,500   **b.** 2.79   **c.** 3

13. a. discrete   **b.** continuous   **c.** discrete   **d.** continuous

14.

	Rosa $\left(\frac{1}{4}\right)$	Juan $\left(\frac{1}{4}\right)$	Carlos $\left(\frac{1}{4}\right)$	Luis $\left(\frac{1}{4}\right)$
Bid on house	$250,000	$240,000	$260,000	$200,000
Fair share of estate	$62,500	$60,000	$65,000	$50,000
Item obtained with highest bid			House	

15.

	Rosa $\left(\frac{1}{4}\right)$	Juan $\left(\frac{1}{4}\right)$	Carlos $\left(\frac{1}{4}\right)$	Luis $\left(\frac{1}{4}\right)$
Pays to estate (+) or receives from estate (−)	$62,500 (−)	$60,000 (−)	$195,000 (+)	$50,000 (−)
Division of estate balance ($22,500)	$5,625 (−)	$5,625 (−)	$5,625 (−)	$5,625 (−)
Summary of cash	Receives $68,125	Receives $65,625	Pays $189,375	Receives $55,625

Chapter 12

1. A wins with 78 points.   **2.** C wins.

3. N wins with 2 points. B gets 1 point, and T gets none.

4. The winner is A with 67 points. Because C has the majority of first-place votes, the majority criterion is not satisfied.

5.

1st	H	H	H	I	I	T	T
2nd	I	I	I	T	T	I	I
3rd	T	T	T	H	H	H	H

6. Using the pairwise comparison method, the winner is W. However, if X and Y are removed, then Z wins the election. The independence-of-irrelevant-alternatives criterion is not satisfied.

7. a. 4 (Let's call them A, B, C and D.)   **b.** 5
   c. {A, B}, {A, C}, {A, B, C}, {A, B, D}, {A, C, D}, {B, C, D}, {A, B, C, D}

8. a.

	Winning Coalitions	Critical Members
1	{A, B}	A, B
2	{A, C}	A, C
3	{A, B, C}	A
4	{A, B, D}	A, B
5	{A, C, D}	A, C
6	{B, C, D}	B, C, D
7	{A, B, C, D}	None

b. A, $\frac{5}{12}$; B, $\frac{3}{12}$; C, $\frac{3}{12}$; D, $\frac{1}{12}$

9. $5! = 120$ **10. a.** B   **b.** A

Chapter 13

1. $7 \times 6 = 42$ **2.** 16   **3.** 24

4. $2 \times 8 \times 3 \times 2 = 96$ different cars   **5.** 2,116,800

6. a. $P(8, 3)$ is the number of ways we can select three different objects from a set of eight and arrange them in a straight line.
   b. 336

7. 792   **8. a.** 1 7 21 35 35 21 7 1   **b.** 35

9. $C(3, 0)$ $C(3, 1)$ $C(3, 2)$ $C(3, 3)$

10. $13 \times C(4, 2) = 13 \times 6 = 78$

Chapter 14

1. a. $\{(1, 5), (2, 4), (3, 3), (4, 2), (5, 1)\}$
   b. $\{bbb, bbg, bgb, gbb\}$

2. 0.72   **3.** 0.808   **4.** 16

5. a. $\dfrac{5}{36}$ **b.** $\dfrac{C(12, 2)}{C(52, 2)} = \dfrac{66}{1,326} = \dfrac{11}{221}$

6. $1 - \dfrac{3}{36} = \dfrac{33}{36} = \dfrac{11}{12}$ **7.** 0.45   **8.** $\dfrac{0.04}{0.09} = 0.44$

9. $\dfrac{12}{52} \cdot \dfrac{11}{51} \approx 0.05$ **10.** 0.18

11. Because $P(O) = \frac{1}{2} = P(O \mid F)$, the events are independent.

12. a. This is the probability that she will be assigned to dormitory X.
   b. 0.40

13. \$126   **14.** $+\frac{1}{9}$; it is now to the student's advantage to guess.

15. \$12.00   **16.** $B\left(8, 2; \frac{1}{6}\right) = 0.2605$

Chapter 15

1.

Score	Frequency	Relative Frequency
1	3	0.15
2	4	0.20
3	2	0.10
4	0	0.00
5	2	0.10
6	1	0.05
7	4	0.20
8	2	0.10
9	1	0.05
10	1	0.05

2.

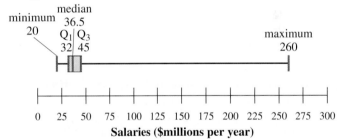

3.

```
6 | 2 2 3 6 7 8
7 | 1 3 5 7
8 | 1 2 2 3 4 8
9 | 0 1 2 8
```

4. a. $\Sigma x = 106$ **b.** $n = 8$ **c.** $\bar{x} = 13.25$

5. The mean number of calls is $\dfrac{\Sigma(x \cdot f)}{\Sigma f} = \dfrac{96}{14} \approx 6.86$.

6. 34.75   **7.** 95

8. a. 20, 32, 36.5, 45, 260
   b.

9. 2.29   **10.** $s = \sqrt{\dfrac{\Sigma(x - 42)^2 \cdot f}{n - 1}} = \sqrt{\dfrac{172}{19}} \approx 3.01$

11. a. 950   **b.** 25

12. a. 0.427   **b.** $0.468 - 0.391 = 0.077$ **c.** 0.412

13. a. 1.29   **b.** 0   **c.** -1.71

14. 24 months

15.

We see in this graph that as x increases, y generally decreases.

16. -0.9718

Answers to Exercises

Chapter 1

Section 1.1

1. Understand the problem; devise a plan; carry out your plan; check your answer.

3. Answers will vary. **5.** *How to Solve It*

7.

10%   pure   5%

9.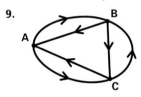

11. Length is l and width is w. **13.** M, C, I, A, and D

15. s for amount invested in stocks; b for amount invested in bonds

17. HH, HT, TH, TT **19.** 16

21. WK, WA, WU, KA, KU, AU

23. $(1, 1), (1, 2), (1, 3), (1, 4), (2, 1), (2, 2), (2, 3), (2, 4), (3, 1), (3, 2), (3, 3), (3, 4), (4, 1), (4, 2), (4, 3), (4, 4)$

25. r_3 is the set of people who are good singers and appeared on *American Idol*. r_4 is the set of people who appeared on *American Idol* and are not good singers.

27. 35, 42, 49, 56, 63 **29.** *bf, cd, ce, cf, cg*

31. 21, 34, 55, 89, 144

33. Possible answer: Three people A, B, C can be lined up in 6 ways; four people can be lined up in 24 ways.

35. Possible answer: With three letters there are 9 ways; with five letters there are 25 ways.

37. Possible answer: For only air conditioning (*A*) and CD player (*C*) there are four possibilities: neither, *A, C, AC*. With three options there are eight possibilities.

39. False; September has 30 days.

41. False; almost any numbers give a counterexample.

43. False; A is the grandfather of C.

45. False; $-9 < -5$, but $(-9)^2 = 81$, which is not less than $(-5)^2 = 25$.

47. false **49.** not the same **51.** the same

53. 5 is a number; $\{5\}$ is a number with braces around it.

55. One is uppercase, the other is lowercase.

57. $\{\ \}$ are different from $(\ )$.

59. The symbol on the right is the number 0; the symbol on the left is not a number.

65. 107, 214 **67.** 9 in the store and 6 giving lessons

69. LaDainian 348, Steven 346, Larry 416

71. $3,500 at 8% and $4,500 at 6% **73.** 3 students

75. Good names help you remember what the unknowns represent.

77. Sometimes when looking at a collection of simpler examples, you will recognize a pattern that leads to a solution.

79. Answers will vary **85.** 6 **87.** 216

89. LCHPL, LCPHL, LHCPL, LHPCL, LPCHL, LPHCL

91. $(41, 43), (59, 61)$ **93.** 55

Section 1.2

1. Example 2

3. We are reasoning using accepted facts and general principles.

7. inductive **9.** deductive **11.** inductive **13.** deductive

15. inductive **17.** 16 **19.** 96 **21.** $\dfrac{1}{64}$ **23.** 21

25.

27.

29. *Hint*: Think of prime numbers.

	X	X		X		X				X		X		

31. $5 + 14$ **33.** $7 + 13$ **35.** 30

37. The total of all the numbers in the square is $1 + 2 + 3 + \cdots + 16 = 136$, so the numbers in each of the four rows is $\frac{136}{4} = 34$. The same is true for the columns and diagonals. From this you can deduce the missing numbers.

7	6	12	9
10	11	5	8
13	16	2	3
4	1	15	14

39. The algebraic interpretation of the steps in the trick are as follows:

   a. Call the number n. **b.** $3n$ **c.** $3n + 9$

   d. $\dfrac{3n + 9}{3} = \dfrac{3n}{3} + \dfrac{9}{3} = n + 3$ **e.** $n + 3 - n = 3$

   In this trick, you will always get the number 3.

41. The algebraic interpretation of the steps in the trick are as follows:

   a. Call the number n. **b.** $8n$ **c.** $8n + 12$

   d. $\dfrac{8n + 12}{4} = \dfrac{8n}{4} + \dfrac{12}{4} = 2n + 3$ **e.** $2n + 3 - 3 = 2n$

   In this trick, you will always get a result that is twice the number that you started with.

43.

45. See definitions in text.

47. If there were six branches, Sharifa would begin by visiting one of the six branches and then visit the remaining five cities in 120 ways. So the total number of ways she could make her visits is $6 \times 120 = 720$ ways.

49. 36644633   **51.** 986763

55. $1 + 2 + 3 + 4 + 5 = \dfrac{5 \times 6}{2}$, $1 + 2 + 3 + 4 + 5 + 6 = \dfrac{6 \times 7}{2}$

57. $1 + 3 + 5 + 7 + 9 = 25$, $1 + 3 + 5 + 7 + 9 + 11 = 36$

59. 20   **61. a.** 60   **b.** 210   **63.** 50

67. If you expand the expression as follows

$$(2n + 5)50 + 1{,}759 - 1{,}991 = 100n + 250 + 1{,}759 - 1{,}991$$
$$= 100n + 2{,}009 - 1{,}991$$
$$= 100n + 18$$

you see that $1{,}759 + 250$ gives a multiple of 100 plus $2{,}009 - 1{,}991$, which is your age. If you have already had your birthday, then you need to add the extra year, which is why we would then add 1,760.

Section 1.3

1. Answers will vary   **3.** Answers will vary

5. Eratosthenes made a remarkably accurate estimate of the circumference of Earth.

7. 37,300   **9.** 8,300   **11.** $20 + 40 + 190 + 40 = 290$

13. $35 - 15 = 20$ **15.** $5 \times 16 = 80$ **17.** $18/3 = 6$

19. $0.1 \times 800 = 80$ **21.** $9\% \times 1{,}000 = 0.09 \times 1{,}000 = 90$

23. $4 \times 5 \times 6 = 120$ **25.** $325/50 = 6.5$ more hours, 7:30 PM

27. $\$80 \times 0.15 = \$12$ **29.** $3 \times \$3 + 4 \times \$1.50 + \$3.00 = \$18$

31. It seems safe. Alicia probably weighs less than 200 pounds, so that leaves $2{,}300 - 200 = 2{,}100$ pounds for the 21 students. They probably do not weigh 100 pounds each.

33. $\$40{,}000 \times 4\% = \$40{,}000 \times 0.04 = \$1{,}600$

35. Her total expenses are about \$100/month; $\dfrac{100}{7} \approx 14$, $12 \times 14 = \$168$

37. dentistry   **39.** between 1980 and 1990

41. category 2   **43.** response 2   **45.** 20.6 million

47. 16 million   **49.** \$866.52 billion   **51.** \$361.05 billion

53. 213,019   **55.** 18.5%

57. Give a rough idea as to whether your estimate is too large or too small.

59. Answers will vary.

61. Mental computations should help you make reliable estimates.

67. If you divide the grassy area into rectangles, you get 10,354 square feet. They need slightly over two bags of fertilizer.

71. a. 153.46   **b.** 8,704

Chapter Review Exercises

1. Understand the problem; devise a plan; carry out your plan; check your answer.

3. 10   **5.** 8 hours as a stockperson; 12 hours as a ski instructor

9. a. deductive   **b.** inductive

11.

		X
	X	

13. a. n **b.** $8n$ **c.** $8n + 12$ **d.** $\dfrac{8n}{4} + \dfrac{12}{4} = 2n + 3$
  e. $2n + 3 - 3 = 2n$

15. a. $210 - 60 = 150$ **b.** $6 \times 15 = 90$

17. a. \$51,700 **b.** \$10,500

Chapter Test

1. Answers will vary   **3.** 404

5. a. 36,000   **b.** 36,500.

11. cde, cdf, cdg   **13.** $23 + 37 = 60$

15. a. Call the number you choose n.   **b.** We get $4n$.
  c. Next we get $4n + 40$.   **d.** Dividing by 2 gives us $2n + 20$.
  e. Subtracting gives us $2n$, or twice the original number.

Chapter 2

Section 2.1

1. Answers will vary.   **3.** Answers will vary.

5. In learning new ideas, babies are classifying objects according to common characteristics similar to what we do in set theory.

7. $\{10, 11, 12, 13, 14, 15\}$

9. $\{17, 18, 19, 20, 21, 22, 23, 24, 25\}$

11. $\{4, 8, 12, 16, 20, 24, 28\}$

13. $\{$Sunday, Monday, Tuesday, Wednesday, Thursday, Friday, Saturday$\}$

15. $\varnothing$ **17.** $\varnothing$

19. $\{x : x$ is a multiple of 3 between 3 and 12 inclusive$\}$

21. $\{y : y$ is a letter in the word *music*$\}$ **23.** $\{-1, -2, -3, \ldots\}$

25. $\varnothing$ **27.** $\{101, 102, 103, \ldots\}$

29. $\{x : x$ is an even natural number between 1 and 101$\}$

31. well defined   **33.** not well defined

35. not well defined   **37.** well defined   **39.** $\notin$ **41.** $\in$

43. $\in$ **45.** $\in$ **47.** $\notin$ **49.** $\in$ **51.** 6   **53.** 0

55. 4   **57.** 2   **59.** 1   **61.** finite   **63.** infinite

65. 4.5   **67.** Sony   **69.** George W. Bush   **71.** Sunday

73. $\{x : x$ is a humanities elective$\}$

75. $\{$History012, History223, Geography115, Anthropology111$\}$

77. $\{$AZ, GA, LA, MN, NJ, TX, VA$\}$

79. $\{x : x$ had an average price of gasoline above 290$\}$

81. $\{$Amazon, Apple, CNET, eBay, Fox, Time Warner, Wikipedia$\}$

83. $\{x : x$ had less than 155 million visitors$\}$

85. when the set is too large or too complicated to list its elements

87. $\{\varnothing\}$ is not empty; it contains one element, $\varnothing$.

93. Answers will vary.

Section 2.2

1. The sets have exactly the same elements.

3. Each time we add an element to the set, the number of subsets doubles.

5. equal   **7.** not equal   **9.** equal   **11.** equal

13. equal   **15.** true   **17.** false   **19.** false

21. true   **23.** false   **25.** equivalent   **27.** equivalent

29. not equivalent   **31.** equivalent   **33.** not equivalent

35. $\{1, 2\}, \{1, 3\}, \{2, 3\}$

37. {1, 2, 3}, {1, 2, 4}, {1, 3, 4}, {2, 3, 4}   **39.** 32; 31

41. *T*   **43.** *L*   **45.** $2^7 = 128$ **47.** $2^8 = 256$ **49.** 7

51. the order of the elements and repetition of elements

53. You must pay careful attention to the differences in notation; they do not all mean the same thing.

55. Answers will vary.   **57.** 25 is not a power of 2.   **59.** 5

61. At the first branching in the tree, the "yes" or "no" indicates that in forming a subset of {1, 2}, we will either take the 1 or omit it. The second branchings in the tree indicate whether we are going to take the 2 as a member of the subset that we are forming. The tree shows all possible ways that we can decide to take or not to take 1 and 2 in forming a subset of {1, 2}. The top branch corresponds to the subset {1, 2}, the second branch corresponds to the subset {1}, and so on.

63. Answers will vary.   **65.** 31

69. The fifth line counts the number of subsets of sizes 0, 1, 2, 3, 4, and 5 of a five-element set.

73. The total in the nth row is 2^n.

75. If $A \subseteq B$ and $B \subseteq C$, then $A \subseteq C$.

Section 2.3

1. $\{x : x \in A \text{ or } x \in B\}$

3. The complement of set *A* is the set of elements in the universal set that are not elements of *A*.

5. $(A \cup B)' = A' \cap B'$. We cannot interchange the order of taking a union and taking complements.

7. Words in everyday life that are the same as mathematical terminology have similar meanings.

9. {1, 3, 5}   **11.** {1, 2, 3, 4, 5, 6, 7, 8}

13. {1, 3, 5, 7, 9}   **15.** {1, 2, 3, 4, 5, 6, 7, 8, 9, 10}

17. {1, 3, 5, 7}   **19.** {9}   **21.** {potato chip, bread, pizza}

23. {apple, fish, banana}   **25.** {fish}

27. 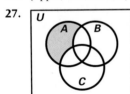   **29.**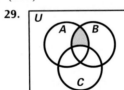

31. 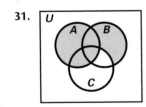   **33.**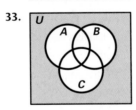

35. $B - A$ **37.** $(A \cup B)'$ **39.** $A \cap B \cap C$

41. $(A \cup C) - B$ **43.** 30   **45.** 28   **47.** 20   **49.** 27

51. *A*   **53.** *B*   **55.** {d, g}   **57.** {c, d, f, g}

59. {b, d, g, h}   **61.** {d}   **63.** {m, mc, hc}

65. {m, mc, bc, c, hc}   **67.** {c, d}   **69.** {f, i}   **71.** {c}

73. *Union* implies "joining together," as in a labor union; *Intersection* implies "overlapping," as in streets.

75. Answers will vary.   **77.** false   **79.** true   **81.** false

87. a. $A \cap B = B \cap A$ **b.** true

89. a. $(A \cap B) \cap C = A \cap (B \cap C)$ **b.** true

Section 2.4

1. $A \cup B = (A - B) \cup (A \cap B) \cup (B - A)$

3. We subtracted the 14 in $A \cap C \cap R$ from the 22 in $C \cap R$.

5. Many of his contemporaries scoffed at his ideas.

7. 2, 3   **9.** 3   **11.** 2   **13.** 2, 3, 5, 6

15. 4, 7   **17.** 7   **19.** 25   **21.** 18

23. 16   **25.** 20   **27.** 19

29. $n(A) = 18, n(B) = 15, n(C) = 14$

31. $n(A) = 5, n(B) = 14, n(C) = 9$ **33.** 59   **35.** 28

37. 30, 49   **39.** 23   **41. a.** 44   **b.** 27   **c.** 8

43. a. 106   **b.** 15   **c.** 40   **45.** 158   **47.** 20   **49.** A^-

51. $A^-, B^-, \text{ or } AB^-$ **53.** $A \cap B \cap Rh$

55. The region r_2 is the part of *A* that is outside *B*. It does not take into account elements in $A \cap B$.

59. 8   **61.** Answers will vary.

Section 2.5

1. Answers will vary.

3. There are an infinite number of infinite cardinal numbers.

5.
1	2	3	4	5	$\cdots$	n	$\cdots$
↕	↕	↕	↕	↕		↕	
4	8	12	16	20	$\cdots$	4n	$\cdots$

7.
1	2	3	4	5	$\cdots$	n	$\cdots$
↕	↕	↕	↕	↕		↕	
8	11	14	17	20	$\cdots$	3n + 5	$\cdots$

9.
1	2	3	4	5	$\cdots$	n	$\cdots$
↕	↕	↕	↕	↕		↕	
2	4	8	16	32	$\cdots$	2^n	$\cdots$

11.
1	2	3	4	5	$\cdots$	n	$\cdots$
↕	↕	↕	↕	↕		↕	
1	1/2	1/3	1/4	1/5	$\cdots$	1/n	$\cdots$

13.
1	2	3	4	5	$\cdots$	n	$\cdots$
↕	↕	↕	↕	↕		↕	
3	6	9	12	15	$\cdots$	3n	$\cdots$

15.
1	2	3	4	5	$\cdots$	n	$\cdots$
↕	↕	↕	↕	↕		↕	
1	4	7	10	13	$\cdots$	3n - 2	$\cdots$

17. Match {2, 4, 6, 8, 10, ...} with {4, 6, 8, 10, 12, ...}; in general, match 2n with 2n + 2.

19. Match {7, 10, 13, 16, 19, ...} with {10, 13, 16, 19, 22, ...}; in general, match 3n + 4 with 3n + 7.

21. Match {2, 4, 8, 16, 32, ...} with {4, 8, 16, 32, 64, ...}; in general, match 2^n with 2^{n+1}.

23. Match {1, 1/2, 1/3, 1/4, 1/5, ...} with {1/2, 1/3, 1/4, 1/5, 1/6, ...}; in general, match $\frac{1}{n}$ with $\frac{1}{n+1}$.

25. Match {1/2, 1/4, 1/6, 1/8, 1/10, ...} with {1/4, 1/6, 1/8, 1/10, 1/12, ...}; in general, match $\frac{1}{2n}$ with $\frac{1}{2n+2}$.

27. 6   **29.** 25

31. We showed that the set of positive rational numbers is in a one-to-one correspondence with the natural numbers.

33. It is impossible to put {1, 2, 3, 4, 5} in a one-to-one correspondence with any of its 31 proper subsets.

35. They already had been matched as 1, 1/2, 1, and 2.   **39.** 6

41. If we take the union of {1}, which has cardinal number 1 and {2, 3, 4, . . .}, which has cardinal number $\aleph_0$, we get {1, 2, 3, 4, . . .}, which has cardinal number $\aleph_0$.

Chapter Review Exercises

1. a. {x : x is an even natural number between 1 and 19}

b. {x : x is one of the 50 states that make up the U.S.}

c. {s, t, u, p, e, n, d, o}

3.

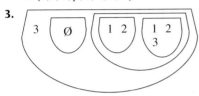

5. a. Yes; order of elements does not matter.

b. Yes; duplication of elements does not matter.

c. No; the first set has elements such as 1,002 that are not present in the second set

7. b. equivalent

9. a. {5, 7, 9}   **b.** {2, 3, 4, 5, 7, 8, 9, 10}

c. {1, 4, 6, 7, 10}   **d.** {7}

11. a.

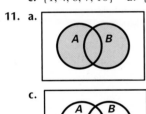

b.

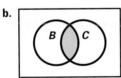

c.

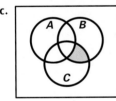

d.

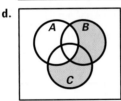

13. a. closure, commutativity, associativity, identity

b. closure, commutativity, associativity, identity

c. union distributes over intersection; intersection distributes over union

15. 83, 58

17. {1, 2, 3, 4, . . .} can be put in a one-to-one correspondence with {2, 4, 6, 8, . . .}.

19. We chose a digit that was different from the third decimal place of the third number in the list.

Chapter 2 Test

1. a. {x : x is a natural number greater than 100}

b. {January, February, March, . . . , December}

3. b. equivalent

5. ∅ contains no elements but {∅} contains one element, namely ∅.

7. 256

9. a. True; every element of the first set is a member of the second.

b. True; every element of the first set is a member of the second.

c. False; ∅ is not one of the numbers 1, 2, 3.

11. a.

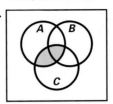

b.

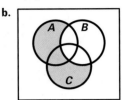

13. 89, 64

15. We can put the set {1, 2, 3, . . .} in a one-to-one correspondence with {2, 4, 6, . . .}.

17. We made the fifth decimal of x to be different from the fifth decimal place of the fifth number in the list.

Chapter 3

Section 3.1

1. not (negation), and (conjunction), or (disjunction), if . . . then (conditional), if and only if (biconditional)

3. A universal quantifier states that all objects have a certain property.

5. "At least one is not."

7. To discover religious truths and lead more perfect lives.

9 statement   **11.** not a statement

13. not a statement   **15.** statement

17. not a statement   **19.** compound; if . . . then, and, or

21. compound; or   **23.** simple

25. compound; if . . . then, or   **27.** compound; if . . . then

29. $g \lor \sim c$ **31.** $(\sim g) \rightarrow (\sim c \land \sim a)$ **33.** $g \leftrightarrow \sim a$

35. The radial tires are included or the sunroof is not extra.

37. It is not true that: the sunroof is extra and the radial tires are included.

39. If the radial tires are included, then the sunroof is extra or the power windows are not optional.

41. There exists at least one snake that is not poisonous.

43. All personal items are covered by this insurance policy.

45. All scientists believe that it is not true that an asteroid collision led to the extinction of the dinosaurs.

47. Some modern art is not difficult to understand.

49. false (Nadia)   **51.** false (all sophomores are athletes)

53. false (Lennox)

55–66. We have not provided answers for these exercises. These statements are complex and sophisticated. You will probably find that you, your classmates, and your instructor do not always agree as to which connectives are present. Nevertheless, it is interesting to try to determine the form of these statements.

67. true   **69.** false   **71.** true

73. In the first expression, we take conjunction and then the negation; in the second expression, we take negations first and then the conjunction.

75. To show "not all objects have a property," we must show that "at least one does not have the property."

Section 3.2

1. The "and" is true only when both components are true; the "or" is false only when both components are false.

3. It makes a difference whether you compute values for "or" first and then negate, or negate first and then compute "or." The situation is similar for "and."

5. Kurt Gödel   **11.** T   **13.** T   **15.** F   **17.** T

19. 32   **21.** yes   **23.** no   **25.** FTFF   **27.** TTFT

29. FFFT   **31.** FFTTFFFF   **33.** TTTTTFFF

35. FFFFTFFF   **37.** exclusive or   **39.** inclusive or

41. Bill is not tall or Bill is not thin.

43. Christian will not apply for a loan and he will not apply for work study.

45. Ken does not qualify for a rebate and he does not qualify for a reduced interest rate.

47. The number x is equal to 5 or s is odd.

49. yes   **51.** no   **53.** yes   **55.** yes

57. true   **59.** false   **61.** true

63. It is not true that: the earned income tax did reduce the tax you owe and gave you a refund.

65. It is not true that: you are single or the head of a household.

67. $p \vee (q \wedge r)$ **69.** $p \vee \sim q$ **71.** line 1

73. Answers will vary.

75. We want the truth table to cover every possibility for truth values for the three variables. There are eight different ways to assign truth values to the three variables.

81. Let the subsets of $\{p, q, r\}$ indicate which of the variables should be true. For example, the subset $\{p, r\}$, specifies that p and r should be true and that q is false. This corresponds to the line labeled TFT in the truth table.

83. The form $\sim(\sim p \vee \sim q)$ is logically equivalent to $p \wedge q$.

85. The stroke connective is logically equivalent to $\sim(p \wedge q)$.

87.

p	q	$(p\vert p)$	$\vert$	$(q\vert q)$
T	T	F	T	F
T	F	F	T	T
F	T	T	T	F
F	F	T	F	T

Section 3.3

1. The only way a statement of the form $p \rightarrow q$ can be false is when p is true and q is false.

3. We can write the original statement symbolically in the form $p \rightarrow q$, then write the contrapositive in the form $\sim q \rightarrow \sim p$, and finally write the contrapositive in words.

5. Once the hypothesis of the conditional of the contract is false, the deal is no longer in effect.

7. true   **9.** true   **11.** true   **13.** false

15. FTTT   **17.** FTFF   **19.** FFTTFTFT

21. TTTTTTTT   **23.** TTTTTFFT   **25.** TTTT

27. If it pours, then it rains.

29. If you do not buy the all-weather radial tires, then they will not last for 80,000 miles.

31. If its sides are not all equal in length, then a geometric figure is not an equilateral triangle.

33. If x does not evenly divide 6, then x does not evenly divide 9.

35. contrapositive   **37.** original conditional

39. original conditional

41. converse: $q \rightarrow (\sim p)$; inverse: $p \rightarrow (\sim q)$; contrapositive: $\sim q \rightarrow p$

43. converse: $\sim(q \wedge r) \rightarrow (\sim p)$; inverse: $p \rightarrow (q \wedge r)$; contrapositive: $(q \wedge r) \rightarrow p$

45. equivalent   **47.** not equivalent

49. If I finish my workout, then I'll take a break.

51. If you qualify for this deduction, then you complete Form 3093.

53. If you receive a free cell phone, then you sign up before March 1.

55. If you remain accident free for three years, then you get a reduction on your auto insurance.

57. If you can be claimed by someone else as a dependent, then your gross income is not more than $2,250.

59. If you decrease the amount being withheld from your pay, then the amount you overpaid is large.

61. true   **63.** true   **67.** true   **69.** false

71. converse: switch clauses; inverse: negate both clauses; contrapositive: switch clauses and negate both clauses

79. Jamie has not been a member for 10 years.

81. $r \rightarrow (p \wedge q)$

83. $p \vee (r \wedge (p \wedge q))$ is logically equivalent to p.

85.

87.

Section 3.4

1. $(p_1 \wedge p_2) \rightarrow c$

3. The final column in the truth table had all trues.

5. Answers will vary.   **7.** law of detachment; valid

9. fallacy of the converse; invalid

11. disjunctive syllogism; valid

13. fallacy of inverse; invalid   **15.** law of syllogism; valid

17. law of contraposition; valid

19. fallacy of the inverse; invalid

21. law of contraposition; valid

23. disjunctive syllogism; valid

25. valid   **27.** invalid   **29.** invalid   **31.** invalid

33. valid   **35.** invalid   **37.** invalid   **39.** valid

41. valid   **43.** invalid

45. The statement p allows us to detach q from $p \rightarrow q$.

47. In the disjunction $p \vee q$, if we don't have p, then we must have q.

49. If we have the premise p, then we conclude q, which is the inverse of $p \rightarrow q$.

59. If $a \wedge b$ is true, then b is true. If b is true and $b \rightarrow c$ is true, by the law of detachment, then c is true. The contrapositive of $d \rightarrow c$ is $c \rightarrow d$. If both c and $c \rightarrow d$ are true, then d is true.

Section 3.5

1. A syllogism is valid if whenever all its premises are true, then the conclusion is also true.

3. Whenever you draw an Euler diagram, you will always draw some extra conditions that were not stated as premises.

5. Jan Lukasiewicz   **7.** valid   **9.** invalid

11. invalid   **13.** invalid   **15.** valid   **17.** invalid

19. invalid   **21.** valid

23. ∴ Some taxes should be abolished.

25. ∴ Some teams that wear red uniforms do not play in a domed stadium.

27. There are many possible diagrams.

29. There are many possible diagrams.

Section 3.6

1. There are an infinite number of values between 0 and 1 inclusive.

3. the minimum of the truth values for p and q

13. 0.05   **15.** 0.25   **17.** 0.85   **19.** 0.15

21. 0.27   **23.** 0.64   **25.** 0.29   **27.** 0.71

29. marketing trainee (0.50)

31. attend Good Old State (0.60)

33. If we only allowed the values 1 (for true) and 0 (for false), then the rules for computing truth values in fuzzy logic are exactly the same as the rules for computing truth tables.

Chapter Review Exercises

1. a. not a statement   **b.** statement   **c.** not a statement

3. a. It is not true that: Antonio is fluent in Spanish and he has not lived in Spain for a semester.

   b. Antonio is not fluent in Spanish or he has not lived in Spain for a semester.

5. a. true   **b.** false   **c.** false   **7. a.** FFTF   **b.** FFFFFTFF

9. a. logically equivalent   **b.** not logically equivalent

11. a. TTTF   **b.** TFTFFFTT

13. a. If the Knicks get to the finals, then they beat the Lakers.

   b. If you are an astronaut, then you have a pilot's license.

15. invalid   **17.** invalid   **19. a.** 0.18   **b.** 0.53   **c.** 0.53

Chapter Test

1. a. statement   **b.** not a statement

3. a. $p \vee (\sim f)$ **b.** $\sim ((\sim p) \wedge f)$ **5.** 64

7. a. 0.25   **b.** 0.62

9. a. If you go to Wikipedia, then you will get enough sources for your research term paper.

   b. If Ticketmaster will mail the concert tickets, then you will pay a fee.

11. a. logically equivalent   **b.** not logically equivalent

13. a. true   **b.** true   **c.** true   **15.** valid

17. $\sim p \vee q$ **19.** invalid

Chapter 4

Section 4.1

1. An odd vertex is connected to an odd number of vertices; an even vertex is connected to an even number of vertices.

3. If we were to traverse it and remove it, we would disconnect the remaining graph and could not finish tracing it.

5. Leonard Euler and Arthur Cayley

7. connected; odd: A, B; even: C, D

9. connected; odd: A, B, E, F; even: C, D

11. not connected; all vertices are even

13. connected; odd: A, B, E, F; even: C, D

15.
A   B
•   •

C   D
•   •

17. impossible

19. **21.**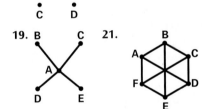

23. yes   **25.** no; four odd vertices

27. no; not connected   **29.** no; four odd vertices

31. ABFHI GCBEF GECDA

33. duplicate edges: GH, GI, FG; ABCDE FDIGF GIBHG HA

35. duplicate edges: CF, DE, EI; ABDEB CEFCF IEIHE DHGDA

37. No; if we represent this map by a graph, it has four odd vertices.

39.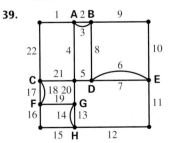

41.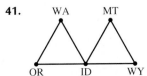

43.

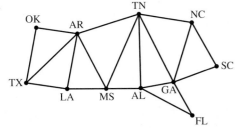

45. The trip is possible because the graph in Exercise 41 has no odd vertices.

47. The trip is not possible because the graph in Exercise 43 has more than two odd vertices.

49. Not possible. In the graph model, B will be an odd vertex, which is neither a beginning nor ending vertex in the attempt to trace the graph.

51. The graph below relates the people who do not get along with each other. Because this graph can be colored with three colors, three tables will be required. One possible seating: Table 1—Peter, Chris, Glenn, and Dianne. Table 2—Carter, Lois, Tom, and Cleveland; Table 3—Barbara, Steve, and Meg.

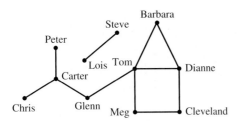

53. Model the committees with vertices. Two vertices are joined if they have common members. Color the graph and then committees that are colored the same can meet at the same time. This graph requires three colors, so three meeting times are needed.

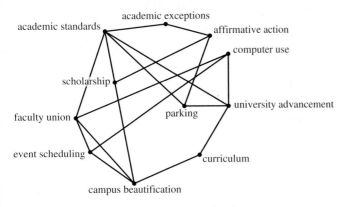

55. Every time we enter the vertex by one edge, we must leave by another.

59. Each edge connects two vertices, so it contributes two to the total of all the degrees. The total of all the degrees of all the edges will be twice the number of edges.

69.

71–72. A graph containing the following cannot be colored with three colors.

73. One sequence is C, F, E, C#, D, A, E, B.

75. One possible schedule: Day one—Anth215, Phy212; Day two—Bio325, Mat311; Day three—Chem264, Fin323, ComD265

Section 4.2

1. two

3. Choose the vertex that is connected to the present vertex by the edge with the smallest weight.

5. He was defeated in an arithmetic contest by Zerah Colburn.

7. a. ADBCEA   **b.** EBCDAE

9. a. ADCBFEA   **b.** EADCBFE

11. ABCDEA, ABCEDA, ABDCEA, ABECDA   **13.** $6! = 720$

15.

17. a. Yes   **b.** Yes   **c.** Yes   **19. a.** No   **b.** No   **c.** No

21. a. 20   **b.** 23   **23.** ACBDA has weight 14.

25. ACBDA has weight 99.   **27.** ABDECA

29. ADCFEBA   **31.** ABEDCA   **33.** ADBECFA

39. a. (We won't count reversals of circuits.) $720/2 = 360$

   b. 6 hours   **c.** PNBRAMCP   **d.** PNBCMRAP

41. a. ABMDHEFA   **b.** ABMDHEFA

43. a. XFEBACDPX   **b.** XFEBACDPX

45. A Hamilton circuit goes through every vertex but does not travel over every edge as an Euler circuit does.

47. They are faster than the brute force algorithm.

49. Answers will vary.   **53.** Answers will vary.

59. without counting reversals, over 1,928 years

61. ABECFDA has weight 42

Section 4.3

1. because the rumors travel in only one direction

3. the total of the direct and two-stage influence of each board member

5. Overnight shipping companies use graph theory to plan the routes that packages travel for efficient delivery.

7. a. ABCE, ABCDE   **b.** ABC   **c.** ABCE

   d. not possible   **e.** ABCEBA

9. a. BCE   **b.** ABCFG   **c.** ACFGCE

   d. not possible   **e.** ABCFGC

11.

From \ To	A	B	C	D	E
A	1	1	1	0	0
B	1	1	1	1	1
C	1	1	0	2	2
D	1	2	0	1	1
E	2	1	1	1	1

13.

From \ To	A	B	C	D	E	F	G
A	0	1	2	0	1	1	0
B	0	0	1	0	1	1	0
C	0	0	0	0	1	1	1
D	0	0	1	0	2	1	0
E	0	0	0	0	0	0	0
F	0	0	1	0	0	0	1
G	0	0	1	0	1	1	0

15. a. Kevin **b.** One way: Reverse the direction on the edge KD.

17. chief financial officer, sales manager, production manager, president, marketing director

19.

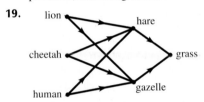

21. Mississippi (5), Florida (6), Alabama and LSU (3), Arkansas (1), Georgia (0)

23. Feinstein (5), Murciano (4), Johnson (3), Cordaro (1), Lee (0)

25. Fuji (8), Dasani (7), Propel (6), Vitaminwater (3), Aquafina (2)

27. when the relationship goes in only one direction

29. I and B are in a block of vertices that do not have any directed edges leaving them.

31. The graph would have to have a directed edge or path from F to I and from F to J. We cannot do this by changing the direction of only one edge.

Section 4.4

1. the time required in months to build the shell

3. a. Begin, A, B, E, I **b.** Begin, A, B, E **c.** day 14
d. day 16 **e.** 17 days **f.** Begin, A, B, E, I, J, End

5. a. Begin, B, E, F, G **b.** Begin, B, E, F, H
c. Begin, B, E, F, G, I, End **d.** to begin on day 6
e. to begin on day 9 **f.** 14 days

7.

Task	Day Task Can Begin
A	1
B	1
C	3
D	3
E	6
F	8
G	8
H	12

9.

Task	Day Task Can Begin
A	1
B	1
C	1
D	3
E	4
F	5
G	7
H	7
I	11
J	11

11.

Get funding 2 → Decide on program 2
Begin 0
Get permits 1
Rent tents 2 → Arrange for speakers 4 → Advertise 2 → Set up festival 1 → 0 End
Set up tents 1

Task	Week Task Can Begin
Get funding	1
Get permits	1
Decide on program	3
Rent tents	5
Arrange for speakers, etc.	7
Advertise	11
Set up tents, booths, etc.	7
Set up festival	13

13.

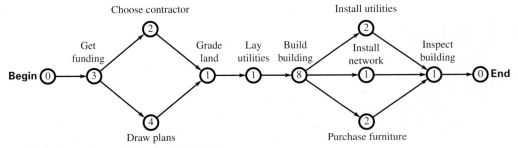

Task	Month Task Can Begin
Get funding	1
Choose contractor	4
Draw plans	4
Grade land	8
Lay utilities	9
Build building	10
Install utilities	18
Install network	18
Purchase furniture	18
Inspect building	20

15.

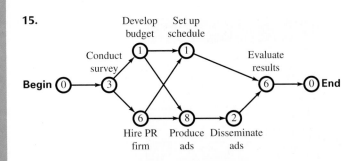

Task	Month Task Can Begin
Conduct survey	1
Develop budget	4
Hire PR firm	4
Set up production schedule	10
Produce ads	10
Disseminate ads	18
Evaluate results	20

17. First find a critical path for T. Then schedule T to begin after the sum of the times along its critical path.

19. No, because "Assemble shell" would no longer be on a critical path.

Chapter Review Exercises

1. a. 12 edges **b.** F, G, H, and I are odd; the rest are even.

   c. yes **d.** FH, GH, and HI are bridges.

3. Graph (a) can be traced because it has only two odd vertices. Graph (b) cannot be traced because it has more than two odd vertices.

5.

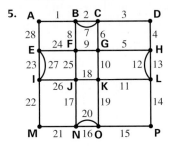

Duplicate edges BC, EI, HL, and NO. The resulting graph will have all even edges, which can be traced using Fleury's algorithm. This tracing will give the indicated route.

7. Four canoes; those who can travel together: A, K, and N; B, C, and M; D and E; J and L

9. ABDECA has weight 13.   **11.** ABDECA

13. Directed graphs are used when the relationship modeled may apply in one direction. For example, object X is related to Y, but Y may not be related to X.

15. a. Begin, A, C, F, H **b.** Begin, A, C, F, I, End

   c. day 10 **d.** 18 days

Chapter Test

1. a. 9 **b.** odd—E, F; even—A, B, C, D, G, H

   c. yes **d.** yes, EF

3. (one way) ABDACEGIHGFEDFA

5.

7. ABDECA and ACEBDA have weight 19. **9.** ACEBDA

11. a. Begin, X, P, B, Z **b.** Begin, X, P, B, Z, End
 c. 16th day **d.** 18 days

Chapter 5

Section 5.1

1. 10, ∩, X, ⌐ **3.** We rewrote one scroll as 10 heel bones.

5. The Rhind papyrus **7.** 324 **9.** 21,324

11. 2,121,210

13. ⊖⊖∩∩∩||||| **15.** [hieroglyphic numeral]

17. [hieroglyphic numeral] **19.** [hieroglyphic numeral]

21. [hieroglyphic numeral] **23.** ⊖⊖∩∩∩∩∩|||||

25. [hieroglyphic numeral]

27. 602 **29.** 1,284 **31.** 4,235 **33.** 59 **35.** 564

37. 1963 **39.** 7,544 **41.** 50,262 **43.** 501,420

45. XLIV **47.** CCLXXVIII **49.** CDXLIV

51. $\overline{\text{IV}}$DCCXCV **53.** $\overline{\text{LXXXIX}}$CDXXIII **55.** 436

57. 5,067 **59.** 9,999 **61.** [Chinese numeral]

63. [Chinese numeral] **65.** [Chinese numeral] **67.** 2,650

69. [hieroglyphic numeral]

71. MCMXXXIX **73.** MCMXCIV

75. [Chinese numeral] ; [Chinese numeral]

77. A number tells us "how many." A numeral is a symbol for writing a number.

79. If we want to omit some power of 10, we simply don't use the symbol for that power.

87. 9,999,999

89. a. $\propto\propto\propto\nabla\otimes\otimes\otimes\otimes\sqrt{}\sqrt{}$; $\propto\propto\propto\propto\propto\propto\propto\nabla\nabla\sqrt{}\sqrt{}$
 b. $\approx\nabla\nabla\nabla\otimes\otimes\otimes\otimes\sqrt{}\sqrt{}\sqrt{}\sqrt{}\sqrt{}$ **c.** $\propto\propto\propto\propto\otimes\otimes\otimes\otimes\otimes\sqrt{}$

Section 5.2

1. Babylonian, Chinese, Mayan, Hindu-Arabic

3. It had a zero. **5.** 20 **7.** 32 **9.** 731 **11.** 8,187

13. [cuneiform numeral] **15.** [cuneiform numeral]

17. [cuneiform numeral]

19. [cuneiform numeral]

21. [cuneiform numeral]

23. [cuneiform numeral]

25. 18,733 **27.** 290,173 **29.** [Mayan numeral] **31.** [Mayan numeral]

33. five hundred **35.** five thousand

37. $2\times10^4+5\times10^3+3\times10^2+8\times10^1+9\times10^0$

39. $2\times10^5+7\times10^4+8\times10^3+0\times10^2+6\times10^1+3\times10^0$

41. $1\times10^6+2\times10^5+0\times10^4+0\times10^3+0\times10^2+4\times10^1+5\times10^0$

43. 5,368 **45.** 370,082 **47.** 3,070,502

49. 3,288 **51.** 6,238 **53.** 142 **55.** 3,655

57. 940 **59.** 20,148 **61.** 136,245

63.

	6	8	5	
3	2/4	3/2	2/0	4
2	4/2	5/6	3/5	7
	1	9	5	

$685\times47=32{,}195$

65. 1,884 **67.** 3,936 **69.** 5,544 **71.** 199,260

73. We can write numerals more efficiently, and we do not need different symbols for different powers of the base.

75. In the galley method, we compute all partial products separately before we add them to get the final product. In our method, we combine several smaller partial products into a single partial product before adding.

77. In writing ▼ ▼▼▼▼▼, it could be difficult to tell whether there is one space or two spaces between the two groups of symbols. If there is one space, the number is $1\times60+5$. If there are two spaces, the number is $1\times60^2+5$.

79. The ancient Chinese system requires a separate symbol for each power of 10, e.g., one billion, 10 billion, and so on. The Hindu-Arabic system uses different powers of 10, 10^9, 10^{10}, and so on. The positioning of symbols in the numeral indicates these powers of 10.

83. [Chinese numeral] **85.** [Chinese numeral] **87.** [Chinese numeral]

Section 5.3

1. 4 **3.** five **5.** It is based on 10 and has place value.

7. 23_5, 30_5 **9.** 1010_2, 1100_2 **11.** EE_{16}, FO_{16}

13. 117 **15.** 184 **17.** 39 **19.** 117

21. 183 **23.** 452 **25.** 756 **27.** 3,336

29. $2{,}314_5$ **31.** $12{,}302_6$ **33.** $1{,}100{,}111_2$

35. $1{,}011{,}110_2$ **37.** $6{,}513_8$ **39.** $AE8_{16}$ **41.** DEA_{16}

43. $4{,}143_5$ **45.** $10{,}310_5$ **47.** $2E53_{16}$ **49.** 110.000_2

51. $2{,}041_5$ **53.** 648_{16} **55.** $100{,}110_2$ **57.** $2{,}043_5$

59. $24{,}041_5$ **61.** 114_5 R11_5 **63.** 44_5 R24_5

65. $1{,}355_8$; $2ED_{16}$ **67.** $1{,}751_8$; $3E9_{16}$ **69.** $10{,}100{,}110_2$

71. $101{,}000{,}111{,}110_2$ **73.** 754_{16} **75.** $1{,}654_8$

77. CANDY **79.** LOVE **81.** 60:5

83. 64; 256. In base 8, we combine the eight symbols, 0 through 7, with themselves; in a base-16 system, we combine the sixteen symbols 0, 1, . . . 9, A, B, . . . , F.

85. The numeral 10 always represents the base, which in this case is 16, so we needed a *single* symbol to represent 10.

91. $4,123_5$ **93.** $3,774_9$

95. 1, 2, 3, 4, 5, 6, 7, 8, 9, A, B, 10_{12}, 11_{12}, 12_{12}, 13_{12}, 14_{12}, 15_{12}, 16_{12}, 17_{12}, 18_{12}, 19_{12}, $1A_{12}$, $1B_{12}$, 20_{12}, 21_{12}

97. 231   **99.** $\neq \oplus @\$$

Section 5.4

1. In doing modulo-12 computations we ignore all multiples of 12.

3. The number -3, which means moving counterclockwise three positions, is the same position as moving clockwise 9. So -3 and 9 are the same on a 12-hour clock.

5. 7   **7.** 4   **9.** true   **11.** true   **13.** false   **15.** 5

17. 2   **19.** 7   **21.** 4   **23.** 0   **25.** 3   **27.** 10

29. 6   **31.** 6   **33.** 3   **35.** 4   **37.** 3, 8   **39.** 24

41. 42   **43.** 3   **45.** 28

47. In 2010 or 2011, 1982; after 2011, 1994.   **49.** 5   **51.** 6

53. Perform the operation as usual and then count off the result on an *m*-hour clock.

55. Solve both congruences separately as in Exercise 54 and then find which solutions are common to both congruences.

61. 17   **63.** 22   **65. a.** 7   **b.** 3, 7

Chapter Review Exercises

1. 1,232,210   **3.** 1,961   **5.** 7,593

7. The Roman system had the subtraction principle, which allowed people to write certain numbers more efficiently. They also had a multiplication principle.

9. 𒐕𒐕𒐕 𒀖𒐕𒐕 𒀖𒐕𒐕

11. 4,738   **13. a.** 375   **b.** 1,440   **15.** $6,513_8$

17. 134_5 $R20_5$ **19.** $1,431_5$ **21. a.** true   **b.** false

23. a. 3, 7, 11   **b.** no solutions

Chapter Test

1. MMMDCLXXXV   **3.** 144; 2622

5. a. 379   **b.** 1,320

7. ꜩꜩꜩꜩꜩꜩꜩꜩ⌒⌒⌒⌒⌒⌒⌒⌒⌒⌒⌒⌒𐆗⋂⋂⋂⋂⋂⋂⋂⋂⋂⋂

9. 3   **11.** $1,100,010_2$ **13.** 15,946

15. a. 8   **b.** 8   **c.** 2

17. They had to use more symbols to write their numerals.

19. 110_5, $R2_5$ **21.** $6,536_8$; $D5E_{16}$

Chapter 6

Section 6.1

1. The square root of 83 is between 9 and 10. If we don't find a prime factor by the time we reach 7, we will not find any.

3. when we reach prime factors

5. He made many contributions to geography.

7. true   **9.** false   **11.** false   **13.** true

15. 53, 59, 61, 67, 71, 73, 79, 83, 89, 97

17. 9   **19.** 12   **21.** 21×11 **23.** prime

25. prime   **27.** 7×17 **29.** 2, 3, 5, 6, 10   **31.** 3, 5, 9

33. 30   **35.** 20   **37.** 4   **39.** 20

41. $2^2 \times 5 \times 7^2$ **43.** $2^2 \times 3^2 \times 5^2 \times 11$ **45.** $3^3 \times 23$

47. 11×29 **49.** 4; 120   **51.** 14; 280   **53.** 72; 864

55. 21; 3,969   **57.** 28   **59.** 45   **61.** 108

63. 360   **65.** 2,231   **67.** 140   **69.** 6

71. 3 feet by 3 feet   **73.** 432   **75.** 17

77. We would multiply the largest powers of all primes that occur in either number.

79. Four will always divide a multiple of 100, so all we have to do to check if 4 divides a number is to determine if 4 divides the number formed by the last two digits. So, 4 automatically divides 36,800. Therefore we only have to check if 4 divides 24, which it does.

81. In each case, the two numbers have common factors greater than 1.

83. not prime   **85.** prime

87. a. 2,369.17 pages   **b.** 26,060.84 inches; 2,171.74 feet
c. 0.41 mile

89. 59 and 61; 71 and 73; 101 and 103

91. No. Both 4 and 6 divide 12, but $4 \times 6 = 24$, which does not divide 12.

93. 15 will divide a number if both 3 and 5 divide the number.

Section 6.2

1. They are balanced on opposite sides of zero.

3. If for the past 3 days, you spent \$5, then 3 days ago, you had \$15 more.

5. They dealt primarily with geometry, which only involved positive quantities.

7. $+3$ **9.** $+10$ **11.** $+9$ **13.** $+19$ **15.** $+23$

17. -13 **19.** -65 **21.** $+50$ **23.** $+35$ **25.** -56

27. $+48$ **29.** -38 **31.** -63 **33.** $+72$

35. $+16 = (-2) \cdot c; c = -8$ **37.** $-14 = (-2) \cdot c; c = +7$

39. -3 **41.** $+15$ **43.** $+5$ **45.** -6 **47.** $+6$

49. -4 **51.** -26 **53.** -12 **55.** -48 **57.** -6

59. true   **61.** false; $(-5) + (+8) = +3$ **63.** false; $\frac{-6}{-3} = +2$

65. true   **67.** 57,260 feet   **69.** 7,201 feet

71. 265 degrees   **73.** 66   **75.** 1,918 years

77. 37 degrees

79. We use the notion of adding signed numbers to define subtraction.

81. When multiplying or dividing signed numbers, if both numbers have the same sign, then the result is positive; if the signs are different, then the result is negative.

83. -2 **85.** 0   **87.** not possible

89. If $\frac{8}{0} = x$, then $8 = 0 \cdot x$, which is not possible.

91. For the next 4 days you spend \$6, so in 4 days, your net change in finances will be $-\$24$. Thus, $(+4)(-6) = -24$.

93. For the past 3 days you have gained \$7, so 3 days ago, you had \$21 less. Thus, $(-3)(+7) = -21$.

97. $a = 2, b = -3, c = -7, d = -8, e = -6$

99. The sum of the nine entries is 9×37.

101. The reasoning is similar to Exercise 100, except now there is no center number. So, put a point, C, in the middle of the 4-by-4 square, and notice that any two numbers that are symmetric with respect to C have the same total. There are eight such pairs in a 4-by-4 box; therefore, add two numbers in the box that are symmetric with respect to the center, C, and multiply this total by 8 to get the total of all 16 numbers.

Section 6.3

1. Cross multiply and see if you get the same result.

3. denominator

5. "Frac" gives an answer in reduced fraction form.

7. equal   **9.** not equal   **11.** not equal   **13.** equal

15. $\dfrac{3}{7}$ **17.** $-\dfrac{1}{3}$ **19.** $\dfrac{9}{14}$ **21.** $\dfrac{13}{14}$

23. $\dfrac{7}{6}$ **25.** $-\dfrac{1}{3}$ **27.** $\dfrac{5}{48}$ **29.** $\dfrac{43}{72}$

31. $\dfrac{59}{24}$ **33.** $-\dfrac{1}{24}$ **35.** $\dfrac{1}{3}$ **37.** $\dfrac{1}{3}$

39. $-\dfrac{21}{5}$ **41.** $-\dfrac{7}{24}$ **43.** $-\dfrac{3}{8}$ **45.** $-\dfrac{33}{4}$

47. 2   **49.** $\dfrac{62}{81}$ **51.** $-\dfrac{49}{10}$ **53.** $\dfrac{79}{5} = 15\dfrac{4}{5}$

55. $\dfrac{29}{3} = 9\dfrac{2}{3}$ **57.** $6\dfrac{3}{4}$ **59.** $8\dfrac{1}{15}$ **61.** $\dfrac{11}{4}$

63. $\dfrac{55}{6}$ **65.** 0.75   **67.** 0.1875   **69.** $0.\overline{81}$

71. $0.\overline{307692}$ **73.** $\dfrac{16}{25}$ **75.** $\dfrac{209}{250}$ **77.** $\dfrac{61}{5}$

79. $\dfrac{4}{9}$ **81.** $\dfrac{7}{37}$ **83.** $\dfrac{7}{22}$ **85.** $\dfrac{1}{4}$ **87.** $\dfrac{5}{48}$

89. $\dfrac{29}{50}$ **91.** 24; no   **93.** $3\dfrac{3}{4}$

95. The small tube costs 49.5 cents per ounce; the large tube costs 50.5 cents per ounce; so the smaller tube is the better buy.

97. $\dfrac{9}{40}$

99. He can get $3\dfrac{5}{9}$ strips from a 12-foot strip and $4\dfrac{4}{9}$ strips from a 15-foot strip. Thus, there is more waste per strip from the 12-foot strips.

101. Cancel common factors from the numerator and denominator.

103. Divide the denominator, d, into the numerator to get a quotient q and a remainder r. Then, write the mixed number as $q\dfrac{r}{d}$.

105. When dividing the denominator into the numerator, the remainders have to repeat, which will cause the digits in the quotient to also repeat.

107. 6   **109.** $\dfrac{10}{9}$ **113.** $2\dfrac{23}{32}$ feet   **115.** $29\dfrac{1}{4}$ inches

Section 6.4

1. Irrational numbers have nonrepeating expansions; rational numbers have repeating expansions.

3. $\sqrt{ab} = \sqrt{a}\sqrt{b}$; $\sqrt{\dfrac{a}{b}} = \dfrac{\sqrt{a}}{\sqrt{b}}$

5. Using approximations can introduce errors, which can eventually lead to an incorrect answer.

7. rational   **9.** irrational   **11.** rational   **13.** irrational

15. $\sqrt{9} = 3$ **17.** $3\sqrt{2}$ **19.** $5\sqrt{3}$

21. $4\sqrt{3}$ **23.** $3\sqrt{21}$ **25.** $11\sqrt{5}$

27. not possible   **29.** $8\sqrt{5}$ **31.** not possible   **33.** 6

35. $6\sqrt{5}$ **37.** $14\sqrt{3}$ **39.** 2   **41.** $\dfrac{4}{3}$ **43.** $\dfrac{2\sqrt{3}}{3}$

45. $\dfrac{3\sqrt{5}}{5}$ **47.** $2\sqrt{6}$ **49.** $\dfrac{5\sqrt{22}}{11}$ **51.** $\dfrac{\sqrt{5}}{2}$

53. 0.435; 0.43512112111211112 . . .

55. 0.45785; 0.45785121121112111112 . . .

57. 0.6; 0.612112111211112 . . .

59. 0.12113; 0.12113121121112111112 . . .   **61.** a, d, b, c

63. a, c, d, b   **65.** distributive   **67.** commutative (addition)

69. associative (addition)   **71.** identity element for addition

73. commutative (addition)   **75.** $4\sqrt{6} \approx 9.8$ miles

77. $20\sqrt{6} \approx 49.0$ mph   **79.** 4.97 seconds

81. 200 pounds   **83.** 10 seconds

85. perfect squares   **87.** the Always Principle

89. False. $\sqrt{2}$ is irrational; 1.414215362 is a rational, decimal approximation of $\sqrt{2}$.

91. False. Counterexample: $\sqrt{25} = 5$, which is rational.

93. True. $\sqrt{3}\sqrt{5} = \sqrt{15}$, and 15 is not a perfect square.

95. True. If a is rational and b is irrational, and $a + b = c$, then if c is rational we could say $b = c - a$. But $c - a$ is the difference of two rational numbers and therefore is rational. This would contradict the fact that b is irrational.

101. a. We can cancel common factors from the numerator and denominator.

 b. Square both sides of a).

 c. Multiply through equation by b^2.

 d. 2 divides the left side of c), so 2 divides the right side of c).

 e. If a were odd, then a^2 would be odd, but it is not.

 f. a is even.

 g. Substitute $2k$ for a in c).

 h. Square $2k$.

 i. Divide both sides of h) by 2.

 j. 2 divides b^2.

 k. If b were odd, then b^2 would be odd.

 l. We assumed that $\dfrac{a}{b}$ was reduced, so a and b cannot both be even.

103. 2

105. These numbers are possible lengths of the three sides of a right triangle. If a right triangle has legs of length 3 and 4 and hypotenuse of length 5, then $3^2 + 4^2 = 5^2$.

Section 6.5

1. $x^2 \cdot x^3 = xx \cdot xxx = xxxxx = x^5$; $x^m \cdot x^n = x^{m+n}$

3. $\dfrac{x^8}{x^3} = \dfrac{\cancel{xxx}xxxxx}{\cancel{xxx}} = xxxxx = x^5$; $\dfrac{x^m}{x^n} = x^{m-n}$

5. growth of money; growth of debt; spread of diseases; growth of interest in a new product

7. 32   **9.** -16 **11.** -9 **13.** 9   **15.** 1   **17.** 0

19. 729   **21.** 117,649   **23.** $\dfrac{1}{25}$ **25.** $\dfrac{1}{729}$ **27.** -3

29. 25   **31.** 36   **33.** $\dfrac{1}{27}$ **35.** 1   **37.** 8

39. 4.356×10^6 **41.** 7.83×10^2 **43.** 2.4×10^{-3}

45. 8.0×10^{-3} **47.** 32,500   **49.** 0.00178   **51.** 63

53. 0.00000045   **55.** 1,000,000   **57.** 2.381×10^7

59. 8.4×10^2 **61.** 6.0×10^{11} **63.** 3.6×10^2

65. 4.0×10^{-5} **67.** 3.36×10^{-1} **69.** 2.6311×10^{-6}

71. 8.076×10^{10} **73.** 1.564×10^{13} **75.** 4.0×10^{-7}

77. 3.736×10^3 **79.** 1.3766×10^{13}

81. a. 5.902×10^4 **b.** 5.902×10^7

    c. The 130-pound person weighs as much as 59 million mosquitoes.

83. 1.459×10^3 **85.** 3.17×10^1 **87.** 8.69×10^1

89. 3.72×10^3 **91.** 3.87×10^1 **93.** 3.24×10^1 seconds

95. See answer 1.   **97.** See answer 3.

99. In evaluating -2^4 we raise 2 to the fourth power to get 16 and then negate that result to get -16. In evaluating $(-2)^4$, we first negate 2 and then raise that negative number to the fourth power to get positive 16.

103. no

Section 6.6

1. We considered the terms of the sequence to all be of the form 3 plus some multiple of 4 to see a pattern.

3. Answers will vary.   **5.** arithmetic; 17; 20

7. geometric; 648; 1,944   **9.** geometric; $\dfrac{1}{16}$; $\dfrac{1}{32}$

11. arithmetic; -10; -15 **13.** geometric; 32; 64

15. arithmetic; 3.5; 4.0   **17.** 35; 220   **19.** 86; 660

21. 10.5; 115   **23.** 436   **25.** 59,049; 88,573

27. $\dfrac{1}{64}$; $\dfrac{127}{64} = 1.984375$ **29.** 0.00002; 2.22222   **31.** 144

33. 377   **35.** $36,810   **37.** $1,475.11   **39.** 4.10 feet

41. 206.24 million   **43.** Answers will vary.

45. Answers will vary.   **49.** $\dfrac{(n-k+1)(a_k+a_n)}{2}$

51. $F_1 + F_2 + F_3 + \cdots + F_n = F_{n+2} - 1$

Chapter Review Exercises

1. 71, 73, 79, 83, 89   **3.** 191 is prime; $441 = 3^2 7^2$.

5. LCM = 7,920; GCD = 264   **7.** yes; yes; no

9. a. 9   **b.** 11   **c.** 72   **d.** -16

11. If $\dfrac{-24}{-8} = c$, then $-24 = -8 \cdot c$. Thus, $c = +3$.

13. If $\dfrac{5}{0} = c$, then $5 = 0 \cdot c$, which is impossible.

15. $67   **17. a.** $\dfrac{5}{27}$ **b.** $\dfrac{18}{37}$ **19.** $14\dfrac{1}{2}$ **21.** $\dfrac{1}{8}$ pound

23. Answers will vary.   **25. a.** $6\sqrt{3}$ **b.** $\sqrt{3}$

27. $\dfrac{8}{\frac{4}{2}}$ does not equal $\dfrac{\frac{8}{4}}{2}$

29. In evaluating -2^4, we first raise 2 to the 4th power and then take its opposite to get -16. In evaluating $(-2)^4$, we raise -2 to the 4th power to get 16.

31. 1,325,000; 0.0000863

33. 6.06×10^1 **35.** 91   **37.** 55 and 610

Chapter Test

1. 101, 103, 107, 109, 113   **3.** 241 is prime; $539 = 7^2 \cdot 11$

5. 156; 10,296   **7. a.** -6 **b.** $+3$ **c.** -54 **d.** $+8$

9. If you spend $5 for the next 7 days, you will have $35 less.

11. 69 degrees   **13.** 377 and 6,765   **15. a.** $\dfrac{1}{60}$ **b.** $\dfrac{16}{3}$

17. $\dfrac{4}{3}$ **19.** 182   **23. a.** $6\sqrt{5}$ **b.** $\dfrac{\sqrt{15}}{5}$

25. a. 8   **b.** $\dfrac{1}{81}$ **c.** 125   **d.** 27

27. In evaluating $(-3)^2$, we square -3 to get 9; in evaluating -3^2, we first square 3 and then take its opposite to get -9.

29. 4.35

Chapter 7

Section 7.1

1. Add, subtract, multiply, and divide both sides by the same quantity.

3. -315; 12.50 and 15

5. a geometry in which we represent geometric objects such as points, lines, and circles by numbers and equations

7. 5   **9.** $-\dfrac{4}{5}$ **11.** 40

13. -12 **15.** $\dfrac{16}{7}$ **17.** 25

19. $w = \dfrac{P - 2l}{2}$ **21.** $\mu = x - \sigma z$ **23.** $r = \dfrac{A - P}{Pt}$

25. $x = \dfrac{-3y + 6}{2}$ **27.** $l = \dfrac{V}{wh}$ **29.** $h = \dfrac{S - 2\pi r^2}{2\pi r}$

31. intercepts: (4, 0), (0, 6)

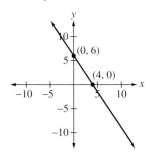

33.

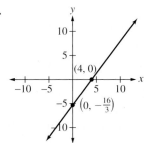

35. intercepts: (9, 0), (0, 6)

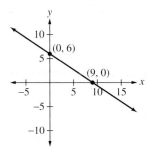

37.

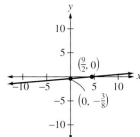

39. intercepts: (8, 0), (0, −0.4)

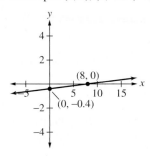

41.

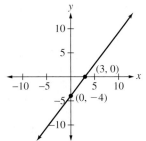

43. $\frac{3}{4}$ **45.** $-\frac{4}{5}$ **47.** $\frac{5}{2}$ **49.** does not exist

51. a, d **53.** f **55.** The rise is zero.

57. y-intercept $= -3$; slope $= 4$

59. y-intercept $= -3$; slope $= -5$

61. 168   **63.** 32   **65.** $d = \left(+\frac{4}{5}\right)t + 64$ **67.** 92

69. $c = 488t + 11{,}900$; $15{,}804$

71. $n = 0.7715g - 27$; $281.60

73. $5.60d + 7.35p = 133$ **75.** $35t + 55c = 14{,}500$

77. $9e + 15c = 342$ **79.** after 60 tokens

81. the y-coordinate   **83.** the x- and y-intercepts

85. 7; If Chuck sells more than seven systems, Buy More is better for him.

91. 3.268%   **93. a.** $y = 105d$ **b.** $d = \frac{y}{105}$

95. $p = 3{,}500e$ **97.** 17.75 feet

Section 7.2

1. Write the equation in standard form; state the x- and y-intercepts; specify the slope and y-intercept.

3. We again used two points to find the the equation.

5. $y = 3x - 5$ **7.** $y = 4x + 14$ **9.** $y = -2x + 11$

11. $y = -5x - 31$ **13.** $y = 2x - 1$ **15.** $y = \frac{1}{4}x + \frac{31}{4}$

17. $y = -\frac{6}{19}x - \frac{10}{19}$ **19.** $y = \frac{7}{2}x + 13$

21. a. $y = 0.3t + 80.97$ **b.** 84.87

23. a. $y = 0.133t + 7.4$ **b.** 9.395 million   **c.** by 2025

25. a. $e = 760t + 11400$ **b.** $19,000

27. a. -14.5 **b.** $v = -14.5t + 495$ **c.** 306.5

29. $d = 6.8t + 87$; 202.6 thousand

31. $y = -0.0154t + 8.5$; 2452

33. a. 1.77 million   **b.** $n = 1.77t + 146$ **c.** 2033

35. the slope of the line

Section 7.3

1. to solve a quadratic equation

3. Even if we use accurate data, if we are using the wrong model, then our predictions are not valid.

5. 2, 8   **7.** 1, $\frac{3}{2}$ **9.** $-3, \frac{2}{3}$ **11.** $-\frac{3}{5}, 4$

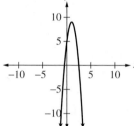

13. opening down; vertex $(3, 1)$; $(2, 0), (4, 0)$; $(0, -8)$

15. opening down; vertex $(1, 9)$; $\left(-\frac{1}{2}, 0\right), \left(\frac{5}{2}, 0\right)$; $(0, 5)$

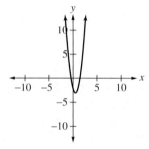

17. opening up; vertex $\left(\frac{1}{2}, -3\right)$; $\left(\frac{1+\sqrt{3}}{2}, 0\right)\left(\frac{1-\sqrt{3}}{2}, 0\right)$; $(0, -2)$

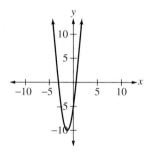

19. opening up; vertex $\left(-\frac{7}{6}, -\frac{121}{12}\right)$ $(-3, 0)$, $\left(\frac{2}{3}, 0\right)$, $(0, -6)$;

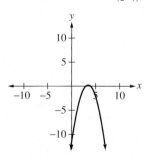

21. opening down; vertex $\left(\frac{7}{2}, \frac{1}{4}\right)$ $(3, 0)$, $(4, 0)$, $(0, -12)$;

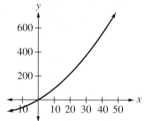

23. **b.** 8.125

25. **a.** week 2 **b.** $4 million **c.** week 3

27. **a.**

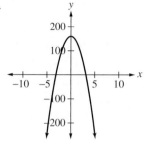

b. t (time) cannot be negative. **c.** 3.16 seconds

29. **b.** 1,468 thousand

31. $b^2 - 4ac$; If it is negative, there are no real roots for the equation; if it is zero, there is one real root, and if it is positive, there are two real roots.

33. The crate begins falling at a point above the ground and picks up speed as it falls. A linear equation would not be appropriate, because the rate of change of the distance of the crate above the ground is not constant.

35. **a.**

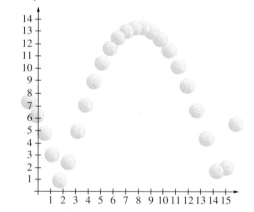

b. t cannot be negative. **c.** The runner is running more slowly when t is closer to 0. **d.** 10.45 seconds.

39.

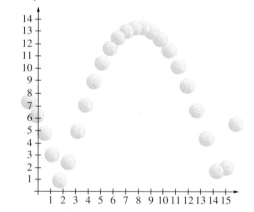

41. **a.** (2004) 305.35; (2006) 378.55; Both predictions are fairly close. **b.** between 2015 and 2016

Section 7.4

1. A is the amount after compounding; P is the principal, r is the interest rate, and n is the number of years the money is compounded.

3. the initial population and the growth rate

5. As the years of compounding increased, neither type of model was very accurate.

7. $1,050   **9.** $4,100   **11.** $6,381.40   **13.** $4,665.60

15. 3.24%   **17.** 3.33%   **19.** 1,389 million

21. 17.5 million   **23.** 0.47%   **25.** 0.97%   **27.** 1.86

29. 2.10   **31.** 0.51   **33.** 1.57   **35.** 0.3669

37. 0.7291   **39.** 35.2%   **41.** 30.9%   **43.** 63.2%

45. 14.21 years   **47.** 2086   **49.** 1,238.6 million

51. 198 years   **53.** 307 mg   **55.** after 8 hours

57. a. $y = 120(1.0299)^x$ **b.** $161.11

59. 80.55%   **61.** 4.5%

63. It reverses the operation of raising 10 to a power.

65. when the rate of growth of a quantity is proportional to the amount present

69. The population will exceed 300 by the end of the fourth year; fishing can begin toward the end of the fourth year.

Section 7.5

1. We compared the ratio of dollars to pesos in the morning with the ratio of dollars to pesos in the gift shop.

3. Step 1: Find the constant of variation, k.

Step 2: Use k to rewrite the variation equation and then substitute values to find the desired quantity.

5. 4   **7.** 62.5   **9.** 18   **11.** 36   **13.** 65

15. 1   **17.** 100   **19.** $\frac{1}{16}$ **21.** 25   **23.** 14

25. $\frac{4}{3}$ **27.** 192   **29.** 2.6 mg   **31.** 35 miles

33. 3.25 hours   **35.** 12,000   **37.** 5 or 6

39. 9.23 minutes   **41.** 126 pounds   **43.** 12 feet

45. 18,000 gal   **47.** 400 feet   **49.** 588 gal

51. 6.4 pounds per square inch

53. y increases.   **55.** inversely

57. In $m:n$, we are comparing m objects with n objects. In $n:(m+n)$, we are comparing the n objects with the total of all the $m+n$ objects.

61. 7:9   **63.** Strength stays the same.

65. Beam is four times as strong.

Section 7.6

1. f is the function that relates c, the cost of gasoline, to n, the number of miles driven.

3. p is the base price of the car. $f(p)$ is the total price of the car after taxes and fees.

5. "y equals f of x."   **7.** "g of x equals 8."   **9.** 10

11. 70   **13.** 6   **15.** does not exist   **17.** {3, 4, 5, 6, 7, 9}

19. 8   **21.** 9   **23.** does not exist   **25.** does not exist

27. −2

29. not a function

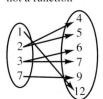

31. function

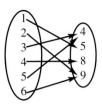

33. function

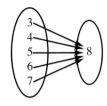

35. yes; 13, 3, −17   **37.** no; (2, 2), (2, −2)

39. yes; $\sqrt{3}$, 3, does not exist   **41.** no; $(1, \sqrt{2})$, $(1, -\sqrt{2})$

43. yes   **45.** no   **47.** $y = 2x^2 - 5x + 3$

49. $y = 3x^2 - 5x$ **51.** yes   **53.** no   **55.** no

57. No; some programs may appear several nights per week.

59. 1,520   **61.** 135.7   **63.** 549.25   **65.** 1,460

67. Domain element means the same thing as independent variable. Range element means the same thing as dependent variable.

69. One domain element, x, has two distinct y's associated with it.

75. −1, −10, −26   **77.** $\frac{1}{3}, \frac{7}{12}, \frac{11}{28}$

Section 7.7

1. All models in both examples are dynamical systems. That is, sequences of numbers of the form $A_0, A_1, A_2, \ldots$, where the value of A_{n+1} depends on the value of A_n.

3. In Example 2, the equilibrium value $\frac{5,000}{6}$ was stable. In Example 3, the equilibrium value $\frac{5}{3}$ was unstable.

5. 5, 9   **7.** 82   **9.** 18.2464   **11.** $a = -3$; unstable

13. $a = \frac{16}{3}$; stable

15. $A_{n+1} = 1.05A_n$, $n = 0, 1, 2, \ldots$, where $A_0 = 1,000$; $1,102.50

17. $P_{n+1} = [1 + (0.08)(1 - P_n)]P_n$, $n = 0, 1, 2, \ldots$, where $P_0 = 0.30$; 33.4%

19. $D_{n+1} = 0.60D_n + 250$, $n = 0, 1, 2, \ldots$, where $D_0 = 0$; 490 mg

21. There are equations relating A_1 to A_0, A_2 to A_1, A_3 to A_2, and so on.

23. If $-1 < m < 1$, then the equilibrium value will be stable.

29. 4,300 years   **31.** 515

Chapter Review Exercises

1. a. 4   **b.** $-\frac{1}{2}$ **3. a.** $s = 200 + 10(h - 40)$ **b.** 260

5. $\frac{3}{4}$ **7.** after 10 months   **9.** $y = \frac{5}{3}x - 1$

11. It is the linear model that best approximates a set of data points.

13. $\frac{1}{2}$, −4

15. a technique by which we find a quadratic equation that best models a set of data

17. $12,641.72   **19.** 0.5075

21. In a logistic model we adjust the rate of growth as the population grows. The larger the population, the smaller the rate of growth.

23. 6.25 gallons   **25.** $\frac{7}{3}$

27. There is no change in the strength of the beam.

29. a. {1, 2, 3, 4, 5, 7}   **b.** {−2, −1, 4, 6, 8}   **c.** −1   **d.** 3 and 7

31. a. not a function   **b.** function   **33.** 27, 111, 447

35. 1,400 mg

Chapter Test

1. a. −12   **b.** 3.36   **3.** $\frac{7}{5}$

5. a. $c = 30 + 0.045(m − 1200)$ **b.** $44.40

7. 1 − b, 2 − d, 3 − a, 4 − c

9. when the rate of change is constant

11. $6.30d + 8.25l = 137.70$

13. $n = 0.3t + 34.7$; 40.1 million

15. $−\frac{2}{3}$, stable   **17.** $6,044.63   **19.** quadratic regression

21. a. {1, 2, 4, 5, 7, 8}   **b.** {−3, −2, 4, 6, 8}
c. −3   **d.** 2 and 7

23. 606.15 mg

25. As the population increases, the environment cannot sustain the same rate of growth, so we must reduce the growth rate.

27. a. function   **b.** not a function

29. 10 or 11   **31.** $\frac{4}{3}$ **33.** b and d

Chapter 8

Section 8.1

1. The lines can intersect in a single point as in Example 1; they can be parallel, with no points in common, as in Example 2; they can be the same line as in Example 3.

3. You have easier numbers to work with.

5. They replace curved surfaces with planar surfaces that are easier to compute.

7.

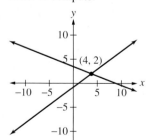

9.

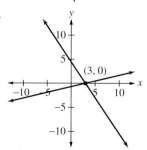

11.

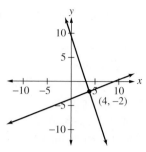

13.

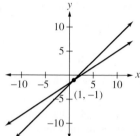

15. (1, 4)   **17.** (−6, −2)   **19.** no solution

21. infinite number of solutions   **23.** no solution

25. $\left(\frac{1}{4}, \frac{1}{2}\right)$ **27.** $\left(\frac{1}{5}, \frac{2}{5}\right)$ **29.** (−1, 3)   **31.** (−1, 4)

33. no solution   **35.** infinite number of solutions

37. 96 wins; 66 losses

39. 4 bagels, 5 ounces of cream cheese

41. After 11 months, WorldCom is the better deal.

43. 20 wins; 14 losses

45. Hartsfield—85 million, O'Hare—77 million

47. 2008   **49.** $11 per hour   **51.** $34

53. In solving, you find a unique value for x or y.

55. You get a statement that is true for any x, such as $0 + 0 = 0$.

63. (4, 0, 1)   **65.** (5, 2, 1)

Section 8.2

1. It is a half-plane.   **3.** We had an infinite number of solutions.

5. a, d   **7.** c, d

9.

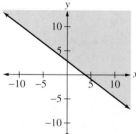

11.

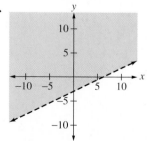

13.

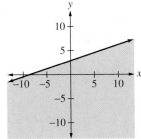

15.

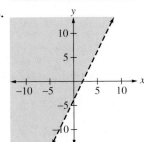

17.

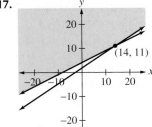

19.

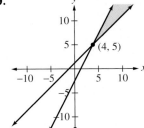

21.

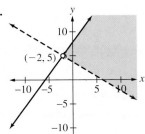

23.

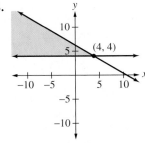

25.

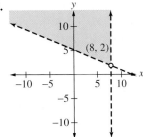

27.

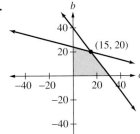

29.

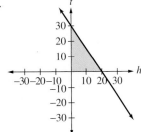

31.

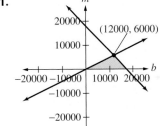

33.

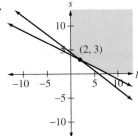

35.

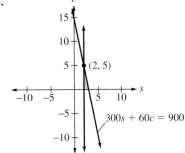

37.

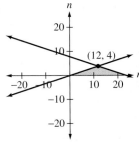

39. to decide which half-plane contains solutions to a linear inequality

41. The x and y may represent real quantities that could not be negative.

43.

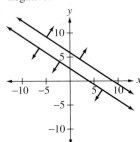

45.

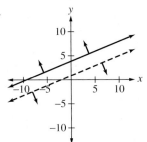

49.

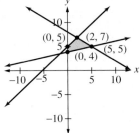

51.

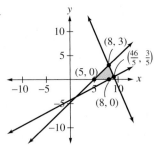

53. $\begin{array}{l} 3y \leq 43 - 5x \\ 11y \geq 27 - 2x \\ 3y \geq 1 + 2x \end{array}$ **55.** $\begin{array}{l} 3y \leq 43 - 5x \\ 11y \geq 27 - 2x \\ 3y \leq 1 + 2x \end{array}$

Section 8.3

1. In Example 1, we are trying to find the maximum value of the quantity $2x - 5y$ within the solution set for the system of inequalities.

3. The constraints are, on copper, $2s + 16c \leq 160$, and on labor, $1s + 3c \leq 40$.

5. $(0, 0)$, $(6, 0)$, $(5, 4)$, $(4, 6)$, $(0, 8)$

7. $(0, 0)$, $\left(\dfrac{22}{3}, 0\right)$, $(6, 2)$, $\left(\dfrac{8}{3}, \dfrac{16}{3}\right)$, $(0, 6)$

9. Maximum is 36 at $(4, 6)$. **11.** Minimum is -3 at $(0, 3)$.

13. Minimum is 4 at $(2, 3)$. **15.** Maximum is 34 at $(2, 2)$.

17. Minimum is 16 at $\left(\dfrac{5}{2}, 3\right)$. **19.** 15 Athens, 20 Barcelona

21. 5 hats, 10 T-shirts **23.** 2 PowerUp, 3 StressTabs

25. 80 lizards, 120 frogs

27. the variables, the constraints, and the objective function

29. Graph the system of constraints; find the corner points; evaluate the objective function at the corner points.

Chapter Review Exercises

1. a. $(3, -5)$ **b.** The lines intersect in a single point.

3. a. no solutions **b.** The lines are parallel.

5. Big Mac 540 calories; McChicken 360 calories

7.

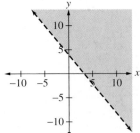

9.

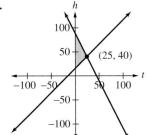

Chapter Test

1. a. no solutions **b.** The lines are parallel.

3. a. $(2, -3)$ **b.** The lines intersect in a single point.

5. 12 hours

7.

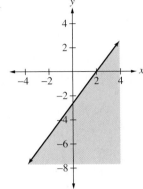

9.

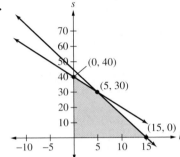

Chapter 9

Section 9.1

1. We represented 19% as 19 hundredths, or 0.19, then positioned the 32 in the two decimal places to the right of 0.19.

3. We found the selling price minus the dealer's cost and divided this difference by the dealer's cost.

5. 0.78   **7.** 0.08   **9.** 0.2735   **11.** 0.0035   **13.** 43%

15. 36.5%   **17.** 145%   **19.** 0.2%   **21.** 75%

23. 31.25%   **25.** 250%   **27.** 1.6%   **29.** 15%

31. 350   **33.** 19.6   **35.** 17.5%   **37.** 128   **39.** 9.4%

41. $397,000   **43.** 26.06%   **45.** 114.41%   **47.** 24.9%

49. 41.3%   **51.** 1,269   **53.** 7.3%   **55.** 27.6 million

57. a. 8%   **b.** She divided by 16,065 instead of 14,875.

59. $37,800   **61.** $680   **63.** 11.5%   **65.** $12,800

67. $35,771.50   **69.** $8,507.50   **75.** per hundred

77. We could have used the dealer's cost for the base, the markup as the amount, and then solved for percent.

79. $10,794.52

81. No. The price after the reduction will be less than the original price.

83. No. It is the same as an increase by 32%.

Section 9.2

1. We substituted values for A, r, and t in the equation $A = P(1 + rt)$, and then solved for P.

3. $\log(3^x) = x \log(3)$

5. $I = \$240$, $P = \$1,000$, $r = 8\%$, $t = 3$ years

7. $I = \$700$, $P = \$3,500$, $r = 5\%$, $t = 4$ years

9. $A = \$3,100$, $P = \$2,500$, $r = 8\%$, $t = 3$ years

11. $A = \$1,770$, $P = \$1,500$, $r = 6\%$, $t = 3$ years

13. $A = \$1,400$, $P = \$1,250$, $r = 6\%$, $t = 2$ years

15. 1.5%   **17.** 12/365%   **19.** $6,381.41

21. $4,686.64   **23.** $23,457.76   **25.** $4,885.48

27. 7.76%   **29.** 6.14%   **31.** 4.95% compounded quarterly

33. $12,278.88   **35.** 2.096   **37.** 14.207   **39.** 2.1544

41. 1.7783   **43.** 4.56%   **45.** 10.34

47. a. $4,896   **b.** $1,296   **49.** $66.67   **51.** 12.5%

53. 75%   **55.** $6.07   **57.** $4.29   **59.** $1,712.15

61. a. 4.74%   **b.** 4.67%   **63.** 10.07 years   **65.** $I = Prt$

67. A is the future amount, P is the principal, r is the yearly interest rate, m is the number of compounding periods per year, and n is the total number of compounding periods.

69. $m = 1$ **73.** 10.5170863%   **75.** 10.5170918%

Section 9.3

1. We added the loan amount and the interest, and then divided the sum by the number of payments.

3. It shows the balance for the loan for each day of September.

5. $46.50   **7.** $37.40   **9.** $243.20; $63.47

11. $3,019.68; $197.91   **13.** 288%   **15.** 384%

17. a. $313.50   **b.** down to $256.50   **c.** $3,982.46

19. a. $536.67   **b.** down to $429.34   **c.** $6,614.91

21. $4.38   **23.** $7.96   **25.** $9.67   **27.** $5.37

29. $4.95   **31.** $6.77   **33.** $5.03

35. add-on method, $87.50; credit card, $82.50   **37.** $61.44

39. We "add on" the interest for the loan before calculating the payments.

41. Charge a large purchase to your card early in the month and then pay off the debt before the end of the month so no interest is charged for that purchase.

Section 9.4

1. The expression $\frac{r}{m}$ is the monthly interest rate; n is the number of compounding periods; R is the monthly payment.

3. $\dfrac{x^8 - 1}{x - 1}$ **5.** $710.59   **7.** $21,669.48

9. $23,008.28   **11.** $85,785.11   **13.** $12,148.68

15. $7,463.67   **17.** $162.14   **19.** $193.75   **21.** 2.7268

23. 1.1073   **25.** 2.3219   **27.** 1.9527   **29.** 42.62

31. 30.91   **33.** 22.43   **35.** $2,435.99   **37.** $6,286.36

39. $121,417.91   **41.** $102.78   **43.** $98.31

45. a. $301,354.51; $226,015.88
   b. $193,354.51; $145,015.88
   c. $247,110.70 ; $199,913.02; tax-deferred earns $47,197.68 more

47. a. $146,709.85; $102,696.90
   b. $50,709.85; $35,496.90
   c. $102,696.90; $92,047.83; tax-deferred earns $10,649.07 more

49. a. $402,627.32; $301,970.49
   b. $192,627.32; $144,470.49
   c. $281,839.12; $258,629.34; tax-deferred earns $23,209.78 more

51. 61 months   **53.** 47 months

55. The exponent property: $\log a^x = x \log a$

59. $141.33   **61.** $197,395.14

Section 9.5

1. The left side represents the amount owed on the loan after 48 months of compounding. The right side represents the amount that must be in the annuity to pay off the loan.

3. $126.82   **5.** $193.44   **7.** $306.64   **9.** $112.37

11.

Payment Number	Amount of Payment	Interest Payment	Applied to Principal	Balance
				$5,000.00
1	$126.82	$41.67	$85.15	$4,914.85
2	$126.82	$40.96	$85.86	$4,828.99
3	$126.82	$40.24	$86.58	$4,742.41

13.

Payment Number	Amount of Payment	Interest Payment	Applied to Principal	Balance
				$12,500.00
1	$306.64	$85.94	$220.70	$12,279.30
2	$306.64	$84.42	$222.22	$12,057.08
3	$306.64	$82.89	$223.75	$11,833.33

15. a.

Payment Number	Amount of Payment	Interest Payment	Applied to Principal	Balance
				$100,000.00
1	$665.31	$583.33	$81.98	$99,918.02
2	$665.31	$582.86	$82.45	$99,835.57
3	$665.31	$582.37	$82.94	$99,752.63

b.

Payment Number	Amount of Payment	Interest Payment	Applied to Principal	Balance
				$100,000.00
1	$765.31	$583.33	$181.98	$99,818.02
2	$765.31	$582.27	$183.04	$99,634.98
3	$765.31	$581.20	$184.11	$99,450.87

c. $1.76

17. a. $276.46   **b.** $1,770.08

19. a. $57.50   **b.** $12.50

21. a. $954.83   **b.** $2,057.23

23. a. $1,101.68   **b.** $2,128.43

25. The present value of the lottery winnings is $425,678.19; this is slightly better than the lump sum of $425,000.

27. $13,593.02

29. The present value of the retirement plan is $38,900.73; this is slightly worse than the lump sum of $40,000.

31. $5,416.17   **33.** $107,389.08   **35.** $4,159.37

37. a. $240.46   **b.** $88.08

39. a. $786.70   **b.** $8,955

41. a. $971.60   **b.** $118.08

43. We saw the left expression in calculating compound interest and the right expression in computing the future value of annuities.

Section 9.6

1. You had paid $300 on an outstanding balance of $1,000, so the interest rate was $\frac{300}{1,000} = 0.30$, or 30%.

3. 12%   **5.** $15   **7.** $13   **9.** 13%   **11.** 14%

13. 12.45%   **15.** 13.67%   **17.** 15%   **19.** 11%

21. 15%   **23.** 15%   **25. b.** (b) is better.

27. b. (b) is better.   **29.** over 16%

31. First calculate the finance charge per $100 on the loan. Next, use the line corresponding to the number of loan payments to locate the amount that is closest to the finance charge per $100. The percent shown at the top of that column is the approximate APR for the loan.

Chapter Review Exercises

1. 12.45%   **3.** 68.75%   **5.** $1,806 million

7. $1,770   **9.** $12,465.25   **11.** 131 months

13. $326.70; $45.75   **15.** $4.57   **17.** $50.85

19. 30 months   **21.** $126.82

23. The present value is $490,907.37; the lump sum is the better option.

25. $912.04; $1,921.07   **27.** 15.2%

Chapter Test

1. 36.24%   **3.** 43.75%   **5.** $3,655   **7.** 11.77%

9. 15%   **11.** 174 months   **13.** $7,052.42

15. 4.55%   **17.** $6.74   **19.** 2.19   **21.** $47.15

23. about 20 months

25. The annuity is the better choice because it has a present value of $811,089.58.

27. a. $347.34   **b.** $10,827.27

Chapter 10

Section 10.1

1. a. acute   **b.** right   **c.** obtuse   **d.** straight

3. The length of arc AB is proportional to the measure of the central angle ACB.

5. 7 and 9   **7.** 7 and 3   **9.** 10   **11.** 9 and 10

13. true   **15.** false   **17.** true   **19.** false   **21.** false

23. e, d **25.** c, g **27.** 60°, 150°

29. No complementary angle; supplementary angle is 60°.

31. 38.8°, 128.8°

33. $m\angle a = 144°$, $m\angle b = 36°$, $m\angle c = 144°$

35. $m\angle a = 45°$, $m\angle b = 135°$, $m\angle c = 45°$

37. $m\angle a = 52°$, $m\angle b = 90°$, $m\angle c = 128°$

39. Arc AB has length 6 ft.   **41.** $m\angle ACB = 120°$

43. circumference = 1,200 mm   **45.** 6 ft   **47.** 144°

49. 60°   **51.** 18°   **53.** 12   **55.** 14.4°

57. The sum of supplementary angles is 180°; the sum of complementary angles is 90°.

59. "Alternate" means that they are on opposite sides of the transversal. "Exterior angles" means that they are outside the parallel lines.

61. yes, if the lines are perpendicular

63. No more than two could be obtuse. Because verticals are equal, if there were a third obtuse angle, then the fourth would also have to be obtuse, giving an angle sum of more than 360°.

65. yes, if the angles are both 45-degree angles

69.

71. 180

Section 10.2

1. Alternate interior angles are equal; $m\angle 1 = m\angle 4$, $m\angle 3 = m\angle 5$.

3. It was the supplement of $\angle WVZ$, which had measure 108°.

5. false   **7.** true   **9.** false   **11.** polygon

13. not a polygon; not made of line segments

15. $m\angle A = 30°$, $m\angle B = 60°$, $m\angle C = 90°$

17. $m\angle A = 45°$, $m\angle B = 55°$, $m\angle C = 80°$

19. 4; 720°   **21.** 162°   **23.** 18   **25.** $\frac{5}{3}$ in.   **27.** 5 in

29. $m\angle E = 40°$, length of EH is 7

31. $m\angle I = 35°$, length of HI is 84

33. 450 ft   **35.** 75 ft   **37.** 192.5 feet

39. 54°   **41.** 20.88 feet

43. Isosceles triangles have two equal sides; equilateral triangles have all three sides equal.

45. No, a scalene triangle has no sides equal.

47. Both have four equal sides. A rhombus does not necessarily have right angles.

51. possible

53. Not possible; the angle sum would be greater than 180°.

57. The measure of the interior angles gets larger. For very large n, the interior angles have measures close to 180°.

59. $\frac{(n-2)90}{n}$ degrees

Section 10.3

1. In both cases, the area is the height times the base.

3. The area of the trapezoid is the sum of the areas of two triangles.

5. 160 ft   **7.** 140 in.^{2} **9.** 108 cm^2 **11.** 72 yd^2

13. 78.5 cm^2 **15.** 50.24 m^2 **17.** 72 yd^2 **19.** 6.88 m^2

21. 0.86 m^2 **23.** 24 in.^{2} **25.** 12 yd^2 **27.** 7 in.^{2}

29. 3.5 in.^{2} **31.** 42.15 cm^2 **33.** 59.81 m^2

35. 6.62 cm   **37.** 13 m   **39.** 6.93 yd   **41.** 69 m^2

43. area   **45.** perimeter   **47.** 127.28 ft   **49.** 2 m

51. 86.13 ft^2

53. The large pizza is the better buy; it costs 4.5 cents per square inch versus 5.3 cents per square inch for the medium.

55. 1,000.96 m^2 **57.** 186.38 m

59. a. 22.61 ft   **b.** 203.47 ft^2

61. rectangle, parallelogram, triangle, trapezoid

63. We had the length of the hypotenuse of ΔTMC and needed the length of another leg to find the height h.

67. 2,779.11 ft^2 **69.** false

71. The area of $WXYZ$ is one-half of the area of $ABCD$.

73. 50 ft by 50 ft   **75.** about 21,195 ft^2

77. $(2r)^2 - \pi r^2 = (4 - \pi)r^2$

Section 10.4

1. Both are the area of the base times the height.

3. The radius of the can that minimized its surface area is between 0.5 and 0.6 of a foot.

5. a. 148 cm^2 **b.** 120 cm^3

7. a. 80.48 in.^{2} **b.** 75.36 in.^{3}

9. a. 408.2 ft^2 **b.** 628 ft^3

11. a. 5,024 cm^2 **b.** 33,493.33 cm^3

13. 75 ft^3 **15.** 180 ft.^{3} **17.** 18.33 m^3 **19.** 1,728

21. 91   **23.** 25   **25.** about 3.7 yd

27. the rectangular cake

29. 2,198; 2,260.8; the radius is being squared, so increasing it contributes greater to the increase.

31. 188.4; 200.96; the radius is being squared, so increasing it contributes greater to the increase.

33. 84 ft^3 **35.** 2,164 mi

37. Its volume is less than the volume of a cylinder with radius r and height h, so it is some fractional part of $\pi r^2 h$, namely, $\frac{1}{3}\pi r^2 h$.

39. The volume is four times as large, because in the formula for the volume of a cone, the radius is squared.

41. Let d be the diameter of the can. The height of the can is $3d$; the circumference of the can is $\pi d \approx 3.14d$, which is larger.

43. radius ≈ 0.68 ft

47. The smaller cubes have more surface area. For example, compare the surface area of one 3-in. cube with the surface area of twenty-seven 1-in. cubes.

49. 6.08 in.

51. $\pi r \sqrt{r^2 + h^2} - \pi \left(\frac{r}{2}\right)\sqrt{\left(\frac{r}{2}\right)^2 + \left(\frac{h}{2}\right)^2} = \frac{3}{4}\pi r\sqrt{r^2 + h^2}$

Section 10.5

1. We realized that moving three columns to the right in the table gave us $10^3 = 1,000$ as many objects.

3. $\dfrac{1.6\ kilometers}{1\ mile}$

5. 1,135 dg   **7.** 4,572 mm   **9.** 15.24 cm

11. 0.02159 dam   **13.** 1.77 dL   **15.** 24,000 dL

17. 2,800 mm   **19.** 350 dL

21. a; a juice glass is about one-eighth of a quart or about one-eighth of a liter.

23. c; an inch is about 2.5 cm, so 5 cm would correspond to a 2-in. nose.

25. b; the dog might be about 2 ft tall, or about $\frac{2}{3}$ of a meter.

27. 59.05 ft   **29.** 105.73 oz   **31.** 554.77 al

33. 1,056.7 qt   **35.** 69.29 in.   **37.** 49.56 tons

39. 7.92 dm   **41.** Don't give him an inch.

43. It is first down and 10 yards to go.

45. **a.** 160 cu m   **b.** 160,000 L   **c.** 160,000 kg

47. 49.72   **49.** 135.93 kL   **51.** $1.25 per pound

53. $8.25   **55.** 60 m   **57.** 12.75 km per liter

59. $86.11   **61.** 65°C   **63.** 140°F   **65.** 68°F

67. 45°C   **69.** 1 hectare = 10,000 sq m   **71.** 90   **73.** 10

75. More decimeters because decimeters are smaller than hectometers.

77. Because dekagrams are much larger than milligrams, we would need fewer dekagrams. Therefore, we would move the decimal point to the left.

Section 10.6

1. There are more rigid motions that we can apply to the star so that its beginning and ending position are the same.

3. The measure of an interior angle for the polygon is greater than 120°, so if we have three or more polygons sharing a vertex, the angle sum at the vertex will exceed 360°.

5–8.

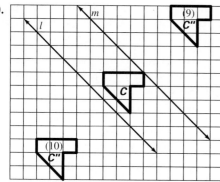

9–10.

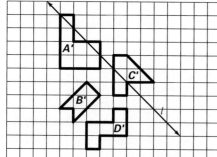

11–12.

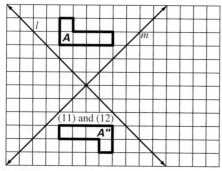

13. Yes; the effect is the same as if we performed a translation.

15–16.

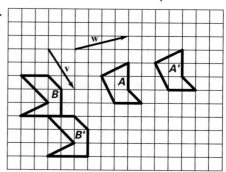

17.

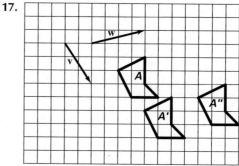

19.

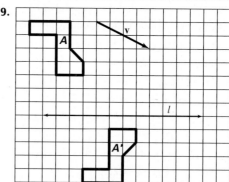

21.

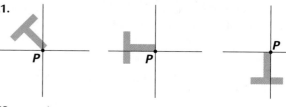

23.

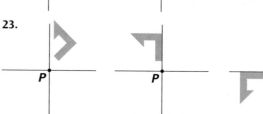

25. 1,800°, 150°

27. Using the interior angles of a regular pentagon, we cannot obtain an angle sum of 360° around a point.

29.

31. (c), (d)   **33.** (a), (e)

35. reflectional symmetries: about a vertical line, a horizontal line, and two diagonal lines; rotational symmetries: 90°, 180°, 270°

37. reflectional symmetries: about a vertical line; no rotational symmetries

39. reflectional symmetries: about a vertical line; no rotational symmetries

41. reflection, translation, glide reflection, rotation

43. With a symmetry, the beginning and ending positions of the object are the same; with a rigid motion, this does not have to be the case.

45. At the points where the vertices of the equilateral triangles and regular hexagons meet, the angle sum is $120° + 60° + 120° + 60° = 360°$.

47. At the points where the vertices of the equilateral triangle, the squares, and the regular hexagon meet, the angle sum is $60° + 90° + 120° + 90° = 360°$.

53.

55. In the figure, we constructed the perpendicular bisectors of segments EE' and CC'. The point where these perpendicular bisectors meet is the center of rotation.

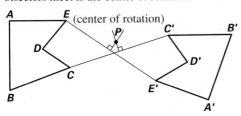

Section 10.7

1. As we magnify the cloud, we keep seeing the same type of pattern.

3. $3^D = 4$; the log function   **5.** $4^5 = 1,024$

7. $\left(\frac{4}{3}\right)^{10} \approx 17.76$

9. a.

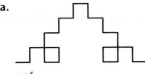

b. $\left(\frac{5}{3}\right)^5$

11. 1.29   **13.** 1.46

15.

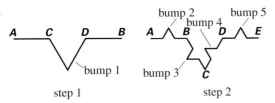

17.

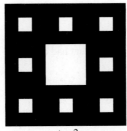

step 2

19. No matter how much we magnify the object, we see patterns that are similar to the patterns seen in the original object.

21. We saw that the area of the gasket at step $n + 1$ was $\frac{3}{4}$ the area at step n, so as n increased, the area of the curve would keep getting smaller and smaller and would be less than any fixed positive number.

25. 5^n **27.** $\left(\frac{3}{4}\right)^{10} \approx 0.06$; $\left(\frac{3}{4}\right)^n$

Chapter Review Exercises

1. a. b, g **b.** e, d **3.** 135°

5. an arc of a great circle

7. $m\angle E = 45°$, length of $EH = \frac{20}{3}$, length of $GF = 4$

9. a. 80 cm² **b.** 20 in.²

11. a. 15.2 cm² **b.** 3.04 cm

13. a. 37.68 in.³ **b.** 330 cm³

15. The volume will be four times as large, because in the formula for the volume of a cone, we square the radius.

17. 56.21 yd   **19.** $118.40   **21. a.** c, e **b.** d

23.

25. zero

Chapter Test

1. a. vertical angles   **b.** corresponding angles
  c. alternate exterior angles   **d.** alternate interior angles

3. $m\angle a = 130°$, $m\angle b = 140°$, $m\angle c = 40°$

5. a. 71.55 cm³ **b.** 48 in.³   **7.** 3 in.

9. a. 27 in.² **b.** 32 cm²

11. $m\angle F = 35°$, length $FG = 8$, length $GH = \frac{20}{3}$

13. 10°   **15. a.** 26.14 ft² **b.** 4.4 ft

17. a. 24 m **b.** 3,460,000 mg **c.** 2,140 cL

19. 164.59 dm

21. reflectional symmetries about lines through, AE, BF, CG, and DH; rotational symmetries through angles of 90°, 180°, and 270°.

Chapter 11

Section 11.1

1. We reassigned the nine board members that had been previously assigned.

3. California: 4; Arizona: 4; Nevada: 3

5. painters: 14; sculptors: 8; weavers: 9

7. Alabama, 7; Mississippi, 5; Louisiana, 7

9. 59,136   **11.** A; 75; 0.045   **13.** 15; 0.016

15. 160,000; 0.364

17. **c.** Paradox occurs when assigning fourteenth seat; business.

19. **a.** Center City, 8; South Street, 3; West Side, 12
    b. In going from 31 to 32, South Street loses a physician.

21. Once representatives have been assigned, do not reassign them at a later time.

23. Absolute unfairness is the difference in the average constituencies; it does not take into account the size of the constituencies as relative unfairness does.

29. **a.** 0.192   **b.** 0.469
    c. Giving the extra representative to Naxxon results in a smaller unfairness.

Section 11.2

1. A smaller relative unfairness occurs if we give the next representative to B.

3. We assigned it to the company with the highest Huntington–Hill number.

5. **a.** 0.162   **b.** 0.434   **c.** musicians

7. **a.** 0.012   **b.** 0.254   **c.** red line

9. New York   **11.** Colorado   **13.** 41,223.2

15. 1,470.9   **17.** 0.362   **19.** 0.330   **21.** Indiana

23. New Hampshire   **25.** musicians   **27.** yellow

29. artists   **31.** yes

33. Assign representatives so as to give the smallest relative unfairness.

35. After we assign one representative to each state, we assign the rest of the representatives to states according to the size of their Huntington–Hill numbers, in decreasing order, starting with the largest.

37. 5.7 million

Section 11.3

1. Naxxon had the largest Huntington–Hill number.

3. row 1, carpenters

5. 104.2, 162.0, 160.2, plumbers   **7.** row 1, Carney

9. 264.5, 181.5, 308.2, Carney   **11.** MTDMTMTMTD

13. **a.** Bronx, 3; Manhattan, 3; Queens, 4
    b. B, M, Q, Q, M, B, Q, M, Q, B

15. UT, ID, OR, OR, UT, OR, ID, OR, UT, OR

17. AR, KS, NE, KS, AR, NE, KS, AR, KS, AR, NE

19. B, B, H; B, 5; H, 4; S, 2

21. region 1, 4, region 2, 1, region 3, 2

23. Calculate a table of Huntington–Hill numbers.

Section 11.4

1. The standard divisor is the total population divided by the number of representatives; the standard quota is the state's population divided by the standard divisor.

3. by trial and error

5. **a.** 32,000; 2.63   **b.** 2   **c.** 3   **d.** 3

7. **a.** 9,272.73; 2.15   **b.** 2   **c.** 3   **d.** 2

9. 13,545.45; CA 4.134, NV 3.027, AZ 3.839

11. CA 4, NV 3, AZ 4

13. 33.2; performers 6.416, food workers 8.223, maintenance 5.361

15. performers 6, food workers 8, maintenance 6

17. 5.15; electricians 4.854, plumbers 3.495, painters 5.631, carpenters 6.019

19. electricians 5, plumbers 4, painters 5, carpenters 6

21. 4.263; fiction 7.037, poetry 4.692, technical 3.988, media 3.284

23. fiction 7, poetry 5, technical 4, media 3

25. 16.667; Pilates 3.36, kick boxing 1.74, yoga 0.660, spinning 0.240

27. Pilates 2, kick boxing 2, yoga 1, spinning 1

29. yes, because A's population has increased faster than B's, but A loses a representative to B in the reapportionment

31. yes, because the number of passengers per week at Bakerstown has increased faster than at Columbia City, but Bakerstown loses a security guard to Columbia City

33. Yes; when C's representatives are added to the commission, B loses a representative to A.

35. No; when D's representatives are added to the commission, there is no change in A's, B's, or C's apportionment.

37. D's standard quota is 87.961; however, using the Jefferson method, D would receive 89 representatives, which is greater than its upper quota.

39. D's standard quota is 240.642; however, using the Adams method, D would receive 239 representatives, which is less than its lower quota.

41. The standard divisor is the average number of people that each representative represents. The standard quota is the number of representatives that a state deserves.

43. Because the Jefferson method rounds the modified quotas down, we often need larger modified quotas and hence smaller modified divisors.

45. She needs to try a larger modified divisor.

49. smallest 926; largest 940

51. smallest 1,058; largest 1,066

Section 11.5

1. In discrete fair division, the objects cannot be divided as they are in continuous fair division.

3. He had the highest bids.

5. **a.** discrete   **b.** continuous   **c.** discrete

7.

	Darnell $\left(\frac{1}{2}\right)$	Joy $\left(\frac{1}{2}\right)$
Bid on car	$110,000	$85,000
Fair share of estate	a. $55,000	b. $42,500
Item obtained with highest bid	c. Car	d.

9.

	Dennis (40%)	Zadie (60%)
Bid on copyright	$20,000	$28,000
Fair share of copyright	a. $8,000	b. $16,800
Item obtained with highest bid	c.	d. copyright

11.

	Ed $\left(\frac{1}{3}\right)$	Al $\left(\frac{1}{3}\right)$	Jerry $\left(\frac{1}{3}\right)$
Bid on painting	$13,000	$8,000	$9,000
Bid on statue	$14,000	$13,000	$15,000
Total value	$27,000	a. $21,000	b. $24,000
Fair share of estate	c. $9,000	d. $7,000	e. $8,000

13.

	Darnell $\left(\frac{1}{2}\right)$	Joy $\left(\frac{1}{2}\right)$
Pays to estate (+) or receives from estate (−)	$55,000 (+)	$42,500 (−)
Division of estate balance ($12,500)	$6,250 (−)	$6,250 (−)
Summary of cash	Pays $48,750	Receives $48,750

15.

	Dennis (40%)	Zadie (60%)
Pays to pool (+) or receives from pool (−)	$8,000 (−)	$11,200 (+)
Division of pool balance ($3,200)	$1,280 (−)	$1,920 (−)
Summary of cash	Receives $9,280	Pays $9,280

17.

	Ed $\left(\frac{1}{3}\right)$	Al $\left(\frac{1}{3}\right)$	Jerry $\left(\frac{1}{3}\right)$
Items obtained with highest bid	Painting		Statue
Pays to estate (+) or receives from estate (−)	$4,000 (+)	$7,000 (−)	$7,000 (+)
Division of estate balance ($4,000)	$1,333.33 (−)	$1,333.33 (−)	$1,333.33 (−)
Summary of cash	Pays $2,666.67	Receives $8,333.33	Pays $5,666.67

19. Betty receives the books, Dennis receives the ring and desk; Dennis pays $9,125 to Betty.

21. Each person has a different estimate of the value of the estate and does not know the others' estimates.

23. Those who think that items in the estate are worth more receive the items and then must contribute cash to the estate that is then (partially) distributed to those who made low estimates of the items in the estate.

Chapter Review Exercises

1. When adding a member to the legislature and without changing populations, a state loses a representative.

3. If the number of representatives increases, we reassign representatives that have been assigned earlier.

5. **a.** Florida, 0.395; Texas, 0.437

7. HBQTHBQHBHQBT

11. pilates, 1; kick boxing, 4; yoga, 1; spinning, 2

13. The new states paradox occurs when a new state is added, and its share of seats is added to the legislature causing a change in the allocation of seats previously given to another state.

15.

	Tito $\left(\frac{1}{3}\right)$	Omarosa $\left(\frac{1}{3}\right)$	Piers $\left(\frac{1}{3}\right)$
Bid on rifle	$12,000	$17,000	$10,000
Bid on sword	$15,000	$13,000	$11,000
Estimated total value of estate	$27,000	$30,000	$21,000
Fair share of estate	$9,000	$10,000	$7,000

Chapter Test

1. When adding a member to the legislature and without changing populations, a state loses a representative.

3. art 2, music 4, theater 2

5. **a.** Arizona, 0.380; Oregon, 0.434

7. Mystery Mountain 4, Jungle Village 3, Great Frontier 3, City Sidewalks 4

9. No, because when Devon is added, no mall has a change in the number of personnel assigned to it.

11. massage 3, aromatherapy 2, yoga 3, meditation 1

13. massage 3, aromatherapy 2, yoga 2, meditation 2

15. Larry gets Branch A and pays $1,666,666.67; Moe gets Branch B and pays $7,666,666.67; Curly gets $9,333,333.33

Chapter 12

Section 12.1

1. Her 18 second-place votes gave her more points than Carim's 1 extra first-place vote.

3. Each received $\frac{1}{2}$ point because the number that preferred T over B was the same as the number that preferred B over T.

5. **a.** no **b.** Edelson   **7.** A (33 votes)   **9.** D

11. C (23 votes)   **13.** M   **15.** T (15 votes)

17. G   **19.** L (17 votes)   **21.** L

23. The Borda count method is being used, with 3 points for first, 2 for second, and 1 for third. **a.** 229 **b.** 291 **c.** 242

25. The first eliminated is in last place, the second eliminated is in second-to-last place, and so on.

27. A, D, C, S   **29.** D, A, C, S   **31.** T, G, E, P, F

33. G, T, E, P, F   **35.** 200   **37.** 110   **39.** 6   **41.** 4

43. Voters can state their second, third, and fourth preferences.

45. A candidate who wins an election should be able to beat the other candidates head-to-head.

47. There is no third candidate to eliminate, so whoever wins is the candidate with the majority of first-place votes.

51. S (92 votes)   **53.** E (37 votes)

Section 12.2

1. Example 1; the Borda method violates the majority criterion.

3. the plurality and pairwise comparison methods

5. B   **7.** Answers will vary.

9. C wins; B defeats everyone else head-to-head.

11. B wins; if we remove A, then C defeats B 105 to 104.

13. C wins; however, A defeats all other options head-to-head.

15. B wins; no one has a majority of first-place votes.

17. There are many correct answers; one is (a) 5; (b) 4.

19. Answers will vary.

21. There are many correct answers; one is 30.   **23.** yes

25. majority, Condorcet's, independence-of-irrelevant alternatives, monotonicity

27. In any election involving more than two candidates, no voting method will satisfy the four fairness criteria.

33. Answers will vary.

Section 12.3

1. 51 was the quota; 26, 26, 12, 12, 12, and 12 were the weights of the six voters.

3. They were critical members in all 11 winning coalitions.

5. a. 5   **b.** Each voter has 1 vote.
c. no dictator   **d.** Each voter has veto power.

7. a. 11   **b.** A, 10 votes; B, 3; C, 4; D, 5
c. no dictator   **d.** No voter has veto power.

9. a. 15   **b.** A, 1 vote; B, 2; C, 3; D, 3; E, 4
c. no dictator
d. No voter has veto power (no resolutions can be passed).

11. a. 12   **b.** A, 1 vote; B, 3; C, 5; D, 7
c. no dictator   **d.** C and D have veto power.

13. a. 25   **b.** A, 4 votes; B, 4; C, 6; D, 7; E, 9
c. no dictator   **d.** C, D, and E have veto power.

15. a. 51   **b.** A, 20 votes; B, 20; C, 20; D, 10; E, 10
c. no dictator   **d.** No voter has veto power.

17. {C, D}, {A, C, D}, {B, C, D}, {A, B, C, D}

19. {A, C, D, E}, {B, C, D, E}, {A, B, C, D, E}

21. C and D are critical in all coalitions.

23. All are critical in {A, C, D, E} and {B, C, D, E}; C, D, and E are critical in {A, B, C, D, E}.

25.

Coalition	Weight	
{P}	5	
{T}	4	
{S}	2	
{P, T}	9	Winning
{P, S}	7	Winning
{T, S}	6	Winning
{P, T, S}	11	Winning

27.

Coalition	Weight	
{A}	3	
{C}	4	
{T}	3	
{N}	2	
{A, C}	7	
{A, T}	6	
{A, N}	5	
{C, T}	7	
{C, N}	6	
{T, N}	5	
{A, C, T}	10	Winning
{A, C, N}	9	Winning
{A, T, N}	8	Winning
{C, T, N}	9	Winning
{A, C, T, N}	12	Winning

29. All voters are critical in {P, T}, {P, S}, and {T, S}; none are critical in {P, T, S}.

31. All voters are critical in all coalitions, with the exception of {A, C, T, N}, which has no critical voters.

33. All have index $\frac{1}{5}$.

35. A and B have index $\frac{1}{11}$; all others have index $\frac{3}{11}$.

37. A and B have index 0; C and D have index $\frac{1}{2}$.

39. K and C have index $\frac{13}{36}$; all others have index $\frac{1}{18}$.

41. a. All have index $\frac{1}{6}$.

43. Without that voter's support, a resolution cannot pass.

45. a. All have index $\frac{1}{5}$.

47. a. A has index 1; others have index 0.

Section 12.4

1. With the notation {A, B, C}, the order of the voters does not matter; with the notation (A, B, C), the order of the voters does matter.

3. We divided the number of times that A was pivotal in some permutation of the voters by total number of permutations of the voters.

5. (X, Y, Z), (X, Z, Y), (Y, X, Z), (Y, Z, X), (Z, X, Y), (Z, Y, X)

7. There are 24 permutations: (W, X, Y, Z), (W, X, Z, Y), (W, Y, X, Z), . . . , (Z, Y, X, W).

9. 720

11. 479,001,600

13.

Sum of Weights of Coalition Members Until We Reach Quota of 5	Pivotal Voter
(A, B, C) 3 + 2	B
(A, C, B) 3 + 2	C
(B, A, C) 2 + 3	A
(B, C, A) 2 + 2 + 3	A
(C, A, B) 2 + 3	A
(C, B, A) 2 + 2 + 3	A

15. An and B have index $\frac{1}{2}$; C has index 0.

17. A and B have index $\frac{5}{12}$; C and D have index $\frac{1}{12}$.

19. A has index $\frac{1}{2}$; the others have index $\frac{1}{6}$.

21. a. Each has index $\frac{1}{5}$.

23. a. A has index 1; the others have index 0.

25. A has index $\frac{3}{4}$; the others have index $\frac{1}{12}$.

27. c. Each has index $\frac{1}{4}$.   **29.** See definition in text.

31. Before B can be pivotal, the managing editor and one other member of the board must already have been chosen.

Chapter Review Exercises

1. a. no **b.** Myers   **3.** social justice

5. D wins; no; R has the majority of first-place votes.

7. B wins; no; if A is removed, then C wins with 80 points.

9. Quota, 17; A, 1 vote; B, 5; C, 7; D, 8; no dictator; B, C, and D have veto power.

11. A and B have index $\frac{1}{10}$, C has index $\frac{3}{10}$, and D has index $\frac{1}{2}$.

13. A, $\frac{7}{10}$; B, $\frac{1}{10}$; C, $\frac{1}{10}$; D, $\frac{1}{10}$

15.

Sum of Weights of Coalition Members Until We Reach Quota of 6	Pivotal Voter
(A, B, C) 4 + 3	B
(A, C, B) 4 + 2	C
(B, A, C) 3 + 4	A
(B, C, A) 3 + 2 + 4	A
(C, A, B) 2 + 4	A
(C, B, A) 2 + 3 + 4	A

17. $\frac{1}{6}$

Chapter Test

1. a. no **b.** Molina

3. Yes; A wins the election and can defeat B, C, and D head-to-head.

5. 6! = 720

7. B wins; however, A has majority of first-place votes. Majority criterion is not satisfied.

9. No; A wins the election; however, if we remove C, then B wins.

11.

Sum of Weights of Coalition Members Until We Reach Quota of 7	Pivotal Voter
(A, B, C) 5 + 3	B
(A, C, B) 5 + 3	C
(B, A, C) 3 + 5	A
(B, C, A) 3 + 3 + 5	A
(C, A, B) 3 + 5	A
(C, B, A) 3 + 3 + 5	A

13. A is winner; however, if we remove D, then B wins. Independence-of-irrelevant-alternatives criterion is not met.

15. A $\frac{1}{2}$; B $\frac{1}{2}$; C 0

Chapter 13

Section 13.1

1. We first listed all pairs that began with C, then all pairs that began with O, then all pairs that began with F, and finally all pairs that began with T.

3. A mathematician who applies mathematics to reveal the organization of terrorist cells

5. JT, JC, JF, JR, TC, TF, TR, CF, CR, FR

7. JT, JC, JF, JR, TJ, TC, TF, TR, CJ, CT, CF, CR, FJ, FT, FC, FR, RJ, RT, RC, RF

9. larger if repetition is allowed   **11.** 4   **13.** 6

15. 30   **17.** 120   **19.** (1, 4), (2, 3), (3, 2), (4, 1)

21. (1, 1), (2, 2), (3, 3), (4, 4), (5, 5), (6, 6)

23. (1, 1), (1, 2), (2, 1), (1, 3), (2, 2), (3, 1), (1, 4), (2, 3), (3, 2), (4, 1)

25. 120   **27.** 24   **29.** 32   **31.** 16

33. 16   **35.** 9   **37.** 16   **39.** 36

41. a. $2^5 = 32$ **b.** 1; 5; 10 **c.** 16 out of 32

43. 24   **45.** 12   **47.** 55   **49.** 27   **51.** 18

53. We thought of flipping the coins one at a time. We also thought of rolling the dice one at a time.

57. Follow the branches in the tree diagram to make the schedules.

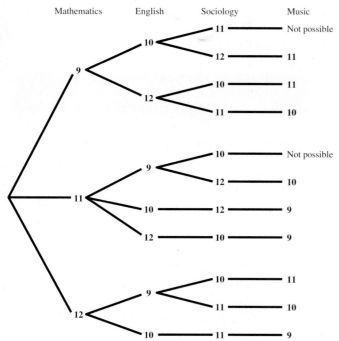

59. RRBG, RBRG, BRRG

61. RGBR, RBGR, GRBR, GBRR, BRGR, BGRR

Section 13.2

1. In a), we thought of flipping the four coins one at a time instead of flipping them all at once. In b), we thought of rolling the red die, then the blue die, and then the green die.

3. A skilled player can win at blackjack.

5. 42 **7.** 336 **9.** 480 **11.** 1,560 **13.** 64

15. 96 **17.** 9,000 **19.** 2,500 **21.** 17,576

23. 479,001,600 **25.** 243

27. a. 20 **b.** 6 **c.** 1 **d.** 10 **e.** 1,200

29. 6 **31.** 1,058,400 **33.** 720 **35.** 72

37. In Example 5, we considered the condition that Louise and her tutor must sit next to each other first.

39. We can use a tree diagram to visualize the total number of possibilities that we count when we use the fundamental counting principle.

Section 13.3

1. Objects cannot be repeated in a permutation.

3. One combination of three people corresponds to $3! = 6$ permutations.

5. 24 **7.** 6 **9.** 720 **11.** 120 **13.** 30

15. 120 **17.** 336 **19.** 1,814,400 **21.** 56

23. 45 **27.** 1 8 28 56 70 56 28 8 1

29. 21 **31.** the second entry in the 18th row

33. the sixth entry in the 20th row

35. $P(8, 8)$ **37.** $C(17, 3)$ **39.** $P(5, 5)$ **41.** $C(10, 3)$

43. $C(9, 5)$ **45.** $C(17, 8)$ **47.** 177,100 **49.** 1,865

51. $P(15, 5) = 360,360$

53. $P(15, 5) \times P(15, 5) \times P(15, 4) \times P(15, 5) \times P(15, 5) = $ a VERY large number

55. $C(6, 2) \times C(8, 3) = 15 \times 56 = 840$

57. 57,750 **59.** 60 **61.** 5,940 **63.** 3,300 **65.** 70

67. In a permutation problem, we are choosing and arranging objects in order; in a combinations problem, we are choosing, but not arranging the objects.

69. the rth entry of the nth row

71. Because of the special conditions on the numbers, we cannot use all digits at each position in the number.

73. The number of ways to do something must be an integer.

79. $C(13, 2) \times C(13, 3) = 78 \times 286 = 22,308$

81. ⊜ **83.** 6 **85.** $C(n - 1, r - 1)$

87. $C(n - 1, r - 1) + C(n - 1, r) = C(n, r)$

89. The number of ways that we can choose an r-element set from n elements is the same as the number of ways that we can choose $n - r$ elements from n elements; for example, the number of ways that we can choose two elements from five elements is the same as the number of ways that we can choose three elements from five elements.

Section 13.4

1. We calculated the number of ways the first wheel could stop multiplied by the number of ways the second wheel could stop multiplied by the number of ways the third wheel could stop.

3. 120 **5.** 20 **7.** 10 **9.** 4

11. a. 4 **b.** 1,287 **c.** $5,148 - 40 = 5,108$ **13.** 123,552

15. We are not simply choosing three cards from a possible 52.

17. three oranges

Chapter Review Exercises

1. PQ, PR, PS, QP, QR, QS, RP, RQ, RS, SP, SQ, SR

3. 240 **5.** 288 **7.** 120

9. In a permutation, order is important; in a combination, it is not.

11. 680 **13.** $C(n, r) = \dfrac{P(n, r)}{r!}$

15. the second entry in row 18 **17.** 5,108

Chapter Test

1. AB, AC, AD, BA, BC, BD, CA, CB, CD, DA, DB, DC

3. 48 **5.** 10 **7.** $P(15, 4) = 32,760$ **9.** 72

11. 1,320 **13.** $C(n, r) = \dfrac{P(n, r)}{r!}$ **15.** 108

Chapter 14

Section 14.1

1. We drew a tree diagram.

3. We know from Chapter 13 that there are $C(52, 5)$ ways to choose 5 cards from 52.

5. The bread could only make one-half of a revolution before hitting the floor. With three lines, there is only a one-third chance that you are in the fastest line.

7. $\{(1, 6), (2, 5), (3, 4), (4, 3), (5, 2), (6, 1)\}$

9. {HHH, HHT, HTH, THH}

11. {(r, b), (r, y) (b, r), (y, r)}

13. {(b, b, r), (b, r, b), (r, b, b), (b, b, y), (b, y, b), (y, b, b)}

15. a. 16
  b. {(1, 1), (1, 3), (2, 2), (2, 4), (3, 1), (3, 3), (4, 2), (4, 4)}
  c. $\frac{1}{2}$ **d.** $\frac{3}{16}$

17. a. 24
  b. {CKMJ, CKJM, CJMK, CJKM, MKCJ, MKJC, MJCK, MJKC, KCJM, KMJC, JCKM, JMKC}
  c. $\frac{1}{2}$ **d.** $\frac{1}{4}$

19. a. 20
  b. {(s, c), (s, w), (s, d), (s, h), (c, s), (w, s), (d, s), (h, s)}
  c. $\frac{2}{5}$ **d.** $\frac{3}{5}$

21. a) $\frac{1}{9}$ b) 8 to 1   **23.** a) $\frac{1}{4}$ b) 3 to 1   **25.** 0.000495

27. 0.002   **29. a.** $\frac{5}{9}$ **b.** 5 to 4   **31.** 0.49   **33.** 0.26

35. The probability that the child is a carrier is $\frac{1}{2}$.

		Second Parent	
		s	**n**
First Parent	**s**	ss	sn
	s	ss	sn

37. a.

		First-Generation Plant	
		r	**r**
First-Generation Plant	**w**	wr	wr
	w	wr	wr

  b. All flowers will be pink. P(pink) = 1; all other probabilities are 0.

39. a.

		Second Parent	
		N	**c**
First Parent	**N**	NN	Nc
	c	cN	cc

  b. P(disease) $= \frac{1}{4}$

41. 0.43   **43.** $\frac{22}{36} = \frac{11}{18}$ **45.** $\frac{2}{36} = \frac{1}{18}$ **47.** 0.57

49. 0.54   **51.** $\frac{1}{336}$ **53.** $\frac{2}{7}$ **55.** $\frac{5}{12}$

57. a. 3 to 7 **b.** 7 to 3   **59.** 45,057,474 to 1

61. $P(E) = \dfrac{\text{number of times } E \text{ occurs}}{\text{number of times experiment is performed}}$

63. All plants had one yellow and one green gene, and yellow was dominant.

65. An outcome is an element in a sample space; an event is a subset of a sample space.

67. Rolling a total less than 13 when rolling two dice

73. 0.000027

Section 14.2

1. It was easier to compute $P(A')$ and then subtract it from 1 instead of computing $P(A)$ directly.

3. Your chances of winning a big lottery are very small compared to many other real-life events.

5. 0.985   **7.** $\frac{999}{1,000}$ **9.** $\frac{25}{36}$ **11.** $\frac{31}{32}$

13. $\frac{7}{13}$ **15.** $\frac{11}{18}$ **17.** 0.10   **19.** 0.55

21. $\frac{15}{26}$ **23.** 0.923   **25.** 0.54   **27.** 0.08

29. 0.92   **31.** 0.84   **33.** 0.65   **35.** 0.19

37. Because $E \cup E' = S$ and $P(S) = 1$.

39. false   **41.** true

43. If E and F are disjoint, then $P(E \cap F) = 0$.

45. $P(A) + P(B) + P(C) - P(A \cap B) - P(B \cap C)$

47. 0.96   **49.** 0.14

Section 14.3

1. the probability that the total was odd, given that the total was greater than nine

3. We found that $P(G \mid F) \neq P(G)$, which meant that knowing that event F occurred had an effect on the probability of G.

5. $\frac{1}{6}$; $\frac{1}{3}$ **7.** $\frac{1}{6}$; $\frac{2}{11}$ **9.** $\frac{1}{2}$ **11.** $\frac{1}{10}$ **13.** $\frac{2}{4}$ **15.** $\frac{2}{3}$

17. 0   **19.** $\frac{50}{61}$ **21.** $\frac{2}{61}$ **23.** 0.947   **25.** 0.90

27. a) $\frac{1}{221}$ b) $\frac{1}{169}$ **29.** a) $\frac{40}{221}$ b) $\frac{30}{169}$ **31.** a) $\frac{8}{663}$ b) $\frac{2}{169}$

33. $\frac{7}{17}$ **35.** $\frac{11}{1.105}$ **37.** $\frac{117}{850}$ **39.** $\frac{1}{729}$ **41.** 0.74

43. 0.63   **45.** dependent   **47.** independent   **49.** $\frac{1}{3}$

51. $\frac{8}{15}$ **53.** 0.1029   **55.** 0.2646

57. Forty percent of the available spaces in dorm Y are rooms.

59. 0.50

61.

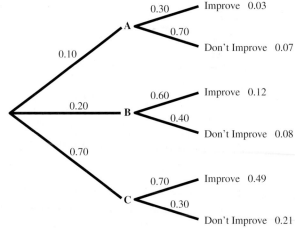

63. 0.64   **65.** 0.375   **67.** 0.336   **69.** 0.736

71. 0.424

73. Multiply both sides of the equation by $P(E)$ to get $P(E \cap F) = P(F \mid E) \cdot P(E)$.

75. Answers will vary.   **77.** Answers will vary.

Section 14.4

1. We multiplied the value of the laptop by the probability of it being stolen.

3. −$0.50; no   **5.** $1.39; $2.39   **7.** −$2.50; $2.50

9. −$0.40; $0.60   **11.** −$2.00; $3.00   **13.** $328

15. −$0.08   **17.** −$0.29

19. Student should guess; expected value is $\frac{1}{16}$.

21. two options   **23.** −$7.50   **25.** $44   **27.** $25,000

33. a. $29.80   **b.** $27.35   **35. a.** 7   **b.** 7   **37.** no

Section 14.5

1. We rolled the dice a fixed number of times; the experiment had two outcomes: seven and nonseven; the probability of rolling a seven was the same from roll to roll; the results of the rolls were independent of each other.

3. yes   **5.** No; the number of trials is not fixed.   **7.** yes

9. 0.2634   **11.** 0.0879   **13.** k is larger than n.

15. 0.0537   **17.** 0.0730   **19.** 0.6778   **21.** 0.2637

23. 0.7472   **25.** $\frac{3}{4}$ **27.** 14.7

29. The experiment is performed a fixed number of trials. The experiment has two outcomes: "success" and "failure." The probability of success is the same from trial to trial. The trials are independent of each other.

33. 0.3770

Chapter Review Exercises

1. a. {HHT, HTH, THH}

   b. {(2, 6), (3, 5), (4, 4), (5, 3), (6, 2)}

3. To calculate the empirical of an event E, we divided the number of the times the event occurs by the number of the times the experiment is performed. To calculate theoretical probability we use known mathematical techniques such as counting formulas.

5. a. $\frac{2}{19}$ **b.** 45 to 55 or 9 to 11

7. $\frac{8}{13}$ **9.** Answers will vary   **11. a.** $\frac{1}{17}$ **b.** $\frac{4}{663}$

13. 0.385   **15.** $0.38   **17.** 0.219

Chapter Test

1. a. {(4, 6), (5, 5), (6, 4), (5, 6), (6, 5), (6, 6)}

   b. {HHHT, HHTH, HTHH, THHH, HHHH}

3. a. $\frac{3}{31}$ **b.** 17 to 3

5. When calculating $P(B\,|\,A)$, we are calculating the probability of B assuming that event A has occurred. In calculating $P(A\,|\,B)$, we are assuming that B has occurred and then calculating $P(A)$.

7. $-\frac{1}{6}$ **9.** 0.0879   **11.** $\frac{1}{2}$ **13.** −$0.40

15. a. $\frac{11}{221}$ **b.** $\frac{8}{663}$

Chapter 15

Section 15.1

1. selection bias and leading-question bias

3. Example 4 dealt with weight, which is a continuous variable, so a bar graph would not be appropriate.

5. Gallup's sampling method was better.

7.

x	Frequency	Relative Frequency
2	1	0.05
3	0	0.00
4	0	0.00
5	3	0.15
6	2	0.10
7	4	0.20
8	4	0.20
9	3	0.15
10	3	0.15

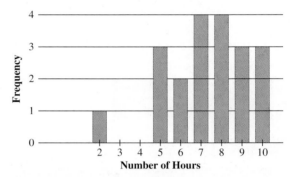

9.

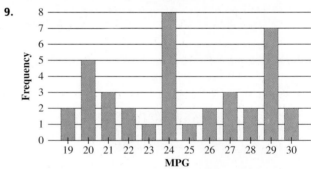

11.

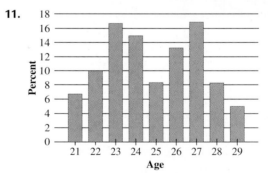

13.

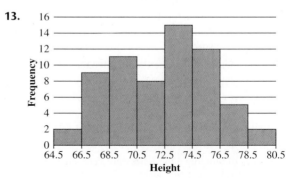

15.

A		B
8 8	1	3 6 8
9 6 2 2 1	2	0 1 1 2 2 2 3 4 6 8 9
9 9 8 7 4 3 2	3	2 3 3 8 9
7 3 3 3 2 2	4	7

17.

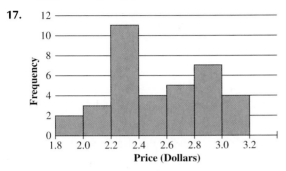

19. a. 0 occurred 13 times   **b.** 12 occurred two times
  c. 6   **d.** 71   **e.** $\frac{22}{71}$

21. a. 8   **b.** 3   **c.** 12   **23.** 10.2%

25. 25 and older, earning at least $10.00 per hour

27. 16 to 19, earning less than $7.51 per hour

29.

No Training		With Training
	9	1	
9 8 8 7 6 6 3 2	2	1 8 8 9 9
4 2	3	2 2 3 6 6 7 9
3 3 2 1 0	4	1 3 4 5

31. The population is the set of all objects being studied; a sample is a subset of the population.

33. A continuous variable can take on arbitrary values; a discrete variable cannot.

35. Answers will vary   **37.** Answers will vary

Section 15.2

1. They ignore the frequencies and divide by 6, which is the number of different scores.

3. A median is the middle strip between a divided highway. This helps us remember that the median is the middle score after the scores have been arranged in order.

5. mean, 6.3; median, 6; mode, 4

7. mean, 6.9; median, 6.5; no mode

9. mean, 5; median, 5; mode, 5

11. mean, 7.3; median, 7.5; mode, 7, 8

13. mean, 7.15; median, 7; mode, 9

15. mean, 11.29; median, 11; modes, 11 and 15

17. mean, 6.75; median, 6.5; mode, 4

19. mean, 12.6; median, 13; mode, 11

21. a. min, 11; Q_1, 23; median, 31; Q_3, 44; max, 53

  b.

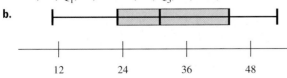

23. a. min, 25; Q_1, 30; median, 33; Q_3, 41; max, 49

  b.

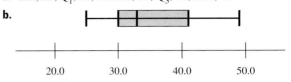

25. mean, 22.85; median, 23; modes, 25 and 26

27. mean, 22.3; median, 16.5; no mode

29. mean, 118.25; median 43.5; no mode

31. 2.8   **33.** 73

35. mean, 24.74; median, 25; modes, 24, 25

37. a. 62   **b.** 112   **39.** $130.32

41. a.

  b.

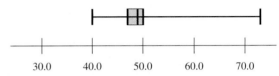

43. liberal arts   **45.** education; no

47. The maximum and minimum salaries are not too far from the middle 50% of the salaries.

49. With Σx, we are adding the individual scores in the distribution; with $\Sigma x \cdot f$, we are multiplying each score x by its frequency f before adding.

51. by a box-and-whisker plot   **53.** mean; mean

Section 15.3

1. One outlier can make the range meaningless.

3. It allows us to compare variation between different sets of data.

5. range, 5; mean, 19.2; standard deviation, 1.55

7. range, 6; mean, 7; standard deviation, 2

9. range, 12; mean, 8; standard deviation, 4.34

11. range, 15; mean, 9; standard deviation, 4.57

13. range, 0; mean, 3; standard deviation, 0

15. 7; 2.13   **17.** mean, 4; standard deviation, 1.60

19. mean, 7; standard deviation, 1.73   **21. a.** set a   **b.** set c

23. 79; 6.59   **25.** 5.46; 1.76   **27.** 4.41; 3.38

29. Family H's income is 1.25 standard deviations above the mean.

31. D

33. CV for Apple Computer is 9.9%; CV for Dell is 9.2%; Apple is more volatile.

35. DJIA has a CV of 0.6%; WebMaster has a CV of 4.5%; WebMaster is more volatile.

37. coffee, 0.111; gasoline, 0.275

39. Family G went from the mean income to 0.94 standard deviations above the mean.

41. In calculating a sample's standard deviation, we divide by $n - 1$; for a population, we divide by n.

43. true

Section 15.4

1. See page 752.

3. The normal curve is symmetric about the mean.

5. 34%   **7.** 2.5%   **9.** 16%   **11.** 16%   **13.** 97.5%

15. 13.5%   **17.** 39.1%   **19.** 11.1%   **21.** 27.3%

23. 8%   **25.** 12.8%   **27.** 7.3%   **29.** 8.1%

31. 90.8%   **33.** 80%   **35.** 1.28   **37.** −1.175

39. 0.2525   **41.** −0.673   **43.** 1.4   **45.** −1.75

47. 0.97   **49.** 64.2   **51.** 33.65   **53.** 31.69

55. a. 50% **b.** 16%   **57. a.** 34% **b.** 16%   **59.** 1.67

61. $0.092 \times 324 = 29.808$ (we expect about 29 or 30)

63. $69.2\% \times 200 = 138.4$ **65.** 9 days; 21.8%   **67.** 53.5

69. Robinson was 2.3 standard deviations above the mean batting average for the 1940s, and Carew was 2.81 standard deviations above the mean for the 1970s. Carew was more dominant.

71. about 10 months

73. a. 2.685 (2 or 3 years) **b.** 1.26 (1 or 2 years)

75. 42.1%

77. The score is almost 2 standard deviations above the mean.

79. By looking at the graph of the normal curve, we see that there is clearly more area under the curve between 0.0 and 0.7 than there is between 1.3 and 2.0.

81. Mean is 20 in both distributions; standard deviation is about 2 or 3 in distribution 1; standard deviation is about 4 or 5 in distribution 2.

87. 0.84   **89.** 42.692   **91.** 96th

Section 15.5

1. Because the grades generally increased as the number of tutoring sessions increased, there was some relationship between the two variables.

3. positive correlation

5. $r = 0.99$

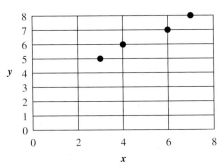

7. $r = 0.74$

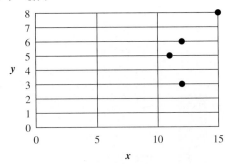

9. We can be 95% confident, but not 99% confident.

11. neither   **13.** $y = 0.7x + 3$ **15.** $y = 0.89x - 5.61$

17. 0.96; we can be 99% confident that there is positive linear correlation.

19. 0.00; neither   **21.** $y = 2.64x + 21.71$

23. $y = 0 \cdot x + 29.4$

25. We can be 99% confident that there is significant negative linear correlation between the variables of car weight and gas mileage.

31. 1

Chapter Review Exercises

1.

Number of Accidents	Frequency
4	1
5	3
6	6
7	2
8	10
9	4
10	5

3.

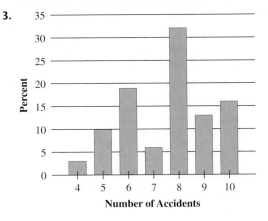

5. a.

	F			M
9 9 8 6 6 5 1		2		9
9 8 5 5 5 4 3 3 3 2 1 0 0		3		1 2 5 6 7 7 7 7 8 8 9
9 9 5 2 1		4		0 2 3 3 5 5 5 6 7
		5		0 1 2 2 4
1 1		6		0 1
4		7		6
0		8		

b. The actors generally seem to be older than the actresses.

7. The mean is the arithmetic average; the median is the middle score; the mode is the most frequent score.

9.

    3   6   8   12   20

11. the spread of the distribution

13. 0.71   **15.** 221   **17.** 14 months

19. a. 0.977   **b.** $y = 2.5x - 2$

Chapter Test

1.

Number of Visits	Frequency
4	1
5	4
6	8
7	9
8	7
9	6
10	2
11	2
12	1

3.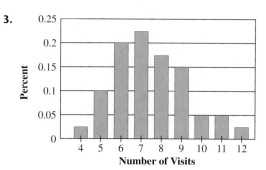

5. 2

7.

Ruth			Aaron
5 2		2	
5 4		3	0 2 4 4 8 9 9
9 7 6 6 6 1 1		4	0 0 4 4 4 4 5 7
9 4 4		5	
0		6	

9. the statistics exam   **11. a.** none   **b.** negative

13. mean, 7.85; median, 8.5; mode, 9

15. 1.40   **17.** 18 months

Photo Credits

p.1: Shutterstock; p. 4: Getty RF; p. 16: Shutterstock; p. 23: Getty Ed; p. 28: Getty RF; p. 30: Creatas RF; p. 38: Shutterstock; p. 41: AFP/Getty Images; p. 44: Library of Congress; p. 49: AP Wideworld Photos; p. 55: Shutterstock; p. 65: Shutterstock; p. 70: Shutterstock; p. 81: Getty Images; p. 83: Getty Images Entertainment; p. 90: Getty Images Entertainment; p. 92: AP Wideworld Photos; p. 115: Stockbyte Action Sports EP002; p. 117: The Image Bank/Getty Rights Ready; p. 118: The Image Bank/Getty Rights Ready; p. 125: Shutterstock; p. 127: PhotoFest; p. 131: NASA Headquarters; p. 138: Getty Images; p. 140: Digital Vision; p. 145: Digital Vision; p. 148: Shutterstock; p. 151: Photographer's Choice/GettyRF; p. 159: Shutterstock; p. 164: Photodisc/Getty RF; p. 166: David Bergman/Corbis; p. 178: Shutterstock; p. 185: Fox/Photofest; p. 186: Courtesy Prindle, Weber, & Schmidt, Inc.; p. 189: Shutterstock; p. 193: © Bettmann/Corbis; p. 195: British Museum; p. 197: public domain; p. 202: © Bettmann/Corbis; p. 207: Shutterstock; p. 213: © Archivo Iconografico, S.A./Corbis, The Granger Collection, New York; p. 221: Courtesy General Mills; p. 222: Shutterstock; p. 227: Shutterstock; p. 238: Shutterstock; p. 245: Shutterstock; p. 253: © Catherine Karnow/Corbis; p. 258: Shutterstock; p. 259: The Granger Collection, New York; p. 277: NASA; p. 279: Shutterstock; p. 287: Shutterstock; p. 288: Shutterstock; p. 296: Eric Nguyen/Corbis; p. 309: California Department of Health Services; p. 322: Stockbyte/Getty RF; p. 333: Shutterstock; p. 335: Photodisc; p. 358: National Geographic/Getty RF; p. 364: Shutterstock; p. 370: Shutterstock; p. 383: Shutterstock; p. 384: Shutterstock; p. 390: Shutterstock; p. 395: Shutterstock; p. 405: Shutterstock; p. 409: Shutterstock; p. 413: Shutterstock; p. 421: Shutterstock; p. 429: Shutterstock; p. 438: Shutterstock; p. 444: Shutterstock; p. 450: Shutterstock; p. 452: Shutterstock; p. 468: Shutterstock; p. 475: Shutterstock; p. 480: Shutterstock; p. 500: Shutterstock; p. 506: Shutterstock; p. 515: Brand X Pictures; p. 520: Courtesy of Anne M. Burns, Professor of Mathematics, Long Island University; p. 518: PhotoDisc; p. 530: iStockphoto; p. 531: Shutterstock; p. 536: Getty Editorial; p. 537: Shutterstock; p. 542: Shutterstock; p. 543: Shutterstock; p. 550: Shutterstock; p. 551: Shutterstock, © Royalty-Free/Corbis; p. 579: Getty Images; p. 589: Getty Images Sport; p. 594: Shutterstock; p. 599: Shutterstock; p. 601: AP/Wide World Photos; p. 606: Digital Vision; p. 610: Shutterstock; p. 616: Shutterstock; p. 622: Getty Images; p. 623: Shutterstock; p. 627: Getty Images Entertainment; p. 628: Jonathan Farley; p. 632: Shutterstock; p. 636: Shutterstock; p. 637: Shutterstock; p. 642: AP/Wide World Photos; p. 648: Beth Anderson, NASA; p. 650: Shutterstock; p. 658: Jupiter; p. 660: Shutterstock; p. 662: Shutterstock; p. 665: © Bettmann/Corbis; p. 668: © Bettmann/Corbis; p. 671: Stockbyte/Getty RF; p. 672: AP/Wide World Photos; p. 676: Shutterstock; p. 680: Shutterstock; p. 695: Digital Vision/Getty RF; p. 697: Photodisc; p. 714: Corbis; p. 720: NOAA; p. 722: Getty Images; p. 723: AFP/Getty Images; p. 730: WireImage/Getty Images; p. 740: Photodisc/Getty RF; p. 745: Digital Vision; p. 748: AP/Wide World Photos; p. 758: Hulton Archive.

Index

HISTORICAL INDEX

Topics

Abacists, 190
Abacus, 190, 213
Algorists, 190
Analytical geometry, 299
Apportionment in U.S. history, 542, 557
Approval voting, 583
Balinski and Young's impossibility
     theorem, 542
Banzhaf power index, 607
Black Plague, 741
Blocking coalition, 607
Boolean algebra, 59
Calculating devices, 213
Circumference of earth, 30
Combinations, 642
Computers, 213
Continuum hypothesis, 76
Counting, 642
Credit, 419
demotic script, 188
Egyptian hieroglyphics, 188
Electoral college, 607
Euclid's Fifth Postulate, 457
Gödel's incompleteness theorem, 96
Hamiltonian circuits, 157
Hindu-Arabic numerals, 190
Huntington-Hill apportionment principle, 542
Inductive reasoning, 19
Interest, 419
IQ tests, 19
Koenigsburg bridge problem, 140
Linear programming, 387
Logic, 85
Lottery, 699
Mayan mathematics and astronomy, 198
Napier's rods, 200
Negative numbers, 241
Non-Euclidean geometry, 457, 466
Permutations, 642
Presidential polls, 717
Probability theory, 668, 679
Problem solving, 14
pseudosphere, 457
Random walk problem, 14

Rhind papyrus, 188
Riemannian geometry, 466
Roman numerals, 190
Rosetta stone, 188
Set theory, 59, 69
Solving equations, 320
Statistics, 741
tractrix, 457

People

Abel, Niels Henrik, 320
Adelman, Leonard, 218
Ahmes, 188
al-Khowarizmi, Mohammed ibn Musal, 320
Aristotle, 85
Babbage, Charles, 213
Balinski, Michael, 542
Bolyai, Janos, 457
Boole, George, 59
Brahmagupta, 241, 642
Cantor, Georg, 69, 76
Cardano, Girolamo, 241, 320, 668
Cayley, Arthur, 141
Champollion, Jean Francois, 188
Charles II (king of England), 741
Cohen, Paul, 76
Colburn, Zerah, 157
Condorcet, the Marquis de, 592
Dantzig, George, 387
DeMorgan, Augustus, 59
Descartes, René, 241, 299
Dewey, Thomas, 717
Eckert, J. Presper, 213
Einstein, Albert, 466
Eratosthenes, 30, 233
Euclid, 457, 466, 642
Euler, Leonhard, 141
Fermat, Pierre, 679
Fibonacci, Leonardo, 286
Fourier, Jean-Baptiste, 188
Franklin, Benjamin, 699
Gallup, George, 717
Galois, Evariste, 320
Gardner, Howard, 19
Gauss, Karl Friedrich, 457

Germain, Sophie, 254
Gödel, Kurt, 96
Gombauld, Antoine, 679
Graunt, John, 741
Halley, Edmund, 741
Hamilton, William Rowan, 157
Henry VII (king of England), 741
Hilbert, David, 69, 76
Hill, Joseph, 542
Huntington, Edward, 542
Hypatia, 476
James I (king of England), 699
Kronecker, Leopold, 69
Lagrange, Joseph, 254
Landon, Alfred, 717
Laplace, Pierre de, 200, 679
Leibniz, Gottfried, 85, 96, 213
Lobachevsky, Nicolai, 457, 466
Lukasiewicz, Jan, 121
Mauchly, John, 213
Napier, John, 200
Napoleon, 188
Nero, 699
Nightingale, Florence, 722
Pascal, Blaise, 213, 241, 647, 679
Pearson, Karl, 722
Poincaré, Henri, 69
Polya, George, 14
Pythagoras, 264
Rayburn, Sam, 557
Rhind, A. Henry, 188
Riemann, Bernhard, 457, 466
Rivest, Ronald, 218
Roosevelt, Franklin D., 542, 717
Russell, Bertrand, 96
Salamucha, Jan, 121
Shamir, Adi, 218
Stern, Wilhelm, 19
Sylvester, James, 722
Tarski, Alfred, 121
Thomas, Albert, 557
Truman, Harry S., 717
Washington, George, 699
Young, H. Peyton, 542

INDEX OF APPLICATIONS

Astronomy

Astronomical distances, 279
Circumference of Earth, 34, 456, 459
Distance light travels, 280
Experiments for the space shuttle, 656
planets in solar system, 43
Selecting astronauts, 648
Space colony project, 172–75, 179
Space flight, 279

Automotive

Automobile accidents, 70, 773
Calculating gas mileage, 403, 499
Car insurance, 702
Car repair *vs.* purchase, 130–31
car sales, 402–403
car speed *vs.* mileage, 770
Car weight *vs.* mileage, 766–67
comparing car features, 171
dealer markup on car, 403
depreciation of a car, 403
EPA mileage ratings, 723, 738
financing purchase of a car, 413
License plates, 629, 637
price of a car, 306, 345–46, 398
purchasing a car (*See under* Consumer
  information)
resale value, 129
saving for a car, 429
seat belt use, 713
Speed limit, 341–42
speed of a vehicle, 269, 498–99
time saved by speeding, 339–40, 342
Tire warranty, 32
vehicle production, 374
Volume of a car's gas tank, 496
voting for car of the year, 591

Biology and Life Sciences

Animal population, 334, 335, 337–38, 342, 357,
  358, 361
Atoms in a person's body, 277
cicada emergence patterns, 237
Cross-breeding plants, 671, 711
Endangered species, 624, 632
fish kill, 729
Food chain, 169
Growth of rabbits, 286
human weight *vs.* mosquito weight, 279
Life expectancy, 314
separating animals from each other, 151
Weights of dinosaurs, 9

Business

Advertising, 179, 257, 307, 383, 626–27,
  632, 673
assigning responsibilities and positions, 636
Award for the best restaurant, 600
Commission on business climate, 565
Communications network, 7–8, 169
Customers at a coffee shop, 724, 748
Customer service, 762
dealer markups, 403–404
Department store introducing a game called
  "Register Roulette," 692
displaying products in a pyramid, 289
Displaying store merchandise, 238, 289
Dissolving a partnership, 572–73, 574
DVD sales, 320–21, 323
Electronics company, 537–38
employee training, 725
Equipment for a business, 410–11, 420, 429–30

Family Services Agency, 543, 544
Fast-food chain, 403, 586–87
Flow of paperwork through a bureaucracy, 169
Governing committee, 645
Law firm, 605–606, 609, 616
life cycle of DVD, 320
Manufacturing, 383, 389
Mergers, 588
music sales on Internet, 312–13, 403
Newspapers, 629, 636
Packaging candy, 222
Pastry business on the Internet, 311
Profit, 322–23, 387–88, 393, 403, 700, 701,
  702, 703
Publishing a textbook, 259
Sales routes, 17, 19–21
Selecting conference attendees, 644, 664–65
Small business, 383, 389–90
Starting a business, 429, 430
Textbook ISBNs, 223
Vending machines, 761
Volunteers, 723
Voting weight of stockholders, 52, 602–603
Wheelchair accessibility, 308

Chemistry

Carbon dating, 358
Gas pressure, 342–43
Radioactive decay, 334

Construction

Age of a building, 16
building a bridge, 465
Building a hummingbird aviary, 489
Building a new factory, 599
Building a student union, 178
Building scaffolding, 470
Ceramic tiles, 508
Constructing a concrete patio, 489
cost of flooring, 499
Energy-efficient house, 179
Furniture construction, 307, 372–73, 469, 470
Gazebo, 469
Landscaping, 253
Making a bookshelf, 259
minimizing waste in posts, 481
Real estate development, 390
Refurbishing a church, 497
Tiling a floor and wall, 238, 257

Consumer Information

Apartment expenses, 32
Better buy, 480
bottled water preferences, 170
buying art supplies, 258
Calculating a tip, 32
Cell phone plans, 307
Charges for word processing, 306
Consumer preferences, 171, 586–87
dealer markups, 403–404
depreciation, 403
DVD warranty, 679
electronic devices sales, 32–33
estimating a shopping bill, 27–28
food service fees, 307
Grocery bill, 27
Health club charges, 306
Insurance policy, 696, 702, 712
Legal documents, 98
Meal possibilities at a restaurant, 35, 636, 656
movie club charges, 306
Nutritional information at a fast-food
  restaurant, 62

Obtaining a complete set of collectibles,
  707–708, 709
original *vs.* sale price, 403
packaging of milk, 732
Paying off a consumer debt, 438
price of a grill, 403
Price of a new home, 402
Price of gasoline, 32, 45, 499, 724
Product reliability, 694, 761
purchasing a car, 49–50, 62, 80, 413, 429
  annual percentage rate and, 441–42, 444
  car sales, 402–403
  financing, 435
  paying off, 438
  present value, 435, 438
  Price of a car, 306, 345–46, 398
purchasing a computer, 32, 403
purchasing plants, 32
purchasing school supplies, 32
Rental costs, 304–305, 307
Renting an apartment *vs.* buying a house, 132
Rent-to-own agreement, 442, 444
Return policy, 104
Satellite dishes, 315, 360, 374
Shopping on the Internet, 677–79, 680
text messaging plan, 338
unit pricing, 258
Universal product code, 220–21
Warranty, 32, 679, 759–60, 762, 774

Economics. *See also* Finance

Calculating tax deductions, 32
comparing stocks, 749
Converting currency, 308, 336–37
Cost of living, 750
Federal income tax, 89, 100, 447
health care spending, 279
income tax, 401, 447, 748
Inflation, 330, 335, 361, 413
interpreting tax forms, 109–10
Medicare expenditures, 324
Poverty data, 315
Price stability, 749
Salaries for men *vs.* women, 725
State income tax, 748
Stock market, 242, 246, 293, 307, 403, 670
Stock prices, 743–44
Supply and demand, 370–72, 374, 393,
  700, 702

Education

allocating resident council, 537
American History Roundtable Society, 576
Assigning grades, 749
Assignment possibilities, 649
Choosing a college, 79, 132
Class schedules, 630
Class seating arrangements, 634–35, 637–38
college athletics committee, 609, 616
college expenses, 306, 314
College tuition, 398
college tuition account, 409–10, 412, 447
Commercial pilot's exam, 763
Comparing salaries to majors, 684–85, 693–94,
  738–39
Comparing years of education to weight, 771
Computer labs on campus, 538, 565
course electives, 45
Dean's list, 343
exam questions, 686, 687–88, 694
exam schedule, 152–53
Fellowships, 550, 564
female enrollment in colleges, 314

grade point average, 738
Grading, 307, 738
Graduate assistantships, 555–56
Guessing on exams, 702, 706–707, 709, 712
intelligence tests, 752–53
Living arrangements *vs.* grade point average, 671
male enrollment in colleges, 314
Number of years of education *vs.* salary, 770
parental involvement in college application process, 110
Perceptions of students, 70
President of the International Students' Organization, 596–97
President of the Undergraduate Labor Council, 581, 582, 584–85
quiz possibilities, 629, 647
Reducing a school budget, 599–600
Relationship between satisfaction with academic advisement and academic success, 680–81
Salaries for college graduates, 684–85, 693–94, 738–39
SAT scores, 762, 770
Scholarship applicants, 70
scholarships allocation, 257
Scores on language aptitude test, 724
Selecting a dormitory room, 689–91
selecting courses, 636
Senior group project, 178
Standardized tests, 699–700, 752–53, 758, 762
student government meetings, 151–52
student loan debt, 400
student loans, 413
Students' data, 52, 89
Students on planning committee, 17
Support services survey, 71
Teach for America research facility, 599
Test scores, 738, 748, 758, 774
Tutoring, 374, 763–64
Voting on improving college life, 588, 589
women earning professional degrees, 32, 34

Engineering

Electrical circuits, 100–101, 111
Strength of a beam, 339, 340–41, 342, 343, 361

Environment

Hurricanes, 720–21
Nuclear power plant, 729
Oil use and average monthly temperature, 338, 342
Pollution, 238, 737
Protesting the destruction of rain forests, 70
Rain, 679, 709, 711
Temperature, 245, 293, 298
Water usage, 342

Finance. *See also* Economics

Add-on interest loan, 415, 420, 421, 444, 447, 448
amortized loans
    Amortization schedule, 432–33, 437, 448
    payments on, 431–32, 435, 437, 438–439, 448
Annual percentage rate, 441–44, 448
Annual yield for an account, 412, 414
annuities, tax-deferred, 429, 447
bank account, 289
bank communications network, 7–8
bond value, 413
college tuition account, 409–10

Compound interest, 326, 333–34, 353, 357, 361, 406–407, 407–408, 409–10, 410–11, 412, 414
Consolidating loans, 444
Credit cards, 110, 417, 447, 738
determining interest rate, 335, 411, 413, 447
Dividing an estate and inheritance, 569–71, 572, 573, 574, 577
Finance charges, 416, 417–18, 419–20, 421, 447
Financing a purchase, 408, 413, 420–21, 422
finding unpaid balance, 436–37, 448
Future value of an account, 405, 412, 414, 447
Future value of an ordinary annuity, 425, 428–29, 429, 447
interest on a student loan, 413
Investments, 10, 16, 17, 132–33, 307, 353, 383, 389, 413, 414, 762
    comparing, 412, 413
Living expenses, 257, 300–301
mortgages
    balance, 436–37
    monthly payments, 448
    paying off, 437, 438, 439
negotiating a basketball contract, 411
Paying interest on late taxes, 413
Paying off a debt, 437–39
paying off a loan early, 438
Payments for a sinking fund, 426–27, 428, 447
Present value of a car, 435, 438
Present value of an account, 405–406, 409–10
Present value of an annuity, 434–35
property taxes, 8–9
Refinancing at a lower interest rate, 439
saving for a business, 410–11
saving for college, 409–10, 412, 447
Saving for living expenses, 300–301
Saving for retirement, 427–28, 429, 430, 438, 447
Saving money for a purchase, 429, 430
Simple interest, 405
zero-down loans, 408

General Interest

ages of volunteers, 723
Ages of World War II veterans, 748
Analyzing logic of advertisements, 93
arts advisory board, 543, 549–50
assigning booths at art show, 537
Beatles' autographs, 403
birth order, 659–60, 661
Buying canned food, 499
calories in fast food, 393
candy sales, 447
Capacity of an elevator, 32
Chain letters, 289
Change in historical dates, 245
Chinese zodiac, 222
choosing an evaluation team, 649
Choosing an outfit, 627–28, 629, 632–33, 656
choosing articles for magazine, 648
Choosing a speaker, 588, 589, 618
Choosing a team for a competition, 649
choosing a team for a seminar, 648
choosing contestants, 53, 670
choosing tasks, 647
Collection of objects, 642
committees
    influence on members, 165–66, 170, 182
    number of, 642
    number of ways of forming, 645, 648
    power of members, 614, 615, 619
    scheduling meetings of, 148, 151–52
    voting and, 609

comparing hamburgers, 490
Cooling a drink with ice, 490
cost of carpet, 342
Counting keys, 637
Counting on a clock, 215–17
Dance lessons, 5
designing a logo, 464
dividing a cheesecake, 293
Drug enforcement, 538
Duplication of birthdays, 695
Earnings of celebrities, 730, 737
editorial board, 614
estimating bags of fertilizer, 33, 342
estimating gallons of paint, 33–34
Estimating gallons of sealant, 361
Estimating number of M&Ms, 30–31
Explaining a number trick, 22–23, 25, 26, 36
Filling glasses with punch, 489
Fire hydrants, 701
Fires per month in a city, 746–47
Flow of information, 169
food sales, 402
Forty-foot-long sandwich, 16
Gambling strategy, 289
global warming survey, 65–66
hamburger toppings, 52
Handshake counting problem, 3–4
Hanging a mirror, 259
hate crimes, 670–71
heights of people, 468, 723, 740, 748, 761
housework, 28–29
Ice cream flavors, 630
Important issues survey, 71
Japanese garden, 253, 489
journalism awards, 648
Lawsuits against tobacco companies, 309–10
legal statements, rewriting, 106
length of statue's shadow, 468
lightning strikes, 672
location of family reunion, 619
Lock patterns, 633–34, 648
Lottery, 285–86, 438, 448, 643, 672, 679, 689–91, 698–99, 708
    expected value, 701, 703
marital status, 100, 662, 672
meal preferences, 586–87
Minimizing waste on a remodeling project, 258
musical patterns, 637
news media survey, 70
noncitizens living in U.S., 43
Nutritional requirements, 369, 373, 374, 381–82, 383, 390
painting a gymnasium, 498
Pizza toppings, 52, 637
planning a wedding, 182
position of surveillance camera, 468–69
Purchasing fruit in Europe, 499
Radio stations, 402, 636
raffle, 701
Recipes, 258, 293
research facility, 599
Sandwich combinations, 637
scheduling committee meetings, 148, 151–52
Seating arrangements, 151, 630, 634–35, 637–38, 656
Selecting officers of a club, 599, 636, 656
Selecting roommates, 648
six degrees of separation, 184
Sleep data, 315
Spread of rumors, 164, 169
Stacking baseballs, 24, 26
Stacking oranges, 630
surveying Mason-Dixon line, 469–70
surveys, 32, 65–68, 70, 71, 80, 110, 676–77, 677–79

tall people, set of, 41
test for extrasensory perception, 670
text message counting problem, 4
time to mow grass, 342
U.S. citizenship, 47
Viewing distance to horizon, 269
weather, 679, 709, 711
web browsers survey, 67–68
women in armed services, 257
writing a list, 648
Zip codes, 636

Geometry

Area of a circular lot, 482
Area of a playground, 472
Area of a stained-glass window, 480
Area of a triangular flower bed, 472–73
Area of trapezoids in air traffic control
    tower, 481
Area of trapezoids in art museum, 481
Area of trapezoids in the base of a statue, 473
area of walls in a room, 258
Capacity of water tanks, 488, 498
diameter of pizzas, 480
Fencing a dog pen, 499
Filling cylindrical glasses, 489
Gallons of juice in a barrel, 489
height of basketball player, 468
length of pond, 468
length of statue's shadow, 468
measuring perimeter *vs.* area, 479
Most efficient shape of a can, 485–86
most efficient shape of right circular cone, 490
position of surveillance camera, 468–69
pyramid height and area, 475–76, 480
Radius of a flower bed, 480
Radius of a tin can, 485–86
Surface area of a cylindrical can, 484
Surface area of a running track, 480
Surface area of the side of a frustum, 490
Volume of a frustum, 490
Volume of a punch bowl, 489
Volume of a swimming pool, 499
Volume of blocks of cheese, 483–84
Volume of cakes, 489
Volume of commercial products, 484–85
Volume of conical storage buildings, 487–88
Volume of ice cream scoops, 489
volume of spherical containers, 489
width of river, 468

Government and Politics

Ages of Supreme Court justices, 737
Arms race, 356–57
Assigning police officers, 548, 550, 551
Budget for snow removal, 724–25
City council, 587–88, 618
Condominium association board, 549
Critical voters, 605, 608, 609, 613
Defense budget, 397–98
Determining the legal drinking age, 599
Elections and voting, 17, 581, 582, 584–85,
    587–88, 589, 590, 591, 592–93,
    594–95, 596–97, 598–601
electoral college, 609–10
Influence on committee members, 165–66,
    170, 182
Jury system, 615
National debt, 279
Oil consortium board, 533–34, 537, 538,
    542–43, 545–47, 553–54,
    559–61, 562
"one person one vote" system, 615
Pivotal voter, 612–13

Political party affiliation, 674
Presidential popularity ratings, 110, 323
presidential preferences, 693
Presidential vetoes, 737
Presidents' ages, 732, 733–34
representative council, 551
scheduling committees, 148
State committee, 609, 616
Tax revenue for federal government, 33
Unfair apportionment, 535–36, 537
U.S. House of Representatives, 535, 537,
    539–41, 543, 544, 550, 554–55, 564,
    565, 566, 576
voters, 674
Voting (*See* Elections and voting (above))
voting on taxes, 594
water authority apportionment, 537, 564
Weighted voting systems, 602–3, 604, 605–608,
    609, 612–13, 615
Winning coalitions, 604, 608, 609, 612, 619

Labor

Calculating earnings, 307
Computing overtime, 306
employee drug testing, 691, 694
Employees negotiating a new contract, 537,
    564, 734–35
Employment data, 315
Family incomes, 749
Hours worked, 16, 36, 738
increasing workload, 403
International Brotherhood of Electrical
    Workers, 537
Interview schedule, 647
Job offers, 132, 307, 374, 747
Job training program, 725
Minimum wage, 680
Net pay, 306
safety in factories, 728
Salary, 283, 289, 307, 403, 684–85, 693–94,
    730, 734–35, 738–39, 749, 763, 770
sales commission, 680
Scheduling lawn service, 238
Staff assignments, 629
Unions, 535, 537, 543, 544, 549, 550, 564, 581,
    582, 584–85
U.S. labor force data, 315, 316
women's average earnings, 36
Work days missed, 742–43

Medical

AIDS awareness test, 718
AIDS cases, 330–31
AIDS deaths, 737
Antibiotics, 354–55, 357, 362
Birth statistics, 761, 762
blood types, 60, 72
cold vaccine, 709
Cystic fibrosis, 671
Doctor's appointment schedules, 630
Donating blood, 709
drug combinations, 646
Drug concentration, 334–35
Drug dosage, 341
Drug testing, 662, 694, 706
emergency clinics, 538
Flu vaccine, 495, 662, 679
Health care program, 179
Heart rates, 761
HIV testing, 712
Measuring a medication, 499
Mononucleosis, 692–93
nursing association, 537
Nursing committee, 542

Psychological test, 656
Scheduling nurses' shifts, 237
Sickle-cell anemia, 667, 671
Smoking rates, 309–310
Spread of disease, 167–68, 169
Storing medical supplies, 238
success rate of new procedure, 709
Survey regarding health concerns, 676–77
testing a cold medication, 694
Testing a flu vaccine, 662
Vaccinations, 495, 662, 709
Weight loss data, 719–20, 723, 725

Physics

Free fall of a sky diver, 342
free fall of MP3 player, 269
Height of a bouncing ball, 289
Path of a bouncing ball, 324
pendulums, 269
Rocket flight, 323, 324
Strength of a spring, 342

Probability

Drawing cards, 643–44, 649, 651–53, 654,
    659–60, 663, 669, 670, 673
    expected value, 701, 712
    probability of events, 664–65, 679, 680, 692,
        693, 711
    probability of union of two events, 676
Flipping a coin, 669, 670, 673, 679, 711,
    712, 713
    expected value, 697
    fundamental counting principle and, 632
    listing combinations, 15
    sample space and, 661, 663
    tree diagram and, 4–5, 15, 625, 629
keno, 657
male *vs.* female children, 664–65, 670
randomly drawing an object from a container,
    663, 692
Rolling dice, 17, 664–65, 669, 670, 673, 679,
    692, 693, 711, 713
    binomial experiment, 704–706
    conditional probability, 683
    Dungeons and Dragons dice, 15, 629,
        636, 656
    expected value, 701, 703
    fundamental counting principle and, 632
    independent *vs.* dependent events, 689, 712
    sample space and, 659–60, 661
    systematic listing and, 629
    tree diagram and, 15, 625–26
Roulette wheel, 668, 697–98, 701–702
Sliding object off a table, 673
spinning a spinner, 670, 681, 711
Tossing an irregular object, 673

Sports and Entertainment

Academy Award winners, 41
Actors' Coalition, 544
ages of actors, 774
American's Cup yacht race, 648
Amusement park ride, 773–74
Arranging a marching band, 222
Assigning tasks for a party, 629
Awards, 588–89
Baseball, 16, 373, 648, 672, 694, 709, 711,
    721–22, 737, 738, 739, 758–59,
    762, 763
Basketball, 373, 374, 393, 399, 709, 774
Basketball court, 477, 480–81
Bingo cards, 648
college athletics committee, 609, 616

Community theater, 640–41
Cy Young award, 589
distances on a baseball diamond, 480
dividing a purse in a poker tournament, 257
DVD sales, 320–22
Earth Day festival, 178
Emmy Award winners, 48
favorite tv programs, 726–27
fitness
    activities and calories burned, 72
    activities survey, 71
    designing a fitness program, 648–49
    scheduling fitness classes, 551, 564, 577
    sets of activities, 55, 57
Football, 16, 166–67, 169, 669, 739
Game show possibilities, 630
Grammy Awards hosts, 15
Great Alaska Challenge boat race, 30
Heights and weights of basketball players, 468,
    723, 748, 761
Heisman trophy, 589
home theater system, 636, 656
Horse racing, 672
life cycle of DVD, 320
Monopoly game, 671–72
movie attendance, 322
movie scheduling, 238
Movies, top-grossing, 63, 726
Music sales on Internet, 312–13, 403
Olympic athletes, 49
Olympic committee, 592–93
Online music channels survey, 71
Organizing a concert, 175–76
Playing cards, 649
Poker, 257, 651, 654, 656
ranking American Gladiators, 170
Ranking teams, 166–67, 169, 170
Recording artists, 626–27
Running a race, 324, 746
Selecting a stage play, 588
selecting singers in competition, 628, 670
Slot machines, 651, 653–54, 656
Sprint car racetrack, 670
stadium taxes, 594

Statistics, 399
strong man competition, 269
The Super Bowl, 739
Theater guild, 608–609, 615
Training for cross country, 32
Trumpet valves, 152
TV programs evaluations, 717, 718–19
TV viewer preferences survey, 66–67
volleyball team, 672
voting for a film, 618–19
waterpark attendance, 342
Women's 100-meter race *vs.* the men's
    marathon, 746
Women's soccer team, 672
YouTube viewership, 32

Statistics and Demographics

Crime statistics, 314–15, 670–71
deception using statistics, 726
ethnic makeup of population, 258
Immigration into the U.S., 33
Land per person in the U.S., 279
minimum wage, 680
Population, 28, 279, 289, 294, 328, 329–30,
    332–33, 334, 361, 403
population density, 306
prisoner growth rate, 323, 324

Technology

Cellular phone communications network, 709
computer animation, 280
computer circuits, 100–101, 111
Computer passwords, 648, 656, 673
computer storage capacity, 277, 280
Digital devices, 637, 670
Electronic devices, 32–33
iMacs *vs.* PCs, 128–129, 343
Internet searches, 46
Internet use, 762
iPhone, 6, 659–60
iTouch warranty, 759–60, 774
most visited Internet sites, 45–46, 402
Photographer's lighting, 342

survey of media users, 68, 71
Web browsers survey, 67–68
Wii Fit warranty, 762

Travel

Airport security, 565, 577
Airport traffic, 374
Air-safety review board, 606–608
Boat's position, 460
Borrowing money for a trip, 413
departure of bullet trains, 236
Distance an elevator travels, 245
Estimating distance, 30, 342
Extreme distances in geography, 245
Hours of flight time for amateur pilots, 723
Koenigsberg bridge problem, 140
Map coloring, 147–48, 150
Mass transit survey, 71
Metro City Transit System, 543, 544
Mishandled baggage, 724
Plane dropping emergency food supplies,
    323, 324
planning a whitewater rafting trip, 181
routes
    for armored car, 637, 649
    Bus routes, 656
    Cheapest, 162
    Efficient, 145–46, 150–51, 163
    number of possible, 17
    sales, 17, 19–21
68-95-99.7 rule, 751, 752–53
Scheduling flights, 237
time based on 7-hour clock, 222
toll road costs, 307
Tourist travel, 315, 316, 361, 374
train passengers, 723
train service at Olympics, 576–77
Transportation authority, 550, 564–65
Traveling salesperson problem (TSP), 19,
    153–54, 184
Travel time, 32, 36, 339–40, 342
Vacation survey, 70
voting for location of reunion, 619

SUBJECT INDEX

Note: *n* designates notes

68-95-99.7 rule, 751, 752–53
Absolute unfairness, 535–36
Acute angle, 453
Adams' apportionment method, 560–61, 563
Addition
  expanded notation and, 197–98
  identity element for, 266
  integers, 240
  rational numbers, 249–50
Additive inverse, 266, 267–68
Add-on interest method, 415–16
  APR and, 440
Adelman, Leonard, 218
Adibi, Jafar, 628
adjustable rate mortgage (ARM), 433
AI (artificial intelligence), 108, 692
AIDS, modeling growth of, 330–31
Alabama paradox, 531, 533–34
Alar, 769
Algebra, Boolean, 93
Algebraic models
  dynamical systems, 352–58
  exponential equations and growth, 325–36
  functions, 343–52
  linear equations, 297–308
  modeling with linear equations, 308–16
  modeling with quadratic equations, 316–24
  proportions and variation, 336–43
Algorithm(s)
  addition, 197–98
  best-edge, 159–60
  brute force, 157–58
  definition, 145
  Fleury's, 145–46
  nearest-neighbor, 158–59
  RSA, 218
  subtraction, 199
al-Khowarizmi, Mohammed, 197
All
  in statements, 86
  in syllogisms, 120–22
Alternate exterior angles, 454
Alternate interior angles, 454
Always Principle, 11, 19, 266
Amortization, 430–37
  adjustable rate mortgage (ARM), 433
  finding unpaid balance, 435–37
  monthly payments, 431–32
  present value of annuity, 434–35
  refinancing, 435–36
  schedules, 432–34
Analogies Principle, 13, 48, 58, 207, 244, 276,
        397, 454, 486, 491, 562, 668, 674,
        689, 732
Analytical geometry. *See* Geometry
And. *See* Conjunction
Angle of rotation, 504
Angle(s), 452–55
  acute, 453
  central, 455
  complementary, 453
  corresponding, 454
  exterior, 454
  interior, 454, 464
  measures of, 454–55
  obtuse, 453
  right, 453
  straight, 453
  sum of, 454–55
    in polygons, 463–64
    of sphere, 466
  supplementary, 453
  types of, 453

vertical, 453
Annual percentage rate (APR), 439–44
  add-on interest method, 440
  estimating, 443
  payday loans and, 442
  rent-to-own, 442
  table, 440–42
Annuity(ies), 422–30
  accumulation time, 427–28
  future value of, 423, 425
  and logarithmic function, 409–10
  present value of, 434–35
  sinking funds, 425–27
  tax deferred, 426
Ants and swarm intelligence, 159
anxiety, 753
Appel, Kenneth, 147
Apportionment, 530–78
  Adams' method, 560–61, 563
  Alabama paradox, 531–32
  applications, 545–51
  criterion, 539–40
  fair division, 567–74
  Hamilton method, 532–34, 563
  House of Representatives, 534
  Huntington-Hill numbers, 541, 547
  Huntington-Hill principle, 541–44
  Jefferson's method, 558–60, 563
  measuring fairness, 534–36
  paradoxes, 531–32, 553–58
  standard divisor and quotient, 552–53
  unfairness in, 535–36, 539–41
  Webster's method, 561–62, 563
Approval voting, 583
Approximations, in traveling salesperson
        problem, 158–59
APR. *See* Annual percentage rate (APR)
Arabic numbers. *See* Hindu-Arabic numbers
Arc of great circle, 457, 466
Area, 470–82. *See also* Surface area
  of circle, 476–77
  definition, 470
  formulas, 471–74
  parallelogram, 471
  perimeter and, 470–74
  rectangle, 470–71
  Sierpinski gasket, 516–17
  trapezoid, 473–74
  triangle, 471–73
Area under the curve, 754–56
Argument(s), 111–19
  common forms, 114–16
  complex, 115–16
  contraposition, law of, 114
  converse, fallacy of, 114
  definition, 112
  detachment, law of, 112, 114
  disjunctive syllogism, 114
  invalid, 113–14
  inverse, fallacy of, 114
  steps of verifying, 112
  syllogism, law of, 114
  valid, 112–13, 114
Aristotelian logic. *See* Logic
arithmetic growth *vs.* geometry growth, 285–86
Arithmetic sequence, 281–83
Arrow, Kenneth, 598
arrow diagrams, 347
arrowhead, 500–501, 506
Arrow's impossibility theorem, 598
Artificial intelligence (AI), 108, 692
ASCII code, 214
Associative property, 266, 267
Average. *See* Central tendency, measures of;
        Mean; Median; Mode

Average constituency, 534–35
Average daily balance method, 417–18, 419
axis of reflection, 502
axis of symmetry, 319

Babylonian numbers, 195–97
Balinski and Young's impossibility
        theorem, 542
Ballot, preference, 581–82
Banzhaf power index, 605–608
  vs. Shapley-Shubik index, 613–14
Bar graphs, 718–19, 720
Base, 271
Base 2 system. *See* Binary system
Base 5 system, 204, 205, 206–12
  operations in, 207–12
Base 8 system, 204, 205, 212
Base 16 system, 204, 212
Base amount, 398
Bell curve. *See* Normal distribution
Bernoulli trials. *See* Binomial experiments
Best-edge algorithm, 159–60
Best fit
  line of, 311–13, 767–69
  parabola of, 322
Bias, 715–16
Biconditional statements, 85–86, 107–108
Bids, sealed, 567–71
Binary digit, 212
Binary system, 203–204, 205, 212
  music and, 206
Binomial probability experiments, 703–709
  application of probability, 706–708
  probability, 703–706
  properties, 704
Bits, 206, 212
Blocking coalition, 607
blood alcohol content (BAC), 496
Bonabeau, Eric, 159
Boole, George, 82
Boolean algebras, 93
Borda count method, 581–83
  defects in, 591, 598
Box-and-whisker plot, 733–34
Bridges
  of connected graphs, 141
  of Koenigsberg, 140, 143
Brute force algorithm, 157–58

calculator(s)
  annual percentage rate (APR), 443
  apportionment computations, 532
  financial calculations, 409
  graphs, 301, 721
  linear correlation coefficient, 767
  linear programming problems, 388
  line of best fit, 313
  permutations and combinations, 644
  standard deviation, 745
  statistics, 731
Cancer, and health scares, 769
Cantor, Georg, 73, 74, 75
capture-recapture method of determining
        population, 337–38
Cardinal numbers, 42–43, 60, 75
casinos and card counting, 635
Cellular phones, and health, 769
Center of a circle, 455
Center of rotation, 504
centimeter, 493
Central angle, 455
Central tendency, measures of, 727–39
  comparing measures, 734–35
  mean, 727–30
  median, 731–33

mode, 734
Chain letters, 285
Check digit, 220–21
Chinese numbers, 191–92
Cicadas, 231
Circle, 455, 476–77
    dividing into regions, 21–22
Circuits
    Euler, 144
    Hamilton, 154–56
Circular cone, 486–87
Circular cylinder, 484–86
Circumference
    of a circle, 455, 476–77
    of Earth, 456
Clocks, modulo *m* system and, 216–17
Closed-ended credit, 415
Closed plane figure, 460
Closure property, 265–66
Coalitions, 603–605
    blocking, 607
    winning, 603–608, 613
Coefficient(s)
    linear correlation, 764–67
    opposite, 366
    of variation, 745–46
coffee, 769
Cognitive development, 42
Coin flipping
    expected value, 697
    tree diagrams and, 4–5, 625
combinations, 641–50
combined variation, 340–41
commas, importance of, 267
Common difference of sequence, 281
Common logarithmic function. *See* logarithmic
    function
Common ratio of geometric sequence, 283
Commutative property, 266, 267
Compatible numbers, 28
Complement
    of event, 673–74, 677–81
    of sets, 56–57, 59
Complementary angles, 453
Complete graph, 155
Complex argument, 115–16
Composite number, 229
Compound interest, 325–27, 404, 406–11
Compound statements, 83
    DeMorgan's Laws for, 97–98
    in fuzzy logic, 128–29
    truth tables, 94–96
Computer graphing programs, 721
Computer science
    artificial intelligence (AI), 108
    expert systems, 692
    national defense and, 116
Conclusion
    in argument, 111, 112
    in conditional statements, 103
    in syllogism, 120
Conditional probability, 681–95. *See also*
        Probability
    computing by counting, 683
    definition, 682
    dependent and independent events, 688–91
    general rule for computing, 684–85
    intersection of events, 685–86
    tree diagrams, 686–88
    Venn diagrams, 683, 685
Conditional statements, 85, 102–107
    alternate wording, 107
    contrapositive, 105–106
    converse, 105–106
    derived forms of, 105–106

in fuzzy logic, 129–30
    inverse, 105–106
    truth table for, 103–104
Condorcet's criterion, 592–93, 598
Cone, volume and surface area, 486–87
Congress. *See* Apportionment
Congruence, in modulo *m* system, 216–17,
        219–20
Congruent numbers, 217
Conjunction, 84
    in fuzzy logic, 128–29
    truth table for, 92
Connected graph, 141
Connectives, 83–86. *See also* Truth table(s)
    in fuzzy logic, 128–29
Constant of variation (proportionality), 338
Constraints, linear, 385
Constraints-objective table, 387
Consumer loans, 414–22
    add-on interest method, 415–16
    annual percentage rate, 439–44
    average daily balance method, 417–18, 419
    comparing finance methods, 419–20
    finding unpaid balance, 435–37
    refinancing, 435–36
    unpaid balance method, 416–17, 419
    zero-down loans, 104, 408
Continuous fair division, 567
continuum, 75
Continuum hypothesis, 76
Contraposition, law of, 113, 114
Contrapositive statements, 105–106
Converse, 105–106
    fallacy of, 114
converting a new problem to older problem,
        10–11, 310, 463
Convex polygon, 461
Corner point, 381, 385
Correlation, 763. *See also* Linear correlation
Corresponding angles, 454
Countable set, 74–75
Counterexample Principle, 11–12, 124
Counting, 623–57
    combinations, 641–644
    combining methods, 644–46
    conditional probability, 683
    fundamental principle, 631–38
    gambling, 635, 650–54
    handshakes, 3–4
    methods, 623–30
    Pascal's triangle, 645–46
    permutations and combinations, 638–50
    and probability, 662–65
    slot diagrams, 633–34
    systematic, 623–24
    tree diagrams, 624–28
    with and without repetition, 626–27
Counting numbers, 74, 228, 232–33
Credit
    closed-ended, 415
    open-ended, 416
Credit cards, 416–20. *See also* Interest
Critical path, in PERT diagram, 173–74
Critical voter, 604–608, 613
Cross multiplication, 247–48
Cross product, 247–48, 336
cube, 482–83
Custer, Gen. George, 668
Cylinder, volume and surface area, 484–86

Data
    bar graphs, 718–19, 720
    box-and-whisker plots, 733–34
    coefficient of variation, 745–46
    definition, 716

five-number summary, 733–34
    frequency table, 716–18
    graphical, and estimating, 28–31
    graphing with calculator, 313
    histogram, 719–20
    linear correlation, 763–71
    line of best fit, 311–13
    mean, 727–30
    measures of central tendency, 727–39 (*See
        also* Central tendency, measures of)
    measures of dispersion, 740–50 (*See also*
        Dispersion, measures of)
    median, 731–33
    mode, 734
    modeling
        with exponential equations, 325–35
        with linear equations, 308–16
        with quadratic equations, 316–24
    normal distribution, 750–63 (*See also*
        Normal distribution)
    organizing, 715–27
    parabola of best fit, 322
    range, 740–41
    scatterplots, 763–64
    standard deviation, 741–44
    stem-and-leaf, 721–22
    terrorism and, 628
    visual representation, 718–22
Data mining, 54, 628
Data set. *See* Data
debt
    credit card, 417
    student, 418
Decimal expansion, 255–56
Decimal form of rational numbers, 255–56
Decision making, and fuzzy logic, 130–31
Deductive reasoning, 22–23
Degree (of angle), 452–53
delivery of packages, 167
Demand, law of, 370–72
DeMorgan, Augustus, 147
DeMorgan's Laws
    for logic, 97–98
    for set theory, 59
Dempsey, James X., 635
Denominator, 247
    rationalizing, 263–64
Dependent events, 688–91
Dependent solution, 368
Dependent variable, 344
Detachment, law of, 112, 114
Deviation from the mean. *See* Standard
        deviation
Diagrams. *See also* Euler diagrams; Graph(s);
        Tree diagram(s); Venn diagrams
    arrow, 347
    PERT, 171–79
    probability trees, 686–88, 690, 691
    slot, 633–34
Diameter, 455
Dice rolling, tree diagrams and,
        625–26, 705
Dictator, 603
Difference, of sets, 57
Digital music, 206
Dimensional analysis, 493–500
Dimensions of fractals, 517–20
Diophantus, 239
Directed edge, 164
Directed graphs, 164–68. *See also* PERT
        diagrams
    modeling disease, 167–68
    modeling influence, 165–67
Directed path, 164–65
Direct variation, 338–39

Discrete fair division, 567
discriminant of quadratic equation, 328
Disjoint sets, 56
Disjunction, 84
    in fuzzy logic, 128–29
    in syllogism, 114
    truth table for, 92
Disjunctive syllogism, 114
Dispersion, measures of, 740–50
    coefficient of variation, 745–46
    range, 740–41
    standard deviation, 741–44
Distribution. *See also* Frequency distribution
    definition, 716
    frequency, 716
    normal, 750–63
    relative frequency, 716, 717, 718
Distributive property, 266, 268
Divisibility tests, 19, 231–32
Division
    integers, 243
    rational numbers, 252–53
Divisor
    modified, 558–62
    standard, 552–53
Domain, of function, 345
drawing strategy, 3–5, 143, 632
Dynamical system(s), 352–58
    definition, 353
    equilibrium values and stability, 355–57

e, 260, 261
Earth, circumference of, 456
Edge(s)
    directed, 164
    of graph, 139–40
    polygon, 461
Egyptian numbers, 186–89
El-Baz, Farouk, 31
Elections. *See* Voting methods
Electoral college, 580
electron charge, 274
Electronic Frontier Foundation, 236
Elements, of set, 39
Element symbol, 42
Elimination method, 366–68
Empirical assignment of probabilities, 63,
       661–62
Empty set, 41
Equal interests, 571
Equality of rational numbers, 247–49
Equal sets, 47
Equation(s)
    exponential, 327–33
    linear, 297–308
    percent, 398–400
    quadratic, 316–24
    recursive, 332
Equilateral triangle, 462
    in Sierpinski gasket, 515, 516–17
Equilibrium point, 370–72
Equilibrium value, in dynamic system, 355
Equivalent equations, 298
Equivalent sets, 51
Equivalent statements, 96–97
Eratosthenes, Sieve of, 229–30
estate, division of, 567–74
Estimation, 26–31
    compatible numbers, 28
    of graphical data, 28–31
    rounding, 26–27
Euclid, 232, 451, 457
Euler, Leonhard, 139, 140, 260
Euler circuits, 144
Euler diagrams, 120
    extra conditions in, 124–25

    invalid syllogism, 121–24
    with quantifier *no*, 122–23
    with quantifier *some*, 123–24
    syllogisms, 120–26
Eulerian graph, 144
Eulerizing a graph, 146
Euler path, 144
Euler's theorem, 141–44
Event(s)
    complement and union, 677–81
    complement of, 673–74
    definition, 660
    dependent and independent, 688–91
    intersection of, 685–86
    odds against, 667–69
    probability of, 661–62
    union of, 675–77
Every, 86
Exclusive or, 92
Existential quantifiers, 86
Expanded notation
    and addition algorithm, 197–98
    and subtraction algorithm, 199
Expected value, 695–703
    definition, 696
    games of chance, 697–99
    other applications, 699–700
Experiment(s), 659
    events in, 661–62
    sample space, 659–61
Expert systems, 692
Exponential equation(s), 325–36
    compound interest, 325–27
    definition, 327
    log function, 329, 410
    logistic models, 330–33
    models, 328–30
Exponential function, 349
exponential growth, 275, 325–28
Exponent property, of log function, 329, 410
Exponent(s), 270–74
    definition, 271
    rules for, 271, 272, 273
Exterior angles, 454

Factorial notation
    in counting, 640–41
    n factorial, 612
Factorization, prime, 228–29, 232–33, 234,
       235, 236
Fair division, 567–74
    method of sealed bids, 567–71
    fairness in game of chance, 698
Fair share, 567
Fallacy of the converse, 114
Fallacy of the inverse, 114
False statement, 91–96. *See also* Truth table(s)
Farley, Jonathan, 628
Farrakhan, Louis, 31
FCP. *See* Fundamental counting principle
Feasible solutions, in linear inequalities, 385
Fibonacci, Leonardo, 197, 286
Fibonacci sequence, 286–88
Fifth postulate, of Euclid, 457
Finance charge, 416–20, 441. *See also*
       Consumer loans; Interest
Finite set, 42
First-order system. *See* Dynamical system(s)
First quartile, in data set, 733
Five-number summary, 733–34
Fleury's algorithm, 145–46
Four-color problem, 147–48
Fourth dimension, 380
Fractal(s), 513–22
    applications of, 518
    dimension, 517–20

    Koch curve, 514–15, 516, 519–20
    self-similarity, 514
    Sierpinski gasket, 515
    tree, 520
Fractions. *See also* Rational numbers
    converting to percent, 397
    improper, 253–54
    reducing, 248–49
Frequency distribution. *See also* Normal
       distribution
    bar graph of, 718–19, 720
    definition, 716
    mean of, 727–30
    median of, 731–33
    mode of, 734
    relative, 716, 717, 718
    standard deviation of, 741–44
Frequency tables, 716–18
    and mean, 729, 730
    and median, 732, 733
    standard deviation, 741–44
Function notation, 344
Function(s), 343–52
    definition, 345
    modeling with, 349
    representing, 346–49
Fundamental counting principle (FCP), 631–38
    and gambling, 635
    slot diagrams, 633–34
    special conditions, 634–35
Fundamental theorem of arithmetic, 232
Future value
    of annuity, 423, 425
    compound interest, 325–27
    simple interest, 405
Fuzzy logic, 127–34
    compound statements in, 128–29
    conditionals, 129–30
    conjunction in, 128–29
    connectives, 128–30
    decision-making with, 130–31
    negation in, 128
    truth tables for, 131
    truth values, 127–28, 129–30

Galley method, 199–201
Gambling, 635, 650–54
    expected value, 697–99
    fundamental counting principle, 650–51
    lotteries, 635, 698–99
    odds against winning, 675
    poker, 651–53
    roulette, 668
    slot machines, 650–51
gamma knife surgery, 262
Gaudi, Antoni, 508
Gauss, Karl Friedrich, 232
GCD. *See* Greatest common divisor
Genetic theory, and probability, 665–67
Geometric growth *vs.* arithmetic growth,
       285–86
Geometric sequence(s), 283–86
    and chain letters, 285
    definition, 283
    Fibonacci sequence, 286–88
Geometry, 450–29
    angles, 452–55
    area, 470, 471–74, 476–77 (*See also* Area)
    circles, 455, 476–77
    circumference, 456, 476–77
    in clothing design, 474
    dimensional analysis, 493–500
    Euclid's principles, 457
    in everyday things, 462
    fractals, 513–22 (*See also* Fractals)
    lines, 451–52

lines, angles, and circles, 451–60
metric system, 490–500 (*See also* Metric system)
nonconvex polygons, 461
non-Euclidean, 457, 466
perimeter, 470–71
polygons, 460–70 (*See also* Polygon(s))
Pythagorean theorem, 259–60, 474–76
surface area, 482–90 (*See also* Surface area)
symmetry and tesselations, 500–13 (*See also* Symmetry)
tessellations, 506–509
three-dimensional, 482–90
volume and surface area, 482–90 (*See also* Volume)
Glide reflections, 503–504
Goldbach's conjecture, 18
Golden ratio (phi), 260, 261, 287
Golden rectangle, 287–88
Google, 39
Gould, Stephen, 231
Gram, 496
Graphics. *See* Computer graphics
Graphing calculator, 301, 313, 532, 721, 731, 745
Graph(s)
complete, 155
connected, 141
of data
bar graphs, 718–19, 720
box-and-whisker plot, 733–34
with calculator, 313
with computer, 721
histograms, 719–20
scatterplot, 763–64
stem-and-leaf, 721–22
definition, 139
directed, 164–68 (*See also* PERT diagrams)
drawing with calculators, 301
edge, 139–40
Eulerizing, 146
Euler's theorem, 141–44
Fleury's algorithm, 145–46
four-color problem, 147–48
functions, 348
Hamiltonian, 154–56
Koenigsberg bridge problem, 140, 143
linear equations, 299–305
of linear inequality, 377–81
misleading, 735
and operations research, 167
parabola of best fit, 322
PERT diagrams, 171–79
quadratic equations, 319–21
and scheduling, 148, 171–79
tracing, 140–41, 142
traveling salesperson problem (TSP), 19–21, 153–63
vertex, 139–40, 141
weighted, 157
Graph theory, 138–84
gravity and weight, 496
Great circle, 457, 466
Greatest common divisor (GCD), 233–34, 235–36
growth model. *See* Logistic models
Guessing, 8–9
Guthrie, Francis, 147

Hajratwala, Nayan, 236
Haken, Wolfgang, 147
half line, 451, 452
Hamilton, William R., 147
Hamilton apportionment method, 531–38, 563
paradoxes, 531–32, 553–58
Hamilton circuit(s), 154–56

Hamilton path, 154–56
Health scares, 769
Heron's formula, 472–73
Hexadecimal system, 204, 212
hexagons, tessellation with, 507
Hindu-Arabic numbers, 197–201
Histogram(s), 719–20
House of Representatives, 531, 534–35, 536. *See also* Apportionment
Huntington-Hill apportionment method, 541–44
Huntington-Hill number(s), 541, 547
Hypothesis in conditional statements, 103

Identity element, 266
identity theft, 218
If . . . then statements, 107
If and only if statements, 85–86, 107–108
Impossibility theorems
Arrow, 598
Balinski and Young, 542
Improper fraction, 253–54
Inclusive or, 92
income tax, 401
Inconsistent system, 368
Independence-of-irrelevant-alternatives criterion, 593–95, 598
Independent events, 688–91
Independent variable, 344
Inductive reasoning, 18–22
incorrect, 21–22
and intelligence tests, 19
Inequalities, linear
solving, 376–79
systems of, 379–84
Infinite sets, 73–77
definition of, 42
natural numbers, 74
Inflection point, 751
Inheritance, division of, 567–74
Initial side of angle, 452
Installment loans. *See* Consumer loans
instant runoff voting (IRV), 583
insurance, expected value of, 696
Integers, 239–46
addition of, 240
definition, 239
division of, 243
multiple operations, 243–44
multiplication of, 242
opposite, 241
set of, 74
subtraction of, 241
Intelligence tests, 19, 752–53
Intercepts
linear equation graphs, 300–302
of parabola, 319
Interest. *See also* Consumer loans
add-on, 415–16
on amortized loan (*See* Amortization)
annual percentage rate, and, 439–44
on annuity (*See* Annuity(ies))
average daily balance method, 417–18, 419
calculating with dynamical system, 353–54
compound, 325–27, 404, 406–11
credit card, 416–20
future value, 325–27, 405, 423, 425
on installment loans (*See* Consumer loans)
on rent-to-own contract, 442
simple, 404–406
unpaid balance method, 416–17, 419
Interest rate, 404
Interior angles, 454, 464
Intersecting lines, 451, 452
Intersection
of event, 685–86 (*See also* Probability)
of sets, 56

Invalid argument, 113–14
Euler diagrams, 121–22
Invalid syllogism, 120, 121–24
inverses, 105–106, 266, 267–68
Inverse statements, 105–106
Inverse variation, 338, 339–40
Investments. *See also* Interest
exponential model and, 326
Irrational numbers, 260–61
Isosceles triangle, 462

Jefferson, Thomas, 534*n*
Jefferson's apportionment method, 558–60, 563
Joint variation, 339

Karinthy, Frigyes, 144
Koch curve, 514–15, 516, 519–20
Koenigsberg bridge problem, 140, 143

Landsteiner, Karl, 60
Langlois, Judith, 500
Laplace, Pierre-Simon de, 200
Law of contraposition, 113, 114
Law of detachment, 112, 114
Law of syllogism, 114
Laws of supply and demand, 370–72
Leading-question bias, 716
Least common multiple (LCM), 234–236
Legislative apportionment. *See* Apportionment
Length
of directed path, 164
of graph, 144
of Koch curve, 516
metric, 491
light-year, 274
Lindley, Dennis, 688
Linear constraints, 385
Linear correlation, 763–71
coefficient, 764–67
health scares and, 769
line of best fit, 767–69
scatterplot, 763–64
Linear equations, 297–308
definition, 297
elimination method, 366–68
equivalent, 298
finding with two points, 310–11
graphing, 299–305
intercepts, 300–302
linear inequalities, 376–79
systems of, 379–84
modeling with, 308–16
point and slope, 309–10
systems of equations, 368–73
systems of inequalities, 381–82
two points, 310–11
programming (*See* Linear programming)
quadratic (*See* Quadratic equations)
rate of change, 301–302
slope, 301–302
slope-intercept form, 303–304
solution(s), 297, 365–66
solving, 297–99
standard form, 297
substitution method, 374–75
supply and demand, 370–72
systems of, 365–75
video games and, 369
Linear function, 349
Linear inequalities
definition, 376
graphing, 377–78
intersection of solution sets, 381
one-point test, 378–79

Linear inequalities (*Continued*)
  solving, 376–79
  systems of, 379–84
    modeling with, 381–82
    two-variable solution, 378
Linear programming, 384–91
  allocation of resources, 386–88
  corner points, 385
  problems, 385–86
  theorem, 385
Linear programming problem, 385
Linear regression, 313
Line of best fit
  for data points, 767–69
  in linear equations, 311–13
  using graphing calculator, 313
Line of symmetry, 505
Line(s)
  Euclid's definition, 451, 452
  intersecting, 451, 452
  parallel, 451, 452
  slope of, 301–302
  transversal, 454
Line segment, 451, 452
  of Koch curve, 514, 516
Listing, systematic, 5–6
Liter, 491, 494–96
Loans. *See* Amortization; Consumer loans
Logarithmic function, 328–30
  and annuities, 409–10
  exponent property of, 329, 410
Logic. *See also* Truth table(s)
  arguments, 111–19
  and artificial intelligence (AI), 108
  biconditional, 107–108
  conditional, 102–107
  connectives, 83–86
  cultural differences, 114
  deductive reasoning, 22–23
  DeMorgan's laws for, 97–98
  Euler diagrams, 120–26
  fuzzy, 127–33
  inductive reasoning, 18–22
  national defense and, 116
  quantifiers, 86–88
  set theory and, 93
  statements, 82–83
    compound, 83
    equivalent, 96–97
    in fuzzy logic, 127–28
    true or false, 91–96
  syllogisms, 120–126
  symbolic, 82
  truth tables, 91–102
Logically equivalent statements, 96–97
Logistic models, 330–33
  growth, 331, 332–33
Lottery
  expected value of, 698–99
  vs. investing, 635
Lower half, in data set, 733
Lower quota, 554
Lowest terms, 248–49

Majority criterion, 590–91, 598
Mandelbrot, Benoit, 514
Map coloring, 147–48
Markup, 398
Mass, metric, 496–97
Mayan numbers, 198
Mean, 727–30
Measurements. *See also* Metric system
  conversion, 493–97
Measures of data. *See* Central tendency,
    measures of; Dispersion, measures of
Median, 731–33

Members, of set, 39
Mendel, Gregor, 665–66
Mersenne prime, 236
Meter, 492–93
Metric system, 490–500
  blood alcohol content (BAC), 496
  common prefixes, 491
  dimensional analysis, 493–97
  gram, 491
  kilogram, 491
  liter, 491, 494–96
  meter, 491, 492–93
  metric conversions, 491–93
Meyer, Mary C., 688
Milgram, Stanley, 144
millimeter, 493
Million Man March, 31
MIT blackjack team, 635
Mixed numbers, 253–54
Mode, 734
Model(ing). *See also* Algebraic models
  dynamical systems, 352–57
  with exponential equations, 328–30
  as functions, 343–52
  with graphs (*See* Graph(s))
  with linear equations, 308–16
  logistic models, 330–33
  with point and slope, 309–10
  with quadratic equations, 320–21
  systems of equations, 368–75
Modified divisor, 558–62
Modified quota, 558–62
Modulo-*m* system, 215–23
  congruences in, 216–17, 219–20
  definition, 216
  operations in, 218–19
  UPC codes and, 220–21
Modulus, 216
Monotonicity criterion, 595–98
mortgages, adjustable rate (ARM), 433
Multiplication
  cross, 247–48
  galley method, 199–201
  identity element for, 266
  integers, 242
  rational numbers, 251–52
Multiplicative inverse, 266, 267–68
Multiplicative system, 191
Murphy's Law, 664
Music, digital, 206

naming strategy, 5
Napier, John, 200, 201
Napier's rods (bones), 200, 201
Nash, John, 628
National defense, logic and, 116
Natural numbers, 74, 228, 232–33
Nearest-neighbor algorithm, 158–59
Negation
  in fuzzy logic, 128
  of quantified statements, 87–88
  of statements, 83
  truth table and, 91–92
Negative numbers, 239–44
Nelson, Gary, 628
Networks. *See* Graph(s)
New amount, 398
New states paradox, 557–58
n factorial, 612
No, in syllogisms, 122–23
Nonconvex polygon, 461
None, 120
Non-Euclidean geometry, 457, 466
Non-polygons, 461
Nonregular tessellations, 508
Normal curve. *See* Normal distribution

Normal distribution, 750–63
  applications, 758–60
  area under the curve, 754–56
  converting raw scores to z-scores, 756–57
  of intelligence tests, 752–53
  percentage of value (68-95-99.7 rule), 751,
    752–53
  properties, 752
  z-scores, 753–55
  table of values, 754
Not. See Negation
Null set, 41
Number line, and addition of integers, 240
Number(s)
  cardinal, 42–43, 60, 75
  composite, 229
  congruent, 217
  definition, 186
  irrational, 260–61
  mixed, 253–54
  natural, 74, 228, 232–33
  negative, 239–44
  prime, 228–30, 231, 236
  rational, 74–75, 247–59
  very large, 236, 274
  whole, 239
  zero, 239
Number system. *See* Real number system
Number theory, 228–39
  divisibility tests, 231–32
  factoring, 232–33
  greatest common divisor (GCD), 233–34,
    235–36
  least common multiple (LCM), 234–236
  prime numbers, 228–30, 231
number tricks, 22–23
Numeral, definition of, 186
Numeration system(s). *See also* Place value
    system(s)
  Babylonian, 195–97
  base 5, 204, 205, 206–12
  binary, 203–204, 205, 206, 212
  Chinese, 191–92
  Egyptian, 186–89
  evolution of, 186–94
  hexadecimal, 204, 212
  Hindu-Arabic, 197–201
  Mayan, 198
  modular, 215–23
  multiplicative
  octal, 204, 205, 212
  other bases, 203–15
  properties of real numbers, 265–68
  Roman, 189–91
  sexagesimal, 195
  simple grouping, 186
Numerator, 247

Objective function, 385
Obtuse angle, 453
Octal system, 204, 205, 212
Odds, 667–69
One man, one vote, 535, 557
One-point test, 378–79
One-to-one correspondence, 51, 73
Open-ended credit, 416
Operations research, 167
Opposite integers, 241
Or. See Disjunction
Ordered pair, 297–98, 365
Order of operations, 12
  in set theory, 58–60
Order Principle, 12–13, 84, 504, 684
Ordinary annuity, 423, 425–28
Outcome(s)
  definition, 659

equally likely, 663–65
probability of, 661
outliers, 731

Pairwise comparison method, 585–87
defect in, 594–95
Parabola of best fit, 322
Parabolas, 317, 319–21
Paradoxes
of apportionment, 531–32, 553–58
of set theory, 46
Parallelepiped, rectangular, 483
Parallel lines, 451, 452
Parallelogram, 462, 471
Pascal's triangle, 6–7, 645–46
Path
critical, 173–74
directed, 164–65
Euler, 144
of graph, 144
Hamilton, 154–56
weight of, 157
Patterns, 6–7
Payday loans, 442
Pentagon(s), 463, 464
tessellations with, 509
Percent(s), 396–404
of change, 397–98
conversions, 396–97
as decimals, 396
equation, 398–400
salary increase and, 400
taxes and, 401
Perimeter, 470–82
Permutations
and combinations, 638–50
counting, 639
definition, 639
factorial notation, 640–41
of a set, 610–12
Perpendicular lines, 453
PERT diagrams, 171–79
phi (φ) (Golden ratio), 260, 261, 287
Pi, 260, 261, 476–77
Pivotal voter, 612
Place value system(s), 194–203. *See also*
Numeration system(s)
Plane(s)
Euclid's definition, 451
figures, 460–61
tessellating, 507–509
Plurality method, 580–81, 587
defects in, 592–93, 594, 598
Plurality-with-elimination method, 583–85, 587
defects in, 596–97, 598
Pluto, 43
point, 451
Poker, and counting, 651–53
Polls, inaccuracy of, 717. *See also* Survey(s)
Polya, George, problem-solving method of, 2–3
Polygon(s), 460–70
angle sum of, 463–64
area, 470–74
classification of, 461
convex, 461
definition, 461
nonconvex, 461
perimeter, 470–74
reflections of, 502–503
regular, 461, 464
similar, 465
tessellations, 506–13
Population paradox, 555–56
Population(s). *See also* Sample
capture-recapture method, 337–38
and samples, 715–16

Positional system. *See* Place value system(s)
Power rule for exponents, 272
Preference ballot, 581–82
Preference table, 581–82, 596
Premises
in arguments, 111, 112
in syllogisms, 120
Present value
of annuity, 434–35
of compound interest account, 409
of simple interest account, 405–406
Prime factorization, 228–29, 232–33, 234,
235, 236
Prime numbers, 228–30, 231
factoring, 228–29
Goldbach's conjecture, 18
very large, 236
Principal, 325, 404
principles, mathematical, 11–14
Probability, 658–713
basics, 659–73
binomial experiments, 703–709
complement and unions of events, 673–81
conditional, 681–95 (*See also* Conditional
probability)
counting and, 662–65
of event, 661–62
expected value, 695–703
expert systems, 692
formulas for calculating
complement and union of events, 677–79
counting, 662–65
empirical information and, 63, 661–62
equally likely outcomes, 663–65
intersection of events, 685–86
not equally likely outcomes, 665
odds against an event *E,* 667–69
probability of complement of event, 673–74
probability of union of two events, 675–76
unions of events, 675–81
using empirical information, 63, 661–62
using theoretical information, 662–65
genetic theory, 665–67
missing, 676–77
and Murphy's Law, 664
outcome(s), 659, 661, 663–65
probability trees, 686–88, 690, 691
in real-life situations, 664, 688
sample space, 659–61
Problem solving, 2–14
careers and, 3
principles of, 11–14
steps, 2–3
strategies, 3–11
Product rule for exponents, 271–72
Proper subset, 48–49
properties of real numbers, 265–68
Proportionality, constant of, 338
Proportions and variation, 336–43
Protractor, 453
Pseudosphere, 457, 466
public demonstrations, 31
punctuation, importance of, 267
Punnett square, 666, 667
puzzles, 3
Pyramid, Pythagorean theorem and, 475–76
Pythagoras, 474
Pythagorean theorem, 259–60, 474–76

Quadratic equations, 316–24
graphing, 319
modeling with, 320–21
parabola of best fit, 322
quadratic formula, 318
Quadratic function, 349
Quadratic regression, 322

Quadrilaterals, 462, 464
Quantifiers, 86–88
in syllogisms, 122–24
Quartiles, in data set, 733
Quota rule, 542, 554
Quota(s)
lower, 554–55
modified, 558–62
standard, 553–55
upper, 554
in weighted voting, 602
Quota sampling, 717
Quotient rule for exponents, 273–74

Radicals, 261–65
Radical sign, 261
Radicand, 261
Radius
of circle, 455
of sphere, 487
and surface area of cylinder, 485–86
Random phenomena, 659
Range, 740–41
of function, 345
Ratio
definition, 336
golden (phi), 287
Rationalizing denominators, 263–64
Rational numbers, 74–75, 247–59
addition and subtraction, 249–50
decimal expansion, 255–56
decimal form, 255–56
definition, 247
equality of, 247–49
mixed numbers, 253–54
multiple operations, 253
multiplication and division, 251–53
reducing, 248–49
repeating decimals, 255–56
set of, 74–75
Raw scores, 756–57
Ray(s), 451, 452
Real number system(s), 259–70
computing with radicals, 261–65
irrational numbers, 260–61
properties, 265–68
properties of real numbers, 265–68
Rectangle, 462, 470–71
Rectangular solid (rectangular parallelepiped),
483
Recursive equation, 332
Refinancing, 435–36
Reflections, 501–503
Regression
linear, 313
quadratic, 322
Regular polygons, 461, 464
tessellation with, 507
Regular tessellations, 506
relation, 346–47
Relative frequency distribution and table, 716,
717, 718
Relative unfairness, 536–37, 539–41
Rent-to-own contract, 442
Repeating decimal, 255–56
Representation. *See* Apportionment
Resources, allocating with linear programming,
384–91
Rhombus, 462
Rice, Marjorie, 509
Richardson, L. F., 356
Right angle, 453
Right circular cone, 486–87
Right circular cylinder, 484–86
Right triangles, Pythagorean theorem and,
259–60, 474–76

Rigid motion(s), 500–504. *See also* Symmetry;
   Tessellations
   glide reflection, 503–504
   reflection, 501–503
   rotation, 504
   translation, 503
rise, in linear equations, 301–302
Rivest, Ronald, 218
Roman numerals, 189–91
Rotations, 504
Roulette
   calculating odds, 668
   expected value of, 697
Rounding, 26–27
RSA algorithm, 218
run, in linear equations, 301–302

Sample, 715
   biased, 715–16
Sample space, 659–61
Scalene triangle, 462
Scatterplot, 763–64
Scheduling tasks, and graphs, 173–79
Scientific notation, 274–77
   applications, 277
   converting to and from, 276
   definition, 274
   rules of converting to, 275
Sealed bids, 567–71
Second-order systems. *See* Dynamical
   system(s)
Selection bias, 715–16
Self-similarity of fractals, 514
Sentences. *See* Statements
Sequence(s), 280–90
   arithmetic, 281–83
   definition, 281
   Fibonacci, 286–88
   geometric, 283–86
Set(s)
   Cantor's set theory, 69, 73
   cardinal numbers of, 42–43, 60
   commonly used, 40
   comparing, 47–51
   comparing sets, 47–53
   complement of, 56–57, 59
   countable, 74–75
   data and, 54
   definition, 39
   DeMorgan's laws, 59
   difference, 57
   disjoint, 56
   elements of, 39, 42
   empty, 41
   equal, 47
   equivalent, 51
   finite, 42
   infinite, 42, 73–77
   infinite sets, 73–77
   intersection, 56, 59–60
   language of, 39–43
   language of sets, 39–43
   and logic, 93
   members of, 39
   null, 41
   operations, 54–63
      order of, 58–60
   paradoxes, 46
   representing, 39–40
   set-builder notation, 13, 40, 41, 42
   set operations, 54–63
   subsets of, 48–50
   survey problems, 64–72, 65–68
   union, 54–56, 59–60
   universal, 41
   Venn diagrams of, 48, 54–57, 59–60, 64–68

well-defined, 40–41
Sets of feasible solutions, 385
Sexagesimal system, 195–97
Shamir, Adi, 218
Shapley, Lloyd, 610
Shapley-Shubik index, 610–16
   definition, 612
   vs. Banzhaf power index, 613–14
Shubik, Martin, 610
Sierpinski gasket, 515, 516–17
Sieve of Eratosthenes, 229–30
Similar polygons, 465
Simple grouping system, 186
Simple interest, 404–406
   future value, 405
   present value, 405–406
Simple plane figure, 460
Simple statement, 83
simplifying a problem, 7–8
Sinking fund, 425–27
SI system. *See* Metric system
Sitting Bull, 668
Slope, 301–302. *See also* Linear equations,
   modeling with
Slope-intercept form, 303–304
Slot diagrams, 633–34
Slot machines, 650–51
small-world problem, 144
social security, 426
Socrates syllogism, 120
Solids, 483–90
Solutions for linear equation, 297, 365–66
Some, in syllogisms, 123–24
Sophie Germain prime, 236
Sphere, volume and surface area, 487–88
Splitting-Hairs Principle, 13, 51, 154, 217, 611
Spreadsheet(s)
   amortization schedule, 434
   Huntington-Hill numbers, 547
   preference table, 596
Square(s), 462
   tessellation with, 507, 508
Square units, 470, 482
Standard deviation, 741–44
   definition, 741
Standard divisor and quotient, 552–54
Standardized tests, expected value of, 699–700
Standard notation, converting to scientific, 276
Standard quota, 553–55
Star
   five-pointed, 464
   symmetries of, 500–501, 505–506
Statements. *See also* Truth table(s)
   algebraic, 11–12
   biconditional, 85–86
   compound, 83
   conditional, 85, 102–107
   conjunction in, 84
   connectives in, 83–86
   disjunction in, 84
   in fuzzy logic, 127–28
   in logic, 82–83
   logically equivalent, 96–97
   negation of, 83, 87–88
   quantifiers in, 86–88
   simple, 83
   true or false, 91–96
Statistics, 714–775. *See also* Data; Population;
   Sample
   definitions, 716
   and health scares, 769
   linear correlation, 763–71
   measures of central tendency, 727–39
   measures of dispersion, 740–50
   normal distribution, 750–63
   organizing and visualizing data, 715–27

Steinhaus, Hugo, 567
Stem-and-leaf display, 721–22
Straight angle, 453
Strategies, problem-solving, 3–11
Subsets, 48–50
Substitution method, in linear equations,
   374–75
Subtraction
   expanded notation and, 199
   integers, 241
   rational numbers, 250
Supplementary angles, 453
Supply, law of, 370–72
Supply and demand, and linear equations,
   370–72
Surface area, 482–90. *See also* Area
   cone, 486–87
   cylinder, 484–86
   rectangular solid, 483
   sphere, 486–87
   of trapezoidal block, 483–84
Survey bias, 715–16
Survey problems, 64–68
   contradictions in, 67–68
Swarm intelligence, 159
Syllogism(s)
   disjunctive, 114
   extra conditions in, 124–25
   invalid, 120, 121–24
   law of, 114
   with quantifier *no,* 122–23
   with quantifier *some,* 123–24
   valid, 120–21
Symbolic logic, 82. *See also* Logic
Symmetry, 500–13
   of arrowhead, 500–501, 506
   axis of (quadratic equations), 319
   definition, 505
   glide reflections, 503–504
   line of, 505
   reflection, 501–503
   rigid motions, 500–504
   rotations, 504
   of star, 500, 505–506
   tesselations, 506–509
   translations, 503
Systematic listing, 5–6, 604, 623–24
Systèm International d'Unités. *See* Metric
   system
Systems of linear equations, 365–75
   elimination method, 366–368
   inequalities, 379–84
   modeling with, 368–73

Tables, frequency. *See* Frequency tables
Tautology, 95
tax, income, 401
Terminal side of angle, 452
Term of sequence, 281–83, 283–85
Terrorism, 628
Tessellations, 506–513
   definition, 506
   nonregular, 508
   with regular polygons, 507
Tests
   binomial experiments, 703–709
   guessing on, 8–9
   intelligence, 19, 752–53
   standardized, 699–700
Theoretical probability, 662–65
Third quartile, in data set, 733
Three-dimensional geometry, 482–90
Three-Way Principle, 14, 88, 234, 249, 272, 755
Tiling. *See* Tessellations
Trace, 140
Tractrix, 457

Translations, 503
Translation vector, 503
Transversal, 454
Trapezoid, 462
    area, 473–74
Traveling salesperson problem (TSP), 19–21,
        153–63
Tree diagrams
    of counting problems, 4–5, 10–11, 625–28
    for factorization, 232–33
    of Hamilton circuits, 155
    of permutations of a set, 611
    of probabilities, 705
    as problem-solving strategy, 4–5
    of sample space, 660
Triangle(s)
    angle sum, 463, 464
        of sphere, 466
    area, 471–73
    classification, 462
    right, 259–60, 474–76
    in Sierpinski gasket, 515, 516–17
    tessellation with, 507, 508
Truth table(s)
    alternative method, 98–99
    arguments, 112, 113, 115–16
    biconditional statements, 107–108
    compound statements, 94–96
    conditional statements, 103–104
    conjunctions, 92
    contrapositive, 105–106
    converse, 105–106
    disjunction, 92
    in fuzzy logic, 131
    inverse, 105–106
    logically equivalent statements, 96–97
    negative statements and, 91–92
    number of lines in, 95
    with three variables, 95–96
Truth values, 127–28. See also Truth table(s)
TSP. See Traveling salesperson problem (TSP)
Tukey, John, 721
TV ratings data, 321

Unfairness
    absolute, 535–36
    in game of chance, 698
    relative, 536–37, 539–41
Union
    of event, 675–81
    of sets, 54–56, 59–60
United Nations Security Council, 601

Unit fraction, 493–94
Universal quantifiers, 86
Universal set, 41
Unknowns, names for, 5
Unpaid balance method, 416–17, 419
UPC codes, and modular m system, 220–21
Upper half, in data set, 733
Upper quota, 554
U.S. customary system, 490
    conversions to and from metric units, 493–97
U.S. House of Representatives. See under
        Apportionment

Valid argument, 112–13, 114
Valid syllogisms, 120–21
Variable, dependent and independent, 344
Variance, 742
Variation
    coefficient of, 745–46
    combined, 340–41
    constant of, 338
    direct, 338–39
    inverse, 338, 339–40
    joint, 339
Vector, translation, 503
Venn diagram(s), 48, 54
    of complement of set, 57
    conditional probability, 683, 685
    of intersection, 56, 59–60
    naming, 64–65
    of set difference, 57
    of survey problems, 64–68
    of union, 54–56
Verification, in problem solving, 219, 407,
        582, 613
Vertex (vertices)
    of angles, 452
    of graph, 139–40, 141, 317
    of parabola, 317
    polygon, 461
    of tessellation, 507, 508
Vertical angles, 453
Vertical line test, 348–49
Veto power, 603
Video games and linear equations, 369
visualizing strategy, 3–5
Volume, 482–90
    base area times height, 483
    cone, 486–87
    cylinder, 484–86
    metric, 491
    of rectangular solid, 483

sphere, 486–87
    of trapezoidal block, 483–84
Voters, critical, 604–608, 613
Voting methods, 580–90
    approval voting, 583
    Borda count, 581–83, 591, 598
    defects in, 590–601
    fairness criteria
        de Condorcet's, 592–93, 598
        independence-of-irrelevant-alternatives,
            593–95, 598
        majority, 590–91, 598
        monotonicity, 595–98
    instant runoff voting (IRV), 583
    pairwise comparison, 585–87, 594–95
    Pareto condition
    plurality method, 580–81, 587, 592–93, 594,
        598
    plurality-with-elimination method, 583–85,
        587, 596–97, 598
    Shapley-Shubik index, 610–16
    weighted, 601–10 (See also Weighted voting
        systems)

Washington, George, 534n
Wealth, accumulating, 427–28
Webb, Glenn, 231
Webster, Daniel, 530
Webster's apportionment method, 561–62, 563
Weighted graph, 157
Weighted voting systems, 601–10
    Banzhaf power index, 605–608
    coalitions, 603–605
    definition, 602
    quotas and voter's weight, 602
    Shapley-Shubik index, 610–16
Weight(s)
    of coalitions, 603–605
    of graphs, 157
    metric, 491
    of path, 157
    of weighted voting system, 602
Whole numbers, 239
Winning coalition, 603–608, 613

x-intercept, 300

y-intercept, 300

z-scores, 753–57